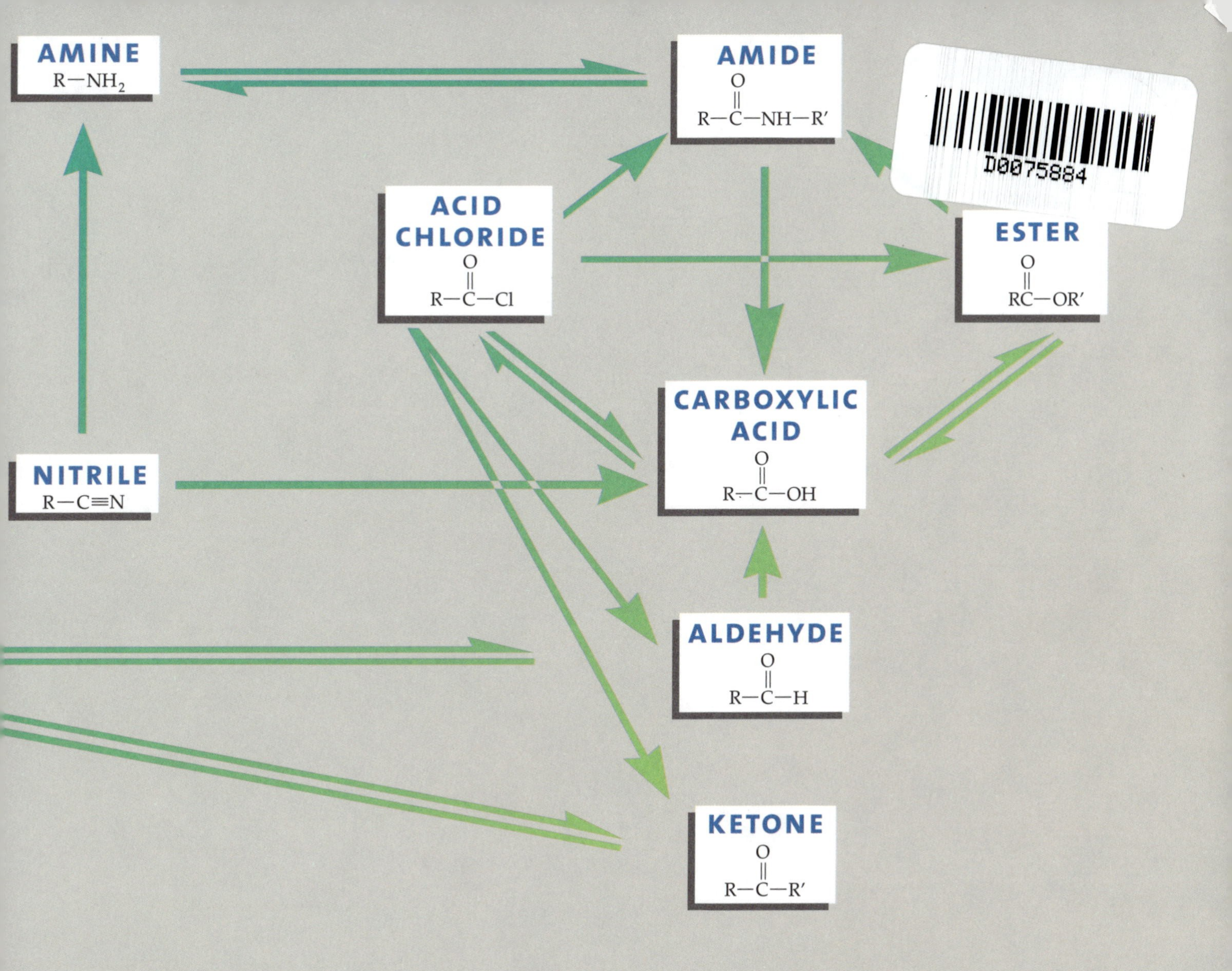

Family of Compounds	General Formula*	Functional Group†	Name of Functional Group
Amine	RNR′R″	$—NH_2$	amino
Aldehyde	RCHO	H–C=O	formyl
Ketone	R—CO—R′	C=O	keto
Carboxylic Acid	RCOOH	C(=O)—OH	carboxyl
Acid Chloride	RCOCl	C(=O)—Cl	acyl halide
Acid Anhydride	$(RCO)_2O$	C(=O)—O—C(=O)	anhydride
Ester	RCOOR′	C(=O)—O	ester
Amide	$RCONH_2$	C(=O)—NH_2	amide
Nitrile	R—CN	C≡N	cyano

*Any R in this table represents an alkyl group or aryl group.
†Carbon is assumed to be tetravalent, with other bonds not shown.

YOUR KEY TO LEARNING ORGANIC CHEMISTRY

The study of organic chemistry is important to anyone who endeavors to understand the complex activities of living systems and the material world. Bill Johnson's **Invitation to Organic Chemistry** is written to help you learn the core concepts of organic chemistry as well as discover how those concepts apply to our everyday world. Here is just a sneak preview of what it has to offer.

> "... the text is very well written and I am particularly impressed with the way in which the synthetic strategy is developed throughout the text."
>
> **ROBERT HILL**
> *University of Glasgow, Scotland*

Main features of the text

- Key to Transformations The Key will help guide you in understanding the relationships between families of organic compounds, and so help you develop strategies for problem solving in organic chemistry.
- How to Solve a Problem This special feature is designed to walk you through the process of analyzing and solving problems typical of those found at the end of the chapters. It will appear as though the author is thinking out loud as a problem is tackled.
- Chemistry at Work Highlights a wide range of applications of organic chemistry in many different fields of industry and research.
- Conceptual Problem Many chapters end with a problem set in a "real-world" context intended to give you a sense of how you might apply what you've learned.

A Complete Teaching and Learning Package

- CHEM TV: Choices by Betty Luceigh. This powerful interactive CD-ROM offers students an entertaining and useful tutorial on introductory topics of organic chemistry. It includes interactive Self-Tests and Concentration Drills.
- Organic Online www.jbpub.com/organic-online. Open to students and instructors alike, this web site features a Gallery of Chime™ Molecules, Mechanism Movies, Chem Web Explorations for exploring the Internet, an Online Quiz, and a Career Resource Center.
- Instructor's ToolKit CD-ROM The ToolKit contains lecture outlines (both Microsoft PowerPoint and browser-based), Chime Molecules, Mechanism Movies, an Image Bank of the text's illustrations, and a Computerized Test Bank.
- Study Guide and Solutions Manual Written by Bill Johnson, this printed supplement provides solutions to problems given in the text as well as amplifies the material covered.
- Models Kit Gives students a tangible experience with the three-dimensional structure of molecules.

> "What I personally liked most about this book was the clear explanations provided in the worked example-problems and solutions. It takes some of the load off the instructor in trying to explain the fundamental principles and at the same time work complicated problems during a lecture period. The excellent How to Solve a Problem analyses are very valuable!"
>
> **GITA SATHIANATHAN**
> *Pennsylvania State University*

Key to Transformations

Organic chemistry is a "large" subject, but one that is readily subdivided into the study of a small number of organic families. **The Key to Transformations** is your guide to these families and how they interrelate. The Key can work as a map to help you determine a strategy for working a problem that involves the synthesis of one compound from or to another.

"This [Key to Transformations] would be a very useful tool for students in relating the functional groups to each other. I think that students in general have a very difficult time seeing the relationship between various functionalities."

RITA MAJERLE
South Dakota State

Alkyl halides are the one family of organic compounds that can be derived from alkanes.

IF YOU REFER to the Key to Transformations (inside front cover), you can see that the alkanes we just studied stand somewhat apart from all other families of compounds. Generally inert, unreactive, and fully saturated with hydrogen, the alkanes are linked to the other families of organic compounds through the alkyl halides. **Alkyl halides,** known in IUPAC nomenclature as **haloalkanes,** can be derived from alkanes (including cycloalkanes) by replacing one or more hydrogen atoms with a *halogen* atom (fluorine, chlorine, bromine, or iodine)—a reaction we will study in Chapter 10. The symbol R—X is used as a general representation of an alkyl halide, with R representing an alkyl group and X representing one of the four halogens.

Alkyl halides are the first family of organic compounds we will study that contain a **functional group,** that part of a molecular structure that is especially susceptible to chemical reaction. In contrast, as you recall, the carbon-hydrogen bonds present in alkanes are notoriously unreactive. In alkyl halides, the halogen atom serves as the functional group, being readily replaceable by nucleophiles in a major type of reaction known as a *nucleophilic substitution*. Important in its own right, nucleophilic substitution also serves here as an introduction to reaction mechanisms—the means by which one compound is transformed into another.

Haloalkanes range from the very simple to the very complex. For example, ethyl chloride (chloroethane), used as a topical "freezing agent" on skin, includes only a single chlorine atom. The compound 1,1,1,2-tetrafluoroethane, one of a new type of refrigerant, has four fluorines. Dichlorodifluoromethane, a refrigerant formerly used in air-conditioning systems, contains four halogens of two different types. And in the compound perfluorodecalin, which is an artificial blood substitute, all of the hydrogens have been replaced by fluorine.

How to Solve a Problem

Even more critical to your problem-solving success are the worked examples called **How to Solve a Problem.** Here the author walks you through a sample problem, showing you how to analyze the situation, getting you to think about what's important and what's not, then he takes you through the solution.

A series of practice problems follow that give you a chance to practice what you've learned. The chapter ends with a large number of **ADDITIONAL PROBLEMS** that are designed to reinforce your understanding of the material.

"The end of chapter problems are superb. I was particularly pleased with the design of problems to progressively lead students toward how to think about the topics covered."

JOHN LEITZEL
Chicago State University

How to Solve a Problem

What is the product of the reaction between sodium cyanide and isopropyl bromide?

PROBLEM ANALYSIS

The first step in solving the problem is to write the information we are given in the form o reaction:

$$\mathrm{Na\,CN} + \mathrm{H{-}C(CH_3)_2{-}Br} \xrightarrow{\text{DMSO}} ?$$

Sodium cyanide — Isopropyl bromide

t we have an alkyl bromide, and the usual reaction is for the bromine atom nucleophile. Can we identify a nucleophile? Yes, because sodium cyanide salt (Na^+ CN^-), and cyanide ion is an effective nucleophile (see Table 3. e S_N2 reaction in which the nucleophile (cyanide) displaces the bromine.

S_N2 reaction are sodium bromide and a nitrile (also known as an alkyl cyani ropyl cyanide). There is no need to worry about stereochemistry in this re arting material has no stereocenter—there are no enantiomers. Do not wr less y asked for it. Instead ur knowledge of the mec

INVITATION TO ORGANIC CHEMISTRY

A SENSE OF RELEVANCE

Chemistry at Work

The **Chemistry at Work** boxes present interesting asides about the application of organic chemistry in a wide variety of career fields.

CHEMISTRY AT WORK

IN THE MARKETPLACE

Plexiglas

Plexiglas, a polymer of methyl methacrylate, is used for windshields in aircraft and "glass" walls around ice hockey rinks, as well as for other applications that require glasslike transparency and decidedly unglasslike strength. Plexiglas was the original bulletproof glass. Its strength and durability are also valuable in more mundane uses: Lucite paint uses the same polymer as a base. The raw material costs for producing poly(methyl methacrylate) are very low because all the ingredients are mass-produced very economically. The commercial preparation relies on the cyanohydrin reaction.

Acetone is converted to the cyanohydrin, which is then dehydrated. The unsaturated nitrile is hydrolyzed to the carboxylic acid and esterified with methanol in a single step to form the monomer, methyl methacrylate. This monomer is polymerized to the linear polymer, which is then formulated into the desired product.

Plexiglass is used to surround hockey rinks in order to protect the spectators from the heavy action and yet provide a clear view. On the ice are the Fighting Sioux of the University of North Dakota at Engelstad Arena.

$(CH_3)_2C{=}O$ (Acetone) $\xrightarrow[HCN]{NaCN}$ $(CH_3)_2C(OH)CN$ $\xrightarrow[\Delta]{H_2SO_4}$ $CH_2{=}C$…

…$\xrightarrow{\text{(polymerization)}}$ $\left[-CH_2-C(CH_3)(COOCH_3)- \right]_n$

Poly(methyl methacrylate)

"I am personally very taken with the style of writing—it matches very well with my own lecture style. The use of contractions and vivid imagery really breathe life into the material. It is very conversational and engaging—as I was reading, I felt like I was sitting down and talking to someone one-on-one."

Brenda Kesler
San Jose State University

Conceptual Problem

Many chapters end with a conceptual problem designed to test your mastery of in-chapter material, as well as suggest some of the broader applications of organic chemistry.

"The [Conceptual Problems] are excellent and so neccessary! In my opinion, a non-science major's chemistry course should show the student the chemistry in the world around them, not the chemistry in the researcher's laboratory."

Jim Burns
Utah State University

CONCEPTUAL PROBLEM

Seek, Identify, and Synthesize: Paclitaxel

In the early 1960s, the National Cancer Institute asked researchers to collect and analyze samples of indigenous plants in the hope of isolating substances that might some day prove effective in the fight against cancer. Researchers cast a wide net and among their catches was a chemical compound isolated from the bark of the Pacific yew. To test its medicinal properties, they placed the compound into some artificially preserved cancer cells—it killed the cells. They quickly set about to analyze this cancer-fighting substance: paclitaxel.

- Can you determine the molecular formula of paclitaxel from the structure?

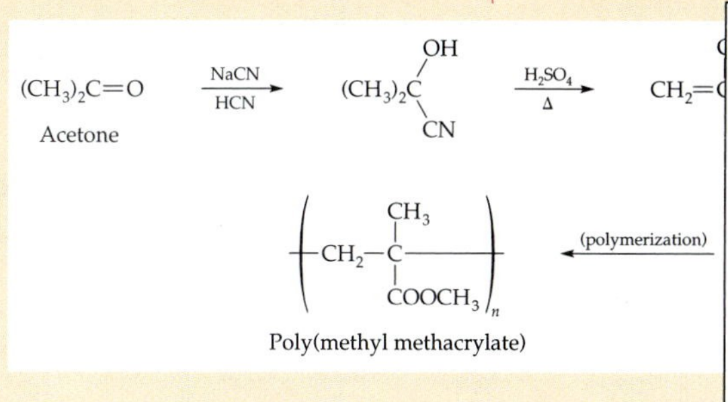

Paclitaxel

…of paclitaxel has 113 atoms in it. Does that …hing to you about how difficult it might …size in the laboratory?
…at you do about organic compounds in …described near the end of this chapter), …xpect paclitaxel to be water-soluble? If …fficulties might that pose for doctors who …his substance to treat their cancer

…biggest difficulties facing the research …orking with paclitaxel was how little of it

Harvesting the bark of the Pacific yew tree yields very small amounts of the naturally occurring cancer-fighting compound we now call paclitaxel. A single 100-year-old tree yields less than a gram, which is not enough for even a single treatment of a cancer patient.

could be extracted from the bark of one yew tree—barely a gram. Not only that, but the tree was killed in the process of stripping the bark. It was clear that chemists would have to find a way to synthesize this compound in the lab. Researchers in France, working with the European yew, discovered in the needles of that kind of tree a compound that was closely related to paclitaxel. This substance could be used as the starting point of a "semi-synthesis" that required just a few steps.

- In addition to providing a relatively easy way to synthesize paclitaxel, what other advantage does use of the compound from the European yew have?

Home Page

Organic Online connects users of **Invitation to Organic Chemistry** to an extensive web site developed by Jones and Bartlett Publishers. The site offers a variety of activities designed to enhance the learning process and to give students access to some of the best organic chemistry and career-related web sites available. You can reach the Organic Online home page by entering the URL **www.jbpub.com/organic-online** into a web browser such as Netscape Navigator or Microsoft Internet Explorer.

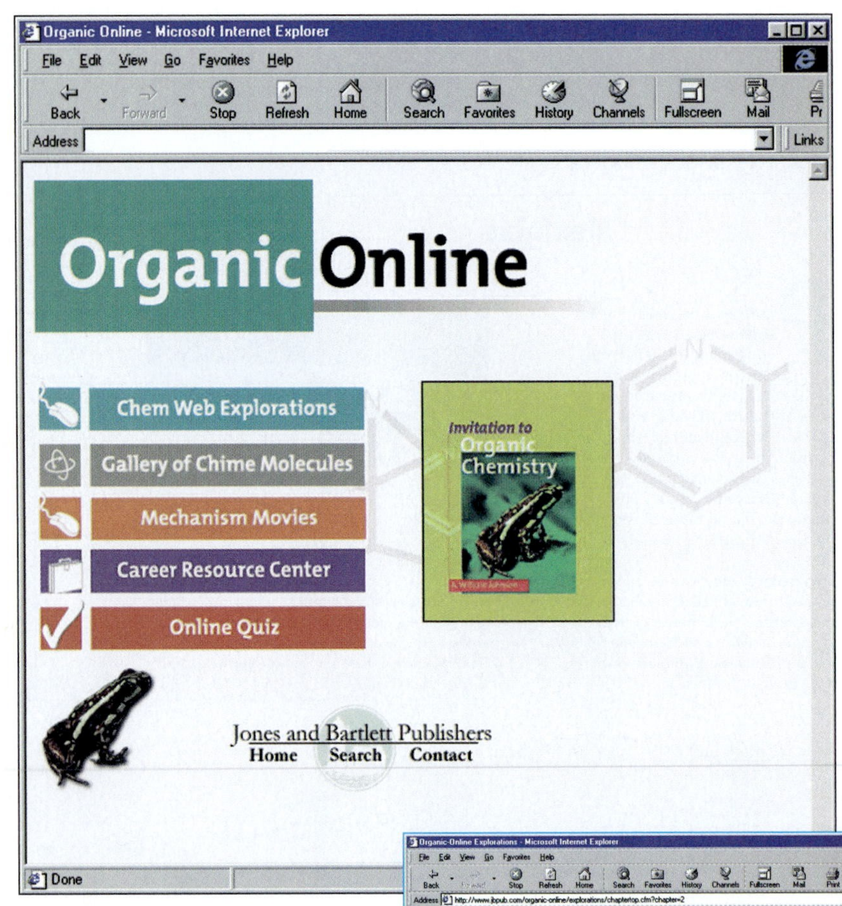

Explorations

The **Explorations** exercises give the student an opportunity to see the kind of information available on the Internet. Prior to searching a site, students are asked a question, and expected to explore the site to formulate a reasonable response to the question.

ORGANIC ONLINE

The Gallery of Chime™ Molecules

The **Gallery of Chime Molecules** features over 300 molecules introduced in the text. These molecules range from simple ones, such as water and methane, to more complex ones, such as buckminsterfullerene and paclitaxel (taxol). Using the **Chime plug-in,** students and lecturers alike can interactively observe each molecule by displaying several model formats such as wire frame, ball and stick and more; by changing from 2- to 3-dimensional representations; by viewing stereo displays and dot surfaces; and by rotating the structure either manually or automatically. Look for the Chime icon in the margin of the text.

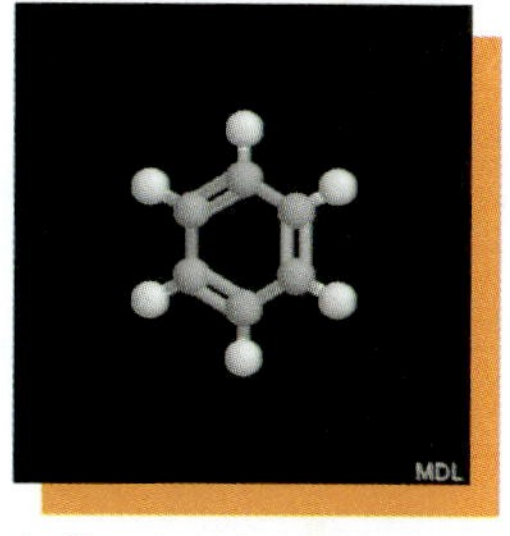

ball and stick

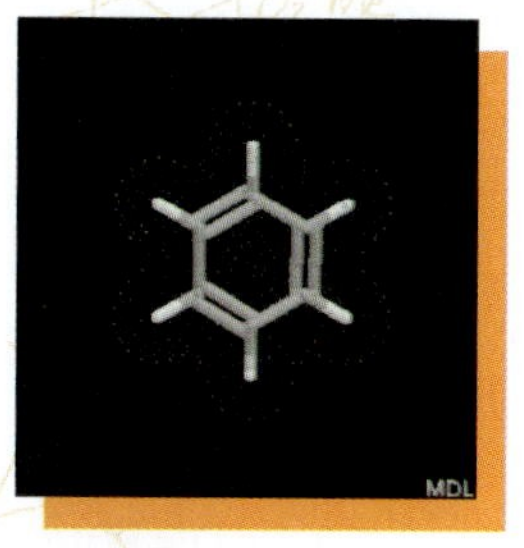

dot surface

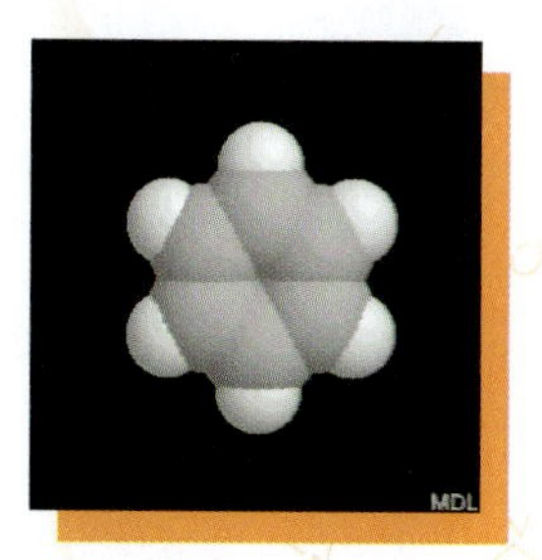

space filling

Mechanism Movies

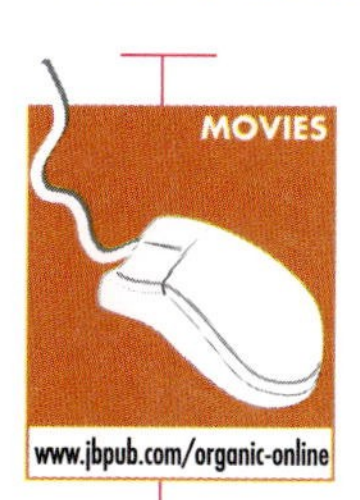

Twelve of the most important reaction mechanisms covered in the text are animated as Quicktime™ movies. Look for the Movies icon in the margin of the text.

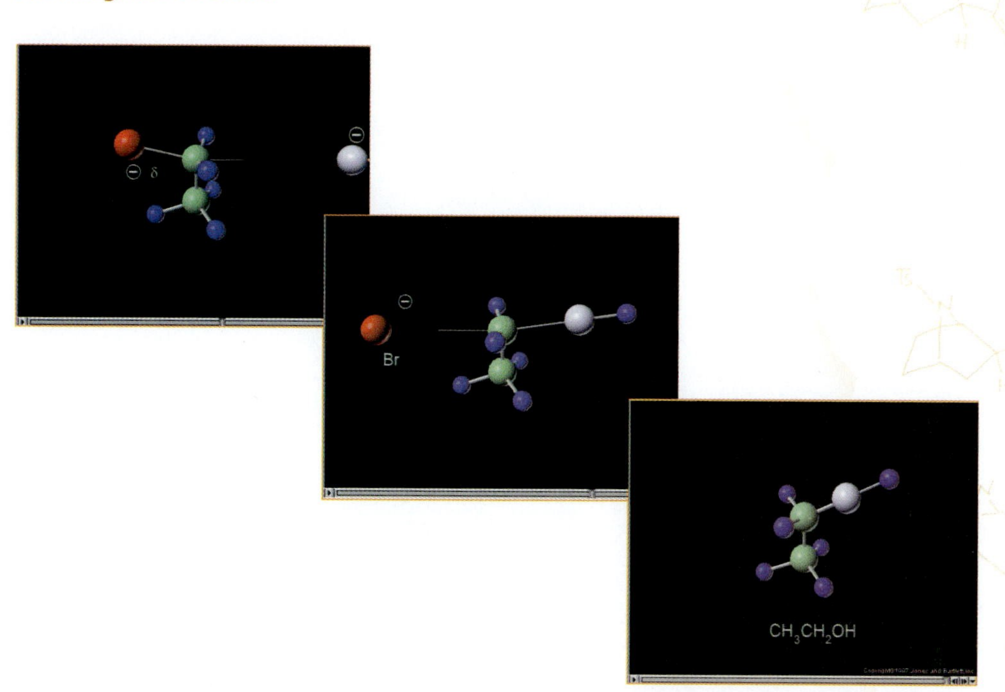

Invitation to Organic Chemistry

PROFESSOR EMERITUS
UNIVERSITY OF NORTH DAKOTA

JONES AND BARTLETT PUBLISHERS
SUDBURY, MASSACHUSETTS
BOSTON TORONTO LONDON SINGAPORE

To my wife Joan
for 42 years of love, company, and patience

WORLD HEADQUARTERS

Jones and Bartlett Publishers
40 Tall Pine Drive
Sudbury, MA 01776
978-443-5000
info@jbpub.com
www.jbpub.com

Jones and Bartlett Publishers Canada
P.O. Box 19020
Toronto, ON M5S 1X1
CANADA

Jones and Bartlett Publishers International
Barb House, Barb Mews
London W6 7PA
UK

Chief Executive Officer: Clayton Jones
Chief Operating Officer: Don Jones, Jr.
Publisher: Tom Walker
V.P., Sales and Marketing: Tom Manning
V.P., Senior Managing Editor: Judith H. Hauck
Marketing Director: Rich Pirozzi
Interactive Technology Director: Mike Campbell
Production Director: Anne Spencer
Manufacturing Director: Therese Bräuer
Executive Editor: Brian L. McKean
Special Projects Editor: Mary Hill
Assistant Production Editor: Ivee Wong
Web Designer: Mike DeFronzo
Interactive Technology Project Editor: W. Scott Smith
Cover & Walkthrough Design: Anne Spencer
Text Design: Seventeenth Street Studios
Text Artwork: HRS Studios; Mark Rodrigues, Stephanie Torta, Anne Spencer
Web Site Design: Stephanie Torta, Mark Rodrigues
Production Service: Jane Hoover, Lifland et al., Bookmakers
Composition: Monotype Composition Company, Inc.
Cover Manufacture: Coral Graphic Services, Inc.
Book Manufacture: World Color Book Services

ABOUT THE AUTHOR

Born in Calgary, Canada, Bill Johnson graduated at the top of his class at the University of Alberta, where he received his Bachelor of Science degree. He received his Ph.D. in organic chemistry from Cornell University and was appointed the first Fundamental Research Fellow at the Mellon Institute in Pittsburgh. He has taught organic chemistry at the University of Saskatchewan (Regina), Cornell University, the University of Massachusetts (Amherst), and the U.S. Military Academy at West Point. Most of his professional career he has spent at the University of North Dakota where, in addition to teaching organic chemistry, he served as Director of Research and Development and Dean of the Graduate School. He is a Fellow of the Chemical Institute of Canada and of the American Association for the Advancement of Science, which cited him for his pioneering research in the chemistry of ylides. Dr. Johnson's outside activities have included curling, hockey, golf, Nordic skiing, camping, fishing, and photography. For ten years he produced and announced a weekly program on grand opera for KFJM-FM in Grand Forks.

ON THE COVER

Epipedobates tricolor, sometimes called the "phantasmal" poison frog, © Gail Shumway. (See back cover for details.)

CREDITS

Credits appear on page C-1, which constitutes a continuation of this copyright page.

Library of Congress Cataloging-in-Publication Data

Johnson, A. William (Alyn William)
Invitation to organic chemistry / A. William Johnson.
p. cm.
Includes index.
ISBN 0-7637-0432-6 (hardcover)
1. Chemistry, Organic. I. Title.
QD253.J63 1999
547—dc21 97-46099
CIP

Copyright © 1999 by Jones and Bartlett Publishers, Inc.

All rights reserved. No part of the material protected by this copyright notice may be reproduced or utilized in any form, electronic or mechanical, including photocopying, recording, or by any information storage and retrieval system, without written permission from the copyright owner.

Printed in the United States.

03 02 01 00 99 98 9 8 7 6 5 4 3 2 1

Brief Contents

Contents

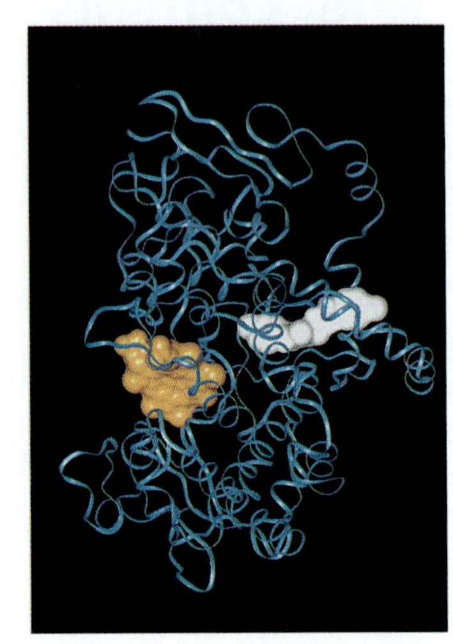

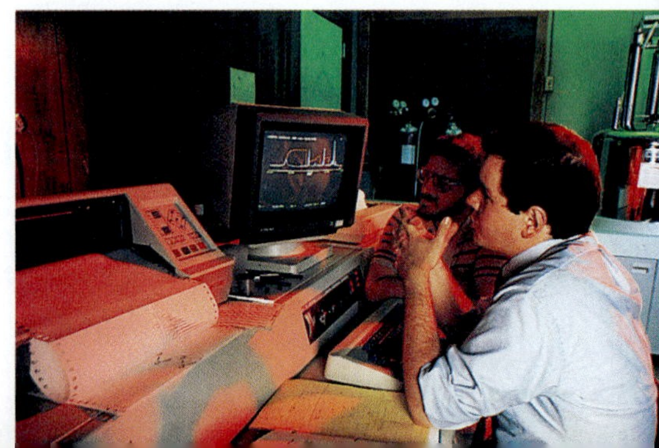

CHAPTER 15 Derivatives of Carboxylic Acids 528

CHAPTER 16 Proteins 577

To the Instructor

Essential Understanding

I wrote ***Invitation to Organic Chemistry*** with one goal in mind—to facilitate the learning of organic chemistry by students who need an **elementary understanding** of the subject to support study in their chosen fields. The paths they take might lead them toward careers in nutrition, nursing, agricultural sciences, biology, psychology, home economics, physical education, allied health sciences, engineering, pharmacy, environmental studies, and others. This text is designed to support a one-semester, one-quarter, or two-quarter course at the college level. Students taking this course will presumably have had a semester or quarter of college-level general chemistry.

Organic chemistry is a large subject. The approach I've taken in this text is to limit the content and present only that which is **essential** for the student's understanding of the elementary workings of organic chemistry. This approach is designed to encourage a thorough **understanding** of a smaller body of knowledge as opposed to memorization of a larger body of knowledge. I seek these outcomes for students using *Invitation to Organic Chemistry*:

- A limited but effective knowledge of organic chemistry and the relationships between different families of organic compounds
- An ability to comprehend basic information about organic compounds and reactions as it is communicated by experts
- An appreciation of the application of organic chemistry to specific fields and society in general

It has been my experience that students taking a short course in organic chemistry struggle for these reasons:

- Too wide a range of material
- Overreliance on memorization
- Discontinuity with their experience of problem solving in general chemistry
- Too little text-based guidance in developing a strategy for problem solving in organic chemistry
- A lack of perspective on the value of organic chemistry in their chosen field

Invitation to Organic Chemistry *addresses all of these concerns.* Students will learn what makes organic chemistry so interesting, as well as gain an appreciation of its relevance. The text moves them logically and gradually from one topic to the next, easing them into new concepts. It is designed to support instruction that encourages students to minimize memorization and to concentrate on understanding basic concepts and relationships as techniques for achieving success in their study. The thorough explanations make this textbook especially suitable for student self-study.

A Logical Organization

The chapter sequence is the result of careful consideration of how students most effectively learn this subject. In my experience, students find it easiest to study organic chemistry through developing familiarity with functional groups, so I have used that organization. Furthermore, becoming familiar with one kind of reactive intermediate and its behavior before being exposed to other reactive intermediates allows students to achieve a comfort level with the subject. Therefore, the text focuses only on carbocation chemistry in the first eight chapters. By the time radical chemistry is introduced in Chapter 10, students will have become comfortable with mechanisms, including issues of reactivity and stabilization. Carbanion chemistry is explored starting in Chapter 13. Even more critical to students' success, I believe, is to focus first on compounds with only sp^3-hybridized carbons. These are the subject of the first five chapters. Once the student becomes comfortable with that content, compounds with sp^2- and sp-hybridized carbons are introduced.

This approach of determining when best to have a student take on a new concept can be seen in other aspects of this book's organization. For example:

- Radical chemistry is concentrated in one chapter (Chapter 10), where the relation between mechanisms is demonstrated, enhancing students' understanding.
- Coverage of spectroscopy (Chapter 11) comes after aromatics have been introduced so that this important family of compounds can be included in the discussion. This also enables spectroscopy to be used in subsequent chapters.
- Amines are covered earlier in this textbook than in many others. Coverage of amines in Chapter 12, before carbonyl compounds, means that they can be considered as simply another nucleophile for the standard additions to carbonyls (Chapter 13) or the nucleophilic acyl substitutions of carboxylic acid derivatives (Chapter 15).
- Chapter 17 (Optical Activity) contains a brief review of the isomerism concepts introduced earlier, with amplification that is made possible because students have covered all reaction chemistry. I have always found such a reiteration useful, considering that much stereochemistry is introduced quite early (Chapters 2 and 3), before its importance is readily appreciated. The concepts of configuration, enantiomers, and racemates are covered thoroughly in Chapter 3. At that early stage the additional details of optical activity, resolution, and multiple stereocenters are unnecessary. Deferring their coverage until Chapter 17 allows concentration on the essential aspects of substitution mechanisms (which are daunting enough for most students). However, Chapter 17 is a stand-alone chapter and so can be taught elsewhere in the sequence if preferred.
- Each functional-group chapter includes a section on important and interesting examples of compounds containing that functional group. Rather than present such compounds at the beginning of each chapter, before the student has an appreciation of the chemistry involved, I have placed all such sections at the ends of the chapters.

Bioorganic Integration

Invitation to Organic Chemistry takes a special approach to biological chemistry. Based on my discussions with faculty and students from around the country, I find that very little time, if any, is given to the separate, often lengthy discussions of lipids, carbohydrates, and proteins that occupy the last few chapters of most textbooks. Many instructors leave the reading of these chapters to the individual initiative of students. With the goal of encouraging meaningful coverage of bioorganic chemistry in this course, I have taken a different approach to the material.

Essential coverage of the standard biological chemicals has been incorporated in those chapters that include the actual functional groups involved:

- Carbohydrates in Chapter 4 (alcohols)
- Nucleic acid bases in Chapter 9 (heterocycles)
- Fats in Chapter 15 (esters)
- Amino acids and proteins in Chapter 16 (immediately following amides)

This approach has the advantage of presenting interesting topics ("the fun stuff") relatively early, mixed in with the regular functional-group chemistry. This not only creates interest, but it has the added advantage of not making these vitally important biological compounds seem separate and possibly even the province of another discipline! Further, by the time Chapter 19 (Introduction to Biological Chemistry) is reached, students are already familiar with the simple amino acids, proteins, fats, carbohydrates, and nucleic acids. The purpose of this final, single chapter is not to give an overview of all of biological chemistry—that would be simply too much. Instead, I have chosen three interesting topics (nucleic acids, metabolism, and medicinal chemistry) as examples of biological chemistry, hoping to incite curiosity and instill appreciation.

The Issue of Relevance

It is fair to say that properly motivating a student to study organic chemistry is half the battle. Organic chemistry has its own rewards, as we in the field are very much aware. To try to convey some of the excitement I feel about organic chemistry and to demonstrate its relevance, I have begun *Invitation to Organic Chemistry* with the essay **What Is Organic Chemistry?** In addition, interspersed throughout the text are 53 vignettes called **Chemistry at Work.** These place chemistry in several contexts, including *In the Marketplace*, *In the Body*, *In Nature*, and *In the Environment*. This special feature is designed to help students connect chemistry with the real world. To solidify this connection, most chapters end with a **Conceptual Problem** designed to give students a sense of how what they've just learned might benefit them in a real-world situation.

Problem-Solving Emphasis

Perhaps one of the greatest challenges facing an instructor of organic chemistry is to prepare students to approach problem solving in a way somewhat different from what they became accustomed to in the study of general chemistry. To help students visualize the various paths organic syntheses can take, I refer them to the **Key to Transformations** that appears on the inside front cover of the book. Then in the text itself, I introduce students to the problem-solving strategies needed to help them think through an organic synthesis. Strategy sessions called **How to Solve a Problem** walk the student through the process of analyzing and solving problems typical of those found in the chapter. It appears as though the author is thinking out loud as a problem is tackled.

It is my hope that students will study this material to the point of being able to use it—to apply it to problems. This textbook has 2490 separate problems of varying degrees of difficulty. The **Problems** within the chapter generally involve single steps in applying the reaction being discussed. The **Additional Problems** found at the end of the chapter are subdivided into sections that test specific concepts from the chapter, as well as a section of *Mixed Problems* of varying degrees of complexity. To make the instructor's selection of homework assignments easier, the relatively more challenging problems are denoted by an orange block. As mentioned earlier, most chapters end with an interesting **Conceptual Problem**, contributed by a colleague, that offers the student an opportunity to see how organic chemistry might be used in a real-world setting. (Solutions to all the problems are included in the ***Study Guide and Solutions Manual*** that accompanies the text.)

A Comprehensive Teaching Package

The text is supported by an impressive package of versatile ancillary materials. The ***Instructor's ToolKit CD-ROM*** offers you all the tools you will need to develop your own customized presentations. The ToolKit can serve as your single resource for preparing new lecture presentations, or its materials can be integrated into the presentations you use now. It offers lecture outlines in both an editable PowerPoint format and in a browser-based WebCD format. Available for use in conjunction with either set of outlines are twelve QuickTime Mechanism Movies as well as an Image Bank of the text's illustrations. The Image Bank art can be printed on acetates to create your own overhead transparencies. The WebCD outlines are also linked to the Gallery of Chime™ Molecules. **All of the materials available on this *Instructor's ToolKit CD-ROM* can be posted directly to a web site so that your students can have access to your lecture outlines and notes.**

PowerPoint Lecture Outlines

The Toolkit includes over 250 PowerPoint slides prepared by Gita Sathianathan (Pennsylvania State University). The slides outline all the major topics of the text and have their own illustrations. You can edit these slides in PowerPoint to better fit your individual needs, adding in your own PowerPoint materials or connecting to the electronic images from the Image Bank or the twelve Mechanism Movies.

WebCD Lecture Outlines

Prepared by Jason Stenzel (Southern Connecticut State University), these complete, ready-to-use outlines cover all the topics covered in the text. Because the WebCD outlines are browser-based, you can simply click on the embedded icons and be linked directly to figures from the Image Bank, the Gallery of Chime™ Molecules, and the Mechanism Movies.

Gallery of Chime™ Molecules

Developed by David Woodcock (Okanagan University College of British Columbia), the Gallery showcases more than 300 interesting and relevant molecules. They range from simple molecules, such as water and methane, to more complex ones, such as buckminsterfullerene and paclitaxel (taxol). Icons placed next to the chemistry in text let you know which molecules are included in the Gallery. Using the free browser plug-in, Chemscape Chime™ (included on the CD-ROM), you can display these virtual molecules for your students, manipulating them in a variety of ways:

- Display the molecules in wire frame, stick, ball-and-stick, and space-filling model formats
- Switch from two-dimensional to three-dimensional representations
- View stereo displays and dot surfaces
- Freely rotate a molecule, either manually or automatically, so as to view it from all angles

- **Mechanism Movies**

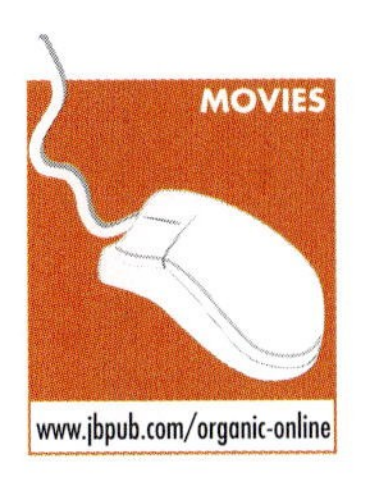

This collection of twelve Mechanism Movies were developed by Metec Ltd. exclusively for *Invitation to Organic Chemistry.* You can include these QuickTime animations as part of your lecture, to help your students visualize certain key reaction mechanisms. The Movie icon appears next to the discussion of these mechanisms in text. Reactions include the chlorination of methane, elimination reactions, nucleophilic substitution reactions, carbonyl addition, and several others.

- **Lecture Demonstrations**

Owen Priest (Hobart and William Smith Colleges) has written a series of 48 easy-to-do, fun, and conceptually interesting classroom demonstrations. These include the preparation of guncotton, the reduction of methyl violet, the combustion of hydrogen, as well as many others. Each demonstration is correlated to a specific numbered section of the text and includes a list of equipment and materials, a quick outline for carrying out the demonstration, cautionary notes on the procedure where appropriate, and questions to stimulate class discussion.

- **Image Bank**

The Image Bank makes the text's illustration program available to you for your electronic presentations. You can also use this beautiful collection of molecules, orbitals, graphs, and reaction diagrams to create overhead transparencies.

- **Computerized Test Bank**

Jason Stenzel (Southern Connecticut State University) wrote more than 800 questions for the Test Bank. The software of Brownstone's Diploma Computerized Test Bank offers test-generation, grading, and online testing capabilities. Use this extensive Test Bank as the basis for your own tests or to provide to your students as self-tests to measure their progress. The software can also be used to complement your own customized quizzes or tests. The network-testing component allows you to create a test that is available to the students online, either through the Internet or an existing intranet.

A Comprehensive Learning Package for Your Students

Invitation to Organic Chemistry also offers your students a full complement of resources to help ensure success in their study. This includes a ***Study Guide and Solutions Manual,*** as well as a free web site, Organic Online (www.jbpub.com/organic-online), which comes complete with web exercises, online quizzes, and a career resource center, plus the Gallery of Chime™ Molecules and Mechanism Movies. Also available through Jones and Bartlett are the tutorial CD-ROM ***CHEM TV: Choices*** by Betty Luceigh (UCLA) and molecular model kits. A complete description of the package is included in "To the Student."

To order any of these products, please call your Jones and Bartlett Publishers' Representative at 1-800-832-0034 or contact us through our home page at www.jbpub.com.

To the Student

An Invitation to Organic Chemistry

Welcome to organic chemistry—a subject I find fascinating, as have thousands of students I have taught. Many of them chose career paths that required them to take a course in organic chemistry. It's hard to know just what you will find yourself doing after you complete your college education. But the fact that you're here, taking an introductory course in organic chemistry, suggests to me that in some way your effectiveness in your career will be tied to an understanding of the essentials of organic chemistry. So you need to give careful thought and attention to the study you now undertake. I also hope that you will come to see what makes the subject so fascinating. For it to be fascinating to you means that the material needs to be presented in a careful manner, progressively building upon a solid foundation while remaining understandable and interesting. I believe that ***Invitation to Organic Chemistry*** has achieved that goal. Your success in learning organic chemistry lies primarily in your hands and is mainly determined by how you approach the subject right from the beginning. Please read this Preface carefully—I want to help you get started on the right foot.

There will appear to be a lot of ground to cover, and at first glance, your task may seem daunting. The task can be made easier, however, when you focus on relationships—on families of organic compounds that share common traits. Being able to recognize a compound as belonging to a certain family will significantly minimize the amount of memory work required. You know that every subject requires some memorization—certainly of the basic terminology, concepts, and major principles—but developing an understanding of why something occurs is the key to success in organic chemistry.

My recipe for your success in this subject, which has been well tested, includes the following four specific suggestions:

- **Keep up.** This is a subject that builds on what you learn each step of the way. Therefore, your cumulative understanding in this course will depend on your comprehension of the material covered earlier. Most students have discovered that organic chemistry is ill suited for last-minute cramming. Leaving study and problem-solving practice until just before examinations is a recipe for failure for many students.

- **Develop problem-solving skills.** You probably have become accustomed to problems of the type associated with general chemistry or physics. Such problems are often mathematical and involve the application of formulas. Problems in organic chemistry are very different and, at this level, are generally nonmathematical. Instead, they call for reasoning and deduction. What makes the process even more interesting is that there may be more than one appropriate solution. Starting in Chapter 3, you will be introduced to problems that involve converting one compound into another. I suggest you consider the following when faced with such problems:

1. *Analyze the problem* before searching for solutions. Do this so that you clearly understand what is at the heart of the problem. Identify just what portion of each chemical structure changes as one compound is converted to another, and what remains unchanged. The text includes many examples of problem analysis in a feature called **How to Solve a Problem.** Work with these to help develop your own approach.
2. Once you have identified what changes in the chemical reaction, *determine what overall transformation must occur*. Do this initially without worrying about the particular reagent(s) required. In other words, ask yourself *what* must change in the structure, not *how* it must happen.
3. *Look for similarities* to other transformations with which you have already become familiar.
4. Now try to *visualize a hypothetical reagent* that can bring about the change identified in steps 1 and 2.
5. *Write the reaction* as you now envisage it. Would the chosen reagent really be expected to produce the desired product? Would you really expect that reagent to react with the starting material? Why?
6. If you get stuck, *don't just look up the answer*. Instead, first struggle with the problem awhile. For instance, jot down the kinds of reactions the starting compound typically undergoes. Alternatively, jot down the known preparations of the kind of compound you are trying to prepare. You will learn more from struggling for awhile than from simply looking up the answer.
7. When doing a multistep synthesis (that is, a problem that asks you to prepare compound C starting with compound A), *think backwards* from your product C (a process called *retrosynthesis*). Think of the different means of preparation you have studied for C and the kinds of compounds used as starting materials. Do you recognize compound A as a starting material? If not, then you may have more than one step to consider. Apply the same thinking going from your desired product C to compound B, then from B to starting material A.
8. Finally, *reread this problem-solving section* many times during your study. You will see many concrete examples of the steps just described in the example problems of *How to Solve a Problem.*

■ **Practice.** This textbook provides many opportunities for you to test yourself by working on problems—there are 2490 individual problems! Only by tackling problems can you really determine for yourself whether you have a command of the material. Don't come out of an exam saying "I thought I knew that stuff" when you really had not tested yourself beforehand. Use the many examples provided in the *How to Solve a Problem* feature of this textbook to guide you—they not only include answers, but will teach you how to develop the thinking skills needed to solve real problems. Use them, understand them, and then try that approach yourself.

The only way to determine if you have learned organic chemistry is to try to use it in solving problems. The **Additional Problems** at the ends of chapters are of varying degrees of difficulty to give you practice. When a problem number has an orange block, that signifies that the problem is a more challenging one.

■ **Focus on principles.** This textbook and your instructor will present you with numerous examples of reactions—the total number may seem mind-boggling. The importance of any given reaction is not its specific example, but rather its *type*. I strongly urge you not to try to memorize examples—there are simply too many of them. Instead, learn the principles and then practice applying them to specific examples, including those in the problems. This textbook will assist you in learning how to *use* the reactions you will study in circumstances you have not seen before, including those that may appear on examinations.

When you have completed your study of organic chemistry using *Invitation to Organic Chemistry*, you should be able to do the following:

- Given an organic structure, derive its name, or, given the name of an organic compound, deduce its three-dimensional structure
- List the major families of organic compounds
- Understand the electronic structure of each family of organic compounds and the influence of that structure on the chemical and physical properties of the compounds
- Write the elementary reactions that permit transformations between the families of organic compounds
- Use these transformations to design elementary organic syntheses
- Determine the structure of new or unknown compounds from chemical, instrumental, and spectroscopic information

If you can work through and understand the problems incorporated into the text, then all of these goals will be within your reach.

To help you keep the details in perspective and not miss the "big picture," I suggest you use the **Key to Transformations** printed on the inside front cover of this textbook as a guide. You will then see how the specific chemistry you are learning at the moment fits into a more global scheme of transformations that connect one organic family to another. You will see that the lefthand portion of the Key applies to the content covered in Chapters 3–7, while the righthand portion applies to the content covered in Chapters 12–15. Although it may look daunting now, you will come to appreciate how much this overview can help as you move through the course. It does, in fact, summarize most of the transformations covered in this book.

Finally, so that you won't lose sight of the value of your study, I have included special features entitled **Chemistry at Work.** In each chapter, these vignettes introduce applications of organic chemistry—*In the Marketplace, In the Body, In Nature,* and *In the Environment.* The **Conceptual Problems** at the ends of most chapters will let you try your hand at some problem solving in a real-world context. In time, you will come to appreciate how important the organic chemistry you are learning now is to an understanding of the world around you and to the career opportunities available to you.

A Comprehensive Learning Program

Invitation to Organic Chemistry is accompanied by a learning program that contains all the resources you will need to make your time with organic chemistry exciting and successful. This includes a powerful web site, where you will find links that allow you to view a molecule in three dimensions, see animations of key reactions, take a test online, explore other web sites that relate to what you are learning in the text, and search through a career resource center. Also available is a comprehensive printed ***Study Guide and Solutions Manual*** that further amplifies the material covered in the text. Finally, enhance your grasp of the three-dimensional nature of organic chemistry with the **Organic Chemistry Models Kit**, an inexpensive and portable way to help you visualize molecular structures. Jones and Bartlett truly ***invites*** you to experience a successful semester of organic chemistry.

Organic-Online

Hosted by the publisher at www.jbpub.com/organic-online, the free web site lets you learn about organic chemistry and explore its possibilities in a variety of ways:

- The **Online Quiz** written by Bill Johnson will help you to prepare for your tests and to review some of the more important concepts in each chapter. Each chapter's quiz contains a set of questions that you can take as a self-test to check your progress. Any incorrect response will refer you to the text page you need to read over before you try that question again.
- The **Career Resource Center** created by Brenda Kesler (San Jose State University) will help you find out more about the career(s) you are interested in or about the kinds of careers available for those taking a short course in organic chemistry. You will be linked to current, specific information provided by real-life employers.
- **Chem Web Explorations** prepared by Brenda Kesler (San Jose State University) and Karen Downey (Kenyon College) are exercises keyed to specific topics covered in the text. These are signaled by the appearance of the Explorations icon in the text's margin. At the Organic-Online web site, you will be asked a question and then provided with a link to an outside web site that will help you answer that question.

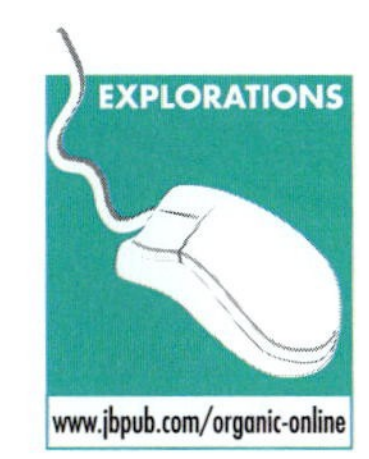

- The **Gallery of Chime™ Molecules** developed by David Woodcock (Okanagan University College of British Columbia) showcases more than 300 interesting and relevant molecules. They range from simple molecules, such as water and methane, to more complex ones, such as buckminsterfullerene and paclitaxel (taxol). Icons placed next to the chemistry in text let you know which molecules are included in the Gallery. Using the free browser plug-in, Chemscape Chime™ (downloaded off the publisher's web site), you can interactively observe these virtual molecules in a variety of ways:

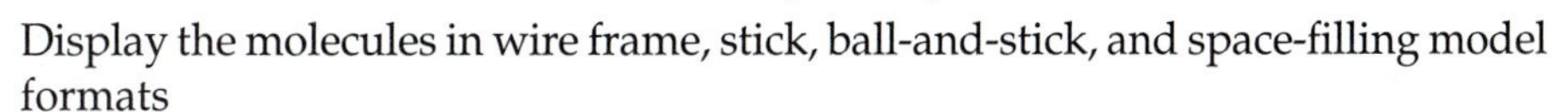

  Display the molecules in wire frame, stick, ball-and-stick, and space-filling model formats

  Switch from two-dimensional to three-dimensional representations

  View stereo displays and dot surfaces

  Freely rotate a molecule, either manually or automatically, so as to view it from all angles

- The **Mechanism Movies** on the web site were developed exclusively for this textbook. These QuickTime animations will help you visualize some of the important reaction mechanisms covered in the text. Look for the Movies icon in the text margin. Some of the reactions featured are the elimination and substitution reactions, the chlorination of methane, and the addition of HX to an alkene.

MOVIES
www.jbpub.com/organic-online

CHEM TV: Choices

Created by Betty Luceigh (UCLA), ***CHEM TV: Choices*** is a powerful interactive CD-ROM that offers a tutorial on various topics of organic chemistry in a fun, inviting atmosphere. Here, participate in a memory game to reinforce the concepts you have learned, or take a trip to a virtual chemistry lab, with a toolbox full of reactions that you have chosen, and perform the reactions in complete safety. Other features include interactive self-tests, concentration drills (matching), structure surveys, selection drills, and a unique example of organic chemistry as it relates to the human body. To learn more about these features and the availability of this CD-ROM, visit the publisher's web site at www.jbpub.com.

- **Study Guide and Solutions Manual**

Written by the text author Bill Johnson, the comprehensive ***Study Guide and Solutions Manual*** completes the organic chemistry learning program. It provides an outline of all the headings in each chapter of the text, states the learning objectives for each chapter, includes a glossary of all the boldfaced terms found in the text, and provides the solutions to all problems from the text, as well as answers to the Conceptual Problems.

- **Organic Chemistry Models Kit**

Create the three-dimensional structures of the molecules you learn about in class and from the text. The model kits are available as an inexpensive supplement packaged with the text or can be purchased separately.

To order any of the last three items, please call the Jones and Bartlett customer service department at 1-800-832-0034 or contact us through our home page at www.jbpub.com.

Acknowledgments

Text Review

To build a successful text and teaching tools, it takes extensive collaboration. I'd like to thank the following reviewers for their suggestions and constructive criticism:

Albert Burgstahler, *University of Kansas*
Allan R. Burkett, *Dillard University*
James W. Burns, *Utah State University*
H. Keith Chenault, *University of Delaware*
Stella Elakovitch, *University of Southern Mississippi*
David Erwin, *Rose Hulman Institute of Technology*
JoAnne Grant, *Middlesex Community Technical College*
Sarah A. Green, *Michigan Technological University*
Robert A. Hill, *Glasgow University, Scotland*
William J. Kerr, *Strathclyde University, Scotland*
Brenda Kesler, *San Jose State University*
Dorothy Kurland, *West Virginia Institute of Technology*
John Leitzel, *Chicago State University*
Anita Maguire, *University College, Cork, Ireland*
Rita Majerle, *South Dakota State University*
Jim Maxka, *Northern Arizona University*
Jimmie Mays, *University of Alabama, Birmingham*
Patricia Moyer, *Phoenix Community College*
Roger Murray, *University of Delaware*
Ken Musker, *University of California at Davis*
David Nelson, *University of Wyoming*
Elva Mae Nicholson, *Eastern Michigan State University*
John Olson, *Augustana University College*
Michael Paton, *Edinburg University, Scotland*
Gita Sathianathan, *Pennsylvania State University*
John Sevenair, *Xavier University of Louisiana*
Jason Stenzel, *Southern Connecticut State University*
Laura Stultz, *Birmingham Southern College*
Eric Trump, *Emporia State University*
Charles Wandler, *Western Washington University*

Ancillary Development and Accuracy Checks

I'd like to acknowledge in particular the efforts of those individuals who have made direct contributions to the teaching/learning package that accompanies *Invitation to Organic Chemistry*:

- Owen Priest, Gita Sathianathan, Jason Stenzel, and David Woodcock for their contributions to the *Instructor's ToolKit*
- Brenda Kesler and Karen Downey for their contributions to Organic-Online
- Bruce Banks (UNC-Greensboro), Jim Burns, Loretta Dorn (Fort Hays State University), David Erwin, James Keefe (San Francisco State University), Brenda Kesler, Rita Majerle, Ken Musker, Gita Sathianathan, and John Sevenair for their contributions to the Conceptual Problems
- Ken Andersen (Professor Emeritus, University of New Hampshire) and Dorothy Kurland for their work in helping me assure accuracy of the text, and Ann Moody (Truman State University) for her work in helping me assure accuracy of the *Study Guide and Solutions Manual*

Last, But Not Least

I want to thank many faculty and students, in numbers too great to recount, who have influenced my thinking about the teaching of organic chemistry. They are part of the 41 years of experience that have shaped this textbook. I offer special thanks to the University of North Dakota and its Chemistry Department and to the Chemistry Department of the United States Military Academy. They provided me with the environment, the encouragement, and the time needed to develop the approach used in this textbook.

I extend sincere thanks to the many people at Jones and Bartlett who, over the past two years, had a hand in making this book a reality. Special acknowledgment is owed to Mary Hill, who applied her expertise in guiding this project from beginning to end. I am also grateful to Jane Hoover, of Lifland et al., Bookmakers, who efficiently shepherded the book through copy-editing and production.

Finally, I acknowledge the support of my wife, Joan. She put up with long periods of my attention being given first to the writing of the manuscript and then to the proofing of the book, even when we were supposed to have more time for each other!

I encourage users of *Invitation to Organic Chemistry* to contact me, through the publisher. I welcome your comments, but most especially your suggestions for its improvement in subsequent editions to better support the teaching of organic chemistry.

What Is Organic Chemistry?

- What is organic chemistry?
- What do organic chemists do?
- Why is organic chemistry important in this society?
- Why should you study organic chemistry?

AS YOU START this course, you are no doubt asking yourself questions such as the ones above. This introductory essay will begin to answer these questions, specifically addressing the first two. Within each chapter, you will find shorter essays entitled "Chemistry at Work" that will further stimulate your interest and curiosity and provide context for your study of organic chemistry. These essays will demonstrate the importance and relevance of organic chemistry—not just to society, but to your chosen field of study.

Chemistry is the branch of science devoted to the study of matter at the molecular level, focusing on the interactions of molecules and their atoms. All matter is chemical: the air we breathe, the water we drink, and the food we eat; all forms of life, whether bacteria, viruses, plants, or animals; and the clothing and building materials we use every day, whether natural (stone, wood), processed (rubber, steel, aluminum), or synthetic (plastics).

Chemists study the composition, properties, and transformations of matter, in both its naturally occurring and synthetic (human-made) forms. Richard Lerner, president of the Scripps Research Institute, describes chemistry as "the central science. Everything that goes on in biology or medicine has a chemical basis." There are four major fields of chemistry: *organic chemistry*, *inorganic chemistry*, *analytical chemistry*, and *physical chemistry*. There are also numerous subfields (for example, *organometallic chemistry*) and interdisciplinary fields (for example, *biochemistry*).

Organic chemistry is, simply put, the study of the compounds of one element, carbon (atomic number 6). Why has carbon alone been singled out as the basis of a separate field of study, while the other 111 known elements are studied in the field of inorganic chemistry? The first reason is historical: *Organ*ic chemistry initially involved the study of compounds that could be obtained from living *organ*isms (the "vitalism" theory), while inorganic chemistry was the study of compounds that originated from nonliving (nonorganic) matter. This distinction and theory became invalid in 1828, when Frederick Wohler demonstrated that urea, an "organic compound" excreted in the urine of mammals, could be produced by heating ammonium cyanate, an inorganic compound.

A second reason why organic chemistry is a separate field has to do with sheer numbers: Approximately 7 million different organic compounds are known at present, while there are "only" about 1.5 million known inorganic compounds. This large number of organic compounds arises from a

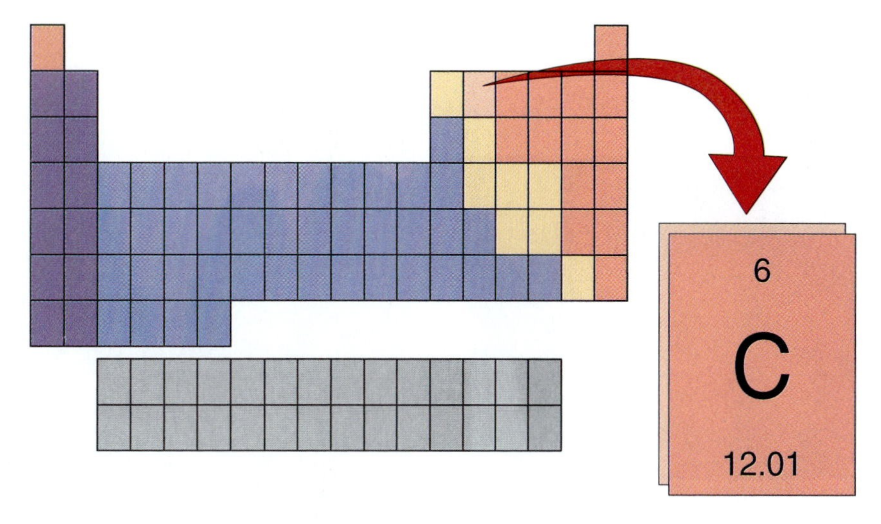

Although only one of 112 elements, carbon is the basis for the entire field of organic chemistry. Over 80% of all known compounds are classified as organic compounds.

unique property of carbon—its atoms can bond to one another in a virtually limitless number of arrangements. Organic compounds are typically rich in carbon and hydrogen, but many also contain one or more additional elements, usually oxygen or nitrogen but occasionally sulfur, phosphorus, or one of the halogens (fluorine, chlorine, bromine, and iodine). The subfield of organometallic chemistry focuses on compounds in which carbon is bonded to one of the metallic elements (for instance, lithium or magnesium).

Organic chemistry is a relatively "young" science, only about 200 years old. However, many activities known to ancient civilizations actually involved the practice of organic chemistry, for example:

- Soap making involves the base-catalyzed hydrolysis of fats (esters of carboxylic acids).
- The fermentation of sugars into wine involves chemical reactions of carbohydrates.
- The production of dyes depends on organic chemical processes and compounds.

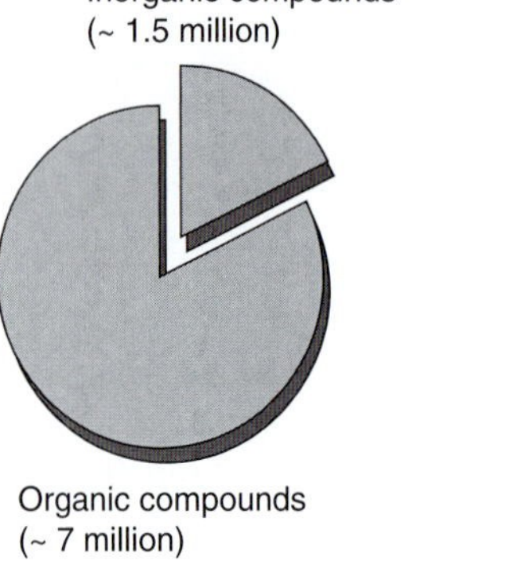

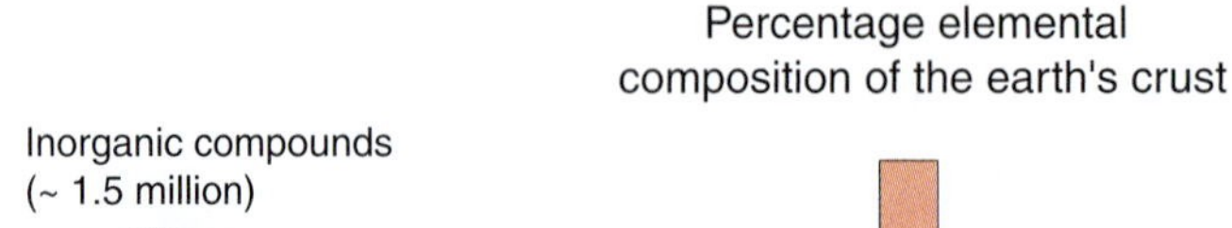

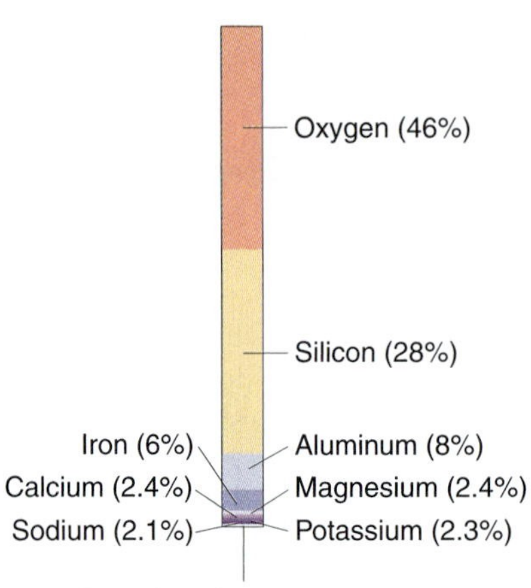

Of the 112 elements known, carbon is far from being the most abundant. It is lumped together with 103 other elements that comprise 1% of all elements in the earth's crust.

The use of certain organic chemicals also dates well back into history, for example:

- An extract from willow bark, a precursor to modern aspirin (acetylsalicylic acid), was used by country folk as a pain reliever in early England.

O
‖
C—OH

O—C—CH_3
‖
O

Acetylsalicylic acid
(aspirin)

The fermentation of grapes to produce wine, a chemical process, is almost as old as recorded history.

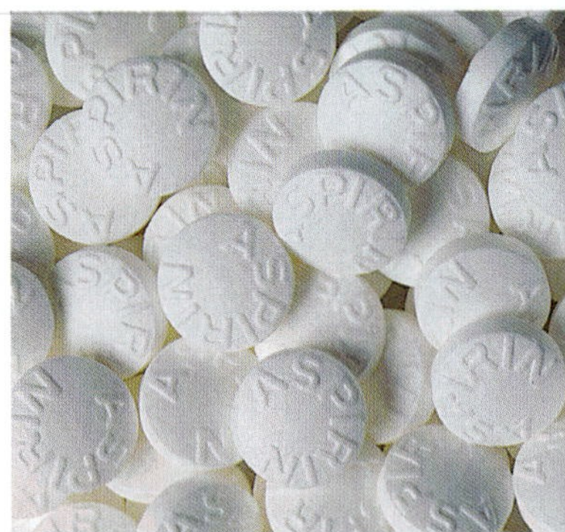

Millions of aspirin tablets are sold in the United States each year. The bark of a single willow tree might have supplied the pain-killing needs of a family in the 1800s.

- The chewing of coca leaves was known by South American tribes to yield an anesthetic effect—an effect now attributed to the presence of cocaine.

Cocaine

- Poisons were applied by various tribes to the tips of arrows to immobilize or kill prey. Epibatidine is one ingredient in the "cocktail" of chemicals excreted from glands on the back of the frog *Epipedobates tricolor* and used for this purpose by Ecuadoran tribes.

Epibatidine

Thus, it's safe to say that organic chemistry, while a relatively new science, is a very old subject.

Organic chemistry is a part of everything material that affects your daily life. Organic chemicals are involved in the biological processes that keep you alive. Aside from water and bone, your physical body consists mainly of organic chemicals: muscles and tissues (protein), blood constituents, hair and fingernails, fat, the glycogen in the liver and cells that provides quick energy, hormones, enzymes (such as those involved in metabolism), the substances responsible for vision, the chemicals that create nerve impulses, and the DNA that constitutes the genetic code in genes and chromosomes. Unnoticed, chemical reactions necessary to life are occurring all the time.

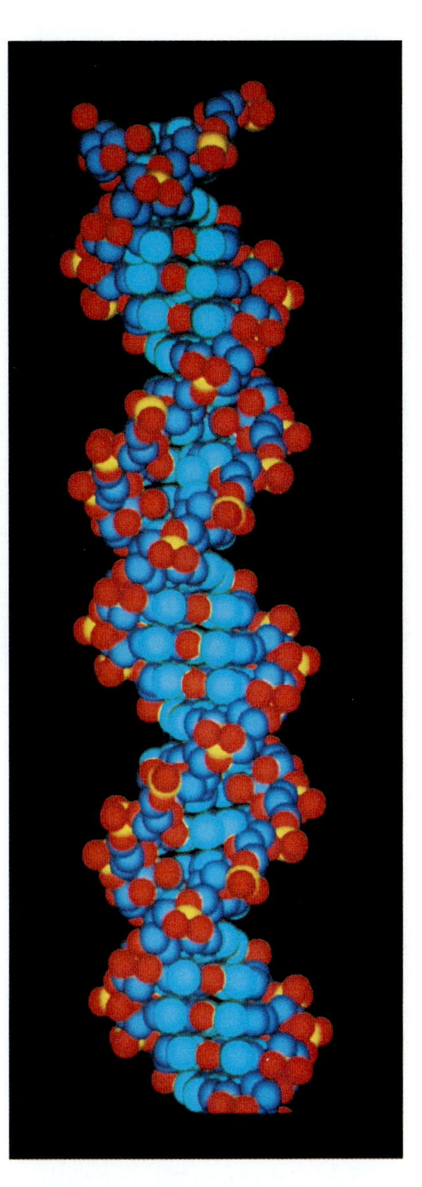

A computer rendering of a strand of DNA, the chemical composition of which contains the genetic code.

Organic materials also provide shelter. The clothes you wear are made from organic chemicals; some (cotton and silk) are naturally occurring, and some (polyesters and nylon) are synthetic. The materials used in the construction of houses are organic as well, and again some are naturally occurring (for example, wood) and some are synthetic (PVC piping, vinyl siding, plastic electrical boxes and wire covering, floor tile, and carpet). Automobiles are increasingly made of organic plastics rather than steel and aluminum, and the gasoline that propels them is organic.

The food you eat is organic (whether or not it carries the popular label "organically grown"), as are many of the substances that preserve food, that kill the insects that would destroy it, and that inhibit the growth of fungi and bacteria. Many of the containers in which food is sold are also made of organic materials, many of them synthetic. Medicines are virtually all organic chemicals, most of them synthetic. Even many vitamin supplements, although identical to the vitamins occurring in nature, were synthesized first in a chemical laboratory and then manufactured in a chemical plant.

The knowledge of chemical compounds and principles of organic reactions that you will aquire applies to understanding a wide variety of complex chemical processes, for example:

- The creation of new synthetic plastics such as Kevlar, which is used in motorcycle helmets and bulletproof vests

- The production of proteins by normal genes, the knowledge of which may prevent diseases such as cystic fibrosis

- The mechanism by which DNA mutations occur
- The synthesis and production of new additives for gasoline, such as MTBE (methyl *t*-butyl ether), which may help reduce pollutants in automobile exhaust
- The chemical processes involved in vision

Pollution caused by automobile exhaust is a worldwide problem. Newer cars run more efficiently, but for the many older vehicles on the road, newly developed gasoline additives, like MBTE, can help reduce the amount of harmful effluents in the exhaust.

Thus, it is important to understand the behavior of the amazing variety of organic chemicals, as well as the processes available for their synthesis and manufacture.

When faced with such a wide-ranging and far-reaching subject, it is fair to ask where and even how to begin. You must remember that science is a systematic study of the material universe, which for the most part has been found to be orderly, not chaotic. This means that the behavior of matter is quite predictable, based on a number of principles. Organic chemistry is not simply an aggregation of a lot of unrelated facts. Those 7 million known organic compounds can be categorized into a small number of general classes. The transformations that occur in organic chemistry as one compound is converted into another via a chemical reaction can be distinguished in terms of a limited number of functional groups and a handful of reaction mechanisms. From great multiplicity comes simplicity. You only need to learn the basic principles involved and recognize when to apply them in a variety of situations. You can make the principles of organic chemistry work for you.

It is not the purpose of this book to prepare you to be an organic chemist, although students who take this introductory course occasionally become so interested in the field that they pursue that career path. More likely, in ways you cannot yet anticipate, you will be using your knowledge of organic chemistry in your chosen field (perhaps dietetics, exercise physiology, petroleum geology, forestry, or wildlife management). Throughout this book, in features titled "Chemistry at Work," you will be introduced to some interesting applications of organic chemistry in daily life—organic chemistry at work in the marketplace, in the body, in nature, and in the environment. These, along with chapter-ending Conceptual Problems, will make it clear how organic chemistry and organic chemists may impact your chosen career. Any way you look at it, your study of organic chemistry will be easier if you understand how organic chemists approach their work.

What do organic chemists do? Organic chemists carry out three major kinds of tasks. One task of organic chemists is *determining the chemical structure of newly discovered compounds*. An organic chemist assumes the role of a "chemical detective," looking for proof of the molecular structure of a compound right down to the three-dimensional details. The compound in question may have been prepared in the laboratory or newly isolated from nature. Naturally occurring compounds are particularly interesting because of their enormous variety and phenomenal structural complexity. Many of these compounds are associated with some physiological activity; some find use as medicines.

In the early days of organic chemistry, the process for determining structure was very laborious and required substantial amounts of the compound on which to perform test reactions. Morphine, for example, obtained initially as a crude extract from the opium poppy, has a long history as a pain reliever. Although it was first isolated and purified in 1804, its molecular formula remained unknown for another 40 years, and its complete molecular structure was not determined until 1925.

Morphine

In contrast, a few milligrams of the relatively simple alkaloid epibatidine were isolated from glands on the back of a poisonous Ecuadoran frog in 1991. Its structure was determined by 1992, and several syntheses were developed in the next two years. The potentially effective anticancer drug *paclitaxel* (initially known by the name *taxol*) was isolated in minute quantities in 1962 from the bark of the Pacific yew tree, during a major national program to screen plant chemicals as anticancer drugs. Its structure was determined in 1971, and a total synthesis was achieved in early 1994.

Paclitaxel (taxol)

A second major task undertaken by organic chemists is *to determine how chemical structure relates to physical, chemical, and biological properties.* What is it, for example, that makes a compound such as aspirin or morphine an effective painkiller and penicillin an effective antibiotic? Or, what is it that makes one compound red and another yellow? Finally, what is it that makes one plastic suitable for soft materials, such as those used in clothing or plastic packaging, while another plastic is hard enough to use in a bulletproof vest or an aircraft windshield? The answer is the same in every case—it is the chemical structure. As knowledge accumulates about the relationship of the structures of organic compounds to various properties, chemists are better able to produce compounds tailored to have certain desired properties.

Commercial and societal needs drive much of the chemical research that focuses on structure and properties. For example, automobile manufacturers have utilized new plastics to replace the metals once used for structural components in cars and trucks, thereby reducing weight and sometimes cost. Some new automobile engines even have plastic manifolds. Another example is the design of drugs to minimize undesirable side effects. For instance, a variety of penicillin antibiotics are now available, permitting their use by people who are allergic to the original peni-

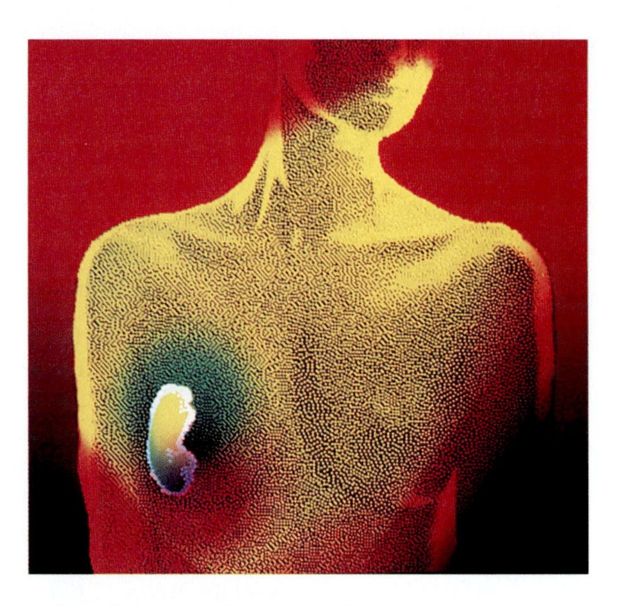

A three-dimensional MRI shows the presence of cancer in the right breast. Paclitaxel is one of the more promising drugs now available for the treatment of advanced breast and ovarian cancers.

cillin G. Finally, in response to the need for replacing the banned CFCs formerly used in air-conditioning units, chemists have designed and synthesized compounds with the very specific properties required of safe and effective refrigerants.

A third task undertaken by organic chemists is the *synthesis of organic compounds, starting with simple, readily available chemicals.* Such synthetic routes are often developed in chemical laboratories but then are "scaled up" to permit large-scale production in chemical manufacturing plants. There are several reasons why organic compounds are synthesized:

- Synthesis makes available reasonable quantities of compounds that are extremely scarce in nature (examples are the painkiller epibatidine and the anticancer drug paclitaxel). This enables researchers to conduct appropriate testing to determine the effectiveness of the compound, and later ensures its market availability, if warranted.
- Synthesis of non–naturally occurring organic compounds with useful physical or chemical properties makes them available for specific commercial purposes (a recent example is the synthesis of new refrigerants).
- Chemists may synthesize a previously unknown compound simply to study its properties or to test a particular chemical theory.
- Synthesis may even be attempted purely as an intellectual challenge—to make a compound that theory predicts cannot exist, or that has resisted synthesis to date, or that has a unique structure or shape.

Snowboard makers rely on modern synthetic plastics (polymers) for fabrication of snowboards that are strong yet flexible at low temperatures.

Some organic syntheses produce compounds that are relatively small and destined for mass production. A recent example is the compound designed to replace Freon-12 (CF_2Cl_2, a CFC, or chlorofluorocarbon), which has been used for decades as a refrigerant in many air-conditioning systems. Its replacement, a new compound that contains no chlorine and is called HFC-134a ($C_2H_2F_4$, a hydrofluorocarbon), was first synthesized in the laboratory and is now produced in large chemical plants.

In contrast, some organic syntheses produce very large and complex compounds originally isolated from plants or animals. The amount available from the natural source is often too small to test the compound's usefulness, let alone to supply potential users. Such compounds are usually destined for smaller-scale production. The antibiotic *penicillin G*, first isolated from a mold found in an old petri dish in 1928 by Sir Alexander Fleming, is now readily available through synthesis.

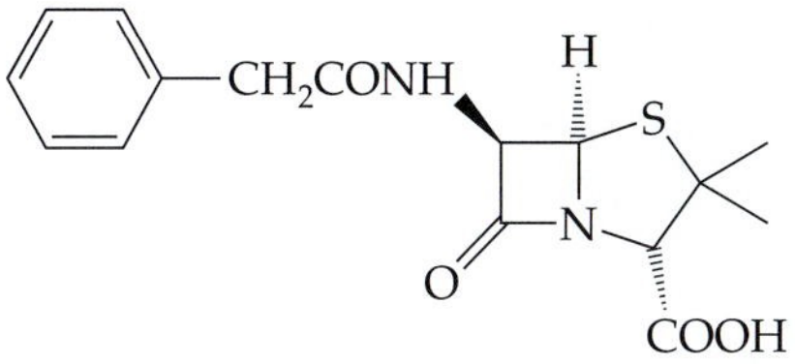

CH_2CONH H S N O COOH

Penicillin G

Cortisone, the first of the medically applied steroid drugs, was initially available in only small quantities from bovine brains, but it, too, became widely available as a result of synthesis.

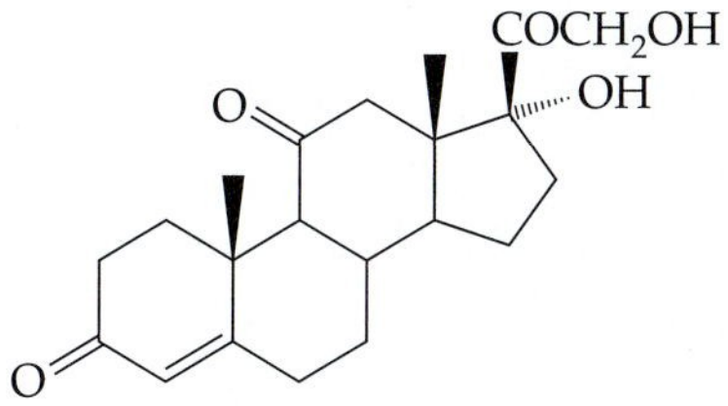

$COCH_2OH$ O OH O

Cortisone

More recently, the anticancer compound *paclitaxel* (shown earlier) has been synthesized in quantities large enough for medical testing. Originally, just milligrams were obtained from the bark of a single Pacific yew tree, and peeling the bark to harvest the paclitaxel killed the tree.

The synthesis of a compound such as paclitaxel is an extraordinarily complex task, requiring numerous steps, each using a chemical reaction whose outcome is absolutely predictable. Before this very careful laboratory work can even begin, a design for the synthesis must be formulated, based on a full knowledge of all the possible reactions that might be employed. You could liken such planning to the strategic thinking of a chess master, thinking through potential moves as a game unfolds. A command of chemical knowledge and a certain wizardry of conception are essential for such a project to succeed.

Organic chemists are very skilled at conceptualizing solutions to problems. The problem may be puzzling out a chemical structure from observed properties; it may be creating from scratch a compound that has specific desired properties not known to exist in naturally occurring compounds; it may be determining the most cost-effective or environmentally sound means of manufacturing a naturally occurring compound. Understanding how organic chemists approach these problems will assist you in understanding the involvement of organic chemistry in your career field. Some of the interesting problems that organic chemists have tackled over the years are described in the short essays throughout this textbook; some of these problems and their solutions may surprise you.

Knowledge of the fundamental characteristics of families of organic compounds is essential no matter what task an organic chemist undertakes. While no one can possibly know every single detail about every organic compound, general principles can be learned, and they facilitate the use of sources such as reference books and chemical journals, which can provide the missing details. This book is meant to provide you with enough of the essentials of organic chemistry to guide you in applying it to your chosen field. You will see how knowledge of chemical families and reactions is organized, sample some of these reactions, learn how to deduce chemical structure from experimental observations, and develop the ability to conceptualize simple organic syntheses. Most important, you will be able to communicate with others about organic chemicals and their behavior.

CHAPTER 1

Carbon Compounds: Bonding and Structure

FUNDAMENTAL TO CHEMISTRY is the nature of the chemical bond. Bonds are the forces that hold atoms together in specific arrangements to form molecules. Under appropriate experimental conditions, bonds break and remake—that is, a chemical reaction occurs. You already know something about chemical bonds from your study of general chemistry, so the early sections of this chapter offer only a brief refresher.

Because this book's focus is on the chemistry of organic compounds, it is important to understand the chemistry of carbon. First, you need to understand the molecular structure of organic molecules—the ways carbon atoms bond to other atoms, including other carbon atoms. Second, you need to know how, during a chemical reaction and in the presence of various reagents, the bonding arrangements of carbon and other atoms in an organic molecule can change. Finally, you must come to appreciate how molecular shape can be critical to the ways molecules interact, especially in biological chemistry. You will learn to represent the structures of organic compounds and to recognize the transformations they undergo. All of these aspects relate directly to the distribution of electrons around carbon atoms and ultimately to the kinds of bonds that these atoms can form.

1.1 Chemical Bonding

1.1.1 The Atom and Its Structure

We begin our review with the *atom*, the fundamental unit of any element. In terms of overall structure, the atom can be divided into two parts (Figure 1.1).

At the center is the *nucleus*, which consists of positively charged *protons* as well as uncharged *neutrons*. (The hydrogen atom is an exception, with a nucleus containing only a single proton.) The number of protons in the nucleus of an element is described by its **atomic number (*Z*),** which is given in the periodic table of the elements (see the inside back cover of this book). Together, the protons and neutrons account for most of an atom's mass, even though they occupy only a small part of the atom's volume. The nucleus has a diameter of about 10^{-15} m; the entire atom itself has a diameter of about 2×10^{-10} m, or 2 Å (1 Å = 10^{-10} m). Given the tiny size of the nucleus, the atom consists largely of open space.

In the space surrounding an atom's nucleus are located negatively charged *electrons*. Because the number of electrons in an atom is equal to the number of protons, an atom is electrically neutral. The bonding of atoms to each other is implemented by the elec-

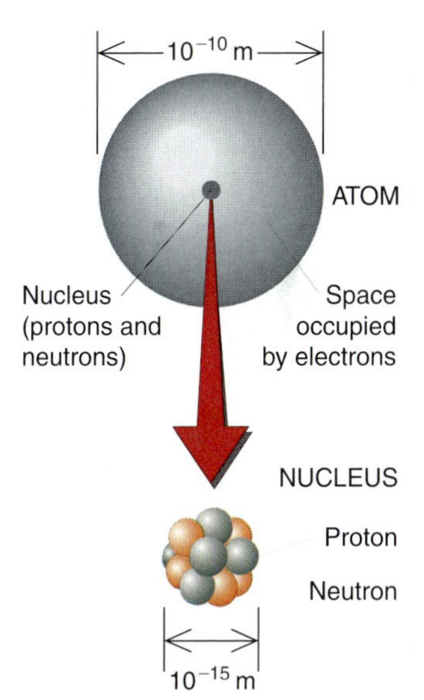

FIGURE 1.1

A schematic view of an atom, showing its small, dense nucleus surrounded by the much larger "cloud" of electrons.

◀ Cultivating a seedling of a Pacific yew tree—whose bark was the original source of the anticancer agent taxol, now known as paclitaxel. Synthesizing this important organic compound proved to be a challenge—see the Conceptual Problem at the end of the chapter.

trons, but only by those in the outer (valence) shell, sometimes referred to as **valence electrons.** It is the attractive force between the positively charged protons and the negatively charged electrons that holds an atom together.

The *mass number* of an atom is the number of protons plus the number of neutrons. Also characteristic of a particular element is its *atomic weight,* an average of the mass of many atoms of that element. The atomic weight is slightly different from the mass number because some atoms of an element contain different numbers of neutrons—these are *isotopes*. For example, hydrogen has an atomic number of 1 and a mass number of 1 but an atomic weight of 1.008, because atoms of a naturally occurring isotope of hydrogen, known as deuterium, have a neutron in the nucleus. Similarly, carbon has an atomic number of 6 and a mass number of 12 but an atomic weight of 12.01, because of the natural presence of some carbon atoms with a mass number of 13 (7 neutrons instead of 6).

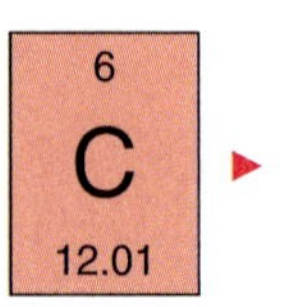

Electrons: A Closer Look

Electrons are positioned around the nucleus but not in a random fashion. Instead, they occupy distinct *energy levels,* called **electron shells,** which are numbered from the inside out with integers called **principal quantum numbers.** The first shell, that closest to the nucleus, represents the lowest energy level. The closer an electron is to the nucleus, the stronger the attraction exerted on it by the protons and the lower its energy. Shells located farther from the nucleus have higher principal quantum numbers; they increase in size, and the electrons within them are at higher energies.

The shells themselves are further subdivided into distinct regions of space called **atomic orbitals.** The first and smallest shell has only one orbital, the spherically shaped *s* orbital. The second shell contains an *s* orbital plus three *p* orbitals.

The *p* orbitals are dumbbell-shaped and are oriented at 90° (orthogonally) to each other, as though along *x*, *y*, and *z* axes. They are designated p_x, p_y, and p_z (Figure 1.2) and have equal energy, which is higher than that of the *s* orbital. There are also *d* orbitals, which first appear in the third shell, and *f* orbitals, which first appear in the fourth shell. We will be concerned only with the first and second shells, which can hold two and eight electrons, respectively, because the *s* and *p* orbitals play the major role in organic chemistry.

Electrons occupy orbitals in a predictable pattern (Figure 1.3), according to these three rules:

Rule 1 Orbitals of lowest energy fill first (the *Aufbau principle*).

Rule 2 A maximum of two electrons (of opposite spin) may occupy a single orbital (the *Pauli exclusion principle*).

Rule 3 When more than one orbital of the same energy (for example, p_x, p_y, and p_z orbitals) is empty, each of these orbitals will first acquire one electron before any orbital acquires a second electron (*Hund's rule*).

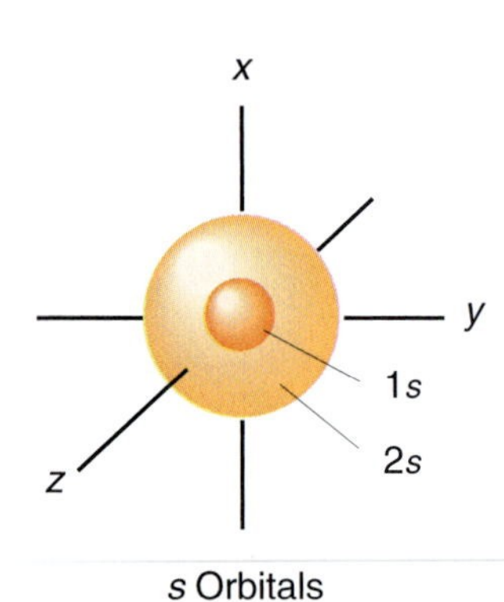

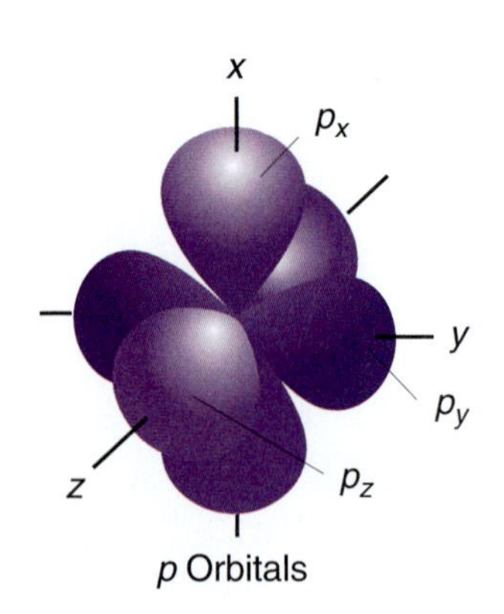

FIGURE 1.2

The shapes of the 1*s*, 2*s*, and 2*p* orbitals.

By applying these rules, you can deduce the ground-state arrangement of electrons about any atomic nucleus, the topic of the next section.

Electron Configurations

Each element has a distinctive configuration of electrons in orbitals—its **electron configuration.** The lowest energy state an atom can have is described as its *ground state.* The electron configuration of the ground state of atoms of any element is determined by applying the three rules listed earlier. The ground-state electron configurations for the first 10 elements (Z = 1, . . . , 10) are given in Table 1.1.

Table 1.1 includes a form of shorthand used to describe the electron configuration of any element. The electron configuration for carbon (Z = 6) is $1s^22s^22p^2$. The number preceding each letter is the principal quantum number (indicating the shell), the letters represent the orbital shape, and the superscript numbers indicate how many electrons are in those orbitals. In the case of carbon, each of the two *p* electrons is located in one of two different (but equal-energy) *p* orbitals. The electron configuration of sodium (Z = 11) is written as $1s^22s^22p^63s^1$ and that of argon (Z = 18) is $1s^22s^22p^63s^23p^6$.

The orderly progression of energy levels and the way in which they are filled in different atoms (see Table 1.1) are reflected in the structure of the periodic table (on the inside back cover of this book). Elements are placed in *rows,* with the row number (or *period*) being the same as the principal quantum number of the outermost shell. For example, boron, found in the second row, has two shells, as reflected by its electron configuration $1s^22s^22p^1$. The vertical columns of the periodic table are designated as *groups.* The elements in any group share similar chemical properties and reactivities because they have exactly the same outer-shell configuration. For example, carbon (Z = 6), with electron configuration $1s^22s^22p^2$, and silicon (Z = 14), with electron configuration $1s^22s^22p^63s^23p^2$, are in the same group, both having four outer-shell (valence) electrons. These outer-shell electrons enter into the various bonding arrangements typical of these elements.

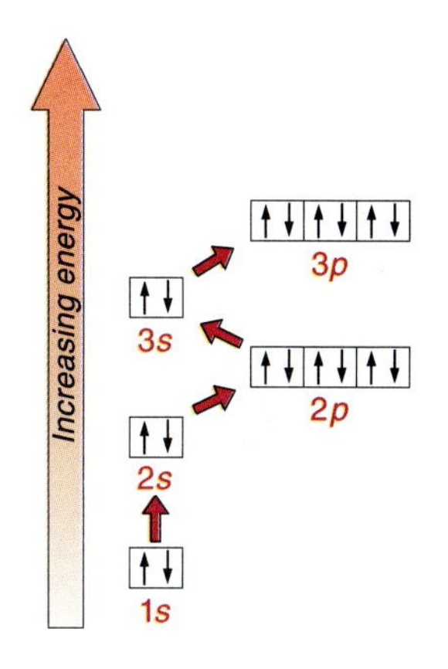

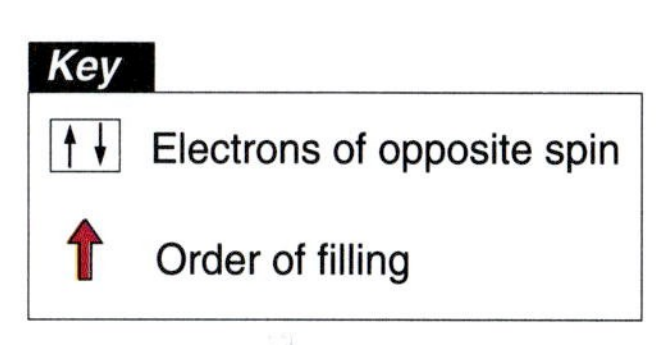

FIGURE 1.3

The order in which atomic orbitals are filled for the first three principal quantum numbers.

When you see this icon, link to the Web Explorations site of Organic-Online to find exercises and information on specific topics related to the material presented in text.

TABLE 1.1

Electron Configurations of the First Ten Elements

Element	Atomic Number	Electron Configuration	Number of Electrons in Each Orbital 1s	2s	$2p_x$	$2p_y$	$2p_z$
H	1	$1s^1$	1				
He	**2**	$1s^2$	**2**				
Li	3	$1s^22s^1$	2	1			
Be	4	$1s^22s^2$	2	2			
B	5	$1s^22s^22p^1$	2	2	1		
C	6	$1s^22s^22p^2$	2	2	1	1	
N	7	$1s^22s^22p^3$	2	2	1	1	1
O	8	$1s^22s^22p^4$	2	2	2	1	1
F	9	$1s^22s^22p^5$	2	2	2	2	1
Ne	**10**	$1s^22s^22p^6$	**2**	**2**	**2**	**2**	**2**

IIA
12
Mg

IVA
14
Si

VA
15
P

VIA
16
S

VIIA
17
Cl

VIII
18
Ar

How to Solve a Problem

Write the ground-state electron configuration for magnesium.

SOLUTION

Magnesium is in the third row of the periodic table, so the principal quantum number of its outermost electron shell is 3. The atomic number of magnesium is 12, indicating that it has 12 protons and 12 electrons. The first two of its electrons fill the first *s* orbital, which completes the first electron shell. The next eight of its electrons fill the 2*s* orbital and the three 2*p* orbitals, completing the second shell. Two electrons are left for the third electron shell, and they fill the 3*s* orbital. Thus, the ground-state electron configuration for magnesium is $1s^22s^22p^63s^2$.

PROBLEM 1.1

Write the ground-state electron configuration for (a) silicon, (b) phosphorus, (c) sulfur, (d) chlorine, and (e) argon.

It was in the configurations of Group VIII, the noble gases, that the key to chemical bonding was first discerned. These elements are *inert*—that is, generally chemically unreactive—because their outer shells are completely filled with electrons, a very stable configuration (see the configurations for helium and neon in Table 1.1). In 1916, based on this observation, the chemist G. N. Lewis proposed the **octet rule** (so named because it usually is concerned with just the eight outer-shell electrons).

Octet rule In forming bonds, atoms tend to acquire the electron configuration of the noble gas nearest to them in the periodic table.

As you will see in the next section, atoms bond with each other by transferring or sharing electrons to achieve filled outer shells.

1.1.2 The Chemical Bond

There are two kinds of chemical bonds: ionic bonds and covalent bonds. Critical to an understanding of all chemical bonds is the concept of **electronegativity,** the power of an atom to attract electrons. Linus Pauling developed a relative scale of electronegativity in which elements are assigned values from 0.7 to 4.0, with a higher value indicating a greater attractive force (Figure 1.4). Elements with high electronegativity values are described as *electronegative,* whereas elements with low electronegativity values are described as *electropositive.* You can use this scale together with the octet rule to understand why and how atoms bond the way they do.

Ionic Bonding

An **ionic bond** is the electrostatic attraction between a cation and an anion. An ionic bond forms when an atom with a high electronegativity meets one with a low electronegativity. If there is sufficient difference in the electronegativities, one atom completely loses its valence electron(s) to the other atom, resulting in the formation of a **cation** (an ion with a positive charge) and an **anion** (an ion with a negative charge). In accordance with the octet rule, the electropositive element tends to give up electrons so that a complete outer shell remains; the electronegative element tends to acquire enough electrons to complete its outer shell.

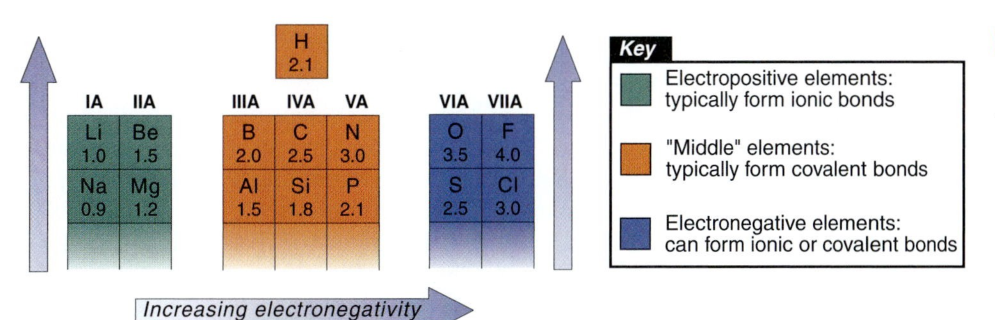

FIGURE 1.4

Pauling electronegativity scale for selected elements.

As indicated in Figure 1.4, the electronegativities of elements increase from left to right across the rows of the periodic table and increase from the bottom up within a group. The strongly electronegative atoms in Groups VIA and VIIA (nonmetals) readily form ionic bonds with the strongly electropositive atoms from Groups IA and IIA (metals). Ionic bonds are typical of inorganic salts, such as lithium fluoride (LiF). In the solid (crystalline) state, the ionic bonding in such salts is not really between any two specific atoms. Instead, each lithium cation is surrounded by six fluoride anions, and each fluoride anion is surrounded by six lithium cations, forming a typical salt crystal lattice (Figure 1.5). Salts such as lithium fluoride dissolve readily in water, existing in aqueous solution as separate cations and anions, each surrounded by water molecules of solvation.

FIGURE 1.5

Crystal lattice of lithium fluoride.

The reaction between elemental lithium and elemental fluorine to form the salt lithium fluoride occurs by the transfer of one electron from an atom of lithium to an atom of fluorine. This transfer can be depicted by using dots to represent electrons and circles for electron shells or by writing electron configurations (Figure 1.6). The neutral lithium atom, in losing one electron and thereby changing to a cationic state (Li^+), achieves an electron configuration identical to that of the noble gas helium. In gaining one electron, the fluorine atom goes from a neutral state to an anionic state (F^-) and achieves an electron configuration identical to that of the noble gas neon. Thus, both of these ions have a filled outer shell and a noble-gas electron configuration.

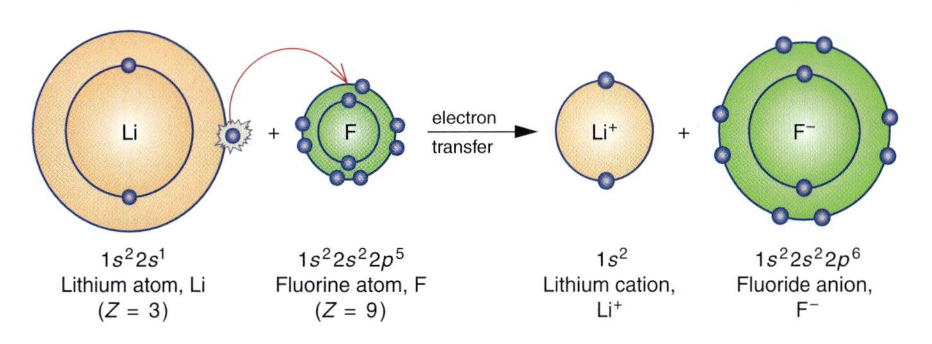

FIGURE 1.6

Electron transfer in the formation of lithium fluoride.

In summary, ionic bonding involves completing the outer (valence) shells of atoms by transfer of one or more electrons to the more electronegative atom(s) followed by electrostatic attraction between the resultant cation(s) and anion(s).

PROBLEM 1.2

Write the ground-state electron configurations for elemental sodium and chlorine. Then write the ground-state electron configurations for sodium and chlorine ions.

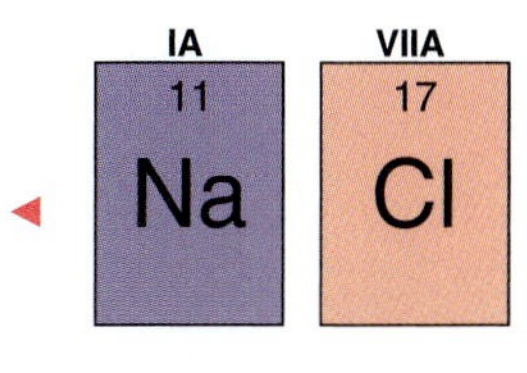

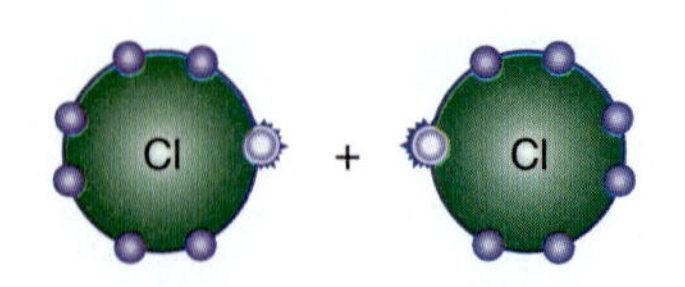

$1s^2 2s^2 2p^6 3s^2 3p^5$
Two chlorine atoms, Cl
(Z = 17)

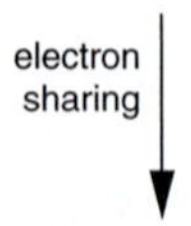

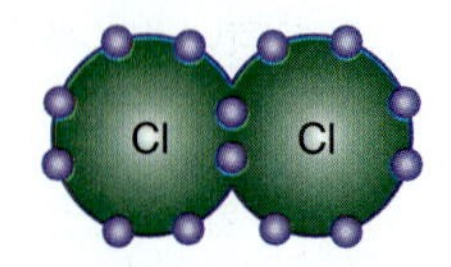

Chlorine molecule

FIGURE 1.7

Electron sharing in the formation of the covalent bond of the chlorine molecule.

Covalent Bonding

Whereas an ionic bond forms because of a significant difference in electronegativities between atoms of two elements, a covalent bond forms between atoms of elements in the middle of the electronegativity scale. These atoms do not gain or lose electrons; instead, they *share* valence electrons. A **covalent bond** is a bond that involves the sharing of a single pair of electrons between two atoms, which completes the outer shell of both. Usually, each atom contributes one electron, as illustrated for chlorine in Figure 1.7.

Covalent bond

$$2\,{:}\ddot{\underset{..}{Cl}}\cdot \xrightarrow[\text{sharing}]{\text{electron}} {:}\ddot{\underset{..}{Cl}}{:}\ddot{\underset{..}{Cl}}{:} \equiv \text{Cl—Cl} \equiv Cl_2$$

Chlorine molecule

In contrast to an ionic bond, a covalent bond forms between two specific atoms, such as two chlorine atoms. Typically, covalent bonding occurs between atoms that are not strongly electronegative or electropositive (those near the middle of the Pauling electronegativity scale) or that have similar electronegativities. Examples are the bonds in molecular hydrogen (H_2), molecular chlorine (Cl_2), water (H_2O), ammonia (NH_3), and methane (CH_4). In each instance, both atoms involved in each covalent bond have achieved a noble-gas configuration. These molecules may be represented by **Lewis structures,** in which only the valence-shell electrons are shown as dots around the element symbol:

H:H	:C̤̈l:C̤̈l:	H:Ö̤:H	H:N̤̈:H with H below	H:C̤̈:H with H above and below
H—H	:C̤̈l—C̤̈l:	H—Ö̤—H	H—N̈—H with —H below	H—C—H with —H above and below
Molecular hydrogen	Molecular chlorine	Water	Ammonia	Methane

Alternatively, the covalent bond between two atoms may be represented by a single line between the atomic symbols. *A line between two atomic symbols always represents two electrons in a covalent bond.* Sometimes it is convenient to mix the two representations, using lines for covalent bonds (shared electrons) and dots for unshared electrons.

A covalent bond is formed by the overlap in space of an atomic orbital of each atom, which creates a new orbital called a **molecular orbital.** Like an atomic orbital, a molecular orbital contains only two electrons, which must be paired (that is, have opposite spins). In the case of molecular hydrogen, the two hydrogen atoms approach sufficiently closely for their individual 1*s* orbitals to overlap, thus forming a molecular orbital in which the two electrons reside and are paired (Figure 1.8). The area of space where the electron is most likely to be found in an individual hydrogen atom is spherical in shape (the 1*s* orbital). In the hydrogen molecule, the highest electron density lies between the two nuclei, where the covalent bond forms. The resulting bond is called a **sigma (σ) bond,** and the shape the molecular orbital takes is cylindrically symmetrical about a line between the two nuclei. (If you imagine standing at the end of an H—H bond and looking from one hydrogen atom along the bond to the second atom, the orbital would appear to surround the bond evenly in the shape of a cylinder.)

Covalent bonds can form only because the bonding results in a molecule that is energetically more stable than are the two separate atoms. That is, there is a net loss of energy (gain of stability) that accompanies the formation of the bond. This amount of energy (called the *bond energy*) can be measured, and the same amount of energy is required to break the covalent bond.

The stability of any covalent bond is the result of a delicate balance between opposing forces. The pairing of the two electrons in the molecular orbital formed by the overlap of two atomic orbitals is energetically favorable (an attractive force). However, bringing the two atoms close enough together for the orbitals to overlap also sets up a repulsive force between their positively charged nuclei. The situation can best be described in terms of potential energy, as diagrammed in Figure 1.9. The original energy of the atoms is represented by 0 on the vertical (energy) axis and point *A* on the energy diagram. As the atoms get closer to one another, their atomic orbitals begin to overlap, and the energy decreases (point *B*). Then, at a certain distance, characteristic of each pair of covalently bonded atoms, a point of optimal balance between the attractive and repulsive forces is reached (point *C*). For two hydrogen atoms, this distance is 0.74 Å—the standard H—H bond length. If the atoms approached any closer (point *D*), the force of repulsion would outweigh the force of attraction. For the hydrogen molecule, the bond energy is 105 kcal/mol (point *E*) when the atoms are 0.74 Å apart (point *C*). This energy value is the net stabilization of the molecule over the two separate nonbonded hydrogen atoms. This value also means that 105 kcal/mol of energy is required to break the H—H bond in molecular hydrogen (quite a strong bond). This same amount of energy has to be supplied to a hydrogen molecule (H_2) for it to react with another atom or molecule—for example, with oxygen (O_2) to form water (H_2O).

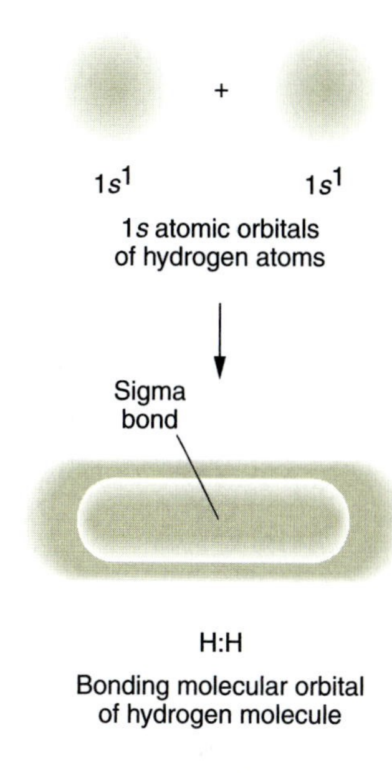

FIGURE 1.8

Atomic and molecular hydrogen.

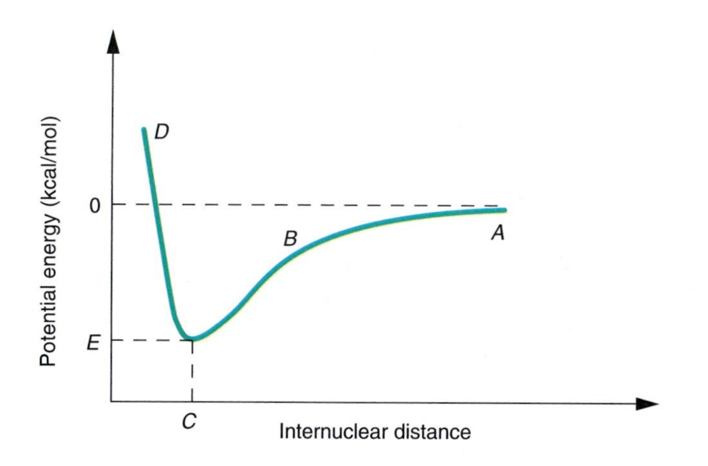

Key
A: no orbital overlap; no stabilization
B: some overlap; little repulsion, some stabilization
C: bond distance at which there is maximum net stability
D: nuclei are too close; repulsion outweighs attraction
E: energy of stabilization at optimal bond distance

FIGURE 1.9

Potential energy of a covalent bond as a function of internuclear distance.

1.1.3 Carbon Covalent Bonding

Carbon is located in the very middle of the electronegativity scale and generally undergoes only covalent bonding. (The same is true of some other elements, such as nitrogen and silicon, that are close to carbon in the periodic table.) The ground-state electron configuration of carbon ($Z = 6$), $1s^22s^22p^2$, indicates the presence of two unpaired electrons, one in each of two *p* orbitals. However, carbon normally forms bonds with four other atoms at a time (*tetravalent* bonding), such as in methane (CH_4), which has four identical C—H bonds. Such tetravalent bonding of carbon cannot be explained by simple sharing of the two unpaired *p* electrons. Instead, the bonding in carbon compounds is accounted for by a process known as **orbital hybridization.** In a typical carbon compound, several different orbitals *hybridize,* or mix their characteristics (including their shapes) to provide the most energetically stable molecular structure. Carbon can form three types of hybrid orbitals: sp^3, sp^2, and *sp*. We will start with structurally simple molecules, those involving just sigma bonds and the sp^3 hybrid orbital. (We'll consider the others in later chapters, when you need them.)

sp^3 Hybridization

To understand how a carbon atom can become tetravalent, you can envision the process as occurring in three sequential steps. In reality, bond formation involves the electrons spontaneously and simultaneously rearranging themselves to produce the most stable electron configuration and spatial arrangement (that is, molecular structure).

In the first step, one of carbon's two 2*s* electrons is promoted to the vacant higher-energy 2*p* orbital. This step requires an initial input of energy:

$$\underset{\text{Atomic carbon}}{1s^2 2s^2 2p^2} \xrightarrow[\text{(requires energy)}]{\text{promotion}} 1s^2 2s^1 2p^3$$

In the second step, the single 2*s* orbital and the three 2*p* orbitals (a total of four orbitals) *hybridize* to form four new identically shaped orbitals. These are called sp^3 hybrid orbitals because they are derived by mixing the characteristics of one *s* orbital and three *p* orbitals. (Note that the superscript 3 in the orbital designation refers to the number of *p* orbitals; this is different from an electron configuration, where a superscript number refers to the number of electrons in the orbital.)

$$1s^2 2s^1 2p^3 \xrightarrow[\text{(no energy change)}]{\text{hybridization}} \underset{\text{Tetrahedral carbon}}{1s^2 2(sp^3)^4}$$

The sp^3 hybrid orbitals have a shape characteristic of a mixture of one part *s* (sphere) and three parts *p* (dumbbell), as shown in Figure 1.10. The hybridization step occurs without much energy change.

Each sp^3 orbital is oriented around the carbon nucleus at 109.5° from any of the other three sp^3 orbitals. In other words, sp^3 hybridization produces orbitals that are oriented

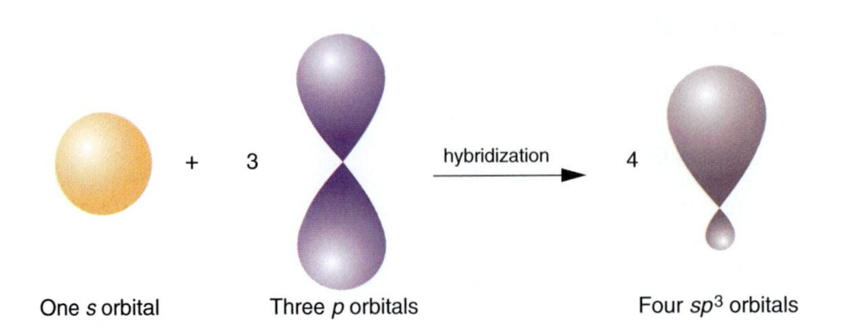

FIGURE 1.10 Tetrahedral (sp^3) hybridization of atomic orbitals of carbon.

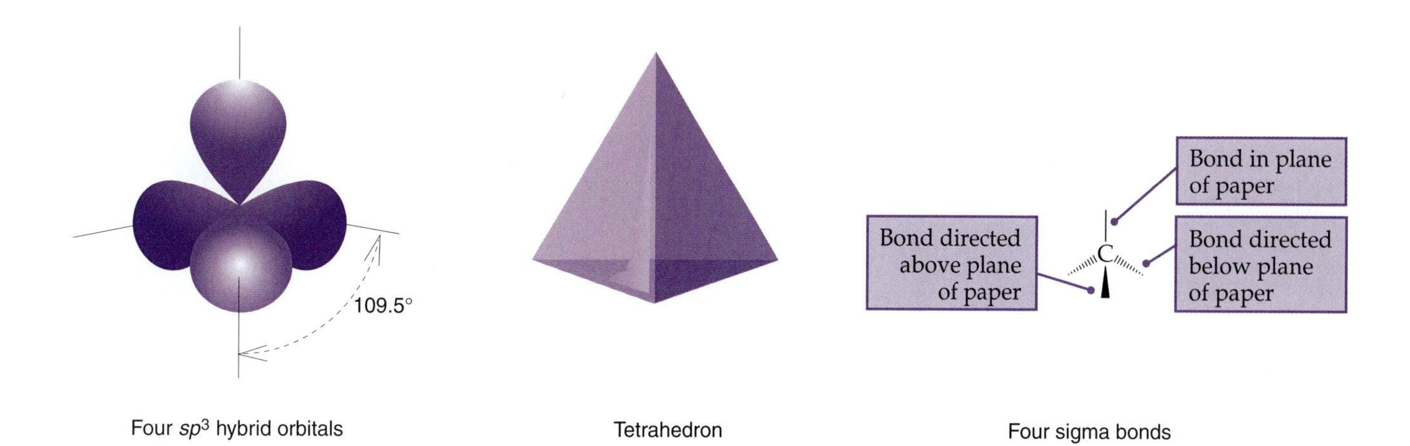

FIGURE 1.11

Three representations of the tetrahedral shape of carbon.

in space as if they pointed from the nucleus to the corners of an imaginary tetrahedron. This process is called **tetrahedral hybridization** (Figure 1.11). The four electrons in the valence shell of carbon occupy these four identical orbitals, one in each orbital. This arrangement allows the four hybridized orbitals to be as far apart from one another as possible, minimizing repulsive forces between electrons.

The third and final step in the covalent bonding of carbon is the overlap of each sp^3 hybrid orbital with an atomic or hybrid orbital of another atom, forming four covalent bonds. This step is energy-releasing. (The next subsection describes the formation of the covalent bond.)

Single Covalent Bonds to Carbon

Single covalent bonds are formed to carbon by overlap of an sp^3 hybrid orbital of a carbon with an orbital of another bonding atom. For example, in methane (CH_4), each sp^3 hybrid orbital of the carbon atom overlaps the 1*s* orbital of a different hydrogen atom to form four covalent bonds, each one a sigma (σ) bond (Figure 1.12). Each of the resulting four carbon-hydrogen sigma bonds is a covalent bond (called a *single bond*) formed of a single pair of electrons, one contributed by the hydrogen atom and one contributed by the carbon atom. In the process, both the carbon atom and the hydrogen atoms have achieved a stable noble-gas electron configuration through electron sharing.

This bonding model explains why the shape of the methane molecule is that of a tetrahedron, with the carbon atom in the center and a hydrogen atom at each point. This tetrahedral shape places the four single bonds from carbon (each with two electrons) as far apart as possible—energetically and spatially the most favorable geometry. It is important to realize that even though chemists usually write methane as CH_4, what is implied by that representation is a three-dimensional molecule with a carbon atom surrounded symmetrically by four hydrogen atoms bonded covalently to it (Figure 1.13). The molecular shape is tetrahedral (that is, the carbon-hydrogen sigma bonds are 109.5° from each other).

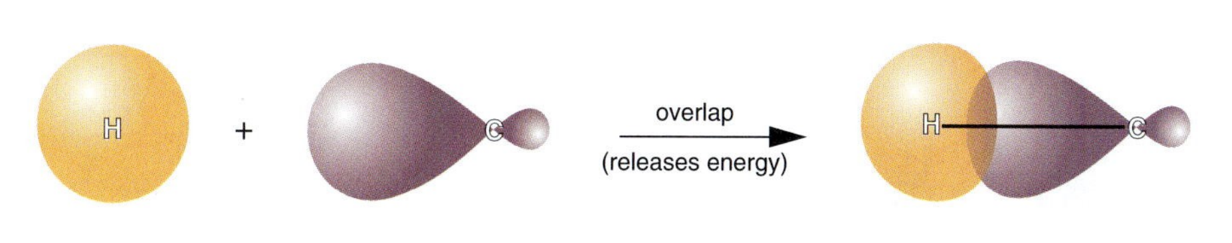

FIGURE 1.12

Sigma bond formation between hydrogen and carbon.

FIGURE 1.13

Representations of methane (CH_4), a compound with four sigma bonds from carbon to four hydrogens (the symbol ≡ means "equivalent to").

When you see this icon, go to the Gallery of Chime™ Molecules at www.jbpub.com/organic-online. You'll be able to interactively view the molecule(s) shown in virtual three-dimensional space.

CH_4 ≡ H—C(H)(H)—H ≡ [three-dimensional formula] ≡

Condensed formula | Two-dimensional structural formula | Formula indicating three-dimensional structure

When representing the three-dimensional structures of organic compounds, it is customary to use the symbolism shown in the third formula in Figure 1.13. In this kind of structural formula, a solid line represents a bond lying in the plane of the paper, a dashed wedge represents a bond directed below the plane of the paper, and a solid wedge represents a bond directed above the plane of the paper, toward the viewer. This book will generally not use a formula indicating three-dimensional structure unless the spatial orientation of a molecule is being considered.

Carbon-Carbon Bonds

As was mentioned in the introductory essay, organic chemistry is a separate field because of the ability of carbon atoms to bond to each other in a seemingly endless number of combinations. A single (sigma) bond can form between two carbon atoms by overlap of an sp^3 hybrid orbital of one of the carbons with the sp^3 hybrid orbital of the other (Figure 1.14). Keep in mind that each carbon atom is tetravalent; each still has three unshared electrons in its remaining sp^3 hybrid orbitals. If these are then used to covalently bond each carbon to three hydrogen atoms through the process of sigma bonding just described, the molecule known as ethane (C_2H_6) results (Figure 1.15).

The process of one carbon linking to another can continue virtually without limit, with each carbon atom potentially bonding to as many as four other carbon atoms. Chemists often show a chain of carbon atoms as a straight row when drawing a structural formula. But in three dimensions, carbon chains actually consist of tetrahedrons linked together at the corners. Most organic molecules have a "backbone," or "skeleton," formed from carbon atoms bonded to one another. Such backbones may be straight chains, branched chains, or even rings of carbon atoms, whose remaining sp^3 orbitals are filled by bonding mainly to hydrogen atoms. These compounds of carbon and hydrogen, when formed by single covalent bonds, are called *alkanes* (the subject of Chapter 2). Alkanes range widely in size and complexity: from methane (CH_4) and ethane (C_2H_6), with only one and two carbon atoms, to such complex long-chain alkanes as the one that forms the wax found on apple skins ($C_{27}H_{56}$).

Often, one or more sp^3 orbitals of carbon overlap with orbitals from one or more noncarbon atoms called **heteroatoms** (oxygen, nitrogen, sulfur, or one of the halogens).

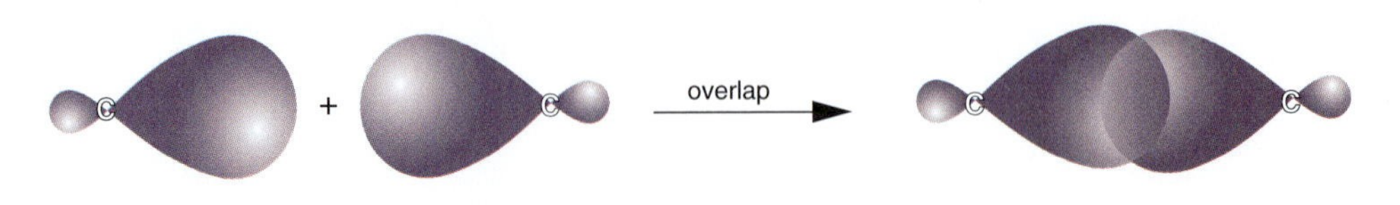

An sp^3 orbital of one carbon atom | An sp^3 orbital of a second carbon atom | A carbon-carbon sigma bond (sp^3-sp^3)

FIGURE 1.14

Sigma bond formation between two carbon atoms.

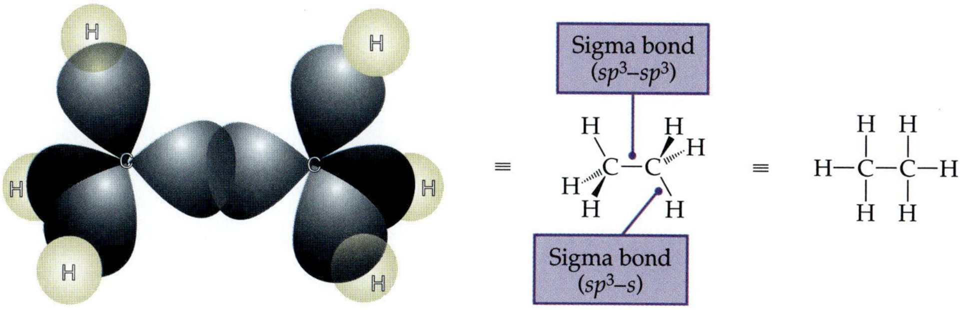

Sigma bond (sp^3–sp^3)

Sigma bond (sp^3–s)

FIGURE 1.15

Representations of ethane (C_2H_6).

For example, the molecular structure of chloroform ($CHCl_3$) is similar to that of methane (CH_4), except that three of the hydrogen atoms in methane have been replaced by chlorine atoms single-bonded to the carbon atom. As you learn to draw the structures of alkanes, remember to make sure that all sp^3 orbitals are filled and that each carbon has four groups attached.

How to Solve a Problem

Draw all possible structures for the alkanes containing only four carbons. Be sure to show the appropriate numbers of hydrogens in the correct locations.

SOLUTION

The simplest way to connect four carbons is in a continuous single row (called a *straight chain*):

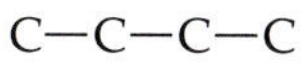

```
C—C—C—C
```

Adding the hydrogens gives us

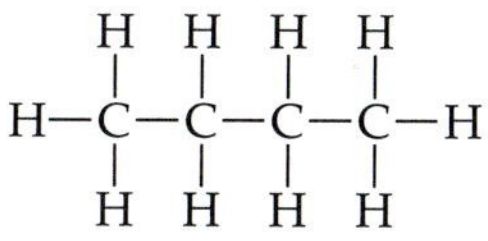

```
   H  H  H  H
   |  |  |  |
H—C—C—C—C—H
   |  |  |  |
   H  H  H  H
```

The only possible variation on this structure has three carbons in a continuous row and the fourth carbon attached to the middle carbon (called a *branched chain*):

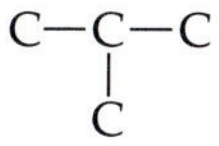

```
C—C—C
  |
  C
```

Adding the hydrogens, we get

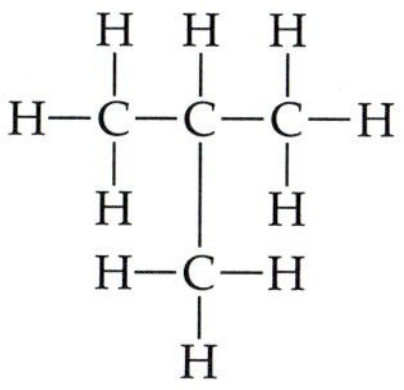

```
   H  H  H
   |  |  |
H—C—C—C—H
   |  |  |
   H  |  H
   H—C—H
      |
      H
```

What about connecting the fourth carbon to the end carbon of a three-carbon chain?

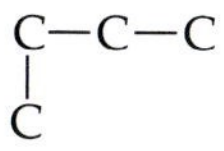

```
C—C—C
|
C
```

That gives the same structure as the first one; all four carbons are still in a continuous row.

PROBLEM 1.3

Draw the structure for the alkane containing only three carbons. Be sure to show the appropriate numbers of hydrogens in the correct locations.

PROBLEM 1.4

Draw a structure for a compound with formula C_5H_{12}.

PROBLEM 1.5

Draw all possible structures with the formula $C_3H_6Cl_2$.

Finally, it is important to mention that carbon atoms can bond to fewer than four other atoms in some molecules. A carbon atom can readily bond to just three other atoms or only two (and even to one atom in the unique molecule carbon monoxide, CO). When this occurs, carbon has formed different kinds of covalent bonds with the other atom(s): *double bonds* or *triple bonds*. These create different orbital geometries, different molecular shapes, and, most important, quite different chemistry. Compounds containing these kinds of bonds will be introduced in Chapters 6 and 7.

1.1.4 Formal Charge

Most organic molecules are neutral, which means that the sum of any charges on atoms in the molecule must be zero. However, some organic molecules are charged, which is what makes them active participants in bringing about chemical reactions. To follow a reaction, it is useful to know which atom is carrying the charge. To do this, you need to determine the **formal charge** on each atom, using the Lewis structure.

After drawing the Lewis structure, you count the number of electrons each atom contributes to the structure and compare it with the number of valence electrons in the neutral atom:

- If the two values are equal, the atom has no net charge.
- If the atom has an excess of electrons, it is negatively charged.
- If the atom is deficient in electrons, it is positively charged.

The formal charge on any atom can be calculated using the following formula:

Formal charge = number of valence electrons − number of unshared electrons − number of covalent bonds

Or, in symbols,

$$FC = VE - UE - CB$$

The formal charge on each atom is calculated in Table 1.2 for water, the hydronium ion (H_3O^+), and two important reactive forms of carbon (to be discussed later)—a *carbocation* and a *carbanion*.

Table 1.3 presents a summary of the formal charges for charged and neutral states of four atoms regularly encountered in organic chemistry.

TABLE 1.2

Calculation of Formal Charges

Species	Structure	Calculation of Formal Charges
Water	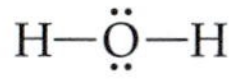 $H-\ddot{\underset{..}{O}}-H$	Hydrogen = 1 − 0 − 1 = 0 Oxygen = 6 − 4 − 2 = 0
Hydronium ion	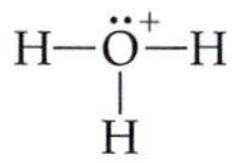 $H-\ddot{O}^{+}(-H)-H$	Hydrogen = 1 − 0 − 1 = 0 Oxygen = 6 − 2 − 3 = +1
Carbocation	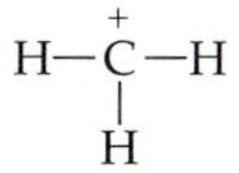 $H-C^{+}(-H)-H$	Hydrogen = 1 − 0 − 1 = 0 Carbon = 4 − 0 − 3 = +1
Carbanion	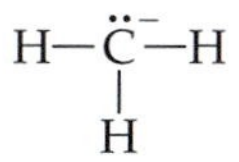 $H-\ddot{C}^{-}(-H)-H$	Hydrogen = 1 − 0 − 1 = 0 Carbon = 4 − 2 − 3 = −1

TABLE 1.3

Summary of Formal Charges

Atom	Number of Valence-Shell Electrons	Formal Charge of +1	Formal Charge of 0	Formal Charge of −1
Carbon	4	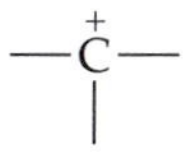 $-C^{+}-$ (three bonds)	 $-C-$ (four bonds)	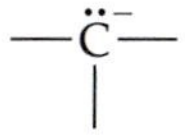 $-\ddot{C}^{-}-$ (three bonds)
Nitrogen	5	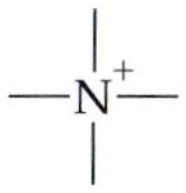 $-N^{+}-$ (four bonds)	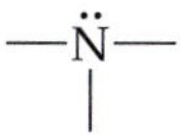 $-\ddot{N}-$ (three bonds)	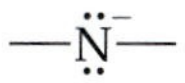 $-\ddot{N}^{-}-$ (two bonds)
Oxygen	6	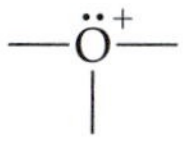 $-\ddot{O}^{+}-$ (three bonds)	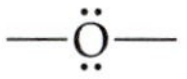 $-\ddot{O}-$ (two bonds)	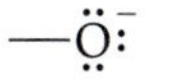 $-\ddot{O}:^{-}$ (one bond)
Halogen	7	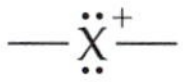 $-\ddot{X}^{+}-$ (two bonds)	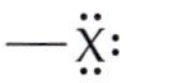 $-\ddot{X}:$ (one bond)	 $:\ddot{X}:^{-}$ (no bonds)

PROBLEM 1.6

Calculate the formal charge on each atom except hydrogen in each of the following structures:

(a) the hydroxide anion (HO^{-}) in sodium hydroxide

(b) ammonia (NH_3)

(c) the ammonium ion (NH_4^{+}) in ammonium chloride

(d) the bicarbonate anion (HCO_3^{-})

(e) the amide anion (NH_2^{-}) from sodium amide

(f) nitric acid $\left(H-\ddot{O}-N(=\ddot{O})-\ddot{O}:\right)$

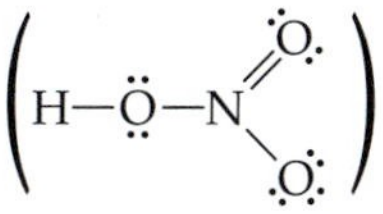

1.2

Organic Compounds

1.2.1 Formulas for Organic Compounds

It is essential that you learn the symbolism of organic chemistry, and initially this involves learning how to write the structures of organic compounds. Because of the large size and complexity of some organic compounds, different techniques have evolved over time to simplify their representation and yet still convey detailed structural information. Thus, you need to become familiar with several kinds of formulas used to represent organic compounds.

Empirical Formulas

An **empirical formula** indicates two things: which elements are present in a compound and the *ratio* of one element to another. It is not meant to show the absolute numbers of atoms present. For example, ethane, C_2H_6, has the empirical formula CH_3. You know from this formula that there are three hydrogens for every carbon present in ethane. However, you also know that the formula CH_3 cannot represent a complete molecule because it indicates that there are only three atoms attached to a tetravalent carbon. Similarly, the compound dichlorobutane, $C_4H_8Cl_2$, has the empirical formula C_2H_4Cl. In Chapter 11, we will look at a limited but important use for empirical formulas.

Molecular Formulas

The **molecular formula** of a compound indicates which elements are present and the *absolute number* of atoms of each. Thus, the molecular formula of ethane, C_2H_6, indicates that there are two carbon atoms and six hydrogen atoms in each molecule of that compound. The molecular formula, which is a common representation for a compound, is always a whole-number multiple of the empirical formula (C_2H_6 is $2 \times CH_3$). In some cases, the molecular and empirical formulas are the same, as for cholesterol, $C_{27}H_{46}O$ (in this case, the whole-number multiple is 1).

The molecular formula is also the basis for calculating the **molecular weight** of a compound (the combined atomic weights of the constituent elements). Ethane (C_2H_6) has a molecular weight of 30: $(2 \times 12) + (6 \times 1) = 30$. Cholesterol ($C_{27}H_{46}O$) has a molecular weight of 386: $(27 \times 12) + (46 \times 1) + (1 \times 16) = 386$. Molecular weights are often rounded off to the nearest whole number, but modern analytical techniques allow them to be determined experimentally to four decimal places. Molecular formulas of organic compounds are always written with carbon first, then hydrogen, and then all other elements in alphabetical order. For example, morphine is $C_{17}H_{19}NO_3$.

Structural Formulas

The molecular formula of a compound tells what atoms are present, but does not reveal molecular *structure*—that is, which atoms are attached to which and how they are connected (their *connectivities*). There are several forms of **structural formulas,** containing varying amounts of detail.

Dot Formulas (Lewis Structures) In a **dot formula,** each single bond is represented by a pair of dots between the element symbols of the two atoms. These are the shared pair of electrons comprising the covalent bond. Each unshared pair of electrons is also shown, next to its element symbol. The dot formula is usually used for careful tracking of the movement of electrons in a reaction, but it is too cumbersome for general use. The dot formula for ethyl alcohol (C_2H_6O) is

$$\begin{array}{c} \mathrm{H\;H} \\ \mathrm{H:\ddot{C}:\ddot{C}:\ddot{O}:H} \\ \mathrm{\ddot{H}\;\ddot{H}} \end{array}$$

Ethyl alcohol

From this formula, you can readily see that each atom has a full outer shell of electrons (a noble-gas configuration).

Dash Formulas In a **dash formula,** each covalent bond is drawn as a single dash, representing the two shared electrons. Such a drawing shows which atoms are attached to which in a molecule, although it does not show the molecule's three-dimensional shape. For example, ethyl alcohol, whose molecular formula is C_2H_6O, can be drawn in several ways as a dash formula. Three of those ways are shown here:

$$\begin{array}{ccccc} & \mathrm{H} & \mathrm{H} & & \\ & | & | & & \\ \mathrm{H-} & \mathrm{C} & \mathrm{-C} & \mathrm{-O-H} \\ & | & | & & \\ & \mathrm{H} & \mathrm{H} & & \end{array} \equiv \begin{array}{cccc} & \mathrm{H} & \mathrm{H} & \\ & | & | & \\ \mathrm{H-} & \mathrm{C} & \mathrm{-C} & \mathrm{-H} \\ & | & | & \\ & \mathrm{H} & \mathrm{O} & \\ & & | & \\ & & \mathrm{H} & \end{array} \equiv \begin{array}{cccc} & \mathrm{H} & & \\ & | & & \\ & \mathrm{O} & \mathrm{H} & \\ & | & | & \\ \mathrm{H-} & \mathrm{C} & \mathrm{-C} & \mathrm{-H} \\ & | & | & \\ & \mathrm{H} & \mathrm{H} & \end{array}$$

Ethyl alcohol

Note that the three formulas are equivalent because a dash formula is a two-dimensional drawing with no three-dimensional implications.

Condensed Formulas Because dot and dash formulas become unwieldy for large molecules, a condensed form of the structural formula is most commonly used in organic chemistry. This is the form you should master as quickly as possible, and it is also the form that is most conveniently used in text. In a **condensed formula,** the atoms single-bonded to any carbon are written immediately following that carbon, with no bonds showing. Thus, the condensed formula for ethane is CH_3CH_3, which indicates that the left carbon atom is single-bonded to three hydrogen atoms as well as to the next (right) carbon atom. The right carbon also has four single bonds, one to the left carbon and three to hydrogen atoms. Ethyl alcohol is written as CH_3CH_2OH, indicating that the right carbon is attached to the left carbon and only two hydrogens, with a fourth covalent bond to oxygen. The oxygen is attached to the right carbon and to the rightmost hydrogen. In writing condensed formulas, you need to consciously check that each carbon atom has the right number of groups or atoms attached to it; this process will eventually become automatic.

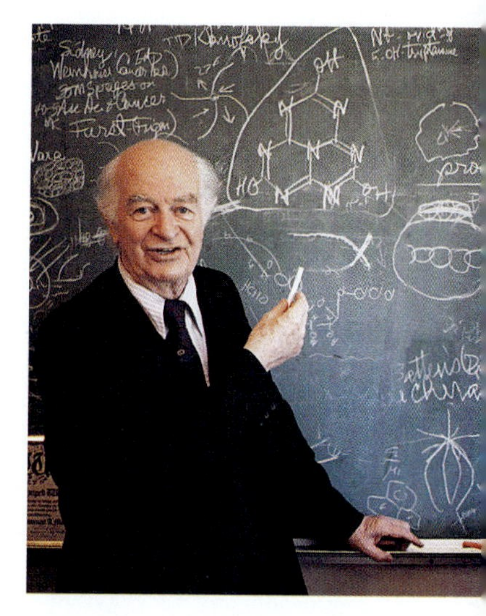

Linus Pauling and a structural formula.

Chemists also draw structures of compounds using a combination of dash and condensed formulas:

$$CH_3{-}CH_3 \qquad\qquad CH_3{-}CH_2{-}OH$$

Ethane — Ethyl alcohol

This type of *combination formula* is frequently used to help visualize some molecular detail while still saving space. This representation condenses only carbon-hydrogen bonds, leaving all others to be represented by a dash. Sometimes dots are added to a combination formula to indicate unshared electrons:

$$CH_3{-}CH_2{-}\ddot{\underset{\cdot\cdot}{O}}H$$

Ethyl alcohol

TABLE 1.4

Condensed, Combination, and Bond-Line Formulas for Three Compounds

Compound	Condensed Formula	Combination Formula	Bond-Line Formula
Butane	$CH_3CH_2CH_2CH_3$	$CH_3{-}CH_2{-}CH_2{-}CH_3$	
n-Butyl chloride	$CH_3CH_2CH_2CH_2Cl$	$CH_3{-}CH_2{-}CH_2{-}CH_2{-}Cl$	Cl
Cyclobutane	$(CH_2)_4$	$CH_2{-}CH_2$ / $CH_2{-}CH_2$	

The rationale for this kind of representation is that most of the chemistry of organic molecules does not take place at carbon-hydrogen bonds. You will see these formulas quite often throughout this book.

Bond-Line Formulas A further simplification in representing compounds is the **bond-line formula,** in which lines are used to represent *all bonds other than carbon-hydrogen bonds.* The lines represent the carbon skeleton, or backbone, of the compound. The carbon-hydrogen bonds and hydrogen atoms are not shown at all; the appropriate numbers of hydrogens are assumed to be present. Thus, for any carbon atom (represented by an apex or the end of a line), the number of attached hydrogen atoms is four minus the number of bonds shown (remember that carbon is tetravalent). In drawing bond-line formulas, you explicitly show any heteroatoms and hydrogens bonded to them. Table 1.4 shows the condensed formulas, the combination formulas, and the bond-line formulas for three compounds: butane (C_4H_{10}), *n*-butyl chloride (C_4H_9Cl), and cyclobutane (C_4H_8).

Bond-line formulas are especially useful to show structures for cyclic compounds, as illustrated by the general carbon skeleton for steroids and the formula for the steroidal compound cholesterol:

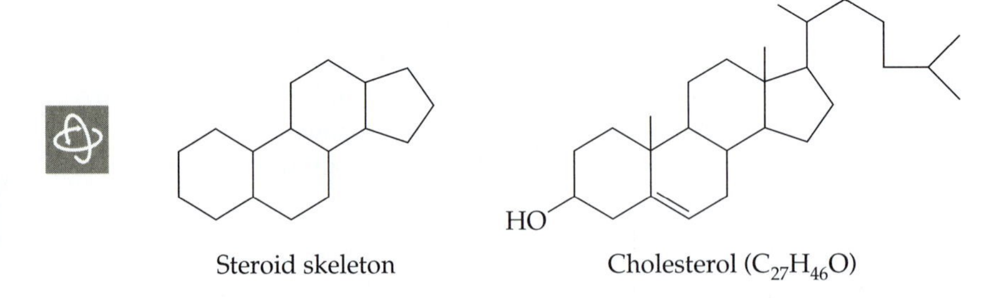

Steroid skeleton

Cholesterol ($C_{27}H_{46}O$)

How to Solve a Problem

(a) Write the molecular formula for $CH_3CH_2CH_2CH_2CH_2Br$.

(b) What is the molecular weight of this compound?

(c) Write the bond-line formula for this compound.

SOLUTION

(a) The molecular formula indicates the total number of each type of atom in the molecule. Counting from the given condensed formula, we see that the molecule has 5 carbons, 11 hydrogens, and 1 bromine. Therefore, the molecular formula is $C_5H_{11}Br$.

(b) The molecular weight of a molecule is the sum of the atomic weights of its constituent elements. For multiples of the same atom, we multiply the atomic weight by the number of times that atom appears in the molecule. The molecular weight of the given molecule is thus

$$\underbrace{(5 \times 12)}_{\text{Five carbons}} + \underbrace{(11 \times 1)}_{\text{Eleven hydrogens}} + \underbrace{(1 \times 80)}_{\text{One bromine}} = 60 + 11 + 80 = 151$$

(c) The bond-line formula is [bond-line structure]Br

PROBLEM 1.7

Write combination formulas for (a) C_3H_8, (b) C_4H_{10}, and (c) C_5H_{12}.

PROBLEM 1.8

Write the molecular formulas corresponding to these structural formulas:

(a) [structure] (b) [structure] (c) $CH_3-CH_2-CH(CH_2-CH_3)-CH_3$

PROBLEM 1.9

What is the molecular formula for each of the following compounds?

(a) The empirical formula is C_2H_5, and the molecular weight is 58.

(b) The empirical formula is C_4H_8O, and the molecular weight is 144.

1.2.2 Classification of Organic Compounds

Given the millions of organic compounds already known and the thousands being discovered or synthesized each year, several means of classifying organic compounds have developed over time. This section will introduce several classification systems, leaving extensive discussion for later chapters.

One classification system considers the *molecular framework* (sometimes called the skeleton or backbone) of an organic compound, resulting in all compounds being classified in one of three types:

- **Acyclic compounds** are those with a linear or branched skeleton, usually containing only carbon atoms and with no closed chain (that is, a ring) of atoms present.
- **Carbocyclic compounds** have only carbon atoms in the skeleton but with some or all of those carbon atoms joined together in one or more rings.

- **Heterocyclic compounds** are ring compounds with a skeleton containing one or more heteroatoms (usually oxygen or nitrogen) in place of ring carbons.

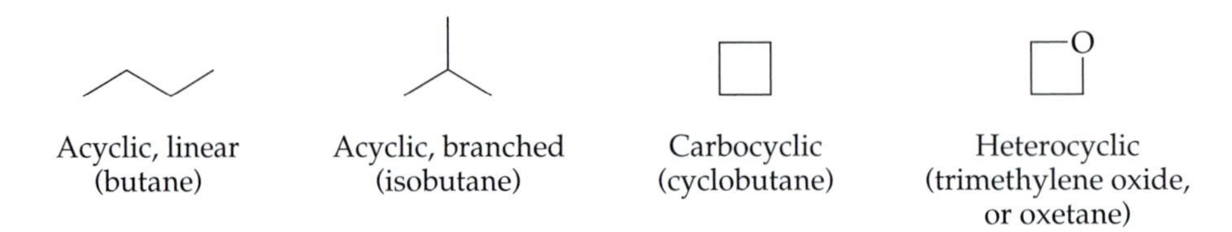

A second classification system is based on compounds being either **saturated** or **unsaturated.** The preceding examples are all saturated compounds, meaning that all the carbon atoms have four sigma bonds. Unsaturated compounds—those with fewer than four sigma bonds attached to some carbons and with multiple bonds between carbons—will be discussed in Chapters 6 through 8.

All of the compounds considered to this point are classified as **aliphatic,** a term arising from a Greek word meaning "fat." Chemists loosely define these as compounds derived from alkanes. The best definition of aliphatic compounds is "compounds that are not aromatic." **Aromatic compounds** are those derived from benzene and related compounds. Although this distinction may not seem helpful at this time, all compounds discussed in Chapters 2 through 8 are aliphatic. It will become apparent when we discuss aromatic compounds in Chapter 9 what the differences are.

The most important classification of organic compounds is by functional group. A **family** of compounds is made up of compounds containing the same functional group. Organic compounds can undergo a wide variety of chemical reactions (transformations) in which their molecular structures can be altered in a controlled manner. However, only a certain portion of an organic molecule is subject to most chemical reactions; the rest of the molecule, usually that part consisting of just carbon-hydrogen bonds, is relatively unreactive. The part of a molecule that is particularly reactive and subject to change is known as a **functional group.** Classification of organic compounds by functional group into families is important because almost all compounds containing a particular functional group undergo the same type of reaction under similar reaction conditions. The structure of the relatively unreactive hydrocarbon part of the molecule usually matters little because typically only the functional group is altered. Thus, learning organic chemistry is simplified immensely by concentrating on functional groups. Shown here are representatives of three families of compounds, with the functional group highlighted in each.

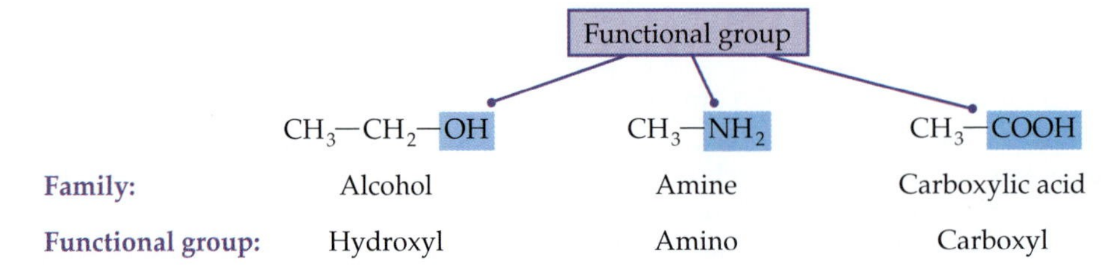

This book is organized by families of compounds, with the first family, alkanes, covered in Chapter 2 and the other 10 major families and their functional groups discussed in subsequent chapters. On the inside front cover is a table of all the families of compounds and their functional groups, along with additional useful information about each. At this point, it is not necessary to memorize the functional groups or families; you will become more familiar with them in due course. For now, simply be aware of the concept of families of compounds and their functional groups. The Key to Transformations located on the inside front cover will help you keep track of the possible interconversions between functional groups.

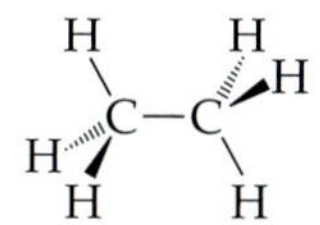

Structural formula

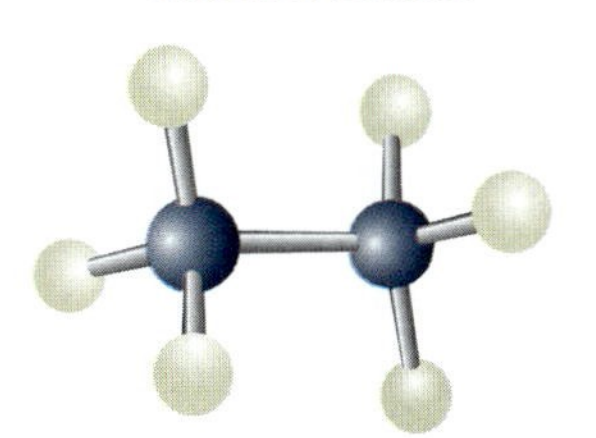

Ball-and-stick model

Space-filling model

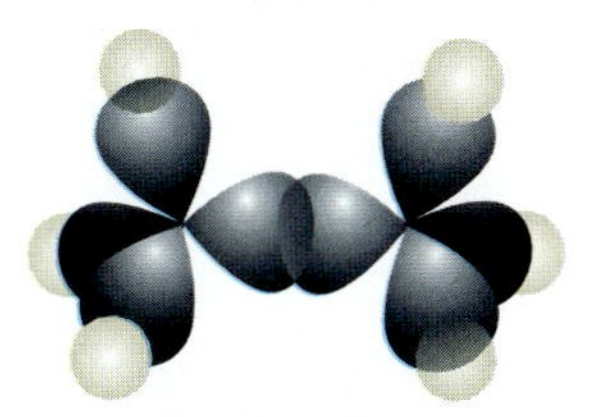

Orbital diagram

FIGURE 1.16

Three-dimensional representations of ethane (C_2H_6).

1.2.3 Physical Properties of Organic Compounds

As each family of organic compounds is discussed in the upcoming chapters, you will discover that the physical properties of those compounds are a logical extension of their bonding and structure. This section is a brief summary of a few of the general properties associated with most organic compounds. These properties often have a significant effect on the chemical behavior of a compound.

Molecular Shape

The nature of covalent bonding means that organic molecules have very specific three-dimensional shapes, which can be described and drawn. It is often helpful to use *molecular models* (ball-and-stick models or space-filling models) to envision these shapes. It is especially important to learn to draw formulas for molecules on a sheet of paper (in two dimensions) in a manner that conveys the actual three-dimensional structure:

- Solid lines represent bonds *in the plane* of the paper.
- Dashed wedges represent bonds directed *below the plane* of the paper.
- Solid wedges represent bonds directed *above the plane* of the paper.

Figure 1.16 shows such a structural formula for ethane as well as a ball-and-stick model, a space-filling model, and an orbital diagram for comparison. (Recall that all of the bond angles in ethane are 109.5°.)

Organic compounds have characteristic three-dimensional molecular structures because of two important properties of the covalent bond: *specific direction* and *specific length.* You learned that the bond angles around a carbon with four sigma bonds are 109.5°; this results from the fact that the carbon is tetrahedrally hybridized (that is, the valence-shell orbitals are sp^3-hybridized). In Figure 1.9 you saw that a covalent bond has a specific length (internuclear distance) for maximum stability. Bond length is usually measured in angstroms, symbolized by Å (1 Å = 10^{-10} m). A few typical and useful bond lengths are listed in Table 1.5.

A covalent bond also has a specific strength. This means that when such a bond is formed, the molecule releases that much energy. However, it also indicates the amount

TABLE 1.5

Typical Bond Lengths and Approximate Bond Energies

Bond Participants	Bond Type	Bond Length (Å)	Bond Energy (kcal/mol)
Two hydrogen atoms	Single	0.74	105
A carbon and a hydrogen atom	Single	1.08	95
Two sp^3 carbon atoms	Single	1.54	85
Two sp^2 carbon atoms	Double	1.32	150
Two sp carbon atoms	Triple	1.18	200

CHEMISTRY AT WORK

IN THE BODY

The Impact of Molecular Shape

Molecular shape is important because of its role in chemical reactions. This is especially true when considering the effects of chemicals on biological systems. How a molecule functions depends on how it "fits" into the structural environment.

Organic chemists discovered that the compound 18-crown-6 (a member of a unique family known as *crown ethers* because of their shape) has a molecular structure that allows it to complex specifically with a potassium ion. The unshared electrons of the six oxygen atoms of 18-crown-6 are shared with the potassium ion as it fits perfectly into the interior of the "crown." The resulting complex has a hydrophobic character on the exterior, which masks the ionic (hydrophilic) character of the potassium ion. A sodium ion is too small to fit snugly inside the crown and be complexed with the oxygen atoms.

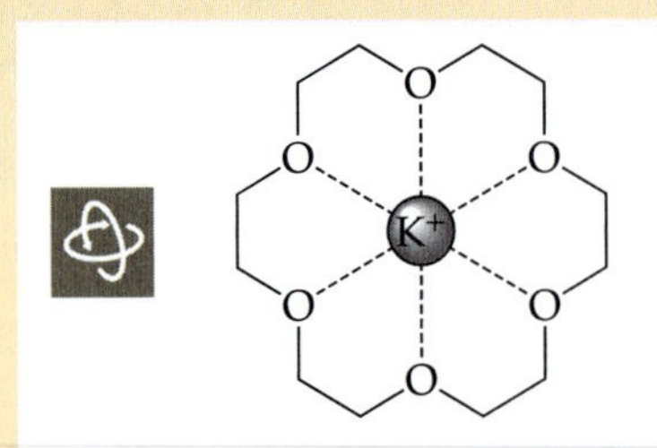

18-Crown-6 ether with potassium ion complexed in the cavity

The human body must maintain a delicate balance of potassium ions outside and inside its cells. An imbalance leads to diseases that disrupt normal nerve and muscle function. The *macrolide antibiotics*, of which nonactin is an example, are thought to affect the potassium ion balance in part by complexing with those ions like the crown ethers do and facilitating their migration into cells.

Molecular shape also explains why you can take two aspirins (or another analgesic) and call the doctor in the morning. These non-steroidal anti-inflammatory drugs (NSAIDs) act in the body to treat symptoms of inflammatory diseases (for example, arthritis) and to alleviate general aches and pains. It has recently been discovered that such drugs bind with, and thereby inhibit the normal activity of, two closely related *cyclo-oxygenase enzymes*, known as COX-1 and COX-2. Inhibition of COX-2 is desirable because it triggers the production of *prostaglandins,* substances that initiate the processes responsible for inflammation and pain. However, inhibition of COX-1 is not desirable because it results in adverse side effects, such as upset stomach and even kidney failure. Existing NSAIDs inhibit both enzymes.

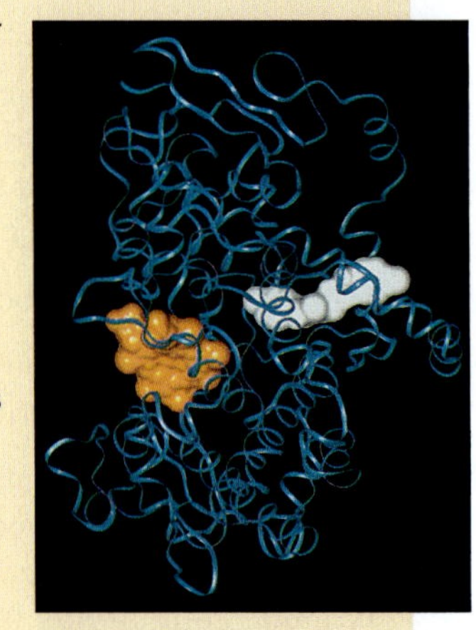

Ribbon structure of the COX-2 enzyme (blue) bound to an inhibitor molecule (white).

Although the chemical compositions and structures of COX-1 and COX-2 are very similar (both are proteins), X-ray analyses revealed that their binding sites differ in a *single* molecular component, making the shapes of the binding cavities in the two enzymes slightly different. As a result of this 1996 discovery, major research efforts are under way to develop new NSAIDs of a specific shape that will bind to COX-2 but not to COX-1. Candidate compounds have been prepared that show up to 200 times greater affinity for COX-2 than for COX-1. New drugs called *COX-2 inhibitors* may be marketed in the near future.

Nonactin
(a macrolide antibiotic)

of energy that must be supplied from an external source in order to break that bond. Thus, this energy is referred to as *bond energy*. In general, a shorter bond is a stronger bond.

Bond Polarity

Recall that a covalent bond involves the sharing of a pair of electrons in a molecular orbital between two atoms. If the bond is formed between two identical atoms with identical groups attached, the electrons are shared equally. Such is the case with the C—C bond in ethane, for example. However, if the two atomic participants in the covalent bond are different and therefore have different electronegativities, the sharing is unequal. In other words, the bond is slightly polarized toward the more electronegative atom. Such is the case with a C—O bond. Often, polarity in a single bond in a molecule (a *bond moment*) is reflected in the entire molecule being polarized. The polarity of molecules can be measured and is represented by a value called a **dipole moment**, expressed in terms of Debye units. (The polarity of a bond cannot be measured directly; that of a molecule can.) A more polar molecule has a larger dipole moment. The dipole moment of a molecule is a vector quantity, meaning that it has both magnitude and direction.

The polarization of a bond may be represented on a structural formula in one of two ways (or sometimes a combination of both), as shown for methane:

1. Place a small arrow near the covalent bond, with the head toward the more electronegative atom (the negative end of the dipole) and a cross at the other end of the arrow to indicate the more electropositive atom.
2. Place a small $\delta+$ or $\delta-$ near each of the atoms being considered, to indicate that there is a very small partial positive or negative charge associated with the atom as a result of the electronegativity difference.

Methane

(1)

(2)

Some organic compounds are nonpolar, some are slightly polar, and some are quite polar. Methane, CH_4, is a nonpolar molecule with a zero dipole moment, even though hydrogen (2.1) is slightly less electronegative than carbon (2.5) (see Figure 1.4). Each of the C—H bonds is slightly and equally polarized from hydrogen toward carbon, but the symmetrical orientation of the four bond moments means that their effects cancel out. Similarly, the linear molecule carbon dioxide, CO_2, has two polar carbon-oxygen double bonds because carbon and oxygen have a significant electronegativity difference (2.5 versus 3.5). However, CO_2 has a zero dipole moment because the two bond moments point in opposite directions and cancel each other exactly.

O=C=O

Carbon dioxide

In contrast, methyl chloride, CH_3Cl, in which a strongly electronegative chlorine atom has replaced a single hydrogen atom of methane, has a dipole moment of 1.9, resulting from the significant C—Cl bond moment. However, carbon tetrachloride, CCl_4, with four polar C—Cl bonds, has a dipole moment of zero because of the canceling directions of the four equal bond moments.

Dipole moment of CH_3Cl (1.9 Debye units)

C—Cl bond moment

CCl_4 has zero dipole moment

Chloromethane (methyl chloride)

Tetrachloromethane (carbon tetrachloride)

Bond polarity will be important as we consider the relative reactivities of various covalent bonds throughout this book. Normally, the C—H bond polarities in a molecule are ignored because the electronegativity difference for that bond is small compared to the differences for the various functional groups (often involving carbon-oxygen, carbon-nitrogen, or carbon-halogen bonds).

PROBLEM 1.10

Draw the bond moment and the dipole moment for HCl.

PROBLEM 1.11

Draw the bond moments for all bonds in CH_3OH and CH_3F.

PROBLEM 1.12

Given that the dipole moment for water (H_2O) is 1.8 and that for ammonia (NH_3) is 1.5, speculate about the three-dimensional structures of these two compounds.

PROBLEM 1.13

Show the bond moments and the dipole moment for chloroform ($CHCl_3$). Would you expect the dipole moment to be larger or smaller than that for CH_3Cl? Why?

Solubility

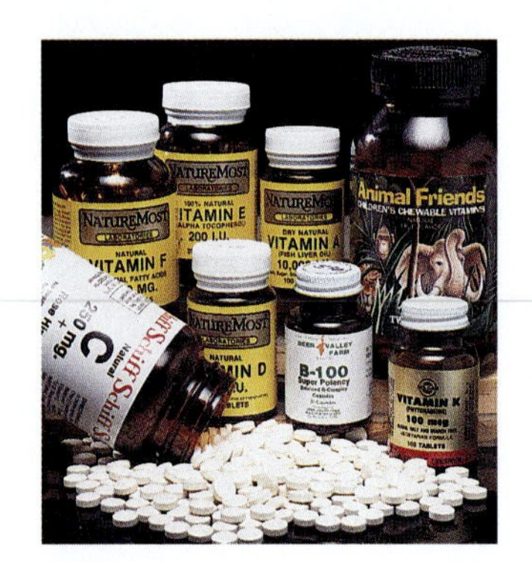

Unlike most organic compounds, vitamin C is hydrophilic and thus passes quickly through the watery environment of the human body. The recommended daily allowance (RDA) of vitamin C for an adult is 60 milligrams (mg) a day. Compare this to the RDAs for vitamins A (0.8–1.0 mg), D (0.005–0.010 mg), and E (8–10 mg). These three vitamins are hydrophobic, which is more typical of organic compounds, and are thus retained longer in the body.

Organic compounds are relatively nonpolar chemicals. From general chemistry, you will recall that it is not unusual for inorganic compounds to dissolve readily in water, which is a very polar solvent (recall how easily table salt, sodium chloride, dissolves). Such salts with ionic bonding are **hydrophilic;** they are attracted to water. In contrast, it is relatively unusual for organic compounds to dissolve in water because of their nonpolarity. Organic compounds are generally **hydrophobic;** they repel water. A drop of gasoline or oil does not dissolve in water to any appreciable extent. There are exceptions, of course, such as the organic compounds sucrose (table sugar) and ascorbic acid (vitamin C), which are readily soluble in water. (Later we will discuss the reasons for these exceptions to the general rule.)

Organic compounds do, however, dissolve readily in nonpolar organic solvents such as chloroform and alkanes (compounds similar to gasoline). Inorganic salts generally do not dissolve in such solvents. The general axiom "like dissolves like" applies to inorganic and organic compounds and solvents.

These differences in solubility are reflected in the chemistry taking place within your body. Vitamin C, one of the organic compounds that is unusually soluble in water, must be regularly ingested. It is hydrophilic and therefore passes quickly through the body in aqueous solution. Failure to consume sufficient vitamin C can result in the deficiency disease known as *scurvy,* which affected many sailors in earlier times. In contrast, vitamin E is hydrophobic; daily intake requirements are much smaller because it is insoluble in water and is therefore retained in the body for longer periods of time.

Intermolecular Attractive Forces

Organic compounds have much weaker intermolecular (molecule-to-molecule) attractive forces than do inorganic compounds. This means that organic compounds have much lower melting points (mp) and boiling points (bp). Small organic compounds (those with low molecular weights), such as methane (bp −164°C), are gases at room temperature. Moderate-sized compounds, such as the various components of commercial gasoline, are usually liquids at room temperature. Generally, only larger compounds, such as cholesterol (mp 149°C), are solids at room temperature, but even their melting points are usually less than 200°C. (We'll discuss specific cases when we consider the different families of compounds.)

- The strongest intermolecular attractive forces typically are *electrostatic forces between oppositely charged species*—for example, between cations and anions in salts (for example, sodium chloride melts at 801°C and boils at 1413°C).
- Next in strength are *electrostatic forces between polar molecules*, such as water and ethyl alcohol, called *dipole-dipole forces*.
- The weakest intermolecular attractive forces are those between nonpolar molecules, such as hydrocarbons, called *van der Waals forces*.

In general, compounds with the same attractive forces and of similar molecular weights have similar boiling points, and higher-molecular-weight compounds usually have higher boiling points. However, as you will see in Chapter 2, molecular surface area is also a factor. The boiling point of a substance is a reflection of the energy required to break the intermolecular attractive forces in the liquid state, allowing the substance to vaporize. Many, but not all, organic compounds have only van der Waals attractive forces and so are vaporized at relatively low temperatures. Those compounds with other, stronger attractive forces, such as alcohols, have comparatively high boiling points.

PROBLEM 1.14

Two compounds with similar molecular weights—water (18) and methane (16)—have very different boiling points: 100°C and −164°C, respectively. Explain this difference.

1.3 Reactions

1.3.1 Acids and Bases

Many organic reaction mechanisms depend on proton transfers. Here we will briefly review certain aspects of acid-base chemistry, which can also be called proton-transfer chemistry.

A **Brønsted acid** is a proton donor, and a **Brønsted base** is a proton acceptor. When hydrogen chloride dissolves in water, HCl serves as the acid and H_2O serves as the base:

$$H_2\ddot{O}: + H{-}\ddot{Cl}: \rightleftharpoons \left[H_2\overset{\delta+}{O}{\cdots}H{\cdots}\overset{\delta-}{Cl} \right] \rightleftharpoons H_2\overset{+}{O}{-}H + :\ddot{Cl}:^-$$

Base Acid Transition state Hydronium ion

Note the use of curved arrows to show electron movement in reactions. These arrows help you keep track of electrons and the charges associated with atoms.

In this reaction, the base (H_2O) attacks the hydrogen of the acid (HCl), removing it to form the *conjugate acid* (H_3O^+) and leaving behind the *conjugate base* (Cl^-). Hydro-

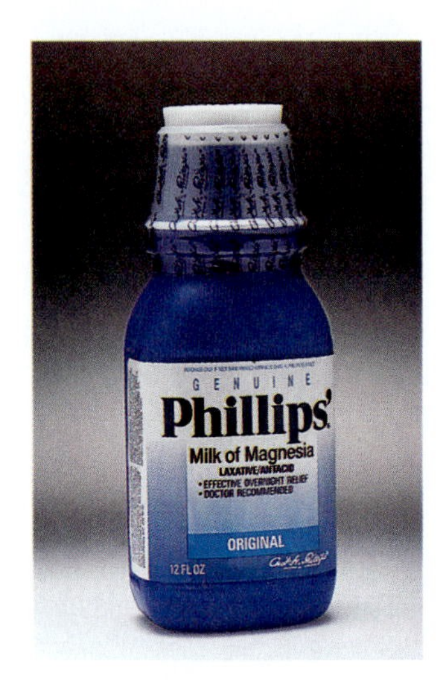

Digestion begins in the acidic environment of the stomach, with the secretion of hydrochloric acid (HCl). Too much food or food that is too spicy can cause overproduction of stomach acid, which leads to acid indigestion. Antacids, such as milk of magnesia [$Mg(OH)_2$], are simply bases that neutralize the acid: $Mg(OH)_2 + 2HCl \rightarrow MgCl_2 + 2H_2O$.

gen can participate in only one covalent bond, so when the unshared electrons of oxygen approach hydrogen (indicated by the longer curved arrow), the existing covalent bond from hydrogen to chlorine begins to break (indicated by the shorter, curvier arrow), with those shared electrons moving toward chlorine. As the base completes the removal of the proton from chlorine, it establishes a new covalent bond to hydrogen, resulting in the hydronium ion (H_3O^+). The originally neutral oxygen atom in water has given up exclusive control of a pair of its nonbonding electrons by sharing them with hydrogen, thereby becoming electron-deficient and positively charged. Simultaneously, the chlorine atom, which formerly shared an electron pair with hydrogen and was neutral, acquires full possession of that pair of electrons and becomes negatively charged. (A neutral chlorine atom has seven electrons in its valence shell. The chloride anion has one more electron, for a total of eight, and is therefore negatively charged.) The charge on the hydrogen atom does not change because it consistently has only one covalent bond to another atom, first chlorine and then oxygen. The structure shown between the reactants and the product is called a **transition state,** and it portrays progress partway through the transfer of hydrogen from chlorine to oxygen.

Acid-base reactions are *equilibria.* That is, the rates of both the forward and the reverse reactions reach a point where the concentrations of the reagents on the two sides of the equation become fixed. In aqueous solution, strong acids—such as HCl, H_2SO_4, and HNO_3—are completely dissociated, and the equilibrium lies completely to the right. Weak acids, however, are not completely dissociated in water.

Acid strength is represented by the **acidity constant, K_a.** For the general acid HA, where A represents the potential anion, we have

$$HA + H_2O \rightleftharpoons H_3O^+ + A^-$$

$$K_a = \frac{[H_3O^+][A^-]}{[HA]} \qquad pK_a = -\log K_a$$

Thus, a larger K_a represents an equilibrium lying more to the right (that is, the magnitude of the denominator, [HA], decreases as the reaction lies more to the right). Therefore, a larger K_a indicates a more dissociated acid and thus a stronger acid. The symbol typically used to represent acid strength is **pK_a**, the negative logarithm of the K_a value. *The larger the K_a, the smaller the pK_a and the stronger the acid.* For example, the K_a for the relatively weak acid acetic acid (found in household vinegar) is 1.8×10^{-5}; the pK_a is 4.8. In later chapters, we will use pK_a values to compare the strengths of organic acids, most of which are relatively weak acids.

The general acid-base reaction for an acid HA (where A represents any potential anion) reacting with a base B^- can also be viewed as a competition between two bases (B^- and A^-) for an available proton:

$$\underset{\text{Base}}{B:^-} + \underset{\text{Acid}}{H{-}A} \rightleftharpoons \underset{\text{Conjugate acid}}{B{-}H} + \underset{\text{Conjugate base}}{:A^-}$$

It is important to be able to predict the direction in which the equilibrium will lie for such an acid-base reaction. The outcome is that the stronger base will acquire the proton. Put another way, the stronger acid will give up its proton.

Whether A^- or B^- is the stronger base can be determined from the pK_a values of their conjugate acids. For example, in the reaction of sodium hydroxide (NaOH) with acetic acid ($C_2H_4O_2$), consider the pK_a values of the two acids present in the equilibrium, acetic acid (pK_a 4.8) and water (pK_a 15.7):

$$\underset{\substack{\text{Acetic acid} \\ (\text{p}K_a\ 4.8)}}{CH_3{-}COOH} + \underset{\substack{\text{Sodium} \\ \text{hydroxide}}}{NaOH} \rightleftharpoons \underset{\substack{\text{Sodium} \\ \text{acetate}}}{Na^+\ CH_3COO^-} + \underset{\substack{\text{Water} \\ (\text{p}K_a\ 15.7)}}{HOH}$$

Because acetic acid is the stronger acid, acetate ion is the weaker of the two bases (acetate anion and hydroxide anion) that are competing for the proton. Thus, hydroxide anion is the stronger base and will acquire the proton. (The larger the pK_a of the conjugate acid, the stronger the base.) Therefore, this reaction equilibrium lies to the right, as indicated by the longer arrow pointing to the right. Put another way, the weaker acid is more likely to continue to hold onto its proton (water is the weaker of the two acids in this equilibrium), and the stronger acid is more likely to lose its proton. This analysis is useful for any acid-base reaction, no matter how strange or unfamiliar it may initially appear.

How to Solve a Problem

Explain why acetylene (C_2H_2, pK_a 25) does not react with sodium hydroxide (pK_a of water is 15.7) but does react with sodium amide ($NaNH_2$), forming sodium acetylide (NaC_2H) and ammonia (NH_3, pK_a 33).

PROBLEM ANALYSIS

The first step is always to write the reaction given:

$$\underset{(\text{p}K_a\ 25)}{C_2H_2} + NaNH_2 \rightleftharpoons NaC_2H + \underset{(\text{p}K_a\ 33)}{NH_3}$$

We recognize that the reaction is an acid-base reaction because a proton has been transferred from acetylene to the amide anion. Next, we identify which species is the acid (the proton donor) on each side of the equation. On the left side, the species giving up a proton as the reaction proceeds to the right is acetylene. On the right side, the species giving up a proton as the reaction proceeds to the left is ammonia. We look up the pK_a values for the two "acids" (although they are given here, they may also be found in Appendix D) and write them below the formulas. Finally, we compare the strengths of the two acids and determine which is more likely to hold onto its proton. Alternatively, we can deduce the relative strength of the two bases. (The left-side base is NH_2^-, which is stronger than the right-side base, C_2H^-.)

SOLUTION

Ammonia has the larger pK_a, so it is the weaker acid and therefore less likely to give up its proton. Also, the stronger base (NH_2^-) will win the battle for the proton, to form ammonia. Thus, the equilibrium must be to the right.

Doing the same analysis, but using sodium hydroxide in place of sodium amide, we find that if the stated reaction were to proceed, it would be written as follows:

$$\underset{(\text{p}K_a\ 25)}{C_2H_2} + NaOH \rightleftharpoons NaC_2H + \underset{(\text{p}K_a\ 15.7)}{H_2O}$$

Acetylene is the weaker acid of the two; so acetylide anion is the stronger base and will acquire the proton. Thus, this equilibrium lies to the left, and acetylene will not react with sodium hydroxide because it is too weak a base. We can look at this reaction another way: Water is the stronger of the two acids and therefore more likely to give up its proton, again making the reaction proceed to the left.

PROBLEM 1.15

Carbonic acid (H_2CO_3) has a pK_a of 6.4, and acetic acid ($C_2H_4O_2$) has a pK_a of 4.8. Write the acid-base reaction between acetic acid and sodium bicarbonate ($NaHCO_3$). In which direction will the equilibrium lie? Why?

PROBLEM 1.16

The organic compound phenol (C_6H_6O), with a pK_a of 10, reacts with aqueous sodium hydroxide to form sodium phenoxide (Na^+ $C_6H_5O^-$) and water. Explain why, in terms of the pK_a values, this reaction occurs.

In many reactions a proton must be removed from an organic compound, and one major issue is determining how strong a base is required for the reaction. You can analyze prospective reactions by using the approach just outlined. Since the pK_a of typical carbon-hydrogen bonds, such as those in ethane (C_2H_6), is extremely high (pK_a 40–60), these bonds are seldom broken by normal acid-base reactions. Therefore, as you contemplate various reactions and mechanisms, remember that the bases you are probably familiar with, such as sodium hydroxide, are not nearly strong enough to remove a hydrogen bonded to carbon.

1.3.2 Electrophiles and Nucleophiles

In the usual acid-base reaction, discussed in Section 1.3.1, a Brønsted acid is described as a proton donor. In such a reaction, the proton acts as an electron-pair acceptor (that is, it gains a share of a pair of electrons previously "owned" by the base). Any species that serves as an electron-pair acceptor is called an **electrophile** ("electron seeker"). Many species other than protons are electrophiles, but the principle is the same. Electrophiles are also referred to as *Lewis acids.*

$$H{-}\ddot{O}{-}H \; + \; H{-}\ddot{Cl}: \;\rightleftharpoons\; H{-}\ddot{O}^{+}(H){-}H \; + \; :\ddot{Cl}:^{-}$$

Brønsted base / Proton acceptor / Nucleophile / Lewis base (H—Ö—H)

Brønsted acid / Proton donor / Electrophile / Lewis acid (H—Cl)

In the same reaction, the species that attacks the proton is referred to as a Brønsted base and is a proton acceptor. In other terms, such a species is an electron-pair donor, with the implication that it is "electron-rich." In other words, it has a pair of electrons it is able to share with the proton it attacks. Such an attacking species is referred to as a **nucleophile** ("nucleus seeker," or seeker of a positive charge center). Nucleophiles are also referred to as *Lewis bases.*

Many organic reactions are initiated by nucleophiles that are seeking out centers of low electron density with which to share a pair of electrons. Nucleophiles may be negatively charged or neutral; the key is that they have an unshared pair of electrons, which they can use to form a new covalent bond. A curved arrow indicates where a nucleophile attacks another species. For example, in Chapter 3, we will discuss nucleophilic substitution reactions, such as this one:

$$HO\!:^- + CH_3\!-\!Cl\!: \longrightarrow HO\!-\!CH_3 + \,:\!Cl\!:^-$$

Hydroxide anion (nucleophile) — Methyl chloride — Methyl alcohol

The left curved arrow indicates that a nucleophile initiates the reaction by attacking with and sharing a pair of electrons with a susceptible carbon atom.

1.3.3 Introduction to Reaction Mechanisms

Organic compounds undergo reactions and structural transformations as a result of bonds breaking and forming under certain physical conditions and/or in the presence of specific chemical reagents. One of the major goals of organic chemists is to be able to discover the conditions and reagents necessary to bring about a predictable reaction and produce a desired product. This involves understanding the mechanisms (the detailed steps and processes) by which reactions occur.

Many reactions involve an initial bond-breaking step that forms a highly reactive and short-lived species often called a **reactive intermediate.** There are three possible ways for a covalent bond between a carbon atom and another atom to break in terms of the fate of the two shared electrons:

1. Both electrons could end up with the carbon.

$$A\!-\!C \xrightarrow{-A^+} \,^-\!:\!C \quad \text{Carbanion}$$

2. Both electrons could end up with the attached atom (represented here as A).

$$A\!-\!C \xrightarrow{-A\!:^-} \,^+C \quad \text{Carbocation}$$

3. One electron could end up with each of the two atoms.

$$A\!-\!C \xrightarrow{-A\cdot} \cdot C \quad \text{Carbon radical}$$

The reactive intermediates formed in these three instances all contain a trivalent carbon atom (that is, with only three groups covalently bonded to it) and are called, respectively, (1) *carbanions*, (2) *carbocations*, and (3) *carbon radicals* (or simply *radicals*, or *free radicals*). Calculating the formal charge on carbon in the intermediates yields −1 for the carbanion, +1 for the carbocation, and zero for the carbon radical (see Table 1.3).

You will learn more about the important role of these reactive intermediates in this text. To make your learning easier, we will first cover reactions involving only carbocations (Chapters 3 through 8), then introduce reactions involving radicals (Chapter 10), and finally look at reactions involving carbanions (Chapters 13 through 15).

PROBLEM 1.17

Explain the formal charges on the three reactive intermediates—the carbocation, the carbanion, and the radical—in terms of the electron count for carbon. (*Hint:* See Section 1.1.4.)

Finally, you should be aware that covalent bonds don't simply fall apart instantaneously to form a reactive intermediate. Instead, the bond dissociation process is gradual: A breaking covalent bond gradually lengthens as energy is applied, until the two atoms are far enough apart that the bond can no longer hold them together. This is an energy-consuming process and invariably involves a *transition state* (a midpoint structure of highest energy), in which the bond has been partially broken. Partial charges may develop in such a transition state as electron density is shifted. For example, in the reaction of hydroxide anion with *t*-butyl chloride to form *t*-butyl alcohol and chloride anion, the transition states are shown with dashed lines representing the partially broken or formed covalent bonds. The curved arrows indicate the shifting of electrons that is to occur in the *following* step.

$$(CH_3)_3C{-}\ddot{\underset{..}{Cl}}{:} \longrightarrow \left[(CH_3)_3C^{\delta+}\cdots\cdots Cl^{\delta-}\right] \xrightarrow{-\,:\ddot{Cl}:^-} \left[(CH_3)_3C^{+}\right]$$

t-Butyl chloride — Transition state — Reactive intermediate (carbocation)

$$\left[(CH_3)_3C^{+}\right] \xrightarrow{^-:\ddot{O}{-}H} \left[(CH_3)_3C^{\delta+}\cdots\cdots \ddot{O}^{\delta-}{-}H\right] \longrightarrow \left[(CH_3)_3C{-}\ddot{O}{-}H\right]$$

Transition state — *t*-Butyl alcohol

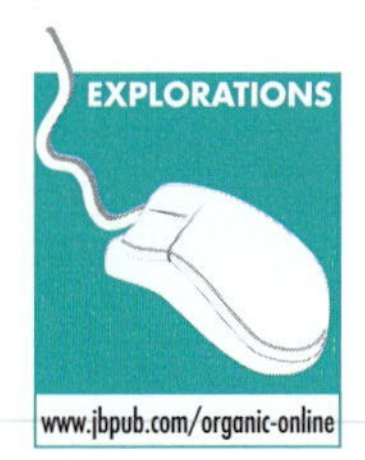

As the carbon-chlorine bond stretches and begins to break, the chlorine atom (the more electronegative of the two atoms forming the bond) gradually acquires the shared electrons, building a partial negative charge (symbolized by $\delta-$ in the transition state). The carbon is losing its share of the bonding electrons and becoming electron-deficient; so a partial positive charge forms (symbolized by $\delta+$ in the transition state). Completion of the bond breaking results in the carbocation reactive intermediate (full positive charge on carbon) and chloride anion (full negative charge). Similar but reverse steps occur in the formation of the new oxygen-carbon bond of the alcohol—the hydroxide ion gradually loses its negative charge as its electrons are shared with carbon.

With this background on the covalent bond to carbon and its implications for the properties of organic compounds, you are now equipped to launch into a study of various families of organic compounds, organized by functional group. Refer occasionally to the Key to Transformations on the inside front cover to visualize the relationships among the families (the "big picture") while you are immersed in the details about each functional group.

Chapter Summary

An atom consists of a *nucleus* surrounded by *electrons* (equal in number to the **atomic number,** Z), which are organized first into **electron shells** designated by **principal quantum numbers** (1,2, . . .) for the element. The electrons in each shell are further organized into **atomic orbitals** (characteristically shaped regions of space with specific energies) designated by lowercase letters (s, p, . . .). The **electron configuration** of an atom of an element can be determined by following three rules that assign the electrons into orbitals, with a maximum of two electrons per orbital. Electrons in the outer shell of an atom are called **valence electrons. Lewis structures** are representations of molecules in which the valence electrons are represented by dots surrounding the symbols for the elements. The **formal charge** on an atom in a molecule may be calculated using the Lewis structure.

An **ionic bond** involves the transfer of electrons from the valence shell of an *electropositive* atom to the valence shell of an *electronegative* atom. The result of the transfer is the completion of an **octet** of electrons in the valence shell of both atoms. In addition, the atoms acquire positive or negative charges, depending on whether they acquired or lost electrons. A **covalent bond** involves two atoms sharing a pair of valence-shell electrons in order for both to achieve an octet.

Carbon ($Z = 6$; $1s^2 2s^2 2p^2$) undergoes mainly covalent bonding. In alkanes, four atomic orbitals of carbon are **hybridized** to form four sp^3 hybrid orbitals that are oriented toward the corners of a *tetrahedron*. Overlap of one of these sp^3 orbitals with an orbital of another atom results in a covalent bond (called a **sigma bond**). As a result, each carbon in an alkane is tetravalent with four *single bonds*.

The **empirical formula** indicates which elements are present in a compound and the ratio of the numbers of atoms of the elements. The **molecular formula** is a whole-number multiple of the empirical formula and indicates the exact number of atoms of each element present. The **structural formula** of a compound is a drawing that indicates which atoms are connected to which (the *connectivities*). Several kinds of structural formulas are used to represent organic compounds: **dot formulas** (Lewis structures), **dash formulas, condensed formulas,** combination (of dash and condensed) formulas, and **bond-line formulas.**

Organic compounds may be classified in several ways: (1) as **acyclic, carbocyclic,** or **heterocyclic;** (2) as **saturated** or **unsaturated;** (3) as **aliphatic** or **aromatic;** or (4) by functional groups or families. All compounds in a **family** have, as part of their structures, the same **functional group**—the part of the molecule that determines the overall properties of the compound and is especially susceptible to chemical reaction.

The nature of the covalent bonds and the orbital hybridization involved determine the specific molecular shape of an organic compound. Carbon-carbon and carbon-hydrogen single bonds are generally **nonpolar** because of the similar **electronegativities** of the atoms involved. However, covalent bonds between carbon and a more electronegative atom, such as a halogen, may be quite polarized, leading to considerable charge separation and a resultant tendency to undergo chemical reactions.

Some organic compounds are weak acids. **Brønsted acids** are proton donors. A *Lewis acid* is an electron-pair acceptor (also called an **electrophile**), and a *Lewis base* is an electron-pair donor (also called a **nucleophile**). The acid strength of weak organic acids is represented by the **acidity constant (K_a)** or the **pK_a.**

Organic compounds undergo chemical reactions through the breaking of one or more bonds. This often involves the formation of a **reactive intermediate** resulting from the breaking of a covalent bond to carbon to form a positively charged carbocation, a negatively charged carbanion, or a neutral carbon radical. The pathway a reaction follows from initiation to product formation is described by its *mechanism*.

Additional Problems

Atomic Structure

1.18 Write the ground-state electron configurations for carbon, nitrogen, and oxygen.

1.19 Write the ground-state electron configurations for potassium and fluorine. Then write the configurations for the two ions in potassium fluoride.

1.20 Indicate whether each of the following elements is electropositive or electronegative:

(a) magnesium (b) sulfur

(c) calcium (d) selenium

1.21 Place the following elements in order of *decreasing* electronegativity: sulfur, oxygen, selenium, tellurium.

1.22 For each of the following atoms, indicate the number of valence electrons: oxygen, carbon, nitrogen, hydrogen, fluorine, sulfur, phosphorus.

Chemical Structure

1.23 Draw all possible structures for compounds with the molecular formula C_4H_9F. Remember that the fluorine can replace any hydrogen in compounds of formula C_4H_{10}.

1.24 Draw a dot formula (Lewis structure) for each of the following compounds:

(a) methyl chloride, CH_3Cl

(b) ethyl alcohol, C_2H_5OH

(c) propyl fluoride, C_3H_7F

1.25 Draw a combination formula for a saturated compound with each of the following molecular formulas, and label each as acyclic, carbocyclic, or heterocyclic:

(a) C_2H_4O (b) C_4H_8 (c) C_5H_{12}

1.26 Write the molecular formula for each of the following compounds:

(a) $CH_3CH_2CHClCH(CH_3)_2$ (b) 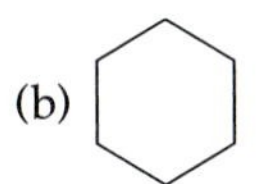

(c) 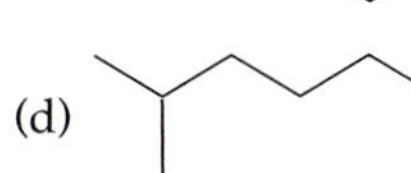

(d)

CONCEPTUAL PROBLEM

Seek, Identify, and Synthesize: Paclitaxel

In the early 1960s, the National Cancer Institute asked researchers to collect and analyze samples of indigenous plants in the hope of isolating substances that might some day prove effective in the fight against cancer. Researchers cast a wide net and among their catches was a chemical compound isolated from the bark of the Pacific yew. To test its medicinal properties, they placed the compound into some artificially preserved cancer cells—it killed the cells. They quickly set about to analyze this cancer-fighting substance: paclitaxel.

- Can you determine the molecular formula of paclitaxel from the structure?

Paclitaxel

- A molecule of paclitaxel has 113 atoms in it. Does that suggest anything to you about how difficult it might be to synthesize in the laboratory?
- Knowing what you do about organic compounds in general (as described near the end of this chapter), would you expect paclitaxel to be water-soluble? If not, what difficulties might that pose for doctors who wish to use this substance to treat their cancer patients?

One of the biggest difficulties facing the research community working with paclitaxel was how little of it could be extracted from the bark of one yew tree—barely a gram. Not only that, but the tree was killed in the process of stripping the bark. It was clear that chemists would have to find a way to synthesize this compound in the lab. Researchers in France, working with the European yew, discovered in the needles of that kind of tree a compound that was closely related to paclitaxel. This substance could be used as the starting point of a "semi-synthesis" that required just a few steps.

Harvesting the bark of the Pacific yew tree yields very small amounts of the naturally occurring cancer-fighting compound we now call paclitaxel. A single 100-year-old tree yields less than a gram, which is not enough for even a single treatment of a cancer patient.

- In addition to providing a relatively easy way to synthesize paclitaxel, what other advantage does use of the compound from the European yew have?

1.27 Write condensed formulas corresponding to the following bond-line formulas:

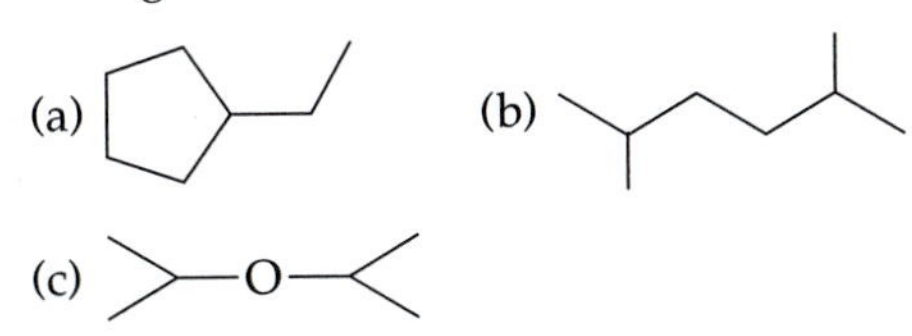

1.28 Draw a dash formula for each of the following compounds, showing the three-dimensional structure:

(a) CH_2Cl_2 (b) CH_3F (c) CH_3CH_2OH

▪ Chemical and Physical Properties

1.29 Arrange the partial structures in each of the following groups in order of *decreasing* bond polarity:

(a) C—H, O—H, N—H

(b) O—H, S—H, Se—H

(c) C—Cl, C—F, C—Br, C—I

(d) C—Li, C—Na, C—K

1.30 Which of the following compounds have a dipole moment? Show each dipole using a structural formula.

(a) CH_3F (b) Cl_3CCCl_3

(c) CH_2Cl_2 (d) $CH_3CH_2CH_2OH$

1.31 Ammonia has a pK_a of about 33, and methanol has a pK_a of about 16. Write the acid-base reaction between the amide anion (NH_2^-) and methanol (CH_3OH), and indicate the direction of the equilibrium.

1.32 Compound A has a K_a of 10^{-7}, and compound B has a K_a of 10^{-4}. Which compound is more acidic? What are the compounds' pK_a values?

1.33 Using arrows to indicate the movement of electron pairs, show the proton-transfer reaction between methanol (CH_3OH, pK_a 16) and sodium hydride [NaH, whose conjugate acid (H_2) has pK_a 35]. Will the equilibrium lie to the right or the left? Why?

▪ Mixed Problems

1.34 You were briefly introduced to reactive intermediates of carbon, two of which were carbanions and carbocations. Indicate the charges associated with the carbanion and the carbocation derived from methane (CH_4), and calculate the formal charges on each atom. Account for these charges in terms of valence electrons. Designate each of these reactive intermediates as an acid or base, and explain your designation.

1.35 The empirical formula of a compound was determined to be C_2H_5O. Its molecular weight was found to be 90. What is its molecular formula?

1.36 Ethers are compounds in which an oxygen atom is bonded to two carbons. Draw all of the ethers with the molecular formula $C_4H_{10}O$.

1.37 Write the electron configuration for atomic boron and aluminum.

1.38 The following four naturally occurring compounds are carboxylic acids and are therefore weakly acidic, with the pK_a values shown. Rank them in order of *decreasing* acidity: formic acid (the substance that makes ant bites sting), pK_a 3.8; ascorbic acid (vitamin C), pK_a 4.2; oxalic acid (found in rhubarb), pK_a 1.3; acetic acid (in vinegar), pK_a 4.8.

1.39 Keeping in mind that carbon, oxygen, nitrogen, and fluorine all tend to maintain an octet of valence electrons, add unshared electron pairs to each of the following formulas to complete all the octets:

(a) H—C(H)(H)—C(H)(H)—O—H (b) CH_3—CH(F)—CH_3

(c) cyclopropyl—NH_2 (d) CH_3—C(=O)—CH_3

(e) CH_3CH_2—O—CH_2CH_3

1.40 Which hydrogen should have the higher electron density, one of those in methane (CH_4) or one of those in silane (SiH_4)? Why?

1.41 Account for the molecular structure and shape of silane (SiH_4) on the basis of the electron configurations of silicon and hydrogen.

1.42 For each of the following molecular formulas, write two different bond-line formulas, one corresponding to a carbocyclic compound and the other to a heterocyclic compound:

(a) C_4H_8O (b) C_3H_7N

1.43 Account for the molecular structure and pyramidal shape of ammonia (NH_3) on the basis of the electron configurations of nitrogen and hydrogen.

CHAPTER 2

Alkanes

THE FIRST FAMILY of organic compounds we will discuss is indeed the "First Family" of organic chemistry. **Alkanes** serve as the parent of all other organic compounds in the sense that all of those compounds can be viewed as being derived from them. Alkanes consist of only carbon and hydrogen and are described as *aliphatic saturated hydrocarbons:*

- **Hydrocarbons**—the only atoms present are carbon and hydrogen
- **Saturated**—all the bonds to carbon are single (sigma) bonds (that is, each carbon has a full complement of four individual bonds)
- **Aliphatic**—alkanes exhibit properties similar to those of fats and oils (*aleiphar* is the Greek word for "fat")

Alkanes can also be classified as **acyclic compounds,** if no carbon rings are present, or **cyclic compounds**, if rings are present (cyclic alkanes are usually referred to as **cycloalkanes**).

Alkanes are commercially important as raw materials for the chemical and petroleum industries. The simplest alkane, methane, is the main constituent of the natural gas used in heating and power systems. Alkanes are important in the study of organic chemistry because they provide a vehicle for learning the terminology of organic chemistry, including *nomenclature* (the assignment of names). However, the reaction chemistry of alkanes is relatively insignificant.

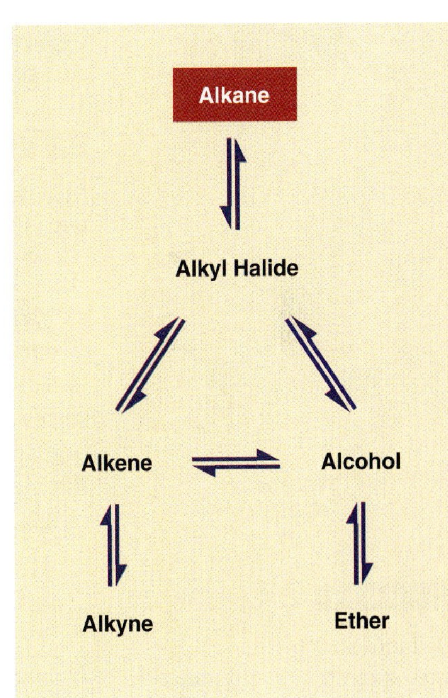

Alkanes are the first family in the Key to Transformations.

2.1

Structure of Alkanes

The simplest member of the alkane family is the one-carbon compound **methane** (CH_4). You learned in Section 1.1.3 that a saturated carbon atom is tetrahedrally hybridized, with four sp^3 hybrid orbitals directed toward the corners of a tetrahedron. Four single bonds are formed to carbon using these orbitals—in the case of methane, the four bonds are to hydrogen atoms.

You also learned in Section 1.1.3 that carbon atoms readily bond to each other. Chemists often use the notation C_1, C_2, C_3, and so on, to designate the number of carbons in an organic compound. Thus, if one hydrogen of methane (CH_4) is replaced by a carbon atom, and its remaining valence orbitals are filled by electrons shared with hydrogens, the C_2 alkane **ethane** (C_2H_6) results. Furthermore, because all four carbon-hydrogen bonds of methane are *equivalent,* it doesn't matter which hydrogen is replaced. The same structure results: There is only one ethane. Replacement of

◀ Methane gas is a natural component of petroleum. Here, at an offshore drilling site, the methane cannot be stored or shipped, and so it is simply burned off (or "flared").

one hydrogen of ethane (all six hydrogens are equivalent) with another carbon (and three hydrogens bonded to it) results in a C_3 alkane, **propane** (C_3H_8). These compounds are the three simplest alkanes.

Methane (CH_4)

Ethane (CH_3CH_3)

Propane ($CH_3CH_2CH_3$)

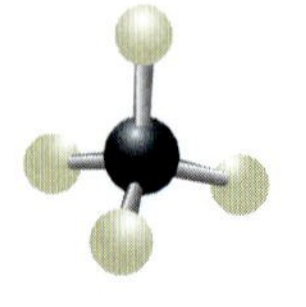

Methane (CH_4)

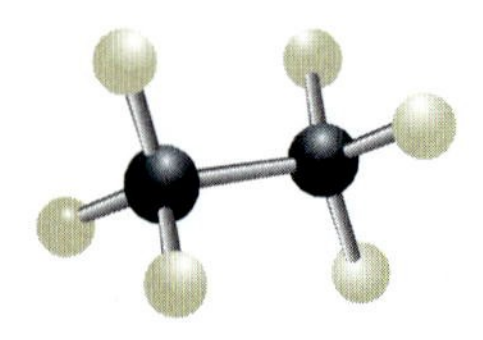

Ethane (C_2H_6)

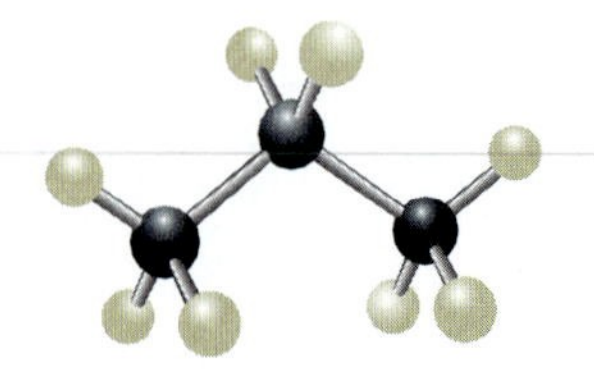

Propane (C_3H_8)

FIGURE 2.1

Ball-and-stick models of methane, ethane, and propane.

At this point, take careful note of the difference between two terms frequently used in organic chemistry: *equivalent* and *identical.* They are *not* the same. Saying that two or more atoms are *equivalent* means that their surroundings and attachments are identical. When either of two equivalent atoms is replaced by any other atom or group, identical new compounds result. For example, all of the hydrogens in ethane are equivalent, since replacement of any one of them with a —CH_3 (methyl) group produces propane. Compounds are *identical* when they have the same composition and structure. This means that there is absolutely no difference between the two; the structure of one can be superimposed on that of the other and every feature will coincide. Their physical and chemical properties are totally indistinguishable. For example, there is no difference between natural vitamin A and synthetic vitamin A—they are identical. By analogy, two dinner forks from the same set of silverware are identical, but the two shoes of a pair are not identical (how do you verify this?).

PROBLEM 2.1

Make a ball-and-stick model of methane. Pick any two hydrogens and replace each, one at a time, with a methyl group (—CH_3). Were the hydrogens equivalent? Are the two products identical?

Methane, ethane, and propane represent part of a **homologous series,** that is, a set of compounds that differ from one another by a common increment. In this case, the increment is a —CH_2— (*methylene*) group. You can picture each member of the alkane series as the same as the one before it, but with an extra CH_2 inserted between a carbon and a hydrogen or another carbon (Figure 2.1). This increment is also apparent in the molecular formulas of the series members: CH_4, C_2H_6, C_3H_8, and so on. These, too, indicate a common difference of CH_2. Thus, alkanes are defined in two additional ways:

1. All alkanes are members of a homologous series whose structural formulas can be pictured as having been "constructed" by inserting a methylene group (—CH_2—) into the formula of the preceding member.
2. The molecular formulas of all alkanes are represented by the general formula C_nH_{2n+2}, where n is any whole number. This formula can be remembered by considering alkanes as long chains of methylene groups (represented by C_nH_{2n}) that must be "capped" at each end by a hydrogen (the source of the +2 in the subscript for H in the general formula).

How to Solve a Problem

What is the molecular formula of a 10-carbon (C_{10}) alkane?

SOLUTION

The general formula for alkanes is C_nH_{2n+2}. If $n = 10$, then the number of hydrogens must be $(2 \times 10) + 2 = 22$. The formula is $C_{10}H_{22}$.

PROBLEM 2.2

Write the molecular formulas for alkanes with (a) 7 carbons, (b) 16 carbons, (c) 30 carbons, and (d) 100 carbons.

Using the general formula C_nH_{2n+2}, you can easily see that the *molecular* formula of a C_4 alkane is C_4H_{10}. Depicting the *structural* formula is more complicated, however, because of the number of possibilities. The fundamental challenge in drawing any structural formula is to depict how the carbon atoms (the backbone, or skeleton) of the molecule are connected. It turns out that there are two, and only two, different ways in which four carbon atoms can be connected, as these dash and bond-line formulas show:

You can arrive at these two structures by starting with the structural formula of propane (C_3H_8) and adding one methylene (CH_2) group in each of the possible positions, and then eliminating duplicate structures. By filling each remaining valence orbital of the C_4 carbon skeletons shown above with electrons shared with hydrogen, you get two structures, each with the molecular formula C_4H_{10}. The obvious conclusion is that two different compounds (that is, two alkanes, each with a unique structural formula) are represented by the molecular formula C_4H_{10}. There are in fact two such compounds, each with distinct physical and chemical properties. Their historic names are *n*-butane and isobutane. We will explore the relationship between these two similar compounds in the next section.

Butane torches (and cigarette lighters) use a mixture of *n*-butane and isobutane.

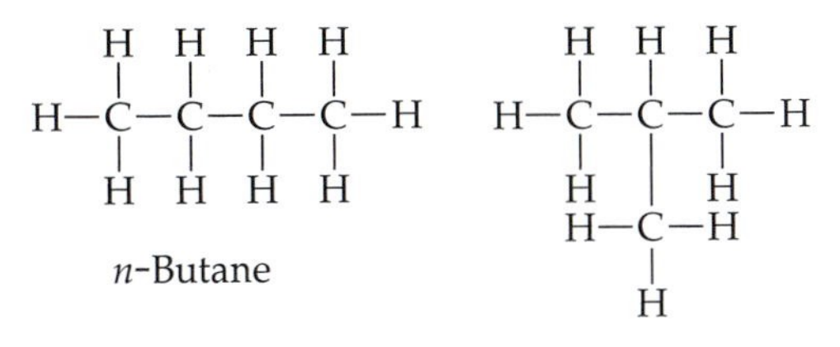
n-Butane Isobutane

How to Solve a Problem

Derive the structures of the two possible C_5 alkanes (C_5H_{12}) that result from adding a methylene group to *n*-butane (C_4H_{10}). (*Note:* The processes described in the following analysis and solution are mental processes, not chemical processes.)

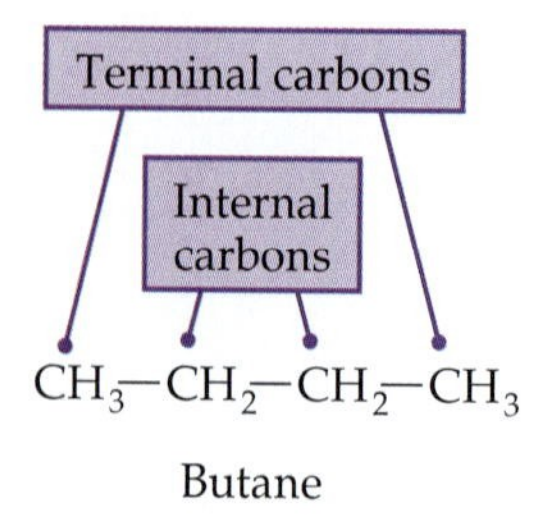

PROBLEM ANALYSIS

To go from a C_4 alkane to a C_5 alkane, we envision adding one carbon at all possible locations and then test our results. Inspection of *n*-butane's structural formula reveals two structurally different kinds of carbon atoms. At each end (or *terminus*) of the structure, there are equivalent carbons (*terminal* carbons) that have only one carbon attached to them. The other two *internal* carbon atoms are also equivalent but are positionally and structurally different from the terminal carbons; they are located in the middle of the chain and each has two carbons attached to it.

SOLUTION

The fifth carbon can be added to the carbon skeleton of *n*-butane at *either end* (that is, at either terminal carbon) of the four-carbon chain. This yields a linear (nonbranched) chain of five carbon atoms. If we add the fifth carbon to an *internal* carbon of *n*-butane, we obtain a branched chain of carbon atoms. The same branched chain results from attaching the fifth carbon to either internal carbon. Thus, two structurally different C_5 alkanes (pentanes) can be formed by adding a fifth carbon to *n*-butane: *n*-pentane and isopentane.

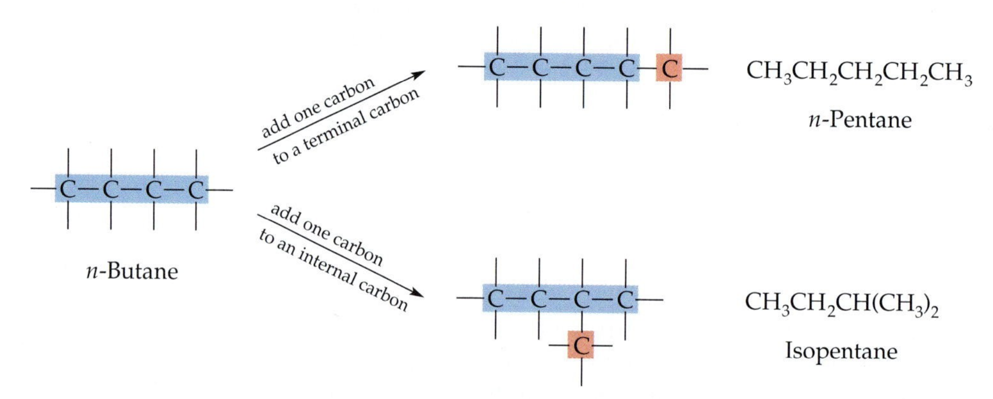

We could have approached the problem by simply drawing as many combinations of five carbons as we could envision and then eliminating duplicates. However, the process we chose to use here is more systematic, making it less likely that we would miss one of the structures.

PROBLEM 2.3

Write the condensed structural formulas for all three possible pentanes.

PROBLEM 2.4

Write the bond-line and the condensed structural formulas for the five possible C_6 alkanes (hexanes).

2.2

Isomerism

You have just seen that the same molecular formula can represent more than one compound. Compounds that have the same molecular formula but different structures are called **isomers.** There are many different kinds of isomers of organic compounds,

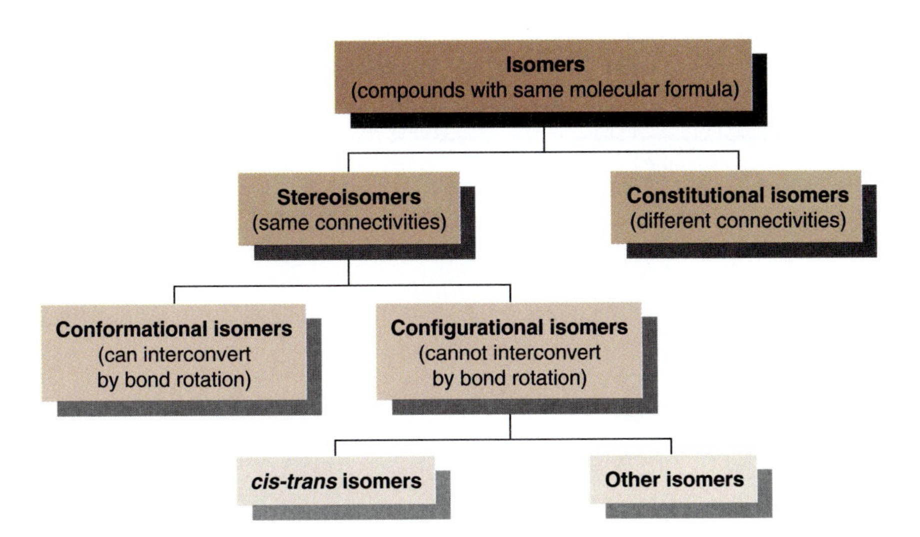

FIGURE 2.2

Selected types of isomers.

and you will be gradually introduced to them in this book. Figure 2.2 is a chart showing the kinds of isomers we will discuss in this chapter. (The inside back cover contains a chart of all the different kinds of isomers discussed in this book; they will be reviewed together in Chapter 17. The chart can serve as a handy reference as you progress through the book.)

2.2.1 Constitutional Isomers

You encountered **constitutional isomers** in Section 2.1 in the form of *n*-butane and isobutane. As you can see from Figure 2.2, constitutional isomers *have the same molecular formula, but different connectivities.* The difference between the two butanes lies in which atoms are connected to which (the connectivities), and this is best seen by examining the carbon skeletons.

n-Butane Isobutane

Note that *n*-butane is a straight chain, whereas isobutane has one branch. The two isomers are different compounds and have different physical and chemical properties, including quite different molecular shapes, as shown by molecular models (Figure 2.3). A mixture of the two isomers can be physically separated into two pure components. The only common feature they share is their molecular formula, C_4H_{10}.

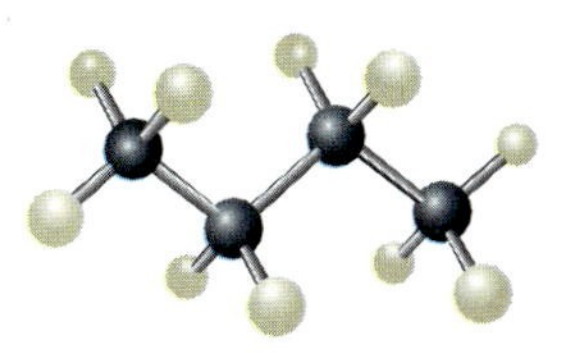

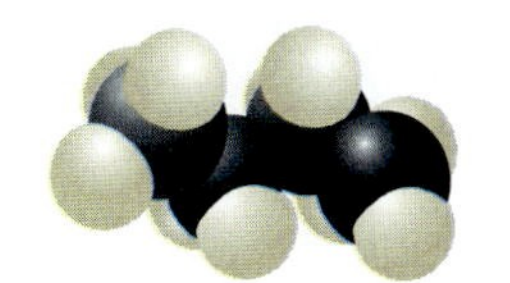

n-Butane

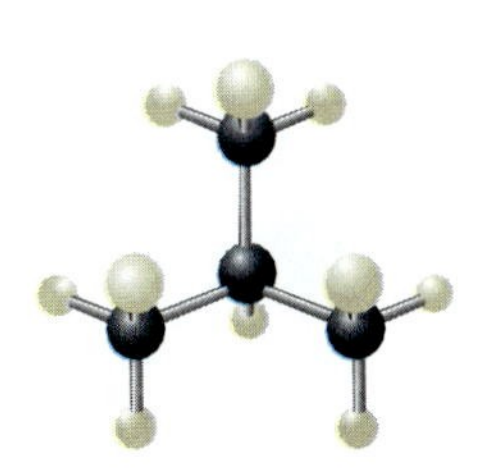

Isobutane

FIGURE 2.3

Space-filling and ball-and-stick models of *n*-butane and isobutane. Constitutional isomers can be distinguished by examining their two-dimensional structural formulas, but their differences are better reflected by models showing the relative sizes and spatial arrangement of the atoms.

TABLE 2.1

Alkanes: Formulas and Isomers

Number of Carbons	Molecular Formula	Number of Constitutional Isomers
1	CH_4	0
2	C_2H_6	0
3	C_3H_8	0
4	C_4H_{10}	2
5	C_5H_{12}	3
6	C_6H_{14}	5
7	C_7H_{16}	9
8	C_8H_{18}	18
9	C_9H_{20}	35
10	$C_{10}H_{22}$	75
15	$C_{15}H_{32}$	4,347
20	$C_{20}H_{42}$	366,319
40	$C_{40}H_{82}$	>60 trillion

As the number of carbons in an alkane increases, the number of possible constitutional isomers grows dramatically, as shown in Table 2.1.

PROBLEM 2.5

The following formulas represent four pairs of constitutional isomers. Identify each pair.

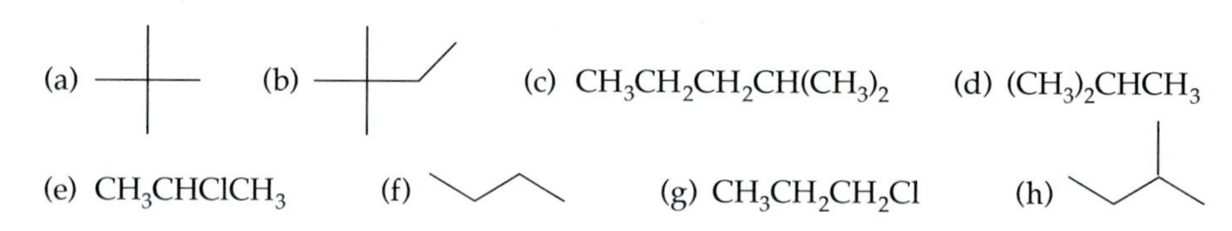

2.2.2 Conformational Isomers

A major category of isomers consists of **stereoisomers.** Two stereoisomers have the same molecular formula and the same connectivities, but *differ in their three-dimensional orientation.*

This section introduces only one kind of stereoisomer: **conformational isomers.** A single compound may exist in two or more conformations. The difference between these isomers, or **conformers** as they are sometimes called, is the result of rotation about a single bond (Figure 2.4).

A.

Axis of rotation

rotate 180°

B.

A.

rotate 180°

B.

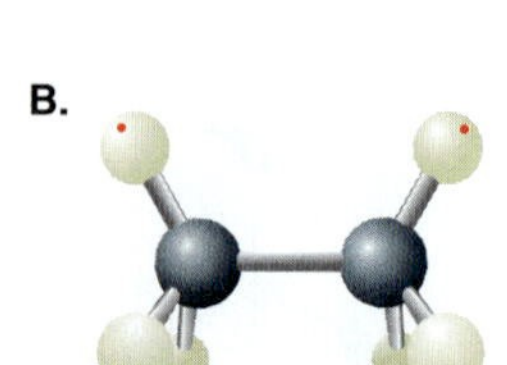

FIGURE 2.4

Two views of ball-and-stick models of two conformers of ethane; conformer B results from a 180° rotation about the C—C bond of A.

Two conformers cannot be physically separated because the energy difference between them is so small that most readily *interconvert* at room temperature. However, at very low temperatures the existence of conformational isomers can be proved using spectroscopic methods (more on this in Chapter 11). The difference between two conformers is best seen using molecular models. However, it is also essential to learn to use two-dimensional drawings to represent their three-dimensional structures.

Conformational isomers owe their existence to the tetrahedral nature of carbon bonding and the fact that the sigma bond is cylindrically symmetrical (see Section 1.1.3). The conformations of ethane, for example, can best be envisioned by approaching the molecule from one end of the central sigma bond. From this perspective, we see only the closest carbon atom; the other is hidden from view behind the closest carbon. We can use *Newman projections* to represent this view, as shown in Figure 2.5. Imagine that the carbon facing us and its three attached hydrogens are fixed in space, while the second (hidden) carbon and its hydrogens are rotated about the axis of the carbon-carbon single bond. All we can see of the hidden carbon are the hydrogen atoms sticking out and rotating. If the rotation is stopped when the hydrogens on each carbon are directly aligned parallel to each other, what results is an **eclipsed conformation.** Each hydrogen on the first carbon "eclipses" a hydrogen on the second carbon. If the rotation is instead stopped when all six hydrogens are visible and equally spaced, what results is a **staggered conformation.** These two conformations are shown in Figure 2.5, using Newman projections, dash-wedge structural formulas, and sawhorse structural formulas.

The staggered conformation is more stable (of lower energy) than the eclipsed conformation because it minimizes the interactions between hydrogens—that is, in the staggered conformation, the hydrogen atoms and their bonding electrons are farther apart from each other across the central bond and so there is less repulsion between them. (Remember that the carbon-hydrogen bonds to each carbon are angled at 109.5° to each other as well as to the carbon-carbon bond and do not lie in the same plane, at 120°, as the Newman projection might suggest.) Repulsive forces come into play when any two atoms or groups are too close together. These forces usually result from both electronic repulsion and steric repulsion (which relates to spatial volume, especially a factor when a group of atoms is bonded to carbon).

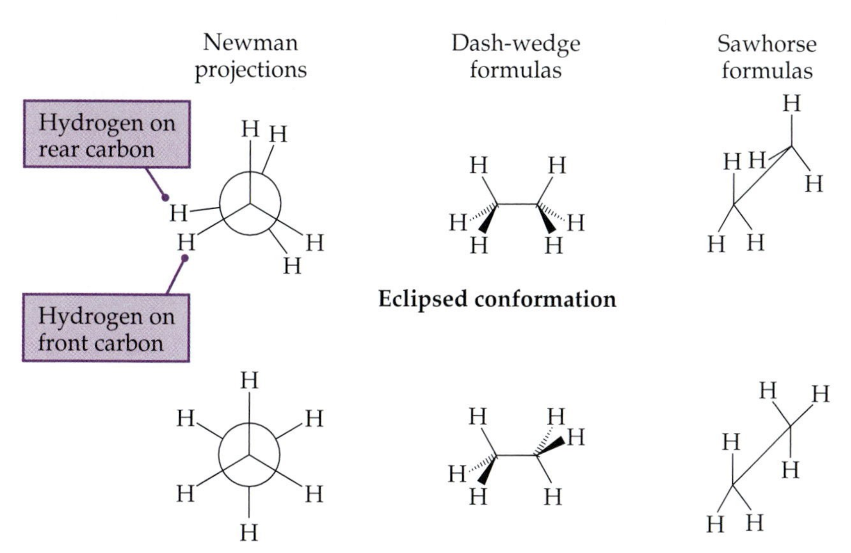

FIGURE 2.5

Three representations of the eclipsed and staggered conformations of ethane.

FIGURE 2.6

Conformations of *n*-butane (C_4H_{10}) at carbons 2 and 3. The gold arrows indicate the direction of rotation of the rear methyl group to arrive at the next conformation.

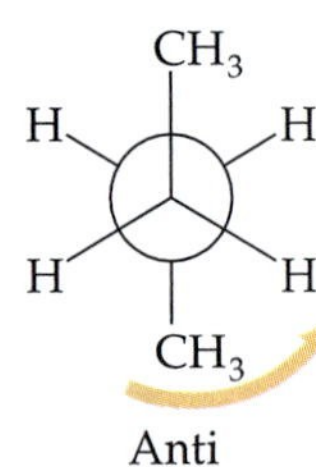

Examination of the conformations of larger alkanes, such as *n*-butane ($CH_3CH_2CH_2CH_3$), reveals larger repulsive effects and the possibility of additional conformations. Figure 2.6 shows Newman projections of several conformations of *n*-butane about the middle carbon-carbon bond, with the front carbon remaining fixed. The **anti** and **gauche conformations** are two variants of a staggered conformation: An *anti conformation* has the two bulkiest groups oriented at 180° to each other; a *gauche conformation* has the two bulkiest groups oriented at 60° to each other. The order of stability for these four conformations of *n*-butane is

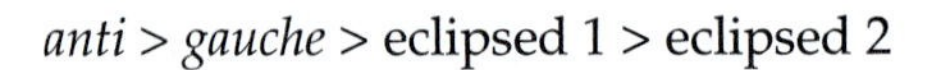

anti > *gauche* > eclipsed 1 > eclipsed 2

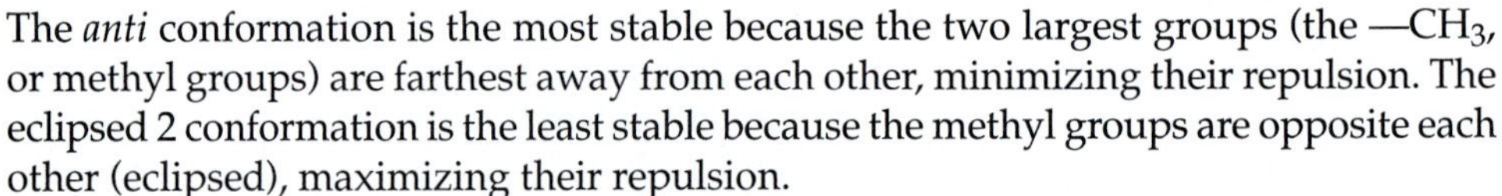

The *anti* conformation is the most stable because the two largest groups (the —CH_3, or methyl groups) are farthest away from each other, minimizing their repulsion. The eclipsed 2 conformation is the least stable because the methyl groups are opposite each other (eclipsed), maximizing their repulsion.

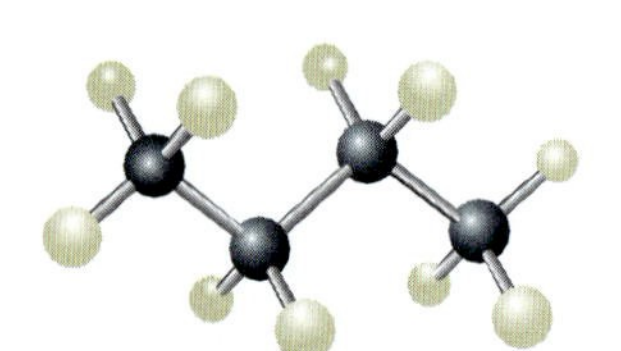

Ball-and-stick model of *n*-butane

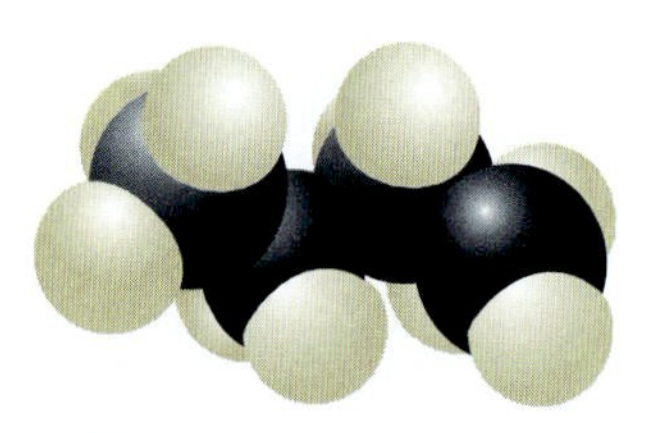

Space-filling model of *n*-butane

FIGURE 2.7

The most stable (*anti*) conformation of *n*-butane (C_4H_{10}).

Of all the possible conformers about each of the three C—C bonds of *n*-butane, the most stable conformation is that with *anti* orientations about all three C—C bonds, as shown in Figure 2.7.

When writing the structures of alkanes and other compounds, you will not usually have to be concerned with conformational isomers. However, in some circumstances, judging the relative stability of specific conformations is critical to your understanding of a chemical reaction. Conformations of any organic molecule can be analyzed using stereochemical considerations, as we just did. Remember that conformational isomers are only one kind of stereoisomer, but they are unique in being mutually interconvertible simply by rotation about a carbon-carbon single bond. There are usually only small energy differences between conformers, but the larger the substituents on the single-bonded carbons, the larger these energy differences will be.

PROBLEM 2.6

Draw the most and the least stable conformations of 1,2-dichloroethane ($ClCH_2CH_2Cl$).

PROBLEM 2.7

(a) Draw the combination formula for the straight-chain C_5 compound pentane (C_5H_{12}).

(b) Draw the most stable conformation for this compound about the bond between carbon 1 and carbon 2.

(c) Draw the most stable conformation about the bond between carbon 2 and carbon 3.

2.3

Nomenclature of Alkanes

Nomenclature is the systematized assignment of names to compounds. Every organic compound must have a unique name so that chemists and nonchemists alike can communicate about compounds without any ambiguity. The International Union of Pure and Applied Chemistry (IUPAC) developed the current nomenclature system, which relies on a number of "rules" that must be learned, not just memorized. The system starts with alkanes as its foundation; the nomenclature of other families of organic compounds is derived from that for the alkanes. A summary of the rules of the IUPAC system as applied to all families of organic compounds appears in Appendix A.

2.3.1 Unbranched Alkanes

The names we have been using so far in this chapter are historical names; *n*-butane, for example, was given that name because it was considered one of the *normal* alkanes. Because of serious limitations with other than simple historical names, you need to learn the IUPAC system of nomenclature. Names of unbranched (straight- or continuous-chain) alkanes contain two components: the stem and the suffix. A Latin or Greek term is used for the **stem** to indicate the number of carbons in the chain. For example, the name of a six-carbon alkane has *hex-* (derived from the Greek word for "six") as the stem. However, for the first four compounds, historical stems were adopted: *meth-*, *eth-*, *prop-*, and *but-*. The name of the alkane is completed by adding *-ane* as the **suffix.** (It is the *-ane* ending on any name that indicates that the compound is an alkane. Other endings are used for other families of compounds, as we will see in later chapters.) The name that results from combining the stem and the suffix is the **parent name** for carbon chains of that particular length. Table 2.2 contains the parent names of the first 20 alkanes.

TABLE 2.2

IUPAC Parent Names of Unbranched Alkanes

Number of Carbon Atoms	Parent Name	Number of Carbon Atoms	Parent Name
1	Methane	11	Undecane
2	Ethane	12	Dodecane
3	Propane	13	Tridecane
4	Butane	14	Tetradecane
5	Pentane	15	Pentadecane
6	Hexane	16	Hexadecane
7	Heptane	17	Heptadecane
8	Octane	18	Octadecane
9	Nonane	19	Nonadecane
10	Decane	20	Icosane

2.3.2 Alkyl Groups

With the nomenclature of alkanes as a foundation, we can now consider derivatives. An **alkyl group** results when a hydrogen is removed from an alkane. An alkyl group is *not* a compound but is simply a structural segment that exists only as a substituent of a larger structure. The name of an alkyl group is formed by deleting the *-ane* ending of the alkane name and replacing it with *-yl*. Thus, methane (CH_4) gives rise to the *methyl* group (CH_3—), and propane ($CH_3CH_2CH_3$) gives rise to the *propyl* group ($CH_3CH_2CH_2$—). In IUPAC nomenclature, only simple alkyl groups (methyl, ethyl, propyl, and butyl) are generally referred to, and the hydrogen removed from the alkane in these instances is always a terminal hydrogen (from a terminal CH_3 group of the alkane).

CH_3—	CH_3CH_2— or C_2H_5—	$CH_3CH_2CH_2$— or C_3H_7—	$CH_3CH_2CH_2CH_2$— or C_4H_9—
Methyl	Ethyl	Propyl	Butyl

An alkyl group can be written as a condensed formula or as a molecular formula, as shown above. The symbol R is frequently used to represent a generic alkyl group attached to another structural unit. For example, RCl represents any alkyl chloride.

2.3.3 Branched Alkanes

As we have seen, not all alkanes consist of straight chains of carbon. Branched chains provide great diversity in alkane structure. Names of branched alkanes are constructed step by step using the following IUPAC rules:

1. *Identify the longest continuous unbranched chain of carbon atoms in the molecule and assign the appropriate parent name (from Table 2.2).* Thus, compound **A** is named as a derivative of the parent compound *butane* (four carbons in the longest chain).

$$CH_3CH_2CH(CH_3)_2 \equiv CH_3{-}CH_2{-}\underset{\displaystyle CH_3}{\underset{|}{CH}}{-}CH_3 \equiv$$

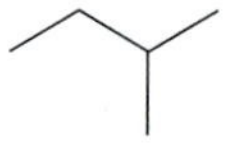

Compound **A** (2-methylbutane)

Compound **B** is named as a derivative of the parent *octane* (eight, not seven, carbons in the longest continuous chain). (Don't be fooled by the way in which a formula is written!)

$$CH_3CH_2CH_2CH_2CH(CH_2CH_2CH_3)CH_2CH_3 \equiv CH_3{-}CH_2{-}CH_2{-}CH_2{-}\underset{\displaystyle CH_2{-}CH_2{-}CH_3}{\underset{|}{CH}}{-}CH_2{-}CH_3 \equiv$$

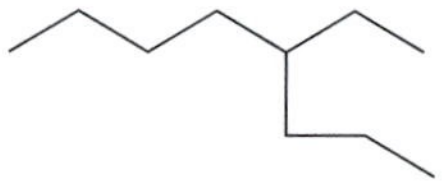

Compound **B** (4-ethyloctane)

2. *Number the* longest *chain, starting from the end closest to any branch (that is, so that the position to which any substituent is attached has the lowest possible number).* Thus, the longest chain of compound **A** is numbered from the right, and that of compound **B** is numbered from the lower carbon.

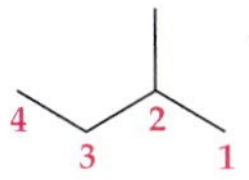

Compound A

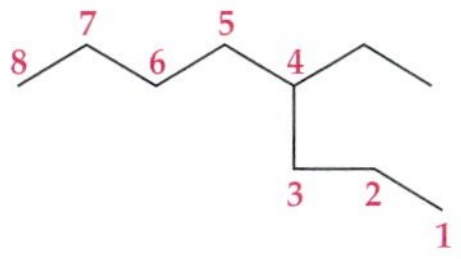

Compound B

3. *Designate the number of the carbon atom to which a substituent is attached (that is, at which branching occurs).* In compound **A,** the branch is at carbon 2, and in compound **B,** it is at carbon 4. These numbers appear before the name of the substituent.

Compound **A:** 2-"substituent"butane Compound **B:** 4-"substituent"octane

Note that in both **A** and **B** the numbers are smaller than if the carbon chain had been numbered from the other direction.

4. *Establish the name of the substituent (that is, the group attached to the main chain).* In compound **A,** the substituent is a methyl (CH_3—) group; in compound **B,** it is an ethyl (CH_3CH_2—) group.

5. *Write the full name.* Compound **A** is 2-methylbutane. (Check that you have the right total number of carbon atoms in the name—in this instance, a total of five carbons.) Compound **B** is 4-ethyloctane (a total of 2 + 8 = 10 carbons). In learning to use the IUPAC system, you will find it easier initially to develop a name from right to left. In other words, name the parent chain, then name any substituent, and finally determine the location (the number) of the substituent. Once you have fully named a compound, you should check the name by verifying that it accounts for each part of the compound's structure.

6. *If there are two or more substituents on the longest chain, name each and place the names in alphabetical order before the parent name; then precede each substituent with the number designating its point of attachment to the chain.* Examples are 4-ethyl-2-methyloctane and 2,3-dimethylpentane:

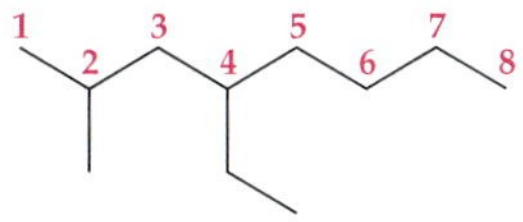

4-Ethyl-2-methyloctane

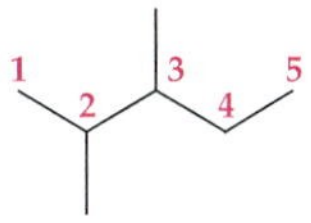

2,3-Dimethylpentane

If there are identical substituents at two or more positions on the longest chain, the prefixes *di-*, *tri-*, and so on, are used, but there must be a number for each individual substituent. (The prefixes do not affect the alphabetization.) Thus, the name for the five-carbon alkane shown to the right is 2,2-dimethylpropane, *not* 2-dimethylpropane. Note that the numbers are separated from one another by a comma and separated from the substituent name by a hyphen. There is no hyphen between the substituent name and the parent name.

2,2-Dimethylpropane

Chemical structures must frequently be represented in a line of text, such as this, and in this case condensed formulas are used. In a condensed formula, parentheses may be used to enclose substituents following the carbon to which they are attached. Thus, 2-methylpentane is written $CH_3CH(CH_3)CH_2CH_2CH_3$, and 2,2-dimethylpropane is written $C(CH_3)_4$.

$$CH_3{-}\underset{\displaystyle CH_3}{\underset{|}{CH}}{-}CH_2{-}CH_2{-}CH_3 \equiv (CH_3)_2CHCH_2CH_2CH_3 \qquad CH_3{-}\overset{\displaystyle CH_3}{\overset{|}{\underset{\displaystyle CH_3}{\underset{|}{C}}}}{-}CH_3 \equiv C(CH_3)_4$$

2-Methylpentane 2,2-Dimethylpropane

The exception to this rule is that when two or three identical groups are attached to the second or second-to-last carbon, they may be written together in parentheses with a subscript, as in the following examples: 4-ethyl-2-methyloctane is written $(CH_3)_2CHCH_2CH(C_2H_5)CH_2CH_2CH_2CH_3$, and 2,2,3-trimethylbutane is written $(CH_3)_3CCH(CH_3)_2$.

How to Solve a Problem

(a) Assign an IUPAC name to $CH_3CH_2C(C_2H_5)_2CH_2CH_3$.

SOLUTION

We determine the longest straight chain of carbon atoms. If necessary, we draw a bond-line structure or a dash formula first.

The longest chain has five carbons, so the parent name is *pentane*. Numbering from right to left or from left to right produces the same position to which the substituents are attached: carbon 3. The substituents, of which there are two, are identical ethyl groups. Thus, the name is 3,3-diethylpentane. We can check by counting the number of carbons in the drawn structure (9) and the number of carbons represented by the various parts of the name (two ethyl groups = 4 and a pentane chain = 5, for a total of 9).

(b) Write a structural formula representing 2,2,4-trimethylpentane (a compound used as a reference in determining the octane number of gasoline).

SOLUTION

The first step is to write the parent chain of five carbons and then number it from one end. The next step is to add two methyl groups at carbon 2 and one methyl group at carbon 4 to produce the bond-line formula for 2,2,4-trimethylpentane:

At first you may have to begin with a dash or bond-line formula, but you should quickly learn to derive the condensed formula directly from the name—for example, 2,2,4-trimethylpentane is $(CH_3)_3CCH_2CH(CH_3)_2$. Check each formula to be sure that each carbon atom has four groups attached to it.

PROBLEM 2.8

Assign IUPAC names to the following compounds:

(a) [bond-line structure] (b) $CH_3CH_2CH_2CH(CH_3)_2$ (c) $(CH_3)_2CHCH_3$

PROBLEM 2.9

Draw bond-line formulas for (a) 2,2-dimethylpropane, (b) 3-ethylhexane, and (c) 2,4-dimethyloctane.

PROBLEM 2.10

Write condensed formulas for (a) 3-methylpentane and (b) 2,2-dimethylpentane.

2.3.4 Historical/Common Names

As was mentioned in the introductory essay, organic chemistry is a young science but an old subject. Historically, the name for an organic compound was assigned by its discoverer long before the structure was known and often in reference to the compound's source. For example, acetic acid, the acid present in vinegar, was derived from *acetum*, the Latin word for "vinegar." And the name for formic acid, which is present in some ants, was derived from *formicae*, the Latin word for "ants." However, with the tremendous increase in the number of known organic compounds, the need arose for a systematic approach to nomenclature. Thus, the IUPAC system was developed. Still, many historical names remain in common use, especially those of some very complex natural products, such as morphine, cholesterol, and paclitaxel (taxol). Some historical names—such as those of the simplest alkanes, methane, ethane, propane, and butane—were even incorporated into the IUPAC system (see Table 2.2).

The historical/common names of C_3 and C_4 compounds—propane, *n*-butane, and isobutane—have given rise to alkyl group names that remain in active use (Figure 2.8). Propane yields two different alkyl groups, the *n*-propyl and isopropyl groups, depending on which hydrogen is removed. The prefix *n*- signifies that a hydrogen has been removed from a terminal carbon; the prefix *iso*- signifies that a hydrogen has been removed from an internal carbon, thus forming a branch. Likewise, *n*-butane gives rise to the *n*-butyl and *sec*-butyl groups, while isobutane gives an isobutyl or *tert*-butyl group. The prefixes *sec*- (for secondary) and *tert*-, or *t*- (for tertiary), are also indicators of the location of an internal carbon from which hydrogen has been removed and of the presence of branches (see Figure 2.8).

2.3.5 Classification of Carbons, Hydrogens, and Alkyl Groups

Carbon atoms in organic molecules are classified by the number of other carbons attached directly to them, using the terms *primary* (one other carbon attached, symbolized 1°), *secondary* (two other carbons attached, symbolized 2°), *tertiary* (three other carbons attached, symbolized 3°), and *quaternary* (four other carbons attached, symbolized 4°). Thus, propane ($CH_3CH_2CH_3$) has two primary carbons (the terminal carbons) and one secondary carbon (the internal carbon). The compound 2-methylpropane, $(CH_3)_3CH$ (common name, isobutane), has three primary carbons and one tertiary car-

FIGURE 2.8

Derivation of names of common alkyl groups.

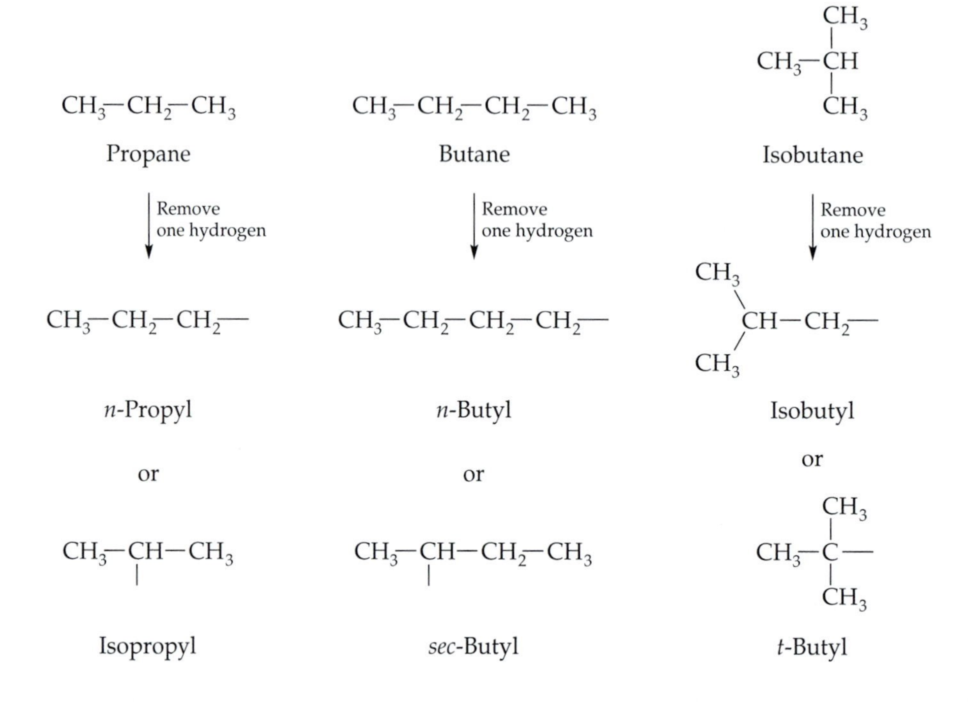

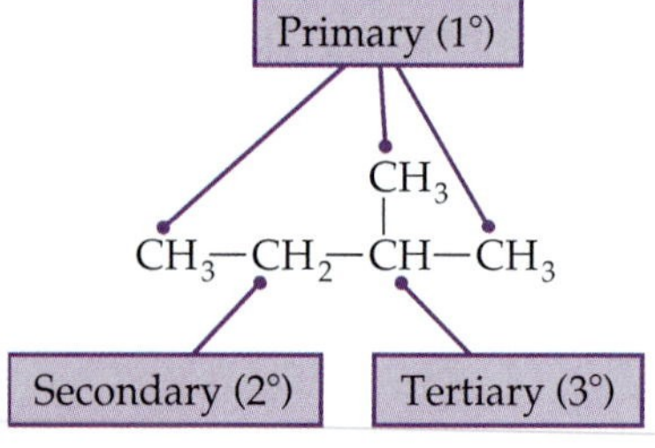

2-Methylbutane (isopentane)

bon; 2-methylbutane, $CH_3CH_2CH(CH_3)_2$ (common name, isopentane), has three primary, one secondary, and one tertiary carbon, as shown to the left.

Hydrogens are also classified as primary, secondary, or tertiary, based on the kind of carbons to which they are attached. Therefore, isobutane (2-methylpropane), $(CH_3)_3CH$, has nine primary hydrogens and one tertiary hydrogen; 2-methylbutane (isopentane), $CH_3CH_2CH(CH_3)_2$, has nine primary hydrogens, two secondary hydrogens, and one tertiary hydrogen.

The terms *primary*, *secondary*, and *tertiary* are also applied to alkyl groups (see Figure 2.8). Thus, the *n*-propyl ($CH_3CH_2CH_2$—) group is primary, the isopropyl [$(CH_3)_2CH$—] group is secondary, and the *t*-butyl [$(CH_3)_3C$—] group is tertiary. Alkyl groups, which differ greatly in size, are often named as substituents in other molecular structures (Figure 2.9).

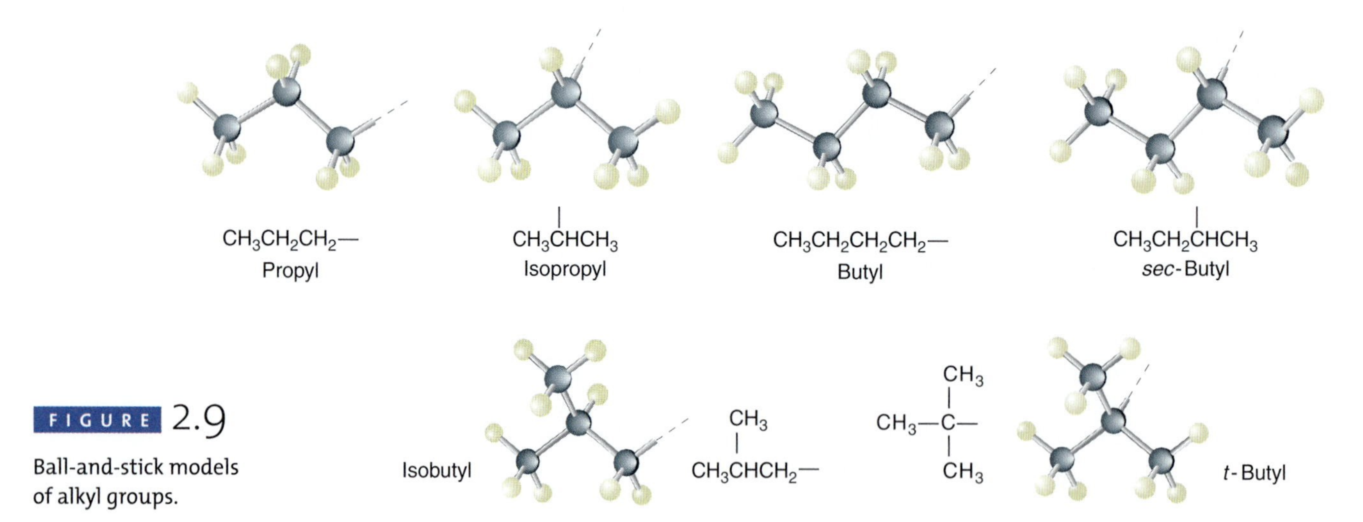

FIGURE 2.9

Ball-and-stick models of alkyl groups.

PROBLEM 2.11

Draw the structure of a primary, a secondary, and a tertiary alkyl group derived from 2-methylpentane.

2.4

Properties, Sources, and Uses of Alkanes

2.4.1 Properties of Alkanes

The physical state of straight-chain alkanes at room temperature varies with their molecular size: The C_1–C_4 compounds are gases, the C_5–C_{16} compounds are liquids, and the C_{17} and higher compounds are solids. For example, methane (C_1) boils at −164°C, pentane (C_5) boils at 36°C, hexadecane (C_{16}) boils at 287°C, and icosane (C_{20}) boils at 343°C and melts at 37°C. Long-chain alkanes are known as **waxes.** Some examples are $C_{27}H_{56}$, the wax found in apple skin; $C_{31}H_{64}$ found in tobacco leaves; and beeswax (a mixture). Most long-chain alkanes have a waxy or oily texture. (Aliphatic compounds, of which alkanes are the parent family, also are referred to as *paraffinic* compounds.) Although you are not expected to memorize specific boiling and melting points, you should develop a feel for general patterns. Generally speaking, higher molecular weight leads to higher boiling points.

Apple skin contains a natural wax, an alkane of formula $C_{27}H_{56}$.

For straight-chain alkanes and for other families of straight-chain compounds, boiling points increase regularly with an increase in chain length. This is a result of the increase in molecule size, leading to larger surface areas for the molecules. The larger the surface area, the greater are the attractive van der Waals forces between molecules. This means that more energy is needed to separate the individual molecules and thus permit vaporization. Alkanes with increased branching have lower boiling points (that is, less energy is required for vaporization) than do the straight-chain isomers. Branched molecules are more compact, with smaller surface areas.

How to Solve a Problem

Arrange the three C_5H_{12} isomers—pentane, 2,2-dimethylpropane, and 2-methylbutane—in order of *decreasing* boiling point.

SOLUTION

First, let's write a structural formula for each isomer:

$$CH_3{-}CH_2{-}CH_2{-}CH_2{-}CH_3$$

Pentane

$$CH_3{-}\overset{\displaystyle CH_3}{\underset{\displaystyle CH_3}{\overset{|}{\underset{|}{C}}}}{-}CH_3$$

2,2-Dimethylpropane
(neopentane)

$$CH_3{-}CH_2{-}\overset{\displaystyle CH_3}{\overset{|}{C}H}{-}CH_3$$

2-Methylbutane
(isopentane)

We see that pentane has the least branching of these three isomers, 2,2-dimethylpropane has the most, and 2-methylbutane is in between. Since the boiling points should decrease as the amount of branching increases, the sequence of boiling points should be

pentane > 2-methylbutane > 2,2-dimethylpropane

In fact, the boiling points of these three compounds are 36°C, 28°C, and 10°C, respectively.

PROBLEM 2.12

Place the following compounds in order of decreasing boiling point: hexane, 2-methylpentane, and 2,3-dimethylbutane.

Alkanes are the least dense of all organic compounds (about 0.6–0.8 g/mL), much less dense than water (1.0 g/mL). This is why gasoline and petroleum products, which are mainly alkanes, float on water. Think of crude oil floating on the surface of the ocean anytime an oil tanker spills its cargo.

Alkanes are insoluble in water because they are nonpolar compounds (see Section 1.2.3); there is very little difference in the electronegativities of carbon and hydrogen, the sole elemental constituents of alkanes. Water molecules are polar and are associated with one another through hydrogen bonding (as we will see in Chapter 4), whereas alkanes cannot participate in hydrogen bonding. This again underscores the importance of the general maxim "like dissolves like." In other words, polar solvents (such as water) dissolve polar compounds (such as salts), but not nonpolar compounds (such as alkanes). Thus, the naturally occurring wax coatings found on leaves and many fruits prevent water loss; water within the plant cannot readily pass through this layer. It also follows that alkanes dissolve most organic compounds, which are nonpolar, but do not dissolve most inorganic compounds, which are polar. So, alkanes are often used as solvents, especially for dissolving tough-to-clean greases (grease is made up of many different organic compounds).

The bulk of the crude petroleum from an oil spill remains on the surface of the ocean because it is less dense than water. Sooner or later, it washes ashore.

The overriding property of alkanes is their general chemical inertness. They simply do not undergo chemical reactions under normal ionic conditions. Thus, they may be stored almost indefinitely and may even be used to protect other materials from chemical reaction. It is for this reason that paraffin wax is often used to seal jars after home canning of fruits and vegetables.

2.4.2 Sources and Uses of Alkanes

EXPLORATIONS

www.jbpub.com/organic-online

Natural Gas

Natural gas, obtained from gas and oil wells throughout the world, is a major source of alkanes. Its composition varies from field to field, but generally it is about 80% methane and 10% ethane, with the remaining portion consisting of higher alkanes, mainly propane. Natural gas is collected by pipeline and then separated into its components (on the basis of their different boiling points).

Much of the methane is transported by pipeline and ships (in a liquefied form) and used for heating fuel. Some is burned in internal combustion engines to power automobiles, buses, and trucks. Some is burned as fuel in gas-fired steam power plants to produce electricity. The ethane is usually converted by an industrial process called *cracking* into ethylene, which is either used as a petrochemical raw material or polymerized into the plastic called polyethylene, found in so many products.

$$\underset{\text{Ethane}}{CH_3{-}CH_3} \xrightarrow[\text{catalyst}]{900^\circ C} \underset{\substack{\text{Ethene} \\ \text{(ethylene)}}}{CH_2{=}CH_2} + H_2$$

Propane is often compressed to liquid form and sold as LP (liquefied propane) fuel. Because of their chemical inertness, propane, butane, and isobutane are used to pressurize some aerosol cans.

Petroleum

Petroleum (crude oil), which is obtained from underground rock formations by drilling oil wells, is a very complex mixture of organic compounds, many of which are alkanes. Some natural gas often exists along with the petroleum, which is formed from the anaerobic decomposition of plant material that was deposited under inland seas millions of years ago. Such decomposition continues today in marshes, bogs, and even garbage landfills, all in the absence of oxygen. Methane is the prinicipal product of the anaerobic decomposition of all organic material.

Petroleum refineries take advantage of the different boiling points of the alkanes to separate petroleum into its major components (which are still mixtures of compounds). This is accomplished with large distillation units called *fractionating towers* that separate the petroleum into seven major "fractions" (Figure 2.10).

1. *Gases boiling below 20°C:* low-molecular-weight alkanes (C_1–C_5), similar to natural gas but including branched alkanes. They may be refined further or may be liquefied and sold as *LPG (liquefied petroleum gas)* fuel.
2. *Liquids boiling from 20 to 100°C:* C_5–C_7 alkanes, branched and unbranched. Sometimes called *naphtha,* this fraction is used for solvents and as raw materials for the chemical industry.
3. *Liquids boiling up to 200°C:* C_5–C_{10} alkanes and aromatics known as *straight-run gasoline* (this fraction accounts for only about 25% of the crude petroleum). It is not sold directly to consumers as fuel because of its low octane rating (about 70), but is reformulated to produce commercial gasoline.
4. *Liquids boiling up to 300°C:* C_{11}–C_{16} alkanes, which serve as a source of kerosene and jet fuel.
5. *Liquids boiling up to 400°C:* C_{15} and higher hydrocarbons, which serve as fuel oil and diesel fuel.
6. *Remaining liquids:* C_{20} and higher hydrocarbons, which serve as the source of lubricating oils.
7. *Solid residue:* asphalt, tar, and paraffin wax.

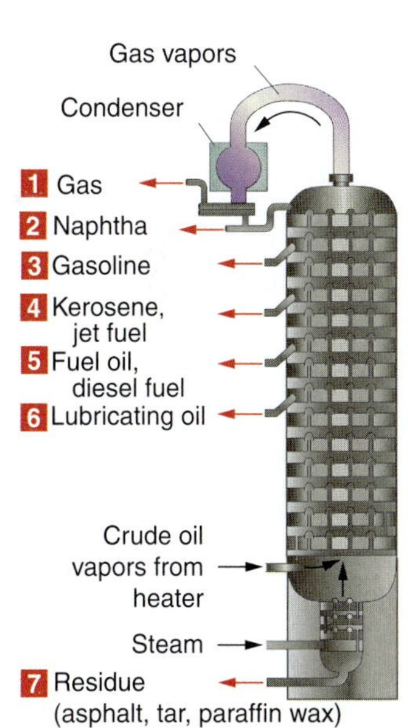

FIGURE 2.10

Fractional distillation towers are used to separate petroleum into seven major fractions. The lower the boiling range of the fraction, the higher up the column it rises before being removed.

Petroleum refineries have two major functions. The first is to carry out the fractionation (distillation) process. However, once those separations are complete, much of the raw material needs to be further processed for commercial use. This involves the use of chemical reactions, usually under very severe conditions (high temperature

CHEMISTRY AT WORK

IN THE MARKETPLACE

Octane Ratings of Gasoline

You have probably seen hundreds of ads for various brands of gasoline. Many of them aim to sell you premium "high-octane" gasoline as the best gas you can buy for your car or truck. But exactly what are these ads talking about? Why is high-octane gasoline better than low-octane gasoline? And what is the connection between the word *octane* used in this way and the eight-carbon straight-chain alkane called *octane?*

The octane rating of any gasoline is determined experimentally by comparing its behavior in an internal combustion engine with that of a uniform reference standard. In particular, the amount of engine "knocking" is examined for each type of fuel. (*Knocking* is the premature, pressure-caused, spontaneous ignition of the fuel in the engine cylinders. The ignition causes counterpressure on the cylinder during the upstroke of the piston. The result is a loud "knock" and inefficient engine performance.) A higher-octane gasoline results in less engine knocking. Higher-compression internal combustion engines require higher-octane fuel for satisfactory performance.

The straight-chain alkane *heptane* is assigned an octane rating of zero in this reference system, and the highly branched alkane 2,2,4-trimethylpentane (historically known as *isooctane*) is assigned an octane rating of 100.

$CH_3CH_2CH_2CH_2CH_2CH_2CH_3$

Heptane

$(CH_3)_3CCH_2CH(CH_3)_2$

Isooctane
(2,2,4-trimethylpentane)

The knocking behavior of a gasoline sample is compared with that of various mixtures of heptane and isooctane. For example, a gasoline that causes knocking similar to that of a mixture of 13% heptane and 87% isooctane is labeled an 87-octane gasoline.

Straight-chain alkanes have low octane ratings, and branched alkanes have higher octane ratings. Straight-run gasoline, obtained directly from the fractionation process

Octane ratings are prominently displayed on gasoline pumps.

in an oil refinery, has an average octane rating of about 70. This is so low that a modern automobile engine burning it would sound terrible. Oil companies have used several approaches to increase the octane ratings of straight-run gasoline. They used to add tetraethyl lead to low-octane gasolines to increase the octane rating. The lead was emitted into the atmosphere from exhaust systems. So-called leaded gasoline was banned by law after it was shown that lead concentrations in the atmosphere were increasing and causing serious environmental and health problems. Today, oil refineries alter the chemical structures of components of straight-run gasoline through extensive cracking and reformulating to produce fuel of sufficiently high octane rating to avoid knocking. Gasoline sold today generally has an octane rating ranging from 85 to 93.

It is interesting to note that a low-octane fuel makes a good diesel fuel. Diesel engines do not use electrical ignition systems; instead, the fuel ignites on its own when sufficient pressure is reached in the engine cylinder. Because the straight-chain alkane *cetane* ($C_{16}H_{34}$) readily ignites under pressures typical in a diesel engine, it is used as the reference compound. This has given rise to the cetane rating system for diesel fuel.

and pressure and often in the presence of special proprietary catalysts), to convert the raw materials into useful commercial products—a process referred to as *cracking* and carried out in tall metal towers called *cracking towers.* Cracking rearranges the structures of alkanes to produce a mixture rich in branched alkanes, which can be sold as commercial gasoline.

2.5

Cycloalkanes

The family of alkanes consists of saturated hydrocarbons—compounds in which all the bonds to carbons are single (sigma) bonds. We have sampled the wide variety of compounds possible when carbon atoms are joined together in straight and branched chains. Also in the alkane family are the cycloalkanes—alkanes in which some of the carbon atoms have formed a closed ring. The simplest cycloalkane has three carbons. The three-carbon acyclic compound is propane; the corresponding cycloalkane has the two terminal carbons joined together, creating *cyclopropane* (Figure 2.11).

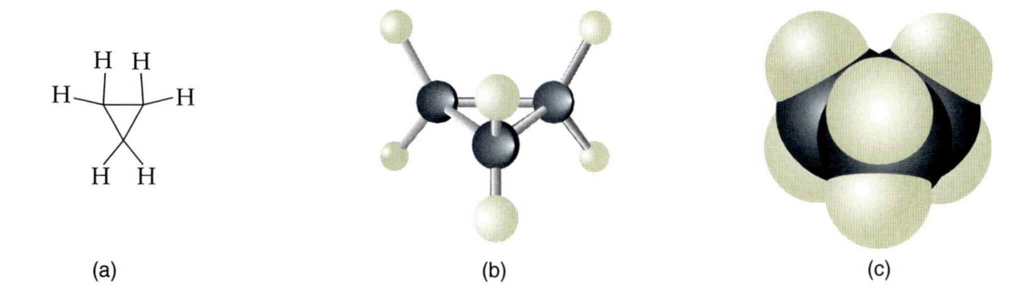

FIGURE 2.11

Three representations of cyclopropane: (a) bond-line formula, (b) ball-and-stick model, and (c) space-filling model.

To imagine forming a ring from a straight-chain alkane, you must envision removing a hydrogen from both of the terminal carbons so that they can form a new C—C single bond. Whereas alkanes have a general molecular formula of C_nH_{2n+2}, cycloalkanes have the general formula C_nH_{2n}, indicating the "loss" of the two terminal hydrogens. Therefore, cycloalkanes can be distinguished from acyclic alkanes by molecular formula. For example, a saturated hydrocarbon with a molecular formula C_6H_{14} must be an acyclic alkane, whereas one with a formula C_6H_{12} must be a cycloalkane.

2.5.1 Nomenclature of Cycloalkanes

Cycloalkanes are named using IUPAC rules similar to those for alkanes. Whereas alkanes are named according to the length of the longest chain, cycloalkanes are named according to the size of the ring. We start with the number of carbons forming the ring to determine the appropriate alkane parent name (see Table 2.2), then add *cyclo-* as a prefix. We then number the ring carbons so as to have substituents on the lowest-numbered carbons possible. The substituents on the ring are designated by their names and numeric locations.

In writing the structures of cycloalkanes, it is standard practice to use only the bond-line representation, occasionally showing the substituents written in condensed form. Further, when only a single substituent is present, the location need not be specified in the name; an example is methylcyclohexane:

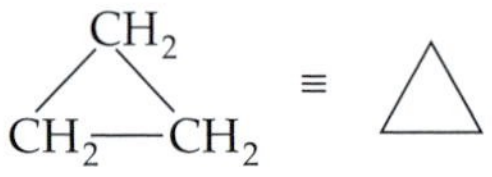

Cyclopropane
(two equivalent representations)

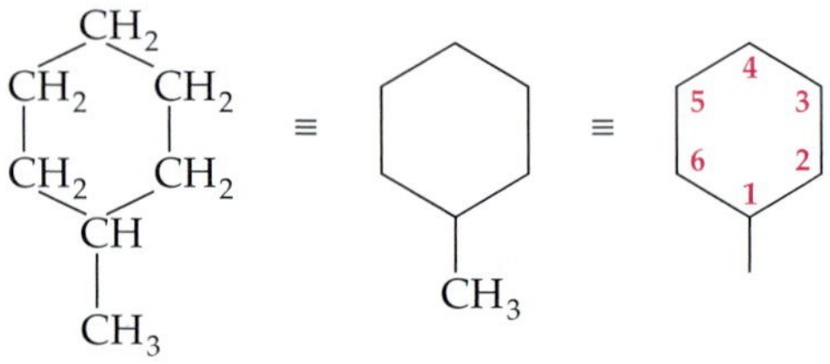

Methylcyclohexane
(three equivalent representations)

The order of substituents in a name is alphabetical. If the ring carries an alkyl substituent containing more carbons than the ring does, the ring is treated as a *cycloalkyl* substituent on an alkane chain. The cycloalkyl name is derived from the cycloalkane in the same way as an alkyl group name is derived from the corresponding alkane (see Section 2.3.2).

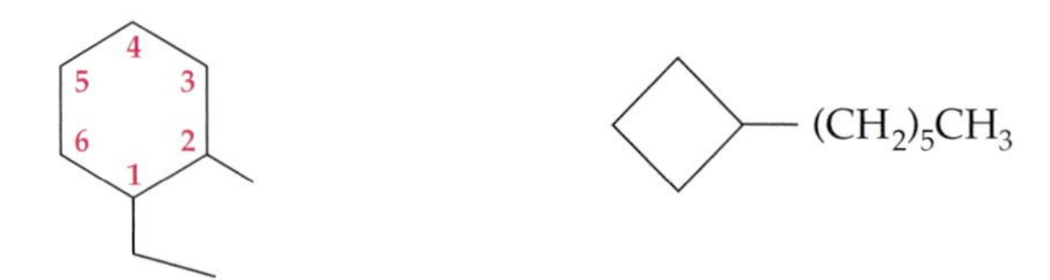

1-Ethyl-2-methylcyclohexane
(*not* 1-methyl-2-ethylcyclohexane)

1-Cyclobutylhexane

How to Solve a Problem

Name the following compound:

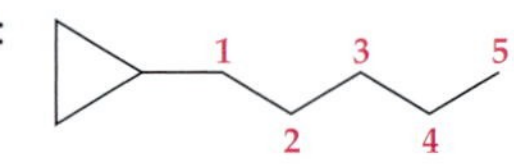

SOLUTION

The ring has three carbons, and the straight chain has five carbons. Therefore, the parent name is pentane, and the substituent is the cyclopropyl group attached to carbon 1. The correct name is 1-cyclopropylpentane (not 1-pentylcyclopropane).

PROBLEM 2.13

Name the following compounds:

(a)  (b) 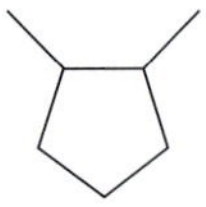 (c) 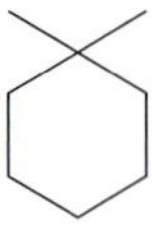 (d)

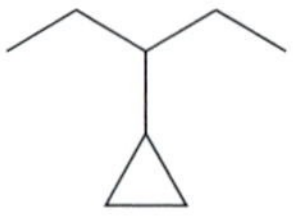

PROBLEM 2.14

Write structural formulas for (a) 1,2-dimethylcyclobutane, (b) 1,2,3,4,5-pentaethylcyclopentane, and (c) cyclopropylcyclopropane.

PROBLEM 2.15

Write structural formulas for all cyclic compounds that fit the molecular formula C_6H_{12}.

Constitutional isomerism also occurs in cyclic compounds. For example, the molecular formula C_5H_{10} for a cycloalkane could represent any one of four compounds with different connectivities:

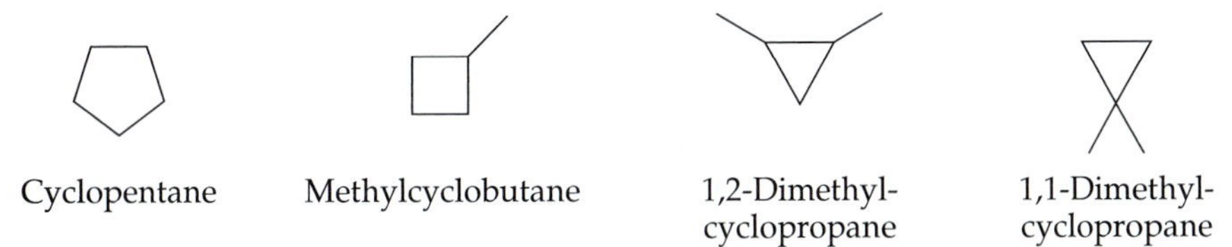

2.5.2 *Cis-Trans* Isomerism in Cycloalkanes

You may recall from the chart of isomers (Figure 2.2) that stereoisomers form a major category with several subcategories, another of which we will consider now. All cycloalkanes that have two or more substituents not attached to the same carbon exhibit a type of **configurational stereoisomerism** known as ***cis-trans* isomerism.** The connectivities in these isomers are the same, but different orientations are possible because the substituents may be attached either to the same side or to opposite sides of a ring. (The ring constitutes an approximate fixed plane of reference for the molecules.) Such configurational isomers are different compounds that can be physically separated and identified, in contrast to conformational isomers, which cannot be separated (see Section 2.2.2). For example, when two methyl groups are attached to two different carbons of a cyclopropane ring, they can be attached either on the same side (both above or both below the plane of the ring) or on different sides (one below and one above the ring or one above and one below, which is the same thing) (Figure 2.12). (Note that if there is only a single methyl group attached to the ring, as in methylcyclopropane, the structure with the group drawn above the plane is identical to the structure with the group drawn below.)

In naming such isomers, *cis-* is used to signify "same side" and *trans-* is used to signify "opposite sides." For example, the cycloalkane 1,2-dimethylcyclopropane exists as two different isomers: *cis*-1,2-dimethylcyclopropane and *trans*-1,2-dimethylcyclopropane. These are two different compounds having totally different physical properties. The *cis* isomer can be drawn with both substituents either above or below the ring; these are identical because rotation of one drawing by 180° produces the other. When the ring is drawn in the plane of the paper, solid and dashed wedges can be used to represent bonds above and below the plane of the paper, respectively.

Two methyl groups on same side of cyclopropane ring

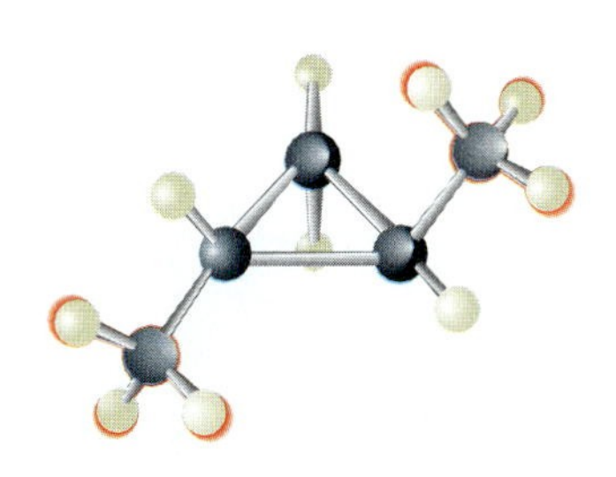

Two methyl groups on opposite sides of cyclopropane ring

FIGURE 2.12

Ball-and-stick models of two isomers of 1,2-dimethylcyclopropane.

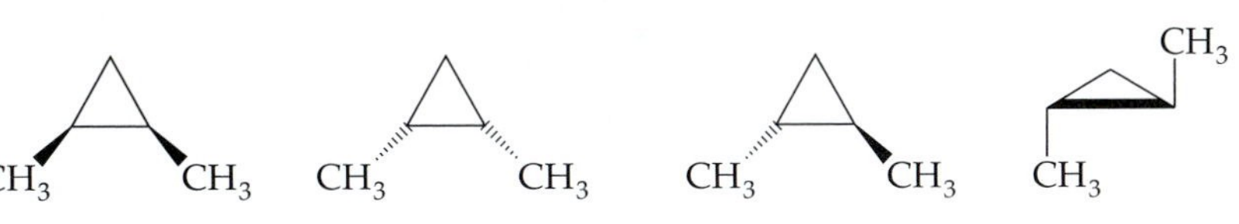

cis-1,2-Dimethylcyclopropane (identical structures)

trans-1,2-Dimethylcyclopropane (two different representations of the same structure)

Alternatively, the ring can be drawn in perspective, as if positioned perpendicular to the plane of the paper; in this case, the heavy-lined bond is closer to the viewer than the other ring bonds, and the substituents lie in the plane of the paper above and below the ring.

FIGURE 2.13

Strains present in cyclopropane.

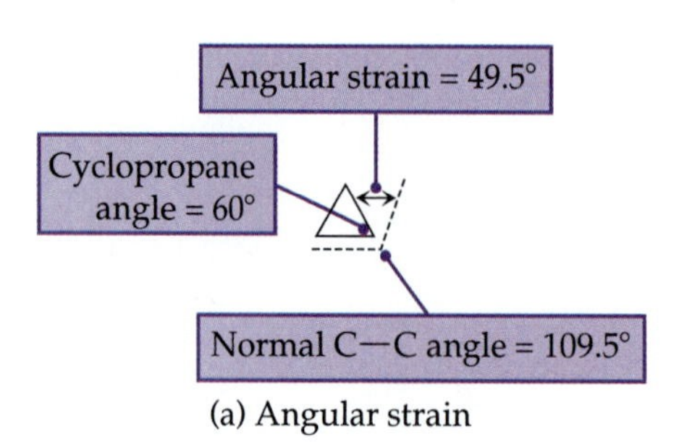

(a) Angular strain

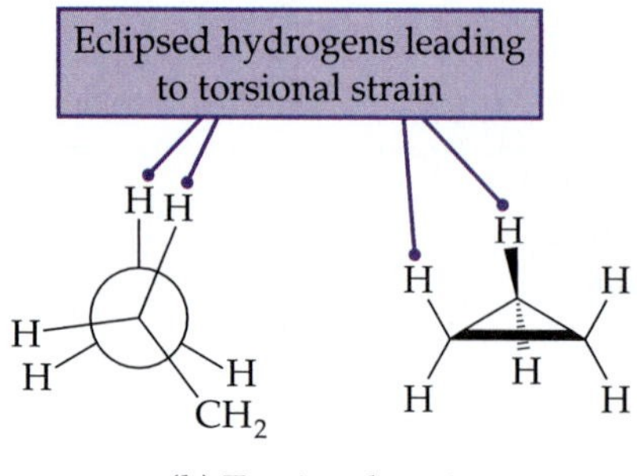

(b) Torsional strain

The C_3 and C_4 rings are under **angular strain** because their internal C—C—C bond angles deviate considerably from the ideal tetrahedral bond angle of 109.5°. For example, the C—C—C angle in cyclopropane is 60°, 49.5° smaller than the ideal (Figure 2.13a). These rings are also under **torsional strain** because the hydrogen atoms attached to the ring carbons are forced to be eclipsed all around the rings (consider the view along each carbon-carbon bond; see Figure 2.13b). This accumulation of strain energy results in compounds that are of higher energy (as we will see later), more reactive, and difficult to synthesize. The C_5 (cyclopentane) ring could be planar, given that each angle within a pentagon is 108°, close to the ideal of 109.5°; however, it is slightly nonplanar to relieve the torsional strain. The C_4 ring is also slightly nonplanar, which increases the angular strain but relieves the torsional strain.

PROBLEM 2.16

Draw structural formulas for (a) *cis*-1,2-dimethylcyclobutane, (b) *trans*-1,3-dimethylcyclopentane, and (c) *trans*-1,3-diethylcyclobutane.

2.5.3 Polycyclic Alkanes

We have been discussing *monocyclic* alkanes, those with only one ring. However, some of the more interesting cycloalkanes contain more than one ring. Some, called *bicyclics*, result from the simple fusion of two rings of varying size; *cis*-decalin has two rings sharing a single C—C bond. Other bicyclics, such as 1,1,1-propellane and bicyclo[3,2,1]-octane, have two rings that share more than one C—C bond.

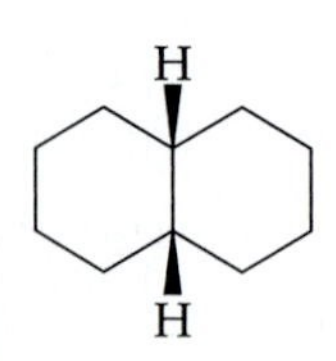

cis-Decalin

Still other cycloalkanes are *polycyclics* (*poly* means "many"). Some of these structures are very highly strained, given the C—C—C bond angles; many are chemical curiosities synthesized in the laboratory as an intellectual challenge. The study of such compounds provides new insights into organic chemistry. Chemists appreciate the beauty of the symmetry present in such unique structures and occasionally assign interesting, although perhaps "trivial," names to these compounds (often in reference to their shapes):

1,1,1-Propellane

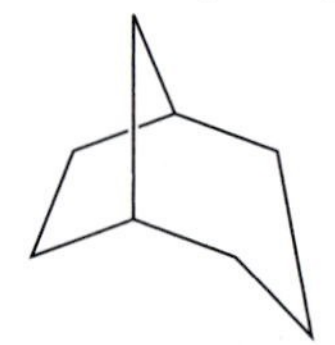

Bicyclo[3,2,1]octane

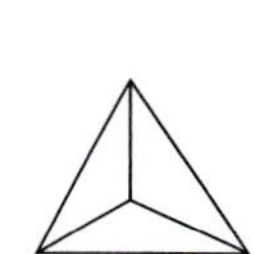

Tetrahedrane

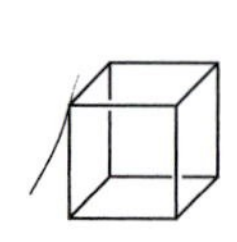

Cubane

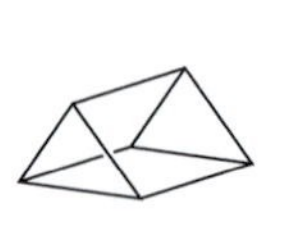

Prismane

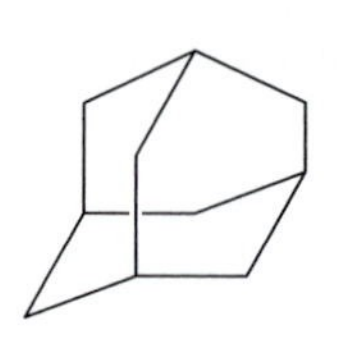

Adamantane

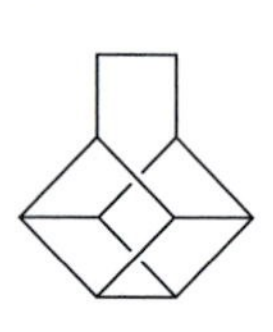

Basketane

2.6

Conformations of Cyclohexane

The most commonly occurring saturated ring in organic compounds is the six-membered ring of *cyclohexane*. If planar, the ring would have the C—C—C bond angles of a hexagon, 120° (creating angular strain), and all hydrogens would be eclipsed (creating torsional strain). The most stable form of the cyclohexane ring is not planar, but instead is a **chair conformation,** which has C—C—C bond angles of 109.5° and staggered C—H bonds around the entire ring. This form eliminates both kinds of strain energy. (Other conformations of cyclohexane, such as the boat and twist conformations, are of higher energy and are generally less important, so we will not discuss them further.) Figure 2.14 shows two different representations of the chair conformation of cyclohexane. Carbons 2, 3, 5, and 6 form the seat of the chair; carbons 1, 2, and 6 form a leg rest angling down from the seat; and carbons 3, 4, and 5 form the back of the chair.

If you look closely at the three-dimensional structure of cyclohexane (best done using a set of molecular models), you can see that there are two distinct orientations for the hydrogen atoms around the ring. Each carbon has one C—H bond directed approximately perpendicular to the plane ("seat") defined by carbons 2, 3, 5, and 6. These C—H bonds are called **axial bonds** because they are parallel to a vertical axis drawn through the center of the ring (Figure 2.15a). The direction in which the axial bonds point is opposite for adjacent carbons (that is, down for C_1, up for C_2, down for C_3, etc., around the ring, as shown in the axial-only drawing, Figure 2.15b).

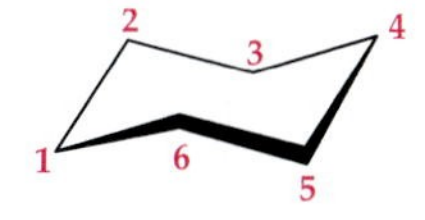

Bond-line formula

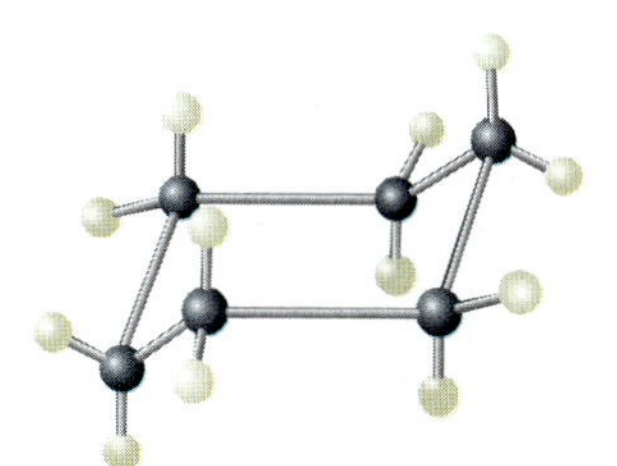

Ball-and-stick model

FIGURE 2.14

Two representations of the chair form of cyclohexane.

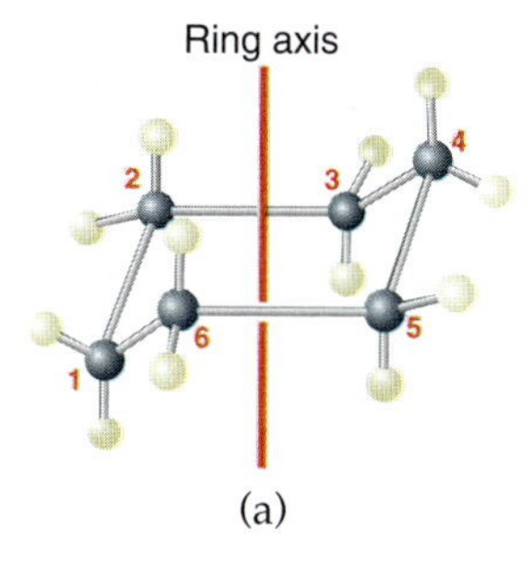

(a)

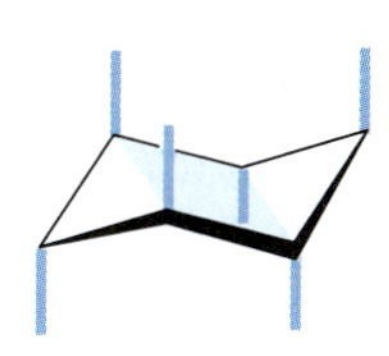

(b) Axial C—H bonds only

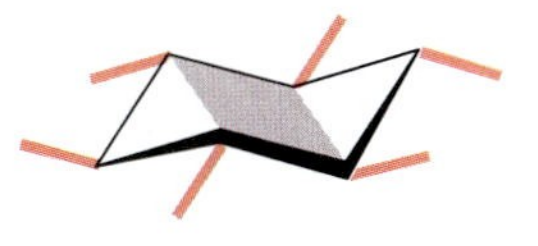

(c) Equatorial C—H bonds only

FIGURE 2.15

Axial and equatorial bonds in the chair conformation of cyclohexane.

The second C—H bond on each carbon is oriented away from the center of the ring. These are called **equatorial bonds.** With all bond angles 109.5°, these equatorial bonds are also directed slightly above or slightly below the plane of carbons 2, 3, 5, and 6, alternating around the ring as shown in the equatorial-only drawing (Figure 2.15c). Put another way, when the axial bond is down, the equatorial bond to the same carbon is pointed slightly above the plane of the ring, and when the axial bond is pointed up, the equatorial bond is oriented slightly below the plane of the ring. Examining each C—C bond using a Newman projection reveals the perfect staggering of all the C—H bonds, shown here for the bonds between carbons 2 and 3 and between carbons 5 and 6:

H H
H CH_2 H
H CH_2 H
H H

PROBLEM 2.17

Draw a Newman projection of the bond between carbon 1 and carbon 2 in methylcyclohexane. There are two possible structures. Which one do you think might be more stable? Why?

Step 1

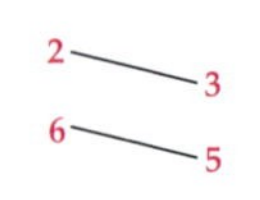

Step 2

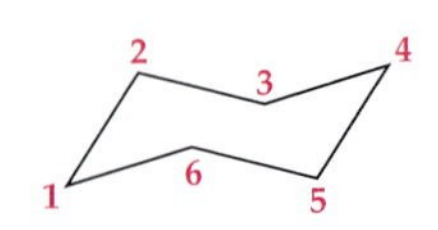

Step 3

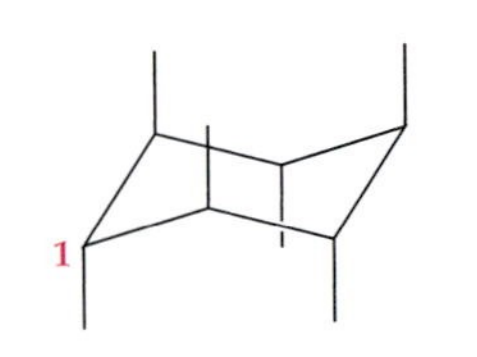

Step 4

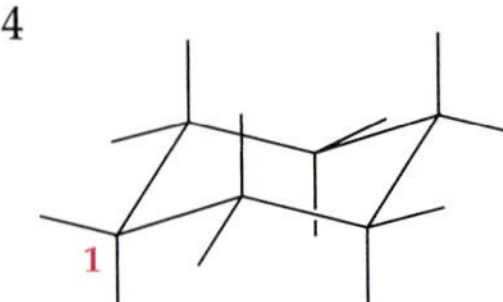

It is important for you to become comfortable drawing cyclohexane rings. Use the following steps for drawing the bond-line chair:

1. Draw the two parallel lines connecting carbon 2 to carbon 3 and carbon 5 to carbon 6 (see Figure 2.14). Imagine that these represent a plane that cuts through the paper (thus, the plane is perpendicular to the paper and the 5-6 bond is closer to you than the 2-3 bond).
2. Draw bonds from carbon 2 and carbon 6 to connect to carbon 1; carbon 1 is below the plane. Draw bonds to connect carbons 3 and 5 to carbon 4, which is above the plane.
3. Start at carbon 1 and draw the axial bond down, then proceed around the ring, drawing the axial bonds alternately up and down, vertical and parallel to each other.
4. Draw the equatorial bond on carbon 1, which goes up slightly because the axial bond goes down. Draw the angle to appear slightly greater than a right angle (90°) with the axial bond on carbon 1. Proceed around the ring, drawing the equatorial bonds, each going in a direction away from the axial bond on the same carbon, at an angle just greater than 90°.

As you become more comfortable drawing the rings, refrain from drawing all the C—H bonds (just use a bond-line formula). Then draw only those bonds to substituents or bonds that are important to a particular reaction.

2.6.1 Conversion of Cyclohexane Conformations

You know from our discussion of the conformations of butane (Section 2.2.2) that conformations result from rotation about carbon-carbon single bonds. Such rotation also occurs in cyclohexane. Through simultaneous rotation about the six C—C bonds of cyclohexane, the chair conformer can be converted into another chair conformer, a process called **conversion of conformation** or chair-chair interconversion. The result of this process is that all axial bonds are converted into equatorial bonds, and all equatorial bonds are converted into axial bonds (Figure 2.16). No bonds are broken in this conversion process.

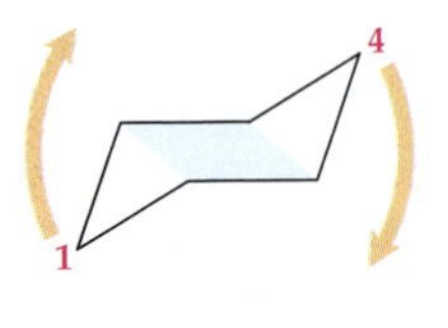

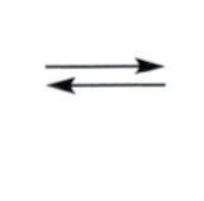

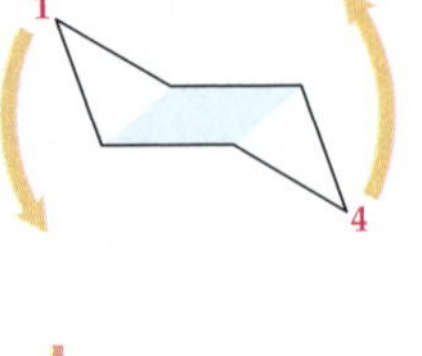

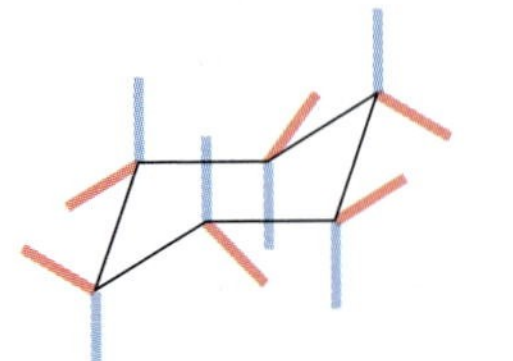

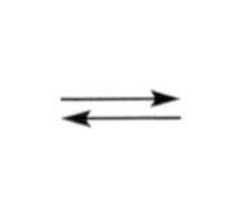

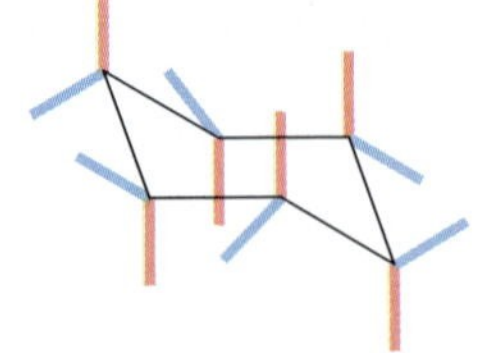

(a) Before conversion (b) After conversion

FIGURE 2.16

Conversion of conformation of cyclohexane (hydrogen atoms omitted for clarity).

To carry out this conversion using a molecular model, you need to twist (rotate) each C—C bond. You can picture it on paper, as shown in Figure 2.16: Carbons 2, 3, 5, and 6 remain as before, carbon 1 moves from below the plane to above the plane, and carbon 4 moves from above the plane to below the plane. Each starting axial hydrogen in (a) (represented by a blue line) becomes equatorial in (b), and each equatorial hydrogen in (a) (represented by a red line) becomes axial in (b).

This conversion occurs extremely rapidly at room temperature (in other words, there is a small energy barrier to the conversion). Thus, the two conformations exist in equilibrium and cannot be physically separated. However, the equilibrium can be slowed by very low temperatures (about −90°C), and spectroscopic signals (see Section 11.6) can be detected for each kind of hydrogen (axial and equatorial). There is no difference in stability between these two conformations of cyclohexane, but there are stability differences when one or more hydrogens has been replaced by a substituent, as we will see in the next section.

2.6.2 Monosubstituted Cyclohexanes

Whereas the two conformations of cyclohexane are equivalent, those of substituted cyclohexanes are not. Conformers of substituted cyclohexanes are of lower energy when substituents are located in the equatorial position, compared to the axial position. An axial substituent X on carbon 1 is crowded by axial hydrogens on carbons 3 and 5 (a form of **steric strain** called 1,3-diaxial interactions), as shown in Figure 2.17. The conformation in which the substituent is equatorial is more stable; it has no 1,3-diaxial interactions.

The size of the substituent X (and therefore the extent of steric repulsion) affects the position of the equilibrium between two chair conformers. The larger the substituent (X), the larger the repulsive forces when it is in the axial position and the larger the difference in stability of the axial and equatorial conformers. The *t*-butyl group is much bulkier than the methyl group. This difference is illustrated by the fact that methylcyclohexane (X = methyl = CH_3—) at equilibrium consists of 95% equatorial conformer and 5% axial conformer, whereas *t*-butylcyclohexane [X = *t*-butyl = $(CH_3)_3C$—] is more than 99% equatorial conformer. Highly branched alkyl groups are much more space-demanding than unbranched alkyl groups, no matter how long the chain (Figure 2.18).

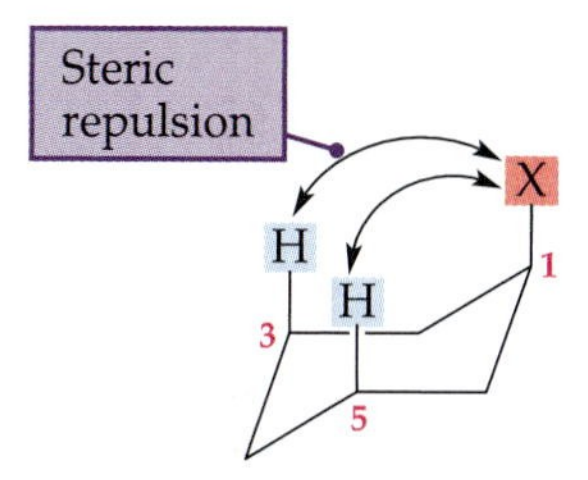

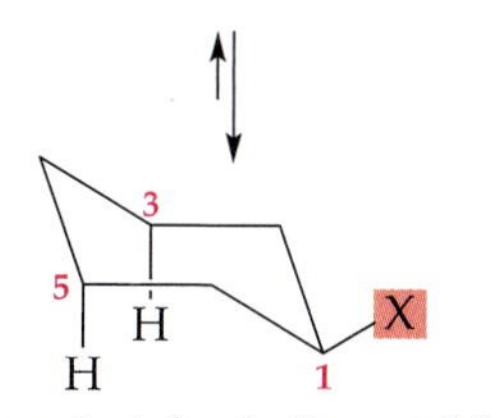

FIGURE 2.17
1,3-Diaxial interactions of a cyclohexane substituent, X.

PROBLEM 2.18

Which compound in each pair will have more of the equatorial conformation at equilibrium?

(a) *n*-propylcyclohexane or isopropylcyclohexane

(b) isopropylcyclohexane or *t*-butylcyclohexane

(c) isopropylcyclohexane or methylcyclohexane

2.6.3 Disubstituted Cyclohexanes

When a cyclohexane ring carries two or more substituents, you need to consider two kinds of isomerism at once—conformational and *cis-trans*. Two issues always arise: (1) identifying each isomer as *cis* or *trans* and (2) determining the relative stability of the two possible conformations of each isomer.

Trans-1,2-dimethylcyclohexane must have the two methyl groups on opposite sides of the ring (remember, *trans* means "opposite"), implying that they must both be in

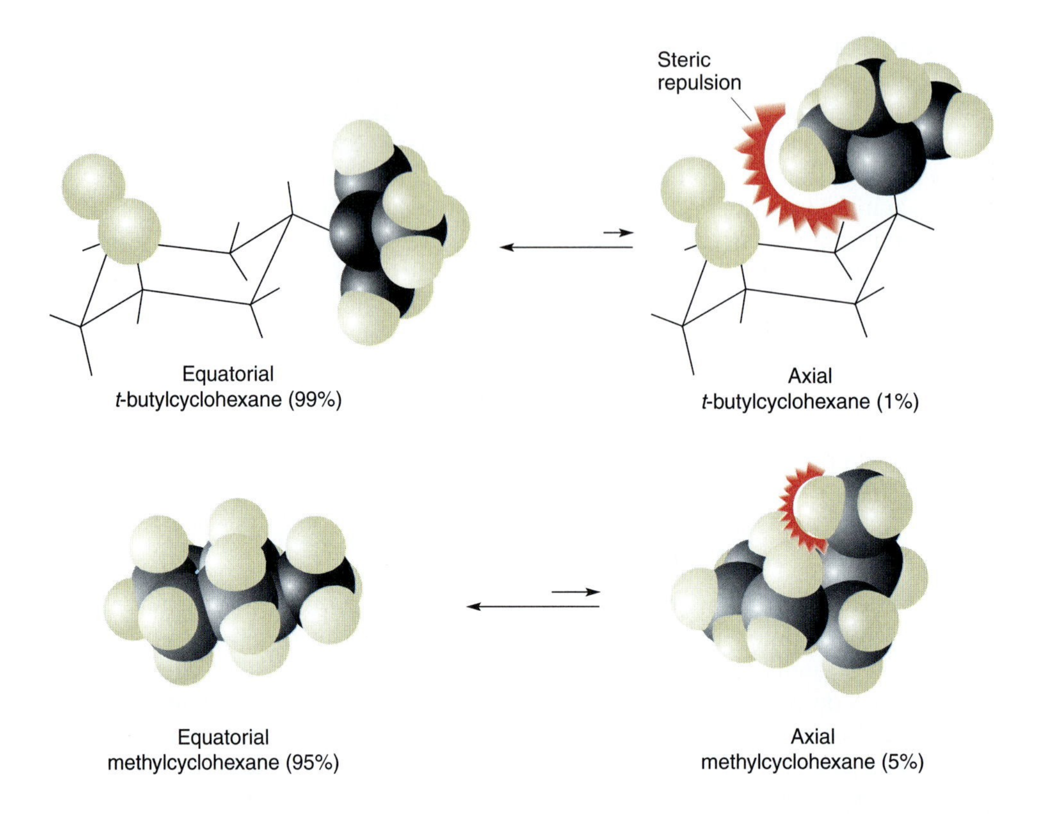

FIGURE 2.18

Space-filling models of axial and equatorial conformers of methylcyclohexane and *t*-butylcyclohexane. The size of the substituent X on the cyclohexane ring affects the position of the equilibrium.

an axial position (Figure 2.19, structure 1). However, since conversion of conformation leads to a *diequatorial* form (structure 2), it also must be *trans*, even though the diequatorial form may not look like it. (Note that no bonds were broken; only a rotation occurred.)

Which *trans* conformer is more stable? If you consider the repulsive effects resulting from 1,3-diaxial interactions, discussed in Section 2.6.2, it becomes apparent that

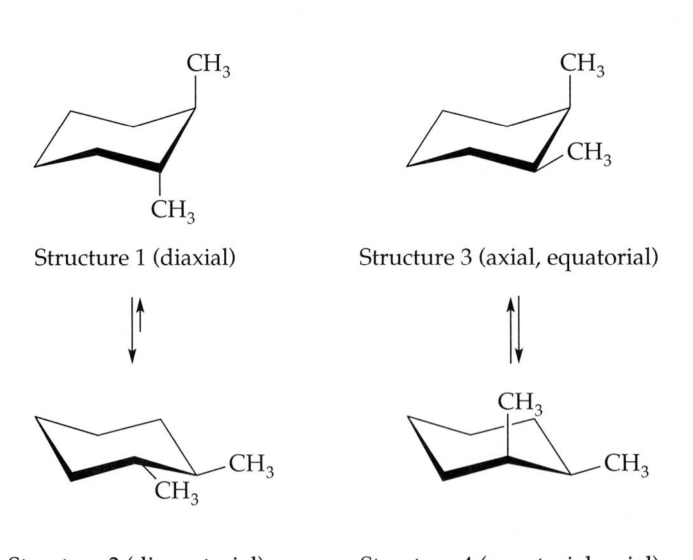

FIGURE 2.19

Conformers of *trans*- and *cis*-1,2-dimethylcyclohexanes.

trans-1,2-dimethylcyclohexane will exist mainly as the more stable diequatorial conformation (Figure 2.20).

If the diaxial and diequatorial forms of 1,2-dimethylcyclohexane (structures 1 and 2 in Figure 2.19) are both *trans*, then the *cis* form must have one methyl group equatorial and one axial (structure 3). Conversion of conformation of this *cis* conformer produces structure 4, which also has one methyl group axial and one equatorial. Therefore, in *cis*-1,2-dimethylcyclohexane, there is no difference in stability between the two possible conformations. However, in a compound such as *cis*-1-isopropyl-2-methylcyclohexane, the conformation with the bulkier group (the isopropyl group) in the equatorial position will be more stable.

Diequatorial conformation
Most stable

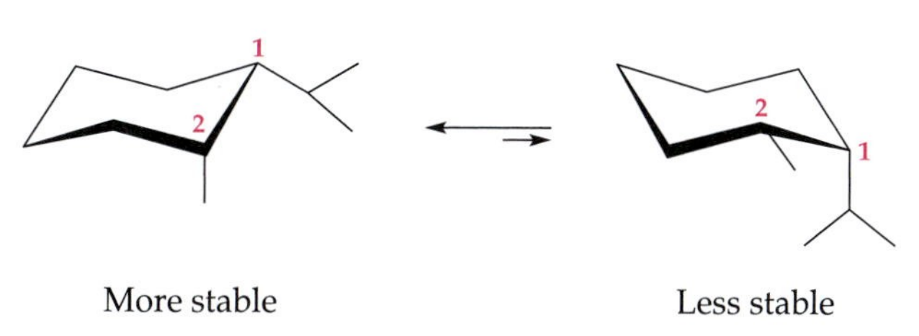

More stable Less stable

Conformers of *cis*-1-isopropyl-2-methylcyclohexane

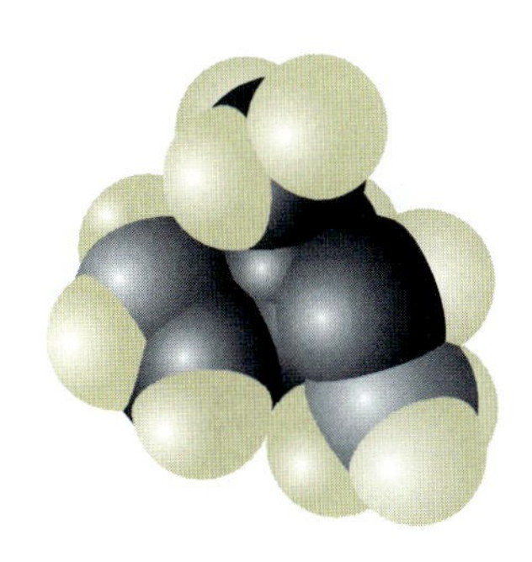

Diaxial conformation

FIGURE 2.20

Space-filling models of conformers of *trans*-1,2-dimethylcyclohexane.

A similar analysis can be performed with other disubstituted and more highly substituted cyclohexanes. The best approach is to sort out which isomers are which by first deciding what the diaxial isomer is (*cis* or *trans*), based on whether the substituents clearly are on the same or opposite sides of the general plane of the ring. Remember that the conformers are in equilibrium, but the configurational *cis* and *trans* isomers are not interconvertible. We can sum up the situation this way: *cis* and *trans* isomers of cyclohexanes have the same molecular formula, have the same connectivities, and cannot be interconverted by rotation about a single bond; however; individual *cis* and *trans* isomers have axial and equatorial conformations that are interconvertible.

How to Solve a Problem

Draw the structures of both conformers of (a) *cis*-1,3-dimethylcyclohexane and (b) *trans*-1,4-dimethylcyclohexane. Which conformation is more stable for each isomer?

SOLUTION

(a) We first draw a diaxial isomer of 1,3-dimethylcyclohexane, which means both methyl groups are on the same side of the ring. It is therefore the *cis* isomer. Then we can draw the converted conformation, which has both methyl groups equatorial. The diequatorial conformation is more stable than the diaxial conformation because both substituents are in the favored equatorial position (lower 1,3-diaxial interactions).

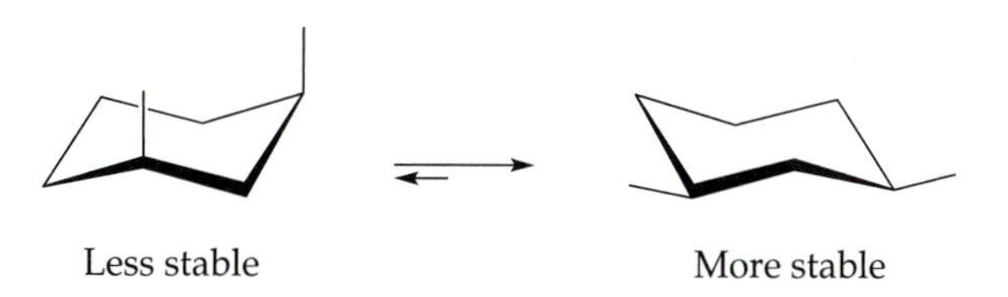

Less stable More stable

cis-1,3-Dimethylcyclohexane

CHEMISTRY AT WORK

IN NATURE

From Cyclohexane to Diamond

You now have some sense of how versatile carbon is in bonding to other carbons to create a great variety of organic molecules. Carbon atoms may bond together into geometric shapes—chains, rings, and even "chairs"—that then become building blocks themselves. Knowing the shape of cyclohexane, you can deduce the shape of molecules incorporating more than one cyclohexane ring. For example, two such rings fused together in parallel make up the bicyclic molecule *trans*-decalin ($C_{10}H_{18}$). The naturally occurring symmetrical tetracyclic hydrocarbon adamantane ($C_{10}H_{16}$), which presents a cyclohexane ring on all four faces, is actually a single cyclohexane ring bridged with four other carbons. Diamantane ($C_{14}H_{22}$) is even more complex, but you can still see the cyclohexane ring on all six of its faces.

The ultimate product created from the fusion of chair cyclohexane rings is the extensive three-dimensional network of one of the elemental forms of carbon—*diamond*. Careful examination of its structure reveals that every face presents a chair form of cyclohexane. It is also possible to spot the adamantane units within the framework. Diamond is essentially pure carbon, with each carbon atom bonded to four other carbon atoms (except for the edge carbons, which are bonded to three other carbon atoms and one hydrogen atom).

Diamond is known for its gem quality—its brilliant reflection of light from its many facets. It is also very important industrially because it is the hardest substance known. This makes it useful in drills and saws, for instance, to cut and fabricate other materials. The hardness of diamond is due to its abundance of strong carbon-carbon bonds, which must be broken in order to break down any part of the structure. However, once these bonds are broken—for example, by a diamond cutter's chisel—they are unlikely to form again, making diamond brittle as well as hard.

In Chapter 9, we will look at two other forms of elemental carbon—graphite and buckminsterfullerene.

An assortment of natural and synthetic diamonds.

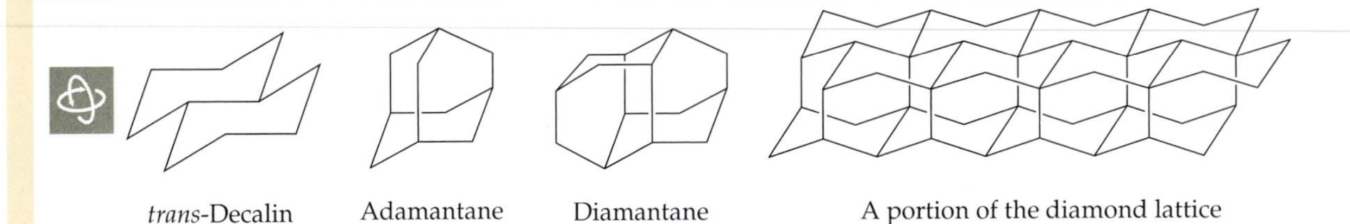

trans-Decalin Adamantane Diamantane A portion of the diamond lattice

(b) Again, we draw a 1,4-diaxial isomer first, which means the two methyl groups are on opposite sides of the ring. Therefore, this is the *trans* isomer. Then we draw the converted conformation, which has both methyl groups in equatorial positions. The diequatorial conformation is more stable than the diaxial conformation because of lower 1,3-diaxial interactions.

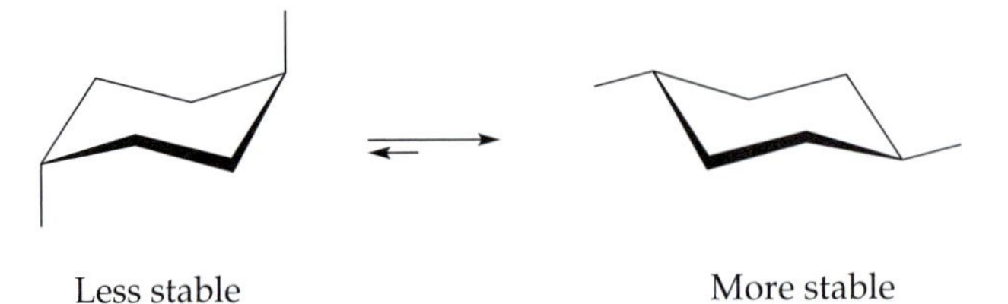

Less stable More stable

trans-1,4-Dimethylcyclohexane

PROBLEM 2.19

Draw both conformations of (a) *trans*-1,3-dimethylcyclohexane and (b) *cis*-1,4-dimethylcyclohexane.

PROBLEM 2.20

Which is the more stable configurational isomer, *cis*- or *trans*-1,4-dimethylcyclohexane? Why?

2.7

Reactions of Alkanes and Cycloalkanes

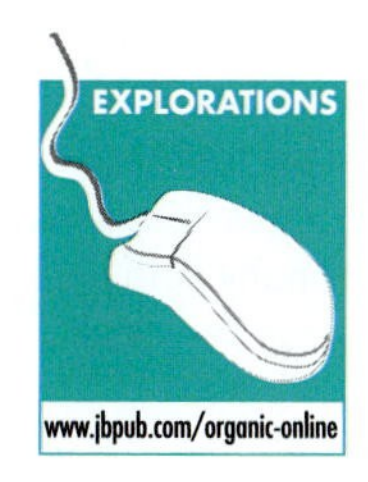

The overriding chemical property of alkanes (here cycloalkanes are included as part of the family) is their relative inertness—in other words, their resistance to chemical reactions. Alkanes do not react with strong acids or bases, with chemical oxidants or reductants, or with any of a host of "normal" chemical reagents that we will study in later chapters. However, they do undergo radical-initiated reactions, such as halogenation and combustion. Under the severe conditions (high temperature and pressure and usually the presence of catalysts) often generated in industrial processes, useful reactions of alkanes do occur, but even then the course of these reactions may be difficult to control precisely.

2.7.1 Halogenation of Alkanes

An important chemical reaction of alkanes is *halogenation*, the substitution of a halogen atom (usually chlorine or bromine) for a hydrogen atom. For example, methane (CH_4) can be converted to methyl chloride (CH_3Cl), also known as chloromethane:

$$\underset{\text{Methane}}{CH_4} + Cl_2 \xrightarrow{\text{heat}} \underset{\substack{\text{Chloromethane}\\ \text{(methyl chloride)}}}{CH_3Cl} + HCl$$

Note in the Key to Transformations on the inside front cover that halogenation is the only transformation shown for alkanes. It is a means for converting an unreactive alkane into a reactive haloalkane (alkyl halide). We will look more closely at this substitution reaction in Chapter 10, when the mechanism of the reaction will be more meaningful to you.

2.7.2 Combustion (Oxidation) of Alkanes

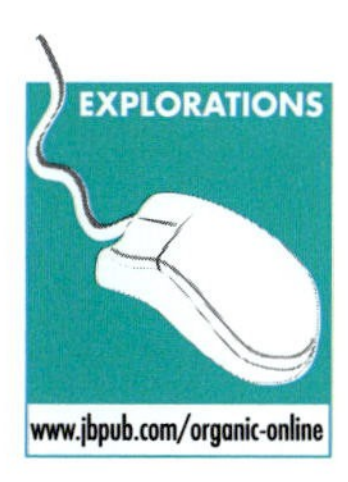

Alkanes are inert to strong chemical oxidants, such as potassium permanganate, under typical laboratory conditions. However, alkanes can be oxidized under more severe conditions. It is well known that gasoline and related materials (natural gas, LP gas, butane in lighters, etc.) burn—that is, they are combustible. **Combustion** is an oxidation reaction initiated with oxygen at a high temperature.

One of the two major ways alkanes are used is as fuels (the other is as raw materials to produce other chemicals). Alkanes are burned to provide energy in the form of

Near the end of the 1991 Gulf War, the retreating Iraqi army set fire to Kuwaiti oil wells. The fires burned for months—emitting tremendous heat and smoke—before finally being extinguished by specialists.

heat. In the presence of an unlimited amount of oxygen, this reaction converts any alkane into carbon dioxide and water, as shown in the following balanced equations for methane (the major component of natural gas), propane (the major component of LP fuel), and cyclohexane:

$$\underset{\text{Methane}}{CH_4} + 2\,O_2 \xrightarrow{\text{heat}} CO_2 + 2\,H_2O$$

$$\underset{\text{Propane}}{C_3H_8} + 5\,O_2 \xrightarrow{\text{heat}} 3\,CO_2 + 4\,H_2O$$

$$\underset{\text{Cyclohexane}}{C_6H_{12}} + 9\,O_2 \xrightarrow{\text{heat}} 6\,CO_2 + 6\,H_2O$$

These oxidation reactions occur only in the presence of externally applied heat, which initiates them. However, the reactions are exothermic (that is, once under way, they produce heat), and the heat produced from oxidation of one molecule is sufficient to initiate the reaction of other molecules. The process thus becomes a *chain reaction*, which proceeds with tremendous speed (see Chapter 10). In some instances, the reaction is too fast and an explosion results. Because the combustion reaction requires oxygen, denial of oxygen and reduction of the temperature are the key techniques used to extinguish a fire.

PROBLEM 2.21

Write a balanced equation for the complete combustion of 2-methylbutane.

The amount of heat produced in the combustion of an alkane (or any compound, for that matter) can be measured experimentally. This provides a means of determining the relative energy levels (stabilities) of compounds. For example, combustion of *n*-butane ($CH_3CH_2CH_2CH_3$) produces 687.5 kcal/mol of energy, while combustion of 2-methylpropane, or isobutane [$(CH_3)_3CH$] produces 685.5 kcal/mol (Figure 2.21). The difference in energy produced while forming identical products demonstrates that isobutane is 2 kcal/mol more stable than *n*-butane (that is, isobutane is at a lower energy level and therefore will emit less energy than *n*-butane). Combustion of cycloalkanes reveals differences in their stabilities as a function of ring size. For instance, cyclohexane emits 157.4 kcal/mol *per methylene* (CH_2) *group*, whereas cyclopropane emits 166.6

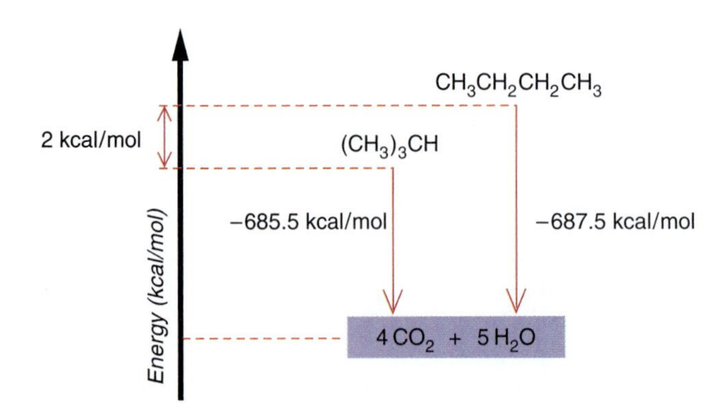

FIGURE 2.21

Heats of combustion of *n*-butane and isobutane.

kcal/mol. This demonstrates cyclopropane's higher energy level, which results from the angular and torsional strain discussed in Section 2.5.2.

In the absence of sufficient oxygen (either in a furnace or an engine), combustion of alkanes produces partially oxidized compounds. These include the deadly carbon monoxide (CO), carbon soot (such as emitted from diesel engines), aldehydes, and carboxylic acids. All of these compounds resulting from incomplete combustion are classified as atmospheric pollutants. The federal government and some state governments have enacted laws to control the emission of such compounds, which means that society must find more efficient methods of combustion for energy-consuming systems such as vehicles and power generators. Smaller hydrocarbons, such as those found in natural gas and LP gas, seem to be "cleaner-burning" as fuels, implying in part that they undergo a more complete combustion.

Chapter Summary

Alkanes are saturated hydrocarbons with the general formula C_nH_{2n+2}; they contain one or more carbon atoms connected in continuous (straight) or branched chains. A **hydrocarbon** is a compound that contains only carbon and hydrogen, and a **saturated hydrocarbon** contains only single bonds between carbon and carbon and between carbon and hydrogen. Alkanes with the same molecular formula but different connectivities are called **constitutional isomers.**

The carbon atoms in alkanes have undergone tetrahedral hybridization to form four sp^3 hybrid orbitals. These overlap with orbitals from hydrogens or other carbon atoms to form molecular orbitals that constitute sigma bonds. The single bond from one saturated carbon to another saturated carbon allows rotation, leading to an equilibrium among various **conformations,** which are best described by Newman projections.

Alkanes are named using the IUPAC system of **nomenclature,** which starts with a **stem name** (indicating the number of carbons in the chain), to which is added the **suffix** *-ane* (to indicate that the compound is an alk*ane*). The result is a **parent name** for an unbranched alkane. Branched alkanes are named as substituted derivatives of the parent alkane. **Alkyl groups** can be thought of as alkanes in which a hydrogen has been removed. They are generalized using the symbol R and named by replacing the suffix *-ane* of the alkane name with the suffix *-yl*. Individual carbon atoms as well as alkyl groups can be classified as *primary, secondary, tertiary,* or *quaternary,* depending on the number of carbon atoms to which they are attached. Hydrogens are classified to match the classification of the carbon to which they are bonded.

Alkanes are nonpolar compounds, which implies that there are only small attractive forces between molecules. Therefore, alkanes are relatively low-boiling compounds and are very insoluble in water. The major sources of alkanes are natural gas and petroleum. The major uses of both petroleum and natural gas are as fuels for combustion and as raw materials for chemical processes.

Cycloalkanes are alkanes in which three or more carbons have formed a ring. Cycloalkanes with substituents on two or more carbons can exist as two separable **configurational isomers,** called ***cis*** and ***trans*** **isomers.** Three-, four-, and five-membered rings have varying degrees of **angular strain** and **torsional strain.** Cyclohexane exists mainly in a nonplanar **chair conformation,** which is essentially devoid of any angular or torsional strain. There is an equilibrium between two chair conformations involving rapid **conversion of conformation,** resulting in **axial** substituents being reoriented to **equatorial** positions, and vice versa. Substituted cyclohexanes are most stable in the chair conformation in which their largest substituents are in the equatorial positions.

Alkanes undergo a substitution reaction called halogenation, in which a halogen atom replaces a hydrogen atom. Alkanes are combustible in the presence of oxygen. The oxidation, or **combustion**, reaction produces carbon dioxide and water and releases considerable energy as heat.

Additional Problems

Alkane Structures and Formulas

2.22 Write the molecular formula of an alkane with (a) 30 carbons, (b) 8 carbons, and (c) 22 carbons.

2.23 Write all possible structures for compounds with molecular formula C_5H_{12} using dash formulas, then condensed formulas, and finally bond-line formulas.

2.24 What is the molecular formula corresponding to each of the following structures?

(a)

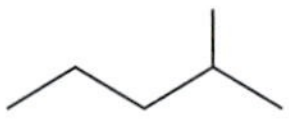

(b) $CH_3CH_2CH(CH_3)_2$

(c) $CH_3-CH(CH_3)-CH(CH_3)-CH_2-CH_3$

2.25 Draw three possible conformations of 1,2-dibromoethane ($BrCH_2CH_2Br$) using both a dash-wedge structural formula and a Newman projection.

2.26 Draw the most stable and the least stable conformations of butane, using both a Newman projection and a dash-wedge formula.

2.27 Draw the *anti* conformation of 1-chloro-2-fluoroethane. Is it the most stable or the least stable of the possible conformations? Why?

2.28 Which, if any, of the following compounds are isomers of each other, and which, if any, are identical?

(a)

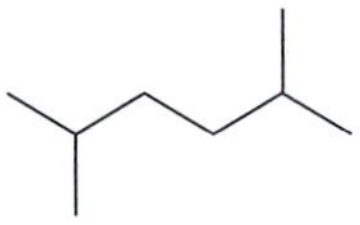

(b) $CH_3CH_2CH(CH_3)CH_2CH(CH_3)_2$

(c) $(CH_3)_2CHCH_2CH_2CH(CH_3)_2$

(d)

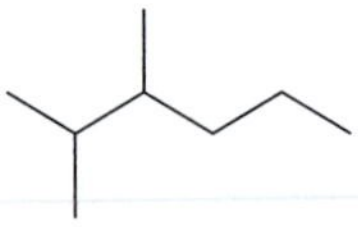

2.29 Draw all possible eclipsed and staggered conformations for 1-chloropropane using Newman projections. Then draw dash-wedge structural formulas for each conformation.

2.30 Identify the primary, secondary, and tertiary hydrogens in 3-methylpentane.

Nomenclature

2.31 Write a bond-line formula for each of the following compounds:

(a) 2,2,4-trimethylheptane

(b) 2,2,3-trimethylbutane

(c) 4-isopropylnonane

(d) 3,3-diethylhexane

(e) 4-*t*-butyloctane

2.32 Write a condensed formula for each of the alkanes named in Problem 2.31.

2.33 Provide full names for the following compounds:

(a) 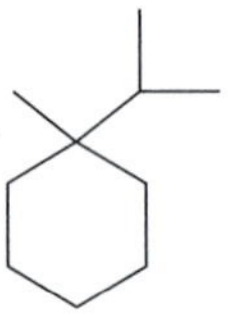 (b)

(c) 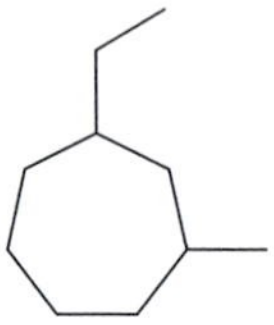 (d)

(e) 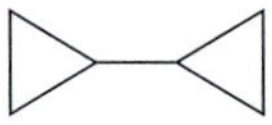 (f) 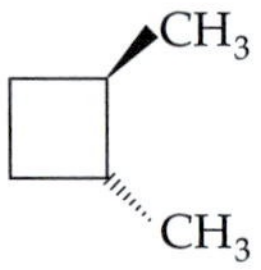 CH_3 / CH_3

2.34 Assign IUPAC names to and write molecular formulas for the following compounds:

(a) 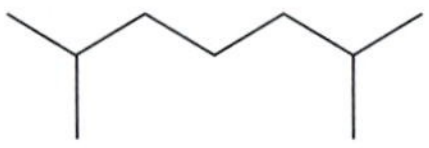 (b)

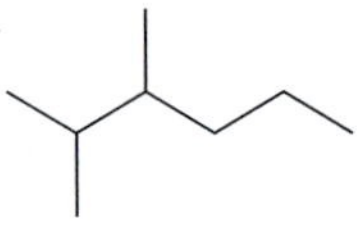

(c) $(CH_3)_3CCH_2CH_3$

(d) $CH_3CH_2CH(C_2H_5)CH_2CH_3$

(e) $CH_3-CH(CH_3)-CH(CH_3)-CH_2-CH_3$

(f)

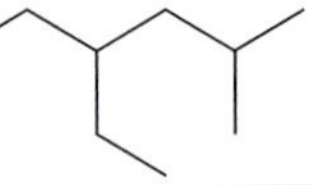

2.35 Draw bond-line formulas for structures (c), (d), and (e) in Problem 2.34.

Cyclohexane Conformations

2.36 Draw structures corresponding to the following names:

(a) *cis*-1,2-dichlorocyclopropane

(b) *trans*-1,3-diethylcyclobutane

(c) *cis*-1,2-dibromocyclohexane

2.37 Draw both chair conformations of each of the following compounds, and indicate which should be more stable:

(a) *cis*-1,2-diethylcyclohexane

(b) *trans*-1,4-dichlorocyclohexane

(c) *cis*-1-*t*-butyl-4-methylcyclohexane

2.38 Give two reasons why cyclohexane exists predominantly in the chair form rather than a planar form.

2.39 Draw both conformations for *cis*-1,3-dimethylcyclohexane and explain which is the more stable conformation and why.

2.40 Of the two possible configurational isomers of each of the following compounds, which should be more stable? Draw the structures.

(a) 1-isopropyl-2-methylcyclohexane

(b) 1-isopropyl-3-methylcyclohexane

(c) 1-isopropyl-4-methylcyclohexane

Mixed Problems

2.41 An alkane has a molecular weight of 86. What is its molecular formula? Draw each of the structural isomers possible for this compound. Which of these isomers would be expected to have the highest boiling point? The lowest boiling point?

2.42 If a bottle of nonane is accidentally dropped and broken in a sink full of water, where will the nonane end up in the mixture of liquids? Explain.

2.43 Why is it not feasible to use straight-run gasoline (that obtained from the simple fractionation process applied to petroleum) in modern automobile engines?

2.44 Draw the structure of, and assign a name to, the compounds that are the immediate higher and lower members of a homologous series involving heptane.

2.45 The important sugar glucose (also known as blood sugar) has the molecular formula $C_6H_{12}O_6$, and its structure is shown here with a planar ring:

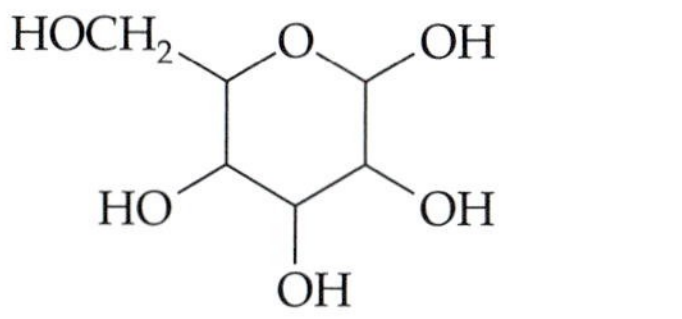

HOCH₂, O, OH, HO, OH, OH

β-Glucose

Note that glucose includes a six-membered heterocyclic ring (an oxygen atom has replaced a methylene group of a cyclohexane ring), which exists mainly in a chair conformation. The most stable form (β-glucose) has all the ring substituents in equatorial positions. Draw this conformational structure of β-glucose.

2.46 For the three dimethylcyclohexanes, explain why the more stable stereoisomer of the 1,3-substituted ring is the *cis* isomer, while for both the 1,2- and 1,4-substituted rings, it is the *trans* isomer.

2.47 Indicate what is incorrect about each of the following names, and provide the correct name.

(a) 1-methyl-2-ethylcyclohexane

(b) 1,3-dimethylbutane

(c) 1-chloroisobutane

(d) 2,3-difluoropropane

(e) 1,5-dimethylcyclopentane

(f) 2,4,4-triethylpentane

2.48 Write a balanced equation for the complete combustion of:

(a) cyclopentane

(b) 2,2,4-trimethylpentane

2.49 Write the condensed and bond-line structures and give the IUPAC name of an alkane that fits each of the following descriptions:

(a) formula C_5H_{12} with only primary hydrogens

(b) formula C_5H_{12} with only one tertiary hydrogen atom

(c) formula C_5H_{12} with only primary and secondary hydrogen atoms

2.50 The *cis* and *trans* relationships (but not the conformational relationships) in cyclohexane rings can be represented using planar hexagonal rings, like those shown for cyclopropane on page 61. Using planar hexagons, show all of the *cis* and *trans* isomers of the compound 1-chloro-2-isopropyl-4-methylcyclohexane.

2.51 For each of the following pairs of compounds, indicate whether the two are identical or are constitutional isomers, configurational isomers, or conformational isomers:

(a) [Newman projections: CH_3, H, H, H, H, CH_3 and H, CH_3, H, H, H, CH_3]

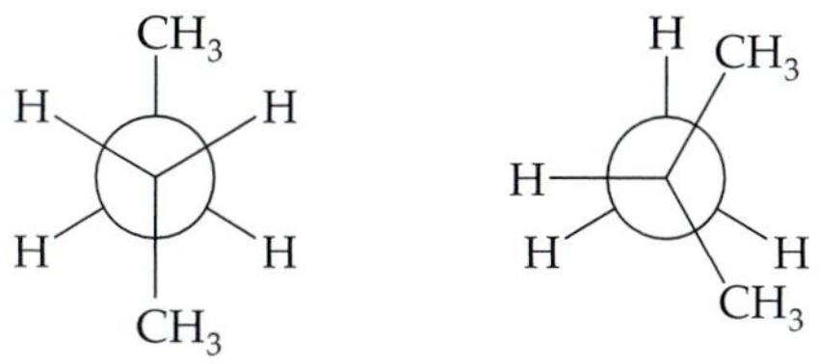

(b) [chair structures]

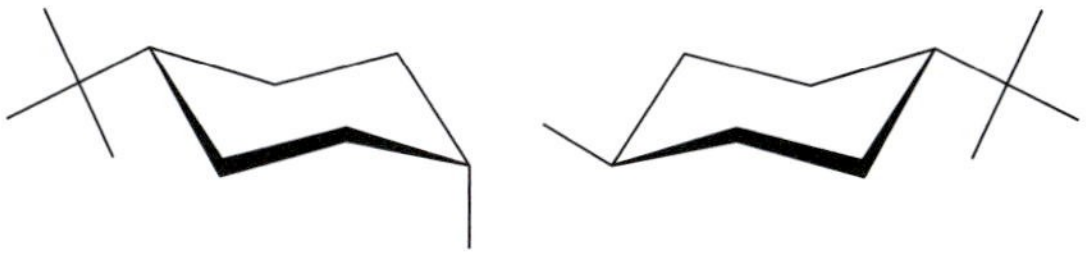

(c) [chair structures]

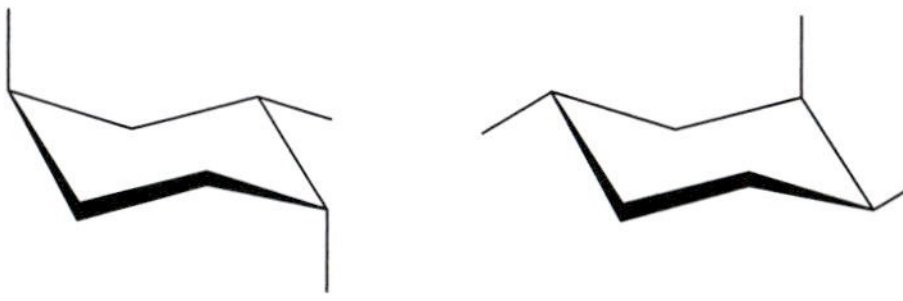

(d) Cl [chair] CH_3 CH_3 [chair] Cl

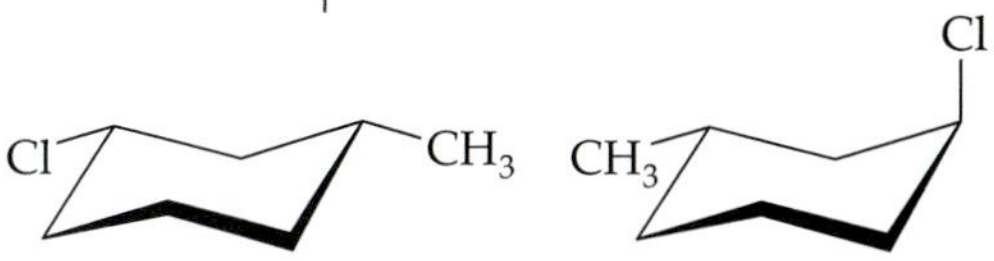

(e) [bond-line structure] $CH_3(CH_2)_4CH_3$

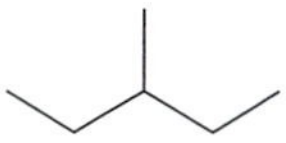

(f) [Newman projections: CH_3, H, H, H, H, CH_3 and H, CH_3, H, CH_3, H, H]

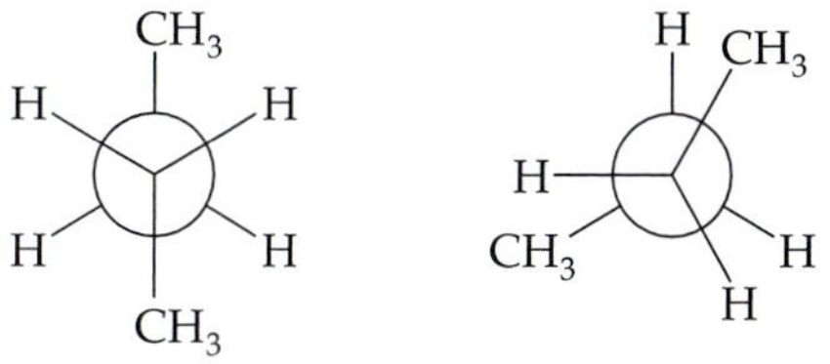

CONCEPTUAL PROBLEM

"Mining" Methane, the Simplest Alkane

Although coal is burned to produce energy, it is more expensive than oil or natural gas to extract and transport. It also emits many by-products, some of which threaten the environment. One way to make the use of coal more environmentally safe is to gasify it to obtain methane, a clean-burning fuel.

The combustion of methane gas in the presence of oxygen gas produces carbon dioxide gas and water—this is the reaction that occurs in a gas furnace. The reaction is exothermic and is a clean and efficient source of energy.

Methane is most readily obtained as part of the natural gas from gas or oil wells, but it can also be produced, along with steam, from the exothermic reaction of a mixture of carbon monoxide gas and hydrogen gas (*synthesis gas*). This mixture of two gases is produced by passing steam over coal (carbon), in an endothermic process called *coal gasification*.

The natural gas company you work for is exploring the possibility of investing in coalbed steam technology. This technology would effectively "mine" the methane that is currently considered just a dangerous by-product of coal mines.

- Starting with steam and coal, write the two balanced chemical reactions that lead to the production of methane. Indicate whether each reaction is endothermic or exothermic.
- What factors would you need to consider to determine the practicality of producing methane in this way?
- What if, in addition to methane, ethane and propane were also produced by the reaction of carbon monoxide gas and hydrogen gas? What could happen to them in the presence of oxygen?
- What happens if the coal used in coal gasification contains an impurity, such as sulfur?

2.52 How many constitutional isomers are possible for C_4H_9Cl, in which a chlorine atom has replaced one hydrogen atom of a C_4 alkane?

2.53 Many chlorinated organic compounds have pesticidal properties. One such compound is 1,2,3,4,5,6-hexachlorocyclohexane. A mixture of all stereoisomers of this compound was sold as an insecticide under the trade name Gammexane. It was later discovered that only one stereoisomer, called the gamma (γ) isomer, possesses the insecticidal properties; this is the *cis*-1,2,4,5-*trans*-3,6-isomer. Draw this isomer using a planar hexagon; then draw its chair conformations and indicate which you would expect to be more stable.

CHAPTER 3

Alkyl Halides (Haloalkanes)

IF YOU REFER to the Key to Transformations (inside front cover), you can see that the alkanes we just studied stand somewhat apart from all other families of compounds. Generally inert, unreactive, and fully saturated with hydrogen, the alkanes are linked to the other families of organic compounds through the alkyl halides. **Alkyl halides,** known in IUPAC nomenclature as **haloalkanes,** can be derived from alkanes (including cycloalkanes) by replacing one or more hydrogen atoms with a *halogen* atom (fluorine, chlorine, bromine, or iodine)—a reaction we will study in Chapter 10. The symbol R—X is used as a general representation of an alkyl halide, with R representing an alkyl group and X representing one of the four halogens.

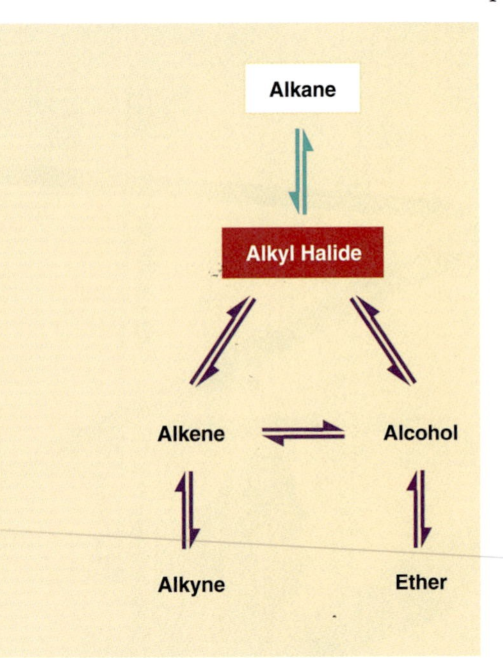

Alkyl halides are the one family of organic compounds that can be derived from alkanes.

Alkyl halides are the first family of organic compounds we will study that contain a **functional group,** that part of a molecular structure that is especially susceptible to chemical reaction. In contrast, as you recall, the carbon-hydrogen bonds present in alkanes are notoriously unreactive. In alkyl halides, the halogen atom serves as the functional group, being readily replaceable by nucleophiles in a major type of reaction known as a *nucleophilic substitution.* Important in its own right, nucleophilic substitution also serves here as an introduction to reaction mechanisms—the means by which one compound is transformed into another.

Haloalkanes range from the very simple to the very complex. For example, ethyl chloride (chloroethane), used as a topical "freezing agent" on skin, includes only a single chlorine atom. The compound 1,1,1,2-tetrafluoroethane, one of a new type of refrigerant, has four fluorines. Dichlorodifluoromethane, a refrigerant formerly used in air-conditioning systems, contains four halogens of two different types. And in the compound perfluorodecalin, which is an artificial blood substitute, all of the hydrogens have been replaced by fluorine.

CH_3CH_2Cl

Ethyl chloride

CH_2FCF_3

1,1,1,2-Tetrafluoroethane

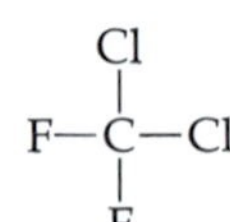

CF_2Cl_2

Dichlorodifluoromethane

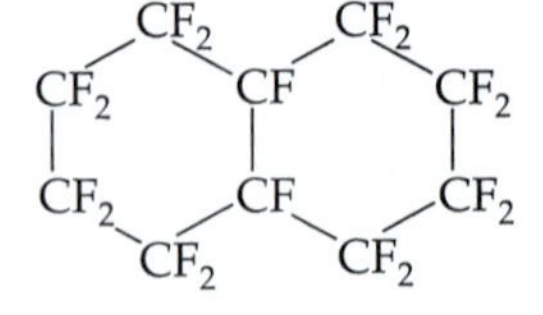

Perfluorodecalin

◀ Overleaf: Milk storage tanks in a cheese-making factory are refrigerated to keep the milk from spoiling. Chlorofluorocarbons (CFCs), which are haloalkanes, were once the refrigerant of choice. In this chapter, you will learn why these useful organic compounds are now banned from production.

3.1

Nomenclature of Alkyl Halides

3.1.1 IUPAC Nomenclature

Haloalkane names are derived from those of the parent alkanes, with the halogen atoms designated as substituent groups (*fluoro-*, *chloro-*, *bromo-*, and *iodo-*) and listed in alphabetical order with other substituents. The position of each halogen is indicated by a number. All of the nomenclature rules introduced in Section 2.3 apply; thus, the first substituent, whether halogen or alkyl group, has the lowest number. Here are some examples of haloalkanes:

$(CH_3)_2CH_2CHBrCH_2CH_3$

3-Bromo-2-methylpentane

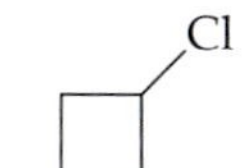

Chlorocyclobutane

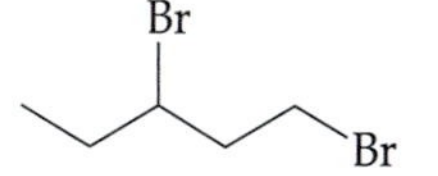

1,3-Dibromopentane

How to Solve a Problem

Give the IUPAC name for each structure:

(a) $CH_3CHFCH_2CH_3$ (b) 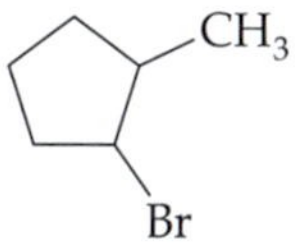 CH_3 Br (c) 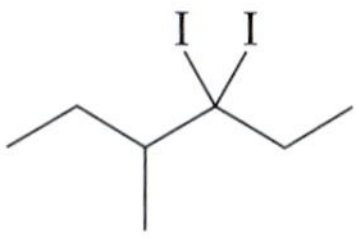 I I

SOLUTION

(a) The compound has four carbons, so the parent name is butane. Fluorine is the only substituent, and so we number the carbons to give fluorine the lowest number. The name is 2-fluorobutane.

(b) The parent name is cyclopentane. Since *bromo* comes before *methyl* alphabetically, we give the lower number to the bromine's position: 1-bromo-2-methylcyclopentane.

(c) The longest carbon chain has six atoms, so the parent name is hexane. This time, whichever way we number the carbons, the substituents appear on carbons 3 and 4. We number them so that the two iodine atoms receive the lower number, rather than the single methyl group, to obtain the lowest possible sum of the numbers involved in the name ($3 + 3 + 4 = 10$ rather than $4 + 4 + 3 = 11$). The name is 3,3-diiodo-4-methylhexane.

PROBLEM 3.1

Draw structures for the following alkyl halides: (a) *cis*-1,4-dichlorocyclohexane, (b) 2,2-difluorobutane, and (c) *trans*-1-bromo-2-chlorocyclobutane.

PROBLEM 3.2

Give the IUPAC names corresponding to the following structures:

(a)  [structure: cyclopropane with two F on one carbon] (b) 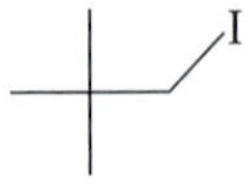 [structure: skeletal formula with I] (c) CHI_3 (d) $CH_3CHBrCH_2CHClCH_2CH(CH_3)_2$

3.1.2 Historical/Common Names

A number of simple haloalkanes have been important for so long that their original (historical) names remain in common use. You need to recognize these. No doubt, some of the common names of the chlorocarbons are familiar to you:

	CH_2Cl_2	$CHCl_3$	Cl_3CCCl_3
Common:	Methylene chloride	Chloroform	Perchloroethane
IUPAC:	Dichloromethane	Trichloromethane	Hexachloroethane

There are analogous common names for other haloalkanes, in which the name of the other halogen replaces chlorine (for example, bromoform for $CHBr_3$). The prefix *per-*, as in *per*chloroethane and *per*fluorodecalin, indicates that all hydrogens in the named alkane have been replaced by the halogen.

Many simple haloalkanes are named as alkyl halides (R—X): The alkyl group (represented by R) is named, and the name of the halide (represented by X) is added as a separate word. Thus, you may see names such as cyclopropyl chloride, isopropyl fluoride, and *t*-butyl chloride.

	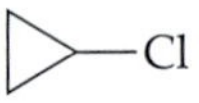 [cyclopropane]—Cl	CH_3CHFCH_3	$(CH_3)_3CCl$
Common:	Cyclopropyl chloride	Isopropyl fluoride	*t*-Butyl chloride
IUPAC:	Chlorocyclopropane	2-Fluoropropane	2-Chloro-2-methylpropane

PROBLEM 3.3

Draw structures for the following compounds: (a) methylene fluoride, (b) isopropyl chloride, (c) isobutyl bromide, (d) iodoform, (e) *n*-pentyl chloride, (f) *t*-butyl iodide, and (g) *t*-pentyl bromide.

3.2 Physical Properties of Alkyl Halides

3.2.1 Bonding

In alkyl halides (R—X), the halogen atom (X) is attached to the carbon atom through a covalent bond that uses an sp^3 hybrid orbital of the carbon. Thus, the molecules are tetrahedral in shape about that carbon, like alkanes. Much bulkier than hydrogen atoms, halogen atoms have sizes that increase down Group VIIA of the periodic table along with their atomic numbers:

Atomic number and atomic weight: F $<$ Cl $<$ Br $<$ I

This is because the larger the atomic number, the more protons there are in the nucleus and the more electrons there are around the nucleus (Figure 3.1). The *length* of the carbon-halogen bond also increases with the size of the halogen atom:

Bond length: C—F < C—Cl < C—Br < C—I

Because the halogens are all considerably more electronegative than carbon, carbon-halogen bonds are polar, with *bond polarity* increasing in the same order as halogen *electronegativity,* from the bottom up within the group (see Figure 1.4):

Electronegativity (and C—X bond polarity): I < Br < Cl < F

All monohalomethanes have a significant dipole moment, ranging from 1.6 to 1.9 Debye units.

The ease with which a carbon-halogen bond is broken in an ionic reaction depends on the polarizability of the bond, not its polarity. **Polarizability** is the ease with which the bonding electrons are shifted, or moved, and is determined in part by the closeness of the atomic nuclei and their protons to the bonding electrons. For any two alkyl halides, the larger the halogen atom, the more electrons there are around the nucleus, and the farther the bonding electrons are from the electrical influence of the protons in the nucleus. Thus, with halogens, polarizability increases with increasing size of the atom:

Polarizability: F < Cl < Br < I

The combined effect of bond polarizability and bond length in the alkyl halides results in this order for *bond strength:*

Bond strength: R—F > R—Cl > R—Br > R—I

Because stronger bonds are more difficult to break in a chemical reaction, the important implication of all of these observations is that the *relative reactivity of alkyl halides* follows this order:

Relative reactivity: R—F < R—Cl < R—Br < R—I

Alkyl fluorides are so inert that we will ignore them in subsequent discussions of reactions. This chemical inertness of fluorocarbons is reflected in the common material Teflon, which is a polyfluoroalkane used as a coating in frying pans and as a flexible roof material on many domed stadiums (more on this in Chapter 18).

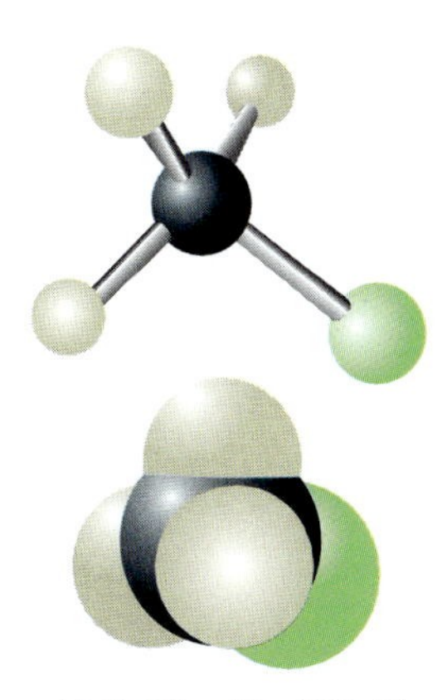

Methyl fluoride (CH_3F)

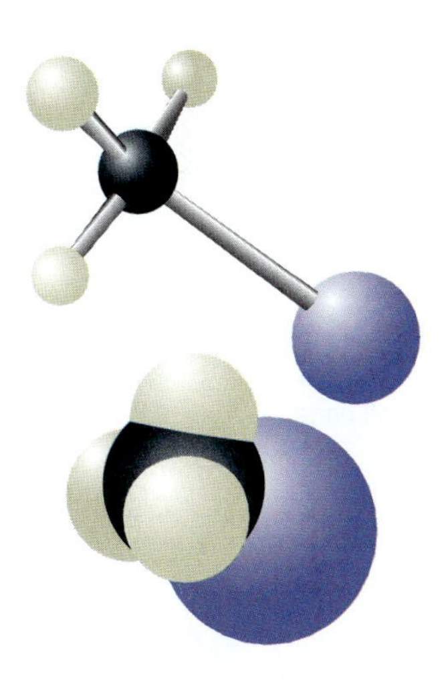

Methyl iodide (CH_3I)

FIGURE 3.1

Ball-and-stick and space-filling models showing the relative sizes of a fluorine atom and an iodine atom attached to a methyl group.

3.2.2 Other Properties

Most alkyl halides are liquids at room temperature and boil at higher temperatures than do the alkanes from which they are derived. The boiling point for a given alkyl halide generally increases from fluoride to chloride to bromide to iodide. Thus, ethane boils at −89°C, and ethyl fluoride boils at −37°C, ethyl chloride at 12°C, ethyl bromide at 38°C, and ethyl iodide at 72°C. This pattern is in the direction expected: Increased molecular weight and increased polarizability (and therefore increased attractive forces in the liquid state) result in higher temperatures being required for vaporization.

Liquid alkyl halides are insoluble in water. Most are more dense than the alkanes from which they were derived, with density increasing with the atomic size of the halogen. Most alkyl halides are more dense than water. For example, the density of dichloromethane (otherwise known as methylene chloride and often used as an organic solvent) is 1.3 g/mL, while that of water is 1.0 g/mL.

PROBLEM 3.4

Arrange each of the following pairs of compounds in *decreasing* order by boiling point:

(a) methyl chloride and ethyl chloride
(b) isopropyl fluoride and isopropyl iodide
(c) cyclopropyl chloride and cyclohexyl chloride

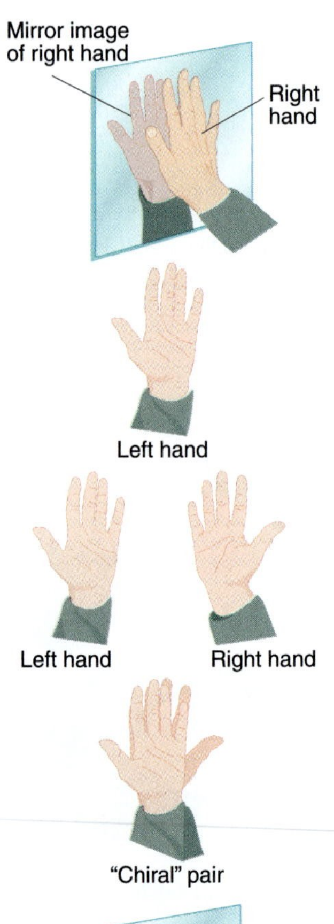

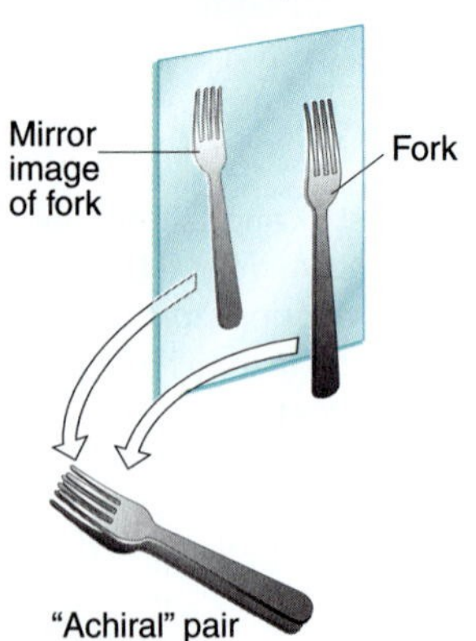

FIGURE 3.2

The mirror image of a left hand is a right hand. They are not identical. The mirror image of a fork is identical to the fork.

3.3

Introduction to Enantiomers

With the alkyl halides, we encounter a second type of configurational isomerism. (We studied *cis-trans* isomers in Chapter 2.) You will recall that isomers are compounds with the same molecular formula but different structural formulas. **Enantiomers** are a kind of stereoisomer, which means that they have the *same connectivities, but differ in their three-dimensional orientation.* They are also a kind of configurational isomer; therefore, they *cannot interconvert by bond rotation.* (Refer to the chart on the inside back cover to see the relationship of enantiomers to other isomers.)

An enantiomer has a unique physical property in that it rotates a beam of plane-polarized light (it is *optically active*). This rotation cannot be seen by the naked eye, but it can be detected and measured by an instrument called a *polarimeter.* Optical activity and related matters will be discussed in detail in Chapter 17.

The great significance of enantiomers lies in the fact that the majority of compounds found in nature exist as a single enantiomer. For example, all but one of the α-amino acids that make up most proteins, whether human or other, exist as a single enantiomer. All natural carbohydrates (sugars) exist as a single enantiomer. Most other natural products, including hydrocarbons (for example, turpentine components) and many alkaloids (such as nicotine and morphine), also exist as a single enantiomer. As we will see, some commercially produced pharmaceuticals are sold as single enantiomers, but these are more the exception than the rule.

3.3.1 Chirality

Chirality is the property of *handedness* (the term is derived from the Greek word for "hand"), which can apply to any object, including a molecule. It refers to the relationship that a right hand has to a left hand. Any object can be described as *chiral* (having this property) or *achiral* (not having this property).

A **chiral object** is one that is not superimposable on its mirror image. If you hold your right hand in front of a mirror, what appears in the mirror (the mirror image) seems to be a left hand. Examination of your two hands (Figure 3.2) reveals that they are very similar, but they are not identical (that is, they are not interchangeable or superimposable). Thus, a hand is said to be *chiral.* In contrast, a table fork is *achiral* because the object (the fork) is superimposable on its mirror image.

Chirality is a characteristic of the three-dimensional shape of a molecule. However, it is important to learn to assess whether an object is chiral or achiral by examining a two-dimensional drawing of it. However, if you are uncertain whether or not any object is chiral, perform the *ultimate test for chirality* by asking: Is the object superimposable on its mirror image?

Consider a pair of leather gloves with stitching on the palm; each glove is clearly chiral (one glove is the object and the other is the mirror image). Now consider a two-dimensional drawing of the right-hand glove (Figure 3.3) and ask yourself whether it appears chiral. Make a drawing of the mirror image (the left-hand glove) and mentally test the superimposability of the drawings. They are not superimposable, and so the gloves are chiral. A simpler way to determine that they are chiral is to look for a plane of symmetry in the drawing of one glove. A **plane of symmetry** exists when an imaginary plane slices an object in half so that one half is the mirror image of the other half. Note that the two-dimensional drawing of either glove *does not* have a plane of symmetry; therefore, the gloves are chiral. This is the predictor of chirality:

- An object *with* a plane of symmetry is *achiral*.
- An object *without* a plane of symmetry is *chiral*.

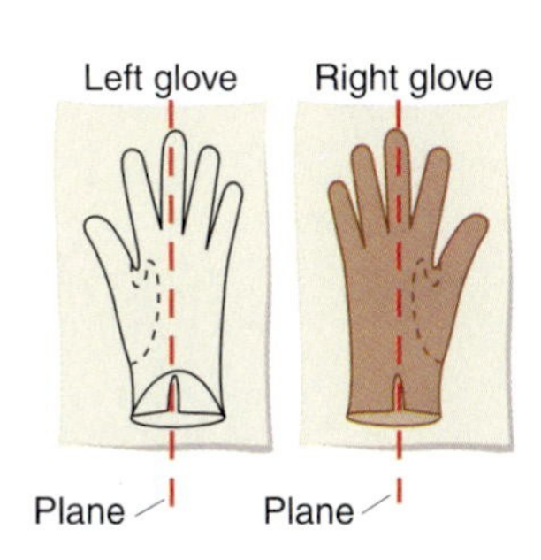

FIGURE 3.3

Lack of a plane of symmetry for either of a pair of gloves.

How to Solve a Problem

Is a single unmarked athletic sock chiral?

SOLUTION

Keeping the pair of gloves in mind, let's now consider a pair of socks. Looked at from the front, it is apparent that each sock has a plane of symmetry; therefore, we can conclude that a sock must be achiral. Practical experience in the real (three-dimensional) world also tells us that there is no right and left sock; unmarked socks are normally interchangeable and therefore identical.

PROBLEM 3.5

Indicate which of the following objects are chiral: (a) automobile, (b) ski (without binding), (c) ice skate, (d) dinner knife, (e) shoe, (f) compact disk, (g) boombox, (h) computer keyboard, (i) computer monitor case.

Enantiomers

Like other three-dimensional objects, organic compounds are either chiral or achiral. In the case of a *chiral compound*, the object and the mirror image are called *enantiomers*. A pair of such enantiomers is called a **racemic mixture,** or a **racemate.**

Consider first the compound 2-iodopropane, represented in two dimensions in part (a) of Figure 3.4. You can see that it is achiral because it has a plane of symmetry (b). This conclusion can be verified by looking at a three-dimensional model (c), together with its mirror image (d). Note that (c) and (d) are identical; the structures are superimposable. You can see this by envisioning (d) rotating 180° about the C—H axis and then setting on top of (c)—all bonds and groups match up. Therefore, 2-iodopropane is achiral, and there are no enantiomers of this compound.

Considering 2-iodobutane (Figure 3.4e) in the same manner reveals that the compound does not have a plane of symmetry (f) and therefore is chiral. Test this conclusion by examining the three-dimensional models of the compound (g) and its mirror

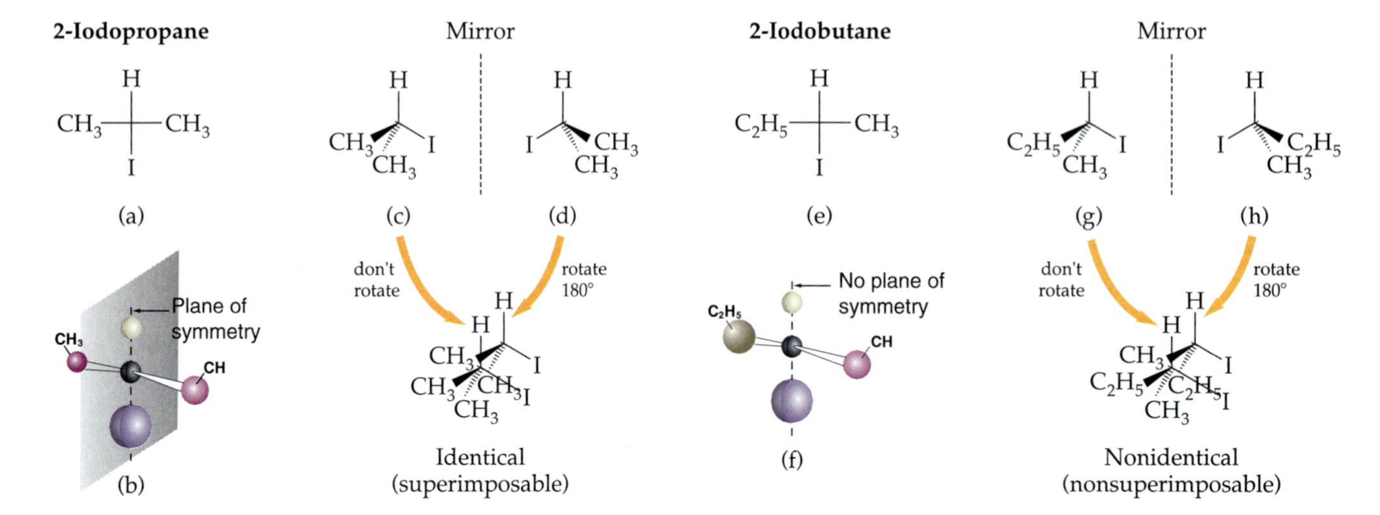

FIGURE 3.4

(a) Two-dimensional formula of 2-iodopropane. (b) Ball-and-stick model showing plane of symmetry. (c) Object. (d) Mirror image. Note that (c) and (d) are identical (superimposable). (e) Two-dimensional formula of 2-iodobutane. (f) Ball-and-stick model testing for a potential plane of symmetry. (g) Object (one enantiomer). (h) Mirror image (other enantiomer).

image (h). They are not identical; the structures are not superimposable. Therefore, 2-iodobutane is chiral, and there are two separable enantiomers of this compound.

A sample of a chiral compound such as 2-iodobutane exists as a 50/50 mixture of the two enantiomers (a *racemic mixture*, or *racemate*) under normal circumstances. Chemically and physically, the sample behaves as though there is only a single compound present. The racemic mixture is *not* optically active. Unusual measures are required to separate it into the two individual enantiomers because the only difference in physical properties between the two is the direction in which they rotate a beam of plane-polarized light. In Chapter 17, the separation (called *resolution*), identification, and properties of enantiomers will be discussed in detail. For now, all you need to know is how to predict whether a compound will exist as enantiomers.

Stereocenters

If a carbon atom has four different substituents attached to it, it is called a **stereocenter.** If you look at 2-iodopropane in Figure 3.4(a), you can see that the central carbon has only three different substituents (methyl group, hydrogen, and iodine); therefore, two of the substituents must be identical (two methyl groups). There is no stereocenter present. Also, having two identical substituents means that the structure has a plane of symmetry and therefore is achiral. In contrast, 2-iodobutane, in Figure 3.4(e) has four different substituents attached to carbon 2 (methyl group, ethyl group, hydrogen, and iodine). Thus, carbon 2 is a stereocenter.

If a compound has one stereocenter, it is chiral and exists in the form of a racemic mixture. In other words, there are two enantiomers of the compound. Thus, you have two means of predicting the existence of enantiomers of a compound and one ultimate test for their existence:

- *Stereocenters.* The presence of a single stereocenter in the structure means that the compound exists in the form of a racemic mixture of two enantiomers.
- *Plane of symmetry.* If a compound's structure has a plane of symmetry, no enantiomers exist.
- *Superimposability.* The ultimate test for the existence of enantiomers (to use as a backup) is the nonsuperimposability of a three-dimensional structure of a compound on its mirror image.

How to Solve a Problem

Determine if bromochlorofluoroiodomethane exists as a pair of enantiomers.

SOLUTION

We draw the two-dimensional structural formula for CBrClFI and search for a stereocenter. The carbon has four different substituents attached to it and therefore is a stereocenter. Two enantiomers of this compound should exist.

For further proof, we can draw a three-dimensional structure or build a molecular model of the compound. Part (a) of Figure 3.5 shows such a drawing. Searching for a plane of symmetry in this structure, we can find no such plane; so the compound should exist as a pair of enantiomers.

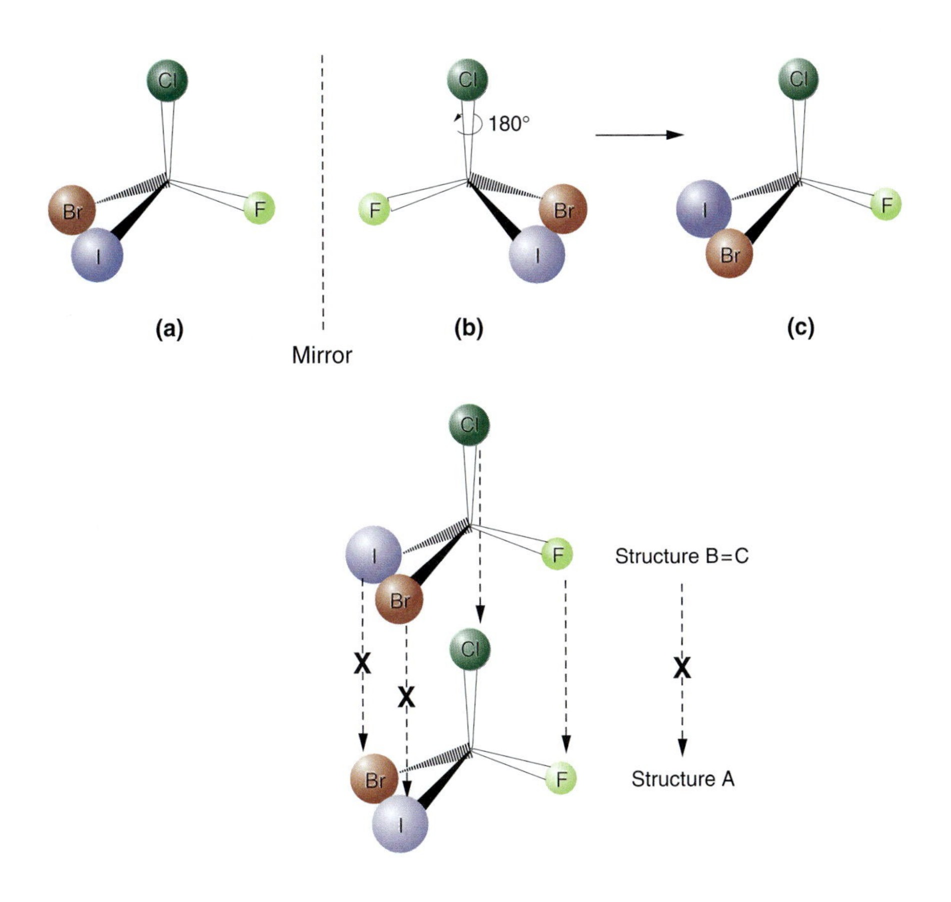

FIGURE 3.5

The structure of bromochlorofluoroiodomethane. (a) is the object, and (b) is the mirror image. Structure (c) is created when (b) is rotated 180° around a vertical axis, the carbon-chlorine bond. When structure (c) is superimposed on the original structure (a), the structures do not match. Therefore (a) and (c) are not identical, and since (c) is the same structure as (b), (a) and (b) are not identical.

Finally, we can draw or build the mirror image, part (b) of Figure 3.5, then test the mirror image for superimposability on the object. To do this, we rotate structure (b) 180° about the C—Cl axis to get structure (c). We lift this and attempt to place it on top of (a). The chlorine and fluorine atoms can be superimposed in the two structures, but the iodine of (c) is lined up over the bromine of (a), and the bromine of (c) is lined up over the iodine of (a). Thus, the mirror image (structure b or c) is not superimposable on the object (a). Therefore, we conclude that the compound exists as two enantiomers, which together form a racemic mixture.

PROBLEM 3.6

For each of the following compounds, draw the structure and mark the stereocenters, if any, with an asterisk: (a) 2-methylpentane, (b) 3-methylpentane, (c) iodocyclohexane, (d) 1-iodo-2-methylcyclohexane, (e) 2,2-difluoro-3-methylpentane, (f) 1,1-dibromocyclopropane.

PROBLEM 3.7

Indicate which of the following compounds exist as a racemic mixture: (a) 1-iodo-2-methylpentane, (b) 2,2-dichlorobutane, (c) 1-bromo-4-fluorocyclohexane, (d) 1-bromo-3-fluorocyclohexane, (e) isopropyl fluoride.

3.3.2 Configurations of Enantiomers

Because a compound with one stereocenter exists as two enantiomers, it is essential that you be able to assign distinguishing names to each. The *R-S* system of configurational nomenclature designates the **absolute configuration** for any enantiomer (in other words, the exact arrangement in space of groups about the stereocenter). The *R-S* system is in corporated into the IUPAC system of nomenclature and is applied as illustrated in the following steps for 2-chlorobutane. (The actual nomenclature rules are steps 1, 2, and 6.)

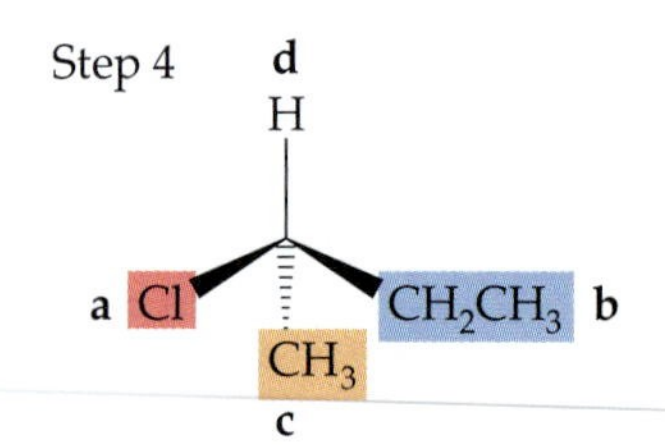

Step 1: Locate the stereocenter of the compound and identify the four substituents. (This can be done using a two-dimensional structural formula.) In 2-chlorobutane (CH_3—CHCl—CH_2—CH_3), carbon 2 is the stereocenter.

Step 2: Assign a priority to each of the four substituents, with **a** *being the highest priority and* **d** *being the lowest.* For 2-chlorobutane, the priority of the substituents is **a** = Cl, **b** = CH_2CH_3, **c** = CH_3, and **d** = H. (The rules for determining priority are given in Section 3.3.3.)

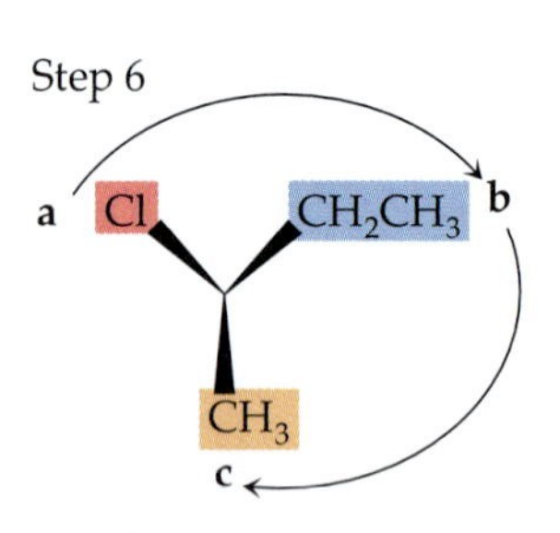

Clockwise route = *R*

Step 3: Draw a three-dimensional structure of the molecule, such that the **d** substituent (lowest priority) is oriented toward the top of the paper and in the plane of the paper.

Step 4: Attach the other three substituents to the stereocenter in any order, labeling them with their priority designations (**a, b,** and **c**).

Step 5: View the molecule from the bottom of the sheet of paper (that is, looking along the plane of the paper and along the bond axis from the stereocenter to the **d** substituent). The three substituents **a, b,** and **c** will appear to be oriented around the edge of an imaginary circle. This is like looking up at an umbrella that has blown inside out, with the umbrella handle being the bond from carbon to substituent **d,** and the substituents **a, b,** and **c** located equidistant from one another around the outer rim of the umbrella.

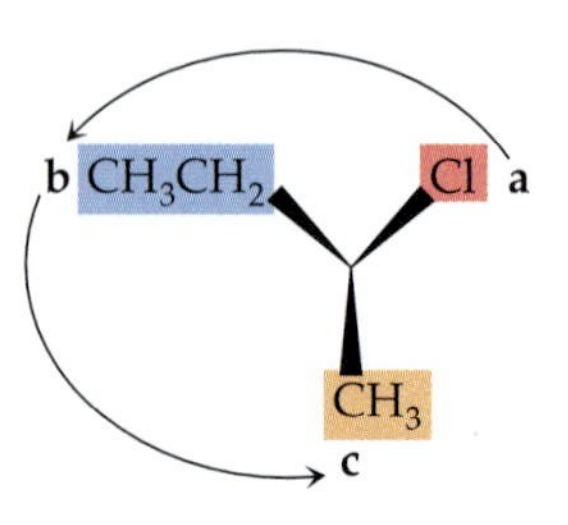

Counterclockwise route = *S*

*Step 6: From the view along the carbon-***d** *bond axis, imagine starting at substituent* **a** *and drawing an arrow to* **b** *and then* **c.** *If this route has a* clockwise *direction, the isomer has the* R *configuration (from the Latin* rectus, *meaning "right"). If this route is* counterclockwise, *the isomer has the* S *configuration (from the Latin* sinister, *meaning "left").* In the case of 2-chlorobutane, the two possible views from the bottom of the sheet of paper (toward the open side of the umbrella) are shown.

The isomer shown in step 4 is (*R*)-2-chlorobutane, which is shown again here together with its mirror image:

d H, a Cl, CH_3 c, CH_2CH_3 b

d H, b CH_3CH_2, CH_3 c, Cl a

(*R*)-2-Chlorobutane (*S*)-2-Chlorobutane

Note that the mirror image is (*S*)-2-chlorobutane. These are the only two configurations possible for 2-chlorobutane. You may try to draw others by interchanging the substituents in every possible order, but all attempts will result in either the *R* or *S* configuration. Try it.

For any pair of enantiomers, one is *R* and the other is *S*, and these designations are incorporated at the beginning of their names, as shown. If the positions of *any two substituents* in one of the enantiomers are interchanged, the resulting configuration will be the other enantiomer. Thus, if the chlorine atom and methyl group are interchanged in the *S* configuration of 2-chlorobutane, what results is structure 1.

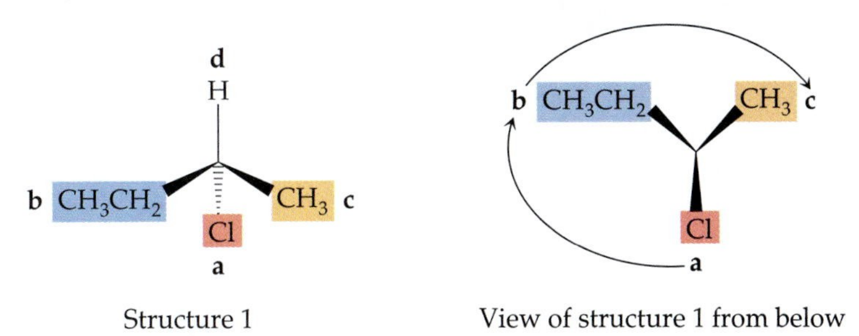

Structure 1 View of structure 1 from below

The view from below structure 1 shows that it has the *R* configuration.

As a final note, many compounds have more than one stereocenter (some examples will be covered in detail in Chapter 17). The approach outlined here can be used in turn for each stereocenter to determine its configuration. The letter *R* or *S* must be included in the compound's name for each stereocenter in the structure.

3.3.3 Determining Substituent Priorities

Application of the *R-S* nomenclature system depends on a systematic assignment of priority to each substituent attached to the stereocenter. Priorities are determined using the following three steps. (This same system of determining substituent priorities will be used again in Chapter 6 to assign configurations to alkenes.)

Step 1: Rank the atoms (not the entire substituents) directly attached to the stereocenter in order of *decreasing* atomic number *Z* (see the periodic table, inside back cover). The highest-numbered atom is assigned the letter **a,** the second-highest is **b,** the third is **c,** and the fourth is **d.** For example, in 2-chlorobutane, chlorine ($Z = 17$) is **a,** and hydrogen ($Z = 1$) is clearly **d,** but the methyl and ethyl groups both have a carbon atom ($Z = 6$) attached to the stereocenter; so these two substituents cannot be distinguished in this step.

Step 1

d H, CH_3—C—CH_2CH_3, Cl a

Step 2

$$\underset{\mathbf{a}}{\text{Cl}}\ \ \mathbf{c}\ CH_3\text{—}\underset{|}{\overset{|}{C}}\text{—}CH_2CH_3\ \mathbf{b}\ \ \overset{\mathbf{d}}{\text{H}}$$

Step 2: If complete assignment of priorities cannot be made in step 1 because two of the atoms directly attached to the stereocenter are the same, then consider the atoms once removed from the stereocenter. For 2-chlorobutane, the methyl group has as its next atom only hydrogen ($Z = 1$), whereas the ethyl group has as its next atom a carbon ($Z = 6$) (as well as two hydrogens, which do not distinguish this group from methyl). Therefore, in this case, the second atom away from the stereocenter provides a difference between the two groups, and the group whose atom has the higher atomic number (the ethyl group) is assigned priority **b.** The remaining group (methyl) is assigned priority **c.**

Evaluate substituent atoms one by one, moving out from the stereocenter, until the first difference is reached. The following examples should be helpful.

Atomic number of *first* atom from stereocenter

1	6	6	6	7	8	9
$-H$	$-CH_3$	$-CH_2CH_3$	$-CH_2(CH_3)_2$	$-NH_2$	$-OH$	$-F$

Increasing priority →

Atomic number of *second* atom from stereocenter

1	6	6	7	8	53
$-CH_3$	$-CH_2CH_3$	$-CH_2CH(CH_3)_2$	$-CH_2NH_2$	$-CH_2OH$	$-CH_2I$

Increasing priority →

Step 3: (*Note:* This step will not come into play until Chapter 6.) If one of the substituents contains a multiple bond, consider the atom at each end of that bond to be attached to a duplicate of the atom in the initial attachment. Several examples illustrate this point:

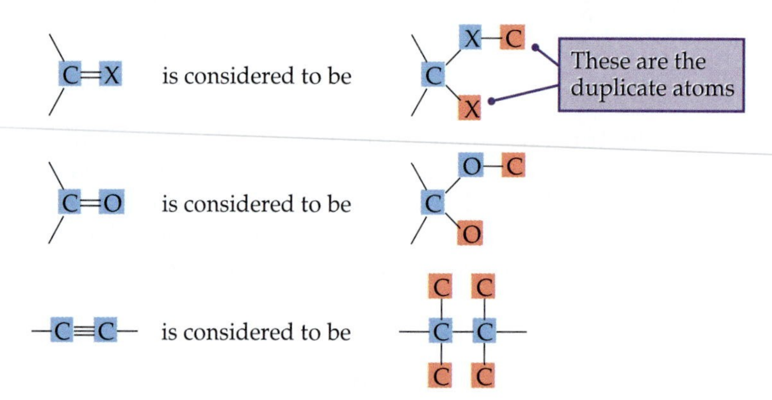

Application of step 3 can be illustrated with the analgesic drug *ibuprofen* (discussed in "Chemistry at Work"). The three-dimensional structure of the physiologically active *S* enantiomer is shown here. From the structure as drawn, can you confirm the configurational assignment?

d H; c CH_3; $CH_2CH(CH_3)_2$ b; HO–C=O a

(*S*)-Ibuprofen

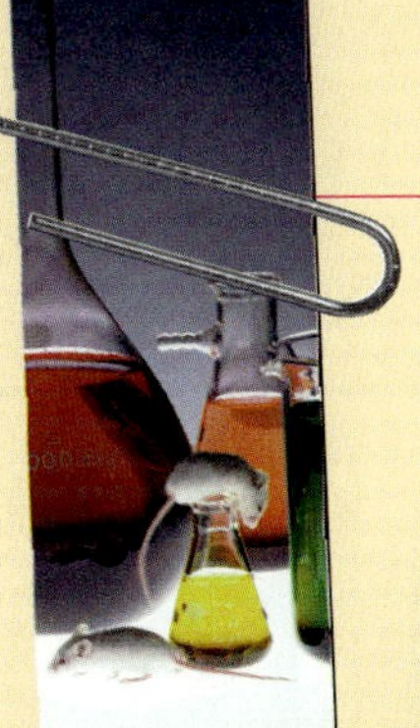

CHEMISTRY AT WORK

IN THE BODY

Chiral Drugs

It sometimes seems as though new drugs appear on the market in a constant stream. How do we know they're really safe? New drugs must be approved by the Food and Drug Administration (FDA) before they can be sold in the United States. An understanding of enantiomers is essential for understanding the action of some drugs and their safety.

Aspirin and acetaminophen are widely used analgesic drugs, available without a prescription. (Acetaminophen is the active ingredient in Tylenol and Excedrin.) They are achiral; that is, they exist as single compounds with no enantiomers.

Aspirin   Acetaminophen

However, the analgesic drug ibuprofen (the active ingredient in Motrin, Nuprin, and Advil) is chiral and exists as two enantiomers (the stereocenter is marked with an asterisk):

$$(CH_3)_2CHCH_2-C_6H_4-\overset{*}{C}H(CH_3)COOH$$

Ibuprofen

One enantiomer (called the *S* enantiomer, described in Section 3.3.3) is physiologically active with the desired analgesic effect, while the other enantiomer (the *R* enantiomer) is inactive. Ibuprofen is sold as a racemic mixture, not as a single enantiomer. Fortunately, the inactive *R* enantiomer is not harmful to the body.

Why is one enantiomer physiologically active while the other is not? For a molecule to have a biological effect, it must interact with a particular receptor site in the body. The receptor site is shaped so that only a molecule with a complementary shape fits correctly, and the receptor sites are chiral. If the wrong enantiomer of the molecule comes along, it won't fit the receptor site. It's like trying to fit your left hand into a right-hand glove.

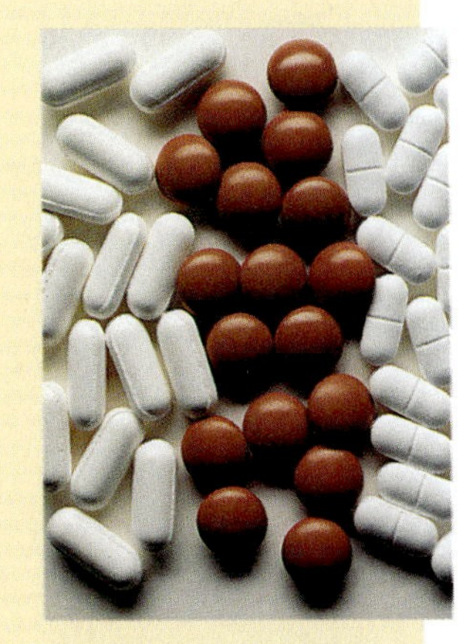

Acetaminophen (left), ibuprofen (center), and aspirin (right).

In 1963, a new drug called *thalidomide* was prescribed for many pregnant women to treat the symptoms of morning sickness. Later, the drug was found to be responsible for horrible birth defects in some of the children born to these women. Thalidomide is chiral, existing as two enantiomers.

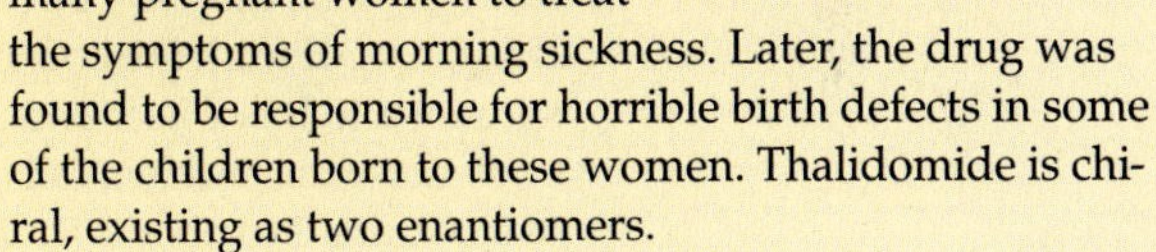

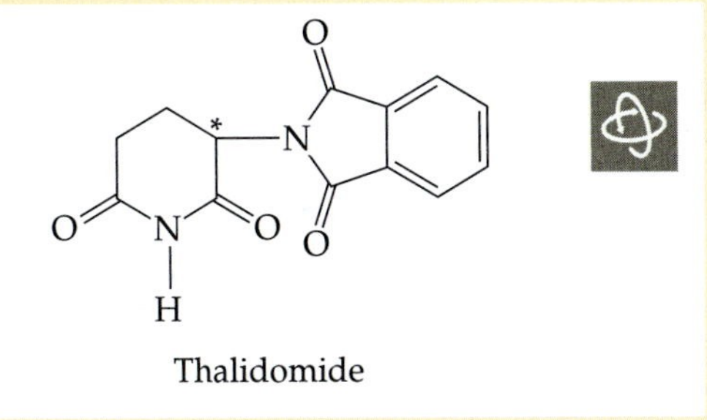

Thalidomide

Researchers subsequently found that one of the two enantiomers had the desired beneficial effect, while the other caused the birth defects. The drug was removed from the market. However, more recent research has revealed substantial beneficial effects of one enantiomer of thalidomide, leading to new requests to the FDA for approval of certain medicinal uses.

Many commercial drugs are produced in the form of racemic mixtures. However, pharmaceutical companies are researching ways to produce pure single enantiomers of drugs rather than mixtures of enantiomers (racemates). These drugs are referred to as *chiral drugs*. Some of these chiral drugs have been marketed abroad and await approval by the FDA for use in the United States.

How to Solve a Problem

Draw the *R* configuration of 3-methylhexane.

PROBLEM ANALYSIS

Solving this problem involves several steps, including checking the structural formula, identifying the stereocenter, identifying the substituents on the stereocenter, and assigning them priorities. Once that is accomplished, we draw *any* three-dimensional structure (without worrying whether it is *R* or *S*) using the steps given in Section 3.3.2. We put the **d** priority group in the plane of the paper and oriented to the top of the paper and then determine the configuration. If we lucked out and happened to draw the *R* configuration, we're done. If we happened to draw the *S* configuration, we can just interchange any two groups to get the *R* configuration. We check the structure at the end.

SOLUTION

The compound is shown as a two-dimensional formula (**A**) with the groups labeled as to priority: **a** = propyl, **b** = ethyl, **c** = methyl, **d** = hydrogen. Then we draw a three-dimensional structure (**B**) with the C—H bond "up" (that is, oriented toward the top of the sheet of paper). Examining **B** from the bottom results in the structure **C,** and the direction of the path from substituent **a** to **b** to **c** is counterclockwise. Therefore, we have drawn the *S* configuration.

Because the problem calls for the *R* configuration, we interchange the ethyl and methyl groups of structure **B** to obtain **D,** the structure of (*R*)-3-methylhexane.

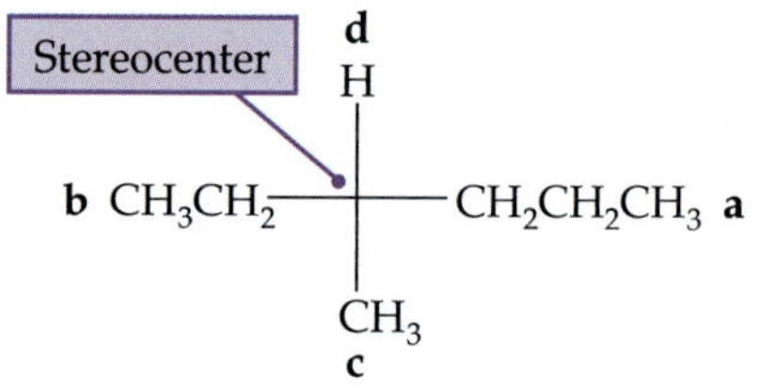

Stereocenter

d H; **b** CH_3CH_2; $CH_2CH_2CH_3$ **a**; CH_3 **c**

A

≡

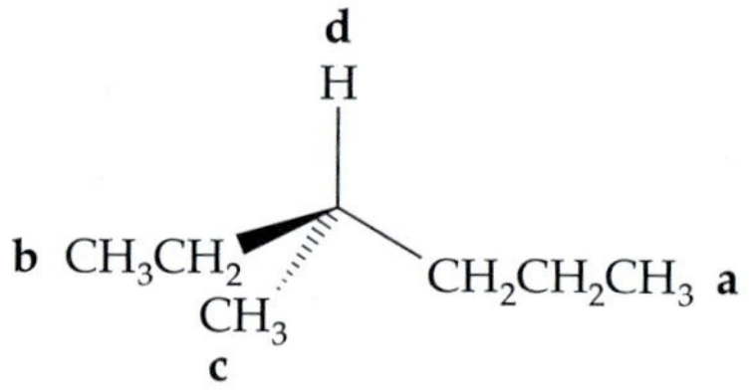

d H; **b** CH_3CH_2; CH_3 **c**; $CH_2CH_2CH_3$ **a**

B

view from below

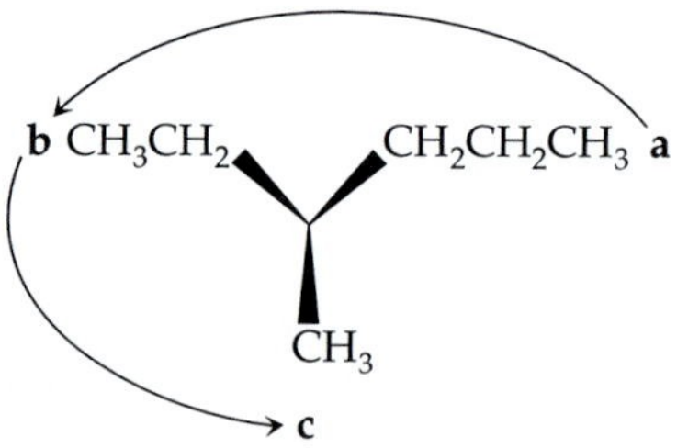

b CH_3CH_2; $CH_2CH_2CH_3$ **a**; CH_3; **c**

C

(*S*)-3-Methylhexane

interchange methyl and ethyl groups in **B**

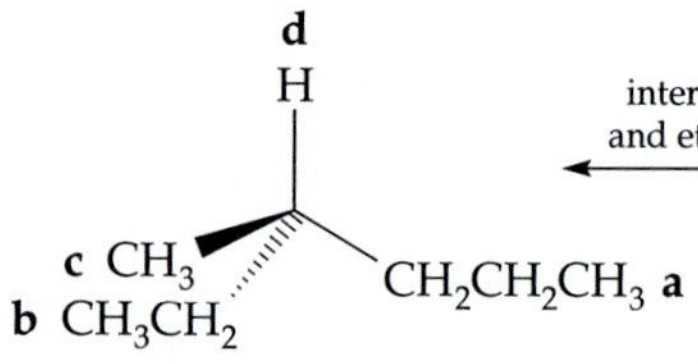

d H; **c** CH_3; **b** CH_3CH_2; $CH_2CH_2CH_3$ **a**

D

(*R*)-3-Methylhexane

PROBLEM 3.8

Draw the *R* configuration of each of the following compounds: (a) 3-bromohexane and (b) 1,3-dichloropentane.

PROBLEM 3.9

Draw the *S* configuration of each of the following compounds:

(a) [skeletal structure with OH] (b) $CH_3CH_2CH(CH_3)CH_2Cl$ (c) [skeletal structure with F and SH]

3.4 Substitution Reactions with Alkyl Halides

Alkyl halides (R—X) are very important in organic chemistry because of the tendency of the polarizable carbon-halogen bond to break under relatively mild reaction conditions. This permits the halogen (X) in an alkyl halide to be replaced by a variety of other substituents (Y), leaving the alkyl group (R) structurally unchanged. This substitution of Y for X is known as a **substitution reaction.**

$$Y\!: \; + \; R\text{—}X \longrightarrow R\text{—}Y \; + \; X\!:$$

Of particular use is **nucleophilic substitution (S_N)**, in which the reagent (Y) is a **nucleophile** (a "nucleus seeker" or electron-pair donor) that displaces and replaces the halide (a **leaving group** or electron-pair acceptor). This reaction can be generalized by the following two equations, where Nu: represents the nucleophile (that is, the electron donor, which may be neutral or negatively charged) and R represents an appropriate alkyl group (for example, ethyl).

General nucleophilic substitution (S_N) reaction:

$$Nu\!:^- \; + \; R\text{—}X \longrightarrow R\text{—}Nu \; + \; X\!:^-$$

$$Nu\!: \; + \; R\text{—}X \longrightarrow R\text{—}Nu^+ \; + \; X\!:^-$$

In both examples, the nucleophile forms a bond with the alkyl group by causing the carbon-halogen bond to break and displacing the halide ion as a leaving group. Since a wide variety of nucleophiles are available, a wide variety of products can be formed by this reaction. Table 3.1 illustrates several of these transformations using ethyl bromide as the **substrate** (that is, the compound undergoing the reaction). The S_N reaction effectively converts one functional group (halide) into another. You will see how the S_N reaction of alkyl halides can be used to synthesize many of the families of compounds listed in Table 3.1 as we move through this book. The S_N reaction can also be used to extend carbon skeletons by making new carbon-carbon bonds.

How to Solve a Problem

What nucleophile would you use to convert 1-bromobutane to 1-butanethiol?

TABLE 3.1

Nucleophiles Reacting with Ethyl Bromide

Nucleophile	Name*	Product	Family Name
$^{-}$:OH	Hydroxide	$C_2H_5{-}OH$	Alcohol
$^{-}$:OR	Alkoxide	$C_2H_5{-}OR$	Ether
$^{-}$:SH	Hydrosulfide	$C_2H_5{-}SH$	Thiol
$^{-}$:SR	Mercaptide	$C_2H_5{-}SR$	Thioether
:NH_3	Ammonia	$C_2H_5{-}NH_3^+$	Alkylammonium ion
$^{-}$:CN	Cyanide	$C_2H_5{-}CN$	Nitrile
$^{-}$:C≡CH	Acetylide	$C_2H_5{-}C{\equiv}CH$	Alkyne
$^{-}$:I	Iodide	$C_2H_5{-}I$	Alkyl iodide
$^{-}$:R	Carbanion	$C_2H_5{-}R$	Alkane

*Each of the anions listed is accompanied by a cation (for example, sodium ion in sodium iodide, NaI) that plays no major role in the reaction.

SOLUTION

The first step should always be to draw the structural formulas of the reactant and product. Because you are not yet familiar with the nomenclature of all the families, examine Table 3.1 for the structure of a compound called a *thiol;* it has an —SH (hydrosulfide) group. Thus, here is 1-butanethiol:

$$\underset{\text{1-Bromobutane}}{CH_3CH_2CH_2CH_2Br} \xrightarrow{?} \underset{\text{1-Butanethiol}}{CH_3CH_2CH_2CH_2SH}$$

Examine the structures of the substrate and the product to determine what has changed from one to the other. There has been no change in the alkyl group during this reaction. The only change is the substitution of the hydrosulfide group for bromine. Therefore, the reagent needed is a nucleophile with an —SH group. We see from Table 3.1 that sodium hydrosulfide (NaSH) should be effective for the desired conversion.

$$\underset{\text{1-Bromobutane}}{CH_3CH_2CH_2CH_2Br} + NaSH \longrightarrow NaBr + \underset{\text{1-Butanethiol}}{CH_3CH_2CH_2CH_2SH}$$

Such a reaction is normally represented by drawing the reagent over the arrow and ignoring the inorganic by-product (in this case, sodium bromide).

$$\underset{\text{1-Bromobutane}}{CH_3CH_2CH_2CH_2Br} \xrightarrow{NaSH} \underset{\text{1-Butanethiol}}{CH_3CH_2CH_2CH_2SH}$$

PROBLEM 3.10

What nucleophile would you use to bring about each of the following conversions?

(a) isopropyl chloride to isopropyl iodide

(b) propyl bromide to propyl alcohol ($CH_3CH_2CH_2OH$)

(c) Cyclohexyl bromide $\xrightarrow{?}$ Cyclohexylacetylene

(Cyclohexyl–Br → Cyclohexyl–C≡CH)

Cyclohexyl bromide Cyclohexylacetylene

PROBLEM 3.11

Show the structure of the product expected from the reaction of each nucleophile with cyclobutyl bromide: (a) sodium hydroxide, (b) potassium cyanide, (c) sodium methyl mercaptide (CH_3SNa), (d) potassium acetylide, (e) methyllithium (CH_3Li).

3.4.1 Mechanism of the S_N2 Reaction

The S_N2 reaction is the first type of chemical reaction considered in this book. We will look at it in some detail in order to illustrate the kind of information that can be acquired about individual reactions. The **mechanism of a reaction** is a detailed description of the individual steps by which the overall transformation occurs. There are two practical advantages to knowing reaction mechanisms:

1. Many reactions occur by the same mechanism. Therefore, instead of learning a host of reactions, you can learn a few mechanisms and apply them to many different reactions.
2. If you know the reaction mechanism, you can predict the effect on a reaction of changing the nature of a reactant or a reagent or changing the reaction conditions.

When carrying out a nucleophilic substitution reaction for the purpose of converting an alkyl halide to another product, chemists choose reaction conditions designed to minimize side reactions. **Side reactions** are secondary reactions that may occur simultaneously with the desired reaction, producing undesirable by-products. (The formation of by-products reduces the yield of the desired product and also means that extra steps must be taken to separate out the by-products and obtain the pure desired product.)

Now let's look at the key substitution reaction known as the **S_N2 reaction,** where S means "substitution," N means "nucleophilic," and 2 means "bimolecular" (that is, two reactants are involved in the slow, key step of the reaction). Understanding the mechanism of the S_N2 reaction is necessary for designing substitution reactions that maximize product yield and minimize by-products. So, let's start with an overview of the reaction mechanism.

S_N2 reactions occur by the mechanism shown in Figure 3.6, which shows the conversion of an alkyl bromide (with three substituents a, b, and c) to an alcohol. An S_N2 reaction is a single-step reaction that occurs without pause. It is sometimes called a **concerted reaction**—a reaction in which the bond changes occur simultaneously (in parallel) rather than sequentially (in series). The overall transformation is a gradual and synchronous shift of electron density from the "attacking" nucleophile (hydroxide ion, in this case) to the leaving group (bromide ion, in this case), as shown by the curved arrows. The **transition state** is a structure that represents how the reactive complex must appear at the point of maximum energy in the reaction—that point when existing bonds are breaking and new ones are forming.

The substitution reaction portrayed in Figure 3.6 requires use of a reagent that "brings in" the hydroxide ion nucleophile; that reagent could be sodium hydroxide or potassium hydroxide. Note that the cation has no role in the mechanism of the reaction.

FIGURE 3.6

Mechanism of the S_N2 reaction of hydroxide ion with an alkyl bromide. (The red curved arrows indicate the flow of electrons needed to accomplish the following step.)

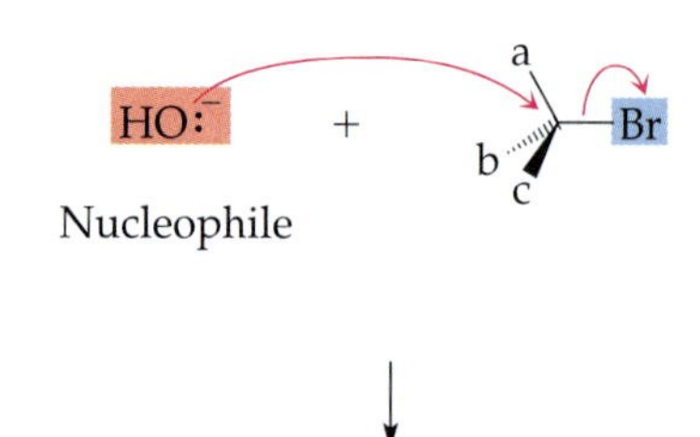

The nucleophilic hydroxide ion attacks the back side of the carbon carrying the bromine. That carbon is relatively electron-deficient because of the high electronegativity of the bromine atom.

The new hydroxyl-carbon bond has partially formed by oxygen donating a pair of its unshared electrons to carbon. Thus, the electron density on oxygen has decreased to a partial negative charge.

The carbon-bromine bond has lengthened and partially broken under the influence of the attack by the hydroxide ion. Thus, bromine has acquired additional electron density, giving it a partial negative charge.

In the transition state, the three noninvolved carbon substituents have moved to a planar arrangement about the central carbon, perpendicular to the hydroxyl-carbon-bromine axis.

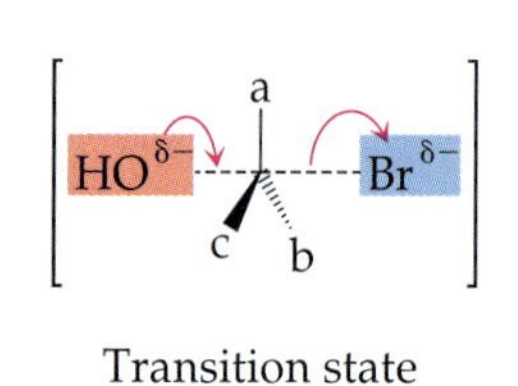

HO—C(a)(b)(c) + Br:⁻

Leaving group

The reaction is completed by the hydroxyl-carbon bond becoming fully established and the carbon-bromine bond fully broken. In the overall reaction, the carbon atom has undergone an *inversion of configuration*.

The reaction is carried out in the lab by dissolving sodium hydroxide and the alkyl bromide in a solvent such as dimethyl sulfoxide (DMSO) and applying heat. The complete reaction is represented in this way:

$$\underset{\text{Sodium hydroxide}}{\text{NaOH}} + \underset{\text{Alkyl bromide}}{\text{R—Br}} \xrightarrow[\text{heat}]{\text{DMSO}} \underset{\text{Alcohol}}{\text{R—OH}} + \underset{\text{Sodium bromide}}{\text{NaBr}}$$

Next, we'll examine several aspects of the S_N2 reaction in detail.

Rate of Reaction

The rate of a reaction is the speed at which it occurs (that is, the rate at which reactant is converted into product). The rate of the S_N2 reaction of sodium hydroxide and an alkyl bromide is represented by a rate equation (shown below), which is determined by measuring the *kinetics* of the reaction. That rate is influenced by the concentrations of *both* reactants. In the example reaction, changing the concentration of either sodium hydroxide or the alkyl bromide will change the rate of the reaction. Therefore, the *rate-determining step of the reaction involves both reactants.* The reaction is therefore described as a second-order reaction; it is first-order in each reactant because, in the rate equation, the concentration of each reactant is raised to the first power. The reaction rate constant is k.

$$\text{Reaction rate} = k[\text{alkyl bromide}][\text{hydroxide ion}]$$

Energetics

The energetics of the S_N2 reaction can be portrayed in a free-energy diagram (Figure 3.7), which describes how the energies of the species involved change as the reaction proceeds. The reaction coordinate indicates the progress of the reaction as bonds are gradually made and broken.

In this reaction, energy must be supplied to the reactants to bring them to the transition state. The transition state has no significant "lifetime"; it is simply a representation of the atomic positions (structure) partway through the reaction. It represents the point of maximum energy in the simultaneous making and breaking of the two bonds. The height of this energy barrier is called the **energy of activation** ($\Delta G^{\ddagger}$), and it determines the rate of the reaction. The higher the energy of activation, the slower the rate of reaction. Once past the transition state, the reaction releases energy as it progresses rapidly to products. This reaction is considered a single-step (concerted) reaction, in which two bonds are broken or formed at approximately the same time. The net energy change of the reaction is called the free-energy change ($\Delta G°$).

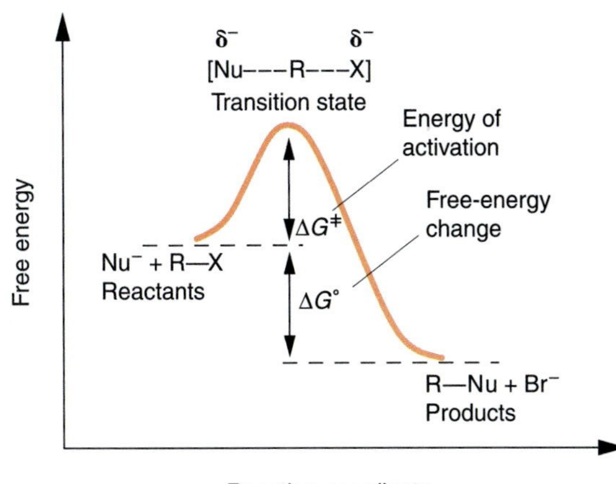

FIGURE 3.7

Free-energy diagram for a hypothetical S_N2 reaction.

Stereochemistry

Examination of the S_N2 mechanism reveals that the nucleophile forms its bond with the carbon on the side opposite that from which the bromine atom (the leaving group) departs. This is called a **back-side attack** on the carbon by the nucleophile. The result is an **inversion of configuration** at the carbon.

The inversion process is like an umbrella being blown inside out by the wind, with the substituents hanging onto their positions around the rim. The substituents start out as if located equidistant from one another around the rim of an umbrella open to the left. However, in the course of the reaction, they are first "blown" to lie in a plane with the central carbon (the transition state structure) and then are "blown" to the rim of the inside-out umbrella (open to the right). This inversion of configuration occurs in all S_N2 reactions of alkyl halides, but it can be detected only when the halogen is attached to a stereocenter. If the starting alkyl halide is in the *R* configuration, the product (with most nucleophiles) will be in the *S* configuration, and vice versa. The *R*-to-*S* or *S*-to-*R* change is the inversion. Thus, (*R*)-2-bromobutane reacts with sodium iodide in acetone solvent to produce (*S*)-2-iodobutane. If the same reaction were conducted with racemic 2-bromobutane, the product would be racemic 2-iodobutane, since each *R* enantiomer would be inverted to the *S* enantiomer, and vice versa.

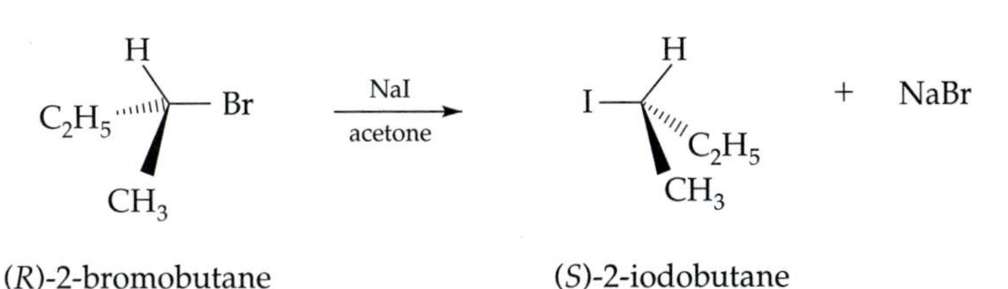

The S_N2 reaction is known as a **stereospecific reaction** because it has a specific stereochemical result; it is not random but produces a specific configuration.

S_N2 reactions of cyclohexyl halides result in an axial halide substituent being replaced by a nucleophile in the equatorial position, and vice versa. For example, *cis*-1-bromo-4-*t*-butylcyclohexane is converted by sodium iodide in acetone solvent to *trans*-1-*t*-butyl-4-iodocyclohexane:

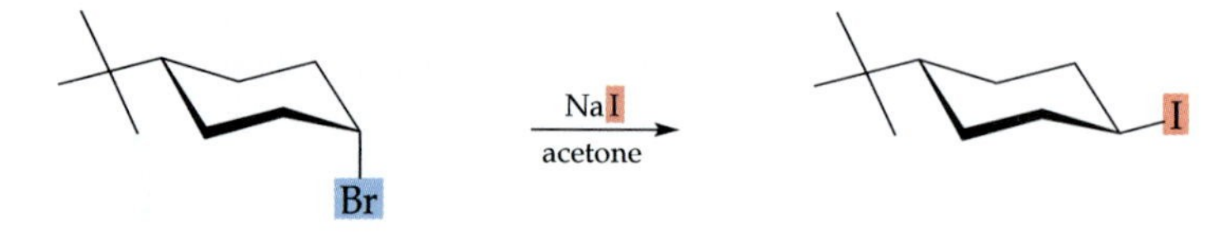

cis-1-Bromo-4-*t*-butylcyclohexane *trans*-1-Butyl-4-iodocyclohexane

How to Solve a Problem

What is the product formed from the reaction of *trans*-2-bromo-1-methylcyclobutane with sodium iodide in acetone solvent?

PROBLEM ANALYSIS

The reaction must be an S_N2 reaction with iodide ion as the nucleophile and bromide ion as the leaving group. We know that such reactions occur as a backside attack by the nucleophile, with inversion of configuration of the carbon under attack. Therefore, the iodide ion must attack the alkyl bromide from the side opposite that from which the bromide ion leaves.

SOLUTION

Since the bromine substituent is "down" in the substrate, the iodide ion must approach from the "top." The new bond is thus directed "up," and the product is *cis*-2-iodo-1-methylcyclobutane.

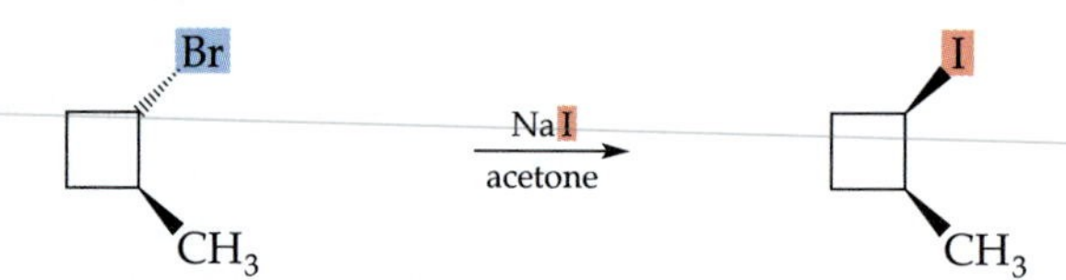

trans-2-Bromo-1-methylcyclobutane *cis*-2-Iodo-1-methylcyclobutane

Alkyl Group Reactivity

Different alkyl groups react with different ease in an S_N2 reaction. The success of the reaction depends on the nucleophile being able to "get at" the back side of the carbon carrying the halogen, in order to form the new bond. Thus, any structural characteristic of the alkyl halide that inhibits approach of the nucleophile makes the reaction more difficult, the energy of activation higher, and the rate slower. Accordingly, there is a **steric effect** in an S_N2 reaction, due to the bulk of the substituents attached to the reacting carbon. The less steric hindrance there is in an alkyl halide, the more reactive it will be in an S_N2 reaction.

The reactivity in an S_N2 reaction of any series of alkyl halides containing the same halogen follows this order:

S_N2 reactivity: methyl > primary (1°) > secondary (2°) > tertiary (3°)

The increased crowding at the back side of the carbon is depicted in Figure 3.8 for ethyl bromide (1°), isopropyl bromide (2°), and *t*-butyl bromide (3°). For example, 1-bromopropane is about 50 times more reactive than 2-bromopropane. The crowding in tertiary halides is so great that S_N2 substitutions of these are virtually nonexistent. Thus, *S_N2 reactions are useful only with methyl, primary, and secondary alkyl halides.* If the nucleophile, the carbon under attack, and the halogen are unable to assume the colinear positions required by the transition state, the reaction does not proceed.

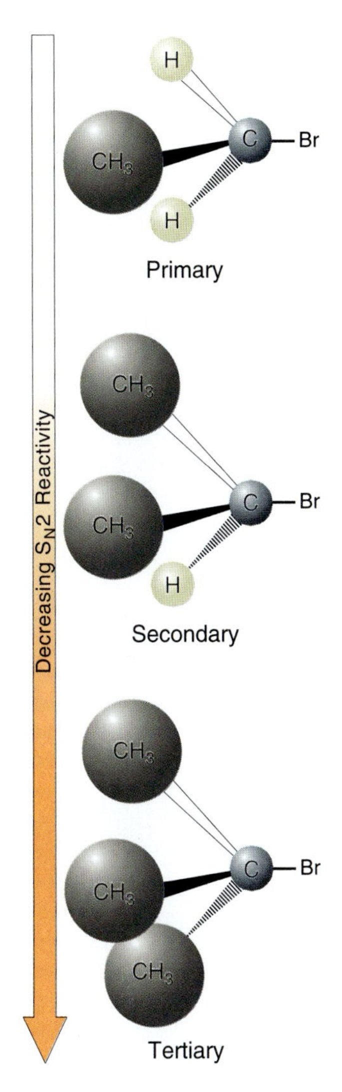

FIGURE 3.8

Steric crowding in ethyl bromide, isopropyl bromide, and *t*-butyl bromide.

Effect of Leaving Group

The more easily broken the carbon-halogen bond in an alkyl halide, the faster the S_N2 reaction will be (because of the lower energy of activation). Recall that the *bond strength* of alkyl halides decreases in the order fluoride > chloride > bromide > iodide (see Section 3.2.1). Therefore, alkyl iodides undergo the fastest S_N2 reactions. In part because of their ease of preparation, alkyl bromides are generally the most convenient to use in S_N2 reactions and for that reason will appear in most examples in this book.

PROBLEM 3.12

Place each of the following pairs of compounds in order by *decreasing* S_N2 reactivity:

(a) cyclohexyl bromide and cyclohexyl fluoride

(b) *n*-butyl bromide and *t*-butyl bromide

(c) 1-iodopentane and 2-iodopentane

PROBLEM 3.13

Using arrows, show the mechanistic and stereochemical details of the reaction of sodium iodide with (a) *n*-butyl chloride and (b) *trans*-1-*t*-butyl-4-chlorocyclohexane.

Effect of Nucleophiles

Because the S_N2 reaction depends on attack by the nucleophile, both the concentration of the nucleophile (recall the rate equation given earlier in this section) and its strength (its **nucleophilicity**) are important factors. The stronger the nucleophile (that is, the more avidly it shares its electrons in forming a new bond to carbon), the faster and more effective the S_N2 reaction will be. You will gradually become familiar with the relative nucleophilicities of many ions and compounds, but for now simply note the following generalizations:

- Anions are always more nucleophilic than the corresponding conjugate acids from which they are derived, for example:

$$HO^- > HOH \quad \text{and} \quad RO^- > ROH$$

- Relative nucleophilicities usually, but not always, are the same as relative basicities, for example:

$$RO^- > HO^-$$

  Recall that basicity can be determined from the pK_a of the conjugate acid (see Section 1.3.1). For example, since water is a stronger acid (pK_a = 15.7) than are alcohols (ROH, p$K_a \approx 17$), hydroxide ion is a weaker base and nucleophile than are alkoxide ions (RO^-).

- Nucleophilicity increases *down* a group in the periodic table, for example:

$$F^- < Cl^- < Br^- < I^- \quad \text{and} \quad HO^- < HS^-$$

- Nucleophilicity increases to the *left* in a row of the periodic table, for example:

$$F^- < O^- < N^- < C^-$$

Note that the cation that accompanies any anionic nucleophile (for instance, sodium in sodium hydroxide) has *no significant role* in the S_N2 reaction mechanism. The cation simply remains in solution and can eventually be isolated, if desired (though this is seldom done), as the salt with the halide anion expelled in the reaction. For example, sodium hydroxide reacts with methyl bromide to yield methyl alcohol and sodium bromide. The sodium bromide is usually not isolated, since the alcohol is the target product.

Effect of Solvent

Finally, the choice of solvent can have an important effect on an S_N2 reaction, affecting various nucleophiles differently. Aprotic polar solvents like dimethyl sulfoxide (DMSO) are preferred. (**Aprotic** means that the solvent contains no polar hydrogen, in contrast to alcohols (ROH) and water (H_2O), which are protic solvents.) However, details of this aspect of the S_N2 reaction are beyond the scope of this book.

How to Solve a Problem

What is the product of the reaction between sodium cyanide and isopropyl bromide?

PROBLEM ANALYSIS

The first step in solving the problem is to write the information we are given in the form of a reaction:

$$\underset{\text{Sodium cyanide}}{NaCN} + \underset{\text{Isopropyl bromide}}{H{-}C(CH_3)_2{-}Br} \xrightarrow{\text{DMSO}} ?$$

We recognize that we have an alkyl bromide, and the usual reaction is for the bromine atom to be displaced by a nucleophile. Can we identify a nucleophile? Yes, because sodium cyanide in solution exists as a salt (Na^+ CN^-), and cyanide ion is an effective nucleophile (see Table 3.1). Now envision the S_N2 reaction in which the nucleophile (cyanide) displaces the bromine.

SOLUTION

The products of the S_N2 reaction are sodium bromide and a nitrile (also known as an alkyl cyanide, in this instance isopropyl cyanide). There is no need to worry about stereochemistry in this reaction because the starting material has no stereocenter—there are no enantiomers. Do not write the reaction mechanism unless you are asked for it. Instead, use your knowledge of the mechanism of the general S_N2 reaction to help you predict the specific reaction outcome.

$$\text{NaCN} + \text{H}-\overset{\overset{\text{CH}_3}{|}}{\underset{\underset{\text{CH}_3}{|}}{\text{C}}}-\text{Br} \xrightarrow{\text{DMSO}} \text{H}-\overset{\overset{\text{CH}_3}{|}}{\underset{\underset{\text{CH}_3}{|}}{\text{C}}}-\text{CN} + \text{NaBr}$$

Sodium cyanide | Isopropyl bromide | Isopropyl cyanide | Sodium bromide

PROBLEM 3.14

Write the structures of the products of the following reactions:

(a) cyclohexyl chloride with sodium hydroxide

(b) sodium cyanide with (*R*)-2-iodobutane

(c) sodium hydrosulfide (NaSH) with *cis*-3-bromo-1-*t*-butylcyclohexane

(d) sodium iodide with (*S*)-2-chloropentane

In summary, the S_N2 reaction of a nucleophile with an alkyl halide is an important tool in organic synthesis.

- The reaction is effective with methyl, primary, and secondary alkyl halides only.
- It occurs stereospecifically with inversion of configuration at a stereocenter.
- Use of alkyl iodides and bromides is preferable, but alkyl chlorides will react slowly.
- It is a second-order reaction, whose rate is affected by the concentrations of the alkyl halide and the nucleophilic agent.
- Stronger nucleophiles are more effective than weaker nucleophiles.

3.4.2 Complications with Nucleophilic Substitution of Tertiary Alkyl Halides

As mentioned earlier, the S_N2 reaction is not useful with tertiary alkyl halides. When tertiary halides react with strong nucleophiles (bases), what occurs instead of nucleophilic substitution is an **elimination reaction,** known as an **E2 reaction.** The nucleophile cannot attack the tertiary carbon from the back side because of steric hindrance (see Section 3.4.1); so, instead, it removes an adjacent hydrogen as part of the E2 process. This converts the alkyl halide to an alkene, one of a family of organic compounds containing a carbon-carbon double bond. Thus, *t*-butyl bromide reacts with sodium cyanide to yield not the S_N2 product (*t*-butyl cyanide), but instead an alkene (isobutylene) from loss of HBr.

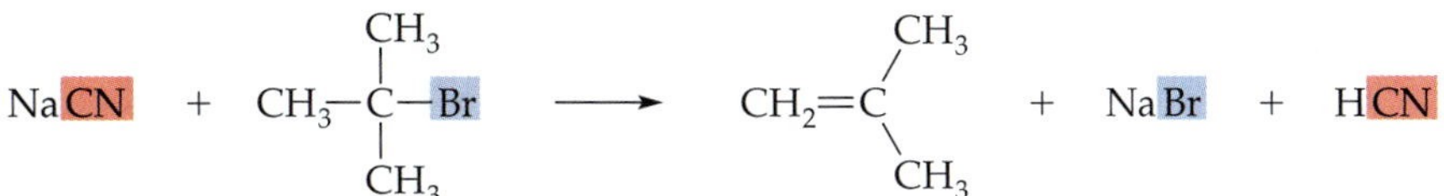

$$\text{NaCN} + \text{CH}_3-\overset{\overset{\text{CH}_3}{|}}{\underset{\underset{\text{CH}_3}{|}}{\text{C}}}-\text{Br} \longrightarrow \text{CH}_2{=}\text{C}(\text{CH}_3)_2 + \text{NaBr} + \text{HCN}$$

Sodium cyanide | *t*-Butyl bromide | Isobutylene

This E2 reaction will be discussed in Chapter 6.

When a weak rather than a strong nucleophile is used with a tertiary halide, a mixture of products results. The products usually include some of the desired substitution product but consist mainly of the elimination product, and it is almost impossible to exert any control over the ratio of the two products. Thus, whereas reaction of *t*-butyl chloride with sodium ethoxide (a strong nucleophile) yields exclusively isobutylene via an E2 elimination, reaction with ethanol (a much weaker nucleophile) results in a mixture of the desired *t*-butyl ethyl ether and isobutylene via two other mechanisms (S_N1 and E1, to be discussed later):

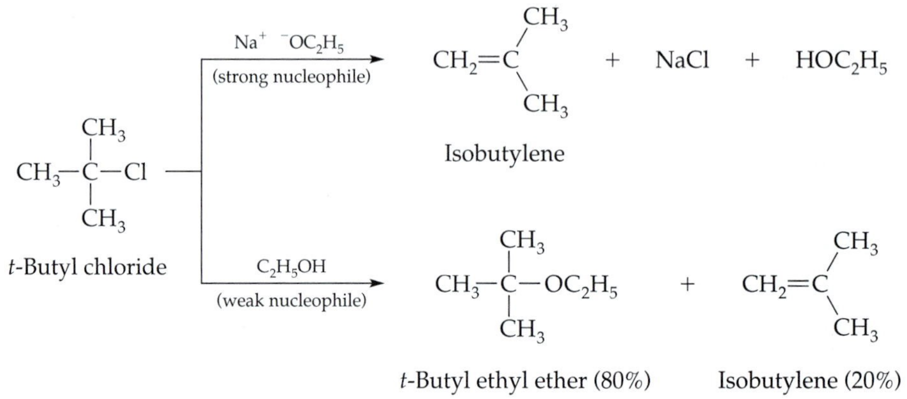

$$(CH_3)_3C{-}Cl \xrightarrow[\text{(strong nucleophile)}]{Na^+\ {}^-OC_2H_5} CH_2{=}C(CH_3)_2 + NaCl + HOC_2H_5$$

t-Butyl chloride → Isobutylene

$$(CH_3)_3C{-}Cl \xrightarrow[\text{(weak nucleophile)}]{C_2H_5OH} (CH_3)_3C{-}OC_2H_5 + CH_2{=}C(CH_3)_2$$

t-Butyl ethyl ether (80%) Isobutylene (20%)

The presence of a mixture of products and the inability to control the ratio of the mixture by varying experimental conditions make substitution reactions with tertiary halides of limited use for organic synthesis.

3.4.3 Mechanism of the S_N1 Reaction

Although of only limited use in synthesis, the S_N1 reaction of alkyl halides is part of organic chemistry and must be understood. Therefore, we will consider the mechanism and compare it with that of the S_N2 reaction.

Note that the overall result is the same as for the S_N2 reaction: the nucleophile replaces the halide in an alkyl halide. For example, a tertiary alkyl bromide can be converted to an alcohol. The experimental conditions for alcohol formation involve warming the alkyl halide in a water-acetone solvent, with water serving as the nucleophile.

General reaction:

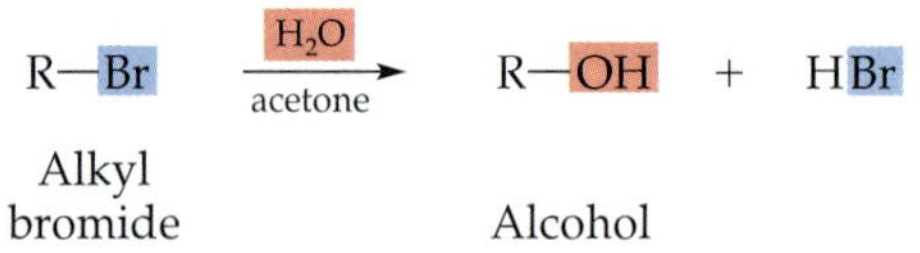

$$R{-}Br \xrightarrow[\text{acetone}]{H_2O} R{-}OH + HBr$$

Alkyl bromide → Alcohol

Mechanism:

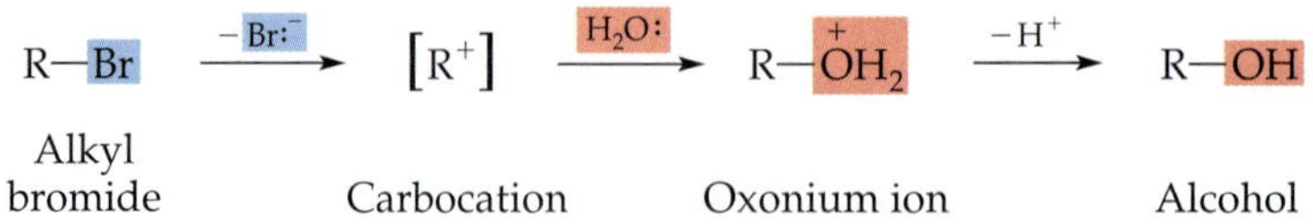

$$R{-}Br \xrightarrow{-Br:^-} [R^+] \xrightarrow{H_2O:} R{-}\overset{+}{O}H_2 \xrightarrow{-H^+} R{-}OH$$

Alkyl bromide → Carbocation → Oxonium ion → Alcohol

The **S_N1 mechanism** (S stands for "substitution," N stands for "nucleophilic," and 1 stands for "unimolecular") is a two-step mechanism summarized above and shown in more detail in Figure 3.9. (The final step shown above is not counted, since it is simply a proton transfer, or acid-base, reaction.) We will discuss the general characteristics of S_N1 reactions in this section. While reading this section, keep in mind how the

FIGURE 3.9

Mechanism of the S_N1 reaction of water with an alkyl bromide.

Under the influence of an ionizing solvent, the carbon-halogen bond in the alkyl halide slowly breaks on its own. The electrons go to the more electronegative halogen, leaving behind an electron-deficient carbon (a carbocation).

The carbocation, although a short-lived species, is a reactive intermediate, not a transition state. There is a transition state to the formation of the carbocation. The carbocation is stabilized by solvation and by hyperconjugation (see p. 101). The carbocation structure has all three substituents lying in a plane around the central carbon. The carbocation therefore has a plane of symmetry. The hybridization of the central carbon is sp^2 (trigonal) with a vacant *p* orbital perpendicular to the plane of atoms (see Figure 3.11).

The nucleophilic water molecule attacks the electron-deficient carbocation from either side (both occur). In both cases, this produces an oxonium ion (a protonated form of an alcohol). The positive charge on the carbon has been neutralized, but the oxygen atom, in giving up a share of its electrons to form the new bond, has become positively charged.

The oxonium ion gives up a proton to the solvent water, forming the hydronium ion and leaving behind the product alcohol. If the carbon bonded to the halogen in the starting alkyl halide was a stereocenter, a pair of enantiomers (a racemate) results.

comparable factors affect an S_N2 reaction. (We will compare S_N1 and S_N2 mechanisms in Section 3.4.4.)

Rate of Reaction

The rate at which the S_N1 reaction occurs depends *only* on the concentration of the alkyl halide. The concentration or strength of the nucleophile has no effect on the reaction rate. Thus, the reaction is first-order in the alkyl halide:

$$\text{Reaction rate} \quad = \quad k[\text{alkyl halide}]$$

Since the rate-determining (slow) step of the S_N1 reaction involves only the alkyl halide, it is a unimolecular reaction. A multistep reaction can proceed no faster than the slowest of its steps. (Consider a water line consisting of pipes with three different diameters: The water can pass through the entire line only as fast as it can pass through the narrowest pipe.) Figure 3.9 reveals that the first step of the reaction, carbocation formation by spontaneous dissociation of the alkyl halide, must be the slow step, since it is the only step that does not involve another species.

Energetics

Figure 3.10 is the free-energy diagram for an S_N1 reaction. The highest-energy point for the reaction is the transition state between alkyl halide and carbocation, a struc-

FIGURE 3.10

Free-energy diagram for a hypothetical S_N1 reaction.

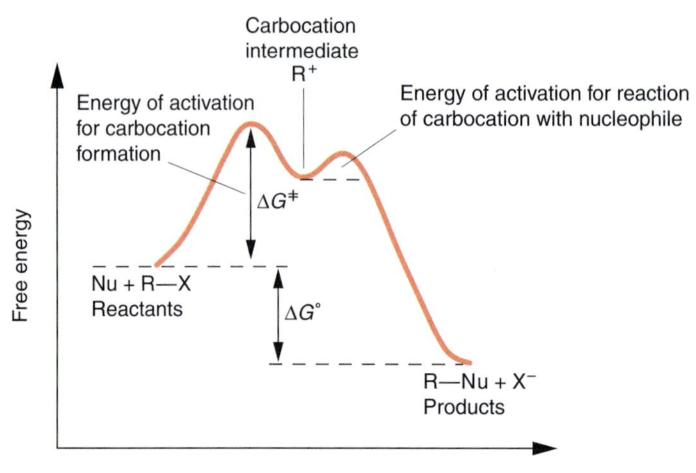

ture in which the carbon-halogen bond has lengthened considerably prior to actually breaking. The formation of this transition state is the slow step of the reaction. The "saddle point" between the two high points of the curve indicates that a structure with a finite lifetime (a **reactive intermediate**) must exist, one of lower energy (more stable) than the structures on either side. This one is a **carbocation**—a species with a positively charged trivalent carbon. Any factors, either structural or electronic, that stabilize the carbocation also stabilize the transition state to its formation, lowering the energy of activation and increasing the rate of the reaction.

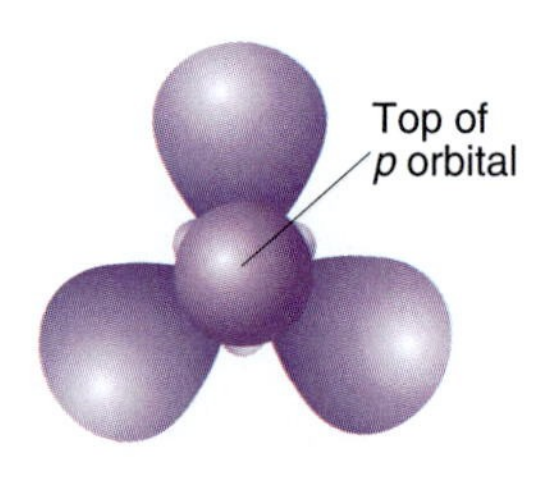

View from above the plane of the sp^2 orbitals (which lie in the plane of the paper)

= sp^2 orbital

Stereochemistry

The carbocation formed in the S_N1 reaction has a plane of symmetry, which means that it is *achiral*. The structure of the carbocation results because the carbon in the alkyl halide is changing from sp^3 hybridization to sp^2 hybridization as the halide anion departs. Instead of the single *s* and three *p* orbitals of the excited state of carbon hybridizing to form four sp^3 hybrid orbitals (tetrahedral hybridization), as in the starting alkyl halide, the single *s* and *two* of the *p* orbitals hybridize to form three **sp^2 hybrid orbitals:**

$$1s^2 2s^2 sp^2 \xrightarrow{\text{promotion}} 1s^2 2s^1 2p^3 \xrightarrow{\text{hybridization}} 1s^2 2(sp^2)^3 + 2p^0$$

Atomic carbon — Trigonal hybridization

View from the edge of the sp^2 plane with the *p* orbital lying in the plane of the paper

FIGURE 3.11

Views of the orbital shapes of a carbocation.

These three orbitals lie in a plane with the central carbon and are directed to the corners of an equilateral triangle with an angle of 120° between them, resulting in what is called **trigonal hybridization** (Figure 3.11). In a carbocation, the three substituents on the carbon form sigma bonds to it using these sp^2 orbitals. Thus, the central carbon and the three substituents lie in a plane—the carbocation has a plane of symmetry. When the three sp^2 hybrid orbitals are formed, a *p* orbital remains unhybridized and is oriented perpendicular to the plane of the three sp^2 hybrid orbitals.

If the halogen in the alkyl halide reactant is attached to a stereocenter and a single enantiomer of that alkyl halide is subjected to S_N1 reaction conditions, the compound's chirality is lost at the same rate as the carbocation is formed. As the carbocation reacts with the nucleophile in the second step of the S_N1 reaction, the nucleophile can approach from either side of the plane of symmetry with equal probability. Thus, a single alkyl halide enantiomer produces equal amounts of both enantiomers of the product (a racemate) in an S_N1 reaction. Because the overall process results in *racemization* of an enantiomer, the S_N1 substitution reaction is *nonstereospecific*. For example, (*R*)-3-bromo-3-methylhexane reacts in water-acetone solution to form a racemic mixture of (*R*)- and (*S*)-3-methyl-3-hexanol:

$$\text{Br}-\text{C}(CH_3)(C_2H_5)(C_3H_7) \xrightarrow[\text{acetone}]{H_2O} (C_2H_5)(C_3H_7)(CH_3)\text{C}-\text{OH} + \text{HO}-\text{C}(CH_3)(C_2H_5)(C_3H_7)$$

(*R*)-3-Bromo-3-methylhexane

(*S*)- and (*R*)-3-Methyl-3-hexanol (racemic 3-methyl-3-hexanol)

Effect of Leaving Group

The more readily the leaving group breaks its bond with carbon (the weaker the bond), the faster the S_N1 reaction will be. Because bond strength decreases in the order fluoride > chloride > bromide > iodide, alkyl iodides undergo the fastest reactions. In practical terms, iodides and bromides are the halides of choice, with bromides the most readily available.

Effect of Solvent

Since the rate-determining step of an S_N1 reaction involves a neutral species being converted to an ionic species, a solvent that supports the ionization facilitates the reaction, stabilizing the carbocation by solvation. Thus, polar protic solvents, such as water and ethanol, are preferred (**protic** means that the solvent contains polar hydrogen). The effect of a solvent on the reaction rate can be very large: There is a difference of a factor of 10^5 in favor of water over ethanol as the solvent in reaction with *t*-butyl chloride. Often, the solvent in an S_N1 reaction is the nucleophile (for example, water or an alcohol), and in these cases, the substitution reaction is referred to as a **solvolysis reaction.** For example, the *methanolysis* of *t*-butyl bromide by gentle heating in methanol solution produces *t*-butyl methyl ether:

$$(CH_3)_3C-Br + CH_3OH \longrightarrow (CH_3)_3C-OCH_3 + HBr$$

t-Butyl bromide | Methanol | *t*-Butyl methyl ether

Effect of Alkyl Group

Because the S_N1 mechanism requires formation of a carbocation intermediate, the alkyl halide must be able to form a reasonably stable carbocation, or the reaction will occur only very slowly or not at all. The carbocation is a highly reactive and electron-deficient species. To the extent that the electron deficiency of the carbocation can be temporarily satisfied, it will be stabilized, the energy of activation for its formation will be lowered, and therefore it will form more readily. The order of stability for carbocations is tertiary (3°) > secondary (2°) > primary (1°) > methyl. In practical terms, carbocations readily form only from tertiary alkyl halides. (This is one reason why only tertiary halides produce complications in the S_N2 reaction.) Carbocations can be stabilized by temporary electron sharing from adjacent C—H bonds in a process called **hyperconjugation** (Figure 3.12).

The effect of hyperconjugation is to lower the concentration of positive charge on carbon by supplying electron density, thereby spreading the charge out among other

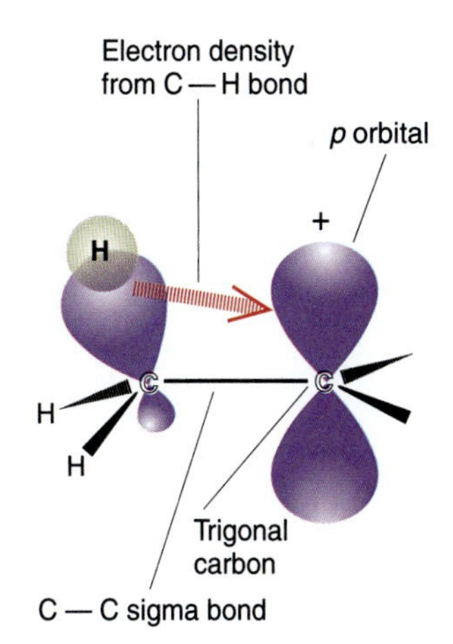

FIGURE 3.12

The sharing of electron density from a C—H bond to an adjacent carbocation.

atoms. Examination of the structures of the *n*-propyl (1°), isopropyl (2°), and *t*-butyl (3°) carbocations reveals that there are, respectively, two, six, and nine equivalent C—H bonds adjacent to the positively charged carbon.

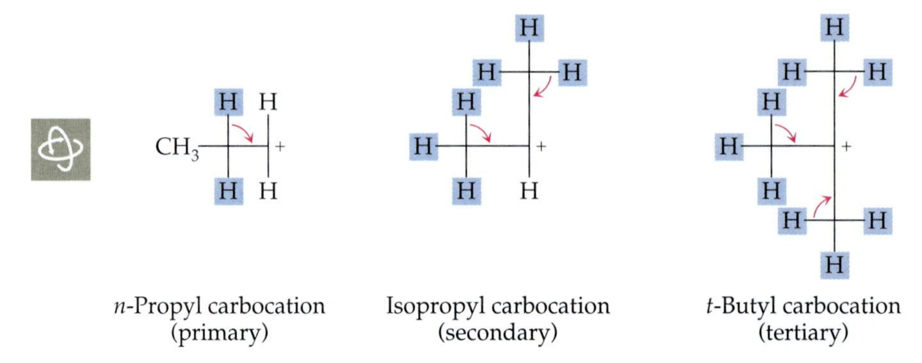

n-Propyl carbocation (primary) | Isopropyl carbocation (secondary) | *t*-Butyl carbocation (tertiary)

The higher level of electron sharing produces the greatest stability for the tertiary carbocation. Therefore, tertiary alkyl halides readily form carbocations, and primary alkyl halides seldom do.

How to Solve a Problem

Which alkyl bromide (R—Br), isopropyl bromide or *t*-butyl bromide, will react faster with water (a solvolysis reaction) to produce the corresponding alcohol (R—OH)? Explain.

SOLUTION

First, we write the given reaction:

$$\mathrm{R{-}Br} \xrightarrow{\mathrm{H_2O}} \mathrm{R{-}OH}$$

Because water is a weak nucleophile, the substitution reaction cannot occur by the S_N2 mechanism and so must occur by the S_N1 mechanism. This means that a carbocation (R^+) must be formed, and we know that the rate of carbocation formation will determine the rate of the reaction. Drawing the structures of the two possible carbocations, the isopropyl and *t*-butyl carbocations, we can see that the positively charged carbon in the *t*-butyl carbocation is surrounded by more C—H bonds (nine) than in the isopropyl carbocation (six). Therefore, there should be more hyperconjugative stabilization of the *t*-butyl carbocation. It follows that *t*-butyl bromide should solvolyze in water faster than isopropyl bromide.

$$\mathrm{H{-}C(CH_3)_2{-}Br} \xrightarrow{-\mathrm{Br{:}^-}} \left[\mathrm{H{-}C^{+}(CH_3)_2}\right] \xrightarrow[-\mathrm{H^+}]{\mathrm{H_2O{:}}} \mathrm{H{-}C(CH_3)_2{-}OH}$$

$$\mathrm{CH_3{-}C(CH_3)_2{-}Br} \xrightarrow{-\mathrm{Br{:}^-}} \left[\mathrm{CH_3{-}C^{+}(CH_3)_2}\right] \xrightarrow[-\mathrm{H^+}]{\mathrm{H_2O{:}}} \mathrm{CH_3{-}C(CH_3)_2{-}OH}$$

Alkyl bromide | Carbocation | Alcohol

PROBLEM 3.15

For each of the following pairs of alkyl halides, which will produce the more stable carbocation, and why?

(a) isopropyl bromide and *n*-butyl bromide

(b) cyclopentyl bromide and *t*-pentyl bromide

(c) 2-chloro-2-methylpentane and 2-chloropentane

(d) 1-bromobutane and 2-bromobutane

PROBLEM 3.16

For each of the following pairs of compounds, which will show the faster rate of formation of a carbocation, and why?

(a) *t*-butyl bromide and *t*-butyl chloride

(b) 1-iodo-1-methylcyclopentane and iodocyclopentane

3.4.4 Comparison of S_N1 and S_N2 Mechanisms

Because nucleophilic substitution reactions are used throughout organic chemistry and throughout this book, it is essential to understand these mechanisms. It will help to have the characteristics of the S_N1 and S_N2 reactions compared succinctly—see Table 3.2.

Substitution reactions are most effective using primary and secondary alkyl halides under conditions that favor the S_N2 mechanism; side reactions are minimized. Tertiary alkyl halides can be substituted using S_N1 conditions, but by-products will form.

Primary, secondary, and tertiary alkyl halides can be distinguished experimentally in simple qualitative tests that rely on their different tendencies toward S_N1 and S_N2 reactions. An unknown alkyl halide is reacted separately with two solutions:

1. sodium iodide dissolved in acetone (iodide is a strong nucleophile that displaces bromine or chlorine in an S_N2 reaction)
2. silver nitrate dissolved in ethanol (silver ion will react with a halide ion formed by dissociation of an alkyl halide into a carbocation and a halide ion).

TABLE 3.2

Comparison of S_N1 and S_N2 Mechanisms

	S_N2	S_N1
Rate of reaction	Second-order	First-order
Nucleophile concentration	Affects rate	No effect on rate
Nucleophile strength	Modest to strong	Irrelevant (weak)
Stereochemistry	Stereospecific, inversion	Nonstereospecific, racemization
Best solvent	Aprotic, polar	Protic, polar
Favored structure	Methyl $> 1° > 2°$	3°
Best leaving group	I $>$ Br $>$ Cl	I $>$ Br $>$ Cl

If the unknown alkyl halide reacts quickly with NaI, it is probably a primary halide (NaBr precipitates, indicating a rapid S_N2 reaction) and will not react with the $AgNO_3$ solution. If the halide reacts quickly with $AgNO_3$, it is probably a tertiary halide (AgX precipitates, indicating a rapid S_N1 reaction) and will not react with the NaI solution. If the reaction is sluggish with both reagents, the alkyl halide is probably a secondary halide.

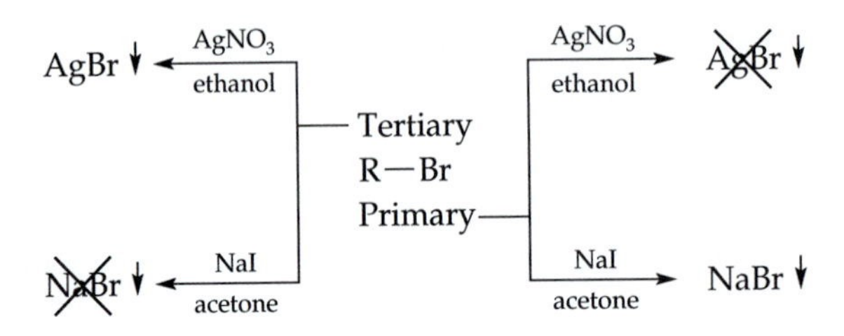

PROBLEM 3.17

Write the structure(s) of the substitution products of the following reactions. If other products are formed under these conditions, indicate their structures as well.

(a) 2-bromopentane with dilute sodium hydroxide

(b) (*R*)-2-bromobutane with sodium iodide in acetone, followed by reaction of the product with dilute potassium hydroxide

(c) *trans*-1-chloro-4-methylcyclohexane with dilute sodium hydroxide

(d) 2-bromo-2-methylpentane with sodium cyanide

3.5 Other Reactions of Alkyl Halides

The most widely used reaction of alkyl halides is the nucleophilic substitution reaction we just discussed. You will see it used many times in the chapters to come. Another important reaction is the elimination reaction (E1 and E2 mechanisms), by which the elements of HX are removed from an alkyl halide. The products that result are alkenes, another kind of hydrocarbon. We will look at this reaction in Chapter 6. One other important reaction of alkyl halides will be discussed in this section—their conversion into organometallic compounds. Although these compounds are almost never isolated, they are very important reagents in organic synthesis (as will be discussed briefly in this section but also in later chapters, especially Chapter 13).

3.5.1 Synthesis of Organometallics

Organometallic compounds are organic compounds containing a carbon-metal bond. Until the middle of the twentieth century, relatively few such compounds were known. Since then, large numbers of these compounds have been prepared, and some play a major role in organic and other chemistry. This section discusses the preparation of three kinds of organometallic compounds that vary in their reactivity. Their most important reactions and uses will be discussed in later chapters.

Any bond between a carbon atom and a metal is bound to be highly polarized because of the electronegativity differences between the two atoms (see Figure 1.4). The metal is always strongly electropositive, resulting in the carbon being the negative end of the bond moment. The carbon-metal bond can be envisioned as covalent, associated (that is, partially dissociated but still complexing the two atoms), or completely dissociated (ionic).

$\overset{\delta-}{\text{C}}$—$\overset{\delta+}{\text{Metal}}$	$\overset{\delta-}{\text{C}}$------$\overset{\delta+}{\text{Metal}}$	$\overset{-}{\text{C}}$: $\overset{+}{\text{Metal}}$
Covalent bond	Associated	Dissociated

The exact structure of any particular organometallic compound may be complex and may even vary with the solvent used. For our purposes, however, we can consider the organometallic reagents (often simply called *organometallics*) as reacting as though they were dissociated, even though most are tightly associated. Their chemistry is typically like that of a dissociated species known as a **carbanion** (an anionic carbon carrying three substituents and an unshared pair of electrons).

The major significance of organometallic reagents is that they provide a means for making carbon into a nucleophile. Recalling the discussion on substitution reactions in Section 3.4 and referring to Table 3.1, in which carbanions are included, it should be apparent that organometallic reagents are key to making new carbon-carbon bonds, and thereby building up complex carbon skeletons. Carbanions are highly reactive and highly nucleophilic species, as we shall discover.

The organometallic reagents we will discuss are organolithium compounds, organomagnesium compounds (known as *Grignard reagents*), and lithium dialkylcuprate compounds. Organometallics are seldom isolated and are usually prepared and reacted in anhydrous diethyl ether (ether) or tetrahydrofuran (THF) as solvent. The nucleophilic reactivity sequence for these compounds is

Nucleophilic reactivity: organolithium > organomagnesium > organocuprate

Later, we will discuss reactions where these differences in reactivity are important.

These three types of organometallic reagents react readily with acids. They must be handled in the absolute absence of moisture because they are also strong enough bases to react with water in an acid-base reaction. (The absence of moisture really means the absence of air, which almost always contains some moisture.) In all instances, reaction with water produces an alkane and the metal hydroxide. The reaction can be generalized as follows and can be explained on the basis of the pK_a values of the two conjugate acids, water (pK_a 15.7) and the alkane (pK_a ~40). (Review Section 1.3.1: The stronger acid will lose its proton; the stronger base will get the proton.)

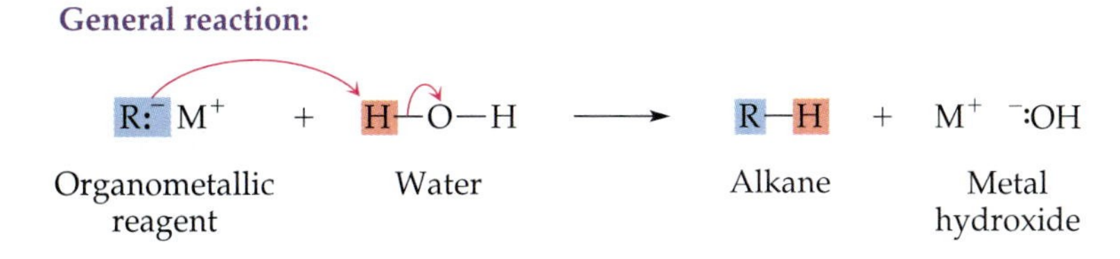

Since organometallics are prepared from alkyl halides (as we will see shortly), this reaction presents a means of converting an alkyl halide into an alkane, a reaction included in the Key to Transformations (inside front cover). For example, cyclohexyl bromide can be converted into an organolithium compound (as described in the next subsection), which, in the presence of water or dilute hydrochloric acid, will produce cyclohexane:

$$\text{C}_6\text{H}_{11}\text{–Br} \dashrightarrow \text{C}_6\text{H}_{11}\text{–Li} \xrightarrow[\text{H}_2\text{O}]{\text{HCl}} \text{C}_6\text{H}_{12} + \text{LiCl}$$

Cyclohexyl bromide — Cyclohexyl lithium — Cyclohexane

Organolithium Compounds

In a reaction that is amazing to watch, treatment of an anhydrous ether solution of an alkyl bromide with metallic lithium (simply metal shavings) results in an exothermic reaction (the solvent actually starts to boil on its own) and the gradual disappearance of the lithium metal. This is an oxidation-reduction reaction, with the lithium being oxidized to Li^+ and the carbon being reduced to a carbanion.

$$CH_3CH_2Br + 2\,Li \xrightarrow{\text{ether}} CH_3CH_2Li + LiBr$$

Ethyl bromide — Lithium metal — Ethyllithium — Lithium bromide

In this example, the ether solution contains ethyllithium and lithium bromide. (The ether solution must be rigorously protected from air because of the moisture.) Bromides are the halogen of choice. The reaction is equally effective with methyl, primary, secondary, and tertiary halides.

General reaction:

$$R{-}Br + Li \xrightarrow{\text{ether}} R{-}Li$$

Grignard Reagents (Organomagnesium Compounds)

Named after the French chemist Victor Grignard, who was awarded the Nobel Prize in chemistry in 1912 for discovering them, Grignard reagents are formed by treating alkyl halides (usually bromides) in anhydrous ether solution with metallic magnesium. Just as with lithium, the metal gradually disappears in an exothermic reaction, producing a solution of the Grignard reagent. For example, cyclopropyl bromide yields cyclopropylmagnesium bromide:

$$\text{C}_3\text{H}_5\text{–Br} + \text{Mg} \xrightarrow{\text{ether}} \text{C}_3\text{H}_5\text{–MgBr}$$

Cyclopropyl bromide — Magnesium metal — Cyclopropylmagnesium bromide

The structure of the Grignard reagent in solution is more complex than is shown here, but the basic chemistry we will study can be explained using this structure. Formation of Grignard reagents occurs readily with methyl, primary, secondary, and tertiary alkyl bromides or iodides.

General reaction:

$$RBr + Mg \xrightarrow{\text{ether}} RMgBr$$

PROBLEM 3.18

Write the products of the following reactions:

(a) 1-bromobutane with magnesium in ether

(b) cyclohexyl bromide with lithium in ether

(c) cyclopropyl bromide with lithium followed by the addition of a small amount of water

---

Lithium Dialkylcuprates

Reaction of an organolithium compound with cuprous iodide produces an organocuprate reagent.

General reaction:

$$2\,RLi + CuI \xrightarrow{\text{ether}} R_2CuLi + LiI$$

Organolithium, Cuprous iodide, Lithium dialkylcuprate, Lithium iodide

Organocuprate formation occurs with most alkyllithium compounds. For example, ethyl bromide can be converted into ethyllithium and then treated with cuprous iodide to produce a solution of lithium diethylcuprate.

$$2\,CH_3CH_2Li + CuI \xrightarrow{\text{ether}} (CH_3CH_2)_2CuLi + LiI$$

Ethyllithium, Lithium diethylcuprate

3.5.2 Alkanes from Organocuprate Reagents

One of the uses of the organocuprate reagents is their reaction with other alkyl halides to produce alkanes. Because such a reagent can be considered as a source of a carbanion, which is a strong nucleophile, it can be used in a substitution reaction with another alkyl halide (methyl, primary, or secondary). What occurs is an S_N2 substitution by the carbanion of the organocuprate reagent on the alkyl halide, thereby forming a new bond between R and R′, a new carbon-carbon bond.

General reaction:

$$R_2CuLi + R'\!-\!X \xrightarrow[S_N2]{\text{ether}} R\!-\!R' + LiX + RCu$$

Lithium dialkylcuprate, Alkyl halide, Alkane

For example, reaction of ethyllithium with cuprous iodide in ether, and then addition of bromocyclopentane, produces ethylcyclopentane:

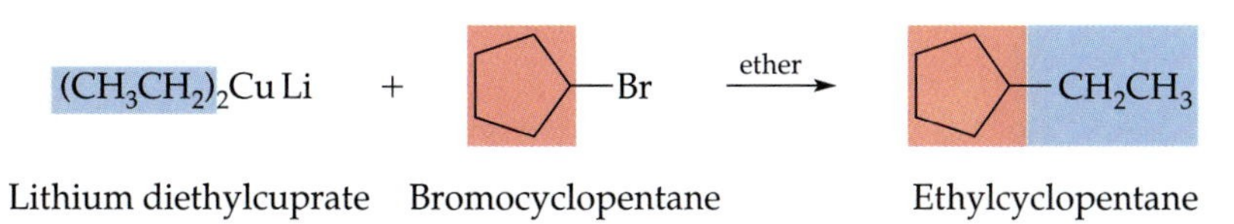

Lithium diethylcuprate   Bromocyclopentane   Ethylcyclopentane

By this means, a five-carbon compound (cyclopentane) has been converted into a seven-carbon compound. It is through reactions such as this that carbon skeletons are elaborated.

This reaction is limited by the fact that the organocuprate reagent is a strong nucleophile and, consistent with the complication cited in Section 3.4.2, cannot be used with tertiary alkyl halides because of the competing elimination reaction. A synthesis of 2,2-dimethylbutane, $CH_3CH_2C(CH_3)_3$, can be accomplished by joining an ethyl group, $—CH_2CH_3$, to a *t*-butyl group, $(CH_3)_3C—$, using an S_N2 reaction employing an organocuprate reagent. There are two possible approaches. Reaction of lithium diethylcuprate with *t*-butyl bromide will not produce 2,2-dimethylbutane by substitution because this reaction involves a carbanion attacking a tertiary alkyl halide, which produces an alkene via elimination instead. However, if lithium di(*t*-butyl)cuprate is reacted with ethyl bromide, the desired alkane is obtained. Here a carbanion attacks a primary alkyl halide.

$$(CH_3)_3C—Br \xrightarrow[\text{ether}]{\text{Li}} (CH_3)_3C—Li \xrightarrow[\text{ether}]{\text{CuI}} [(CH_3)_3C]_2—CuLi \xrightarrow{C_2H_5—Br} CH_3—C(CH_3)_2—CH_2—CH_3$$

t-Butyl bromide

2,2-Dimethylbutane

$$C_2H_5—Br \xrightarrow[\text{ether}]{\text{Li}} C_2H_5—Li \xrightarrow[\text{ether}]{\text{CuI}} [C_2H_5]_2—CuLi \xrightarrow[(CH_3)_3C—Br]{} \times$$

Ethyl bromide

This is an excellent example of how knowledge of the mechanism of the S_N2 reaction permits an appropriate choice of reagents to accomplish a synthesis.

How to Solve a Problem

Design a synthesis of *t*-butylcyclobutane using any four-carbon starting materials.

PROBLEM ANALYSIS

As always, we first write the structure of the desired product:

$$\text{cyclobutyl}—C(CH_3)_3$$

t-Butylcyclobutane

Then we consider the other information we are given: The desired product must be synthesized from four-carbon fragments. We draw a dashed line to split the molecule and indicate possible four-carbon fragments. There are two in this case, the cyclobutyl group and the *t*-butyl group:

Cyclobutyl *t*-Butyl

Next, we realize that our task is to make a new carbon-carbon bond between these two four-carbon fragments. This normally is accomplished by an S_N2 substitution reaction. To bring about such a reaction, one fragment must be a nucleophile, and in this case it must be a carbon nucleophile (a carbanion). The other fragment must have a leaving group, usually a bromine atom. We have two possible choices: (1) to make the cyclobutyl group a nucleophile and have the *t*-butyl group carry a bromine atom, or (2) to make the *t*-butyl group a nucleophile and have a bromine atom on the cyclobutyl ring:

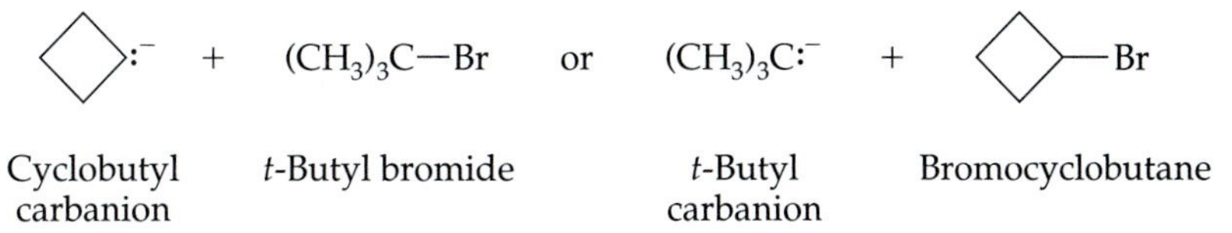

Cyclobutyl carbanion + $(CH_3)_3C{-}Br$ (*t*-Butyl bromide) or $(CH_3)_3C{:}^-$ (*t*-Butyl carbanion) + Bromocyclobutane

How to choose? Recall that the S_N2 substitution reaction is effective on primary or secondary alkyl halides but is not effective with a tertiary halide. That rules out the first option.

The next task is to get carbon to act as a nucleophile, and making an organocuprate reagent is the method of choice. Grignard reagents and organolithium compounds, though they, too, act as carbanions, are simply too nucleophilic.

SOLUTION

We react *t*-butyl bromide with lithium to form *t*-butyllithium. We treat the resulting solution with cuprous iodide to produce lithium di(*t*-butyl)cuprate. Finally, we add bromocyclobutane as the substrate for the S_N2 substitution reaction.

$$(CH_3)_3C{-}Br \xrightarrow[\text{ether}]{\text{Li}} (CH_3)_3C{-}Li \xrightarrow[\text{ether}]{\text{CuI}} [(CH_3)_3C]_2{-}CuLi$$

t-Butyl bromide; *t*-Butyllithium

$[(CH_3)_3C]_2{-}CuLi$ + Br-cyclobutane (Bromocyclobutane) → cyclobutyl$-C(CH_3)_3$

t-Butylcyclobutane

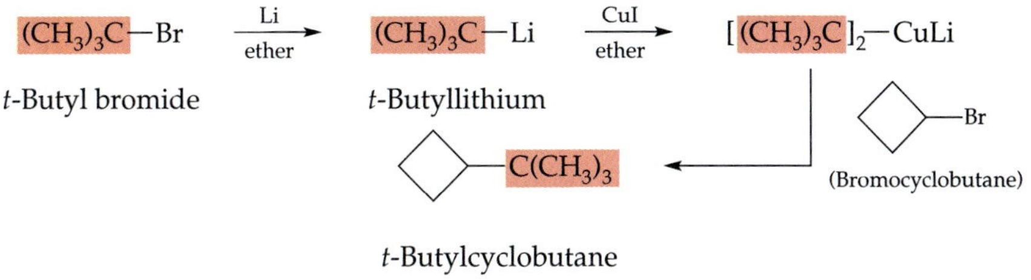

Note that in synthesis problems such as this, you need not write a separate single-line equation for each step (unless you are asked to), and you need not write balanced equations for each step (unless you are asked to). Instead, write the key reactants and any conditions and reagents necessary. This is standard practice in organic chemistry, and you should adopt it.

PROBLEM 3.19

Draw the products of the following reactions:

(a) lithium dimethylcuprate with cyclohexyl bromide
(b) lithium diethylcuprate with *t*-butyl bromide
(c) lithium dicyclohexylcuprate with *n*-butyl bromide

PROBLEM 3.20

Devise syntheses of the following compounds using organocuprate reagents. Show every step required, including the reagent and the product.

(a) methylcyclohexane
(b) 2,4-dimethylhexane
(c) 2,2-dimethylpentane

Krakatoa, a famous and still active volcano in Indonesia. Erupting volcanoes emit large quantities of halogens and halides.

3.6 Important Alkyl Halides

In contrast to alkanes, of which large quantities in significant concentrations exist naturally, there are no comparable natural sources for alkyl halides. Therefore, most halogen-containing organic compounds used in bulk in industry are produced synthetically. However, contrary to the general opinion of only a few years ago, there are literally thousands of naturally occurring organic halogen compounds, some of which are produced continually. It is not surprising that organohalogen compounds exist in nature since there is large-scale and widespread occurrence of chlorine (in the form NaCl in the earth and oceans and HCl emitted from volcanoes), bromine (mainly in the form of ocean salts), fluorine (emitted from volcanoes as HF and in minerals), and iodine (as brine in deep wells and in minerals). These elements react with organic compounds with which they come in contact.

One of the major uses of alkyl halides is as chemical intermediates. In other words, they are used in reactions leading to other important end products, often in substitution reactions. You will see them used in this way throughout this book. For now, however, we will focus on a number of important and interesting alkyl halides familiar to the general public.

In contrast to alkyl halides with a single halogen, organic halogen compounds containing *several* halogen atoms often are especially stable compounds (a carbon-halogen bond in such a compound is not readily broken). This means that they resist chemical change and are thus persistent. In other words, they resist nature's tendency to break down (degrade or detoxify) chemicals in the environment and in the body. However, the stability of organohalogen compounds provides a significant advantage in making them useful in many ways. Here lies society's dilemma: How to balance the significant beneficial use of these compounds against the negative impact of their persistence in the biosphere? This issue has been the source of much controversy over the last several decades.

3.6.1 Anesthetics

In 1847, chloroform (trichloromethane, $CHCl_3$) became the first widely used general anesthetic, rendering patients unconscious and relaxed. However, it has been replaced by other low-boiling haloalkanes, one of which has the trade name Halothane (CF_3—$CHClBr$). Ethyl chloride (C_2H_5Cl) is used as a numbing agent applied to the surface of skin when minor surgery or stitches are required.

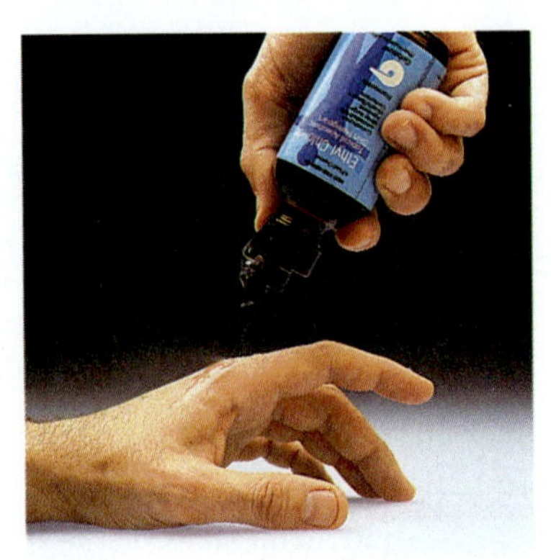

Ethyl chloride spray, a skin "refrigerant," is used as a topical anesthetic.

3.6.2 Solvents

Low-molecular-weight organic halogen compounds are excellent solvents, both for conducting chemical reactions and as cleaners and degreasers. Carbon tetrachloride (CCl_4) was the first major dry-cleaning solvent, but it is toxic and carcinogenic upon continued exposure. Thus, 1,1,1-trichloroethane (Cl_3CCH_3, also known as methyl chloroform), trichloroethylene ($Cl_2C{=}CHCl$), and perchloroethylene ($Cl_2C{=}CCl_2$) are now used instead. Methylene chloride is widely used in laboratories as an extraction and reaction solvent, but only in fume hoods, which protect the user.

3.6.3 Refrigerants

Much of our way of life would not be possible were it not for efficient refrigeration systems, providing for the storage of food for long periods, free of spoilage. The discovery of **Freons** (usually referred to as CFCs, for chlorofluorocarbons) led to safe home freezers, refrigerators, and then air-conditioning systems in buildings and automobiles. Production of CFCs totaled billions of pounds per year.

The most common CFC was discovered in 1930 and is trade-named Freon-12 (dichlorodifluoromethane, CCl_2F_2). Its significant advantages include superb heat-transfer properties, nontoxicity, nonflammability, and such high stability that it never needs to be replaced (unless there is leakage). Although its use was originally limited to refrigeration and air-conditioning systems, Freon-12 was later also used as a cleaning agent for semiconductors and computer parts, as the propellant agent in aerosol cans, and as a foaming agent in insulation board, foam cups, and packing materials. These uses, along with some leakage from air-conditioning systems, led to the release of large quantities of Freon-12 into the atmosphere.

As a result of measurements started in the 1970s, scientists began to realize that the CFCs were so stable that they kept rising through the atmosphere into the stratosphere (about 15 miles high) without being destroyed by natural degradation processes. Laboratory research, some of which resulted in a 1995 Nobel Prize in chemistry for Sherwood Rowland and Mario Molina, proved that in the polar regions of the earth, CFCs reacted with and depleted ozone. (We will discuss this chemistry in Chapter 10.) Atmospheric ozone is essential to protect the earth from harmful ultraviolet radiation (UV-C) from the sun. In 1987, an international agreement called the Montreal Protocol committed the industrialized nations to a total phaseout of CFC production, actually accomplished in 1996 (developing countries were granted an additional 10 years).

Urgent research and development programs have resulted in a series of replacement refrigerants that are already being incorporated in new refrigeration systems, but they

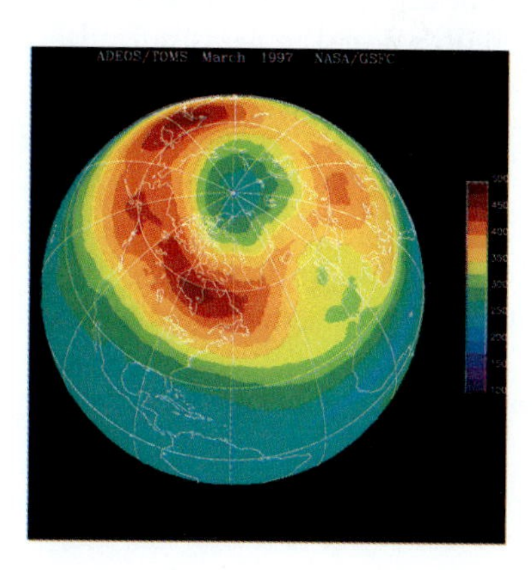

The concentration of ozone in the atmosphere is measured in Dobson units (DU). Here, the highest concentrations are shown in red, the lowest in blue and purple. Twenty years ago, the North Pole had an ozone concentration over 500 DU; shown here as measured in March 1997, the ozone concentration was in the 100–200 DU range.

cannot be used in existing systems (including 1994 and earlier-model automobiles). The most prominent materials are HFC-134a ($CF_3—CH_2F$) and HCFC-141b ($CH_3—CCl_2F$).

3.6.4 Fire Retardants

Closely related to CFCs are the **halons,** materials that have critical uses as fire retardants and fire-fighting chemicals. The major halons are halon 1301 (CF_3Br) and halon 1211 (CF_2BrCl). Their use in fire-extinguishing systems is critical in "clean" environments (those that cannot stand water), such as libraries, computer rooms, and telephone equipment. The halons leave no residue on evaporation and do not conduct electricity, making them important in applications involving electrical equipment, military equipment, and civilian aircraft.

3.6.5 Insecticides

It is safe to say that millions of people have died as a result of insects transmitting disease and destroying crops. The battle of humankind against insects changed dramatically in 1939 with the discovery that DDT (dichlorodiphenyl-trichloroethane) could kill a wide variety of harmful insects, including mosquitoes that carry malaria and locusts that ravage crops.

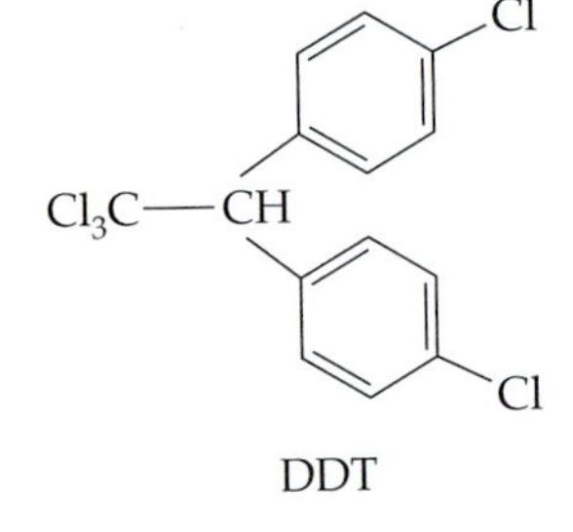

DDT

These peregrine falcon eggs have weakened shells, due to the presence of DDT in the birds' diet. Affected shells can be so thin that they break when incubated by an adult. The egg on the right is badly dented.

A major advantage of DDT was its persistence: It was hydrophobic, so it clung to insects and plants and did not immediately need to be reapplied after rains; and it was not readily degraded in the environment—another example of the stability of polychlorinated compounds. However, this same persistence led to its accumulation in the environment. One result, as documented in Rachel Carson's 1962 influential book *Silent Spring,* was the serious decline in the population of eagles and some other raptors. DDT was accumulating in their bodies and affecting calcium deposition in egg shells. The shells were so fragile that they simply collapsed in the nests during incubation.

DDT is now banned for agricultural uses, although occasional reuse is permitted to fight serious insect infestations, like that of the gypsy moth in the northeastern United States in the 1980s.

3.6.6 Medicinal Compounds

Many synthetic medicinal compounds that contain halogens are used to treat a wide variety of human afflictions, including depression, arthritis, pain, fungal diseases, malaria, and cancer. Most of these, however, are not aliphatic halides but aromatic halides, which we will discuss in Chapter 9.

Perfluorocarbon compounds have an unusual application due to their unique ability to dissolve more oxygen than does human blood itself (60% by volume versus about

20%). Fluosol-DA, an aqueous emulsion of perfluorodecalin, has been used as a short-term blood substitute.

Supercomputer cooled by perfluorocarbon.

Perfluorodecalin
(Fluosol-DA)

It is effective when whole blood is unavailable or when a blood match is not available, such as in trauma cases. Its use is temporary because it is gradually excreted from the body, but it can ensure patient survival until whole blood is available. Fluosol-DA is nontoxic and can be stored frozen for several years before use. Other related compounds are being evaluated for similar uses.

One of the more interesting uses for perfluorocarbons is as cooling fluids that dissipate the large amount of heat produced in supercomputers. The fluid actually flows through the computer and bathes the working parts. This is possible because the fluorocarbon is inert, does not conduct electricity, and has effective heat-conducting properties.

3.6.7 Naturally Occurring Alkyl Halides

Among many recent environmental issues is a call by some for a total ban on the production and use of chlorine. This concern stems from the persistence of chlorocarbons and implications for problems related to health and the atmosphere. Those who argue against the ban point out that nature itself produces many chlorocarbons, some of which are beneficial to animal life and some of which are very toxic. The production of many simple alkyl halides in nature has been well documented, and some, such as methyl chloride, appear to be produced in far larger quantities than are those produced by humans. Marine organisms account for a tremendous variety of simple chlorocarbons, as well as some very complex compounds.

$Cl_3C-C(=O)-CCl_3$

Chlorocarbons produced by marine algae

Volcanoes emit huge quantities of halocarbons, and the burning of biomass (which contains halides), as in a forest fire, also results in halocarbon production.

It is clear that completely removing halocarbons from the environment is impossible, and even undesirable. But reasonable controls on unnecessary release of synthetic halocarbons has already been shown to be effective in reducing negative environmental impacts.

Chapter Summary

Haloalkanes contain a halogen substituent (fluoro-, chloro-, bromo-, or iodo-) attached to a parent alkane. These substituents are named in alphabetical order and located by numbers on the alkane chain. Simple haloalkanes can also be named as **alkyl halides** and generalized by the structure R—X, where R is an alkyl group and X is a halogen.

A second kind of configurational isomerism involves chirality. **Chirality** is the property of *handedness* in any object. A compound (or object) is **chiral** if its mirror image *cannot* be superimposed on it; in that instance, there exist two isomers of the compound. If the mirror image of a compound can be superimposed on it, it is **achiral.** Achirality is usually due to the presence of a **plane of symmetry** in the compound.

A carbon atom with four *different* substituents attached is a **stereocenter.** Any compound with a single stereocenter is chiral, and the object and the mirror image (the two isomers) are called **enantiomers.** Each enantiomer has the property of optical activity. All other physical properties of the two enantiomers are identical.The **absolute configuration** of an enantiomer, the precise spatial location of the substituents about the stereocenter, is specified using the ***R-S* system** of configurational nomenclature. A 50/50 mixture of the two enantiomers (*R* and *S*) of a compound is a **racemic mixture** (also called a **racemate**). Chiral compounds produced by laboratory synthesis normally exist as a racemic mixture unless unusual steps have been taken to separate them into the two enantiomers (a process called *resolution*).

The most important reaction of alkyl halides is the replacement of the halide by another nucleophile—a **nucleophilic substitution.** The mechanism of an **S_N2 substitution reaction** involves a single step, the **back-side attack** of a nucleophile on the carbon to form a new bond, with *concerted* expulsion of the halide anion. The reaction is **stereospecific;** reaction at a stereocenter results in **inversion of configuration** of the stereocenter. It is a second-order reaction involving a **transition state,** and the size of the **energy of activation** determines the rate of the reaction. The relative rates of S_N2 reactions are methyl halide > primary halide > secondary halide. Tertiary halides normally will not undergo an S_N2 substitution reaction; instead, they eliminate HX to produce alkenes. Iodide ion is the best **leaving group,** and stronger nucleophiles make the S_N2 reaction go faster.

Tertiary halides undergo an **S_N1 substitution reaction** with weak nucleophiles. The rate of the reaction depends on the stability of the **carbocation (a reactive intermediate)** formed in the two-step reaction. The carbocation is stabilized by **hyperconjugation** from adjacent C—H bonds. The S_N1 reaction is nonstereospecific, resulting in the formation of a racemate. Primary and secondary halides do not undergo significant S_N1 substitution.

A major use of alkyl bromides is to form *organometallic reagents,* which serve as sources of carbanions. *Organolithium compounds* are formed by reaction with metallic lithium, while *Grignard reagents* (organomagnesium halides) are formed with metallic magnesium. Organolithium reagents react with cuprous iodide to form *lithium dialkylcuprates.* All three organometallics are strong nucleophiles, with the order of nucleophilicity being organolithium > Grignard reagent > lithium dialkylcuprate. Lithium dialkylcuprates react with primary or secondary alkyl halides to effect an S_N2 reaction, thereby forming new carbon-carbon bonds.

Summary of Reactions

1. **Nucleophilic substitution (Section 3.4).** Alkyl halides react with a wide variety of nucleophiles, resulting in the halogen atom being displaced by the nucleophile. This reaction therefore serves as a means of synthesizing a wide variety of families of compounds (R—Y). (Note that X = halogen and Y = −OH, −OR, −SH, etc.; see Table 3.1.)

$$\text{Y:} + \text{R—X} \longrightarrow \text{R—Y} + \text{X:}$$

2. **Formation of organometallics (Section 3.5.1).** Alkyl halides react with lithium or magnesium to form organometallic reagents. Organolithium reagents react further with cuprous iodide to form lithium dialkylcuprates.

$$\text{R—X} + 2\,\text{Li} \xrightarrow{\text{ether}} \text{R—Li} + \text{LiX}$$

Organolithium reagent

$$\text{R—X} + \text{Mg} \xrightarrow{\text{ether}} \text{R—Mg—X}$$

Grignard reagent

$$2\,\text{R—Li} + \text{CuI} \xrightarrow{\text{ether}} \text{R}_2\text{CuLi} + \text{LiI}$$

Lithium dialkylcuprate

3. **Reaction of organometallics with water or acids (Section 3.5.1).** Organometallic reagents are very strong bases. They react in an acid-base, or proton transfer, reaction with a proton source such as water, a weak acid, or a strong acid to form an alkane.

$$R{-}M \quad + \quad H{-}OH \longrightarrow R{-}H \quad + \quad M{-}OH$$

4. **Alkanes from cuprates and alkyl halides (Section 3.5.2).** Lithium dialkylcuprates react via an S_N2 mechanism with primary and secondary alkyl halides to form alkanes.

$$R_2CuLi \quad + \quad R'{-}X \xrightarrow{\text{ether}} R{-}R' \quad + \quad LiX \quad + \quad RCu$$

Additional Problems

Nomenclature

3.21 Name the following compounds:

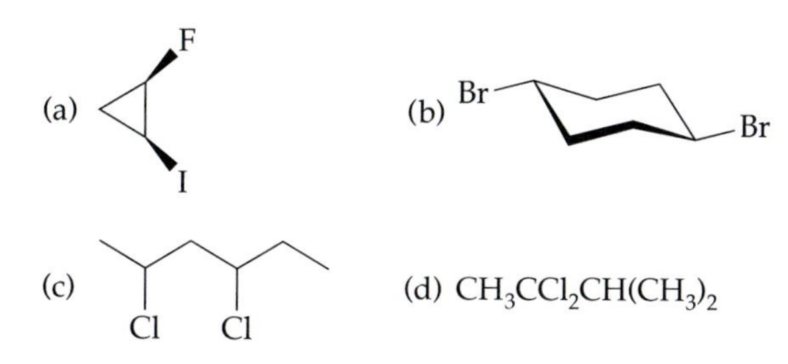

(d) $CH_3CCl_2CH(CH_3)_2$

3.22 Draw structures for the following compounds:

(a) 3-bromo-1,1-dichloro-1-fluoropropane
(b) *trans*-1,3-diiodocyclobutane
(c) *cis*-1,4-difluorocyclohexane
(d) 1,3,5-trichlorohexane
(e) isopropyl chloride
(f) *t*-butyl fluoride

Stereochemistry

3.23 Which of the following compounds has a stereocenter? Identify it by its position number (for example, carbon 2).

(a) 3-chloropentane
(b) 2-iodohexane
(c) bromocyclohexane
(d) 1,2-dichlorocyclohexane
(e) 3-fluoro-2-methylpentane
(f) iodocyclobutane

(g)

(h)

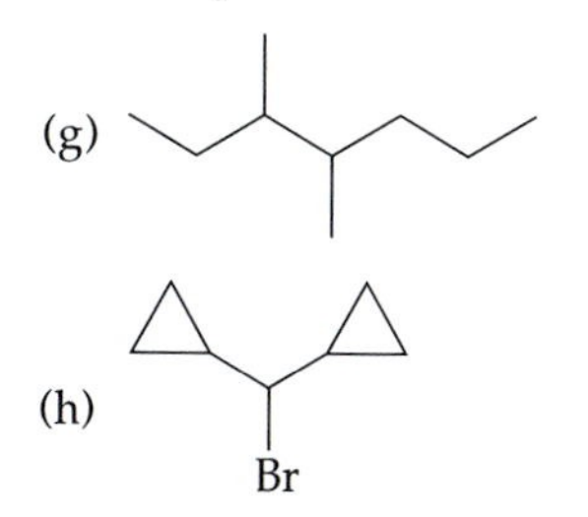

3.24 There are several isomers of a compound with molecular formula C_4H_9Br. Draw all those that are chiral.

3.25 Using the *R-S* system for indicating absolute configuration, place the substituents in each of the following series in *descending* order of priority:

(a) CH_3—, H—, Br—, CH_3O—
(b) CH_3—, $ClCH_2$—, CH_3CH_2—, NH_2—
(c) HO—, $HOCH_2$—, —CH=O, CH_3CH_2—.

Substitution Reactions

3.26 Write the major product to be expected from reacting each of the following reagents with 1-bromobutane:

(a) sodium iodide
(b) sodium hydroxide
(c) sodium cyanide
(d) sodium methoxide (CH_3O^- Na^+)

3.27 Write the reaction and show the product(s), including their stereochemistry, of the following reactions:

(a) (*R*)-2-bromopentane with sodium methoxide
(b) *trans*-1-*t*-butyl-4-iodocyclohexane with sodium cyanide

3.28 Draw the structure of the alkyl halide and the reactants necessary to produce each of the following products via an S_N2 substitution:

(a) $CH_3CH_2CH_2OH$
(b) $(CH_3)_2CHCH_2CH_2SCH_3$
(c) $CH_3CH_2OCH_3$

3.29 Which of the two reactions in each of the following sets would occur more rapidly?

(a) *n*-butyl iodide with sodium hydroxide or with water to produce $CH_3CH_2CH_2CH_2OH$
(b) *n*-butyl chloride with sodium iodide or sodium bromide
(c) ethyl iodide with one equivalent of sodium methoxide (CH_3ONa) or with two equivalents
(d) sodium iodide in acetone with *n*-butyl bromide or with *t*-butyl bromide

(e) 1-bromobutane or 2-bromobutane with sodium hydroxide

(f) 1-chlorobutane or 1-iodobutane with sodium hydroxide

(g) 1-bromobutane with sodium hydroxide or with sodium methoxide

3.30 Solvolysis of *t*-butyl bromide with water produces a mixture of *t*-butyl alcohol, $(CH_3)_3COH$, and isobutylene, $(CH_3)_2C{=}CH_2$. Exactly the same ratio is obtained when *t*-butyl chloride or *t*-butyl iodide is employed in the solvolysis reaction. Explain.

3.31 Indicate the structure and name of the starting bromoalkane, as well as the necessary reagents, that you would use to synthesize each of the following products by a substitution reaction:

(a) 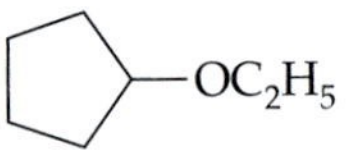 $-OC_2H_5$

(b) 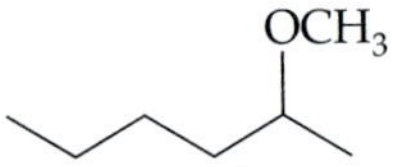 OCH_3

(c) $CH_3CH_2CH(CH_3)CH_2CN$

(d)  OH

Organometallics

3.32 Write the organic products of the following reactions:

(a) cyclohexyl bromide with magnesium in ether

(b) *t*-butyl iodide with lithium in ether

(c) isobutyl bromide with lithium in ether, followed by treatment with cuprous iodide

(d) isopropyl bromide with magnesium in ether, followed by a small amount of water

(e) cyclopropyl bromide with lithium in ether, then addition of cuprous iodide, and finally addition of ethyl bromide

Mixed Problems

3.33 Indicate the relationship between the compounds in each of the following pairs:

(a)

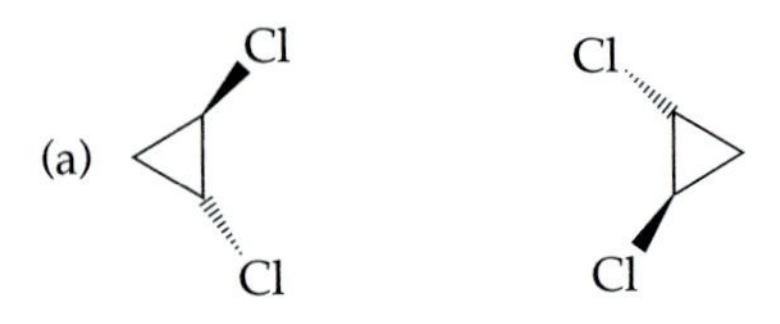

(b) 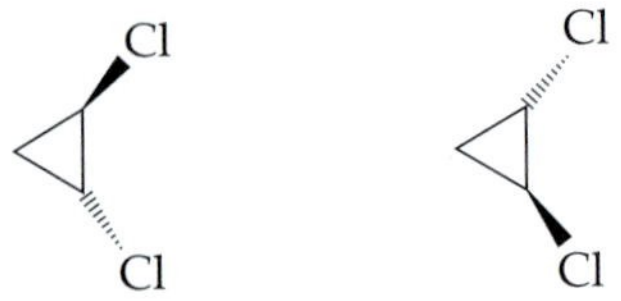 Cl, Cl — Cl, Cl

(c)  C_2H_5, I, H, CH_3 — C_2H_5, H, I, CH_3

(d) C_2H_5, H, I, CH_3 — C_2H_5, I, H, CH_3

(e) 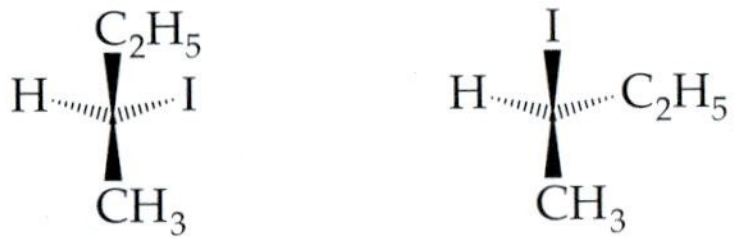 C_2H_5, H, I, CH_3 — I, H, C_2H_5, CH_3

3.34 For the structures shown in Problem 3.33(c)–(e), draw the *R* configuration.

3.35 Predict the product(s) of each of the following reactions. If more than one is formed, which is the major product?

(a) $CH_3CH_2CH_2I \ + \ CH_3O^- \longrightarrow$

(b) $(CH_3)_3CO^- \ + \ CH_3CH_2Br \longrightarrow$

(c) 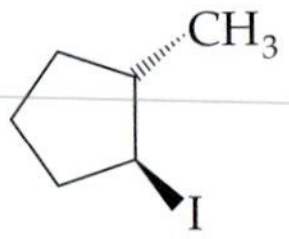 CH_3, I $+$ NaSH $\longrightarrow$

(d) (*S*)-2-iodopentane $+$ NaOH $\longrightarrow$

(e) $(CH_3)_3C{-}I \ + \ NaOH \longrightarrow$

3.36 Heating 2-bromo-2-methylpentane with hot sodium hydroxide produces a mixture of two organic products. Draw their structures and explain their formation.

3.37 Two compounds of molecular formula C_4H_9Br were found in separate unlabeled bottles. To help determine their structure, each was treated with sodium iodide in acetone. Compound A did not react, but compound B produced a precipitate almost immediately. What conclusions can you draw about the structures of A and B?

3.38 Explain why the first of the following pairs of reactions is the preferred synthetic route to the indicated product:

CONCEPTUAL PROBLEM

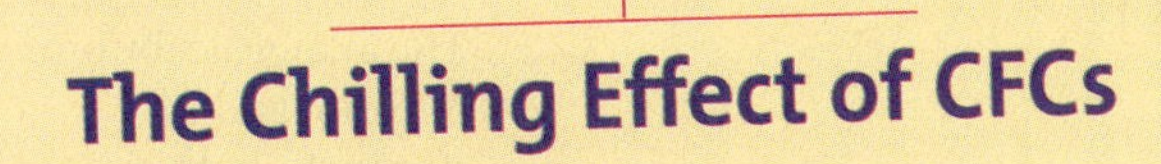

The Chilling Effect of CFCs

You work in a large university's physical plant department, which oversees the day-to-day operations of the buildings on campus. The university currently maintains 14 large-scale air-conditioning units (or "chillers") that still use CFCs as the refrigerant. These units were installed well before the 1996 ban on CFC production. Some are as little as 10 years old; others are much older. It is part of your job to develop a plan to either retrofit or replace all the CFC units, as well as to establish guidelines for future purchases of air-conditioning equipment. A member of the budget committee comes to you because she is concerned about the potential costs of replacing so many units. She asks you these questions:

- Since a cooling unit is sealed (the refrigerant is kept under pressure in a sealed system), why does the continued use of CFCs pose any risk to the environment?
- Can't the university simply buy a different refrigerant to replace the CFCs and use it in the existing units? What would be the concern associated with such refrigerant replacement?
- Environmental issues aside, why can't the university just run the units it has until they are no longer operational and replace them on an as-needed basis? Wouldn't this be the most cost-efficient plan?

After answering these questions, you outline your plan to replace the oldest units first, maintaining some of the younger CFC units for a longer time.

- What can you do to keep these CFC units as environmentally safe as possible?

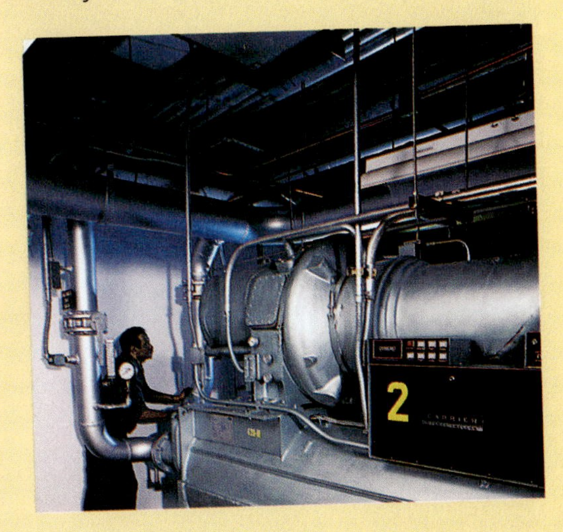

An industrial-sized "chiller."

(a) 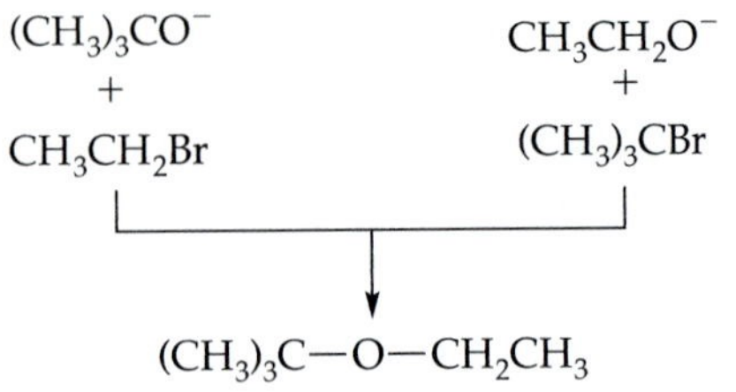 $(CH_3)_3CO^-$ + CH_3CH_2Br ; $CH_3CH_2O^-$ + $(CH_3)_3CBr$

→ $(CH_3)_3C{-}O{-}CH_2CH_3$

(b) 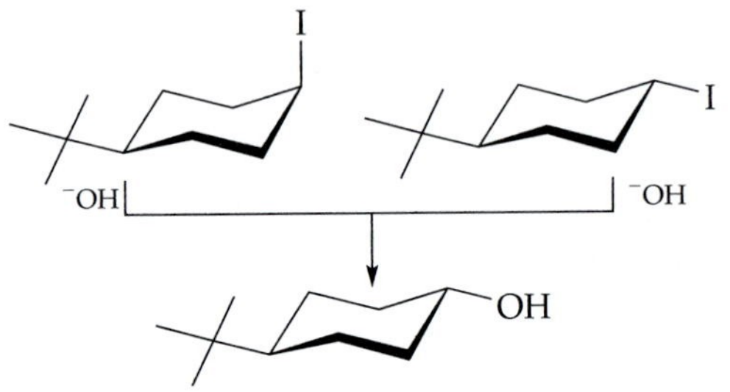 [iodo-tert-butylcyclohexane, I axial] + ^-OH ; [iodo-tert-butylcyclohexane, I equatorial] + ^-OH

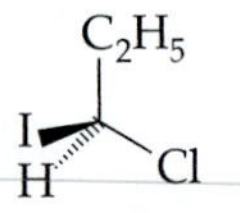

→ [tert-butylcyclohexanol, OH equatorial]

(c) 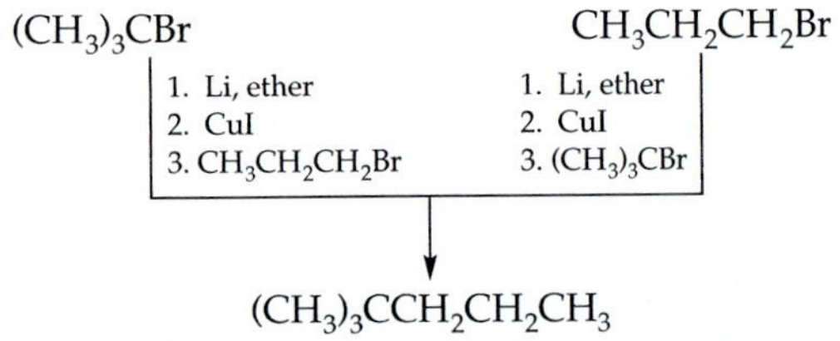 $(CH_3)_3CBr$: 1. Li, ether 2. CuI 3. $CH_3CH_2CH_2Br$; $CH_3CH_2CH_2Br$: 1. Li, ether 2. CuI 3. $(CH_3)_3CBr$

→ $(CH_3)_3CCH_2CH_2CH_3$

3.39 Label each of the following compounds with the proper absolute configuration:

(a) 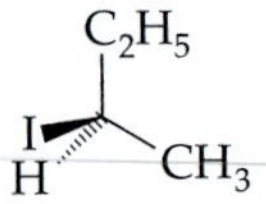 C_2H_5, I, H, CH_3

(b) C_2H_5, I, H, Cl

(c) CH_3, C_3H_7, H, NH_2

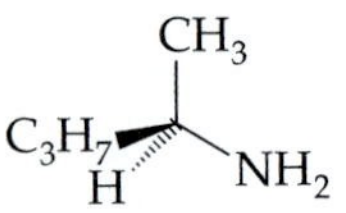

(d) 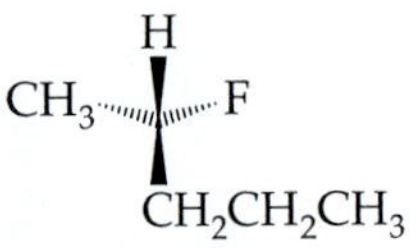 H, CH_3, F, $CH_2CH_2CH_3$

3.40 Describe the reaction—conditions and reagents—to most effectively carry out the following conversions:

(a) 1-bromopentane to 1-pentanol ($CH_3CH_2CH_2CH_2CH_2OH$)

(b) 2-bromo-2-methylpropane to 2-methyl-2-propanol [$(CH_3)_3COH$]

(c) 1-bromo-1-methylcyclohexane to 1-methoxy-1-methylcyclohexane

3.41 When cyclohexyl bromide ($C_6H_{11}Br$) is reacted with methanol (CH_3OH), a weak nucleophile, the major product is cyclohexyl methyl ether ($C_6H_{11}{-}O{-}CH_3$), but when cyclohexyl bromide is reacted with hot sodium methoxide ($NaOCH_3$), a strong nucleophile, the major product is cyclohexene (C_6H_{10}). Explain.

3.42 The analgesic ibuprofen (which is sold as Advil, Nuprin, and Motrin) is chiral. In the commercial products, both enantiomers are present but only the *S* enantiomer is physiologically active. Draw the absolute configuration of the inactive form, (*R*)-ibuprofen.

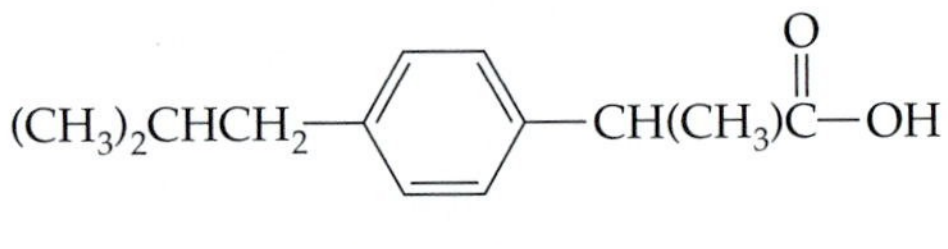

$(CH_3)_2CHCH_2{-}C_6H_4{-}CH(CH_3)C(=O){-}OH$

Ibuprofen

3.43 When a solution of (*R*)-2-iodobutane in acetone is treated with sodium iodide, the solution loses its optical activity and a racemic mixture of (*R*)- and (*S*)-2-iodobutane results. Explain.

3.44 Write syntheses of the following compounds from alkyl bromides:

(a) 2,3-dimethylbutane

(b) cyclohexyl iodide

(c) methylcyclohexane

(d) 2-pentanol [$CH_3CH(OH)CH_2CH_2CH_3$]

(e) isopropylcyclobutane

3.45 There are four dichlorocyclopropane isomers. Draw and name each, and identify the kind of isomer it is.

3.46 Show the single reaction needed to convert (*R*)-2-bromobutane to (*S*)-2-butanol [$CH_3CH(OH)CH_2CH_3$].

C_2H_5, Br, H, CH_3

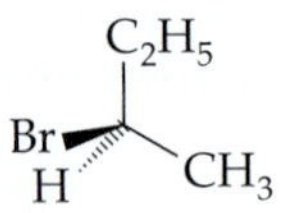

(*R*)-2-Bromobutane

3.47 Show the two reactions needed to convert (*R*)-2-bromobutane to (*R*)-2-butanol [$CH_3CH(OH)CH_2CH_3$].

CHAPTER 4

Alcohols

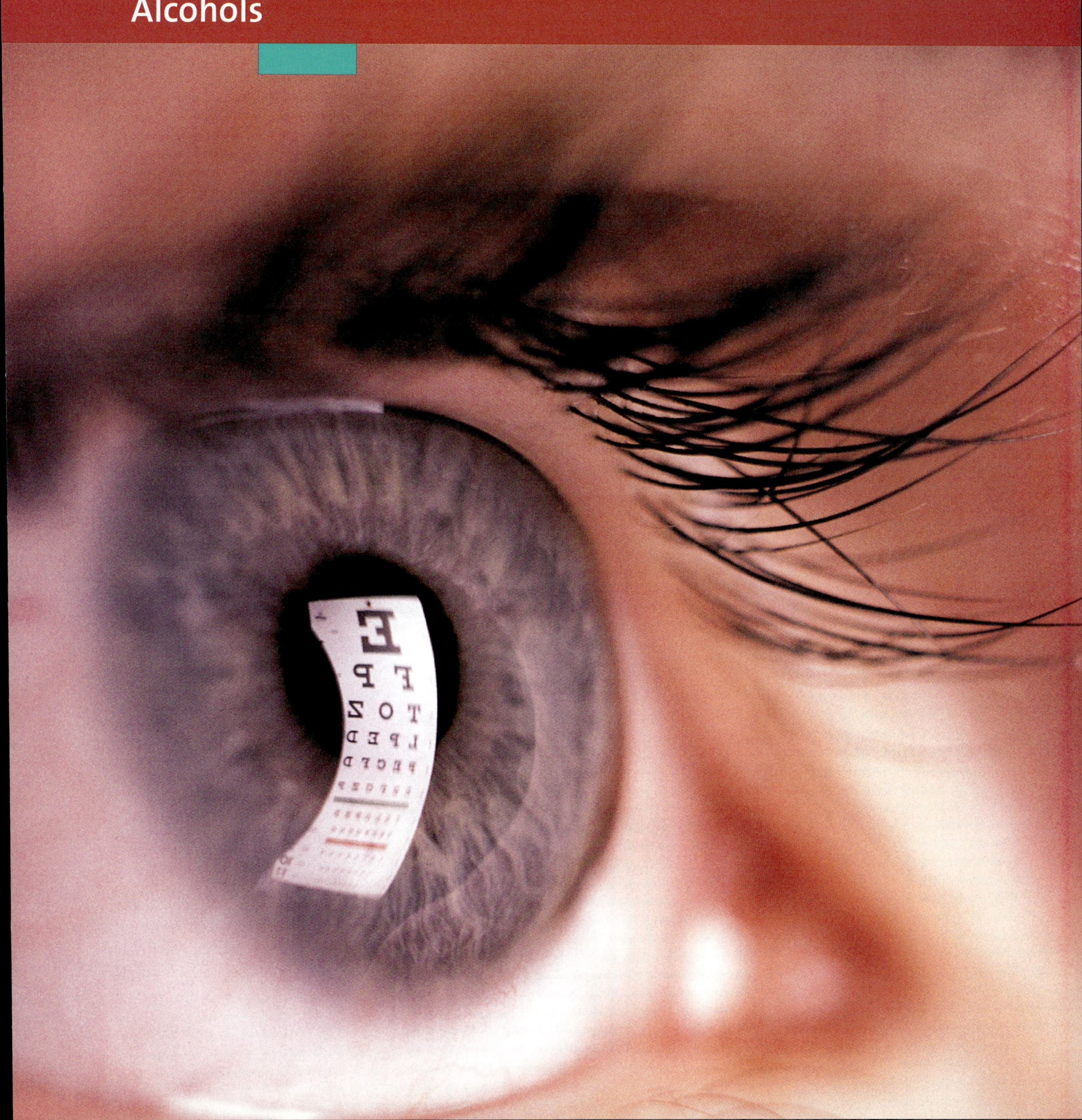

ALCOHOLS ARE a family of compounds whose defining characteristic is the attachment of a single **hydroxyl group** (—OH) to a tetrahedrally hybridized carbon. Alcohols can be considered to be derivatives of alkanes in which one (or more, but on different carbons) of the hydrogen atoms of an alkane has been replaced by an —OH functional group. If just one hydroxyl group is present, the general formula for an alcohol is that of an alkane plus an oxygen: $C_nH_{2n+2}O$. Alcohols are important because of their direct commercial applications, their use as chemical intermediates in the production of other useful end products, their biological significance, and their widespread natural occurrence. This chapter will use relatively simple alcohols to show the properties and chemistry of the family but will close with a brief description of some more complex alcohols.

The hydroxyl group is a key functional group, as you can see from the central location of alcohols in the Key to Transformations (inside front cover). Alcohols can be converted into and derived from alkyl halides, alkenes, ethers, and carbonyl compounds (aldehydes and ketones).This chapter will introduce you to various methods used to achieve some of these important transformations. Utilizing these two-way conversions, you will begin to learn of the many possibilities for organic synthesis.

Alcohols react with a wide variety of reagents, only a few of which we will consider in this chapter. However, all such reactions fall into two general patterns: The first pattern involves cleavage of the bond between the oxygen of the hydroxyl group and carbon. Such reactions include dehydration of alcohols to form alkenes and conversion of alcohols to alkyl halides. The second pattern involves cleavage of the bond between the hydrogen and the oxygen of the hydroxyl group. These reactions include formation of alkoxide anions, formation of esters and ethers, and oxidation of alcohols to carbonyl compounds. The possibilities are summarized in Figure 4.1.

Alcohols are a family of organic compounds that are central to many syntheses.

4.1 Nomenclature of Alcohols

4.1.1 IUPAC Nomenclature

Alcohols provide the first example of how the parent alkane name serves as the starting point for the nomenclature of all families of compounds: The *-ane* suffix is modified in a systematic manner to indicate the presence of a functional group. An alcohol is named by identifying the longest straight carbon chain (parent alkane) containing the

Overleaf: There are many biologically active alcohols involved in chemical reactions in the human body. Once such alcohol is retinol, otherwise known as vitamin A. It is essential in chemical processes occurring in the eye that enable humans and other animals to see.

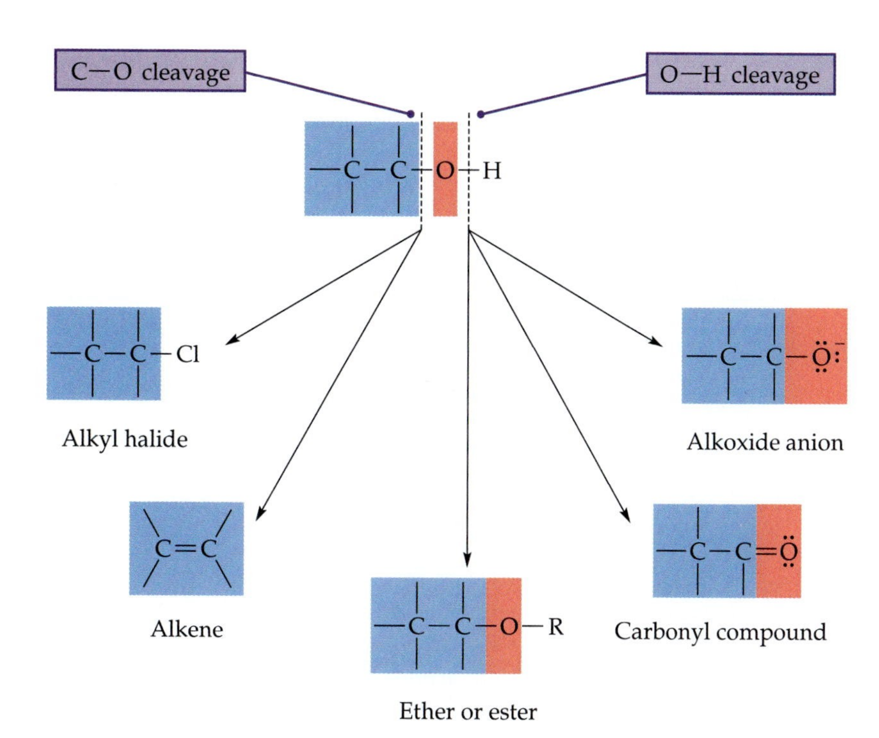

FIGURE 4.1

Patterns of bond cleavage for reactions of alcohols.

—OH group. The *-ane* suffix is replaced with *-anol,* and the location of the —OH group on the chain is designated by a number. For example, CH_3OH is methanol, C_2H_5OH is ethanol, and $C_6H_{11}OH$ is cyclohexanol. Numbers are not used in these cases because there is only a single possible location for the —OH group. However, there are two propanols and three pentanols, requiring the use of a number to designate the position of the —OH group. The number designating that location is kept as small as possible.

$CH_3CH_2CH_2OH$ — 1-Propanol

$(CH_3)_2CHOH$ — 2-Propanol

1-Pentanol

2-Pentanol

3-Pentanol

The generic IUPAC name for alcohols is **alkanols,** and they are represented in reactions by the general formula R—OH.

The presence of other substituents in an alcohol is indicated by their names and numerical positions, always keeping the lowest possible number for the hydroxyl group:

2-Methylcyclopentanol

$(CH_3)_2C(OH)CH_2CH(CH_3)_2$ — 2,4-Dimethyl-2-pentanol

HO~~~OH — 1,3-Propanediol

The presence of more than one hydroxyl group is indicated by the suffixes *-diol*, *-triol*, and so on, and numbers to indicate the position of *each* group. In cyclic alcohols, the —OH group is assigned position 1, so usually the number is not even included in the name.

How to Solve a Problem

Name the following compounds:

(a) $CH_2-CH-CH_2$ with OH on each carbon (OH OH OH)   (b) $CH_3CH(OH)CH(Cl)CH_3$

SOLUTION

(a) This compound has a three-carbon chain, so the parent alkane name is propane. Because the compound contains three hydroxyl groups, the suffix must be *-triol.* Therefore, the name is 1,2,3-propanetriol. The common name of this compound is glycerol (also called glycerine), and it is an important by-product in the manufacture of soaps.

(b) This compound has a four-carbon chain carrying the functional group, so the parent name is butane. The functional group is a hydroxyl group, so the compound is named as a butanol. The location of the hydroxyl group takes precedence in numbering over the chlorine substituent. Therefore, numbering from the left gives the name 3-chloro-2-butanol, which is sometimes written as 3-chlorobutan-2-ol.

PROBLEM 4.1

Draw the structures corresponding to the following names: (a) 3-cyclopropyl-1-propanol, (b) 4-methyl-1,4-pentanediol, (c) 1,3,5-cyclohexanetriol, and (d) dicyclopropylmethanol.

PROBLEM 4.2

Write the names corresponding to the following formulas:

(a) OH   (b) OH OH   (c) OH

(d) $(CH_3)_2CHOH$ (e) $(CH_3)_2CHCH_2CH(OH)CH_2CH(CH_3)_2$

4.1.2 Historical/Common Names

Historically, alcohols were named using the alkyl group name followed by the word *alcohol:* **alkyl alcohol.** Historical names are commonly used for simple alcohols. Here are a few examples (the IUPAC name is given in parentheses): methyl alcohol (methanol), ethyl alcohol (ethanol), and cyclohexyl alcohol (cyclohexanol). However, as the alkyl group becomes more complex, this approach becomes useless. For some very complex structures, historical names come into play, with the suffix *-ol* indicating an alcohol. Examples are menthol and cholesterol.

Alcohols may be classified as primary (1°), secondary (2°), or tertiary (3°), depending on the classification of the carbon to which the hydroxyl group is attached. Thus, a primary alcohol has the —OH group attached to a carbon also bonded to one other carbon (as in ethanol), a secondary alcohol has the —OH group attached to a carbon also bonded to two other carbons (as in *sec*-butyl alcohol), and a tertiary alcohol has the —OH group attached to a carbon bonded to three other carbons (as in *tert*-butyl alcohol):

CH_3CH_2OH — Ethyl alcohol (ethanol)

OH — *sec*-Butyl alcohol (2-butanol)

OH — *tert*-Butyl alcohol (2-methyl-2-propanol)

The abbreviations *n*- for normal (straight chain) and *t*- for *tert*- (tertiary) are standard. The prefix *iso*- is sometimes used for compounds containing the isopropyl unit, as illustrated for isopropyl alcohol and isobutyl alcohol.

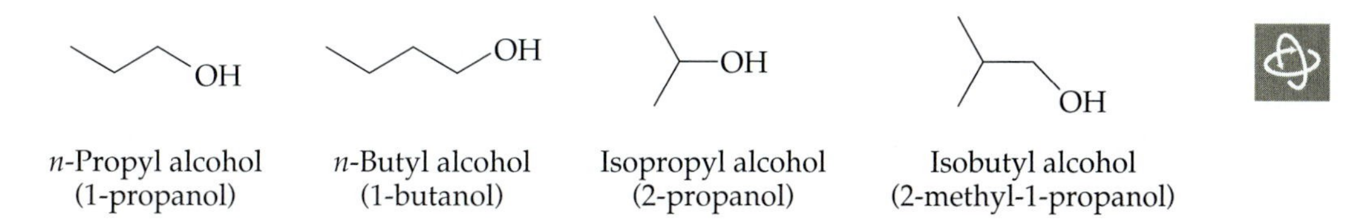

n-Propyl alcohol (1-propanol)

n-Butyl alcohol (1-butanol)

Isopropyl alcohol (2-propanol)

Isobutyl alcohol (2-methyl-1-propanol)

PROBLEM 4.3

Draw structural formulas for the following C_5 alcohols: (a) *n*-pentyl alcohol, (b) *sec*-pentyl alcohol, (c) *t*-pentyl alcohol, and (d) isopentyl alcohol.

4.2

Properties of Alcohols

You can think of alcohols as being hydroxyl derivatives of alkanes, but you can also view them as derivatives of water, in which one of the hydrogens of the water molecule has been replaced by an alkyl group. As this might lead you to expect, alcohols and water share many similarities in physical and chemical properties. The molecules have similar shapes, with the H—O—H bond angle being about 105° and the R—O—H bond angle being about 109° (Figure 4.2).

Most common alcohols (those through C_{10}) are liquids at room temperature. Methanol boils at 65°C and ethanol at 78°C. However, large-molecular-weight alcohols can be solid at room temperature; cholesterol ($C_{27}H_{46}O$, mp 149°C) is one example (Figure 4.3).

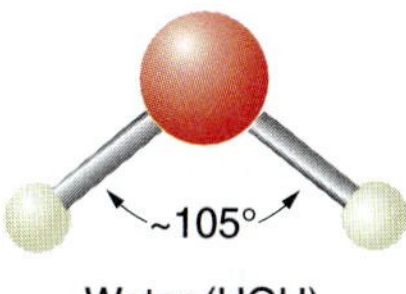

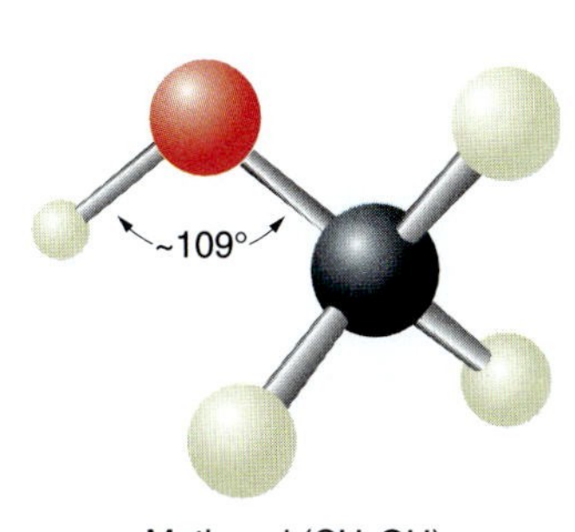

FIGURE 4.2

The similar shapes of water and methanol.

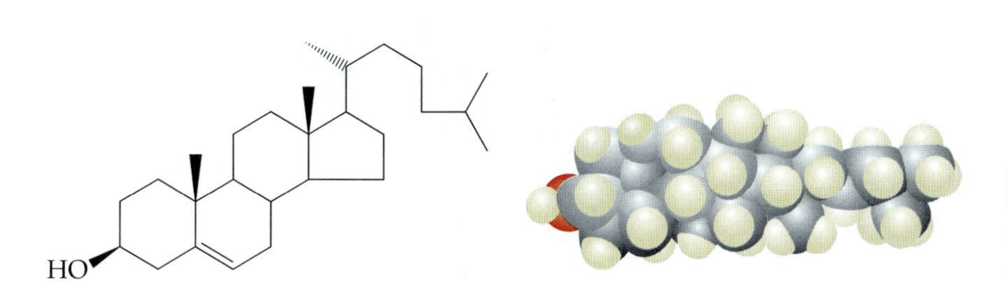

FIGURE 4.3

Cholesterol is a large-molecular-weight alcohol.

CHEMISTRY AT WORK

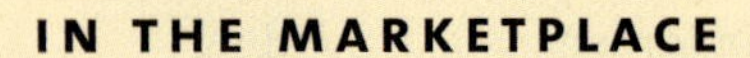

IN THE MARKETPLACE

Thiols: What Hair, Onions, and Skunks Have in Common

Sulfur is not a common element in organic compounds, but when it does appear in a compound, it can make a strong impression. The powerful odors of onions, garlic, and skunks are due to the presence of sulfur compounds, some of which are known as *thiols.*

Thiols are sulfur analogs of alcohols, compounds containing the —SH functional group. (The letters *thia* or *thio* in any name imply the presence of sulfur.) Thiols are named in the same way as alcohols, but with the suffix *-thiol* instead of *-ol* and with retention of the last *e* of the alkane parent name. Some examples are ethanethiol (CH_3CH_2SH), 2-propanethiol [$CH_3CH(SH)CH_3$], and cyclohexanethiol ($C_6H_{11}SH$); the generic name is *alkanethiol.*

Thiols undergo many of the same reactions as alcohols, but they also have some unique chemistry. Thiol groups are important in maintaining the three-dimensional structure of many proteins, including the hormone *insulin* and the protein *keratin* in your hair. Keratin exists as multiple long, parallel chains of connected α-amino acids, including cysteine, which has a thiol group (see Chapter 16):

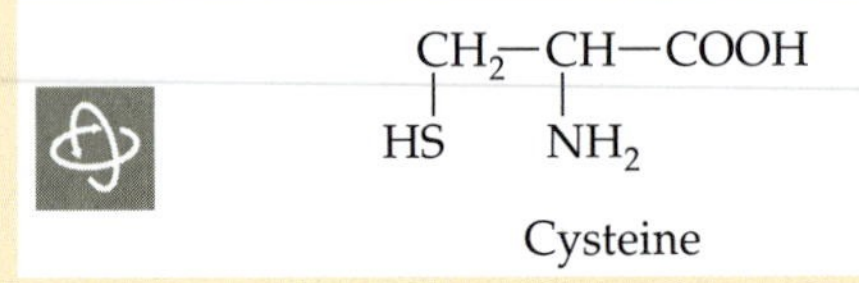

Cysteine

The chemistry involved in hair styling consists of oxidation and reduction reactions affecting the thiol groups. In normal hair, the keratin chains are joined together by *disulfide linkages* (—S—S—) between two cysteine units, formed by removal of two hydrogens. These cross-links hold the hair in a specific shape. The disulfides can be reduced easily, resulting in the formation of thiol groups and the unlinking of the keratin chains.

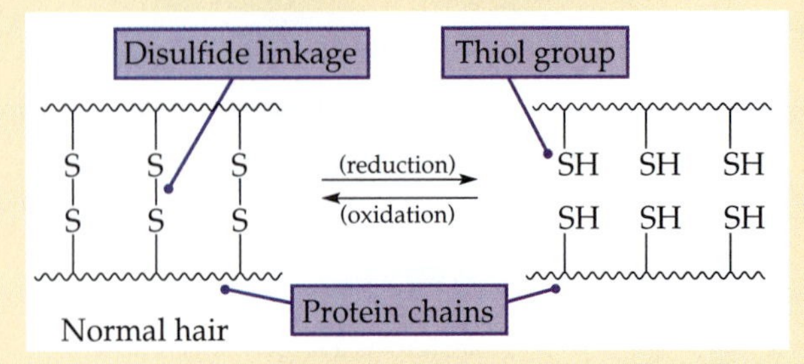

If the hair is then reshaped, with rollers, for example, and then gently oxidized, as with a mild solution of hydrogen peroxide, the thiols are oxidized to new disulfide linkages. These hold the keratin chains in place again, giving the hair a new shape. This is an example of how a chemical reaction can affect the structure and function of larger molecules and even parts of the body.

Sulfur confers a very strong odor in certain organic compounds, not unlike the odors of the inorganic compounds hydrogen sulfide (H_2S) and sulfur dioxide (SO_2). Organosulfur compounds (thiols and thioethers) are responsible for the pungent odors of onions and garlic. Several thiols give the skunk its signature odor. This property of thiols is used in a positive way to make natural gas and LP gas safe for household use. The hydrocarbons (such as methane and propane) in these gases are odorless and extremely flammable. Because the presence of these hydrocarbons cannot be detected by the human nose, suppliers add trace amounts of low-molecular-weight thiols to their gas products. Leaks can then be detected by smell before there is a concentration large enough to ignite and cause a fire or explosion.

Water is a nucleophile (a Lewis base) and is subject to protonation by strong acids to form the hydronium ion (H_3O^+). In a similar rapidly reversible process, an alcohol can be protonated by a strong acid to form an **oxonium ion** (ROH_2^+) in solution:

$$\underset{\text{Alcohol}}{R{-}\ddot{\underset{..}{O}}{-}H} \quad \xrightleftharpoons{H^+} \quad \underset{\text{Oxonium ion}}{R{-}\overset{\overset{\displaystyle H}{|}}{\underset{..}{O^+}}{-}H}$$

The chemistry of alcohols often involves replacing the —OH group with other groups. However, the —OH group is among the poorest of leaving groups and must first be converted into a good leaving group in order to be replaced. This is often accomplished by protonation of the hydroxyl group to form an oxonium ion. Water is an excellent leaving group. It can break away spontaneously from a tertiary carbon, leaving behind a *carbocation* that reacts further (in an E1 or S_N1 reaction). Alternatively, water can be displaced from a primary or secondary carbon by attack of a nucleophile (in an S_N2 reaction). Examples of these kinds of reactions will be discussed in Section 4.3.

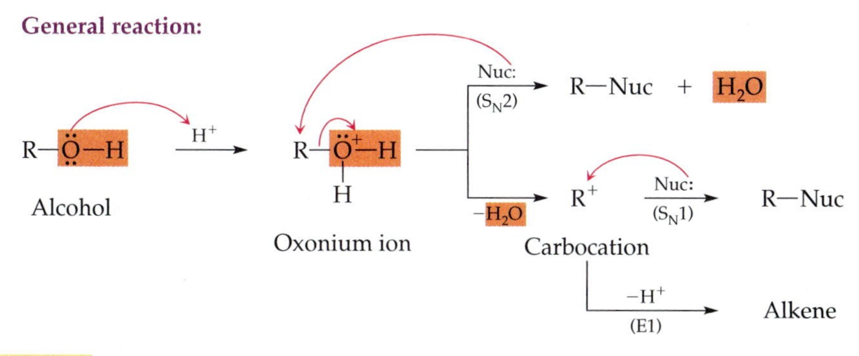

4.2.1 Hydrogen Bonding

Water boils at a much higher temperature (100°C) than expected, compared to the similar but higher-molecular-weight compounds H_2S (bp −61°C) and H_2Se (bp −43°C). Likewise, 1-butanol (bp 118°C) boils much higher than alkanes with nearly identical molecular weights, such as pentane (bp 36°C). The reason behind these observations is the same: the presence of the —OH group in water and in alcohols and its polarity due to the high electronegativity of oxygen. The polarization of a hydroxyl group is $C{—}O^{\delta-}{—}H^{\delta+}$. The result of this polarization is a **hydrogen bond**—a weak electrostatic bond to another molecule (a dipole-dipole attractive force of about 5 kcal/mol). Hydrogen bonding occurs between a site of low electron density (the hydrogen) and a site of high electron density (the oxygen). It can occur between two separate molecules (*inter*molecular hydrogen bonding) or within a single large molecule containing the appropriate functional groups (*intra*molecular hydrogen bonding). A hydrogen bond is often represented by a dotted line, as shown in Figure 4.4.

Water in the liquid form exists as associated complexes of several molecules held together by hydrogen bonds (see Figure 4.4). These hydrogen bonds must be broken before vaporization can occur. The same holds true for alcohols, in which an alkyl group has replaced one hydrogen of water. Low-molecular-weight alcohols (up to and including C_5 alcohols) readily dissolve in water because the alcohol molecule can hydrogen-bond with water, in effect replacing one water molecule in an associated complex. When the alkyl group of an alcohol is too large (larger than C_5), the **hydrophilic** ("water-seeking") character of the hydroxyl group is outweighed by the **hydrophobic** ("water-repelling") character of the larger alkyl (hydrocarbon) group, and the alcohol is insoluble in water.

FIGURE 4.4

Hydrogen bonding of water, an alcohol, and ethanol in water. (The dotted lines represent the hydrogen bonds.)

Compounds with more than one hydroxyl group, such as glycerol (1,2,3-propanetriol), ethylene glycol (1,2-ethanediol), and the common sugars like sucrose ($C_{12}H_{22}O_{11}$, containing eight —OH groups), are infinitely soluble in water because each hydroxyl group can form a hydrogen bond.

Hydrogen bonding is a very important phenomenon that occurs with other families of compounds. Most important, hydrogen bonding is crucial in maintaining the structures and shapes of proteins (see Section 16.3) and DNA (see Section 19.1).

PROBLEM 4.4

Place each set of compounds in order of increasing boiling point:

(a) ethyl alcohol, ethane, ethyl chloride

(b) cyclohexane, cyclohexanol, cyclohexane-1,2-diol

4.2.2 Acidity of Alcohols

Given the electronegativity difference between hydrogen and oxygen and the ability of oxygen to carry a negative charge (that is, to be anionic, as in the hydroxide ion), alcohols are susceptible to removal of the proton from oxygen. In other words, they are weak acids. Alcohols are generally weaker acids than water (pK_a 15.7), ranging from pK_a = 15.9 for ethanol to pK_a = 18 for *t*-butyl alcohol. Thus, *a base stronger than hydroxide is required to remove a proton from an alcohol.*

$$\text{R—O—H} + \text{B:}^- \rightleftharpoons \text{R—O}^- + \text{BH}$$

Alcohol, Base, Alkoxide anion, Conjugate acid of base

$$C_2H_5\text{—OH} + \text{NaOH} \rightleftharpoons C_2H_5\text{—O}^-\ \text{Na}^+ + \text{HOH}$$

Ethanol (pK_a 15.9), Sodium ethoxide, Water (pK_a 15.7)

The relative acidities of alcohols are affected by the substituents attached to the carbon carrying the hydroxyl group, often by an electronic effect called the **inductive effect.** The inductive effect is the relative electron-donating or electron-withdrawing effect exerted by a substituent as the result of electronegativity differences. An example is the lower acidity of *t*-butyl alcohol (pK_a 18) compared to methyl alcohol (pK_a 15.5), due in part to the electron-donating effects of the three methyl groups in *t*-butyl alcohol, which increase the relative electron density on oxygen. In contrast, 2,2,2-trifluoroethanol (pK_a 12.4) is more acidic than methyl alcohol because of the strong electron-withdrawing effects of the three fluorine atoms in the trifluoromethyl group, delocalizing the negative charge on oxygen.

Methyl alcohol (pK_a 15.5)

$$H{-}\underset{H}{\overset{H}{C}}{-}OH \rightleftharpoons H{-}\underset{H}{\overset{H}{C}}{-}O^- + H^+$$

t-Butyl alcohol (pK_a 18)

$$CH_3 \dashrightarrow \underset{CH_3}{\overset{CH_3}{C}}{-}OH \rightleftharpoons CH_3 \dashrightarrow \underset{CH_3}{\overset{CH_3}{C}}{-}O^- + H^+$$

Electron-donating

2,2,2-Trifluoroethanol (pK_a 12.4)

$$CF_3 \dashleftarrow \underset{H}{\overset{H}{C}}{-}OH \rightleftharpoons CF_3 \dashleftarrow \underset{H}{\overset{H}{C}}{-}O^- + H^+$$

Electron-withdrawing

In other words, any structural effect that disperses the negative charge on oxygen stabilizes the anion, making it form more readily. Any structural effect that concentrates charge on oxygen destabilizes the anion, making it more difficult to form.

The relative acidity of any two alcohols can be predicted by considering the relative inductive effects of the substituents on their alkoxide anions, and these inductive effects can be predicted by comparing relative electronegativities.

PROBLEM 4.5

Place each of the following pairs of compounds in order of increasing acidity:

(a) cyclohexanol and 2,2,6,6-tetrachlorocyclohexanol

(b) methanol and isopropyl alcohol

4.2.3 Alkoxides

When a proton is removed from the hydroxyl group of an alcohol (alkanol), the resulting anion is called an **alkoxide anion,** often referred to as an *alkoxide.* Thus, *meth*anol produces *meth*oxide anion, *eth*anol produces *eth*oxide anion, and *t-but*yl alcohol produces *t-but*oxide anion. Alkoxides are widely used in organic chemistry as bases and nucleophiles that are stronger than hydroxide and are soluble in organic solvents. The three alkoxides just mentioned are the most common. The relative

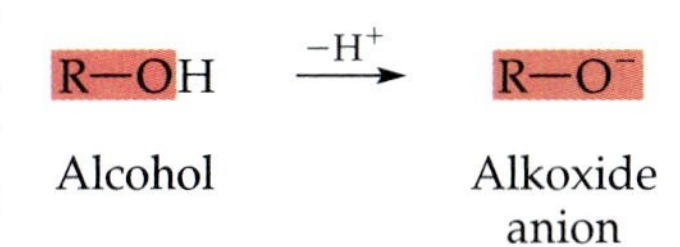

strength of the three bases is the reverse of the acidity of their corresponding alcohols (recall Section 1.3.1):

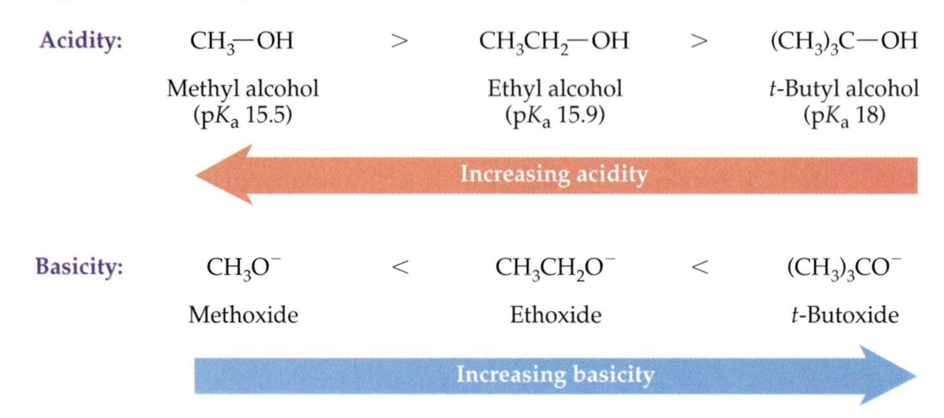

An alcohol can be converted to an alkoxide by treatment with a stronger base, one whose conjugate acid has a pK_a of about 20 or larger. We will encounter such bases later in this book (some examples are sodium amide and various organometallics). However, these reactions are seldom used in the lab because there is a much easier method for preparing alkoxides. This method consists of carefully reacting methanol with metallic sodium to form sodium methoxide ($NaOCH_3$ or NaOMe), ethanol with metallic sodium to form sodium ethoxide ($NaOC_2H_5$ or NaOEt), or *t*-butyl alcohol with metallic potassium to form potassium *t*-butoxide ($KOBu^t$ or *t*-BuOK). In these oxidation-reduction reactions, the metal is oxidized to its cation and the alcohol is reduced in a one-electron transfer. The reaction is analogous to that which occurs when sodium is added to water (sodium hydroxide and hydrogen result, with the hydrogen occasionally catching fire); in the case of alcohols, however, the reaction is somewhat more subdued, though still exothermic. The alkoxide forms in the liquid alcohol and typically is used as a reagent dissolved in that alcohol as a solvent.

$$\underset{\text{Ethanol}}{CH_3CH_2{-}OH} + \underset{\text{Sodium metal}}{Na} \longrightarrow \underset{\text{Sodium ethoxide}}{CH_3CH_2O^-\ Na^+} + \tfrac{1}{2}H_2$$

$$\underset{t\text{-Butyl alcohol}}{(CH_3)_3C{-}OH} + \underset{\text{Potassium metal}}{K} \longrightarrow \underset{\text{Potassium } t\text{-butoxide}}{(CH_3)_3CO^-\ K^+} + \tfrac{1}{2}H_2$$

How to Solve a Problem

Sodium methoxide is treated with acetic acid (pK_a 4.8). Predict the direction in which the equilibrium of this reaction will lie.

SOLUTION

First, we write out the equilibrium reaction, together with the pK_a of the acid (acetic acid) and of the conjugate acid (methanol):

$$\underset{\substack{\text{Sodium}\\\text{methoxide}}}{CH_3O^- \ Na^+} + \underset{\substack{\text{Acetic acid}\\(pK_a\ 4.8)}}{CH_3COOH} \rightleftharpoons \underset{\substack{\text{Methanol}\\(pK_a\ 15.5)}}{CH_3OH} + \underset{\substack{\text{Sodium}\\\text{acetate}}}{CH_3COO^- \ Na^+}$$

From the pK_a values, it is clear that acetic acid is the stronger of the two acids and that sodium methoxide is the stronger of the two bases. Therefore, the equilibrium will lie far to the right, with the methoxide anion removing the proton from acetic acid.

PROBLEM 4.6

Write the equation for the reaction of each of the following alcohols with metallic potassium: (a) isopropyl alcohol, (b) 1-propanol, and (c) *t*-pentyl alcohol.

PROBLEM 4.7

Place the following alcohols in order of increasing acidity: ethanol, 2-propanol, 2-methyl-2-propanol.

PROBLEM 4.8

Place the alkoxides of the alcohols in Problem 4.7 in order of increasing basicity.

PROBLEM 4.9

Show the reaction and the product you would expect when a solution of sodium ethoxide is treated with methyl bromide. (*Hint:* Recall the S_N2 reaction mechanism and identify a substrate and a nucleophile.)

4.3

Conversion of Alcohols to Alkyl Halides

Chapter 3 discussed the conversion of alkyl halides to alcohols using an S_N2 substitution reaction of hydroxide ion with an alkyl bromide:

$$\underset{\substack{\text{Alkyl}\\\text{bromide}}}{R{-}Br} + \underset{\substack{\text{Sodium}\\\text{hydroxide}}}{NaOH} \xrightarrow{(S_N2)} \underset{\text{Alcohol}}{R{-}OH} + NaBr$$

This reaction is successful for several reasons, but especially because the bromide ion is a good leaving group and the hydroxide ion is a strong nucleophile. It seems logical that a substitution reaction could be employed to convert an alcohol back to an alkyl halide if these same criteria could be met.

$$\underset{\text{Alcohol}}{R{-}OH} + \ ? \longrightarrow \underset{\text{Alkyl bromide}}{R{-}Br}$$

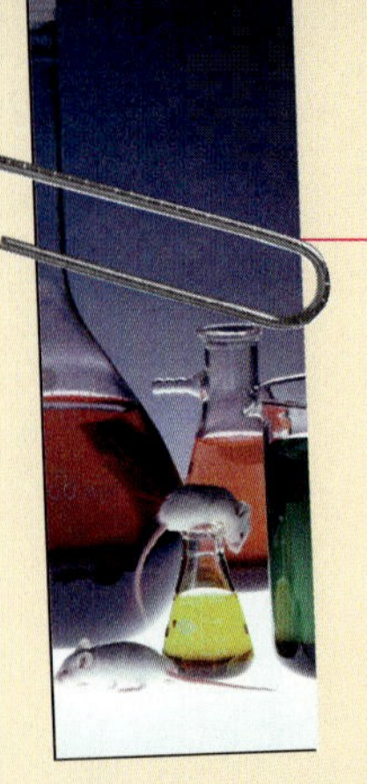

CHEMISTRY AT WORK

IN THE MARKETPLACE

Which Alcohol Is "Alcohol"?

The word *alcohol* is widely used for a class of beverages, once referred to as "spirits." As you now know, *alcohol* is the scientific term for an entire family of organic compounds. What most people really mean when they refer to "alcohol" is the organic compound ethanol (CH_3CH_2OH), also known as ethyl alcohol or grain alcohol.

About 2 billion pounds of ethanol are produced commercially every year. Much of it is used in industrial processes for producing other chemicals, such as acetaldehyde, acetic acid, and ethyl ether. Ethanol is also used as a solvent in chemical reactions, as an ingredient in lotions and perfumes, and as an additive to raise the octane rating in a type of fuel known as *gasohol*. Ethanol for these purposes is produced by the acid-catalyzed hydration of readily available ethylene (a reaction to be described in Section 6.4):

$$\underset{\text{Ethylene}}{CH_2{=}CH_2} + H_2O \xrightarrow{H^+} \underset{\text{Ethanol}}{CH_3CH_2OH}$$

The wort from which beer is made consists of mashed malted grain and hot water. As it ferments, carbon dioxide is formed, creating bubbles.

The ethanol used in alcoholic beverages is produced by fermentation of agricultural products, such as grapes (for wine), barley (for beer), wheat or rye (for whiskey), corn (for bourbon), potatoes (for vodka), or rice (for saki). The raw material is first hydrolyzed to the simple C_6 sugar glucose. The fermentation is carried out by enzymes in yeast, which first convert a glucose molecule into two molecules of pyruvate. (This aerobic metabolic process is identical to the human metabolism of carbohydrates; see Section 19.2.) The enzymes then break down the pyruvate via anaerobic processes to form acetaldehyde and carbon dioxide (the gas that causes bubbles in beer and some wines). Finally, they convert the acetaldehyde to ethanol.

$$\underset{\text{Glucose}}{C_6H_{12}O_6} \xrightarrow{\text{yeast}} \underset{\text{Pyruvate anion}}{\left[CH_3{-}\overset{\overset{\displaystyle O}{\|}}{C}{-}\overset{\overset{\displaystyle O}{\|}}{C}{-}O^-\right]}$$

$$\xrightarrow{-CO_2} \underset{\text{Acetaldehyde}}{\left[CH_3{-}\overset{\overset{\displaystyle O}{\|}}{C}{-}H\right]} \xrightarrow{\text{(reduction)}} \underset{\text{Ethanol}}{CH_3CH_2OH}$$

In recent years, a great deal of industrial ethanol has been produced by large-scale fermentation plants using farm products such as corn and barley as raw materials. Blackstrap molasses left over from sugar-refining processes is also fermented to produce industrial-grade ethanol.

In the human body, ethanol acts as a food, as a drug, and as a poison. It is a food because the body is able to metabolize it through normal processes, commencing with oxidation (see Section 4.4), and ultimately convert it to carbon dioxide, water, and energy. As a drug, ethanol is a depressant (not a stimulant, as many think), significantly depressing brain functions. Long-term consumption of ethanol frequently leads to cirrhosis of the liver, which can be fatal. When consumed in large amounts over a short time, ethanol is lethal.

However, the —OH group almost never functions as a leaving group—it is too strong a base and is not very polarizable. The only viable approach, then, is to convert the —OH group into another group that is an effective leaving group, able to be displaced by bromide anion, which is a moderate nucleophile. There are two reactions, discussed in this section, that accomplish the desired transformation.

4.3.1 Reaction of Alcohols with Hydrogen Halides

Alcohols react with hydrogen halides (HCl, HBr, and HI) to form alkyl halides:

$$CH_3CH_2CH_2{-}OH + HBr \longrightarrow CH_3CH_2CH_2Br + H_2O$$

1-Propanol → 1-Bromopropane

$$C_5H_9{-}OH + HI \longrightarrow C_5H_9{-}I + H_2O$$

Cyclopentanol (Cyclopentyl alcohol) → Iodocyclopentane (Cyclopentyl iodide)

Primary and secondary alcohols require heating, but tertiary alcohols react readily at room temperature. Thus, this is the reactivity order for alcohols:

Reactivity of alcohols: $3° > 2° > 1°$

The reactivity order for the halogen acids is HI > HBr > HCl, for two reasons. First, this is the acidity order, which affects the initial protonation step. Second, iodide is the most nucleophilic anion because of its polarizability. However, HBr is usually the acid chosen because of its reactivity and ease of handling and because the resulting bromide has convenient reactivity for subsequent use. This general reaction applies to all alcohols.

General reaction:

$$ROH + HBr \longrightarrow RBr + H_2O$$

Alcohol + Hydrogen bromide → Alkyl bromide + Water

The mechanism of this transformation is readily understandable based on your knowledge of the S_N1 and S_N2 mechanisms discussed in Chapter 3. The first step of the mechanism for these reactions involves protonation of the hydroxyl group to form an oxonium ion. Water is a very good leaving group, able to be replaced by a halide anion in a substitution reaction. The actual mechanism for this replacement depends on the nature of the alcohol, as shown in Figure 4.5. When a stable carbocation can be formed, such as from a tertiary alcohol, the S_N1 mechanism operates. When the original alcohol is primary, the resulting carbocation would not be sufficiently stable to form spontaneously, so the halide anion initiates an S_N2 substitution. A secondary alcohol reacts mainly by the S_N2 mechanism, but some S_N1 substitution occurs concurrently.

The difference in how readily the three classes of water-soluble alcohols (1°, 2°, and 3°) form carbocations provides a basis for distinguishing them quickly via a qualitative test called the *Lucas test*, which is conducted in a test tube. The reagent is $ZnCl_2$ dissolved in hydrochloric acid. The $ZnCl_2$ is a Lewis acid, which complexes effectively

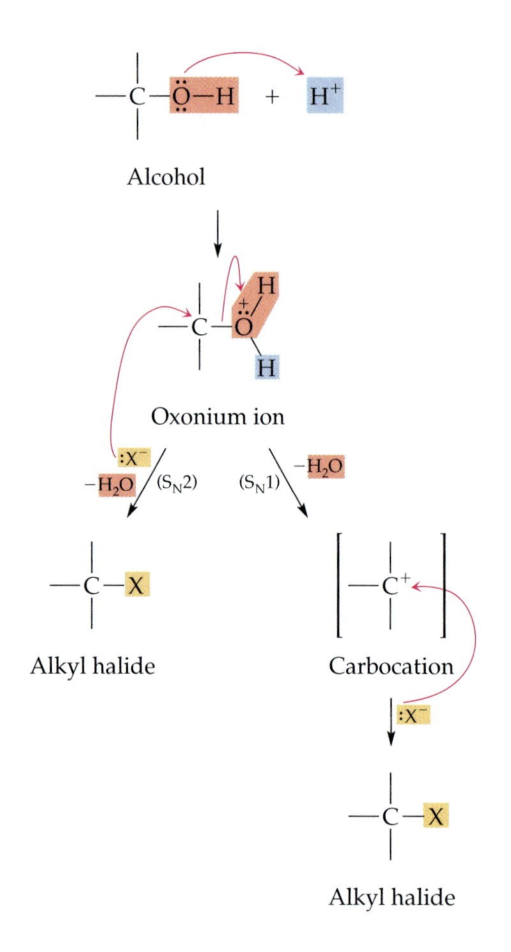

FIGURE 4.5

Mechanisms for conversion of alcohols to alkyl halides.

The alcohol is a nucleophile and so can be protonated by a strong halogen acid to form an oxonium ion.

Water is a good leaving group. It can leave spontaneously when the alcohol is tertiary, to form a stable carbocation (via an S_N1 mechanism). When a stable carbocation cannot be formed (the alcohol is primary or secondary), the oxonium ion undergoes an S_N2 substitution by halide ion.

The carbocation is trapped quickly by the nucleophilic halide anion.

with the unshared electrons of the alcohol to form an oxonium ion, compensating for the lower acidity of HCl (relative to HI or HBr) and the weak nucleophilicity of chloride anion. A tertiary alcohol reacts immediately because it rapidly forms a carbocation and then forms the alkyl chloride. The alkyl chloride is insoluble in the aqueous solution and is visible as cloudiness or as a separate layer. A secondary alcohol reacts after being heated for a few minutes to form an insoluble secondary alkyl halide. Primary alcohols do not react in a reasonable time, requiring extensive heating with the reagent because of the instability of the required carbocation.

Stereochemistry

Because an alcohol may be converted into an alkyl halide by the S_N1 and S_N2 mechanisms operating concurrently, this conversion is *nonstereospecific* if the —OH group is attached to a stereocenter. The reason for this lack of stereospecificity is that any carbocation that may be formed has a plane of symmetry, and in the final step of the S_N1 mechanism, both enantiomers (a racemic mixture) will form by reaction of the halide anion at either side of the symmetrical carbocation (see Section 3.4.3). The overall stereochemical process is called **racemization**—the conversion of an enantiomer to a racemate. Any enantiomeric alkyl halide formed by the S_N2 mechanism results from inversion of configuration (see Section 3.4.1). Therefore, reaction of any enantiomeric alcohol with a hydrogen halide may lead to a racemic alkyl halide, and an enantiomeric tertiary alcohol is certain to yield a racemic alkyl halide.

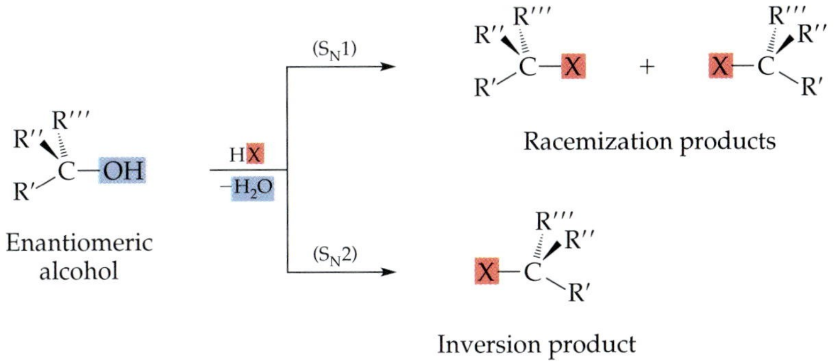

R'''
R''
C—OH
R'
Enantiomeric alcohol

HX
$-H_2O$

(S_N1) → R'''
R''
C—X
R'
+
R'''
R''
X—C
R'

Racemization products

(S_N2) → R'''
R''
X—C
R'

Inversion product

As a result of the lack of stereospecificity in the conversion of alcohols to alkyl halides using hydrogen halides, alternative reactions have been devised to permit stereospecific conversions. These alternatives will be described next.

How to Solve a Problem

Starting with only isopropyl alcohol (2-propanol) and any necessary inorganic reagents, produce isopropyl ether, $(CH_3)_2CHOCH(CH_3)_2$. (The nomenclature of ethers, which have the general structure R—O—R′, will be covered in Chapter 5.)

PROBLEM ANALYSIS

First, we always write the overall reaction:

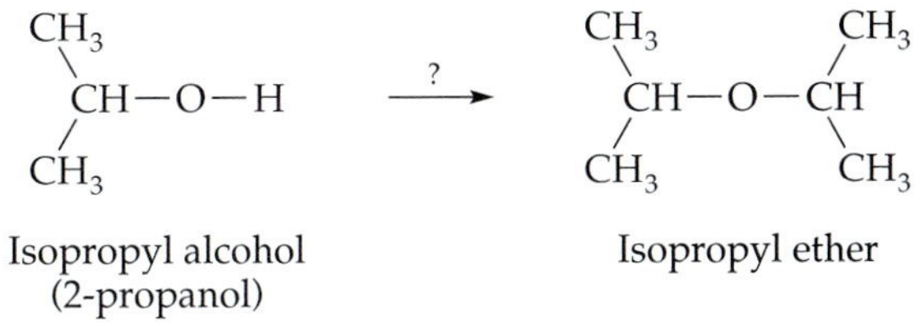

$$(CH_3)_2CH\text{—}O\text{—}H \xrightarrow{?} (CH_3)_2CH\text{—}O\text{—}CH(CH_3)_2$$

Isopropyl alcohol (2-propanol) — Isopropyl ether

Then we examine the product to locate where the structural elements of the starting material may be found. Finally, we use a dashed line to indicate which bond(s) must be formed. The product in this instance is structurally symmetrical, so the dashed line can be placed on either side of the oxygen:

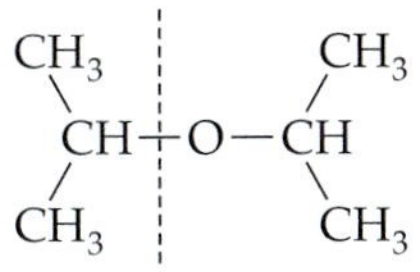

$$(CH_3)_2CH \,\vdots\, O\text{—}CH(CH_3)_2$$

Conversion of the starting alcohol to an isopropoxide ion will produce the right-hand fragment. We can use this alkoxide as a nucleophile to carry out an S_N2 reaction on the left-hand C_3 fragment. However, accomplishing such a reaction requires that the left-hand fragment originally carry a leaving group. Therefore, a molecule such as isopropyl bromide (2-bromopropane) must be available. Since the only organic starting material we are permitted to use is isopropyl alcohol, the challenge is to convert isopropyl alcohol to isopropyl bromide (that is, R—OH to R—Br).

SOLUTION

Reaction of isopropyl alcohol with hot HBr yields isopropyl bromide:

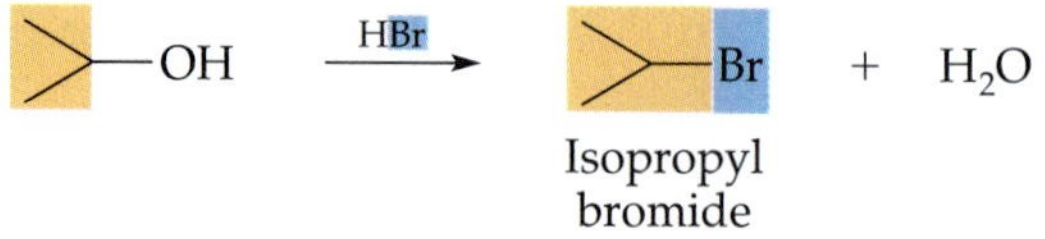

$$\text{(CH}_3)_2\text{CH–OH} \xrightarrow{\text{HBr}} \text{(CH}_3)_2\text{CH–Br} + H_2O$$

Isopropyl bromide

Reaction of isopropyl alcohol with metallic sodium produces sodium isopropoxide in solution:

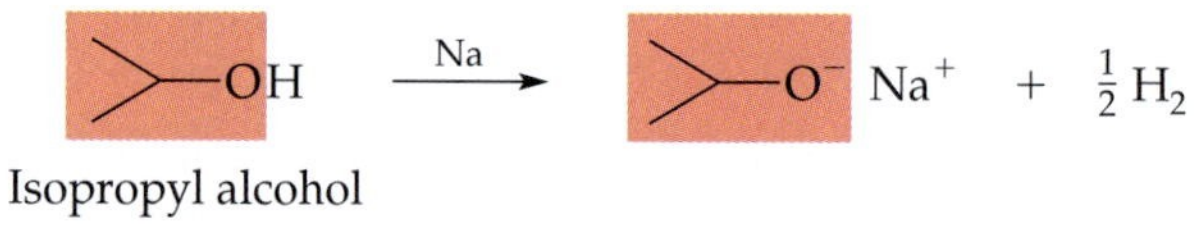

$$\text{(CH}_3)_2\text{CH–OH} \xrightarrow{\text{Na}} \text{(CH}_3)_2\text{CH–O}^- \text{Na}^+ + \tfrac{1}{2} H_2$$

Isopropyl alcohol

Addition of isopropyl bromide to that solution results in an S_N2 substitution reaction, producing the desired ether product (as you will learn in Chapter 5, this reaction is simply an example of the standard Williamson ether synthesis):

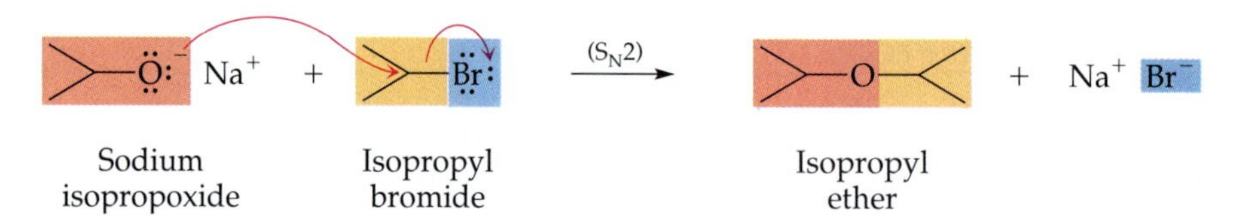

This synthesis is the first example we have considered of putting two reactions together in sequence to bring about a multistep organic synthesis. The analysis of synthesis problems is best accomplished using the approach outlined in this example.

PROBLEM 4.10

Why does 2-methyl-2-butanol react at the same rate with the three different acids HI, HCl, and HBr to form an alkyl halide?

PROBLEM 4.11

Why does 2-methyl-1-butanol react with the three acids HI, HCl, and HBr at different rates (HI > HBr > HCl)?

PROBLEM 4.12

Write a balanced equation for the reaction of each of the following alcohols with HBr: (a) cyclohexanol, (b) *t*-butyl alcohol, and (c) (*R*)-2-butanol.

PROBLEM 4.13

Write a balanced equation for the formation of each of the following alkyl halides from an alcohol:

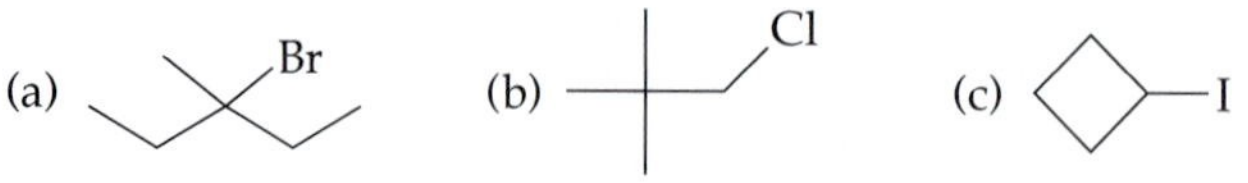

(a) Br (b) Cl (c) I

4.3.2 Stereospecific Conversion of Alcohols to Alkyl Halides

The key to achieving stereospecific conversion of an enantiomeric alcohol to an enantiomeric alkyl halide is to avoid the formation of a carbocation intermediate. A carbocation has lost the chirality of its precursor alcohol because it has a plane of symmetry. Two techniques have been developed to permit stereospecific alkyl halide formation from a starting enantiomeric alcohol. Of course, these reactions can also be used with compounds that are not chiral.

Inversion of Configuration

Primary and secondary alcohols react with the reagent thionyl chloride ($SOCl_2$) in the presence of pyridine to form alkyl chlorides in a stereospecific reaction. Pyridine is a base used to trap the HCl by-product (as pyridinium hydrochloride) and ensure the presence of chloride anion. The reaction involves the initial formation of a sulfite ester, which serves as an effective leaving group in an S_N2 reaction. The by-product sulfur dioxide (SO_2) is a gas. For example, cyclohexanol is converted to cyclohexyl chloride (chlorocyclohexane) in excellent yield:

$$\text{C}_6\text{H}_{11}\text{—OH} + \text{SOCl}_2 \xrightarrow{\text{pyridine}} \text{C}_6\text{H}_{11}\text{—Cl} + \text{SO}_2(g) + \text{pyridine}\cdot\text{HCl}$$

Cyclohexanol · Thionyl chloride · Cyclohexyl chloride

This reaction is effective for three key reasons:

1. The —OH group (a poor leaving group) is converted to a sulfite ester (a good leaving group).
2. The C—O bond is not broken, preventing racemization at the stereocenter.
3. The stereospecific S_N2 substitution step produces inversion.

Therefore, the reaction is *stereospecific* and occurs with *inversion* of configuration at a stereocenter.

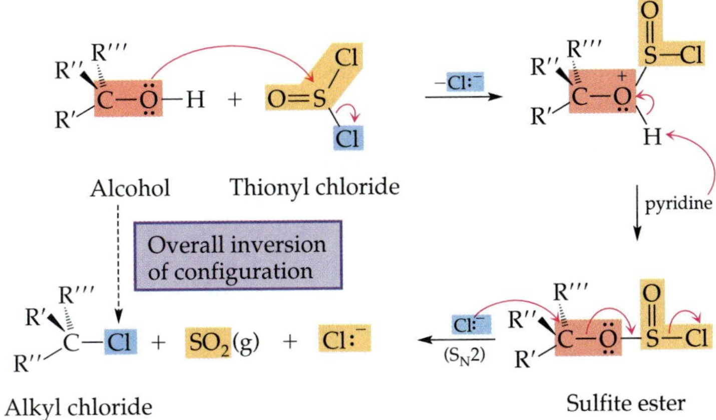

A similar stereospecific conversion of alcohols to alkyl bromides using the reagent phosphorus tribromide (PBr_3) follows an analogous mechanism:

$$\text{R—OH} \xrightarrow{\text{PBr}_3} \text{R—Br}$$

Alcohol · Alkyl halide (with inversion of configuration)

Retention of Configuration

There are only two ways to convert one compound into another with retention of configuration: use a single reaction whose mechanism forces retention, or use a sequence of two reactions that both occur with inversion of configuration. Through an ingenious use of a sequence of three reactions (one with retention and two with inversion), it is possible to create a reaction sequence that converts an enantiomeric alcohol into an enantiomeric alkyl halide with overall retention of configuration. Two of the reactions needed have already been discussed.

The first step is reaction of the enantiomeric alcohol with a reagent called *p*-toluenesulfonyl chloride (p-$CH_3C_6H_4SO_2Cl$), usually shortened to *tosyl chloride* and abbreviated as TsCl. The alcohol serves as a nucleophile to displace chloride from the reactive TsCl in a manner similar to the reaction with $SOCl_2$. The C—O bond of the alcohol does not break, allowing the stereocenter to retain its configuration in the product tosyl ester (tosylate).

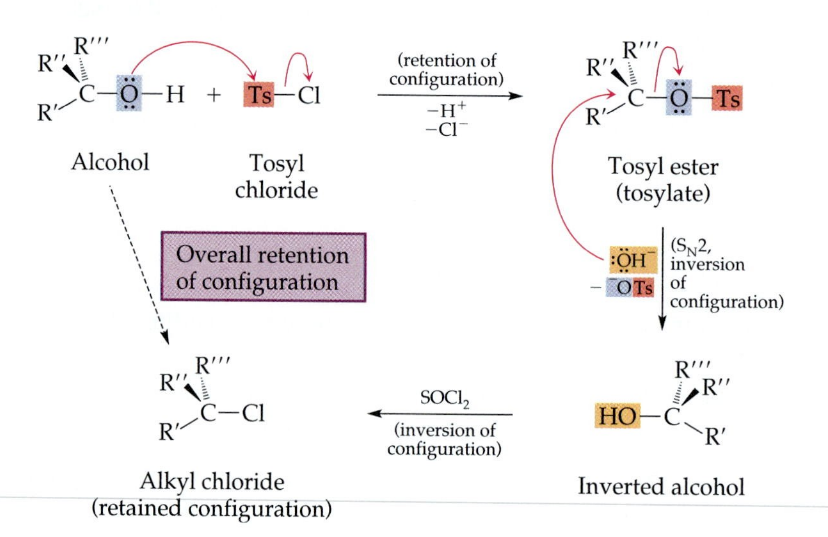

Next, the tosyl ester is subjected to an S_N2 substitution reaction with hydroxide (a strong nucleophile), in which the tosylate group serves as a good leaving group. The result is inversion of configuration of the stereocenter, yielding the inverted alcohol. In the final step, reaction with thionyl chloride produces the usual inversion of configuration. The overall result of the three steps (retention, inversion, and inversion) is an alkyl chloride returned to the original configuration of the alcohol.

In summary, alcohols can be converted to alkyl chlorides using thionyl chloride and to alkyl bromides using phosphorus tribromide. Both reactions are stereospecific and occur with inversion of configuration. Nonstereospecific conversion of alcohols to alkyl halides using hydrogen halides is also effective when stereochemistry is not an issue.

How to Solve a Problem

Show the reagents and reactions necessary to convert *cis*-2-methylcyclobutanol to *trans*-1-bromo-2-methylcyclobutane.

PROBLEM ANALYSIS

First we draw the structures of the reactant and product to clearly envision the required change:

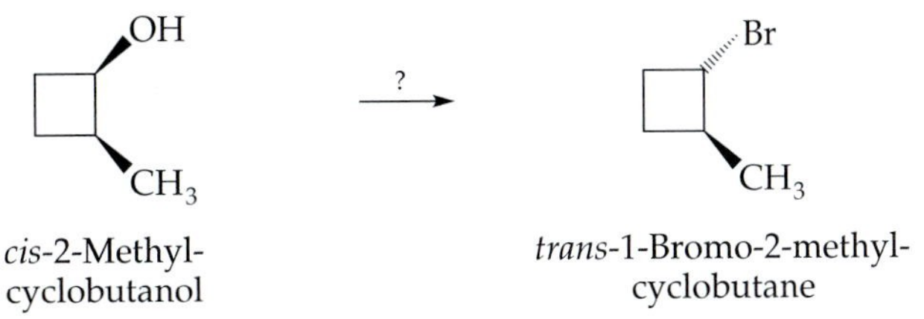

cis-2-Methyl-cyclobutanol → *trans*-1-Bromo-2-methyl-cyclobutane

What is required is replacement of the hydroxyl group with a bromine and inversion of configuration at that carbon. Accordingly, a stereospecific conversion is required.

There are two possible reagents for conversion of a hydroxyl group to a bromine substituent. Hydrogen bromide (HBr) is effective; it works by forming an oxonium ion and then replacing it with a bromide ion. The replacement step can be S_N2 or S_N1 or a mixture of the two mechanisms occurring concurrently. To the extent that the mechanism is S_N1, the reaction is nonstereospecific because a carbocation is involved. Because the stereochemistry of this reaction cannot be controlled, this route is not appropriate for this problem.

The second possible reagent is phosphorous tribromide (PBr_3), which converts alcohols to alkyl bromides via a stereospecific inversion process, analogous to that shown earlier for $SOCl_2$.

SOLUTION

Reaction of *cis*-2-methylcyclobutanol with PBr_3 produces, stereospecifically, *trans*-1-bromo-2-methylcyclobutane. The intermediate phosphorus ester group, oriented above the ring, is displaced in an S_N2 reaction by the bromide attacking from below the ring. Therefore, the new carbon-bromine bond is oriented below the ring, resulting in the *trans* isomer.

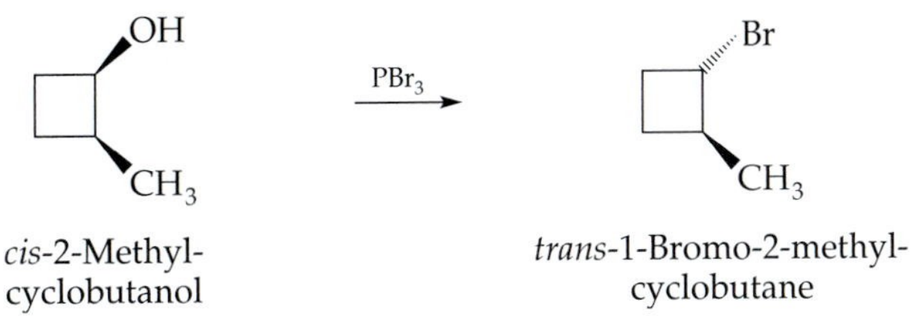

cis-2-Methyl-cyclobutanol —PBr_3→ *trans*-1-Bromo-2-methyl-cyclobutane

PROBLEM 4.14

Show the products of each of the following reactions:

(a) *trans*-2-methylcyclohexanol with $SOCl_2$

(b) cyclopentanol with HBr

(c) (*R*)-2-butanol with PBr_3

PROBLEM 4.15

Show how to prepare (*R*)-2-chlorobutane from each of the following:

(a) (*R*)-2-butanol

(b) (*S*)-2-butanol

4.4

Oxidation of Alcohols

Oxidation reactions are important and prevalent in organic chemistry. A number of different reagents have been developed to permit *selective oxidation*—that is, oxidation of one functional group without affecting another susceptible functional group. It is important to gradually develop a feel for the different oxidizing reagents and their strengths. As you will see, most of these reagents are inorganic compounds. Remember that when any compound (often generically called a *substrate*) is oxidized, the reagent is reduced (and vice versa). The detailed mechanisms of oxidation reactions will not be discussed in this book.

When an organic compound is oxidized, one of two things happens: There is an increase in the oxygen content of the substrate, *or* there is a decrease in the hydrogen content of the substrate. Families of organic compounds can be placed in a hierarchy of oxidation states, as reflected in the Key to Transformations (inside front cover) and specifically illustrated in Table 4.1. These relationships are important. You can see that the effective laboratory oxidations are those from alcohol (alkanol) to aldehyde (alkanal) to carboxylic acid (alkanoic acid). Note also that only C—H bonds are broken in these oxidations, not C—C bonds. No carbon atoms are removed.

We will discuss three oxidation reagents here:

- A solution of **potassium permanganate ($KMnO_4$).** Deep purple in color, such a solution is a strong oxidant. In the course of reaction, the purple Mn(VII) is reduced to Mn(IV), which precipitates as brown manganese dioxide (MnO_2).
- **Chromic acid (H_2CrO_4).** A strong oxidant usually used with alcohols, chromic acid can be produced in solution by two methods: (1) from sodium dichromate ($Na_2Cr_2O_7$) and sulfuric acid, or (2) by dissolving chromic anhydride (CrO_3) in concentrated sulfuric acid and water (this version is called **Jones' reagent**). During an oxidation reaction, the orange-colored Cr(VI) in this reagent forms greenish-blue Cr(III), which remains in solution.

TABLE 4.1

Oxidation-State Hierarchy of Some Organic Compounds

	Family Name (IUPAC)	Structure	Ease of Oxidation
Highest oxidation state of carbon	Carbon dioxide	O=C=O	
↑	Carboxylic acid (Alkanoic acid)	R—C(=O)—OH	If R = alkyl, only by combustion If R = H, very easy
	Aldehyde (Alkanal)	R—C(=O)—H	Very easy
	Alcohol (Alkanol)	$R—CH_2—OH$	Readily
Lowest oxidation state of carbon	Alkane	$R—CH_3$	Extremely difficult

- **Pyridinium chlorochromate ($C_5H_6NCrO_3Cl$, usually abbreviated PCC).** A mild oxidizing reagent, PCC is a soluble complex of chromic anhydride (CrO_3) and pyridine in dilute HCl.

Chemists normally write an organic oxidation reaction showing the reagent used and the organic product(s), but not the inorganic products. Occasionally, they indicate an oxidation process by writing [O] over the reaction arrow, without specifying the actual reagent.

In a primary alcohol, two hydrogens can be removed from the carbon being oxidized. The alcohol is oxidized by $KMnO_4$ or H_2CrO_4, initially to an aldehyde. However, it is very difficult to obtain good yields because the aldehyde is more easily oxidized than the alcohol (see Table 4.1) and so is oxidized rapidly to the carboxylic acid. Therefore, the normal product of oxidation of a primary alcohol with a strong oxidant is a carboxylic acid:

General reaction:

$$R{-}CH_2{-}OH \xrightarrow[H_2SO_4]{Na_2Cr_2O_7} \left[R{-}\overset{O}{\overset{\|}{C}}{-}H\right] \longrightarrow R{-}\overset{O}{\overset{\|}{C}}{-}OH$$

1° Alcohol — Aldehyde — Carboxylic acid

For example, oxidation of isobutyl alcohol (2-methyl-1-propanol) yields isobutyric acid (2-methylpropanoic acid):

$$(CH_3)_2CHCH_2OH \xrightarrow[H_2SO_4]{Na_2Cr_2O_7} (CH_3)_2CH{-}COOH$$

Isobutyl alcohol (2-methyl-1-propanol) — Isobutyric acid (2-methylpropanoic acid)

An aldehyde (RCHO) can be obtained from a primary alcohol (RCH_2OH) if PCC (pyridinium chlorochromate) is used. PCC is a much milder oxidizing agent and so does not oxidize the aldehyde to the carboxylic acid. For example, isobutyl alcohol can be converted to isobutyraldehyde (2-methylpropanal) using this reagent:

$$(CH_3)_2CHCH_2OH \xrightarrow{PCC} (CH_3)_2CH{-}CHO$$

Isobutyl alcohol (2-methyl-1-propanol) — Isobutyraldehyde (2-methylpropanal)

Secondary alcohols, which have only one hydrogen bonded to the carbon carrying the hydroxyl group, are oxidized by chromic acid or permanganate to ketones. The oxidation can proceed no further because the carbon double-bonded to the oxygen has no more hydrogens.

General reaction:

$$R'R''C(H)OH \xrightarrow[H_2SO_4]{Na_2Cr_2O_7} R'R''C{=}O$$

2° Alcohol — Ketone

For example, cyclopentanol produces cyclopentanone:

$$\text{Cyclopentanol (cyclopentyl–OH)} \xrightarrow[H_2SO_4]{Na_2Cr_2O_7} \text{Cyclopentanone (cyclopentyl=O)}$$

Cyclopentanol Cyclopentanone

Note that the overall change produced in the oxidation of primary and secondary alcohols is removal of a hydrogen from the hydroxyl group and from the carbon to which it is attached:

$$-\underset{H}{\overset{|}{C}}-O-H \longrightarrow \;\rangle C=O$$

1° or 2° Alcohol

$$R''-\underset{R'''}{\overset{R'}{C}}-OH \xrightarrow[H_2SO_4]{Na_2Cr_2O_7} \text{No reaction}$$

3° Alcohol

On this basis, you can see why tertiary alcohols cannot be oxidized, even with strong reagents—there is no hydrogen on the hydroxyl carbon.

Jones' reagent can be used in a qualitative test with alcohols of unknown structure to distinguish primary and secondary alcohols, which will undergo oxidation, from tertiary alcohols, which will not. A positive test result is the change in color from orange Cr(VI) to greenish-blue Cr(III), indicating that the reagent has been reduced and the substrate has been oxidized.

The oxidation of alcohols is a very important reaction in biological chemistry, as is covered in some detail in Section 19.2. Oxidation reactions metabolize food and also provide much of the body's energy supply. Although the particular reagents may differ from those we have considered, the end result is the same—primary alcohols are oxidized to aldehydes and carboxylic acids, and secondary alcohols are oxidized to ketones. For example, during strenuous exercise, body cells cannot get enough oxygen to complete metabolism, thereby creating an oxygen "debt." As a result, lactic acid builds up in the muscles—the source of muscle ache. When the body returns to rest and its oxygen supply is restored, the alcohol group in lactic acid is oxidized to the ketone group in pyruvic acid, which then reenters the normal metabolic cycle.

$$CH_3-\underset{OH}{\underset{|}{CH}}-COOH \xrightarrow{[O]} CH_3-\underset{O}{\underset{\|}{C}}-COOH$$

Lactic acid Pyruvic acid

PROBLEM 4.16

Write the following reactions and their products:

(a) cyclohexanol with potassium permanganate

(b) 1-butanol with PCC

(c) 1-pentanol with potassium permanganate

CHEMISTRY AT WORK

IN THE BODY

Breathalyzer Tests and the Metabolism of Alcohol

Oxidation reactions are the means by which our bodies metabolize ingested substances for the production of energy. Carbohydrates and fats are our main sources of energy, and their oxidation produces the carbon dioxide we exhale. Enzymes in our bodies bring about the oxidation reactions under very mild conditions (at the normal body temperature of about 37°C, compared to the temperatures of over 100°C often required in the laboratory). Biological oxidations are described in greater detail in Chapter 19.

Ethanol, the alcohol in alcoholic beverages, is oxidized at a fixed rate by an enzyme in the liver, producing acetaldehyde. The acetaldehyde in turn is rapidly oxidized by another enzyme to acetate anion (the anion from acetic acid).

Acetate is also the end product of the direct metabolism of fats and carbohydrates. It is either further metabolized through the Krebs cycle (see Section 19.2.1) to carbon dioxide (which is exhaled through the lungs) or used by the body to synthesize many needed substances, including fats, amino acids, and cholesterol.

If the concentration of ethanol in the blood is too high, some remains in circulation and results in intoxication. This unoxidized ethanol, which eventually returns to the liver for oxidation, is what a Breathalyzer test measures. The amount of alcohol in a person's breath can be determined because ethanol is readily oxidized by sodium dichromate. Exhaled air containing ethanol is forced through a tube of orange-colored sodium dichromate ($Na_2Cr_2O_7$). Oxidation of the alcohol and reduction of the dichromate occur. When dichromate is reduced, orange Cr(VI) is converted to greenish-blue Cr(III). The extent to which the color in the tube changes from orange to greenish-blue indicates the amount of ethanol in the exhaled air. Because the ethanol in the bloodstream is in rapid equilibrium with the air deep in the lungs, this simple test reflects the amount of ethanol in the bloodstream. Calibration of the device permits fairly accurate determination of the blood alcohol content.

Portable gas chromatographs are now being used by police to measure blood alcohol levels.

Although Breathalyzers are still manufactured and used, most police departments now use small gas chromatographs for determining blood alcohol content. These devices are relatively inexpensive and are much more accurate in determining the amount of alcohol present in breath.

Oxidation of ethanol by enzymes in the liver:

$$\underset{\text{Ethanol}}{CH_3{-}CH_2{-}OH} \xrightarrow{\text{enzyme 1}} \underset{\text{Acetaldehyde}}{CH_3{-}CH{=}O} \xrightarrow{\text{enzyme 2}} \underset{\text{Acetate anion}}{CH_3{-}CO_2^-}$$

PROBLEM 4.17

Draw the structures and provide the names for the alcohols that must be used as the starting materials to prepare the following carbonyl compounds using an oxidation reaction. In addition, indicate the reagent(s) used for each transformation.

(a) cyclopentanone (b) $(CH_3)_2CHCOCH_2CH_3$ (c) [skeletal structure: ketone, C=O with O] (d) [skeletal structure] CHO

4.5 Preparation of Alcohols from Alkyl Halides

Alcohols (R—OH) can be prepared from other organic compounds, but they cannot be easily prepared in the laboratory by simply inserting an oxygen atom between a carbon and a hydrogen in an alkane (R—H). Therefore, the approach must be indirect, as is evident in the Key to Transformations (inside front cover).

Although there are four important methods for alcohol preparation, only one is covered here—conversion of alkyl halides to alcohols. The other three methods (conversion of alkenes, Section 6.4.2; conversion of carbonyl compounds, Section 13.3.1; and conversion of carboxylic acids, Section 14.3.1) will be described in detail once you have been introduced to those families of compounds and their functional group behavior.

Section 4.3 described how alcohols can be transformed into alkyl halides (by a substitution reaction). We will see how this transformation can be reversed—how alcohols can be obtained *from* alkyl halides. This is the first instance we have encountered where an interconversion is possible between two families of compounds.

For preparation of an alcohol (R—OH) from an alkyl halide (R—X), it is clear that what is required is a *substitution* reaction. The halide (the leaving group) must be replaced by a hydroxyl group (the nucleophile), a reaction discussed in detail earlier. Recall from Section 3.4.1 that in an S_N2 reaction, a primary or secondary alkyl halide reacts with sodium hydroxide to produce an alcohol. Starting with an enantiomeric alkyl halide results in inversion of configuration at the stereocenter. Recall from Section 3.4.3 that conversion of a tertiary halide to a tertiary alcohol occurs under S_N1 conditions (with water as the nucleophile), but may be accompanied by formation of an alkene, an elimination product.

Primary and secondary:

$$R—X \xrightarrow[(S_N2)]{NaOH} R—OH$$

Tertiary:

$$R—X \xrightarrow[(S_N1)]{H_2O} R—OH \text{ (plus alkene)}$$

Chapter 10 will describe the conversion of alkanes to alkyl halides. That reaction, followed by conversion of the alkyl halide to an alcohol as described here, provides a means to bring about the overall transformation of an alkane (R—H) to an alcohol (R—OH), a transformation not readily accomplished in a single step (as the Key to Transformations, inside front cover, shows).

How to Solve a Problem

Show how to synthesize cyclohexanone starting with bromocyclohexane.

Bromocyclohexane $\xrightarrow{?}$ Cyclohexanone

PROBLEM ANALYSIS

We examine the structures to determine precisely what has changed, and what has not changed, from reactant to product. In this synthesis, the carbon skeleton does not change; only the substituent on the cyclohexane ring changes. Note also that we have not discussed any one-step conversion from an alkyl halide to a ketone. Therefore, more than one step is required.

In analyzing synthesis problems, it is useful to apply an approach known as **retrosynthesis**—the mental process of working backward from product to starting compound on a step-by-step basis. Chemists consider the possible reactions by which a compound could be obtained from an immediate precursor without worrying initially about the detailed chemistry. They use a double-lined arrow to point to the potential precursor. Learning the transformations outlined in the Key to Transformations is essential for using this approach.

We have seen that ketones can be obtained by oxidation of secondary alcohols. But how can we obtain a secondary alcohol from the specified starting material? The answer is by carrying out an S_N2 substitution reaction, replacing the bromine with a hydroxyl group. Therefore, our retrosynthetic scheme is as follows:

Cyclohexanone $\Longrightarrow$ Cyclohexanol $\Longrightarrow$ Bromocyclohexane

All that remains is to specify the reagents needed for each step of the synthesis.

SOLUTION

Reaction of the starting compound bromocyclohexane with aqueous sodium hydroxide will result in the S_N2 substitution of bromine, producing cyclohexanol. Oxidation of the secondary alcohol cyclohexanol with sodium dichromate and sulfuric acid will produce the desired ketone, cyclohexanone. Therefore, the overall synthesis is

Bromocyclohexane $\xrightarrow{NaOH}$ Cyclohexanol $\xrightarrow[H_2SO_4]{Na_2Cr_2O_7}$ Cyclohexanone

PROBLEM 4.18

Indicate the reagents necessary for each step involved in the following transformations:

(a) 1-bromobutane to butanal ($CH_3CH_2CH_2CHO$)

(b) 2-bromobutane to 2-butanone ($CH_3CH_2COCH_3$)

Formula 1 racing cars burn methanol in their engines.

4.6

Important Alcohols

Because the hydroxyl group is so widely occurring, there are many important alcohols. The hydroxyl group is important in many industrial chemicals, most of which have relatively simple structures. Such alcohols will be discussed in this section. However, the hydroxyl group is also present in many compounds of biological significance, and some of these have quite complex structures, often including more than one functional group. Section 4.7 will present an overview of carbohydrates, also known as sugars, a biologically important class of compounds whose properties reflect the presence of several hydroxyl groups.

Methanol (CH_3OH), one of the industrial alcohols, is also called *wood alcohol* because, in the past, it was obtained from the destructive distillation of wood in the presence of a limited amount of oxygen. It is now produced on a very large scale (billions of pounds per year) in industrial plants. The process involves catalytic reduction of carbon monoxide:

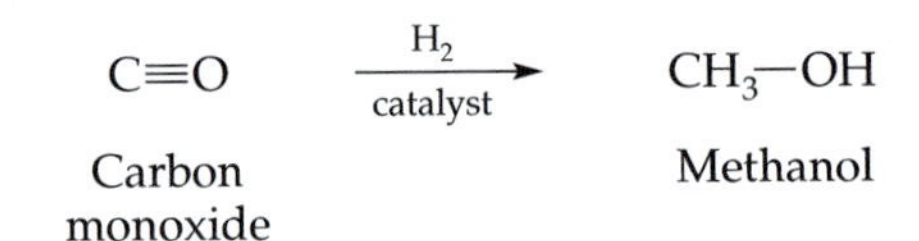

Large quantities of industrial-grade ethanol are produced by fermentation of crops such as corn.

Methanol is very toxic to humans, leading to blindness and even death. Methanol is often used as a premium fuel to power racing car engines, in which it is oxidized (combusted) to carbon dioxide and water. Controlled oxidation of methanol is a primary route to formaldehyde ($H_2C{=}O$), which is used to make a wide variety of plastics and insulation.

Ethanol (C_2H_5OH), found in alcoholic beverages, results from the fermentation of the carbohydrates (sugars) present in a wide variety of vegetable matter (see "Chemistry at Work" on page 130). The flavor of any alcoholic beverage comes not from the ethanol, which is colorless and tasteless, but from the vegetable matter that was fermented. Ethanol is also used as a topical antiseptic.

Very large industrial plants have been built in recent years to ferment agricultural products (mainly corn, barley, and molasses obtained from sugar refining) to produce large quantities of ethanol for use in automotive fuel. Ethanol is mixed with petroleum products in various proportions (5–20%) to produce nonleaded gasoline for use in internal combustion engines (one such mixture is called *gasohol*). The presence of ethanol raises the octane number of the gasoline, eliminating the need for lead and thereby reducing pollutants emitted from internal combustion engines. Ethanol is one of the group of compounds called *oxygenates* (oxygen-containing compounds) that are used in gasolines to cause more complete combustion. Domestic manufacturing of processed fuels such as ethanol reduces the demand for imported petroleum. Such manufacturing has been important in South Africa and Brazil for many years and is now assuming greater importance in the United States.

Isopropyl alcohol is often used as the bulk solvent for cosmetics, lotions, perfumes, and other products applied externally. When content labels for such products list "alcohol," they usually mean isopropyl alcohol or ethanol.

The alcohol ethylene glycol (1,2-ethanediol) is the major constituent in automobile engine coolants. It has both a low freezing point (−12°C) and a high boiling point (199°C), making it ideal as both antifreeze and coolant.

Glycerol (1,2,3-propanetriol) is produced from the hydrolysis of fats. Also produced in the reaction are fatty acids, used to make soap and detergents (see Section 14.2.3):

$$\begin{array}{l} CH_2{-}O{-}CO{-}R \\ | \\ CH{-}O{-}CO{-}R \\ | \\ CH_2{-}O{-}CO{-}R \end{array} \xrightarrow[2.\ H^+]{1.\ NaOH} \begin{array}{l} CH_2{-}OH \\ | \\ CH{-}OH \\ | \\ CH_2{-}OH \end{array} + 3\ R{-}COOH$$

Fat — Glycerol (1,2,3-propanetriol) — Fatty acid

Nitration of glycerol produces glycerol trinitrate (an ester, also known as nitroglycerine), a shock-sensitive liquid explosive:

$$\begin{array}{l} CH_2{-}OH \\ | \\ CH{-}OH \\ | \\ CH_2{-}OH \end{array} + 3\ HNO_3 \xrightarrow{H_2SO_4} \begin{array}{l} CH_2{-}O{-}NO_2 \\ | \\ CH{-}O{-}NO_2 \\ | \\ CH_2{-}O{-}NO_2 \end{array} + 3\ H_2O$$

Glycerol — Nitroglycerine

Alfred Nobel discovered that mixing the very sensitive nitroglycerine with a porous material produces a stable explosive, known as dynamite. Nitroglycerine is also used in medicine to treat angina.

There are many important biologically active alcohols in addition to the carbohydrates discussed in the next section. Examples are cholesterol, found in blood and gallstones (see Figure 4.3); menthol, obtained from the oil of peppermint; and retinol, also known as vitamin A and found in fish oils, liver, and dairy products. Retinol is a key precursor to the chemicals that interact with light entering the eye, enabling us to see (see Chapter 8).

CH_3, HO, $CH(CH_3)_2$

Menthol

OH

Retinol (vitamin A)

4·7 Carbohydrates

Carbohydrates are a large and very important class of naturally occurring compounds that contain two functional groups: the hydroxyl group we have been studying and a carbonyl group (aldehyde or ketone) to be described in Chapter 13. Carbohydrates occur in all plants and animals and serve many vital functions:

- As glucose, glycogen, and starch, they are the storehouse for energy in living systems.
- As cellulose, they provide structural support in plants and trees.
- As deoxyribose and ribose, they are one of three components of the genetic material DNA and RNA (D = deoxyribose and R = ribose).

- They comprise all or part of many biologically active compounds, such as the antibiotic streptomycin and the compounds that determine blood type.
- They are major components of foods—for example, flour is mainly starch, and milk products contain lactose.

4.7.1 Terminology

The term *carbohydrate* arose from the early observation that the chemical formula for many simple carbohydrates could be written as though they were "hydrates of carbon"—that is, as $C_x(H_2O)_y$. For example, glucose (also known as blood sugar) has the formula $C_6H_{12}O_6$, which was represented as $C_6(H_2O)_6$; sucrose (common table sugar) has the formula $C_{12}H_{22}O_{11}$, which was represented as $C_{12}\,(H_2O)_{11}$. **Carbohydrates** can best be described as *polyhydroxy aldehydes* or *polyhydroxy ketones.* As already mentioned, they contain two functional groups. In addition to the carbonyl group, which makes it an aldehyde or ketone, a carbohydrate has two or more hydroxyl groups (thus "polyhydroxy"), which makes it also an alcohol. The most common simple carbohydrates are C_5 and C_6 compounds.

Carbohydrates are also known as **saccharides,** a term derived from the Latin word *saccharum,* meaning "sweet" (many of the simple carbohydrates have a very sweet taste). They are also commonly referred to as **sugars.** Examples are glucose (also known as dextrose or blood sugar and found in corn syrup), fructose (also known as levulose and found in honey), and sucrose (table sugar, made from sugar cane or sugar beets). The *-ose* ending on a chemical name invariably indicates that the compound is a carbohydrate or a derivative of one. All of the names for common carbohydrates are historical in derivation.

Carbohydrates are classified as monosaccharides, disaccharides, or polysaccharides.

- **Monosaccharides** are the simplest sugars since they cannot be hydrolyzed (cleaved by water). Glucose, fructose, and ribose are monosaccharides.
- **Disaccharides,** as the name suggests, are made up of two simple sugars (monosaccharides) bonded together; these may be cleaved by hydrolysis. Examples are sucrose (which can be hydrolyzed to one molecule of fructose and one of glucose), lactose (which can be hydrolyzed to one molecule of glucose and one of galactose), and maltose (which can be hydrolyzed to two molecules of glucose).
- **Polysaccharides** contain many monosaccharide units—they are polymeric saccharides. The most common examples are starch (hydrolyzed to glucose), glycogen (hydrolyzed to glucose), and cellulose (also hydrolyzed to glucose). Starch may contain as many as several thousand units of glucose chemically joined together, and glycogen may have a molecular weight as high as 1 million.

A few examples of the three classes of saccharides are presented in the following sections.

4.7.2 Structures of Monosaccharides

The best-known monosaccharide is **glucose,** or blood sugar. The structure of glucose is shown here in three different representations, each of which makes clear that glucose has four stereocenters:

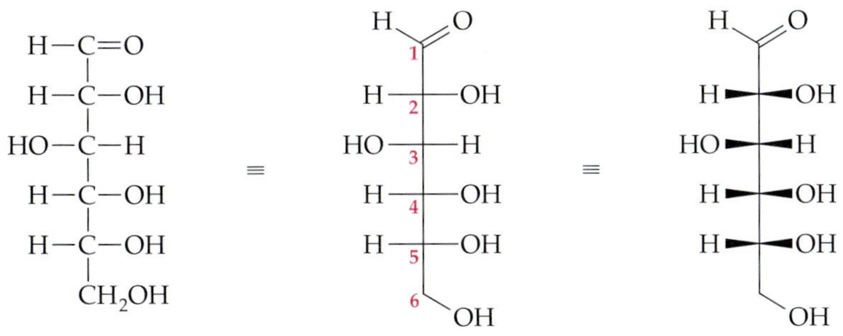

D-Glucose

The left-hand structure is seldom used; it is simplified to the bond-line formula in the middle. This bond-line formula is known as a **Fischer projection** because, although drawn in two dimensions, it is understood to represent (project) the three-dimensional structure on the right. In a Fischer projection, all horizontal bond lines represent substituents directed toward the observer (above the plane of the paper), while all vertical lines represent substituents directed away from the observer (below the plane of the paper). The small capital letter D indicates the absolute configuration at carbon 5; it is an old symbol still used to identify what is now known as the *R* configuration for that carbon.

Note that glucose has an *ald*ehyde group (—CHO) at carbon 1. All such monosaccharides are referred to as **aldoses.** The alternative is for a monosaccharide to have a *ket*one group, usually on carbon 2, as in fructose; in that case, the monosaccharide is referred to as a **ketose.** Monosaccharides are very soluble in water because of the presence of many hydroxyl groups and the resulting hydrogen bonding.

PROBLEM 4.19

Draw the absolute configuration of carbon 5 of D-glucose (shown above). Using the letters **a, b, c,** and **d** to assign priorities, prove that the configuration is *R*.

PROBLEM 4.20

The symbol L designates the configuration opposite to D, and in glucose these symbols refer to carbon 5. Draw the Fischer projection for glucose with the L configuration at carbon 5.

PROBLEM 4.21

The monosaccharide D-galactose is part of the disaccharide lactose. D-Galactose an isomer of D-glucose that varies in configuration at only one stereocenter: It has the opposite configuration at carbon 4. Draw the bond-line formula for D-galactose.

The structure of glucose is actually more complicated than shown above because in solution there is an equilibrium among three structures. Because of the presence of carbonyl and hydroxyl functional groups in the same molecule and at appropriate distances from one another, the hydroxyl group at carbon 5 can "add across" the carbonyl group (that is, the hydroxyl hydrogen goes to the carbonyl oxygen, and the hydroxyl oxygen goes to the carbonyl carbon) to form a **hemiacetal,** which has a six-membered ring with all single bonds. (This chemistry will be discussed in Section 13.4.2.) Such six-membered cyclic sugars are called **pyranoses,** after the six-membered oxygen-containing heterocycle *pyran*, and they are shaped like the chair form of cyclohexane.

Open-chain D-glucose　　Hemiacetal D-glucose

The hydroxyl addition across the carbonyl group in glucose can occur in either of two ways with respect to the chair ring: The newly formed hydroxyl group on carbon 1 may be equatorial (in β-D-glucose) or axial (in α-D-glucose). Carbon 1 is therefore called an **anomeric carbon,** a carbon that has become a stereocenter because of hemiacetal formation. This anomeric carbon can have two possible configurations, α (alpha) and β (beta):

α-D-Glucose　　Open-chain D-glucose　　β-D-Glucose

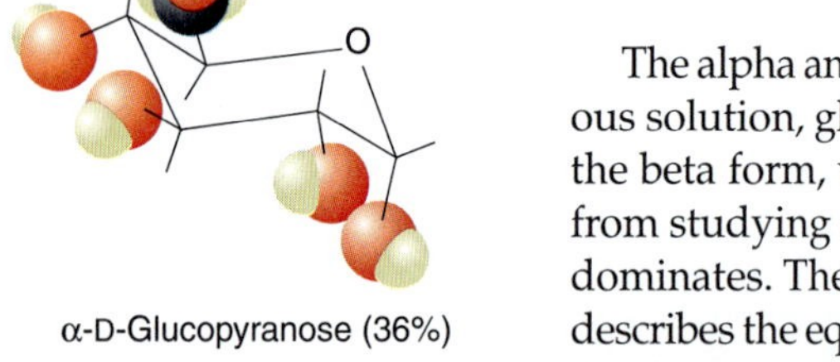

α-D-Glucopyranose (36%)

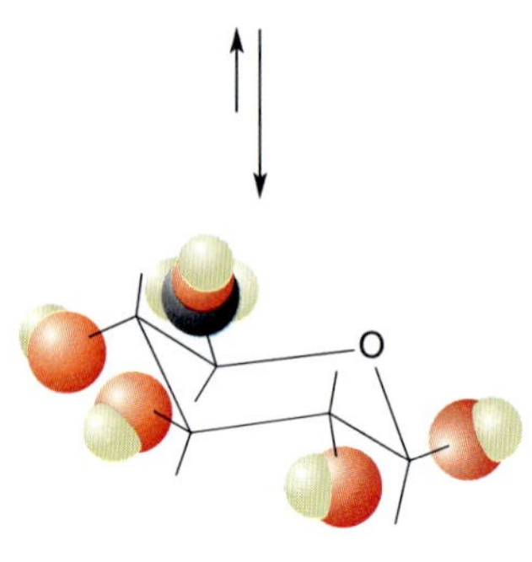

β-D-Glucopyranose (64%)

The alpha and beta forms of glucose can be isolated and crystallized. However, in aqueous solution, glucose exists in an equilibrium, about 36% of the alpha form and 64% of the beta form, with only a trace of the open-chain form present. As you would expect from studying cyclohexane conformations, the more stable equatorial beta isomer predominates. The full name of the predominant form of glucose is **β-D-glucopyranose:** β describes the equatorial stereochemistry about the anomeric carbon (carbon 1); D describes the absolute configuration (*R*) about carbon 5; *gluco* arises from the historical name of the compound (glucose), and *pyranose* identifies the six-membered hemiacetal ring.

An aldehyde group (—CHO) is easily oxidized to a carboxylic acid group (—COOH), which means that the oxidizing agent is reduced. (You learned this fact in Section 4.4 when we discussed the oxidation of primary alcohols, and we will discuss it further in Section 13.2.2.) Carbohydrates with aldehyde groups (aldoses) are therefore called *reducing sugars*—they reduce the oxidizing reagent—and they are readily detected. Glucose is a reducing sugar. If the aldehyde group of a reducing sugar is converted into another group, the new sugar is no longer a reducing sugar. This is the case with glycosides (discussed in the next section). However, if the aldehyde group remains in equilibrium with a hemiacetal form, even in low concentration as in glucose solutions, it can still be detected as a reducing sugar.

A second important C_6 monosaccharide is **D-fructose,** with the same molecular formula as glucose ($C_6H_{12}O_6$). D-Fructose is a *ketose* whose open-chain form exists in an equilibrium with four cyclic forms. The forms called **furanoses** contain five-membered hemiacetal rings (after the five-membered oxygen-containing heterocycle called *furan;* see Section 9.7.3). In solution, D-fructose exists as an equilibrium mixture of its five forms: about 3% of the α-pyranose form, 57% of the β-pyranose form, 9% of the α-furanose form, 31% of the β-furanose form, and a trace of the open-chain form. The alpha stereochemistry about the anomeric carbon (carbon 2) in the pyranose form puts the largest ring substituent (—CH_2OH) in the less stable axial position, whereas the beta stereochemistry places the large group in the more stable equatorial position.

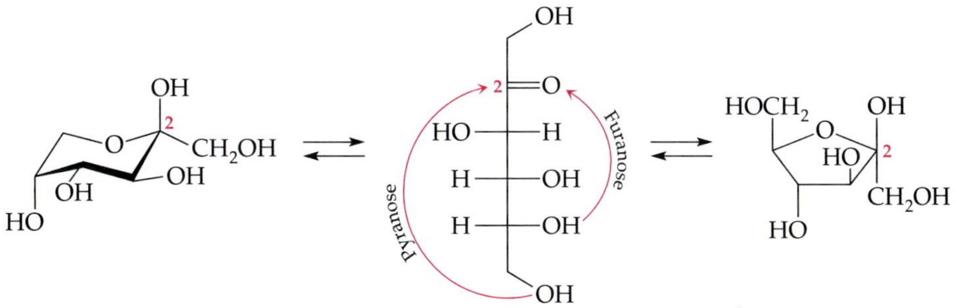

β-D-Fructopyranose

D-Fructose
(open-chain form)

β-D-Fructofuranose

Two important naturally occurring monosaccharides are D-ribose and 2-deoxy-D-ribose. These are C_5 monosaccharides (aldofuranoses) that are part of the structure of the nucleotides that make up RNA (ribonucleic acid) and DNA (deoxyribonucleic acid), respectively. The nucleic acids have a backbone structure of alternating carbohydrate and phosphate units, with a nucleic acid base attached to each carbohydrate at carbon 1 (see Section 19.2). The two carbohydrates have the cyclic structures shown, with the term *2-deoxy* implying a ribose structure without the oxygen on carbon 2.

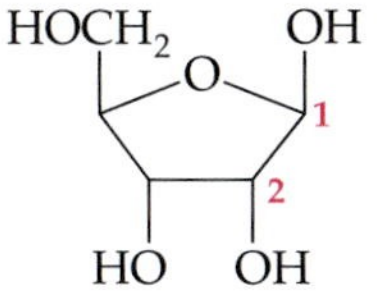

β-D-Ribofuranose
(D-ribose)

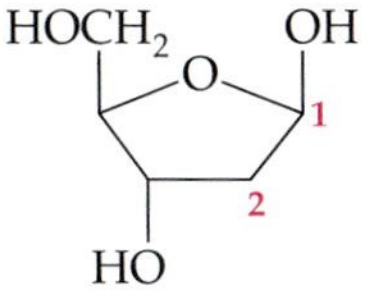

β-D-Deoxy-D-ribofuranose
(2-deoxy-D-ribose)

PROBLEM 4.22

2-Deoxyribose is the five-carbon sugar just shown. Draw the conformational structure of β-D-2,3-dideoxyglucose.

4.7.3 Structures of Disaccharides

Disaccharides are carbohydrates that contain two monosaccharide molecules, each in the hemiacetal form, joined together by the elimination of a molecule of water between two hydroxyl groups. The elimination of water (dehydration) involves the anomeric carbon of one monosaccharide and may or may not involve the anomeric carbon of the other monosaccharide. When the hemiacetal hydroxyl group on an anomeric carbon is involved in a dehydration, the resulting product is an **acetal.** In carbohydrate chemistry, this product is called a **glycoside:**

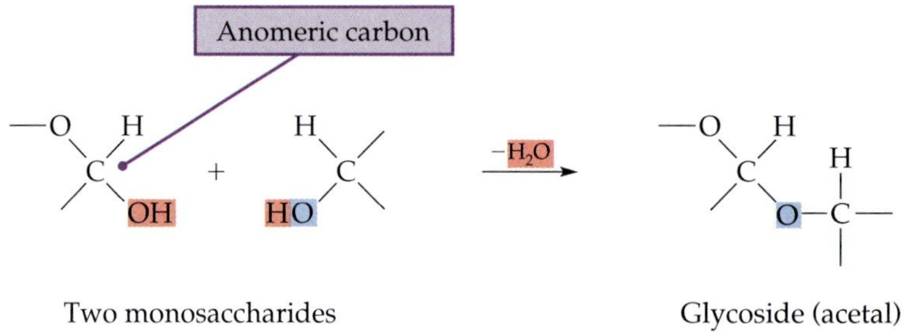

Anomeric carbon

$-H_2O$

Two monosaccharides

Glycoside (acetal)

As we will see in Chapter 13, acetals are stable in basic solution but hydrolyze readily in acidic solution. Thus, a disaccharide is readily hydrolyzed to its two monosaccharide units in dilute aqueous acid. Disaccharides are also very soluble in water because of hydrogen bonding of the many hydroxyl groups.

The disaccharide **sucrose** ($C_{12}H_{22}O_{11}$) consists of two different monosaccharides, a molecule of α-D-glucose joined to a molecule of β-D-fructose. The elimination of a molecule of water occurs between the hemiacetal hydroxyl on carbon 1 of α-D-glucose and the hemiacetal hydroxyl on carbon 2 of β-D-fructose. Thus, a common shorthand notation for sucrose is α1G-β2F:

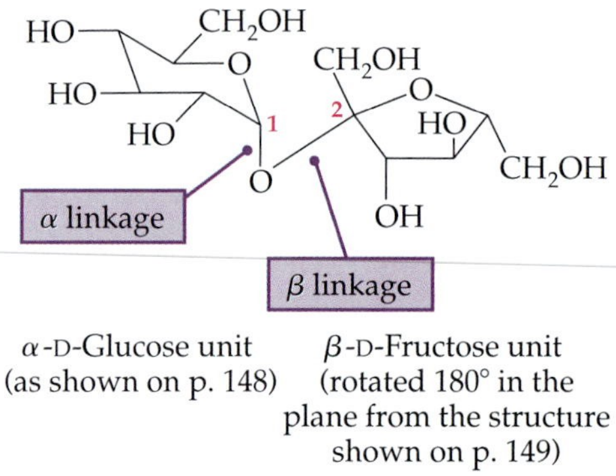

α linkage

β linkage

α-D-Glucose unit (as shown on p. 148)

β-D-Fructose unit (rotated 180° in the plane from the structure shown on p. 149)

Sucrose (α-D-glucopyranosyl-β-D-fructofuranoside)

Table sugar is sucrose, which is produced on a large scale from sugar beets (shown here) or sugar cane.

The formal chemical name of sucrose is α-D-glucopyranosyl-β-D-fructofuranoside, and it describes completely the stereochemistry of the junction as being between an alpha form of glucose and a beta form of fructose. The terms *pyranosyl* (six-membered ring) and *furanosyl* (five-membered ring) describe the ring size of the component monosaccharides. Sucrose is a nonreducing sugar because the aldehyde group of the glucose component is masked as a glycoside linkage.

Enzymes in the human metabolic system are able to hydrolyze sucrose to the monosaccharides glucose and fructose, which then undergo further metabolism in the cells to CO_2 and H_2O, producing energy as a by-product. Alternatively, the glucose may be carried to the liver or to the muscles (as blood sugar), where it is converted into glycogen, a polysaccharide consisting only of glucose units joined together by alpha linkages, and stored.

Another common disaccharide is **maltose,** which is formed from two molecules of glucose connected between carbon 1 of one molecule and carbon 4 of the second molecule. In this instance, the glycoside formation involves the anomeric hydroxyl (a hemiacetal hydroxyl) of one glucose and the hydroxyl on carbon 4 (an alcohol hydroxyl) of the second glucose. Therefore, in this case, the notation α1G-β4G indicates the bonding between the two monosaccharides. Note also that in maltose, the hemiacetal functional group on carbon 1 of the β-D-glucose unit can be in equilibrium with its aldehyde form. As a result, maltose is a reducing sugar. Maltose is an intermediate hydrolysis product of starch and is formed by the enzyme maltase in the malting step of beer production.

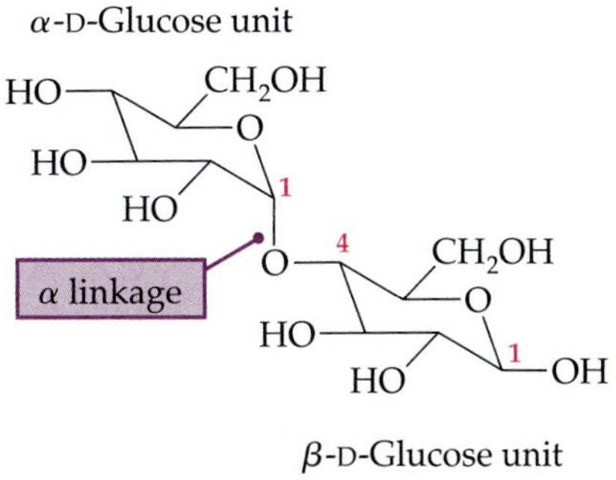

α-D-Glucose unit

α linkage

β-D-Glucose unit

Maltose [4-*O*-(α-D-glucopyranosyl)-β-D-glucopyranose]

Lactose, the common disaccharide in milk, is formed from one molecule of glucose and one molecule of galactose; it can be represented as β1GA-α4G. Galactose is called a carbon 4 *epimer* of glucose. An **epimer** is one of a pair of configurational isomers that have the same absolute configurations except at one designated stereocenter, where their configurations are opposite. Thus, galactose has the same three-dimensional structure as glucose except that the absolute configuration is opposite at carbon 4 (note that the hydroxyl on that carbon is axial in galactose but equatorial in glucose).

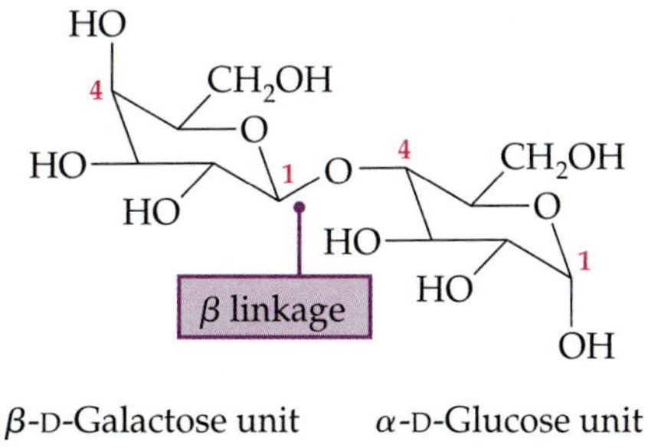

β linkage

β-D-Galactose unit α-D-Glucose unit

Lactose [4-*O*-(β-D-galactopyranosyl)-α-D-glucopyranose]

Some individuals cannot digest lactose because they lack a key enzyme (lactase), a condition known as *lactose intolerance.* They may choose to avoid milk products or may add lactase to their diet to facilitate the metabolism of lactose.

PROBLEM 4.23

Compare the structures of sucrose and maltose. Explain why maltose is a reducing sugar in aqueous solution, whereas sucrose is not.

4.7.4 Structures of Polysaccharides

Polysaccharides are **polymers** of monosaccharides—that is, large molecules containing many ("poly") monosaccharide units joined together by intramolecular dehydration (glycoside formation).

Starch and cellulose are both polymers of glucose, proved by the fact that both are hydrolyzed in dilute acid to produce glucose as the only product.

$$\text{Starch} \xrightarrow{H^+/H_2O} \text{D-Glucose} \xleftarrow{H^+/H_2O} \text{Cellulose}$$

Starch is a basic food for all people. It is interesting to note that the human metabolic system readily hydrolyzes starch to glucose, but cannot hydrolyze cellulose. In other words, plant fiber ingested by humans does not get metabolized; it proceeds through the digestive system relatively unchanged and is excreted in the feces. Cellulose is the substance promoted as "dietary" fiber in high-fiber products. Foraging animals such as cattle, on the other hand, have enzymes that can metabolize both cellulose and starch, enabling them to survive on feeds such as hay. The fundamental reason for this difference is that *starch has alpha linkages* connecting the glucose units, while *cellulose has beta linkages*. Human enzymes cannot hydrolyze the beta linkages of cellulose.

The dietary fiber in bran muffins is not digestible by humans. Cows, however, can digest the cellulose in hay.

Starch is a mixture of two different polymers of glucose, each containing up to 4000 glucose units per molecule. Starch consists of about 20% of a water-soluble component called *amylose* and 80% of a water-insoluble component called *amylopectin*. **Amylose** is a linear polymer containing alpha linkages between carbons 1 and 4 of adjacent glucose units, as shown here, represented as α1G-α4G.

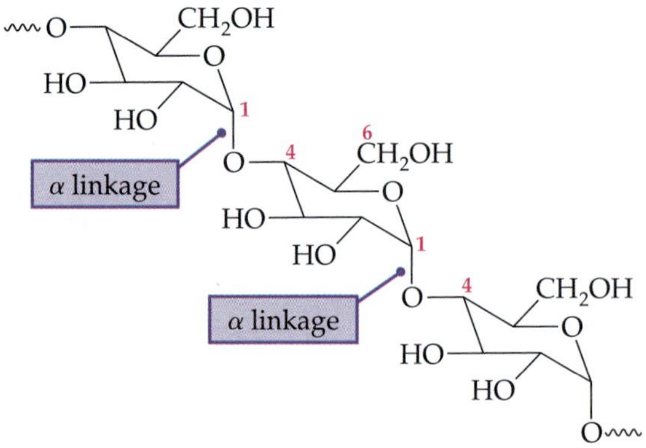

A portion of the amylose molecule

Amylopectin contains the same basic backbone of α1G-α4G units but also contains some α1G-α6G linkages. Thus, amylopectin is not a linear polymer, but has considerable branching from the long chains of connected glucose units.

Cellulose is a structural material found in all plant life, forming the basic framework of stalks, trunks, bark, and leaves. Cotton, which is a fibrous material found in the seed pods of plants of the genus *Gossypium*, is almost pure cellulose. Like amylose and amylopectin, cellulose is a polymer of glucose (about 3000 glucose units per molecule). It is linear and contains only linkages between carbons 1 and 4. These are beta linkages and can be represented as β1G-β4G.

Cotton is almost pure cellulose.

A portion of the cellulose molecule

A cellulose fiber is a bundle of cellulose chains held together by extensive hydrogen bonding, possible because of the many hydroxyl groups present. Cellulose is insoluble in water.

Another important polysaccharide, **glycogen,** is not found in nature but is synthesized in the human body and stored mainly in the liver and muscle tissue. Glycogen is the major source of energy in the body. The energy from glycogen is rapidly available, in contrast to the energy from fat, which is metabolized much more slowly (see Section 19.2). When marathon runners "carb up" (eat large amounts of carbohydrates) before a competition, their bodies produce and store a large supply of glycogen, which is then readily available to support strenuous muscular activity during the race. Glycogen is also a polymer of α-D-glucose but has a much higher molecular weight than starch. It has the usual 1,4-alpha linkages between glucose units (α1G-α4G), but also has a high degree of branching due to additional 1,6-alpha linkages (α1G-α6G).

4.7.5 Chemistry of Carbohydrates

As you have seen, even simple carbohydrates contain many hydroxyl groups, each of which behaves chemically like any hydroxyl group. Thus, carbohydrates can be converted to alkyl halides, can be oxidized, and can be converted to ethers (see Chapter 5) and esters (see Chapter 15). The obvious problem facing carbohydrate chemists is how to bring about a reaction with one specific hydroxyl group without affecting the others. Many special approaches have been developed over the years to accomplish this goal, but they are beyond the scope of this book. The chemistry of carbohydrate metabolism is discussed in Section 19.2.

CHEMISTRY AT WORK

IN THE MARKETPLACE

Artificial Sweeteners

When you add sugar to coffee or tea or use it to make cookies, you are using the disaccharide sucrose—that is, plain table sugar. Sucrose is readily available from sugar cane and sugar beets and has been used for centuries as a sweetener. Annual worldwide production exceeds 100 million tons. The monosaccharides glucose and fructose are even sweeter. Fructose, marketed as corn syrup, is about 1.6 times as sweet as sucrose and is widely used in making soft drinks.

Although it has long been of interest and is still a subject of serious study, the fundamental cause of sweetness is not known. No acceptable general theory can explain what makes certain compounds taste sweet and precisely how they interact with human taste-sensing systems. As a result, most synthetic sweeteners have been discovered by accident. The low calorie count of some synthetic sweeteners has led to their use in diet foods and created a major industry.

Sorbitol (1,2,3,4,5,6-hexahydroxyhexane) is a naturally occurring compound that is closely related to glucose. It is produced commercially by reduction of the aldehyde group of glucose to a hydroxyl group. It is as sweet as sucrose but is metabolized more slowly. Xylitol is a pentahydroxypentane similar to sorbitol. Xylitol and sorbitol are mainly used in "sugarless" candies and gums.

Sorbitol   Xylitol

The first widely used synthetic low-calorie sweetener was saccharin, best known to the public as Sweet 'n' Low. It is about 110 times sweeter than sucrose, but leaves a bitter, metallic aftertaste for some users. The next major discovery was cyclamate, about 20 times sweeter than sucrose and having no aftertaste. Both were serendipitous discoveries.

Saccharin   Cyclamate

In 1970, the U.S. Food and Drug Administration (FDA) banned the sale of saccharin because of suspicions that it was carcinogenic; however, Congress permitted its continued use under certain conditions and with a warning label. Cyclamate was banned soon after that for the same reason.

Today, the most widely used low-calorie artificial sweetener is aspartame, marketed as NutraSweet:

Aspartame (NutraSweet)

Aspartame, discovered serendipitously in 1965, is about 110 times as sweet as sucrose. It has been very successfully marketed since the removal of saccharin and cyclamate by the FDA. The major disadvantage of aspartame is that it hydrolyzes upon heating, resulting in a bitter taste that makes it unsuitable for use in baking. It also cannot be used by people with the condition called PKU (phenylketonuria) because it is metabolized to produce the α-amino acid phenylalanine [$C_6H_5CH_2CH(NH_2)COOH$], which these individuals cannot further metabolize. The original patent for aspartame has expired, so there is now widespread research activity to develop and market competing artificial sweeteners.

Chapter Summary

Alcohols (R—OH) are a family of compounds in which the **hydroxyl** (—OH) functional group is bonded to an sp^3 hybridized carbon of an alkyl group. Alcohols are named as **alkanols** to indicate that they are derived from parent alkanes; the alcohol suffix (*-anol*) has replaced the alkane suffix (*-ane*). The parent chain must contain the carbon to which the hydroxyl group is bonded, and its location on the chain is indicated by a position number. Historical names for simple alcohols consist of the name of the alkyl group followed by the word *alcohol*, as in ethyl alcohol. Alcohols are classified as primary, secondary, or tertiary, depending on the classification of the carbon to which the hydroxyl group is attached. The sulfur analog of an alcohol is a **thiol** (R—SH).

Alcohols are polar compounds as a result of the polarity of the carbon-oxygen and oxygen-hydrogen bonds. Alcohols are weak Lewis bases because of the unshared electrons on oxygen; an alcohol can be protonated by a strong acid to form an **oxonium ion** (ROH_2^+). An alcohol molecule can form a **hydrogen bond** with water or another alcohol molecule. Alcohols are weak acids with pK_a values ranging from 16 to 18. The proton of an alcohol can be removed by a very strong base or, more commonly, by reaction with metallic sodium or potassium. The products of the latter reaction are molecular hydrogen and a soluble metal **alkoxide** ($RO^- M^+$), a very strong base.

Alcohols can be converted to alkyl halides by reaction with halogen acids (HCl, HBr, and HI). The reaction involves initial formation of an oxonium ion, from which water is displaced by halide anion in an S_N1 or S_N2 reaction.

Alcohols are converted into alkyl chlorides by reaction with thionyl chloride and pyridine. This reaction occurs with inversion of configuration at a stereocenter, an (*R*)-alcohol producing an (*S*)-alkyl chloride. Alcohols react with *p*-toluenesulfonyl chloride (tosyl chloride) to form tosyl esters, the alcohol serving as a nucleophile. This reaction occurs with retention of configuration at a stereocenter—that is, an (*R*)-alcohol produces an (*R*)-tosylate. An enantiomeric tosylate undergoes an S_N2 reaction with sodium hydroxide with inversion of configuration (that is, the *S*-enantiomer of the alcohol is formed). Reaction of the latter with thionyl chloride and pyridine produces the inverted alkyl halide (*R* enantiomer). These transformations permit conversion of an enantiomeric alcohol into an enantiomeric alkyl halide with or without inversion of configuration. Phosphorus tribromide (PBr_3) converts alcohols into alkyl bromides with inversion of configuration.

Primary and secondary alcohols can be oxidized. Primary alcohols (RCH_2OH) are oxidized to aldehydes (RCHO) using the moderate oxidant pyridinium chlorochromate (PCC). Stronger oxidants, such as $Na_2Cr_2O_7/H_2SO_4$ or $KMnO_4$, initially form an aldehyde, which cannot be isolated but is rapidly oxidized further to a carboxylic acid (RCOOH). Secondary alcohols (R_2CHOH) are oxidized to ketones (R_2CO). Tertiary alcohols cannot be oxidized.

Alcohols are prepared from three other families of compounds (see the Key to Transformations). One method is by S_N2 substitution by hydroxide on primary and secondary alkyl halides. Tertiary halides are converted to alcohols by S_N1 solvolysis with water.

A major family of compounds containing hydroxyl groups are the **carbohydrates,** also called **saccharides** or **sugars.** They are polyhydroxy aldehydes (**aldoses**) or polyhydroxy ketones (**ketoses**). The simplest sugars are **monosaccharides,** typically C_5 and C_6 compounds that cannot be hydrolyzed to simpler compounds. A typical aldose monosaccharide is **glucose,** $C_6H_{12}O_6$, whose most stable form is β-D-glucose, with all substituents in equatorial positions on the six-membered ring. **Fructose** is a common C_6 monosaccharide ketose. Ribose and 2-deoxyribose are C_5 monosaccharides that are important constituents of the nucleic acids RNA and DNA, respectively.

Disaccharides contain two monosaccharide units joined together by a **glycoside (acetal) linkage** that is susceptible to enzymatic or acid-catalyzed hydrolysis. **Sucrose** (glucose-fructose, table sugar), **maltose** (glucose-glucose, malt sugar), and **lactose** (glucose-galactose, milk sugar) are common disaccharides.

Polysaccharides contain many monosaccharide units joined by glycoside linkages; they are **polymers. Starch** is a high-molecular-weight polymer of α-D-glucose. **Cellulose** is a high-molecular-weight polymer of β-D-glucose. **Glycogen,** also a high-molecular-weight polymer of α-D-glucose, is synthesized and stored in the body as a source of energy.

Summary of Reactions

1. **Alkoxide formation (Section 4.2.2).** Alcohols are converted to alkoxides by reaction with metalic sodium or potassium. The alkoxides serve as strong nucleophiles and bases.

$$\underset{\text{Alcohol}}{R{-}OH} + Na \longrightarrow \underset{\text{Sodium alkoxide}}{R{-}O^- \ Na^+} + \tfrac{1}{2}H_2$$

2. **Conversion of alcohols to alkyl halides (Section 4.3).** Alcohols can be converted into alkyl chlorides, bromides, or iodides via four types of reactions.
   a. **Reaction with hydrogen halides** is a nonstereospecific reaction.

$$\underset{\text{Alcohol}}{R{-}OH} \xrightarrow[(X = Cl, Br, I)]{HX} \underset{\text{Alkyl halide}}{R{-}X} + H_2O$$

b. **Reaction with thionyl chloride and pyridine** produces an alkyl chloride with inversion of configuration at a stereocenter.

$$\text{R—OH} \xrightarrow[\text{pyridine}]{SOCl_2} \text{R—Cl} + SO_2(g) + \text{HCl}$$

Alcohol Alkyl chloride

c. **Reaction with tosyl chloride** produces a tosyl ester with retention of configuration. An S_N2 reaction with sodium hydroxide produces the original alcohol but with an inverted configuration. Reaction of that alcohol with thionyl chloride produces the alkyl chloride with the same configuration as the starting alcohol.

$$\text{R—OH} \xrightarrow{\text{TsCl}} \text{R—OTs} \xrightarrow{\text{NaOH}}$$

Alcohol Tosyl ester

$$\text{R—OH} \xrightarrow[\text{pyridine}]{SOCl_2} \text{R—Cl}$$

Inverted alcohol Alkyl chloride

d. **Reaction with phosphorus tribromide** produces an alkyl bromide, and a stereocenter is inverted in the process.

$$\text{R—OH} \xrightarrow{PBr_3} \text{R—Br}$$

Alcohol Alkyl bromide

3. **Oxidation of alcohols (Section 4.4).** Primary alcohols can be oxidized to aldehydes (using PCC) or to carboxylic acids (using $KMnO_4$ or $Na_2Cr_2O_7/H_2SO_4$). Secondary alcohols are oxidized to ketones. Tertiary alcohols cannot be oxidized.

$$RCH_2\text{—OH} \xrightarrow{\text{PCC}} \text{R—C(=O)—H}$$

1° Alcohol → Aldehyde

$$RCH_2\text{—OH} \xrightarrow[\text{or } KMnO_4]{Na_2Cr_2O_7/H_2SO_4} \text{R—C(=O)—OH}$$

1° Alcohol → Carboxylic acid

$$\text{R—CH(OH)—R'} \xrightarrow[\text{or } KMnO_4]{Na_2Cr_2O_7/H_2SO_4} \text{R—C(=O)—R'}$$

2° Alcohol → Ketone

4. **Alcohols from alkyl halides (Section 4.5.3).** A standard means of preparing alcohols is the S_N2 reaction of primary or secondary alkyl halides with sodium hydroxide or the S_N1 reaction of tertiary alkyl halides with water.

Primary and secondary:

$$\text{R—X} \xrightarrow[(S_N2)]{\text{NaOH}} \text{R—OH}$$

Tertiary:

$$\text{R—X} \xrightarrow[(S_N1)]{H_2O} \text{R—OH}$$

Additional Problems

Nomenclature

4.24 Name the following compounds:

(a) HO–(cyclohexane chair)–OH

(b) $CH_3CH_2CH(OH)CH(CH_3)_2$

(c) HO OH HO OH (skeletal structure)

(d) H OH (skeletal structure)

(e) $HOCH_2CH(CH_3)CH_2CH_2OH$

(f) SH (skeletal structure)

(g) OH OH (cyclobutane)

(h) OH (skeletal structure)

(i) (cyclohexane chair)–OH

4.25 Draw a structure for each compound:

(a) isobutyl alcohol

(b) *cis*-2-methylcyclopentanol

(c) 3,3-dimethyl-4-heptanol

(d) (*R*)-2-pentanol

(e) *trans*-4-chlorocyclohexanol

(f) potassium ethoxide

(g) *cis*-cyclobutane-1,3-diol

(h) 2,2,6,6-tetramethylcyclohexanethiol

(i) *t*-butyl alcohol

(j) sodium isopropoxide

(k) butane-2,3-diol

4.26 Classify each of the following alcohols as primary, secondary, or tertiary:

(a) 2-butanol

(b) 1-methylcyclohexanol

(c) 3-methyl-3-hexanol

(d) tricyclopropylmethyl alcohol

(e) (*R*)-2-pentanol

Properties of Alcohols

4.27 Arrange each of the following sets of compounds in order of decreasing solubility in water, and explain your reasoning:

(a) 1-hexanol; hexane; hexane-1,3,5-triol

(b) ethanol; 1-hexanol; 1-decanol

4.28 Explain why hydrogen bonding occurs with alcohols but not with alkanes.

4.29 Write the reaction of each of the following compounds with metallic sodium, and name the product:

(a) methanol

(b) isopropyl alcohol

(c) 1-hexanol

4.30 Place the following alcohols in order of increasing acidity: ethanol; 2,2,2-trichloroethanol; *t*-butyl alcohol.

4.31 Explain why *n*-butyl alcohol is less soluble in water but has a higher boiling point than *t*-butyl alcohol.

4.32 Show the hydrogen bonding between methanol and water in two ways: (a) with methanol as the hydrogen donor and (b) with methanol as the hydrogen acceptor.

Reactions of Alcohols

4.33 Show the major product of each of the following reactions:

(a) cyclohexanol with hydrobromic acid

(b) (*R*)-2-butanol with hydroiodic acid

(c) cyclopropanol with thionyl chloride and pyridine

(d) 3-pentanol with sodium dichromate and sulfuric acid

(e) (*S*)-2-pentanol with metallic sodium

4.34 Explain why the reaction of an enantiomeric alcohol with tosyl chloride to form a tosylate does not change the configuration at the stereocenter, while reaction of the same alcohol with thionyl chloride and pyridine to form an alkyl chloride inverts the configuration at the stereocenter.

4.35 Show how to convert (*S*)-2-pentanol to (*S*)-2-chloropentane. Then show how to convert the same starting material to (*R*)-2-chloropentane.

4.36 Show how you would convert (*R*)-2-pentanol to each of the following:

(a) 2-pentanone ($CH_3COCH_2CH_2CH_3$)

(b) (*S*)-2-pentanol

(c) (*S*)-2-chloropentane

4.37 In the reaction of *t*-butyl alcohol with hydrochloric acid, an 80% yield of *t*-butyl chloride is obtained. However, in the reaction of 1-butanol with the same reagent, the yield of 1-chlorobutane is nearly 100%. Explain the difference in the reactions, and indicate what might have happened to the other 20% of *t*-butyl alcohol.

4.38 Draw the structure for the product of each of the following oxidation reactions:

(a) cyclopentanol with sodium dichromate and sulfuric acid

(b) 1-butanol with potassium permanganate

(c) 1-pentanol with pyridinium chlorochromate (PCC)

(d) 1-pentanol with sodium dichromate and sulfuric acid

4.39 Write the major products of the reaction of each of the following reagents with 1-hexanol:

(a) pyridinium chlorochromate (PCC)

(b) phosphorus tribromide (PBr_3)

(c) potassium permanganate

(d) hydrobromic acid

(e) sodium dichromate and sulfuric acid

(f) thionyl chloride and pyridine

Carbohydrate Chemistry

4.40 The configurational reference compound for carbohydrates is D-glyceraldehyde, the three-carbon aldose [$HOCH_2CH(OH)CHO$], which has the *R* configuration (the same configuration as carbon 5 in D-glucose). Draw the Fischer projection for D-glyceraldehyde and then identify each of the following structures as either D- or L-glyceraldehyde.

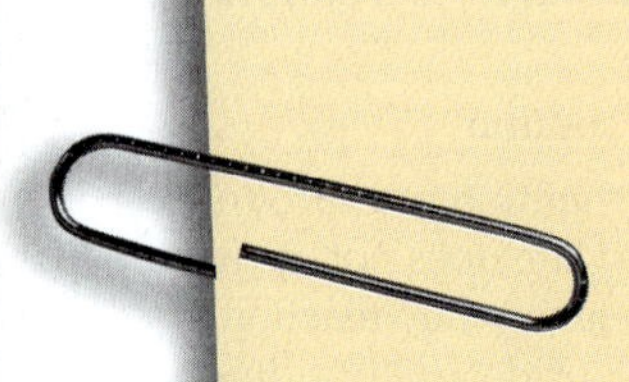

CONCEPTUAL PROBLEM

A Case of Mistaken Identity or Murder?

A famous mycologist, Sir Eugene Francisco, was found dead in his kitchen one Sunday, ostensibly from eating an omelette freshly prepared with mushrooms he had collected just that morning. The police initially classified it as an accidental death, due to mushroom poisoning. Then a close friend of the deceased notified them that Sir Eugene was an expert on mushroom identification and therefore highly unlikely to mistakenly pick and eat the wrong species. Furthermore, he was embroiled in a highly political battle for the presidency of an international mycological society. At this point, you, as the county forensic expert, are contacted by the police to help in the investigation. It is also part of your job to explain the scientific evidence to the police and a jury.

The mushroom presumed to be the culprit is *Amanita muscaria*, which grows in Sir Eugene's locality and is known to contain the deadly poisonous compound muscarine.

Muscarine is a bifunctional compound containing a secondary alcohol and a quaternary ammonium salt (you'll learn more about this in Chapter 12). A sample of muscarine chloride isolated from the mushroom has a melting point of 180–181°C and is optically active (you may want to review Section 3.3). A sample obtained from Sir Eugene's body has the same melting point, but shows absolutely no optical activity.

H_3C … O … $CH_2\overset{+}{N}(CH_3)_3\ Cl^-$ … HO

Muscarine chloride

Amanita muscaria.

During the course of the trial, you are asked a series of questions, some on basic chemistry and some that require you to provide your professional opinion. Based on the facts you've collected and your knowledge of organic chemistry, answer these questions:

- How would you characterize the structure of muscarine? (Do this by identifying the number of stereocenters and assigning each of them a configuration.)
- Given that the naturally occurring muscarine has the configuration shown, draw the other enantiomer. What would the properties of this enantiomer be? What would the optical activity of a 50/50 mixture of the two enantiomers be?
- What can you tell the jury concerning why the muscarine found in the victim's body does not show optical activity like that from the actual mushroom?
- Can you give an explanation for the presence of racemic muscarine in Sir Eugene's body?
- What is your conclusion? Is it murder, or a case of mistaken identity (of the mushroom, that is)?

(a)
CHO
HO—┼—H
CH_2OH

(b)
CHO
H—┼—OH
CH_2OH

(c)
OH
H—┼—CH_2OH
CHO

(d)
OH
$HOCH_2$—┼—H
CHO

(e)
CH_2OH
HO—┼—H
CHO

4.41 β-D-glucose has the hydroxyl group on the anomeric carbon in the equatorial position, while α-D-glucose has this hydroxyl group in the axial position. Explain why the beta form is more stable than the alpha form.

4.42 Explain the meaning of each term: (a) deoxy, (b) epimer, (c) anomeric carbon.

4.43 Starch and cellulose are both polymers of D-glucose. What is the key structural difference between these two polymers?

4.44 Draw the Fischer projections for two D-aldotetroses and for two L-2-ketopentoses.

4.45 L-Fucose is an aldohexose that is present in the cell walls of many bacteria. It can also be described as L-6-deoxygalactose (see Section 4.7.3 for D-galactose, the carbon 4 epimer of glucose). Applying the fact that L-galactose is the mirror image of D-galactose, draw a Fischer projection for L-fucose.

Mixed Problems

4.46 Why are alcohols Lewis bases? Show a step of a reaction that involves an alcohol acting as a Lewis base.

4.47 Show how to carry out the following conversions (they may require more than one step):

(a) 1-butanol to 1-methoxybutane

(b) cyclohexanol to cyclohexyl chloride

(c) cyclohexanol to methylcyclohexane

(d) 2-bromobutane to 2-butanone ($CH_3COCH_2CH_3$)

(e) 1-butanol to butanal ($CH_3CH_2CH_2CHO$)

(f) ethanol to ethyl tosylate

4.48 Explain why each of the following names involves incorrect usage, and supply a proper name:

(a) 2-ethyl-2-butanol

(b) 2,5-pentanediol

(c) isobutanol

(d) 3-hydroxy-1-chlorocyclohexane

4.49 Based on your chemical knowledge to this point, show the expected product if D-glucose is oxidized with the strong oxidant sodium dichromate/sulfuric acid.

4.50 Show the products from each of the following steps of a typical synthesis:

(a) Cyclohexanol reacts with sodium hydride to form A (NaH, a source of hydride anion, is a very strong base; the pK_a of its conjugate acid, H_2, is larger than 20).

(b) Cyclohexanol reacts with PBr_3 to form B.

(c) A and B react to form C.

4.51 Draw a possible structure for a compound of molecular formula $C_4H_{10}O$ that reacts instantly with the Lucas reagent (HCl and $ZnCl_2$).

4.52 Show the reaction and the products when *trans*-4-*t*-butylcyclohexanol reacts with each of the following reagents:

(a) metallic sodium

(b) hydrobromic acid

(c) potassium permanganate

(d) metallic sodium, then methyl iodide

(e) tosyl chloride

(f) tosyl chloride, then sodium hydroxide

(g) thionyl chloride and pyridine

(h) sodium dichromate/sulfuric acid

4.53 Hydroxyl groups in carbohydrates can be "protected" by converting them to acetoxy groups (—OH converted to —$OCOCH_3$). Draw 1,2,3,4-tetraacetoxy-β-D-glucose, and write its reaction with PCC.

4.54 Explain why potassium *t*-butoxide is a stronger base than sodium methoxide, the most common alkoxide base.

4.55 Cyclohexanone can be synthesized from cyclohexane using three separate reactions in sequence. The first reaction is a conversion of cyclohexane to cyclohexyl bromide by reaction with liquid bromine (to be discussed in Chapter 10). Write all three reactions required to complete the synthesis.

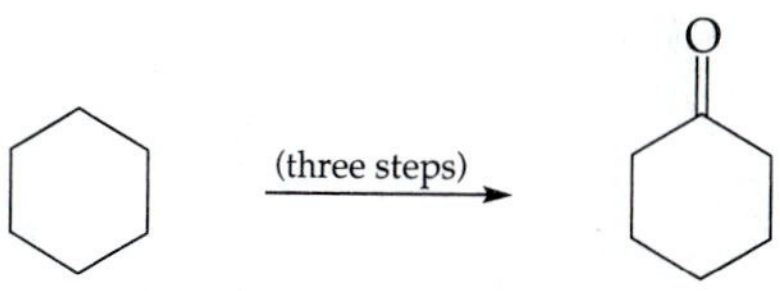

4.56 Write a possible structure for a compound of molecular formula $C_4H_{10}O$ that is oxidized by sodium dichromate/sulfuric acid to produce a ketone.

4.57 Many recipes for "sweets" call for boiling table sugar in water with a trace of acid, such as citric acid (derived from lemons or oranges). The resulting solution has an even sweeter taste after this treatment. Explain why in terms of carbohydrate chemistry.

CHAPTER 5

Ethers

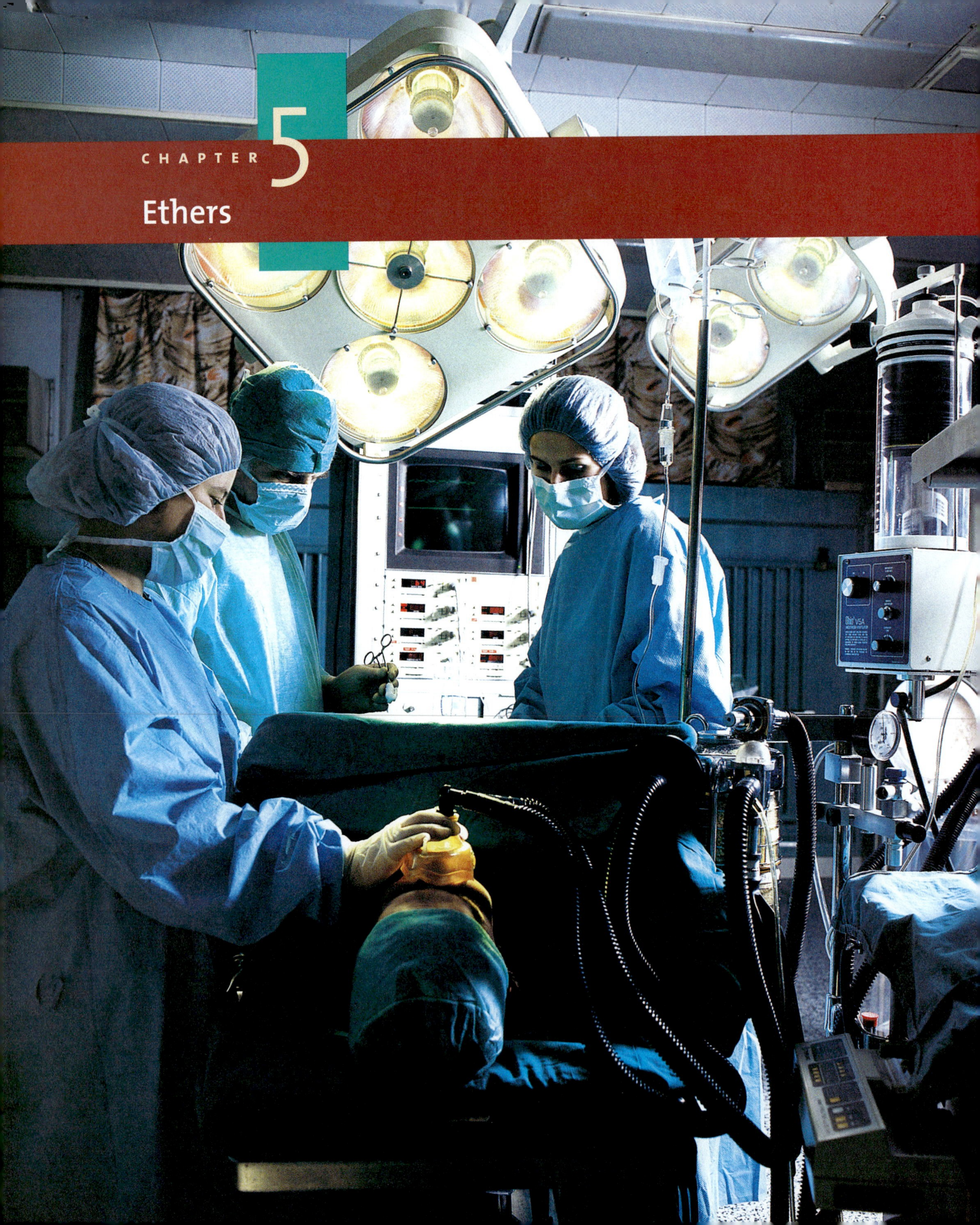

WE BEGAN our study of the families of organic compounds with the parent family of saturated hydrocarbons (alkanes, R—H). Their structure is based on tetrahedral carbon (sp^3 hybridization) and sigma bonding (single bonds). They are distinguished by their chemical inertness. We have also examined two families of compounds in which a functional group has replaced a hydrogen in a parent alkane. We looked first at alkyl halides (haloalkanes, R—X), in which a halogen (—X) is substituted for hydrogen. Next, we saw that a hydroxyl group (—OH) could replace a hydrogen, resulting in alcohols (R—OH). Both of these families of compounds are chemically reactive. Finally, we come to ethers (R—O—R′), which can be viewed as a family in which an alkoxy group (—OR′) replaces a hydrogen of an alkane. Ethers may also be viewed as derivatives of alcohols (R—OH), in which an alkyl group (R′) has replaced the hydrogen of the hydroxyl group. Note also that we are working our way through one segment of the Key to Transformations (see inside cover), but ethers are a "dead end." They can be obtained from alcohols but can be converted only back to alcohols. Thus, their chemistry is limited.

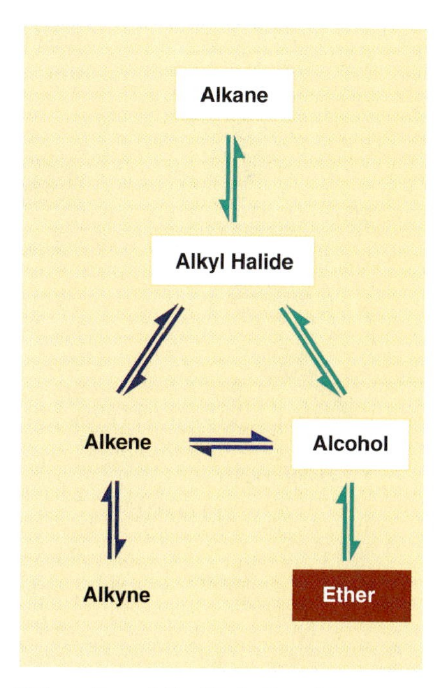

This portion of the Key to Transformations shows the interconnections of families and functional groups having single bonds. Ethers can be viewed as a "dead end" in that they do not serve as a route to other families of organic compounds.

Ethers are represented by the general formula R—O—R′, representing two alkyl groups (R and R′) joined together by an oxygen atom. (Later, we will encounter compounds in which aromatic groups replace one or more alkyl groups.) The functional group of an ether consists of the oxygen atom with two carbon groups attached to it (C—O—C), as if a water molecule had both hydrogens replaced by alkyl groups. Compounds containing only the ether functional group form a relatively small family with only a few significant members. The methyl ether group (CH_3—O—R) is the most common, appearing in a wide variety of polyfunctional natural products.

The general formula for an alkyl ether is $C_nH_{2n+2}O$, apparently an alkane plus an inserted oxygen atom. For example, methyl ether (CH_3—O—CH_3) can be viewed as ethane with an oxygen fitted in between the two methyl groups. Note that the general formula is the same as that for a simple alcohol, which can be viewed as an alkane with an oxygen fitted in between a carbon and a hydrogen. Thus, an ether and an alcohol may be constitutional isomers of each other: They have the same molecular formula but different connectivities. Compare methoxyethane with 1-propanol in Figure 5.1. Both have the molecular formula C_3H_8O.

When you are confronted with the molecular formula of a saturated compound containing a single oxygen, the compound may be an alcohol or an ether. These can be readily distinguished chemically: An alcohol reacts exothermically with sodium (to form an alkoxide and hydrogen gas), whereas an ether

◀ The discovery that inhaling ether could make a patient insensitive to pain revolutionized the practice of medicine.

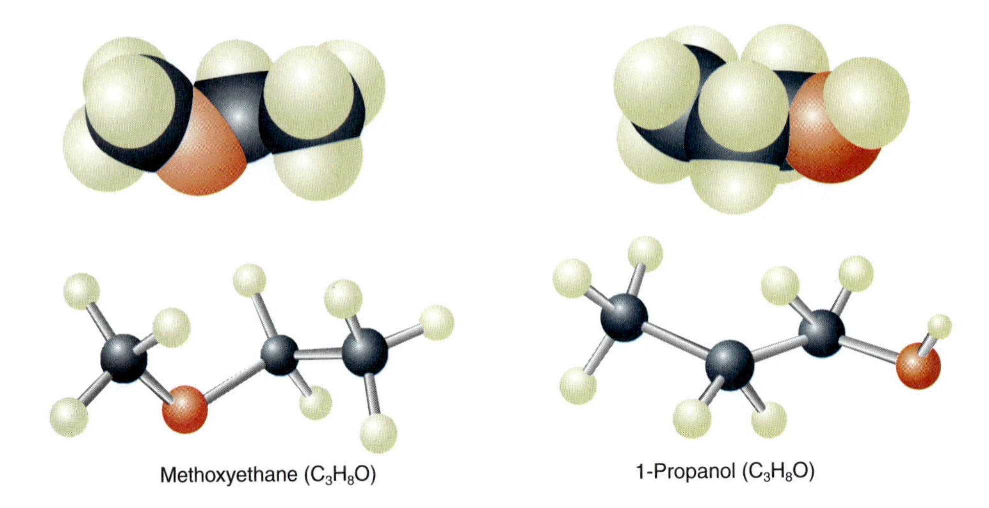

FIGURE 5.1

A pair of constitutional isomers: methoxyethane and 1-propanol.

is inert to sodium. Both will dissolve in concentrated sulfuric acid because of protonation of the oxygen atom to form oxonium salts.

There are straight-chain ethers (**acyclic ethers**) and ethers in which the oxygen is part of a ring system (**cyclic ethers**). Acyclic ethers may contain hydrocarbon rings as part of the alkyl group, as we will see in Section 5.1. The two classes of ethers will be discussed separately.

5.1

Acyclic Ethers

Simple alkyl ethers are relatively low-boiling, colorless liquids. They boil at about the same temperatures as alkanes of similar molecular weights. Ethers are not capable of hydrogen-bonding with each other because of the absence of an —OH group. However, they can hydrogen-bond with other molecules that have —OH groups, such as alcohols. Thus, ethers are relatively soluble in alcohols, and vice versa. Low-molecular-weight ethers can hydrogen-bond to water molecules and so are very slightly soluble in water. However, ethyl ether and water are immiscible—they form two separate layers.

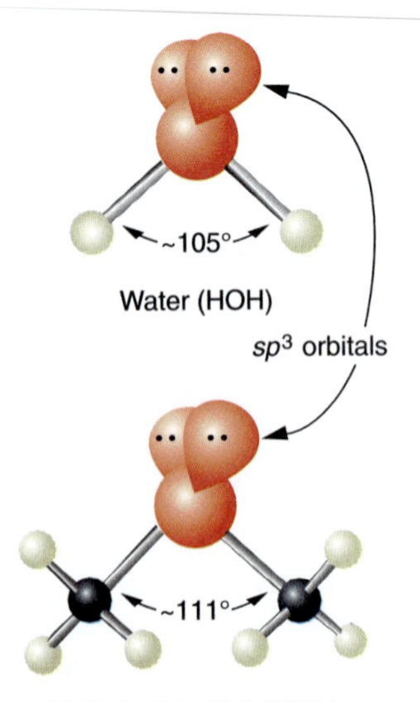

FIGURE 5.2

Models of water and methyl ether.

An ether molecule is shaped like water because the oxygen is sp^3 hybridized in both. The H—O—H bond angle in water (only 105°) is less than the ideal 109° tetrahedral angle because of repulsion between the two lone pairs of electrons and also the small size of the hydrogens. However, the larger methyl groups in dimethyl ether force the C—O—C angle to be wider, about 111° (Figure 5.2).

Ethers are chemically quite unreactive. Acyclic ethers are inert to aqueous acids, bases, and reactive metals such as sodium. Because they are inert and yet slightly polar, ethers are superb solvents. Ethyl ether has been used for this purpose for decades. However, its low boiling point (35°C), high vapor pressure, extreme flammability, and tendency to form explosive peroxides upon extended exposure to air have led to a marked decrease in its use. Other ethers, such as tetrahydrofuran (THF), are more suitable (see Section 5.2).

Although they do not dissolve in water and aqueous acids, ethers dissolve in concentrated sulfuric acid by protonation of the oxygen atom and formation of an oxonium ion. However, no further reaction occurs. This fact allows alkanes (also chemically inert) to be distinguished from ethers, since alkanes are insoluble in concentrated sulfuric acid.

$$\underset{\text{Ether}}{R{-}\ddot{\underset{\cdot\cdot}{O}}{-}R'} + H_2SO_4 \rightleftharpoons \underset{\text{Oxonium ion}}{R{-}\overset{H}{\overset{|}{\underset{\cdot\cdot}{O^+}}}{-}R'} + HSO_4^-$$

5.1.1 Nomenclature of Acyclic Ethers

The IUPAC system of nomenclature treats the RO— group of an acyclic ether R—O—R′ as an **alkoxy** substituent on the alkane R′—H. The name of the alkoxy substituent is derived from that of the alkoxide by replacing *-ide* with *-y*. For example, methoxide, CH_3O—, becomes methoxy, and butoxide, C_4H_9O—, becomes butoxy. This system works well because most ethers have at least one simple, easily named alkyl group (R or R′).

$C_2H_5{-}O{-}C_2H_5$

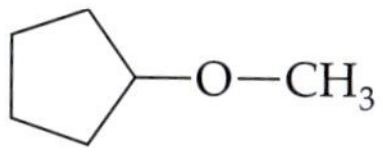

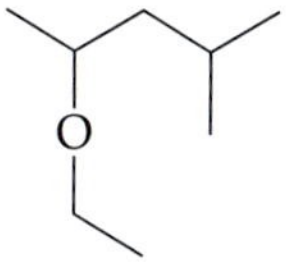

IUPAC:	Ethoxyethane	Methoxycyclopentane	2-Ethoxy-4-methylpentane
Common:	(ethyl ether)	(methyl cyclopentyl ether)	

Because most common ethers are relatively simple compounds, their historical names are still used. In these historical/common names, shown in parentheses below the IUPAC names, the two alkyl groups that are attached to the oxygen of the ether are simply named as alkyl groups. The two alkyl groups may be different (an unsymmetrical ether, such as ethyl methyl ether) or the same (a symmetrical ether, such as ethyl ether). If the two groups are the same, as in ethyl ether, the expected *di-* prefix is often not used because the name remains unambiguous. Thus, since only one group is named in isopropyl ether, there must be two isopropyl groups to make the compound an ether.

How to Solve a Problem

Assign a name to the following compound:

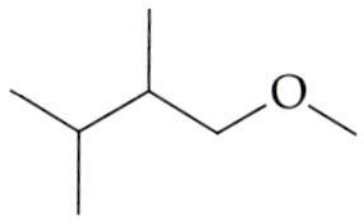

SOLUTION

First we determine if a common name is applicable, one in which two simple (generally C_1 through C_4) alkyl groups can be named. In this case, the methyl group is obvious, but the other group, a C_6 group, is not a simple alkyl group. Therefore, we turn to the IUPAC system, which requires identifying an alkoxy group as a substituent on a parent alkane. The longest alkane chain has four carbons, so the parent name is butane. The substituents are two methyl groups and a methoxy group. The methoxy group is assigned the lowest number, because it precedes the methyl groups in alphabetical order. Therefore, the name is 1-methoxy-2,3-dimethylbutane.

PROBLEM 5.1

Draw the structural formulas for the following compounds: (a) cyclohexyl methyl ether, (b) *n*-propyl ether, (c) 2-ethoxybutane, (d) isopropoxycyclopropane, and (e) ethyl propyl ether.

PROBLEM 5.2

Assign a name to the following compounds:

(a) ethyl–O–propyl (skeletal) (b) cyclobutyl–OC_2H_5 (c) cyclopropyl–O–cyclopropyl

---

5.1.2 Williamson Synthesis of Ethers

A single reaction is used to prepare most ethers. It is called the **Williamson synthesis,** after its discoverer. The reaction is a normal S_N2 substitution by a nucleophilic alkoxide on an alkyl halide. (An alkoxide, you remember from Section 4.2.3, is an anion prepared from an alcohol.) This direct application of the S_N2 reaction studied in Section 3.4.1 is subject only to the normal limitations on that type of reaction.

General reaction for the Williamson synthesis:

$$\underset{\text{Alcohol}}{R{-}O{-}H} \xrightarrow{Na} \underset{\text{Alkoxide}}{R{-}O^- \; Na^+} \xrightarrow[\text{(alkyl halide)}]{R'{-}Br} \underset{\text{Ether}}{R{-}O{-}R'} + NaBr$$

Because the Williamson synthesis requires an alcohol and an alkyl halide as starting materials, it is possible to conceptualize two routes for the synthesis of any unsymmetrical ether. For example, in the case of the synthesis of ethoxycyclopentane, routes A and B are possible:

Route A: $\underset{\text{Cyclopentanol}}{C_5H_9{-}OH} \xrightarrow{Na} C_5H_9{-}O^- \xrightarrow[\text{(bromoethane)}]{C_2H_5{-}Br} \underset{\text{Ethoxycyclopentane}}{C_5H_9{-}OC_2H_5}$

Route B: $\underset{\text{Ethanol}}{C_2H_5{-}OH} \xrightarrow{Na} C_2H_5{-}O^- \xrightarrow[\text{(bromocyclopentane)}]{C_5H_9{-}Br} \underset{\text{Ethoxycyclopentane}}{C_5H_9{-}OC_2H_5}$

Route A starts with cyclopentanol, forms its alkoxide, and then substitutes it into bromoethane (the source of the second alkyl group). Route B starts with ethanol, forms its ethoxide, and then substitutes it into bromocyclopentane. Route A is the better choice because it involves an S_N2 substitution on a primary halide, bromoethane. Bromocyclopentane is a secondary halide, and with a secondary halide, there is always a slight possibility of a competing elimination reaction. Therefore, the yield of the desired ether is maximized in route A, because of minimal by-product formation.

The prominence of the methyl ether group ($—OCH_3$) was mentioned at the beginning of this chapter. One way to "protect," or "mask," a hydroxyl group and keep it from being affected by another reagent is to convert it into a methyl ether. In these cases, the alcohol is converted to its alkoxide, which is then reacted with methyl iodide to produce the methyl ether:

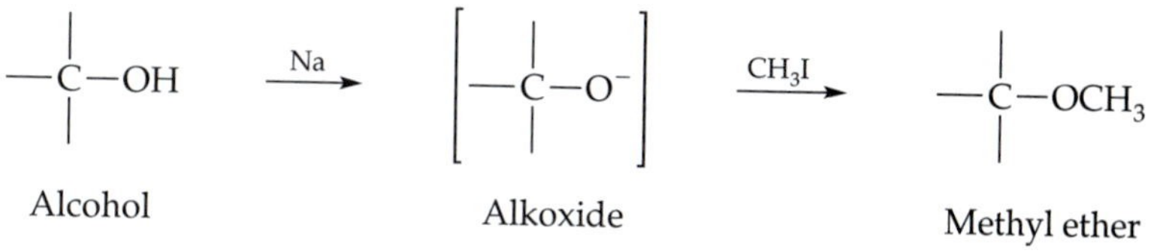

$$\text{—C—OH} \xrightarrow{\text{Na}} [\text{—C—O}^-] \xrightarrow{\text{CH}_3\text{I}} \text{—C—OCH}_3$$

Alcohol — Alkoxide — Methyl ether

How to Solve a Problem

Design a synthesis of *t*-butyl ethyl ether (2-ethoxy-2-methylpropane).

PROBLEM ANALYSIS

As in Chapter 4, we apply the approach known as *retrosynthesis*. Ethers are prepared by the Williamson synthesis. Using retrosynthetic analysis, we find that we could, in principle, obtain *t*-butyl ethyl ether using either of two possible sets of reactants, designated on the bond-line formula by the dashed lines A and B. The bond at A could be made from ethoxide displacing bromide from *t*-butyl bromide. Alternatively, the bond at B could be made from *t*-butoxide displacing bromide from ethyl bromide.

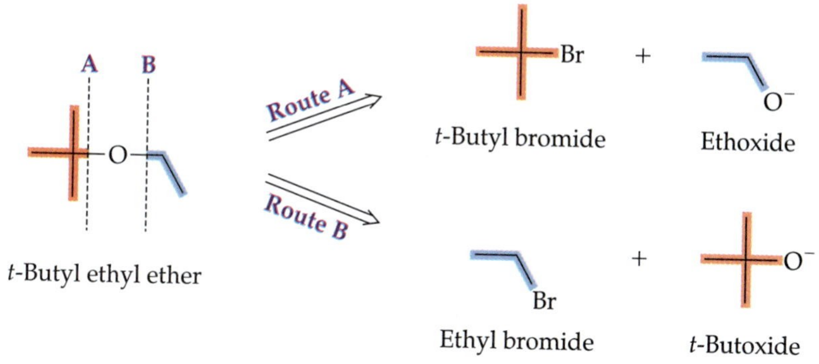

A B

O

t-Butyl ethyl ether

Route A

Br + O^-

t-Butyl bromide — Ethoxide

Route B

Br + O^-

Ethyl bromide — *t*-Butoxide

SOLUTION

For a practical synthesis of *t*-butyl ethyl ether, only route B will be successful. Route B involves forming *t*-butoxide anion from *t*-butyl alcohol and sodium, then bringing about a substitution reaction on the primary bromide, ethyl bromide. Recall that S_N2 substitutions are most effective on primary alkyl halides.

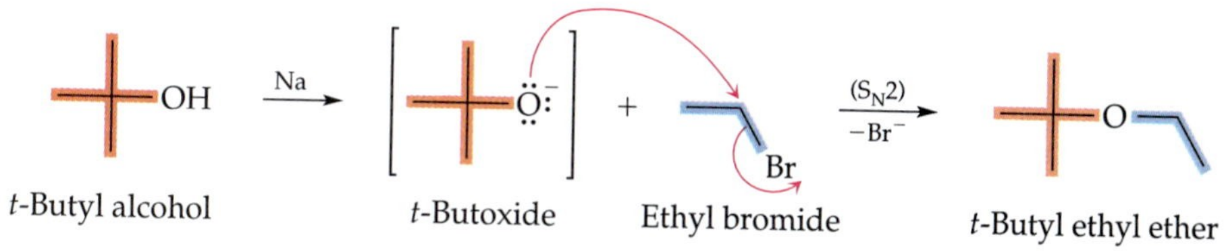

OH $\xrightarrow{\text{Na}}$ [$\ddot{\text{O}}\!:^-$] + Br $\xrightarrow[-\text{Br}^-]{(S_N2)}$ O

t-Butyl alcohol — *t*-Butoxide — Ethyl bromide — *t*-Butyl ethyl ether

Route A is not a practical synthesis of *t*-butyl ethyl ether. The major product will be the alkene isobutylene because the strong base (sodium ethoxide) carries out an E2 elimination reaction on the tertiary alkyl halide *t*-butyl bromide, rather than an S_N2 substitution reaction:

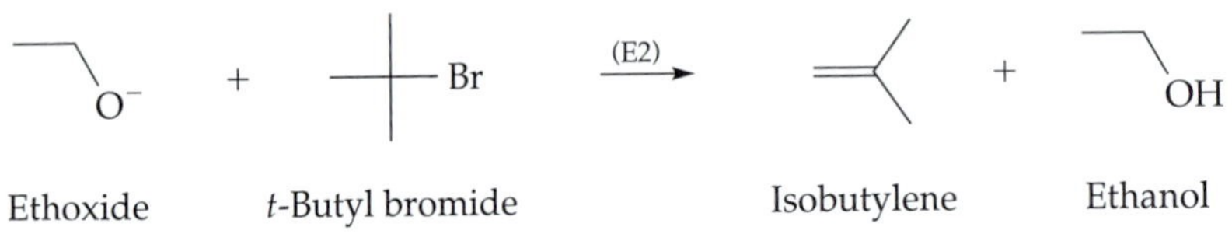

Ethoxide + *t*-Butyl bromide $\xrightarrow{(E2)}$ Isobutylene + Ethanol

Recall that tertiary alkyl halides are the least useful in S_N2 substitutions because of steric hindrance to attack of the nucleophile at the carbon carrying the halogen (see Section 3.4.2) and predominant E2 elimination.

PROBLEM 5.3

Show the products of the following reactions:

(a)  Cyclopropanol (—OH) $\xrightarrow[2.\ C_2H_5I]{1.\ Na}$

(b)  1-Butanol (OH) $\xrightarrow[2.\ CH_3Br]{1.\ Na}$

PROBLEM 5.4

Show the reactions required to synthesize each of the following compounds, starting with any alcohol and any other required reagent: (a) methoxycyclohexane, (b) isopropyl ether, (c) 2-methoxy-2-methylbutane.

5.1.3 Cleavage of Methyl Ethers

As mentioned earlier, hydroxyl groups are often temporarily converted to methyl ethers in a process called *protection*, which masks the hydroxyl group to keep it from being affected by another reagent. Protection is effective because the ether functional group is so chemically unreactive. However, for such protection to be useful, there must be a means of reconverting the methyl ether group back to the hydroxyl group. Such reconversion is accomplished by cleaving the ether with one equivalent of a very strong acid, usually hydroiodic acid (HI). (Cleavage by HI is effective with other ethers, not just methyl ethers, but these will not be discussed in this book.)

In this reaction, the ether is first protonated to form an oxonium ion. Then the iodide anion (an excellent nucleophile) attacks the methyl group of the oxonium ion via an S_N2 substitution. The remainder of the oxonium ion (an alcohol) serves as an excellent leaving group. The methyl group is where the S_N2 substitution invariably occurs because of the minimal steric hindrance there (see Section 3.4.1).

General cleavage reaction:

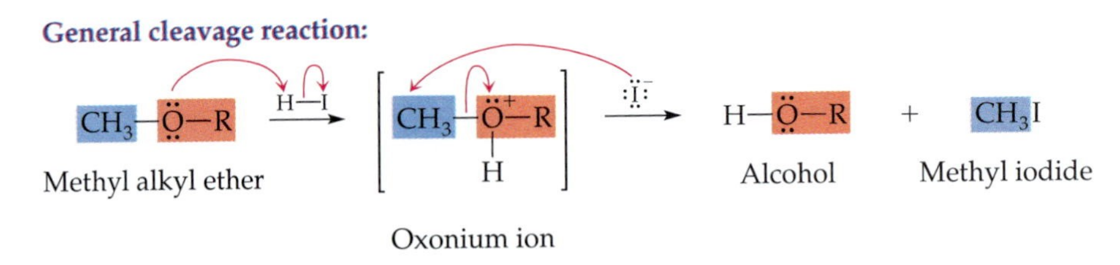

With this cleavage reaction, a starting alcohol can be recovered after being protected. For example, cyclohexanol can be protected as its methyl ether and then recovered after HI cleavage:

Cyclohexanol $\xrightarrow[\text{2. } CH_3I]{\text{1. Na}}$ Cyclohexyl methyl ether $\xrightarrow{HI}$ Cyclohexanol

PROBLEM 5.5

Show the products of the following reactions:

(a) cyclobutyl–OCH_3 $\xrightarrow{HI}$

(b) butyl–OCH_3 $\xrightarrow{HI}$

(c) tert-butyl–OCH_3 $\xrightarrow{HI}$

5.2 Cyclic Ethers

Cyclic ethers are compounds that fit the general ether definition, having a C—O—C linkage, but with that linkage incorporated in a ring. Cyclic ethers have widely varying ring sizes, but the most useful and common cyclic ethers are those with a three-membered ring. Known as epoxides, these compounds are by far the most significant cyclic ethers, so we will focus on their chemistry. Cyclic ethers with larger rings are much less prevalent. Many larger cyclic ethers are unique and naturally occurring—products of bacterial or fungal metabolism, for instance. However, two larger cyclic ethers, best known by their historical names tetrahydrofuran (THF) and dioxane, are especially effective reaction solvents.

Cyclic ethers are one of the main components of epoxy glues. Such a glue is strong, but also lightweight. It is used as a component of the Stealth bomber.

5.2.1 Nomenclature of Cyclic Ethers

Cyclic ethers with three-membered rings are called **epoxides** (historical name) or **oxiranes** (IUPAC name). You will see both family names used regularly. The IUPAC names for individual compounds are based on the name of the unsubstituted parent compound, *oxirane*. Substituents are located on the ring by numbering, starting with the oxygen atom. The historical name for a three-membered cyclic ether is derived from the name of the alkene from which it is most commonly prepared, followed by the word *oxide*, for an *alkene oxide*. Examples are ethylene oxide and cyclohexene oxide. Note that *cis-trans* isomerism is possible with these cyclic compounds, just as with cycloalkanes.

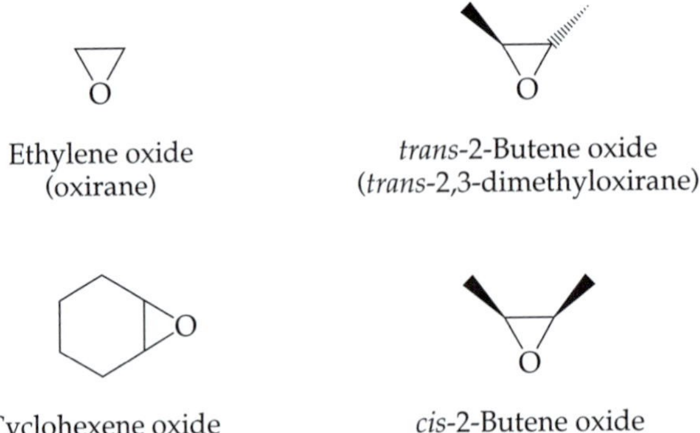

Ethylene oxide (oxirane)

trans-2-Butene oxide (*trans*-2,3-dimethyloxirane)

Cyclohexene oxide

cis-2-Butene oxide (*cis*-2,3-dimethyloxirane)

5.2.2 Preparation of Epoxides (Oxiranes)

Most epoxides (oxiranes) are made from an alkene, an unsaturated hydrocarbon that contains a double bond (to be described in Chapter 6). The reaction is called **epoxidation,** and it involves mild oxidation of the alkene using a **peracid,** also called a *peroxyacid*, which is a carboxylic acid with an "extra" oxygen. You haven't studied carboxylic acids (RCOOH) yet, but all you need to know at this time is that peracids have an extra oxygen (R—CO—**O**—OH), analogous to hydrogen peroxide (H—**O**—O—H), which may be regarded as water with an extra oxygen. In this reaction, the peracid transfers its extra oxygen directly to the alkene, forming an epoxide and a carboxylic acid by-product. The stereochemistry of the alkene is unchanged in this reaction. Peracetic acid is frequently used as the reagent for this reaction.

General epoxidation reaction:

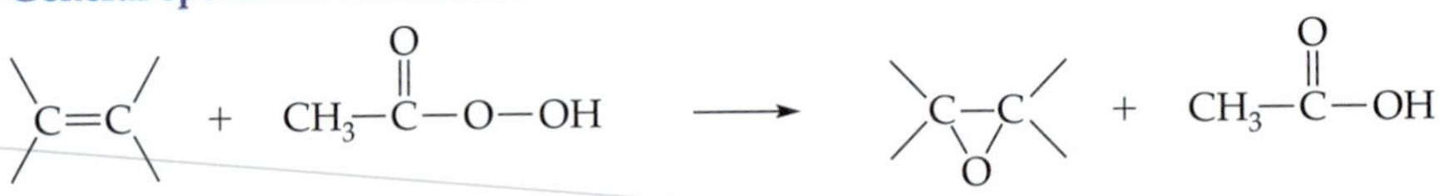

$$>C{=}C< \; + \; CH_3{-}\overset{O}{\overset{\|}{C}}{-}O{-}OH \longrightarrow >C{-}C< \text{(epoxide, O bridging)} \; + \; CH_3{-}\overset{O}{\overset{\|}{C}}{-}OH$$

Alkene Peracetic acid Epoxide Acetic acid

As an example, cyclopentene is converted to cyclopentene oxide in high yield:

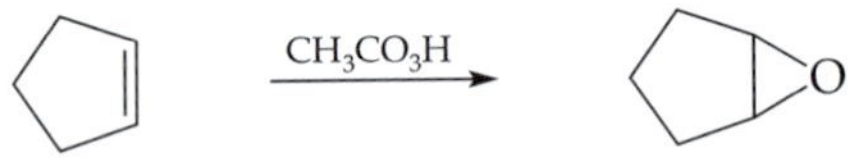

$$\text{Cyclopentene} \xrightarrow{CH_3CO_3H} \text{Cyclopentene oxide}$$

Cyclopentene Cyclopentene oxide

Of course, preparing an epoxide requires the precursor alkene. We will cover alkenes in the next chapter.

PROBLEM 5.6

Although cyclopentene oxide can be readily prepared by the epoxidation reaction just shown, it can also be prepared from *trans*-2-chlorocyclopentanol. Can you envision such a reaction? Show the reagents and steps required. (*Hint:* Think about the S_N2 reaction and the Williamson ether synthesis.)

PROBLEM 5.7

Show the products from reacting peracetic acid with (a) 2,3-dimethyl-2-butene, $(CH_3)_2C{=}C(CH_3)_2$, and (b) propene, $CH_3CH{=}CH_2$.

5.2.3 Acid-Catalyzed Ring Opening of Epoxides

Ethers are known for their lack of chemical reactivity. This inertness is also characteristic of the cyclic ethers except for those with a highly strained ring, the epoxides (oxiranes). The bond angles at carbon and oxygen in an epoxide are only about 60°, much smaller than the optimal 109.5° that occurs with tetrahedral hybridization. The resulting ring strain is reflected in unusual reactivity.

Epoxides are readily cleaved by acids in aqueous or alcoholic solution. The epoxide is protonated by a strong acid to form a cyclic oxonium ion. The positively charged oxygen is highly electronegative, weakening the carbon-oxygen bond. (This is analogous to the oxonium ion, ROH_2^+, formed from an alcohol whose carbon-oxygen bond is easily broken; see Section 4.3.1.) The positively charged oxygen is an excellent leaving group when a carbon is attacked by a nucleophile. If hydroiodic acid (HI) is used, iodide serves as the nucleophile to produce an *iodohydrin:*

Ethylene oxide —HI→ [Oxonium ion] —I^-→ Iodohydrin

The full mechanism of the acid-catalyzed ring opening with water or methanol is shown in Figure 5.3.

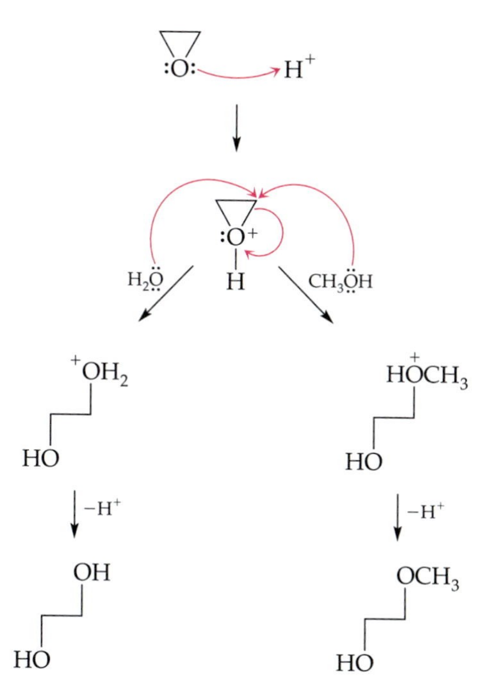

The nucleophilic oxirane (ethylene oxide) is protonated, forming a cyclic oxonium ion.

The carbon-oxygen bonds are weak. Attack of a nucleophile (water or methanol) in an S_N2-type reaction opens the ring. Attack can occur at either carbon, but is a back-side attack in both cases.

The product of ring opening is an oxonium ion that loses its proton to the solvent (water or methanol). The overall reaction is acid-catalyzed (the acid is consumed in the first step but regenerated in the last step).

FIGURE 5.3
Acid-catalyzed reaction of ethylene oxide (oxirane) with water or methanol.

If the epoxide is unsymmetrical, as in propylene oxide, S_N2 attack occurs mainly at the primary (less substituted) carbon rather than at the secondary carbon, mainly because of lower steric hindrance.

S_N2 attack at primary carbon

Propylene oxide (2-methyloxirane) $\xrightarrow{HI}$ 1-Iodo-2-propanol

However, if one of the carbons is tertiary, as in isobutylene oxide, the cyclic oxonium ion opens spontaneously to form a tertiary carbocation, which then is attacked by the nucleophile.

S_N1 reaction at tertiary carbon

Isobutylene oxide (2,2-dimethyloxirane) $\xrightarrow[CH_3OH]{H^+}$ 2-Methoxy-2-methyl-1-propanol

If the epoxide ring is fused to a cycloalkane ring, a **stereospecific ring opening** results. For example, cyclopentene oxide reacts with acidic methanol to produce only *trans*-2-methoxycyclopentanol (Figure 5.4). This is because the S_N2 attack is a back-side attack on the oxonium ion. The methanol nucleophile attacks the cyclic oxonium ion from above the plane of the ring, while the C—O bond breaks on the other side of the ring. This is called a *trans* ring opening. The incoming nucleophile and the hydroxyl group are on opposite sides of the cyclopentane ring.

Coincidentally, cleavage of the resulting *trans*-2-methoxycyclopentanol (an acyclic ether) with HI produces the *trans* diol, which can also be produced from cyclopentene

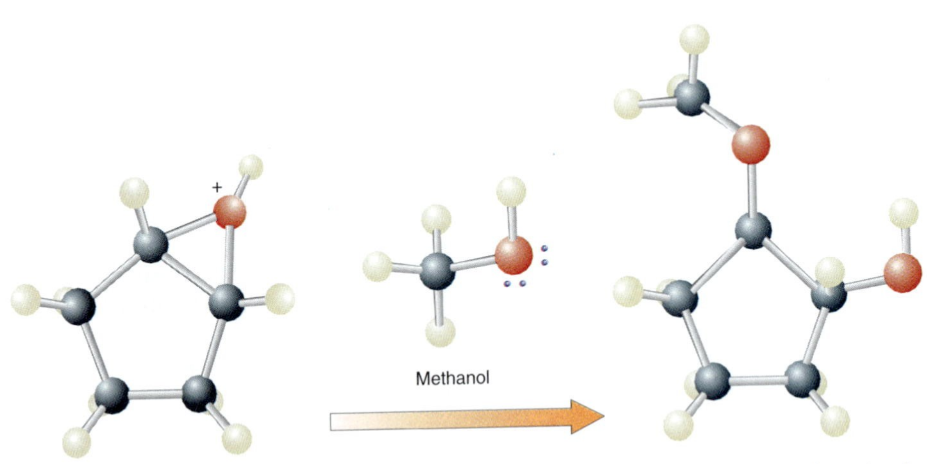

FIGURE 5.4 Ball-and-stick models showing the acid-catalyzed methanolysis of cyclopentene oxide.

oxide by ring opening with dilute acid and water (in place of methanol). In Chapter 6, we will see how to produce the *cis* diol.

Cyclopentene oxide —(H^+ / CH_3OH)→ *trans*-2-Methoxycyclopentanol —(HI)→ *trans*-1,2-Cyclopentanediol

Cyclopentene oxide —(H^+/H_2O)→ *trans*-1,2-Cyclopentanediol

PROBLEM 5.8

Show the product(s) and stereochemistry of the reaction of cyclohexene oxide with dilute aqueous acid.

PROBLEM 5.9

Show the reactions and reagents necessary to convert ethylene ($CH_2{=}CH_2$) into 2-ethoxyethanol. (*Hint:* Use retrosynthetic analysis.)

5.2.4 Nucleophilic Ring Opening of Epoxides

The epoxide (oxirane) ring is reactive enough to be opened by a strong nucleophile in an S_N2 substitution without initial formation of an oxonium ion intermediate. The ring strain is relieved by this opening. For example, propylene oxide reacts with hydroxide or methoxide to produce oxyanions. These anions, which are stable in solution, can be protonated during reaction work-up to yield the same products obtained from the acid-catalyzed ring openings. Note that the nucleophile preferentially attacks the *least substituted carbon,* consistent with an S_N2 mechanism.

Propylene oxide (2-methyloxirane) + Na^+ $^-$:ÖH → [$^-$O … OH] —(H^+/H_2O)→ HO … OH (1,2-Propanediol)

Propylene oxide (2-methyloxirane) + Na^+ $^-$:ÖCH$_3$ → [$^-$O … OCH_3] —(H^+/H_2O)→ HO … OCH_3 (1-Methoxy-2-propanol)

The most important application of this type of nucleophilic substitution reaction involving ring opening relies on the use of either organolithium compounds (R—Li) or Grignard reagents (R—MgBr). Because these are both very strong carbanionic nucleophiles (see Section 3.5.1), they attack the carbons of the epoxide ring in similar S_N2 substitutions and follow the preferred S_N2 order of position for such attack (that is, $1° > 2° > 3°$ carbon). The reactions are always "worked up" by quenching with dilute acid (H^+/H_2O), which protonates the oxyanion (alkoxide) to form the corresponding alcohol.

General reaction:

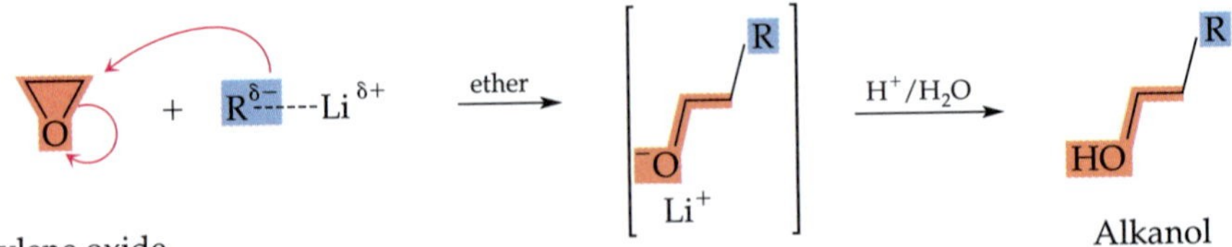

The significance of this reaction is that it provides an important means of making new carbon-carbon bonds. For example, reacting ethylene oxide with ethyllithium converts a C_2 compound into a C_4 compound, an example of how carbon skeletons can be built. Starting with an alcohol, converting it to the alkyl bromide, making the organolithium reagent, and finally reacting this with ethylene oxide yields a new alcohol with a carbon chain two carbons longer:

$$\text{R—OH} \xrightarrow{\text{HBr}} \text{R—Br} \xrightarrow[\text{ether}]{\text{Li}} \text{R—Li} \xrightarrow[\text{2. H}^+/\text{H}_2\text{O}]{\text{1. ethylene oxide}} \text{R—CH}_2\text{CH}_2\text{—OH}$$

(two-carbon homologation)

This is called a **two-carbon homologation reaction**—adding two carbons at once to a compound. The reaction of an organometallic reagent with ethylene oxide is a superb means of adding a two-carbon unit to a structure. For example, cyclohexanol is converted to 2-cyclohexylethanol using ethylene oxide, and cyclopentene oxide is converted to 2-propylcyclopentanol using propylmagnesium bromide:

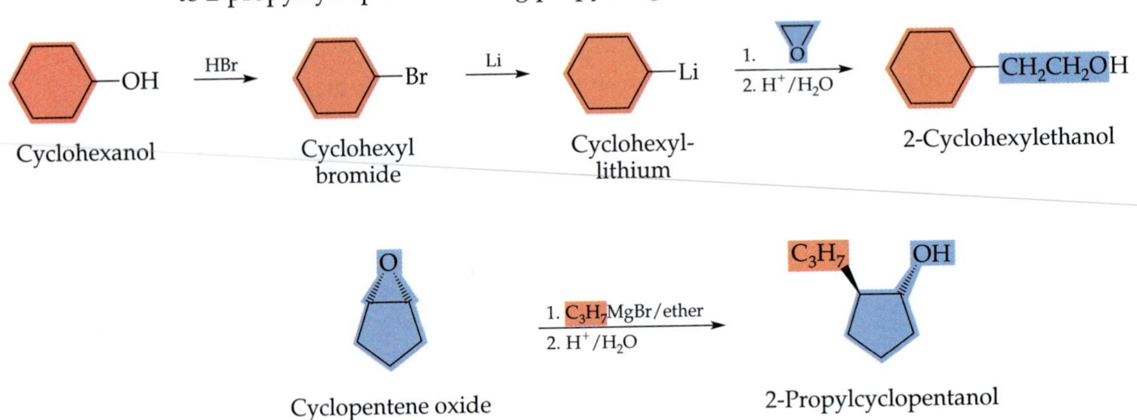

Note that each nucleophilic ring opening of an epoxide (oxirane) involves S_N2 attack on the ring, requiring that the nucleophile attach itself to the side of the ring opposite the hydroxyl group. Therefore, these are *trans* additions, as exemplified by the formation of *trans*-2-methylcyclopentanol from the reaction of methyllithium with cyclopentene oxide:

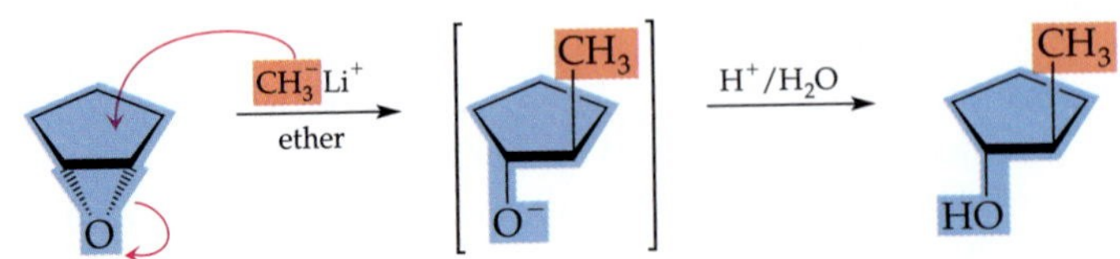

How to Solve a Problem

How can we convert cyclopropyl bromide to 1-bromo-2-cyclopropylethane?

PROBLEM ANALYSIS

First, we draw the structures for the given compounds and identify by a dashed line the key bond(s) that must be constructed.

Cyclopropyl bromide $\xrightarrow{?}$ 1-Bromo-2-cyclopropylethane ($\triangleright$–CH_2CH_2Br)

Then we inspect the structures and note the key changes. In this case, what is required is a two-carbon homologation reaction.

The best advice in synthesis is to worry first about getting the proper number of carbons joined and worry later about installing the right functional group. Thus, the first challenge is how to join two carbons to the cyclopropyl ring. (*Note:* Carbon-carbon bonds frequently are formed by S_N2 reactions.) There are two possible approaches:

(1) The bromine on the starting cyclopropyl ring is a leaving group, so substitution (via S_N2) of cyclopropyl bromide with a nucleophile can be envisioned. The nucleophile would have to be a carbanion, but would have to have a bromine attached to the carbon next to the negatively charged carbon. Thus, there would be two alkyl halide functional groups available for reaction, and that complicates matters. So let's consider the alternative approach.

(2) Since there is a bromine already on the cyclopropyl ring, the starting compound could be converted into an organolithium (or Grignard) reagent, making it a nucleophile. Needing a C_2 unit to react with the organometallic compound, we think of ethylene oxide as a substrate for an S_N2 reaction. That would form the new carbon-carbon bond, would add the necessary two carbons, and would leave a hydroxyl group on the terminal carbon. It is easy to convert a primary alcohol to a primary alkyl bromide (see the Key to Transformations).

SOLUTION

We react cyclopropyl bromide with lithium in ether, then add ethylene oxide. We quench the reaction with acid and water and isolate 2-cyclopropylethanol. Finally, we react the alcohol with HBr (or PBr_3) to produce the desired product.

Cyclopropyl bromide $\xrightarrow[\text{ether}]{\text{Li}}$ Cyclopropyllithium $\xrightarrow{\text{ethylene oxide}}$ [$\triangleright$–$CH_2CH_2O^-$] $\xrightarrow{H^+/H_2O}$ 2-Cyclopropylethanol ($\triangleright$–CH_2CH_2OH) $\xrightarrow{\text{HBr}}$ 1-Bromo-2-cyclopropylethane ($\triangleright$–CH_2CH_2Br)

Overall transformation desired

PROBLEM 5.10

Show the products of the following reactions.

(a) Cyclohexene oxide $\xrightarrow[\text{2. } H^+/H_2O]{\text{1. } CH_3MgBr/\text{ether}}$

(b) 2-Hexene (skeletal structure) $\xrightarrow[\text{2. } C_2H_5Li \text{ 3. } H^+/H_2O]{\text{1. } CH_3COOOH}$

PROBLEM 5.11

Show the reagents necessary to complete the following transformations—more than one step will be required. (Remember to apply your retrosynthetic skills; show each step with its products.)

(a) Cyclobutyl–OH $\xrightarrow{?}$ Cyclobutyl–CH_2CH_2OH

(b) Cyclopentyl–Br $\xrightarrow{?}$ Cyclopentyl–$CH_2CH_2CH_2$Br

5.3 Important Ethers

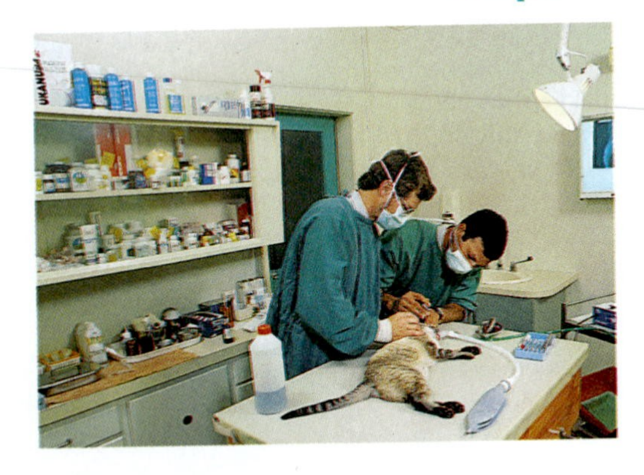

Anesthesia being used in veterinary surgery.

Probably one of the most common associations people have for the term *ether* is its use as an anesthetic. Ethyl ether (called "ether" in hospitals) was an important general anesthetic first used in surgery in the 1840s. It is seldom used today because it has a tendency to form explosive peroxides in containers left exposed to air. Ethyl ether is made commercially by intermolecular dehydration of ethanol, a reaction not widely applicable to the production of other ethers:

$$\underset{\text{Ethanol}}{CH_3CH_2{-}OH} \xrightarrow[140^\circ C]{H_2SO_4} H_2O + \underset{\text{Ethyl ether}}{CH_3CH_2{-}O{-}CH_2CH}$$

As was mentioned earlier, the cyclic ethers tetrahydrofuran (THF) and dioxane are very important as solvents for chemical reactions. They are effective solvents because their inertness keeps them from entering into reactions while their slight polarity and the unshared electrons on oxygen make them good dissolving and complexing agents. Also of practical importance are the ethers called epoxy glues, introduced in "Chemistry at Work."

CHEMISTRY AT WORK

IN THE MARKETPLACE

Epoxy Glues

If the word *epoxy* is familiar to you, it's probably in connection with glue. At this very minute, you are probably sitting near an object manufactured using an epoxy glue: a chair, a table, a pair of skis or in-line skates, maybe even the binding of this book. Epoxy glues are mainstays of home repair and hobby kits because they are easy to use and bond a wide variety of materials very strongly, including metals, glass, wood, and plastics. Epoxies are used in aircraft manufacturing and other industries to bond certain structural parts together, replacing the need for rivets or screws. The advantages of epoxies are strength, low weight, toughness, and resistance to chemical change.

The bonding property of epoxy glues depends on the nucleophilic ring opening of epoxides. All such glues have two components: the epoxide (called the *resin*) is mixed with a hardener, or *curing agent,* which is a di- or trinucleophile. The two components are mixed just before the epoxy glue is applied. When mixed with the resin, the curing agent opens the ring of the epoxide in an S_N2 reaction, thereby joining the two components in a chain. The key to the strength of the glue, however, is the *cross-linking* between polymer chains, forming a three-dimensional web. The compounds generalized in the following reaction may have any of several structures for groups **A** and **B;** the common formulation illustrated here uses an amine group (—NH_2) as the nucleophile.

The ring-opening reaction of epoxy glues is an important example of how organic reactions can provide useful products with desired physical properties.

$$\overset{O}{\overbrace{CH_2{-}CH}}{-}CH_2{-}\mathbf{A}{-}CH_2{-}\overset{O}{\overbrace{CH{-}CH_2}} \quad + \quad H_2\ddot{N}{-}\mathbf{B}{-}\ddot{N}H_2$$

Epoxy resin

Curing agent (triamine)

$$\downarrow$$

$$\sim H\ddot{N}{-}\mathbf{B}{-}\ddot{N}H{-}CH_2{-}\underset{}{\overset{OH}{CH}}{-}CH_2{-}\mathbf{A}{-}CH_2{-}\overset{OH}{CH}{-}CH_2{-}H\ddot{N}{-}\mathbf{B}{-}\ddot{N}H\sim$$

Polymer chain

$$\downarrow$$

Crossed-linked polymer chains

$$\mathbf{A} = {-}O{-}C_6H_4{-}\underset{CH_3}{\overset{CH_3}{C}}{-}C_6H_4{-}O{-} \qquad \mathbf{B} = {-}CH_2CH_2NHCH_2CH_2{-}$$

PROBLEM 5.12

Ethyl ether is often prepared from ethanol by an intermolecular dehydration reaction under acid catalysis, as already shown. Although the details of the reaction were not discussed, write a mechanism for this reaction. (*Hint:* There is only one molecule for the acid to protonate. After that, there is only one nucleophile present to react with the resulting oxonium ion.)

Because automobile emissions contribute to air pollution, government agencies have mandated the use of "oxygenate" additives in gasolines in some metropolitan areas. One major additive is called MTBE (methyl *t*-butyl ether). It is made commercially by adding methanol to isobutylene in an acid-catalyzed process:

$$\text{Isobutylene} + CH_3OH \xrightarrow{H^+} \text{Methyl } t\text{-butyl ether (MTBE)}$$

Isobutylene — Methanol — Methyl *t*-butyl ether (MTBE)

In this reaction, isobutylene is protonated to form the *t*-butyl cation, which is attacked by the nucleophile, methanol. The additive MTBE accounts for about 11% of the volume of reformulated gasoline. The oxygen atom in MTBE facilitates more complete combustion of the gasoline, thereby reducing harmful emissions. MTBE also raises the octane level of the gasoline (the octane rating of pure MTBE is about 110).

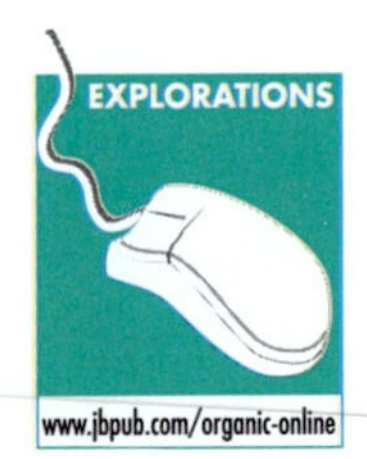

You may recall being introduced to a special type of ether in Section 1.2.3 (see "Chemistry at Work," p. 28) The cyclic polyether shown there belongs to a family of very large cyclic polyethers called *crown ethers* because of their shape. They have become important in recent years because of their ability to form complexes with metal cations. The diameter of the central cavity of 18-crown-6 is just right to hold a potassium ion, which forms a *complex* with the unshared electrons on the six ether oxygen atoms. (The name 18-crown-6 is a shorthand notation that indicates the size of the ring, 18 atoms, and the number of ether functional groups—that is, the number of oxygens, 6.)

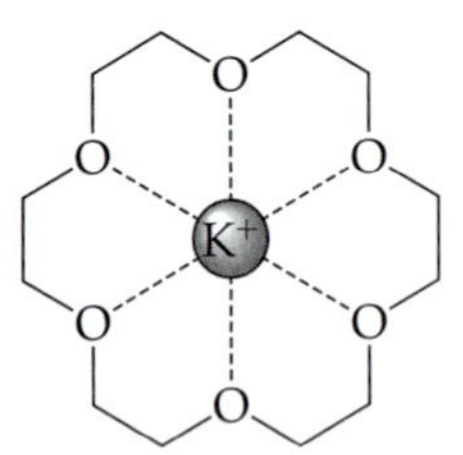

18-Crown-6-ether with potassium ion complexed in the cavity

The cavity in this crown ether is too big to hold the smaller sodium or lithium ion and too small to hold a larger cation. This property has been taken advantage of to bring about separations of ions and to dissolve inorganic salts in organic solvents. Potassium salts are normally insoluble in organic solvents, but the addition of the soluble 18-crown-6 brings them into the solution. Some naturally occurring antibiotics are cyclic polyethers that transport potassium ions through cell membranes by similar complexation, leaving behind sodium ions.

As mentioned in the introduction to this chapter, methyl ether groups are widespread in natural products. Numerous examples will be encountered later (see especially Chapters 9, 12, and 13). Note that codeine (used in some cough syrups) is the methyl ether of morphine, the well-known narcotic painkiller.

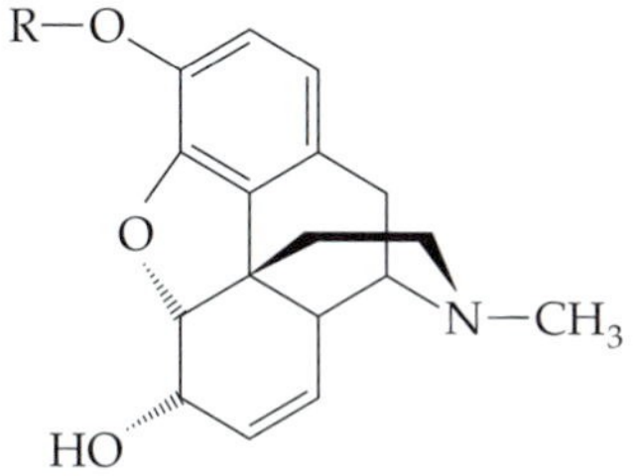

Morphine (R = H)
Codeine (R = CH_3)

Chapter Summary

Ethers are a family of compounds whose members contain an oxygen atom attached to two carbons. Ethers are represented by the general structure R—O—R′. In alkyl ethers, R and R′ represent alkyl groups. Ethers are quite inert chemically but are excellent solvents for organic reactions.

Acyclic ethers are named by designating an **alkoxy group** as a substituent on a parent alkane. Historical/common names are often used, in which the two alkyl groups attached to the oxygen are named, followed by the term *ether.*

Ethers are usually prepared by the **Williamson synthesis,** in which an alcohol (ROH) is converted to its alkoxide anion (RO^-). This anion then carries out an S_N2 reaction with a primary or secondary alkyl halide (R′X) to produce the ether (ROR′). A methyl ether ($ROCH_3$) is often formed to protect a hydroxyl group and then cleaved by one equivalent of HI, to re-form the alcohol and methyl iodide.

Cyclic ethers, with the exception of three-membered compounds, are very similar in behavior to **acyclic ethers.** Three-membered cyclic ethers are known as **epoxides** (historical name) or **oxiranes** (IUPAC). Oxiranes (epoxides) are prepared by the **epoxidation** of alkenes, which results in the cis addition of oxygen to the double bond.

Oxiranes undergo a **ring-opening reaction** under the influence of acid catalysts (S_N1 or S_N2 reaction) or with nucleophilic reagents (S_N2 reaction). With aqueous acid, they produce 1,2-diols. Reaction of an epoxide with a nucleophilic reagent such as hydroxide or alkoxide forms a diol or an ether-alcohol, respectively. Reaction of an epoxide with a Grignard reagent or an organolithium reagent also results in ring opening and is a useful **two-carbon homologation reaction.**

Summary of Reactions

1. **Williamson synthesis of acyclic ethers (Section 5.1.2).** An alcohol is converted to its alkoxide anion with sodium metal. Addition of an alkyl halide (other than a tertiary one) results in an S_N2 substitution by the nucleophilic alkoxide.

$$\underset{\text{Alcohol}}{R{-}OH} \xrightarrow{Na} \underset{\text{Alkoxide}}{R{-}ONa} \xrightarrow{R'X} \underset{\text{Ether}}{R{-}O{-}R'}$$

2. **Cleavage of methyl ethers with HI (Section 5.1.3).** HI cleaves methyl ethers by protonating the ether oxygen to form an oxonium ion. Iodide carries out an S_N2 attack on the methyl group, displacing an alcohol and forming an alkyl iodide.

$$\underset{\text{Ether}}{R{-}O{-}CH_3} \xrightarrow{HI} \underset{\text{Alcohol}}{ROH} + \underset{\text{Methyl iodide}}{CH_3I}$$

3. **Epoxidation of alkenes to form epoxides (oxiranes) (Section 5.2.2).** Peracetic acid (or another peracid) transfers an oxygen atom to an alkene in a *cis* addition to form an epoxide (oxirane).

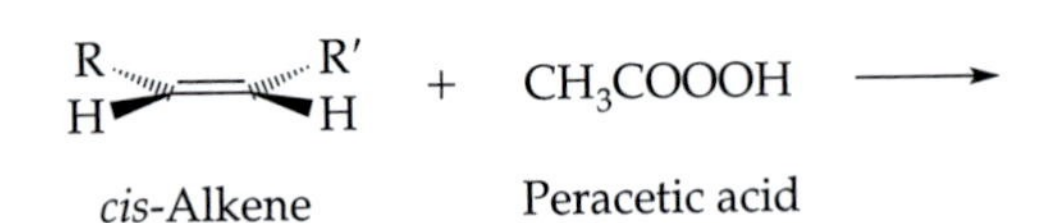

R R′ H H + CH_3COOH

O

cis-Oxirane

Acetic acid

4. **Acid-catalyzed ring opening of epoxides (oxiranes) (Section 5.2.3).** An acid protonates an oxirane/epoxide to form an oxonium ion, which can be opened by attack of a nucleophile through an S_N1 or S_N2 mechanism. The reaction is commonly used to produce diols.

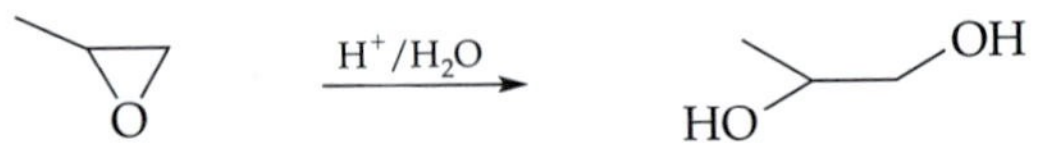

O $\xrightarrow{H^+/H_2O}$ HO OH

Propylene oxide 1,2-Propanediol

5. **Nucleophilic ring opening of epoxides (oxiranes) (Section 5.2.4).** The epoxide ring can be opened by direct attack of any strong nucleophile. The reaction using an organolithium or Grignard reagent is a very effective two-carbon homologation technique.

O $\xrightarrow[2.\ H_2O]{1.\ Nuc}$ HO Nuc

Ethylene oxide

O $\xrightarrow[2.\ H_2O]{1.\ R-metal}$ $R-CH_2CH_2-OH$

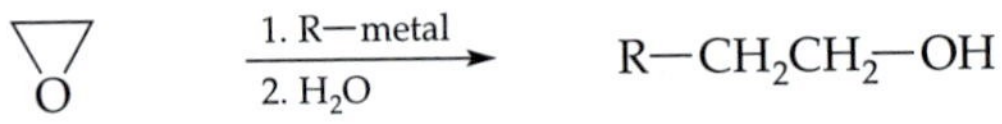

Ethylene oxide

Additional Problems

Nomenclature

5.13 Name the following compounds:

(a) $CH_3OCH_2CH_2CH_3$ (b) $CH_3OCH(CH_3)_2$

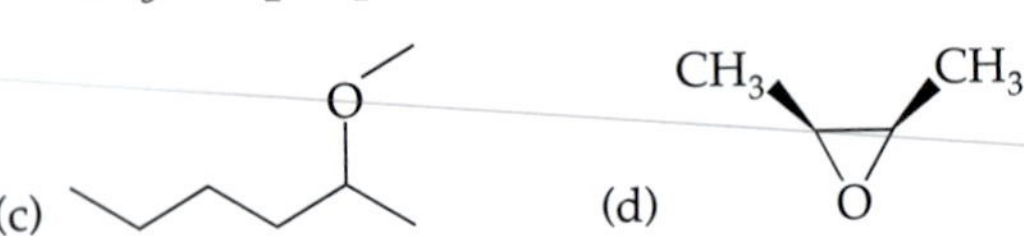

(e) 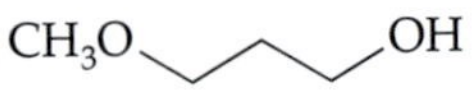 CH_3O OH

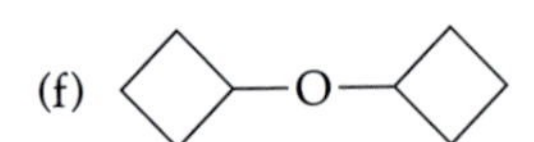

(g) 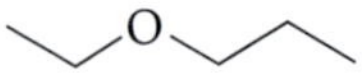 O

5.14 Draw a structure for each of the following compounds:

(a) (*R*)-2-methoxybutane

(b) *cis*-1,4-diethoxycyclohexane

(c) isopropyl ether

(d) *trans*-2-ethyl-3-methoxyoxirane

(e) 2,4-dimethyl-1-ethoxypentane

(f) *cis*-2-butene oxide

5.15 Draw all possible constitutional isomers having the molecular formula $C_4H_{10}O$.

5.16 *t*-Butyl ethyl ether, a gasoline oxygenate additive, is referred to in the press as ETBE. Draw its structure.

Synthesis of Ethers

5.17 Show a synthesis of each of the following compounds, using an alcohol as the starting material:

(a) methoxycyclobutane

(b) isopropyl ether

(c) methyl *t*-pentyl ether

(d) (*R*)-2-methoxypentane

5.18 Show the structure of the product of epoxidation with peracetic acid for each of the following alkenes:

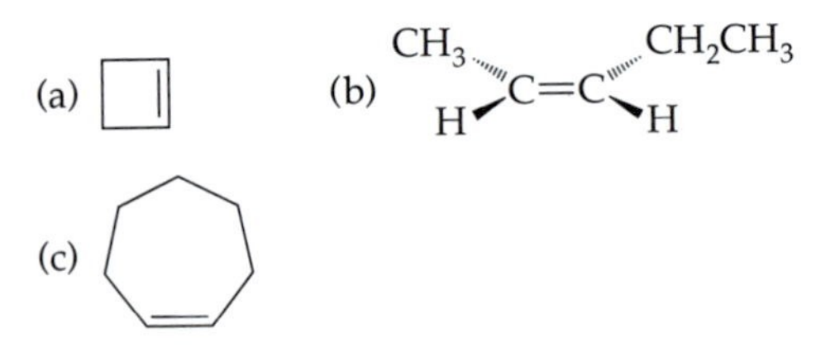

Cleavage of Ethers

5.19 Show the products of the following reactions:

(a) cyclopentyl methyl ether and HI

(b) tetrahydrofuran with excess hot HBr

(c) oxirane with dilute acid and water

(d) oxirane with sodium methoxide, followed by H^+/H_2O work-up

(e) (*S*)-2-chlorobutane with sodium ethoxide

CONCEPTUAL PROBLEM

In an Ethereal Mood

You are employed as the patient liaison in the surgical group of a large hospital. You sit in on a preoperative meeting between a college athlete and her anesthesiologist, Dr. Dunn. Dr. Dunn discusses the pros and cons of various possible anesthetics with the athlete and then concludes that isoflurane (1-chloro-2,2,2-difluoroethyl difluoromethyl ether, $CF_3CHCl—O—CHF_2$) would be the most appropriate choice, given the circumstances and her medical history. He sets a bottle of the compound on the table for all to see—the label reads simply "isoflurane." After the anesthesiologist leaves, the athlete, who has just completed a course in organic chemistry, decides to pass the time by testing your knowledge of organic chemistry.

She asks you to draw structures for the two enantiomers of isoflurane, showing the correct tetrahedral geometry and configurations about the stereocenter. She wonders whether the anesthetic is used as an enantiomer or as a racemate?

She mentions that she knows that halothane ($CF_3CHBrCl$) is another widely used anesthetic. She asks you what benefits it might have over ether compounds in a normal surgical environment, reminding you that operating rooms usually contain pure oxygen supplies. (You might want to refer to Section 5.3.)

This monument to ether stands in the Boston Public Garden. Ethyl ether was first used as a general anesthetic at the Massachusetts General Hospital in the 1840s.

5.20 Show the product(s) of the reaction of cyclohexene oxide with aqueous sodium hydroxide. Account for the stereochemistry of the product(s).

5.21 Show the products formed from reaction of each of the following ethers with one equivalent of HI:

(a) *n*-propyl ether

(b) cyclobutyl methyl ether

(c) (*R*)-2-methoxyhexane

5.22 Show the reaction and products when ethylene oxide is treated with each of the following reagents:

(a) one equivalent of HI

(b) excess HBr

(c) dilute H_2SO_4

(d) methanol and acid

Mixed Problems

5.23 Draw all constitutionally isomeric ethers with the molecular formula $C_5H_{12}O$ and name them.

5.24 The following compounds have similar molecular weights but different physical properties. Name each and place them in order of (1) increasing boiling point and (2) increasing solubility in water.

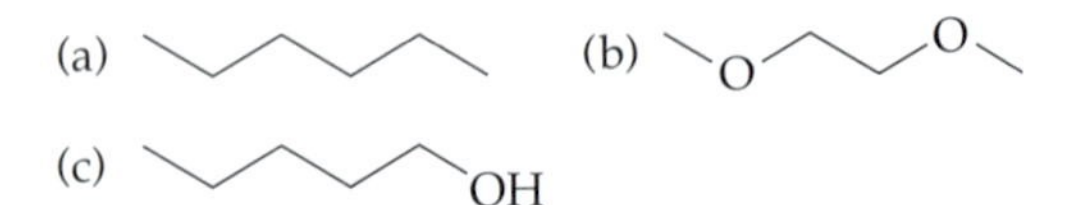

5.25 Propylene oxide (prepared by epoxidation of propylene, $CH_3CH{=}CH_2$) reacts with sodium methoxide in methanol to produce a compound with two functional groups, an alcohol-ether. The product can have two possible structures. What are they? What test would you perform to distinguish between the two possible structures? Which product should predominate in such a reaction?

5.26 Indicate two qualitative tests that can be used to distinguish between the three classes of compounds represented in Problem 5.24.

5.27 Compound **A,** with molecular formula $C_4H_{10}O$, was insoluble in water and did not react with metallic sodium. It did react with hydroiodic acid and produced methyl iodide and compound **B,** which reacted with metallic sodium. Compound **B** reacted with concentrated hydrobromic acid to produce compound **C.** In turn, compound **C** reacted slowly with sodium iodide in acetone and slowly with silver nitrate in ethanol solution. What are the structures of compounds **A, B,** and **C?**

5.28 Starting with 2-butene ($CH_3CH{=}CHCH_3$), write the reactions necessary to produce 2,3-butanediol.

5.29 Show the products formed from each step in the following conversion of cyclohexene to 1,2-cyclohexanediol:

(a) Reaction of cyclohexene with peracetic acid produces compound **Y.**

(b) Reaction of **Y** with water and dilute acid produces the desired product. This reaction is not stereospecific, however.

(c) Reaction of **Y** with sodium hydroxide produces the desired product stereospecifically. Draw its structure and name it.

5.30 Show the products of each of the following steps:

(a) Reaction of 1-butene ($CH_3CH_2CH{=}CH_2$) with peracetic acid forms **A.**

(b) Reaction of 1-butanol with PBr_3 forms **B.**

(c) Reaction of **B** with magnesium in ether forms **C.**

(d) Reaction of **C** with **A,** followed by acidification of the reaction mixture, produces the desired product **D,** $C_8H_{18}O$.

5.31 A series of important commercial solvents prepared from ethylene oxide are called *cellosolves.* An example is methyl cellosolve, $CH_3OCH_2CH_2OH$. Show a possible low-cost, one-step synthesis of methyl cellosolve from ethylene oxide, and write the mechanism for the reaction.

5.32 1,2-Dimethoxyethane can be prepared from methyl cellosolve ($CH_3OCH_2CH_2OH$). Show each step of the necessary reactions.

5.33 Show the product(s) of the following reactions:

(a) $CH_3OCH_2CH_2Br$ with Mg in ether, followed by addition of ethylene oxide

(b) $CH_3OCH_2CH_2Br$ with Mg in ether, followed by addition of water

5.34 Write the product(s) of the following reactions. If no reaction will occur, say so.

(a) cyclopentanol with sodium

(b) methyl cyclopentyl ether with sodium

(c) methyl cyclopentyl ether with magnesium

(d) ethyl ether with HI

(e) *n*-propyl ether with concentrated sulfuric acid

(f) cyclopropyl ether with hot sodium hydroxide

5.35 Complete the following transformations using any necessary reagents. More than one step may be required.

(a) ethyl bromide to 1-bromobutane

(b) cyclohexanol to cyclohexyl ether

(c) cyclobutene to *trans*-1,2-cyclobutanediol

(d) cyclopropyl alcohol to 2-cyclopropylethanol

5.36 What simple chemical tests could you use to distinguish between the members of each of the following pairs of compounds?

(a) pentane and diethyl ether

(b) ethanol and methyl ethyl ether

(c) pentane and pentanol

5.37 A compound of molecular formula $C_4H_{10}O_3$ exhibited chemical behavior typical of both an alcohol and an ether. Its reaction with *excess* hot hydrogen iodide produced 1,2-diiodoethane as the *only* organic product. What is its structure?

CHAPTER 6

Alkenes

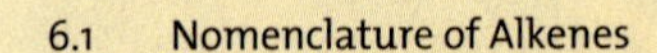

SO FAR, this book has focused primarily on the reactions and configurations of single-bonded, saturated, sp^3-hybridized carbon compounds. In this chapter, the scope broadens to include a large family of hydrocarbons, called **alkenes,** in which two carbon atoms are joined by a **double bond.** In a carbon-carbon double bond, two of a carbon atom's four valence electrons are shared with another carbon atom. This four-electron double bond enables each carbon to maintain its eight-electron valence shell (Figure 6.1).

The double-bonded carbons comprise the functional group of alkenes, which have distinctive chemistry. Rather than having four attached groups, as does each carbon in alkanes, a double-bonded carbon can have only three groups attached to it. Also, while a double bond is drawn as though the two bonds are identical, they are quite different. Finally, the double bond is very reactive, in striking contrast to the single bond.

A pair of lines represents the double bond in structural formulas:

$$(H)_2C{=}C(H)(CH_3) \equiv CH_2{=}CHCH_3 \equiv$$

Propene (propylene)

Alkane ⇌ Alkyl Halide; Alkyl Halide ⇌ Alkene; Alkyl Halide ⇌ Alcohol; Alkene ⇌ Alcohol; Alkene ⇌ Alkyne; Alcohol ⇌ Ether

Alkenes are the first family of organic compounds we examine that have a carbon-carbon multiple bond—a double bond.

Double bonds are also found in cyclic compounds, such as cyclohexene. There are many compounds that contain more than one double bond—for example, 1,3-cyclopentadiene and vitamin A alcohol:

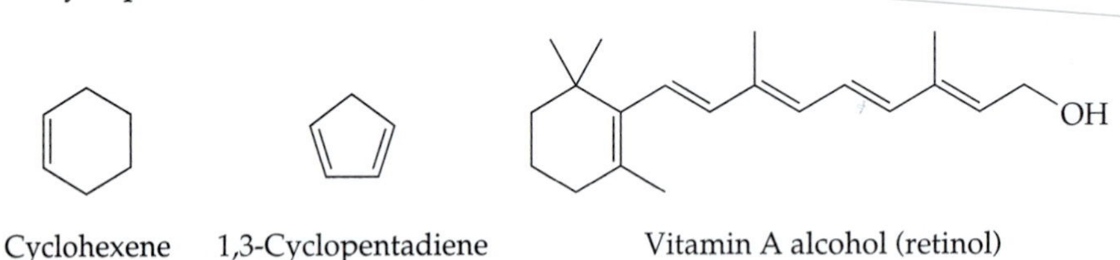

Cyclohexene   1,3-Cyclopentadiene   Vitamin A alcohol (retinol)

Alkenes are often described by their degree of substitution, meaning the number (not position) of substituents *other than hydrogen* attached to the carbon atoms sharing the double bond. If all hydrogens are replaced by substituents, the alkene is tetrasub-

FIGURE 6.1

Lewis structure, dash and condensed formulas, and ball-and-stick model of the simplest alkene, ethene.

$H_2C::CH_2$ $(H)_2C{=}C(H)_2$ $CH_2{=}CH_2$

◀ *Overleaf:* Ethylene (ethene), the simplest alkene, is produced naturally by plants. The Ethylene gas produced by apples themselves promotes their ripening. Commercial growers often store unripe fruit in an ethylene-free environment to extend its shelf life or expose unripe fruit to ethylene to ripen it quickly.

stituted. The double bond in ethene has no substituents, propene is monosubstituted, and cyclohexene is disubstituted. Cyclopentadiene has two disubstituted double bonds. Vitamin A alcohol has two disubstituted double bonds, two trisubstituted double bonds, and one tetrasubstituted double bond.

Alkenes are this book's first example of compounds classified as **unsaturated** (containing one or more double or triple bonds). All of the families of compounds you've studied to this point have been **saturated** (that is, their carbon atoms have all been sp^3 hybridized, with four sigma bonds). Acyclic alkenes with one double bond have the general formula C_nH_{2n}, containing two less hydrogens than alkanes. The addition of two hydrogens to an alkene produces an alkane (C_nH_{2n+2}); this process is called *hydrogenation* (see Section 6.3), and it results in saturation of the double bond. Note that C_nH_{2n} is the same general formula as for cycloalkanes. Therefore, an acyclic alkene and a cycloalkane cannot be distinguished by molecular formula alone.

Because alkenes are hydrocarbons, they have physical properties similar to those of other hydrocarbons, such as alkanes. They are nonpolar compounds and therefore low-boiling, with C_2 to C_4 alkenes being gases at room temperature and most higher alkenes being liquids. Alkenes are insoluble in water but soluble in organic solvents such as other hydrocarbons, halocarbons, and ethers.

6.1

Nomenclature of Alkenes

6.1.1 IUPAC Nomenclature

Alkene nomenclature is based on the general principles of the IUPAC system, with the *-ane* suffix of an alkane being replaced by *-ene*. For example, ethene ($CH_2{=}CH_2$) and propene ($CH_3CH{=}CH_2$) are names of common alkenes. When there is more than one possible location for the double bond in a molecule, its location in the longest straight chain (parent alkene) is indicated by a number. The number assigned is that of the *lower-numbered carbon* involved in the double bond, which means that the double bond is between that carbon and the next-higher-numbered carbon. The numbering is done so as to give the double bond functional group the lowest number possible, with other substituent positions numbered accordingly. Thus, compound **A** is 2-pentene (*not* 3-pentene).

1 2 3 4 5

2-Pentene

A

3 2 4 1 5 6

3-Methylcyclohexene

B

OH

3-Penten-1-ol

C

In cycloalkenes, the double bond is assigned position number 1, so compound **B** is 3-methylcyclohexene (*not* 6-methylcyclohexene). Alkenes that also contain the hydroxyl group (a second functional group) are named with both *-ene* and *-ol* suffixes. The location of the hydroxyl group takes precedence in determining the numbering scheme. (Oxygenated functional groups take precedence over other functional groups; see Appendix A for a complete table of functional group precedence.) Thus, compound

C is 3-penten-1-ol rather than 2-penten-5-ol. Note that the number precedes the suffix for the functional group whose position is being designated and that the final *-e* of *-ene* is dropped.

If there is more than one double bond in a compound, the location of each is designated using the same principles. A number designates the first carbon involved in each double bond. The number of double bonds is indicated by inserting *di-*, *tri-*, and so on, before the *-ene* suffix is added to the alkane stem name. Thus, the alkene with structural formula $CH_2{=}CH{-}CH_2{-}CH{=}CH_2$ is 1,4-pentadiene.

PROBLEM 6.1

Draw the structure corresponding to each of the following names, and indicate how many substituents are on each double bond: (a) 3-methylcyclobutene, (b) 2,3-dimethyl-2-butene, and (c) 4-chloro-2,4-pentadien-1-ol.

PROBLEM 6.2

Assign names to the following compounds, and indicate the degree of substitution of each double bond:

(a) 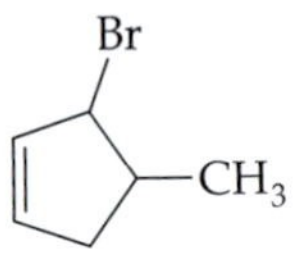 Br, CH_3 (b) 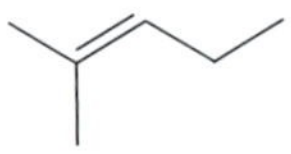 (c) 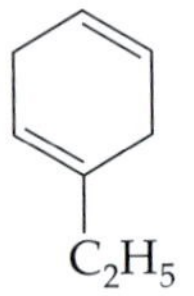 C_2H_5

6.1.2 Historical/Common Names

Historical names for some simple alkenes remain in common use—for example, ethylene for ethene and propylene for propene. In the C_4 series of alkenes, isobutylene is commonly used instead of 2-methylpropene, but the IUPAC names 1-butene and 2-butene are used for the other two C_4 isomers.

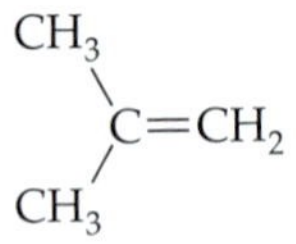

$(CH_3)_2C{=}CH_2$ $CH_3CH_2CH{=}CH_2$ $CH_3CH{=}CHCH_3$

Isobutylene (2-methyl-1-propene) 1-Butene 2-Butene

The $CH_2{=}CH{-}$ group is called the *vinyl group*, and $CH_2{=}CH{-}CH_2{-}$ is called the *allyl group*. The $CH_2{=}$ group is called the *methylene group* when attached to a ring.

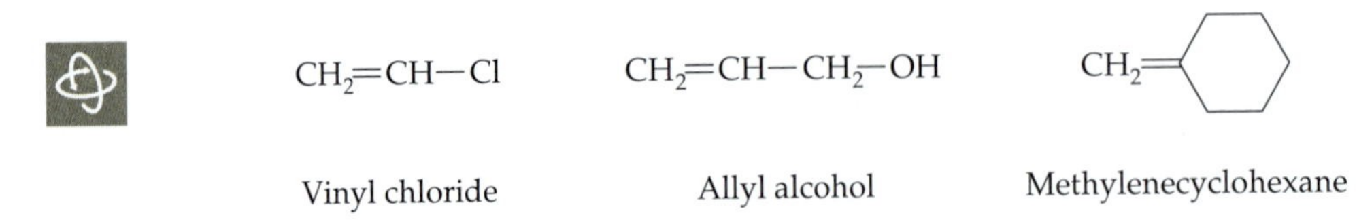

Vinyl chloride Allyl alcohol Methylenecyclohexane

PROBLEM 6.3

Draw the structures corresponding to the following names, and indicate the degree of substitution of each double bond: (a) 1,1-divinylcyclopropane, (b) allyl bromide, (c) 2,4-cyclohexadien-1-ol, and (d) 1-vinylcyclohexene.

6.2 Structure of Alkenes

6.2.1 Electronic Structure of Double Bonds: sp^2 Hybridization

Each carbon atom involved in a double bond is **trigonally hybridized,** having three sp^2 hybrid orbitals arranged about the nucleus in the same plane and at an angle of 120° to each other (Figure 6.2). The process of visualizing trigonal hybridization of a carbon atom is similar to that for tetrahedral (sp^3) hybridization (see Section 1.1.3). The first step, promotion of a 2*s* electron to a 2*p* orbital, is the same. In the second step, only two of the three 2*p* orbitals hybridize with the single 2*s* orbital, forming three sp^2 hybrids.

$$1s^2 2s^2 2p^2 \xrightarrow[\text{(requires energy)}]{\text{promotion}} 1s^2 2s^1 2p^3 \xrightarrow[\text{(energy-neutral)}]{\text{hybridization}} 1s^2 2(sp^2)^3 + 2p^1$$

Atomic carbon — Trigonal carbon

This hybridization process leaves one remaining unhybridized 2*p* orbital, which lies perpendicular to the plane of the sp^2 hybrid orbitals (Figure 6.3).

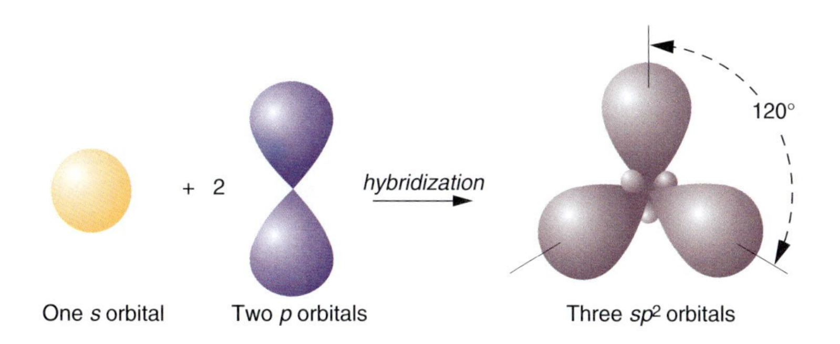

FIGURE 6.2

Trigonal (sp^2) hybridization of carbon.

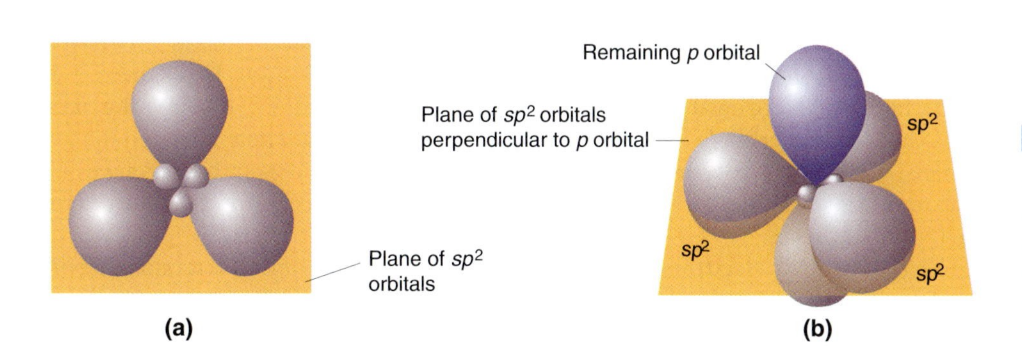

FIGURE 6.3

(a) Three sp^2 hybrid orbitals lying in the plane of the page. (b) The *p* orbital lying in the plane of the page and oriented perpendicular to the plane of the three sp^2 hybrid orbitals.

FIGURE 6.4

Representation of the formation of the ethene molecule from sp^2-hybridized carbons.

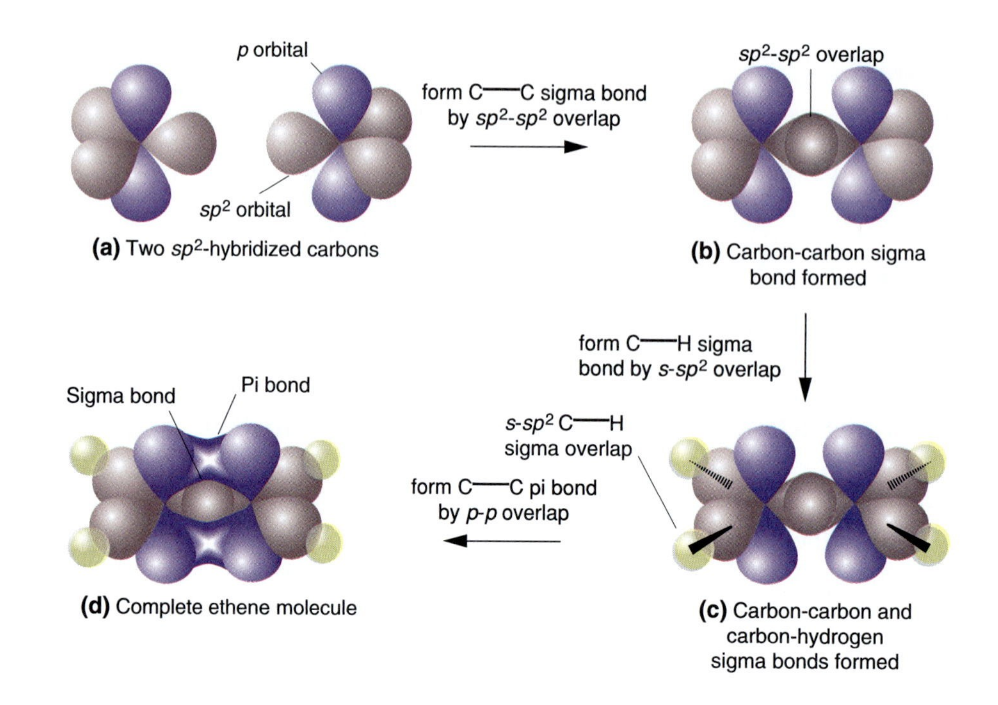

The formation of the double bond in ethene can be pictured as occurring in three steps. Single sp^2 hybrid orbitals from each of the two trigonally hybridized carbons (Figure 6.4a) overlap end to end to form a molecular orbital, resulting in a carbon-carbon *sigma* bond (sp^2-sp^2) (Figure 6.4b); it is a sigma bond because its shape is cylindrical about the C—C axis. The remaining two sp^2 hybrid orbitals of each carbon form two covalent single (sigma) bonds by overlapping with the *s* orbitals of two hydrogen atoms (s-sp^2) (Figure 6.4c). The atomic positions in ethene (all six atoms lying in a plane with 120° bond angles) are determined by the sigma bond framework. The bonding in ethene is completed by the overlap of the remaining unhybridized and parallel *p* orbitals from each trigonal carbon. This overlap of two *p* orbitals creates a new molecular orbital with its own distinctive shape, called a **pi (π) orbital,** and a new covalent bond called a **pi (π) bond** (Figure 6.4d). There are two electrons in this orbital, shared by the two trigonal carbons; therefore, each carbon has a filled valence shell of eight electrons (six involved in three sigma bonds and two involved in a single pi bond).

From this description, it is evident that a carbon-carbon double bond consists of two very different bonds, a sigma bond and a pi bond, between the two trigonal carbons. In contrast to the molecular orbital for a sigma bond, in which overlap is "end to end" between the two atomic orbitals, a pi orbital is formed by "side-by-side" overlap of two parallel *p* orbitals (above and below the plane of the molecule). This results in an area of high electron density between the two carbons and above and below the plane of the three hybrid orbitals. The sigma bond electrons are tightly held between the two carbons, whereas the electrons of the pi bond are relatively loosely held. As we will see later in this chapter, the pi bond is chemically reactive, accounting for the overall reactivity of alkenes.

The higher total electron density between two double-bonded trigonal carbons results in these atoms being held closer and more tightly. Consider the bond between the two carbons in ethene. It has a length of 1.34 Å, considerably shorter than the length of the sigma bond in ethane (1.54 Å). The bond strength for a double bond is about 150 kcal/mol, compared to about 90 kcal/mol for the C—C sigma bond in ethane. This means that the pi bond accounts for approximately 60 kcal/mol and thus is weaker than the sigma bond. The pi bond is more readily broken in chemical reactions.

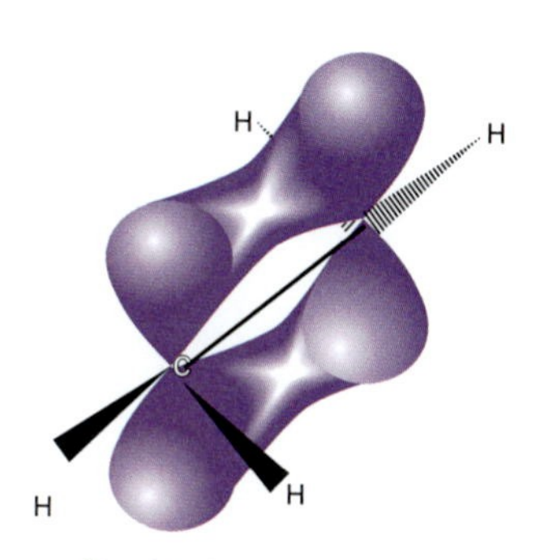

Overlap between *p* orbitals

rotate C—C bond 90°

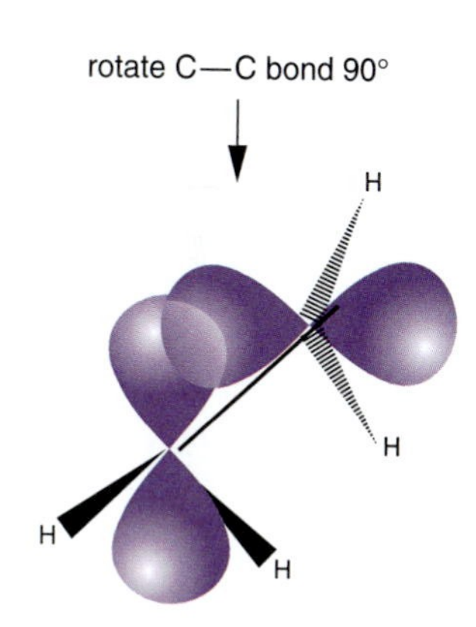

No overlap between *p* orbitals

FIGURE 6.5

Rotation does not occur about a carbon-carbon double bond because the pi bond would be broken.

6.2.2 *Cis-Trans* Isomerism

An important implication of sp^2 hybridization in alkenes is that in order for the pi bond to be maintained, the two *p* orbitals must overlap and thus must be parallel. If one carbon and its *p* orbital were rotated 90°, this *p* orbital overlap would be impossible (Figure 6.5). Because the energy of the molecule would be increased by breaking the pi bond, the two double-bonded carbon atoms do not rotate, but are fixed in position relative to each other. Thus, there is no free rotation about the double bond in ethene (or other alkenes) as there is about the single bond in ethane (and other alkanes).

The restricted rotation at a double bond accounts for *cis-trans* isomerism in appropriately substituted alkenes. For example, 2-butene can exist in two isomeric forms: *cis*-2-butene and *trans*-2-butene. These are *configurational isomers* (same molecular formula, same connectivities, but different configurations that are not interchangeable). They are drawn here with the double-bonded carbons and all their substituents lying in the plane of the paper.

$$(CH_3)(H)C{=}C(CH_3)(H) \not\rightleftharpoons (CH_3)(H)C{=}C(H)(CH_3)$$

cis-2-Butene   *trans*-2-Butene

cis-2-Butene and *trans*-2-butene are two different compounds; they have different physical and chemical properties and are capable of being separated by standard chemical means. In contrast, 1-butene (CH_3—CH_2—CH=CH_2) does not exist in *cis* and *trans* forms because one of the double-bonded carbons carries two identical substituents (two hydrogens), making the following two structures identical:

$$(C_2H_5)(H)C{=}C(H)(H) \xrightarrow{180^\circ} (H)(C_2H_5)C{=}C(H)(H)$$

1-Butene   1-Butene

Rotate the entire molecule 180° (top to bottom) to test superimposability; they are identical structures!

Cis-trans nomenclature works well for any alkene, like 2-butene, that carries the same substituent on each of the two trigonal carbons. In the *cis* configuration, the two substituents are on the same side of the double bond. In the *trans* configuration, the two substituents are on opposite sides of the double bond. Thus, this nomenclature is similar to that used for cycloalkanes, whose substituents can be found on the same (*cis*) or opposite (*trans*) sides of the ring (see Section 2.5.2).

The *cis-trans* terminology is historical and remains useful for simple alkenes, generally those with a single substituent on each trigonal carbon. However, the question of which isomer to label *cis* and which to label *trans* becomes confusing when two or more substituents on the double-bonded carbons are different. To avoid confusion, chemists apply a different isomer designation system, using the symbols *Z* and *E*. The *Z-E* system is based on the same substituent priority system used to determine the absolute configuration of stereocenters (review Section 3.3.3).

Consider, for example, the alkene 1-chloro-2-fluoroethene (CHCl=CHF):

Cl, F / C1=C2 / H, H — (*Z*)-1-Chloro-2-fluoroethene

H, F / C1=C2 / Cl, H — (*E*)-1-Chloro-2-fluoroethene

In naming these compounds, you use substituent priorities as follows:

- Determine the priority of the substituents on each trigonal carbon separately, using the atomic numbers as in the system employed for *R* and *S* configurations. (The higher priority is assigned to the substituent whose atom that is attached to the trigonal carbon has the higher atomic number.) In this case, chlorine has a higher priority than hydrogen on carbon 1, and fluorine has a higher priority than hydrogen on carbon 2.
- Determine whether the higher-priority groups on the two carbons are on the same or opposite sides of the double bond. If they are on the same side, use the prefix *Z*- (from *Zusammen*, the German word for "together"); if they are on opposite sides, use *E*- (from *Entgegen;* German for "opposite").

Higher, Higher / C=C / Lower, Lower

Z isomer
(higher-priority groups on *same* side of double bond)

Higher, Lower / C=C / Lower, Higher

E isomer
(higher-priority groups on *opposite* sides of double bond)

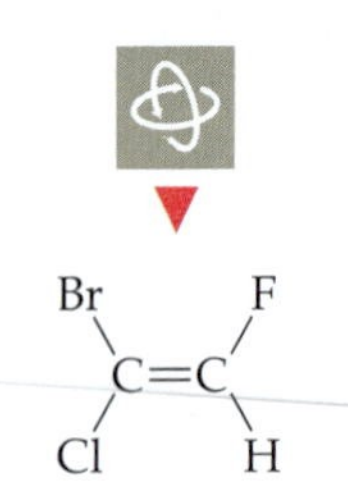

(Z)-1-Bromo-1-chloro-2-fluoroethene

Using the same approach, you see that (Z)-1-bromo-1-chloro-2-fluoroethene has this configuration: Bromine and fluorine are the two higher-priority groups on the trigonal carbon atoms, and they are on the same (Z) side of the double bond.

How to Solve a Problem

Draw the structures and assign full names to both configurational isomers of 3-methyl-3-heptene.

SOLUTION

First, we draw the structure in either configuration to clearly identify the substituents on the two trigonal carbons:

CH_3, H / C3=C4 / CH_3CH_2, $CH_2CH_2CH_3$

3-Methyl-3-heptene

Then we consider each trigonal carbon separately and assign relative priorities to the substituents on each. On carbon 3, the ethyl group is of higher priority than the methyl group: The directly attached atoms are both carbons, but the second atom away from the trigonal carbon is a hydrogen (Z = 1) in the case of the methyl group but a carbon (Z = 6) in the case of the ethyl group. On carbon 4, the propyl group has a higher priority than hydrogen.

Finally, we determine whether the higher-priority groups are on the same or opposite sides of the double bond. In the structure as originally drawn, the two higher-priority groups are on the same side of the double bond, indicating that the structure is the Z isomer. The structure with the two higher-priority groups on opposite sides of the double bond is the E isomer.

CH_3 / H, C=C, CH_3CH_2 / $CH_2CH_2CH_3$

(Z)-3-Methyl-3-heptene

CH_3CH_2 / H, C=C, CH_3 / $CH_2CH_2CH_3$

(E)-3-Methyl-3-heptene

PROBLEM 6.4

For which of the following compounds are *cis* and *trans* isomers possible? Draw the isomers for each compound for which they exist.

(a) 1-pentene
(b) 2-hexene
(c) 1-chloroethene
(d) 1-chloropropene

PROBLEM 6.5

Draw the structures of (a) (E)-1-fluoropropene, (b) (Z)-2-chloro-2-butene, and (c) (E)-3-methyl-2-pentene.

PROBLEM 6.6

Assign full names to the following compounds:

(a)　(b) Br　(c) OH

6.2.3 Relative Stabilities of Alkenes

Trans alkenes are more stable than *cis* alkenes, but usually by only a few kilocalories per mole. *Cis* isomers typically exhibit more steric hindrance (repulsion between groups positioned on the same side of a double bond) (Figure 6.6). The larger the groups, the greater the repulsion and the greater the energy difference between the *cis* and *trans* isomers. Under certain special conditions, the less stable *cis* alkenes can be converted to *trans* alkenes, but this involves breaking and remaking of a pi bond. As we will see

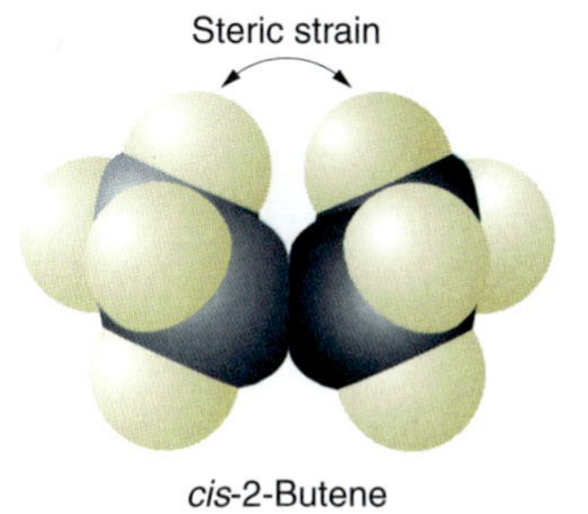

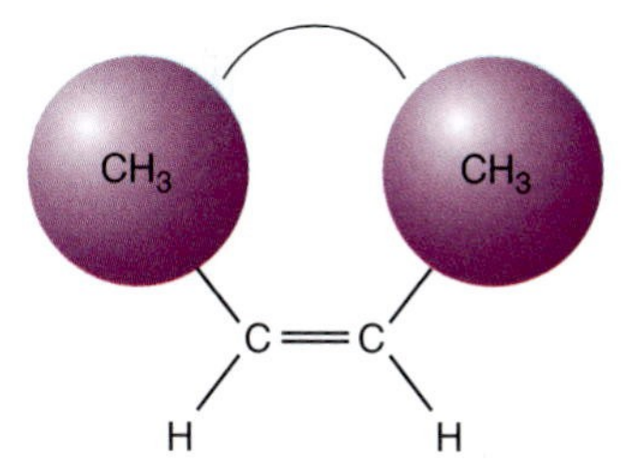

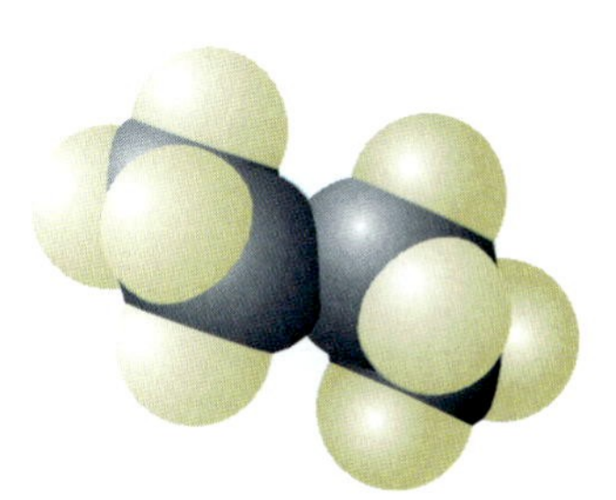

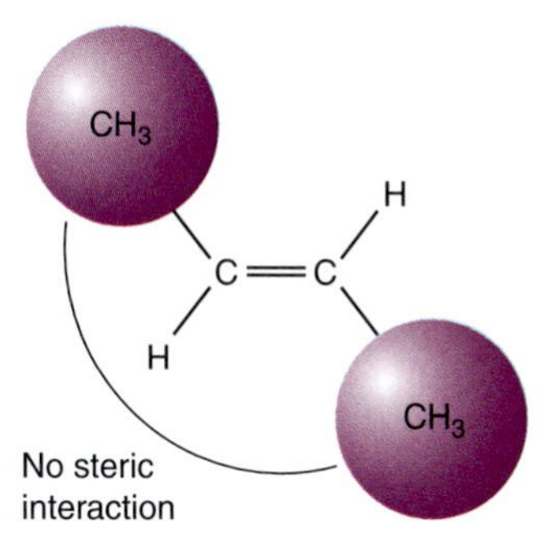

FIGURE 6.6 Steric interactions in *cis*- and *trans*-2-butene.

in the following sections, reactions that form alkenes typically produce a predominance of the *trans* alkene if the reaction mechanism permits both isomers to form.

It will also become evident that more highly substituted alkenes are more stable than less highly substituted alkenes. For example, in a reaction that produces butenes, 2-butene (disubstituted) normally forms in preference to 1-butene (monosubstituted), and the 2-butene is mainly the more stable *trans* isomer.

Smaller cycloalkenes are always the *cis* isomers. These cycloalkenes, especially cyclopropene and cyclobutene, are difficult to make and very reactive because of the high degree of ring strain. (Normal alkene bond angles are 120°, while in cyclopropene the C—C—C bond angles must be 60°.)

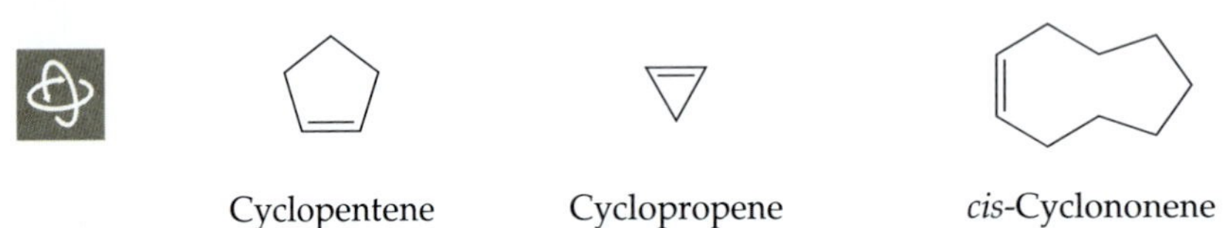

Cyclopentene Cyclopropene *cis*-Cyclononene

6.3

Hydrogenation (Reduction) of Alkenes

Alkenes typically undergo **addition reactions,** in which both parts of a reagent A—B, such as H—H, H—Cl, or H—OH, *add across the double bond:*

$$\mathrm{C{=}C} + \mathrm{A{-}B} \xrightarrow{\text{(addition)}} \mathrm{-\overset{A}{\overset{|}{C}}-\overset{B}{\overset{|}{C}}-}$$

The overall result is that the pi bond of the alkene (~60 kcal/mol) and the sigma bond of the reagent (~80–100 kcal/mol) are broken. This requires an energy input totaling approximately 140–160 kcal/mol. However, two new sigma bonds form (~80–100 kcal/mol each), releasing a total of about 160–200 kcal/mol. Thus, addition reactions are energetically favorable.

The simplest addition reaction of alkenes is **hydrogenation,** the addition of molecular hydrogen to a double bond. In the presence of a slurry of a metal catalyst (usually finely powdered platinum or palladium) in an inert solvent, the addition of hydrogen takes place rapidly at room temperature. This reaction readily converts an alkene to an alkane.

General reaction:

$$\underset{\text{Alkene}}{\mathrm{C{=}C}} \xrightarrow[\mathrm{Pt}]{\mathrm{H_2}} \underset{\text{Alkane}}{\mathrm{CH{-}CH}}$$

This reaction is referred to as *catalytic hydrogenation* or **reduction.** Reduction means that the product has gained electrons (and protons in this case) and therefore been reduced.

FIGURE 6.7

Schematic view of hydrogenation of an alkene.

Occasionally, the process is called *saturating a double bond*, or converting an unsaturated compound to a saturated compound. For example, ethene is converted to ethane, and cyclohexene is converted to cyclohexane:

$$CH_2{=}CH_2 \xrightarrow[Pt]{H_2} CH_3{-}CH_3$$

Ethene → Ethane

Cyclohexene → Cyclohexane (H_2, Pt)

The hydrogenation reaction begins with the powdered metal catalyst adsorbing molecular hydrogen onto its surface, splitting the hydrogen molecule into two atoms. The hydrogen atoms are then delivered, one at a time, to both carbons of the double bond while the alkene is also adsorbed to the catalyst's surface (Figure 6.7).

Catalytic hydrogenation is usually a *stereospecific* reaction, called a ***syn*** **addition** because both hydrogen atoms add to the *same* side of the alkene (*syn* means "together"). This is demonstrated by the fact that 1,2-dimethylcyclohex*ene* yields mainly *cis*-1,2-dimethylcyclohex*ane* upon catalytic hydrogenation:

1,2-Dimethylcyclohexene $\xrightarrow[Pt]{H_2}$ *cis*-1,2-Dimethylcyclohexane

PROBLEM 6.7

Show the products of hydrogenation of each of the following alkenes, including their stereochemistry, if relevant: (a) *cis*-2-butene, (b) 1,2-dichlorocyclopentene, (c) 1,3-butadiene, and (d) (Z)-3-methyl-2-pentene.

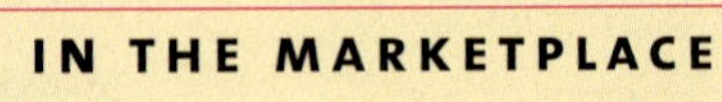

CHEMISTRY AT WORK

IN THE MARKETPLACE

Margarine

Does the word *hydrogenated* sound familiar to you? If so, you may have come across it in an advertisement for margarine or vegetable oil or in connection with the amount of saturated or unsaturated fat in a product. What do these terms really mean?

We will discuss fats in more detail in Chapter 15, but briefly, fats belong to the family of compounds called esters. A fat is a triester derived from a molecule of glycerol and three molecules of fatty acid. The fatty acids (and therefore the groups R, R′, and R″ in any particular fat) may be identical or different. These R groups are long, unbranched hydrocarbon chains, generally C_{11} to C_{17}, which may be either saturated or unsaturated (having one or more double bonds).

Glycerol: $CH_2{-}OH$ / $CH{-}OH$ / $CH_2{-}OH$

Fatty acid: $HO{-}C({=}O){-}R$

Fat (lipid): $CH_2{-}O{-}C({=}O){-}R$ / $CH{-}O{-}C({=}O){-}R'$ / $CH_2{-}O{-}C({=}O){-}R''$

A major component of butter is butterfat, which contains a large proportion of **saturated fats;** that is, the fats have no double bonds in their hydrocarbon chains. By contrast, most vegetable oils—such as soybean, corn, coconut, and olive oils—contain mainly **unsaturated fats;** that is, the fats have one or more double bonds in their hydrocarbon chains. Saturated fats are generally solid at room temperature, whereas unsaturated fats are generally liquid.

Because consumers seem to prefer solid spreads to liquids, food companies began to hydrogenate some of the double bonds in vegetable oils, producing the solid known as margarine. This is the same chemical process as converting an alkene to an alkane. Often, the labels on margarine containers specify "partially hydrogenated vegetable oils" as the major ingredient. Margarine is hydrogenated enough to be a solid, but still has some unsaturation.

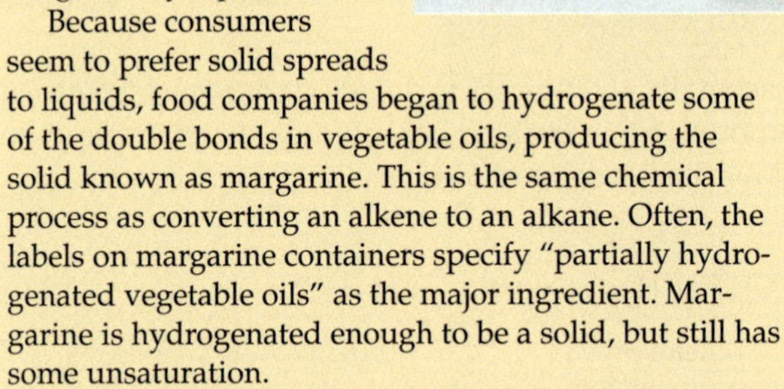

Three major factors have motivated consumers to replace butter with margarine. First, butter was in short supply during World War II, and margarine eventually became an acceptable substitute. Second, the shelf life of margarine is longer than that of butter. Finally, reducing the amount of saturated fats in a diet appears to result in significant health benefits.

6.4 Electrophilic Addition to Alkenes

The most important reaction of alkenes is the addition of electrophilic reagents to the electron-rich double bond. Electrophiles are electron-deficient species and so are "seeking" sites of high electron density. The general mechanism for such ionic reactions involves the initial attack of the nucleophilic alkene on the polarized electrophilic reagent (Figure 6.8).

In the first step, the pi bond of the alkene serves as a nucleophile, with one carbon forming a sigma bond with the incoming electrophile (A). In the process, this carbon rehybridizes to tetrahedral (sp^3) hybridization. The other carbon becomes electron-deficient, forming a carbocation (sp^2 hybridization). The reactive intermediate (the carbocation) is formed in this first (slow) step of the addition reaction. In the second (rapid)

(slow)

Carbocation

(fast)

The electron-rich pi bond attacks the positive end of the dipolar reagent (that is, the reagent acts as an electrophile, or electron seeker). As the C—A bond is formed, the A—B bond must break. The transition state has the C—A and A—B bonds partially formed or broken. The intermediate is a carbocation.

The carbocation seeks to form a covalent bond to return the positively charged carbon to the stable electron configuration with eight electrons in the valence shell. The anion B^- is the strongest nucleophile in the reaction.

FIGURE 6.8

Mechanism of electrophilic addition to alkenes.

step of the reaction, the electron-deficient carbocation is attacked by the electron-rich nucleophile (B), which was released in the first step. Thus, the overall reaction is initiated by the attack of the alkene on the electron-deficient portion of the reagent (A), but it results in the addition of the entire reagent (A—B) across the double bond.

In the first step of such additions, the electrophile (A) could, in principle, attach itself to either of the two double-bonded carbon atoms of the alkene. Thus, if these two carbons are different (that is, have different substituents), an issue of **regiochemistry** arises: To which carbon does the electrophile attach itself, and what is the determining factor? *The first step of electrophilic addition proceeds so as to form the most stable carbocation intermediate.* This **regioselectivity,** the preferential formation of one isomer over another, determines the entire course of the reaction. Recall that because of differences in stabilization due to hyperconjugation, the order of carbocation stability is 3° > 2° > 1° (see Section 3.4.3). The carbocation-forming step is the slowest step in electrophilic addition, the step with the highest energy of activation, so carbocation stability determines the height of that energy barrier. The reaction proceeds so as to minimize the energy barrier in that first step by forming the most stable carbocation. Note that in the second step of the reaction, the negative portion of the adding reagent (B:⁻) ends up attached to what was the positively charged carbon in the intermediate carbocation, as shown here for propene and 1-methylcyclopentene:

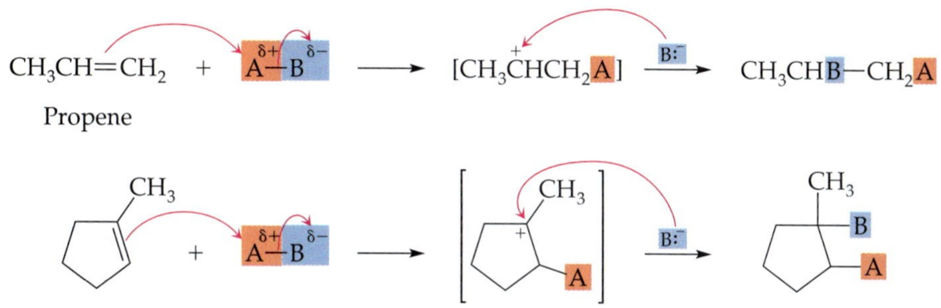

Long before this mechanism was known, the Russian chemist Vladimir Markovnikov enunciated a rule, now bearing his name, to predict the regiochemistry of such additions. **Markovnikov's rule** can be stated as follows: *The major product will be that in which the negative portion of the adding reagent (B:−) adds to the most highly substituted carbon.* The outcome of applying this rule is exactly the same as that obtained from the mechanistic explanation because the carbon with the greatest number of alkyl groups makes the most stable carbocation. However, it is more helpful to understand the carbocation chemistry, work out each addition reaction, and predict the products rather than simply memorize Markovnikov's rule. The mechanism of these additions explains their outcome, whereas Markovnikov's rule simply predicts the major product.

Electrophilic additions are normally not stereospecific because of the intermediacy of the carbocation. Recall that carbocations are sp^2 hybridized and therefore are planar (that is, have a plane of symmetry). Thus, even though a stereocenter may be created during the second step of an addition, such as with 1-butene, the nucleophile may attack from either side of the carbocation plane. Therefore, both enantiomers form, and the product is a racemic mixture.

$C_2H_5{-}CH{=}CH_2 \xrightarrow{A-B}$ [carbocation: C_2H_5, $^+$, ACH_2, H] $\xrightarrow{B:^-}$ C_2H_5, B, H, CH_2A + C_2H_5, H, B, ACH_2

1-Butene — Carbocation — Racemic mixture

The following subsections apply these general principles of all electrophilic additions to a number of key addition reactions of alkenes.

6.4.1 Addition of Hydrogen Halides

The addition of hydrogen halides (HCl, HBr, HI) to alkenes produces alkyl halides (see Key to Transformations, on inside front cover). The reaction is usually rapid and may occur simply by bubbling the gaseous acids through the alkene or adding the acids in a solvent.

General reaction:

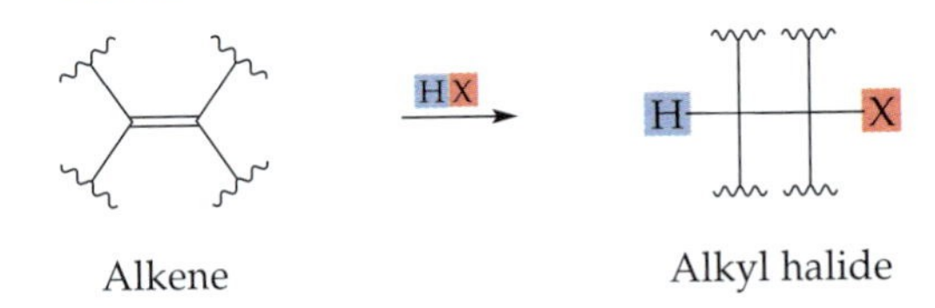

For example, cyclopentene reacts with HCl to produce chlorocyclopentane, and ethene adds HBr to form ethyl bromide:

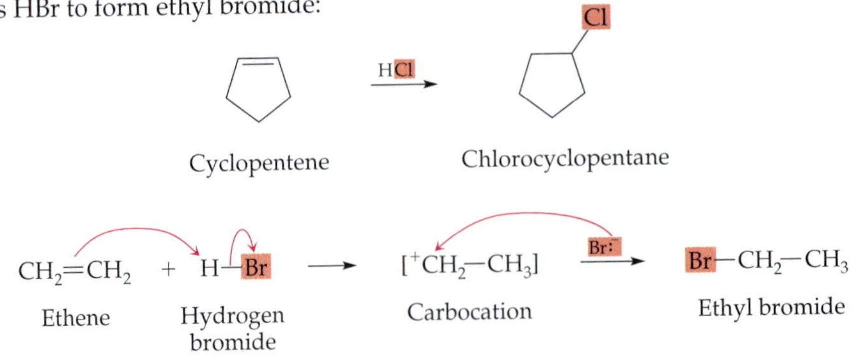

The addition reaction proceeds via initial attack of the alkene on the proton of the acid, forming a carbocation, as illustrated for ethene. In the final step, the carbocation is attacked by the nucleophilic bromide ion released in the first step.

The addition of a hydrogen halide is *regioselective* in that the more stable of the two possible carbocations predominates if the alkene is unsymmetrical. For example, 2-methyl-1-propene yields 2-bromo-2-methylpropane with HBr, and 1-butene yields 2-iodobutane with HI (these are Markovnikov-oriented additions):

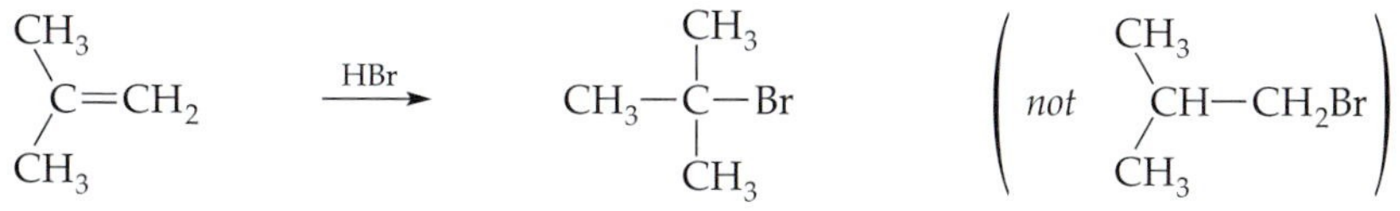

$$(CH_3)_2C{=}CH_2 \xrightarrow{HBr} CH_3{-}C(CH_3)_2{-}Br \quad \left(not\ (CH_3)_2CH{-}CH_2Br\right)$$

2-Methyl-1-propene → 2-Bromo-2-methylpropane

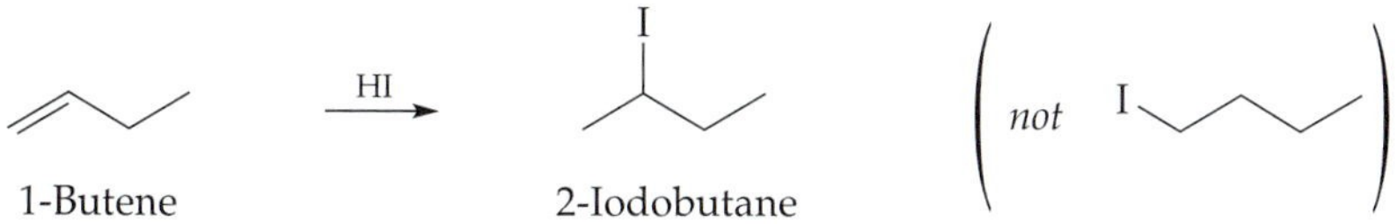

1-Butene → (HI) 2-Iodobutane (*not* 1-iodobutane)

In the case of HI addition to 1-butene, a stereocenter is created at carbon 2. However, the intermediate carbocation is achiral (has a plane of symmetry), so in the final step attack of the iodide ion can occur from either side of the planar carbocation. Thus, a 50/50 mixture of (*R*)- and (*S*)-2-iodobutane is formed—a racemate.

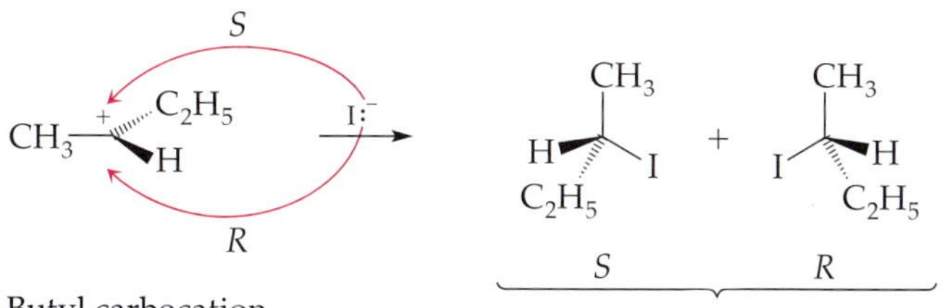

2-Butyl carbocation (CH_3, C_2H_5, H; attack by I^- from either face: *S*, *R*) → *S* + *R*

Racemic 2-iodobutane

In summary, the addition of a hydrogen halide *is not stereospecific* (it produces both enantiomers), but it *is regioselective* (the halide anion attaches to the carbon that results in the most stable carbocation).

PROBLEM 6.8

Draw the structure of the major product of each reaction:

(a) 1-pentene with hydrogen chloride

(b) *cis*-2-pentene with hydrogen bromide

(c) cyclobutene with hydrogen chloride

PROBLEM 6.9

Name the alkene you would start with to produce each alkyl halide via hydrogen halide addition: (a) *t*-butyl chloride, (b) 1-bromo-1-methylcyclobutane, (c) 2-bromobutane.

6.4.2 Addition of Water

The addition of water to alkenes (**hydration** of alkenes) produces alcohols, permitting another of the conversions outlined in the Key to Transformations.

General reaction:

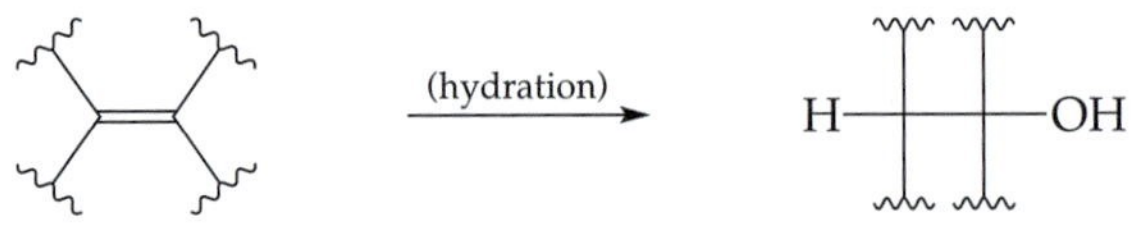

Alkene $\xrightarrow{\text{(hydration)}}$ Alcohol (H–C–C–OH)

Two different processes for the addition of water are illustrated here for isobutylene (2-methyl-1-propene). The normal acid-catalyzed addition of water follows Markovnikov's rule (producing *t*-butyl alcohol), but another reaction was developed to accomplish *anti-Markovnikov* addition (producing isobutyl alcohol, in this example).

$(CH_3)_2C{=}CH_2$ (Isobutylene) $\xrightarrow{\text{(Markovnikov addition)}}$ $CH_3{-}C(CH_3)_2{-}OH$ (*t*-Butyl alcohol)

$(CH_3)_2C{=}CH_2$ (Isobutylene) $\xrightarrow{\text{(anti-Markovnikov addition)}}$ $(CH_3)_2CH{-}CH_2OH$ (Isobutyl alcohol)

Acid-Catalyzed Hydration of Alkenes

Treating an alkene with water in the presence of dilute sulfuric acid results in the formation of an alcohol. For example, 1-butene produces 2-butanol, and cyclobutene produces cyclobutanol:

$$\underset{\text{1-Butene}}{CH_3CH_2CH{=}CH_2} \xrightarrow[H_2O]{H_2SO_4} \underset{\text{2-Butanol}}{CH_3CH_2CHOHCH_3}$$

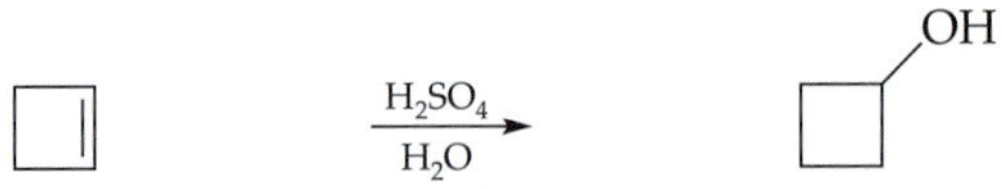

$$\text{Cyclobutene} \xrightarrow[H_2O]{H_2SO_4} \text{Cyclobutanol (OH)}$$

The acid-catalyzed addition of water is regioselective, shown by the fact that 1-butene produces mainly 2-butanol, as Markovnikov's rule predicts, rather than 1-butanol. Again, the reason for this observed regioselectivity is the relative stability of the two possible carbocations that could form in the first step. The mechanism involves initial protonation of 1-butene at the terminal carbon to form a secondary carbocation:

$$CH_3{-}CH_2{-}CH{=}CH_2 \xrightarrow{H^+} CH_3{-}CH_2{-}\overset{+}{C}H{-}CH_3 \quad (\textit{not}\ CH_3CH_2CH_2\overset{+}{C}H_2)$$

1-Butene — 2-Butyl carbocation (a 2° carbocation) — 1-Butyl carbocation (a 1° carbocation)

The secondary carbon is then attacked by nucleophilic water to form an oxonium ion:

$$CH_3{-}CH_2{-}\overset{+}{C}H{-}CH_3 \xrightarrow{H_2\ddot{O}:} CH_3{-}CH_2{-}\underset{}{\overset{^{+}\ddot{O}H_2}{\overset{|}{C}H}}{-}CH_3$$

2-Butyl carbocation — Oxonium ion

Finally, the resulting oxonium ion releases a proton (which is picked up by water or bisulfate ion) to re-form the original acid:

$$CH_3{-}CH_2{-}\overset{H\diagdown\overset{+}{\ddot{O}}\diagup H}{\overset{|}{C}H}{-}CH_3 \xrightarrow{H_2\ddot{O}:} CH_3{-}CH_2{-}\overset{OH}{\overset{|}{C}H}{-}CH_3 + H_3O^+$$

Oxonium ion — 2-Butanol

Thus, the reaction is *acid-catalyzed*—the acid is not consumed.

PROBLEM 6.10

Draw the product(s) obtained by treating each alkene with dilute sulfuric acid: (a) cyclohexene, (b) 1-pentene, (c) *trans*-2-pentene, and (d) 1-methylcyclopentene.

Anti-Markovnikov Addition of Water

In the general mechanism for the addition of water just shown, it is apparent that the regiochemistry is determined by the first step of the reaction, the addition of the electrophilic proton. The only way to achieve the opposite regiochemistry, known as *anti-Markovnikov addition*, is to somehow avoid proton addition as the first step. This means avoiding direct addition of H_2O because it is not possible to have OH addition occur in the first step (the hydroxyl group is never an electrophile; it is always a nucleophile).

The solution to this dilemma was discovered by H. C. Brown, who won the 1979 Nobel Prize in chemistry for this research. The key to the process is the chemistry of the element boron. Boron has only six electrons in its outer shell and so is able to accommodate two more. Thus, trivalent boron (for example, in borane, BH_3) is an electrophile (a Lewis acid) and adds to an alkene with the same regiochemistry as an electrophilic proton. The boron-hydrogen bond in borane is polarized, with the boron atom as the positive end of the dipole ($B^{\delta+}{-}H^{\delta-}$). Borane normally exists as a dimer (diborane, B_2H_6, two BH_3 units joined together), but it reacts as the monomer BH_3.

Reaction of an alkene with diborane produces a trialkylborane from addition of three B—H units across the double bonds of three molecules of alkene:

$$CH_3{-}CH_2{-}CH{=}CH_2 \;(\text{1-Butene}) + \overset{\delta-}{H}{-}\overset{\delta+}{B}H_2 \longrightarrow \left[CH_3{-}CH_2{-}\underset{H}{CH}{-}CH_2{-}BH_2\right]$$

(repeat with two more molecules of 1-butene)

$$CH_3{-}CH_2{-}CH_2{-}CH_2{-}B(CH_2{-}CH_2{-}CH_2{-}CH_3)_2 \;(\text{Tributylborane}) \xrightarrow[NaOH]{H_2O_2} 3\,CH_3{-}CH_2{-}CH_2{-}CH_2{-}OH \;(\text{1-Butanol}) + Na_3BO_3 \;(\text{Sodium borate})$$

The significant fact is that the alkene's pi electrons attack the electrophilic boron, not the hydrogen (the negative end of the B—H dipole in borane). Note that this step by itself results in a hydrogen being added to the secondary carbon of 1-butene, whereas in acid-catalyzed hydration, a hydrogen (in the form of a proton) adds to the primary carbon. All that remains is to remove the boron atom (break the carbon-boron bond) and replace it with an —OH group. This is accomplished by using the mild oxidant hydrogen peroxide (H_2O_2) in basic solution. The C—B bond is oxidized and cleaved, with a hydroxyl group being attached to the carbon and to the boron. Thus, the final products are three equivalents of the desired alcohol (in which water has appeared to have been added with anti-Markovnikov regiospecificity) and boric acid (neutralized in the basic solution to the borate anion).

Using the hydroboration route, the typical regiochemistry of the alkane hydration reaction can be reversed. For example, 1-methyl-cyclopentene can be converted to the two isomeric alcohols using the two hydration methods:

$$\text{1-Methylcyclopentene} \xrightarrow[H_2O]{H_2SO_4} \text{1-Methylcyclopentanol (Markovnikov addition)}$$

$$\text{1-Methylcyclopentene} \xrightarrow[2.\ H_2O_2/NaOH]{1.\ B_2H_6} \text{2-Methylcyclopentanol (anti-Markovnikov addition)}$$

There is another important outcome of the hydroboration-oxidation route to alcohols: The reaction is stereospecific. Examining the mechanism for the reaction reveals that the boron and the hydrogen must add to the double bond from the *same* side at the *same* time (a *concerted reaction*). This is a *syn* addition, just like catalytic hydrogenation.

When the carbon-boron bond is oxidized, the stereochemistry is not changed. Thus, the overall effect is for the hydroboration-oxidation of the alkene to be a regioselective and stereospecific *syn* addition, as shown for 1-methylcyclopentene:

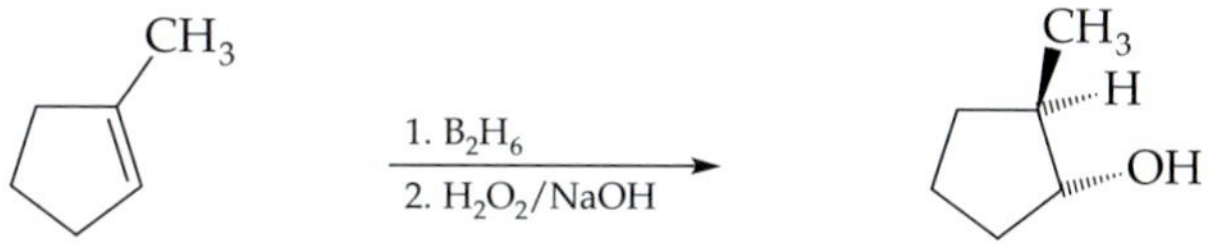

CH_3 — 1. B_2H_6 / 2. H_2O_2/NaOH → CH_3, H, OH

1-Methylcyclopentene — *trans*-2-Methylcyclopentanol

In contrast, the acid-catalyzed hydration of alkenes is regioselective but not stereospecific (because it involves a planar symmetric carbocation). Thus, the availability of both reactions provides regiochemical choice in converting alkenes to alcohols.

How to Solve a Problem

Show how to convert (*E*)-3-methyl-2-pentene into 3-methyl-3-pentanol and into 3-methyl-2-pentanol.

PROBLEM ANALYSIS

We draw the structures of the reactant and the desired products (two constitutional isomers):

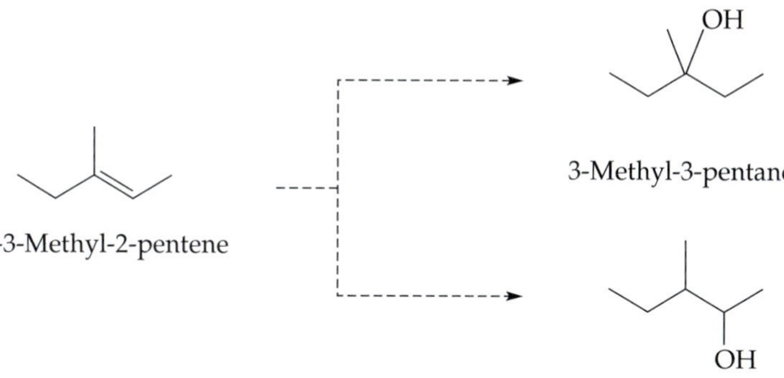

(*E*)-3-Methyl-2-pentene → 3-Methyl-3-pentanol (OH); 3-Methyl-2-pentanol (OH)

The overall change is that water must be added to the double bond with two different regioselectivities. Two different reactions will accomplish the desired transformations. We first consider the Markovnikov-oriented addition of water using acid catalysis. Protonation of the alkene occurs preferentially to produce the most stable carbocation, which is the tertiary carbocation. Its reaction with water followed by loss of a proton produces 3-methyl-3-pentanol:

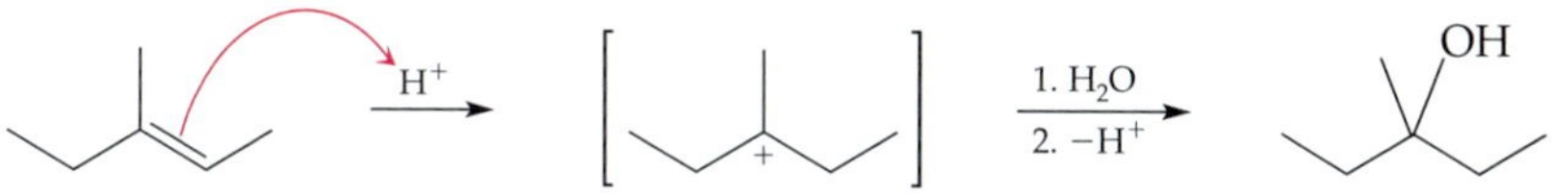

(*E*)-3-Methyl-2-pentene —H^+→ [3° Carbocation] —1. H_2O / 2. $-H^+$→ 3-Methyl-3-pentanol (OH)

This is the Markovnikov-oriented addition. Therefore, in order to obtain anti-Markovnikov addition, we resort to the hydroboration reaction.

SOLUTION

Reaction of (*E*)-3-methyl-2-pentene with dilute sulfuric acid produces 3-methyl-3-pentanol as the major product. Reaction of the same alkene with diborane, followed by treatment with hydrogen peroxide and sodium hydroxide, produces 3-methyl-2-pentanol:

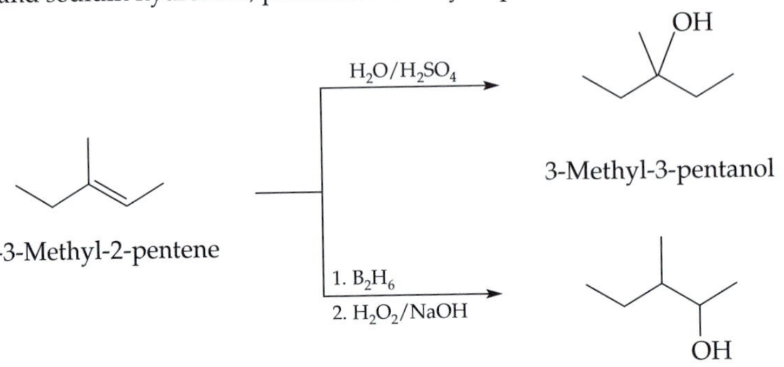

(*E*)-3-Methyl-2-pentene — H_2O/H_2SO_4 → 3-Methyl-3-pentanol

(*E*)-3-Methyl-2-pentene — 1. B_2H_6 2. H_2O_2/NaOH → 3-Methyl-2-pentanol

PROBLEM 6.11

Draw the product obtained by treating each alkene with diborane and then hydrogen peroxide and base: (a) cyclohexene, (b) 1-pentene, (c) vinylcyclopentane, and (d) 1-methylcyclopentene.

PROBLEM 6.12

Name the alkene and reagent you would employ to produce each alcohol: (a) *t*-butyl alcohol, (b) 1-methylcyclohexanol, (c) 2-methylcyclohexanol, and (d) 3-methyl-2-pentanol.

6.4.3 Addition of Halogens

The halogens chlorine (a gas) and bromine (a liquid) react rapidly with an alkene in an addition reaction that forms a **vicinal dihalide.** *Vicinal* (*vic-*) means that the two halogen substituents are attached to *adjacent* carbons; the alternative is a *geminal* (*gem-*) *dihalide*, where the two halogens are on the same carbon.

General halogenation reaction:

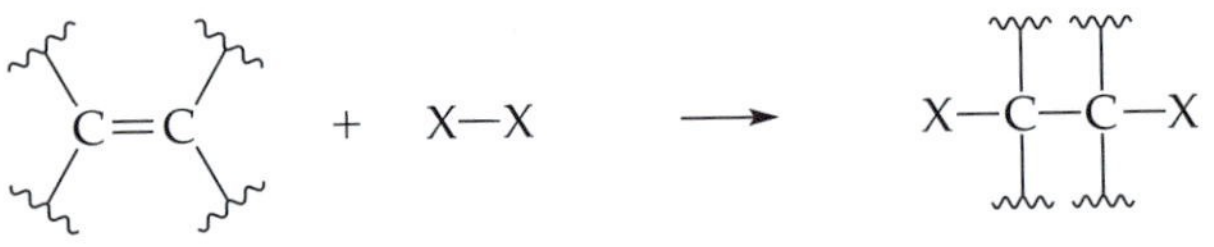

C=C + X—X → X—C—C—X

Alkene Halogen *vic*-Dihalide

For example, cyclohexene produces 1,2-dibromocyclohexane:

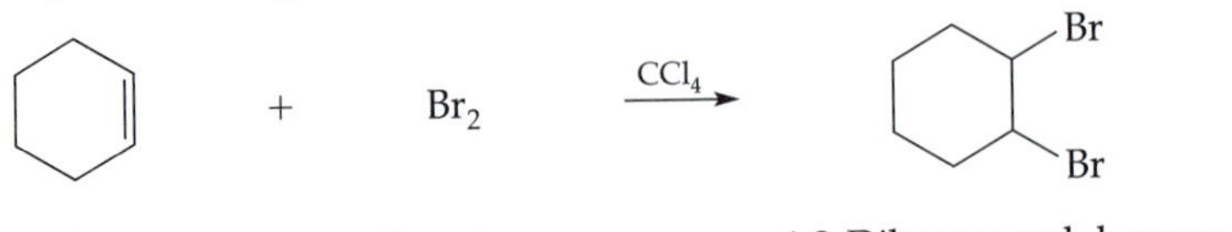

Cyclohexene (colorless) + Br_2 Bromine (reddish brown) —CCl_4→ 1,2-Dibromocyclohexane (colorless)

Bromine Cyclobutene

The nucleophilic alkene attacks molecular bromine, displacing bromide anion. This occurs in a single step to produce the reactive intermediate, a bromonium ion.

Bromonium ion

The nucleophilic bromide anion attacks the back side of a carbon attached to the bromine, forming a new sigma bond and returning the shared electrons to the other bromine atom. The bromide anion can attack either carbon with equal ease. Therefore, both products result, as a 50/50 racemic mixture.

Racemic mixture of 1,2-dibromocyclobutane

FIGURE 6.9

Mechanism of the bromination of cyclobutene.

The reaction of reddish-brown bromine with alkenes is used as a qualitative test to distinguish readily between two families of hydrocarbons: alkenes (which make the bromine's color disappear immediately as the addition reaction proceeds) and alkanes (with which bromine does not react in the absence of light).

The outcome of this addition reaction seems parallel to the addition to alkenes of another symmetrical reagent, molecular hydrogen (see Section 6.3). However, the mechanisms are entirely different. Halogen addition involves an electrophilic addition mechanism. Chlorine and bromine are electrophilic reagents in that they are susceptible to attack by nucleophiles—and an alkene is a nucleophile, as is apparent from the reactions of alkenes with hydrogen halides, water and acid, and diborane. The mechanism of the halogenation of alkenes is presented in Figure 6.9, using cyclobutene as the substrate.

Note that the addition of halogens is stereospecific because of the necessity of backside attack by the bromide ion on the bromonium ion. Therefore, the addition of a halogen to any alkene is an *anti addition* (that is, the two components of the reagent add to opposite sides of the double bond). Knowing the stereochemistry of the starting alkene, you can work out the expected stereochemistry of the product(s).

How to Solve a Problem

What product(s) are formed from the addition of chlorine to (a) cyclohexene and (b) *trans*-2-butene?

PROBLEM ANALYSIS AND SOLUTION

(a) In dealing with a ring system, it is clear from the addition mechanism that the two chlorine atoms must end up on opposite sides of the ring. Further, in view of the stereochemistry of the reaction, the two chlorines should be axial after attack of the chloride ion. However, we know that the equatorial position is more stable for cyclohexane substituents. Conversion of the conformation of the initially formed 1,2-diaxial isomer yields the more stable 1,2-diequatorial isomer as the final product.

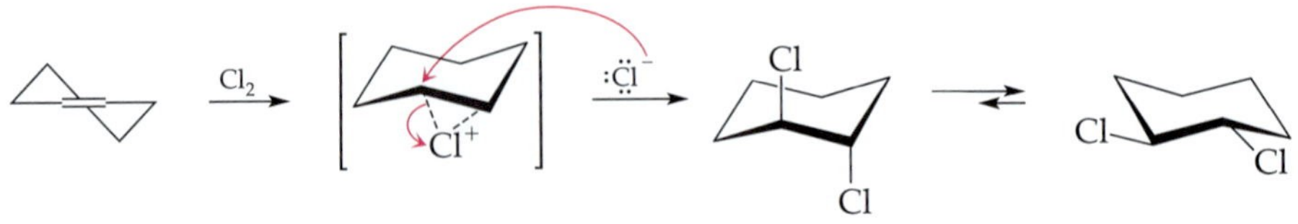

Cyclohexene 1,2-Dichlorocyclohexane

(b) The same approach is required with *trans*-2-butene. The only difference is the absence of the fixed plane of the ring for reference, which makes it necessary to concentrate more on the conformations of the products. A similar chloronium ion intermediate is involved, which undergoes back-side attack by chloride ion, the attack being able to occur equally well at either carbon atom. The product of the reaction is 2,3-dichlorobutane.

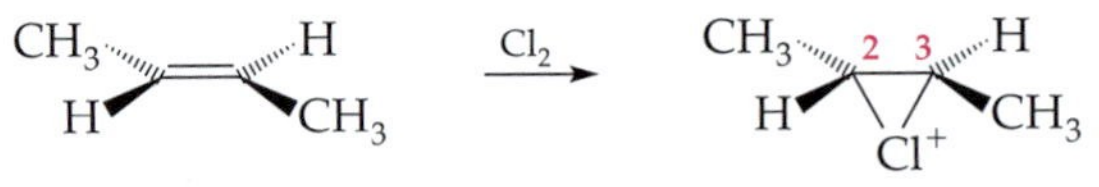

trans-2-Butene Chloronium ion

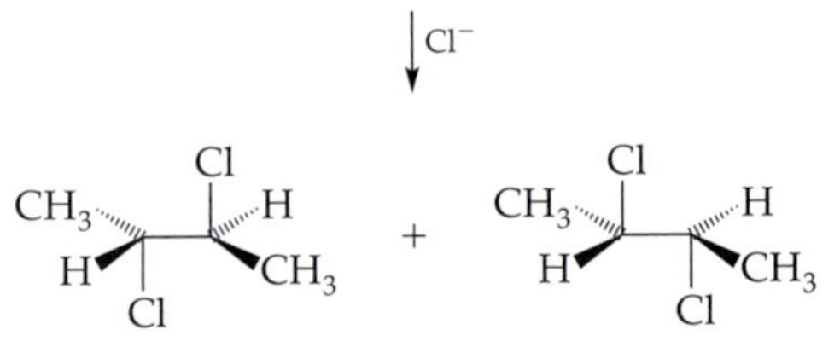

B from attack at position 3 **A** from attack at position 2

Identical structures for 2,3-dichlorobutane

We note that structure **B** is superimposable on structure **A,** and so they are identical. (If we rotate **A** 180° in its plane, it becomes **B.**) Also, the structures have a plane of symmetry between the two carbons carrying chlorine (imagine rotating the right-hand carbon and substituents of **A** or **B** 180°) and so cannot be enantiomers. We also note that carbons 2 and 3 are both stereocenters. Because of their identical pattern of substitution and opposite absolute configurations, these structures are not enantiomers but are instead a *meso form*, a term to be explained in Chapter 17.

PROBLEM 6.13

Write the product(s) (including stereochemistry, if relevant) of the reaction of bromine with each of the following alkenes: (a) cyclopropene, (b) 1,2-dimethylcyclohexene, (c) 1-pentene, and (d) *trans*-2-pentene.

PROBLEM 6.14

Show the stereochemistry of the reaction of chlorine with *cis*-2-butene, indicating how the various isomers arise.

6.5

Oxidation of Alkenes

The carbon-carbon double bond is very reactive because of the ready accessibility of the pi electrons, as you learned in Sections 6.3 and 6.4. In contrast, recall how unreactive alkanes usually are to oxidation, undergoing combustion only under severe conditions. Therefore, it is not too surprising that alkenes readily undergo oxidation. We will discuss four such reactions in this section.

6.5.1 Epoxidation of Alkenes

Epoxidation of alkenes was discussed briefly in Section 5.2.2 as a means of preparing epoxides. Alkenes react readily with a family of compounds called peracids (RCO_3H), which are organic carboxylic acids (see Chapter 14) that have an extra oxygen inserted between the oxygen and hydrogen of the —OH group, usually through reaction with hydrogen peroxide (H_2O_2). These peracids are mild oxidizing agents and readily transfer their "extra" oxygen to alkenes.

General reaction:

$$\text{C=C} \xrightarrow{RCO_3H} \text{C—C (with O bridge)}$$

Alkene Epoxide/oxirane

This transformation is illustrated by the conversion of cyclohexene to cyclohexene oxide using peracetic acid:

$$\text{Cyclohexene} + CH_3\text{—}\overset{O}{\overset{\|}{C}}\text{—O—OH} \longrightarrow \text{Cyclohexene oxide} + CH_3COOH$$

Cyclohexene Peracetic acid Cyclohexene oxide Acetic acid

The preparation of epoxides is important commercially (for example, they are used in epoxy resins), as mentioned in Chapter 5. The conversion of an alkene to an epoxide is also an important biological reaction, appearing in the biosynthetic process for the cyclization of the terpene squalene into the steroid nucleus from which cholesterol is derived (see Section 8.3.2). One of the major laboratory uses of epoxide preparation is as a first step in the overall conversion of alkenes to *trans* diols by reaction of epoxides with hydroxide ion (see Section 5.2.4). This reaction involves a nucleophilic and stereospecific ring opening:

$$\text{Cyclohexene oxide} \xrightarrow[H_2O]{NaOH} \textit{trans}\text{-1,2-Cyclohexanediol (OH, OH)}$$

Cyclohexene oxide *trans*-1,2-Cyclohexanediol

Epoxidation of an alkene is a stereospecific reaction, with the geometry of the alkene retained in the epoxide through a *syn* addition. Thus, *cis*-2-butene produces *cis*-2,3-dimethyloxirane, and *trans*-2-butene produces *trans*-2,3-dimethyloxirane:

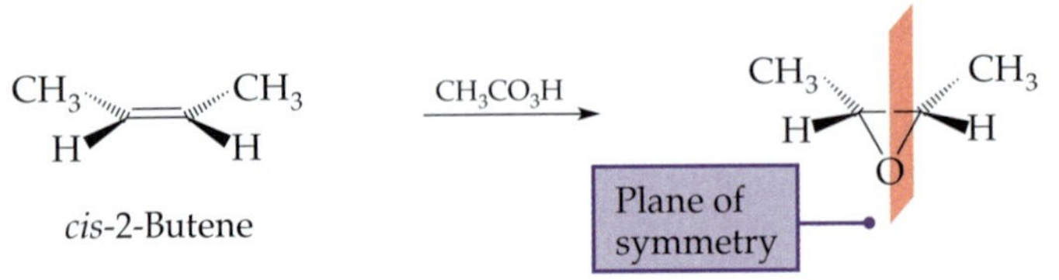

CH_3CO_3H

cis-2-Butene

Plane of symmetry

cis-2,3-Dimethyloxirane (achiral)

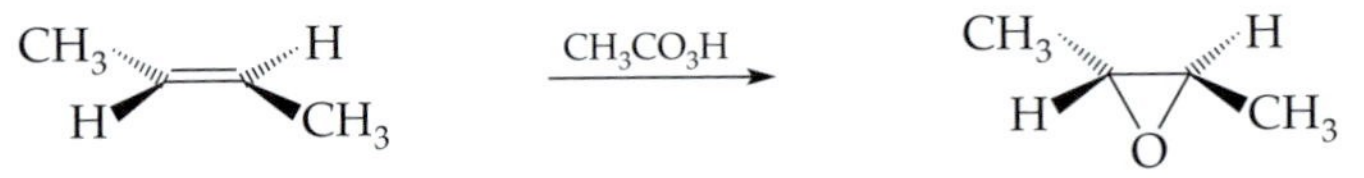

CH_3CO_3H

trans-2-Butene

trans-2,3-Dimethyloxirane (chiral, obtained as a racemic mixture)

Note that although the *cis*-dimethyloxirane has two stereocenters, it has a plane of symmetry and therefore is achiral; a mirror image of the structure will be superimposable. The *trans*-dimethyloxirane also has two stereocenters, but it does not have a plane of symmetry. Therefore, it is chiral; the structure shown is not superimposable on its mirror image, and two trans enantiomers are formed as a racemic mixture.

6.5.2 Oxidation with Potassium Permanganate

Potassium permanganate is a strong oxidizing agent that we first discussed as one reagent for the oxidation of alcohols (see Section 4.4). When used in a cold solvent in the presence of water, its strength is muted and it oxidizes alkenes to vicinal diols (also known as glycols, or 1,2-diols). This is a stereospecific *syn* addition of two hydroxyl groups: The permanganate anion approaches from one side of the alkene and forms two carbon-oxygen bonds simultaneously.

General reaction:

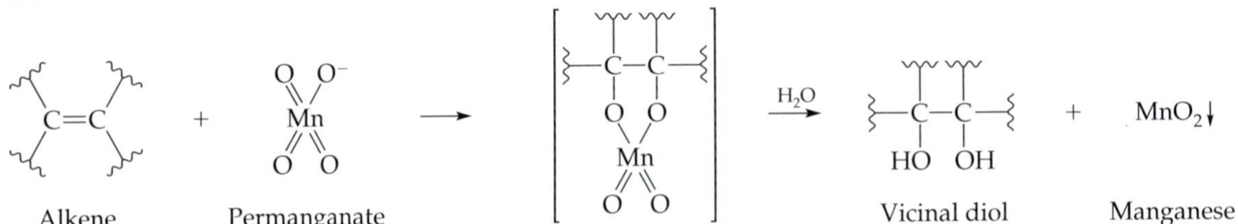

Alkene + Permanganate anion → [Permanganate ester] $\xrightarrow{H_2O}$ Vicinal diol + Manganese dioxide ($MnO_2\downarrow$)

For example, cyclopentene yields *cis*-1,2-cyclopentanediol. Recall that *trans*-1,2-cyclopentanediol can be obtained via epoxidation of cyclopentene, followed by ring opening of the epoxide using hydroxide ion (described in Section 6.5.1):

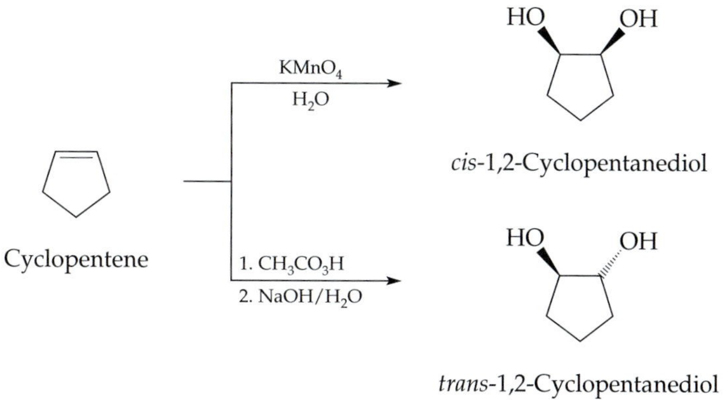

Cyclopentene

$KMnO_4$, H_2O → *cis*-1,2-Cyclopentanediol

1. CH_3CO_3H 2. $NaOH/H_2O$ → *trans*-1,2-Cyclopentanediol

Potassium permanganate in aqueous solution has a deep purple color. When it oxidizes a substrate, the Mn(VII) is reduced to Mn(IV), which precipitates as brown MnO_2. Because alkanes are not susceptible to permanganate oxidation, this reaction is used as a quick qualitative test to distinguish between two families of hydrocarbons: alkanes and alkenes.

PROBLEM 6.15

Give structures and names for the products of the following reactions:

(a) cyclohexene with peracetic acid, followed by treatment with sodium hydroxide
(b) cyclohexene with cold potassium permanganate
(c) *cis*-2-butene with peracetic acid, followed by treatment with sodium methoxide

6.5.3 Ozonolysis of Alkenes

Ozonolysis of alkenes is a reaction that is rarely used for synthesis and mainly used for determining the structure of an unknown alkene. Ozone is readily generated in the laboratory by passing a 10,000-volt electric discharge through oxygen gas. (Ozone is what you smell immediately after a lightning bolt has struck nearby.) Reaction of ozone with an alkene results in the addition of ozone to the double bond. The initially formed ozonide is usually immediately reduced (cleaved) with zinc and water to produce two carbonyl compounds. Overall, the double bond is cleaved, and each of its carbons becomes double-bonded to an oxygen. For example, isobutylene is cleaved to formaldehyde and acetone:

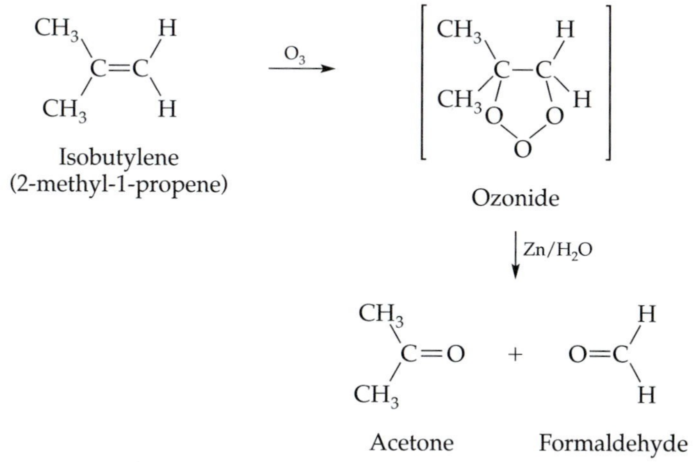

$$(CH_3)_2C{=}CH_2 \xrightarrow{O_3} [\text{Ozonide}] \xrightarrow{Zn/H_2O} (CH_3)_2C{=}O + O{=}CH_2$$

Isobutylene (2-methyl-1-propene) — Ozonide — Acetone + Formaldehyde

The structure of any alkene can be deduced from the known structures of the resulting carbonyl compounds. Simply envision putting the two carbonyl groups together, removing both oxygen atoms, and replacing them with a carbon-carbon double bond. For example, if 2-butanone and acetaldehyde are obtained from the ozonolysis of an unknown alkene, the alkene must have been 3-methyl-2-pentene:

Products obtained by ozonolysis — **Deduced structure of alkene**

$$(CH_3)(C_2H_5)C{=}O \;+\; O{=}C(CH_3)(H) \;\Longrightarrow\; (CH_3)(C_2H_5)C{=}C(CH_3)(H)$$

2-Butanone, Acetaldehyde, 3-Methyl-2-pentene

Likewise, if acetone is the only product obtained by ozonolysis of an alkene, the alkene must have been 2,3-dimethyl-2-butene. (Because two products must be formed, in this case they must be identical in structure.)

$$(CH_3)_2C{=}O \;+\; O{=}C(CH_3)_2 \;\Longrightarrow\; (CH_3)_2C{=}C(CH_3)_2$$

Acetone, Acetone, 2,3-Dimethyl-2-butene

How to Solve a Problem

Bay leaves

After having used bay leaves in preparing a stew for supper, a chemistry student wondered what it was that provided the pleasant flavor in the bay leaf. From a supply of bay leaves, the student obtained a flavorful, colorless liquid with the expected odor and taste (compound **A**). It was shown to be a hydrocarbon, and its molecular formula was determined to be $C_{10}H_{16}$. The student found that compound **A** decolorized bromine and also reacted with cold potassium permanganate. The student subjected compound **A** to catalytic hydrogenation, producing the known compound 2,6-dimethyloctane (compound **B**). Ozonolysis of compound **A** produced three different carbonyl compounds: two equivalents of formaldehyde ($H_2C{=}O$), one equivalent of acetone [$(CH_3)_2C{=}O$], and one equivalent of a tricarbonyl (having three $C{=}O$ groups) compound **C** with formula $C_5H_6O_3$. Compound **C** was shown to contain two aldehyde groups ($—CH{=}O$) and one ketone group ($R_2C{=}O$). Spectroscopic analysis indicated that the ketone group had only one methylene (CH_2) group adjacent to it. What is the structure of the compound in bay leaves (compound **A**)?

PROBLEM ANALYSIS

This is this book's first example of a structural analysis problem, sometimes called a *road map problem.* It places you in the role of a detective, using the minimum necessary amount of chemical information to deduce the structure of an unknown compound. The process involves putting together individual pieces of evidence, either to include or to rule out various structural possibilities.

The best approach to solving road map problems is to read slowly through the problem, jotting down specific conclusions (not observations) as they come to mind, even if the conclusions are a range of possibilities. It is also useful to write down structural fragments as they become evident. Finally, always keep in mind the total number of carbons in the unknown; don't worry about the number of hydrogens in the early stage of analysis. When you get to the end of the problem and your list of conclusions, reverse the direction of your analysis and the pieces should come together.

SOLUTION

Compound **A** is a hydrocarbon; its molecular formula, $C_{10}H_{16}$, indicates that there must be three double bonds and/or carbocyclic rings present. (The formula is "short" six hydrogens from being an alkane, $C_{10}H_{22}$; remember, C_nH_{2n+2} for alkanes and C_nH_{2n} for alkenes and cycloalkanes.) The positive reaction with bromine and permanganate indicates that there is at least one double bond in compound **A**. Hydrogenation saturates all carbon-carbon double bonds, and since 2,6-dimethyloctane (compound **B**) was obtained, we now know the carbon skeleton of **A**. We may conclude that there are no rings in **A**, so there are probably three double bonds.

Ozonolysis is the final key. If there are three double bonds, they will be cleaved to form six carbonyl groups. We are told of all six: two formaldehydes, one acetone, and the remaining three carbonyls in a single C_5 compound, **C**.

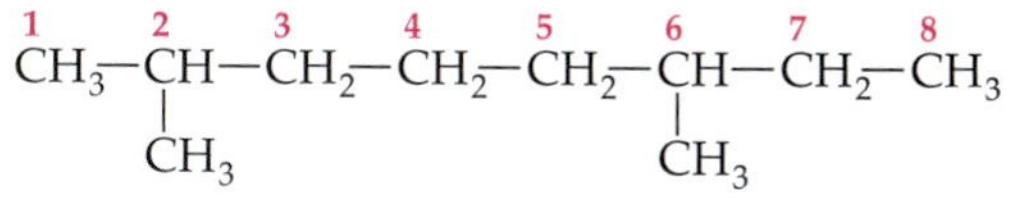

$$\overset{1}{CH_3}-\underset{\displaystyle CH_3}{\underset{|}{\overset{2}{CH}}}-\overset{3}{CH_2}-\overset{4}{CH_2}-\overset{5}{CH_2}-\underset{\displaystyle CH_3}{\underset{|}{\overset{6}{CH}}}-\overset{7}{CH_2}-\overset{8}{CH_3}$$

2,6-Dimethyloctane

B

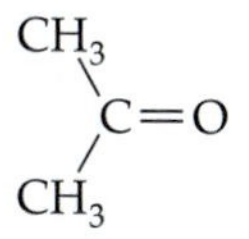

$$(CH_3)_2C{=}O$$

Acetone

$$O{=}CH_2$$

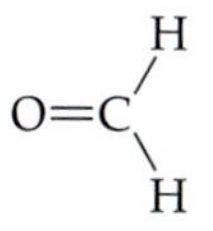

Formaldehyde

Examining the C_{10} carbon skeleton of compound **B**, we see that there is only one location that could yield acetone: a double bond between carbon 2 and carbon 3. The formaldehydes, then, could only come from the other two methyl groups, whose carbons must have been involved in double bonds. Thus, a tentative structure for compound **A** can be written based on the information gathered so far:

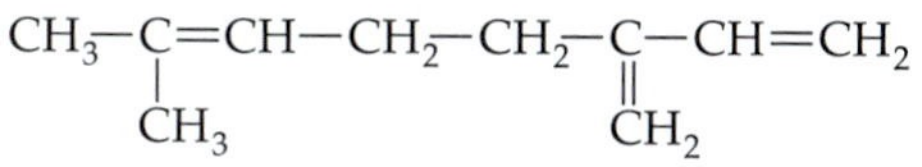

$$CH_3-\underset{\displaystyle CH_3}{\underset{|}{C}}{=}CH-CH_2-CH_2-\underset{\displaystyle CH_2}{\underset{\|}{C}}-CH{=}CH_2$$

Tentative structure for **A**

Checking the expected products from an ozonolysis of **A** indicates that they are all accounted for (imagine cleaving each double bond in **A** with ozone). Thus, the structure of **C** must be

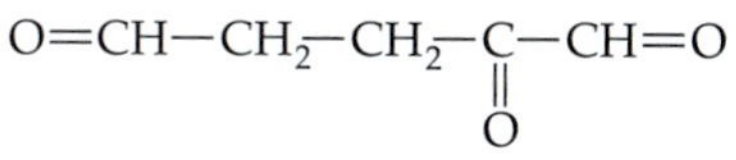

$$O{=}CH-CH_2-CH_2-\underset{\displaystyle O}{\underset{\|}{C}}-CH{=}O$$

C

The structure of the tricarbonyl compound **C** ($C_5H_6O_3$) can be deduced separately. The three functional groups of **C** together account for all of its molecular formula except for a CH_2 group:

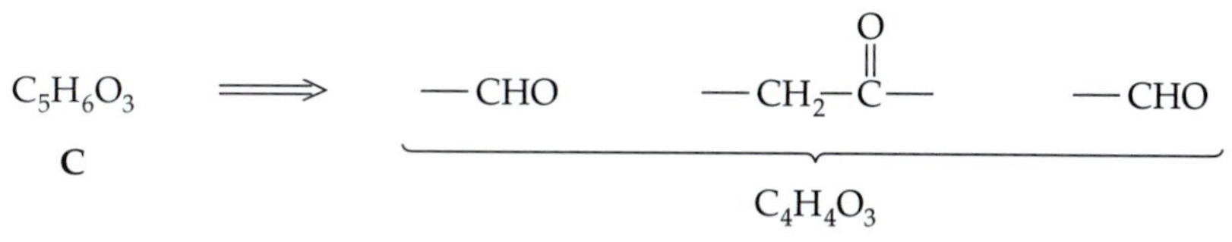

$$\underset{\textbf{C}}{C_5H_6O_3} \Longrightarrow \underbrace{-CHO \quad -CH_2-\overset{\displaystyle O}{\overset{\|}{C}}- \quad -CHO}_{C_4H_4O_3}$$

There is only one location for the CH_2 group that is consistent with all the information provided for the tricarbonyl compound (recall that the ketone group had only *one* adjacent CH_2 group); it must be between the left-hand aldehyde group and the CH_2 adjacent to the ketone group. Thus, the C_5 fragment is confirmed to be compound **C**.

Therefore, the unknown bay leaf liquid has the chemical structure that we adopted as the tentative structure for compound **A**. Its name is myrcene, and it is a member of the terpene family (see Section 8.3.2).

PROBLEM 6.16

The following carbonyl compounds were obtained upon ozonolysis of various alkenes. Show the structure of each starting alkene.

(a) CH_3CHO and $(CH_3)_2CO$
(b) $OHC(CH_2)_4CHO$
(c) OCH—CHO and CH_2O
(d) $CH_3COCH_2CH_2CHO$

PROBLEM 6.17

A compound with formula C_5H_8 decolorized bromine and absorbed one equivalent of hydrogen upon catalytic hydrogenation over platinum. Ozonolysis of the compound yields only a single compound. What is a possible structure for C_5H_8?

6.5.4 Combustion of Alkenes

Like any hydrocarbon, alkenes undergo complete oxidation (combustion in air) when sufficient energy is provided in the presence of excess molecular oxygen. Thus, alkenes are flammable, and the oxidation reaction proceeds completely to carbon dioxide and water, just as for alkanes. Alkenes are generally not used for fuels, however, since they must first be obtained from alkanes, which can be burned directly.

$$\underset{\text{Alkene}}{C_nH_{2n}} \xrightarrow[\Delta]{O_2} n\,CO_2 + n\,H_2O$$

$$\underset{\text{Hexene}}{C_6H_{12}} + 9\,O_2 \xrightarrow{\Delta} 6\,CO_2 + 6\,H_2O$$

6.6 Preparation of Alkenes

Conceptually, the preparation of alkenes can be accomplished through two broad approaches: by converting an existing bond into a double bond or by joining two fragments together and creating the double bond in the process.

In principle, converting an existing bond into a double bond in a structure that has all the other necessary groups and atoms already in place can be accomplished in two ways. You can start with a single bond and eliminate two groups, one from each of the carbons (elimination), or you can start with a triple bond and add back one hydrogen to each carbon (hydrogenation). These are the possible reactions:

1. Elimination of HOH by dehydration of alcohols (see Section 6.6.1).
2. Elimination of HX by dehydrohalogenation of alkyl halides (see Section 6.6.2).

3. Starting with a triple bond and adding back one hydrogen to each carbon (see Section 7.3.1, hydrogenation of alkynes).

These three reactions are shown here and are incorporated in the Key to Transformations (see inside front cover).

$$\text{—CH(A)—CH(B)—} \xrightarrow{-\text{A—B}} \text{—CH=CH—} \xleftarrow{+\text{H—H}} \text{—C}\equiv\text{C—}$$

Alcohol or alkyl halide — Alkene — Alkyne

The other approach for preparation of an alkene is to join two fragments together and create the double bond in the process. This is the approach of the Wittig reaction (to be discussed in Section 13.5.2).

6.6.1 Dehydration of Alcohols

Synthesis of an alkene by dehydration of an alcohol involves creating a double bond in a molecule where a single bond already exists.

General reaction:

$$\text{—C(H)—C(OH)—} \xrightarrow{-\text{HOH}} \text{C=C}$$

Alcohol — Alkene

This **dehydration reaction** is the reverse of the *hydration* of alkenes to form alcohols (see Section 6.4.2 and the Key to Transformations). Both hydration and dehydration are important biological reactions, appearing, for example, in the metabolism of fats and carbohydrates in the Krebs cycle (described in Chapter 19).

$$\text{H—C—C—OH} \underset{\text{(hydration of alkene)}}{\overset{\text{(dehydration of alcohol)}}{\rightleftharpoons}} \text{C=C}$$

Alcohol — Alkene

Heating an alcohol with concentrated sulfuric acid (or 85% phosphoric acid) removes the elements of water to form a double bond. For tertiary alcohols, the temperature required is 100°C, and for primary alcohols, about 200°C (the symbol Δ is often used for heat). Note that what is lost from the alcohol is the —OH group and a hydrogen from an *adjacent* carbon, as indicated in the general reaction by the red box. For example, cyclohexanol yields cyclohexene by loss of water, and both 1-propanol and 2-propanol yield propene:

Cyclohexanol $\xrightarrow[\Delta]{H_3PO_4}$ Cyclohexene + H_2O

1-Propanol $\xrightarrow[\Delta]{H_3PO_4}$ Propene + H_2O

2-Propanol $\xrightarrow[\Delta]{H_3PO_4}$ Propene + H_2O

When there is more than one location from which a hydrogen may be lost in the dehydration process, leading to two constitutionally isomeric alkenes, the issue of regiochemistry arises. Usually, both possible products are formed, but one is usually the major product and the other is a minor product. The *major product* is the alkene with the most substituents on the double-bonded carbons—in other words, the *most highly substituted alkene*. This regiochemistry is predicted by **Zaitzev's rule** (often spelled Saytzeff's), named after the Russian chemist who observed this behavior. The reason for the regiochemical outcome, as you now know, is the relative stability of the possible products: The more highly substituted alkene is the more stable alkene. For example, 1-methylcyclohexanol produces mainly 1-methylcyclohexene, while 2-butanol produces mainly 2-butene:

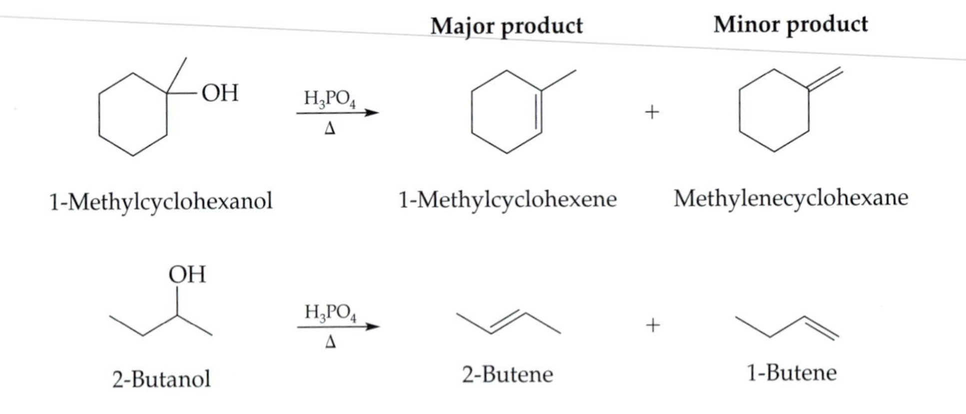

Recall that the way to determine which alkene is the more highly substituted is simply to count how many substituents *other than hydrogen* are attached to the double-bonded carbons. The parent alkene is ethene ($CH_2{=}CH_2$), which has no substituents other than hydrogen on the double-bonded carbons. The maximum possible number of substituents is four; four groups have replaced the four hydrogens of ethene. 1-Methylcyclohexene, the major product in the first example, has three substituents on its double-bonded carbons, while the minor product, methylenecyclohexane, has two. 2-Butene has two substituents, while 1-butene has only one.

How to Solve a Problem

Draw and name the product(s) formed when 3-methyl-2-butanol is heated with concentrated sulfuric acid. If more than one product is expected, which should predominate?

PROBLEM ANALYSIS

As always, we first draw the structure of the starting material.

$$\underset{\text{a}}{CH_3}-\underset{OH}{CH}-\overset{CH_3}{\underset{\text{b}}{CH}}-CH_3$$

3-Methyl-2-butanol

We recognize that sulfuric acid is a strong acid and that the given conditions are those for a dehydration reaction. Such a reaction involves loss of water (the hydroxyl group and an adjacent hydrogen). There are two adjacent carbons (labeled **a** and **b**) carrying hydrogens that could be lost to form an alkene. Loss of a hydrogen from carbon **a** would form one alkene (**A**), while loss of a hydrogen from carbon **b** would produce a different alkene (**B**).

$$CH_2{=}CH-\overset{CH_3}{CH}-CH_3 \qquad CH_3-CH{=}\overset{CH_3}{C}-CH_3$$

A **B**

SOLUTION

The given alcohol undergoes dehydration to form alkenes, and both alkenes are possible products. Alkene **B** predominates because it is the more stable alkene—the more highly substituted alkene. Alkene **B** has three substituents on the carbons of the double bond, while alkene **A** has only one. Zaitzev's rule applies, resulting in 2-methyl-2-butene (alkene **B**) as the major product and 3-methyl-1-butene (alkene A) as the minor product.

The mechanism for the dehydration reaction is called an **E1 mechanism** (E stands for elimination, of water in this instance; 1 means unimolecular). The elimination mechanism involves the formation of a carbocation in the slow (rate-determining) step, as shown in Figure 6.10. Because tertiary carbocations are more stable than secondary or primary carbocations (as a result of hyperconjugation; see Section 3.4.3), they form more readily; thus, the relative ease of dehydration of alcohols is 3° > 2° > 1°. To form a carbocation, the —OH group must leave, but as you already know, it is not a good leaving group. The acid catalysis in the reaction is needed to convert the hydroxyl group to an oxonium group, which allows water to be the leaving group in the slow step. (The reaction is acid-catalyzed because the acid is not consumed but is regenerated in the final step.)

The major disadvantage of the dehydration of alcohols as a means of preparing alkenes is that the reaction is not regiospecific, (it is regioselective, according to Zaitzev's rule), nor is it stereospecific. An alcohol that can form either a *cis* or *trans* alkene forms both, usually with the more stable *trans* alkene as the major product. For example, 2-pentanol produces mainly *trans*-2-pentene (but accompanied by small amounts

FIGURE 6.10

E1 mechanism for the dehydration of alcohols.

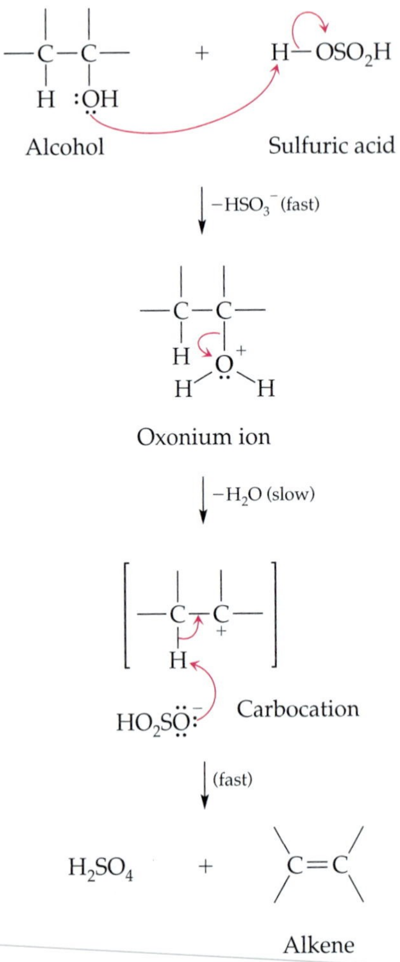

Alcohol + Sulfuric acid ($H{-}OSO_2H$)

$-HSO_3^-$ (fast)

The alcohol is protonated by a strong acid—in this case, sulfuric acid—to form an oxonium ion. Such protonations are very rapid.

Oxonium ion

$-H_2O$ (slow)

Water is a good leaving group, taking the carbon-oxygen bonding electrons with it and leaving behind a carbocation. Because of the relative instability of a carbocation, this is the slowest step (it requires the most energy).

Carbocation

HO_2SO^-

(fast)

H_2SO_4 + Alkene ($C{=}C$)

The positively charged carbon of the carbocation can return to a tetravalent form by acquiring a shared pair of electrons through one of two possible steps: by reacting with a nucleophile (an S_N1 reaction) or by sharing a pair of electrons with the adjacent carbon atom, which obtains the necessary electrons through loss of a proton to bisulfate anion (an E1 reaction, leading to formation of an alkene).

of *cis*-2-pentene and 1-pentene), and 2-methylcyclohexanol produces mainly 1-methylcyclohexene (accompanied by a small amount of 3-methylcyclohexene):

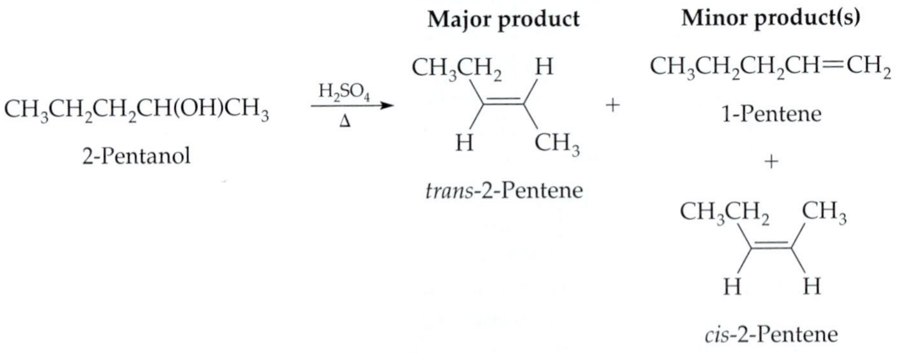

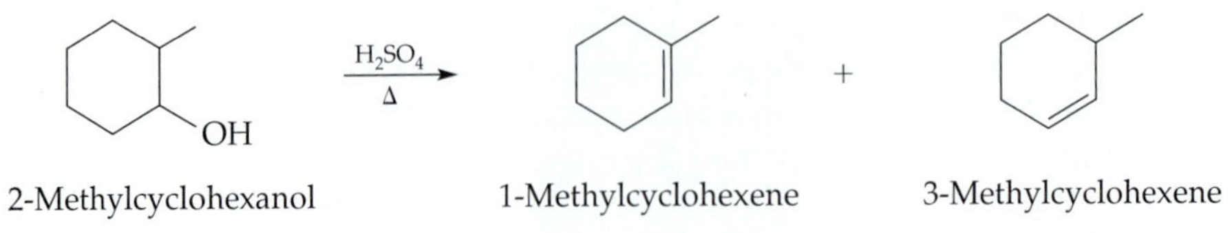

$$CH_3CH_2CH_2CH(OH)CH_3 \xrightarrow[\Delta]{H_2SO_4}$$

2-Pentanol

Major product: *trans*-2-Pentene (CH_3CH_2, H / H, CH_3)

Minor product(s): $CH_3CH_2CH_2CH{=}CH_2$ 1-Pentene + *cis*-2-Pentene (CH_3CH_2, CH_3 / H, H)

2-Methylcyclohexanol (OH) $\xrightarrow[\Delta]{H_2SO_4}$ 1-Methylcyclohexene + 3-Methylcyclohexene

How to Solve a Problem

Show all steps and the reagents required to convert cyclohexanol into cyclohexane.

PROBLEM ANALYSIS

As usual, we write the overall reaction we wish to accomplish:

OH

Cyclohexanol Cyclohexane

Then we consider whether there is any direct single-step means to remove a hydroxyl group. For this to happen, the —OH group would have to be a good leaving group. But —OH is a poor leaving group, so there probably is no single reaction by which to accomplish the desired conversion. Therefore, we start thinking in terms of taking two steps for the conversion.

We can consider two approaches: (1) think of the kinds of reactions alcohols undergo (conversion to alkyl halides, dehydration to alkenes, and oxidation to carbonyl groups), and (2) think of the kinds of reactions from which alkanes are obtained (reaction of organometallics with water and hydrogenation of alkenes). We want to know if there is a kind of compound into which the alcohol could be converted and which in turn could be converted into an alkane. From approaches (1) and (2), it is apparent that an alkene is a possible intermediate compound: It can be obtained from an alcohol and can be hydrogenated to an alkane. Using retrosynthesis, we envision the following scheme:

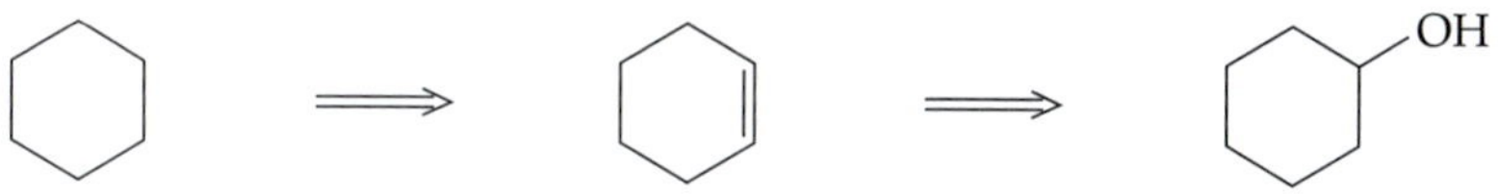

OH

Cyclohexane Cyclohexene Cyclohexanol

SOLUTION

Cyclohexanol is first converted to cyclohexene by acid-catalyzed dehydration. Because —OH is a poor leaving group, treatment with a strong acid (H_2SO_4 or H_3PO_4) is required to protonate the hydroxyl group to form an oxonium ion, making water an effective leaving group. Next, the reaction of cyclohexene with hydrogen over a platinum catalyst produces the desired cyclohexane.

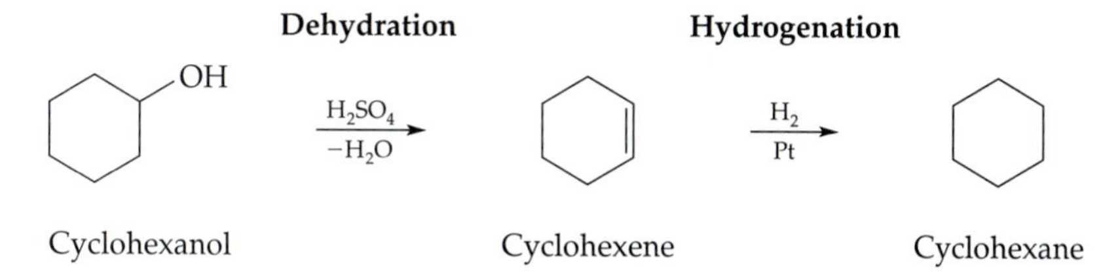

There is another, longer route that we should review. It involves the conversion of cyclohexanol to bromocyclohexane, conversion of bromocyclohexane to cyclohexylmagnesium bromide (a Grignard reagent), and quenching of cyclohexylmagnesium bromide with dilute acid. In the last step, the strongly nucleophilic organometallic carbanion is protonated by water to form cyclohexane.

PROBLEM 6.18

Write the reaction for the dehydration of each of the following alcohols with phosphoric acid, showing all products and indicating which should be the major product:

(a) 2,2-dimethylcyclohexanol

(b) 4-methyl-2-pentanol

(c) 2-methylcyclopentanol

PROBLEM 6.19

Show the alcohol that should be used to produce each of the following alkenes as the major product of a dehydration reaction, while minimizing the minor products:

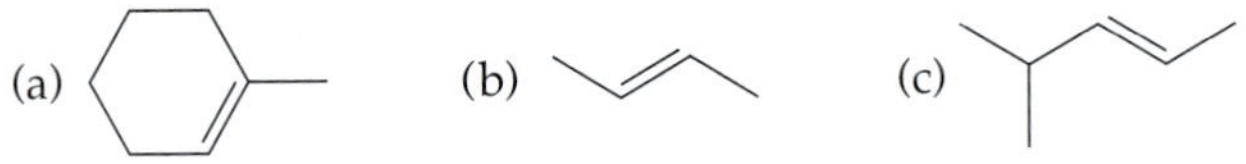

(a) (b) (c)

PROBLEM 6.20

Show the reactions and reagents necessary to convert 2-methyl-2-butanol into 3-methyl-2-butanol.

6.6.2 Dehydrohalogenation of Alkyl Halides

Dehydrohalogenation is the elimination of a hydrogen halide (H—X) from an alkyl halide to form a double bond between two carbons, where a single bond already exists.

General reaction:

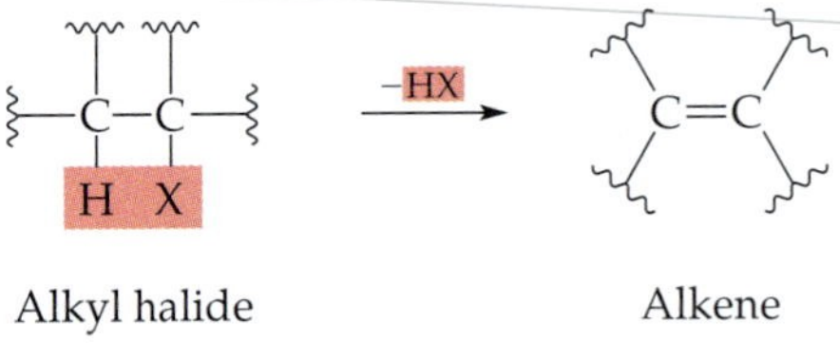

C—C with H and X $\xrightarrow{-HX}$ C=C

Alkyl halide   Alkene

For example, cyclopentyl chloride can be converted to cyclopentene by elimination of hydrogen chloride.

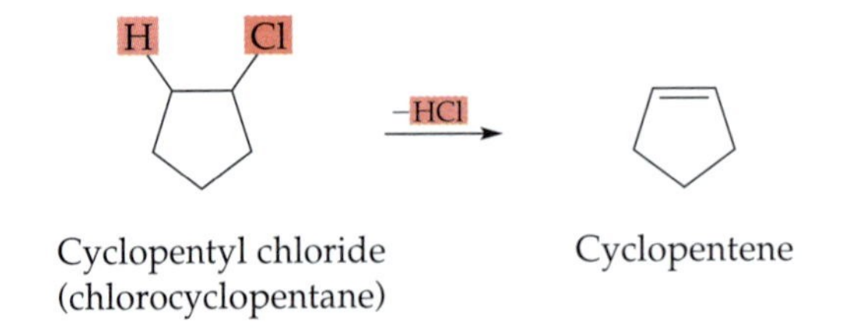

Cyclopentyl chloride (chlorocyclopentane)   Cyclopentene

This dehydrohalogenation reaction is the reverse of the hydrohalogenation reaction (addition of a hydrogen halide) of alkenes to form alkyl halides (see Section 6.4.1 and the Key to Transformations).

H—X (dehydrohalogenation of alkyl halide) ⇄ (hydrohalogenation of alkene)

Alkyl halide Alkene

The removal of a hydrogen halide (usually HCl or HBr) from an alkyl halide requires heating in the presence of a strong base. Most often, the base is sodium hydroxide (NaOH) in concentrated solution, sodium ethoxide ($Na^+\ ^-OC_2H_5$, often abbreviated NaOEt, where Et stands for ethyl) in ethanol solution, or potassium *t*-butoxide ($K^+\ ^-OC_4H_9$, abbreviated $KOBu^t$ or *t*-BuOK) in *t*-butyl alcohol solution.

E2 Elimination

Understanding the mechanism of the dehydrohalogenation reaction (Figure 6.11) will allow you to predict the reaction outcome with any starting alkyl halide. The mechanism is called E2 (E stands for elimination, 2 stands for bimolecular) because the slow step involves two species forming the transition state, the alkyl halide and the base. The rate of the reaction depends on the concentration of both reactants. In other words, this is a second-order reaction, which is first-order in each reactant.

E2 elimination is a stereospecific reaction (*anti* elimination) in that the leaving halogen atom must be oriented at 180° from the hydrogen being removed; all atoms undergoing a bonding change in the transition state must lie in a plane. Proof of this stereospecificity comes from the dehydrochlorination of two derivatives of natural products, menthyl chloride and neomenthyl chloride. Note that elimination occurs only with a hydrogen oriented *anti* to the chlorine.

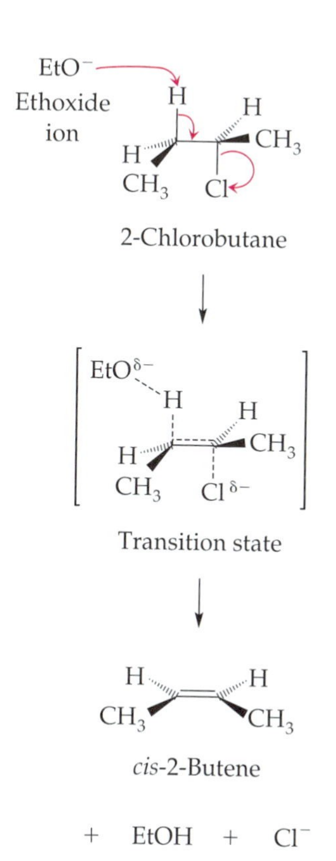

The base attacks a hydrogen on a carbon adjacent to the leaving group (chlorine). The reaction is a single-step reaction, all of the electron shifts portrayed take place approximately simultaneously. Thus, it is often called a *concerted reaction*. The transition state for the reaction has the base, the hydrogen, the two carbons, and the chlorine all lying in a single plane (in this case, the plane of the page). The transition state is depicted with dashed lines representing the bonds that are partially formed or partially broken at that point in the reaction. The ethoxide ion is losing its charge as those electrons are used to form the bond with hydrogen, and the chlorine is acquiring charge as its bond to carbon is being broken.

FIGURE 6.11

Mechanism of the E2 elimination reaction of 2-cholorobutane.

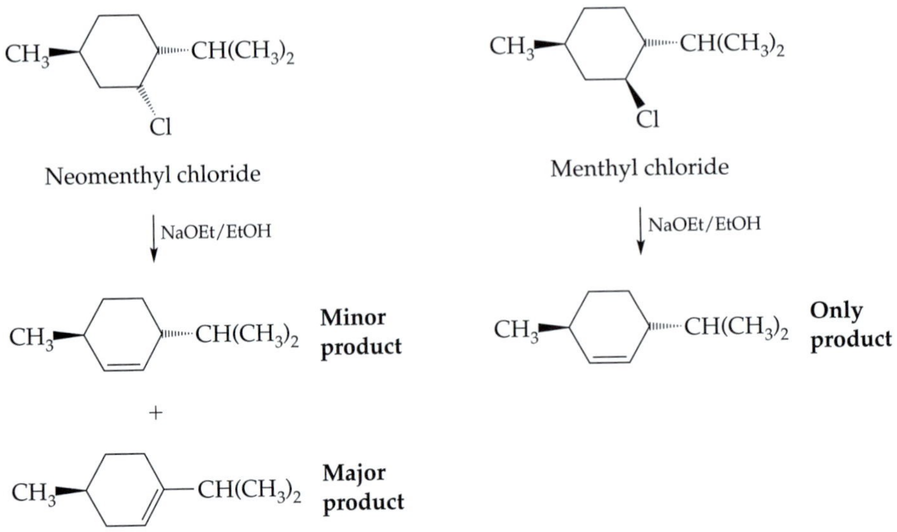

Neomenthyl chloride → (NaOEt/EtOH) → Minor product + Major product

Menthyl chloride → (NaOEt/EtOH) → Only product

The course of these reactions is made even clearer when the conformational structures of the two reactants are considered. (The hydrogens abstracted by the base are shown in red, and the *trans* chlorine is shown in blue.) Note that the menthyl chloride must convert to its less stable conformation (all substituents axial) in order for the hydrogen and chlorine to be positioned *anti* to each other.

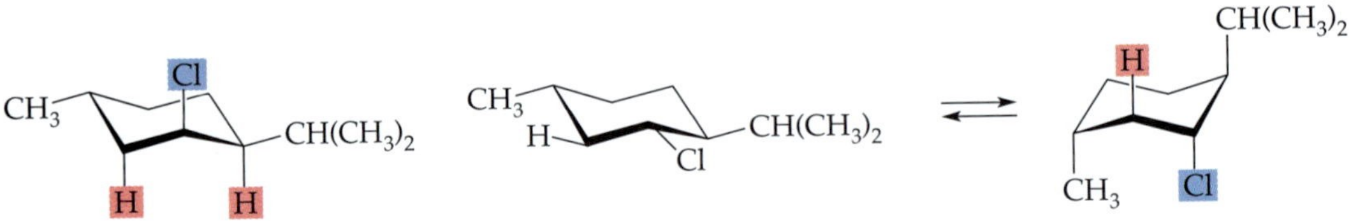

Neomenthyl chloride

Menthyl chloride

The elimination reaction with neomenthyl chloride also indicates that regiochemistry is an issue. Two alkenes can be formed because two *anti* hydrogens are available for elimination. Zaitzev's rule applies (see Section 6.6.1)—the major product is the most highly substituted alkene. The reason for this is found in the transition state, in which the double bond is partially formed (see Figure 6.11). Recall that more highly substituted alkenes are more stable than less highly substituted alkenes, and the same factors that stabilize an alkene come into play in the transition state as the double bond is forming. Thus, the dehydrohalogenation reaction is regioselective.

The relative stability of the two transition states, due to different degrees of substitution, also explains why tertiary alkyl halides eliminate under milder conditions than do primary alkenes. For example, ethyl chloride requires heating with sodium ethoxide, whereas *t*-butyl chloride is converted to isobutylene with only mild sodium hydroxide treatment:

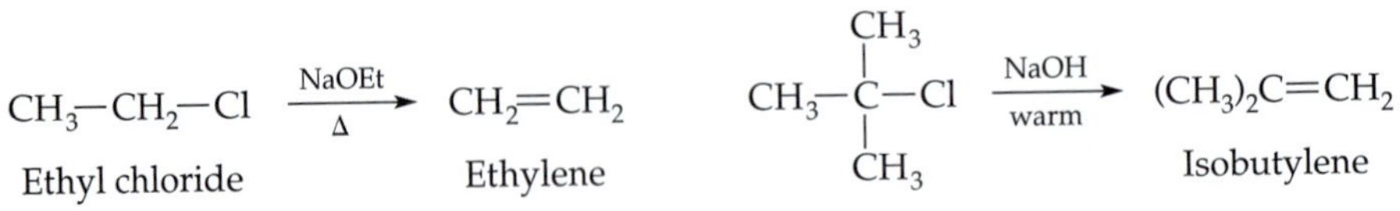

$$CH_3{-}CH_2{-}Cl \xrightarrow[\Delta]{NaOEt} CH_2{=}CH_2$$

Ethyl chloride　　Ethylene

$$CH_3{-}C(CH_3)_2{-}Cl \xrightarrow[warm]{NaOH} (CH_3)_2C{=}CH_2$$

t-Butyl chloride　　Isobutylene

PROBLEM 6.21

Show the possible product(s) of the following elimination reactions. If there is more than one product, indicate which is major and which is minor.

(a) *cis*-1-chloro-2-methylcyclohexane with sodium ethoxide

(b) *trans*-1-chloro-2-methylcyclohexane with sodium ethoxide

(c) 2-methylcyclohexanol with phosphoric acid

PROBLEM 6.22

What starting material and reagent(s) would you employ to produce each of the following alkenes from a saturated compound?

(a) 2-methyl-1-butene (b) 1-methylcyclohexene (c) 3-methyl-1-butene

---

Certain experimental conditions can reverse the regiochemistry of the E2 dehydrohalogenation reaction. Because it is essential for the attacking base to "get at" the hydrogen being eliminated, a very strong base of a large size usually attacks the *least sterically hindered* hydrogen (to yield the least highly substituted alkene) rather than the most hindered hydrogen (which would yield the most highly substituted and most stable alkene). For example, the reaction of 2-bromo-2-methylbutane with sodium ethoxide produces mainly 2-methyl-2-butene, as expected from Zaitsev's rule. However, use of potassium *t*-butoxide as the base produces mainly 2-methyl-1-butene. This change in regiochemistry is due to the large size of the *t*-butoxide ion, which makes it difficult for it to reach the secondary hydrogen that a base normally would attack.

$$CH_3-\underset{CH_3}{\overset{Br}{C}}-CH_2-CH_3 \xrightarrow[\text{heat}]{\text{NaOEt}} CH_3-\underset{CH_3}{C}=CH-CH_3 \quad \textbf{Major product}$$

2-Bromo-2-methylbutane → 2-Methyl-2-butene (more substituted alkene)

$$CH_3-\underset{CH_3}{\overset{Br}{C}}-CH_2-CH_3 \xrightarrow[\text{heat}]{\text{KOBu}^t} CH_2=\underset{CH_3}{C}-CH_2-CH_3 \quad \textbf{Major product}$$

2-Methyl-1-butene (less substituted alkene)

$$\text{NaOEt} \equiv CH_3-CH_2-O^- \; Na^+$$

Sodium ethoxide

$$\text{KOBu}^t \equiv CH_3-\underset{CH_3}{\overset{CH_3}{C}}-O^- \; K^+$$

Potassium *t*-butoxide

Comparison of S_N2 and E2 Reactions

The E2 reaction of an alkyl halide with a base should remind you of the S_N2 reaction of alkyl halides with nucleophiles, which are also bases. In considering the reaction of any alkyl halide with a base/nucleophile, you must always consider two possible outcomes: elimination (E2) to form an alkene, or substitution (S_N2) to form an alcohol

TABLE 6.1

Substitution (S_N2) versus Elimination (E2) Reactions of Alkyl Halides

Reaction	Favoring Conditions
Substitution (S_N2)	1. Low reaction temperatures 2. Modest bases/nucleophiles 3. Sterically small bases 4. Relative reactivity of alkyl groups: $1° > 2° > 3°$
Elimination (E2)	1. Higher reaction temperatures 2. Strong bases (for example, ethoxide) 3. Bulkier bases (reserve *t*-butoxide for reversing Zaitzev's rule) 4. Relative reactivity of alkyl groups: $3° > 2° > 1°$

or ether. In the elimination reaction, the base attacks hydrogen; in the substitution reaction, the base attacks carbon, as illustrated here for 1-bromopropane:

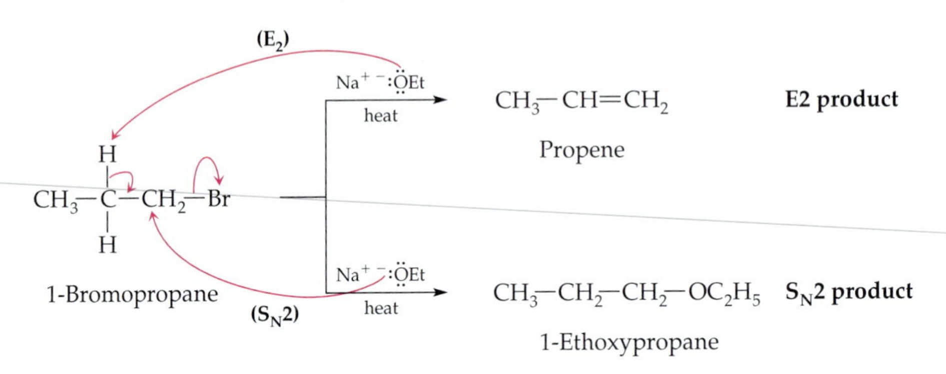

The reaction outcome for any alkyl halide with a nucleophile (base) can be controlled to some extent by experimental conditions but is determined mainly by molecular structure. Table 6.1 summarizes the conditions and structures favoring substitution or elimination. These tabulated reaction characteristics can often be applied to direct the course of a reaction of a base with an alkyl halide. The most difficult reaction to bring about is a nucleophilic substitution on a tertiary halide, because mainly elimination products form. Thus, *t*-butyl chloride with warm sodium hydroxide yields mainly isobutylene, and even with a base as weak as water, the substitution product (*t*-butyl alcohol) is accompanied by 20% isobutylene, the elimination product.

How to Solve a Problem

Predict the major product of the reaction of 2-bromo-4-methylpentane with (a) sodium hydroxide at low temperature, (b) hot sodium ethoxide, and (c) hot potassium *t*-butoxide.

PROBLEM ANALYSIS

First, we draw the structure of the substrate:

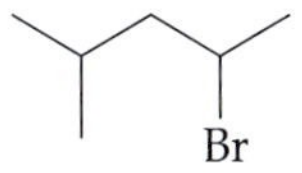

Br

2-Bromo-4-methylpentane

The substrate is an alkyl halide, and we recall that a halogen atom (bromine) makes a good leaving group as halide anion. The bromine ion can be made to leave through reaction with a nucleophile, either by a nucleophilic substitution or by an elimination. Because the substrate is a secondary alkyl bromide, both reactions are feasible. Substitution is favored by moderate bases and low temperatures, whereas elimination is favored by strong bases and heat.

(a) Reaction with the moderately strong base sodium hydroxide at low temperature is likely to cause a substitution.

(b) Reaction with the stronger base sodium ethoxide is likely to cause an elimination to form an alkene. Examination of the structure of the starting alkyl halide shows that there are two different hydrogens that could be eliminated, H_a and H_b:

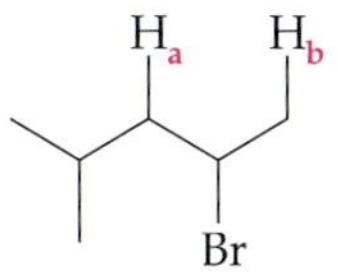

H_a H_b

Br

Elimination should follow Zaitzev's rule and remove H_a, forming the more stable alkene.

(c) Potassium *t*-butoxide is a stronger and much bulkier base and will have difficulty attacking H_a. Therefore, it should attack H_b to form the less stable alkene.

SOLUTION

(a) Reaction of 2-bromo-4-methylpentane with sodium hydroxide will result mainly in an S_N2 substitution and the formation of 4-methyl-2-pentanol.

(b) Reaction with sodium ethoxide will result in an E2 elimination and the formation of mainly 4-methyl-2-pentene.

(c) Reaction with potassium *t*-butoxide will result in an E2 elimination and the formation of mainly 4-methyl-1-pentene.

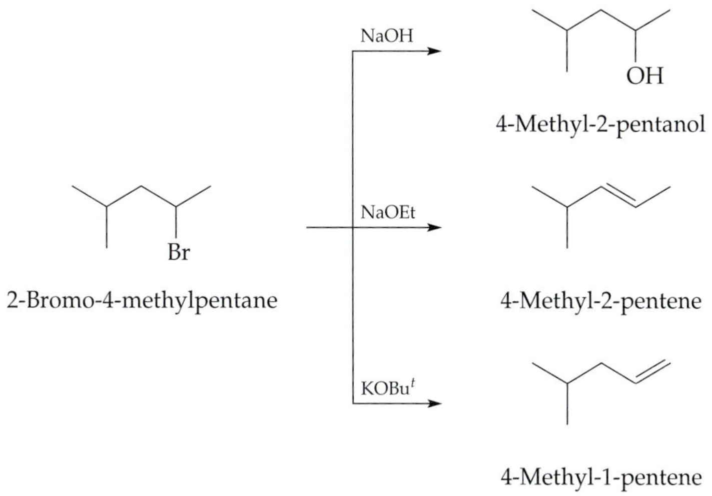

NaOH

OH

4-Methyl-2-pentanol

Br

2-Bromo-4-methylpentane

NaOEt

4-Methyl-2-pentene

$KOBu^t$

4-Methyl-1-pentene

PROBLEM 6.23

Predict the major product of each of the following reactions:

(a) 2-bromo-3-methylbutane with hot potassium *t*-butoxide
(b) 2-bromo-3-methylbutane with dilute sodium hydroxide
(c) 2-bromo-3-methylbutane with hot sodium ethoxide

PROBLEM 6.24

Designate the reaction conditions and reagents you would employ to effect each transformation:

(a) 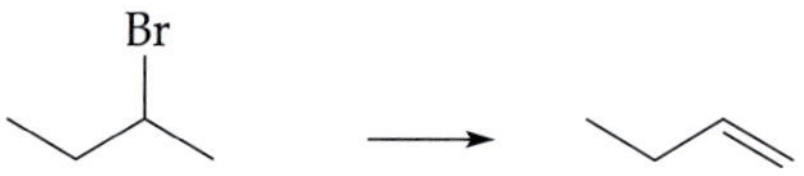 Br
(b) 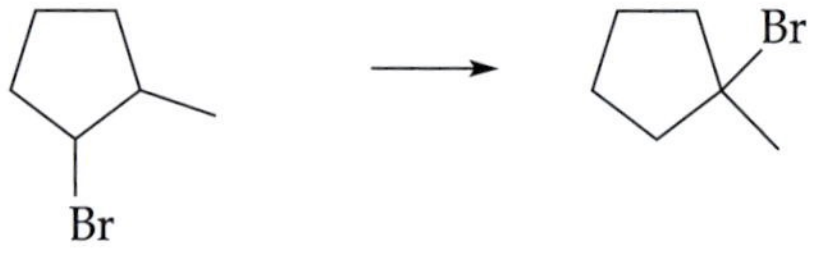 Br Br
(c) 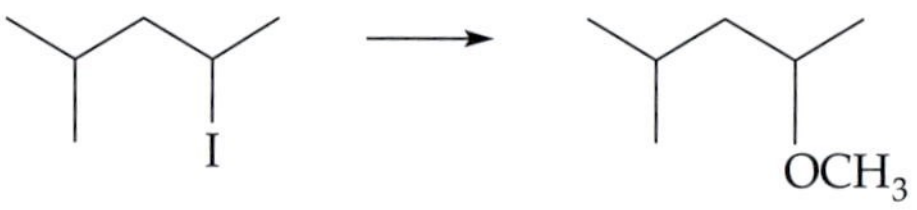 I OCH_3
(d)  I

How to Solve a Problem

Convert 2-methylpropene into 1-chloro-2-methylpropane.

PROBLEM ANALYSIS

Examining the structures of the given compounds suggests that the problem is one of addition of the elements of HCl to the alkene.

Cl

2-Methylpropene 1-Chloro-2-methylpropane

Knowing that alkenes react readily with hydrogen chloride, we should consider a straightforward addition reaction. However, the mechanism and regiochemistry of such an addition (as codified in Markovnikov's rule) predict that it will produce 2-chloro-2-methylpropane rather than the desired 1-chloro isomer:

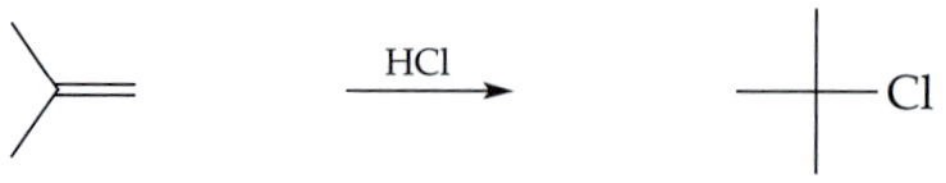

HCl Cl

2-Methylpropene 2-Chloro-2-methylpropane

This mental exercise has now identified the real problem—how to add HCl in an anti-Markovnikov manner. It cannot be done directly by the reactions we have considered so far. Is there any reaction we have covered that results in anti-Markovnikov addition to an alkene? How about hydroboration? That reaction with the given alkene in this case will result in an alcohol with the —OH group on the terminal carbon. We can take advantage of that outcome because an alcohol can readily be converted to an alkyl chloride (see Key to Transformations).

SOLUTION

First, we hydroborate 2-methylpropene to bring about an anti-Markovnikov addition and obtain 2-methyl-1-propanol. Then we replace the hydroxyl functional group with a chlorine. This can be done using hydrogen chloride or thionyl chloride/pyridine:

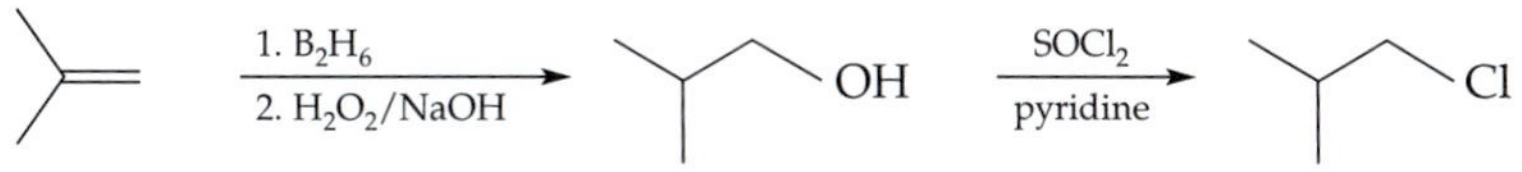

1. B_2H_6 2. H_2O_2/NaOH ; OH ; $SOCl_2$ pyridine ; Cl

2-Methylpropene 2-Methyl-1-propanol 1-Chloro-2-methylpropane

This is another example of putting together a sequence of two reactions to accomplish an organic synthesis.

PROBLEM 6.25

Show the reagents required for and the products obtained from each step of the following synthesis. (*Hint:* Apply your retrosynthetic skills to work backward from product to reactant.)

(a)  →

(b) 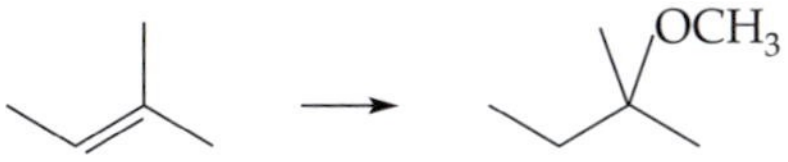 → OCH_3

(c)  OCH_3 →

6.7 Important Alkenes

The carbon-carbon double bond is very widespread and frequently occurs in compounds along with other functional groups. It is especially prevalent in naturally occurring compounds. This section focuses on a few compounds in which the double bond is the only functional group.

The simplest alkene, ethene, most often referred to by its common name ethylene, is produced in greater quantity than any other organic chemical in the United States, with annual production in billions of pounds. Ethylene is produced commercially via

CHEMISTRY AT WORK

IN THE MARKETPLACE

Ethylene, a Plant Hormone

Although ethylene is one of the most widely used industrial chemicals in the world, this simple and common alkene is also an important plant hormone, essential in small quantities for the ripening of many fruits and also involved in the response of a plant to being wounded.

Ethylene is synthesized in fruit-bearing plants from the amino acid 1-aminocyclopropanecarboxylic acid:

$$\text{1-Aminocyclopropanecarboxylic acid (cyclopropane bearing } NH_2 \text{ and } COOH) \xrightarrow{\text{enzymes}} \underset{\text{Ethylene}}{CH_2{=}CH_2}$$

The mechanism by which ethylene interacts with the plant and stimulates fruit ripening and growth of flowers is still being investigated.

Commercially, many fruits are picked before they become ripe and are shipped and stored in an ethylene-free atmosphere. When the unripe fruit reaches a packing plant, small amounts of ethylene vapor (bp −104°C) are passed over it, completing the ripening process just before the fruit is brought to market. Alternatively, a solution of a chemical called *ethrel* ($ClCH_2CH_2PO_3H_2$) can be sprayed over the fruit, either before it is picked or while in storage. Plants absorb ethrel and convert it into ethylene, hastening the ripening process.

A technician tests for the presence of ethylene gas in a genetically engineered melon. This fruit would normally ripen in 3–4 days but has been engineered to block its production of ethylene gas. Thus, it remains fresh for 50 days.

the *thermal dehydrogenation* of ethane in large petrochemical plants. That is, heating ethane to high temperatures over special catalysts produces ethylene and hydrogen:

$$\underset{\text{Ethane}}{CH_3{-}CH_3} \xrightarrow[\text{catalyst}]{\Delta} \underset{\text{Ethylene}}{CH_2{=}CH_2} + H_2$$

(Recall that ethane is the second most abundant component of natural gas. It is separated from methane in gas plants; then most of the ethane is converted to ethylene.) The hydrogen is used for many purposes, including as fuel for space shuttle launches and for reduction of molecular nitrogen to ammonia for use in fertilizer.

The most widespread use of ethylene is as a *monomer,* which is polymerized to the *polymer* polyethylene. *Polymerization,* a process we will discuss in Chapter 18, is a reaction in which large numbers of molecules of a simple compound are chemically joined together to form a long chain of repeating monomer units. In the case of ethylene, the repeating unit is $—CH_2—CH_2—$. Polyethylene takes many forms; it is the common plastic used for bottles (milk bottles and "squeeze" bottles), other containers, and wrapping material.

Propylene (propene) is another major industrial chemical, prepared by thermal dehydrogenation of propane. Propene can be polymerized to form the common plastic polypropylene. Polypropylene can be formulated to be much harder than polyethylene, making it especially useful in automobile parts. Isobutylene can be polymerized to make what is called butyl rubber, one of the first synthetic rubbers:

$$(CH_3)_2C{=}CH_2 \xrightarrow{\text{(polymerization)}} \left(-C(CH_3)_2-CH_2- \right)_n$$

Isobutylene — Polyisobutylene (butyl rubber)

A large family of naturally occurring alkenes are the terpenes (to be discussed in detail in Chapter 8).

Chapter Summary

Alkenes comprise a family of **unsaturated hydrocarbons** with the general formula C_nH_{2n}. Alkenes contain a **double bond** between two carbon atoms, each of which is attached to two other atoms or groups (that is, these carbons are trivalent). The double bond consists of a sigma (σ) bond (sp^2–sp^2 overlap) and a **pi (π) bond** (p–p overlap) and is formed between two carbons that are **trigonally hybridized.** Because rotation does not normally occur around the double bond, substituted alkenes can exist as *cis* and *trans* isomers (configurational isomers), also designated as *E* and *Z*. As a generalization, *trans* isomers are more stable than *cis* isomers of alkenes, and more substituted alkenes are more stable than less substituted alkenes.

According to the IUPAC system, alkenes are named by changing the alkane suffix *-ane* to *-ene* for the corresponding alkene. The double bond must be included in the carbon chain from which the parent alkane name is derived. The location is indicated by a single number designating the *first* carbon of the double bond. This number is chosen to be as small as possible. Historical/common names are often used for simple alkenes, such as ethylene and propylene.

The characteristic reactions of alkenes are **addition reactions** to the double bond. The simplest addition reaction is **hydrogenation,** in which hydrogen adds to one face of the double bond in the presence of a catalyst in a stereospecific *syn* addition to form an alkane.

The mechanism of most addition reactions involves *electrophilic addition* in a regioselective manner. Hydrogen halides add to alkenes to form alkyl halides, and water (using acid catalysis) adds to alkenes to form alcohols. The orientations of these additions are in accord with **Markovnikov's rule.** The addition of diborane (B_2H_6) followed by oxidation results in alcohol formation with the opposite regioselectivity (*anti-Markovnikov addition*).

Molecular halogens (bromine and chlorine) are symmetrical reagents that add to alkenes in a stereospecific *anti* addition to form **vicinal dihalides.**

Alkenes are very susceptible to oxidation. Cold potassium permanganate oxidizes alkenes to 1,2-diols (glycols or vicinal diols) in a stereospecific *syn* addition. Alkenes are oxidized to epoxides/oxiranes in a *syn* addition using peracids. Cleavage at the carbon-carbon double bond by *ozonolysis*, a reaction used to deduce the structure of an unknown alkene, yields carbonyl compounds.

Alkenes are prepared by four general reactions, two of which are described in this chapter. Alcohols are readily **dehydrated** using acid catalysis to form alkenes. The reaction is regioselective but not stereospecific. The most stable alkene (that is, the most highly substituted *trans* alkene) is usually the major product—according to **Zaitzev's rule.** Alkyl halides are **dehydrohalogenated** to form alkenes, using strong bases such as sodium ethoxide. Such elimination reactions occur by an E2 mechanism and are stereospecific (*trans* elimination) and regioselective (producing the most stable alkene). Use of a bulky base such as potassium *t*-butoxide produces the least stable alkene.

The reaction of alkenes with bromine is a qualitative test for the presence of unsaturation; the reddish-colored bromine is converted to colorless alkyl halides. The reaction of alkenes with potassium permanganate is also an indication of the presence of unsaturation; the purple permanganate solution produces a brown precipitate of manganese dioxide.

Summary of Reactions

1. **Hydrogenation of alkenes (Section 6.3).** In the presence of a catalyst, usually finely powdered platinum or palladium, molecular hydrogen adds to an alkene in a stereospecific *syn* addition to form an alkane.

Alkene $\xrightarrow[\text{Pt}]{H_2}$ Alkane (H H)

2. **Halogenation of alkenes (Section 6.4.3).** Molecular halogen (usually chlorine or bromine) adds to an alkene in a stereospecific *anti* addition to produce a vicinal dihalide.

Alkene $\xrightarrow{X_2}$ Vicinal dihalide (X X)

3. **Hydrohalogenation of alkenes (Section 6.4.1).** A hydrogen halide adds to an alkene to form an alkyl halide in a regioselective reaction following Markovnikov's rule.

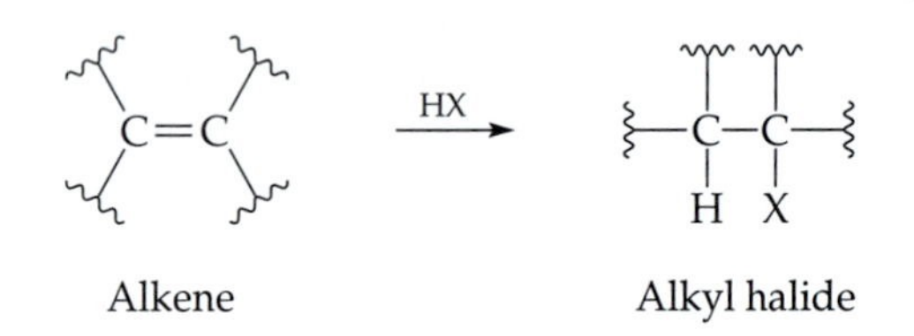

4. **Hydration of alkenes (Section 6.4.2)**

   a. With acid catalysis, water adds to an alkene in a Markovnikov orientation.

Alkene $\xrightarrow[H_2O]{H^+}$ Alcohol (H OH)

   b. An alcohol resulting from apparent anti-Markovnikov addition of water can be produced by hydroboration of an alkene followed by oxidation of the organoborane.

Alkene $\xrightarrow[\text{2. } H_2O_2/\text{NaOH}]{\text{1. } B_2H_6}$ Alcohol (HO H)

5. **Oxidation of alkenes (Section 6.5)**

   a. Alkenes react with peracids to form epoxides/oxiranes via a stereospecific *syn* addition.

Alkene $\xrightarrow{CH_3CO_3H}$ Epoxide/oxirane

   b. Alkenes react with cold potassium permanganate to form vicinal diols via a stereospecific *syn* addition.

Alkene $\xrightarrow{KMnO_4}$ Vicinal diol (HO OH)

   c. Ozonolysis of alkenes yields two carbonyl compounds, aldehydes or ketones.

Alkene (RR'C=CR''R''') $\xrightarrow[\text{2. Zn}/H_2O]{\text{1. } O_3}$ RR'C=O + O=CR''R'''

Carbonyl compounds (aldehydes or ketones)

6. **Preparation of alkenes (Section 6.6)**

   a. **Dehydration of alcohols (Section 6.6.1).** Heating an alcohol with a strong acid (H_3PO_4 or H_2SO_4) eliminates water in a regioselective manner to form the most stable alkene as the major product, according to Zaitzev's rule.

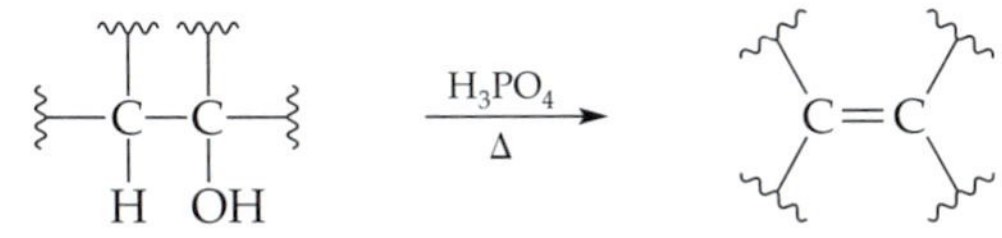

H OH $\xrightarrow[\Delta]{H_3PO_4}$ C=C

Alcohol Alkene

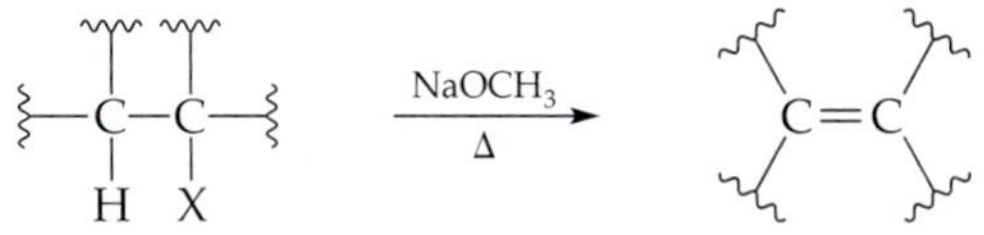

b. Dehydrohalogenation of alkyl halides (Section 6.6.2). Treating an alkyl halide with a strong base (sodium ethoxide) eliminates HX to form the most stable alkene. The reaction is stereospecific (*anti* elimination) and regioselective (the most stable alkene is the major product).

H X $\xrightarrow[\Delta]{NaOCH_3}$ C=C

Alkyl halide Alkene

The least stable alkene can be produced using the bulky base potassium *t*-butoxide.

Additional Problems

Nomenclature and Structure

6.26 Name the following compounds:

(a) $CH_3CH{=}CHCH_2CH_3$ (b) 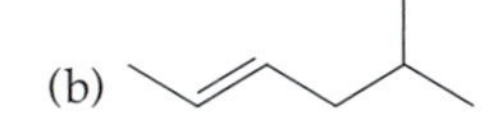

(c) 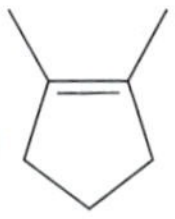 (d) 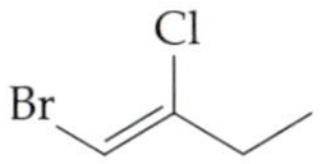 Br Cl

(e) 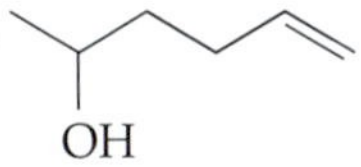 OH (f) $(CH_3)_2C{=}C(CH_3)_2$

(g) 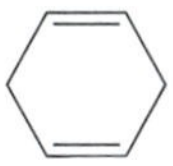 (h) $CH_2{=}CHCH(C_2H_5)CH_3$

6.27 Draw structural formulas corresponding to the following names:

(a) 1-methylcyclobutene

(b) (*E*)-2-hexene

(c) cyclopropylethene

(d) *cis*-3,4-dimethylcyclopentene

(e) 5,5-dimethyl-1,3-cyclopentadiene

(f) cyclohex-3-en-1-ol

(g) allyl alcohol

(h) 3-vinylcyclobutene

(i) (*Z*)-3,4-dimethyl-3-heptene

(j) isobutylene

6.28 Explain why the following names are incorrect. Then provide a correct name.

(a) *cis*-3-pentene

(b) 2-methylpropylene

(c) *trans*-3-methyl-3-hexene

(d) 1,1,2,2-tetramethylethene

(e) 2-methylcyclobutene

(f) 4,5-dibromocyclopentene

(g) 1-methyl-1-pentene

(h) 2-penten-5-ol

(i) 1,1-dimethyl-2,4-cyclohexadiene

(j) 1-vinyl-1-butene

6.29 Draw all possible constitutional isomers and stereoisomers having the molecular formula C_4H_8.

6.30 Determine whether each of the alkenes drawn in Problem 6.29 contains a mono-, di-, tri-, or tetra- substituted double bond.

6.31 Identify the more stable member in the following pairs of isomeric alkenes, and give a reason for your choice:

(a) 2-methyl-1-butene and 3-methyl-1-butene

(b) (*Z*)-2-pentene and (*E*)-2-pentene

(c) 1-methylcyclohexene and methylenecyclohexane

(d) 2,3-dimethyl-2-butene and 2-methyl-2-pentene

(e) *cis*- and *trans*-3,4-dimethylcyclobutene

6.32 Assign *E* or *Z* configurations to the following compounds:

(a) 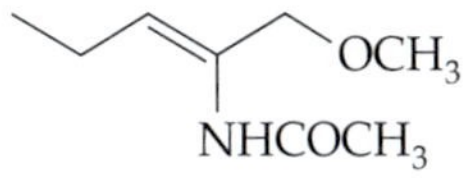 OCH_3 $NHCOCH_3$

(b) 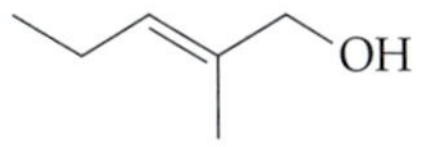 OH

(c) 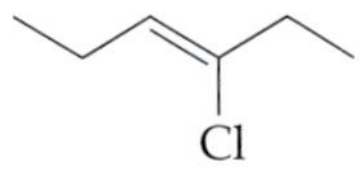 Cl

Addition to Alkenes

6.33 When 1,2-dimethylcyclohexene is hydrogenated, only one isomer of 1,2-dimethylcyclohexane is obtained. Draw its structure and name it. Explain why two isomers are not obtained.

6.34 Caryophyllene, $C_{15}H_{24}$, is a member of the terpene family obtained from oil of cloves. When it is fully hydrogenated over a platinum catalyst, the product is a hydrocarbon with formula $C_{15}H_{28}$. What can you conclude about the structure of caryophyllene from this observation?

6.35 Show the major product(s) of the reaction of 1-pentene with each of the following reagents:

(a) bromine in CCl_4 in the dark

(b) dilute sulfuric acid and heat

(c) hydrogen over a palladium catalyst

(d) hydrogen chloride

(e) ozone, then zinc and water

(f) peracetic acid, then sodium hydroxide

(g) diborane, then hydrogen peroxide and sodium hydroxide

(h) cold potassium permanganate

(i) hydrogen bromide, then sodium iodide in acetone

6.36 Repeat Problem 6.35 using cyclopentene as the starting alkene.

6.37 Repeat Problem 6.35 using 2-methyl-2-butene as the starting alkene.

6.38 How many different alkenes will yield 2,3,4-trimethylpentane upon hydrogenation? Draw their structures.

6.39 Draw the structure of the alkene that produces each set of products from ozonolysis:

(a) CH_2O and $(CH_3)_2CHCHO$

(b) CH_3CH_2CHO

(c) $O{=}CHCH_2CH_2CH{=}O$

6.40 How many different alkenes with the molecular formula C_5H_{10} will produce formaldehyde (CH_2O) as a product of ozonolysis? Draw the structure of each alkene (ignore possible stereoisomers).

6.41 Draw the structure of the alkene that you would employ to produce each of the following compounds, and specify the reagent(s) needed:

(a) Br (b) OH

(c) Cl (d) H Br Br H

(e) OH (f) Br Br

(g) OH OH (h) OH OH

Synthesis of Alkenes

6.42 Show the product(s) of an acid-catalyzed dehydration reaction of each of the following alcohols, indicating major and minor products if both are produced:

(a) 2-methylcyclohexanol

(b) 1-butanol

(c) 2-butanol

(d) 3-methyl-2-butanol

6.43 What alcohol would you dehydrate with concentrated sulfuric acid to produce mainly each of these alkenes?

(a) $CH_3CH{=}C(CH_3)_2$ (b) CH_3

6.44 Using the organic chemistry you've learned so far, show how you would convert 1-butene (a) to 1-chlorobutane and (b) to 2-chlorobutane. More than one step may be involved in each transformation.

6.45 Show the reaction and reagent necessary to accomplish each of the following alkene syntheses:

(a) isobutyl chloride to isobutylene

(b) *t*-butyl alcohol to isobutylene

(c) 1-bromobutane to 1-butene

(d) 2-bromobutane to mainly 2-butene

(e) 2-bromobutane to mainly 1-butene

6.46 Show the detailed mechanism, using curved arrows, for the dehydration of 1-methylcyclopentanol with phosphoric acid. Show all possible products, indicate which is expected to be the major product, and explain why.

6.47 Show the product(s) of the following reactions:

(a) 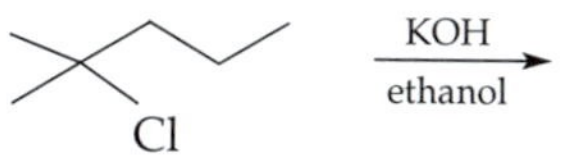 Cl — KOH / ethanol →

(b) 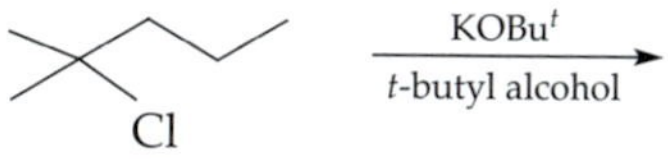 Cl — KOBu^t / t-butyl alcohol →

(c) 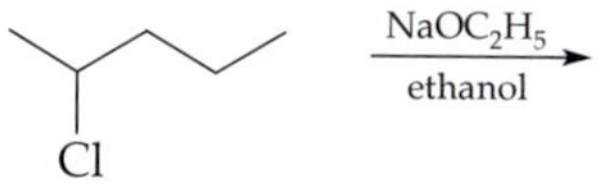 Cl — $NaOC_2H_5$ / ethanol →

6.48 Starting with an alkyl halide, outline a synthesis that will produce each alkene as the major or only product:

(a) 1-pentene

(b) 3-methyl-1-butene

(c) 2,3-dimethyl-2-butene

(d) 1-methylcyclopentene

Mixed Problems

6.49 The compounds known as myrcene (obtained from bay leaves), limonene (from lemons and oranges), and farnesol (from lily-of-the-valley) are odiferous members of the terpene family (see Section 8.3.2). Are there *cis-trans* isomers for each of these compounds? If so, how many?

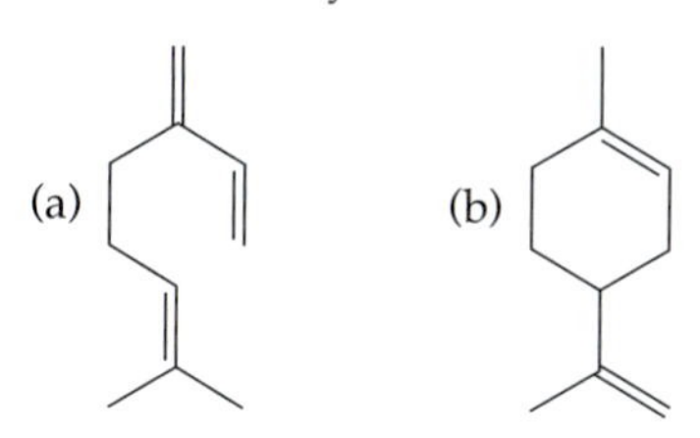

Myrcene Limonene

(c) 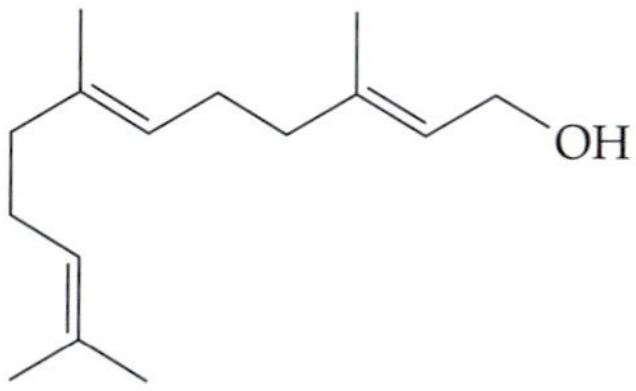 OH

Farnesol

6.50 Show the possible product(s) of dehydration of each of the following alcohols using hot concentrated sulfuric acid, and indicate which is the major product:

(a) 1-methylcyclohexanol

(b) 2-methylcyclohexanol

(c) 2-methyl-3-hexanol

(d) 2,4-dimethyl-3-pentanol

6.51 Three important unsaturated fatty acids (carboxylic acids derived from liquid fats such as vegetable oils) are the C_{18} acids oleic acid, linoleic acid, and linolenic acid. They contain, respectively, one, two, and three carbon-carbon double bonds. How many *cis-trans* isomers exist for each of these fatty acids?

(a) $CH_3(CH_2)_7CH{=}CH(CH_2)_7COOH$
Oleic acid

(b) $CH_3(CH_2)_4(CH{=}CHCH_2)_2(CH_2)_6COOH$
Linoleic acid

(c) $CH_3CH_2(CH{=}CHCH_2)_3(CH_2)_6COOH$
Linolenic acid

6.52 Applying what you've learned about the chemistry of alkyl halides, alcohols, and alkenes, show the reaction(s) that could be employed to effect the following transformations. (*Hint:* Remember to apply the concept of retrosynthesis and use the Key to Transformations.)

(a) 1-methyl-1-bromocyclohexane to 2-methyl-1-bromocyclohexane

(b) 2-methyl-2-butanol to 3-methyl-2-butanol

(c) 1-chlorobutane to 1,2-dichlorobutane

(d) cyclopentanol to *cis*-1,2-cyclopentanediol

(e) cyclopentanol to *trans*-1,2-cyclopentanediol

(f) cyclohexylmethyl chloride to 1-methylcyclohexanol

(g) cyclohexylmethyl chloride to cyclohexylmethanol

(h) 4-methylcyclohexanol to methylcyclohexane

6.53 When reacted with metallic sodium, a compound with formula $C_4H_{10}O$ produced a gas. Further, it reacted with sodium dichromate in sulfuric acid. Treatment with concentrated sulfuric acid produced a compound with formula C_4H_8, and only a single isomer of this product was detected in the reaction. The C_4H_8 compound was converted to C_4H_{10} upon reaction with hydrogen over a platinum catalyst. Propose a structure for each of these compounds that is consistent with the evidence presented.

6.54 Show the alkyl bromide that will yield each of the following compounds as the *sole product* of dehydrobromination:

(a) methylenecyclohexane

(b) 4-methylcyclohexene

(c) 2-methyl-1-butene

6.55 Two isomeric alkenes react with hydrogen over a catalyst to form methylcyclohexane. Ozonolysis of the alkenes affords the dialdehydes shown below. What is the structure of the alkene from which each dialdehyde was derived?

(a) 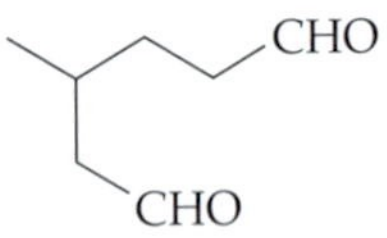 CHO CHO (b) 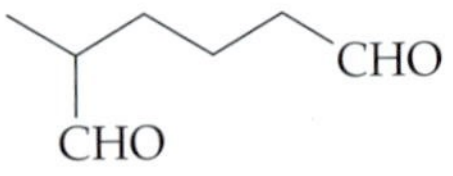 CHO CHO

6.56 A compound **X** was shown to be an alkyl bromide with formula $C_6H_{13}Br$. It reacted slowly with both silver nitrate in ethanol and sodium iodide in acetone. Treatment of **X** with sodium ethoxide and heat produced a mixture of two compounds, **Y** and **Z**, both with formula C_6H_{12}. Ozonolysis of **Y** produced $(CH_3)_2CO$ and CH_3CH_2CHO. Ozonolysis of **Z** produced $(CH_3)_2CHCHO$ and CH_3CHO. What are the structures of **X**, **Y**, and **Z**?

6.57 A sex-attractant pheromone of a female moth contains a hydrocarbon of formula $C_{21}H_{40}$. The compound decolorized bromine and reacted with potassium permanganate. It also reacted with hydrogen over a platinum catalyst, absorbing two equivalents of hydrogen. Ozonolysis of the pheromone produced three carbonyl compounds:
$CH_3(CH_2)_{10}CHO$, $CH_3(CH_2)_4CHO$, and $O{=}CHCH_2CH{=}O$. What is the structure of the pheromone?

6.58 The compound limonene, a member of the terpene family, is obtained from lemons. Its molecular formula is $C_{10}H_{16}$ and it is known to contain a six-membered ring. Upon hydrogenation, it absorbs two equivalents of hydrogen. Ozonolysis of limonene produces two carbonyl compounds: CH_2O and $CH_3COCH_2CH_2CH(CH_2CHO)COCH_3$. What is the structure of limonene?

6.59 Hydration of limonene (see Problem 6.58) with dilute sulfuric acid produces a diol ($C_{10}H_{20}O_2$) called *terpin hydrate,* which is used as an expectorant in over-the-counter cough medicines. Propose a structure for terpin hydrate. How many stereoisomers would you expect to be formed?

6.60 α-Pinene is a terpene component of turpentine with molecular formula $C_{10}H_{16}$. Ozonolysis of α-pinene produces compound **A.** β-Pinene is a constitutional isomer of α-pinene that produces formaldehyde (CH_2O) and compound **B** under the same oxidation conditions. Draw the structures of α- and β-pinene.

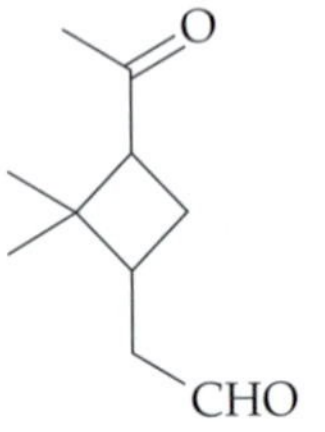

O CHO O
A 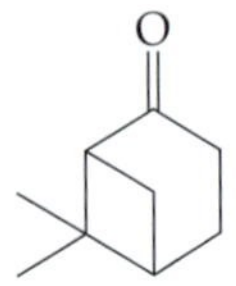 **B**

6.61 Draw all the constitutionally isomeric alkenes with molecular formula C_5H_{10}. Then identify which of these can exist as stereoisomers (ignore conformational isomers), draw the stereoisomers, and name them.

6.62 Using your knowledge of the mechanism of electrophilic additions to alkenes, propose a mechanism for the reaction of isobutylene with methanol in the presence of sulfuric acid, to form *t*-butyl methyl ether (known as MBTE). (*Hint:* Identify an electrophile and a nucleophile.)

6.63 When cyclohexyl bromide is heated with methanol, the major product is cyclohexyl methyl ether, but when cyclohexyl bromide is heated with sodium methoxide, the major product is cyclohexene. Explain these results.

6.64 (a) How many different alkenes (that is, constitutional isomers) can you envision having the same carbon skeleton as the alkane 2,3-dimethylpentane? (b) Write each of them, name each of them, identify those for which stereoisomers exist, and draw the stereoisomers. (c) Which of the constitutional isomers should be the more stable?

6.65 Show the reactions necessary to accomplish each of the following syntheses:

(a) cyclohexyl bromide to *cis*-1,2-cyclohexanediol

(b) cyclohexyl bromide to *trans*-1,2-cyclohexanediol

(c) cyclohexyl bromide to *trans*-1,2-dibromocyclohexane

(d) cyclohexyl bromide to *cis*-1,2-dibromocyclohexane

(e) cyclohexyl bromide to ethylcyclohexane

6.66 An unknown compound **A** has molecular formula C_5H_8O and is slightly soluble in water. It gave off a gas when treated with metallic sodium, and the addition of methyl iodide to the resulting solution produced a new compound **B.** Both **A** and **B** decolorized bromine in CCl_4 and gave a brown precipitate when treated with cold potassium permanganate. Reaction of **A** with hydrogen over a platinum catalyst resulted in the addition of one equivalent of hydrogen and the formation of **C.** Compound **C** did not decolorize bromine but did react slowly with potassium permanganate; it also reacted with Jones' reagent (chromic anhydride in sulfuric acid). Reaction of **C** with metallic sodium and then methyl iodide gave compound **D,** which was also produced by hydrogenation of **B.** Compound **A** reacted with PBr_3 to produce compound **E,** C_5H_7Br, and **E** yielded compound **F,** C_5H_6, upon heating with sodium ethoxide. Ozonolysis of **A** gave a single symmetrical compound **G,** containing two aldehyde (—CHO) groups. **F** was also ozonized and produced two compounds (**H** and **I**), both containing two aldehyde groups. Write structural formulas for compounds **A–I** (*Hint:* Think logically and stepwise! There are only five carbons in compound **A.**)

CONCEPTUAL PROBLEM

Frogs: The New "Mine Canaries"?

You work for your state's wildlife management office. Increasingly, reports of sightings of deformed frogs in local ponds have crossed your desk. Just as canaries were once used to warn miners of poor air quality deep in the mines, so frogs may serve to warn us about environmental degradation.

Your initial research tells you that the first cases of deformed frogs were reported in Minnesota. Similar reports have since been documented from almost every state. The deformities include extra legs, malformed or missing legs, and other severe defects. One of the most common abnormalities is the absence of a hind limb. Your task is to consider hypotheses for "natural" causes, as well as chemically induced causes (that is, those that might result from human intervention in the environment), for such deformities.

- Can you propose any non–chemically induced or "natural," causes for the observed deformities?
- Researchers who are studying this phenomenon are also vigorously pursuing a number of hypotheses as to the possible cause(s). One hypothesis involves the chemical insecticide *trans,trans*-(*S*)-methoprene, which is used in wetlands to kill mosquitoes and can also be used in various sprays to kill fleas.

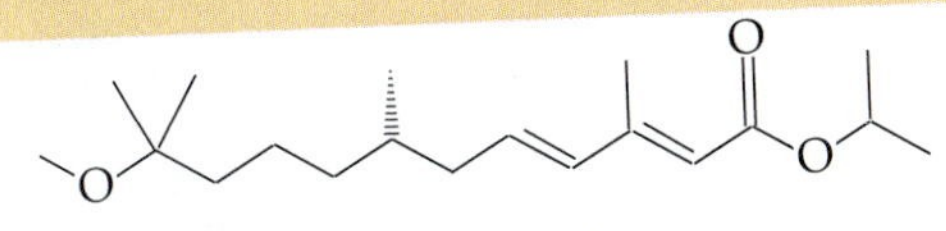

trans,trans-(*S*)-Methoprene

Spraying insecticide to kill mosquitoes.

The suspicion is that the compound binds to a receptor that regulates limb development in frog embryos. However, when frog embroyos in a lab are treated with methoprene, normal development occurs.

- Can you think of any reason why methoprene would not cause abnormalties in the lab, but might in a natural environment?

You learn that methoprene is reactive in the natural environment and that the *trans*-trisubstituted double bond isomerizes to *cis*. The resulting *cis,trans*-isomer may be mutagenic. Draw that isomer.

CHAPTER 7

Alkynes

ONE FAIRLY SIMPLE PATH through the Key to Transformations, starting with alkanes, leads to a relatively small family of unsaturated hydrocarbons, the **alkynes.** They, like the alkenes before them, are distinguished by a functional group consisting of two multiple bonded carbons, but in alkynes the two carbon atoms are attached to each other by a **triple bond.** A triple bond involves six electrons, which means that each of the carbon atoms can be attached to only one other group by a covalent bond. (Recall that carbon can have a maximum of eight electrons in its outer shell.) The triple bond is indicated by three lines between the two carbons. The simplest alkyne is ethyne (commonly called acetylene), for which several representations are shown in Figure 7.1. Other examples of alkynes are propyne, 2-butyne, and ethynylcyclohexane:

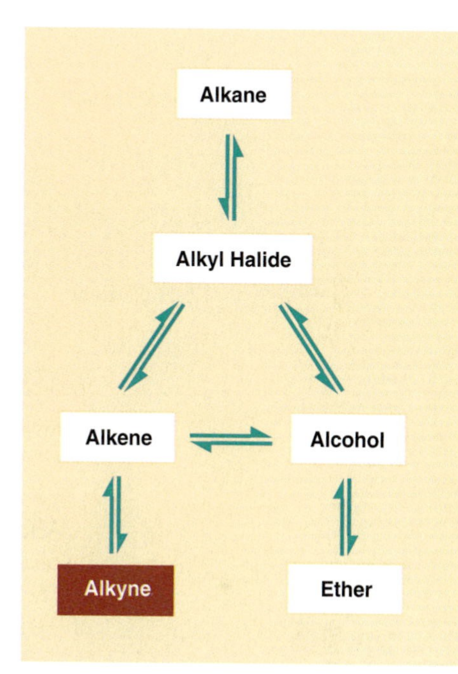

Alkynes are the first family of organic carbons that we study that have a triple bond.

$CH_3C{\equiv}CH$ or (line structure) — Propyne (methylacetylene)

$CH_3C{\equiv}CCH_3$ or (line structure) — 2-Butyne (dimethylacetylene)

(cyclohexane)—$C{\equiv}CH$ or (line structure) — Ethynylcyclohexane (cyclohexylacetylene)

Alkynes are categorized as either **terminal alkynes,** in which the triple bond is located at the end of a chain, or **internal alkynes,** in which the triple bond is *not* located at the end of the chain. Ethyne (see Figure 7.1) and ethynylcyclohexane (shown above) are terminal alkynes, which means that there is at least one hydrogen attached to one of the triple-bonded carbons. 2-Butyne is an internal alkyne, and so there is no hydrogen attached to either of its triple-bonded carbons.

H:C:::C:H

$H{-}C{\equiv}C{-}H$ $HC{\equiv}CH$

FIGURE 7.1

Lewis structure, dash and condensed formulas, and ball-and-stick model of ethyne (acetylene).

Overleaf: As we learn in this chapter, pheromones are chemical signals that allow animals to communicate with each other. Pheromones affect the behavior of small animals, like insects, as well as large. The long nose of this wolf has many signal receptors.

Alkynes have the general formula C_nH_{2n-2}. In other words, alkynes are "short" four hydrogens compared to alkanes having the same number of carbons (C_nH_{2n+2}). Note that an unknown hydrocarbon whose molecular formula is "short" four hydrogens can be one of several types of compounds: an alkyne, a compound with two alkene groups, a compound with two rings, or a compound with one ring and one double bond. For example, all of the following structures (as well as others not shown) have the molecular formula C_6H_{10}:

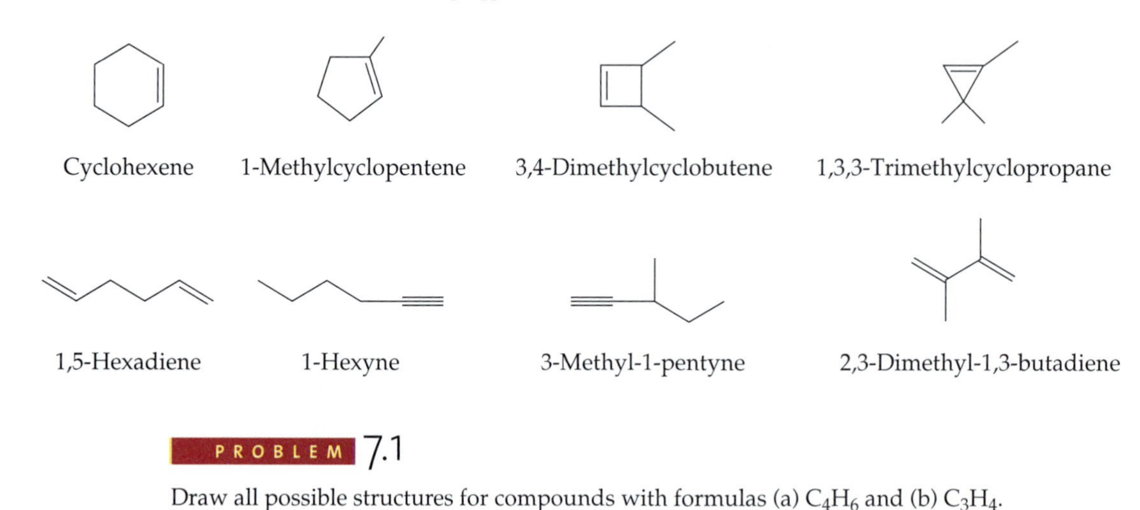

PROBLEM 7.1

Draw all possible structures for compounds with formulas (a) C_4H_6 and (b) C_3H_4.

Alkynes are typical hydrocarbons in that they are nonpolar. They are therefore insoluble in water but soluble in other organic solvents. They are relatively low-boiling; for example, acetylene (C_2H_2) is a gas at room temperature with a sublimation point of −84°C. Alkynes with five or more carbons are liquids at room temperature.

7.1 Nomenclature of Alkynes

7.1.1 IUPAC Nomenclature

Alkynes are named following the IUPAC protocols: Start with the parent alkane name corresponding to the carbon chain containing the triple bond, and replace the suffix *-ane* with the suffix *-yne*. Thus, the C_2 alkane eth*ane* becomes the alkyne eth*yne*. The location of the triple bond in the longest straight chain of carbon atoms is indicated by assigning it the lowest carbon number possible. Thus, the first compound below is 2-pentyne (rather than 3-pentyne). Other substituents and branches on the chain are located using the same numbering sequence, as illustrated for 1-chloro-2-butyne. A compound with two or more alkyne groups is named using the same numbering principles and the prefixes *di-*, *tri-*, and so on. Thus, the third compound below is 1,5-heptadiyne.

$\overset{1}{C}H_3-\overset{2}{C}\equiv\overset{3}{C}-\overset{4}{C}H_2-\overset{5}{C}H_3$

2-Pentyne (*not* 3-pentyne)

$\overset{4}{C}H_3-\overset{3}{C}\equiv\overset{2}{C}-\overset{1}{C}H_2-Cl$

1-Chloro-2-butyne

1 2 3 4 5 6 7

1,5-Heptadiyne (*not* 2,6-heptadiyne)

If both a double bond and a triple bond are present, the compound is an *alkenyne*. The numbering starts from the end containing the lower-numbered multiple bond. If the double bond and the triple bond will have the same numbers from either end of the chain, the numbering of the double bond takes locational priority; that is, the double bond is assigned the lower number. Therefore, the following compound is 1-penten-4-yne, not 4-penten-1-yne:

$$\overset{5}{HC}\equiv\overset{4}{C}-\overset{3}{CH_2}-\overset{2}{CH}=\overset{1}{CH_2}$$

1-Penten-4-yne
(*not* 4-penten-1-yne)

Ethynylcyclohexane

The group obtained by removing a hydrogen from ethyne is called the **ethynyl group** (H—C≡C—). It can be named as a substituent on a carbon skeleton, as in ethynylcyclohexane.

7.1.2 Historical/Common Names

Acetylene is the historical and commonly used name of the simplest alkyne, ethyne (C_2H_2), one of the oldest-known organic compounds. Simple alkynes can be named as derivatives of acetylene, with the derivative name reflecting substitution of one or both hydrogens. Thus, a terminal alkyne (a monosubstituted acetylene) can be named as an alkylacetylene, such as methylacetylene:

$CH_3C\equiv CH$	$CH_3C\equiv CCH_3$	$CH_3C\equiv CCH_2CH_3$
Methylacetylene	Dimethylacetylene	Ethylmethylacetylene

In a similar manner, an internal alkyne can be named as a disubstituted acetylene, such as dimethylacetylene (in which both substituents are the same) or ethylmethylacetylene (in which the two substituents are different).

How to Solve a Problem

Name the following compounds:

(a) $CH_3CH_2CH_2C\equiv CH$ (b) (c)

SOLUTION

(a) The total number of carbons in this compound is five, and all are in a single chain. Therefore, the parent alkane is pentane. The functional group is the triple bond, so the suffix must be *-yne*. Because the triple bond starts at carbon 1, the IUPAC name is 1-pentyne. The common name is propylacetylene.

(b) The total number of carbons in this compound is eight, but the longest chain including the triple bond has six carbons. Therefore, the parent name is hexyne. The triple bond is located between carbons 3 and 4, so the compound is a 3-hexyne. There are two methyl substituents located on carbons 2 and 5. Therefore, the name is 2,5-dimethyl-3-hexyne.

(c) This compound contains two functional groups, an alkene and an alkyne. Both are located in the seven-carbon chain, so it must be named as a heptenyne. Either the double bond or the triple bond can be numbered as located on carbon 2, depending on the direction chosen to number the carbons. Because the alkene group takes priority, the IUPAC name is 2-hepten-5-yne or hept-2-en-5-yne, both forms being acceptable. Note that the number immediately precedes the functional group to which it refers.

PROBLEM 7.2

Draw structural formulas for the following compounds: (a) dicyclopropyl acetylene, (b) 2-pentyne, (c) 2-hepten-4-yne, and (d) 1-ethynylcyclobutene.

PROBLEM 7.3

Assign names to the following structures:

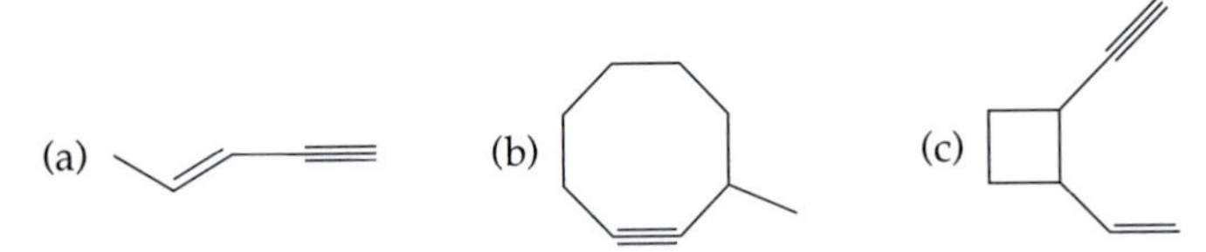

7.2

Structure of Alkynes

7.2.1 Electronic Structure of Triple Bonds: *sp* Hybridization

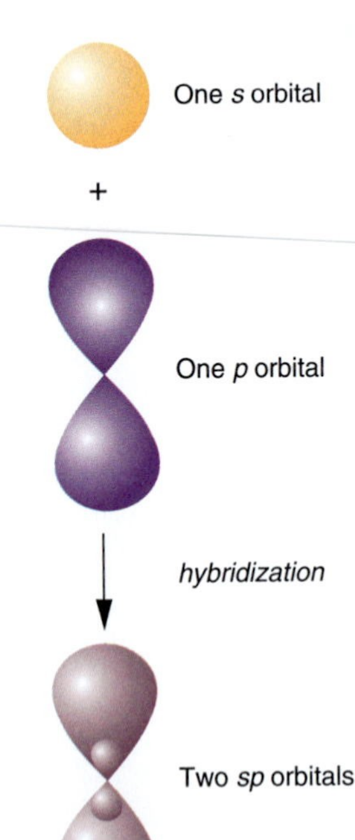

FIGURE 7.2

Linear *sp* hybridization of atomic carbon.

Each carbon atom involved in a triple bond is **linearly hybridized,** having two *sp* hybrid orbitals arranged about the nucleus at an angle of 180° to each other (Figure 7.2). The process of *sp* hybridization of atomic carbon is similar to that of sp^3 (tetrahedral) hybridization (see Section 1.1.3). The first step, promotion of a 2*s* electron to a 2*p* orbital, is the same.

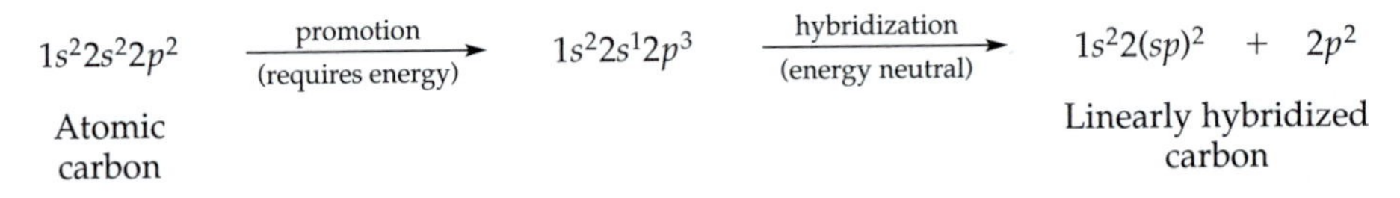

In the second step, only one of the three 2*p* orbitals hybridizes with the single 2*s* orbital, forming two *sp* hybrid orbitals. As a result of this hybridization process, there are two unhybridized *p* orbitals (perpendicular to each other) remaining on each carbon. They lie in planes perpendicular to the axis of the *sp* hybrid orbitals (Figure 7.3).

The formation of a triple bond in ethyne can be pictured as occurring in three steps, as illustrated in Figure 7.4. First, a single *sp* hybrid orbital from each of the two linearly hybridized carbons (Figure 7.4a) overlap end to end to form a molecular orbital, resulting in a carbon-carbon sigma bond (*sp-sp*) (Figure 7.4b); it is a sigma bond because its shape is cylindrical about the C—C axis. Second, the remaining *sp* hybrid orbital of each carbon forms a covalent single (sigma) bond with an *s* orbital of a hydrogen atom (*s-sp*) (Figure 7.4c). The atomic positions in ethyne (all four atoms in a line with 180° bond angles) are determined by the sigma bond framework. The structure

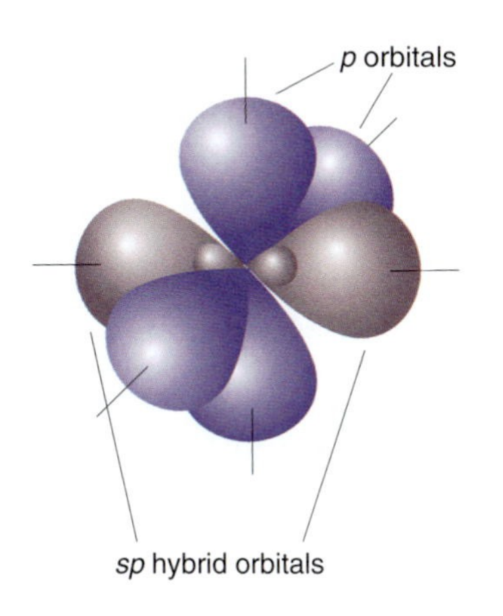

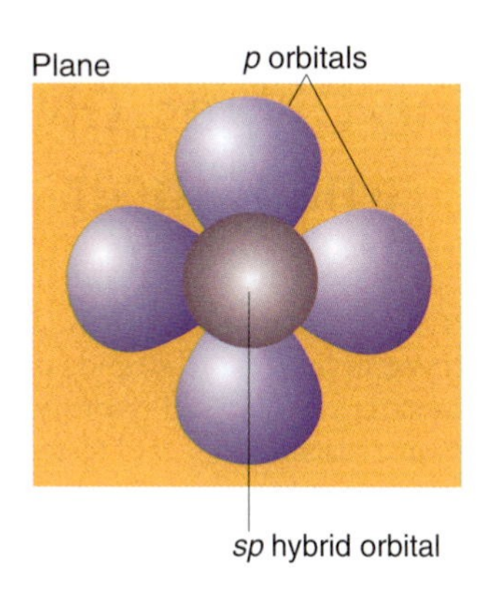

FIGURE 7.3

Two views of an *sp*-hybridized carbon, showing the *sp* hybrid orbitals and the unhybridized *p* orbitals perpendicular to each other.

of ethyne is completed when each of the two unhybridized *p* orbitals on each carbon overlaps with a counterpart parallel *p* orbital on the other carbon to form *two* pi (π) molecular orbitals perpendicular to each other (Figure 7.4d). There are two electrons in each pi orbital, shared by the two carbons; therefore, each carbon has a filled valence shell of eight electrons (four involved in two sigma bonds and four involved in two pi bonds). Thus, the triple bond between two carbons actually consists of a sigma bond and two separate pi bonds, accounting for a total of six electrons shared between the two carbon atoms.

If the ethyne molecule were viewed from one end, there would appear to be electron density above, below, and on both sides of the carbon-carbon sigma bond. Thus, ethyne and other alkynes can serve as nucleophiles and are reactive in addition reactions with electrophiles. The carbon-carbon bond in ethyne has a very high total electron density between the carbons and is even stronger than that in ethene. The stronger

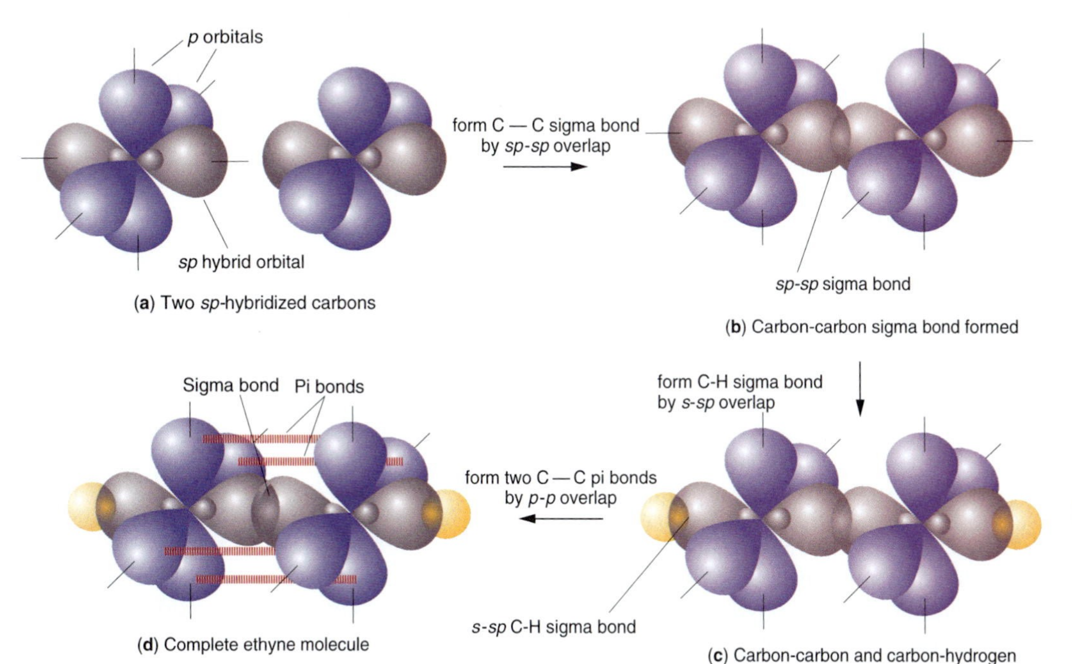

FIGURE 7.4

Representation of the formation of the ethyne molecule from two *sp*-hybridized carbons and two hydrogens.

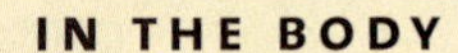

CHEMISTRY AT WORK

IN THE BODY

Ene-Diynes, DNA, and Cancer

Alkynes are relatively rare in nature. However, in the 1980s, several research groups discovered some very complex compounds that contain not just one but two triple bonds (thus, the name *diyne*), in addition to other functional groups. The two triple bonds are separated by a double bond, which led to compounds containing this type of structural feature being called *ene-diynes*.

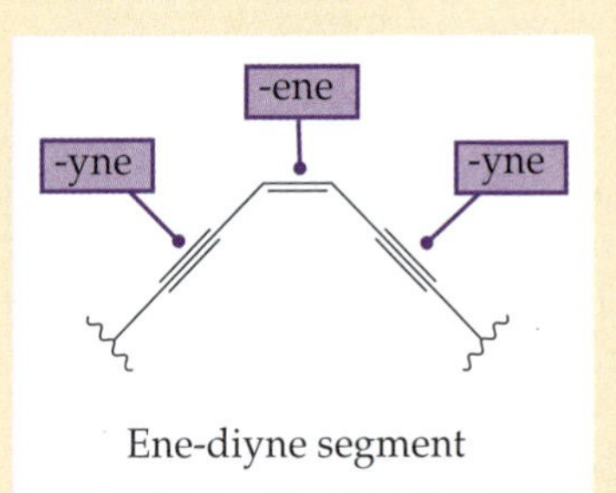

Ene-diyne segment

The ene-diyne compounds were collected in minute quantities from large-volume cultures of several unrelated bacteria, some of which are found in soils. Initially, these compounds were of interest because of their unique and very complex chemical structures. Also of interest was the fact that several of these similar compounds are produced by unrelated organisms. Interest exploded when researchers discovered that, as a family, ene-diynes exhibit potentially important antitumor activity. They exert this effect by causing cleavage of strands of DNA, thereby interrupting cell division. Presently, synthetic ene-diynes are being developed and tested for even greater selectivity in targeting cancer cells.

Although several members of this class of compounds are now known, we will look at only one typical member, *dynemicin A:*

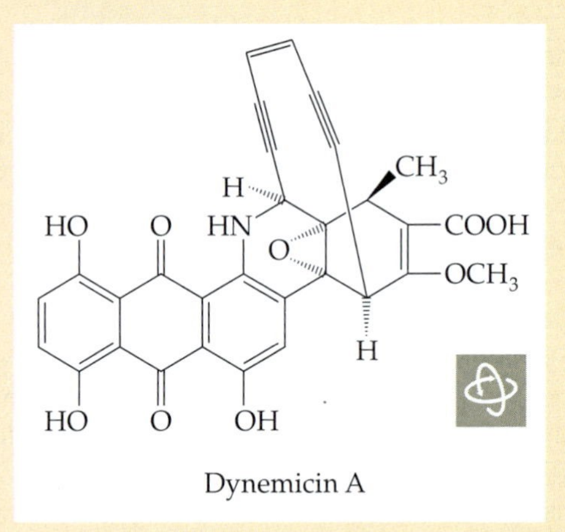

Dynemicin A

Dynemicin A was originally discovered in milligram amounts in fermentation broths of *Micromonospors chersina*, but it has since been synthesized in the laboratory.

The chemical details of the anticancer activity of ene-diynes will be covered in Section 19.3.3.

bond between the two carbons in ethyne has a length of 1.20 Å, shorter than that in ethene (1.34 Å) and that in ethane (1.54 Å).

Since the structure of a triple bond is linear, there can be no *cis-trans* isomerism about the triple bond in alkynes. Also, because of the linear nature of the triple bond, it cannot be incorporated into a carbocyclic ring smaller than C_6—the distance between the ends of the triple bond unit is too long to be closed by sp^3 sigma bonds involving only three other carbons.

7.2.2 Acidity of Terminal Alkynes

Hydrogen atoms attached to carbon are not normally thought of as being acidic—that is, capable of being removed by a base. The reason is that removal of such a hydrogen (without its electron pair) would leave behind a carbanion. This carbanion would be so strongly basic that the hydrogen would be drawn back to it. Thus, there is no ready means of removing a hydrogen from ethane (pK_a ~50) or ethene (pK_a ~44). Alkynes, however, are weakly acidic. Any hydrogen attached to a triple-bonded carbon (that is, on a terminal alkyne such as acetylene, pK_a ~25) can be removed readily by a very strong base, typically one known as sodium amide ($NaNH_2$, also known as sodamide) in liquid ammonia solution.

Sodamide and a terminal alkyne react to form an *alkynide*—in the case of acetylene (ethyne), the alkynide is commonly known as sodium acetylide (sodium ethynide):

$$\underset{\substack{\text{Terminal alkyne} \\ (\text{p}K_a \sim 25)}}{R{-}C{\equiv}C{-}H} + \underset{\text{Sodium amide}}{Na^+ \ {}^-{:}\ddot{N}H_2} \xrightleftharpoons{NH_3} \underset{\text{Sodium alkynide}}{R{-}C{\equiv}C{:}^- \ Na^+} + \underset{\substack{\text{Ammonia} \\ (\text{p}K_a \sim 33)}}{NH_3}$$

Analysis of the pK_a values in the equilibrium indicates that the reaction "works" (proceeds to the right) because the alkyne is a stronger acid than ammonia. Put another way, the amide anion is a stronger base than the acetylide anion and therefore removes a proton from the alkyne. Any base whose conjugate acid has a pK_a larger than 25 will shift the equilibrium to the right, removing a proton from the alkyne and forming an alkynide. However, a weaker base, such as hydroxide (pK_a of water, ~15.7) or even an alkoxide such as *t*-butoxide (pK_a of *t*-butyl alcohol, ~18), will not convert an alkyne to an alkynide.

The higher acidity level of alkynes, as compared to the more inert alkenes and alkanes, can be explained on the basis of the relative stability of the resulting carbanions. Consider the order of acidity of the hydrocarbons ethane, ethene, and ethyne:

$$\underset{\text{Ethane}}{H_3C{-}CH_3} < \underset{\text{Ethene}}{H_2C{=}CH_2} < \underset{\text{Ethyne}}{HC{\equiv}CH}$$

Increasing acid strength →

The significant difference between these hydrocarbons is in the hybridization of the orbital that would carry the unshared pair of electrons in the carbanion. In ethane, the orbital is sp^3 hybridized; in ethene, it is sp^2 hybridized; and in ethyne, it is *sp* hybridized. An *s* orbital is more electronegative than a *p* orbital; a pair of electrons in an *s* orbital is at a lower energy level than a pair of electrons in a *p* orbital. Therefore, as the orbital holding the electrons in the carbanion becomes more like an unhybridized *s* orbital (100% *s* character), the more stable (that is, lower-energy) the carbanion is and the more readily it will form. The *sp* hybrid orbital of triple-bonded carbons has 50% *s* character; the sp^2 hybrid orbital of double-bonded carbons has 33% *s* character; and the sp^3 hybrid orbital of single-bonded carbons has 25% *s* character. Thus, the alkyne carbanion has the lowest-energy electrons of these three carbanions, and forms the most readily, as reflected by the pK_a data in Table 7.1.

TABLE 7.1

The Relationship of the *s* Character of the Carbon-Hydrogen Bond to the pK_a of the Compound

Compound	Resulting Carbanion	Hybridization of Carbanion	Percentage *s* Character	pK_a
Ethane	$CH_3CH_2{:}^-$	sp^3	25%	~50
Ethene	$CH_2CH{:}^-$	sp^2	33%	~44
Ethyne	$HCC{:}^-$	sp	50%	~25

Terminal alkynes can be detected chemically by an acid-base reaction with silver ammonium hydroxide, which produces a silver alkynide precipitate. Because an internal alkyne has no acidic hydrogen, it does not react with silver ammonium hydroxide.

$$RC{\equiv}CH + Ag(NH_3)_2^+ \ {}^-OH \longrightarrow RC{\equiv}CAg\downarrow + H_2O + 2\,NH_3$$

Terminal alkyne — Silver ammonium hydroxide — Silver alkynide (precipitate)

PROBLEM 7.4

Name the following three compounds, and place them in order of decreasing acidity:

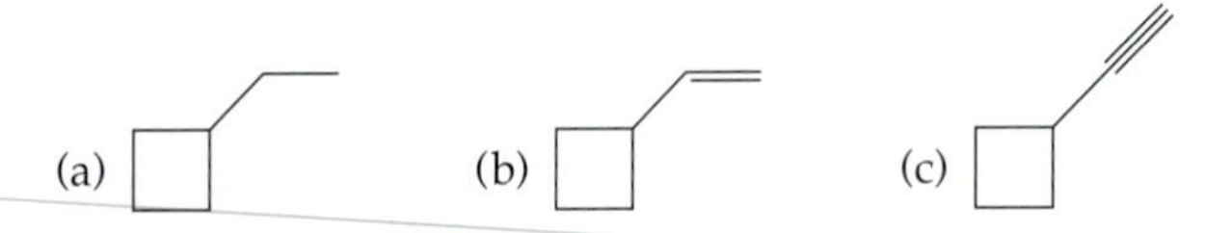

PROBLEM 7.5

Indicate which of the following bases (the pK_a values of their conjugate acids are given in parentheses) will convert 1-propyne (pK_a ~25) into its anion, and show the acid-base reaction for each: (a) NaOH (15.7), (b) $NaOCH_3$ (17), (c) $NaOBu^t$ (18), (d) $NaNH_2$ (33), (e) CH_3Li (50), (f) $CH_2{=}CHLi$ (44).

7.3 Reactions of Alkynes

Like alkenes, alkynes are unsaturated compounds, and therefore typically undergo addition reactions. Furthermore, because alkynes contain pi electrons (as do alkenes), they are electron-rich, and *electrophilic* addition is their typical reaction. The overall mechanism of such addition reactions is similar to that described for alkenes. Usually, the addition reactions of alkynes can be controlled to bring about the addition of one or two equivalents of the reagent. There are many such reactions, including the addition of hydrogen halides, halogens, and water.

General reaction:

$$\text{—C}\equiv\text{C—} \xrightarrow{A-B} \text{AC=CB} \xrightarrow{A-B} \text{—C(A)(A)—C(B)(B)—}$$

One example of this type of reaction is the addition of one equivalent of HCl to acetylene to produce vinyl chloride, a major industrial chemical that serves as a monomer in making the polymer poly(vinyl chloride), or PVC:

$$\underset{\text{Acetylene}}{HC\equiv CH} \xrightarrow{HCl} \underset{\text{Vinyl chloride}}{CH_2{=}CHCl} \xrightarrow{\text{(polymerization)}} \underset{\text{Poly(vinyl chloride) (PVC)}}{\text{—}(CH_2\text{—}CHCl)_n\text{—}}$$

Since the electrophilic addition reactions have relatively little significance for organic synthesis, they will not be discussed further. This section will focus on two other reactions for alkynes: hydrogenation and oxidation.

7.3.1 Hydrogenation (Reduction) of Alkynes

Hydrogen can be added to alkynes, just as it can to alkenes, and the symbol [H] is used to represent this reduction in general. Hydrogenation of alkynes to alkanes results from addition of two equivalents of hydrogen. However, special techniques make it possible to add the two equivalents of hydrogen stepwise: first adding one equivalent of hydrogen to produce an alkene, and then adding a second equivalent of hydrogen to produce an alkane.

General reaction:

$$\underset{\text{Alkyne}}{\text{—C}\equiv\text{C—}} \xrightarrow{[H]} \underset{\text{Alkene}}{\text{—CH=CH—}} \xrightarrow{[H]} \underset{\text{Alkane}}{\text{—}CH_2\text{—}CH_2\text{—}}$$

Most important, it is possible to add the first two atoms of hydrogen in either a *syn* or an *anti* manner, stereoselectively producing a *cis* or *trans* alkene, respectively. In fact, this process is one of the best means of preparing pure *cis* and *trans* isomers of various alkenes, as illustrated for the reduction of 2-butyne to *cis*- or *trans*-2-butene:

$$\underset{\text{2-Butyne}}{CH_3C\equiv CCH_3} \xrightarrow{[H]\ (syn\ \text{addition})} \textit{cis}\text{-2-Butene} \xrightarrow{[H]} \underset{\text{Butane}}{CH_3\text{—}CH_2\text{—}CH_2\text{—}CH_3}$$

$$\underset{\text{2-Butyne}}{CH_3C\equiv CCH_3} \xrightarrow{[H]\ (anti\ \text{addition})} \textit{trans}\text{-2-Butene} \xrightarrow{[H]} \text{Butane}$$

$$\text{2-Butyne} \xrightarrow{[H]} \text{Butane}$$

EXPLORATIONS

www.jbpub.com/organic-online

Catalytic Hydrogenation of Alkynes to Alkanes

Reaction of alkynes with hydrogen over a platinum or palladium catalyst (the same conditions used for hydrogenation of alkenes; see Section 6.3) results in the addition of two equivalents of hydrogen, producing alkanes in excellent yield. Under these conditions, the alkene formed initially by addition of one equivalent of hydrogen is hydrogenated to the alkane as quickly as it is formed, and so cannot be isolated. For example, 2-hexyne produces hexane:

$$\underset{\text{2-Hexyne}}{CH_3CH_2CH_2C{\equiv}CCH_3} \xrightarrow[\text{Pt}]{2\,H_2} \underset{\text{Hexane}}{CH_3CH_2CH_2CH_2CH_2CH_3}$$

Reduction of Alkynes to *Cis* Alkenes

Use of a catalyst that has been *poisoned* (treated with a reagent to reduce its catalytic activity) results in limiting the hydrogenation of an alkyne to the addition of one equivalent of hydrogen. This allows the alkene to be isolated. One poisoned catalyst is *Lindlar's catalyst*, consisting of finely divided palladium metal deposited on a support of solid calcium carbonate. Hydrogenation using Lindlar's catalyst is a *stereospecific syn addition*. The two hydrogen atoms are delivered from the catalyst surface to the alkyne in a manner similar to that described for alkenes (see Section 6.3). For example, 2-butyne is converted to *cis*-2-butene:

$$\underset{\text{2-Butyne}}{CH_3C{\equiv}CCH_3} \xrightarrow[\text{Pd/CaCO}_3]{H_2} \underset{\textit{cis}\text{-2-Butene}}{(CH_3)HC{=}CH(CH_3)}$$

Reduction of Alkynes to *Trans* Alkenes

Alkynes can be reduced by "chemical" means, in addition to the "catalytic" means already described. The process involves reacting the alkyne with sodium or lithium metal in liquid ammonia, a reducing medium. An electron is removed from each of two atoms of the metal and transferred to the alkyne. The resulting carbanion acquires hydrogen from the liquid ammonia. This reaction oxidizes the metal to its cation and reduces the alkyne to an alkene. The reduction is *stereospecific*, producing overall *anti* addition of hydrogen and thus yielding the *trans* alkene. This *anti chemical* reduction therefore complements the *syn catalytic* reduction described earlier. For example, 2-butyne is reduced to *trans*-2-butene:

$$\underset{\text{2-Butyne}}{CH_3C{\equiv}CCH_3} \xrightarrow[NH_3]{\text{Li (or Na)}} \underset{\textit{trans}\text{-2-Butene}}{(H)(CH_3)C{=}C(CH_3)(H)} + LiNH_2\ (\text{or } NaNH_2)$$

How to Solve a Problem

Show all the products that could be obtained by application of all possible hydrogenation or chemical reduction techniques to 3-hexyne.

PROBLEM ANALYSIS

3-Hexyne has a triple bond, which can be reduced completely to an alkane using a metal (platinum) catalyst and hydrogen. Reduction to an alkene requires a partial reduction, necessitating use of a poisoned catalyst (Lindlar's catalyst) and hydrogen or a metal and liquid ammonia. Lindlar's catalyst and hydrogen produces a *cis* alkene, whereas the chemical reduction produces a *trans* alkene.

SOLUTION

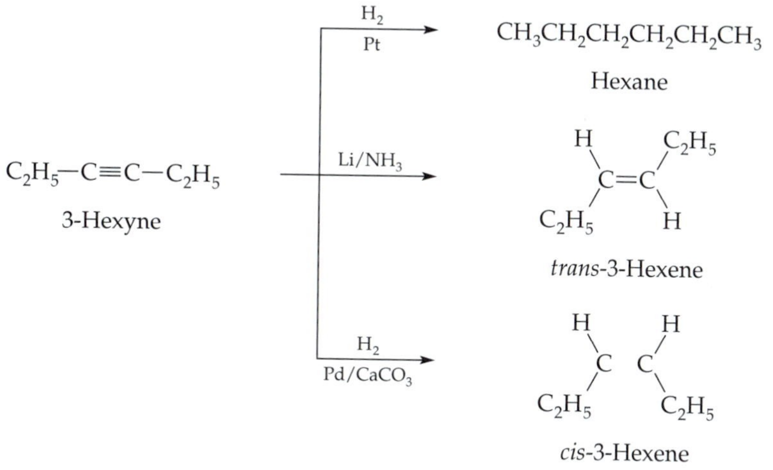

$C_2H_5{-}C{\equiv}C{-}C_2H_5$ (3-Hexyne)

$\xrightarrow[Pt]{H_2}$ $CH_3CH_2CH_2CH_2CH_2CH_3$ (Hexane)

$\xrightarrow{Li/NH_3}$ $(H)(C_2H_5)C{=}C(C_2H_5)(H)$ with H and C_2H_5 trans (*trans*-3-Hexene)

$\xrightarrow[Pd/CaCO_3]{H_2}$ $(H)(C_2H_5)C{=}C(H)(C_2H_5)$ with H atoms on the same side (*cis*-3-Hexene)

PROBLEM 7.6

Show the reagents necessary to accomplish each of the following conversions:

(a) 2-pentyne to pentane
(b) 1-butyne to 1-butene
(c) 2-pentyne to *trans*-2-pentene
(d) cyclooctyne to *cis*-cyclooctene

7.3.2 Oxidation of Alkynes

Alkynes are very readily oxidized by a variety of oxidizing reagents. The result in all cases is complete cleavage of the alkyne at the triple bond, with both resulting fragments being carboxylic acids. Thus, 2-pentyne produces ethanoic acid and propanoic acid, and 1-hexyne produces pentanoic acid and carbonic acid. These same products result from the use of ozonolysis (a weak oxidant), nitric acid (a moderate oxidant), or potassium permanganate (a strong oxidant).

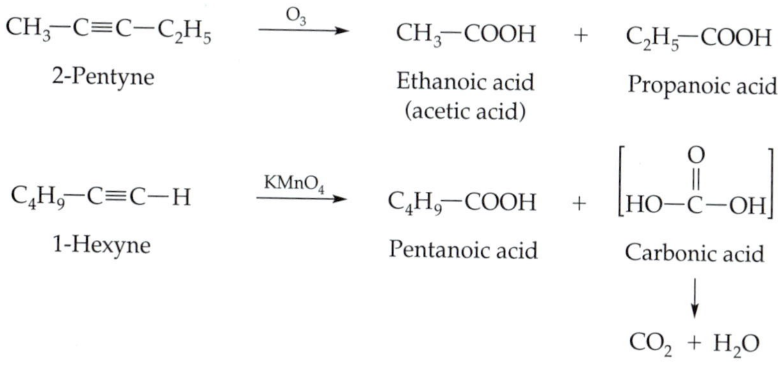

$$\underset{\text{2-Pentyne}}{CH_3{-}C{\equiv}C{-}C_2H_5} \xrightarrow{O_3} \underset{\text{Ethanoic acid (acetic acid)}}{CH_3{-}COOH} + \underset{\text{Propanoic acid}}{C_2H_5{-}COOH}$$

$$\underset{\text{1-Hexyne}}{C_4H_9{-}C{\equiv}C{-}H} \xrightarrow{KMnO_4} \underset{\text{Pentanoic acid}}{C_4H_9{-}COOH} + \underset{\text{Carbonic acid}}{\left[HO{-}\overset{O}{\overset{\|}{C}}{-}OH\right]}$$

$$\text{Carbonic acid} \longrightarrow CO_2 + H_2O$$

The combustion of acetylene provides tremendous amounts of heat, sufficient to cut metal using an oxyacetylene torch.

The oxidation reaction is not generally used in synthesis; its only use is in proving the structure of an unknown alkyne. By determining the structures of the two carboxylic acids that must result from any alkyne oxidation, chemists can reconstruct what must have been the structure of the starting alkyne. For example, if a linear C_6 acid and a linear C_4 acid result from oxidizing an unknown alkyne, the alkyne must have been 4-decyne:

$$\underset{\text{Hexanoic acid}}{C_5H_{11}COOH} + \underset{\text{Butanoic acid}}{HOOCC_3H_7} \Longrightarrow \underset{\text{4-Decyne}}{C_5H_{11}-C{\equiv}C-C_3H_7}$$

Finally, like all hydrocarbons, alkynes are subject to complete combustion in the presence of excess oxygen and heat. The products are carbon dioxide and water. In the case of acetylene, a great deal of heat is also emitted—this is the very hot flame of an oxyacetylene torch, often used to cut metals.

$$\underset{\text{Acetylene}}{H-C{\equiv}C-H} + \tfrac{5}{2}O_2 \longrightarrow 2\,CO_2 + H_2O + \text{heat}$$

PROBLEM 7.7

Show the structure and provide the name for the compound that produces each of the following products from the specified oxidation reaction:

(a) CH_3COOH only, from ozonolysis
(b) CH_3CHO, $OHCCH_2COOH$, and CH_3COOH, from ozonolysis
(c) $HOOC(CH_2)_6COOH$, from ozonolysis
(d) CH_3COOH and CO_2, from $KMnO_4$ oxidation

7.4 Synthesis of Alkynes

Most alkyne syntheses start with acetylene, taking advantage of the fact that a terminal hydrogen in an alkyne is acidic enough to be removed by the strong base sodamide (see Section 7.2.2). This removal produces a carbanion that can serve as the nucleophile in an S_N2 reaction with an alkyl halide, thereby forming a carbon-carbon bond via an alkylation reaction. Starting with acetylene, a larger monosubstituted or disubstituted alkyne (unsymmetrical or symmetrical) can be prepared using two separate S_N2 steps.

General reaction:

$$\underset{\text{Acetylene}}{H-C{\equiv}C-H} \xrightarrow[\text{2. RX}]{\text{1. NaNH}_2} \underset{\text{Monoalkylacetylene}}{R-C{\equiv}C-H} \xrightarrow[\text{2. R'X}]{\text{1. NaNH}_2} \underset{\text{Dialkylacetylene}}{R-C{\equiv}C-R'}$$

A specific example is the preparation of propyne and 2-hexyne:

$$\underset{\text{Acetylene}}{H-C{\equiv}C-H} \xrightarrow{\text{NaNH}_2} \left[H-C{\equiv}C:^-\right] \xrightarrow[(S_N2)]{CH_3-Br} \underset{\text{Propyne}}{CH_3-C{\equiv}C-H}$$

$$\downarrow \text{NaNH}_2$$

$$\underset{\text{2-Hexyne}}{CH_3-C{\equiv}C-C_3H_7} \xleftarrow[(S_N2)]{C_3H_7-Br} \left[CH_3-C{\equiv}C:^-\right]$$

The only significant limitation of this alkylation synthesis is that the alkyl halide cannot be tertiary (because E2 elimination would be the major course of the reaction). It is apparent that almost any alkyne can be prepared using this reaction, and a *cis* or *trans* alkene can be prepared from that alkyne by reduction. In fact, the eventual preparation of the alkene is usually the purpose of the alkyne alkylation reaction.

How to Solve a Problem

Starting with cyclobutene, a three-carbon compound, and any needed inorganic reagents, synthesize *cis*-1-cyclobutylpropene.

PROBLEM ANALYSIS

We write the starting material and end product, and decide where the starting material appears in the structure of the product. This identifies the key bond that must be formed. In this case, a C_3 chain must be joined to the cyclobutyl ring.

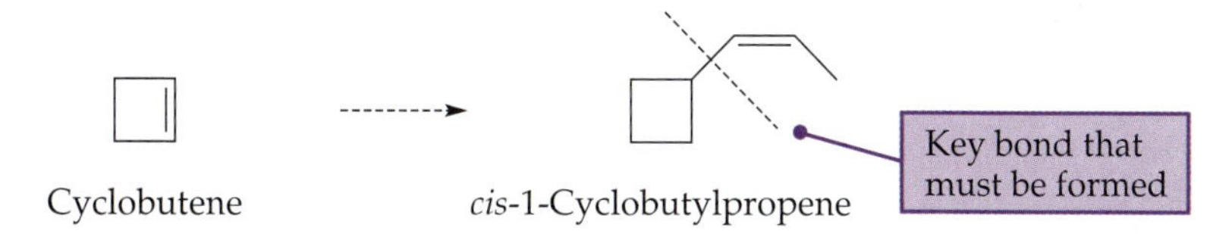

To make a carbon-carbon bond, we could use an S_N2 reaction, which requires a carbanion fragment and an alkyl halide fragment. Using the chemistry covered to this point, it is not feasible to convert cyclobutene into a carbanion. However, it is possible to convert it into an alkyl halide. Recall that alkenes normally undergo electrophilic addition, and so adding HBr to cyclobutene should produce cyclobutyl bromide, which is capable of being a substrate in an S_N2 reaction. The S_N2 reaction also requires a C_3 carbanion, which might take any of three forms; in principle:

(1) an alkyl carbanion, such as an organolithium reagent (but how, then, to introduce the double bond into the C_3 chain?);
(2) an alkenyl carbanion (but these have not yet been discussed); or
(3) an alkynyl carbanion, which is readily obtained from a terminal alkyne. Furthermore, after attachment, a triple bond can be converted stereospecifically into a double bond by reduction.

The approach involving the C_3 alkynyl carbanion is clearly the solution, embodied in this retrosynthetic scheme:

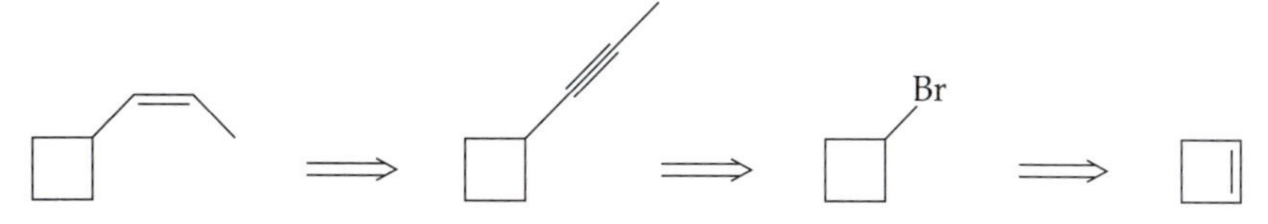

SOLUTION

We convert cyclobutene into cyclobutyl bromide by addition of HBr.

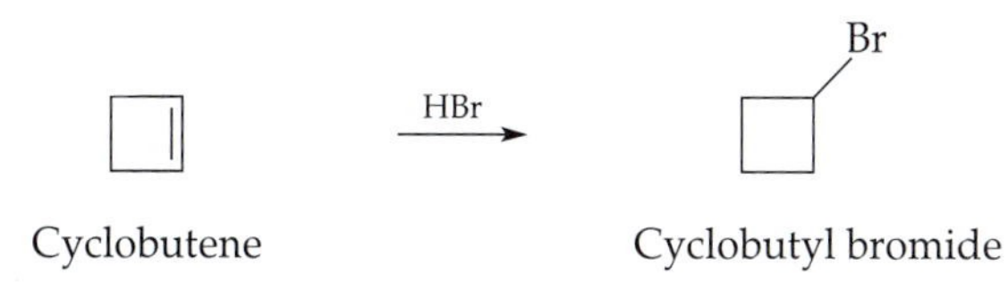

We then convert methylacetylene into its carbanion by removal of the proton with sodamide; we react the carbanion with cyclobutyl bromide to produce cyclobutylmethylacetylene:

$$CH_3-C\equiv C-H \xrightarrow{NaNH_2} [CH_3-C\equiv C:^- \; Na^+] + NH_3$$

Methylacetylene — Sodium methylacetylide

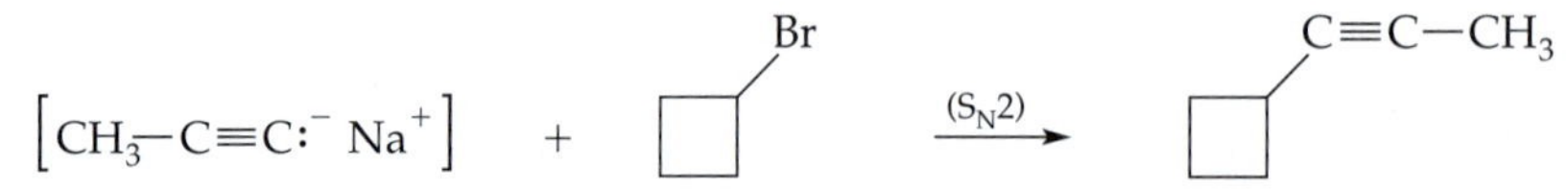

$$[CH_3-C\equiv C:^- \; Na^+] + \text{Cyclobutyl bromide} \xrightarrow{(S_N2)} \text{Cyclobutylmethylacetylene}$$

Sodium methylacetylide — Cyclobutyl bromide — Cyclobutylmethylacetylene

We reduce cyclobutylmethylacetylene stereospecifically with Lindlar's catalyst and hydrogen to *cis*-1-cyclobutylpropene.

$$\text{Cyclobutylmethylacetylene} \xrightarrow[Pd/CaCO_3]{H_2} \textit{cis}\text{-1-Cyclobutylpropene}$$

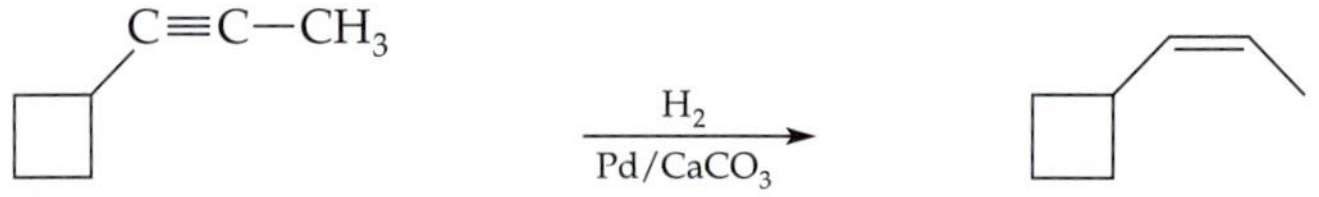

Cyclobutylmethylacetylene — *cis*-1-Cyclobutylpropene

PROBLEM 7.8

Synthesize each of the following unsaturated hydrocarbons from acetylene. (*Hint:* Use retrosynthesis.)

(a) 2-heptyne

(b) dicyclopropylacetylene

(c) *cis*-3-hexene

(d) *trans*-2-pentene

PROBLEM 7.9

Show the steps, reagents, and products involved in each of the following syntheses. (*Hint:* Use retrosynthesis.)

(a) 2-hexanol from acetylene

(b) 2-cyclopropylethanol from cyclopropanol

7.5 Important Alkynes

Few alkynes are naturally occurring. Even acetylene (ethyne, C_2H_2), the simplest alkyne, must be produced synthetically. It was originally produced by heating coal with lime in an electric furnace to produce calcium carbide (CaC_2), which, when mixed with water, forms acetylene gas and calcium hydroxide. Before the days of battery-powered lamps, miners used a type of lamp in which drops of water slowly fell on solid calcium carbide, producing gaseous acetylene, which was then burned as it formed. Acetylene is the fuel used in oxyacetylene torches because of the great heat produced as it burns, resulting in the ability to cut metals.

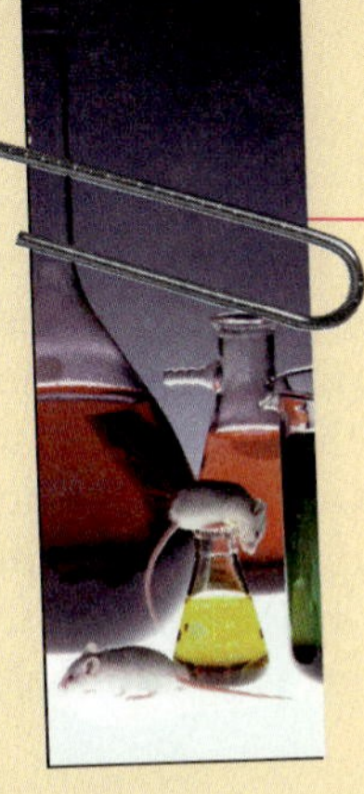

CHEMISTRY AT WORK

IN THE BODY

Alkynes and Pheromones

We humans communicate so regularly by spoken and written language that we sometimes forget that there are other ways to communicate. In fact, one of the most common methods of communication among animals involves secretion and detection of tiny amounts of chemicals called *pheromones*. Most pheromones are species-specific, meaning that the pheromone elicits a response only from animals of the same species. (There are exceptions, however, such as (Z)-7-dodecenyl acetate, which is a sexual pheromone of elephants, butterflies, and moths.) Pheromones are potent; only tiny amounts are involved, and they can often be detected by sensitive chemical receptors over great distances. Much of the research on pheromones involves insects and is done in the hope of finding environmentally benign ways to control insect populations. (Human pheromones are also currently a subject of research.)

Pheromones serve widely different purposes in the insect world. For example, the honeybee secretes a pheromone called *isoamyl acetate* (3-methylbutyl acetate) that acts as an alarm signal to other bees. Other pheromones control the development of the insect from larva to adult. But the most widely known insect pheromones are the sex attractants.

A gypsy moth caterpillar

Sex-attractant pheromones have been identified for many species of insects. For example, the C_{19} epoxide/oxirane known by the common name *disparlure* is the sex attractant secreted by female gypsy moths. A small amount of disparlure was first obtained by extraction from hundreds of thousands of female gypsy moths. The compound's structure was determined using only a few milligrams. Disparlure is now available in significant quantities as a result of a laboratory synthesis that relies on the alkyne chemistry discussed in this chapter. Disparlure is a *cis* epoxide that can be prepared by epoxidation of the precursor *cis* alkene (see Sections 5.2.2 and 6.5.1), which is obtained using alkyne alkylation and reduction. Synthetic disparlure is used as a bait to attract and trap male gypsy moths, thereby diminishing the breeding of the moths and the damage their offspring would do to hardwood forests, especially in the northeastern United States.

A standard method used to synthesize pheromones in the laboratory involves the initial synthesis of an alkyne with the triple bond located in the same position as the desired double bond or epoxide/oxirane group. Preparation of the C_{12} sex pheromone of the cabbage looper illustrates this procedure, which takes advantage of the acidity of terminal alkynes (see Section 7.2.2) to bring about an alkylation. The final step in creating a *cis* alkene is the *syn* addition of hydrogen to the alkyne using Lindlar's catalyst (see Section 7.3.1).

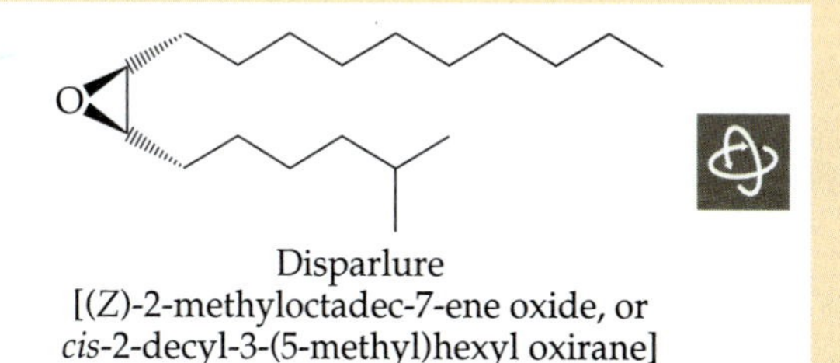

Disparlure
[(Z)-2-methyloctadec-7-ene oxide, or
cis-2-decyl-3-(5-methyl)hexyl oxirane]

Synthesis of cabbage looper sex pheromone:

$HC\equiv C$—(chain)—dioxolane $\xrightarrow[C_4H_9\text{—}Br]{NaNH_2}$ C_4H_9—$C\equiv C$—(chain)—dioxolane

↓ (several steps)

C_4H_9—$C\equiv C$—(chain)—O—C(=O)—CH_3 $\xrightarrow[Pd/CaCO_3]{H_2}$ (Z)-7-Dodecenyl acetate: $C_4H_9(H)C{=}C(H)(CH_2)_6OCOCH_3$

Most organic polymers are electrical insulators (they do not conduct electricity). However, acetylene can be polymerized to a long-chain *polyacetylene* that acts as an insulator. However, if "doped" with a trace of iodine, the polyacetylene can conduct electricity. (Doping is the addition of a small amount of material that will either add a charge to the system or remove electron density from the system.) Alternation of double and single bonds is the key to such conductivity, as will be explained in Chapter 8. Such polymers, called *organic metals,* have an electrical conductivity greater than that of copper when compared in terms of volume, and many times greater than that of copper when the comparison is based on weight.

$$\underset{\text{Acetylene}}{HC{\equiv}CH} \xrightarrow{\text{(polymerization)}} \underset{\text{Polyacetylene}}{-\!\!(HC{=}CH)_n\!\!-}$$

$$\sim\!-HC{=}CH{-}HC{=}CH{-}HC{=}CH-\!\sim$$

A segment of a polyacetylene chain

Recently, long chains of carbons joined by alternating triple and single bonds have been prepared. Called *polyalkynes,* these chains are also sometimes referred to as "molecular wires" because of their linear geometry and electrical conductivity:

$$\sim\!-C{\equiv}C{-}C{\equiv}C{-}C{\equiv}C{-}C{\equiv}C-\!\sim$$

A segment of a polyalkyne chain

Such a material is another form of carbon (the chain has only carbon atoms in it), this one involving *sp*-hybridized carbons. (Other forms of carbon are diamond, with sp^3-hybridized carbons, and graphite, with sp^2-hybridized carbons.) Such polyalkynes have considerable strength as carbon rods because of their rigidity and recently have been shown to have electrical conductivity properties. It is hoped that this property may lead to molecular electronics in the future.

Chapter Summary

Alkynes are a family of unsaturated hydrocarbons with the general formula C_nH_{2n-2}. Alkynes contain a **triple bond** between two carbon atoms, each of which is attached to one other atom or group. The triple bond consists of a sigma (σ) bond (*sp*–*sp* overlap) and two pi (π) bonds (*p*–*p* overlap) and is formed between two carbons that are **linearly *sp* hybridized.** Because of this structure, the two carbons of the triple bond and their substituents all lie in a straight line.

In the IUPAC system, alkynes are named by changing the suffix *-ane* of the parent alkane to *-yne* for the corresponding alkyne. The triple bond must be included in the carbon chain of the named parent. The location is indicated by a single number designating the first carbon of the triple bond, and this number is chosen to be as small as possible. Historical/common names are used for the simple alkynes, which are named as derivatives of **acetylene** (H—C≡C—H).

The characteristic reaction of an alkyne is addition to the triple bond. The simplest addition reaction of alkynes is *hydrogenation (reduction)* to form an alkene or alkane. Normal catalytic hydrogenation of an alkyne over a platinum or palladium catalyst produces an alkane. However, use of Lindlar's catalyst enables hydrogenation to be stopped after *syn* addition of one equivalent of hydrogen; thus, a *cis* alkene is obtained. Chemical reduction of alkynes using a metal and liquid ammonia results in *anti* addition of hydrogen and the production of a *trans* alkene.

Alkynes are readily oxidized by ozone, nitric acid, or potassium permanganate. The result is cleavage of the carbon-carbon triple bond to form two carboxylic acids, a reaction used mainly as a tool for proof of structure. **Terminal alkynes** can be distinguished from **internal alkynes** by their reaction with silver ammonium hydroxide to form precipitates of silver alkynides and by their production of carbon dioxide upon oxidation.

Higher alkynes are synthesized by *alkylation* of acetylene. Removal of the acidic hydrogen of acetylene using sodamide produces a carbanion (alkynide), which is then used as the nucleophile in an S_N2 reaction with a primary or secondary alkyl halide.

Summary of Reactions

1. **Reduction of alkynes (Section 7.3.1)**

   a. Normal hydrogenation over a platinum or palladium catalyst results in complete saturation to form alkanes.

$$\underset{\text{Alkyne}}{R-C\equiv C-R'} \xrightarrow[\text{Pt}]{H_2} \underset{\text{Alkane}}{R-CH_2-CH_2-R'}$$

   b. Hydrogenation using Lindlar's catalyst ($Pd/CaCO_3$) results in *syn* addition of only one equivalent of hydrogen to form a *cis* alkene.

$$\underset{\text{Alkyne}}{R-C\equiv C-R'} \xrightarrow[Pd/CaCO_3]{H_2} \underset{cis\text{-Alkene}}{(H)(R)C=C(H)(R')}$$

   c. Chemical reduction using sodium or lithium in liquid ammonia results in overall *anti* addition of hydrogen to form a *trans* alkene.

$$\underset{\text{Alkyne}}{R-C\equiv C-R'} \xrightarrow{Li/NH_3} \underset{trans\text{-Alkene}}{(H)(R)C=C(R')(H)}$$

2. **Oxidation of alkynes (Section 7.3.2).** Oxidation of an alkyne with most oxidizing agents, including ozone, nitric acid, and potassium permanganate, cleaves the triple bond to form two carboxylic acids.

$$\underset{\text{Alkyne}}{R-C\equiv C-R'} \xrightarrow{[O]} \underset{\text{Carboxylic acids}}{RCOOH + HOOCR'}$$

3. **Preparation of alkynes (Section 7.4).** Alkynes are prepared by forming the acetylide anion using sodamide, then alkylating with a primary or secondary alkyl halide.

$$\underset{\text{Acetylene}}{HC\equiv CH} \xrightarrow[\text{2. RBr}]{\text{1. NaNH}_2} \underset{\text{Alkyne}}{R-C\equiv CH} \xrightarrow[\text{2. R'Br}]{\text{1. NaNH}_2} \underset{\text{Alkyne}}{R-C\equiv C-R'}$$

Additional Problems

Nomenclature and Structure

7.10 Name the following compounds:

(a)

(b)

(c)

(d)

(e)

(f) HO

7.11 Draw structures for the following compounds:

(a) vinylacetylene (b) 4,4-dimethyl-2-pentyne
(c) ethynylcyclobutane (d) isopropylacetylene
(e) 4-pentyn-1-ol (f) 2-methyl-1-penten-4-yne

7.12 Describe the shaded sigma bond in each compound in terms of the hybridization of the carbon atoms involved:

(a) $CH_3-C\equiv CH$ (b) $CH_2=CH-C\equiv CH$

(c) $CH_2=CH-CH=CH_2$

7.13 Explain the fact that the lengths of the highlighted sigma bonds in Problem 7.12 are as follows: (a) 1.49 Å, (b) 1.43 Å, and (c) 1.47 Å.

Reactions of Alkynes

7.14 Show the products of the following reactions:

(a) 2-hexyne with hydrogen and Lindlar's catalyst ($Pd/CaCO_3$)

(b) 1-pentyne with sodium in liquid ammonia

(c) ozonolysis of 2-hexyne

(d) 1-propyne with sodium amide; then 1-bromobutane added

(e) 3-methylcyclooctyne with potassium permanganate

7.15 Show the reaction(s) necessary to bring about each of the following transformations:

(a) 1-pentyne to 2-bromopentane

(b) 1-pentyne to 1-pentanol

(c) 1-propyne to *cis*-2-butene

(d) 2-butyne to 2,3-butanediol

(e) acetylene to vinylcyclohexane

(f) acetylene to *trans*-2-hexene

Preparation of Alkynes

7.16 Show the reactions necessary to prepare each alkyne from acetylene:

(a) 1-butyne

(b) 2-butyne

(c) cyclobutylacetylene

(d) 1-ethynyl-4-methylcyclohexane

Mixed Problems

7.17 Draw all possible acyclic constitutional isomers containing double and/or triple bonds for compounds with the molecular formula C_5H_8.

7.18 Starting with 3-methyl-1-butyne and using any reagents and as many steps as necessary, show how you could synthesize each compound:

(a) 1-bromo-3-methylbutane

(b) 3-methyl-2-butanol

(c) *trans*-4-methyl-2-pentene

(d) *cis*-2-methyl-3-hexene

(e) 2-methyloctane

7.19 Three compounds, **A**, **B**, and **C**, have the same molecular formula, C_5H_8. All three compounds decolorize bromine in carbon tetrachloride and give a positive test with potassium permanganate. When reacted with excess hydrogen over a platinum catalyst, both **A** and **B** produce pentane. Under these same conditions, **C** absorbs one equivalent of hydrogen.

(a) Suggest possible structures for **A**, **B**, and **C**.

(b) From the reaction of **A** with hot potassium permanganate, the carboxylic acid $CH_3CH_2CH_2COOH$ was isolated, while similar treatment of **B** produced CH_3CH_2COOH and CH_3COOH. What are the structures of **A** and **B**?

(c) Ozonolysis of **C** produced a single compound, the dialdehyde $OHC(CH_2)_3CHO$. What is the structure of **C**?

7.20 Show how to synthesize propylcyclohexane from any starting materials and reagents containing no more than six carbon atoms.

7.21 An unknown hydrocarbon **W** has an empirical formula C_3H_4 and a molecular weight of 80. It decolorized bromine and produced manganese dioxide when treated with cold potassium permanganate. **W** absorbed two equivalents of hydrogen over a platinum catalyst. When **W** was oxidized with ozone, acetic acid (CH_3COOH) was isolated from the reaction mixture. Propose a structure for **W**.

7.22 Show the reactions and reagents necessary to accomplish each of the following syntheses:

(a) cyclohexene to cyclohexylacetylene

(b) 1-butene to 3-methyl-1-pentyne

(c) isobutylene to 4-methyl-1-pentyne

(d) ethylene to 2-pentyne

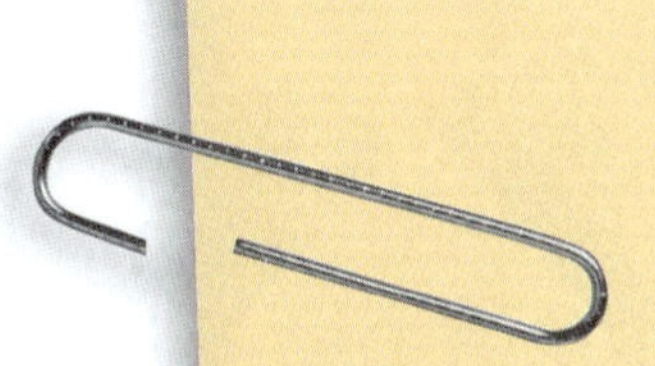

CONCEPTUAL PROBLEM

Pheromones: A Cheap Date

You work for the agricultural extension service of a large university. In increasing numbers, the owners of farms and orchards in the surrounding community have come to you seeking advice because the conventional insecticides that they use for controlling crop-destroying pests are losing their effectiveness. Many of the insects targeted have developed a resistance to the insecticides, and it takes more of an insecticide to control a pest population, which means more time and money spent by the growers. They are also concerned about the long-term environmental impact of increased insecticide use.

The extension service agrees to oversee a community-based pest management program that will involve the integrated use of insecticides and pheromones. The first prong of the attack involves using synthetic pheromones that mimic the sex pheromones produced by the female insects. Pheromone dispensers are put out at the time when mating is known to occur. In the presence of so much pheromone, the males are confused as to the whereabouts of the females and less mating occurs.

- What are some of the advantages of using mating disruption as a means of pest control?
- Some of the growers are concerned that the pheromone will also attract females and increase the likelihood of infestation. How do you allay their fears?
- They also ask whether a pheromone might be designed that would be effective for closely related species of pests. What factors would determine the likelihood of that?
- What are some of the disadvantages of using pheromones in this way, as compared to conventional insecticides?

The codling moth is the most destructive insect pest affecting apples and pears worldwide. Because of growing genetic resistance to conventional insecticides, growers are trying an integrated pest management program that includes pheromone dispensers as well as insecticides. The dispensers contain a synthetic female hormone that disrupts mating by confusing the males. If mating can be prevented, then the females cannot lay the eggs that subsequently develop into the larvae that infest the fruit.

CHAPTER 8

Conjugated Dienes

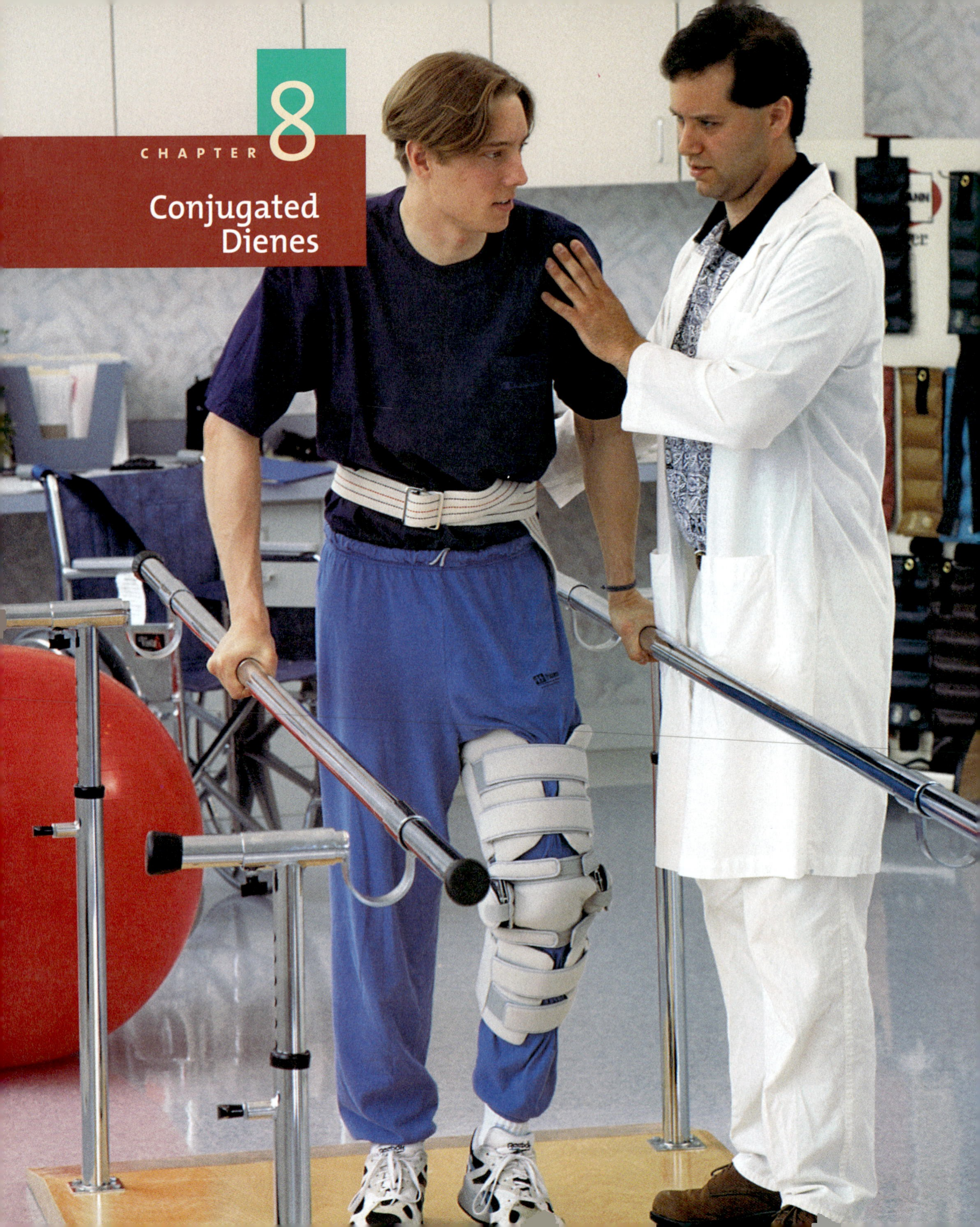

CHAPTER 6 DISCUSSED the chemistry of the alkene functional group but concentrated on alkenes with just one double bond. Alkenes having more than one double bond in their structure are referred to as **dienes** (for two), **trienes** (for three), **tetraenes** (for four), and so on. Those compounds incorporating more than four double bonds are usually referred to as **polyenes.** Nomenclature follows the normal IUPAC rules for alkenes: the suffix *-ene* to indicate the functional group and numbers for locations of the double bonds. Examples are 1,3,5-hexatriene and 1,4-cyclohexadiene:

$$\overset{1}{CH_2}{=}\overset{2}{CH}{-}\overset{3}{CH}{=}\overset{4}{CH}{-}\overset{5}{CH}{=}\overset{6}{CH_2}$$

or

1,3,5-Hexatriene

1,4-Cyclohexadiene

The focus of this chapter is on dienes, alkenes with two double bonds. There are three different classes of dienes:

- **Cumulated dienes** (also called *cumulenes* or *allenes*) are dienes in which the two double bonds are adjacent, as in 1,2-pentadiene. These compounds are relatively rare and will not be discussed in this book.
- **Isolated dienes** are dienes in which the double bonds are separated by more than one single bond, as in 1,4-pentadiene. The double bonds usually behave independently, undergoing separately the kinds of reactions described for alkenes in Chapter 6.
- **Conjugated dienes** are dienes in which the two double bonds are separated by one single bond, as in 1,3-pentadiene. Conjugated dienes, which exhibit some unusual and important chemistry, are the focus of this chapter.

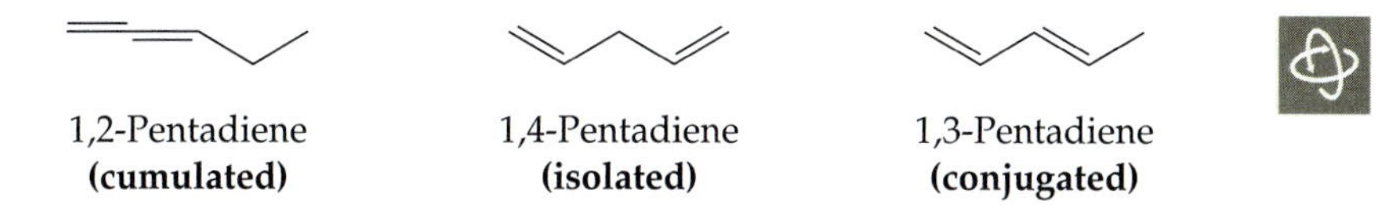

1,2-Pentadiene **(cumulated)** | 1,4-Pentadiene **(isolated)** | 1,3-Pentadiene **(conjugated)**

PROBLEM 8.1

Draw a structural formula for each of the following compounds: (a) 2,3-dimethyl-1,3-butadiene, (b) 1,3,5-cycloheptatriene, (c) 2*Z*,4*E*-hexadiene, and (d) 2*E*,4*E*,7*E*-decatriene.

◀ One important group of conjugated dienes are steroids, which in humans serve principally as hormones. Anabolic steroids are synthetic hormones that promote muscle development . These can be used safely as part of a physical therapy program, but their use in sports training and body building is discouraged because of potentially harmful side effects.

8.1 Uniqueness of Conjugated Dienes

8.1.1 Conjugation

Conjugation means that there is an alternating pattern of double and single bonds within a molecule. We will see many examples of conjugation in later chapters, some involving double bonds and single bonds other than carbon-carbon ones. But this chapter will concentrate on a conjugated system that consists of four carbons containing two double bonds separated by a single bond: —C=C—C=C—. This structural unit is present within many different kinds of molecules, including acyclic ones such as 2,4-heptadiene and cyclic ones such as 1,3-cyclohexadiene:

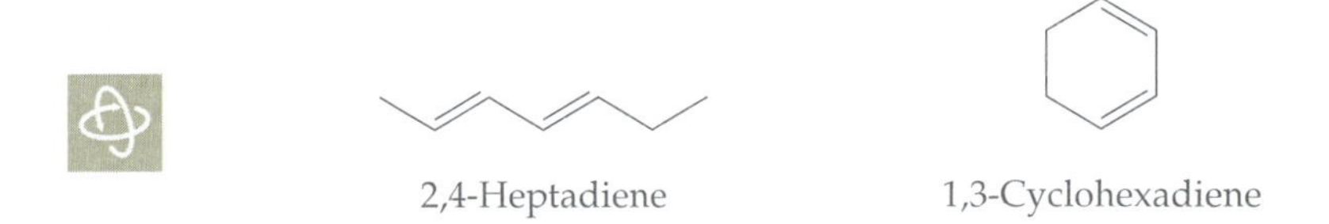

2,4-Heptadiene 1,3-Cyclohexadiene

It is important to be able to recognize this structural unit, regardless of what other functional groups or structures are present.

Conjugated dienes are special in at least two ways that will be discussed in this chapter:

1. They have high stability.
2. They produce two kinds of products from electrophilic addition, rather than just one.

8.1.2 Stability of Conjugated Dienes

By comparing the properties of conjugated dienes with those of isolated dienes, chemists have demonstrated that *conjugated dienes are about 4 kcal/mol more stable than comparable isolated dienes.* While this is a small amount of energy, it is sufficient to dictate the course of some reactions.

There is much evidence for the enhanced stability of conjugated dienes, but the most obvious is the fact that conjugated dienes form preferentially. For example, when standard elimination reactions, such as dehydrohalogenation and dehydration, are conducted to introduce a second double bond into an alkene, and there is a choice as to the regiochemistry of the elimination, the reaction invariably produces a conjugated diene rather than an isolated diene. For example, upon dehydration, 5-hexen-3-ol forms mainly the conjugated diene, 1,3-hexadiene:

OH

$\xrightarrow[\Delta,\ -H_2O]{H_2SO_4}$

Major product **Minor product**

5-Hexen-3-ol 1,3-Hexadiene + 1,4-Hexadiene

The enhanced stability of conjugated dienes is attributed to the existence of a pi (π) electron "cloud" that extends over the four carbon atoms (not just two, as in a simple alkene) (Figure 8.1). The electrons are said to be *delocalized* over the four carbon atoms, which reduces

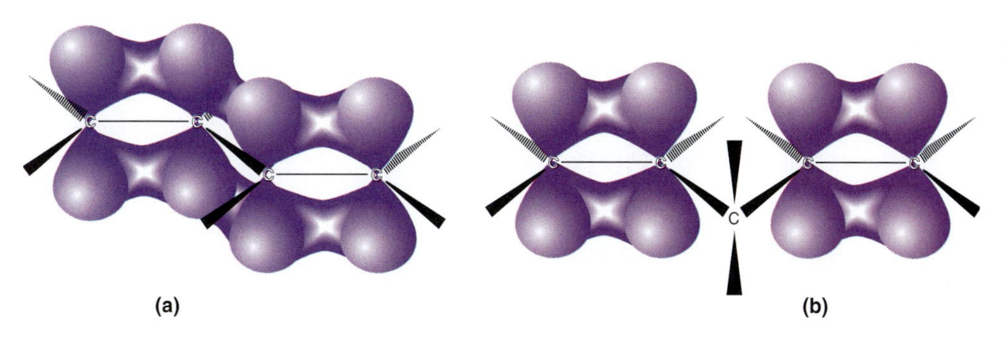

FIGURE 8.1

(a) Extended pi electron cloud in a conjugated diene compared with (b) pi bonds in an isolated diene.

the energy level of the system—in other words, delocalization increases the stability of the system. You can see that the alternating double and single bonds of the conjugated diene structure are essential for such delocalization (conjugation), providing a *p* orbital on each carbon that can overlap with adjacent *p* orbitals to form the extended pi molecular orbital. If there were two single bonds between the double bonds (as in an isolated diene), the two double bonds would be separated by an sp^3-hybridized carbon (having no *p* orbital available) and could not overlap to form the extended pi molecular orbital.

8.2 Electrophilic Addition to Conjugated Dienes

8.2.1 Addition of Bromine (or Chlorine)

Recall that molecular bromine reacts with an alkene in an addition reaction. (It is, effectively, a 1,2-adduct because the two bromines end up on adjacent carbons, forming a 1,2-dibromide, sometimes called a vicinal dibromide.) Reaction of one equivalent of bromine with a conjugated diene starts by the same mechanism (to be discussed in Section 8.3.2), but produces two products. The first is the result of 1,2-addition, and the second is the result of 1,4-addition, as illustrated with 1,3-cyclohexadiene:

1,3-Cyclohexadiene —(1 equiv. Br_2)→ 3,4-Dibromocyclohexene **(1,2-adduct)** + 3,6-Dibromocyclohexene **(1,4-adduct)**

1,3-Cyclohexadiene —(excess Br_2)→ 1,2,3,4-Tetrabromocyclohexane

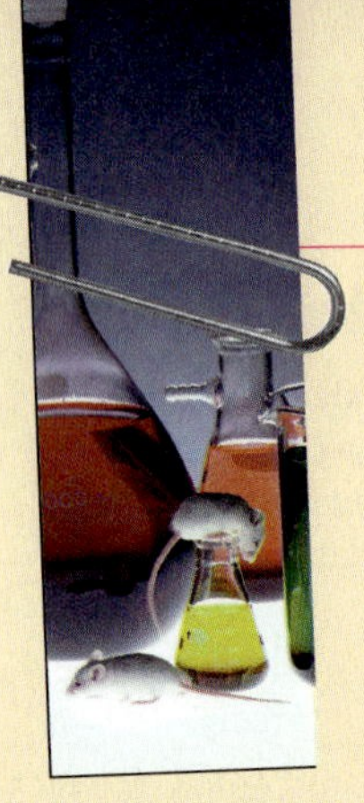

CHEMISTRY AT WORK

IN THE MARKETPLACE

β-Carotene, Vitamin A, and Vision

We all know that vegetables are an essential part of a good diet. One reason is that many vegetables contain the pigment β-carotene, which gives carrots their orange color. As it turns out, β-carotene is an important source of one of the chemicals that enable us to see. β-Carotene is a symmetrical tetraterpene (see Section 8.3.2) that contains 40 carbon atoms. This C_{40} compound can be characterized as a fully conjugated polyene, with a pi molecular orbital that extends over 11 double bonds in conjugation.

After ingestion, β-carotene is oxidatively cleaved by enzymes in the body at its center (the midpoint double bond). This reaction produces two equivalents of vitamin A, which is also called *retinol*. Vitamin A is then further oxidized to *retinal*, the corresponding aldehyde. Retinal is the active chemical involved in the photochemistry of vision.

The chemical mechanism of vision starts with the substance **rhodopsin,** which is a light-sensitive pigment located in the retina of the eye. Rhodopsin is formed by a chemical dehydration involving a protein called opsin and *cis*-11-retinal. The two compounds are joined by an imine (C=N) bond in place of the carbonyl (C=O) bond of retinal (see Section 13.4.1).

Rhodopsin has a *cis* configuration at the double bond between carbons 11 and 12. When rhodopsin is struck by light, it briefly goes into an excited electronic state. This excited state rapidly (in about 2×10^{-13} s) and spontaneously returns to the ground state, but now with a *trans* double bond at carbon 11. You should recall that *trans* double bonds are of lower energy than *cis* double bonds. The excess energy emitted by the molecule in returning to the ground state is eventually detected by the optic nerve, which sends a signal to the brain.

The mechanism that permits the eye to continually detect light is the rapid recycling of the chemicals involved. After energy transfer to the optic nerve, the isomerized rhodopsin hydrolyzes to opsin and *trans*-11-retinal. The *trans*-11-retinal is then enzymatically converted back to *cis*-11-retinal (an energy-consuming process). The newly re-formed *cis*-11-retinal reacts with an amino group ($-NH_2$) of the protein opsin to eliminate water and re-form rhodopsin, ready to start the process over again with another photon of light.

β-Carotene — Center of molecule

(enzymatic oxidation) →

Vitamin A (retinol) $R = CH_2OH$

Retinal $R = CHO$

(11, 12)

cis-11-Retinal (CH=O) — Double bond between carbons 11 and 12

H_2N-opsin / $-H_2O$ →

Rhodopsin (CH=N−opsin) — *cis*-11

(light strikes rhodopsin) → Energy emitted to optic nerve

CH=N−opsin — *trans*-11

← (hydrolysis) −opsin

trans-11-Retinal (CH=O)

↑ (enzymatic isomerization)

The 1,2-adduct (with the bromines added to carbons 1 and 2 of the starting diene) is the result of a normal alkene addition reaction. The 1,4-adduct (with the bromines added to carbons 1 and 4) is the result of **conjugate addition,** in which the two bromines have added to the ends of the conjugated diene unit and a double bond has changed position. (*Note:* When adducts are described as 1,2- or 1,4-adducts, the numbers refer only to the *relative* positions of the four carbons in the diene unit, not necessarily to the positions of the substituents in the products according to IUPAC nomenclature.) If the bromine is present in excess or if additional bromine is added, a second equivalent of bromine reacts with the 1,2- and 1,4-adducts in the normal manner, adding to the remaining double bond to produce the same product: 1,2,3,4-tetrabromocyclohexane.

Any conjugated diene (a 1,3-diene) will react with bromine to produce a mixture of the normal addition product (1,2-adduct) and the conjugate addition product (1,4-adduct). Often, the ratio of the two products can be controlled by the temperature chosen for the addition reaction.

8.2.2 Addition of Hydrogen Bromide (or Chloride)

Recall that hydrogen bromide adds to an alkene in an electrophilic addition, producing an alkyl bromide; the regiochemistry follows Markovnikov's rule (see Section 6.4.1). With a conjugated diene, one equivalent of hydrogen bromide initiates the reaction by the same mechanism (to be discussed in Section 8.2.4), forming a mixture of two products: a 1,2-adduct and a 1,4-adduct. For example, 1,3-butadiene produces 3-bromo-1-butene and 1-bromo-2-butene:

1,3-Butadiene —(1 equiv. HBr)→ 3-Bromo-1-butene **(1,2-adduct)** + 1-Bromo-2-butene **(1,4-adduct)**

The 1,2-adduct appears to result from a normal addition, the proton having added to carbon 1 of the diene and the bromide to the adjacent carbon (carbon 2). However, obtaining the 1,4-adduct requires that the proton add to carbon 1 and bromide add to carbon 4, with the double bond having changed position, once again resulting in conjugate addition.

Any conjugated diene (1,3-diene) will react with HX to produce the products from normal addition (1,2-adduct) and conjugate addition (1,4-adduct). Often, the ratio of the two products can be controlled by the temperature chosen for the reaction.

8.3 Introduction to Resonance

8.3.1 Stabilization of the Allylic Carbocation

To understand conjugate addition, you need to consider the stabilization of the reactive intermediate involved in electrophilic addition reactions—a carbocation, and in this case, an *allylic* carbocation. As you learned in Chapter 3, the relative stability of *alkyl* carbocations is 3° > 2° > 1°. The tertiary carbocation is more stable because of *hyperconjugation*—the availability of nine adjacent C—H bonds to share electron density with the electron-deficient carbon (see Section 3.4.3). In other words, the greater the extent to which the positive charge on carbon can be delocalized—that is, spread slightly over other atoms—the more stable the carbocation is.

Although allyl bromide is a primary halide, in methanol solution it undergoes solvolysis by an S_N1 mechanism to produce allyl methyl ether. (Primary halides are normally substituted only via the S_N2 mechanism and by stronger nucleophiles.) The reason for this unexpected result is that an allylic carbocation forms readily, because of the special stabilization of that cation (it is about as stable as a tertiary carbocation).

$$CH_2{=}CH{-}CH_2{-}Br \xrightarrow[-Br^-]{CH_3OH} \left[CH_2{=}CH{-}\overset{+}{C}H_2\right] \xrightarrow[-H^+]{CH_3OH} CH_2{=}CH{-}CH_2{-}O{-}CH_3$$

Allyl bromide (3-bromo-1-propene) — Allyl carbocation — Allyl methyl ether (3-methoxy-1-propene)

The stabilization of the allylic carbocation is explained by the **resonance theory,** which states that *whenever a molecule or ion can be represented by two or more structures that differ only in the positions of the electrons,* there are three implications:

1. No single structure by itself can adequately represent the actual structure of the molecule or ion.
2. The actual structure of the molecule or ion is a *hybrid* of the individual structures.
3. The molecule or ion is considerably more stable (that is, of lower energy) than would be expected based on any single structure.

The resonance theory is one of the fundamentals of organic chemistry and will be encountered many times in this book.

We can apply the resonance concept to the allylic carbocation. There are two **contributing structures** (**A** and **B**) for this carbocation. Neither represents a structure that actually exists, but both are needed to account for the characteristics of the actual structure. The actual structure of the allylic carbocation is a single structure that is a hybrid of the two contributing structures (**A** and **B**) and is called a **resonance hybrid.**

$$\underset{\mathbf{A}}{CH_2{=}CH{-}\overset{+}{C}H_2} \longleftrightarrow \underset{\mathbf{B}}{\overset{+}{C}H_2{-}CH{=}CH_2}$$

The double-headed arrow between **A** and **B** implies a resonance relationship between the contributing structures. (Note that the double-headed arrow does not imply an equilibrium between **A** and **B;** the typical equilibrium double arrows are not appropriate.) The structure of the allylic carbocation is neither **A** nor **B,** but a more stable hybrid of the two. The electronic relationship between the two contributing structures is represented by the small blue curved arrows, indicating the necessary electron flow to get from structure **A** to structure **B,** and vice versa.

Sometimes a single structure is used to indicate that the electron deficiency of the carbocation is spread over the entire allyl structure (Figure 8.2a). In terms of molecular orbital theory, the allylic carbocation has p orbitals on three adjacent carbons, but there are only two electrons to be shared (Figure 8.2b). This leaves the structure elec-

$$\overset{\delta+}{C}H_2{-}CH{-}\overset{\delta+}{C}H_2$$

(a)

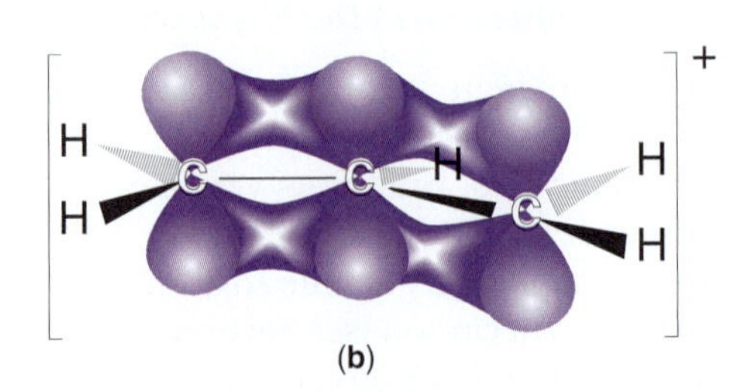

(b)

FIGURE 8.2

(a) A single structure that represents the allyl carbocation. (b) Orbital diagram of the allyl carbocation.

tron-deficient, but with continuous overlap in an extended pi molecular orbital, similar to that shown for conjugated dienes in Section 8.1.2.

A chemical structure in which a positively charged carbon is adjacent to a carbon-carbon double bond is called an *allylic unit*. The fact that an allylic unit has some positive charge (an electron deficiency) located at both ends is demonstrated by the fact that solvolysis of 3-chloro-1-methylcyclopentene produces two isomeric ethers as products of S_N1 substitution:

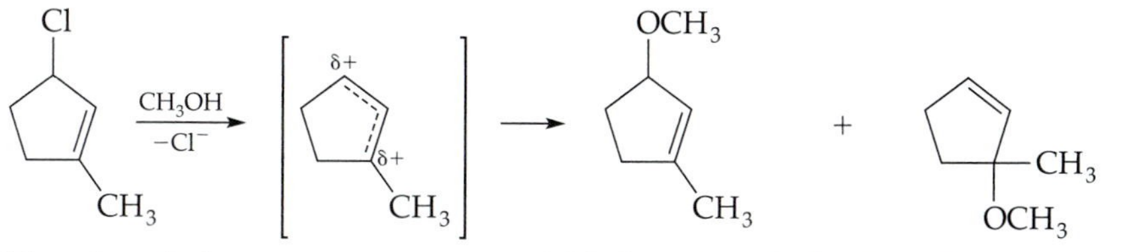

3-Chloro-1-methyl cyclopentene $\xrightarrow[-Cl^-]{CH_3OH}$ [δ+ allylic cation δ+] → 3-Methoxy-1-methyl cyclopentene + 3-Methoxy-3-methyl cyclopentene

Thus, two different carbons (1 and 3) had sufficiently low electron density (sufficient positive charge) to undergo attack by the nucleophile methanol. These were the two ends of the resonance-stabilized allylic carbocation. In summary, resonance-stabilized allylic cations form readily and can be attacked by a nucleophile at either end of the allylic system.

In any structural environment where a cationic site is located adjacent to one or more carbon-carbon double bonds, resonance stabilization of the cation is to be expected. The stabilization is represented by drawing contributing structures, as for the allylic cation, and by indicating shifts of electrons required to get from one structure to another. For example, the carbocation from ionization of 3-bromo-1-methylcycloocta-1,4-diene is resonance-stabilized.

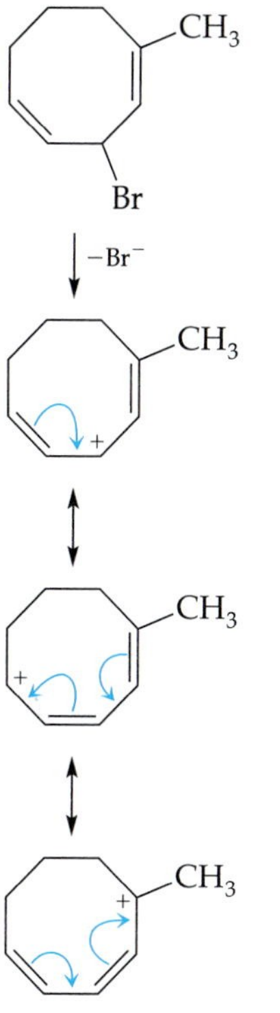

Carbocation from 3-bromo-1-methylcycloocta-1,4-diene

How to Solve a Problem

Draw the products from the S_N1 solvolysis in water of 3-bromo-3-methylcyclohexene. Draw the contributing structures for the carbocation intermediate.

SOLUTION

First, we write the ionization of the substrate into its carbocation, with bromide ion leaving (first step of an S_N1 reaction):

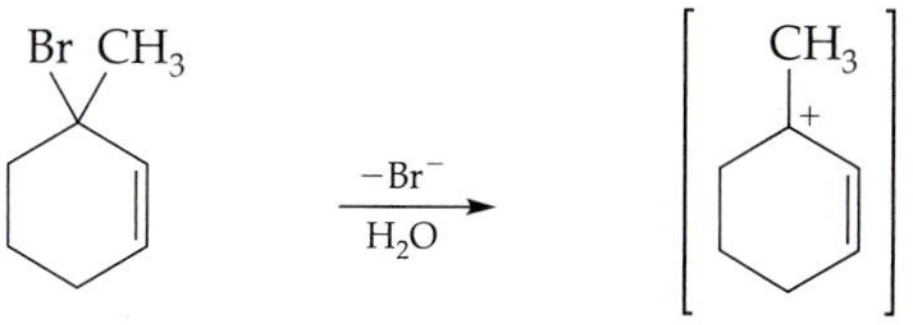

3-Bromo-3-methyl cyclohexene $\xrightarrow[H_2O]{-Br^-}$ [Carbocation]

Next, we examine the structure of the carbocation to determine whether there is conjugation present. The *p* orbital of the carbocation is next to the pi orbital of the double bond. In fact, this is an allylic system with the following two contributing forms:

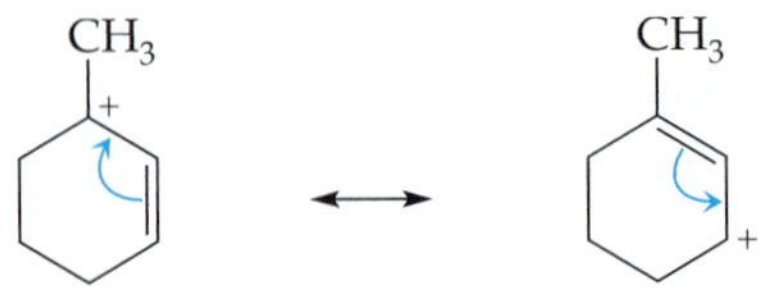

Carbocation

The solvolysis reaction is completed by attack of nucleophilic water on the carbocation (at both positions that carry some positive character), followed by loss of a proton from the oxonium ion:

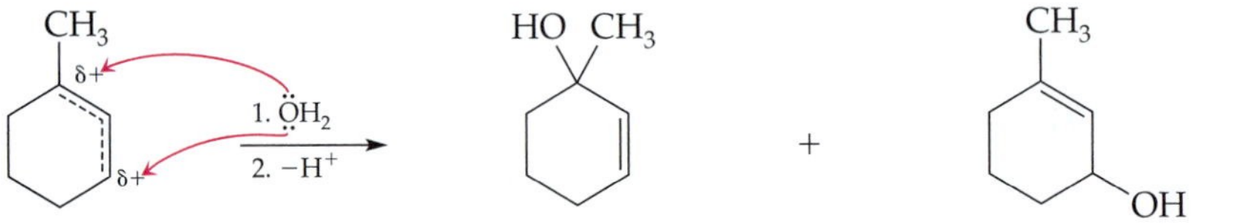

Carbocation 1-Methylcyclohex-2-enol 3-Methylcyclohex-2-enol

PROBLEM 8.2

Draw the contributing structures for the following resonance-stabilized carbocations:

(a) 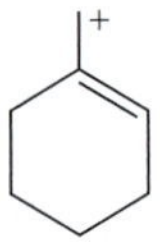 (b) 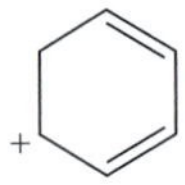 (c)

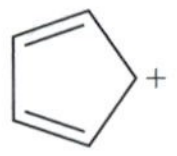

8.3.2 Mechanism of Conjugate Addition

Now that you know about the stabilization and structure of allylic carbocations, you can understand why and how conjugate addition occurs to conjugated dienes. In the addition of hydrogen bromide to 1, 3-butadiene, the usual regioselective initial attack of a proton on the alkene occurs at a terminal carbon to form a resonance-stabilized allylic carbocation. The second step of the addition reaction, attack by bromide anion at center(s) of electron deficiency, can occur at either carbon 2 (to form the 1,2-adduct from normal addition) or carbon 4 (to form the 1,4-adduct from conjugate addition):

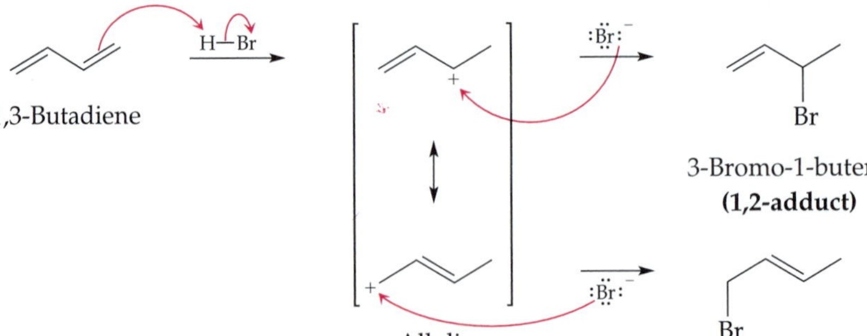

1,3-Butadiene

Allylic carbocation

3-Bromo-1-butene **(1,2-adduct)**

1-Bromo-2-butene **(1,4-adduct)**

This mechanism accounts for the two products obtained from addition of hydrogen bromide to a conjugated diene. They result from normal addition and conjugate addition occurring concurrently.

A similar explanation accounts for the formation of 1,2- and 1,4-adducts from the reaction of bromine with a conjugated diene. In the first step, the diene attacks bromine in a usual electrophilic addition at a terminal carbon to form an intermediate resonance-stabilized allylic carbocation. In the second step, attack of bromide ion at site(s) of electron deficiency occurs at carbons 2 and 4, producing 1,2- and 1,4-adducts, respectively.

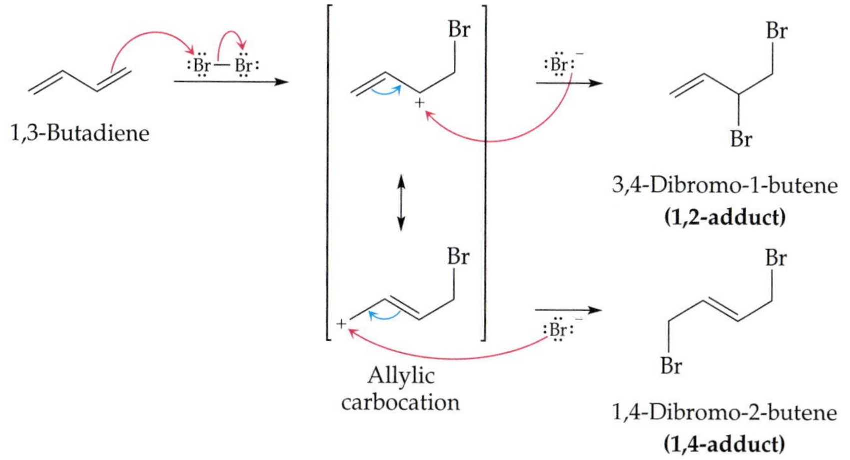

1,3-Butadiene

Allylic carbocation

3,4-Dibromo-1-butene
(1,2-adduct)

1,4-Dibromo-2-butene
(1,4-adduct)

In summary, conjugate addition to a 1,3-diene is the result of the formation of a resonance-stabilized allylic carbocation intermediate (C=C—C+) during the reaction.

How to Solve a Problem

Show the product(s) from the reaction of one equivalent of hydrogen chloride with 6-methyl-1,3-cycloheptadiene.

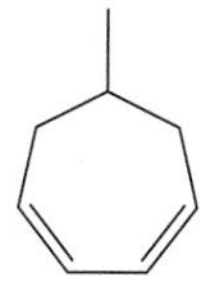

6-Methyl-
1,3-cycloheptadiene

PROBLEM ANALYSIS AND SOLUTION

The "1,3-diene" part of the name of the substrate indicates immediately that it is a conjugated diene, and this is confirmed by a drawing of the structure. The double bonds are located within a seven-membered ring, but the compound will still undergo the usual diene addition reactions, producing both 1,2- and 1,4-adducts.

Hydrogen chloride will add to the diene in two steps. First, the proton adds to one end of the diene structure to form a carbocation:

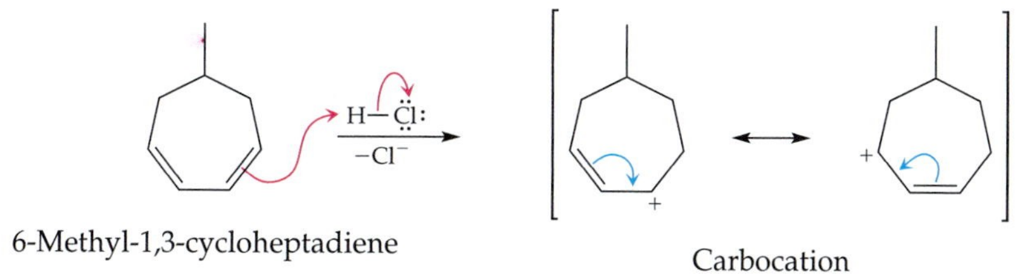

6-Methyl-1,3-cycloheptadiene

Carbocation

Because this carbocation is allylic, it is resonance-stabilized. In the final step, the chloride ion released in the first step attacks either center of positive charge, thereby producing two isomeric chloroalkenes: 3-chloro-6-methylcycloheptene (the 1,2-adduct) and 3-chloro-5-methylcycloheptene (the 1,4-adduct):

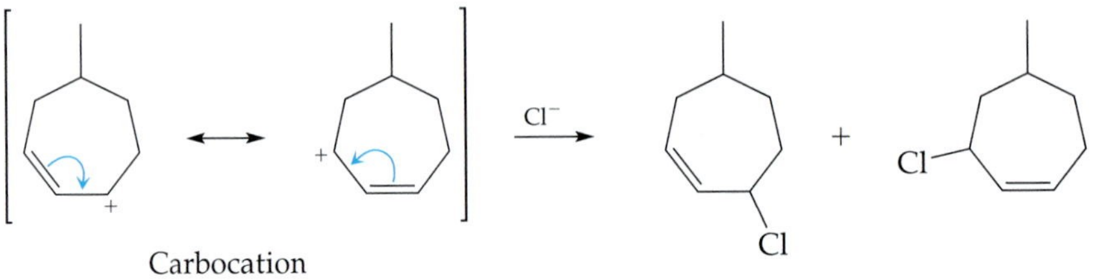

Carbocation

Cl^-

3-Chloro-6-methylcycloheptene

3-Chloro-5-methylcycloheptene

PROBLEM 8.3

Write the products formed from the following reactions:

(a) 1,3-cyclopentadiene with one equivalent of HBr
(b) 2,4-hexadiene with two equivalents of bromine
(c) 2,4-hexadiene with one equivalent of bromine
(d) 1,3-cyclohexadiene with one equivalent of bromine

PROBLEM 8.4

Show the steps and reagents necessary to synthesize 1,4-dimethoxycyclohexane from 1,3-cyclohexadiene.

8.4

Important Conjugated Dienes

8.4.1 Rubber

Conjugated dienes became especially important to the United States during World War II, when the usual supply of natural rubber was cut off. Natural rubber, critical for the war effort, had been imported mainly from Southeast Asia, which was occupied by Japan early in the 1940s. The immediate result was that the United States undertook a massive research program to develop useful synthetic rubbers. An earlier discovery had revealed that slow, gentle heating of natural rubber led to the distillation of a clear liquid, shown to be the conjugated diene known as *isoprene* (2-methyl-1,3-butadiene). In the best tradition of retrosynthesis, chemists surmised that if natural rubber was a polymer of isoprene, then synthetic rubber might be prepared by polymerizing isoprene, which was readily available from other sources. The polymerization reaction did succeed, but it produced a lower-quality rubber. In 1955, discovery of a new polymerization catalyst finally permitted polymerization of isoprene to produce a product that is indistinguishable in physical properties from natural rubber. Polymerization of conjugated dienes will be discussed in detail in Section 18.3.6.

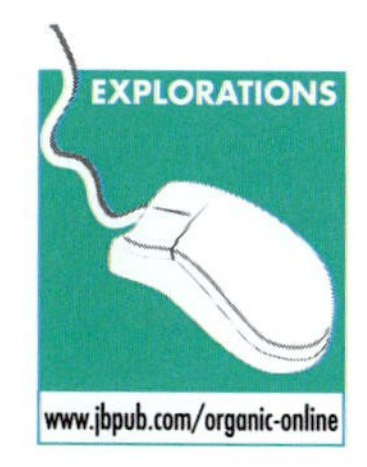

$$CH_2{=}C(CH_3){-}CH{=}CH_2 \xrightarrow{\text{(polymerization)}} \left[-CH_2-C(CH_3){=}CH-CH_2-\right]_n$$

Isoprene (carbons numbered 1–4) | *cis*-Polyisoprene (synthetic "natural" rubber)

8.4.2 Terpenes and Steroids

The very large family of naturally occurring compounds known as **terpenes** was discovered in the "essential oils" of plants. These oils could be obtained from the plants by distillation and were found to be the source of the fragrance of many of those plants. The common chemical feature of terpenes is that their structures are multiples of a five-carbon unit based on the conjugated diene whose common name is isoprene (2-methyl-1,3-butadiene; see Figure 8.3). Thus, terpenes are sometimes referred to as *isoprenoid compounds.* Isoprene itself is emitted into the atmosphere by many plants, especially in warmer climates. It is partly responsible for the haze that hangs over the Blue Ridge Mountains of Virginia and the Blue Mountains of Australia.

The blue haze characteristic of the Blue Ridge Mountains in Virginia is due in part to the emission of isoprene by the vegetation.

Examples of terpenes are myrcene (the oil of bay leaves), limonene (the oil of lemon and orange), menthol (the oil of peppermint), α-pinene (the turpentine from pine trees), farnesol (the oil of the lily-of-the-valley flower), and caryophyllene (the oil of cloves). Although many terpenes are hydrocarbons (alkenes), some have other functional groups present, such as the alcohol group in menthol and farnesol (note the suffix *-ol* on these historical names, indicating that the compounds are alcohols).

Myrcene (bay oil)

Limonene (lemon and orange)

Menthol (peppermint)

α-Pinene (turpentine)

Farnesol (lily-of-the-valley)

Caryophyllene (oil of cloves)

FIGURE 8.3

Structural formula and ball-and-stick model of isoprene (2-methyl-1,3-butadiene).

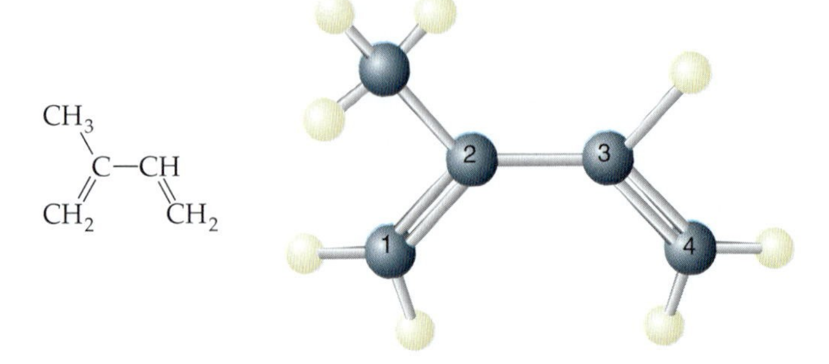

It has been demonstrated that terpenes are formed by the stepwise joining together of isoprenoid (C_5) units, but not of isoprene itself. The steps involved in the biosynthetic pathway to terpenes have been fully elaborated but will not be described here. Each isoprene unit is said to have a "head" (carbon 1 in Figure 8.3) and a "tail" (carbon 2 in Figure 8.3). Monoterpenes involve two isoprene units and are C_{10} compounds (for example, myrcene); sesquiterpenes are C_{15} compounds (for example, farnesol); and diterpenes are C_{20} compounds (for example, abietic acid, from pine-tree rosin). In the structures shown in Figure 8.4, the isoprene units are identified by alternating groups of light and dark spheres. (Each sphere indicates a single carbon, but the hybridization of that carbon is not specified.) Note that these terpenes all have the isoprenoid units connected "head to tail," the most common arrangement.

Triterpenes are C_{30} compounds, and one of the most important is *squalene*. While most polyisoprenoids have all the isoprene units connected "head to tail," squalene has a single "tail to tail" connection at the center of the molecule, making it symmetrical.

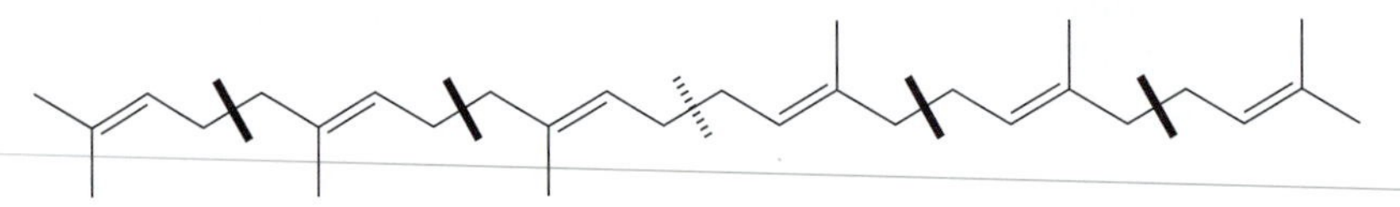

Squalene ($C_{30}H_{50}$)
2,6,10,15,19,23-hexamethyl-2,6*E*,10*E*,14*E*,18*E*,22-tetracosahexaene

— Head-to-tail isoprene connection
''''''''' Tail-to-tail isoprene connection

FIGURE 8.4

The isoprene units of several terpenes.

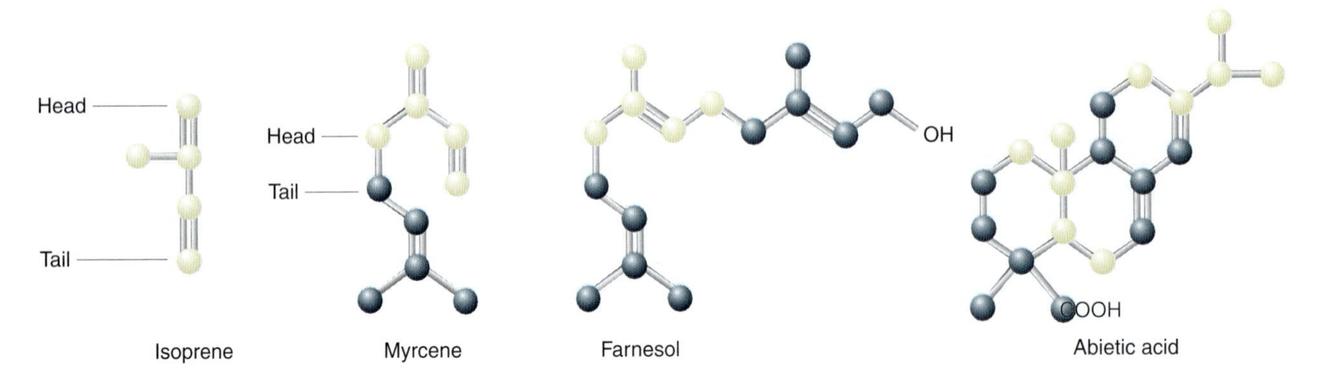

Tetraterpenes are C_{40} compounds; one example is β-carotene. Note that β-carotene also has one tail-to-tail connection of isoprenoid units, marked with the hatched line.

One source of the tetraterpene β-carotene is carrots. This C_{40} compound not only gives the carrot its orange color, but is also essential as a precursor in the chemistry of vision.

β-Carotene

Steroids have in common the tetracyclic ring system, shown in color for ergosterol:

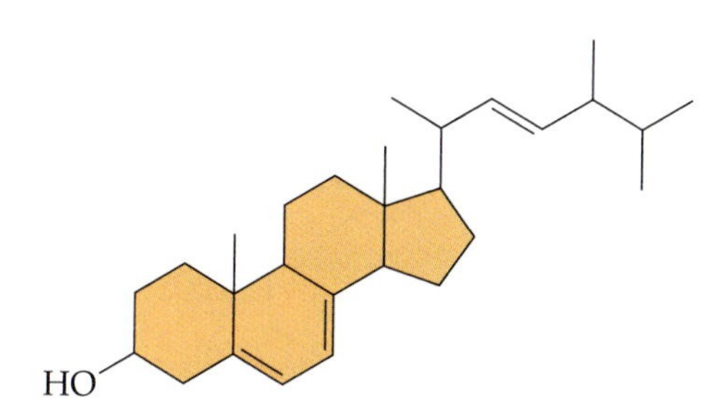

Ergosterol

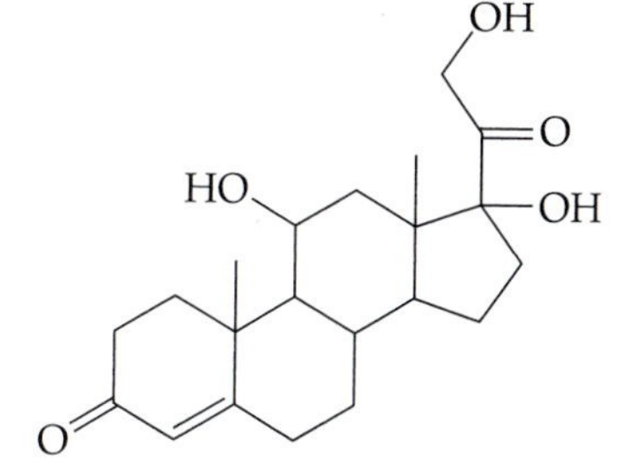

Cortisol (hydrocortisone)

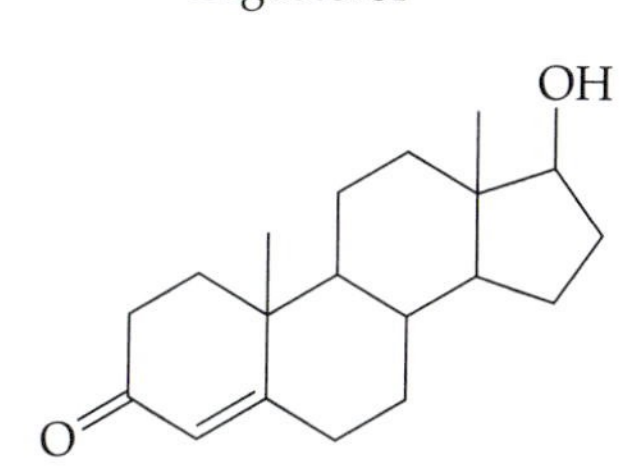

Testosterone

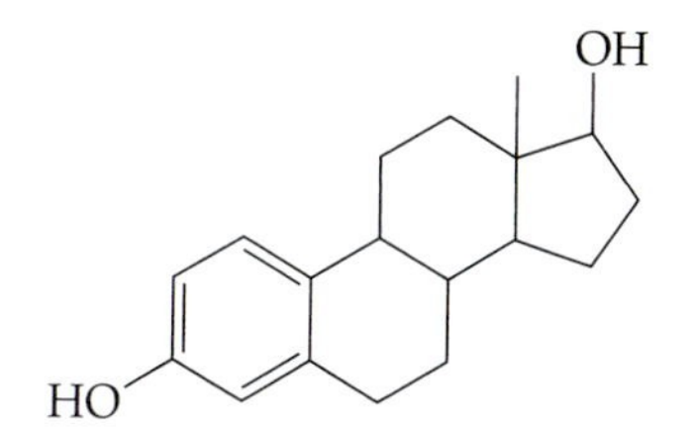

Estradiol

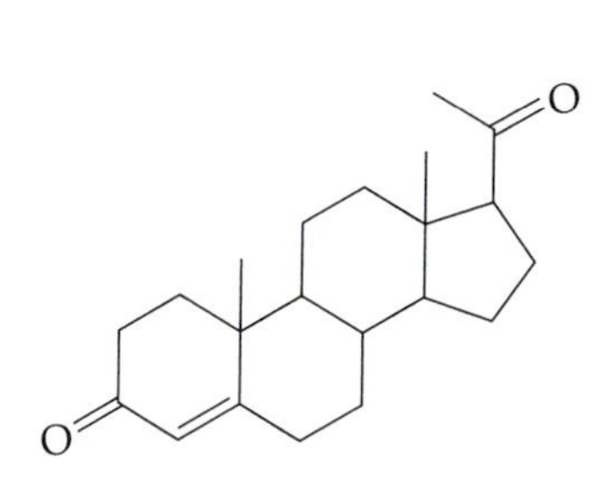

Progesterone

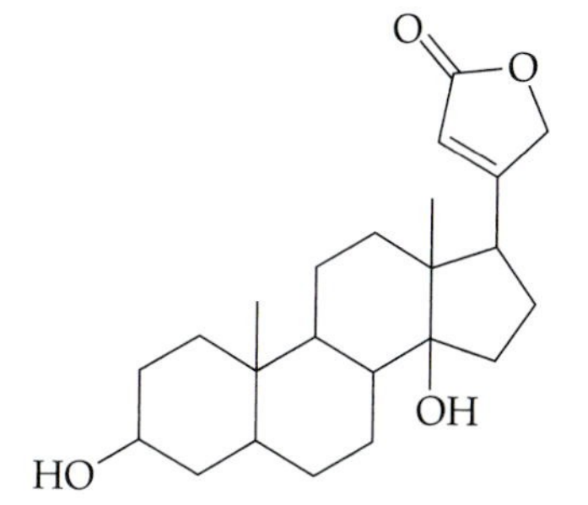

Digitoxigenin

Steroids are synthesized in the human body as well as in many other animals and plants. They play essential roles in many body functions: as vitamins (ergosterol is converted into vitamin D_2), as part of the immune system (cortisol, also known as hydrocortisone), as the sex hormones (testosterone is the principal male sex hormone, and estradiol is the principal female sex hormone), and as key participants in the reproduction process (progesterone is the hormone that prepares the uterus for implantation of

Digitalis, more commonly known as foxglove, produces the plant steroid digitoxigenin.

a fertilized egg and suppresses further ovulation). Plant steroids are widespread. For example, digitoxigenin is a steroid from the digitalis plant and is used to treat certain heart problems; related steroids occur in yams.

All terpenes and steroids are biosynthesized starting with acetate ion (CH_3COO^-), which is converted into biologically active equivalents of isoprene [$CH_2{=}C(CH_3){-}CH{=}CH_2$] known as 3-methyl-2- or 3-methyl-3-butenylpyrophosphate. These are the basic isoprenoid (C_5) building blocks. Their polymerization, first to monoterpenes (C_{10}) and then to sesquiterpenes (C_{15}), followed by dimerization tail to tail, produces squalene (C_{30}). Squalene is converted, through many steps, into lanosterol, from which the other steroids, including cholesterol ($C_{27}H_{46}O$), are produced.

$CH_3{-}COO^-$

Acetate anion

↓

3-methyl-2-butenylpyrophosphate → Squalene → Lanosterol → Cholesterol

Chapter Summary

Conjugated dienes are alkenes with a single bond separating two double bonds. The **conjugation** of the double bonds results in overlap of the *p* orbitals of four adjacent carbons, creating an extended pi system. Conjugated dienes are therefore more stable than similar nonconjugated dienes.

Conjugated dienes undergo electrophilic **conjugate addition,** forming 1,2-adducts and 1,4-adducts. The mechanism of addition reactions involves formation of an intermediate carbocation, an allylic carbocation in this case, which is stabilized by **resonance.** Two contributing structures must be drawn to represent the actual structure of the allylic carbocation, which has two electrons delocalized over three carbons. The final addition products result from nucleophilic attack at both centers of positive charge.

Conjugated dienes are important in the synthetic rubber industry as monomers for polymerization. A conjugated diene unit is also important as a structural feature of some **terpenes.** All terpenes contain two or more **isoprenoid** structural units, usually joined together in a head-to-tail arrangement.

Summary of Reactions

1. **Addition of HBr (Section 8.2).** A conjugated diene reacts with one equivalent of hydrogen bromide to form isomeric allylic bromides, the 1,2- and 1,4-adducts.

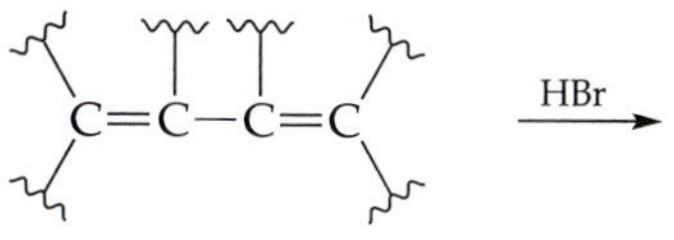

C=C–C=C $\xrightarrow{HBr}$

Diene

C=C–C(Br)–C(H) + C(Br)–C=C–C(H)

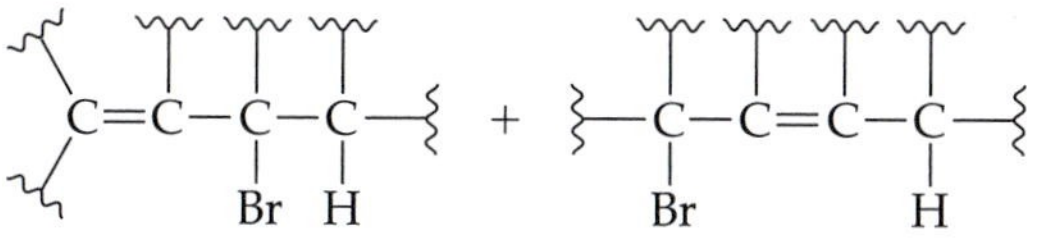

1,2-adduct 1,4-adduct

2. **Addition of bromine (Section 8.2).** A conjugated diene reacts with one equivalent of bromine to produce the isomeric dibromoalkenes, the 1,2- and 1,4-adducts.

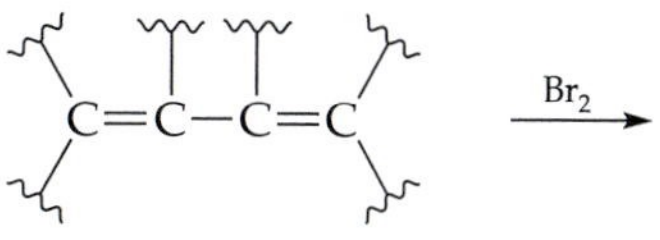

C=C–C=C $\xrightarrow{Br_2}$

Diene

C=C–C(Br)–C(Br) + C(Br)–C=C–C(Br)

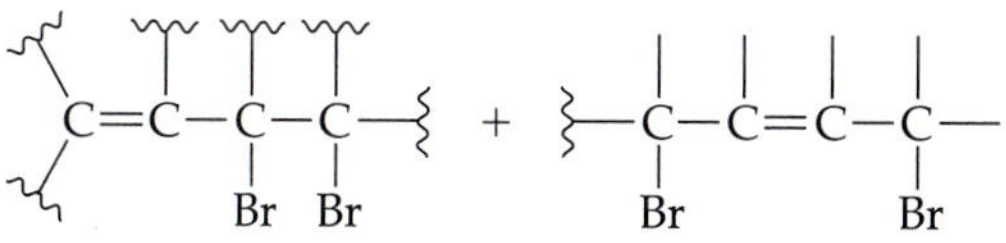

1,2-adduct 1,4-adduct

Additional Problems

Nomenclature and Structure

8.5 Name the following compounds:

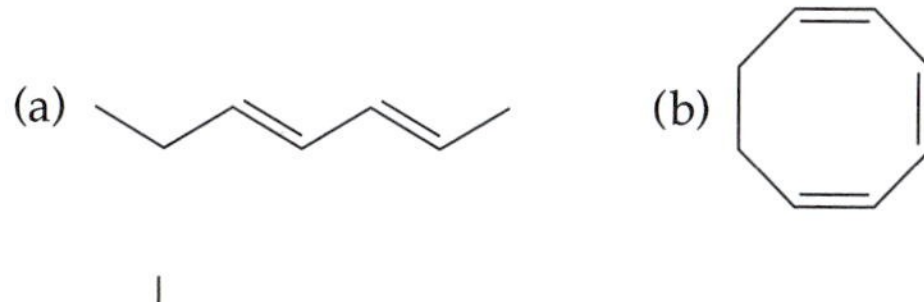

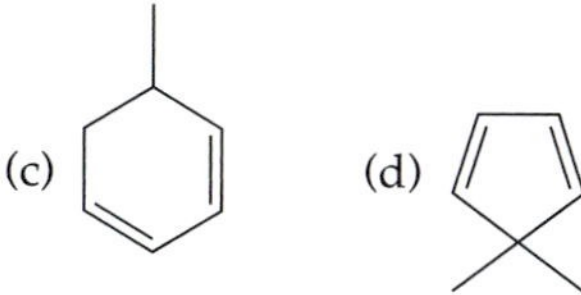

8.6 Draw structures corresponding to the following names:

(a) 2-methyl-1,4-cycloheptadiene

(b) (2*E*,4*E*,6*E*)-octatriene

(c) 2-bromo-1,3-butadiene

(d) 1,3-dimethyl-1,3-cyclobutadiene

8.7 Draw the contributing structures for each resonance-stabilized allylic cation:

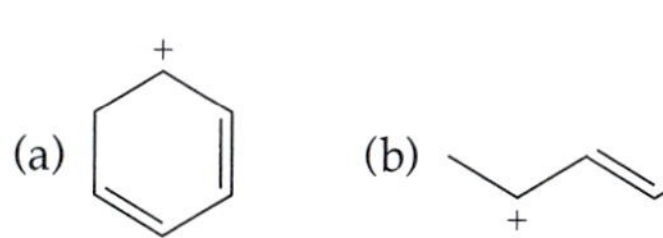

Addition Reactions

8.8 Addition of one equivalent of bromine to 1,3-cyclopentadiene yields two structural isomers with the formula $C_5H_6Br_2$. Write their structures and show the mechanism by which each forms.

8.9 Show the product(s) expected from each of the following reactions:

(a) 1,3-pentadiene with one equivalent of hydrogen chloride

(b) cyclopentadiene with one equivalent of hydrogen bromide

(c) 4-methyl-1,3-pentadiene with hydrogen bromide (Which product should predominate?)

CONCEPTUAL PROBLEM

Testing . . . 1, 2, 3

You have just started working at a laboratory that conducts drug testing for the Federal Drug Administration. Your supervisor decides to test your knowledge of organic chemistry. He places three pills before you and says that one is the birth-control pill norethindrone, one is RU-486 (mifepristone, the "morning after" pill), and the other is progesterone (used in estrogen therapy). He asks whether you know a simple way to distinguish the three compounds using chemical tests. What would you tell him? Explain how you could distinguish the three drugs.

Norethindrone

Mifepristone (RU-486)

Progesterone

(d) 1-methyl-1,3-cyclohexadiene with excess hydrogen over a platinum catalyst

(e) 1,4-cyclohexadiene with one equivalent of bromine

(f) 1,4-cyclohexadiene with excess bromine

(g) 1,3-cyclohexadiene with one equivalent of bromine

(h) 1,3-cyclohexadiene with ozone

Mixed Problems

8.10 Outline a synthesis of 1,3-butadiene from each of the following compounds:

(a) 1,4-dibromobutane (b) 1-buten-3-yne

(c) 3-buten-1-ol (d) 1,4-butanediol

8.11 Explain the observation that 1-bromopropane produces a precipitate with alcoholic silver nitrate only after lengthy heating, whereas allyl bromide (3-bromo-1-propene) produces an immediate precipitate at room temperature with the same reagent. (*Hint:* Remember the S_N1 mechanism.)

8.12 Draw the contributing structures for the resonance-stabilized carbocation that results from protonation of and loss of water from 4-methyl-3-penten-2-ol. Which should be the major contributor, and why?

8.13 Explain the observation that 2-cyclohexen-1-ol reacts readily with concentrated sulfuric acid at room temperature to produce 1,3-cyclohexadiene as the only product.

8.14 Show the structure of the product(s) formed from each of the following elimination reactions:

(a) 2-cyclopenten-1-ol with concentrated sulfuric acid

(b) 2-bromo-4-nonene with hot sodium ethoxide

(c) 3-cyclohepten-1-ol with phosphoric acid

CHAPTER 9 Aromatics

THE COMPOUNDS STUDIED to this point have all been derived from parent alkanes and have been named accordingly, using the IUPAC system. These families fall into the broad classification of **aliphatic compounds.** The term *aliphatic* relates to the derivation of these compounds from fat-like sources (*aleiphar* is Greek for "fat").

This chapter introduces the other broad classification of organic compounds, the **aromatic compounds.** The term *aromatic* originated from the fact that the first of these compounds to be identified were derived from substances with pleasant aromas (such as oil of bitter almonds and oil of wintergreen). As will become evident, aliphatic compounds may be best defined as those which are *not aromatic,* and aromatic compounds are characterized as being *cyclic, planar, highly unsaturated, conjugated,* and *resonance-stabilized.*

The active ingredient associated with the aroma of almonds is benzaldehyde, an aromatic compound.

The parent aromatic hydrocarbon is **benzene** (C_6H_6), and the family of aromatic hydrocarbons is called the **arenes.** However, aromatic compounds include not only hydrocarbons, but also their derivatives, as we will see later in this chapter. Derivatives of benzene—that is, compounds containing the benzene skeleton—are often referred to as **benzenoid compounds.** There is no general molecular formula for arenes, as there is for aliphatic hydrocarbons. It should be noted, however, that arenes have low hydrogen-to-carbon ratios, typically about 1:1, as in benzene, whereas alkanes (C_nH_{2n+2}), alkenes (C_nH_{2n}), and alkynes (C_nH_{2n-2}) have hydrogen-to-carbon ratios around 2:1.

The alkane family provides the parent structures for all the families of compounds studied in the preceding chapters and shown in the left portion of the Key to Transformations (see inside front cover). The various functional groups are either built into an alkane structure (alkenes and alkynes) or attached to that structure (alkyl halides, alcohols, and ethers). Likewise, benzene is the parent structure of benzenoid compounds, and halogens, hydroxyl groups, and alkoxy groups may be attached to the benzene ring. In addition, other functional groups yet to be studied (shown in the right-hand portion of the Key to Transformations) may also be attached to the benzene ring. These include the amines (Chapter 12), carbonyl groups (Chapter 13), carboxylic acids (Chapter 14), and carboxylic acid derivatives (Chapter 15). Thus, the functional groups shown in the Key to Transformations are attachable to either an aliphatic or an aromatic carbon skeleton.

9.1 The Structure of Benzene

Benzene (C_6H_6) is a colorless liquid, boiling at 80°C and melting at 5°C. It was first isolated by Michael Faraday in 1825 and was later prepared from benzoic acid ($C_7H_6O_2$) by forcing the ejection of CO_2. Benzoic acid was obtained by chemical degradation of

◀ *Overleaf:* Polycyclic aromatic hydrocarbons have been found in meteorites of Martian origin, including those bearing what appear to be fossils of primitive life forms (inset).

any of a wide variety of naturally occurring fragrant compounds. The C_6 structural unit of benzene seemed to remain intact through many different kinds of reactions, and benzene itself was produced from many different starting materials. It became clear that benzene has special stability, and its chemical behavior is not typical of aliphatic hydrocarbons.

9.1.1 Experimental Evidence

Determining benzene's structure was a difficult task that was not accomplished until 47 years after its discovery. The essential experimental facts known before the structure was determined were as follows:

1. The molecular formula for benzene is C_6H_6. Therefore, it has eight fewer hydrogens than necessary for a C_6 alkane (C_6H_{14}). Thus, it must have some combination of double bonds, triple bonds, and/or rings.
2. Benzene is not hydrogenated by hydrogen and a platinum catalyst under the usual conditions that hydrogenate an alkene or an alkyne.
3. Under high pressure, benzene is hydrogenated and absorbs three equivalents of hydrogen, producing the known compound cyclohexane (C_6H_{12}).
4. Benzene is not oxidized by hot potassium permanganate.
5. Benzene does not react with bromine, as do alkenes or alkynes, which undergo rapid addition reactions. When bromine is mixed with benzene in the presence of a Lewis acid such as $FeBr_3$, a substitution reaction occurs—producing bromobenzene (C_6H_5Br) and evolving HBr.
6. Further bromination of benzene using bromine and $FeBr_3$ produces a mixture of three isomeric dibromobenzenes ($C_6H_4Br_2$).

The first three items of evidence suggested that benzene should be 1,3,5-cyclohexatriene. However, a cycloalkene structure is inconsistent with the behavior cited in the last three items of evidence; benzene simply could not contain ordinary double bonds. In 1872, August Kekulé sought to reconcile the evidence by proposing that benzene contains three double bonds that shift position so rapidly that the two forms cannot be isolated; Kekulé was proposing an equilibrium. These representations of benzene are now referred to as *Kekulé structures*.

1,3,5-Cyclohexatriene

Kekulé structures for benzene

Three pieces of modern evidence are essential for establishing the actual structure of benzene:

1. X-ray analysis of benzene reveals that it has a symmetrical planar hexagonal structure (C—C—C and C—C—H bond angles are 120°). The six identical carbon-carbon bond lengths are 1.39 Å. (For reference, C—C and C═C bond lengths are 1.54 Å and 1.34 Å, respectively.) The six C—H bond lengths are identical at 1.09 Å (Figure 9.1).
2. Nuclear magnetic resonance (NMR) spectroscopy (see Section 11.6) indicates that benzene's hydrogen atoms are electronically equivalent but are in a very different electronic environment than are the hydrogens attached to a carbon of a carbon-carbon double bond. Benzene possesses what is called a *ring current*, which is unique

FIGURE 9.1

Space-filling and ball-and-stick models showing the shape of benzene.

for aromatic compounds. Finally, NMR spectroscopy provides no evidence for the existence of two equilibrating forms of benzene.

3. The heat of hydrogenation of cyclohexene to cyclohexane is -28.6 kcal/mol. The heat of hydrogenation of 1,3,5-cyclohexatriene would be expected to be about $3 \times -28.6 = -85.8$ kcal/mol. The actual heat of hydrogenation of benzene is only -49.8 kcal/mol, a difference of 36 kcal/mol. Therefore, benzene is much more stable (by 36 kcal/mol) than expected for 1,3,5-cyclohexatriene.

On the basis of all the evidence, it is clear that benzene has a symmetrical, planar, hexagonal structure with carbon-carbon bonds that are between double bonds and single bonds in character; it possesses a kind of unsaturation that is unreactive and that confers special stability. Benzene could have been represented like this:

Possible representations of benzene

However, chemists now have a clear understanding of the electronic structure of benzene, as we will see in the next subsection.

9.1.2 Electronic Structure of Benzene

Benzene as a hybrid of two contributing structures

A single-structure representation of benzene

Benzene is the classic example of a resonance-stabilized molecule whose actual structure is a hybrid of two equivalent contributing structures, the Kekulé structures of 1872. However, these structures are *not* in equilibrium as Kekulé proposed, but are contributing structures connected by the double-headed arrow indicating resonance. Benzene possesses a significant resonance stabilization energy of 36 kcal/mol.

A single structure often used to represent benzene is a hexagon with a circle inside it. In this book, a single contributing (Kekulé) structure will generally be used for benzene, but remember that it represents benzene as it actually is, a hybrid of both contributing structures. (The Kekulé structure is used because it makes it easier to keep track of the electrons when writing reaction mechanisms).

The orbital description of benzene involves formation of the carbon skeleton by the connection of six sp^2-hybridized carbons (the same hybridization found in alkenes) via sigma bonds (sp^2-sp^2 overlap). The carbon-hydrogen bonds are sigma bonds formed by sp^2-s overlap. Each carbon also has an unhybridized p orbital containing a single electron. Overlap of the parallel p orbitals from adjacent carbons provides a continuous pi electron cloud, containing six electrons, above and below the plane of the ring (Figure 9.2). The

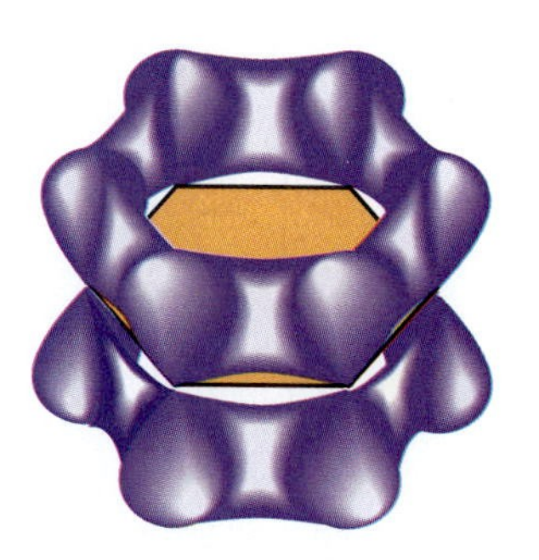

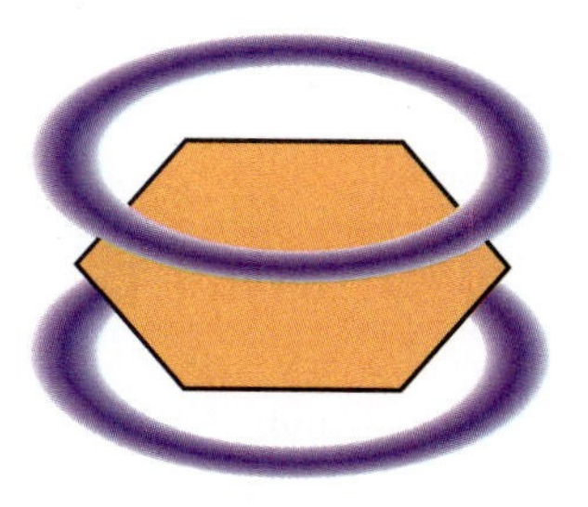

Pi cloud above and below hexagonal sigma bond framework of benzene

FIGURE 9.2

Orbital descriptions of benzene.

six electrons are referred to as the **aromatic sextet**. The delocalization of the six electrons over a molecular orbital, appearing as a "doughnut" of electron density above and below the plane of the ring, provides the unusual stabilization associated with **aromaticity,** the resonance stabilization of benzene.

PROBLEM 9.1

How do the Kekulé structures for benzene account for the fact that there are three different dibromobenzenes ($C_6H_4Br_2$)?

9.1.3 Other Monocyclic Aromatic Compounds

Based on molecular orbital calculations, the theoretical chemical physicist Erich Hückel enunciated the **Hückel rule,** which predicts that cyclic, planar compounds containing $4n + 2$ pi electrons in a continuous overlap system will be aromatic, provided n is a positive integer. In other words, *compounds with 2, 6, 10, 14, 18, . . . pi electrons should be aromatic.* This theory has withstood many tests. For benzene, $n = 1$ and the number of pi electrons is $(4 \times 1) + 2 = 6$, the so-called *aromatic sextet.* Cyclobutadiene (4 pi electrons, $n = \frac{1}{2}$) and cyclooctatetraene (8 pi electrons, $n = \frac{3}{2}$) both react as alkenes and are not aromatic, in accord with the theory. The compound with 10 pi electrons (10-annulene, $n = 2$) is nonplanar and for that reason cannot be aromatic. However, 14-annulene ($n = 3$) and 18-annulene ($n = 4$) are planar and are aromatic, as predicted.

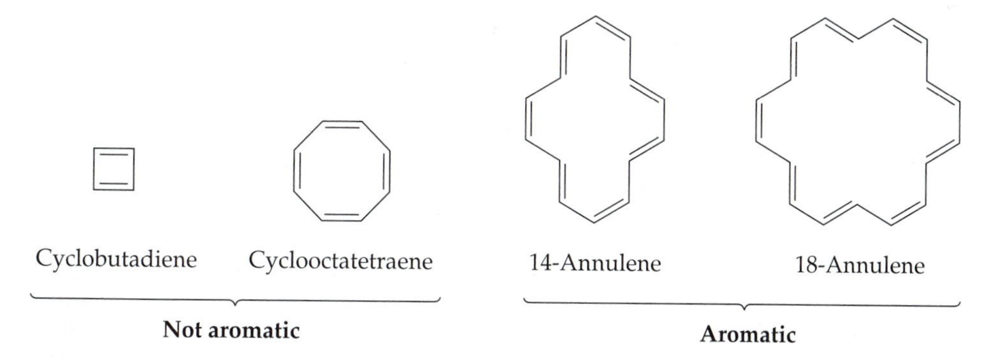

Interestingly, several carbon ions that fit the Hückel criteria are aromatic, making them unusually stable and therefore unusually easy to produce. (Recall that carbon normally tends not to be ionic.) These include the cyclopropenyl cation ($n = 0$), the

cyclopentadienide anion ($n = 1$), and the cycloheptatrienyl cation ($n = 1$); the latter two contain an aromatic sextet.

Cyclopropenyl cation

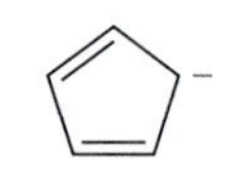

Cyclopentadienide anion

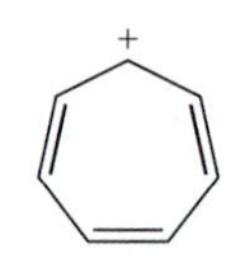

Cycloheptatrienyl cation

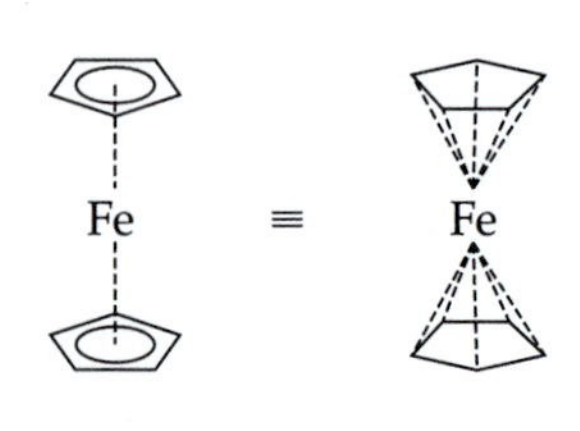

Ferrocene

An interesting family of aromatic organometallic compounds consists of cyclopentadienide anions complexed to metal atoms. The earliest discovered example was ferrocene (dicyclopentadienyl iron), but many similar compounds are now known.

Therefore, *an aromatic system is one in which there are 4n + 2 pi electrons delocalized over a continuous pi system on a cyclic planar skeleton*. As you will see later, the atoms involved in the cyclic system are not restricted to carbon; there are many heterocyclic aromatic compounds. We will also look at examples of polycyclic aromatics.

9.2 Nomenclature of Benzenoid Compounds

9.2.1 Monosubstituted Benzenes

Compounds in which another substituent (X) has replaced a single hydrogen of benzene are called *monosubstituted benzenes* (C_6H_5X). IUPAC nomenclature designates them as substituted benzenes—for example, *bromo*benzene and *nitro*benzene. However, several of the original historical names have been adopted by IUPAC and are routinely used. Shown here are the IUPAC names and the historical names (in parens), if different, for the most common monosubstituted benzenes. The name used more often is printed in bold.

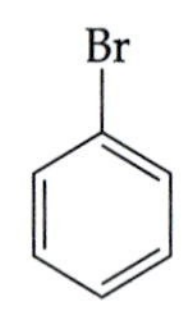

Bromobenzene

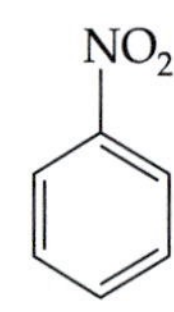

Nitrobenzene

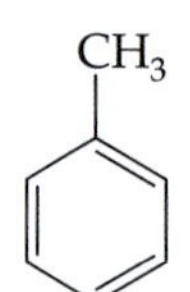

Methylbenzene **(toluene)**

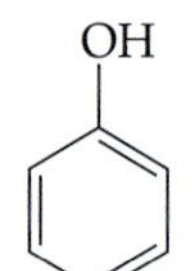

Hydroxybenzene **(phenol)**

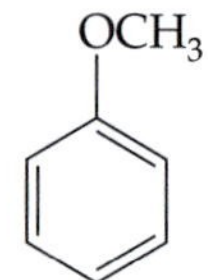
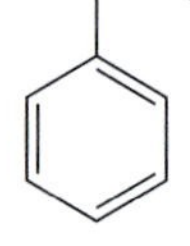

Methoxybenzene **(anisole)**

NH$_2$ — Aminobenzene **(aniline)**

COOH — Benzene-carboxylic acid **(benzoic acid)**

CHO — Benzene-carbaldehyde **(benzaldehyde)**

SO$_3$H — **Benzene-sulfonic acid**

CN — Cyanobenzene **(benzonitrile)**

Any benzene substituted by an aliphatic group may be named by simply naming the substituent followed by the word *benzene*—for example, vinylbenzene, isobutylbenzene, and cyclohexylbenzene:

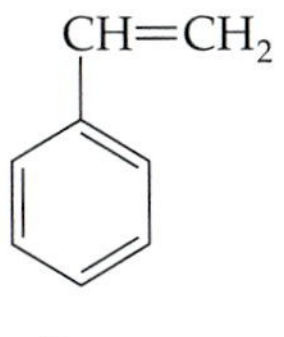

$CH{=}CH_2$

Vinylbenzene
or phenylethene
(styrene)

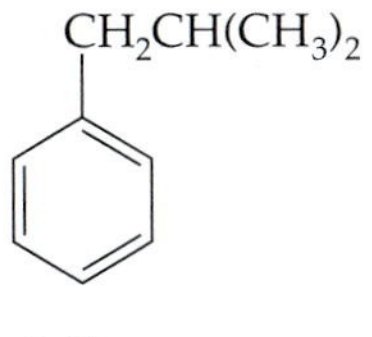

$CH_2CH(CH_3)_2$

Isobutylbenzene
(1-phenyl-2-methylpropane)

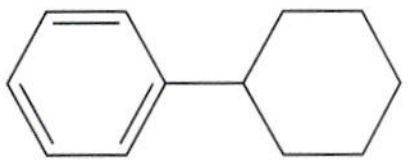

Cyclohexylbenzene
(phenylcyclohexane)

Frequently, an alkyl substituent attached to benzene is too complicated for a simple alkyl name to be used. In those instances, the IUPAC *alkane* parent name is used, and the benzene ring is considered a substituent, called a **phenyl group** (abbreviated C_6H_5— in condensed formulas). The **benzyl group** is also frequently encountered.

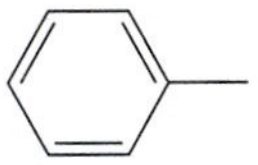

$-CH_2-$

Phenyl group

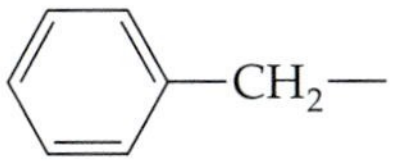

Benzyl group

The generalized designation for an aromatic group is Ar— (short for *aryl*), comparable to the R— used to represent an alkyl group. Thus, Ar—R represents an alkylaromatic compound.

PROBLEM 9.2

Draw the structures corresponding to the following names: (a) isopropylbenzene, (b) 2-phenyl-3-methylpentane, (c) 1-phenylethanol, (d) (*Z*)-2-phenyl-2-butene, (e) 1,4-diphenyl-1,3-butadiene, (f) allylbenzene, and (g) methoxybenzene (anisole).

PROBLEM 9.3

Provide names for the following structures:

(a) 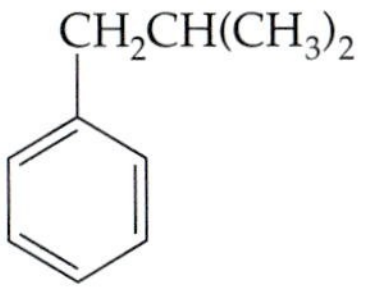 $CH_2CH(CH_3)_2$

(b)

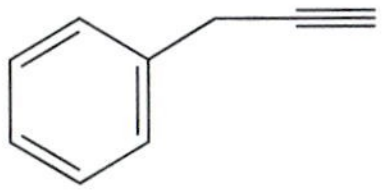

(c) $CH_3CH(C_6H_5)CH_2CH_3$

9.2.2 Disubstituted Benzenes

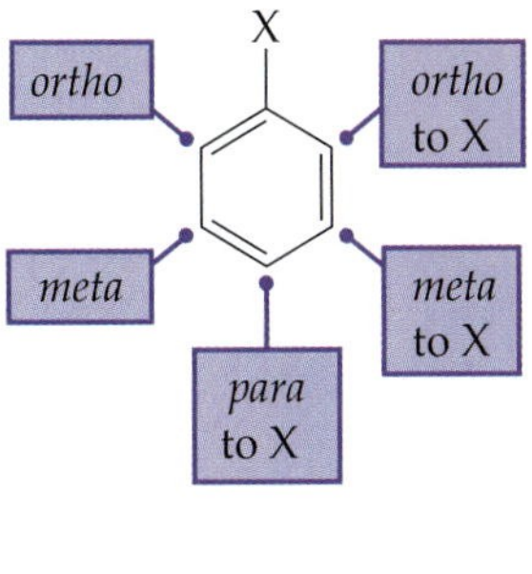

Two substituents on a benzene ring may occupy three different sets of relative positions. The original system for describing and naming disubstituted benzenes is still used today. The terms ***ortho, meta,*** and ***para*** (abbreviated *o-*, *m-*, and *p-*) refer to the positional relationship of the second substituent relative to the first. If the two substituents are identical, the prefix *di-* is used. If the two substituents are different, one is named as a substituent located on a monosubstituted benzene. Occasionally, this gives rise to two reasonable names, both of which may be found in indexes and chemical dictionaries.

X
ortho
ortho to X
meta
meta to X
para to X

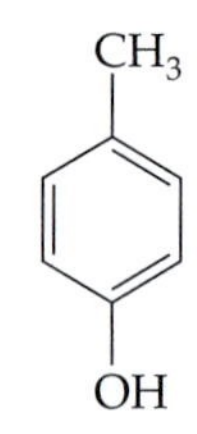

p-Hydroxytoluene (*p*-methylphenol) | *o*-Chloronitrobenzene (*o*-nitrochlorobenzene) | *m*-Iodobenzoic acid 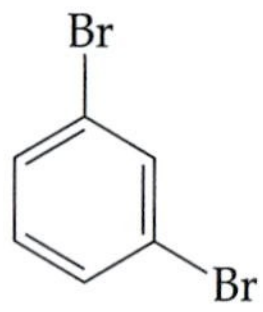 | *m*-Dibromobenzene

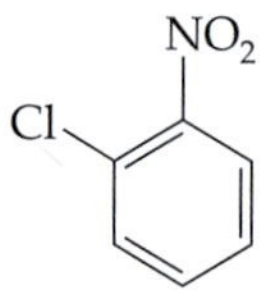

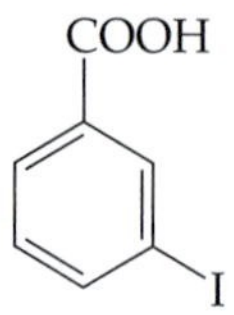

How to Solve a Problem

Draw the structures corresponding to the following names: (a) *m*-chlorotoluene, (b) *p*-fluorobenzaldehyde, and (c) *o*-toluenesulfonic acid.

SOLUTION

(a) The parent compound is toluene, which has a methyl group attached to the benzene ring. The name indicates that there is a chlorine atom attached to the ring at the position *meta* to the methyl group:

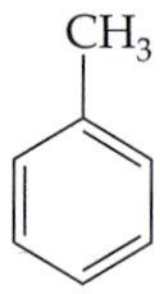

Toluene 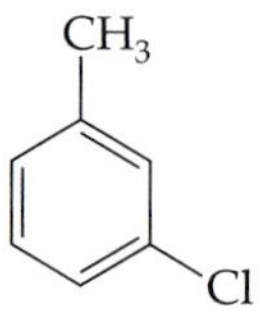 | *m*-Chlorotoluene

(b) Benzaldehyde has the —CHO group attached to a benzene ring. The name indicates that there is a fluorine atom at the *para* position:

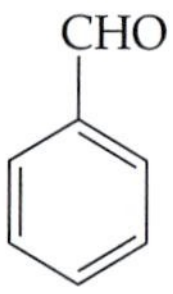

Benzaldehyde 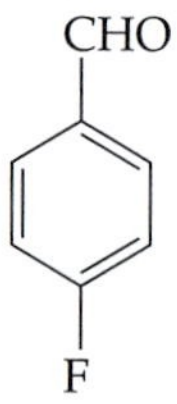 | *p*-Fluorobenzaldehyde

(c) The compound is named as a sulfonic acid, so it must contain the —SO_3H group. Toluene has a methyl group attached to a benzene ring, so the structure must have both a methyl group and a sulfonic acid group on the benzene ring, *ortho* to each other:

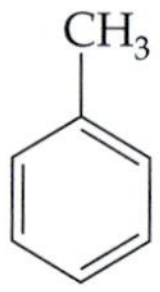

Toluene 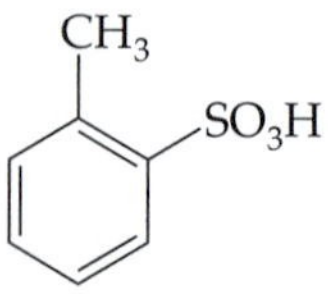 | *o*-Toluenesulfonic acid

PROBLEM 9.4

Draw the structure corresponding to each of the following names: (a) *p*-ethylaniline, (b) *o*-chlorotoluene, (c) *m*-fluorobenzoic acid, (d) 3-(*p*-bromophenyl)propyne, and (e) *m*-diethylbenzene.

PROBLEM 9.5

Provide names corresponding to the following structures:

(a)  Cl, Br (b) 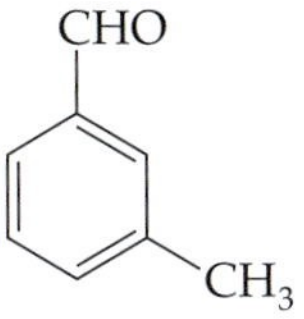 CHO, CH_3 (c) 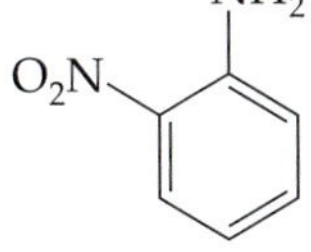 NH_2, O_2N (d) 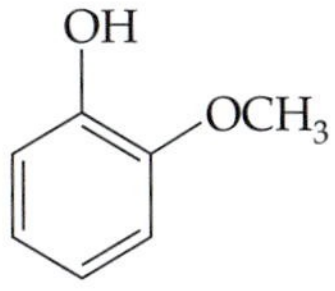 OH, OCH_3

9.2.3 Polysubstituted Benzenes

When more than two substituents are attached to a benzene ring, a numerical positional system is used. First, a parent name is chosen based on one of the substituents (X), which determines position 1:

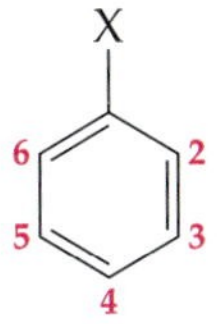

X, 2, 3, 4, 5, 6

Then other substituents are located using the numbers 2 through 6 and keeping the numbers as small as possible. Substituents are mentioned in the name in alphabetical order, not numerical order. For example, 4-ethyl-2-nitrobenzaldehyde is a correct name, whereas 2-nitro-4-ethylbenzaldehyde is not.

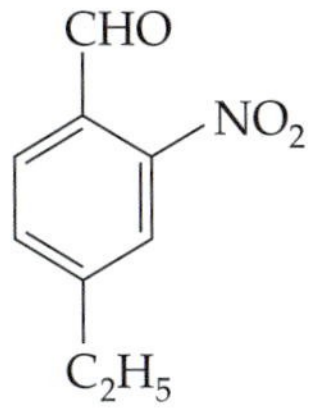

CHO, NO_2, C_2H_5
4-Ethyl-2-nitrobenzaldehyde

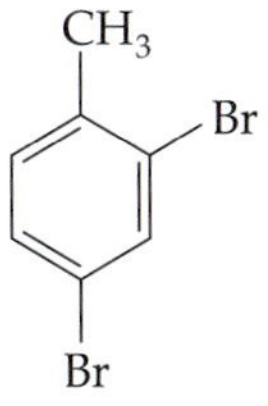

CH_3, Br, Br
2,4-Dibromotoluene

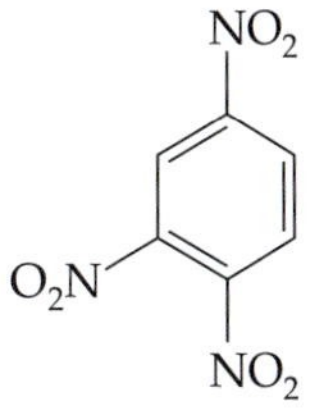

NO_2, O_2N, NO_2
1,2,4-Trinitrobenzene

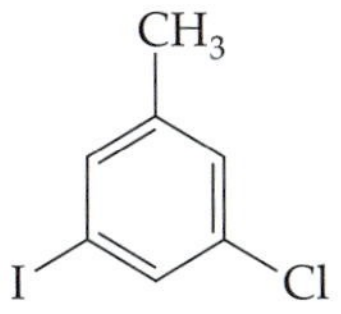

CH_3, I, Cl
3-Chloro-5-iodotoluene

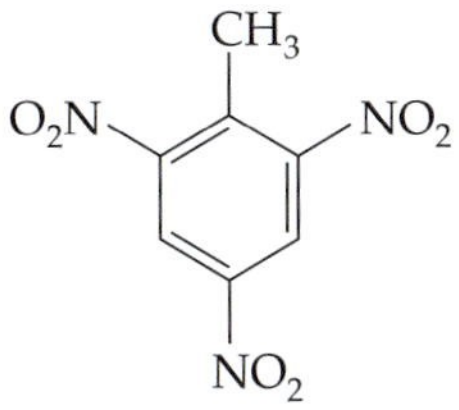

CH_3, O_2N, NO_2, NO_2
2,4,6-Trinitrotoluene (TNT)

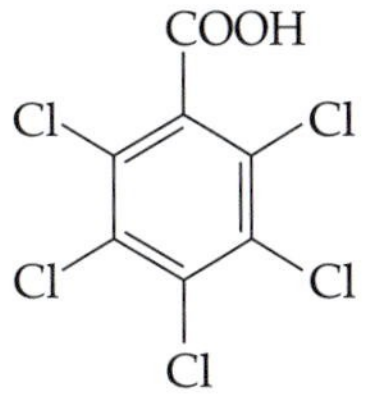

COOH, Cl, Cl, Cl, Cl, Cl
Pentachlorobenzoic acid

PROBLEM 9.6

Draw structures corresponding to the following names: (a) 3,4,5-trifluorotoluene, (b) 2,6-dimethylaniline, (c) 1,3,5-trinitrobenzene, and (d) 3,5-diethoxybenzoic acid.

PROBLEM 9.7

Provide names for the following structures:

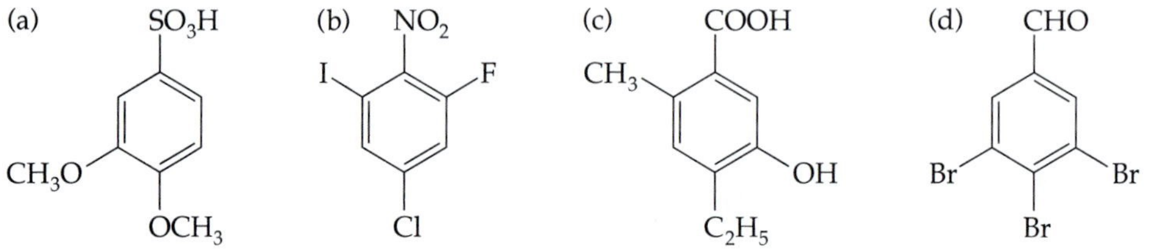

(a) SO_3H, CH_3O, OCH_3

(b) NO_2, I, F, Cl

(c) COOH, CH_3, OH, C_2H_5

(d) CHO, Br, Br, Br

9.3

Electrophilic Aromatic Substitution

The characteristic reaction of aromatic compounds is substitution by a wide variety of electrophilic reagents—**electrophilic aromatic substitution.** The overall effect is replacement of one or more hydrogens on the ring by a substituent while the aromatic system remains intact. Important substitution reactions that will be discussed in this section include placement of the following groups on the ring: Cl or Br (*halogenation*), NO_2 (*nitration*), SO_3H (*sulfonation*), R (*alkylation*), and R—C=O (*acylation*).

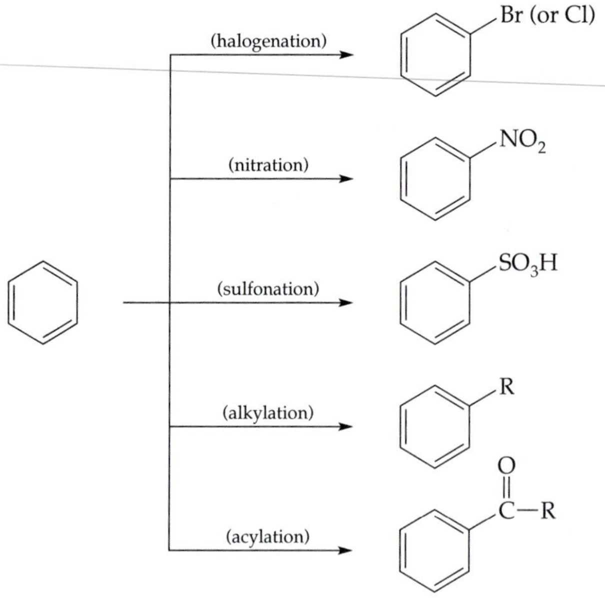

(halogenation) → Br (or Cl)

(nitration) → NO_2

(sulfonation) → SO_3H

(alkylation) → R

(acylation) → C(=O)—R

Although these reactions may appear very different, they all occur by the same fundamental mechanism, described next.

9.3.1 Mechanism of Electrophilic Aromatic Substitution

In Chapter 6, we considered electrophilic addition reactions of alkenes. In the first step, the pi electrons of the alkene attack an electrophile, such as bromine ($Br^{\delta+}$—$Br^{\delta-}$); in the second step, the anionic portion of the reagent adds to the carbocation. The overall result is cleavage of the pi bond and formation of two new sigma bonds.

Electrophilic addition:

C=C + $\overset{\delta+}{E}$—$\overset{\delta-}{A}$ ⟶ [+C—C—E] $\xrightarrow{:A^-}$ A—C—C—E

Alkene | Electrophilic reagent | Carbocation intermediate | Addition product

The initial step of a reaction of benzene with an electrophilic reagent is also the attachment of the electrophile to a carbon of the ring. In this step, two of the pi electrons become localized on the carbon, which then rehybridizes to sp^3. However, the second step is removal of a proton in order to reestablish the aromatic system of the benzene ring, rather than addition of an anion, which would destroy the system.

Electrophilic substitution:

Benzene (C–H) + $\overset{\delta+}{E}$—$\overset{\delta-}{A}$ (Electrophilic reagent)

$\xrightarrow{-A:^-}$ Substitution product (C–E) + H—A

$\xrightarrow{+A:^-}$ (✕) Addition product—*not formed* (H, E, A, H)

Although these two reactions have a similar first step, the reason for the different outcomes has to do with energetics. In the addition reaction with an alkene, one single and one double bond are broken, but two new single bonds are formed, releasing more energy than is consumed in the bond breakings. However, the substitution reaction with benzene occurs because of the release of the resonance stabilization energy when the aromatic ring is reestablished. If addition occurred, the resulting cyclohexadiene would be of much higher energy than the benzenoid substitution product.

Electrophilic aromatic substitution is a two-step process, with the first step being the slower step (that is, its rate determines the rate of the overall reaction). The mechanism is the same for all electrophilic reagents and is generalized in Figure 9.3 for benzene.

Benzene + E^+ (Electrophile)

Step 1

In the first step, the pi electrons of the ring attack the electrophile, creating a carbocation. This arenium cation is stabilized by conjugation with the extended pi system of the cyclohexadiene ring. Three contributing structures can be drawn.

Contributing structures for resonance-stabilized arenium cation

+ $:A^-$

Step 2

In the second step, the anion ($:A^-$) from the original reagent (E—A) removes a proton from the arenium cation, releasing the C—H bonding electrons back to the ring to re-establish the aromatic pi system, and in the process produce the substitution product.

Substituted benzene + H—A

FIGURE 9.3 Mechanism of electrophilic aromatic substitution.

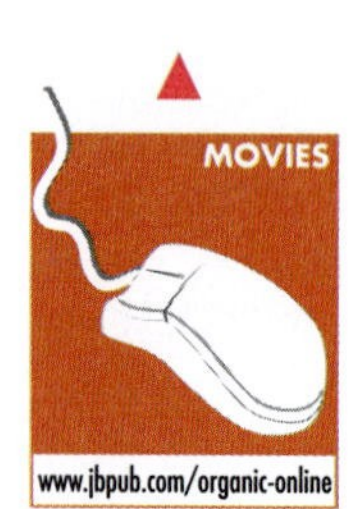

All of the electrophilic aromatic substitution reactions presented in this section proceed via the same general mechanism. The only significant difference is the nature of the electrophile (E^+):

- The electrophile may be a stand-alone cation—for example, $^+NO_2$.
- The electrophile may be part of a complex with an electron-deficient site—for example, $R^{\delta+}$----Cl----$^{\delta-}AlCl_3$.

Both kinds of electrophiles are susceptible to attack by the electron-rich and nucleophilic benzene ring.

The free-energy diagram for electrophilic aromatic substitution is shown in Figure 9.4. The energy of activation for forming the transition state in the first step is large because the resonance stabilization energy of benzene must be overcome. As we will see in Section 9.4, any substituents already on the benzene ring that stabilize the intermediate arenium cation will also stabilize the transition state for its formation. This lowers the energy of activation for the cation's formation and increases the relative rate for the reaction.

This general mechanism for electrophilic aromatic substitution is followed in halogenation, nitration, sulfonation, alkylation, and acylation (described in the following

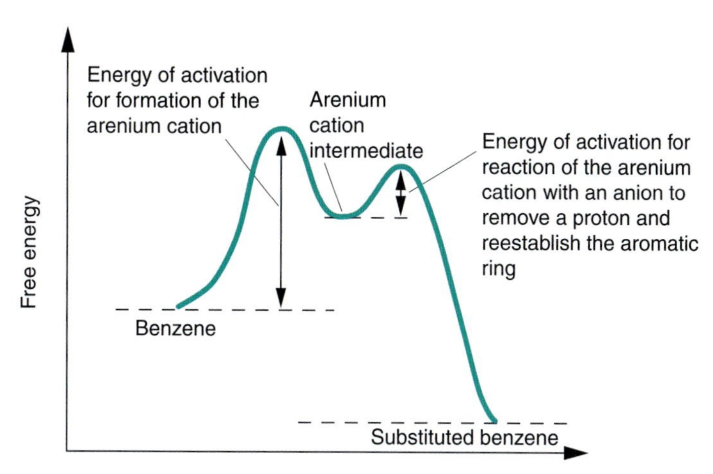

FIGURE 9.4

Free-energy diagram for an electrophilic aromatic substitution.

subsections). The only difference in those reactions is the means by which the electrophile is produced. Once the electrophile is available, the mechanism shown in Figure 9.3 proceeds for all of these reactions.

9.3.2 Halogenation

The halogenation of benzene is useful for attaching chlorine and bromine to an aromatic ring. However, fluorination and iodination are ineffective in attaching fluorine or iodine. Section 12.5.2 will describe how iodine or fluorine can be attached to an aromatic ring via replacement of an amino group (—NH_2).

Halogenation requires a Lewis-acid catalyst to form the electrophile. The reaction of chlorine (a Lewis base), with a Lewis acid, usually $FeCl_3$ or $AlCl_3$, produces a complex containing an electron-deficient chlorine. This electrophile is subject to attack by the nucleophilic aromatic ring.

$$:\ddot{Cl}-\ddot{Cl}: + FeCl_3 \longrightarrow \overset{\delta+}{:\ddot{Cl}}\text{---}\ddot{Cl}\text{---}\overset{\delta-}{FeCl_3}$$

Chlorine-$FeCl_3$ complex

$$\text{C}_6\text{H}_6 + \overset{\delta+}{:\ddot{Cl}}\text{---}\ddot{Cl}\text{---}\overset{\delta-}{FeCl_3} \xrightarrow{-FeCl_4^-} [\text{C}_6\text{H}_6\text{Cl}]^+$$

Arenium cation

The intermediate arenium cation gives up a proton to $FeCl_4^-$, regenerating $FeCl_3$, releasing HCl, and re-forming the aromatic system of benzene, which now carries a chlorine substituent:

$$[\text{C}_6\text{H}_6\text{Cl}]^+ + :\ddot{Cl}-FeCl_3^- \xrightarrow[-FeCl_3]{-HCl} \text{C}_6\text{H}_5\text{Cl}$$

Chlorobenzene

CHEMISTRY AT WORK

IN THE ENVIRONMENT

Biologically Active Chloroaromatics

As noted in Section 3.6, chloroalkanes are highly stable and resist degradation in the environment. This same characteristic is associated with a wide variety of commercially important chloroaromatic chemicals.

One of the earliest insecticides was *p*-dichlorobenzene, the active ingredient in some mothballs. It was prepared by the dichlorination of benzene, using two equivalents of chlorine:

$$\text{Benzene} \xrightarrow[\text{FeCl}_3]{2\ \text{Cl}_2} p\text{-Dichlorobenzene}$$

Mothballs were added to storage chests and closets to kill moths that eat woolen garments, especially winter clothes held in storage over the summer.

DDT (dichlorodiphenyltrichloroethane, not a correct IUPAC name) was first used as an insecticide in 1939, and was extremely successful in ridding large areas of the world of the insects that carried the killer diseases malaria and typhus. DDT was prepared by first chlorinating benzene, then reacting it with trichloroacetaldehyde:

$$\text{Benzene} \xrightarrow[\text{AlCl}_3]{\text{Cl}_2} \text{Chlorobenzene} \xrightarrow[\text{H}_2\text{SO}_4]{\text{Cl}_3\text{CCHO}} \text{Cl}-\text{C}_6\text{H}_4-\underset{\text{CCl}_3}{\text{CH}}-\text{C}_6\text{H}_4-\text{Cl}$$

DDT
(dichlorodiphenyltrichloroethane)

Chemical insecticides and herbicides, sprayed from the air, prevent crops from being destroyed by insects and infested with weeds.

An area does not have to be frequently resprayed with DDT since it is a persistent insecticide. It *resists* degradation, but it does degrade in nature, slowly and in what turns out to be a deadly way. DDT degrades by dehydrochlorination (elimination of HCl), forming the compound DDE [1,1-dichloro-2,2-di(*p*-chlorophenyl)ethene, $(p\text{-ClC}_6\text{H}_4)_2\text{C}{=}\text{CCl}_2$]. It was discovered that when DDE is ingested by birds, it inhibits the enzyme that facilitates the supply of calcium for eggshell formation. As a result, the birds' eggs are too fragile to survive incubation. This led to a marked decline in the population of some birds, especially raptors such as hawks and eagles, which are higher in the food chain. Because of these detrimental effects, use of DDT in the United States was banned in 1970.

2,4-Dichlorophenoxyacetic acid (2,4-D) is one of the best-known herbicides, especially for broad-leaf plants such as dandelions. It is prepared by an S_N2 reaction of the sodium salt of 2,4-dichlorophenol with the sodium salt of chloroacetic acid.

During the Vietnam War, 2,4-D and the related 2,4,5-T (2,4,5-trichlorophenoxyacetic acid) were combined and used as the large-scale defoliant called Agent Orange. It was subsequently discovered that trace amounts of 2,3,7,8-tetrachlorodioxin, one member of a family of compounds generically referred to as *dioxins*, were formed in the 2,4,5-T

Preparation of 2,4–D:

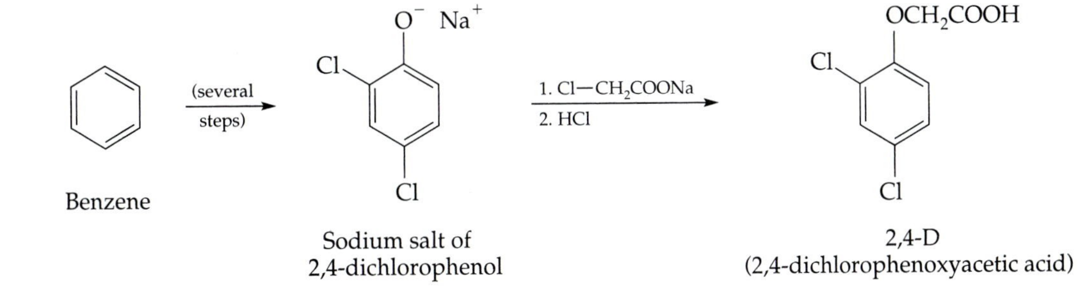

Benzene → (several steps) → Sodium salt of 2,4-dichlorophenol → 1. Cl—CH_2COONa 2. HCl → 2,4-D (2,4-dichlorophenoxyacetic acid)

Preparation of 2,4,5–T and dioxins:

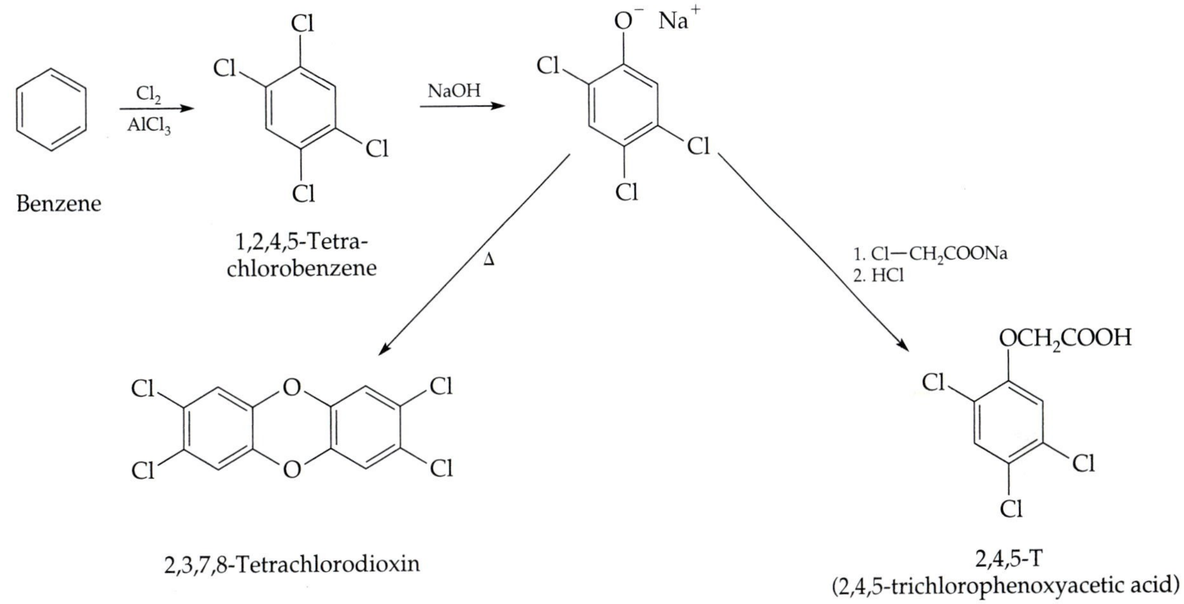

Benzene → Cl_2 / $AlCl_3$ → 1,2,4,5-Tetrachlorobenzene → NaOH → (sodium 2,4,5-trichlorophenoxide)

→ Δ → 2,3,7,8-Tetrachlorodioxin

→ 1. Cl—CH_2COONa 2. HCl → 2,4,5-T (2,4,5-trichlorophenoxyacetic acid)

preparation reaction, especially if the temperature of the reaction was not carefully controlled. Dioxins are also extremely persistent chemicals and have been found as trace by-products from several industrial processes involving chloroaromatics. Dioxins are toxic to humans and have been implicated in a number of serious medical problems.

Polychlorinated biphenyls (PCBs) are a mixture of various chloroaromatics obtained by chlorination of biphenyl (also called diphenyl). PCBs, too, are very stable and persistent compounds, and their release into the environment has led to their accumulation in the food chain. PCBs are very toxic compounds, leading to a ban on their production and use. One of their major uses was as a coolant inside electrical transformers because they are inert to degradation during exposure to electrical currents, continued heating, and confinement. However, even this use has been banned, and all such PCBs are being removed.

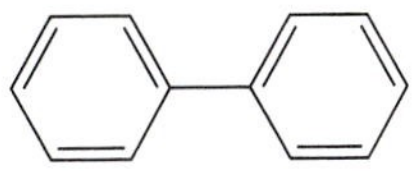

Biphenyl

By the same process, bromine and $FeBr_3$ bring about bromination of benzene:

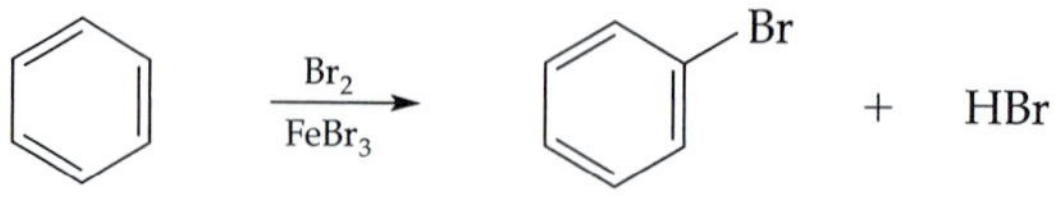

$$\text{Benzene} \xrightarrow[\text{FeBr}_3]{\text{Br}_2} \text{Bromobenzene} + \text{HBr}$$

Bromobenzene

It may be helpful to summarize all of the halogenation reactions studied to this point:

- Halogenation of alkenes results in ionic electrophilic addition.
- Halogenation of aromatics results in ionic electrophilic substitution.

9.3.3 Nitration

The nitro group ($—NO_2$) can be attached to a benzene ring through a process called **nitration,** involving treatment with concentrated nitric and sulfuric acids. The nitration of an aromatic ring is a very important reaction, especially because the nitro group is readily converted into the amino group ($—NH_2$), which is readily converted into a variety of other substituents (see Section 12.5.2). There is no other generally practical means of attaching an amino group to an aromatic ring. An example of the use of nitration to initiate such a series of conversions is as follows:

$$\text{Benzene} \xrightarrow[\text{H}_2\text{SO}_4]{\text{HNO}_3} \text{Nitrobenzene} \dashrightarrow \text{Aniline} \dashrightarrow \text{Iodobenzene}$$

Benzene Nitrobenzene Aniline Iodobenzene

The electrophile essential for nitration is the **nitronium ion** ($^+NO_2$), which is generated from nitric acid by protonation with a stronger acid, usually sulfuric acid:

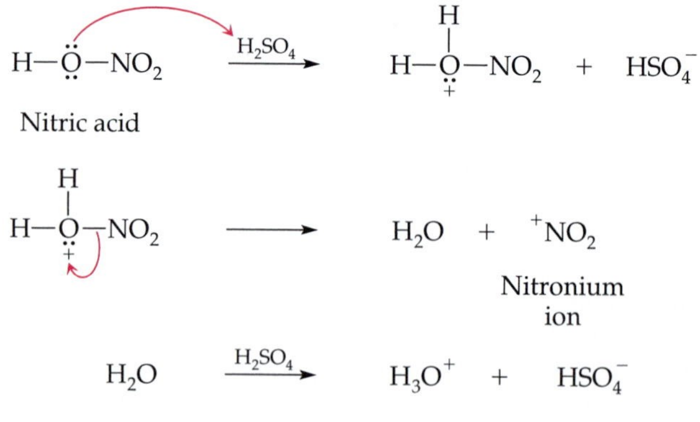

$$\text{H—}\ddot{\text{O}}\text{—NO}_2 \xrightarrow{\text{H}_2\text{SO}_4} \text{H—}\overset{\text{H}}{\overset{|}{\text{O}}}{}^{+}\text{—NO}_2 + \text{HSO}_4^-$$

Nitric acid

$$\text{H—}\overset{\text{H}}{\overset{|}{\text{O}}}{}^{+}\text{—NO}_2 \longrightarrow \text{H}_2\text{O} + {}^+\text{NO}_2$$

Nitronium ion

$$\text{H}_2\text{O} \xrightarrow{\text{H}_2\text{SO}_4} \text{H}_3\text{O}^+ + \text{HSO}_4^-$$

Overall reaction: $HNO_3 + 2\,H_2SO_4 \longrightarrow {}^+NO_2 + 2\,HSO_4^- + H_3O^+$

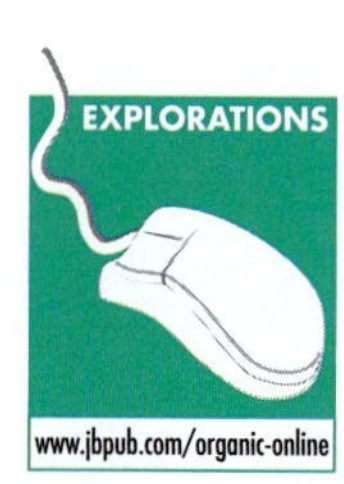

9.3.4 Sulfonation

Aromatic sulfonic acids are related to sulfuric acid ($HO—SO_2—OH$): A single —OH group of sulfuric acid is replaced by an aromatic ring. They are strong acids that have

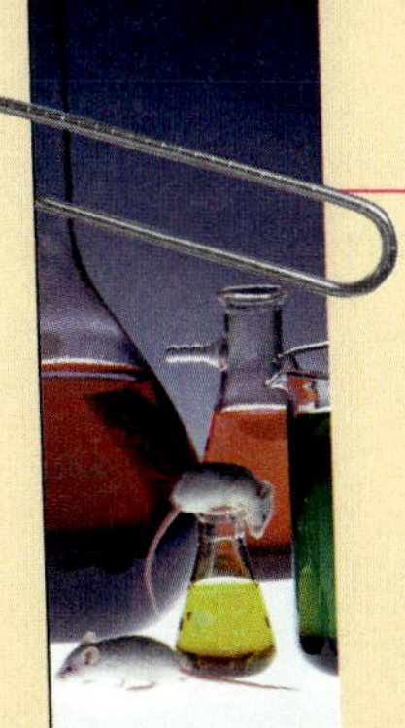

CHEMISTRY AT WORK

IN THE MARKETPLACE

Detergents

Soaps have been used for centuries—for cleaning clothes, skin, dishes, and just about anything you can think of. A soap consists of a long-chain hydrocarbon with a highly polar sodium salt of a carboxylic acid at one end ($R—COO^- Na^+$). Soaps are one kind of **surfactant** (surface-active agent) and will be described in detail in Chapter 14. Another kind of surfactant, the modern synthetic detergent, is often prepared via Friedel-Crafts alkylation and aromatic sulfonation.

All surfactants consist of a long hydrocarbon chain with a highly polar ion (usually an anion) at one end. This structure gives the surfactant molecule a hydrophilic ("water-liking") end (the anion) and a lipophilic ("fat-liking") chain. Generally, dirt has an oily (hydrocarbon) surface that keeps it from dissolving in water (recall that "like dissolves like"). The surfactant molecules act as intermediaries, orienting themselves across the water-oil interface with their polar (ionic) ends dissolved in the water and their nonpolar hydrocarbon chains dissolved in the oily dirt. By this means, dirt droplets are suspended in aqueous solution as an emulsion and can be washed away with water.

Detergents, as cleansers, are an invention of the chemical industry that successfully solve two problems that soaps present. These compounds are neutral in aqueous solution and do not hydrolyze to create irritating basic or acidic solutions. Also, their metal salts formed in hard water do not precipitate, but remain in solution. Although several different synthetic detergents (*syndets*) have been developed, many are alkylbenzenesulfonates ($R—C_6H_4—SO_3^- Na^+$).

Alkylbenzenesulfonates are prepared by a process that starts with a Friedel-Crafts alkylation of benzene using a C_{10}–C_{14} terminal alkene and acid. The resulting alkylbenzene is reacted with fuming sulfuric acid, producing an alkylbenzenesulfonic acid. When this acid is neutralized with sodium hydroxide, the sodium salt of an alkylbenzene sulfonate—the syndet—results.

Detergents purchased in the store contain many chemicals in addition to the active ingredient (the surfactant). Phosphates, and more recently silicates, are added as "builders" to complex the hard water cations. Other additives—including fabric softeners, enzymes to degrade protein stains, fragrances, antistatic agents, and brighteners—are also included to make the detergent product more effective and appealing to the consumer.

Benzene + $R—CH{=}CH_2$ (Terminal alkene) $\xrightarrow[\text{(Friedel-Crafts alkylation)}]{HF}$ $R—CH(CH_3)—C_6H_5$ (Alkylbenzene)

Alkylbenzene $\xrightarrow{H_2SO_4/SO_3}$ $R—CH(CH_3)—C_6H_4—SO_3H$ (Alkylbenzenesulfonic acid)

Alkylbenzenesulfonic acid $\xrightarrow{NaOH}$ $R—CH(CH_3)—C_6H_4—SO_3^-\ Na^+$

Sodium alkylbenzenesulfonate (a syndet)

many uses. The sulfonic acid group ($—SO_3H$) is attached to a benzene ring by direct **sulfonation** using fuming sulfuric aid (H_2SO_4 with SO_3 added):

Benzene $\xrightarrow[H_2SO_4]{SO_3}$ Benzenesulfonic acid

The electrophile in the sulfonation reaction is sulfur trioxide (SO_3, a Lewis acid), which is attacked by the nucleophilic aromatic ring. The proton of the arenium ion is removed by any of a number of bases in the solution (the $—SO_3^-$ group in the arenium ion itself, SO_3, or HSO_4^-).

Benzene + Sulfur trioxide → arenium ion; Anion removes H → $C_6H_5SO_3^-$ $\xrightarrow{H_2SO_4}$ Benzenesulfonic acid

9.3.5 Friedel-Crafts Alkylation

The reaction that attaches an alkyl group to an aromatic ring is known as the **Friedel-Crafts alkylation,** named after Charles Friedel and James Mason Crafts, the French and American chemists responsible for its discovery in 1877. The electrophile in this reaction must be a carbocation, and it is formed from an alkyl halide and aluminum chloride (a Lewis acid):

$$R—\ddot{Cl}: + AlCl_3 \longrightarrow \left[\overset{\delta+}{R}\text{----}\ddot{Cl}\text{----}\overset{\delta-}{AlCl_3}\right] \longrightarrow [R^+]\ AlCl_4^-$$

Alkyl chloride — Molecular complex — Carbocation

$$\text{Benzene} + \left[\overset{\delta+}{R}\text{----}\ddot{Cl}\text{----}\overset{\delta-}{AlCl_3}\right] \text{ or } [R^+]\ AlCl_4^- \xrightarrow[-HCl]{-AlCl_3} \text{Alkylbenzene}$$

The electrophilic species may not be a free carbocation, but instead a complex of a carbocation with tetrachloroaluminate ($AlCl_4^-$). However, a carbocation to alkylate benzene can also be produced by two other reactions already studied—the protonation of an alkene and the protonation of an alcohol:

$$R{-}CH{=}CH_2 \text{ (Alkene)} \xrightarrow{H^+} \left[R{-}\overset{+}{C}H{-}CH_3\right] \text{ (Carbocation)} \xrightarrow{C_6H_6} C_6H_5{-}CH(R){-}CH_3 \text{ (Alkylbenzene)}$$

$$R{-}CH(OH){-}CH_3 \text{ (Alcohol)} \xrightarrow[-H_2O]{H^+} \left[R{-}\overset{+}{C}H{-}CH_3\right]$$

Specific examples of these three alkylation reactions, all of which are called *cyclohexylation*, are the alkylation of benzene with cyclohexene and hydrofluoric acid, with cyclohexanol and phosphoric acid, and with chlorocyclohexane and aluminum chloride:

Cyclohexene —HF→ / Cyclohexanol —H_3PO_4→ / Chlorocyclohexane —$AlCl_3$→ [Cyclohexyl carbocation] —C_6H_6→ Cyclohexylbenzene (phenylcyclohexane)

Although the alkylation reaction as presented here is clear-cut, serious problems are often encountered under many circumstances. These problems are so significant that if the intention is to attach a primary alkyl group to an aromatic ring, it is best to avoid the Friedel-Crafts alkylation and use the alternative approach outlined in the next section. Tertiary carbocations are usually effective in Friedel-Crafts alkylations.

PROBLEM 9.8

Styrene ($C_6H_5CH{=}CH_2$) is an important industrial chemical for the production of the plastic polystyrene. It is made commercially by dehydrogenating ethylbenzene over a special catalyst. Propose three different syntheses of ethylbenzene, starting with benzene in each case.

9.3.6 Friedel-Crafts Acylation

Friedel and Crafts used the same concept as for their alkylation reaction to attach acyl groups ($R{-}C{=}O$) to aromatic rings. This reaction is called a **Friedel-Crafts acylation.** Reaction of benzene with an acyl chloride in the presence of aluminum chloride as catalyst yields an acylbenzene. As an example of Friedel-Crafts acylation, reaction of acetyl chloride with benzene in the presence of aluminum chloride produces acetophenone.

General reaction:

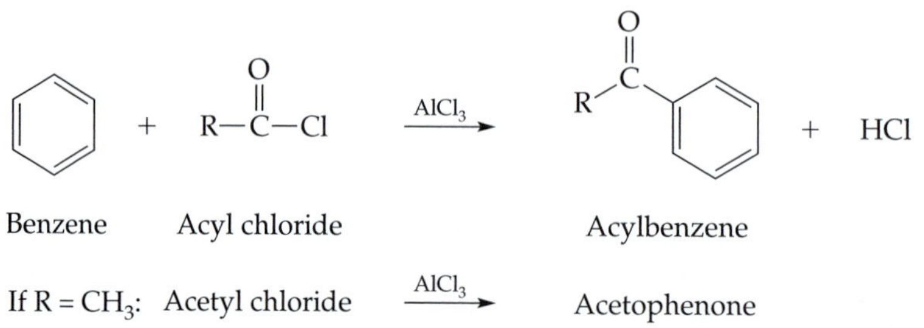

Benzene + Acyl chloride $\xrightarrow{AlCl_3}$ Acylbenzene + HCl

$$\text{Benzene} + R{-}\overset{O}{\overset{\|}{C}}{-}Cl \xrightarrow{AlCl_3} \text{Acylbenzene} + HCl$$

If R = CH_3: Acetyl chloride $\xrightarrow{AlCl_3}$ Acetophenone

This *acylation reaction* does not have the limitations of the alkylation reaction. It is especially useful for two major reasons: (1) The acyl group can be converted into a wide variety of other functional groups (as we will see in Chapter 13), and (2) the acyl group can be converted into the corresponding primary alkyl group, eliminating the need to use the Friedel-Crafts alkylation to attach such substituents.

The mechanism of acylation is similar to that of alkylation. Aluminum chloride complexes with the chlorine of the acyl chloride, weakening the carbon-chlorine bond. The complex, having an electron-deficient carbon, can dissociate to form an **acylium cation.** The acylium cation and the acylium complex are both electrophiles, susceptible to attack by the pi electrons of the benzene ring to form an arenium cation intermediate.

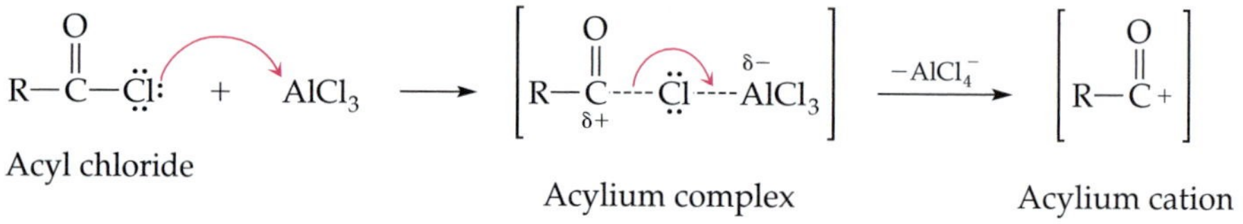

$$R{-}\overset{O}{\overset{\|}{C}}{-}\ddot{Cl}{:} + AlCl_3 \longrightarrow \left[R{-}\underset{\delta+}{\overset{O}{\overset{\|}{C}}}{\cdots}\ddot{Cl}{\cdots}\overset{\delta-}{AlCl_3}\right] \xrightarrow{-AlCl_4^-} \left[R{-}\overset{O}{\overset{\|}{C}}{+}\right]$$

Acyl chloride — Acylium complex — Acylium cation

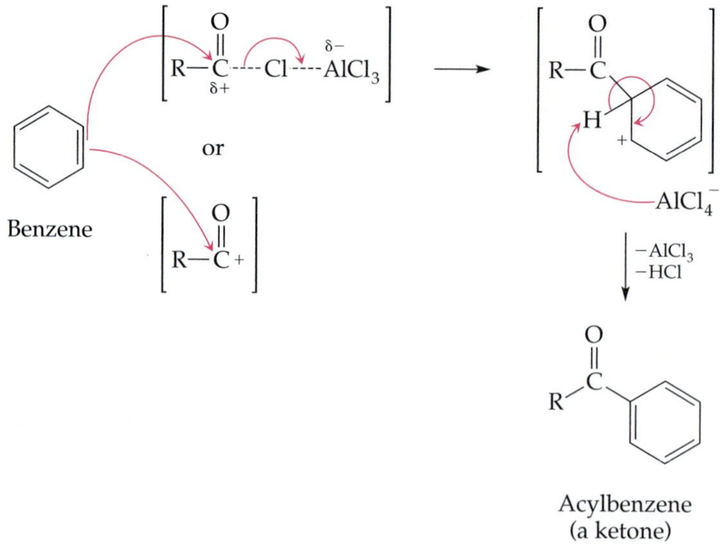

$$\text{Benzene} + \left[R{-}\underset{\delta+}{\overset{O}{\overset{\|}{C}}}{\cdots}Cl{\cdots}\overset{\delta-}{AlCl_3}\right] \text{ or } \left[R{-}\overset{O}{\overset{\|}{C}}{+}\right] \longrightarrow [\text{arenium cation with } H \text{ and } R{-}C{=}O] \xrightarrow[-HCl]{-AlCl_3 \; (AlCl_4^-)} \text{Acylbenzene}$$

Benzene

or

Acylbenzene
(a ketone)

Acylbenzenes can be reduced to primary alkylbenzenes via the **Clemmensen reduction,** a long-known reaction using zinc amalgam (Zn/Hg) and concentrated hydrochloric acid:

Clemmensen reduction:

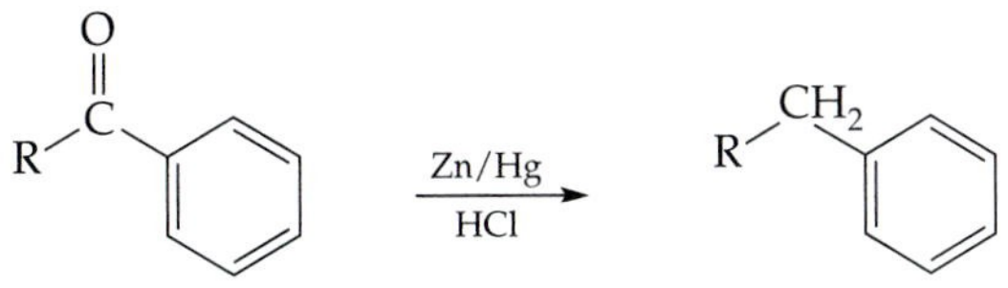

$$R{-}C(=O){-}C_6H_5 \xrightarrow[HCl]{Zn/Hg} R{-}CH_2{-}C_6H_5$$

Acylbenzene → Alkylbenzene

In this reaction, the carbonyl group is reduced to a methylene group. This overall route from benzene to an acylbenzene to a primary alkylbenzene is the best route to a primary alkylbenzene, better than Friedel-Crafts alkylation of benzene. For example, conversion of acetophenone to ethylbenzene under these conditions is the best method for the laboratory preparation of ethylbenzene, since direct Friedel-Crafts alkylation of benzene with ethyl chloride produces some ethylbenzene but also a considerable amount of *p*-diethylbenzene, which then must be separated.

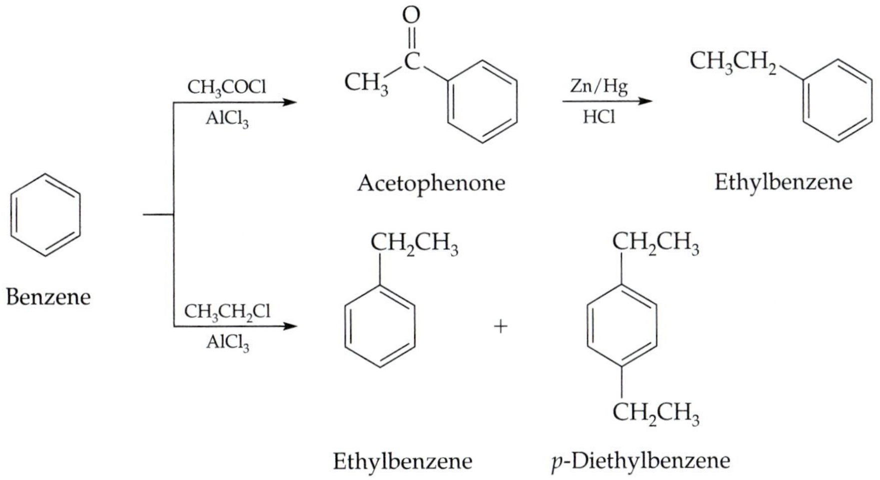

$$C_6H_6 \xrightarrow[AlCl_3]{CH_3COCl} CH_3C(=O){-}C_6H_5 \xrightarrow[HCl]{Zn/Hg} CH_3CH_2{-}C_6H_5$$

Benzene → Acetophenone → Ethylbenzene

$$C_6H_6 \xrightarrow[AlCl_3]{CH_3CH_2Cl} C_6H_5CH_2CH_3 + CH_3CH_2{-}C_6H_4{-}CH_2CH_3$$

Benzene → Ethylbenzene + *p*-Diethylbenzene

How to Solve a Problem

Show the reactions necessary to convert benzene into 1-phenylbutane (butylbenzene).

$$C_6H_6 \dashrightarrow C_6H_5{-}CH_2CH_2CH_2CH_3 \equiv C_6H_5{-}C_4H_9$$

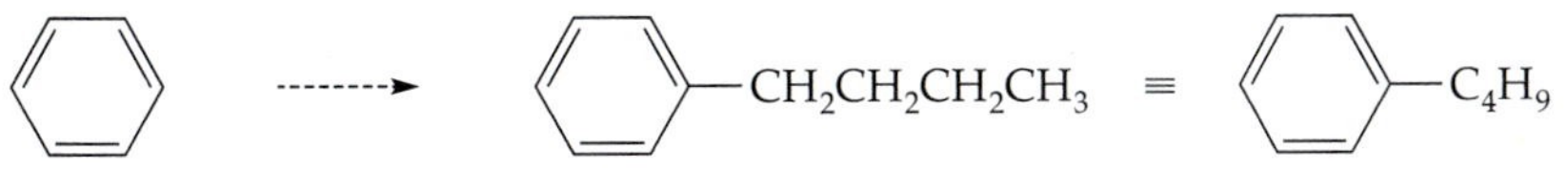

Benzene → 1-Phenylbutane (butylbenzene)

PROBLEM ANALYSIS

It is usually best to assume that the critical step in a synthesis is to establish the necessary carbon-carbon bonds, building the carbon skeleton of the desired product. Therefore, the key reaction in this problem is the attachment of a C_4 group to the benzene ring. Forming such C—C bonds can only be done using the Friedel-Crafts approach. Since the alkylation reaction has limitations, we consider first the acylation reaction using a C_4 acyl halide. The resulting acyl substituent can be converted to an alkyl group using the Clemmensen reduction.

Once again, we use the mental process of retrosynthesis, thinking backward from the desired product to the necessary starting material:

$$C_6H_5-CH_2CH_2CH_2CH_3 \Longrightarrow C_6H_5-\overset{O}{\overset{\|}{C}}-CH_2CH_2CH_3 \Longrightarrow C_6H_6$$

1-Phenylbutane Butanoylbenzene Benzene

SOLUTION

Reaction of butanoyl chloride with benzene in the presence of aluminum chloride produces butanoylbenzene, which is reduced with zinc amalgam and HCl to produce 1-phenylbutane (butylbenzene):

$$C_6H_6 \xrightarrow[AlCl_3]{CH_3CH_2CH_2COCl} C_6H_5-\overset{O}{\overset{\|}{C}}-C_3H_7 \xrightarrow[HCl]{Zn/Hg} C_6H_5-C_4H_9$$

Benzene Butanoylbenzene 1-Phenylbutane

PROBLEM 9.9

Show the steps and reagents you would use to prepare propylbenzene from benzene by two different routes.

9.4

Substituent Effects and Their Use in Synthesis

Section 9.3 described reactions by which different substituents are attached to a benzene ring. These same reactions may be used to place additional substituents on a benzene ring already carrying a substituent. Benzene is sufficiently reactive to undergo all of the electrophilic aromatic substitution reactions described earlier, and all six carbons of benzene are equivalent in terms of the point of attack of the electrophile. However, both of these conditions change when an electrophilic aromatic substitution reaction is conducted on a benzenoid ring already carrying one or more substituents.

The reactivity of a benzenoid ring is affected by a preexisting substituent. Figure 9.5 shows the relative rates of nitration of toluene (about 25 times as fast as benzene) and nitrobenzene (about 10,000 times slower than benzene). The rate of substitution on benzene is the reference rate. Any substituent that causes the rate of electrophilic aromatic substitution on a benzenoid ring to be higher than for benzene, like the methyl group of toluene, is called an **activating group.** Any substituent that causes the rate of electrophilic aromatic substitution on a benzenoid ring to be lower than for benzene, like the nitro group of nitrobenzene, is called a **deactivating group.**

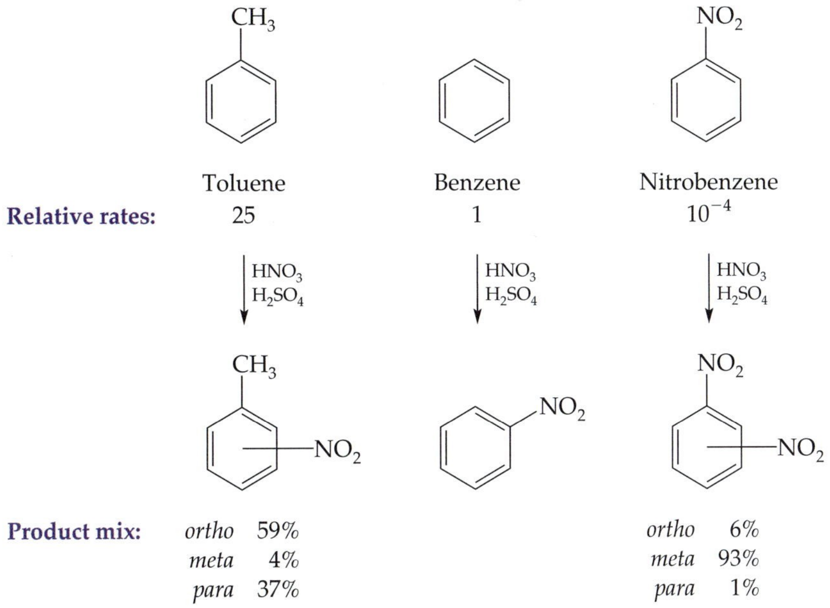

	Toluene	Benzene	Nitrobenzene
Relative rates:	25	1	10^{-4}

HNO_3, H_2SO_4

	Toluene	Nitrobenzene
Product mix:	*ortho* 59%	*ortho* 6%
	meta 4%	*meta* 93%
	para 37%	*para* 1%

FIGURE 9.5

Comparison of activating or deactivating and directing effects during the nitration of toluene, benzene, and nitrobenzene.

Figure 9.5 also indicates the differences in the reactivity of different positions at which the electrophilic aromatic substitution reaction can occur. Toluene is nitrated mainly at the *ortho* and *para* positions, with almost no substitution occurring at the *meta* position. By contrast, nitrobenzene is nitrated almost exclusively at the *meta* position. The influence of a substituent on the location at which an aromatic electrophilic substitution occurs is called the **directing effect** of that substituent. Toluene is said to be *ortho-para* directing; nitrobenzene is said to be *meta* directing.

For any electrophilic aromatic substitution of a substituted benzene, both the *activating or deactivating effect* of the substituent and the *directing effect* of the substituent must be considered when predicting the feasibility and outcome of a reaction. This predictability can be summarized by five rules (see Table 9.1):

1. Electron-donating groups are *usually* activating groups.
2. Electron-donating groups are *always ortho-para* directing groups.
3. Electron-withdrawing groups are *always* deactivating groups.
4. Electron-withdrawing groups are *usually meta* directing groups.
5. Halogens are the single exception and are the reason for the word *usually* in rules 1 and 4; they are electron-withdrawing groups (being highly electronegative) and so are deactivating. However, they can donate electrons on demand and so are *ortho-para* directors.

The next three subsections explain these patterns of behavior in terms of the electrophilic aromatic substitution mechanism considered in Section 9.3.1, as well as illustrating how they can be used to advantage in planning a synthesis.

9.4.1 Reactivity Effects of Substituents

The rate of an electrophilic aromatic substitution reaction depends on the height of the energy barrier to the transition state of the first step—the formation of the arenium cation; in other words, it depends on the energy of activation for the reaction. (Refer to the

TABLE 9.1

Effects of Substituents on Electrophilic Aromatic Substitution Reactions

Ortho-Para Directors		*Meta* Directors	
Strong activators	$-\ddot{N}H_2$ (amino) $-\ddot{O}H$ (hydroxy)	Strong deactivators	$-NO_2$ (nitro) $-\overset{+}{N}R_3$ (ammonium) $-CX_3$ (trihalomethyl)
Moderate activators	$-\ddot{N}HCOCH_3$ (acetamido) $-\ddot{O}COCH_3$ (acetoxy) $-\ddot{O}R$ (alkoxy)	Moderate deactivators	$-CN$ (cyano or nitrile) $-SO_3H$ (sulfonic acid) $-CHO$ (aldehyde) $-COR$ (ketone) $-COOH$ (carboxyl)
Weak activators	$-CH_3$ (methyl) $-C_6H_5$ (phenyl)		

Weakly Deactivating *Ortho-Para* Directors

$-\ddot{F}:$	$-\ddot{Cl}:$	$-\ddot{Br}:$	$-\ddot{I}:$
(fluoro)	(chloro)	(bromo)	(iodo)

electrophilic aromatic substitution mechanism in Figure 9.3 and the free-energy diagram in Figure 9.4.) Any electronic factor that stabilizes the arenium cation intermediate stabilizes the transition state, with its increasing positive character. Because positive charge is stabilized by delocalization due to an infusion of electron density, any substituent on the arenium cation that is electron-donating should stabilize the cation, lower the energy of activation, and thereby make the substitution reaction proceed more rapidly. In other words, an electron-donating substituent is an *activating* substituent. Conversely, any substituent that is electron-withdrawing destabilizes the transition state, increases the energy of activation, and slows the reaction; an electron-withdrawing substituent is therefore a *deactivating* substituent.

For example, the presence of a hydroxyl group on the benzene ring in phenol stabilizes the arenium cation through delocalization, as shown in the following resonance contributing structures:

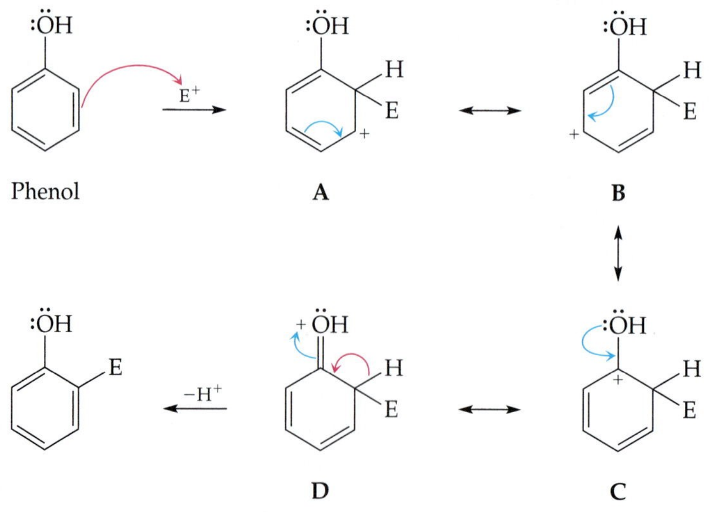

:ÖH E⁺ :ÖH H E ⟷ :ÖH H E

Phenol A B

:ÖH E $-H^+$ +ÖH H E ⟷ :ÖH H E

D C

Contributing structures **A, B,** and **C** are those normally drawn for any arenium cation resulting from *ortho* substitution (see Figure 9.3). However, the unshared electrons on oxygen can be delocalized into the ring, as shown in contributing structure **D.** The extra stabilization resulting from the contribution of **D** lowers the energy of activation significantly. Thus, the hydroxyl group is a strong activator.

Any substituent that can provide such stabilization to the arenium cation intermediate is an activator. The groups listed in Table 9.1 as activators all provide electron density to the arenium cation. Most of those groups activate through the use of unshared electrons on the atom attached directly to the ring. The methyl group and other alkyl groups provide electron density through hyperconjugation (via electrons in the adjacent C—H bonds), just as they do when stabilizing carbocations (see Section 3.4.3).

The presence of a cyano group (—CN) on the benzene ring in benzonitrile (cyanobenzene) destabilizes the arenium cation relative to that from benzene itself:

Benzonitrile

E

F

G

$\overset{\delta+}{—C}\equiv\overset{\delta-}{N}$

Cyano group

The cyano group is polarized toward the nitrogen, making the carbon electron-deficient. Contributing structures **E, F,** and **G** are those normally drawn for any arenium cation from *ortho* substitution (see Figure 9.3). However, note that in structure **G** the arenium cation's positive charge is localized on a ring carbon attached to another electron-deficient carbon, the cyano carbon. This proximity of two electron-deficient sites is energetically unfavorable. Thus, the arenium cation intermediate is destabilized relative to that for unsubstituted benzene. Therefore, the energy of activation is higher for this substitution, and the rate of reaction is slower. Thus, the cyano group is a *deactivator.*

Any substituent that can cause such destabilization of the arenium cation intermediate is a deactivator. The groups listed in Table 9.1 as deactivators are all electron-withdrawing groups because the atom attached directly to the ring is electron-deficient. The nitro (—NO_2) and ammonium (—$\overset{+}{N}R_3$) groups have an actual positive charge on the nitrogen atom, while in the other deactivators, the atom attached to the ring is the positive end of a dipole. The trihalomethyl groups, such as —CF_3 and —CCl_3, are strong deactivators because of the strong cumulative electronegativity effect of three halogens attached to the carbon (recall that this is an *inductive* effect).

The situation is slightly different when a halogen is attached directly to the ring. The dominant influence on the arenium cation is the electronegativity of the halogen atom; the resulting electron withdrawal leads to a slight destabilization and therefore weak deactivation. This effect predominates even though there are unshared electrons on the halogen. The next subsection describes how these electrons come into play in the directing effects of substituents.

In summary, substituents on benzene rings have the following reactivity effects:

- Any substituent that has electron density to share with the intermediate arenium cation is an activator.
- Any substituent that has an electron-deficient atom attached to the ring is a deactivator.

PROBLEM 9.10

Place the compounds in each of the following pairs in decreasing order of reactivity (susceptibility to electrophilic aromatic substitution):

(a) aniline and acetanilide ($C_6H_5NHCOCH_3$)

(b) ethylbenzene and acetophenone ($C_6H_5COCH_3$)

(c) anisole and toluene

(d) aniline and nitrobenzene

(e) bromobenzene and toluene

PROBLEM 9.11

Apply the knowledge you have acquired in the preceding subsections to explain why the Friedel-Crafts alkylation of benzene results in polysubstitution, while the Friedel-Crafts acylation results in monosubstitution.

9.4.2 Directing Effects of Substituents

Keep in mind that in any electrophilic aromatic substitution on a monosubstituted benzene, there is a competition regarding the position of attack—*ortho, para,* or *meta*—of the incoming electrophile. The dominant products will be those with the fastest rate of formation. In other words, the lower the energy of activation, the faster the rate of reaction, leading to the dominant product. Therefore, it is possible to predict the dominant position of attack—the directing effect of the preexisting substituent—by comparing the stability of the three possible arenium intermediates. As we have seen, substituents fall into two categories: *ortho-para* directors and *meta* directors. *Ortho-para* directors direct to both positions, as will be seen shortly, but the *para*-substituted product usually predominates because steric hindrance is involved in making attack at the nearby *ortho* position more difficult. Normally, there is no means to control the ratio of *ortho* to *para* products, so a separation step is inevitably required to obtain either isomer in pure form.

Ortho-Para–Directing Effects

Aniline can serve as an example for illustrating *ortho-para*–directing effects. The following contributing structures can be written for attack of an electrophile (E^+) on aniline at each of the three positions:

***Ortho* attack:**

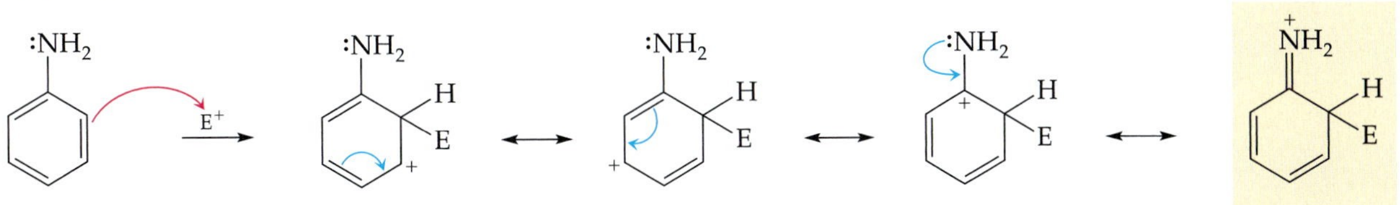

Aniline A B C D

***Meta* attack:**

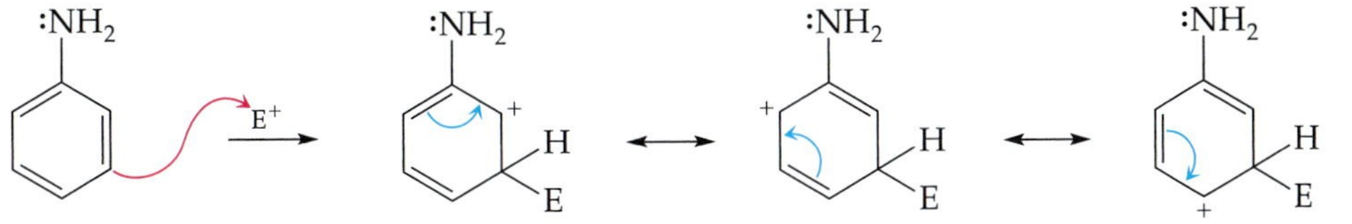

E F G

***Para* attack:**

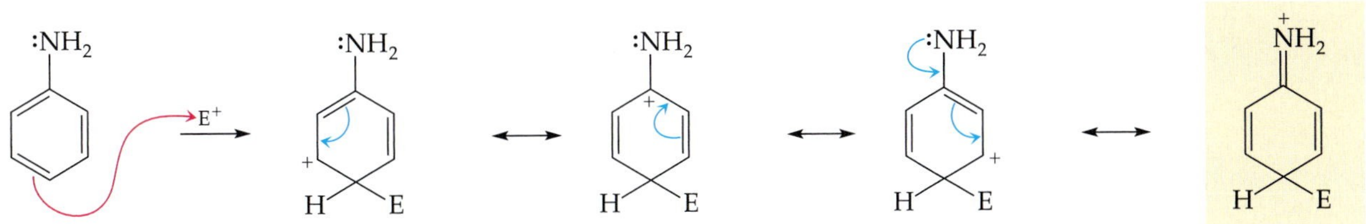

H I J K

The contributing structures **A–C, E–G,** and **H–J** all involve delocalization around the arenium ring. However, the *ortho* and *para* attacks both result in an additional contributing structure, **D** and **K,** respectively, each involving delocalization to the amino substituent. Since the more contributing structures a resonance hybrid has, the more stable it is, the arenium cation should be more stable when it results from *ortho* or *para* attack, leading to lower energies of activation and faster rates of formation. Therefore, the amino group is an *ortho-para* director, meaning that only a minor amount of *meta*-substituted product will be obtained. The same is true of all activating substituents.

The behavior of the halogens in this situation is unique. Although a halogen atom deactivates a benzene ring toward electrophilic aromatic substitution (owing to its electronegativity), the arenium cations resulting from successful attack at the three positions are parallel to those shown for aniline (structures **A–K**). That is, the unshared electrons on the halogen can stabilize the arenium cation resulting from *ortho* and *para* attack, but not that resulting from *meta* attack:

***Ortho* attack:**

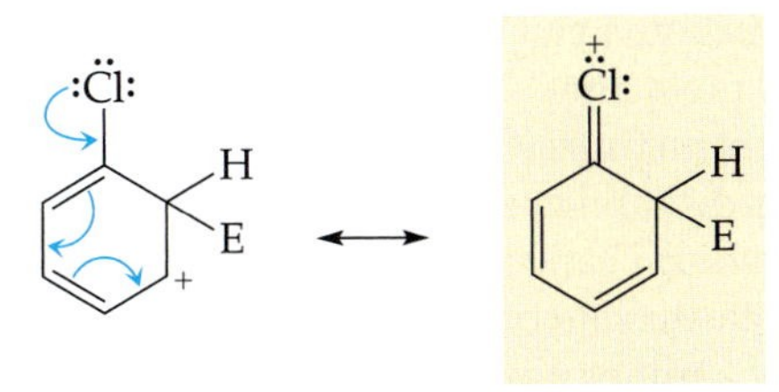

***Para* attack:**

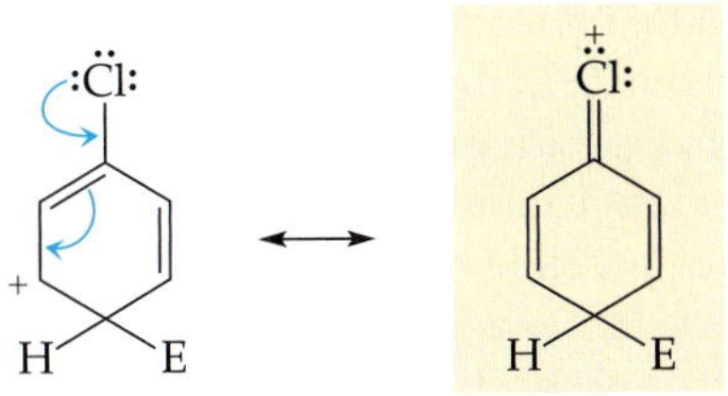

Therefore, even though electrophilic aromatic substitution is less likely to occur on halobenzenes than on benzene itself (the halogen substituents are deactivators), when substitution is forced, it occurs mainly *ortho-para* because of the greater stabilization of the respective arenium cation intermediates.

Meta-Directing Effects

Benzaldehyde is a good example for illustrating *meta*-directing effects. The following contributing structures can be written for the attack of an electrophile (E^+) on benzaldehyde at each of the three positions:

Ortho **attack:**

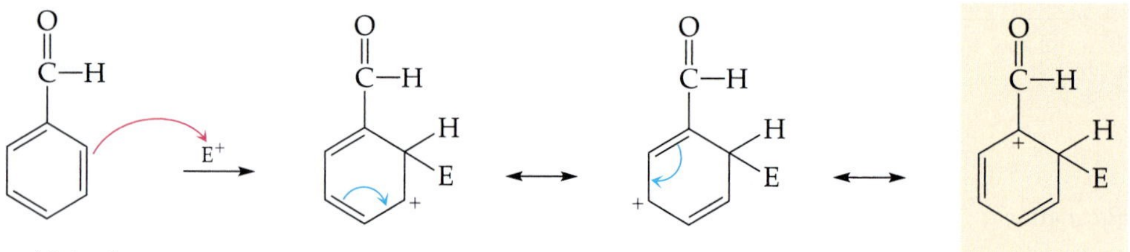

Benzaldehyde **L** **M** **N**

Meta **attack:**

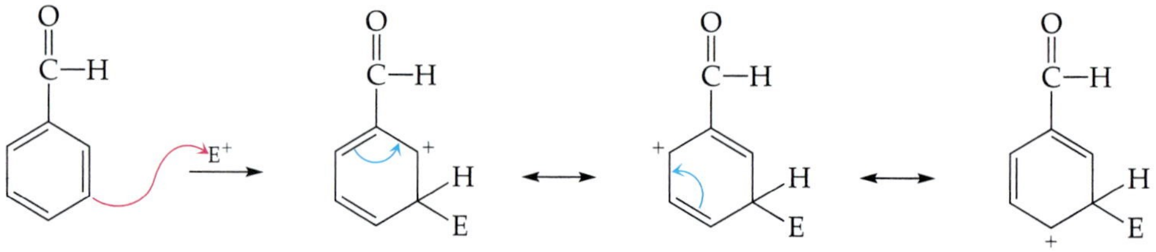

O **P** **Q**

Para **attack:**

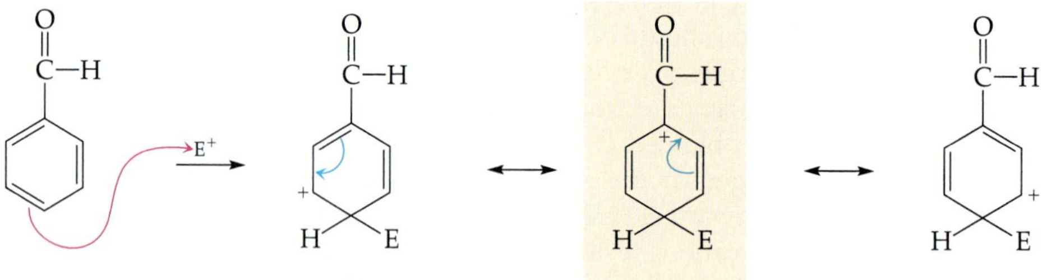

R **S** **T**

The aldehyde group is polarized toward oxygen ($^{\delta+}C{=}O^{\delta-}$), making the carbonyl carbon electron-deficient. In *ortho* and *para* attacks, one of the three contributing structures (structure **N** or **S**) distributes the positive charge of the arenium cation to the ring carbon that already carries the electron-deficient carbonyl carbon. This is a destabilizing influence. In the arenium cation resulting from meta attack, there is no such conjunction of positive charges in the three contributing structures (**O, P,** and **Q**). Therefore, the arenium cation resulting from *meta* attack is less destabilized; it is the *most stable* of the three possible arenium cations. Because, the arenium cation from *meta* attack is more stable, it has a lower energy of activation and a faster rate of formation. Therefore, the carbonyl substituent is a *meta* director, and only a minor amount of *ortho-* or *para*-substituted product will be obtained. The same is true of all substituents with an atom directly attached to the ring that is positively charged or is the positive end of a dipole.

9.4.3 Application of Reactivity and Directing Effects in Synthesis of Disubstituted Aromatics

By considering the directing influence of substituents, chemists can choose the best sequence for attaching substituents to a benzene ring in order to obtain the desired product. You should practice devising syntheses using the information in Table 9.1. However, you do not need to memorize Table 9.1; you just need to understand three aspects of the arenium cation intermediates involved:

1. Electron-donating groups are activators and *ortho-para* directors because they stabilize the arenium cations.
2. Electron-withdrawing groups are deactivators and *meta* directors because they destabilize the arenium cations.
3. Because of their unique electronic character, halogens are inductively withdrawing and therefore deactivating, but they can supply electrons upon demand and so are *ortho-para* directors.

Solutions to several benzenoid synthesis problems will show how to apply this knowledge and reasoning.

How to Solve a Problem

(a) Convert benzene into *m*-bromonitrobenzene.

PROBLEM ANALYSIS AND SOLUTION

The problem is how to attach a nitro group and a bromine atom to a benzene ring. First, we simply reason through one of the possible sequences of electrophilic aromatic substitution and consider the problems that result. Since bromine is an *ortho-para* director, if it is attached first to benzene, nitration as the second step will not result in the *meta* isomer. If nitration is conducted first, bromination as the second step will occur at the *meta* position, since the nitro group is a *meta* director. This sequence leads to the desired product:

NO_2 NO_2

$\xrightarrow[H_2SO_4]{HNO_3}$ $\xrightarrow[FeBr_3]{Br_2}$ Br

Benzene Nitrobenzene *m*-Bromonitrobenzene

(b) What is the major product from the mononitration of *p*-chlorobiphenyl?

PROBLEM ANALYSIS AND SOLUTION

p-Chlorobiphenyl has two aromatic rings potentially available for nitration. The most reactive ring will undergo nitration. Because chlorine is a deactivator for its ring, the nitration occurs on the ring without the attached chlorine. The next consideration is the position of attack on this unsubstituted ring. A phenyl ring as a substituent is a modest activating group, because it can supply electrons. Therefore, the phenyl group is an *ortho-para* director. The major product is *p*-chloro-*p*′-nitrobiphenyl, having been nitrated at the *para* position of the formerly unsubstituted ring (the *ortho* substitution product is minor because of steric hindrance).

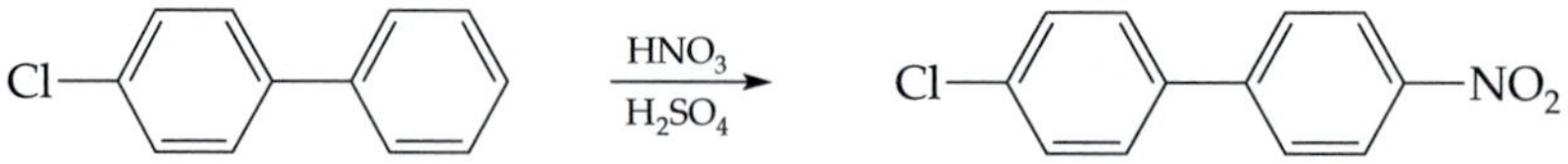

Cl—(biphenyl) $\xrightarrow[H_2SO_4]{HNO_3}$ Cl—(biphenyl)—NO_2

p-Chlorobiphenyl

p-Chloro-*p*′-nitrobiphenyl (major product)

(c) Why does Friedel-Crafts alkylation of benzene frequently lead to polysubstitution, while Friedel-Crafts acylation leads only to monosubstitution?

PROBLEM ANALYSIS AND SOLUTION

In alkylation with ethyl chloride and aluminum chloride, for example, the first reaction is attachment of an ethyl group to form ethylbenzene in the reaction flask. Additional ethyl carbocation complexes form and seek a benzene ring to attack. The aromatic ring of the newly synthesized ethylbenzene is more reactive than benzene itself, owing to the activating effect of the ethyl group. Thus, a second substitution occurs and *p*-diethylbenzene is formed (this process occasionally continues even further, leading to the attachment of third and fourth ethyl groups to the ring):

Friedel-Crafts alkylation:

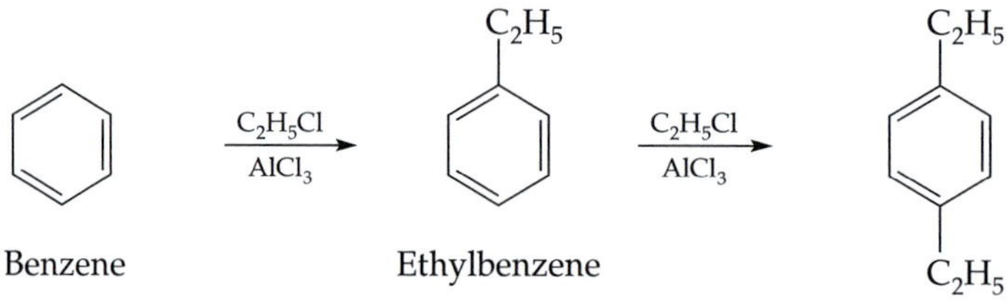

Benzene $\xrightarrow[AlCl_3]{C_2H_5Cl}$ Ethylbenzene (C_2H_5) $\xrightarrow[AlCl_3]{C_2H_5Cl}$ *p*-Diethylbenzene (C_2H_5, C_2H_5)

In acylation with acetyl chloride and aluminum chloride, for example, the first substitution results in the formation of acetophenone, a benzene ring with an acyl group attached. The acyl group, in contrast to the ethyl group, is a deactivator. As each additional acylium cation complex seeks an aromatic ring to attack, it substitutes on benzene. The unsubstituted benzene ring is more reactive than the ring in acetophenone. Thus, no *p*-diacetylbenzene is formed.

Friedel-Crafts acylation:

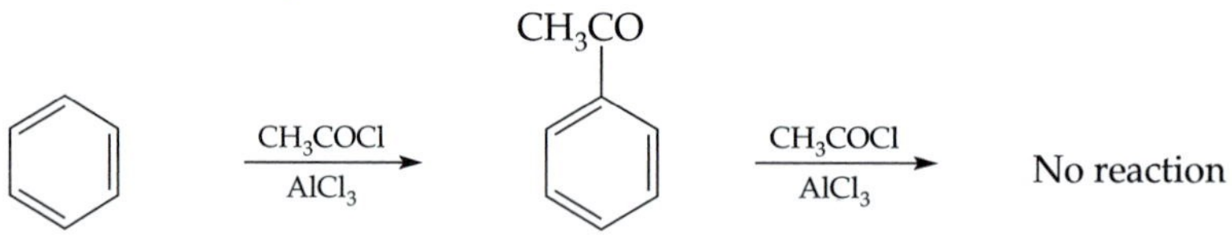

Benzene $\xrightarrow[AlCl_3]{CH_3COCl}$ Acetophenone (CH_3CO) $\xrightarrow[AlCl_3]{CH_3COCl}$ No reaction

PROBLEM 9.12

Show the product(s) of each of the following reactions:

(a) nitration of iodobenzene

(b) sulfonation of toluene
(c) acetylation of anisole
(d) bromination of acetanilide ($CH_3CONHC_6H_5$)

PROBLEM 9.13

Show the reactions and reagents necessary for each transformation:

(a) benzene to *p*-bromoethylbenzene
(b) benzene to *m*-nitroethylbenzene
(c) benzene to 1-(*p*-nitrophenyl)butane
(d) benzene to diphenylmethane

9.4.4 Effect of Two Substituents on Electrophilic Aromatic Substitution

The principles discussed in the preceding section also apply when there are two or more substituents on a benzene ring undergoing electrophilic aromatic substitution. There are two major factors that determine the position of attachment of the third substituent:

- If the directing effects of two substituents are competing, the influence of the *ortho-para*–directing substituent will predominate.
- The stronger of two activators will prevail in directing an attacking substituent.

Application of these "rules" is demonstrated in the following solved problem, where chlorination is used as a typical substitution reaction.

How to Solve a Problem

(a) What is the major product expected from the chlorination of *p*-nitrotoluene?

SOLUTION

The methyl group directs *ortho* to itself; the *para* position is blocked. The nitro group is a *meta* director. Therefore, the influences of the two substituents on the starting compound are complementary; both direct the attacking chlorine to the same positions. Therefore, the product will be 2-chloro-4-nitrotoluene:

CH_3 / NO_2 $\xrightarrow[FeCl_3]{Cl_2}$ CH_3 / Cl / NO_2

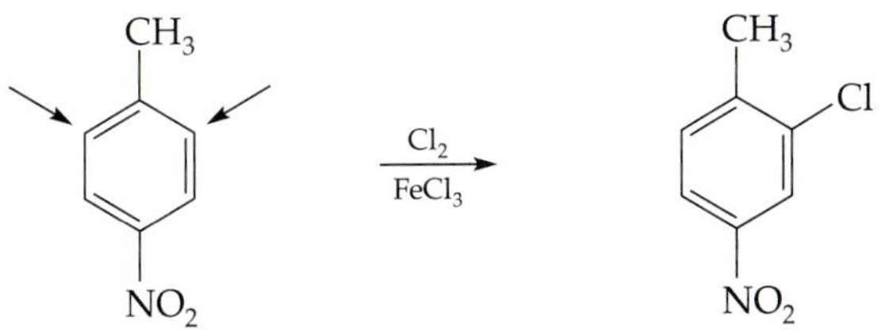

p-Nitrotoluene 2-Chloro-4-nitrotoluene

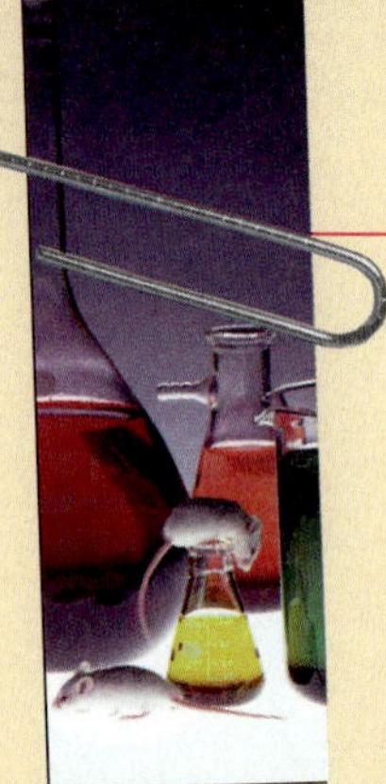

CHEMISTRY AT WORK

IN THE MARKETPLACE

The Explosive Effect of the Nitro Group

Nitrogen is one of the most important elements in organic chemistry, showing up in compounds ranging from agricultural fertilizer to the genetic material essential to life. Nitrogen is present in several different functional groups. One of the most distinctive of these is the nitro group (—NO_2), whose presence often confers explosive properties on organic compounds.

Among the earliest explosives was trinitroglycerin, commonly called *nitro,* a shock-sensitive liquid. Alfred Nobel discovered that mixing nitro with porous inert materials produced a solid explosive material that could be safely handled; he called it *dynamite.* This discovery made his fortune and was the source of the money he endowed for the Nobel Prizes. Guncotton, the major constituent of the explosive known as *smokeless powder,* is the nitrated form of cellulose, produced by nitrating three free hydroxyl groups per glucose unit. Another early explosive was trinitrotoluene (TNT), prepared by the exhaustive nitration of toluene.

$CH_2—O—NO_2$
$CH—O—NO_2$
$CH_2—O—NO_2$

Trinitroglycerin
(nitro)

Guncotton
(a portion of the nitrated cellulose structure)

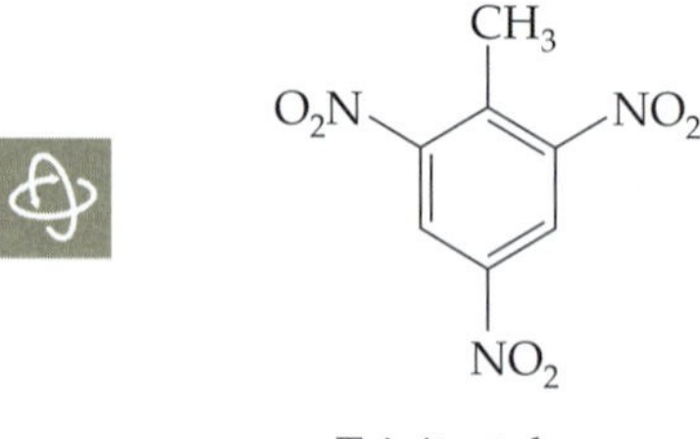

Trinitrotoluene
(TNT)

A number of newer explosives have appeared in recent decades, many of them relying on the presence of multiple nitro groups. The compound *N*-methyl-*N*,2,4,6-tetranitroaniline, commonly called Tetryl or Tetralite, is one example. The infamous Semtex, which has been implicated in numerous terrorist bombings, is a plastic explosive that consists of a mixture of RDX and PETN:

Explosive chemicals, like those applied to the demolition of the Hartford Hilton, invariably have nitro groups as a key part of their structure.

1,3,5-Trinitro-1,3,5-triazacyclohexane
(RDX)

$$O_2N—O—CH_2—C(CH_2—O—NO_2)_2—CH_2—O—NO_2$$

Pentaerythritol nitrate
(PETN)

Several explosives detection systems have been developed in recent years to provide safer travel for airline passengers. Some of these systems detect the presence of explosives in packages and luggage by sensing nitro groups.

It is easy to focus on the more dramatic and cruel uses of explosives. However, you should not forget that they have many constructive uses, such as in clearing rock for the construction of highways and buildings, and in the demolition of decrepit buildings. In fact, nitroglycerin is also a common heart medication.

(b) What is the major product expected from the chlorination of *o*-cyanophenol?

SOLUTION

The directing effects of the two substituents are again complementary—*ortho* and *para* to the hydroxyl group (an *ortho-para* director) and *meta* to the cyano group (a *meta* director). The position marked with an X is less likely because of steric hindrance from the adjacent hydroxyl group. Therefore, the major product will be 4-chloro-2-cyanophenol:

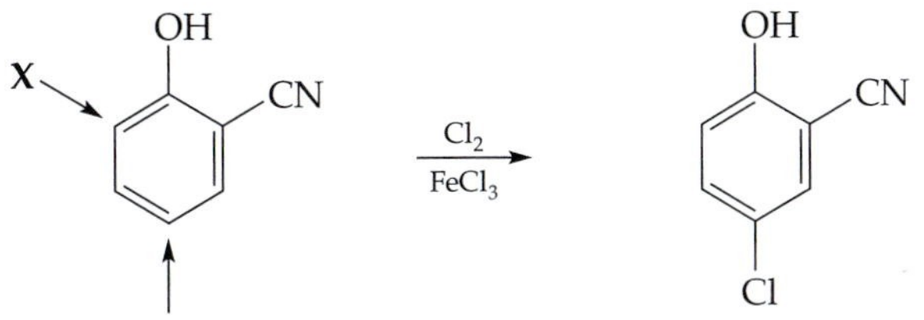

$$\xrightarrow[FeCl_3]{Cl_2}$$

o-Cyanophenol 4-Chloro-2-cyanophenol

(c) What is the major product expected from the chlorination of *m*-nitroanisole?

SOLUTION

The directing effects of the two substituents are not complementary. Therefore, the major product will be determined by the predominant effect of the *ortho-para*–directing and activating methoxy group, with its unshared pair of electrons. Steric hindrance prevents any significant reaction at the position between the methoxy and nitro groups. Both other possible products will form. However, because the methoxy group is smaller than the nitro group, substitution adjacent to the methoxy group will predominate slightly, yielding 2-chloro-5-nitroanisole:

$$\xrightarrow[FeCl_3]{Cl_2}$$

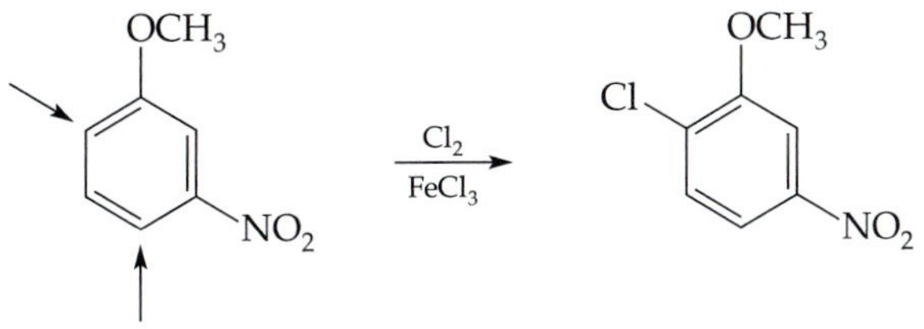

m-Nitroanisole 2-Chloro-5-nitroanisole

(d) What is the major product expected from the chlorination of *p*-methylanisole?

SOLUTION

In this instance, both substituents are *ortho-para*–directors. The stronger directing group is the one whose influence is dominant. The resonance effect of oxygen's unshared electrons in the methoxy group provides more stabilization than does the hyperconjugative effect of the C—H bonds of the methyl group. The methoxy group therefore controls the position of attachment of the incoming electrophile, and the major product will be 2-chloro-4-methylanisole:

$$\xrightarrow[FeCl_3]{Cl_2}$$

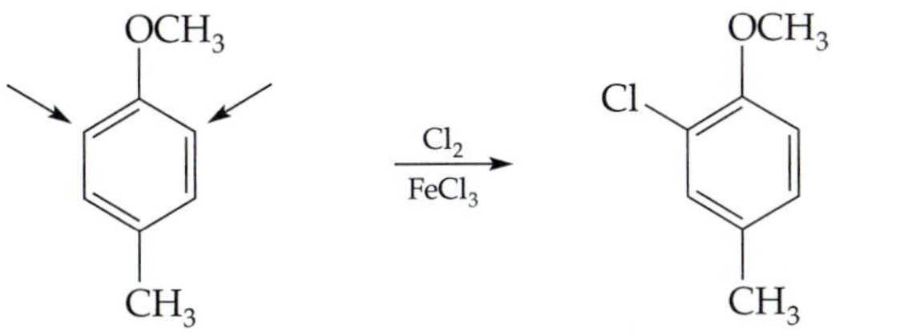

p-Methylanisole 2-Chloro-4-methylanisole

PROBLEM 9.14

Predict the major product formed from each of the following reactions:

(a) nitration of *p*-bromotoluene
(b) sulfonation of *p*-nitroanisole
(c) bromination of *m*-nitrotoluene

PROBLEM 9.15

Show the reactions and reagents required to complete each synthesis:

(a) chlorobenzene to 4-ethyl-3-nitrochlorobenzene
(b) toluene to 2-bromo-4-ethyltoluene
(c) acetophenone to 5-chloro-2-propylbenzenesulfonic acid

9.5 Phenols

The urushiols in poison ivy that cause the allergic reactions are phenols.

Phenols are a family of benzenoid compounds that carry one or more hydroxyl groups attached to the aromatic ring. The sulfur analogs of phenols, with —SH rather than —OH group(s) on the aromatic ring, are called *thiophenols* and have many of the same properties as phenols.

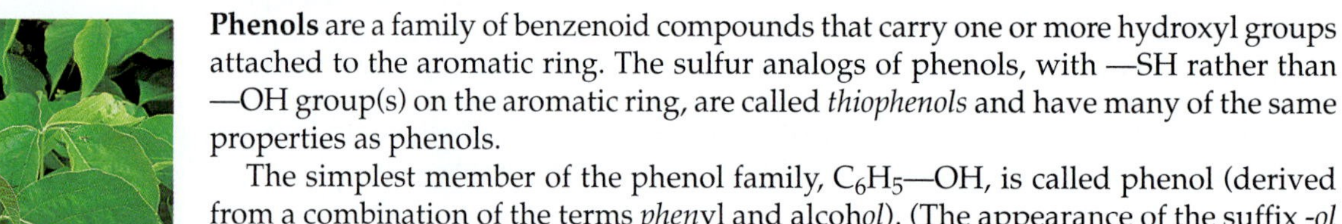

The simplest member of the phenol family, C_6H_5—OH, is called phenol (derived from a combination of the terms *phen*yl and alcoh*ol*). (The appearance of the suffix *-ol* in any name, IUPAC or historical/common, indicates the presence of a hydroxyl group.) The normal system of benzenoid nomenclature applies for substituted phenols, such as 2,4-dichlorophenol. A number of historical names are still used today, such as cresol. (Drawing the CH_3 group with a line to the center of the benzene ring implies all possible positions of attachment.)

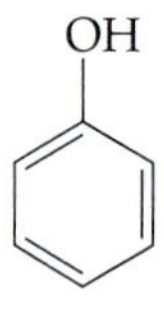

Phenol

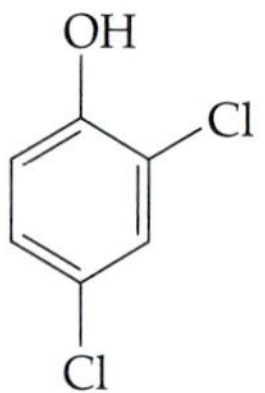

2,4-Dichlorophenol

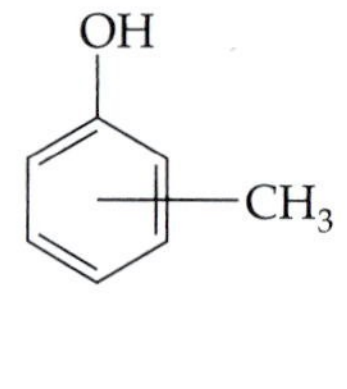

Cresols

The chemistry of phenols illustrates the significant effect that the aromatic ring has on the hydroxyl group, a functional group already studied in the alcohol family, and the reciprocal major effect that the hydroxyl group has on the benzene ring. This section outlines a few of the properties of phenols that illustrate this point.

9.5.1 Acidity of Phenols

One of the important properties of phenols is their acidity. The hydroxyl proton can be readily removed from a phenol to form a phenoxide anion. Phenol is a much stronger acid than cyclohexanol:

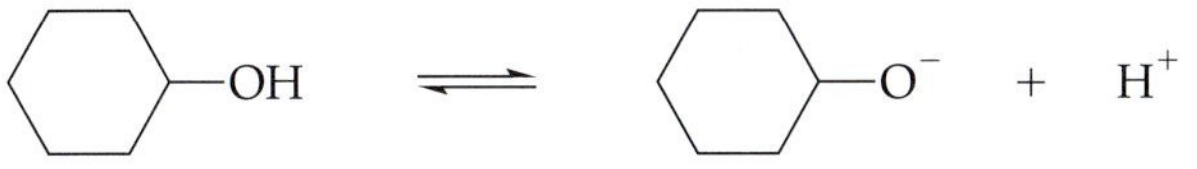

Cyclohexanol ⇌ Cyclohexoxide anion + H^+

$K_a = 1 \times 10^{-17}$
$pK_a = 17$

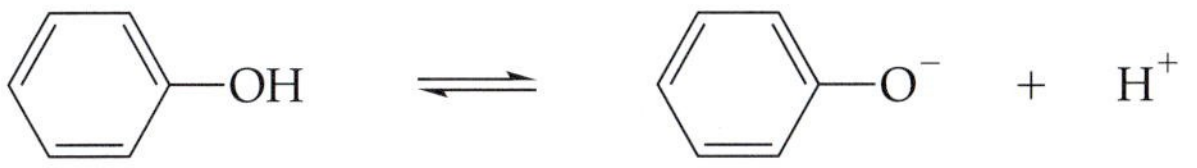

Phenol ⇌ Phenoxide anion + H^+

$K_a = 1 \times 10^{-10}$
$pK_a = 10$

Phenol is the stronger acid (by a factor of 10^7) because the phenoxide anion is stabilized by resonance, as shown here, whereas the cyclohexoxide anion is not. Thus, phenol more readily gives up a proton to form the phenoxide anion, and the ionization equilibrium is shifted to the right. Clearly, attachment of a phenyl group to a hydroxyl group affects the acidity of the latter.

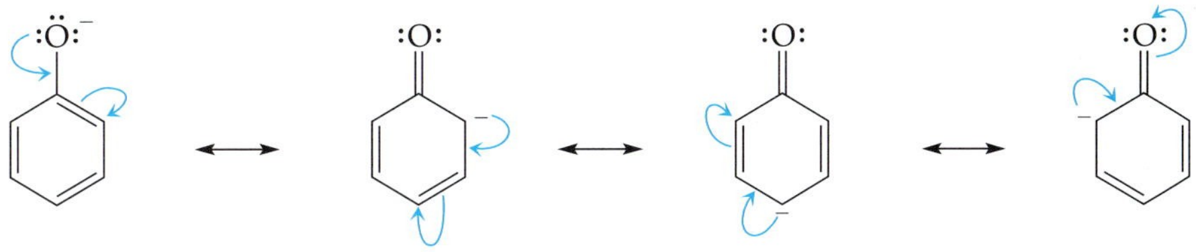

Contributing structures for resonance stabilization of the phenoxide anion

It is possible to modify the acidity of a phenol by attaching substituents that increase or decrease the relative stability of the phenoxide anion. The key is the electronic effect of the substituent: Electron-withdrawing groups (such as chlorine and the nitro group) stabilize the phenoxide anion (they delocalize the negative charge), while electron-donating groups (such as the methyl group) destabilize it. Some examples with their pK_a values are shown here:

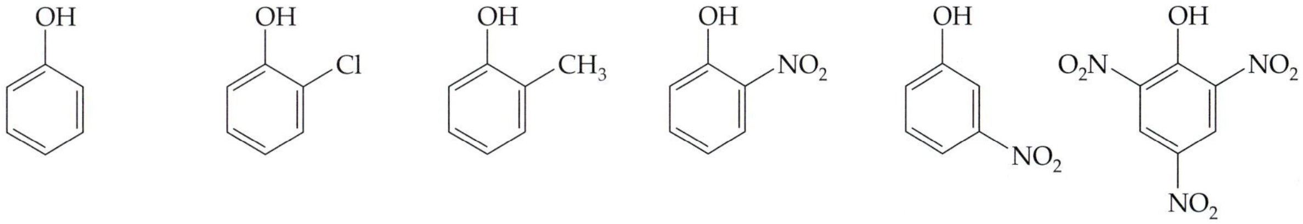

	Phenol	*o*-Chlorophenol	*o*-Cresol	*o*-Nitrophenol	*m*-Nitrophenol	Picric acid
$pK_a =$	10	8.1	10.2	7.2	8.3	0.4

Using *o*-nitrophenol as an example, the following equation shows removal of the proton and formation of the stabilized *o*-nitrophenoxide anion:

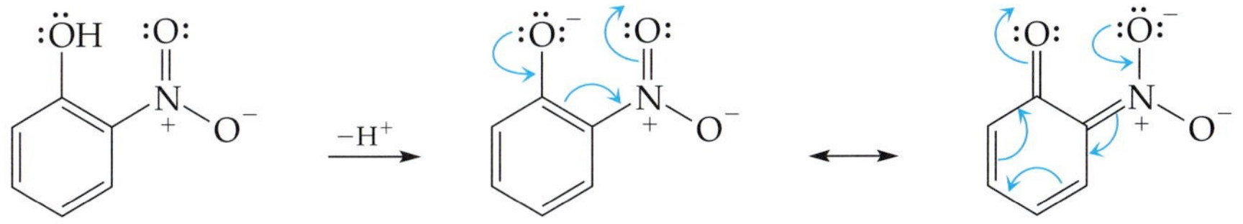

o-Nitrophenol $\xrightarrow{-H^+}$ *o*-Nitrophenoxide anion
(2 of 5 contributing structures)

The inductive effect of the electron-withdrawing nitro group (due to the electronegativity effect of the partial positive charge on nitrogen) stabilizes the resulting *o*-nitrophenoxide anion. Further, this anion has all four of the contributing structures shown previously for the phenoxide anion itself. Most important, however, the negative charge on oxygen can be delocalized by resonance to the nitro group, as shown above. These effects are multiplied nearly 10 million times in picric acid (2,4,6-trinitrophenol)—an organic acid nearly as acidic as hydrochloric acid.

How to Solve a Problem

Explain why it is possible to use aqueous sodium hydroxide extraction but not aqueous sodium bicarbonate extraction to separate phenol from cyclohexanol when they are mixed together in ether solution.

SOLUTION

Consider the acid-base equilibria (see Section 1.3) of cyclohexanol and phenol with aqueous sodium hydroxide:

$$\text{C}_6\text{H}_{11}\text{–OH} + \text{NaOH} \rightleftharpoons \text{C}_6\text{H}_{11}\text{–O}^- \text{Na}^+ + H_2O$$

Cyclohexanol
(pK_a 17)

(pK_a 15.7)

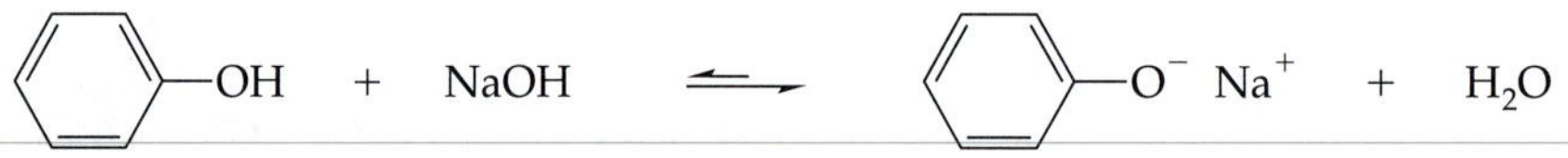

$$\text{C}_6\text{H}_5\text{–OH} + \text{NaOH} \rightleftharpoons \text{C}_6\text{H}_5\text{–O}^- \text{Na}^+ + H_2O$$

Phenol
(pK_a 10)

(pK_a 15.7)

The equilibria lie in the directions shown because the more acidic species is the one that gives up its proton. Water is more acidic than cyclohexanol, but phenol is more acidic than water, as indicated by the pK_a values. Therefore, phenol exists as its phenoxide salt in the presence of hydroxide ion because phenol is more acidic than water, and the phenoxide salt is soluble in water. Cyclohexanol exists as the un-ionized alcohol and therefore remains dissolved in the organic ether layer. Separation of the organic and aqueous layers enables separation of the compounds. Cyclohexanol can be recovered by simple evaporation of the ether solvent. Phenol can be recovered by acidifying the separated aqueous layer with HCl (a stronger acid):

$$\text{C}_6\text{H}_5\text{–O}^- \text{Na}^+ + \text{HCl} \longrightarrow \text{C}_6\text{H}_5\text{–OH} + \text{NaCl}$$

Sodium phenoxide — Phenol

If sodium bicarbonate were used as the base, the comparison of pK_a values would involve carbonic acid, with a pK_a of 6.4. Because this is more acidic than both phenol and cyclohexanol, sodium bicarbonate is not a strong enough base to remove the proton from either compound, so both would remain un-ionized and in the ether solution.

PROBLEM 9.16

In each of the following pairs of compounds, indicate which is more acidic:

(a) *p*-chlorophenol or *p*-nitrophenol
(b) *o*-iodophenol or *p*-iodophenol
(c) *p*-ethylphenol or *p*-acetylphenol

9.5.2 Electrophilic Aromatic Substitution of Phenols

Just as the attachment of a phenyl group to a hydroxyl group significantly affects the hydroxyl group (increasing its acidity compared to that of nonaromatic alcohols, as shown in Section 9.5.1), the hydroxyl group significantly affects the behavior of the phenyl ring. As shown in Table 9.1 (page 290), the electron-donating —OH group is an *ortho-para* director for electrophilic aromatic substitution. More significantly, it is a very strong activator, causing most substitution reactions of phenols to result in polysubstitution. For example, nitration of phenol results in the formation of 2,4,6-trinitrophenol, and bromination produces 2,4,6-tribromophenol.

The only way to monosubstitute phenol is to first convert it into acetoxybenzene (also known as phenyl acetate) by acetylation using acetyl chloride, a reaction to be described in Section 15.1.2. Note from Table 9.1 that the acetoxy group is an *ortho-para* director and a moderate activator; thus, electrophilic aromatic substitution of acetoxybenzene results in monosubstitution, mainly at the *para* position. The acetyl group can be removed by simple hydrolysis to re-form the hydroxyl group. This process is illustrated for the preparation of *p*-bromophenol:

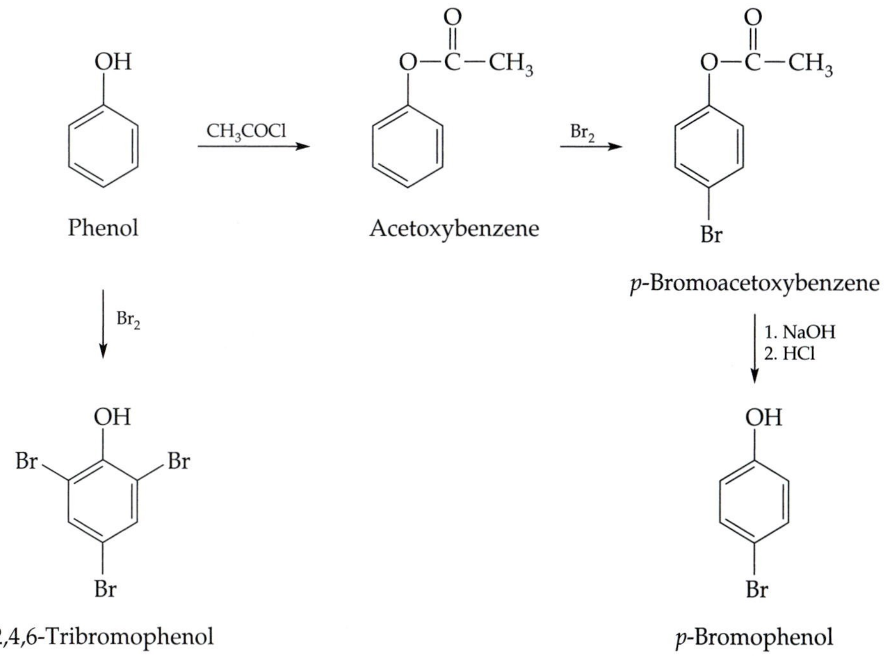

OH (Phenol) $\xrightarrow{CH_3COCl}$ $O-\overset{O}{\overset{\|}{C}}-CH_3$ (Acetoxybenzene) $\xrightarrow{Br_2}$ $O-\overset{O}{\overset{\|}{C}}-CH_3$, Br (*p*-Bromoacetoxybenzene)

Phenol $\xrightarrow{Br_2}$ OH, Br, Br, Br (2,4,6-Tribromophenol)

p-Bromoacetoxybenzene $\xrightarrow{1.\ NaOH\ \ 2.\ HCl}$ OH, Br (*p*-Bromophenol)

Attaching the acetyl group to the phenolic oxygen moderates its electron-donating effect because the acetyl group is electron-withdrawing and competes with the ring for electrons, as shown by its resonance-contributing structures. Thus, oxygen's electrons are less available to the ring, and acetoxybenzene is less reactive toward substitution than is phenol.

Contributing structures for resonance stabilization of acetoxybenzene (phenylacetate)

9.5.3 O-Alkylation of Phenols (Williamson Synthesis)

Because phenols are acidic and can be converted easily into their phenoxide anions, it is very easy to form phenyl alkyl ethers via the Williamson synthesis of ethers (described in Section 5.1.2), usually brought about using methyl iodide for convenience. The methyl group can be readily removed by a typical ether cleavage (see Section 5.1.3).

Phenol —NaOH→ [Phenoxide (O^- Na^+)] —CH_3I (S_N2)→ Anisole (OCH_3)

Anisole —HI (S_N2)→ Phenol

This overall technique is used to protect the hydroxyl group of a phenol. If a reaction is to be conducted on the phenol that might affect the —OH group, that group can first be protected as the methyl ether. Then the desired reaction can be conducted, and finally the methyl ether can be cleaved to "free" the hydroxyl group once again (that is, deprotect it).

Methyl ethers of phenols occur very frequently in nature (see Section 9.5.5 for examples), and the Williamson synthesis is often used in preparing such compounds in the laboratory and in industry.

9.5.4 Synthesis of Phenols

The synthesis of phenols is not straightforward. A number of specialty reactions have been developed, but they will not be considered in this book.

The only widely used laboratory synthesis of phenols is that from the corresponding anilines through a process called *diazotization,* to be described in Section 12.5. This route from benzenoid compounds to phenols starts with the nitration reaction, followed by reduction of the nitro group (—NO_2) to an amino group (—NH_2), diazotization of the amine to a diazonium ion (—N_2^+), and finally displacement of the diazonium group by the hydroxyl group (—OH) upon heating in water:

Benzene —HNO_3/H_2SO_4→ Nitrobenzene (NO_2) —H_2/Pt→ Aniline (NH_2) —HNO_2→ Benzenediazonium ion ($^+N_2$) —Δ/H_2O→ Phenol (OH)

Fortunately, all of these steps are usually very high-yield reactions. It is important just to be aware of the overall process—a phenolic hydroxyl group is attached to an aromatic ring by first introducing a nitro group at that position.

The primary constituent of jalapeno peppers, a key ingredient for Mexican cooking, is capsaicin, a phenol. Capsaicin also is used in topical ointments to relieve arthritis pain and as an additive to bird seed to discourage feeding by squirrels.

PROBLEM 9.17

Show the reactions necessary to complete each transformation:

(a) toluene to *p*-cresol (*p*-methylphenol)
(b) ethylbenzene to *p*-ethylphenol
(c) nitrobenzene to *m*-bromophenol
(d) acetophenone to *m*-ethylphenol

9.5.5 Important and Interesting Phenols

Phenols are important commercial commodities. Phenol and its sodium salt are used as household and institutional disinfectants. The cresols are used as preservatives for wooden poles placed in the earth; for example, the lower portion of a utility pole is "creosoted," which accounts for the black, tarlike material at the bottom of the pole.

Phenols and the closely related methyl ethers occur widely in natural products. Examples include tyrosine (an essential amino acid for humans), methyl salicylate (oil of wintergreen), vanillin (the flavor of vanilla), eugenol (oil of cloves, also used as an antiseptic), and thymol (the flavor of garden thyme).

OH, CH_2CHNH_2, COOH — Tyrosine

OH, $COOCH_3$ — Methyl salicylate

OH, OCH_3, CHO — Vanillin

OH, OCH_3, $CH_2CH{=}CH_2$ — Eugenol

OH, $CH(CH_3)_2$, CH_3 — Thymol

The urushiols are the blistering ingredients in poison ivy, estradiol is a female sex hormone, and epinephrine (also known as adrenaline) is a neurotransmitter in mammals. The active ingredient in hot peppers is capsaicin, which is used in cooking, as a topical agent for arthritis, and as an additive to bird food to discourage squirrels from eating it.

OH, OH, R — Urushiols ($R = C_{15}$ alkyl, alkenyl, or alkadienyl groups)

OH, HO — Estradiol

HO, HO, $CH(OH)CH_2NHCH_3$ — Epinephrine

CH_3O, HO, $CH_2NHCO(CH_2)_4CH{=}CHCH(CH_3)_2$ — Capsaicin

Phenols are key intermediates in the synthesis of commercially important products. Aspirin (acetylsalicylic acid) is the acetate ester of salicylic acid (note its similarity to oil of wintergreen). When ingested, it is quickly hydrolyzed to salicylic acid, which is the active agent. The widely used herbicides 2,4-D and 2,4,5-T are synthesized from phenol and chloroacetic acid (see page 281).

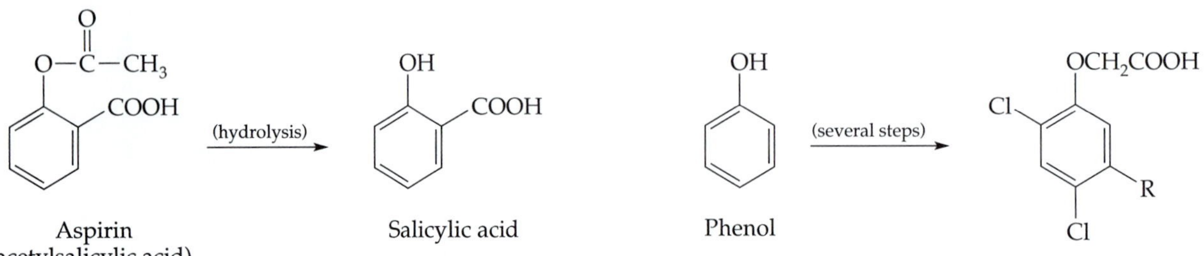

O—C(=O)—CH_3, COOH (hydrolysis) → OH, COOH (several steps) → OCH_2COOH, Cl, Cl, R

Aspirin (acetylsalicylic acid) — Salicylic acid — Phenol — 2,4-D (R = H), 2,4,5-T (R = Cl)

PROBLEM 9.18

Show the products of the following reactions:

(a) phenol with sodium hydroxide and then ethyl bromide
(b) acetoxybenzene with chlorine and $FeCl_3$
(c) phenol with nitric acid and sulfuric acid
(d) anisole with bromine and $FeBr_3$
(e) the product of (d) with hydrogen iodide

9.6 Side-Chain Chemistry of Aromatics

It was clearly demonstrated in Section 9.4 that the substituents on a benzene ring affect the reactions of the ring, but it is also true that the presence of a benzene ring affects the behavior of the substituents attached to the ring. This phenomenon was described for phenols, but it also is true for alkyl groups attached to benzenoid rings, the so-called *side chains*. We will briefly explore such effects and how they can be exploited to do organic syntheses. Most of the chemistry described in this section has been discussed earlier, but will now be applied in an aromatic environment.

9.6.1 Alkenylbenzenes

Alkenylbenzenes are compounds in which a double bond is located adjacent to a benzenoid ring (for example, $C_6H_5—CH{=}CH—R$). The simplest example is styrene (vinylbenzene, $C_6H_5—CH{=}CH_2$). The double bond is conjugated with the ring, and some resonance stabilization results.

Formation of Alkenylbenzenes

There appears to be a significant driving force toward the formation of alkenylbenzenes. When an elimination reaction is conducted on an alkyl side chain and the resulting double bond can be either conjugated or nonconjugated with the ring, the reaction invariably follows a regiochemistry such that the double bond is formed in conjugation with the ring. For example, in the dehydration of both 1-phenyl-1-propanol and 1-phenyl-2-propanol, *only* the conjugated product, 1-phenylpropene, forms—not 3-phenylpropene:

$$C_6H_5-CH_2-CH(OH)-CH_3 \text{ (1-Phenyl-2-propanol)} \quad \text{and} \quad C_6H_5-CH(OH)-CH_2-CH_3 \text{ (1-Phenyl-1-propanol)} \xrightarrow[\Delta]{H_2SO_4} C_6H_5-CH{=}CH-CH_3 \text{ (1-Phenylpropene)}$$

Similarly, dehydrohalogenation of 2-bromo-1-phenylbutane yields *only* the conjugated product, 1-phenyl-1-butene:

$$C_6H_5-CH_2-CH(Br)-CH_2-CH_3 \text{ (2-Bromo-1-phenylbutane)} \xrightarrow[\Delta]{NaOEt} C_6H_5-CH{=}CH-CH_2CH_3 \text{ (1-Phenyl-1-butene)}$$

Additions to Alkenylbenzenes

Addition reactions to alkenylbenzenes proceed in the same manner as for any alkene. Thus, the double bond can be hydrogenated or added to by bromine, water, and hydrogen halides. The only new aspect of these addition reactions is the difference in the regiochemistry caused by the phenyl ring. Ionic additions always occur so that the most stable carbocation is formed. In most cases, the most stable carbocation is that with positive charge located on the benzylic carbon, the carbon next to the ring. The reason is that the reactive intermediate at that position can be stabilized through resonance delocalization with the ring (see Section 9.6.2). This position is often referred to as the alpha (α) position, meaning the first position from the ring. For example, addition of HBr to styrene under ionic conditions (see Section 6.4.1) yields only α-bromoethylbenzene (1-bromo-1-phenylethane) because the benzylic carbocation is by far the most stable possible intermediate.

$$C_6H_5-CH{=}CH_2 \text{ (Styrene)} \xrightarrow{HBr} \left[C_6H_5-\overset{+}{C}H-CH_3\right] \xrightarrow{Br^-} C_6H_5-CH(Br)-CH_3 \text{ (1-Bromo-1-phenylethane)}$$

9.6.2 Benzylic Reactive Intermediates

As just described, the attachment of a phenyl group to an intermediate carbocation adds significantly to the stability of that reactive intermediate. The reason is the delocalization of positive charge over the aromatic ring. For example, the benzylic carbocation has these four contributing structures because the ring is conjugated with the carbocation:

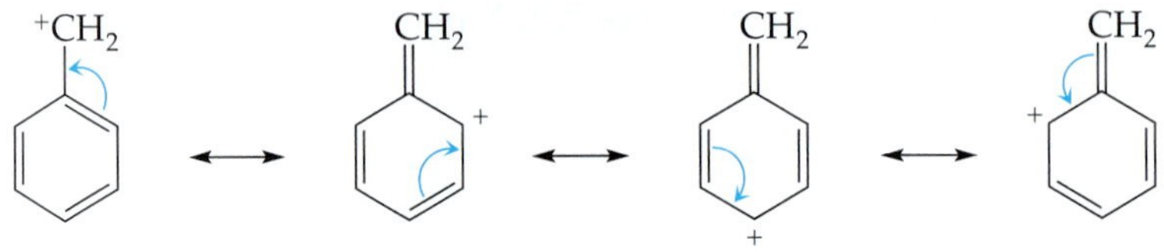

$^+CH_2$ ⟷ CH_2 ⟷ CH_2 ⟷ CH_2

Benzylic carbocation

In general, for any reaction involving a carbocation where one option is for the positive charge to be located at a benzylic position (a position alpha to the ring), that position is preferred and normally determines the regiochemistry of the reaction. This was demonstrated in Section 9.6.1 for the ionic addition of HBr.

The stabilizing effect of the phenyl ring is also indicated by the fact that benzyl halides undergo S_N1 substitution, which involves a carbocation intermediate. This occurs in spite of the fact that the intermediate is a primary carbocation that otherwise would not be expected to be very stable. For example, benzyl bromide is solvolyzed by methanol to benzyl methyl ether in an S_N1 substitution:

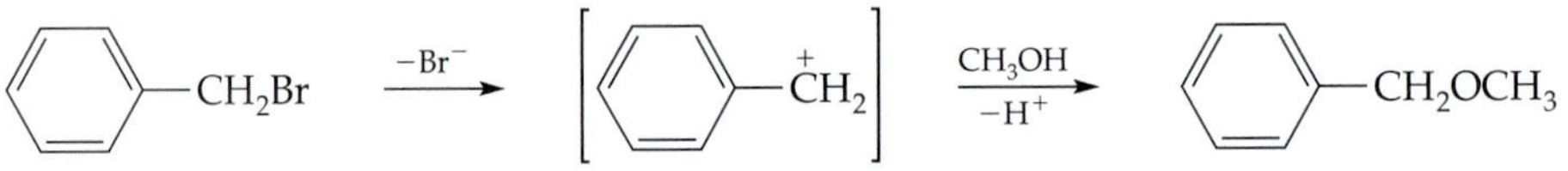

$$C_6H_5CH_2Br \xrightarrow{-Br^-} [C_6H_5\overset{+}{C}H_2] \xrightarrow[-H^+]{CH_3OH} C_6H_5CH_2OCH_3$$

Benzyl bromide — Benzylic carbocation — Benzyl methyl ether

The effect of the phenyl group on the stability of a carbocation can be multiplied by having more than one conjugated phenyl group. For example, triphenylmethyl chloride dissociates in water to form the triphenylmethyl carbocation, known also as the *trityl carbocation*, which can actually be isolated as its perchlorate salt because it is stabilized by three phenyl groups. This is one of the very few carbocations ever to be isolated.

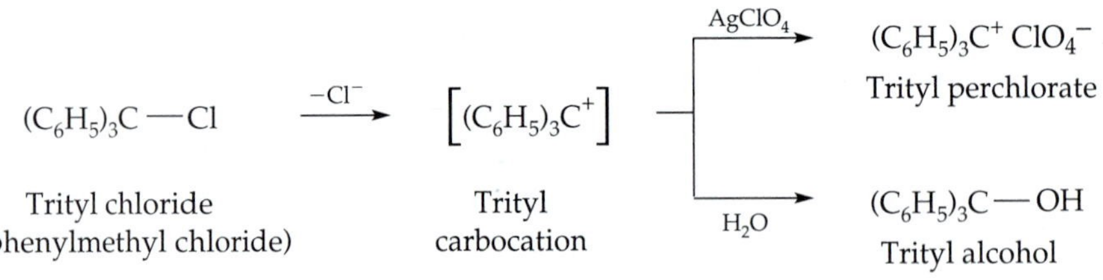

$$(C_6H_5)_3C{-}Cl \xrightarrow{-Cl^-} [(C_6H_5)_3C^+] \xrightarrow{AgClO_4} (C_6H_5)_3C^+\ ClO_4^-$$

$$[(C_6H_5)_3C^+] \xrightarrow{H_2O} (C_6H_5)_3C{-}OH$$

Trityl chloride (triphenylmethyl chloride) — Trityl carbocation — Trityl perchlorate — Trityl alcohol

PROBLEM 9.19

Predict the major product of each of the following reactions:

(a) 1-phenylpropene with HBr

(b) 1-phenylpropene with dilute aqueous sulfuric acid

9.6.3 Oxidation of Alkyl Side Chains

Recall that alkanes are inert to normal chemical oxidations, such as with hot potassium permanganate. They succumb only to the severe oxidation conditions created during combustion, and in that case are converted completely to carbon dioxide and water (see Section 2.7.2). Similarly, benzene rings are not susceptible to normal laboratory oxidation conditions, but also can be "burned" in a combustion (oxidation) reaction. In fact, alkylbenzenes are important octane-raising components of nonleaded gasoline.

$$\underset{\text{Alkane}}{R{-}CH_2{-}R'} \xrightarrow[\Delta]{KMnO_4} \text{No reaction} \xleftarrow[\Delta]{KMnO_4} \text{Benzene}$$

However, when an alkyl group is attached to a benzene ring and there is at least one *benzylic* hydrogen, oxidation with hot potassium permanganate occurs at the benzylic position to cleave the alkyl chain and produce benzoic acid (after final acidification):

$$\underset{\text{Alkylbenzene}}{C_6H_5{-}CH_2{-}R} \xrightarrow[\Delta]{KMnO_4} \underset{\text{Potassium benzoate}}{C_6H_5{-}COO^-\ K^+} \xrightarrow{H^+} \underset{\text{Benzoic acid}}{C_6H_5{-}COOH}$$

This reaction occurs in the presence of other substituents and may occur with more than one alkyl substituent. For example, *p*-chlorotoluene yields *p*-chlorobenzoic acid, and both *o*-xylene and tetralin (tetrahydronaphthalene) produce phthalic acid:

$$\underset{p\text{-Chlorotoluene}}{CH_3{-}C_6H_4{-}Cl} \xrightarrow[2.\ H^+]{1.\ KMnO_4} \underset{p\text{-Chlorobenzoic acid}}{HOOC{-}C_6H_4{-}Cl}$$

$$\text{Tetralin},\ \underset{o\text{-Xylene}}{C_6H_4(CH_3)_2} \xrightarrow[2.\ H^+]{1.\ KMnO_4} \underset{\text{Phthalic acid}}{C_6H_4(COOH)_2}$$

Any primary or secondary alkyl group attached to the benzene ring, even if part of a carbocyclic ring, will end up as a carboxyl group (—COOH) under such oxidation conditions.

The oxidation reaction is frequently used to determine the position(s) of attachment of alkyl (and alkenyl) substituents on benzene rings. This is effective because the three benzene dicarboxylic acids—the *ortho* isomer (phthalic acid), the *meta* isomer (isophthalic acid), and the *para* isomer (terephthalic acid)—are well known. If an unknown dialkylbenzene is oxidized, one of these dicarboxylic acids is obtained, and its identity can be readily determined simply by measuring its melting point. The structure of the precursor dialkyl benzene can be deduced from the resulting dicarboxylic acid. For example, if an unknown xylene (dimethylbenzene) is oxidized and produces terephthalic acid, then the unknown can only have been *p*-xylene.

How to Solve a Problem

An unknown compound **A** has the molecular formula C_8H_{10}. Oxidation of **A** with hot potassium permanganate, followed by acidification with dilute hydrochloric acid, results in compound **B,** $C_7H_6O_2$. What are the structures of compounds **A** and **B**?

PROBLEM ANALYSIS

To solve a road map problem such as this, we first work systematically "down" the problem, drawing obvious conclusions and deductions from the data in the sequence presented. Only then do we work back "up" the problem, using structures that should be apparent by then. Also, we try to deduce as much as possible from the molecular formula before considering the reactions.

The molecular formula (C_8H_{10}) tells us that the compound is eight hydrogens short of being an alkane (C_8H_{18}). The low hydrogen-to-carbon ratio suggests an aromatic compound, which, with only eight carbons in the molecule, must be benzenoid. Benzene itself (C_6H_6) is eight hydrogens short of being an alkane (C_6H_{14}). This accounts for the ring and three "double" bonds, each requiring the "loss" of two hydrogens. If the compound is benzenoid, there are only two carbons and no additional unsaturation left unaccounted for. These two carbons could be attached to the ring as a single group, which must be an ethyl group, or as two methyl groups. Since strong oxidation kept seven of the eight carbons together in **B** (only one carbon was lost), the system is confirmed to be benzenoid. Further, any such oxidation reaction must produce a carboxyl group (—COOH). Therefore, deducting the partial formula CO_2H from the formula for **B,** $C_7H_6O_2$, leaves C_6H_5. This is consistent only with **B** being benzoic acid—a phenyl group attached to a carboxyl group. (*Important clue:* Always suspect that $C_7H_6O_2$ is benzoic acid.) Because **B** is benzoic acid, there can only have been a single alkyl group attached to the benzene ring in **A.** If the eighth carbon had been attached to the ring, benzoic acid could not have been formed in the oxidation; one of the benzene dicarboxylic acids (phthalic acids) would have been produced instead.

SOLUTION

Compound **B** must be benzoic acid. This means that compound **A** has only a single group attached to the benzene ring, which oxidizes into the carboxyl group. Thus, a partial structure for compound **A** is **A1,** which accounts for seven of the eight carbons. The eighth carbon can only be attached as shown in structure **A2** (it cannot be attached to the ring; see the preceding analysis). Therefore, the carbon skeleton is as shown in **A2,** and adding in the necessary hydrogens gives structure **A,** ethylbenzene.

—COOH ⟹ —C— ⟹ —C—C ⟹ $—CH_2CH_3$

B Benzoic acid ($C_7H_6O_2$) | **A1** | **A2** | **A** Ethylbenzene (C_8H_{10})

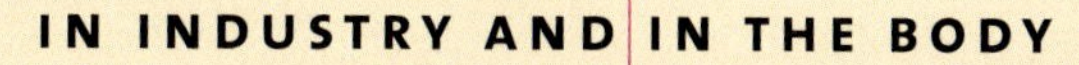

Industrial Oxidation and Detoxification

Oxidation of organic compounds is perhaps the most common reaction known—occurring in nature, in the laboratory, in the home, in industrial plants, and in our bodies. The ease with which families of compounds are oxidized varies widely. Of the families discussed so far in this book, alkanes and alkyl halides are inert to oxidation (other than combustion); alcohols oxidize under moderate conditions; alkenes and alkynes oxidize under mild conditions; aromatic rings are inert to oxidation; and alkylbenzenes oxidize under rather strong conditions. Quite a range of reaction conditions!

Two oxidation reactions are used in industry to produce the raw materials for the popular fabric with the trade name *Dacron.* Dacron was first prepared as an alternative fiber to cotton and is manufactured in very large quantities all over the world. It is a polymer, a *polyester,* which can be produced from the repeated ester-forming reaction between terephthalic acid and ethylene glycol (see Sections 15.3.4 and 18.4.2):

$$\underset{\text{Terephthalic acid}}{HOOC{-}C_6H_4{-}COOH} + \underset{\text{Ethylene glycol}}{HOCH_2CH_2OH} \xrightarrow{-H_2O} \underset{\text{Dacron (a polyester)}}{\left[\overset{O}{\overset{\|}{C}}{-}C_6H_4{-}\overset{O}{\overset{\|}{C}}{-}OCH_2CH_2O\right]_n}$$

Ethylene glycol (the compound used in automobile coolants) is produced by the oxidation of ethylene. Ethylene, in turn, is obtained by thermal catalytic dehydrogenation of ethane. (These reactions were discussed in Sections 5.2.3 and 6.6.1.)

$$\underset{\text{Ethane}}{CH_3{-}CH_3} \xrightarrow[\text{catalyst}\ \Delta]{-H_2} \underset{\text{Ethylene}}{CH_2{=}CH_2} \xrightarrow[2.\ H_2O]{1.\ O_2} \underset{\text{Ethylene glycol}}{HO{-}CH_2CH_2{-}OH}$$

Terephthalic acid is produced by catalytic air oxidation of *p*-xylene, which is obtained from coal tar or by catalytic conversion from petroleum. The oxidation of *p*-xylene requires very severe reaction conditions.

$$\underset{p\text{-Xylene}}{CH_3{-}C_6H_4{-}CH_3} \xrightarrow[\text{catalyst}]{O_2} \underset{\text{Terephthalic acid}}{HOOC{-}C_6H_4{-}COOH}$$

In contrast to the very severe conditions required for oxidizing an arylalkane to an arylcarboxylic acid, the same overall conversion is accomplished in the body at normal body temperature (~37°C). For example, toluene, a constituent of unleaded gasoline and a major industrial chemical, is toxic to humans, though not necessarily fatal. If accidentally ingested, toluene is detoxified by an enzymatic oxidation in the liver, forming benzoic acid, which is converted to its water-soluble salt and excreted from the body in the urine.

$$\underset{\text{Toluene}}{C_6H_5{-}CH_3} \xrightarrow[\text{oxidation)}]{\text{(enzymatic}} \underset{\text{Benzoic acid}}{C_6H_5{-}COOH}$$

In contrast, benzene cannot be detoxified in this manner because it has no alkyl group for oxidative attack, and the ring is too resistant to oxidation. Therefore, benzene remains in the body much longer than toluene, eventually acting as a carcinogen (a cancer-causing substance).

PROBLEM 9.20

Show the expected product of oxidation of each of the following compounds with hot potassium permanganate followed by acidification:

(a) *p*-nitrotoluene
(b) *p*-isopropyliodobenzene
(c) 1,3,5-trimethylbenzene (mesitylene)
(d) 1-phenyl-1-butene
(e) cyclobutylbenzene

PROBLEM 9.21

An unknown compound **A** has molecular formula $C_{10}H_{12}$. It does not decolorize bromine or react with cold potassium permanganate. Compound **A** was converted into a dicarboxylic acid ($C_8H_6O_4$) upon oxidation with hot potassium permanganate followed by acidification of the filtrate remaining after the removal of manganese dioxide. This diacid is shown to be phthalic acid. What is a possible structure for compound **A**?

9.7 Important Aromatic Compounds

9.7.1 Benzenoid Compounds

Benzene is a widely used commercial chemical. Most benzene used to be obtained from coal tar (a by-product of the formation of coke from coal), but now most is obtained from petroleum. Modern refineries have the ability, through patented catalytic processes, to convert straight-chain alkanes into cycloalkanes in a dehydrogenation process and then to convert cycloalkanes into benzenoid compounds in a second dehydrogenation process.

Benzenoid compounds are important to several major industries. Some compounds are used in gasoline; the addition or inclusion of aromatics is a major means of increasing the octane rating of unleaded gasoline. A large amount of benzene is converted into styrene, using the Friedel-Crafts alkylation followed by dehydrogenation. The styrene is then polymerized to polystyrene or, together with butadiene, polymerized to synthetic rubber.

$$\text{Benzene} + CH_2{=}CH_2 \text{ (Ethylene)} \xrightarrow{H_3PO_4} C_6H_5{-}CH_2CH_3 \text{ (Ethylbenzene)}$$

$$C_6H_5{-}CH_2CH_3 \xrightarrow[\text{catalyst, } \Delta]{-H_2} C_6H_5{-}CH{=}CH_2 \text{ (Styrene)} \xrightarrow{\text{(radical polymerization)}} \left[CH(C_6H_5){-}CH_2 \right]_n \text{ (Polystyrene)}$$

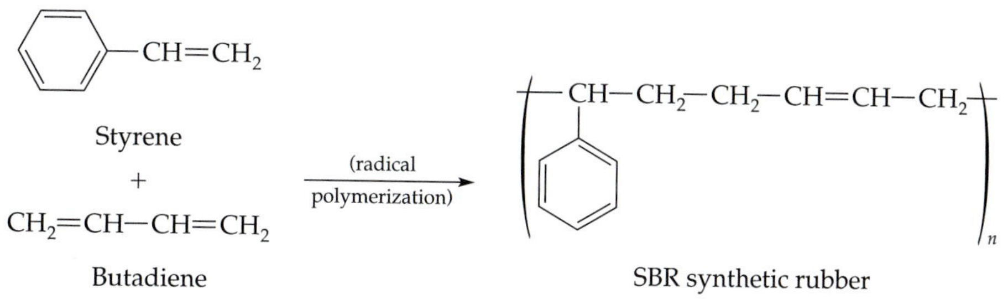

Styrene (C_6H_5—CH=CH$_2$) + Butadiene (CH$_2$=CH—CH=CH$_2$) $\xrightarrow{\text{(radical polymerization)}}$ SBR synthetic rubber [—CH(C_6H_5)—CH$_2$—CH$_2$—CH=CH—CH$_2$—]$_n$

Benzene is also important as a raw material, used to produce other chemicals important in manufacturing processes. Some benzene is hydrogenated under strong catalytic conditions to cyclohexane, which is converted to chemicals used in the manufacture of nylon (see Chapter 18). Benzene can be alkylated with propylene to form isopropylbenzene (cumene), which is oxidized to form phenol and acetone. Benzene also can be nitrated to nitrobenzene, which is reduced to aniline, a very important raw material for many chemical processes, including the manufacture of many dyes.

9.7.2 Polynuclear Aromatic Hydrocarbons

Benzene is a monocyclic aromatic hydrocarbon. A large family of compounds contain two or more benzene rings "fused" together to form polycyclic compounds known as *polynuclear aromatic hydrocarbons* (sometimes abbreviated PAHs). The fusion of two or more benzenoid rings fulfills the Hückel rule requirement for $4n + 2$ pi electrons. Polynuclear aromatic hydrocarbons are normally planar. Naphthalene is a bicyclic PAH, with 10 pi electrons, while anthracene (from linear fusion to naphthalene) and phenanthrene (from angular fusion to naphthalene) are tricyclic PAHs with 14 pi electrons.

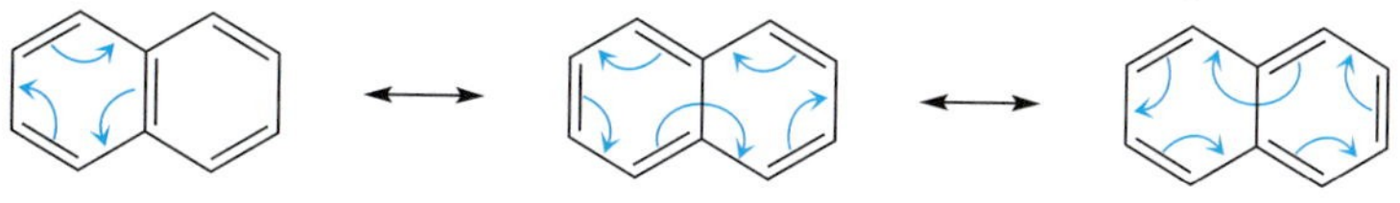

Naphthalene

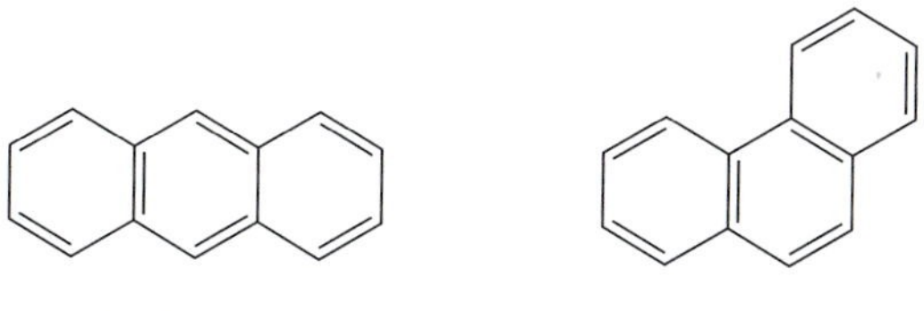

Anthracene

Phenanthrene

Contributing structures can be drawn for polynuclear aromatic hydrocarbons (note the three shown for naphthalene). PAHs undergo electrophilic aromatic substitution, just as benzene does, although the reaction outcomes are complicated by the possibility of substitution at various nonequivalent positions. While naphthalene has only two nonequivalent positions with substitutable hydrogens, there are many more such positions in the larger polynuclear aromatic hydrocarbons. Naphthalene can be substituted at position 1 or 2, depending on the electrophilic reagent used and the reaction conditions:

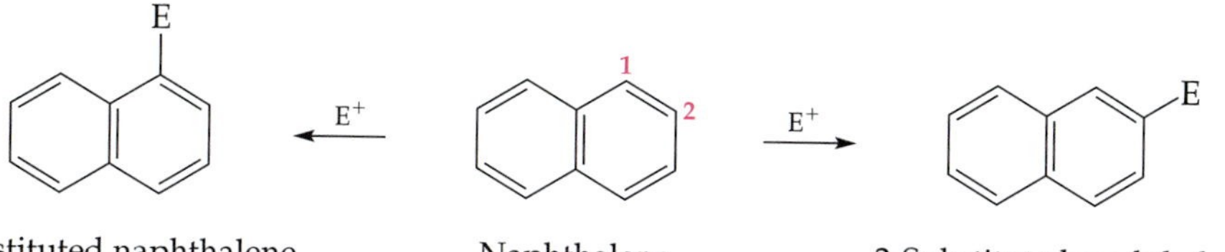

CHEMISTRY AT WORK

IN NATURE

Elemental Forms of Carbon

You may think that knowledge of the chemical elements is complete, especially for such a common element as carbon. And much of what is known about carbon *was* discovered over 100 years ago. But it may surprise you to learn that one of carbon's natural forms was discovered only in 1985.

As noted in Chapter 2, *diamond* is one of the elemental (pure) forms of carbon. Diamond is a continuous network of chair-form cyclohexane rings fused together. The carbon atoms use only sp^3 hybrid orbitals, with each carbon sigma-bonded to four others. This "architecture" creates a particularly rigid structure that makes diamond the hardest substance known.

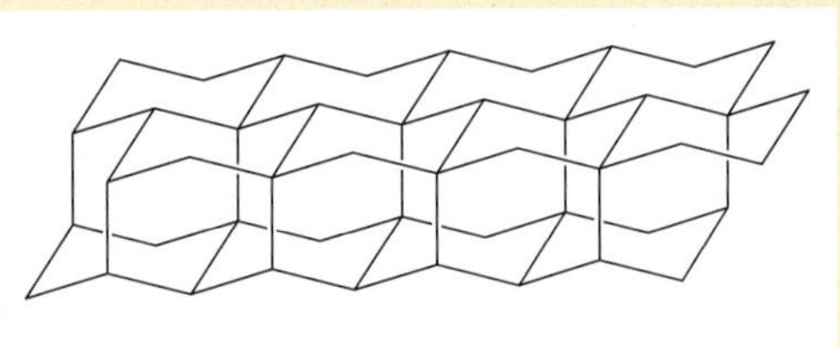

A portion of the diamond lattice

A second elemental form of carbon is *graphite*, which consists of layers made up of benzene rings fused together in a continuous planar network. The architecture here is that of a giant polynuclear aromatic hydrocarbon, with continuous angular and linear fusion of benzene rings. Each carbon atom is sp^2 hybridized and connected to three other carbon atoms. Graphite is, therefore, a kind of aromatic polymer. The C—C bond lengths are 1.42 Å, compared to 1.39 Å for benzene rings, and the individual layers are 3.4 Å apart. These layers are held together by relatively weak electrical forces, not by chemical bonds, so they are relatively free to slide past each other. This accounts for the slippery feel and lubricating qualities of graphite, which also is the "lead" in pencils.

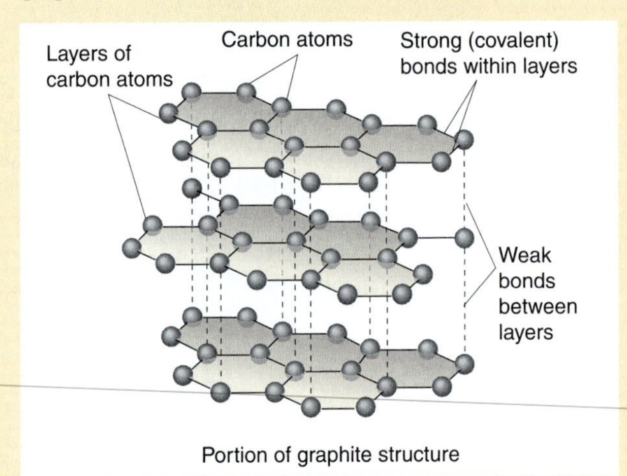

Portion of graphite structure

Some polynuclear aromatic hydrocarbons have a large number of fused rings, and the variety of combinations are too numerous to show here. However, one interesting example is coronene, which contains six angularly fused rings in a planar array. (Is the center ring a hole or a benzenoid ring?) The fusion of five benzene rings results in the nonplanar compound corranulene. In corranulene, each ring fusion junction is slightly bent. The molecule is saucer-shaped (it looks like a third of a soccer ball) because the interior pentagon would have to have very long benzenoid bonds if the molecule were planar. It is as if a drawstring were placed around five planar fused benzene rings and drawn so tight that the outer circumference was forced to shrink to the point where it buckled.

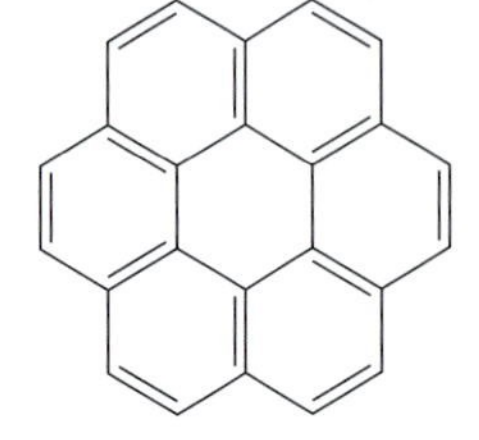

Coronene

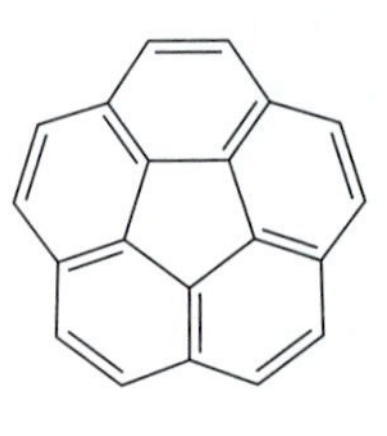
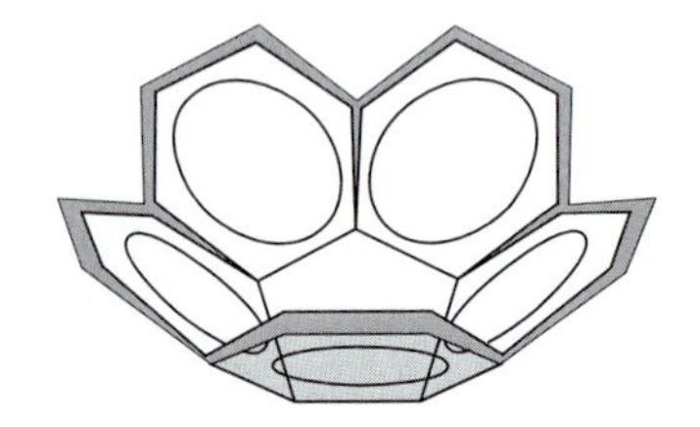

Corranulene

Yet a third form of elemental carbon was discovered in 1985 by Robert Curl, Jr., Richard E. Smalley, and Harold W. Kroto, who shared the 1996 Nobel Prize in chemistry for their work. They obtained only very small quantities of this new elemental form from soot; larger quantities are now produced by vaporization of carbon electrodes in a high vacuum. Unlike diamond and graphite, the new form is not a network, but a defined structural unit. It has the molecular formula C_{60} and possesses a molecular architecture of incredible symmetry—a sphere whose surface contains 20 hexagonal rings and 12 pentagonal rings. Each of the 12 pentagons is surrounded by 5 hexagons, and each hexagon is surrounded by 3 pentagons and 3 hexagons, exactly like a soccer ball (specifically, it is a truncated icosahedron).

The similarity in shape to that of the geodesic domes first created by the architect Buckminster Fuller led to the name *buckminsterfullerene,* frequently shortened to *buckyball.* It was named "Molecule of the Year" by Science magazine in 1991!

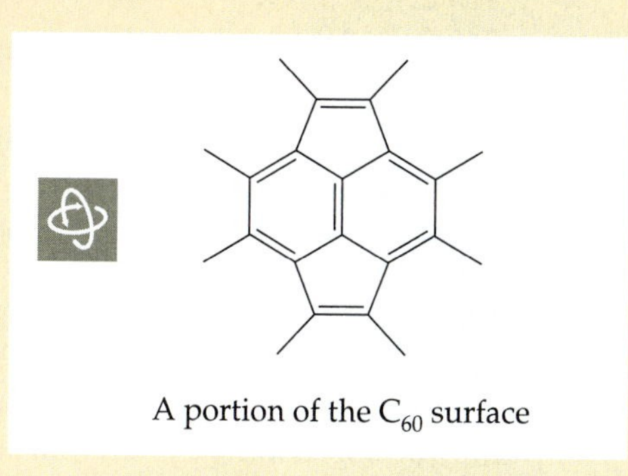

A portion of the C_{60} surface

C_{60} has a very stable structure with aromatic character. Each carbon is *sp*^{2} hybridized, and there are no "edges" to the molecule. Because each benzenoid ring is not perfectly planar, C_{60} undergoes a number of addition reactions. Further, buckyballs are large enough to encapsulate other atoms, and many new examples of "atoms in a sphere" are being synthesized, including the potassium ion, the rubidium ion (Rb_3C_{60} is a superconductor), and the helium atom.

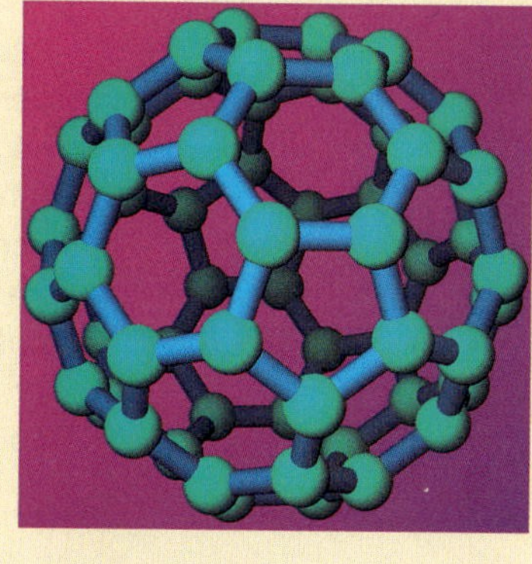

Computer model of buckminsterfullerene.

Traces of C_{60} with helium trapped inside were discovered in 1994 in the 2-billion-year-old meteorite impact crater near Sudbury, Ontario, Canada. Very interesting questions arise. For example, was the C_{60} formed in outer space or upon impact? And was the helium encapsulated on earth after impact or in outer space before impact? The isotopic ratio of helium-3 to helium-4 is consistent with that of helium found in interplanetary space and not with earth-bound helium.

A great deal of research is under way on the chemistry and potential applications of C_{60}. Additional stable cycloaromatic molecules, such as C_{70}, have also been discovered, and the whole family is now called *fullerenes.*

Some PAHs are known to be carcinogenic (cancer-causing). Workers with extended exposure to soot and coal tar, like coal miners, have unusually high incidences of cancer, linked to exposure to polynuclear aromatic compounds. Early medical tests showed that simply painting mouse skin with solutions of these compounds caused tumors to form. Polynuclear aromatic hydrocarbons are also present in cigarette smoke and are blamed for the high incidence of cancer among heavy smokers. For example, the carcinogen benzo[a]pyrene, present in cigarette smoke, has been shown to cause lung cancer.

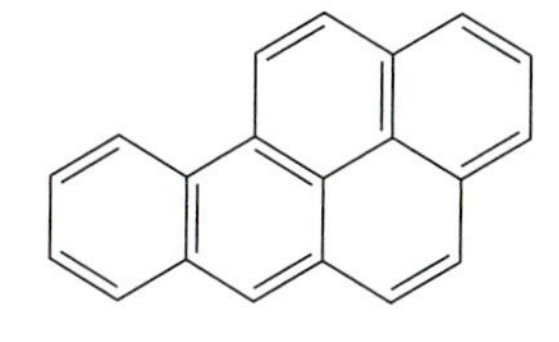

Benzo[a]pyrene

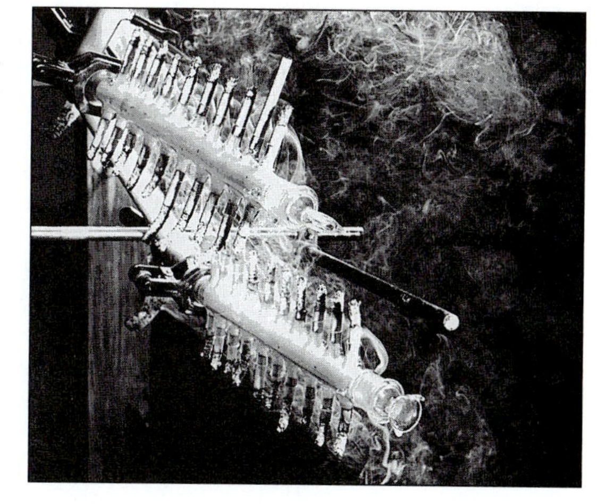

Smoking machines are used to collect cigarette smoke for analysis of the products, one of which is the carcinogen benzo[a]pyrene.

9.7.3 Heterocyclic Aromatic Compounds

The Hückel rule (Section 9.1.3) predicts that any planar closed cyclic conjugated system having $4n + 2$ pi electrons will be aromatic. To this point, all of the compounds described as aromatic have had ring systems containing only carbon atoms. However, an entire field of study focuses on aromatic compounds whose rings contain one or more atoms other than carbon (heteroatoms). Most often, these atoms are oxygen, nitrogen, or sulfur, but there are also more exotic ones, such as phosphorus, silicon, and boron. Here we can only briefly consider such compounds; their synthesis and reactions will not be presented. Heterocyclic aromatic compounds, which may be monocyclic or polycyclic, are referred to as **heteroaromatic compounds.** They are widely dispersed in nature and are part of some very important and complex chemical structures.

Monocyclic Heteroaromatics

The four most common monocyclic aromatic ring systems containing a single heteroatom are pyridine, pyrrole, furan, and thiophene. Their ring positions are numbered starting with the heteroatom:

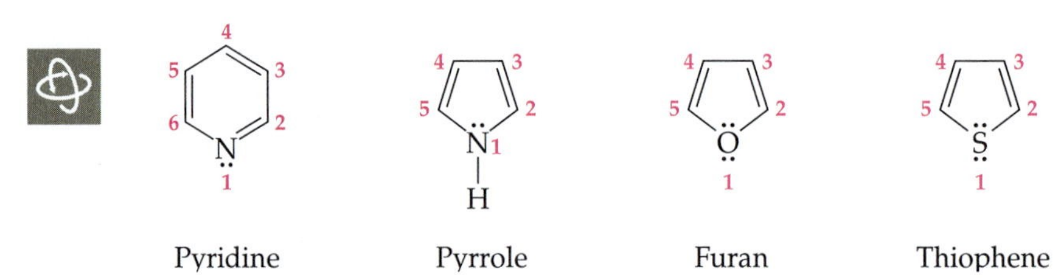

Pyridine is aromatic because it has six pi electrons, all located in p orbitals (Figure 9.6). One electron is contributed by each of the five sp^2-hybridized carbons and one by the sp^2-hybridized nitrogen. The nitrogen is *divalent;* the two nitrogen-carbon sigma bonds arise from sp^2-sp^2 overlap. The third sp^2 orbital contains nitrogen's lone pair of electrons, which are not involved in the aromatic sextet. Thus, these electrons are available to share, making nitrogen a base, just as it is in ammonia or amines (see Section 12.2.2). Therefore, pyridine is basic and is protonated by dilute acids, forming pyridinium salts.

Pyrrole is a different matter, however. The four sp^2-hybridized carbon atoms provide four pi electrons. Note that this nitrogen is *trivalent,* with the sigma bonds involving three sp^2 hybrid orbitals. Completion of an aromatic sextet requires both of the nitrogen's unshared electrons (in a p orbital). Therefore, pyrrole is not basic, since the unshared electrons are not free to be protonated without destruction of the aromatic character of the molecule (Figure 9.7).

FIGURE 9.6

The electronic structure of pyridine.

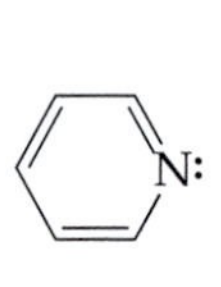

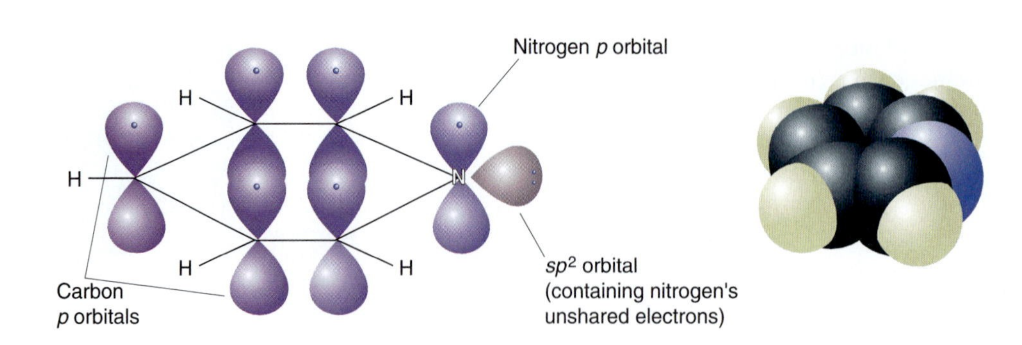

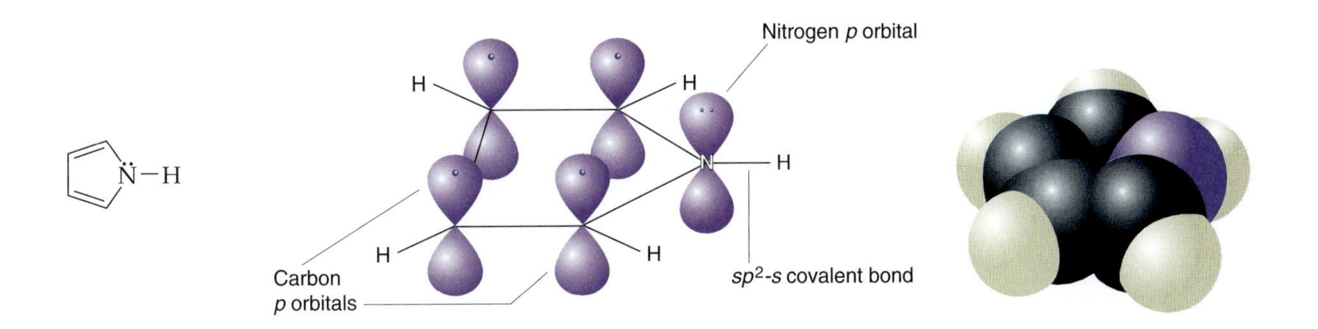

FIGURE 9.7

The electronic structure of pyrrole.

In both furan and thiophene, completion of the aromatic sextet requires four electrons from the carbons and two electrons from a *p* orbital of the respective heteroatom. Pyrrole, furan, and thiophene can be considered to be isoelectronic with the cyclopentadienide anion (that is, they have the same electron distribution), in which four *p* electrons are contributed by the four carbons involved in the double bonds and two are contributed by the negatively charged carbon:

H

Cyclopentadienide anion

There are many heteroaromatics that contain more than one heteroatom. The heteroatoms may be the same or different, and they may be separated by one or more carbons or be adjacent to each other. The compound imidazole contains two nitrogen atoms. One nitrogen atom is basic (similar to the nitrogen of pyridine), and one is not because its two unshared electrons are involved in the aromatic sextet (similar to the nitrogen of pyrrole). Other examples of five-membered heterocyclic aromatic compounds, which are isoelectronic with the cyclopentadienide anion, are oxazole and isoxazole. Pyrimidine is isoelectronic with benzene.

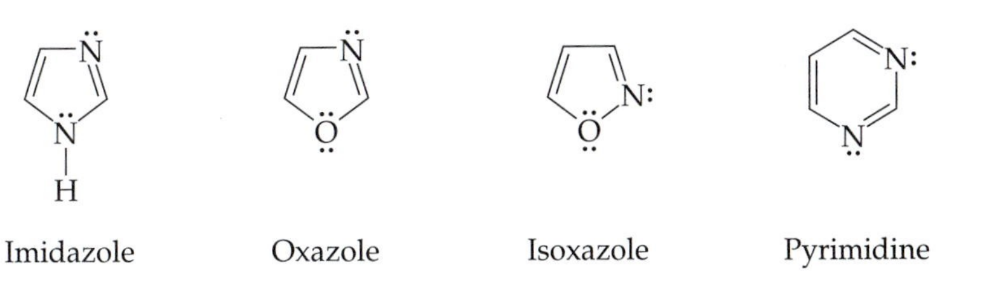

Heterocyclic aromatic compounds exhibit chemical behavior similar to benzene's, but the presence of the heteroatom occasionally permits additional kinds of reactions. Heterocyclic aromatic compounds undergo electrophilic aromatic substitution reactions analogous to those of benzene, but the heteroatom exerts directing effects on the substitutions.

Heterocyclic aromatic compounds have fully hydrogenated counterparts, referred to as **heterocycles** (as opposed to *carbocycles*, such as cyclohexane). Many of these heterocycles are important commercial and biological chemicals. Examples include piperidine (hexahydropyridine, a cyclic amine), tetrahydrofuran (a cyclic ether), and tetrahydrothiophene (a cyclic thioether).

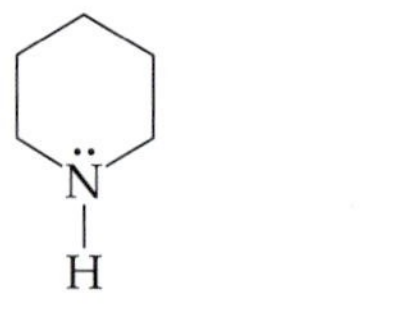

Piperidine

Tetrahydrofuran (THF)

Tetrahydrothiophene

Polycyclic Heteroaromatics

Just as benzene rings fuse to form polynuclear aromatic hydrocarbons, benzene and heteroaromatic rings can be fused to form polynuclear heteroaromatic compounds. There are hundreds of such ring combinations known; here are just a few examples:

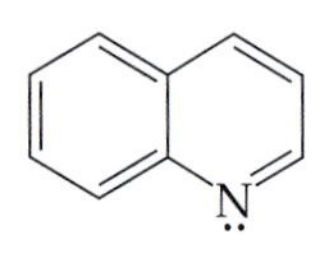

Quinoline

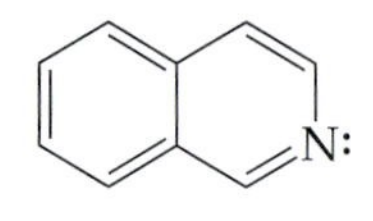

Isoquinoline

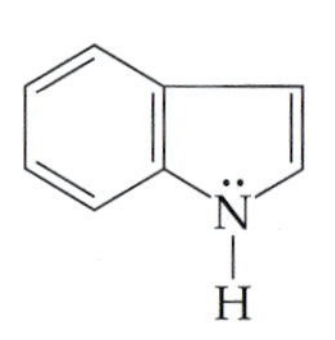

Indole

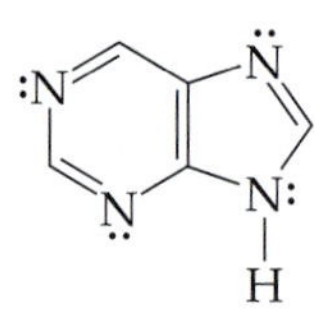

Purine

The pyridine ring is incorporated in the structure of nicotine (obtained from the tobacco plant) and niacin (vitamin B3).

The incorporation of these ring systems in biologically important compounds will be discussed in later chapters.

Important Heteroaromatics

There are countless interesting and biologically important heteroaromatics, but only a few examples will be considered here. Many heteroaromatics are also members of the family of compounds known as *alkaloids*, which are naturally occurring nitrogenous bases, to be described in Section 12.7.1.

Monocyclic heteroaromatics include nicotine, an alkaloid from the tobacco plant, whose ingestion from cigarette smoke is a public health issue. Oxidation of nicotine produces nicotinic acid, the amide of which is known as niacin. Niacin is an essential vitamin (B_3) that becomes part of the coenzyme known as NAD, a catalyst for oxidation reactions in the body. Histamine is a toxic substance formed in the body by decarboxylation of the amino acid histidine; it is released in response to inflammation and allergic reactions. Thiamin (Vitamin B_1) is essential for carbohydrate metabolism.

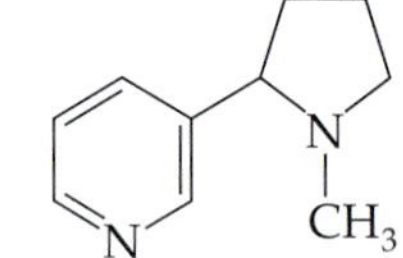

Nicotine

Nicotinic acid amide (niacin, vitamin B_3)

Histamine

Thiamin (vitamin B_1)

Polynuclear heteroaromatics include serotonin, an important neurotransmitter. Quinine was the original antimalarial drug. Caffeine is obtained from coffee beans, tea leaves, and cola beans; it can be removed from coffee beans by extraction with liquid carbon dioxide, producing decaffeinated coffee. The compound 6-mercaptopurine has been used in the treatment of cancer; structurally similar to the purine bases involved in cell replication, 6-mercaptopurine can disrupt the replication process.

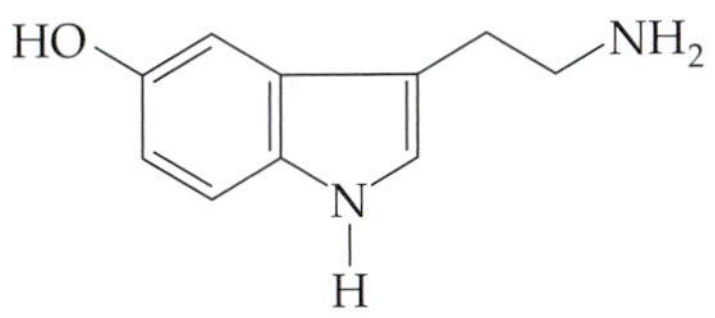

Serotonin 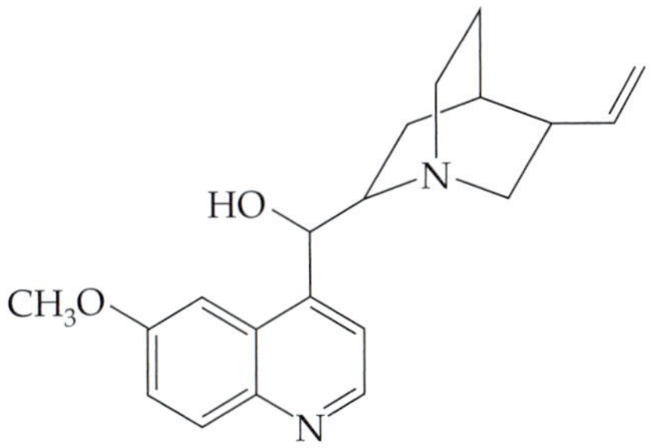 Quinine 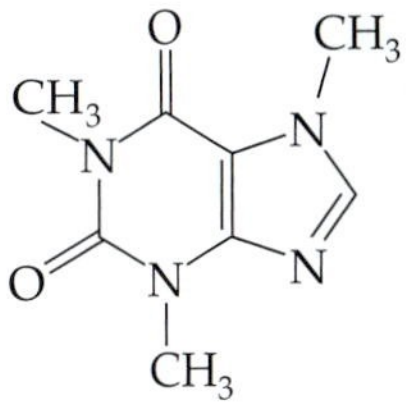 Caffeine 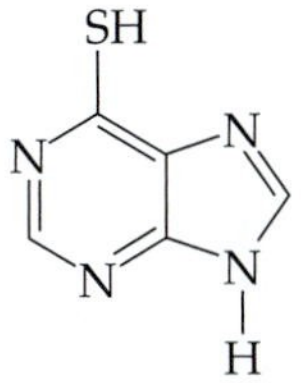 6-Mercaptopurine

The heme molecule contains four pyrrole units, bonded to four of the six coordination sites of iron(II). Four of these heme units attach to the protein *globin* by bonding between the fifth coordination site of iron and a histidine unit of globin. The result is hemoglobin, the red blood cell constituent responsible for the transport of oxygen and carbon dioxide. During their transport to and from the lungs, carbon dioxide and oxygen are temporarily bonded to the sixth coordination site of the iron atom, located in the middle of the ring system.

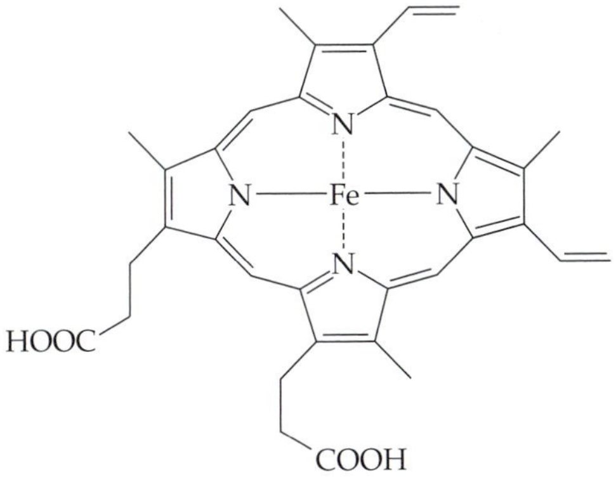

Heme

EXPLORATIONS

www.jbpub.com/organic-online

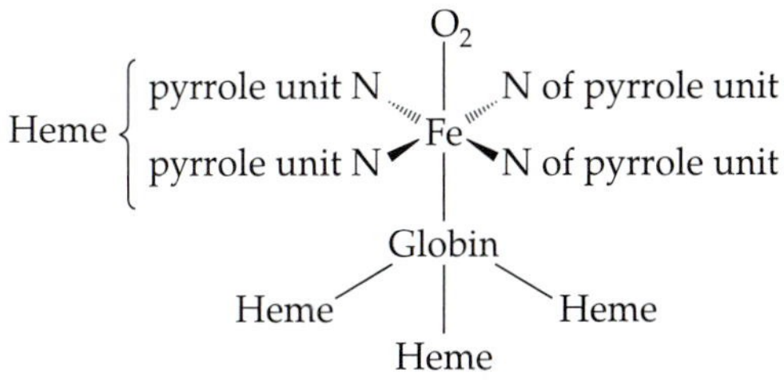

Schematic representation of an oxygenated heme unit. Four heme units are complexed to one globin protein to make up the hemoglobin molecule.

DNA and RNA Bases

DNA (deoxyribonucleic acid) is the molecule responsible for the storage of genetic information in the cells (that is, the genetic code). RNA (ribonucleic acid) is a similar molecule, responsible for transferring the genetic information within a cell so that it can be used in the synthesis of proteins. More discussion of the structure and role of DNA and RNA will appear in Chapter 19; here we focus on the heteroaromatic bases involved.

DNA (or RNA) and its components can be represented using a shorthand scheme in which B represents the heteroaromatic base, R represents ribose (a carbohydrate), and P represents phosphate ion. The nucleic acids are long chains of alternating carbohydrate (R) and phosphate units (P), with a heteroaromatic base (B) attached to each carbohydrate:

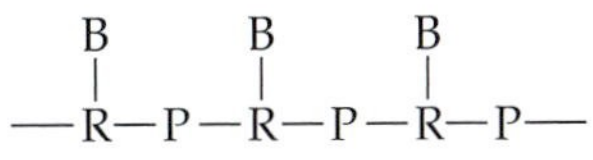

```
    B     B     B
    |     |     |
 —R—P—R—P—R—P—
```

Section of DNA or RNA
(a nucleic acid)

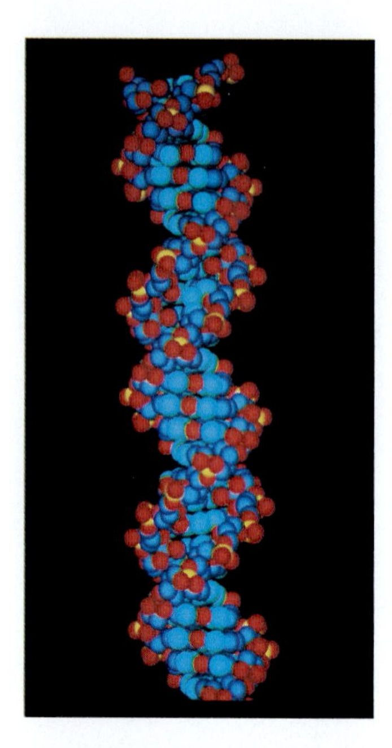

A computer rendering of the DNA molecule.

Nucleic acids can be sequentially hydrolyzed, being cleaved first into smaller molecules called *nucleotides*. Each nucleotide contains a carbohydrate molecule, a molecule of phosphoric acid, and a heteroaromatic base. An example is adenine deoxyribonucleotide:

NH_2 … Base (B)

$HOCH_2$ … Sugar (R)

OPO_3H_2 … Phosphate (P)

Adenine deoxyribonucleotide

Removal of the phosphoric acid unit from the nucleotide yields a *nucleoside*. Further hydrolysis of a nucleoside produces a carbohydrate (ribose from RNA and deoxyribose from DNA) and the free DNA base.

DNA or RNA ⟶ Nucleotide (B—R—P) ⟶ Nucleoside (B—R) + H_3PO_4

Nucleoside ⟶ DNA (or RNA) base + Ribose or Deoxyribose

Chemists break down nucleic acids in order to determine what bases occur in the original long chain and in what sequence they occur.

Attached to each carbohydrate of DNA or RNA is a heteroaromatic base. DNA, although containing thousands of nucleotide units, contains only four bases: abbreviated A (for adenine), G (for guanine), C (for cytosine), and T (for thymine). The *genetic code* consists of the particular sequence of the four bases in DNA. RNA also contains only four bases, three of which are identical to the DNA bases (A, G, and C). The fourth RNA base is abbreviated U (for uracil). Note that the RNA base uracil (U) differs from the DNA base thymine (T) only by the absence of a methyl group.

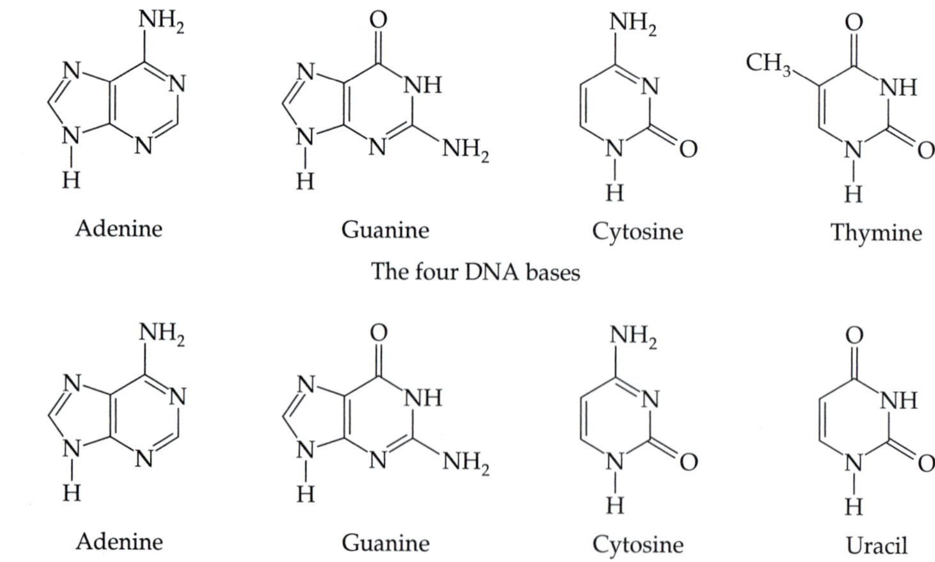

The four DNA bases

The four RNA bases

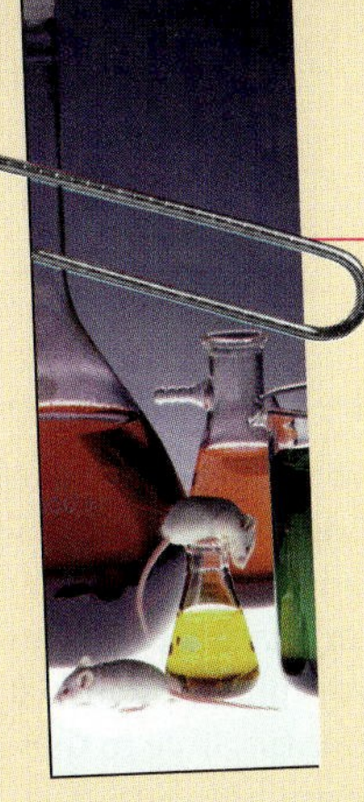

CHEMISTRY AT WORK

IN THE BODY

Drugs to Treat AIDS

Acquired immune deficiency syndrome (AIDS) is a twentieth-century disease that has become a major medical problem in the United States and around the world. It has ravaged worldwide populations as badly as any disease since the bubonic plague of the seventeenth century. This dreaded disease is caused by the human immunodeficiency virus (HIV), which can remain inactive in the body for a considerable time before becoming active and resulting in full-blown AIDS. The fatality rate for those infected with AIDS has been extremely high. No cure has yet been found. Research has focused on treatment, on preventing HIV from developing into AIDS, and on developing an AIDS vaccine.

The number of drugs that are effective in treating HIV infection and AIDS is small, but increasing. These drugs are not cures, but are designed to delay the onset of AIDS symptoms. Many of these drugs are nucleosides; that is, they contain heteroaromatic bases. They are designed to be similar enough to the nucleosides in the RNA of the HIV virus that they block further replication of the virus.

One family of AIDS drugs, the reverse transcriptase inhibitors (RTIs), inhibits an essential HIV-replicating enzyme called reverse transcriptase. AZT was the first example of this family of drugs. A completely different family of drugs, the protease inhibitors, blocks the protease enzyme of HIV. A member of this family of drugs is Indinavir. Note the heteroaromatic bases in both molecules.

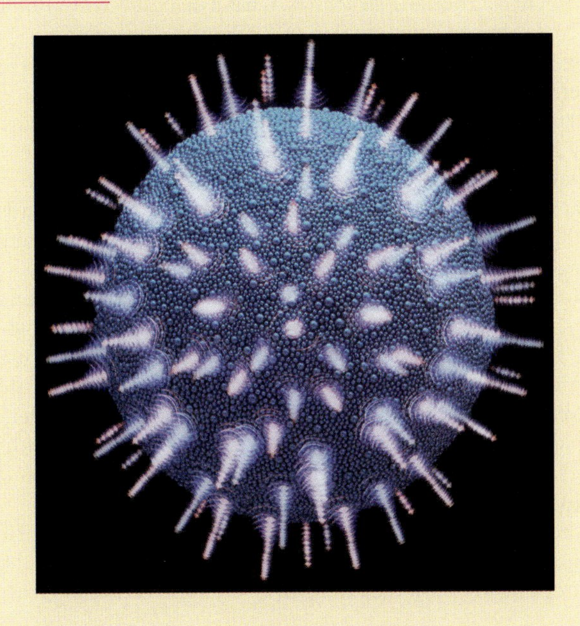

The HIV virus.

A more detailed discussion of HIV, AIDS, and the mechanisms of various treatments appears in Section 19.3.4.

AZT
(azidothymine
or zidovudine)

Indinavir

Chapter Summary

Aromatic compounds are structurally related to **benzene** and are referred to as **arenes.** Benzene is a resonance hybrid of two equivalent contributing structures. It has a delocalized electronic structure formed from overlap of six pi orbitals containing six pi electrons. As a result, benzene, and aromatic compounds in general, are stabilized by resonance energy, amounting to 36 kcal/mol in the case of benzene. To be aromatic, a compound must be planar, each of the ring atoms must have a parallel *p* orbital, and the number of electrons in the *p* orbitals must be equal to $4n + 2$ (the **Hückel rule**). **Heteroaromatic compounds** are those in which one or more heteroatoms, such as oxygen, nitrogen, or sulfur, are incorporated into an aromatic ring.

Benzenoid compounds are named by the IUPAC system, but many historical names have been incorporated into that system. Disubstituted benzenoid compounds are named using the prefixes ***ortho-***, ***meta-***, and ***para-*** (*o-*, *m-*, and *p-*) to indicate the relative positions of substituents. Polysubstituted benzenoid compounds are named using numbers as position indicators. The C_6H_5— group is known as the **phenyl group,** and $C_6H_5CH_2$— is known as the **benzyl group.**

The major reaction of aromatic compounds is **electrophilic aromatic substitution.** The pi electrons of the ring attack the electrophile in a slow step to form an arenium cation intermediate, which then loses a proton to restore the **aromatic sextet.** The major substitution reactions are **halogenation, nitration, sulfonation,** and **Friedel-Crafts alkylation** and **acylation.**

Substituents already on the benzene ring have a major influence on the course of subsequent electrophilic aromatic substitutions, affecting the rate of substitutions (relative to that for benzene) and directing incoming substituents to particular positions. Electron-donating substituents are generally **activators** and direct to the *ortho* and *para* positions. Electron-withdrawing substituents are **deactivators** and generally direct to the *meta* position. Halogen substituents are deactivators but direct substituents to the *ortho* and *para* positions.

Phenols, such as phenol itself (C_6H_5OH), are very easily substituted because of the strong activating effect of the hydroxyl group. They are also more acidic ($pK_a \sim 10$) than alcohols.

Carbocations adjacent to the phenyl ring (benzylic carbocations) are highly stabilized by resonance delocalization through the ring. Also, **alkenyl benzenes,** compounds in which a double bond is conjugated with the ring, are especially readily formed.

Primary or secondary alkyl side chains attached to a benzene ring are oxidized by hot potassium permanganate to carboxyl groups.

Summary of Reactions

1. **Electrophilic aromatic substitution (Section 9.3).** Benzene and its derivatives can be substituted by appropriate electrophilic reagents.

a. Halogenation (bromination and chlorination) (Section 9.3.2)

$$\text{Benzene} \xrightarrow[\text{FeX}_3]{\text{X}_2} \text{Halobenzene} + \text{HX}$$

Benzene → Halobenzene (X = Br or Cl)

b. Nitration (Section 9.3.3)

$$\text{Benzene} \xrightarrow[\text{H}_2\text{SO}_4]{\text{HNO}_3} \text{Nitrobenzene } (C_6H_5NO_2) + H_2O$$

c. Sulfonation (Section 9.3.4)

$$\text{Benzene} \xrightarrow[\text{H}_2\text{SO}_4]{\text{SO}_3} \text{Benzenesulfonic acid } (C_6H_5SO_3H) + H_2O$$

d. Friedel-Crafts alkylation (Section 9.3.5)

$$\text{Benzene} \xrightarrow[\text{AlCl}_3]{\text{RCl}} \text{Alkylbenzene } (C_6H_5R) + HCl$$

e. Friedel-Crafts acylation (Section 9.3.6)

$$\text{Benzene} \xrightarrow[\text{AlCl}_3]{\text{RCOCl}} \text{Acylbenzene } (C_6H_5CO\text{—}R) + HCl$$

2. O-Alkylation and O-acylation of phenols (Sections 9.5.2 and 9.5.3). O-Alkylation is employed to synthesize ethers or to protect the hydroxyl group. O-Acetylation is used to lower the activating effect of the hydroxyl group to permit monosubstitution of the ring.

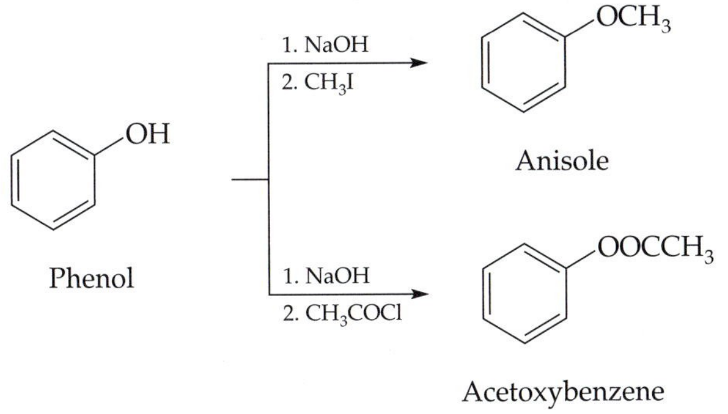

Phenol → (1. NaOH, 2. CH_3I) → Anisole ($C_6H_5OCH_3$)

Phenol → (1. NaOH, 2. CH_3COCl) → Acetoxybenzene ($C_6H_5OOCCH_3$)

3. Oxidation of alkylbenzenes (Section 9.6.3)

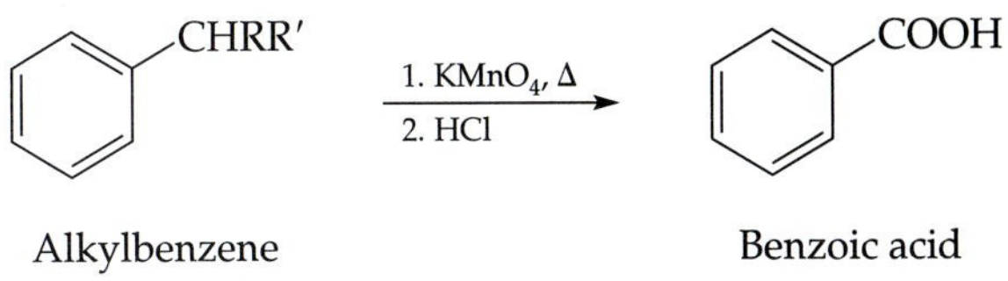

Alkylbenzene (C_6H_5CHRR') → (1. $KMnO_4$, Δ; 2. HCl) → Benzoic acid (C_6H_5COOH)

Additional Problems

Nomenclature and Structure

9.22 Write structural formulas for the following compounds:

(a) *o*-chlorotoluene
(b) *p*-dinitrobenzene
(c) 2,5-dimethylpyrrole
(d) 3,5-dinitrophenol
(e) 2,4,6-tribromoaniline
(f) *p*-isopropylbenzenesulfonic acid
(g) 2,4-difluoronitrobenzene
(h) 1,3,5-trimethylbenzene
(i) 2,4-dimethylbenzaldehyde
(j) phenylacetylene
(k) (*Z*)-2,3-diphenyl-2-butene
(l) 2-chloropyridine
(m) 3-nitrofuran
(n) *p*-iodobenzoic acid
(o) 3-hydroxythiophene
(p) vinylbenzene (styrene)
(q) pentachlorophenol
(r) benzyl alcohol
(s) (*E*)-2-phenyl-2-pentene
(t) *p*-toluenesulfonic acid
(u) diphenyl
(v) (*R*)-1-phenylethanol
(w) *p*-methoxyanisole

9.23 Assign names to the following compounds:

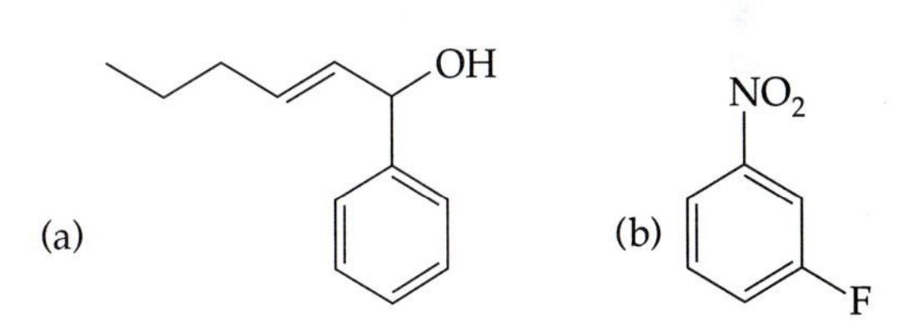

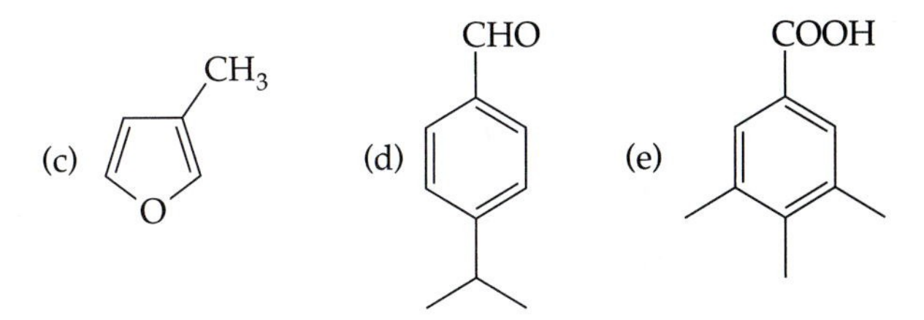

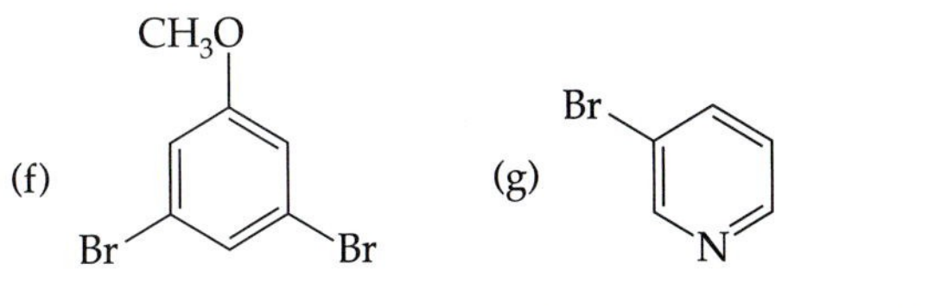

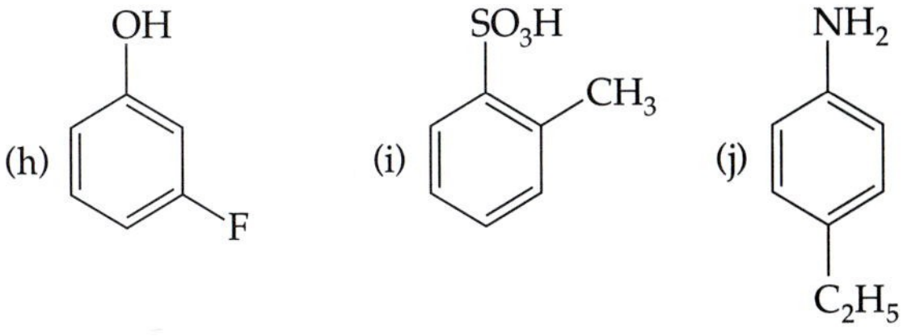

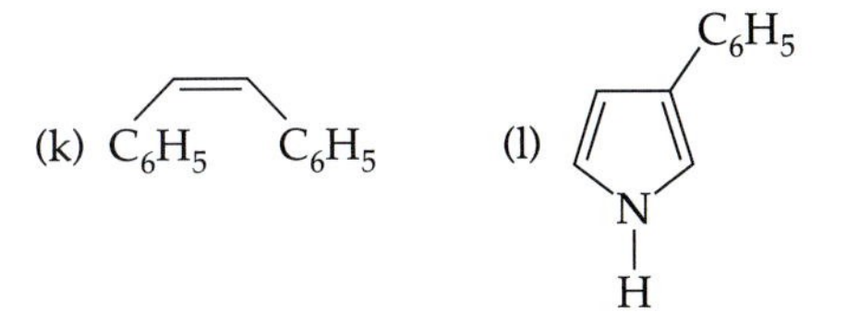

9.24 (a) Draw the three contributing structures for resonance stabilization of naphthalene, using curved arrows to show how one is converted into the other.

(b) Draw the five contributing structures for phenanthrene.

(c) Draw the four contributing structures for diphenyl.

9.25 How many different mononitro derivatives can you envisage for (a) naphthalene and (b) anthracene?

9.26 Predict whether each of the following compounds or ions is expected to be aromatic (assume a planar structure), and explain each answer:

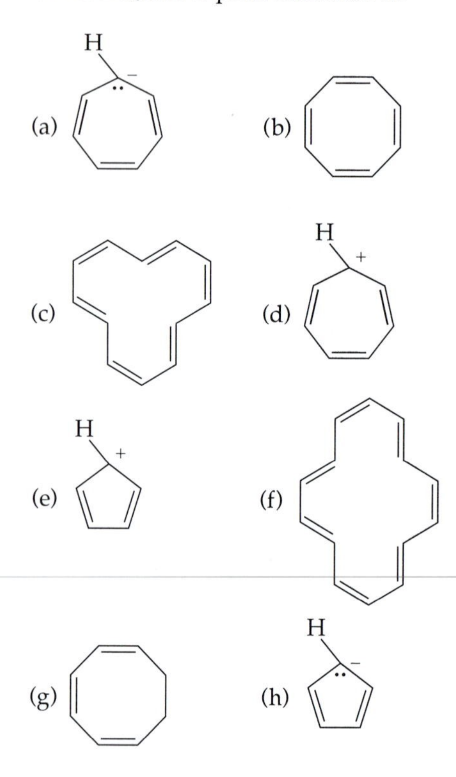

9.27 Draw the contributing structures for resonance stabilization of aniline, reflecting the delocalization of the lone pair of electrons on nitrogen into the benzene ring.

9.28 Draw the contributing structures reflecting the delocalization of the ring electrons into the nitro group of nitrobenzene.

Substitution Reactions

9.29 Show the contributing structures for resonance stabilization of the arenium cation that results from the attack of the bromous ion (Br^+) on benzene. Use curved arrows to indicate the electron shifts in converting one form into another.

9.30 For each of the following, write the mechanism for the production of the electrophile and its reaction with the other compound:

(a) toluene with chlorine in the presence of $AlCl_3$

(b) chlorobenzene with nitric acid in the presence of H_2SO_4

(c) bromobenzene with acetyl chloride in the presence of $AlCl_3$

(d) isopropyl benzene with sulfur trioxide and sulfuric acid

(e) iodobenzene with isopropyl chloride in the presence of $AlCl_3$

(f) bromobenzene with *t*-butyl alcohol in the presence of H_3PO_4

9.31 Each of the following substituents is an *ortho-para* director when attached to the benzene ring. Show how the substituent interacts electronically with the arenium cation formed from reaction of the nitronium ion (NO_2^+) with the substituted benzene at the *para* position:

(a) amino (—NH_2) (b) methoxy (—OCH_3)

(c) vinyl (—CH=CH_2) (d) phenyl (—C_6H_5)

(e) acetamido (—$NHCOCH_3$)

9.32 Using benzene or toluene as the starting material, devise a synthesis of each of these compounds:

(a) *p*-toluenesulfonic acid

(b) 2,6-dibromo-4-nitrotoluene

(c) *p*-bromonitrobenzene

(d) *m*-bromonitrobenzene

(e) *m*-chlorobenzoic acid

(f) *p*-bromobenzoic acid

(g) *p*-cyclohexyltoluene

(h) 4-chloro-2-nitrobenzoic acid

(i) *p*-methylacetophenone (*p*-$CH_3C_6H_4COCH_3$)

9.33 Place the compounds in each set in increasing order of reactivity toward electrophilic aromatic substitution:

(a) benzene, —NH_2, —$COCH_3$, —$NHCOCH_3$

(b) —OCH_3, —CH_3, —CHO

(c) 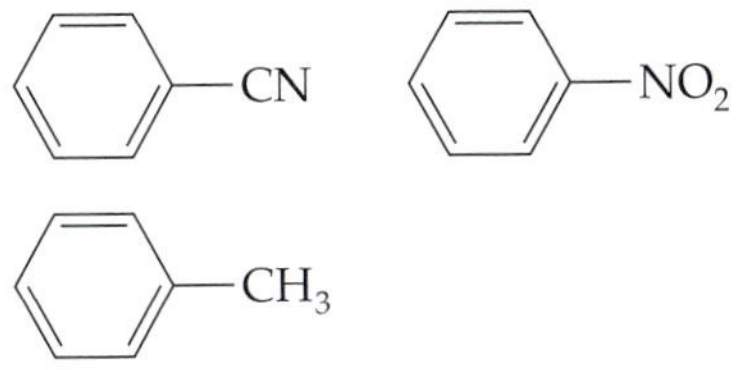 —CN —NO_2

—CH_3

9.34 For each of the following disubstituted arenes, circle the substituent that is more activating. Then draw the product of chlorination of each compound.

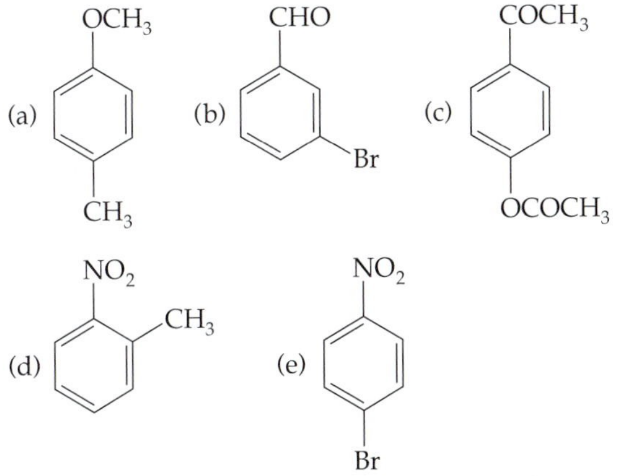

(a) OCH_3, CH_3 (b) CHO, Br (c) $COCH_3$, $OCOCH_3$

(d) NO_2, CH_3 (e) NO_2, Br

9.35 The sodium salt of 4-hexylbenzenesulfonic acid is to be tested for its suitability as a detergent. Show how it could be prepared from benzene.

Phenols

9.36 Explain why *p*-nitrophenol is more acidic than *p*-methoxyphenol.

9.37 Show the major product of each of the following reactions:

(a) anisole with $Br_2/FeBr_3$

(b) *p*-methylphenol (*p*-cresol) with $Cl_2/FeCl_3$

(c) phenol with HNO_3/H_2SO_4

(d) acetoxybenzene with HNO_3/H_2SO_4

9.38 Show the scheme you would use to separate the constitutional isomers benzyl alcohol and *p*-methylphenol (*p*-cresol) if they were dissolved together in an ether solution.

9.39 Show how to accomplish the following syntheses:

(a) benzene to phenol

(b) chlorobenzene to *p*-chlorophenol

(c) nitrobenzene to *m*-chlorophenol

(d) toluene to *p*-methylanisole

Mixed Problems

9.40 BHT (butylated hydroxytoluene; see p. 342) is a commercial antioxidant added to many foods as a preservative. It can be synthesized by a reaction between 2-methylpropene and *p*-cresol. What kind of catalyst would you use for this reaction? Show the mechanism of the reaction.

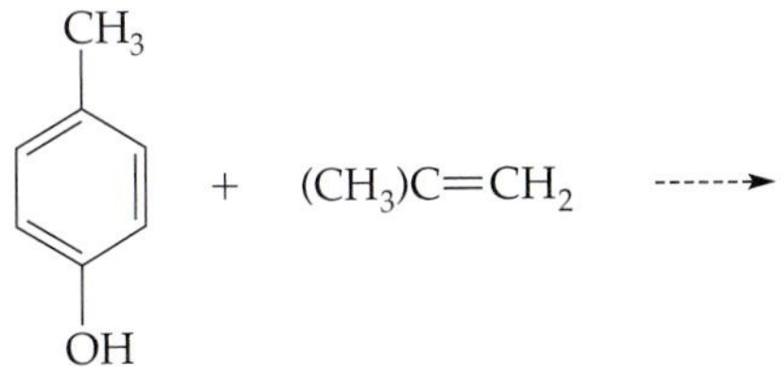

CH_3, OH + $(CH_3)C{=}CH_2$ ⟶

p-Cresol 2-Methylpropene

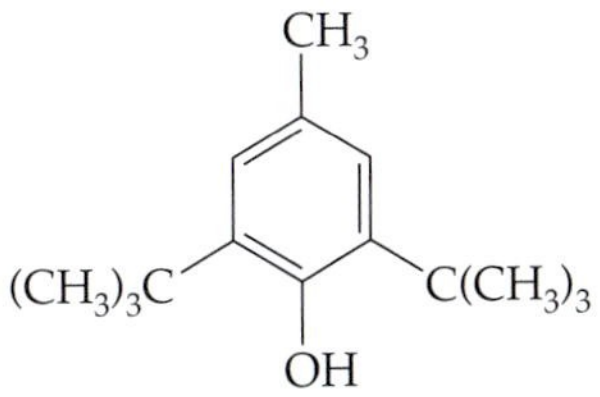

CH_3, $(CH_3)_3C$, $C(CH_3)_3$, OH

BHT

9.41 Show the products of each of the following reactions:

(a) styrene with HBr

(b) 2-chloro-1-phenylpropane with hot sodium ethoxide

(c) 1-phenyl-2-butanol with concentrated H_2SO_4 and heat

(d) product of part (c) with H_2O/H^+

(e) product of part (c) with diborane, then H_2O_2 and NaOH

(f) product of part (c) with H_2/Pt

(g) product of part (f) with hot $KMnO_4$, then H^+

9.42 To synthesize 3-bromo-5-nitrobenzoic acid, which would you choose as the starting material—3-bromobenzoic acid or 3-nitrobenzoic acid? Explain and show the reaction.

9.43 Give the structures and names for all possible trichlorobenzenes.

9.44 Show the reactions necessary to complete each transformation. (*Hint:* Use retrosynthesis.)

(a) benzyl bromide to *cis*-1-phenyl-2-butene

(b) 1-bromo-1-phenylethane to *trans*-4-phenyl-2-pentene

(c) 1-phenylpropene to 1-phenyl-1-propanol

(d) 1-phenylpropene to 1-phenyl-2-propanol

9.45 Although not discussed in Section 9.3.6, acid anhydrides can be used in the Friedel-Crafts acylation in place of acid chlorides to generate acylium cations. For example, succinic anhydride with aluminum chloride produces a ketocarboxylic acid ($C_{10}H_{10}O_3$).

Draw the structures of compounds **A–D**, formed by reactions in the sequence resulting in the overall conversion of benzene into naphthalene:

COOH

($C_{10}H_{10}O_3$)
A ketocarboxylic acid

$\xrightarrow[HCl]{Zn/Hg}$ **A** $\xrightarrow[2.\ AlCl_3]{1.\ SOCl_2}$ **B**

($C_{10}H_{12}O_2$) ($C_{10}H_{10}O$)

Zn/Hg, HCl

C ($C_{10}H_{12}$) $\xrightarrow[peroxides]{Br_2}$ Br ($C_{10}H_{11}Br$) $\xrightarrow[ethanol]{NaOEt}$ **D** ($C_{10}H_{10}$)

$\xrightarrow[peroxides]{Br_2}$ Br ($C_{10}H_9Br$) $\xrightarrow[ethanol]{NaOEt}$ Naphthalene ($C_{10}H_8$)

9.46 Show the two possible monobromination products of the reaction of $Br_2/FeBr_3$ with benzanilide ($C_6H_5NH—COC_6H_5$). Which product should predominate in the reaction mixture?

9.47 Using toluene as the starting material, show the reactions necessary to produce (a) benzoic acid and (b) *p*-nitrobenzoic acid.

9.48 (a) Although primary halides usually do not undergo S_N1 reactions, *p*-methylbenzyl bromide is solvolyzed rapidly by methanol to produce *p*-methylbenzyl methyl ether. Explain by writing the mechanism for the reaction.

(b) If the *p*-methyl group of *p*-methylbenzyl bromide is replaced by a *p*-methoxy group, the reaction described in (a) proceeds even faster, while if it is replaced by a *p*-nitro group, the reaction slows considerably. Explain.

9.49 The heterocyclic compounds pyrrole and pyridine are both nitrogen-containing aromatic compounds. A key difference between them is that pyridine is basic (as are most amines; see Chapter 12), whereas pyrrole is not. Explain in terms of the aromatic sextet.

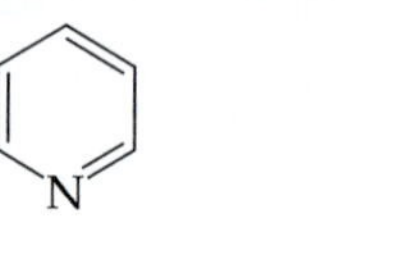

Pyridine

N
H

Pyrrole

9.50 Show how to convert toluene to 2,4-dinitrobenzoic acid and 3,5-dinitrobenzoic acid.

9.51 Compound **A** (C_9H_{10}) readily reacts with Br_2/CCl_4 in the dark and can be oxidized with cold potassium permanganate. Reaction of **A** with H_2/Pt leads to the absorption of one equivalent of hydrogen and the formation of compound **B**, which is inert to Br_2/CCl_4 and cold $KMnO_4$. Vigorous oxidation of either **A** or **B** with hot potassium permanganate followed by removal of MnO_2 leaves a colorless aqueous solution that, upon acidification, yields a colorless precipitate (**C**), with formula $C_7H_6O_2$. Ozonolysis of **A** produces formaldehyde ($H_2C{=}O$) and compound **D**. Compound **D** can be synthesized by Friedel-Crafts acetylation of benzene with acetyl chloride (CH_3COCl) and aluminum chloride. What are the structures of compounds **A–D**?

9.52 Show the reactions necessary to complete each of the following syntheses:

(a) phenol to *p*-bromoanisole

(b) benzene to *m*-nitroethylbenzene

(c) benzene to phenylcyclohexane

9.53 *p*-Bromodiphenyl undergoes nitration on the unsubstituted ring to produce 4-bromo-4′-nitrodiphenyl ($Br—C_6H_4—C_6H_4—NO_2$). Explain this outcome.

9.54 Explain why aniline ($C_6H_5NH_2$) undergoes rapid tribromination when treated with bromine in water, whereas the acetyl amide of aniline (acetanilide, $C_6H_5—NHCOCH_3$) undergoes only monobromination.

9.55 Show the reactions necessary to convert aniline into 2-bromo-4-nitroaniline.

9.56 A compound **W** is shown to have an empirical formula of C_4H_5 and a molecular weight of 106. It does not react with bromine in the dark or with cold $KMnO_4$. Oxidation of **W** with hot $KMnO_4$, removal of the MnO_2 precipitate, and acidification of the colorless filtrate produces a colorless precipitate **X** of formula $C_7H_6O_2$. What are the structures of compounds **W** and **X**?

9.57 Show the reaction you would employ to convert morphine to codeine. Write a reaction that would convert codeine back to morphine. Finally, show a reaction that would convert both hydroxyl groups of morphine to methoxy groups.

R—O
O
N—CH$_3$
HO

Morphine (R = H)
Codeine (R = CH_3)

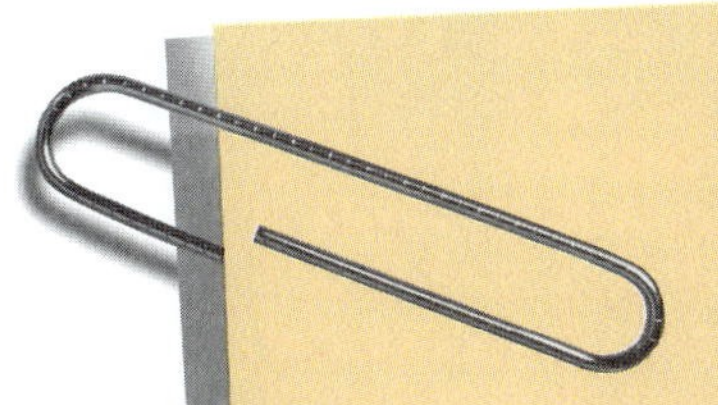

CONCEPTUAL PROBLEM

Dioxins: Up in Smoke

You have recently been hired by the state environmental office as an environmental engineer. The major focus of your job is to establish tougher but practicable standards to ensure a substantial decrease in the dioxin levels in the state, particularly in the larger cities. You soon realize that your job will also entail some public relations, since a series of small chemical accidents in the last year has put dioxins in the news. You attend a local hearing to discuss one such chemical accident. The citizens in attendance are justifiably upset and demand that *all* dioxins be eliminated. As a citizen, you share their concern; as an engineer, you wonder how realistic such a goal would be.

What many people do not realize is that the general term *dioxins* applies not just to a single compound, but to a family of chlorinated compounds derived from the parent compound dioxin (compound **A**). The most prominent member of that family is2,3,7,8-tetrachlorodioxin (compound **B**).

A B

Any or all of the hydrogens on the parent dioxin ring can be replaced by chlorines, which means there are a huge number of different compounds in the dioxin family. Many halogenated compounds are produced natually by a variety of organisms (such as red algae and plants) and by natural physical events (such as volcanic eruptions and forest fires). The carbon-chlorine bond is not innately lethal. Indeed, plastics that get used on a daily basis, such as Saran Wrap and PVC, are not considered dangerous at all.

- Since dioxins are produced naturally during any burning of organic chlorine-containing material, what are some of the sources you must consider as potential sources of dioxins in your state?
- Why must you seek cooperation from the environmental offices of neighboring states?
- Are dioxins and the family of organochlorines water-soluble? Are dioxins found in food? Which foods might contain higher levels of these compounds, fruits and vegetables or fish?

Dioxins are produced when organic materials burn. Forest fires are therefore one source of dioxins.

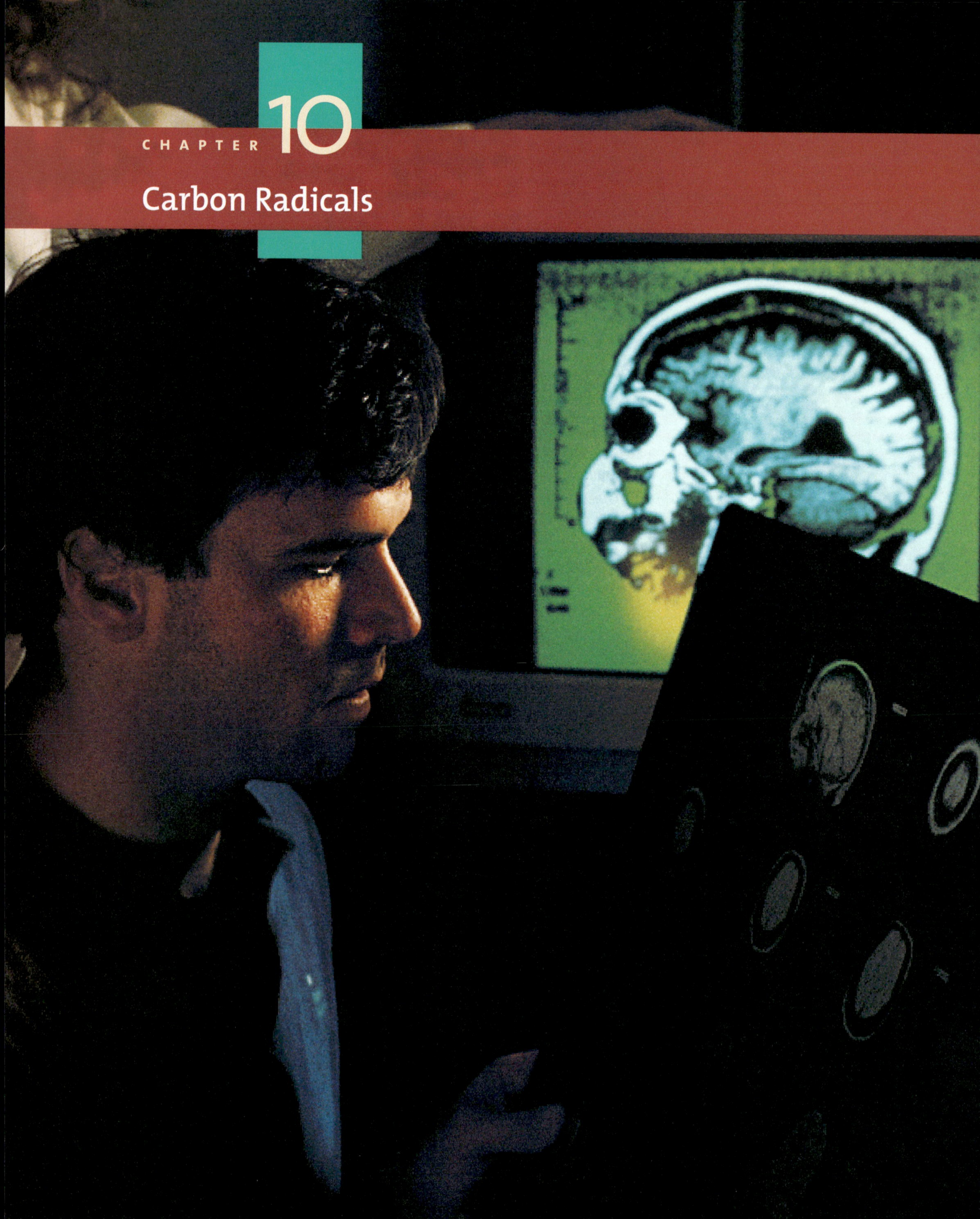

CHAPTER 10

Carbon Radicals

YOU ARE now familiar with the two common parent families of organic compounds: aliphatic hydrocarbons and aromatic hydrocarbons. You have learned about their structures, their nomenclature, their electronic configurations, and their molecular shapes. You have already encountered over half of the functional groups that can be attached to or incorporated into these parent structures. These functional groups (halides, alcohols, ethers, and the unsaturated groups, alkenes and alkynes) can be attached to both aliphatic and aromatic carbon skeletons.

In spite of the significant amount of organic chemistry covered so far, all of the reactions studied in detail have had one feature in common—the nature of the reactive intermediate in the mechanism by which they occur. All of the organic chemistry studied so far has involved carbocations as the only reactive intermediates. Recalling the discussion in Section 1.3.3, you should remain aware that there are two other kinds of reactive carbon intermediates: carbon radicals and carbanions. In this chapter, we will look at radical chemistry. And since you are familiar with the general issues regarding reaction mechanisms, this will be a brief topic.

10.1 Introduction to Carbon Radicals

The **carbon radical** is a carbon atom that is sharing three of its four valence electrons in covalent bonds to three other atoms while its fourth electron remains as a single unshared electron. This carbon therefore has seven electrons about it—not eight. A carbon radical is electrically neutral but exists as a very reactive intermediate that generally cannot be isolated. The simplest example is the methyl radical ($\cdot CH_3$). The presence of the unshared electron in the radical is indicated by a single dot next to the formula, in contrast to the bond line used to indicate a shared pair ($—CH_3$). Any of the three hydrogens of the methyl radical can be replaced by other covalently bonded groups, creating a more complex radical. When one radical reacts with another, each contributes its single electron to make a pair of electrons, producing a covalent bond that joins the two radicals together. Thus, two methyl radicals react to form ethane, and a single methyl radical reacts with a hydrogen atom ($H\cdot$, which is a radical) to form methane:

$$\underset{\text{Ethane}}{CH_3{-}CH_3} \xleftarrow{\cdot CH_3} \underset{\text{Methyl radical}}{\cdot CH_3} \xrightarrow{H\cdot} \underset{\text{Methane}}{CH_3{-}H}$$

◀ Researchers study brain scans of an Alzheimer patient. Highly reactive free radicals are thought to be involved in the development of degenerative diseases such as Alzheimer's disease.

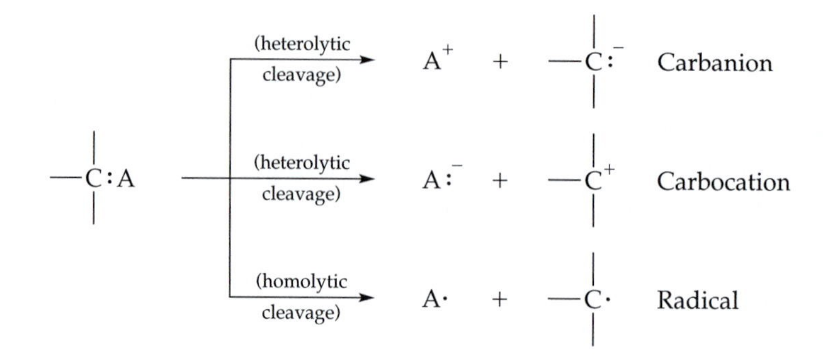

FIGURE 10.1

Cleavage options for a C—A bond.

Radicals do not form spontaneously. The process starts with a covalent bond between a saturated carbon C and another atom A (Figure 10.1). The two share a pair of electrons in a molecular orbital formed by overlap of an sp^3 (tetrahedral) hybrid orbital from carbon and an orbital from atom A. The C—A covalent bond can be broken, and the pair of electrons forming that bond can, in principle, be distributed in three possible ways:

1. In the first type of bond cleavage, both bonding electrons remain with the carbon; in this **heterolytic cleavage,** the two electrons are unequally distributed. This results in the carbon acquiring excess electron density and forming a *carbanion intermediate.* Carbanions were briefly encountered in the form of the organometallic reagents (Grignard reagents, organolithium reagents, organocuprates) in Section 3.5.1. They also are formed when a proton is removed from a terminal alkyne by a strong base (see Section 7.2.2). Their chemistry will be discussed in depth beginning in Chapter 13.
2. In the second type of possible cleavage (also *heterolytic*), both bonding electrons remain with the atom A, resulting in the carbon becoming electron-deficient and forming a *carbocation intermediate.* Carbocations were encountered in several reactions, including the S_N1 reaction of alkyl halides (see Section 3.4.3), the addition of a proton to alkenes (see Section 6.4.1), and the dehydration of alcohols (see Section 6.6.1).
3. In the third type of cleavage, the two bonding electrons are divided equally between C and A; this is, therefore, a **homolytic cleavage,** with the two electrons equally distributed. One electron of the pair remains with atom A and the other with the carbon. Both of these resulting species are called *radicals;* they have no charge but are very reactive. A carbon radical intermediate is sometimes called a *free radical.*

This chapter will describe the chemistry of carbon radicals, both their formation through homolytic cleavage and their subsequent reactions.

10.1.1 Generation of Carbon Radicals

EXPLORATIONS www.jbpub.com/organic-online

The very reactive carbon radical is usually produced by reaction of a noncarbon radical with a carbon-hydrogen bond. The hydrogen atom (hydrogen radical) is abstracted, leaving behind a carbon radical.

Radical reactions are often started by the addition of an **initiator,** which is a compound that readily dissociates into a radical that will in turn remove a hydrogen atom from an organic compound. Peroxides (represented as R—O—O—R) are often used as initiators because they readily fragment with mild heating into alkoxy radicals (R—O·). The alkoxy radical removes a hydrogen atom from a C—H bond to form an alcohol. Remaining behind is a carbon radical:

$$\text{R—O—O—R} \xrightarrow{\Delta} [2\,\text{R—O}\cdot]$$

Peroxide — Alkoxy radical

$$\text{—C:H} + [\cdot\text{O—R}] \longrightarrow \text{H—O—R} + [\text{—C}\cdot]$$

Alcohol — Carbon radical

Note the convention of using a *half-headed curved arrow* to designate an electron shift involving *only one electron*. Such shifts occur in all radical mechanisms.

A second way to initiate a radical reaction is to expose the reactants to a source of ultraviolet (UV) light. Light of this frequency can transmit sufficient energy to break bonds homolytically and create radicals. Chemists indicate the use of UV light as a reaction condition by placing the symbols $h\nu$ over the reaction arrow, where h stands for Planck's constant and ν indicates the frequency of light. As we will see in Chapter 11, the symbol $h\nu$ represents the application of an amount of energy.

The final means of producing radicals is simply the application of heat. This method is relatively indiscriminate and therefore not useful, except in a few cases (discussed later in this chapter).

10.1.2 Structure of Carbon Radicals

With seven electrons around it—four from atomic carbon and three from attached substituents—the radical carbon is sp^2 (trigonally) hybridized. This means that a p orbital is left unhybridized and perpendicular to the plane of the three sp^2 hybrid orbitals. Six of the seven electrons occupy the three sp^2 hybrid orbitals, which become part of the sigma bonds to the attached substituents (Figure 10.2).

A carbon radical has a seventh (odd or unpaired) electron in the unhybridized p orbital. The trigonal hybridization in a carbon radical means that the carbon atom and its three substituents lie in a plane, just as in a carbocation (Figure 10.3). The only difference is that the carbocation has an empty unhybridized p orbital, whereas the radical has one electron in the unhybridized p orbital.

$$1s^22s^22p^2 \xrightarrow[\text{(requires energy)}]{\text{promotion}} 1s^22s^12p^3 \xrightarrow[\text{(no energy change)}]{\text{hybridization}} 1s^22(sp^2)^3 + 2p^1$$

Atomic carbon — Trigonal carbon

s orbital + 2 p orbital —hybridization→ Three sp^2 orbitals

FIGURE 10.2

Trigonal (sp^2) hybridization of carbon.

FIGURE 10.3

Views of the orbital structure of a carbon radical.

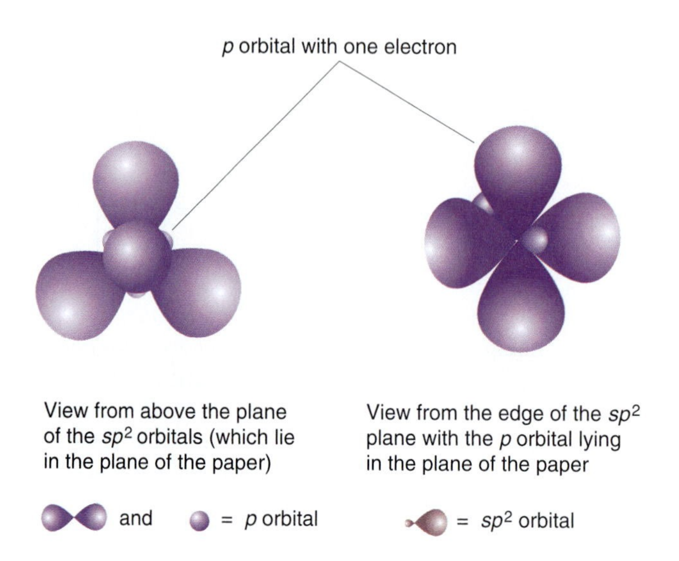

In a formal manner, then (but not necessarily representing actual reaction processes), carbon radicals can be viewed as being formed from a carbon atom and three organic groups, where each covalent bond is formed by the contribution of one electron by each of the bonding partners. This leaves the fourth carbon electron unpaired—a carbon radical. Alternatively, a carbon radical can be viewed as being formed by removal of atomic hydrogen from an alkane, taking its single electron with it.

$$\cdot\dot{\text{C}}\cdot \; + \; 3\,\text{R}\cdot \longrightarrow \text{R}_3\text{C}\cdot \xleftarrow{-\text{H}\cdot} \text{R}_3\text{C{:}H}$$

Carbon atom — Carbon radical — Alkane

10.1.3 Stability of Carbon Radicals

Carbon radicals are very reactive intermediates with very short lifetimes. However, their presence can be detected and their relative stability determined using a number of experimental techniques.

Carbon radicals can be classified in a manner similar to carbocations—on the basis of the number of alkyl groups attached to the carbon atom. A *primary radical* (for example, an ethyl radical) has a single alkyl group attached to the radical carbon; a *secondary radical* (for example, isopropyl radical) has two alkyl groups attached; and a *tertiary radical* (for example, *t*-butyl radical) has three alkyl groups attached.

$CH_3CH_2\cdot$	$(CH_3)_2CH\cdot$	$(CH_3)_3C\cdot$
Ethyl radical (1°)	Isopropyl radical (2°)	*t*-Butyl radical (3°)

Increasing radical stability →

Radical stability follows the same order as carbocation stability: tertiary (3°) > secondary (2°) > primary (1°). This is because the radical carbon is slightly electron-deficient, so the sharing of electron density from adjacent groups helps to stabilize the radical.

Two special cases must be noted: the allyl radical and the benzyl radical. Both of these carbon radicals are especially stable because of resonance delocalization throughout their conjugated systems. The allyl radical is stabilized by having two equivalent contributing structures:

$$\cdot CH_2{-}CH{=}CH_2 \longleftrightarrow CH_2{=}CH{-}CH_2\cdot$$

Allyl radical delocalization

Structural units that can form allyl radicals normally do so in radical reactions.

The benzylic radical (often formed by loss of a hydrogen from a carbon alpha to the ring) has four contributing structures, with the unshared electron being delocalized throughout the benzene ring:

Benzyl radical delocalization

In any reaction involving a radical intermediate where one option is for a radical carbon to form at a benzylic position, that position is preferred and normally determines the regiochemistry of the reaction.

10.1.4 Importance of Carbon Radicals

Reactions of radicals with organic compounds and reactions of carbon radicals are often not very selective. Therefore, radical chemistry is not widely used in laboratory chemical reactions, and deliberate steps are often taken to preclude radical formation. However, there are a few instances where radical reactions are very useful.

The radical reaction of halogens with alkanes is one that is chemically useful. (This reaction was mentioned briefly in Section 2.7.1, and we will look at it in detail in Section 10.2.) It is the only means of replacing a hydrogen atom in an alkane with a halogen (alkanes, as you recall, tend to be chemically inert). We studied the ionic addition of hydrogen bromide to alkenes to produce alkyl bromides in Chapter 6, but the same addition also occurs under radical conditions, leading to a useful reversal of the regioselectivity. (We will discuss this reaction in Section 10.3.)

A very important use of radical chemistry is in the polymerization of alkenes to produce commercially important addition polymers, such as polyethylene, polypropylene, and polystyrene. (We will discuss these in Chapter 18.)

Radical chemistry plays an important part in the polymerization process that produces plastics like this polyethylene.

Finally, the combustion reaction of alkanes (R—H) that forms carbon dioxide and water (see Section 2.7.2) is known to proceed via a radical mechanism. While the details of the entire mechanism are unclear, it is known that conversion of an alkane to a carbon radical by molecular oxygen (which is a diradical itself) is a key beginning step. This reaction does not occur spontaneously but must be initiated, often by the application of heat. Once alkyl radicals (R·) are formed, peroxy radicals (ROO·), hydroperoxides (ROOH), alkoxy radicals (RO·), and hydroxy radicals (HO·) are produced one after the other by reaction with oxygen. These new radicals also attack the alkane and so feed the **chain reaction,** a reaction that is self-propagating once initiated. The steps involved are *initiation*, *propagation*, and *termination*. Parts of the chain reaction of alkane combustion are shown here. (*Note:* Every step that produces an alkyl radical starts another chain.)

Initiation:

$$\underset{\text{Alkane}}{R{-}H} + \underset{\text{Oxygen}}{O_2} \xrightarrow{\Delta} \underset{\text{Alkyl radical}}{R\cdot} + \underset{\text{Hydroperoxy radical}}{H{-}O{-}O\cdot}$$

Propagation:

$$R\cdot + O_2 \longrightarrow \underset{\text{Alkyl peroxy radical}}{R{-}O{-}O\cdot}$$

$$R{-}O{-}O\cdot + RH \longrightarrow \underset{\text{Alkyl hydroperoxide}}{R{-}O{-}O{-}H} + R\cdot$$

$$R{-}O{-}O{-}H \xrightarrow{\Delta} \underset{\text{Alkoxy radical}}{R{-}O\cdot} + \underset{\text{Hydroxy radical}}{\cdot O{-}H}$$

$$R{-}O\cdot + R{-}H \longrightarrow R\cdot + ROH$$

$$\cdot O{-}H + R{-}H \longrightarrow R\cdot + H_2O$$

The chain reaction generates the chemical intermediates and the energy necessary to initiate a new propagation step and speed up the reaction. While the initiating step requires energy, subsequent steps *produce* energy, only some of which is needed to initiate new chains. This excess energy is the source of the heat emitted in a combustion reaction. Therefore, combustion reactions are self-perpetuating.

The amount of energy given off by combustion of an alkane (the heat of combustion) can be measured very precisely. For methane, it is 213 kcal/mol, and for *n*-butane, it is 687 kcal/mol.

$$\underset{\text{Butane}}{C_4H_{10}} + \tfrac{13}{2}\,O_2 \longrightarrow 4\,CO_2 + 5\,H_2O + 687\text{ kcal/mole}$$

For all alkanes, the heat of combustion averages about 160 kcal/mol *per CH_2 group*. By comparison, the heat of combustion for acetylene, used in oxyacetylene torches, is 310 kcal/mol.

When the concentrations of alkane and oxygen are just right, the cumulative rate of the combustion chain reactions becomes so fast and the amount of energy released so large that an explosion occurs. The rate of a combustion reaction is controlled by the amount of fuel supplied—this accounts for the difference between a controlled combustion (such as in a gas furnace) and an explosive combustion (such as when a gas pipeline breaks). A combustion reaction ceases either when the fuel is expended or when the necessary oxygen is excluded or fully consumed. The same combustion reaction that occurs in a gas furnace to release heat also occurs in the internal combustion engine of an automobile to provide an explosive force in the cylinder. The fundamental difference between the two reactions is their rates.

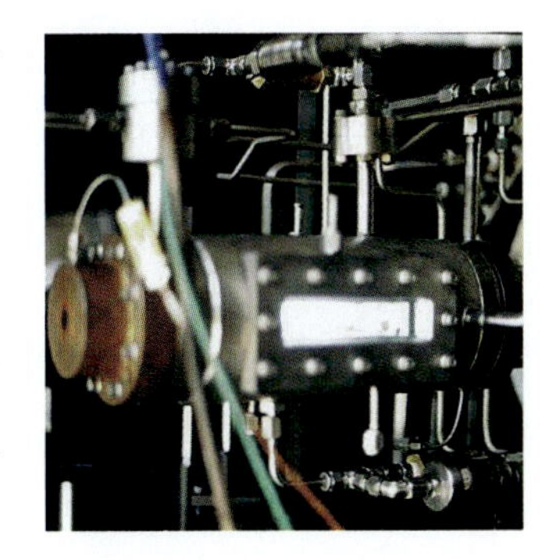

The combustion of gasoline in an internal combustion engine (shown here in a testing facility) proceeds via an oxidation reaction with carbon radical intermediates. The rate of this reaction is so high that it creates an explosion.

PROBLEM 10.1

Write balanced equations for the complete combustion of (a) methane, (b) ethene, (c) acetylene, and (d) octane.

10.2

Reaction of Alkanes with Halogens

When an alkane is mixed with chlorine or bromine in the dark at room temperature, there is no reaction. However, in the presence of heat (~100°C), ultraviolet light (*hv*), or a radical initiator (such as a peroxide, ROOR), a reaction occurs. For example, chlorine reacts with methane to produce a mixture consisting mainly of methyl chloride but also including some methylene chloride, chloroform, and carbon tetrachloride:

$$\underset{\text{Methane}}{CH_4} \xrightarrow[Cl_2]{100°C} \underset{\text{Methyl chloride}}{CH_3Cl} + \underset{\text{Methylene chloride}}{CH_2Cl_2} + \underset{\text{Chloroform}}{CHCl_3} + \underset{\text{Carbon tetrachloride}}{CCl_4} + HCl$$

Similarly, propane produces mainly isopropyl chloride and *n*-propyl chloride, plus some polychloropropanes:

$$\underset{\text{Propane}}{CH_3CH_2CH_3} \xrightarrow[Cl_2]{hv} \underset{\text{Isopropyl chloride}}{CH_3CHClCH_3} + \underset{n\text{-Propyl chloride}}{CH_3CH_2CH_2Cl} + HCl$$

These are radical substitution reactions: Chlorine replaces hydrogen, with hydrogen chloride being evolved.

Radical halogenation is useful for the conversion of the very inert alkanes into alkyl bromides and chlorides, compounds that can be used in a variety of ionic reactions, including the substitutions and eliminations discussed in Chapters 3 through 8. Note the place of this reaction in the Key to Transformations (inside front cover). The only reaction shown for alkanes, it is the standard means of "functionalizing" them.

10.2.1 Mechanism of the Chlorination of Methane

Methane is chlorinated by molecular chlorine (chlorine gas) in the presence of heat (Δ) or light (*hv*) to produce, initially, methyl chloride and hydrogen chloride:

$$\underset{\text{Methane}}{CH_4} + Cl_2 \xrightarrow{100°C} \underset{\text{Methyl chloride}}{CH_3Cl} + HCl$$

The overall result is a substitution of chlorine for hydrogen. The mechanism of this chlorination (or bromination) of methane involves a radical chain reaction:

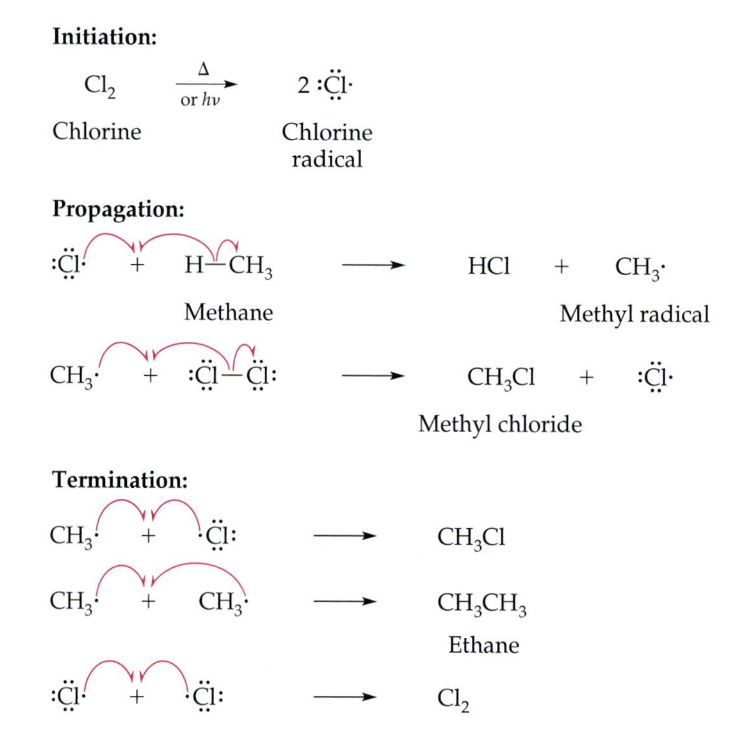

The reaction is initiated by the formation of chlorine radicals, either by heating or by using ultraviolet light. The chlorine radicals form the methyl radical by removing a hydrogen atom from methane; HCl is a reaction by-product. The methyl radical then attacks a chlorine molecule to produce methyl chloride and release another chlorine radical. Thus, the products of the two propagation steps are the substituted alkane (methyl chloride) and a regenerated chlorine radical needed to produce another methyl radical. *This is the essence of a chain reaction: One of the products (chlorine radical) is also a reactant.* This process can continue until all of the chlorine or methane is consumed. (In theory, a single chain is sufficient, but in practice, many chains are operating concurrently.) The cleavage of molecular chlorine (Cl_2) into two chlorine atoms (Cl·) consumes relatively little energy. Once the reaction is initiated, subsequent steps release more than enough energy for it to continue without additional external energy being supplied; so the overall reaction is exothermic.

The chain reaction will be interrupted (*terminated*) if one of the radical intermediates is removed and is thereby unable to continue the propagation. For example, any two radicals may react and thereby fail to produce a new radical to continue the reaction, leading to termination of the chain. Any other process that consumes a radical without producing another one will also stop the chain reaction. This is the secret behind radical inhibitors, as we will see later in this chapter.

The products of polychlorination (methylene chloride, chloroform, and carbon tetrachloride) are formed mainly in the presence of excess chlorine. For example, if the chlorine radical attacks methyl chloride, it removes a hydrogen atom and forms a *chloromethyl radical* ($ClCH_2·$) that reacts with more chlorine to produce methylene chloride (CH_2Cl_2):

$$:\ddot{Cl}\cdot + H{-}CH_2Cl \longrightarrow HCl + ClCH_2\cdot$$

Methyl chloride — Chloromethyl radical

$$ClCH_2\cdot + :\ddot{Cl}{-}\ddot{Cl}: \longrightarrow CH_2Cl_2 + :\ddot{Cl}\cdot$$

Methylene chloride

Likewise, methylene chloride produces the *dichloromethyl radical* ($Cl_2CH\cdot$), which then reacts with chlorine to produce chloroform. Chloroform produces the *trichloromethyl radical* ($Cl_3C\cdot$), which then reacts with chlorine to produce carbon tetrachloride.

PROBLEM 10.2

Write the steps involved in the mechanism for the bromination of propane to form 1-bromopropane (*n*-propyl bromide).

10.2.2 Halogenation of Alkanes in Synthesis

Higher alkanes can be chlorinated or brominated using the same reaction and mechanism shown for methane. The only additional question is one of regioselectivity: Which hydrogen will be initially abstracted by the chlorine or bromine radical? For example, monochlorination of isobutane could produce isobutyl chloride or *t*-butyl chloride. What happens is that some of both is produced. This fact, together with the inevitable formation of polychlorinated products, makes the halogenation reaction of alkanes of somewhat limited synthetic use.

In the chlorination of isobutane, two factors work in opposite directions. The first is simply *probability*. The chlorine radical can remove any one of nine primary hydrogens (those on the three methyl groups), but there is only one tertiary hydrogen. Probability dictates that the reaction will produce a 9:1 product ratio of isobutyl chloride to *t*-butyl chloride. The second factor is *radical stability*. Recall that a tertiary radical is more stable than a primary radical. This factor leads to the prediction that *t*-butyl chloride will be the predominant product. Actual experiments show the product ratio to be 63% isobutyl to 37% *t*-butyl chloride. Therefore, the higher stability of the *t*-butyl radical intermediate forces the product mix in the direction of *t*-butyl chloride. Nonetheless, the need to separate these two products, as well as the products of polychlorination, make this a useless synthetic reaction.

Isobutane: $(CH_3)_3C{-}H$

$$\xrightarrow[\Delta]{Cl_2} (CH_3)_3C{-}Cl + ClCH_2{-}CH(CH_3)_2$$

37% *t*-Butyl chloride — 63% Isobutyl chloride

$$\xrightarrow[\Delta]{Br_2} (CH_3)_3C{-}Br + BrCH_2{-}CH(CH_3)_2$$

99% *t*-Butyl bromide — 1% Isobutyl bromide

Thus, radical chlorination is of practical synthetic use only in those instances in which the starting alkane is symmetrical and has only a single kind of hydrogen. Examples of such alkanes are 2,2-dimethylpropane (only primary hydrogens are present) and any cycloalkane (only secondary hydrogens are present). Chlorination of 2,2-dimethylpropane yields 1-chloro-2,2-dimethylpropane, and chlorocyclohexane can be effectively synthesized by chlorination of cyclohexane:

$$\underset{\text{2,2-Dimethylpropane (neopentane)}}{C(CH_3)_4} \xrightarrow[\Delta]{Cl_2} \underset{\text{1-Chloro-2,2-dimethylpropane (neopentyl chloride)}}{(CH_3)_3C{-}CH_2Cl}$$

Cyclohexane $\xrightarrow[\Delta]{Cl_2}$ Chlorocyclohexane

Radical bromination is much more regioselective than chlorination because of the overall energetics of the reaction. For example, monobromination of isobutane results in a mixture of 99% *t*-butyl bromide and 1% isobutyl bromide. Clearly, the higher stability of the tertiary radical is the dominant factor in product formation. However, even bromination of higher alkanes is not synthetically useful unless there is either a single class of hydrogens or a single hydrogen of a specific class, usually tertiary. For example, cyclohexane (with a single class of hydrogens—all secondary) produces good yields of bromocyclohexane (cyclohexyl bromide), but bromination of pentane is useless because it yields a diverse mixture of three monobrominated and several polybrominated products. Bromination of 2-methylbutane produces mainly 2-bromo-2-methylbutane because there is a single tertiary hydrogen that can be selectively removed by the bromine radical.

Cyclohexane $\xrightarrow[75°C]{Br_2}$ Bromocyclohexane

Pentane $\xrightarrow[75°C]{Br_2}$ 1-Bromopentane + 2-Bromopentane + 3-Bromopentane

2-Methylbutane $\xrightarrow[75°C]{Br_2}$ 2-Bromo-2-methylbutane

How to Solve a Problem

Show the reactions necessary to convert cyclohexane into cyclohexanol.

PROBLEM ANALYSIS

Cyclohexane ⇢ Cyclohexanol

The required transformation—replacing H by OH—seems quite straightforward on the surface, but there is no single-step reaction that we have encountered that will convert an alkane into an alkanol. (Recall the general unreactivity of alkanes.) There is only one reaction of alkanes we have studied, and that is radical halogenation to produce an alkyl halide. If bromocyclohexane can be obtained, we know that we will have a good leaving group in the bromide ion. We can then use a substitution reaction (S_N2) of hydroxide on bromocyclohexane to form the alcohol.

SOLUTION

We brominate cyclohexane under free-radical conditions to produce bromocyclohexane. Then we carry out a substitution reaction using sodium hydroxide to produce cyclohexanol.

Cyclohexane $\xrightarrow[h\nu]{Br_2}$ Bromocyclohexane $\xrightarrow{NaOH}$ Cyclohexanol

Important clue: In most syntheses that require starting with an alkane, there is almost nothing chemically that can be done with the alkane other than halogenate it. Further, once you have a halogen on the alkane, you can use all of the reactions now in our chemical tool kit—the functional group interconversions outlined in the Key to Transformations. Finally, many carbon-carbon bonds are made by substitution reactions, which means that you need an alkyl halide as one substrate to build up a more complex carbon skeleton.

PROBLEM 10.3

Show all possible monobromination products of each of the following compounds: (a) cyclobutane, (b) butane, (c) 2,3-dimethylbutane, and (d) methylcyclohexane.

PROBLEM 10.4

Show the reactions and reagents necessary to carry out each transformation:

(a) cyclopentane to ethylcyclopentane

(b) 2,2-dimethylpropane to 1-methoxy-2,2-dimethylpropane

(c) cyclohexane to vinylcyclohexane

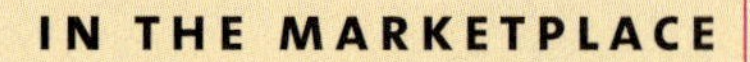

CHEMISTRY AT WORK

IN THE MARKETPLACE | AND IN THE BODY

The Fight against Autoxidation

There is no doubt about it. Rancid butter smells awful. In fact, it is usually the taste and smell of spoiled food, rather than its color or texture, that tell us it has gone bad. But chemically, what has happened to the food to make it spoil?

The chemical process that makes butter rancid is called **autoxidation.** It converts an R—H group to an R—O—O—H group, called a hydroperoxide. The process occurs by initial abstraction of hydrogen, usually by molecular oxygen, which is itself a diradical. This leaves behind a carbon radical, which reacts with additional oxygen to form a peroxy radical (R-O-O·). The peroxy radical then removes a hydrogen from another substrate, forming a hydroperoxide (ROOH) and another carbon radical.

$$\text{R—H} + O_2 \longrightarrow \underset{\text{Carbon radical}}{\text{R}\cdot} + \text{H—O—O}\cdot$$

$$\text{R}\cdot + O_2 \longrightarrow \underset{\text{Peroxy radical}}{\text{R—O—O}\cdot}$$

$$\text{R—H} + \text{R—O—O}\cdot \longrightarrow \text{R}\cdot + \underset{\text{Hydroperoxide}}{\text{R—O—O—H}}$$

Hydroperoxides degrade and rearrange to other compounds, including carbonyl compounds and carboxylic acids.

Autoxidation occurs readily in the many foods that contain polyunsaturated fats, such as margarine, butter, and food oils (for example, corn oil and olive oil). These fats contain at least one double bond in the hydrocarbon chain, which means that there is always at least one allylic position that is prone to giving up a hydrogen atom and forming a stable allylic radical. With time, oxygen removes an allylic hydrogen to form an allylic radical, which in turn produces hydroperoxides and related products. It is the taste and odor of these compounds, especially the carboxylic acids, that are associated with the spoilage (rancidity) of such foods.

A well-known example of autoxidation occurs with the fatty acid linoleic acid. Linoleic acid has two double bonds (at carbons 9 and 12), making the methylene group at carbon 11 doubly allylic and very susceptible to autoxidation.

$CH_3(CH_2)_4$—CH=CH—CH_2—CH=CH—$(CH_2)_7COOH$

Linoleic acid (9*Z*,12*Z*-octadecadienoic acid)

↓ O_2

$CH_3(CH_2)_4$—CH=CH—$\dot{C}H$—CH=CH—$(CH_2)_7COOH$

Allylic intermediate (resonance-stabilized)

↓ O_2

$CH_3(CH_2)_4$—CH(OOH)—CH=CH—CH=CH—$(CH_2)_7COOH$

13-Hydroperoxy-9*Z*,11*E*-octadecadienoic acid

Today, *antioxidants* are added to many foods to prevent autoxidation and spoilage and allow long-term

10.2.3 Special Cases: Allylic and Benzylic Bromination

Propene reacts with bromine at room temperature in the dark to yield 1,2-dibromopropane, the product of normal electrophilic addition of a halogen to a double bond (see Section 6.4.3). However, when the reaction is conducted at high temperature, the product is allyl bromide (3-bromopropene), the result of a substitution reaction. The use of high temperature as a reaction condition causes the bromine to dissociate into bromine radicals, which then initiate a radical substitution reaction by the same mechanism shown for chlorination of methane.

storage. Two widely used additives, called *radical inhibitors*, are BHT (butylated hydroxytoluene) and BHA (butylated hydroxyanisole):

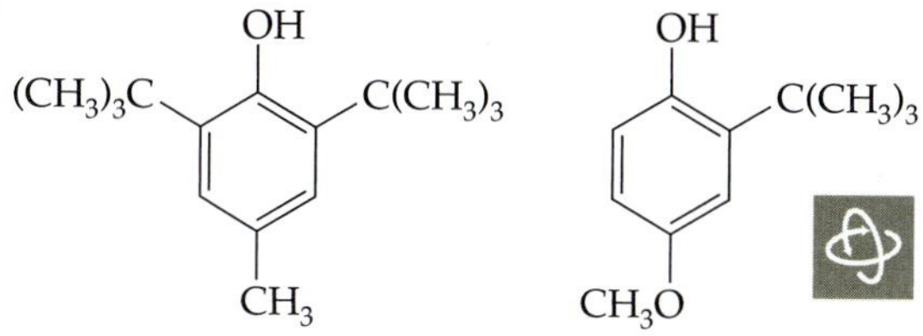

OH
$(CH_3)_3C$ — $C(CH_3)_3$
CH_3
BHT

OH
$C(CH_3)_3$
CH_3O
BHA

These protective agents function by interrupting the chain reaction of the autoxidation process. The carbon radical produced in the first step of the autoxidation reaction removes a hydrogen atom from the hydroxyl group of BHA to produce a new radical (a phenoxy radical). This radical is relatively unreactive because of extra stability due to resonance with the benzene ring and because of steric hindrance from the *ortho t*-butyl group. Because the BHA oxyradical is too unreactive to attack the unsaturated fat and form more carbon radicals, it quenches the chain reaction.

The body contains many polyunsaturated fats, and it also has a means of protecting itself against autoxidation of those fats. (Remember that oxygen is transported throughout the body; also, some enzymatic reactions proceed via radical mechanisms.) The essential vitamin E, also known as α-tocopherol, is an antioxidant, a radical inhibitor whose role is to prevent deterioration of body fats by autoxidation:

Vitamin E pills.

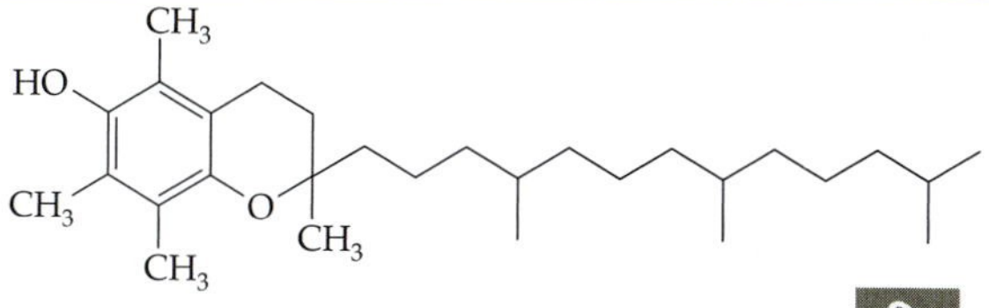

HO, CH_3, CH_3, CH_3, O, CH_3
Vitamin E
(α-tocopherol)

There has recently been some speculation that human aging is partially the result of autoxidation processes. You may see ads for skin creams or vitamin supplements containing extra amounts of vitamin E or other antioxidants.

Radical inhibition by BHA:

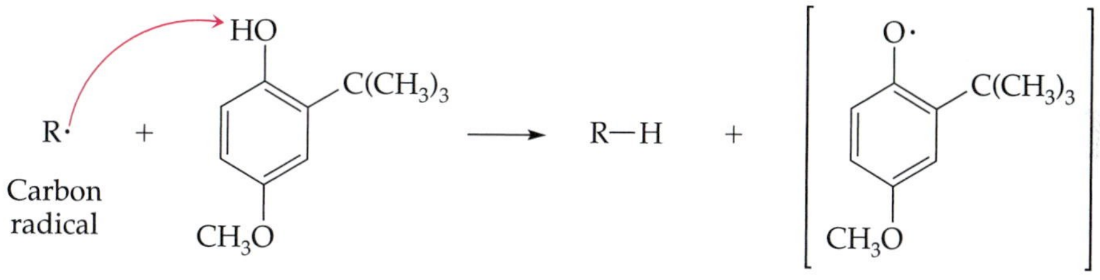

$$R\cdot + \text{BHA} \longrightarrow R{-}H + \text{Resonance-stabilized phenoxy radical}$$

R· — Carbon radical
HO, $C(CH_3)_3$, CH_3O — BHA
O·, $C(CH_3)_3$, CH_3O — Resonance-stabilized phenoxy radical

$$CH_2{=}CH{-}CH_3 \xrightarrow[\text{low temp.}]{Br_2} \underset{Br}{CH_2}{-}\underset{Br}{CH}{-}CH_3 \quad \textbf{Addition}$$

Propene → 1,2-Dibromopropane

$$CH_2{=}CH{-}CH_3 \xrightarrow[\text{high temp.}]{Br_2} CH_2{=}CH{-}CH_2{-}Br + HBr \quad \textbf{Substitution}$$

Propene → Allyl bromide (3-bromopropene)

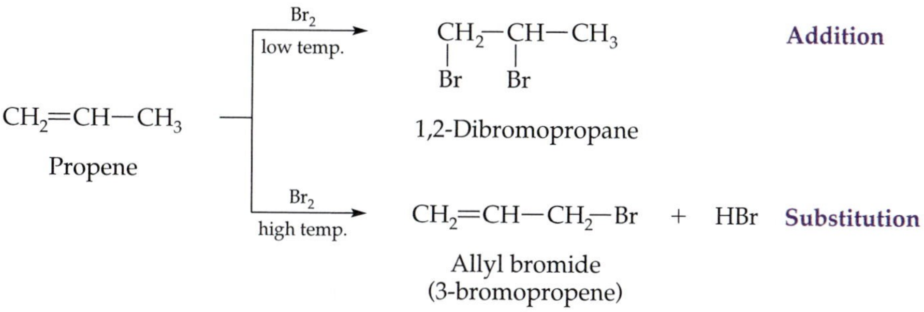

In the case of propene, the most stable radical that can form by loss of a hydrogen is an allyl radical, which is stabilized by resonance (see Section 10.1.3). This radical then attacks bromine to complete the substitution.

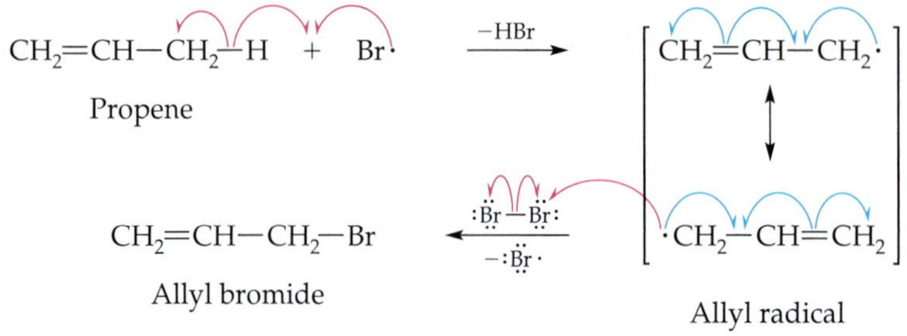

$$CH_2{=}CH{-}CH_2{-}H \; + \; Br\cdot \xrightarrow{-HBr} \left[CH_2{=}CH{-}CH_2\cdot \longleftrightarrow \cdot CH_2{-}CH{=}CH_2 \right]$$

Propene — Allyl radical

$$\text{Allyl radical} \xrightarrow[-:\ddot{Br}\cdot]{:\ddot{Br}-\ddot{Br}:} CH_2{=}CH{-}CH_2{-}Br$$

Allyl bromide

The radical bromination (or chlorination) of a compound with a potential allylic radical always results in substitution at the allylic position. This is a regiospecific and very useful reaction. For example, radical bromination of cyclohexene produces 3-bromocyclohexene:

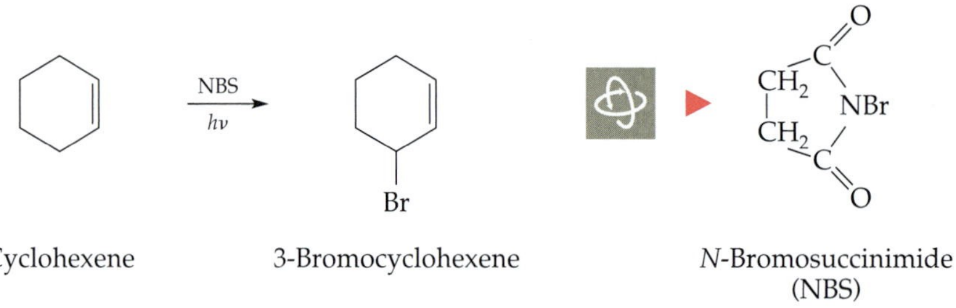

$$\text{Cyclohexene} \xrightarrow[h\nu]{NBS} \text{3-Bromocyclohexene}$$

Cyclohexene — 3-Bromocyclohexene — *N*-Bromosuccinimide (NBS)

In place of liquid bromine and high temperatures, a more convenient bromination reagent is the compound *N*-bromosuccinimide, abbreviated NBS, which provides a source of bromine radicals at room temperature in CCl_4 solution when exposed to UV light or a peroxide initiator.

Radical bromination of an alkylaromatic carrying one or more benzylic hydrogens occurs at the benzylic position. A bromine atom will remove a benzylic hydrogen in preference to any other hydrogen because the result is a resonance-stabilized benzyl radical (see Section 10.1.3). For example, reaction of ethylbenzene with bromine at high temperatures or with NBS and light at room temperature produces 1-bromo-1-phenylethane, not 1-bromo-2-phenylethane:

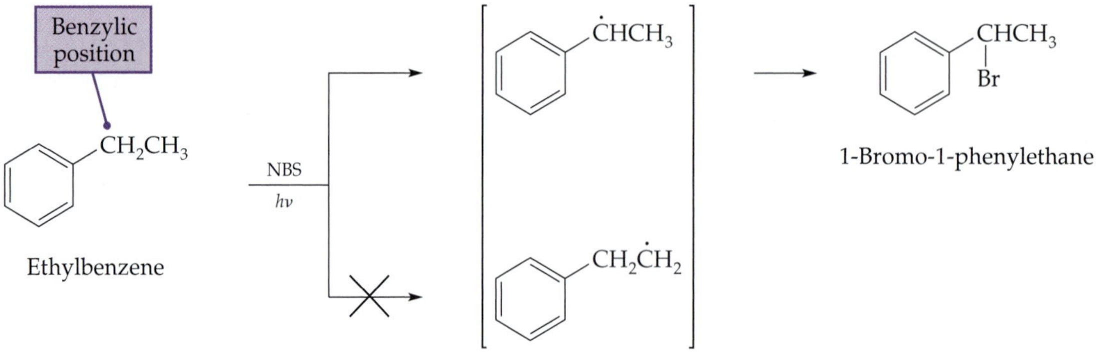

Benzylic position

$$C_6H_5CH_2CH_3 \xrightarrow[h\nu]{NBS} \left[C_6H_5\dot{C}HCH_3 \right] \longrightarrow C_6H_5CH(Br)CH_3$$

(not $C_6H_5CH_2\dot{C}H_2$)

Ethylbenzene — 1-Bromo-1-phenylethane

When the bromine radical removes a hydrogen from the ethyl group, it chooses the hydrogen that will result in the most stable radical, since the energy of activation will be lower. The most stable radical is the one at the benzylic position (the methylene group), not the one at the methyl group.

PROBLEM 10.5

Show the reactions and reagents necessary to produce each of the following transformations. (*Hint:* Use retrosynthesis.)

(a) propylbenzene ------> 1-ethoxy-1-phenylpropane (OC_2H_5)

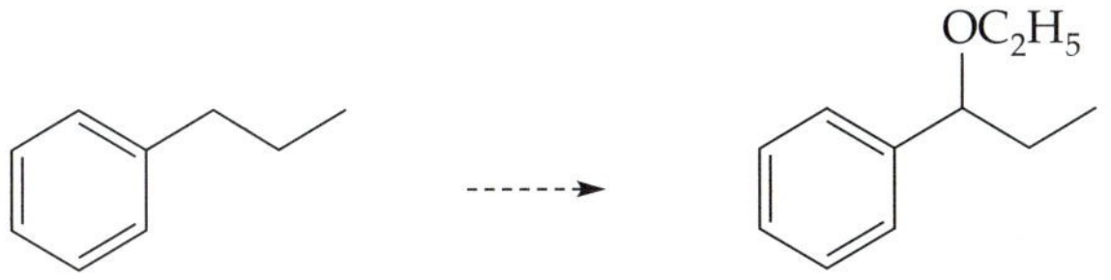

(b) tetralin ------> 1,2-dihydronaphthalene

10.3 Reaction of Alkenes with Hydrogen Bromide

Chapter 6 included a description of the reaction of hydrogen bromide with alkenes, an ionic electrophilic addition resulting in the formation of alkyl bromides. The reaction begins with regioselective attack on the alkene by a proton to form a carbocation, which dictates Markovnikov orientation of the addition product. This reaction must be scrupulously free of radical initiators (including oxygen) and is usually conducted in the dark because hydrogen bromide also adds to alkenes via a radical mechanism.

Reaction of propene with HBr under ionic conditions produces 2-bromopropane. However, the same reaction, when conducted in the presence of ultraviolet light or with the addition of peroxide initiator, yields 1-bromopropane:

$$CH_3{-}CH{=}CH_2 \text{ (Propene)} \xrightarrow[\text{dark}]{\text{HBr}} CH_3{-}CHBr{-}CH_3 \text{ (2-Bromopropane, **(Markovnikov addition)**)}$$

$$CH_3{-}CH{=}CH_2 \text{ (Propene)} \xrightarrow[h\nu]{\text{HBr}} CH_3{-}CH_2{-}CH_2{-}Br \text{ (1-Bromopropane, **(anti-Markovnikov addition)**)}$$

The reason for the opposing regioselectivity of the two reactions is that the first proceeds via ionic electrophilic addition and the second proceeds via radical addition. However, both produce the most stable reactive intermediate.

CHEMISTRY AT WORK

IN THE ENVIRONMENT

Radicals in the Ozone Layer

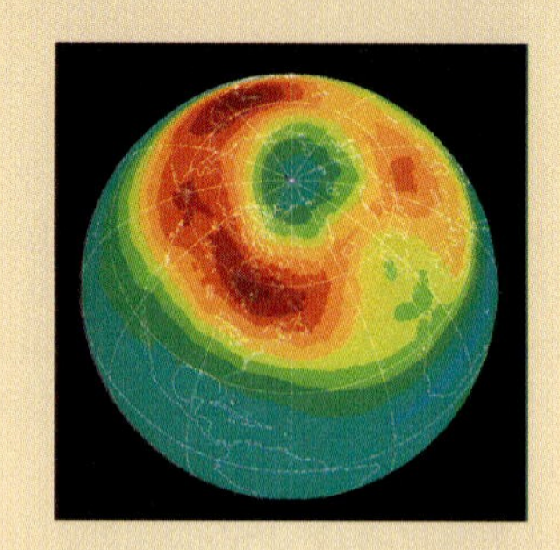

Knowing that ultraviolet light can initiate free-radical reactions, you might think that would make sunlight rather dangerous—and in fact it is. Scientists believe that too much UV light can cause skin cancer and cataracts. The earth's surface has been protected from too much UV light by a layer of ozone, a triatomic form of oxygen (O_3), in the atmosphere enveloping the planet. Ozone exists in the stratosphere, the region of the atmosphere about 24 km (15 mi) above the earth. You can imagine the concern when, in 1985, researchers found that the concentration of ozone in the atmosphere over the Antarctic was being diminished. What could possibly be affecting this balance so high above the earth?

In 1974, Sherwood Rowland and Mario Molina proved that chlorofluorocarbons (CFCs) can attack ozone under a very specific set of conditions found in the high stratosphere over the Antarctic in the winter season. The reactions involve chlorine radicals initially formed by the breaking of a carbon-chlorine bond in CFCs.

The chlorine radicals catalyze the overall conversion of two molecules of ozone into three molecules of oxygen, thereby diminishing the ozone concentration.The chlorine radical is not consumed, but can continue to catalyze the conversion of ozone to oxygen. This research led to the award of the 1995 Nobel Prize in chemistry to Rowland and Molina (along with Paul Crutzen for related atmospheric research).

How do the CFCs get up so high? As described in Section 3.6.3, CFCs have been widely used commercial products. One of their attributes is that they are very stable compounds, not subject to normal degradation processes in the atmosphere. Thus, they have a long lifetime and, through general air circulation patterns, rise to much greater heights than do most earth-generated compounds, even into the stratosphere.

As a result of the widespread concern over this phenomenon, the international community agreed, in the Montreal Protocol of 1987, to gradually phase out the production of CFCs. Because of the major importance of CFCs to society, a crash program was undertaken to develop replacement compounds. Clearly, these compounds must have the same physical properties that have made CFCs so useful, yet be unable to produce chlorine radicals in the stratosphere.

Two approaches were followed in seeking replacement compounds. The first was to develop compounds that do not have carbon-chlorine bonds and so cannot produce chlorine radicals. The compound $CH_2F—CF_3$, called HFC-134a (HFC for hydrofluorocarbon), is now used in refrigerants. The compound $CH_3—CHF_2$, called HFC-152a, is widely used as an aerosol propellant. The second approach was to develop compounds that are more quickly degraded in the atmosphere, even though they may have carbon-chlorine bonds. This was achieved by ensuring the presence in the compounds of carbon-hydrogen bonds, which are susceptible to cleavage by oxygen before the compounds reach the stratosphere. The compound $CH_3—CCl_2F$, called HCFC-141b (HCFC for hydrochlorofluorocarbon), is one example now used in foam insulation.

These two approaches seem to be working, because the atmospheric chlorine concentrations appear to have peaked and may now be slightly declining. Forecasts are that ozone could return to 1979 levels by the year 2050.

Conversion of ozone into oxygen:

$$Cl_2CF_2 \xrightarrow[-65^\circ C]{h\nu} ClCF_2\cdot + Cl\cdot$$

$$O_3 \xrightarrow[-65^\circ C]{h\nu} O_2 + O\cdot$$

$$Cl\cdot + O_3 \longrightarrow ClO\cdot + O_2$$

$$ClO\cdot + O\cdot \longrightarrow Cl\cdot + O_2$$

Net equation:

$$2\,O_3 \longrightarrow 3\,O_2$$

10.3.1 Mechanism of Radical Addition of Hydrogen Bromide

Radical addition of HBr proceeds first by conversion of HBr into a bromine radical—the *initiation* process. This step is brought about by light or by addition of a radical initiator, such as a peroxide:

$$\left.\begin{array}{l} R{-}O{-}O{-}R \xrightarrow{\Delta} 2\,R{-}O\cdot \\ R{-}O\cdot + H{-}Br \longrightarrow ROH + :\ddot{Br}\cdot \end{array}\right\}\ \textbf{Initiation}$$

The next step involves addition of the bromine radical to the alkene, which could, in principle, occur at either end of the double bond. In fact, this addition is regioselective in producing the more stable radical intermediate—in the case of propene, a secondary rather than a primary radical:

$$\underset{\text{Propene}}{CH_3CH{=}CH_2} + :\ddot{Br}\cdot \longrightarrow \underset{2^\circ\ \text{radical}}{CH_3\dot{C}H{-}CH_2{-}Br} \qquad \not\longrightarrow \underset{1^\circ\ \text{radical}}{CH_3CHBr{-}\dot{C}H_2}$$

Therefore, in the *propagation* steps, the bromine radical adds to the terminal methylene carbon of propene to form a secondary carbon radical. This radical abstracts a hydrogen from hydrogen bromide to produce the product (1-bromopropane) and a new bromine radical:

$$\left.\begin{array}{l} CH_3CH{=}CH_2 + :\ddot{Br}\cdot \longrightarrow CH_3\dot{C}H{-}CH_2{-}Br \\ CH_3\dot{C}H{-}CH_2{-}Br + H{-}Br \xrightarrow{-:\ddot{Br}\cdot} CH_3CH_2CH_2Br \end{array}\right\}\ \textbf{Propagation}$$

The overall reaction is a chain reaction because the bromine radical that is produced starts the chain over again by attacking another molecule of propene.

Note that the radical addition results in anti-Markovnikov regiochemistry because the bromine adds first, producing the more stable (radical) intermediate. In an electrophilic addition, the hydrogen adds first, also producing the more stable (carbocation) intermediate. In both cases, the first step of the addition determines the regiochemistry of the reaction, which is directed to produce the more stable secondary intermediate rather than the less stable primary intermediate.

10.3.2 Application in Synthesis

The ability to control the addition of HBr to alkenes to produce either Markovnikov or anti-Markovnikov regiochemistry is an extremely important tool in organic chemistry. It is relatively easy to obtain or synthesize alkenes from a variety of sources: alcohols, alkyl halides, alkynes, and, as we will see later, carbonyl compounds (see Key to Transformations). Subsequent reactions can be targeted to either end of the double bond, depending on where a bromide is needed. For example, styrene can be separately converted into the two isomeric bromophenylethanes, 1-bromo-1-phenylethane and 1-bromo-2-phenylethane:

$$\underset{\text{1-Bromo-2-phenylethane}}{C_6H_5{-}CH_2{-}CH_2Br} \xleftarrow[h\nu]{HBr} \underset{\text{Styrene}}{C_6H_5{-}CH{=}CH_2} \xrightarrow[\text{dark}]{HBr} \underset{\text{1-Bromo-1-phenylethane}}{C_6H_5{-}\overset{\overset{\displaystyle Br}{|}}{C}H{-}CH_3}$$

How to Solve a Problem

Show the reactions necessary to convert styrene into two isomeric phenylbutenes: 4-phenyl-1-butene and 3-phenyl-1-butene.

PROBLEM ANALYSIS

First, we write the structures of the reactant and products:

$$\underset{\text{Styrene}}{C_6H_5{-}CH{=}CH_2} \dashrightarrow \underset{\text{4-Phenyl-1-butene}}{C_6H_5{-}CH_2CH_2CH{=}CH_2}$$

$$\underset{\text{Styrene}}{C_6H_5{-}CH{=}CH_2} \dashrightarrow \underset{\text{3-Phenyl-1-butene}}{C_6H_5{-}\underset{\underset{\displaystyle CH_3}{|}}{C}HCH{=}CH_2}$$

Examination of the products indicates that in both a two-carbon unit has been added to the carbons present in styrene. To make carbon-carbon bonds, an S_N2 reaction is frequently employed, which means that a leaving group and a nucleophilic carbon are required. Therefore, we have to convert styrene either into a compound with a leaving group (typically a bromide) or into a nucleophile (so far, the only carbon nucleophiles we have encountered have been organometallic reagents or acetylides). Recall that we usually obtain alkenes by either elimination reactions (dehydration or dehydrohalogenation) or reduction of alkynes. With these thoughts in mind, especially regarding acetylide, and using a retrosynthetic approach, we get the following possibility for the synthetic scheme:

$$\underset{\text{Alkene}}{R{-}CH{=}CH_2} \Longrightarrow \underset{\text{Alkyne}}{R{-}C{\equiv}CH} \Longrightarrow \underset{\text{Alkyl halide}}{R{-}Br} + \underset{\text{Acetylide}}{HC{\equiv}C{:}^-}$$

This shows how to add a two-carbon unit to an alkyl halide and convert it to an alkene.

All that remains is to obtain the proper alkyl halide on which acetylide can carry out an S_N2 substitution. Note that the desired products can be obtained by attaching the acetylide unit either to the methylene group ($=CH_2$) or to the methinyl group ($-CH=$) of the styrene unit. Therefore, it is essential to be able to place a bromine on each of the two carbons separately. This can be done using the ionic and radical addition of HBr to styrene, as described earlier.

SOLUTION

The solution involves (1) adding HBr to styrene with the desired regiochemistry, (2) carrying out an S_N2 substitution with sodium acetylide, and (3) reducing the triple bond to a double bond.

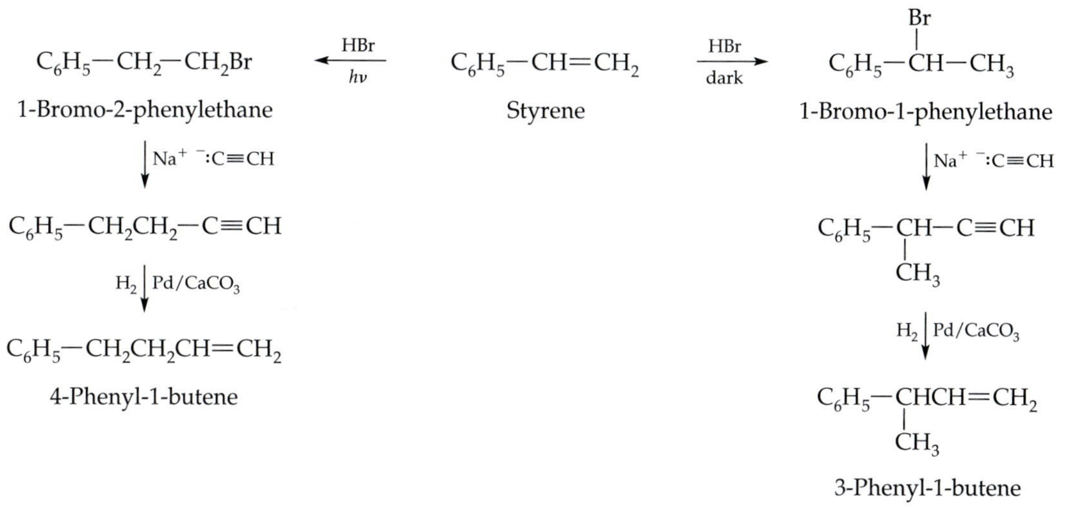

$$C_6H_5{-}CH_2{-}CH_2Br \xleftarrow[h\nu]{HBr} C_6H_5{-}CH{=}CH_2 \xrightarrow[dark]{HBr} C_6H_5{-}CH(Br){-}CH_3$$

1-Bromo-2-phenylethane — Styrene — 1-Bromo-1-phenylethane

$$C_6H_5{-}CH_2{-}CH_2Br \xrightarrow{Na^+ \ ^-:C{\equiv}CH} C_6H_5{-}CH_2CH_2{-}C{\equiv}CH \xrightarrow[H_2]{Pd/CaCO_3} C_6H_5{-}CH_2CH_2CH{=}CH_2$$

4-Phenyl-1-butene

$$C_6H_5{-}CH(Br){-}CH_3 \xrightarrow{Na^+ \ ^-:C{\equiv}CH} C_6H_5{-}CH(CH_3){-}C{\equiv}CH \xrightarrow[H_2]{Pd/CaCO_3} C_6H_5{-}CH(CH_3)CH{=}CH_2$$

3-Phenyl-1-butene

PROBLEM 10.6

Show the major product formed from each of the following reactions:

(a) 1-butene with HBr in the dark

(b) 1-butene with HBr in the presence of peroxides

(c) methylenecyclohexane with HBr in the dark

(d) methylenecyclohexane with HBr under ultraviolet light

PROBLEM 10.7

Show the reactions and reagents necessary for each synthesis:

(a) bromocyclohexane to 1-bromo-2-cyclohexylethane

(b) 1-butene to 1-methoxybutane

(c) 1-butene to 2-methoxybutane

Chapter Summary

Carbon radicals are a type of reactive intermediate in which a carbon is trigonally hybridized. The radical is a neutral species with a carbon having three groups attached by sigma bonds (requiring six electrons, three from the carbon) and having a seventh (unshared) electron in a *p* orbital. The stability order for alkyl radicals is tertiary > secondary > primary. Allylic radicals and benzylic radicals are primary radicals that have enhanced stability due to resonance delocalization of the unshared electron. Radicals can be generated by heat, light, or an **initiator.** The combustion of organic compounds occurs by a radical chain reaction.

Alkanes undergo radical halogenation resulting in substitution of halogen for hydrogen and the formation of alkyl halides. The mechanism is a **chain reaction** involving *initiation, propagation,* and *termination* steps. The reaction is relatively unselective and is useful mainly for alkanes with either a single class of hydrogens (as in cycloalkanes, which have only secondary hydrogens) or a single tertiary hydrogen. Bromination is much more selective than chlorination. Compounds with allylic or benzylic hydrogens undergo preferential bromination at the allylic or benzylic positions.

Hydrogen bromide adds to alkenes under radical conditions to form alkyl halides in a chain reaction. The addition follows anti-Markovnikov regiochemistry, the opposite of ionic addition of hydrogen bromide. The availability of the two mechanisms of addition provides a choice of regiochemistry in HBr addition to alkenes.

Summary of Reactions

1. Halogenation of alkanes (Section 10.2)

$$\text{R—H} + \text{X}_2 \xrightarrow[\text{or } h\nu]{\Delta} \text{R—X} + \text{HX}$$

2. Bromination of benzylic and allylic hydrocarbons with *N*-bromosuccinimide (NBS) (Section 10.2.3)

$$\text{C}_6\text{H}_5\text{CH}_2\text{—R} \xrightarrow[\Delta]{\text{NBS}} \text{C}_6\text{H}_5\text{CHBr—R}$$

$$\text{CH}_2\text{=CHCH}_3 \xrightarrow[\Delta]{\text{NBS}} \text{CH}_2\text{=CHCH}_2\text{Br}$$

3. Addition of HBr to alkenes (Section 10.3)

$$\text{R—CH=CH}_2 \xrightarrow[\text{ROOR or } h\nu]{\text{HBr}} \text{R—CH}_2\text{—CH}_2\text{Br}$$

Additional Problems

Radical Reactions

10.8 Using an alkoxy radical (RO•) from a peroxide (R—O—O—R) as the initiator, show the mechanism by which hydrogen is abstracted from each of the following compounds to form the most stable radical:

(a) cyclohexane (b) 2-methylbutane
(c) 1-butene (d) 2,2-dimethylpropane
(e) isobutane (f) ethylbenzene

10.9 What are the three experimental techniques for initiating a radical reaction?

10.10 Show all of the steps in the mechanism for the heat-initiated radical monochlorination of ethane.

10.11 Show all of the steps in the mechanism for the peroxide-initiated radical monobromination of cyclohexene.

10.12 Show all of the possible alkyl radicals that could be formed by abstraction of a hydrogen from 2-pentene. Which is the most stable, and why?

10.13 In addition to polychlorinated products, the chlorination of butane yields a number of isomers with the formula C_4H_9Cl. Show the structure of all of the different isomers you would expect to be produced.

10.14 Explain why the synthesis of cyclohexyl bromide by radical bromination of cyclohexane is useful, but the formation of 1-bromobutane by the same reaction is not especially useful.

10.15 Show the mechanism for the radical addition of HBr to 1-butene using a peroxide as initiator.

10.16 Show the major product of the monobromination (using bromine and UV light) of each of the following compounds:

(a) cycloheptane (b) isobutane
(c) propane (d) 3-methylpentane
(e) propylbenzene

10.17 Radical chlorination of ethane yields a single monochlorinated product but two different dichlorinated products. What are these products? Show a mechanism for the formation of both dichloroethanes.

10.18 The radical bromination of propane is not regioselective, but the radical bromination of propylbenzene is very regioselective. Explain.

10.19 Show the major product of each of the following reactions:

(a) 1-pentene with HBr in the dark
(b) 1-pentene with HBr in UV light
(c) cyclopentene with bromine in the dark
(d) cyclopentene with bromine in the presence of a peroxide
(e) cyclopentene with *N*-bromosuccinimide (NBS)
(f) cyclopentane with bromine and heat

10.20 What product(s) would you expect if propene labeled with ^{14}C at carbon 1 ($^{14}CH_2{=}CH{-}CH_3$) were subjected to reaction with NBS? Explain.

10.21 Show the product of the reaction of 1-phenyl-1-butene with HBr (a) in the dark and (b) in the presence of UV light.

Organic Synthesis

10.22 Show the reactions necessary to convert toluene to the two constitutional isomers *p*-bromotoluene and benzyl bromide.

10.23 Show the reactions necessary to convert benzene to styrene, starting with a Friedel-Crafts acetylation. (*Hint:* Use retrosynthesis.)

10.24 Show the reactions necessary to bring about each conversion:

(a) 2-bromo-2-methylpropane to 1-bromo-2-methylpropane

(b) 1-butene to 1,3-butadiene

(c) cyclohexane to 1,3-cyclohexadiene

(d) cyclohexene to *cis*-1,2-dimethoxycyclohexane

(e) cyclohexene to *trans*-1,2-dimethoxycyclohexane

(f) isobutane to 3,3-dimethyl-1-butanol

10.25 Show the reactions necessary to complete each of the following syntheses, using the designated starting material and any other necessary reagents:

(a) cyclopentane to cyclopentanol

(b) cyclobutane to cyclobutylacetylene

(c) cyclopentane to 2-cyclopentylethanol

(d) cycloheptane to ethylcycloheptane

(e) bromocyclohexane to 1-cyclohexylethanol

(f) cyclohexene to 3-vinylcyclohexene

10.26 Complete the following syntheses, showing each reaction and its products:

(a) acetophenone to 1-methoxy-1-phenylethane

(b) ethylbenzene to *p*-bromostyrene

10.27 Show how each of the following compounds can be obtained. For part (a), start with toluene, and for parts (b) and (c), start with any compound prepared in the preceding part(s).

(a) propylbenzene

(b) 1,2-diphenylethane

(c) 1,2-diphenylethene

10.28 Show how to accomplish these conversions:

(a) toluene to benzyl alcohol

(b) 4-pentylanisole to 1-(4-methoxyphenyl)-1-pentene

10.29 Show how to obtain each of the following products, starting with ethylbenzene for part (a) and starting with any compound prepared in the preceding part(s) for parts (b) and (c).

(a) 2-phenylethanol

(b) 4-phenyl-1-butanol

(c) 1-phenyl-1,3-butadiene

Mixed Problems

10.30 Alkenes are known for undergoing addition reactions. Some additions are stereospecific, and some are regioselective.

(a) Explain why the addition of HBr to an alkene (in the dark or light) is regioselective.

(b) Explain why the addition of bromine is stereospecific, but the addition of HBr (in the dark) is not.

10.31 Compounds **W** and **X** have the molecular formula C_5H_8. Both decolorize bromine and react with cold potassium permanganate. Reaction with hydrogen over platinum results in the uptake of two molar equivalents of hydrogen by each isomer. Oxidation of **W** with ozone or hot potassium permanganate produces two carboxylic acids (RCOOH) and no carbon dioxide. Oxidation of **X** with hot potassium permanganate produces carbon dioxide, a dicarboxylic acid, and acetic acid (CH_3COOH). What are the structural formulas of **W** and **X**?

10.32 The compound known as epichlorohydrin is an important commercial chemical used in polymer chemistry. Show how it can be synthesized from propene.

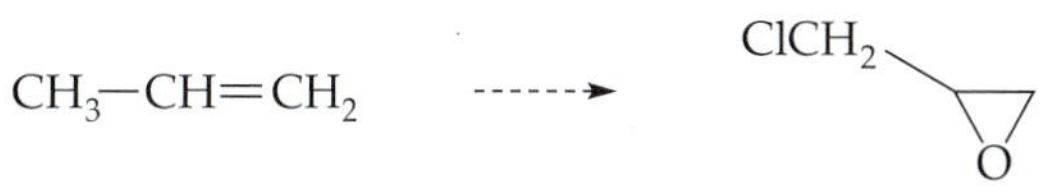

$CH_3{-}CH{=}CH_2$ - - - - - → $ClCH_2$ (epoxide)

Epichlorohydrin

10.33 Show the product of each of the following reactions:

(a) 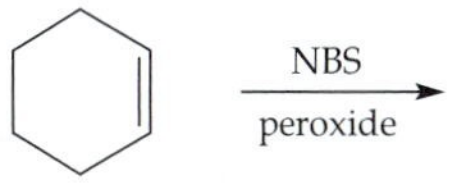 cyclohexene $\xrightarrow[\text{peroxide}]{\text{NBS}}$

(b) product of part (a) with sodium acetylide

(c) product of part (b) with H_2 and $Pd/CaCO_3$

10.34 Benadryl is an over-the-counter antihistamine that can be synthesized from toluene. What reagents are necessary to bring about each of the following steps?

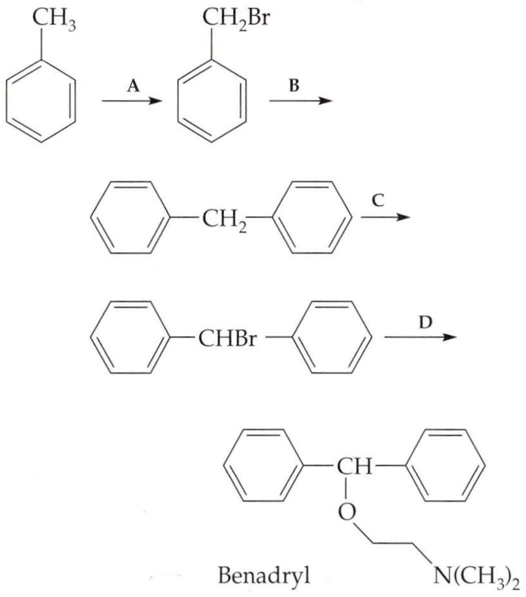

Toluene (CH_3) $\xrightarrow{\text{A}}$ benzyl bromide (CH_2Br) $\xrightarrow{\text{B}}$

Ph–CH_2–Ph $\xrightarrow{\text{C}}$

Ph–CHBr–Ph $\xrightarrow{\text{D}}$

Ph–CH(–O–CH_2CH_2–$N(CH_3)_2$)–Ph

Benadryl

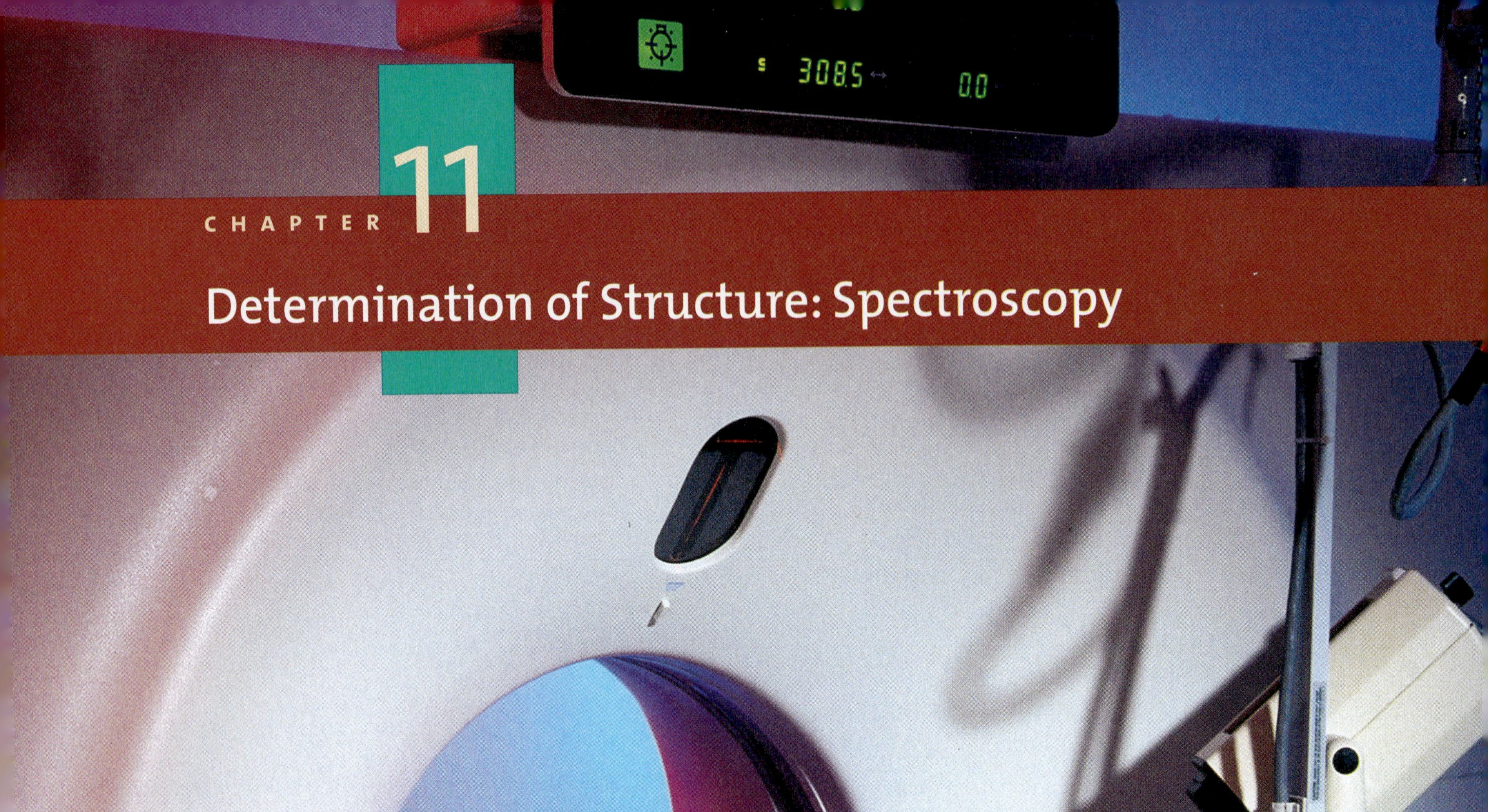

CHAPTER 11

Determination of Structure: Spectroscopy

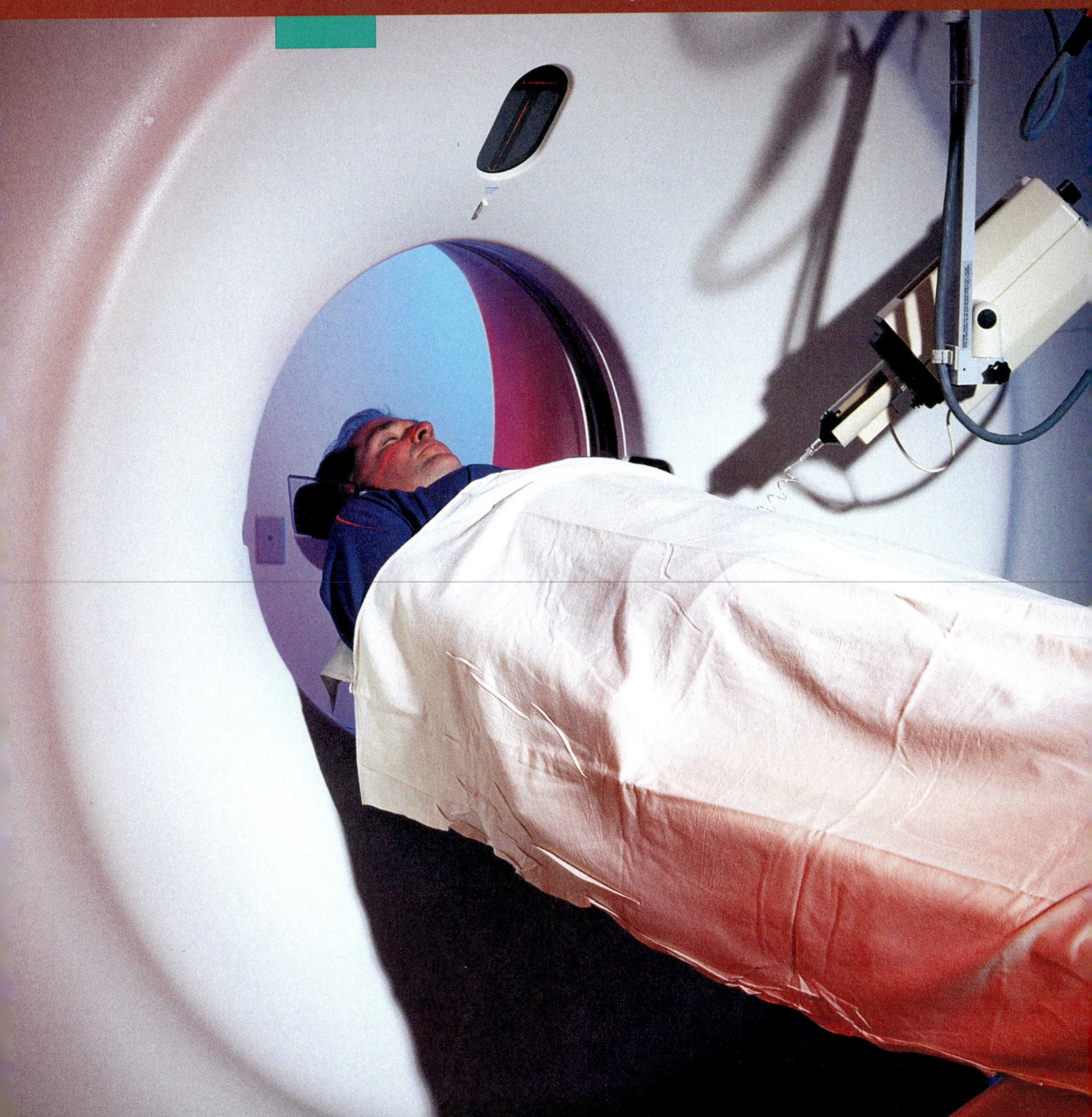

THE INTRODUCTORY ESSAY in this book pointed out that one of the three tasks undertaken by organic chemists is to determine the chemical structures of unknown compounds. In this task, the organic chemist functions as a chemical detective, gathering critical evidence that will lead to an irrefutable conclusion regarding the chemical structure of an "unknown." Such unknowns arise from the isolation of new compounds from natural sources, such as from bacteria or plants; from syntheses of new compounds in the laboratory; and from chemical reactions that occasionally produce unexpected products or by-products.

Here we will review the means available for determining the structure of an unknown, giving special attention to several instrumental techniques that have revolutionized the process. Determining the structure of an unknown (sometimes called a *structure proof*) can now be done using much smaller quantities (milligrams) of a sample and can be accomplished much more quickly as well.

11.1 Structure Proof

When chemists try to identify an unknown chemical, their first task must be to ensure that it is pure. Once an unknown compound is pure, chemists can begin to determine its structure—the structure proof. They start by determining the molecular formula. Historically, the next step was to study the chemical behavior of the compound by testing for the presence or absence of functional groups. Such procedures remain a part of organic chemistry today (and will be summarized in Section 11.1.3). However, much of the work done in the lab now involves special instruments that can detect different structural features of a compound. The final step is often to *degrade* the unknown into a known compound—that is, to carry out on the unknown chemical reactions that eventually produce and establish a structural relationship with known compounds.

The next three subsections summarize the noninstrumental approaches to structure proof.

11.1.1 Determination of Molecular Formula

The **empirical formula** of a pure organic compound is determined by *microanalysis.* This process is normally accomplished by commercial analytical laboratories that combust a few milligrams of the compound and determine the amounts of the various combustion products, principally carbon dioxide and water. The results are reported as the elemental composition of the compound (the *percentage* of each element

◀ A patient being diagnosed in an MRI device.

present). For example, a compound may be reported to contain 53.30% carbon, 11.10% hydrogen, and, by difference, 35.60% oxygen. (Oxygen composition is always calculated as the difference between the sum of the percentages of other elements and 100%.) These percentages are transformed into a *ratio* of the numbers of atoms of each kind present by dividing each percentage by the atomic weight of the corresponding element. In the example given, the result is a ratio of 4.44 carbon to 10.99 hydrogen to 2.22 oxygen. Finally, dividing all three numbers by the smallest number reduces the ratio to the smallest possible numbers, with one number being 1.0. In the example, division by the smallest number (2.22) gives 2.0 carbon to 4.99 hydrogen to 1.0 oxygen. The empirical formula must be written in whole numbers (to represent whole atoms). Thus, the empirical formula for this example, which is the ratio of the number of atoms of each element present, is C_2H_5O.

The empirical formula is simply a ratio, while the desired **molecular formula** gives the actual number of each type of atom present in a single molecule. The molecular formula is always some multiple of the empirical formula. This multiple is derived from the molecular weight of the unknown. As will be described in Section 11.2, the molecular weight of an unknown is usually determined by the technique known as mass spectrometry, and it can be done to an accuracy of four decimal places, if needed.

Dividing the experimental molecular weight by the sum of the atomic weights of the empirical formula produces approximately a whole number, which is the multiplier to be applied to the empirical formula to convert it into the molecular formula. For our example, the molecular weight is determined to be 90.09, while the sum of the atomic weights of the empirical formula, C_2H_5O, is

$$(2 \times 12.01) + (5 \times 1.01) + (1 \times 16) = 45.06$$

Dividing the molecular weight (90.09) by the sum of the atomic weights (45.06) produces a multiplier of 1.99, which is rounded off to the whole number 2. Thus, the molecular formula in our example is the empirical formula, C_2H_5O, multiplied by 2, or $C_4H_{10}O_2$.

PROBLEM 11.1

(a) Calculate the molecular formula for an unknown compound of molecular weight 86 whose microanalysis indicates the following percentage composition: 83.7% carbon and 16.3% hydrogen.

(b) Calculate the molecular formula for an unknown compound of molecular weight 93 whose microanalysis indicates the following percentage composition: 77.4% carbon, 7.5% hydrogen, and 15.0% nitrogen.

(c) Calculate the molecular formula for an unknown compound of molecular weight 136 whose microanalysis indicates the following percentage composition: 70.6% carbon and 5.9% hydrogen.

11.1.2 Hydrogen Deficiency Index (HDI)

Because alkanes have the general formula C_nH_{2n+2}, any compound having fewer hydrogens must either be unsaturated (have double or triple bonds or aromatic rings) or contain one or more rings. The number of hydrogens *fewer* than $2n + 2$ is called the **hydrogen deficiency** (**HD**) and is always an even number. The general formulas of alkenes (C_nH_{2n}), cycloalkanes (C_nH_{2n}), and alkynes (C_nH_{2n-2}), indicate that an HD of 2 could be caused by the presence of a ring or double bond. Similarly, an HD of 4 could be caused by two rings, two double bonds, one ring and one double bond, or one triple bond. Dividing the HD by 2 gives the **hydrogen deficiency index (HDI)**, which represents the number of double bonds or rings in a compound. Be aware that a triple bond (which is the equivalent of two double bonds) is also a possibility if HDI ≥ 2. In addition, a carbonyl group (C=O) or an imine group (C=N) also has a double bond. A benzene ring has an HDI of 4 (equivalent to three double bonds and a ring).

Once chemists know the molecular formula and HDI of an unknown compound, it is possible to arrive at some early conclusions about its structure. Here are some common HDI values and the structural conclusions that may follow:

HDI = 1 A ring or a double bond

HDI = 2 Two rings, two double bonds, one ring and one double bond, or one triple bond

HDI = 4 Frequently accounted for by a benzene ring (one ring and the equivalent of three double bonds)

How to Solve a Problem

What is the HDI for a compound with the molecular formula C_6H_8? What can we deduce about the structure of this compound?

SOLUTION

The formula for an alkane of six carbons is C_6H_{14}. Therefore, the HD of the unknown is 6 ($14 - 8 = 6$), and its HDI is 3. This means that there must be some combination of rings, double bonds, or triple bonds, totaling 3. If the compound decolorizes bromine and reacts with potassium permanganate, there must be at least one multiple bond present. If, upon catalytic hydrogenation, two equivalents of hydrogen are absorbed, the product must have at least one ring and either two double bonds or one triple bond. These conclusions can then be used to suggest subsequent experiments to eliminate one or more of these possibilities.

PROBLEM 11.2

Calculate the HDI for the following compounds: (a) C_6H_{12}, (b) C_8H_{10}, and (c) $C_7H_{12}O$.

PROBLEM 11.3

A compound with molecular formula C_8H_{14} reacts with bromine in CCl_4 and is oxidized by cold potassium permanganate. It absorbs one equivalent of hydrogen over a platinum catalyst. What conclusions can you draw about the structure of this compound?

11.1.3 Functional Group Tests

This book presents many characteristic reactions of the various functional groups. For most families of compounds, there are one or more characteristic qualitative tests. These tests can be used to determine the presence or absence of certain functional groups in an unknown compound. (Recall that alkanes are very inert, and there are no unique tests for them; they are usually identified by excluding other possibilities.) These functional group tests are listed in Table 11.1 for easy reference. (Refer to the appropriate chapter for a detailed discussion.) Essential spectroscopic information is also included in Table 11.1 but will be discussed in detail in later sections.

TABLE 11.1

Functional Group Tests

Functional Group (Chapter reference)	Chemical Test	Spectral Keys
Alkene (2)	Decolorizes bromine/CCl_4; reacts with cold $KMnO_4$	IR: ~1650 cm^{-1}
Alkyne (7)	Same tests as for alkenes; $Ag(NH_3)_2OH$ gives precipitate with terminal alkynes	IR: ~2150 cm^{-1}
Alcohol (4)	Na metal produces H_2 bubbles; for Lucas test ($HCl/ZnCl_2$), 3° > 2°; CrO_3 oxidation of 1° and 2°	IR: ~3300 cm^{-1}
Phenol (9)	Soluble in NaOH, not in $NaHCO_3$	IR: ~3300 cm^{-1}
Ether (5)	The only oxygenated inert compound	
Amine (12)	Soluble in dilute HCl	IR: ~3400 cm^{-1}
Aldehyde (13)	Reacts with DNPH; oxidized by Tollens reagent	NMR: 9.5 ppm IR: ~1730 cm^{-1}
Ketone (13)	Reacts with DNPH	IR: ~1730 cm^{-1}
Carboxylic acid (14)	Soluble in $NaHCO_3$; CO_2 evolved	IR: ~1720 cm^{-1}, ~3300 cm^{-1} (broad) NMR: 10–13 ppm
Ester (15)	Hydrolyzed to acid and alcohol	IR: ~1700 cm^{-1}
Amide (15)	Insoluble in HCl; hydrolyzed to acid and amine	IR: ~1700 cm^{-1} and ~3400 cm^{-1}

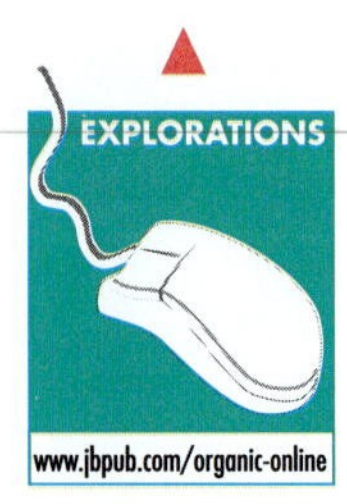

How to Solve a Problem

Compound **A** analyzes as having molecular formula C_8H_8 and readily reacts with Br_2/CCl_4 in the dark and with cold $KMnO_4$. Reaction with hydrogen over a platinum catalyst leads to absorption of one equivalent of hydrogen and the formation of compound **B.** Compound **B** does not decolorize bromine or react with cold $KMnO_4$. Vigorous oxidation of **A** and **B** with hot $KMnO_4$ in water, followed by filtering off of the brown MnO_2 precipitate, yields a colorless filtrate. Acidification of the filtrate produces a colorless precipitate (compound **C**), which has molecular formula $C_7H_6O_2$. Compound **C** is soluble in aqueous $NaHCO_3$ solution. Write structural formulas for compounds **A, B,** and **C.**

PROBLEM ANALYSIS

The molecular formula of compound **A** indicates an HDI of 5, and such a low ratio of hydrogen to carbon frequently indicates an aromatic compound. With a total of eight carbons present, a likely possibility is the presence of a benzenoid ring (C_6), which itself requires an HDI of 4 (three double bonds and a ring). If this is so, then only two other carbons can be present. The bromine and cold $KMnO_4$ tests indicate that an alkene or alkyne is present. Since only one equivalent of hydrogen is absorbed, there is only one double bond present, and this requires two carbon atoms (C_2). We can deduce that there is a two-carbon fragment attached to a benzene ring (C_6H_5—C—C). Benzenoid compounds with a single carbon chain attached, no matter how long the chain, are oxidized by hot $KMnO_4$ to benzoic acid, so compound **C** is probably benzoic acid. Therefore, compound **A** is styrene, and compound **B** is ethylbenzene.

SOLUTION

$CH{=}CH_2$ (on benzene ring) **A** Styrene —(H_2/Pt)→ $CH_2{-}CH_3$ (on benzene ring) **B** Ethylbenzene —(1. $KMnO_4$, Δ; 2. HCl)→ COOH (on benzene ring) **C** Benzoic acid

PROBLEM 11.4

What conclusions can you draw about each of the following unknowns? (*Note:* In some cases, the complete structure or a very few possibilities can be deduced; in other cases, only partial structural conclusions can be made. Consider only the structural possibilities of families studied to this point.)

(a) A compound with formula C_7H_8 does not decolorize bromine and does not absorb hydrogen over a platinum catalyst.

(b) A compound with formula $C_6H_{12}O$ does not decolorize bromine. It does not react with sodium metal.

(c) A compound with formula $C_6H_{12}O$ decolorizes bromine in CCl_4 and absorbs one equivalent of hydrogen over a platinum catalyst. It gives off a gas when reacted with sodium metal.

(d) A compound with formula C_5H_8 decolorizes bromine in CCl_4 and absorbs two equivalents of hydrogen over a platinum catalyst. It gives a precipitate when reacted with silver ammonium nitrate.

PROBLEM 11.5

Propose structures for the compounds for which the following experimental observations have been made:

(a) A compound of molecular formula C_3H_8O is soluble in water and treating it with metallic sodium produces bubbles. In the Lucas test, the compound fails to react with $HCl/ZnCl_2$ in a reasonable time.

(b) A compound with molecular formula C_7H_8O is insoluble in water. It does not decolorize bromine or react with cold potassium permanganate. It does not react with metallic sodium.

(c) A compound with molecular formula C_8H_6 decolorizes bromine and reacts with cold potassium permanganate. It absorbs two equivalents of hydrogen over a platinum catalyst to produce C_8H_{10}. Upon oxidation with hot potassium permanganate, removal of the brown MnO_2 precipitate, and acidification of the colorless filtrate, a colorless precipitate with formula $C_7H_6O_2$ is obtained.

11.2

Mass Spectrometry

Mass spectrometry (abbreviated MS) is a highly accurate technique for analyzing very small quantities (1–2 mg) of liquids or solids. Mass spectrometers are computer-controlled, bench-top instruments. Their results can be compared with stored computerized libraries of spectra to assist in identification. Specifically, mass spectrometry can provide information on molecular weight and the presence of heteroatoms and isotopes. Further, this technique can identify "fragments" of organic structure that are formed in the spectrometer.

11.2.1 The Mass Spectrometry Process

In mass spectrometry, the sample to be analyzed is vaporized at very low pressure, injected into the spectrometer, and then bombarded by a 70-eV (~1600-kcal) electron beam in an ionization chamber. When an electron having that much energy impacts a molecule (symbolized as M), another electron is ejected from the molecule. The molecule thereby acquires a positive charge and becomes a **molecular ion** (symbolized $M^{+\cdot}$).

$$\underset{\text{Neutral molecule}}{M} + \underset{\text{Electron}}{e^-} \longrightarrow \underset{\text{Cation radical (molecular ion)}}{M^{+\cdot}} + \underset{\text{Two electrons}}{2\,e^-}$$

This molecular ion is a cation radical—*cation* because it is positively charged and *radical* because it has an odd number of electrons. Its mass (molecular weight) is essentially unchanged from that of the parent molecule, because the mass of the ejected electron is negligible compared to the remaining mass of protons plus neutrons. For example, methane produces a molecular ion with a **mass-to-charge ratio** (m/z) of $16/1 = 16$.

$$\underset{\text{Methane}}{CH_4} \xrightarrow{-e^-} \underset{\text{Methane radical cation } (m/z = 16)}{[CH_4]^{+\cdot}}$$

Once formed, the molecular ion is accelerated by a series of electrical plates to a very high speed and enters the mass analyzer portion of the spectrometer. This analyzer contains a strong magnetic field that deflects the cations into a circular path. The extent to which the cation beam is deflected depends on the mass of the cation: Larger-mass cations are deflected less than cations with smaller masses. Therefore, if the beam contains cations of different masses, the beam will be dispersed. The cations then strike a detector, which measures the current created by their impact and records its intensity on a recorder, producing a kind of graph called a *mass spectrum*. Figure 11.1 shows a schematic and photograph of a mass spectrometer.

The spectrum produced by a mass spectrometer (shown on a computer screen and/or printed on a recorder) shows the m/z ratio of each cation detected and the intensity of each (a *peak*). The simple mass spectrum of methane is shown in Figure 11.2.

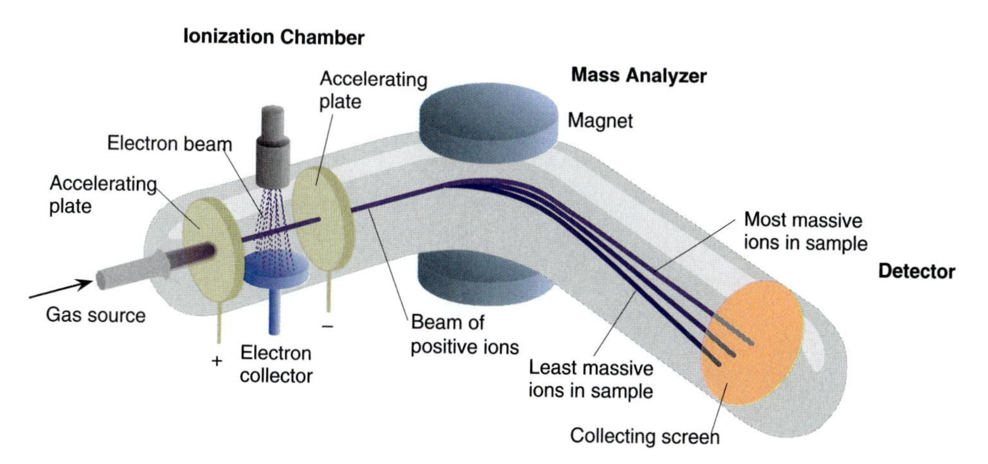

FIGURE 11.1

Schematic diagram and photograph of a mass spectrometer.

11.2.2 Molecular Ion

The strongest peak in a mass spectrum is called the *base peak*. Its intensity is arbitrarily set to 100 on a vertical intensity scale, and the relative intensity of other peaks is calibrated against that of the base peak. The horizontal axis is the m/z ratio, and the molecular ion is usually a strong peak at the second highest m/z ratio. Since the charge (z) in the m/z ratio is invariably 1, *the m/z ratio of the molecular ion is equal to the molecular weight of the sample.* It can be determined to four decimal places on a high-resolution mass spectrometer, but usually a low-resolution instrument that determines the molecular weight to a whole number only suffices.

It should be noted that the molecular weights of most organic compounds (all hydrocarbons, organic halides, and oxygen-containing compounds) are even numbers. (This follows from the general formulas—for example, C_nH_{2n}—check this yourself.) The major exceptions are compounds containing nitrogen. Therefore, when an MS analysis is done for an unknown compound and the molecular ion mass is an odd number, the presence of an odd number of nitrogen atoms is indicated.

Because isotopes of many elements are naturally present to a significant degree, other peaks are observed in mass spectra. For example, natural carbon contains 1.1% ^{13}C along with the usual ^{12}C. Therefore, the mass spectrum of an organic compound usually contains a small $M^{\ddagger} + 1$ peak reflecting the ^{13}C content. For example, the mass spectrum of methane in Figure 11.2 shows, in addition to the base peak at $m/z = 16$, a small peak

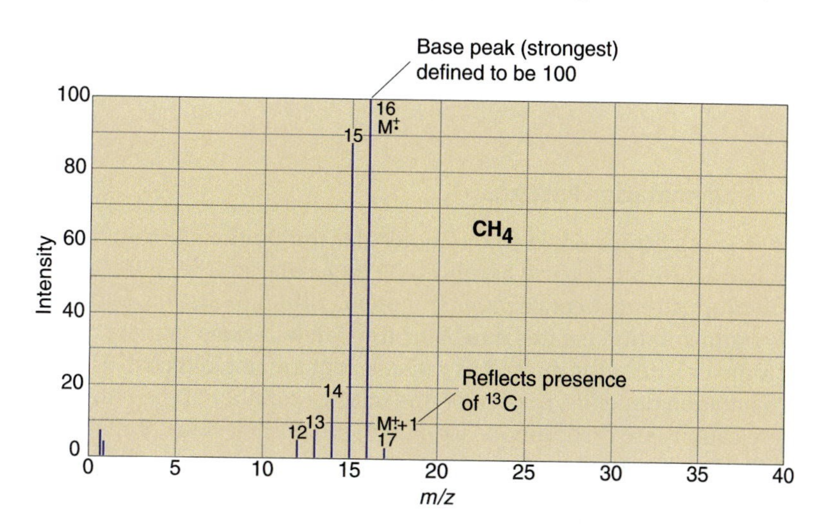

FIGURE 11.2

Mass spectrum of methane.

FIGURE 11.3

Mass spectrum of 1-bromopropane.

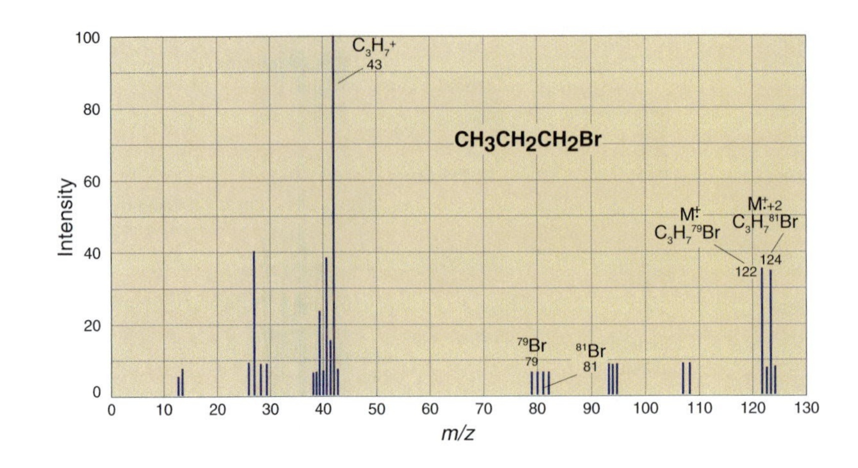

at $m/z = 17$, reflecting the presence of ^{13}C. Of the common elements present in organic compounds, those with significant concentrations of isotopes are chlorine (76% ^{35}Cl and 24% ^{37}Cl) and bromine (51% ^{79}Br and 49% ^{81}Br). Therefore, an alkyl bromide or chloride will show a peak at $M^{+\cdot}$ and $M^{+\cdot} + 2$. A mass spectrum for 1-bromopropane containing such isotopes is shown in Figure 11.3 as an example.

PROBLEM 11.6

What m/z value does the molecular ion peak have for each of the following compounds?

(a) C_6H_{12} (b) C_3H_6BrCl (c) C_2H_7N

PROBLEM 11.7

For each of the following compounds, indicate the m/z value of the molecular ion peak and any other peaks expected:

(a) Br (b) Cl (c) OH (d) NH_2

11.2.3 Fragmentation Patterns

The amount of energy transferred to the molecular ion by the electron beam is more than enough to fracture any and all bonds in the molecular ion. Such fracturing occurs extensively to produce every conceivable fragment, although distinct patterns usually prevail. This fragmentation creates **daughter ions,** new cations that are also detected by the spectrometer's detector, as well as radicals that are not detected. The likelihood of such fragmentation depends, in part, on the relative stability of the potential daughter ions. These cations are carbocations for which the normal rules of relative stability apply (3° > allylic ≈ 2° > 1°).

$$\text{A:B} \xrightarrow{-e^-} [\text{A}\cdot\text{B}]^{+} \xrightarrow{\text{(fragmentation)}} \text{A}\cdot + \text{B}^{+} \text{ (Daughter ion)} \quad \text{and/or} \quad \text{A}^{+} \text{ (Daughter ion)} + \text{B}\cdot$$

Molecule — Molecular ion

The mass spectrum of methane in Figure 11.2 shows the expected peaks at 16 (the $M^{+\cdot}$ peak) and 17 (the $M^{+\cdot} + 1$ peak), which originate from the molecular ion and its ^{13}C isotope, respectively. The peaks at 15, 14, 13, and 12, however, are those of daughter ions. The sequential loss of a hydrogen atom (with its electron, therefore a radical) results in daughter ions with m/z values of 15 (CH_3), 14 (CH_2), 13 (CH), and 12 (C), which are detected. Note that the intensities of these peaks decrease regularly as the number of steps required for formation of the daughter ion increases and the likelihood of its occurrence therefore diminishes.

$$\underset{\text{Methane}}{CH_4} \xrightarrow{-e^-} \underset{M^{+\cdot}}{[CH_4]^{+\cdot}} \xrightarrow{-H\cdot} \underset{M^{+\cdot}-1}{[CH_3]^{+}} \xrightarrow{-H\cdot} \underset{M^{+\cdot}-2}{[CH_2]^{+}} \xrightarrow{-H\cdot} \underset{M^{+\cdot}-3}{[CH]^{+}} \xrightarrow{-H\cdot} \underset{M^{+\cdot}-4}{[C]^{+}}$$

Mass spectra are normally quite complex, showing many peaks due to extensive fragmentation. However, an organic chemist seldom attempts to account for most peaks in a mass spectrum, but instead looks for those of greatest likelihood and prominence (Table 11.2). Certain groups tend to break away readily from a molecular ion, and they become recognizable in two ways:

1. In some cases, the group leaves the parent molecular ion as a cation, and its existence is indicated by an m/z peak. (The remaining parent is a neutral radical and so is not detected.)

TABLE 11.2

Mass-to-Charge Ratios for Common Fragments in Mass Spectroscopy

Group	Group Formula	m/z Value(s)
Methyl	CH_3—	15
Ethyl	CH_3CH_2—	29
Isopropyl	$(CH_3)_2CH$—	43
Allyl	CH_2=CH—CH_2—	41
Phenyl	C_6H_5—	77
Water	H_2O	18
Acetyl	CH_3CO—	43
Chlorine	Cl—	35 and 37
Bromine	Br—	79 and 81

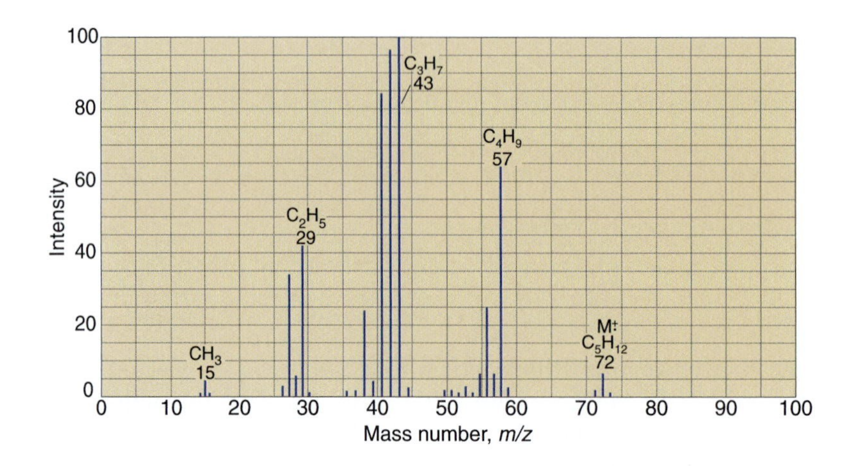

FIGURE 11.4

Mass spectrum of 2-methylbutane.

2. In some cases, the group leaves as a neutral radical (it is not detected), leaving behind a cation that is detected, and an m/z peak results. The remaining cationic fragment has an m/z value that has been reduced by that of the leaving group (that is, an $M^{\overset{+}{\cdot}} - m/z$ value).

In the mass spectrum of 1-bromopropane (n-propyl bromide) (see Figure 11.3), the molecular ion ($m/z = 122$) can be identified. But there is also a peak at $m/z = 43$. This peak corresponds to the propyl carbocation that remains after a bromine radical ($m/z = 79$) is lost ($122 - 79 = 43$). Note that the spectrum indicates weaker peaks at $m/z = 79$ for ^{79}Br and at $m/z = 81$ for ^{81}Br.

$$CH_3CH_2CH_2Br^{\overset{+}{\cdot}} \xrightarrow{\text{(fragmentation)}} \begin{cases} CH_3CH_2CH_2^{+} \text{ and } Br\cdot \\ CH_3CH_2CH_2\cdot \text{ and } Br^{+} \end{cases}$$

1-Bromopropane molecular ion ($m/z = 122$)

Propyl carbocation ($m/z = 43$)

Bromous ion ($m/z = 79$)

$$CH_3 \,\vdots\, CH_2 \,\vdots\, \underset{}{\overset{CH_3}{\overset{|}{CH}}} \,\vdots\, CH_3$$

2-Methylbutane

The mass spectrum for 2-methylbutane (C_5H_{12}, molecular weight 72) (Figure 11.4) shows peaks for the following groups: the molecular ion (72), the fragments formed by loss of a methyl group ($72 - 15 = 57$), loss of an ethyl group ($72 - 29 = 43$), or loss of an isopropyl group ($79 - 43 = 29$), and the methyl group fragment ($m/z = 15$). These groups can be obtained by cleaving every possible C—C bond in the molecular ion, as illustrated.

PROBLEM 11.8

Figure 11.5 shows the mass spectrum of 2-methyl-2-butanol. Account for the peaks at m/z values of 88 ($M^{\overset{+}{\cdot}}$), 73, 70, 59, and 55.

PROBLEM 11.9

Propose a structure that is consistent with each set of MS data. (*Hint:* Remember that for any molecular weight, there can be only so many carbons present.)

(a) peaks at 59, 58 ($M^{\overset{+}{\cdot}}$), 43, and 29 (b) peaks at 95, 94, 93, 92 ($M^{\overset{+}{\cdot}}$), 77, and 57

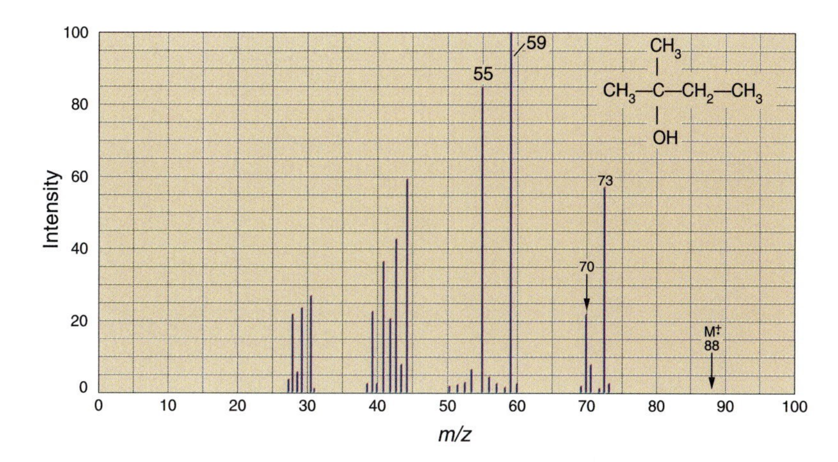

FIGURE 11.5

Mass spectrum of 2-methyl-2-butanol.

11.2.4 How to Use Mass Spectra

It is important not to overinterpret data contained in a mass spectrum. The information about an unknown compound that is readily obtained from its mass spectrum usually includes the following:

- The molecular weight (from the molecular ion peak, $M^{\dot{+}}$)
- The presence of nitrogen (an odd-number m/z value for the $M^{\dot{+}}$)
- The presence of major isotopes, most of which are $M^{\dot{+}} + 2$ ions (but remember that carbon has a small $M^{\dot{+}} + 1$ peak)
- The loss of any obvious groups (methyl, phenyl, and so on), based on their $M^{\dot{+}} - m/z$ values. (*Note:* The presence of such groups is frequently confirmed by the NMR spectrum; see Section 11.6.)

When a tentative structure has been assigned to the unknown, based on all experimental data, the mass spectrum of the unknown can be compared with spectra contained in various MS libraries for comparison. Two different pure samples of any one compound, no matter what their source, should have mass spectra that are identical in all respects. If the spectra are not identical, then the compounds are different. Mass spectroscopy is a tool widely used for proving or disproving the identity of two samples, such as drugs or additives.

How to Solve a Problem

A compound with empirical formula C_3H_8O is infinitely soluble in water. Its mass spectrum is shown in Figure 11.6. When a piece of sodium metal is added to the pure compound, gas bubbles appear. What is the structure of the compound? Account for the major (numbered) peaks shown in the mass spectrum.

PROBLEM ANALYSIS

The mass spectrum indicates that the molecular weight is 60. Since the formula weight for C_3H_8O is also 60, the molecular formula for the compound must be C_3H_8O. The compound is an alcohol

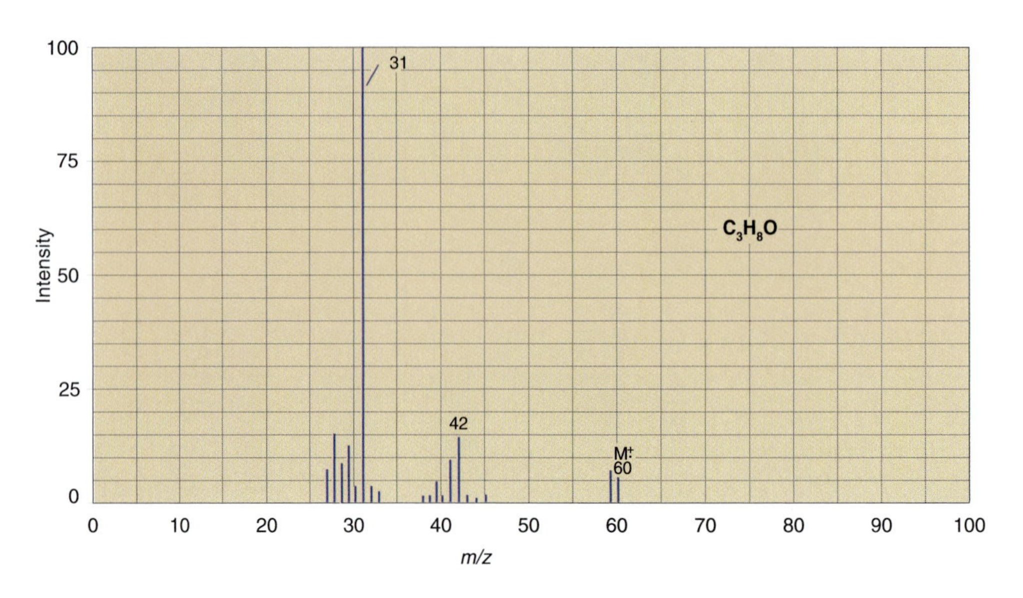

FIGURE 11.6

Mass spectrum of C_3H_8O.

because it gives off hydrogen when treated with sodium metal, forming an alkoxide in solution. If it is an alcohol, there should be a peak at $M^{+\cdot} - 18 = 42$, and there is. The peak at $m/z = 31$ indicates the loss of a group with $m/z = 29$, which corresponds to an ethyl group. Based on the molecular formula, there are only two possible alcohols, *n*-propyl alcohol and isopropyl alcohol:

$$CH_3{-}CH_2{-}CH_2{-}OH \qquad (CH_3)_2CH{-}OH$$

n-Propyl alcohol — Isopropyl alcohol

SOLUTION

It is not possible to lose an ethyl group from a simple fragmentation of isopropyl alcohol, so the compound C_3H_8O is *n*-propyl alcohol. This can be confirmed by comparing the mass spectrum with that of a known sample of *n*-propyl alcohol or with the known spectrum in a library of mass spectra.

PROBLEM 11.10

Account for the following peaks in the mass spectrum of 2-octanone, $CH_3CO(CH_2)_5CH_3$: 129, 128 ($M^{+\cdot}$), 113, 85, and 43.

PROBLEM 11.11

An unknown hydrocarbon produces a molecular ion at $m/z = 92$ with a small $M^{+\cdot} + 1$ peak. The mass spectrum also had a peak at $m/z = 77$. The compound does not react with bromine, with cold potassium permanganate, or with hydrogen over a platinum catalyst. What is its structure?

11.3

Introduction to Molecular Spectroscopy

One of the marvels of twentieth-century organic chemistry is the revolutionary impact of molecular spectroscopy in determining molecular structure. Structure is determined by interpreting the frequencies of radiation absorbed by a substance. There are three major advantages to molecular spectroscopic techniques:

- Spectroscopy is nondestructive (the sample is not destroyed or transformed).
- Structure can be determined from very small samples (usually a few milligrams).
- The results of spectroscopy are highly reproducible.

Molecular spectroscopy is the passing of electromagnetic radiation of various frequencies through a sample of a compound and determining the absorption pattern of that radiation. The result is a **spectrum** (a graph), viewed on a computer screen or printed on paper. The spectrum is a plot of the intensity of the energy absorbed or transmitted against the frequency of the radiation. The kinds of molecular spectroscopy most widely used in structure determination are ultraviolet-visible (UV-Vis) spectroscopy, infrared (IR) spectroscopy, and nuclear magnetic resonance (NMR) spectroscopy. This introductory section describes some of the characteristics common to all of these spectroscopic methods.

11.3.1 Electromagnetic Radiation

Electromagnetic radiation, which is a form of radiant energy, includes the familiar visible light (white light) as well as radiation of higher energy (for example, ultraviolet rays, X-rays, and gamma rays) and radiation of lower energy (for example, microwaves, infrared radiation, and radio waves). All of these forms of radiation are represented in the electromagnetic spectrum shown in Figure 11.7.

Electromagnetic radiation has the properties of both a wave and a particle. When electromagnetic radiation is viewed as a wave traveling at the speed of light, the unit of reference is the **wavelength**, λ (lambda), the distance from crest to crest of one complete cycle of the wave (Figure 11.8). Wavelength is reported in units of length (meters), with radio waves about 1–300 m long, infrared radiation about 10^{-5} m (2–16 μm),

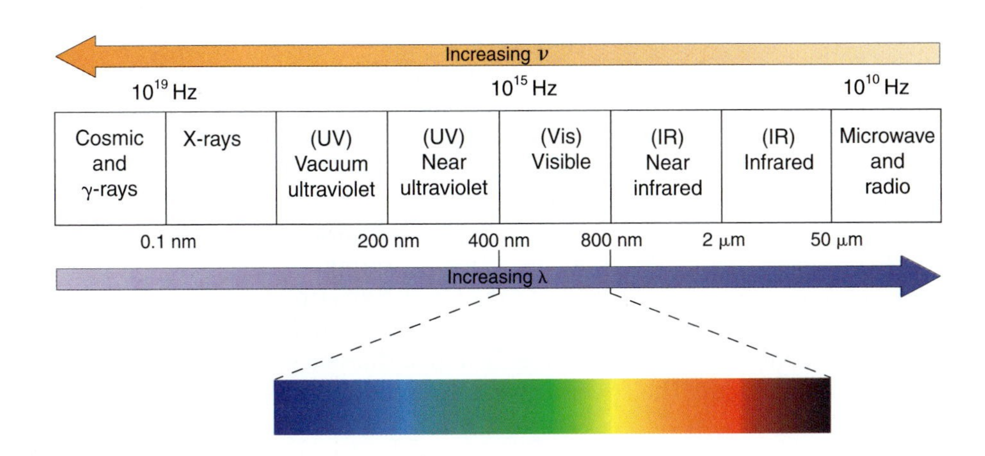

FIGURE 11.7
The electromagnetic spectrum.

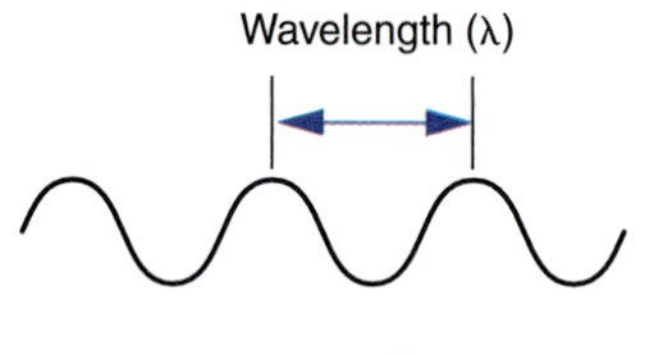

FIGURE 11.8

A wave representing electromagnetic radiation.

visible light about 10^{-6} m (400–800 nm), ultraviolet radiation about 10^{-7} m (200–400 nm), and gamma rays about 10^{-11} m. The shorter the wavelength, the higher the energy transmitted by the radiation.

An alternative means of characterizing electromagnetic radiation is by its **frequency,** ν (nu), which is the number of wave crests that pass any point per second. The standard frequency unit is cycles per second, now referred to as hertz (Hz). Higher-frequency radiation is higher-energy radiation.

Wavelength and frequency are inversely related by these equations:

$$\nu = c/\lambda \quad \text{and} \quad \lambda\nu = c$$

where λ is wavelength (in meters), ν is frequency (in Hz or s^{-1}), and c is the speed of light (3.0×10^8 m/s, equivalent to 186,000 mi/s). Thus, higher-frequency radiation is lower-wavelength radiation.

Electromagnetic radiation can also be considered as a stream of particles, small packets of energy called **photons.** The energy of a photon is related to its frequency by the equation

$$E = h\nu$$

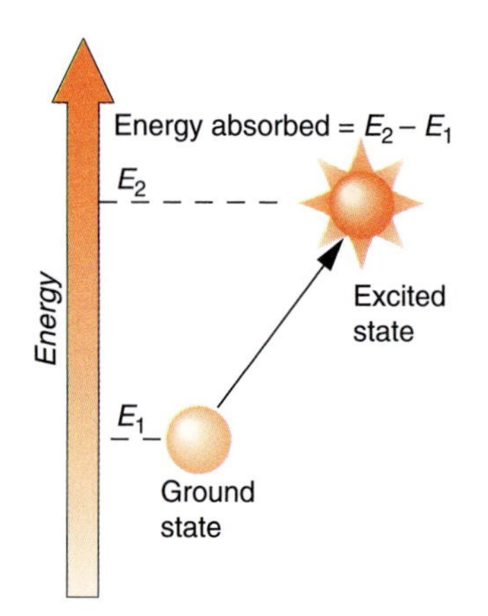

FIGURE 11.9

Absorption of energy leading to a change from a ground state (energy level E_1) to an excited state (energy level E_2).

where E is energy (in kilocalories per mole, kcal/mol), ν is frequency (in Hz), and h is Planck's constant (9.537×10^{-14} kcal-s/mol). This relationship demonstrates that higher-energy radiation has a higher frequency, and thus also a shorter wavelength. The amount of energy transmitted by various forms of radiation is impressive. For example, ultraviolet radiation of 200 nm (2×10^{-7} m) transmits about 143 kcal/mol, which, if totally absorbed, is sufficient to break a C—C or C—H bond.

When radiation is passed into a sample of an organic compound, most of it is transmitted through the sample—that is, there is no absorption of the radiation. However, a molecule can absorb radiation of the right frequency and thereby move from one energy state (E_1) to a higher energy state (E_2) (Figure 11.9). This absorption occurs if the energy difference of the two states ($E_2 - E_1$) is exactly equal to the energy of the impinging radiation.

Molecular spectroscopy is concerned with measuring the precise frequency (or wavelength) of radiation that is absorbed by molecules and correlating the pattern of absorption with molecular structure. The absorption patterns expected for organic molecules are quite predictable, as we will see.

11.3.2 Spectrophotometers

A **spectrophotometer,** often called a *spectrometer*, is the instrument used to produce absorption spectra. Its components are a radiation source, a monochromator (to permit a beam of specific wavelength radiation to pass through the sample), a sample compartment, a detector (to measure the intensity of radiation passing through the sample), and a device for observing the spectrum (computer screen and/or printer) (Figure 11.10). The wavelength of the electromagnetic radiation of the beam is changed gradually during the recording of a spectrum, a process called **scanning.**

When a sample is placed in the spectrometer and absorption of radiation occurs at a specific wavelength, the detector senses a reduction in the intensity of radiation passing through the sample, and the detector signal shows a **peak,** indicating absorption at that wavelength. The spectrum is calibrated either by wavelength or by frequency.

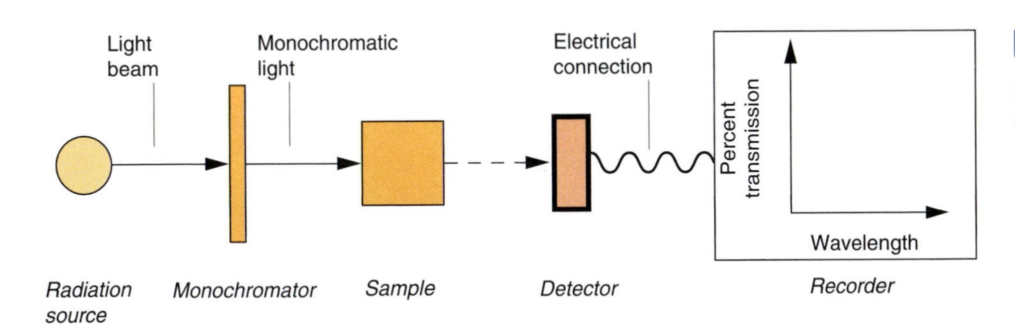

FIGURE 11.10

Schematic drawing of a spectrometer.

11.3.3 Comparison Spectra

IR, UV-Vis, and NMR spectra have been run for most previously isolated or synthesized organic compounds. Libraries typically hold many reference volumes of spectra for common compounds, organized by functional group families. In addition, many computerized spectrometer systems store thousands of reference spectra, so that once a new spectrum is recorded, the computer can find the best match with its stored spectra. The operating principle for comparing spectra and samples is that if two samples are identical in chemical structure, their spectra will be identical. However, depending on the kind of spectroscopy used, two very similar spectra may be obtained from compounds of different chemical structure.

11.4 Ultraviolet-Visible Spectroscopy

Ultraviolet-visible (UV-Vis) spectroscopy uses ultraviolet radiation and visible light to help determine chemical structure. The ultraviolet region of the electromagnetic spectrum corresponds to wavelengths of 200–400 nm (1 nm = 10^{-9} m). The visible region (the light our eyes can perceive) is adjacent, with wavelengths of 400–800 nm. These two regions are analyzed by the same spectrometer, usually called a *UV* or *UV-Vis spectrometer* (Figure 11.11).

11.4.1 Electronic Excitation

Radiation of ultraviolet and visible wavelengths is absorbed by compounds with pi electrons, which, upon absorption of this energy, are raised to an excited state. This is called an **electronic transition,** and the energy of the radiation corresponds to the energy difference between the excited and ground states. UV-Vis spectra generally have very broad peaks (maxima).

For a compound with a single double bond, such as ethylene, the wavelength of maximum absorption is about 171 nm, which is too low to be detected by typical UV-Vis spectrometers; this means that the electronic transition energy is very high. However, conjugation of double bonds lowers the excitation energy, so that absorption appears in the usual UV range. For example, 1,3-butadiene has its maximum at 217 nm, and *trans*-1,3,5-hexatriene absorbs at 268 nm (Figure 11.12). The more conjugated a pi system is, the longer the wavelength of UV radiation absorbed.

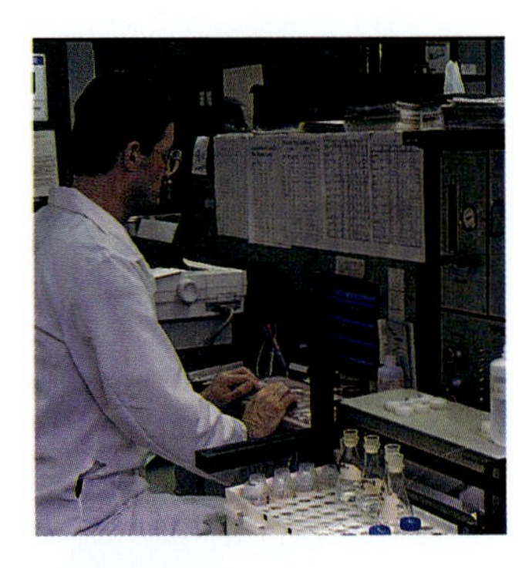

FIGURE 11.11

Ultraviolet-visible spectrometer.

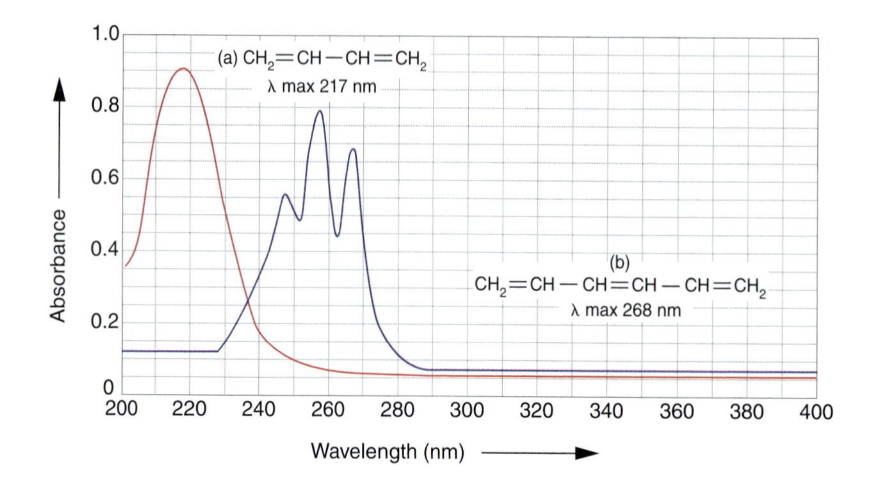

FIGURE 11.12

UV-Vis spectra of (a) 1,3-butadiene and (b) *trans*-1,3,5-hexatriene.

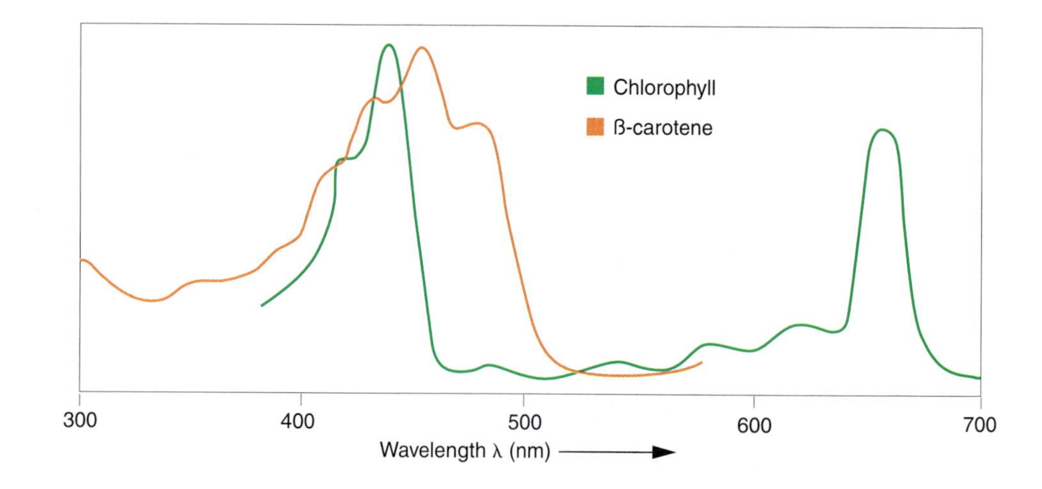

FIGURE 11.13

UV-Vis spectra of β-carotene and chlorophyll.

Extraction of green leafy vegetables, such as spinach, produces two important chemicals whose UV-Vis spectra are markedly different, reflecting different structures and different colors. Yellow β-carotene, which has a very extensive system of 11 conjugated double bonds, absorbs at 497 nm (Figure 11.13). Green chlorophyll, which has a highly conjugated porphyrin system, shows a major peak near 700 nm (Figure 11.13).

UV spectra are especially useful for detecting aromatic systems. Again, the more extensive the conjugated system, the longer the wavelengths of the absorption maxima. (A more conjugated system requires less energy for electronic excitation.) Thus, while benzene absorbs at 204 and 254 nm, styrene absorbs at 248 and 282 nm (Figure 11.14). Polynuclear aromatic compounds, such as phenanthrene, absorb at even longer wavelengths and frequently exhibit multiple maxima (Figure 11.14).

PROBLEM 11.12

Two compounds, **A** and **B**, are isolated from a reaction. Both have molecular formula C_9H_{10}. The mass spectrum of each compound shows a peak at $m/z = 77$. On a UV-Vis spectrum, **A** shows λ_{max} at 258 nm, and **B** shows λ_{max} at 279 nm. **A** does not decolorize bromine, whereas **B** does. What are possible structures for **A** and **B**?

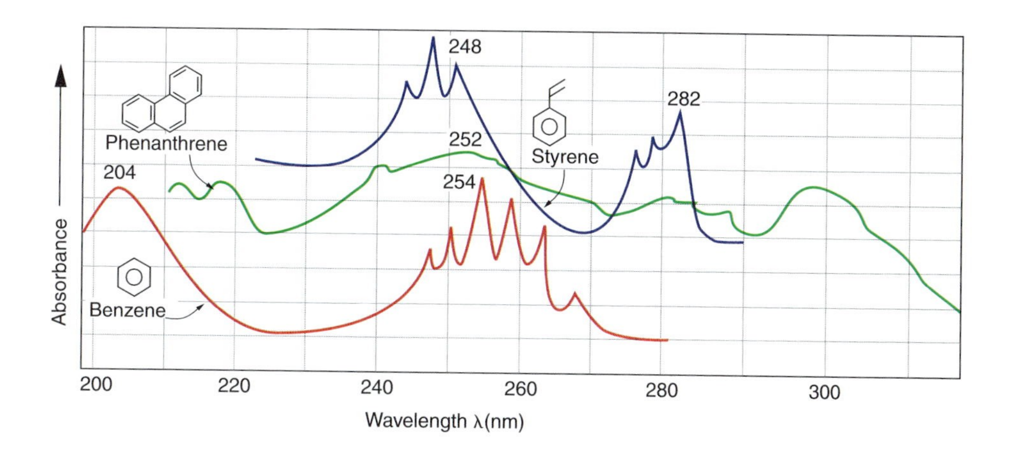

FIGURE 11.14

UV-Vis spectra of styrene, benzene, and phenanthrene.

11.4.2 UV-Vis Spectra and Color

Almost everything we use and even eat is colored by natural or synthetic chemical dyes. Although dyes have been around for thousands of years, most were originally obtained from plants. The dye industry today almost exclusively produces synthetic dyes made by organic chemists, whether new colors for clothing, foods, or for automobile paints.

The color exhibited by a substance depends on the wavelength of light that it absorbs. Recall that white light is composed of a spectrum of colors (remember the rainbow). If white light has a certain wavelength removed by absorption, the remaining light appears as a color. This color is known as the complementary color, the color we see (Figure 11.15). For example, *β*-carotene absorbs light in the 450–500 nm region (the violet end of the visible spectrum); its orange color is the complementary color that remains and is visible. Similarly, chlorophyll absorbs near 700 nm, so its complementary color, which we see, is green.

Organic synthesis is the approach used to make new colors for paints and dyes. The challenge is to find just the right conjugated system to absorb radiation at the right wavelength. Often, simple changes in the structure of a known dye will produce a new color. A blue dye (such as the indigo used for blue jeans) must absorb radiation at 600–700 nm. For this to occur, the dye must have an extensive conjugated system. Blue jeans are vat-dyed by first impregnating them with an aqueous solution of leucoindigo, a colorless compound with a limited conjugated acylbenzene system. Leucoindigo is then oxidized in air to the compound indigo, which has a more extensive conjugated system (from one ring through to the other). Indigo absorbs the appropriate radiation (in the yellow-orange part of the spectrum), creating the complementary color blue.

Blue M&Ms absorb orange light.

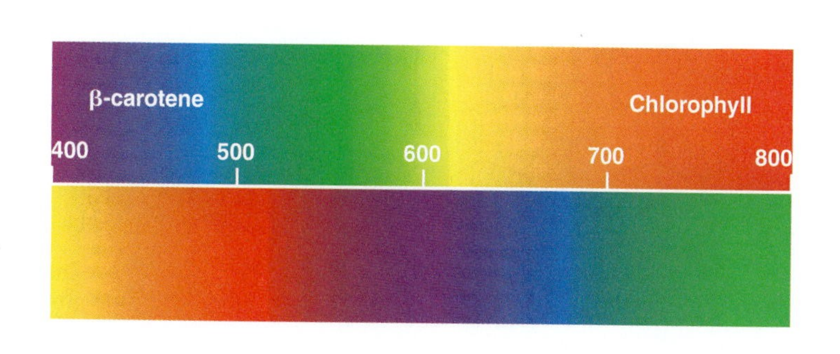

FIGURE 11.15

Complementary colors.

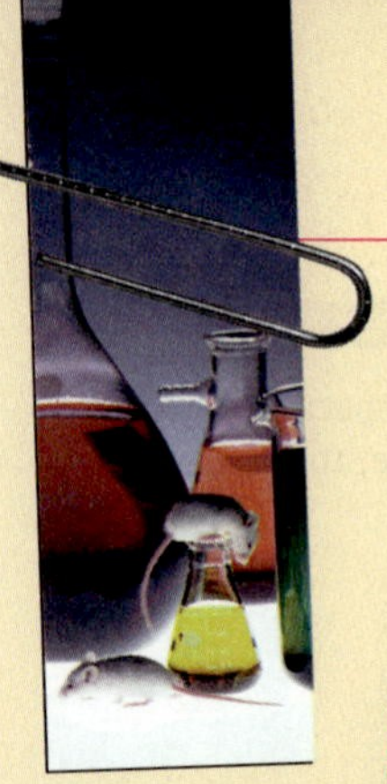

CHEMISTRY AT WORK

IN THE LABORATORY

The Structure of Acacialactam

Many of the most useful drugs for fighting diseases come from plants. In 1989, chemists in Japan and Thailand took part in a program to study medicinal plants in search of new drugs. One plant they examined was *Acacia concinna DC* (*leguminosae*), whose seeds are used in folk medicine for treating skin disease. Chemists collected 200 g of seeds, but obtained only 32 mg of the pure active ingredient, which they named acacialactam. They then faced the challenge of how to determine the chemical structure of this tiny amount of compound—it was accomplished without ever doing a single chemical reaction on the sample, relying totally on spectroscopic techniques.

Acacialactam is an enantiomeric, colorless, oily liquid. Mass spectroscopy indicated a molecular ion at $m/z = 165.1146$, corresponding to a molecular formula of $C_{10}H_{15}NO$. The UV-Vis spectrum showed weak absorption at 210 nm, indicating a conjugated carbonyl system. The IR spectrum had peaks at 1670 cm^{-1} (corresponding to an amide carbonyl group) and 1600 cm^{-1} (an alkene). The ^1H-NMR spectrum showed the usual pattern for a vinyl group ($CH_2{=}CH{-}$) plus an additional alkenic proton. It also showed peaks due to two different methylene groups ($-CH_2-$) and two different methyl groups ($-CH_3$), which had no C—H bonds on adjacent carbons. Finally, the ^{13}C-NMR spectrum showed signals attributable to 10 carbon atoms, including two double bonds, a carbonyl group, a quaternary carbon, two methylene groups, and two methyl groups.

Can you deduce the structure of acacialactam? Probably not, but by the end of this chapter, you should be able to analyze all this information and understand what it indicates about chemical structure. The chemists were able to prove their deductions with this basic information and a few further NMR studies. They determined that the compound's structure is:

$CH_2{=}CH$ CH_3
HN
O
CH_3

Acacialactam

Acacia seeds are the source of acacialactam.

The significance of determining the structure from just 32 mg of a compound is that this structure proof allows the compound to be prepared synthetically if its medicinal value is proved sufficient. The importance of such activity is underscored by the number of plant and animal species rapidly disappearing from the earth, representing a potential loss of as yet undiscovered but medically potent chemical compounds.

O H N N H O [O] → O H N N H O

Leucoindigo (colorless)

Indigo (blue)

A wide selection of dyes is available for industrial use, but only eight are approved by the Food and Drug Administration for use in foods (they are given FDC numbers). For example, FDC No. 6 (Sunset Yellow) is used to color margarine, and Allura Red (FDC No. 40) is used to dye maraschino cherries.

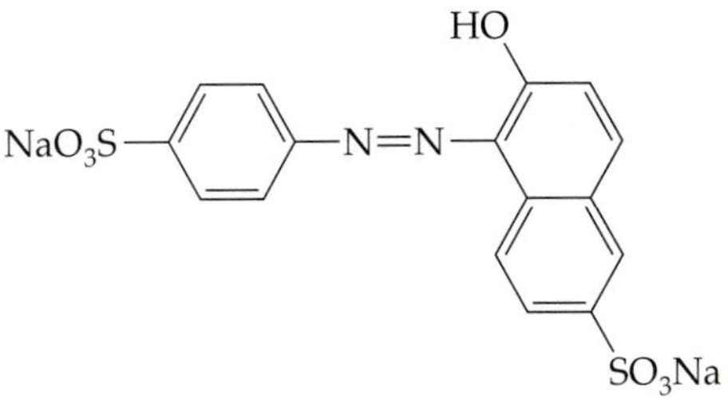

Sunset Yellow
(FDC No. 6)

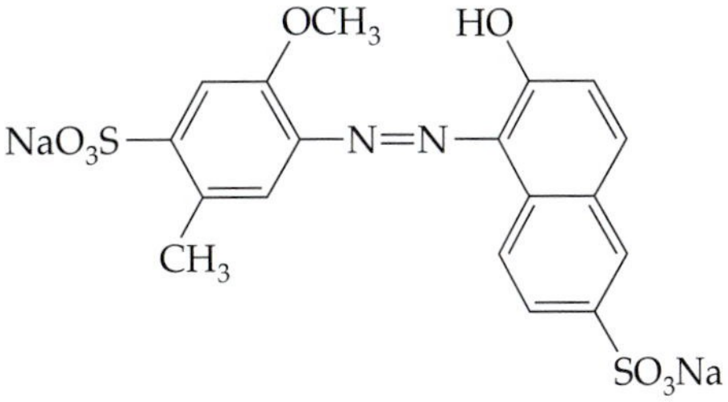

Allura Red
(FDC No. 40)

11.4.3 Uses of UV-Vis Spectroscopy

UV-Vis spectroscopy has limited application in the determination of chemical structure because it detects "systems" rather than specific structural features. It is used mainly to detect the presence of conjugation. The longer the wavelength of absorption by a compound, the more extensive is the conjugated system. UV-Vis spectroscopy is especially useful in distinguishing between two similar compounds with a difference in conjugation. For example, the conjugated 1-phenylpropene (λ_{max} = 279 nm) can readily be distinguished from the nonconjugated 3-phenylpropene (λ_{max} = 258 nm) because 1-phenylpropene absorbs at a longer wavelength.

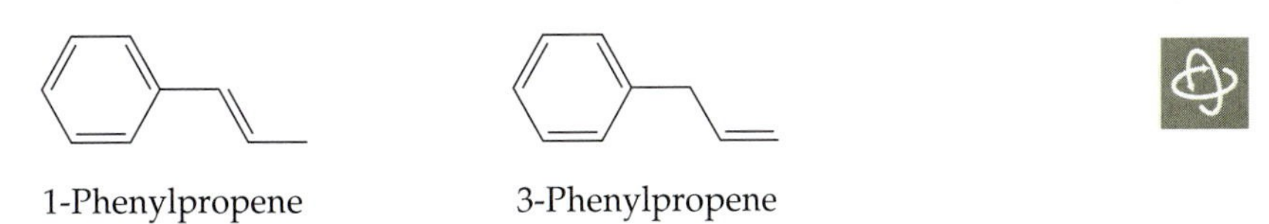

1-Phenylpropene   3-Phenylpropene

Also, if an unknown compound is suspected of having a cyclohexadiene-type structure, a conjugated 1,3-cyclohexadiene unit can be readily distinguished from a nonconjugated 1,4-cyclohexadiene.

PROBLEM 11.13

Which of the following compounds are likely to be colored?

(a) 1,4-pentadiene (b) 1-phenyl-1,3,5-hexatriene
(c) *p*-nitrophenol (d) vinylcyclohexane

PROBLEM 11.14

If you wanted to create (a) red paint, (b) blue paint, (c) yellow paint, and (d) green paint, approximately what wavelength of absorption would you look for in candidate compounds to be used as dyes?

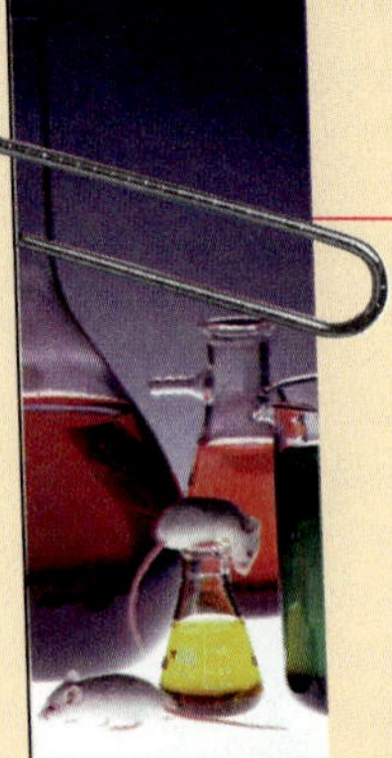

CHEMISTRY AT WORK

IN THE MARKETPLACE

Sunscreens

Chapter 10 discussed the ozone hole in the atmosphere and the dangers of more UV radiation reaching the earth's surface from the sun. UV radiation can cause dermatological problems, including skin cancer. How can you avoid such problems?

Staying out of the sun is one remedy, but not a very practical one. What you can do is augment nature's sunscreen (the ozone layer) by using sunscreens and sunblocks, now commonly found in skincare products. Such ointments applied to the skin limit the amount of UV radiation reaching the skin. They are "rated" by a skin protection factor (SPF), which is the ratio of the time required to get a sunburn with and without the protection applied. Thus, SPF 15 means that it takes 15 times as long to get a burn with the protection applied.

For dermatological purposes, ultraviolet radiation is broken into three wavelength classifications: UV-C (100–290 nm, dangerous but absorbed by the atmosphere), UV-B (290–315 nm, causing sunburn), and UV-A (315–400 nm, less dangerous but causing some changes in the skin). To be effective, the active ingredient in sunscreens and sunblocks should (1) absorb UV-B radiation, rather than letting it pass through to the skin, (2) recycle itself (that is, not be destroyed by the radiation), and (3) not cause other dermatological problems, such as skin rashes and other allergic reactions.

Sunscreens provide protection against UV-A and UV-B radiation.

The recycling ability of sunscreens comes from the active ingredients, organic chemicals that absorb UV radiation (that is, absorb energy). These chemicals are thereby raised to an excited state, but then quickly return to the ground state. Effectiveness as a sunscreen requires that the excited state of the compound lose its acquired energy as thermal energy through low-energy vibrational motions.

One of the earliest compounds used as a sunscreen was *p*-aminobenzoic acid (PABA), which has UV absorption maxima (denoted by λ_{max}) at 285 and 310 nm. However, a small percentage of the population has a photoallergic reaction to PABA derivatives. Oxybenzone (λ_{max} 289 and 325 nm) and various ester derivatives of *p*-methoxycinnamic acid (λ_{max} 310 nm) are now the most widely used ingredients. Both have strong UV-B absorption, but also absorb into the UV-A range. All of these are relatively ordinary and inexpensive organic compounds, but they are formulated with a wide variety of additives, such as moisturizers and antioxidants.

COOH, NH_2 — PABA

O, CH_3O, OH — Oxybenzone

CH=CHCOOR, CH_3O — *p*-Methoxycinnamate

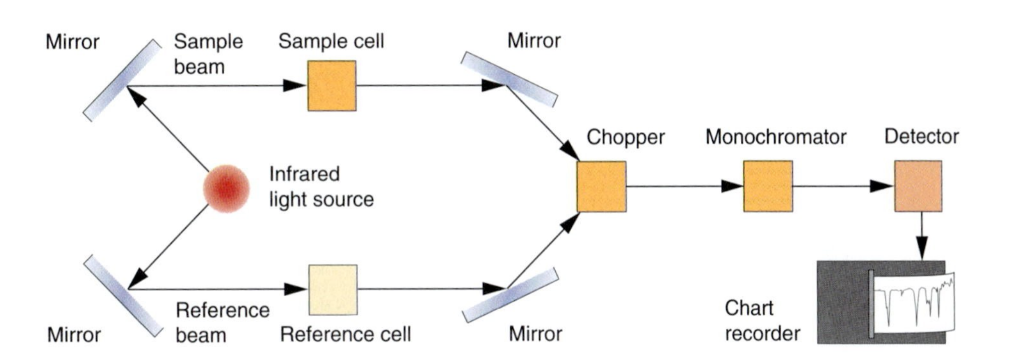

FIGURE 11.16
Infrared spectrometer.

11.5

Infrared Spectroscopy

In **infrared (IR) spectroscopy,** the absorption of infrared radiation is recorded (Figure 11.16). The infrared region used in most IR spectroscopy is from 2.5 μm to 16 μm (1 μm = 10^{-6} m). Infrared absorption peaks are usually reported in terms of wavenumber $\bar{\nu}$ (cm^{-1}), which is the number of wave crests per centimeter. The conversion from wavelength (in μm) to wavenumber (in cm^{-1}) is

$$\bar{\nu} = 10{,}000 \div \lambda$$

Therefore, the usual IR region used is 4000 cm^{-1} to 625 cm^{-1}. The higher the wavenumber is, the shorter the wavelength, the higher the frequency, and the higher the energy involved. Figure 11.17 shows the relationships. Infrared radiation is relatively low-energy radiation, having about 2–12 kcal/mol. This energy is too little to cause electronic excitation, but is just right to match the energy of bond stretching and bending, called *bond vibrations.*

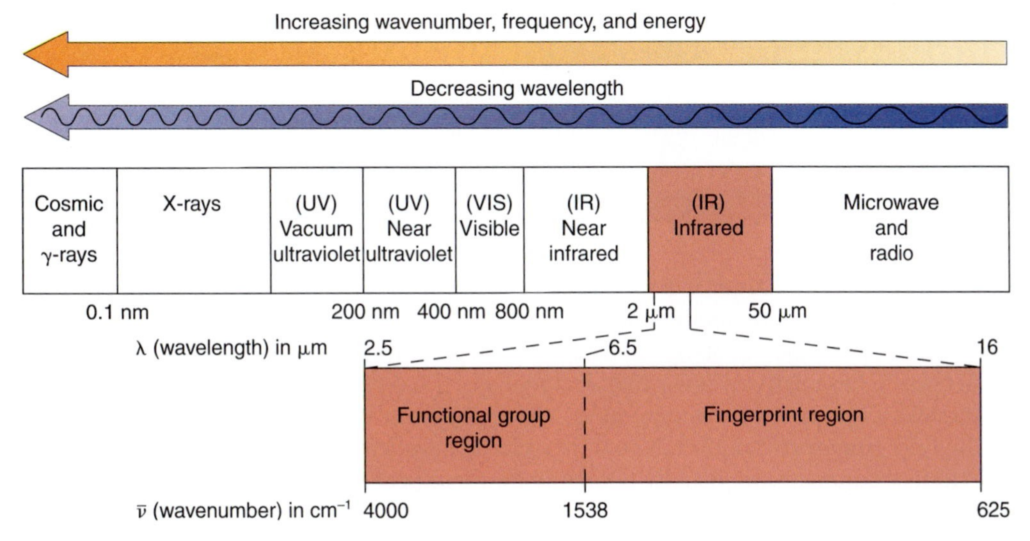

FIGURE 11.17
Infrared region of the spectrum.

FIGURE 11.18

(a) Bending and (b) stretching of C—H bond.

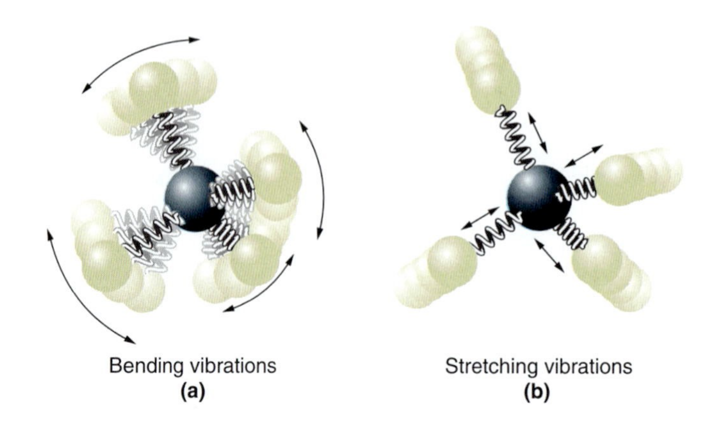

11.5.1 Vibrational States

The covalent bond between two atoms is not rigid or static. You know from working with stereoisomers that rotation can occur about some bonds. A bond can also bend and stretch, somewhat like a small spring (Figure 11.18). The energy required for each stretching and bending motion of a given type of bond is specific, and only the radiation with the specific wavelength corresponding to that energy will be absorbed. This absorption is what an IR spectrum records. For a nonlinear molecule containing n atoms, there are $3n - 6$ possible vibrational states. The IR spectrum for methane ($n = 5$) is relatively simple, with 9 fundamental vibrations. As the number of atoms increases, however, the number of fundamental vibrations grows dramatically; for 1-hexanol, with $n = 21$, the number is 57. Because the number of vibrations is so large for most molecules, IR spectra usually contain a large number of peaks. Spectroscopists analyze IR spectra in detail, but organic chemists interested in structure analysis generally do not.

The energy difference between vibrational states corresponds to the energy level of infrared radiation. The actual energy of vibration of a particular bond depends on the mass of the atoms involved and the bond strength. The order of C—C bond strength is triple bond > double bond > single bond, and the IR absorption order for C—C bond stretching is the same:

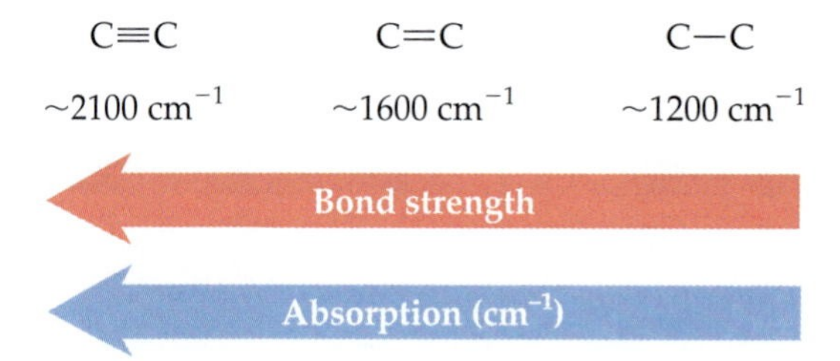

It takes more energy (larger wavenumber) to stretch a stronger bond.

There are two major regions of interest in an infrared spectrum. The **functional group region,** from 4000 cm^{-1} to about 1550 cm^{-1}, is the region reflecting most bond stretching. It usually has relatively few peaks, but many of those peaks are characteristic of the functional groups present in a compound. The region from about 1550 cm^{-1} to 625 cm^{-1} is called the **fingerprint region.** It contains a large number of peaks; however, specific structural assignments are usually not made for these peaks. Figure 11.19 shows the IR spectrum of hexanoic acid, with the functional group and fingerprint regions marked.

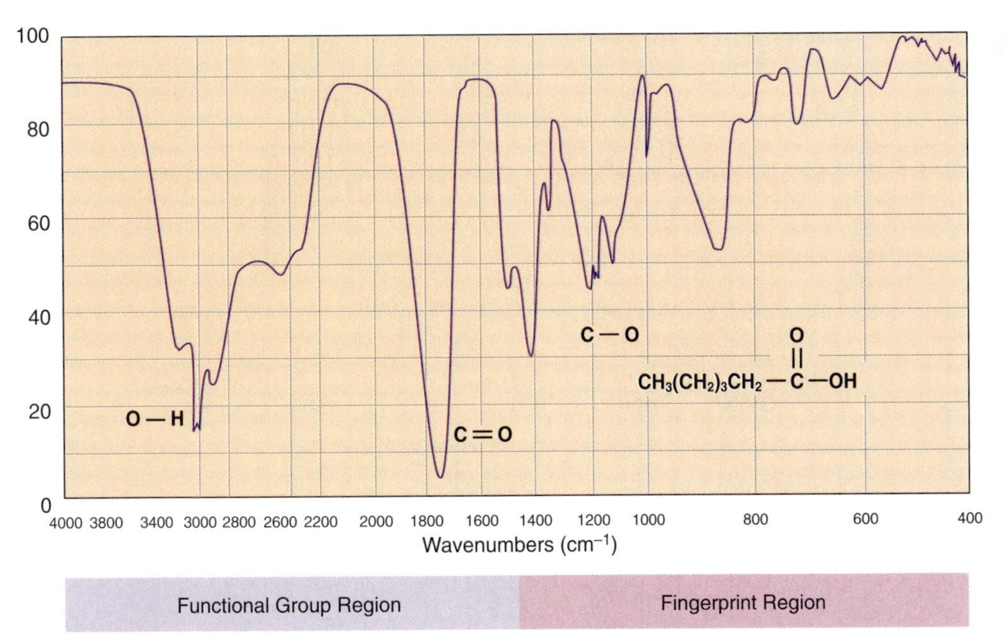

FIGURE 11.19

IR spectrum of hexanoic acid.

11.5.2 The Fingerprint Region

Peaks in the fingerprint region of an IR spectrum frequently are not considered when IR spectroscopy is used to analyze the structure of an unknown. However, there is an important use for this region because of its richness of peaks. If two samples are really identical, their IR spectra must be identical. This means that all of the peaks and subtle inflections must match exactly. Therefore, the major use of the fingerprint region is to prove or disprove identity between samples, much as human fingerprints are compared for identity purposes. The comparison between *o*-toluic acid and *p*-toluic acid in Figure 11.20 illustrates how two similar compounds with identical functional groups produce IR spectra with slightly different "fingerprints."

11.5.3 The Functional Group Region

Most of the functional groups discussed in this book are readily identified through IR spectroscopy. Therefore, for determining the structure of an unknown, this type of spectroscopy is indispensable early in the process. The various functional groups exhibit infrared absorptions corresponding to the energies of their bond stretchings over rather narrow and predictable regions, regardless of the rest of the molecular environment. Figure 11.21 shows where the absorption peaks of functional groups and of key C—H bonds are located in an IR spectrum. Table 11.3 correlates the same functional groups with regions of absorption in the infrared. Both Figure 11.21 and Table 11.3 are reproduced in Appendix B for ease of reference in later chapters.

Here is a summary of specific data from Figure 11.21 and Table 11.3 that may be helpful in IR spectral interpretation:

- O—H and N—H peaks are both near 3300 cm^{-1}.
- C—H peaks are near 3000 cm^{-1}.
- C≡C and C≡N peaks are near 2200 cm^{-1}.

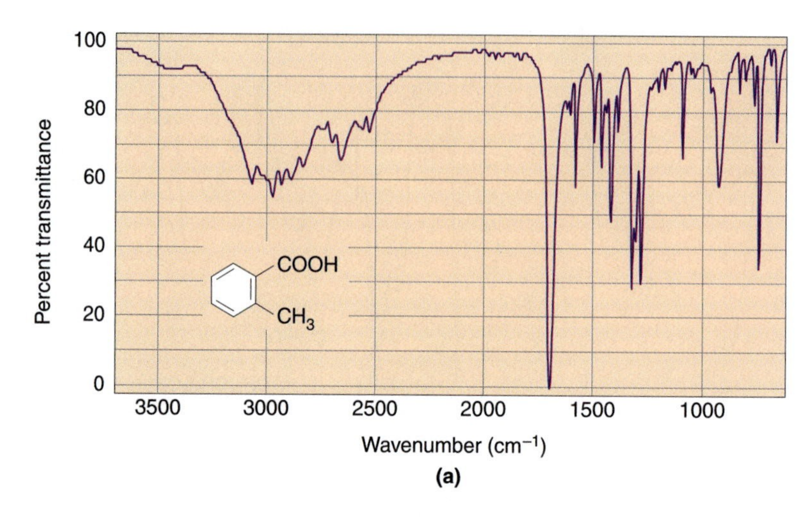

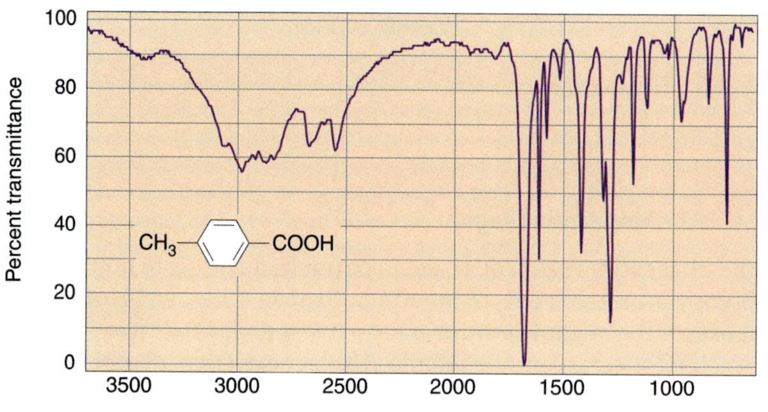

FIGURE 11.20

IR spectra of (a) *o*-toluic acid and (b) *p*-toluic acid.

- C=O and C=N peaks are near 1700 cm^{-1}.
- Carboxylic acids have a sharp C=O peak near 1700 cm^{-1} and a very broad strong O—H peak centered near 3300 cm^{-1}.
- C=C and aromatic peaks are near 1625 cm^{-1}.

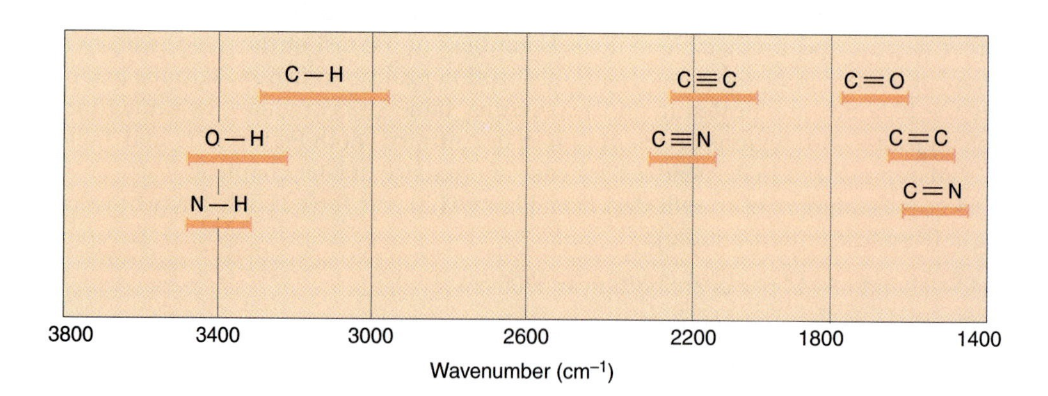

FIGURE 11.21

IR peaks for some functional groups.

TABLE 11.3

Correlation of Bond Stretching and IR Absorption

Type of Bond	Group	Family of Compounds	Wavenumber Range (cm^{-1})
Single bonds	—C—H	Alkanes	2850–3300
	=C—H	Alkenes, aromatics	3000–3100
	≡C—H	Alkynes	3300–3320
	O—H	Alcohols	3200–3600
	N—H	Amines	3300–3500
Double bonds	C=C	Alkenes, aromatics	1600–1680
	C=O	Carbonyls	1680–1750
		Aldehydes, ketones	1710–1750
		Carboxylic acids	1700–1725
		Esters, amides	1680–1750
	C=N	Imines	1500–1650
Triple bonds	C≡C	Alkynes	2100–2200
	C≡N	Nitriles	2200–2300

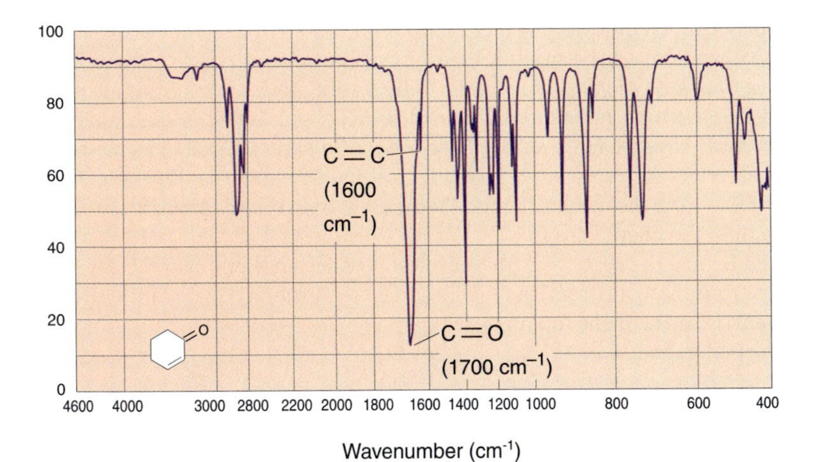

FIGURE 11.22

IR spectrum of cyclohex-2-enone.

IR spectra can be quickly analyzed using these generalizations. For example, Figure 11.22 shows the IR spectrum of cyclohex-2-enone, with the alkene group absorbing at 1600 cm^{-1} and the carbonyl group absorbing at 1700 cm^{-1}.

How to Solve a Problem

Compound **A** has molecular formula $C_9H_{12}O$, and its mass spectrum shows a molecular ion at $m/z = 136$ and major peaks at 121, 77, and 59. The UV spectrum shows a maximum at 265 nm, and the IR spectrum has peaks at 3000 and 1625 cm^{-1}. Treatment of **A** with hydriodic acid produces compound **B,** which shows a new peak in the IR spectrum at 3300 cm^{-1}. Compound **B** is oxidized with Jones' reagent ($Na_2Cr_2O_7/H_2SO_4$) to yield a ketone, identified by its IR peak at 1720 cm^{-1}. Heating **B** with concentrated sulfuric acid leads to a new compound **C** (C_8H_8), which has lost the IR peak at 3300 cm^{-1}, but which decolorizes bromine in CCl_4. Compound **C** reacts with hydrogen over a platinum catalyst, leading to the absorption of one equivalent of hydrogen and producing compound **D.** Oxidation of **A** and **D** with hot potassium permanganate, followed by acidification of the resultant filtrate, produces a colorless precipitate, compound **E.** Compound **E** is analyzed as having molecular formula $C_7H_6O_2$ and shows infrared absorption at 3300 (very broad peak), 3000, 1700, and 1625 cm^{-1}. Write structures for compounds **A–E.**

PROBLEM ANALYSIS

In working a road map problem such as this, we work "down" the problem, noting any specific conclusions we can make. Finally, we work back "up" the problem, at which time we should be able to deduce the structure of each compound in turn. In the following analysis, specific conclusions are in italics.

Analysis of the molecular formula indicates that the HDI is 4. The hydrogen-to-carbon ratio is also nearer 1 than 2, leading to initial suspicion of an aromatic system (*aromaticity* requires an HDI of 4, with no other double bonds or rings). The UV spectrum indicates a conjugated system, consistent with a benzenoid ring. The mass spectrum indicates the loss of a group with $m/z = 15$ (*methyl group*) and the loss of a group of $m/z = 77$ (*phenyl group*). The IR peak at 1625 cm^{-1} is consistent with a benzenoid compound, and that at 3000 cm^{-1} indicates aliphatic hydrogen. Compound **A** has one oxygen, but the IR spectrum indicates that it cannot be a carbonyl compound (no peak at ~1700 cm^{-1} region), nor can it be an alcohol (no peak at ~3300 cm^{-1} region). Thus, it is probably an *ether.* Therefore, just by using spectroscopic information, we have accounted for a C_6 group (phenyl) and a C—O—C (ether) group: eight carbons out of the nine in the unknown.

Reaction of **A** with HI produced **B,** an alcohol, as indicated by the IR peak at 3300 cm^{-1}, which dehydrated to **C,** which has to be an *alkene* (decolorized bromine and could be hydrogenated). Thus, **B** must have a *C(OH)—CH group* to undergo dehydration. Oxidation of this alcohol with Jones' reagent produced a ketone, not a carboxylic acid, indicating that it was a secondary alcohol. This means that **A** must have a *CH(OCH₃)—CH* group for three reasons: (1) The methyl ether group (—O—CH_3) accounts for the methyl group lost in the MS analysis; (2) the methyl ether was cleaved with HI to form the alcohol, **B;** and (3) there are only nine carbons in the unknown, these three and six from the phenyl group.

Compound **E** must be *benzoic acid* (the simplest possible aromatic acid) because of its formula and its IR spectrum showing aromatic, —OH, and C=O groups. This indicates that there is only one substituent on the benzene ring in **A,** which was oxidized down to the carboxyl group with potassium permanganate.

SOLUTION

Compound **E** is benzoic acid. Compounds **B, C,** and **D** must have a two-carbon side chain, and **B** must be the secondary alcohol 1-phenylethanol. Dehydration produces styrene (**C**) and its reduction yields ethylbenzene (**D**). Therefore **A** must be 1-methoxy-1-phenylethane.

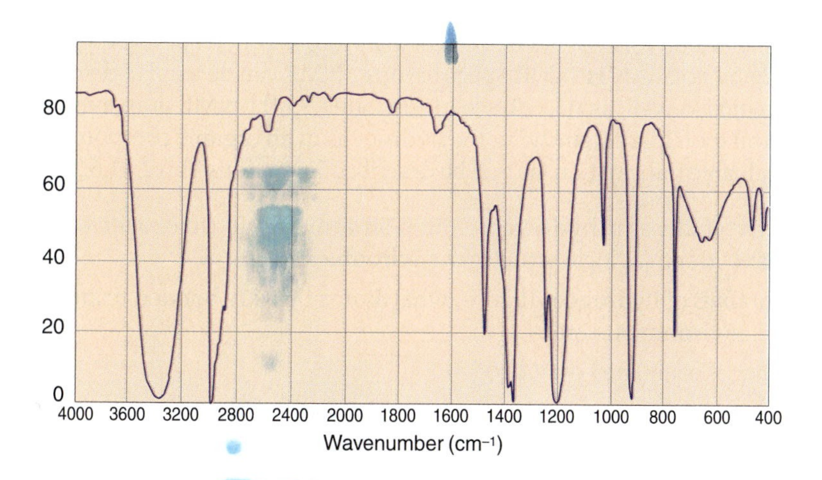

FIGURE 11.23 IR spectrum of $C_4H_{10}O$ (Problem 11.15).

PROBLEM 11.15

A sample is known to have the molecular formula $C_4H_{10}O$ and to be one of two constitutional isomers, either *t*-butyl alcohol or isopropyl methyl ether. The IR spectrum is shown in Figure 11.23. What is the structure of the unknown?

PROBLEM 11.16

Indicate how the following pairs of compounds could be distinguished using characteristic IR peaks:

(a) benzaldehyde (C_6H_5CHO) and benzoic acid (C_6H_5COOH)

(b) cyclobutene and 2-pentyne

(c) cyclohexanol and cyclopentyl methyl ether

11.6 Proton Nuclear Magnetic Resonance Spectroscopy

Nuclear magnetic resonance (NMR) spectroscopy is the newest of the spectroscopic techniques we will discuss. It was discovered by Felix Bloch and Edward Mills Purcell in 1946, and they shared a Nobel Prize in physics in 1952. An NMR spectrometer applies radiofrequency (RF) radiation ranging from 60 to 700 MHz to a sample in the presence of a very strong external magnetic field (Figure 11.24). Typically, the magnetic field measures about 14,000 gauss (1.4 tesla, or T) to 70,000 gauss (7.0 T); as a point of reference, the magnetic field of the earth measures about 0.5 gauss. The radiation applied in the presence of a magnetic field can change the orientation of protons in the nucleus of certain elements, and thus their energy levels, in a way that can be measured. The energy levels involved in such nuclear transitions are extremely small, only about 0.3 cal/mol, compared to the IR vibrational energies of 2–12 kcal/mol and the UV-Vis electronic transition energies of 40–150 kcal/mol.

FIGURE 11.24 NMR spectrometer.

The most common form of NMR detects the protons (nuclei of hydrogen atoms) in a compound and is called **^{1}H-NMR spectroscopy.** (NMR can be used to detect the nuclei of other elements as well, as we will see in Section 11.7.) ^{1}H-NMR is an extremely powerful analytical tool. Just by studying the hydrogens in an organic compound, chemists can obtain valuable information (as will be described in Sections 11.6.2, 11.6.3, and 11.6.4):

1. The *number of kinds* of hydrogens found in a molecule (hydrogens of the same "kind" have identical local structural environments)
2. The actual *kinds* of hydrogens (that is, what atoms they are bonded to and what their electronic environments are)
3. The *number* of each kind of hydrogen
4. For any kind of hydrogen, the nature of the local structural environment (how many hydrogens are on each adjacent carbon)

These four categories of information are obtained, respectively, by observing the following four characteristics of a ^{1}H-NMR spectrum:

1. The number of major peaks in the spectrum
2. The locations of these peaks in the spectrum
3. The relative areas under these peaks
4. The "splitting," if any, of the major peaks into doublets, triplets, quartets, or multiplets by adjacent hydrogens

With this kind of detailed information available from ^{1}H-NMR spectra, it is possible to determine a great deal about the structure of unknown organic compounds. With simple organic compounds, the ^{1}H-NMR spectra are fully interpretable, and every peak can be assigned to a structural feature, as we will see in this section. However, for very complex structures, the spectra can become quite complex, with overlapping of many peaks. Interpretation of these spectra often requires computer analysis—a topic beyond the scope of this book.

11.6.1 The NMR Phenomenon

A proton, located in the nucleus of a hydrogen atom, can have one of two nuclear spin states. These spin states are referred to as $+\frac{1}{2}$ and $-\frac{1}{2}$ states and are of equal energy. The spin states have small magnetic moments associated with them. Ordinarily, the magnetic moments of hydrogen nuclei are randomly oriented (Figure 11.25a).

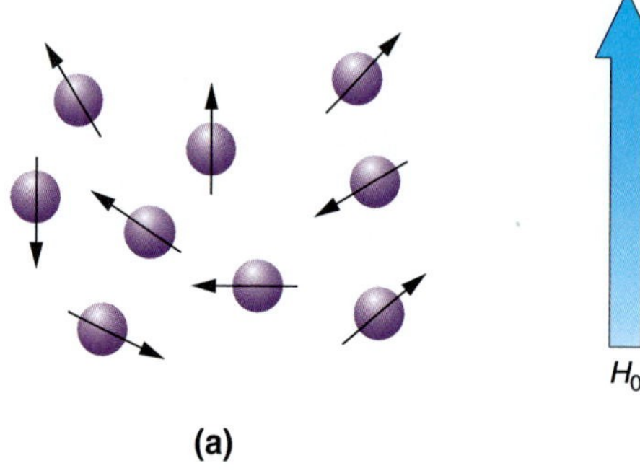

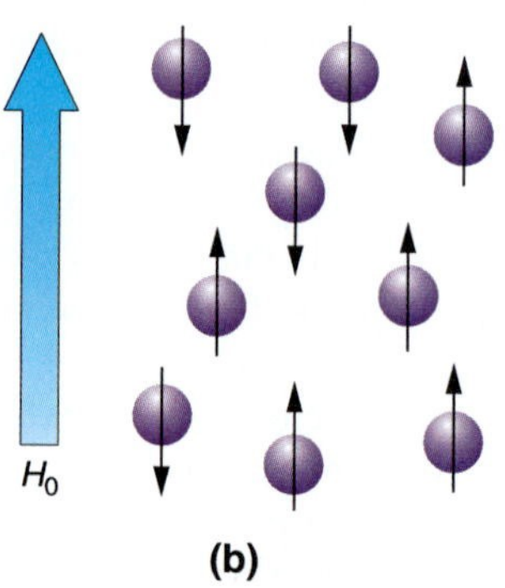

FIGURE 11.25

(a) Random orientation of nuclear spins in the absence of a magnetic field. (b) Orientation of nuclear spins with or against an external magnetic field (H_0).

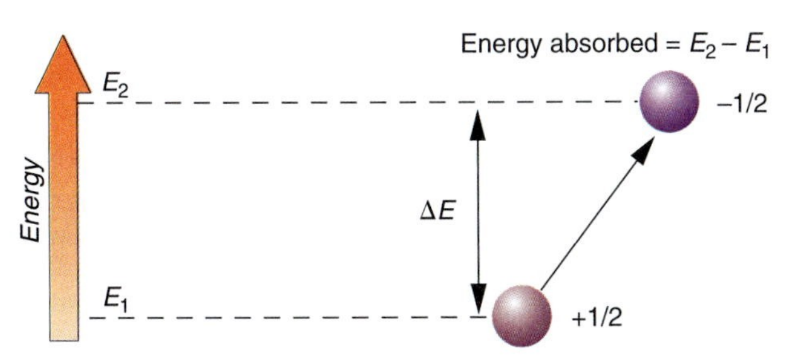

FIGURE 11.26

Energy absorption in changing a nucleus from spin state $+\frac{1}{2}$ to spin state $-\frac{1}{2}$.

When the hydrogen nuclei are placed in a strong magnetic field, the spin states become of unequal energy and the magnetic moment of each proton is forced to align either with the external field ($+\frac{1}{2}$) or against it ($-\frac{1}{2}$) (Figure 11.25b). It takes more energy to align a spin state against the external field than with it. The spin state whose magnetic moment is aligned with the external magnetic field ($+\frac{1}{2}$) is of lower energy, and the state whose magnetic moment is aligned against the external magnetic field ($-\frac{1}{2}$) is of higher energy. Therefore, energy is required for a spin state to change from $+\frac{1}{2}$ to $-\frac{1}{2}$ (Figure 11.26).

The energy difference between the two spin states is about 10^{-5} kcal/mol, approximately corresponding to radiation of 60–700 MHz, which lies in the radiofrequency (RF) range. When the combination of the applied external magnetic field and the applied RF radiation corresponds to the energy difference between the two spin states, the nuclei in the lower ($+\frac{1}{2}$) spin state absorb the energy and "flip" to the higher state, aligned against the magnetic field. The nuclei are then said to be **in resonance** with the radiation. (Be careful not to confuse this with resonance stabilization of aromatic and other conjugated systems.) When a proton is in resonance, RF energy is absorbed, and a peak is observed in the NMR spectrum.

For example, the ^{1}H-NMR spectrum of *p*-xylene shows two major peaks, indicating that two different kinds of hydrogens are present. This means that the hydrogens in *p*-xylene are in two different local structural environments: the methyl group hydrogens and the hydrogens on the aromatic ring (Figure 11.27).

The scale used to measure the energy of absorption is a relative one. The reference compound used for ^{1}H-NMR spectroscopy is tetramethylsilane, $(CH_3)_4Si$, abbreviated

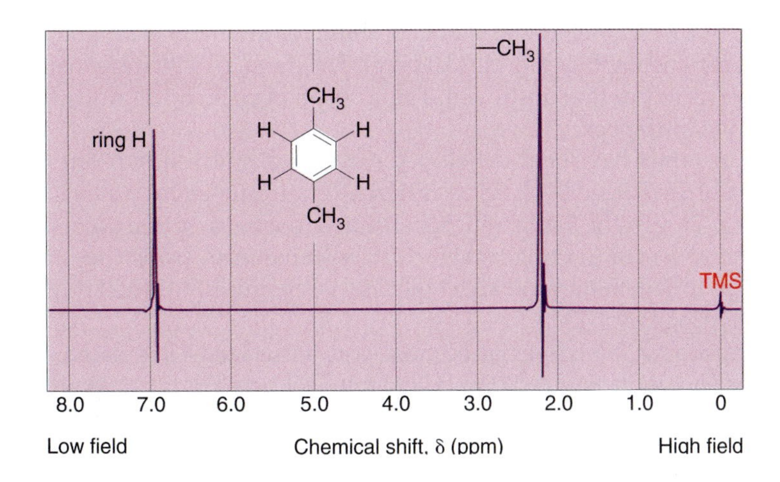

FIGURE 11.27

^{1}H-NMR spectrum of *p*-xylene.

TMS. The spectrometer is calibrated with TMS, whose absorption is electronically set at 0 parts per million (ppm) on the scale of the spectrum. Thus, the ^{1}H-NMR spectrum of a sample shows peaks that are shifted from 0 ppm, usually to the left (*downfield*) of the standard position of the TMS peak. This is described as a **chemical shift** downfield, for which the symbol is δ and the units are ppm. (TMS was chosen as the standard because it absorbs radiation at the far right of the NMR spectrum.) The unit ppm measures the extent of the shift of any absorption peak from that of TMS.

11.6.2 The Chemical Shift

All equivalent hydrogens in a molecule absorb the same RF energy and show a single peak, called a **singlet.** For example, all six methyl hydrogens of *p*-xylene (Figure 11.27) are indicated by the same singlet peak at 2.3 ppm, and all four hydrogens on the aromatic ring are indicated by the singlet peak at 7.0 ppm. Clearly, however, something is different about the two sets of hydrogens, because they absorb different amounts of RF energy.

Remember that each hydrogen nucleus is surrounded by negatively charged electrons in its covalent bond. When a molecule is placed in a magnetic field, these electrons also generate their own small magnetic field. Thus, the strength of the magnetic field actually experienced at the nucleus of a hydrogen is slightly different from that applied by the external field. The hydrogen nucleus is **shielded** from that external magnetic field by the electrons in the covalent bond with which it is attached to the molecule. Therefore, hydrogen atoms in different structural environments absorb RF energy at different magnetic field strengths and consequently show different positions of absorption in ^{1}H-NMR spectra. *Different kinds of hydrogens have different chemical shifts.* This is why ^{1}H-NMR spectroscopy can supply information about how many different kinds of hydrogens there are in the molecule (that is, how many different electronic environments there are in the molecule). The fact that the spectrum of *p*-xylene (see Figure 11.27) shows two peaks reveals that the 10 hydrogens in *p*-xylene are located in two, and only two, different electronic (and therefore structural) environments.

The electronegativity of the atom or group attached to a hydrogen determines its chemical shift. In a molecule H—A, the more electronegative A is, the lower is the electron density around the hydrogen and the more **deshielded** (that is, the less shielded) is the hydrogen. The more deshielded a hydrogen is relative to the hydrogens in TMS, the more its peak will be shifted downfield from that of TMS. You can see the effects of deshielding by comparing the chemical shifts for methane (0.2 ppm), methyl bromide (2.7 ppm), and methyl chloride (3.0 ppm): The chemical shift increases with increasing electronegativity of the substituent. The absorption peak moves downfield—farther away from the TMS peak.

Correlation tables have been developed to relate chemical shift positions to the different kinds of hydrogens—those in different electronic environments (Table 11.4). Because the hydrogens of TMS are highly shielded (their absorption occurs at high magnetic field), most organic hydrogens absorb at lower magnetic fields; they are deshielded by the groups to which they are attached (that is, chemically shifted downfield) relative to the hydrogens of TMS.

The information in Table 11.4 may be more easily visualized if it is placed on a ^{1}H-NMR spectrum, as shown in Figure 11.28. Both Table 11.4 and Figure 11.28 are reproduced in Appendix C for ready reference when studying subsequent chapters and doing problems.

TABLE 11.4

Chemical Shifts of Representative Hydrogens

Kind of Hydrogen	Chemical Shift (ppm)	Kind of Hydrogen	Chemical Shift (ppm)
C—CH_3	0.8–1.0	—CH—N—	2.2–2.9
C—CH	1.0–1.6	O=C—CH	2.0–2.6
C=C—C—**H**	1.6–1.9	O=C—**H**	9.5–9.7
Ar—C—**H**	2.2–2.8	O=C—O—**H**	10–13
C=C—**H**	4.6–5.7	halogen—C—**H**	3.1–4.1
C≡C—**H**	2.5–2.7	O—**H**	0.5–6.0*
Ar—**H**	6.5–8.5	N—**H**	0.6–3.0*
—CH—O—	3.3–4.0		

*The shifts of hydroxyl and amino hydrogens are highly variable and cannot be predicted with accuracy. The best advice is to identify those peaks by elimination *or* to shake the sample with D_2O. In that instance, the deuterium will replace the hydrogen and the peak will disappear (deuterium does not show up on NMR spectra).

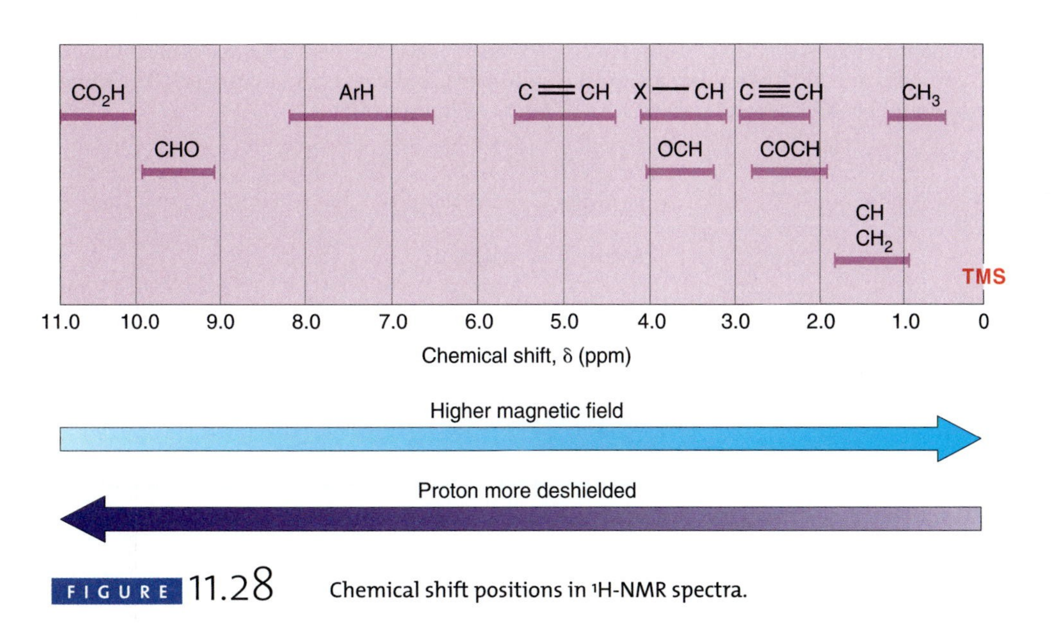

FIGURE 11.28 Chemical shift positions in ¹H-NMR spectra.

As an example of the application of these correlations, let's interpret the NMR spectrum of chloromethyl methyl ether ($ClCH_2OCH_3$) (Figure 11.29). The hydrogens bonded to the carbon with the highly electronegative chlorine and oxygen attached are expected to be strongly deshielded and therefore shifted far downfield from TMS. Their peak is at 5.5 ppm. The methyl hydrogens are deshielded only by the oxygen atom and so should be deshielded less. Their peak appears farther upfield at 3.5 ppm.

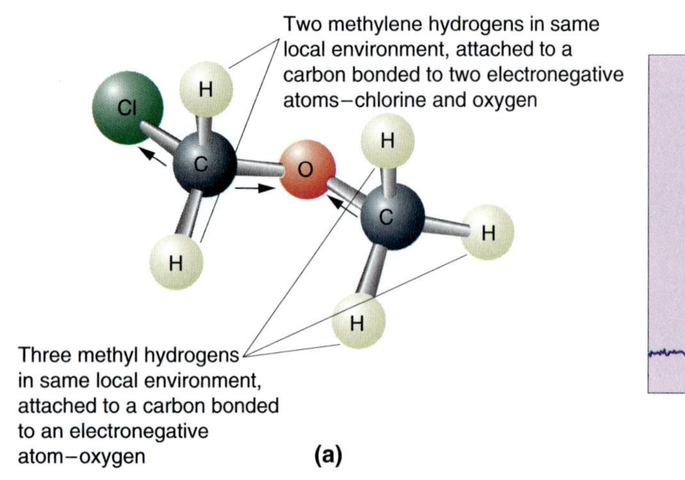

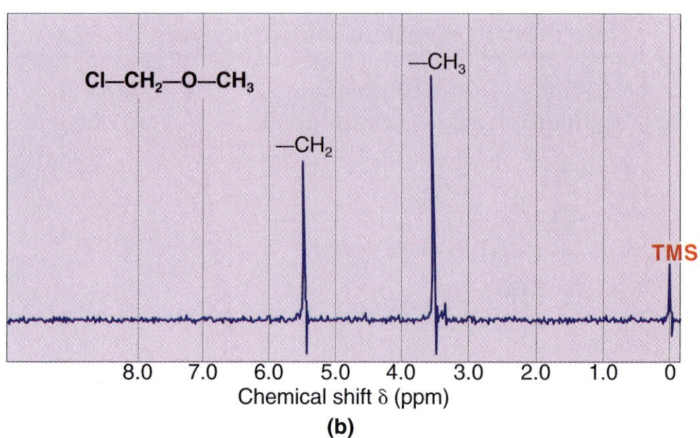

FIGURE 11.29

(a) Structure showing bond moments and (b) NMR spectrum of chloromethyl methyl ether.

How to Solve a Problem

Account for the peaks in the ¹H-NMR spectrum of methyl phenylacetate (Figure 11.30).

SOLUTION

Examination of the structure of methyl phenylacetate reveals that there are three different kinds of hydrogens: methyl hydrogens, methylene hydrogens, and benzenoid hydrogens. Therefore, there should be three separate peaks with different chemical shifts.

$$C_6H_5-CH_2-\overset{\overset{\large O}{\|}}{C}-O-CH_3$$

Methyl phenylacetate

Table 11.4 reveals that methyl protons, normally near 0.9 ppm, are shifted downfield by being attached to an electronegative oxygen atom. The peak at 3.7 ppm is consistent with methoxy protons. The aromatic protons are always near 7.3 ppm. The methylene protons are deshielded by the carbonyl group and the aromatic ring (both of these are sp^2 hybridized and therefore relatively electronegative), and so are downfield from their normal location. The peak at 3.6 ppm is assigned to the methylene protons.

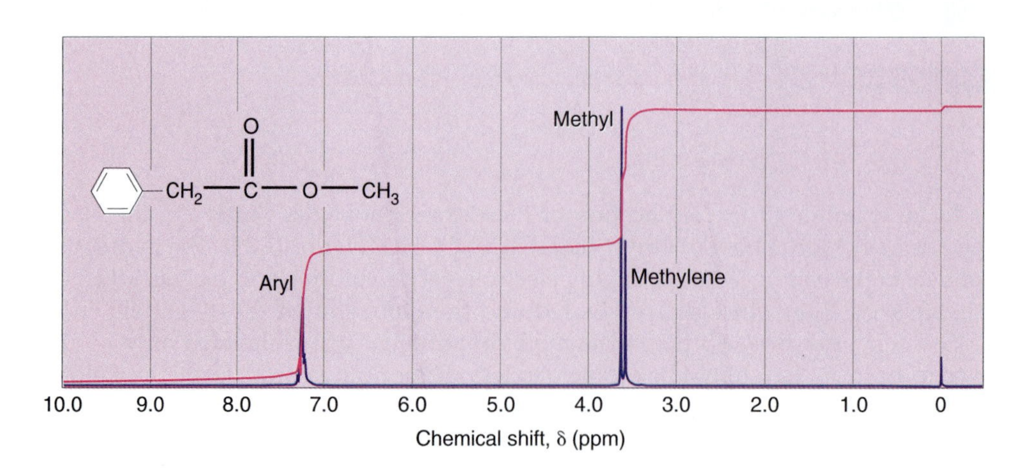

FIGURE 11.30

¹H-NMR spectrum of methyl phenylacetate.

PROBLEM 11.17

A compound with molecular formula $C_5H_{10}O$ shows IR absorption at 1725 cm^{-1} and NMR absorption centered at 0.9 and 2.6 ppm. What is its structure?

PROBLEM 11.18

Write a structural formula for each of the following four compounds, which produce only a single peak in their NMR spectra: (a) C_2H_6O, (b) $C_3H_6Cl_2$, (c) C_5H_{12}, and (d) C_4H_6.

11.6.3 Peak Integration

A second important feature of NMR spectroscopy is the ability to determine the area under each peak electronically. This is important because the area under a peak is proportional to the number of hydrogens causing that peak. The mathematical process used is called **integration.** Modern spectrometers plot a separate integral line above the peaks. The amount by which this line increases in height can be measured for each peak (simply using a ruler), and the ratio of those distances is the ratio of the numbers of hydrogens.

A simple example is the spectrum of *p*-xylene shown in Figure 11.27. The ratio of the peak heights is 1.0:1.5 (1.0 for the aryl hydrogens at 7.0 ppm and 1.5 for the methyl peaks at 2.3 ppm). Since the molecule has a total of 10 hydrogens, the ratio of the absolute numbers of hydrogens must be 4:6. The reasoning goes this way: We determine the whole-number factor needed (4) to multiply the sum of the ratio numbers ($1.0 + 1.5 = 2.5$) to give us the total number of hydrogens in the molecule (10). We use this factor to multiply the original ratio of 1.0:1.5, and we get 4:6 (four aryl hydrogens and six methyl hydrogens in *p*-xylene). Another example is chloromethyl methyl ether (Figure 11.29), in which the peak height ratio is 1:1.5. Using the preceding approach, we find that the ratio of the two kinds of hydrogens is 2:3.

Integration ratios are relatively simple to extrapolate into the absolute numbers of hydrogens needed to propose actual structures. Just remember that a methyl group always has three hydrogens, an aldehyde or carboxylic acid group has only one hydrogen, and a monosubstituted benzenoid compound has five aryl hydrogens.

How to Solve a Problem

An unknown compound with molecular formula $C_5H_{12}O$ has the ^{1}H-NMR spectrum shown in Figure 11.31. What is its structure?

PROBLEM ANALYSIS

The unknown compound cannot be a cyclic or unsaturated compound because of its molecular formula (HDI = 0); it cannot be a carbonyl compound (an aldehyde or ketone) because the double bond would yield an HDI of 1. Because there are only two NMR peaks, the compound has only two kinds of hydrogen. A single oxygen can only be found in an alcohol or an ether. The compound cannot be an alcohol (which would have at least three kinds of hydrogens). Further, since it is an acyclic compound, its structure must have at least two "ends" that must be methyl groups.

The peak at ~1.2 ppm is consistent with the conclusion that one of the "ends" consists of methyl groups bonded to carbon (C-methyl groups). The peak at 3.4 ppm is consistent with hydrogens

FIGURE 11.31

^{1}H-NMR spectrum of $C_5H_{12}O$.

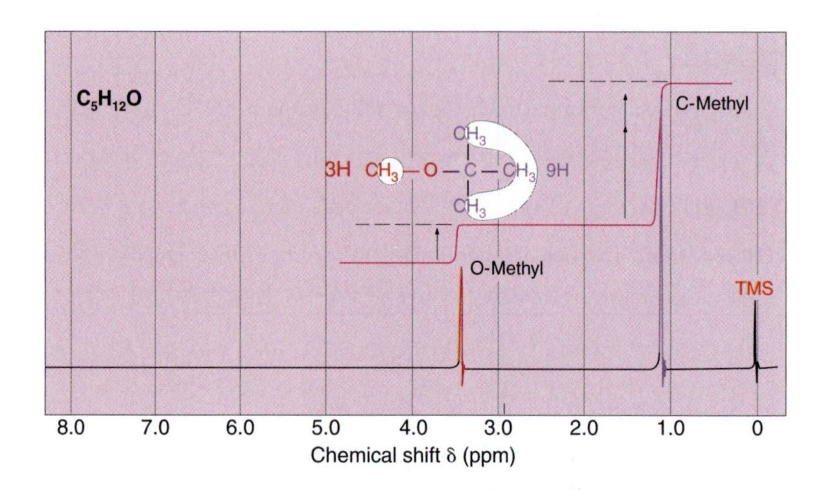

on a carbon bonded to oxygen. The ratio of the two peaks is 1:3, and since there are 12 hydrogens in the compound, the actual numbers of hydrogens must be in a ratio of 3:9. This is consistent with an O-methyl group and three C-methyl groups. By this reasoning, only one carbon is unaccounted for, and it must be the carbon to which the C-methyl groups are attached.

SOLUTION

The compound is *t*-butyl methyl ether, with the 3.4 ppm peak representing the O-methyl group (integration area = 1) and the 1.2 ppm peak representing the three identical C-methyl groups (integration area = 3).

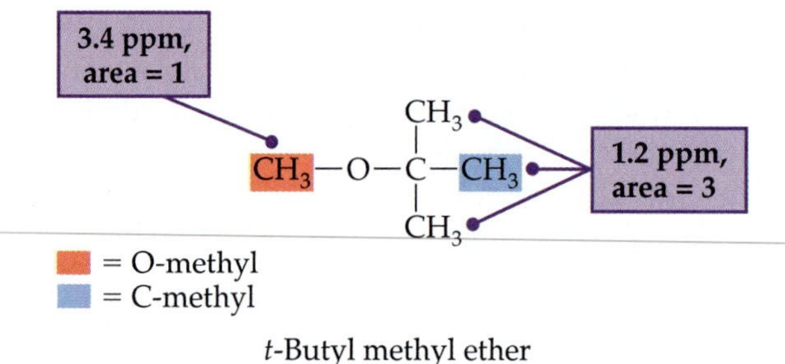

t-Butyl methyl ether

PROBLEM 11.19

For the two isomeric compounds *t*-butyl formate and pivalic acid, predict the number of different peaks in the ^{1}H-NMR spectrum, indicate the ratio of the areas of those peaks, and draw the expected spectrum, using lines for peak positions and heights.

$(CH_3)_3C{-}O{-}C({=}O){-}H$ — *t*-Butyl formate

$(CH_3)_3C{-}C({=}O){-}O{-}H$ — Pivalic acid

11.6.4 Spin-Spin Splitting

So far, we have seen how ^{1}H-NMR spectroscopy reveals (1) the number of different kinds of hydrogens (by the number of peaks in a spectrum), (2) just what kind of hydrogen is represented by each peak (from the chemical shift), and (3) the number of

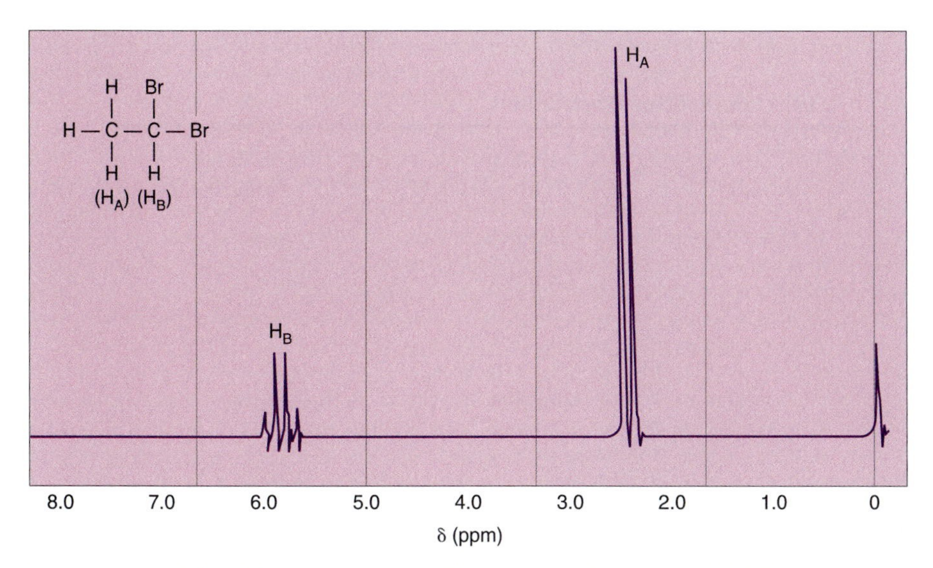

FIGURE 11.32 ^{1}H-NMR spectrum of 1,1-dibromoethane (H_A = methyl hydrogen; H_B = methinyl hydrogen).

hydrogens of each kind (by integration). This section shows how ^{1}H-NMR spectra provide (4) details of the overall structural environment of the compound—what hydrogens go where.

The examples we have considered so far have been compounds that exhibit a series of single peaks in their NMR spectra. Each of these compounds (except the aromatic compounds, a special case) share a common feature: They do not have hydrogens on adjacent carbon atoms. Therefore, the electronic environment of one kind of hydrogen has no effect on the electronic environment of another. The final kind of information obtainable from ^{1}H-NMR spectra comes from considering what happens when nonequivalent (dissimilar) hydrogens are located on adjacent carbons.

We will use as an example 1,1-dibromoethane, CH_3CHBr_2. Its ^{1}H-NMR spectrum (Figure 11.32) shows six peaks, arranged in two separate clusters, but the compound clearly contains only two kinds of hydrogens. What has happened is that the methyl (CH_3) hydrogens have influenced the methinyl (CH) hydrogen, and vice versa.

For two hydrogen atoms, H_A and H_B, in a compound (Figure 11.33a), the chemical shift of H_A is influenced by the presence of a nonequivalent H_B on an adjacent carbon. This interaction between the hydrogens is called **coupling;** H_A is said to be coupled to H_B. The reverse is also true: H_A affects the chemical shift of H_B. Such couplings are routinely observable only when the hydrogens are separated by three bonds (H_A—C—C—H_B)—that is, only for hydrogens on *adjacent* carbons. That is why no coupling effect is seen in the spectrum of chloromethyl methyl ether ($ClCH_2OCH_3$), where an oxygen separates the carbons (see Figure 11.29).

The source of the coupling is the effect of the small magnetic moment of H_B on the intensity of the external magnetic field experienced by H_A. If the moment of H_B is in the same direction as that of the external field, it adds to the internal field surrounding H_A, and so a smaller external magnetic field is required to achieve resonance for H_A. If the moment of H_B is opposite to that of the external magnetic field, a slightly larger external field is required to achieve resonance for H_A. Because H_B is present in both magnetic orientations, H_A will be in resonance at two slightly different RF values. The effect of coupling is not observed if H_A and H_B are equivalent (that is, have identical structural environments), such as in 1,2-dichloroethane ($ClCH_2CH_2Cl$).

In the absence of H_B (or if its effect is electronically *decoupled*), H_A exhibits a single sharp peak, or **singlet;** it is in resonance at a single combination of RF energy and external magnetic field. (This is true of the chloromethyl and methyl groups in chloromethyl

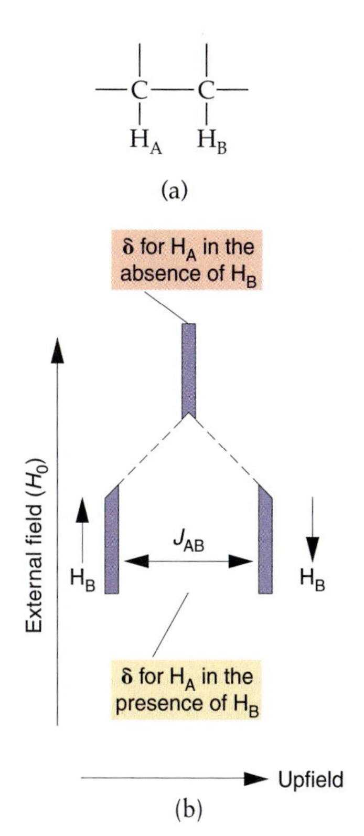

FIGURE 11.33 (a) H_A-H_B coupling and (b) spin-spin splitting.

TABLE 11.5

Spin-Spin Splitting Correlations

Structure	Left-Hand Hydrogen(s), H_A	Right-Hand Hydrogen(s), H_B
CH_3-CH_2-X	Triplet	Quartet
$CH_3-CH(X)-X$	Doublet	Quartet
$(CH_3)_2CH-X$	Doublet	Multiplet
Y_2CH-CH_2-X	Triplet	Doublet

methyl ether; see Figure 11.29.) However, the effect of coupling of H_A to H_B is that the signal *splits* and the spectrum shows a double peak, called a **doublet** (see Figure 11.32). The doublet is caused by absorption of RF energy by H_A at two slightly different magnetic field strengths (Figure 11.33b). The magnetic interaction between coupled nonequivalent hydrogens results in **spin-spin splitting.** The distance between the two peaks of a doublet is called the **coupling constant** (symbolized J_{AB}). The chemical shift of H_A is at the center of the doublet and is reported as δ for $H_A = x$ ppm (d), where d stands for doublet and x is the chemical shift. Further, the chemical shift of H_B is also split by H_A, and the coupling constant is identical ($J_{AB} = J_{BA}$).

As we will see, the fact that the peak for H_A (a methyl hydrogen) in 1,1-dibromoethane is split into a doublet means that there is a *single* nonequivalent hydrogen on the adjacent carbon (the carbon "next door"). We can use a simple formula to predict the number of splittings: A singlet will be split into $n + 1$ peaks, where n represents the number of nonequivalent hydrogens located on adjacent carbons. If there are two H_B's on the adjacent carbon, the singlet of H_A will be split into a **triplet** (t). If there are three H_B's, the singlet will be split into a **quartet** (q). Without special analysis, splittings higher than quartets are difficult to discern; they will be referred to simply as **multiplets** (m) in this book. All these splitting patterns are illustrated in Table 11.5.

Referring back to Figure 11.32 for 1,1-dibromoethane, you can see that there is a doublet representing the methyl hydrogens (H_A) being split by a single methinyl hydrogen (H_B) and a quartet representing the methinyl hydrogen being split by three identical methyl hydrogens. The inference drawn when analyzing the spectrum without knowing the structure beforehand is that because the methyl peak is a doublet, there must be a single adjacent hydrogen, and because the methinyl peak is a quartet, there must be three adjacent hydrogens (remember $n + 1$).

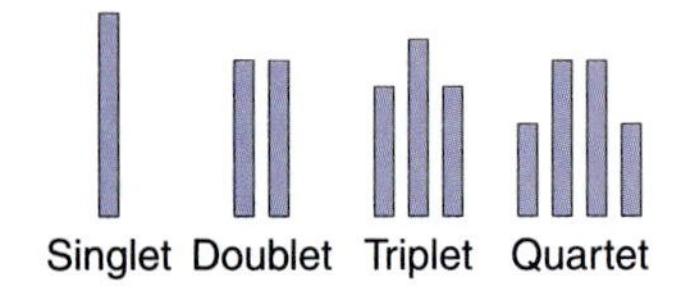

FIGURE 11.34

The shapes of NMR split peaks.

When doublets, triplets, quartets, and multiplets are integrated, the area under all peaks for each kind of hydrogen are added together to obtain the correct integration ratios. The multiple peaks are approximately symmetrical (in both height and spacing), centered about what would have been the center of the singlet position. The ideal shapes of singlets, doublet, triplets, and quartets are represented in Figure 11.34.

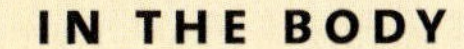

CHEMISTRY AT WORK

IN THE BODY

Magnetic Resonance Imaging (MRI)

NMR spectroscopy was discovered by physicists in the late 1940s and was applied by chemists to structural problems starting in the early 1950s. Not until the 1970s was it realized that NMR spectroscopy could be useful in diagnosing medical problems. It proved to be so useful that it had become a regular weapon in the diagnostic arsenal of most major hospitals by the late 1980s. Because many patients were made nervous by a name that included the word "nuclear" (commonly associated with atomic radiation or nuclear medicine), medical professionals began calling the technique **magnetic resonance imaging (MRI)**.

MRI is usually nothing more than ^{1}H-NMR spectroscopy applied to the human body rather than to a single compound. The magnet used to establish the external magnetic field is made large enough for the patient to be placed directly within the field. Fortunately, the magnetic field does not have to be as strong or as uniform as that required in the NMR spectrometers used in chemical laboratories.

Every cell in the body has protons, mostly in water. MRI analysis relies on the fact that after a proton is excited to a higher energy state by absorbing RF radiation, it eventually returns to its ground state (a process called *relaxation*). The sensitive electronics of MRI machines can detect differences in the relaxation times of protons due to differences in their magnetic environment. MRI instruments apply a gradient magnetic field (that is, a nonuniform field) and, on the basis of the differences in magnetic environment, produce images of cross-sectional "slices" through the body. Combining these images creates a three-dimensional image of a part of the body.

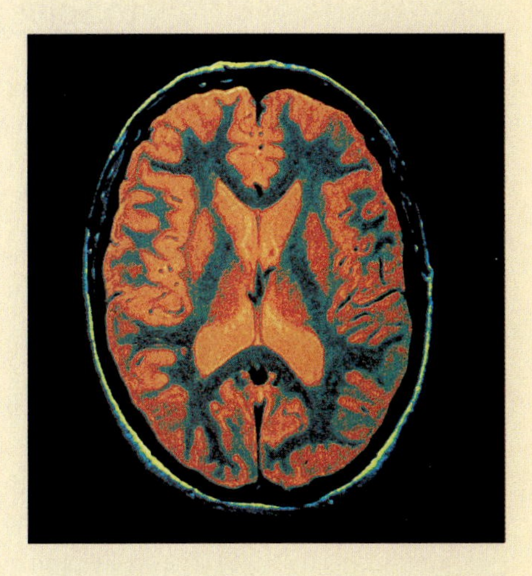

A color-enhanced MRI scan of a normal brain.

Medical researchers discovered that the relaxation times of protons in water within diseased tissue are different from those of protons in water within normal tissue. By compiling a library of MRIs of normal, healthy body parts, doctors can identify abnormal MRIs by comparison. Further diagnosis and treatment can then be focused on those specific areas showing abnormalities.

MRI seems to be replacing computerized tomography (CT) scans for many uses because CT scans rely on much higher-energy X-ray radiation (10^5–10^7 kcal/mol), which can break chemical bonds; thus, CT scans have the potential for causing tissue damage. However, the techniques are complementary, because X-rays are especially useful in diagnosing denser materials, such as bone, whereas MRI is more useful with soft tissue and organs.

Since benzenoid spin-spin splittings can be complex, this book will simply consider all aromatic hydrogens as being represented by multiplets (m), usually centered near 7.5 ppm. Further, hydrogens on nitrogen (in amines, such as RNH_2) and on oxygen (in carboxylic acids, RCO—OH, and alcohols, ROH) generally are not split and do not split other hydrogens because of hydrogen bonding and rapid proton exchange.

In summary, the spin-spin splitting pattern observed for hydrogens provides information about the structural environment on adjacent carbons. Full interpretation of a ^{1}H-NMR spectrum, which may not always be straightforward for complex molecules, can reveal virtually the total structure of a molecule—therein lies its tremendous power in modern organic chemistry.

How to Solve a Problem

(a) A compound with the formula C_3H_7Cl shows the following ^{1}H-NMR peaks: 1.0 ppm (d, area = 6) and 3.4 ppm (m, area = 1). What is its structure?

PROBLEM ANALYSIS

(1) The relative areas under the peaks add up to 7. Since there are seven hydrogens in the compound, all are accounted for, and the ratio of areas is also that of the actual numbers of hydrogens.

(2) There are only two kinds of hydrogens present.

(3) The 1.0 ppm hydrogens must have only a single hydrogen adjacent, since the splitting is into a doublet.

(4) The 3.4 ppm hydrogen must have more than three hydrogens adjacent because of the splitting into a multiplet.

(5) The six 1.0 ppm hydrogens are probably those of two methyl groups attached to carbon, given the value of their chemical shift.

(6) The 3.4 ppm hydrogen is shifted downfield, and the attachment of a chlorine is the only available reason.

SOLUTION

Since HDI = 0, there are only two possible structures: isopropyl chloride and *n*-propyl chloride. *n*-Propyl chloride, $CH_3CH_2CH_2Cl$, has three kinds of hydrogens; isopropyl chloride, $(CH_3)_2CHCl$, has only two. Therefore, the structure is isopropyl chloride.

(b) Predict and draw (using lines as in Figure 11.34) the ^{1}H-NMR spectrum of ethyl acetate, CH_3—CO—O—CH_2CH_3, including spin-spin splittings and approximate chemical shifts.

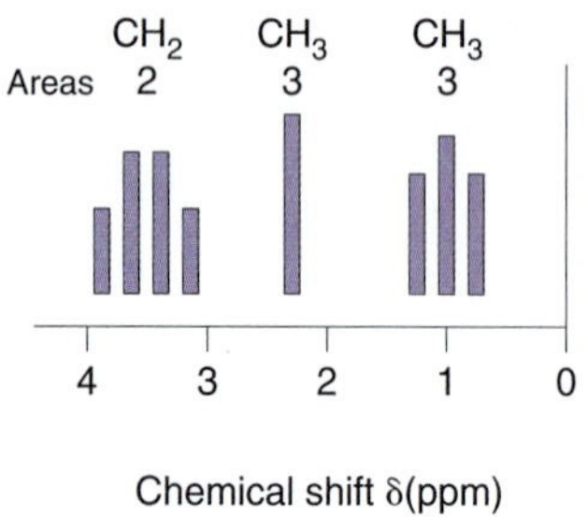

PROBLEM ANALYSIS

(1) The peak for the hydrogens of the methyl group attached to the carbonyl will be shifted downfield owing to the electronegativity of the carbonyl group. Table 11.4 indicates that it should be at about 2.3 ppm. It will be a singlet of relative area 3.

(2) The peak for the methyl hydrogens of the ethyl group should be upfield, near 1.0 ppm, since they are not significantly deshielded. It should be split into a triplet (2 + 1 = 3) by the adjacent CH_2 group. Its relative area is 3.

(3) The CH_2 group should be deshielded by the oxygen, so its position should be about 3.5 ppm. It should appear as a quartet (3 + 1 = 4), having been split by the adjacent methyl group. Its relative area is 2.

SOLUTION

The spectrum should appear schematically as shown in Figure 11.35. Note that a pattern of triplet and quartet (with the same coupling constants) is invariably an indication of the presence of an ethyl group.

FIGURE 11.35

Schematic ^{1}H-NMR spectrum of ethyl acetate,

$$CH_3-\overset{\overset{\displaystyle O}{\|}}{C}-O-CH_2-CH_3.$$

PROBLEM 11.20

How should the ^{1}H-NMR spectra of the two isomers methyl propanoate and ethyl acetate differ? Draw the spectrum for each, using lines for peak positions and heights.

$$CH_3-CH_2-\overset{O}{\overset{\|}{C}}-O-CH_3 \qquad CH_3-\overset{O}{\overset{\|}{C}}-O-CH_2-CH_3$$

Methyl propanoate   Ethyl acetate

PROBLEM 11.21

Draw a structure for each of the following compounds, which exhibit the indicated ^{1}H-NMR peaks. Items in parentheses indicate the splittings (s, d, t, q, m) and the relative areas under the peaks (for example, 9H means a relative area of 9). Remember to consider the HDI for each compound and the limited number of possible carbon skeletons.

(a) $C_5H_{11}Br$: 1.1 (s, 9H), 3.2 (s, 2H)
(b) $C_2H_4Br_2$: 2.5 (d, 3H), 5.9 (q, 1H)
(c) C_4H_8O: 1.0 (t, 3H), 2.1 (s, 3H), 2.4 (q, 2H)
(d) C_4H_9Br: 1.1 (d, 6H), 1.9 (m, 1H), 3.4 (d, 2H)
(e) C_4H_9Br: 1.8 (s)

11.7

Carbon-13 Nuclear Magnetic Resonance Spectroscopy

We have looked at ^{1}H-NMR spectroscopy in some detail because it was the first developed, is the most advanced, and is still the most powerful form of NMR spectroscopy for organic chemistry. However, the properties that make the proton of hydrogen susceptible to NMR also appear in the nuclei of some other atoms as well, most importantly in ^{13}C, ^{31}P, and ^{19}F.

^{13}C-NMR spectroscopy relies on the presence of 1.1% ^{13}C mixed in with the most abundant (98.9%) form of carbon, ^{12}C. Analysis of a ^{13}C-NMR spectrum gives information specific to the carbon skeleton of a molecule, providing the information directly rather than through inference, as in ^{1}H-NMR spectroscopy. An advantage of ^{13}C-NMR spectroscopy is that the chemical shifts are very large (about 200 ppm) compared to the shifts in ^{1}H-NMR spectroscopy (13 ppm). The same reference compound, $(CH_3)_4Si$ (TMS), is used, but as a standard with reference to its carbons, not its protons. Because the hydrogens attached to carbon atoms produce spin-spin splitting, represented by J_{HC} and J_{CH}, ordinary ^{13}C-NMR spectra can be relatively cluttered. A ^{13}C-NMR spectrum can be simplified to the point of being very useful by electronically decoupling the effects of the hydrogens. Thus, most ^{13}C-NMR spectra are **hydrogen-decoupled spectra.** Figure 11.36 shows the coupled and decoupled ^{13}C-NMR spectra of 2-bromobutane.

^{13}C-NMR hydrogen-decoupled spectra reveal the different kinds of carbon atoms present in a compound (that is, those with different local environments) simply by the

FIGURE 11.36

(a) Coupled and
(b) decoupled ^{13}C-NMR spectra of 2-bromobutane.

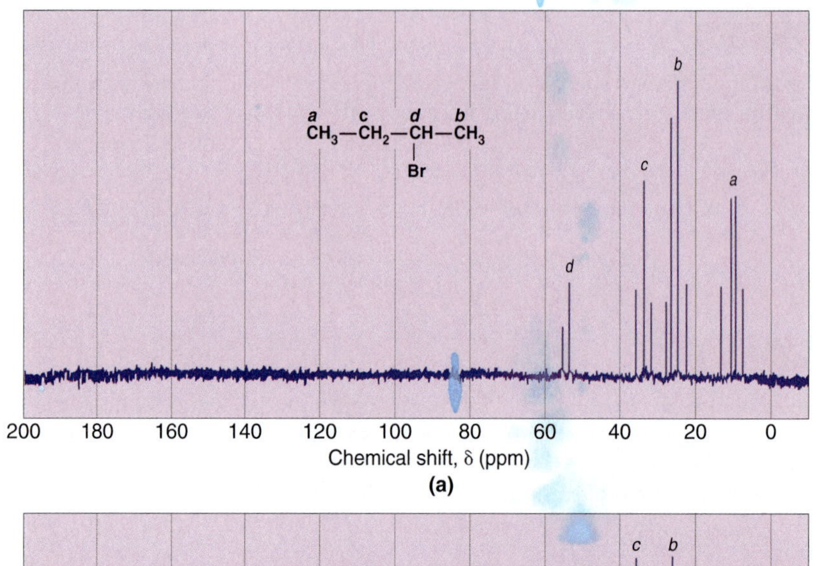

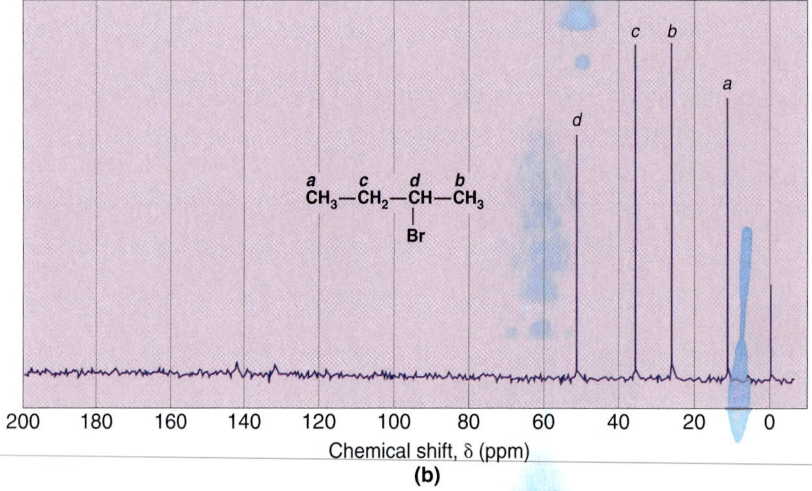

TABLE 11.6

^{13}C-NMR Chemical Shifts

Kind of Carbon	Chemical Shift (ppm)	Kind of Carbon	Chemical Shift (ppm)
R—**C**H$_3$	0–40	**C**≡C	65–85
R—**C**H$_2$—R′	15–50	Benzenoid	110–170
R$_3$**C**H	25–60	R—**C**O—R or R—**C**O—H	200–215
R—**C**H$_2$—O—	40–80		
R—**C**H$_2$—halogen	25–80	R—**C**OO—R or R—**C**OO—H	160–185
C=C	100–150		

number of different peaks (which are not split) and their chemical shifts. Table 11.6 lists the ranges of chemical shifts for a few typical kinds of carbons. This section's elementary analysis of ^{13}C-NMR spectra will not be concerned with these specific chemical shift values, but only with the number of different kinds of carbons present in a compound, as indicated by the number of different peaks in the ^{13}C-NMR spectrum.

How to Solve a Problem

An unknown compound with molecular formula C_3H_5Cl has a ^{13}C-NMR hydrogen-decoupled spectrum containing only two peaks. Draw a structure for the compound.

PROBLEM ANALYSIS

The HDI = 1, indicating the presence of one double bond or a ring. Since the compound has only three carbon atoms, the only possibilities are one of three chloropropenes or chlorocyclopropane:

$CH_2{=}CH{-}CH_2Cl$	$ClCH{=}CH{-}CH_3$	$CH_2{=}CCl{-}CH_3$	▷—Cl
3-Chloropropene (allyl chloride)	1-Chloropropene	2-Chloropropene	Chlorocyclopropane

SOLUTION

We can deduce that the compound must be chlorocyclopropane because it is the only one with only two kinds of carbons: the methylene carbons (CH_2) and the methinyl carbon (CH). The propenes all have three different kinds of carbons—that is, carbons with different atoms or groups attached.

This solution illustrates the tremendous power of ^{13}C-NMR spectroscopy. A structure can be deduced based on only two experiments: determination of the molecular formula and analysis of the ^{13}C-NMR spectrum.

PROBLEM 11.22

How many ^{13}C-NMR peaks will be shown by *p*-xylene and *m*-xylene? Identify them by position.

PROBLEM 11.23

A compound with molecular formula $C_5H_{12}O$ shows no obvious functional groups on the IR spectrum. The decoupled ^{13}C-NMR spectrum shows four different peaks. The ^1H-NMR spectrum shows the following major peaks (plus others): 1.0 ppm (d, 6H) and 3.4 ppm (s, 3H). What is the compound's structure?

PROBLEM 11.24

A hydrocarbon shows a molecular ion at 134 on its mass spectrum and has major peaks at $m/z = 76$ and 29. The ^1H-NMR spectrum shows only three peaks, at 1.2 ppm (t, 3H), 2.6 ppm (q, 2H), and 7.1 ppm (s, 2H). The decoupled ^{13}C-NMR spectrum shows four peaks, two in the aliphatic region and two in the benzenoid region. What is the compound's structure?

Chapter Summary

The structure of an organic compound can be determined using traditional laboratory methods, modern instrumental methods, or a combination of both. The **molecular formula** and **hydrogen deficiency index (HDI)** are critical pieces of information. Functional group tests can be useful, but can consume significant quantities of the compound.

Mass spectrometry (MS) involves bombarding a compound with high-energy electrons, creating molecular cations that can be detected. The technique provides (1) an accurate molecular weight, (2) information on the presence of heteroatoms such as nitrogen, chlorine, bromine, and sulfur, and (3) information on the loss of recognizable alkyl and other groups.

Molecular spectroscopy is the process by which the absorption of particular frequencies of **electromagnetic radiation** can be detected and the absorption patterns can be related to molecular structure. A **spectrometer** produces a **spectrum** containing **peaks** indicating the absorption of specific amounts of electromagnetic radiation.

Ultraviolet-visible (UV-Vis) spectroscopy, operating in the range of 200–800 nm, detects electronic transitions between the ground state and excited states of a compound. It is especially useful for detecting the presence of various patterns of conjugation in compounds.

Infrared (IR) spectroscopy, operating in the range 4000–625 cm^{-1} (2.5–16 μm), detects vibrational transitions in organic molecules (bond stretching and bending). The **functional group region** (4000–1550 cm^{-1}) is most useful for detecting the presence or absence of functional groups. The **fingerprint region** (1550–625 cm^{-1}) is most useful for proving identity or nonidentity of two samples.

Proton nuclear magnetic resonance (^{1}H-NMR) spectroscopy detects the "flipping" of the nuclear spins of protons from a low-energy state to a high-energy state in the presence of a powerful external magnetic field. The energy absorbed during the "flipping" in the presence of a magnetic field of 14,000–70,000 gauss is in the radiofrequency (RF) range of 60–700 MHz. The precise energy absorbed depends on the local structural environment of each proton. The information that may be obtained from a ^{1}H-NMR spectrum of a compound comes from (1) the number of peaks (the number of different kinds of protons), (2) the **chemical shift** (peak position, or the nature of the local structural environment for each kind of proton), (3) the integration of the area under each peak (the number of each kind of proton), and (4) the **splitting pattern** of each major peak (how many equivalent hydrogens there are of each kind).

Carbon-13 nuclear magnetic resonance (^{13}C-NMR) spectroscopy supplies similar information, based on the absorption of RF energy by the 1.1% of ^{13}C atoms present in any carbon compound. When splitting patterns caused by the protons on adjacent carbons are decoupled electronically, such spectra indicate the number of different kinds of carbon atoms present in a compound.

Additional Problems

Noninstrumental Structure Proof

11.25 Determine the molecular formula for each of the following compounds, based on the data given:

(a) Empirical formula is C_5H_{10}, and the mass spectrum shows $M^{+\cdot}$ at $m/z = 210$.

(b) Microanalysis shows a composition of 87.50% carbon and 12.50% hydrogen, and the mass spectrum shows $M^{+\cdot}$ at $m/z = 164$.

(c) Microanalysis shows a composition of 75.95% carbon, 6.33% hydrogen, and 17.72% nitrogen. The molecular ion is at $m/z = 79$.

(d) Microanalysis shows 67.6% carbon and 9.8% hydrogen. The molecular ion is at $m/z = 142$.

11.26 Draw one possible structural formula for each of the following compounds:

(a) The empirical formula is C_5H_8O, and the molecular ion is at $m/z = 84$.

(b) Microanalysis shows 85.7% carbon and 14.3% hydrogen, and the molecular ion is at $m/z = 56$.

11.27 Specify the minimal functional group tests you would perform to determine as much as possible about the structure of each of the following compounds. (You may not be able to prove any single structure with only this evidence.) Indicate the observations and conclusions you would expect to make.

(a) A compound has a molecular formula $C_4H_{10}O$.

(b) A compound has a molecular formula C_5H_{10}.

(c) A compound has a molecular formula $C_5H_{11}Br$.

(d) Microanalysis of a compound shows 87.8% carbon and 12.2% hydrogen; its molecular weight is 82.

11.28 What can you conclude about the structure of each of the following compounds (representative of families studied so far) from the following observations:

(a) An oxygen-containing compound evolves a gas when treated with sodium metal.

(b) A compound decolorizes bromine in the dark.

(c) A compound does not decolorize bromine, but does react with sodium dichromate in sulfuric acid (Jones' reagent).

(d) A compound reacts with cold potassium permanganate to produce a brown precipitate.

(e) An oxygen-containing compound does not evolve a gas when treated with metallic sodium.

(f) A water-insoluble oxygen-containing compound dissolves in aqueous sodium hydroxide.

(g) A hydrocarbon does not decolorize bromine.

11.29 What single qualitative test could you employ to distinguish between the members of the following pairs of compounds?

(a) 1-butanol and 2-methyl-2-butanol

(b) 1-hexene and hexane

(c) *t*-butyl chloride and *n*-butyl chloride

(d) 1-methoxypropane and 1-butanol

(e) cyclohexanol and phenol

(f) benzoic acid and benzaldehyde

11.30 Compound **A** of unknown structure has molecular formula $C_6H_{10}O$. It decolorizes bromine and reacts with cold $KMnO_4$. Reaction with hydrogen over a platinum catalyst leads to the absorption of one equivalent of hydrogen to form compound **B.** Compound **B** does not decolorize bromine, but does evolve a gas when treated with sodium metal. Heating **B** with concentrated sulfuric acid produces a hydrocarbon, **C** (C_6H_{10}), which decolorizes bromine. Heating **A** with concentrated sulfuric acid also produces a hydrocarbon, **D** (C_6H_8), which showed UV absorption at 255 nm. Ozonolysis of **C** produces a dialdehyde, $OCH(CH_2)_4CHO$. What are the structures of compounds **A–D**?

Mass Spectrometry

11.31 Determine the m/z value for the molecular ion of each of the following compounds:

(a) C_7H_{12} (b) C_7H_7Cl (c) $C_{10}H_{17}N$

11.32 For each of the following compounds, indicate the m/z value expected for the molecular ion. Also indicate any expected major peaks due to fragmentation, by name and m/z value:

(a) ethanol (b) isopropyl bromide

(c) 3-methylpentane (d) ethylbenzene

11.33 The molecular ion for 2-methylpentane occurs at $m/z = 86$. Account for the observed strong peaks at $m/z = 71, 57, 43$, and 29.

11.34 An unknown compound with molecular formula C_3H_8O is water-soluble and evolves a gas when treated with sodium metal. Its mass spectrum shows major peaks at $m/z = 60, 45, 42$, and 31. What is the structure of the unknown?

11.35 A compound that contains only carbon, hydrogen, and bromine shows peaks at $M^{+\cdot} = 198$, $M^{+\cdot} + 2 = 200$, and $M^{+\cdot} + 4 = 202$. Propose a molecular formula and a possible structural formula for this compound.

11.36 The labels came off two bottles of alcohols (designated **A** and **B**) that were known to be 1-pentanol and 2-methyl-2-butanol. In order to identify the unlabeled bottles, the mass spectrum was obtained for each alcohol; these spectra are reproduced in Figure 11.37. Identify which alcohol is in each bottle.

11.37 The most intense peak in the mass spectrum of 2,2-dimethylbutane is at $m/z = 57$. Propose a structure for this daughter ion and for the lost fragment.

11.38 The mass spectrum for a hydrocarbon shows major peaks at $m/z = 86, 71, 57$, and 43. Propose a structure that is consistent with this spectrum. (*Hint:* For a given molecular weight, there is a limit to the number of carbons that may be present.)

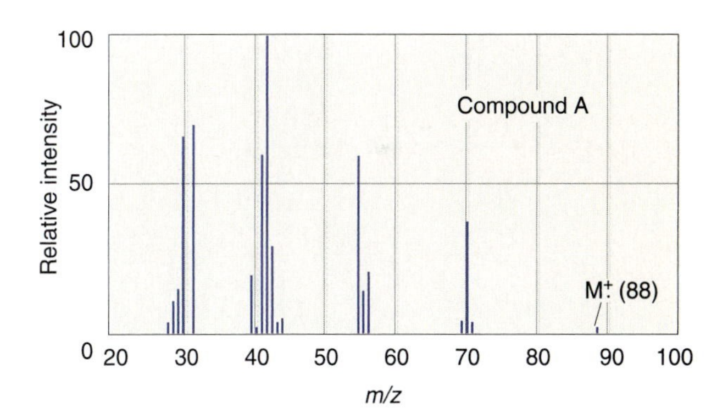

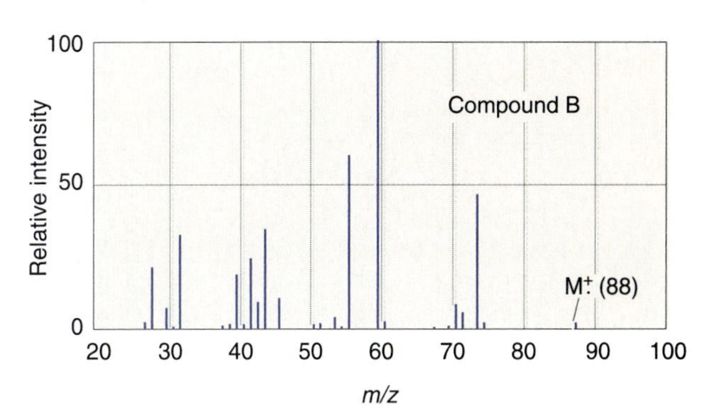

FIGURE 11.37

Mass spectra of alcohols A and B (Problem 11.36).

11.39 Propose a structure for the nonhydrocarbon compound whose mass spectrum shows two equal-intensity peaks at $m/z = 158$ and 156 and another peak at $m/z = 77$.

UV-Vis Spectroscopy

11.40 Two compounds **A** and **B** have the same molecular formula, C_6H_8. Both decolorize bromine and react with cold $KMnO_4$. Both also absorb two equivalents of hydrogen when treated with hydrogen over a platinum catalyst, forming cyclohexane in each instance. Compound **A** shows a UV absorption maximum near 255 nm, while compound **B** shows no absorption in the UV region. Propose structures for **A** and **B.**

11.41 Indicate which of the following compounds is likely to absorb in the UV region:

(a) ethanol
(b) benzyl alcohol
(c) 1,3,5-hexatriene
(d) 1-phenylpropane
(e) cyclopentadiene
(f) divinyl ether
(g) anthracene
(h) 1,4-pentadiene
(i) diphenyl
(j) diphenylmethane

IR Spectroscopy

11.42 Figure 11.38 shows the IR spectra for two structural isomers (**X** and **Y**): *t*-butyl methyl ether and 2-methyl-1-butanol. Assign a structure to each spectrum.

11.43 Indicate the approximate wavenumbers of the IR peaks that could be used to distinguish between the members of each of the following pairs of compounds:

(a) 1-propanol and methyl ethyl ether
(b) benzoic acid and benzaldehyde
(c) 1-pentyne and 1,3-pentadiene
(d) anisole and benzyl alcohol
(e) allyl ethyl ether and 2-pentanone ($CH_3COCH_2CH_2CH_3$)
(f) decane and 1-decene

11.44 An unknown oxygen-containing compound is suspected of being an alcohol, a ketone, *or* a carboxylic acid. Its IR spectrum shows a broad strong peak at 3100–3400 cm^{-1} and a sharp strong peak at 1700 cm^{-1}. What kind of compound is it?

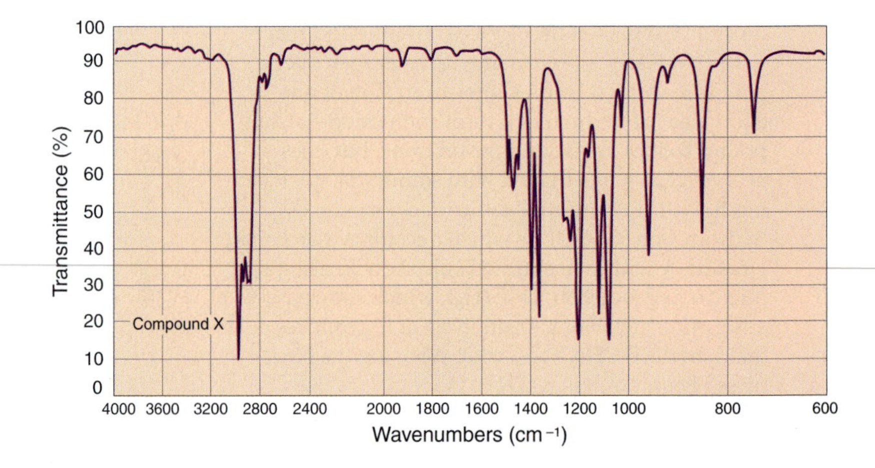

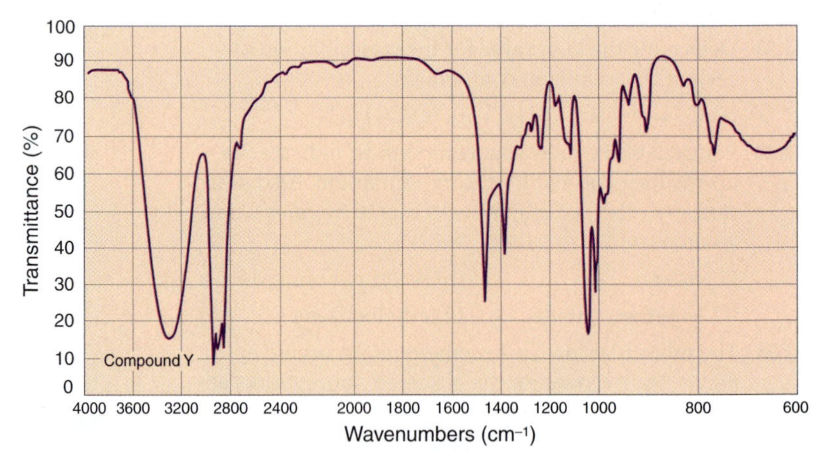

FIGURE 11.38

IR spectra of compound X and compound Y (Problem 11.42).

11.45 A compound with molecular formula C_4H_8O shows no absorption in the IR region near 1700 cm^{-1} or 3400 cm^{-1}. What can you deduce about its structure? Propose one possible structure.

11.46 Indicate the wavenumber of one major peak in the IR spectrum that would permit distinguishing between the two compounds in each of the following pairs:

(a) 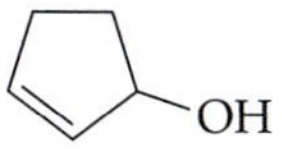 [cyclopentene ring]–OH and [cyclopentane ring]=O

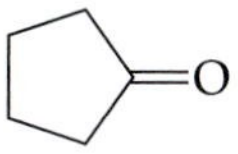

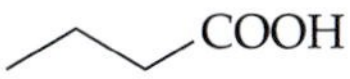

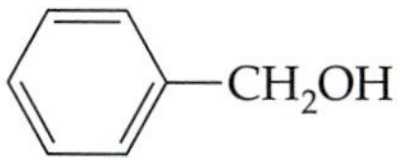

(b) 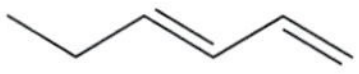 [diene chain] and [alkyne chain]

(c) [chain]–CHO and [chain]–COOH

(d) C_6H_5–CH_2OH and C_6H_5–OCH_3

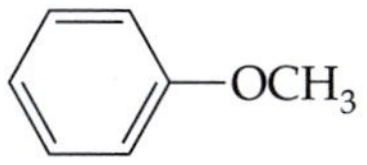

(e) 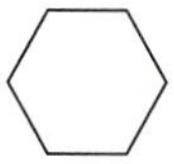 [cyclohexane] and [alkene chain]

11.47 You are assigned the task of oxidizing cyclopentanol to its corresponding ketone (cyclopentanone) with CrO_3 in sulfuric acid. How could you use IR spectroscopy to monitor the progress of the reaction?

^{1}H-NMR Spectroscopy

11.48 Indicate the number of different peaks you would expect to find in the ^{1}H-NMR spectrum of each of the following compounds (ignore spin-spin splitting):

(a) cyclopentane
(b) propane
(c) 1-bromopropane
(d) methylcyclopropane
(e) *p*-methoxytoluene
(f) ethyl methyl ether
(g) ethyl acetate (CH_3CH_2O—CO—CH_3)
(h) methyl ethyl ketone (CH_3CH_2—CO—CH_3)

11.49 Draw the structure of a compound that fits each molecular formula and has a ^{1}H-NMR spectrum showing a single peak. Indicate the approximate peak position you would expect for each compound. (*Hint:* Remember to consider HDI.)

(a) C_2H_6O (b) C_6H_{12}
(c) C_4H_6 (d) C_6H_6
(e) $C_3H_6Br_2$ (f) C_3H_6O
(g) C_4H_9Br (h) C_8H_8

11.50 Place the following compounds in order of the increasing extent to which the methyl group is *deshielded* from TMS:

(a) methylcyclohexane
(b) dimethyl ether
(c) acetone (CH_3—CO—CH_3)
(d) methyl chloride

11.51 Using lines for peaks (as in Figure 11.34), sketch the ^{1}H-NMR spectrum for each of the following compounds, showing all peaks, including those from coupling:

(a) ethyl methyl ether (b) ethyl bromide
(c) propylbenzene (d) acetaldehyde (CH_3CHO)

11.52 What features of the ^{1}H-NMR spectra of the following pairs of compounds could be used to distinguish the two compounds? Sketch the spectrum of each compound using lines for peaks (including splittings as in Figure 11.34) and numbers to indicate relative areas:

(a) $CH_3-C(=O)-O-CH_2CH_3$ and $CH_3CH_2-C(=O)-O-CH_3$

(b) $CH_3CHBrCH_3$ and $CH_3CH_2CH_2Br$

(c) $CH_2BrCHBr_2$ and CH_3CBr_3

(d) 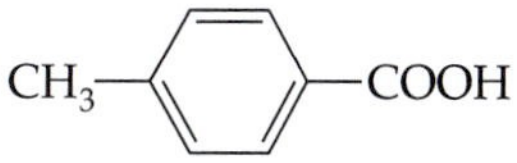 CH_3–C_6H_4–COOH and C_6H_5–CH_2COOH

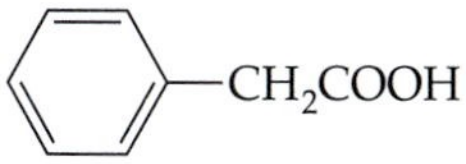

(e)  CH_3–C_6H_4–CH_3 and C_6H_5–CH_2CH_3

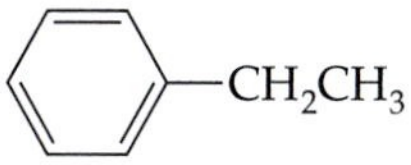

11.53 Propose structures that are consistent with each set of peaks:

(a) $C_4H_{10}O$, 1.28 ppm (s, 9H), 4.35 ppm (s, 1H)
(b) C_3H_7Br, 1.71 ppm (d, 6H), 4.32 ppm (m, 1H)
(c) C_4H_9Cl, 1.04 ppm (d, 6H), 1.95 ppm (m, 1H), 3.35 ppm (d, 2H)
(d) C_8H_{10}, 1.25 ppm (t, 3H), 2.68 ppm (q, 2H), 7.23 ppm (m, 5H)
(e) C_7H_8O, 2.43 ppm (s, 1H), 4.58 ppm (s, 2H), 7.28 ppm (m, 5H); IR peak at 3350 cm^{-1}

(f) $C_3H_6O_2$, 1.27 ppm (t, 3H), 2.66 ppm (q, 2H), 10.95 ppm (s, 1H); IR peaks at 1715 cm^{-1} and 3500–3000 cm^{-1}

(g) $C_5H_{10}O$, 1.10 ppm (d, 6H), 2.10 ppm (s, 3H), 2.50 ppm (m, 1H); IR peak at 1720 cm^{-1}

(h) C_8H_9Br, 2.00 ppm (d, 3H), 5.15 ppm (q, 1H), 7.35 ppm (m, 5H)

11.54 Propose a structure for the compound whose molecular formula is C_4H_9Br and whose ^{1}H-NMR spectrum is shown in Figure 11.39.

11.55 Figure 11.40 shows the ^{1}H-NMR spectra of three compounds with molecular formula $C_4H_8O_2$; their structural formulas are $CH_3CH_2CO—OCH_3$, $CH_3CO—OCH_2CH_3$, and $HCO—OCH_2CH_2CH_3$. Assign a compound to each spectrum, and designate which signals are from which hydrogens.

11.56 The ^{1}H-NMR spectrum of the compound $C_3H_3Cl_5$ shows peaks at 4.5 ppm (t, 1H) and 6.0 ppm (d, 2H). What is the compound's structure?

^{13}C-NMR Spectroscopy

11.57 When 2-chloro-2-methylbutane is treated with a variety of strong bases, the product usually contains a mixture of two isomers (**A** and **B**) with molecular formula C_5H_{10}. When sodium hydroxide is the base, isomer **A** predominates. When potassium *t*-butoxide is used, isomer **B** predominates. The ^{13}C-NMR spectrum of **A** shows four peaks (four kinds of carbons), and that of **B** shows five. Propose structures for **A** and **B**, and explain the different product ratios with different bases.

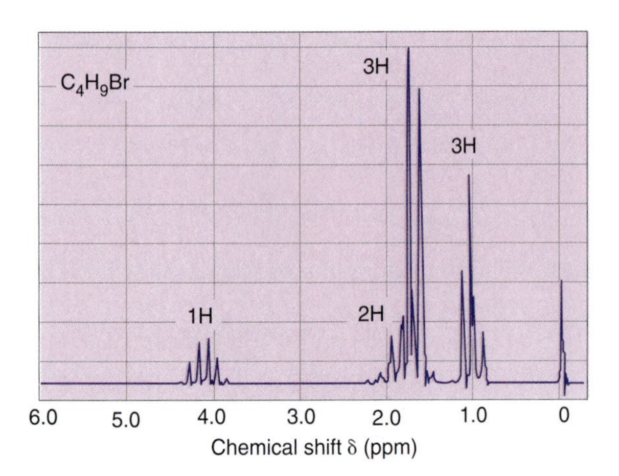

FIGURE 11.39

^{1}H-NMR spectrum of C_4H_9Br (Problem 11.54).

11.58 How many peaks should be observed in the hydrogen-decoupled ^{13}C-NMR spectrum of each of the following compounds?

(a) 2-propanol (b) 2-chloro-2-methylpentane
(c) cyclopentane (d) methylcyclopropane
(e) toluene (f) *p*-chlorobenzoic acid
(g) *m*-dichlorobenzene

11.59 Three unlabeled samples are known to be 1,2-dimethylcyclohexane, 1,3-dimethylcyclohexane, and 1,4-dimethylcyclohexane. The ^{13}C-NMR hydrogen-decoupled spectrum is obtained for each sample. The spectrum for sample **A** contains three peaks, that for sample **B** has four peaks, and that for sample **C** has five peaks. Assign a structural formula to each sample (use a planar cyclohexane ring in your analysis).

11.60 A chemist obtained the ^{13}C-NMR hydrogen-decoupled spectra of three isomeric C_7 alcohols: 3-ethyl-1-pentanol, 2-methyl-2-hexanol, and 1-heptanol. One spectrum has five peaks, one has six peaks, and one has seven peaks. Match the spectra to the compounds.

Mixed Problems

11.61 A compound $C_5H_{12}O$ shows four peaks in its ^{13}C-NMR hydrogen-decoupled spectrum, shows no peak near 3300 cm^{-1} in its IR spectrum, and shows peaks at 1.0 ppm (d, 6H), 3.4 ppm (s, 3H), 3.0 ppm (d, 2H), and 1.2 ppm (m, 1H) in its ^{1}H-NMR spectrum. Propose a structure for the compound.

11.62 A compound with empirical formula $C_8H_{10}O$ has a molecular ion of $m/z = 122$. The mass spectrum also shows peaks at $m/z = 107$, 104, and 77. The IR spectrum and ^{1}H-NMR spectrum are shown in Figure 11.41. Propose a structure for the compound.

11.63 A compound with a molecular ion of $m/z = 86$ has a peak on its IR spectrum at 1720 cm^{-1}. The ^{1}H-NMR spectrum shows peaks at 1.0 ppm (t, 3H) and 2.5 ppm (q, 2H). What is the compound's structure?

11.64 A hydrocarbon decolorizes bromine in CCl_4, but it does not absorb in the UV region. Its mass spectrum shows a molecular ion at $m/z = 82$ and other peaks at $m/z = 67$ and 53. The ^{1}H-NMR spectrum has peaks at 0.9 ppm (t, 3H) and 2.2 ppm (q, 2H). What is its structure? (*Hint:* Determine a molecular formula that fits, calculate the HDI, and then deduce possible structural features. Next analyze the mass spectrum for peaks from fragments. Finally, turn to the NMR spectrum for recognizable positions, splitting patterns, and then groups.)

11.65 A compound analyzes as having molecular formula $C_9H_{12}O$ and has the ^{1}H-NMR and IR spectra shown in Figure 11.42. What is its structure?

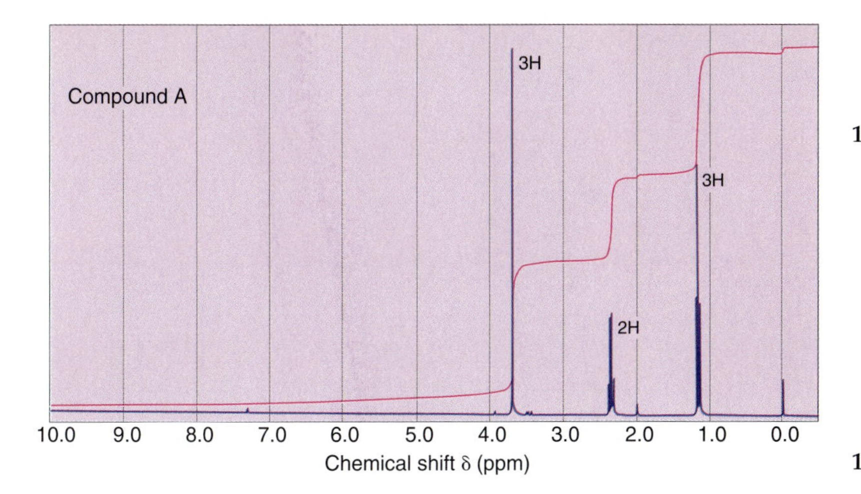

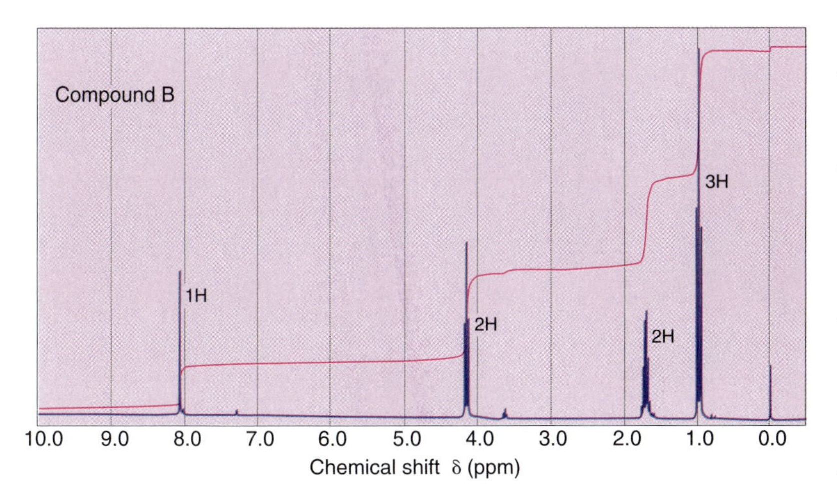

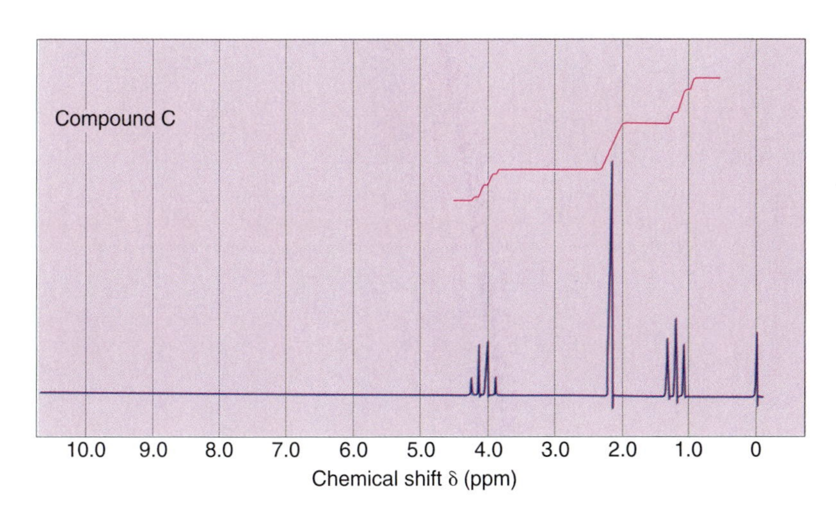

FIGURE 11.40

^{1}H-NMR spectra of three compounds with molecular formula $C_4H_8O_2$ (Problem 11.55).

11.66 A compound has an empirical formula of C_4H_5. The mass spectrum shows a molecular ion at $m/z = 106$ and a major fragment appears at $m/z = 91$. The compound absorbed at 1625 cm^{-1} in the IR, and its ^{1}H-NMR spectrum showed two singlets (with area ratio 3:2) at 1.5 ppm and 7.5 ppm. The hydrogen-decoupled ^{13}C-NMR spectrum shows three different peaks. What is the compound's structure?

11.67 A compound analyzes as 69.8% carbon and 11.6% hydrogen. It produces the mass, IR, and ^{1}H-NMR spectra shown in Figure 11.43. What is its structure?

11.68 An unknown compound produces the mass, IR, and ^{1}H-NMR spectra shown in Figure 11.44. What is its structure?

11.69 A compound has a molecular-ion peak at $m/z = 74$ and other peaks at $m/z = 59$ and 56; there are no peaks at $m/z = 45$ or 31. The IR spectrum shows a peak at 3400 cm^{-1}. What is its structure?

11.70 A compound C_3H_8O shows IR absorption at 3325 cm^{-1}. Its ^{1}H-NMR spectrum shows peaks at 3.5 ppm (s, 1H), 2.9 ppm (m, 1H), and 1.1 ppm (d, 6H). What is its structure?

11.71 A compound has a molecular ion with $m/z = 122$ and fragments at 107, 93, and 77. The IR spectrum shows no peaks in the 3300 or 1700 cm^{-1} region. The ^{1}H-NMR spectrum shows peaks at 7.2 ppm (m, 5H), 3.8 ppm (q, 2H), and 1.2 ppm (t, 3H). What is the compound's structure?

11.72 Methamphetamine is a controlled substance prescribed under a number of trade names; it is also the illegal drug known as "speed" or "meth."

H
N

Methamphetamine

Methamphetamine can be detected by the fragmentation pattern in its mass spectrum, by the fingerprint region in its IR spectrum, and by its characteristic ^{13}C-NMR and ^{1}H-NMR spectra. What is the m/z value for its molecular ion? What peaks should be most readily apparent in the IR spectrum? How many different peaks should be seen in the ^{13}C-NMR spectrum? Sketch all of the peaks—their approximate positions and relative areas—that you would expect to see in the ^{1}H-NMR spectrum. (Use lines for peaks, as in Figure 11.34.)

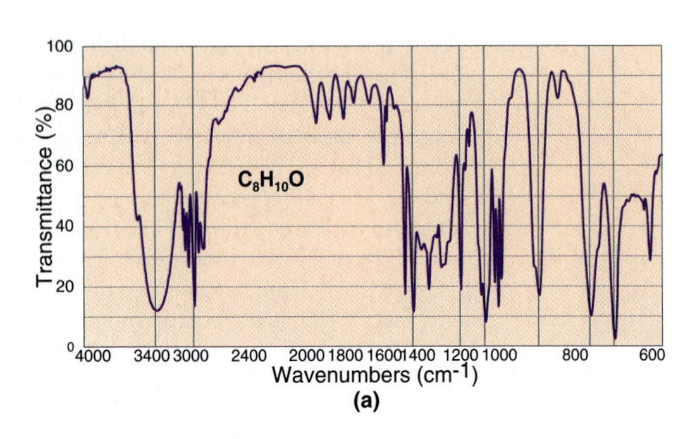

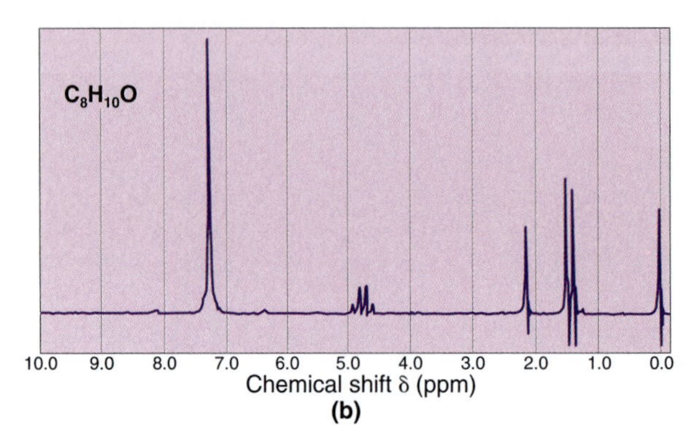

FIGURE 11.41

(a) IR spectrum and (b) ^{1}H-NMR spectrum of a compound with empirical formula $C_8H_{10}O$ (Problem 11.62).

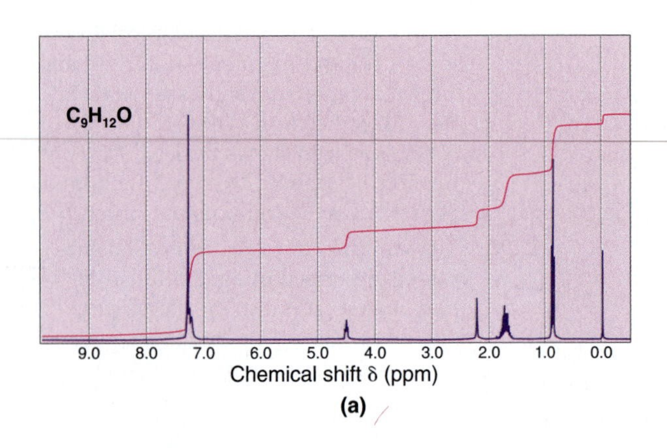

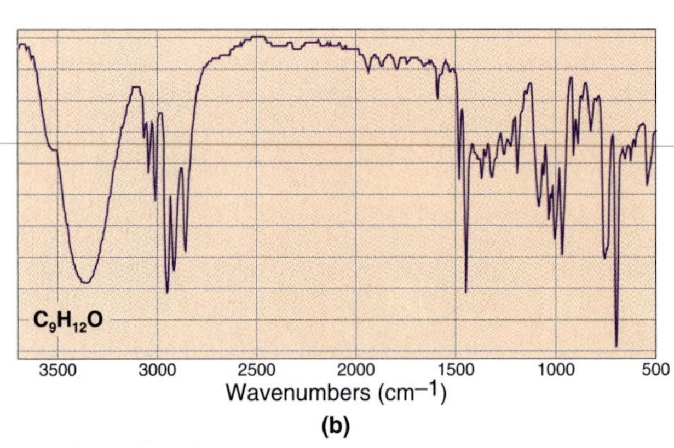

FIGURE 11.42

(a) ^{1}H-NMR spectrum and (b) IR spectrum of a compound with molecular formula $C_9H_{12}O$ (Problem 11.65).

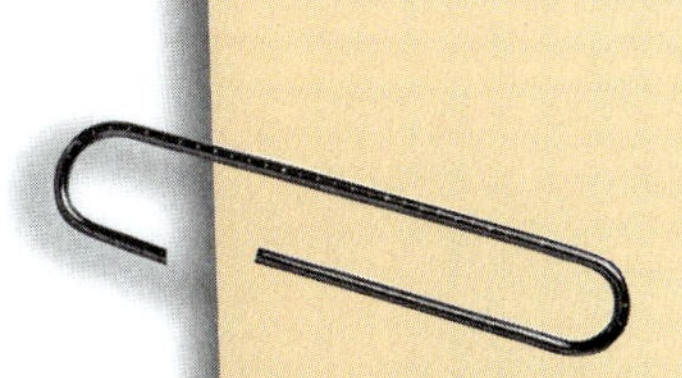

PVC: Under Suspicion

The poly(vinyl chloride) (PVC) monomer, chloroethene, came under suspicion as a carcinogen in 1970, when it was discovered that workers exposed to high levels of chloroethene developed liver cancer. The mode of action involves mammalian P-450 cytochrome catalysts, which cause epoxidation of chloroethene to yield chlorooxirane. At physiological pH (~2), chlorooxirane rearranges to compound **A.** Compound **A** has an HDI of 1, shows IR absorption at 1720 cm^{-1}, and shows a ^{1}H-NMR peak near 9 ppm. When compound **A** is mixed with guanine (one of the five bases in human DNA), a reaction occurs to form compound **B** and expel hydrogen chloride.

Guanine

You work for an environmental "watchdog" committee and are asked by a neighborhood group to analyze air and waste-water samples from a nearby PVC manufacturing plant. You perform ^{1}H-NMR spectroscopy on the samples.

- How many peaks would you observe on an ^{1}H-NMR spectrum of chloroethene? What would the approximate chemical shifts of these peaks be?
- Suggest a laboratory method to convert chloroethene to chlorooxirane.
- What would be the approximate chemical shift observed for the protons in chlorooxirane?
- What are the structures of compounds **A** and **B**?
- Write a simple one-step mechanism for the formation of compound **B** from compound **A** and guanine. Is the reaction S_N1, S_N2, addition, or elimination?

Many people associate poly(vinyl chloride), or PVC, with the plastic piping used in construction. The plastic is valued for being lightweight, as well as durable. Compact discs, shown here on an assembly line, are also made of poly(vinyl chloride). The PVC disc is coated with a fine film of metal, which reflects the laser light that "reads" the depressions made on the surface of the disc.

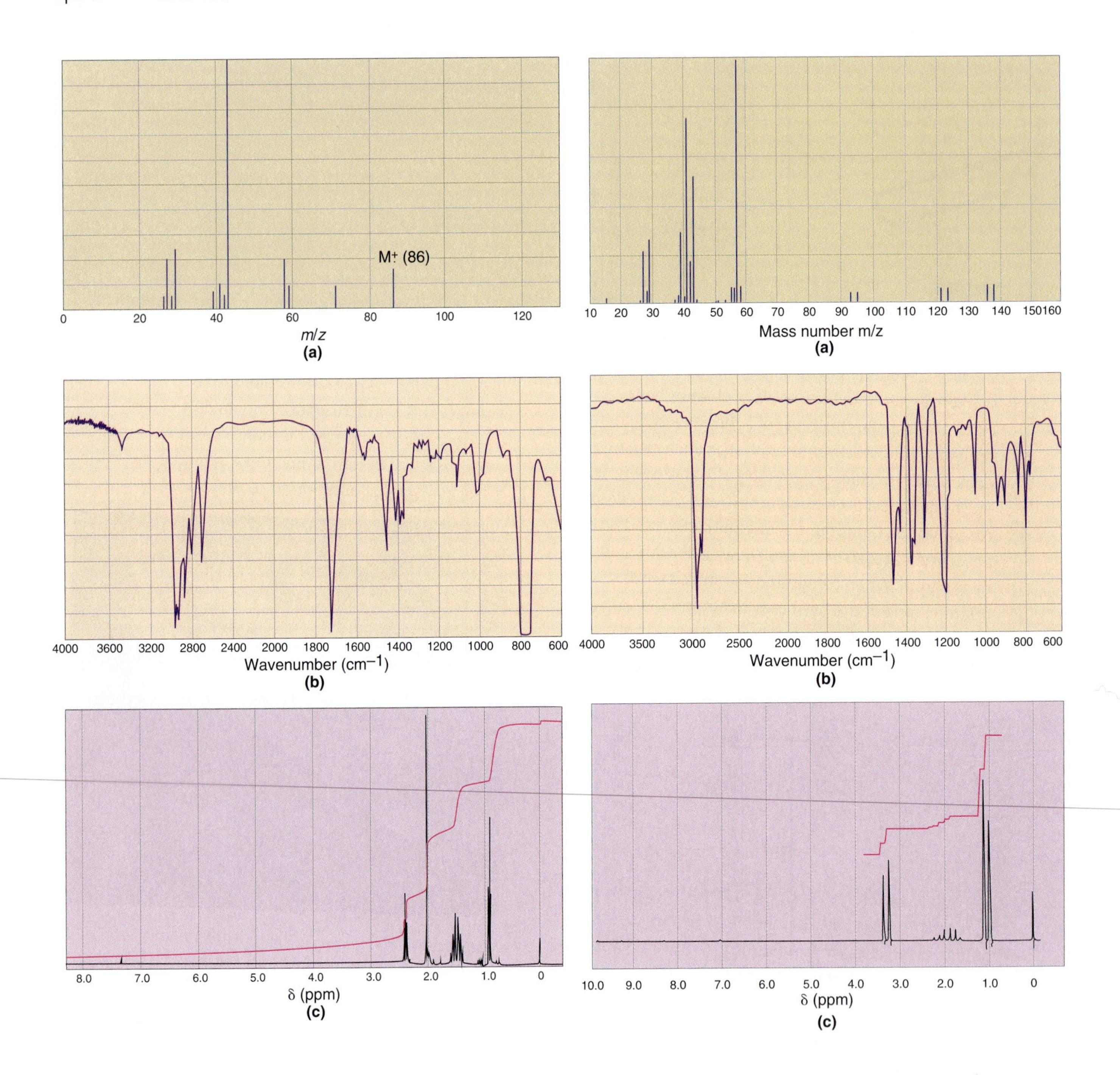

FIGURE 11.43

(a) Mass spectrum, (b) IR spectrum, and (c) ^{1}H-NMR spectrum of an unknown compound (Problem 11.67).

FIGURE 11.44

(a) Mass spectrum, (b) IR spectrum, and (c) ^{1}H-NMR spectrum of an unknown compound (Problem 11.68).

CHAPTER 12

Amines

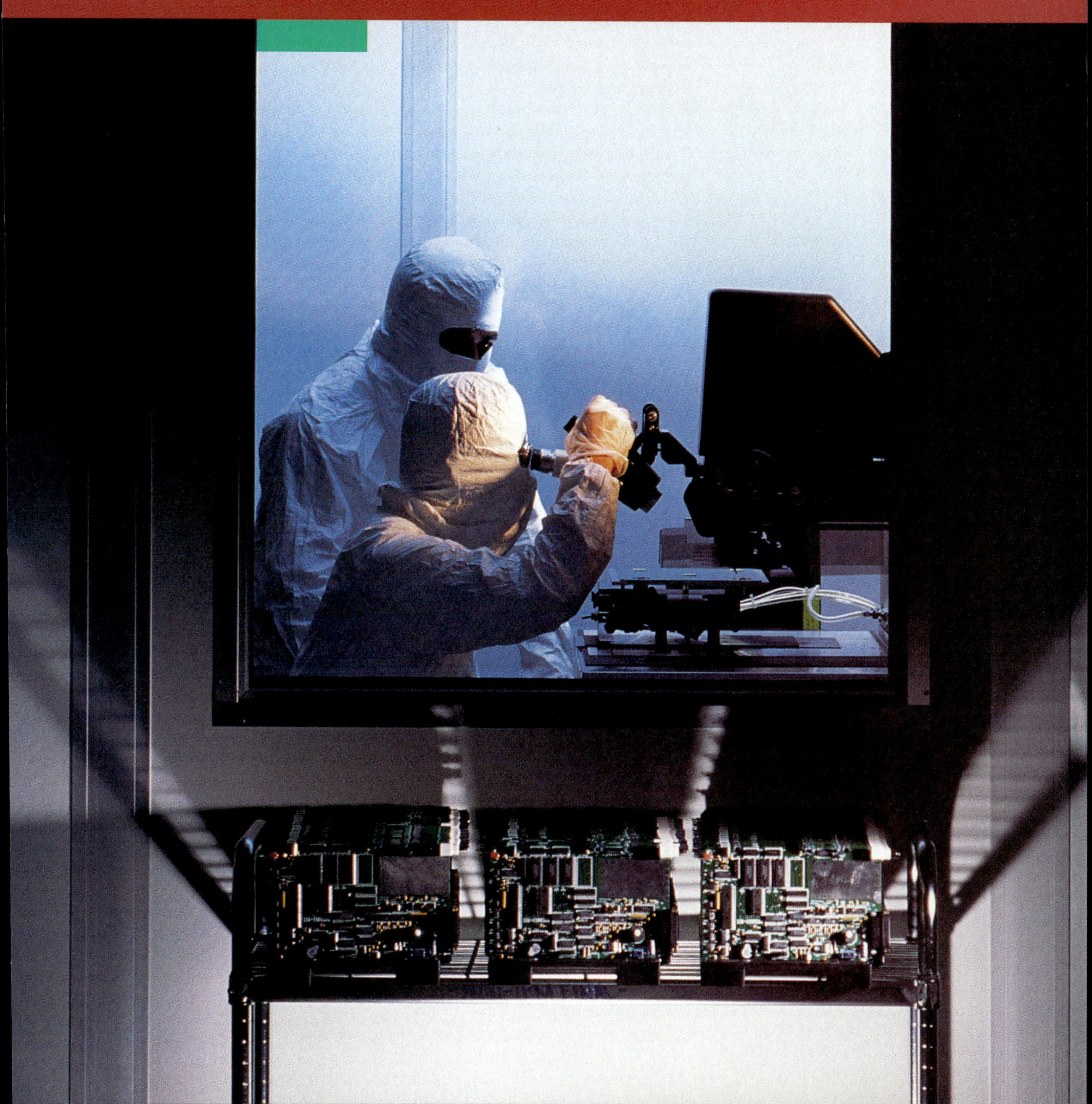

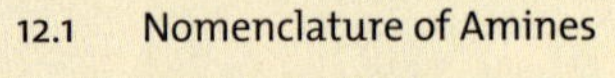

AMINES ARE a large family of organic compounds that contain nitrogen as the heteroatom. They are of special importance because of their role in biological chemistry—the chemistry of living systems. Although molecular nitrogen (N_2) accounts for 78% of the composition of air, it is inaccessible to most biological systems because of its chemical inertness. Thus, the nitrogen that is essential to all forms of life must be incorporated into organic compounds in the form of amines before it can be utilized. Leguminous plants function as chemical processors by carrying out the activity referred to as *nitrogen fixation*—the incorporation of nitrogen into organic amines.

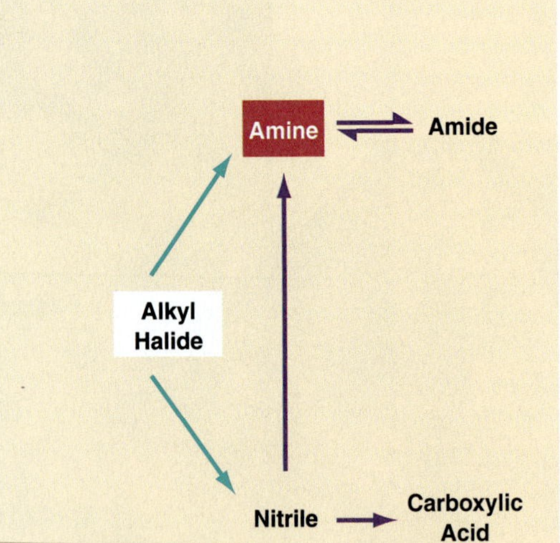

Amines, which are nitrogen-containing organic compounds, are basic.

An important occurrence of the functional group characteristic of amines is in proteins. Proteins are polymers of α-amino acids, which, as the name implies, contain the amino group. Some DNA bases, the key to the genetic code, contain amino groups. Many medicinals, both modern and ancient, are amines whose complex structures have been determined by organic chemists. A wide variety of amines, from simple to extremely complex, are produced by plants and other organisms; many of these have been isolated and then synthesized by organic chemists.

Amines can be viewed in two ways: (1) They can be considered derivatives of alkanes or aromatic hydrocarbons in which a hydrogen atom has been replaced by the **amino group (—NH_2);** examples are aminomethane (CH_3—NH_2) and aniline (C_6H_5—NH_2) (Figure 12.1). This is analogous to considering an alcohol (R—OH) or phenol (Ar—OH) as a derivative of an alkane or benzenoid compound in which a hydroxyl group has replaced a hydrogen. (2) Amines can be considered derivatives of ammonia (NH_3) in which a hydrogen atom has been replaced by an alkyl (R—NH_2) or aromatic (Ar—NH_2) organic group; an example is ethylamine (C_2H_5—NH_2). This is analogous to considering an alcohol or phenol as being derived from water by replacing a hydrogen with an alkyl or aryl group.

While most organic compounds are neutral, and therefore not converted to ions in aqueous acid, amines are the only major family of organic compounds that are basic in aqueous solution. This has a major impact on their properties.

◀ Microchips are produced using a chemical process called photoresist technology, which is based on a process first developed using diazonium salts.

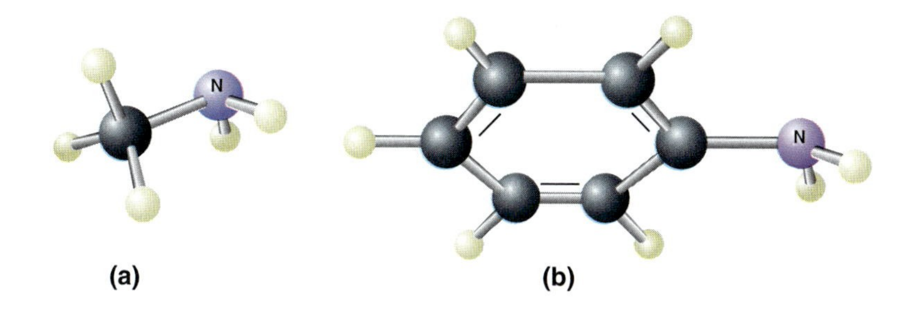

FIGURE 12.1

Ball-and-stick models of (a) aminomethane (methylamine, CH_3NH_2) and (b) aminobenzene (aniline, $C_6H_5NH_2$), viewed from below the plane of the benzene ring.

12.1

Nomenclature of Amines

12.1.1 Classification of Amines

A general classification system (separate from the nomenclature system) for amines designates any amine as *primary* (RNH_2 or $ArNH_2$), *secondary* (R_2NH or Ar_2NH), or *tertiary* (R_3N or Ar_3N), depending on whether one, two, or three hydrogen atoms, respectively, of the "parent" ammonia (NH_3) have been replaced by alkyl or aryl groups. This classification system is useful in understanding the similar chemistry of apparently unrelated amines. The substituents on the nitrogen in a secondary or tertiary amine may be identical or different, and the nitrogen atom may be incorporated into one or more ring systems.

Alfalfa is a leguminous plant that converts molecular nitrogen from the air into organic amines.

$(CH_3)_3C—NH_2$	$C_2H_5—NH—C_6H_5$	$C_6H_5CH_2—N(CH_3)_2$
Primary (1°) amine (*t*-butylamine)	Secondary (2°) amine (ethylphenylamine)	Tertiary (3°) amine (benzyldimethylamine)

PROBLEM 12.1

Classify each amine as primary, secondary, or tertiary: (a) $(C_6H_5)_2NH$, (b) $CH_3CH_2NHCH_3$, (c) $CH_3CH_2N(CH_3)_2$, (d) aniline, (e) $CH_3\,(CH_2)_2NH_2$, and (f) $(CH_2)_4NH$.

12.1.2 Historical/Common Names

Amines are named by two different systems: historical and IUPAC. In both systems, virtually any compound whose name has the suffix *-ine* is an amine, even most commercial products with trade names. For example, caffeine, nicotine, and Benzedrine are all amines.

The historical/common nomenclature for amines is widely used. It involves simply naming the organic groups attached to the nitrogen atom, followed by the suffix *-amine* (for example, methylamine). Up to three different groups may be named in alphabetical order (for example, benzylethylmethylamine), and two or three identical groups are designated by the prefixes *di-* and *tri-* (trimethylamine). Clearly, this system is applicable only when the organic groups are simple and commonly recognized.

$CH_3CH_3NH_2$	$C_6H_5CH_2NHCH_3$	$(CH_3)_3N$
Ethylamine (1°)	Benzylmethylamine (2°)	Trimethylamine (3°)

Many amines, especially naturally occurring ones such as nicotine and histamine, have retained their simple historical names. Derivatives of these amines are named using these as the "parent" names.

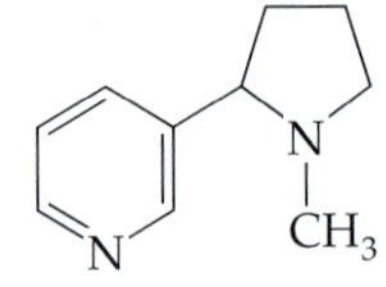

Nicotine

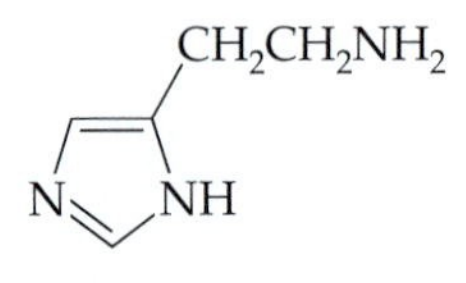

Histamine

Further, a wide variety of heterocyclic amines—for example, pyridine, piperidine, and purine—have been assigned unique names, usually of historical origin. The positional numbering system for these three examples is

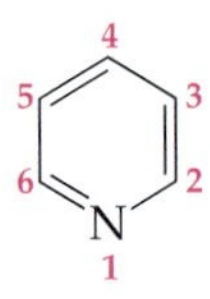

Pyridine

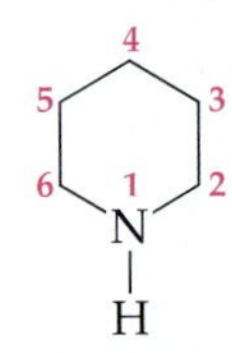

Piperidine

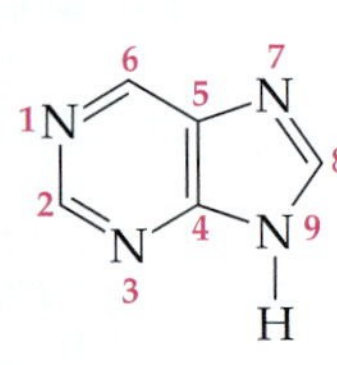

Purine

12.1.3 IUPAC Nomenclature

The second system of nomenclature for amines follows the IUPAC principles used for most other families of organic compounds, such as alcohols. The longest straight chain of carbon atoms carrying the functional group is named as an alkane; then the suffix *-amine* is added. Therefore, alkyl amines are generically called **alkanamines.** Thus, $CH_3CH_2NH_2$ is called ethanamine in the IUPAC system. Other substituents are designated in the usual manner (name of group and position number). Finally, secondary and tertiary amines have their second and third groups designated by *N*-alkyl or *N,N*-dialkyl terminology. Thus, *N*-methyl-2-propanamine has a methyl group on the nitrogen atom (N) of what otherwise would be 2-propanamine.

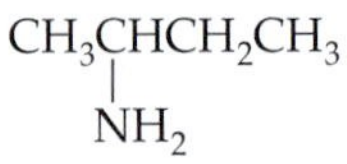

$CH_3CHCH_2CH_3$ (with NH_2 on C-2) — 2-Butanamine

Cyclohexane ring with $-NH_2$ and CH_3 — 3-Methylcyclohexanamine

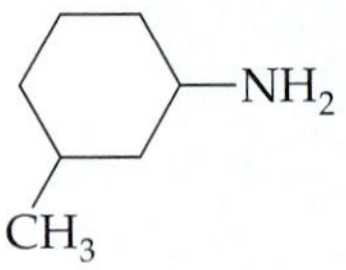

$CH_3NHCH(CH_3)_2$ — *N*-Methyl-2-propanamine

The amino group ($—NH_2$) can be designated as a substituent for compounds carrying higher-priority functional groups. When the amino group is part of a compound containing an alcohol, carbonyl, or carboxylic acid functional group, it is named as a substituent. Examples are 2-aminoethanol and *p*-aminobenzoic acid:

$HO—CH_2CH_2—NH_2$

2-Aminoethanol
(*not* 2-hydroxyethanamine)

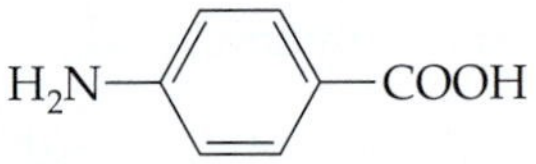

$H_2N—C_6H_4—COOH$

p-Aminobenzoic acid
(*not* *p*-carboxyaniline)

Benzenoid amines are named as derivatives of aniline (C_6H_5—NH_2). For example, *m*-bromoaniline and *N,N*-dimethyl-2,4,6-trichloroaniline have the following structures:

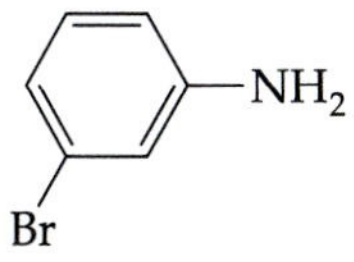

m-Bromoaniline

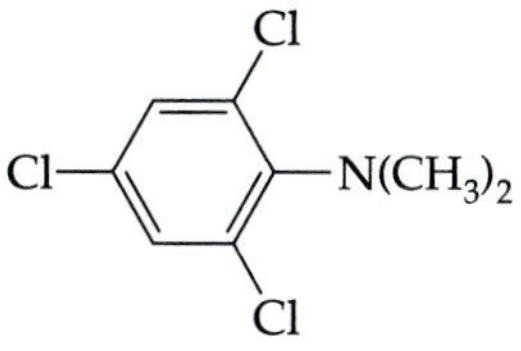

N,N-Dimethyl-2,4,6-trichloroaniline

How to Solve a Problem

Draw the structural formula for each of the following compounds: (a) benzylethylamine, (b) *N,N*-diethylaniline, and (c) 4-amino-2-butanol.

SOLUTION

(a) There are two alkyl group names preceding the suffix *-amine,* so the compound must be a secondary amine. One attached group is benzyl and the other is ethyl, so we have

$C_6H_5CH_2NHCH_2CH_3$

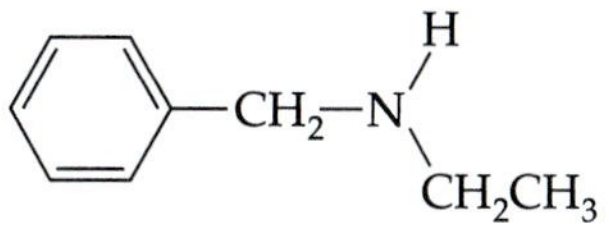

Benzylethylamine

(b) The parent compound is aniline. The two *N*'s indicate that two groups are attached to nitrogen, and they are both ethyl groups.

$C_6H_5N(C_2H_5)_2$

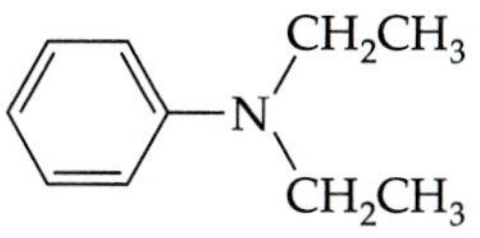

N,N-Diethylaniline

(c) The stem name *butan-* means that the parent chain is the C_4 alkane butane. The suffix *-ol* indicates that the compound is an alcohol, so the parent name is butanol. The hydroxyl group is located on carbon 2, and the amino group is attached to carbon 4:

$H_2NCH_2CH_2CH(OH)CH_3$ $\quad$ $H_2N—CH_2—CH_2—CH(OH)—CH_3$

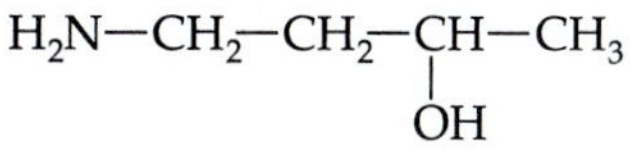

4-Amino-2-butanol

PROBLEM 12.2

Assign names to the following compounds using both nomenclature systems:

(a) 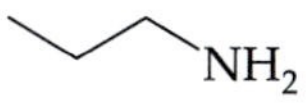 $CH_3CH_2CH_2NH_2$ (b) 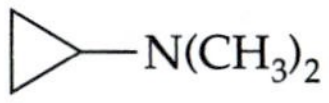 cyclopropyl—$N(CH_3)_2$

(c) 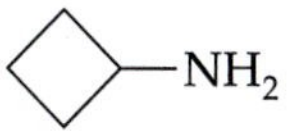 cyclobutyl—NH_2 (d) $(CH_3)_2CHNHCH_3$

PROBLEM 12.3

Draw structures corresponding to the following names: (a) 2-butanamine, (b) 3-methylcyclopentanamine, (c) 3-aminocyclohexanol, (d) (*Z*)-3-(*N*-phenylamino)-2-pentene, (e) (*E*)-2-buten-1-amine, and (f) triisopropylamine.

PROBLEM 12.4

Assign structures to the following compounds: (a) *N*-methyl-*p*-nitroaniline, (b) *N*,*p*-dimethylaniline, and (c) *N*-cyclopropylaniline.

PROBLEM 12.5

Assign names to the following structures:

(a) C_6H_5–N(H)–C_6H_5 (b) (CH_3)(CH_3)C_6H_3–NH_2 (c) $CH_2{=}CH$–C_6H_4–$N(C_2H_5)_2$

12.2

Structure of Amines

The physical and chemical properties of amines arise from their electronic structure, which can be derived from the orbital description of atomic nitrogen. (Review the discussion of atomic orbitals and carbon bonding in Chapter 1.) Nitrogen, with atomic number 7, has seven electrons, and the electron configuration is $1s^22s^22p^3$. In the course of forming covalent bonds, the four outer-shell atomic orbitals (one 2*s* and three 2*p* orbitals) hybridize into four identical sp^3 hybrid orbitals (tetrahedral hybridization). The five outer-shell electrons fill these four sp^3 orbitals according to Hund's rule, resulting in one full orbital and three others with just one electron each:

$$^7N = 1s^22s^22p^3 = \underset{1s}{\uparrow\downarrow}\ \underset{2s}{\uparrow\downarrow}\ \underbrace{\uparrow\ \uparrow\ \uparrow}_{2p} \xrightarrow{\text{hybridization}} \underset{1s}{\uparrow\downarrow}\ \underbrace{\uparrow\downarrow\ \uparrow\ \uparrow\ \uparrow}_{sp^3}$$

Ground state of nitrogen — Bonding orbitals of nitrogen

Pyramidal amine

R⁗–N^+(R′)(R″)(R‴)

Tetrahedral ammonium ion

The three sp^3 orbitals with only one electron overlap with the orbitals from three atoms or groups and form covalent bonds with those three atoms or groups; the nitrogen is therefore trivalent. Thus, all amines have an unshared pair of electrons and three covalent bonds (Figure 12.2). The unshared pair of electrons, which are in a filled sp^3 orbital, are the electrons responsible for the basicity and nucleophilicity of amines.

Because the hybridization is tetrahedral, the bonds in an amine from nitrogen to its three substituents are directed at approximately 109.5° to one another (a tetrahedral angle). Viewed from the nitrogen atom, the four electron pairs in an amine point to the corners of a tetrahedron, but the three covalent bonds radiate to form a pyramidal shape (Figure 12.2). Thus, amines are pyramidal (with nitrogen at the top of the pyramid). It should also be clear, however, that when a fourth group is covalently bonded

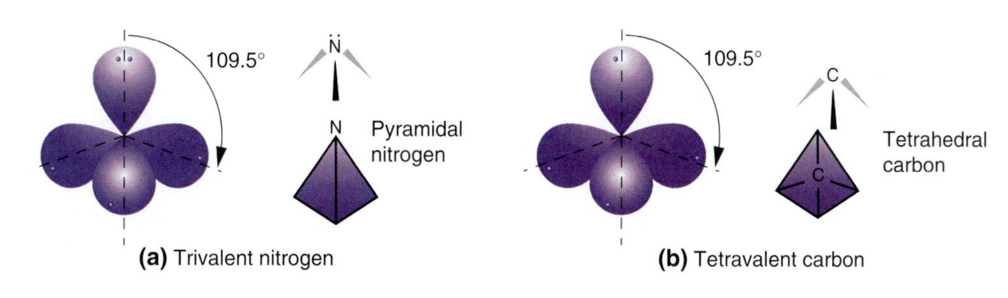

FIGURE 12.2

The hybrid orbitals of (a) trivalent nitrogen and (b) tetravalent carbon.

to nitrogen using the nonbonding lone pair of electrons, the resulting molecule is tetrahedral around nitrogen. This is the case for what are called ammonium compounds ($R_4N^+\ X^-$), such as ammonium chloride ($NH_4^+\ Cl^-$) and tetramethylammonium chloride [$(CH_3)_4N^+\ Cl^-$].

12.3

Properties of Amines

12.3.1 Hydrogen Bonding

Amines are polar compounds because of the difference in electronegativity between nitrogen (3.0) and carbon (2.5) and hydrogen (2.1). Primary and secondary amines, because they have hydrogen attached to nitrogen, are hydrogen-bonding *acceptors*—that is, their hydrogens "accept" electrons from a donor for sharing. The reason for this property is the fact that nitrogen is more electronegative than hydrogen; therefore, the N—H bond is polarized such that hydrogen is relatively electron-deficient (that is, $N^{\delta-}$—$H^{\delta+}$). Because of the higher electronegativity of nitrogen relative to carbon and because of nitrogen's unshared pair of electrons, all three classes of amines are hydrogen-bonding *donors*—that is, they share their unshared electrons with a hydrogen on an acceptor molecule.

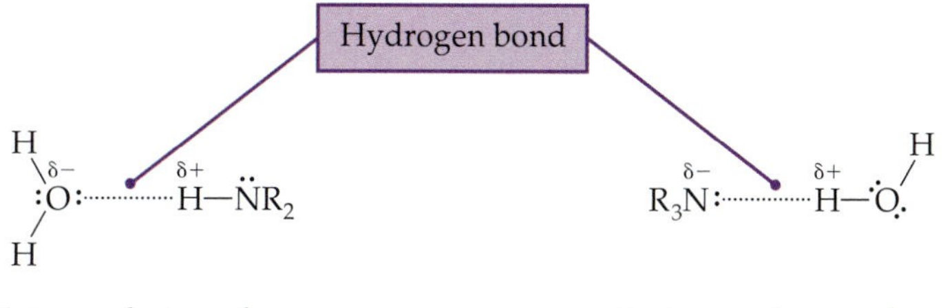

Water as electron donor, amine as electron acceptor

Amine as electron donor, water as electron acceptor

The effects of the hydrogen-bonding capabilities of amines are twofold:

1. Low-molecular-weight amines (up to C_6) are soluble in water as a result of hydrogen bonding with water molecules, whereas hydrocarbons of the same molecular weight are not.
2. Primary and secondary amines have higher boiling points than hydrocarbons of comparable molecular weight because of hydrogen bonding with other amine molecules. For example, propane (molecular weight 44) boils at −42°C, while ethylamine (45) boils at +17°C. Tertiary amines cannot hydrogen-bond with other amine molecules because of the absence of a hydrogen bonded to nitrogen.

12.3.2 Basicity of Amines

The unshared pair of electrons on nitrogen accounts for the most important chemical properties of amines, causing them to be basic (they attack protons) and nucleophilic (they attack carbon). Amines are protonated by aqueous dilute acids, such as hydrochloric acid, to form water-soluble ammonium salts. An ammonium salt can be converted back to the amine by removal of a proton using a stronger base, usually sodium hydroxide.

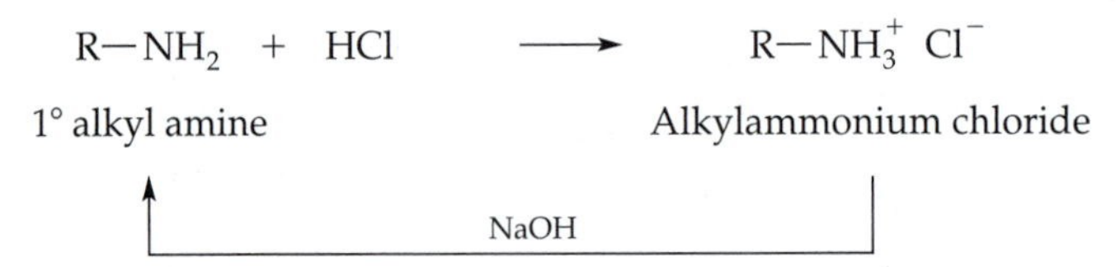

Amines are more basic than water and therefore remove a proton from water, forming an aqueous solution containing the **ammonium cation** and the hydroxide anion. This is similar to ammonia in water forming ammonium hydroxide (NH_4^+ OH^-). An aqueous solution of an amine is basic:

$$R_3N\!: + H{-}OH \rightleftharpoons R_3\overset{+}{N}{-}H + {}^{-}\!:\!OH$$

Amine; Ammonium cation

$$K_b = \frac{[R_3NH^+][{}^{-}OH]}{[R_3N]}$$

The relative base strength of an amine can be determined and is often represented by pK_b, the negative log of the basicity constant, K_b, derived from the above equilibrium. The pK_b values for alkylamines are generally 3–4. However, it is easier to take advantage of the fact that relative base strength can also be determined from the pK_a of the **conjugate acid of the amine**—the ammonium cation—and the knowledge that $pK_a = 14 - pK_b$. The pK_a of an ammonium cation is derived as follows:

$$R_3\overset{+}{N}{-}H + H{-}\ddot{O}{-}H \rightleftharpoons R_3N\!: + H_3O^+$$

Ammonium cation; Amine

$$K_a = \frac{[R_3N][H_3O^+]}{[R_3NH^+]} \qquad pK_a = -\log K_a$$

The more readily an ammonium cation gives up its proton to another base (for example, water), the more acidic it is, the farther the above equilibrium lies to the right, and the less basic is the free amine. Therefore, *smaller* pK_a values are associated with more acidic ammonium salts, which are derived from less basic amines. Put another way, amines that are stronger bases tend to hold on to the proton, making their conjugate acids (ammonium salts) less acidic. The pK_a of the parent ammonium ion is 9.3, that of the methylammonium ion is 10.6, and that of the dimethylammonium ion is 10.7. These data indicate that the basicities of the amines increase from ammonia to methylamine to dimethylamine. The pK_a of the anilinium ion is 4.6, while that of the *p*-chloroanilinium ion is 4.0, indicating that aniline is more basic than *p*-chloroaniline.

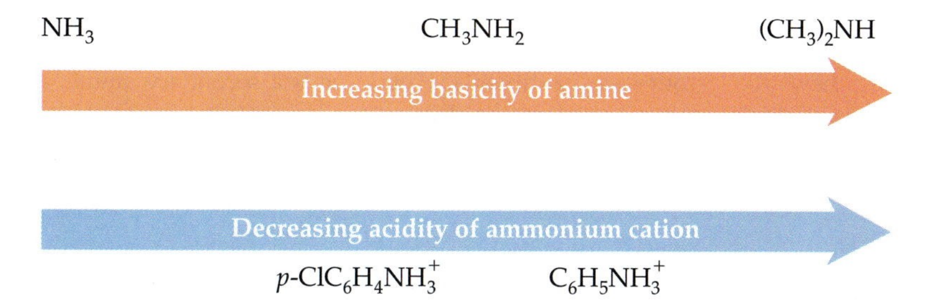

The pattern of relative basicities of amines can be accounted for by considering the electronic effects of the substituents on the nitrogen atom. Recall that alkyl groups more readily share electron density than do hydrogen atoms. (Remember the activating effect of alkyl groups in electrophilic aromatic substitution and the stabilizing effect of alkyl groups on carbocations.) Therefore, an ammonium salt ($R_2NH_2^+$) derived from a secondary amine (with two electron-donating groups) has its positive charge delocalized more effectively than does an ammonium salt (RNH_3^+) from a primary amine (just one electron-donating group). Thus, a secondary amine forms an ammonium salt more readily than a primary amine does, and is therefore more basic. In both instances, the ammonium salt is stabilized relative to the appropriate starting amine, which does not carry a positive charge. Thus, this is generally the *basicity* sequence of alkyl amines (when determined in the gas phase):

Basicity: 3° amine > 2° amine > 1° amine

In the case of aromatic amines, the aromatic ring is relatively electron-withdrawing, delocalizing the lone pair of electrons into the ring (see Section 9.4). Therefore, aromatic amines are generally less basic (their conjugate acids are more acidic and therefore have lower pK_a values) than alkyl amines. Ring substituents that increase the delocalization of the lone pair of electrons on nitrogen have the expected effect of decreasing the basicity of the amine. (Imagine that the electrons on nitrogen, by being more delocalized into the ring, are less available to serve as a proton acceptor.)

The basicity sequence of any set of amines, alkyl or aromatic, relative to ammonia, can be predicted by considering the relative electron-donating and electron-withdrawing effects of the attached groups on nitrogen's unshared electrons: The higher the electron density on nitrogen, the more basic is the amine. Electron-donating groups (such as alkyl groups) increase the basicity of amines, and electron-withdrawing groups decrease the basicity of amines.

To summarize, the general order of basicity for water and amines is

Basicity: alkanamines > arylamines > water

How to Solve a Problem

Which is the stronger base, piperidine or cyclohexylamine? Explain.

N—H　　　　NH_2

Piperidine　　　　Cyclohexylamine

SOLUTION

Both amines have alkyl groups as substituents. Piperidine is a secondary amine, and cyclohexylamine is a primary amine. Therefore, the electron density on nitrogen should be higher in piperidine, making it the more basic of the two. This means that its conjugate acid should be less acidic. The pK_a of piperidine is 11.12, and the pK_a of cyclohexylamine is 10.67.

PROBLEM 12.6

Place each series of amines in order of decreasing basicity:

(a) methylamine, aniline, *p*-methoxyaniline, *p*-nitroaniline

(b) cyclohexylamine, diethylamine

The basicity of amines is frequently the basis for separating them from other organic compounds, especially those found in plant extracts. If a mixture of organic compounds that includes a water-insoluble amine is first dissolved in an organic solvent (for example, dichloromethane) and then mixed with dilute hydrochloric acid, the amine is protonated to form the ionic ammonium salt, which is more soluble in the aqueous phase than in the organic phase. The nonbasic compounds ("neutrals") remain dissolved in the organic solvent phase. Separation of the two phases, followed by reconversion of the ammonium salt to the amine by treatment with dilute sodium hydroxide (a stronger base), permits isolation of the amine (see Figure 12.3). This separation scheme is regularly used to extract complex, naturally occurring amines from plants, often for use as medicinals.

12.3.3 Detection of Amines

Amines can be detected and identified by a number of techniques. The first evidence that a compound of unknown structure is an amine is usually insolubility in water but solubility in dilute hydrochloric acid (as in the case of ammonium salts), indicating that the unknown is a base. (Note that this test is useless for low-molecular-weight amines, which *are* water-soluble.) If fusion of an unknown compound with sodium results in the formation of cyanide ion (^{-}CN), which can be detected, this proves the presence of nitrogen. Quantitative combustion microanalysis can provide the empirical formula. Mass spectrometry can determine the molecular weight and therefore the molecular formula. An odd number for the molecular weight is a likely sign of the presence of nitrogen; remember that compounds with only carbon, hydrogen, oxygen, or an even number of nitrogens always have even-numbered molecular weights.

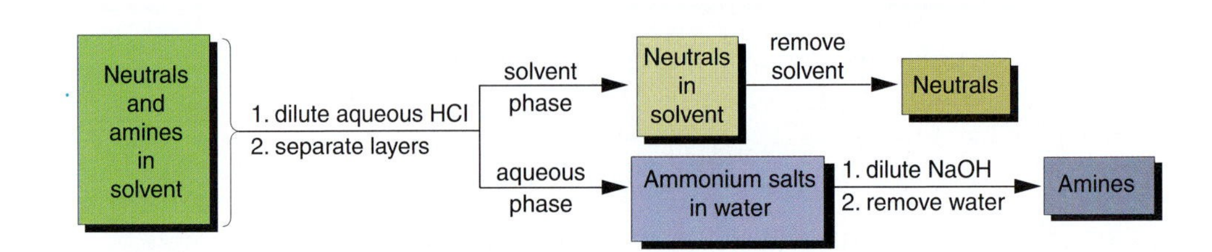

FIGURE 12.3

Separation scheme for the isolation of amines.

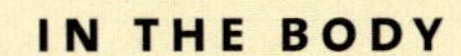

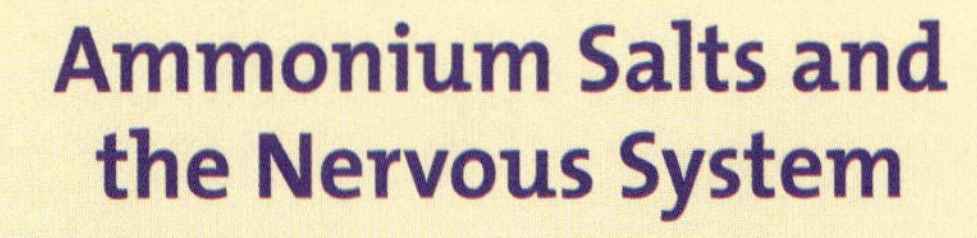

CHEMISTRY AT WORK

IN THE BODY

Ammonium Salts and the Nervous System

Nerve impulses travel very quickly through the human nervous system. If you accidentally touch a very hot object, such as a pie pan right out of the oven, your hand recoils even before your brain registers the message "Hey, this is hot!" What is also interesting about the nervous system is that after it sends a signal to the muscles or brain, it is ready to send the next signal right away. How does that happen? The answer involves biologically active ammonium salts.

Nerve impulses travel from one nerve cell to the next across a space called a *synapse.* What actually moves across the synapse is an ammonium cation called acetylcholine. It travels from the ending of one nerve cell to a receptor site on the next cell, having the same effect as the movement of electrical charge. In motor nerves, the electrical impulse eventually stimulates contraction of the muscle. When acetylcholine reaches a receptor site, it is hydrolyzed to choline by an enzyme called acetylcholine esterase:

$$\underset{\text{Acetylcholine}}{(CH_3)_3\overset{+}{N}CH_2CH_2OAc} \xrightarrow{\text{acetylcholine esterase}} \underset{\text{Choline}}{(CH_3)_3\overset{+}{N}CH_2CH_2OH}$$

The choline is later reesterified to acetylcholine by a compound called acetyl coenzyme A (Sections 15.1.3 and 19.2.1). It is then ready to transmit another nerve impulse. If this cyclic chemical process is interrupted or blocked, the muscles cannot contract. Depending on the severity of the interruption, the result can be muscle relaxation, paralysis, or even death.

Certain synthetic quaternary ammonium salts, such as decamethonium bromide and succinylcholine bromide, have been developed for use as muscle relaxants during surgery. Because they are organic ammonium salts, these compounds are attracted to the active enzyme sites where acetylcholine normally undergoes hydrolysis. Their presence effectively blocks the function of acetylcholine esterase. Therefore, the motor nerve action is neutralized.

Another ammonium salt with powerful effects on the human nervous system is tubocurarine, the active ingredient in the poison known as *curare.* Curare occurs naturally in certain plants and is used by some South American tribes on the tips of arrows to paralyze prey. It works by the same chemical mechanism as the muscle relaxants just described and, in fact, was used in surgery at one time for the same purpose. However, its use has been generally discontinued because an excess can paralyze the respiratory muscles, killing the patient.

$$(CH_3)_3\overset{+}{N}-(CH_2)_{10}-\overset{+}{N}(CH_3)_3 \quad 2\,Br^-$$

Decamethonium bromide

$$(CH_3)_3\overset{+}{N}-(CH_2)_2-O-\overset{O}{\overset{\|}{C}}-(CH_2)_2-\overset{O}{\overset{\|}{C}}-O-(CH_2)_2-\overset{+}{N}(CH_3)_3 \quad 2\,Br^-$$

Succinylcholine bromide

Tubocurarine chloride

The spectroscopic methods described in Chapter 11 also provide evidence regarding the structure of an amine (Figure 12.4). The N—H stretching frequency shows as a strong-to-medium peak at 3300–3500 cm^{-1} in the IR spectrum, with a primary amine showing two peaks and a secondary amine showing a single peak. A ^1H-NMR spectrum also shows the proton(s) bonded to nitrogen but over quite a wide range—at a

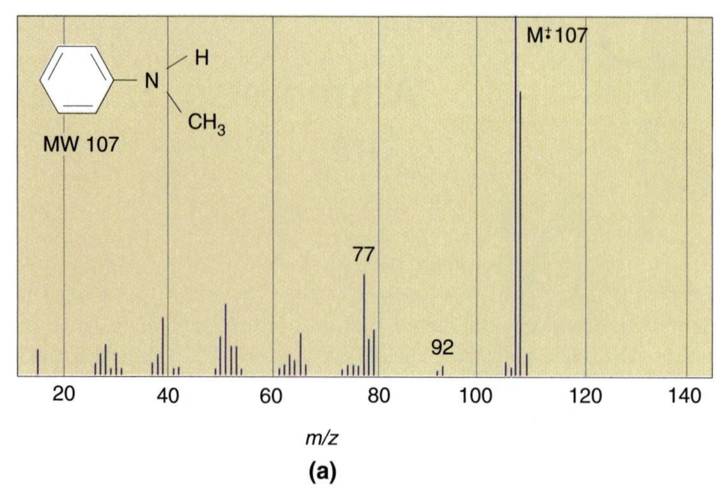

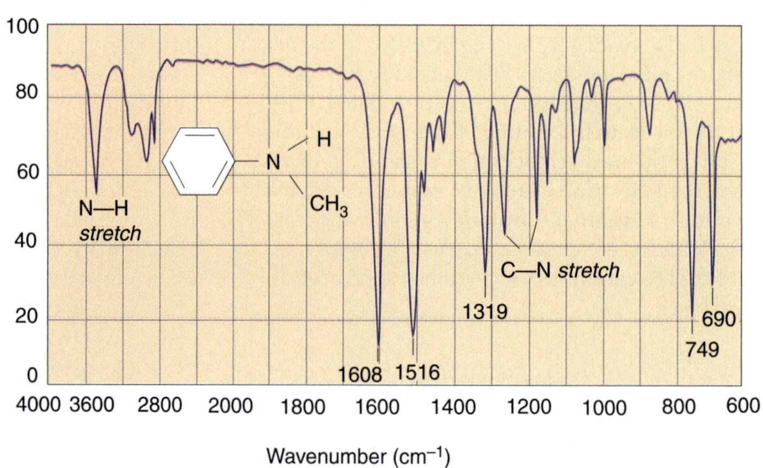

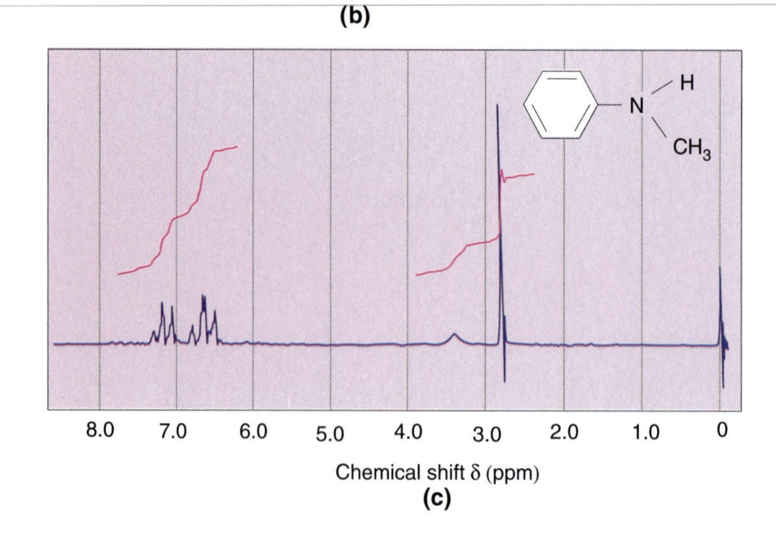

FIGURE 12.4

(a) Mass spectrum, (b) IR spectrum, and (c) ^{1}H-NMR spectrum of *N*-methylaniline.

chemical shift of 0.6–3.0 ppm. The N—H peak disappears when the sample is shaken with D_2O (heavy water) because N—H is replaced by N—D, and deuterium (hydrogen that has one proton and one neutron) does not show up on ^{1}H-NMR spectra. Thus, NMR spectroscopy is not useful for initial proof of the presence of an amine, but is essential for confirmation of tentative structures.

PROBLEM 12.7

An odiferous liquid compound has a molecular ion at $m/z = 73$ on its mass spectrum with a very small $M^{\dot{+}} + 1$ peak and peaks at $m/z = 44$ and 58. IR spectroscopy shows no functional groups present. The decoupled ^{13}C-NMR spectrum shows three peaks. The ^{1}H-NMR spectrum shows peaks at 1.0 ppm (t, 3H), 2.1 ppm (s, 6H), and 2.3 ppm (q, 2H). What is the compound's structure?

12.4 Amines as Nucleophiles

The major chemical properties of amines arise from the presence of the unshared electron pair on nitrogen, making amines effective as nucleophiles. We will look at two kinds of reactions. First, amines can serve as nucleophiles in substitution reactions, such as alkylation and acylation. The second kind of reaction involves the replacement of aromatic primary amine groups by other substituents (see Section 12.5). This complements the electrophilic aromatic substitution reaction by which substituted benzenes are synthesized.

12.4.1 Alkylation of Amines

Alkylation replaces a hydrogen atom on the nitrogen of an amine (or ammonia) with an alkyl group. Reaction of an amine (the nucleophile), such as ethylamine, with an alkyl iodide susceptible to S_N2 substitution, such as methyl iodide, results in a typical S_N2 reaction to produce a more highly substituted amine, such as ethylmethylamine:

$$C_2H_5-\ddot{N}H_2 + CH_3-I \longrightarrow C_2H_5-\overset{+}{N}H_2-CH_3\ I^- \xrightarrow{:Base} C_2H_5-\ddot{N}H-CH_3$$

Ethylamine Methyl iodide Ethylmethylammonium iodide Ethylmethylamine

Because the initially formed species is an ammonium salt, which is acidic, a base is needed to abstract a proton and release the free amine. The base can be excess ethylamine, added sodium hydroxide, or added ammonia. In this example, a primary amine, ethylamine, has been converted to a secondary amine, ethylmethylamine.

Using this alkylation reaction, it is possible, in principle, to convert ammonia to a primary amine, a primary amine to a secondary amine, a secondary amine to a tertiary amine, and a tertiary amine to a quaternary ammonium salt. In the first three cases, a nitrogen-hydrogen bond is replaced by a nitrogen-alkyl bond. In practice, a reaction to produce a secondary amine also produces a tertiary amine and a quaternary ammonium salt as by-products—these result from subsequent alkylations that occur uncontrollably.

The alkylation reaction is practical in only two situations. First, it can be used to form primary amines from alkyl halides and ammonia—the alkylation of ammonia. This reaction is useful because it is economical to use a large excess of ammonia as the nucleophile.

General reaction:

$$\underset{\text{Alkyl bromide}}{R{-}Br} + \underset{\text{Ammonia (excess)}}{NH_3} \longrightarrow \underset{\text{Alkylammonium bromide}}{\left[R{-}\overset{+}{N}H_3\ Br^-\right]} \xrightarrow{NH_3} \underset{1^\circ\text{ amine}}{R{-}NH_2} + \underset{\text{Ammonium bromide}}{NH_4Br}$$

The presence of excess ammonia minimizes the side reaction involving the newly formed primary amine, which is more nucleophilic than the starting ammonia. Ordinarily, the primary amine would serve as a nucleophile in a subsequent S_N2 reaction, reacting with more starting alkyl halide and producing secondary and tertiary amines. The alkylation reaction with ammonia, therefore, offers an effective means of synthesizing primary amines from any primary or secondary alkyl halide. (Remember that most tertiary alkyl halides undergo elimination rather than substitution with many nucleophiles, including some amines.) Refer to the Key to Transformations (on the inside front cover) for this conversion of an alkyl halide to an alkyl amine.

The ammonia alkylation reaction is important in the laboratory and in the commercial synthesis of α-amino acids (such as phenylalanine) from readily available α-haloacids. α-Amino acids are the constituents of proteins, and some are used as supplements in animal diets.

$$\underset{\text{2-Bromo-3-phenylpropanoic acid}}{C_6H_5{-}CH_2{-}\underset{\displaystyle Br}{\underset{|}{C}H}{-}COOH} \xrightarrow{\text{excess } NH_3} \underset{\text{Phenylalanine (2-amino-3-phenylpropanoic acid)}}{C_6H_5{-}CH_2{-}\underset{\displaystyle NH_2}{\underset{|}{C}H}{-}COOH} + NH_4Br$$

The second situation where alkylation is useful is in the conversion of tertiary amines to quaternary ammonium salts. The tertiary amine serves as the nucleophile in an S_N2 reaction, usually with a primary halide, frequently methyl iodide:

$$\underset{N,N\text{-Dimethylaniline}}{C_6H_5{-}\ddot{N}(CH_3)_2} + \underset{\text{Methyl iodide}}{CH_3{-}I} \longrightarrow \underset{N,N,N\text{-Trimethylanilinium iodide}}{C_6H_5{-}\overset{+}{N}(CH_3)_3\ I^-}$$

The formation of the usually crystalline quaternary ammonium salts is a means of characterizing the often liquid amines.

PROBLEM 12.8

Show the major product of each of the following alkylation reactions:

(a) aniline with excess ethyl iodide

(b) 2-bromobutane with excess ammonia

(c) isopropyl bromide with dimethylamine

PROBLEM 12.9

What nonnitrogenous starting material would you employ to produce each of the following compounds? (Use any other reagents necessary.)

(a) cyclohexylamine (b) cyclopropyltrimethylammonium iodide

12.4.2 Acylation of Amines

Acylation of an amine involves the replacement of a hydrogen on the nitrogen atom with an **acyl group,** R—CO— (see Section 15.4), to form compounds called **amides** (R—CO—NH_2). Amides are important commercial and biological compounds. They can be viewed as acyl derivatives of amines, in which the acyl group replaces a hydrogen, or as amine derivatives of carboxylic acids, in which the amino group replaces the —OH of a carboxylic acid (R—CO—OH).

The reagent used to acylate an amine is usually an acyl chloride (also called an acid chloride), which is very reactive because the chloride ion is an effective leaving group. Amines are effective nucleophiles that readily attack the carbonyl carbon, leading to substitution of the chlorine by the amine.

General reaction:

$$\underset{\text{Acyl chloride}}{R{-}CO{-}Cl} + \underset{\text{Alkylamine}}{R'NH_2} \xrightarrow{\text{pyridine}} \underset{\text{Amide}}{R{-}CO{-}NHR'} + \underset{\text{Pyridine hydrochloride}}{\text{pyridine} \cdot HCl}$$

The reaction occurs spontaneously at room temperature, and a base such as pyridine is added simply to react with the HCl released in the reaction. Although the reaction appears to be an S_N2 substitution, it is not, as Chapter 15 will discuss in detail. The initial result of the attack by nucleophilic nitrogen is the formation of an intermediate quaternary ammonium salt, from which a proton is readily abstracted either by excess starting amine or by another added base (such as sodium hydroxide or pyridine):

$$\underset{1^\circ\text{ amine}}{R{-}\ddot{N}H_2} + \underset{\text{Acetyl chloride}}{CH_3{-}C({=}O){-}Cl} \xrightarrow{\text{(two steps)}} \left[R{-}\overset{+}{N}H_2{-}C({=}O){-}CH_3 \; Cl^-\right] \xrightarrow{:\text{Base}} \underset{\text{Amide}}{R{-}\ddot{N}H{-}C({=}O){-}CH_3}$$

Based on the mechanism of the reaction, it is clear that amides can be formed only from primary and secondary amines, the only amines that contain at least one hydrogen atom on nitrogen that can be removed in the final step. Amides can be produced from almost any primary or secondary amine and from almost any acyl chloride.

$$\underset{1^\circ\text{ amine}}{R{-}NH_2} + \underset{\text{Acyl chloride}}{R''COCl} \xrightarrow{\text{pyridine}} \underset{\text{Amide}}{R{-}NH{-}CO{-}R''}$$

$$\underset{2^\circ\text{ amine}}{RR'NH} + \underset{\text{Acyl chloride}}{R''COCl} \xrightarrow{\text{pyridine}} \underset{\text{Amide}}{RR'N{-}CO{-}R''}$$

In contrast to amines, which are basic and therefore soluble in dilute acid, amides are neutral and therefore insoluble in dilute acid. Although the nitrogen in an amide still carries a lone pair of electrons, the strong electron-withdrawing effect of the acyl group delocalizes the lone pair toward the acyl group, rendering it unavailable for protonation (see Chapter 15 for details). Two resonance contributing structures represent the actual structure of an amide:

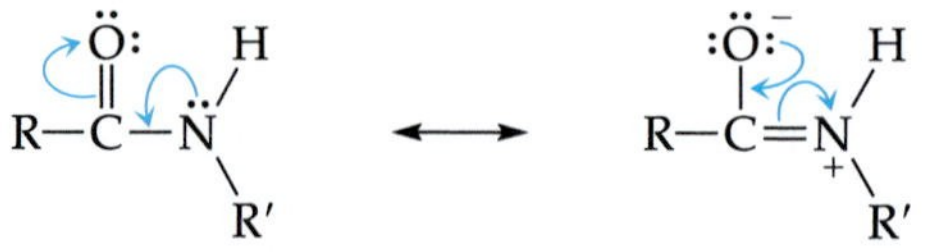

$$R-\overset{\overset{\ddot{O}:}{\|}}{C}-\ddot{N}(H)R' \longleftrightarrow R-\overset{\overset{:\ddot{O}:^-}{|}}{C}=\overset{+}{N}(H)R'$$

Contributing structures for resonance stabilization of an amide

How to Solve a Problem

Show the product of the reaction of benzoyl chloride with aniline in the presence of dilute sodium hydroxide:

$$\underset{\text{Benzoyl chloride}}{C_6H_5-\overset{\overset{O}{\|}}{C}-Cl} + \underset{\text{Aniline}}{C_6H_5-NH_2} \xrightarrow{NaOH} ?$$

PROBLEM ANALYSIS

First, we review the characteristics of each reactant. Benzoyl chloride is an acid chloride, and we know that halides usually function as leaving groups. Aniline is an amine with an unshared pair of electrons; it can function as a nucleophile. Therefore, we have the ingredients for a substitution reaction: a nucleophile and a leaving group.

SOLUTION

The nitrogen of aniline attacks the carbonyl group (C=O) of benzoyl chloride, displacing the chloride ion. The intermediate ammonium salt loses its proton to hydroxide anion. The product is benzanilide, an amide.

$$\underset{\text{Benzoyl chloride}}{C_6H_5-\overset{\overset{O}{\|}}{C}-Cl} + \underset{\text{Aniline}}{C_6H_5-NH_2} \xrightarrow{NaOH} \underset{\text{Benzanilide}}{C_6H_5-\overset{\overset{O}{\|}}{C}-NH-C_6H_5}$$

PROBLEM 12.10

Show the product(s) of the following reactions: (a) diphenylamine with CH_3COCl, (b) cyclopentanamine with C_6H_5COCl, and (c) 1-propanamine with C_2H_5COCl.

PROBLEM 12.11

From what starting materials would you produce each of the following compounds?

(a) $C_6H_5CONHC_6H_5$

(b) 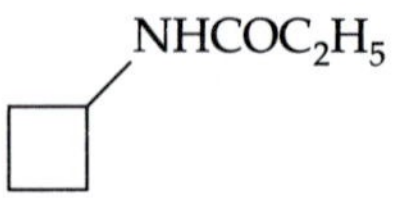 cyclobutane ring bearing $NHCOC_2H_5$

(c) 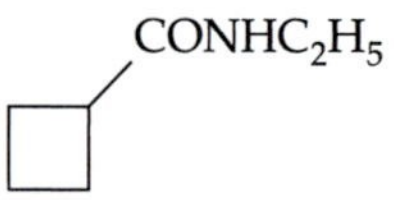 cyclobutane ring bearing $CONHC_2H_5$

(d) $CH_3NHCOCH(CH_3)_2$

CHEMISTRY AT WORK

IN THE MARKETPLACE AND IN THE BODY

Amides in Insect Repellents, Proteins, and Nylon

The amide functional group is very common in commercial products. It is found in things as different as insect repellents, proteins, and nylon stockings.

The active ingredient of the insect repellent sold as Off is an amide, *N,N*-diethyl-*m*-toluamide. It is formed from the acylation of diethylamine with *m*-toluyl chloride:

$$m\text{-}CH_3C_6H_4-\overset{O}{\overset{\|}{C}}-Cl + (CH_3CH_2)_2NH \xrightarrow{NaOH} m\text{-}CH_3C_6H_4-\overset{O}{\overset{\|}{C}}-N(CH_2CH_3)_2$$

m-Toluyl chloride; Diethylamine; *N,N*-Diethyl-*m*-toluamide

The amide bond between an acyl group and an amine (—CO—NH—) is the key linkage in all proteins. It is called a **peptide bond** in biological chemistry. Proteins are polyamides—that is, long-chain compounds containing many amide bonds. A large number of α-amino acids are linked together by peptide bonds, with the amino group of one molecule joining to the acyl group of another molecule.

$$HO-\overset{O}{\overset{\|}{C}}-C(H)(R)-NH_2$$

α-Amino acid

$$\sim\overset{O}{\overset{\|}{C}}-C(H)(R)-\underset{H}{\underset{|}{N}}-\overset{O}{\overset{\|}{C}}-C(H)(R')-\underset{H}{\underset{|}{N}}-\overset{O}{\overset{\|}{C}}-C(H)(R'')-\underset{H}{\underset{|}{N}}\sim$$

A fragment of a protein chain showing three different α-amino acids joined by two peptide bonds, which are colored

The number of α-amino acids in a protein varies from a few (in which case the protein is often called a polypeptide) to many thousands. A total of 22 different α-amino acids are found in proteins, and organisms synthesize all proteins in their cells from α-amino acids. Chemists have learned how to synthesize a large number of proteins in the laboratory; we will discuss protein chemistry in Chapter 19.

Nylon is the material used in parachutes. In World War II, production of nylon was diverted from making nylon stockings to making nylon parachutes.

Closely related to proteins are a group of synthetic polyamides called *nylons.* The first nylon was prepared in 1935 by Wallace Carothers, a chemist employed by the DuPont Company. Different members of the nylon family are used in a wide variety of products, ranging from nylon stockings (all the rage at their introduction at the 1939 World's Fair in New York) and parachutes to automobile tires and bulletproof vests. Nylon-6 may be considered as being derived from the polymerization (polyamide formation reaction) of ϵ-aminocaproic acid. These materials and reactions will be covered in detail in Chapter 18.

$$n\ H_2N-(CH_2)_5-\overset{O}{\overset{\|}{C}}-OH \xrightarrow{-H_2O} \left[NH-(CH_2)_5-\overset{O}{\overset{\|}{C}}\right]_n$$

ϵ-Aminocaproic acid; Nylon-6

12.5

Diazotization of Primary Aromatic Amines

12.5.1 Formation of the Diazonium Ion

The reaction of aromatic primary amines, such as aniline, with nitrous acid is important for commercial and laboratory syntheses. The process is called **diazotization,** and it produces a **diazonium salt,** which is not isolated but is instead utilized immediately in solution. Aryl diazonium salts are important synthetic intermediates whose diazonium group ($—N_2^+$) can (1) be converted into other functional groups and (2) serve as weak electrophiles that react with reactive aromatic compounds to form *azo compounds.*

In a diazotization reaction, the primary aromatic amine serves as a nucleophile that reacts with a nitrosonium ion ($^+N{=}O$). The nitrosonium ion is formed in aqueous solution from hydrochloric acid and sodium nitrite:

$$\underset{\text{Sodium nitrite}}{NaNO_2} + HCl \longrightarrow \underset{\text{Nitrous acid}}{H—O—N{=}O} + NaCl$$

$$\underset{\text{Nitrous acid}}{H—\ddot{O}—\ddot{N}{=}\ddot{O}} + H^+ \longrightarrow \left[H—\overset{+}{\underset{|}{\overset{H}{O}}}—\ddot{N}{=}\ddot{O}\right] \xrightarrow{-H_2\ddot{O}:} \underset{\text{Nitrosonium ion}}{\left[{}^+\ddot{N}{=}\ddot{O}\right]}$$

The mechanism of diazonium salt formation involves attack of a nucleophilic aromatic primary amine on the electrophilic nitrosonium ion, forming an *N*-nitrosamine. The *N*-nitrosamine is not isolated; instead it undergoes protonation in the acidic solution to eventually form an aromatic diazonium salt ($Ar—N_2^+ Cl^-$). Note that both original hydrogens of the amine are lost; therefore, only primary amines can be diazotized.

$$\underset{\text{Aromatic amine}}{Ar—\ddot{N}H_2} + \left[{}^+\ddot{N}{=}\ddot{O}\right] \longrightarrow \left[Ar—\overset{+}{N}H_2—\ddot{N}{=}\ddot{O}\right] \xrightarrow{H\ddot{O}H} Ar—\ddot{N}(H)—\ddot{N}{=}\ddot{O}$$

$$\xrightarrow[(1)\,H^+]{(2)\,H_2\ddot{O}:} Ar—\ddot{N}{=}\ddot{N}—\ddot{O}—H \xrightarrow{H^+} \left[Ar—\ddot{N}{=}\ddot{N}—\overset{+}{O}H_2\right] \xrightarrow{-H_2O} Ar—\ddot{N}{=}\ddot{N}^+ \longleftrightarrow \underset{\text{Aromatic diazonium ion}}{Ar—\overset{+}{N}{\equiv}\ddot{N}}$$

The diazonium ion is stabilized by resonance. Aromatic diazonium salts are stable in ice-cold aqueous solution, but evolve nitrogen gas upon warming. The same reaction is possible with primary aliphatic amines; but the resulting diazonium ions are much less stable, spontaneously losing nitrogen and forming carbocations that in turn form the products expected from nucleophilic substitution and elimination (see Chapter 3). This is, by the way, an effective means of selectively forming a specific alkyl carbocation.

$$\underset{\text{1° alkylamine}}{R{-}NH_2} \xrightarrow{HNO_2} \underset{\text{Diazonium salt}}{\left[R{-}\overset{+}{N}_2\ Cl^-\right]} \xrightarrow{-N_2} \underset{\text{Carbocation}}{\left[R^+\right]}$$

PROBLEM 12.12

Show the reaction and the diazonium ion that will result from diazotization of each amine: (a) aniline and (b) *p*-nitroaniline.

PROBLEM 12.13

Although diazotization of primary alkylamines is not described in detail, use your knowledge of carbocations to work out the diazotization reaction of 2-propanamine in aqueous solution and show the expected products.

12.5.2 Displacement of Nitrogen

One important feature of aromatic diazonium ions is that molecular nitrogen (N_2) is readily displaced from such ions by an appropriate nucleophile. (Note that there is always a strong driving force toward the formation of molecular nitrogen, whose stability is indicated by its persistence in earth's atmosphere.) Treatment of an aryldiazonium ion with any of the following reagents results in replacement of the nitrogen molecule with a nucleophile:

$$\underset{\text{Aryldiazonium ion}}{Ar{-}\overset{+}{N}{\equiv}\ddot{N}} \longrightarrow \begin{cases} \xrightarrow{CuCl} Ar{-}Cl \\ \xrightarrow{CuBr} Ar{-}Br \\ \xrightarrow{KI} Ar{-}I \\ \xrightarrow{HBF_4} Ar{-}F \\ \xrightarrow{CuCN} Ar{-}CN \\ \xrightarrow{HOH} Ar{-}OH \\ \xrightarrow{H_3PO_2} Ar{-}H \end{cases}$$

The significance of these reactions is that they permit the attachment of a variety of substituents to an aromatic ring, some of which could not be attached via direct electrophilic aromatic substitution. (Chapter 9 discusses direct attachment of only Br, Cl, NO_2, SO_3H, acyl (R—CO—), and alkyl groups to aromatic rings.) The nucleophile becomes attached to the ring at the same position from which the nitrogen left. Therefore, aromatic primary amines are important intermediary compounds in the preparation of substituted aromatics. This is the only efficient method of placing some of these substituents (I, F, OH, and CN) on aromatic rings.

The overall process for bringing about substitution on an aromatic ring using diazotization involves four separate steps: (1) nitration of the ring (see Section 9.3.3), followed by (2) reduction of the nitro group to an amino group (see Section 12.6.3, to follow), then (3) diazotization of the amino group, and finally (4) replacement of nitrogen.

Although four steps are involved, each is effective and efficient, and so the overall yield of product can be quite good.

$$\text{Ar—H} \xrightarrow[\text{(nitration)}]{\text{1. HNO}_3\text{, H}_2\text{SO}_4} \text{ArNO}_2 \xrightarrow[\text{(reduction)}]{\text{2. Fe/HCl}} \text{ArNH}_2 \xrightarrow[\text{4. [X}^-\text{], –N}_2\text{ (replacement)}]{\text{3. NaNO}_2\text{/HCl (diazotization)}} \text{Ar—X}$$

The use of diazotization, together with what you learned in Chapter 9, allows a greater range of synthetic sequences involving aromatic compounds. In tackling such syntheses, it is essential to keep in mind that a nitro group is easily attached to a ring and can be replaced by a wide variety of substituents.

How to Solve a Problem

(a) Synthesize *p*-iodonitrobenzene from benzene.

Benzene ------→ I—C₆H₄—NO₂

Benzene   *p*-Iodonitrobenzene

PROBLEM ANALYSIS

Two substituents are to be attached to the ring. Which should be attached first? We remember from Chapter 9 that the nitro group is a *meta* director. If it is attached first, the second substituent will be directed to the *meta* position, rather than to the desired *para* position. On the other hand, iodine is an *ortho-para* director, so if it is attached first, the nitro group will end up in the desired *para* position. The obvious conclusion is to attach the iodine atom first and the nitro group second.

How can each substituent be attached? There is no way to directly attach an iodine atom to an aromatic ring. Thus, there is only one possible approach: Attach another group, then replace it with iodine. That thinking should lead almost automatically to four-step diazotization as the best route, starting with direct nitration of benzene. After formation of iodobenzene, direct nitration is needed to attach the nitro group in the final step.

SOLUTION

$$\text{Benzene} \xrightarrow[\text{H}_2\text{SO}_4]{\text{HNO}_3} \text{Nitrobenzene} \xrightarrow{\text{Fe/HCl}} \text{Aniline} \xrightarrow{\text{1. NaNO}_2\text{, HCl; 2. KI}} \text{Iodobenzene} \xrightarrow[\text{H}_2\text{SO}_4]{\text{HNO}_3} p\text{-Iodonitrobenzene}$$

Benzene   Nitrobenzene   Aniline

p-Iodonitrobenzene   Iodobenzene

(b) Synthesize *p*-cresol (*p*-hydroxytoluene) from toluene.

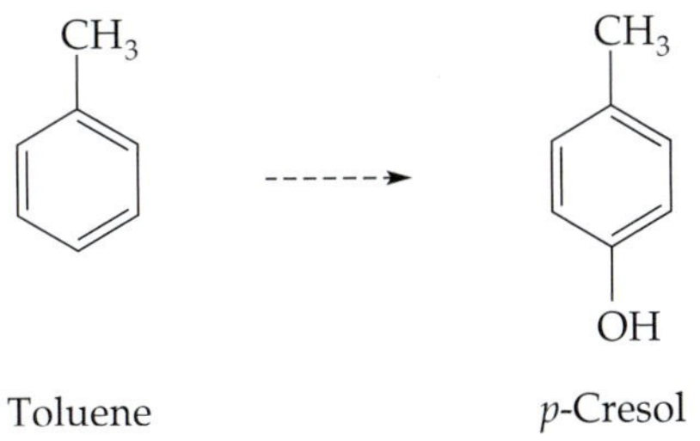

CH_3 ⟶ CH_3 / OH

Toluene *p*-Cresol

PROBLEM ANALYSIS

The key problem is how to attach an —OH group to a substituted aromatic ring and in the *para* position. There is no means to do so via direct electrophilic aromatic substitution, as discussed in Section 9.5. Therefore, the —OH group must replace another substituent. Once again, the idea of replacement should lead to thinking of the diazotization reaction, whereby nitrogen can be displaced by warm water and lead to attachment of the hydroxyl group. The initial step is direct nitration, which occurs *para* to the methyl group in toluene.

SOLUTION

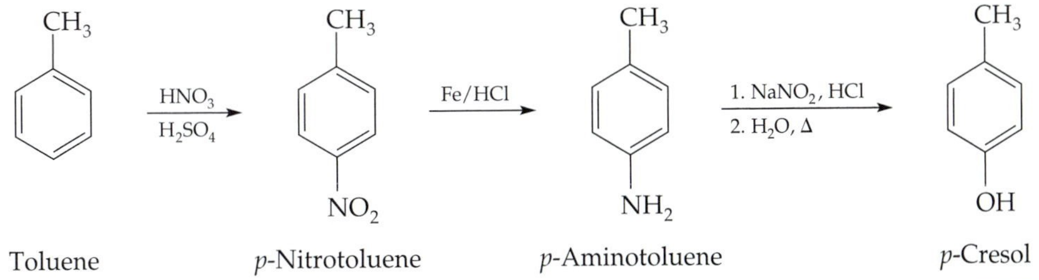

CH_3 —(HNO_3 / H_2SO_4)→ CH_3 / NO_2 —(Fe/HCl)→ CH_3 / NH_2 —(1. $NaNO_2$, HCl; 2. H_2O, Δ)→ CH_3 / OH

Toluene *p*-Nitrotoluene *p*-Aminotoluene *p*-Cresol

PROBLEM 12.14

Use the diazotization reaction to accomplish each of the following transformations:

(a) aniline to phenol

(b) benzene to fluorobenzene

(c) ethylbenzene to *p*-ethyliodobenzene

(d) acetophenone to *m*-cyanoethylbenzene

The reduction of a diazonium ion with hypophosphorous acid (H_3PO_2)—that is, the replacement of molecular nitrogen by hydrogen—has a special application. It effectively removes a nitro or amino substituent from an aromatic ring. But why remove a substituent when most problems involve attaching substituents? The answer is that it is frequently necessary to attach a substituent *temporarily*, usually for the purpose of exerting a directional effect on a subsequent electrophilic aromatic substitution reaction. Once that substitution is made, the substituent needs to be removed.

The overall course of such a reaction is to nitrate a ring, reduce the nitro group to an amino group, diazotize the amino group, and then replace it with hydrogen:

$$\text{Benzene} \xrightarrow[H_2SO_4]{HNO_3} \text{Nitrobenzene } (C_6H_5NO_2) \xrightarrow{Fe/HCl} \text{Aniline } (C_6H_5NH_2) \xrightarrow[HCl]{NaNO_2} \text{Benzenediazonium salt } (C_6H_5N_2^+) \xrightarrow{H_3PO_2} \text{Benzene}$$

This sequence permits the exertion of a *meta*-directing effect via the nitro group or the exertion of an *ortho-para*–directing effect via the amino group. Then the group can be removed by the diazotization sequence, followed by treatment with hypophosphorous acid.

How to Solve a Problem

Convert toluene to *m*-bromotoluene.

$$\text{Toluene } (C_6H_5CH_3) \dashrightarrow m\text{-Bromotoluene } (CH_3, Br)$$

PROBLEM ANALYSIS

The key problem is that a new substituent must be attached to the aromatic ring in a position *meta* to the methyl group, but we know that the methyl group is *ortho-para* directing. Therefore, we cannot use direct bromination of toluene. To get the bromine atom *meta* to the methyl group, we have to introduce another group to direct it to that position. This problem has an almost universal solution: temporarily placing in the *para* position a group that will direct an incoming bromine atom *ortho* to itself, a position which is also *meta* to the methyl group.

The amino group is such an *ortho-para* director. Recall that if there are two substituents on a benzene ring, the more powerful *ortho-para* director dictates the position of the next substitution. In this case, the amino group is more powerful than the methyl group. Note also that the amino group is such a strong activator that its effect must be moderated by temporarily converting it to an amide before bromination. The acetyl group must be removed by sodium hydroxide hydrolysis after bromination; otherwise, dibromination will result.

CH_3 CH_3 / $NHCOCH_3$

Positions of attachment of incoming electrophile

SOLUTION

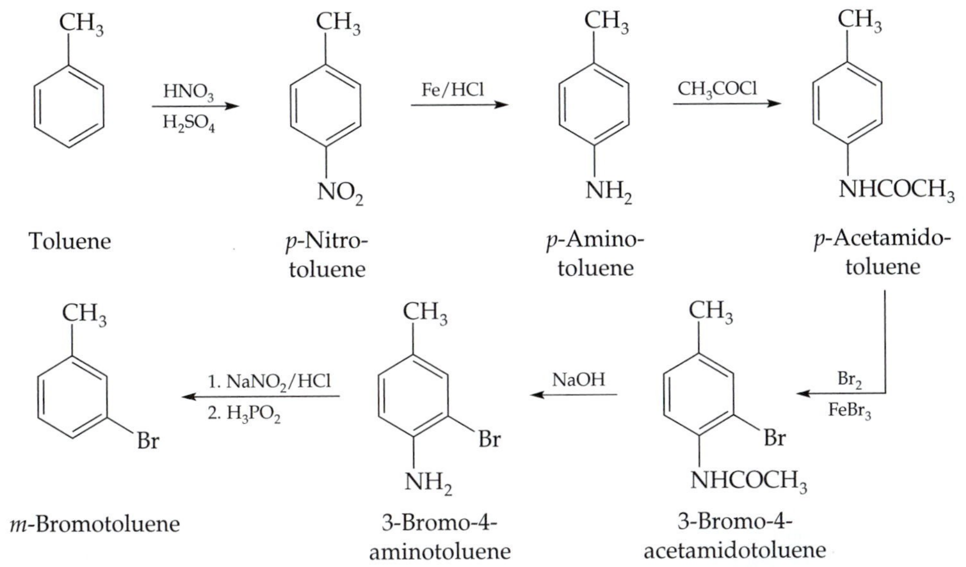

Toluene —(HNO_3, H_2SO_4)→ *p*-Nitrotoluene —(Fe/HCl)→ *p*-Aminotoluene —(CH_3COCl)→ *p*-Acetamidotoluene —(Br_2, $FeBr_3$)→ 3-Bromo-4-acetamidotoluene —(NaOH)→ 3-Bromo-4-aminotoluene —(1. $NaNO_2$/HCl; 2. H_3PO_2)→ *m*-Bromotoluene

Although this is a lengthy synthesis (having six discrete steps), conceptually it involves only three transformations: attaching the new directing group, carrying out the bromination, and removing the directing group.

PROBLEM 12.15

Show the reactions necessary to convert isopropylbenzene to 3,5-dichloroisopropylbenzene.

12.5.3 The Coupling Reaction of Diazonium Ions

The second major use of diazonium ions is in the preparation of dyes for a wide variety of commercial applications. This use depends on the fact that the diazonium ion itself is an electrophile. Thus, in reaction with an aromatic ring system, the diazonium ion brings about an electrophilic aromatic substitution on the ring to form an **azo compound** (a compound with the —N=N— grouping), effectively **coupling** the diazonium ion to the other aromatic ring. Note that nitrogen is not lost in the coupling reaction, as it is in the displacement reactions discussed in Section 12.5.2.

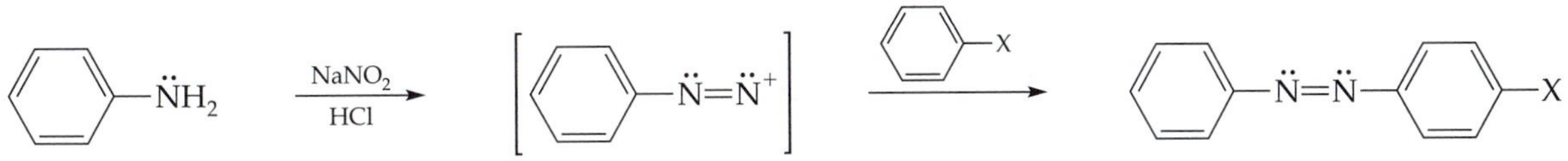

Aniline —($NaNO_2$, HCl)→ [Diazonium ion, C_6H_5—N≡N$^+$] —(C_6H_5—X)→ Azo compound (C_6H_5—N=N—C_6H_4—X)

The coupling reaction is limited by the fact that the diazonium ion is a weak electrophile and therefore reacts only with highly reactive aromatic rings. In practice, this means that the ring being attacked must usually carry a hydroxyl group (be a phenol), an alkoxy group (—OR), or a dialkylamino group [—N(CH_3)$_2$]. The coupling usually occurs only at the *para* position, if it is available; because of the large size of the attacking electrophile, a steric effect blocks the *ortho* position.

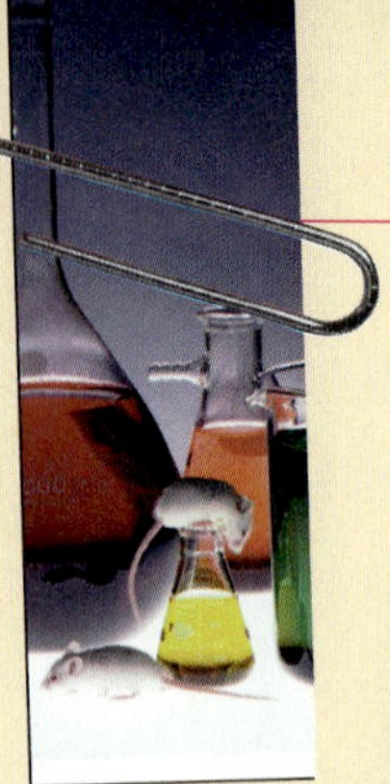

CHEMISTRY AT WORK

IN THE MARKETPLACE

Photoresist Technology

Electronic devices affect every aspect of our existence. Communications, transportation, business, education—everything would be different without computers and solid-state electronics. The key development was the introduction of integrated circuits, which contain thousands of transistors and circuit connections on tiny semiconductor crystals (silicon wafers). The most common way of producing these chips uses a chemical process called *photoresist technology,* which is based on the principles of diazo copying, a process first developed using diazonium salts.

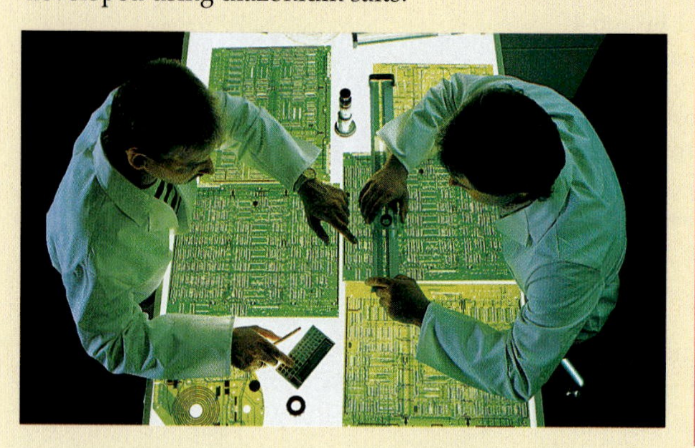

Diazo copying has been used for many years to prepare copies of engineering and architectural drawings. The original drawing is made on clear plastic, such as cellophane or sheets of polyethylene. The copy paper is soaked in a three-part solution of a diazonium salt ($ArN_2^+ \ Cl^-$), 2-naphthol, and an "acid stabilizer" such as *p*-toluenesulfonic acid, which keeps the naphthol (a phenol) in protonated form. Then the original is placed over the copy paper and exposed to ultraviolet light. The UV light passes through the original, exposing the copy paper everywhere except where there is a line in the original drawing. The UV light causes the diazonium salt absorbed on the copy paper to decompose to nitrogen and the aryl chloride. Thus, the exposed copy paper contains a *latent image* (it cannot be seen) of the original, consisting of unchanged diazonium salt.

The latent image on the copy paper is developed by placing the paper in an ammonia solution. This neutralizes the acid stabilizer and removes the proton from the 2-naphthol, producing 2-naphthoxide anion. This anion is susceptible to rapid electrophilic attack by the weakly nucleophilic diazonium ion remaining in the latent image. Thus, everywhere there is a line on the original drawing, a line of azo dye appears on the copy paper. In effect, the original line drawing serves as a "mask"

Compounds containing the azo group are usually highly colored because of the extended conjugated system in their molecules. They are frequently called **azo dyes.**

$$C_6H_5-N{=}N-C_6H_4-\ddot{O}-H \longleftrightarrow {}^{-}C_6H_5{=}N-N{=}C_6H_4{=}\overset{+}{O}-H$$

Contributing structures for resonance stabilization of *p*-hydroxyazobenzene, an azo dye

This extensive delocalization permits the absorption of electromagnetic radiation in the visible region of the spectrum, making the compound appear colored to human eyes (see Section 11.4.2). Changing the nature of other substituents on the aromatic ring, and thereby affecting the extent of conjugation, will alter the frequency of radiation absorbed. Thus, the color can vary from yellow all the way to blue.

Attachment of a sulfonic acid group ($—SO_3H$) to the ring of an azo dye followed by conversion to the sulfonate salt makes such compounds water-soluble and therefore readily usable as dyes for fabrics and in photographic processes. Azo compounds also have been used in a wide variety of other products, such as pH indicators (Methyl

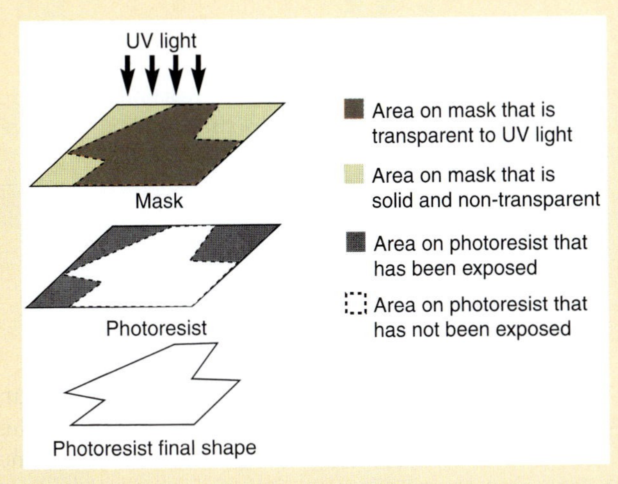

to protect the copy paper in those areas where the lines should appear.

The same principle of masking is used in modern photoresist technology to make patterns on thin silicon wafers. Three layers are involved. The bottom layer is the semiconductor silicon material (analogous to the copy paper of the diazo process). The second layer is made of a polymeric *photoresist*, a material susceptible to decomposition by UV light (this functions in a manner similar to the diazonium salt solution). The top layer is the mask (the original, inscribed with the pattern needed for the final product). The mask can block out large areas or the very fine lines needed in a complex circuit.

In the process called *lithography*, the silicon wafer is exposed to UV light, and the area of photoresist that is exposed decomposes and is removed by chemical processes. What is left is the unexposed design, the "copy." The exposed areas can then be laminated, etched, or whatever is needed to complete the process.

2-Diazonaphthoquinone $\xrightarrow{1.\ h\nu \quad 2.\ H_2O}$ Indenecarboxylic acid (COOH)

A variety of polymers can serve as photoresists. A common example is 2-diazonaphthoquinone (a stable diazonium compound) embedded in a hydrophobic resin. Upon exposure to UV light, the diazo group decomposes by loss of nitrogen; then a molecular rearrangement occurs to form indenecarboxylic acid. This acid is solubilized in sodium hydroxide by formation of its salt, which is water-soluble and can be washed away from the exposed area of the photoresist.

The ability to lithograph ever smaller features on silicon wafers has doubled the number of transistors that can be inscribed on a computer chip every year. These advances have relied on the use of shorter-wavelength radiation to etch the features. The wavelength presently employed is about 248 nm, allowing the etching of lines just 0.25 μm (10^{-6} m) thick. It is expected that wavelengths as short as 193 nm will be used, producing lines as small as 0.13 μm wide.

Orange), coloring for orange skins (Citrus Red), red dye for cotton (Para Red), and coloring in margarine (formerly Butter Yellow but now Sunset Yellow, FDC No. 6).

$(CH_3)_2N$–C_6H_4–N=N–C_6H_4–$SO_3^-\ Na^+$

Methyl Orange

Para Red (OH, N=N, NO_2)

Citrus Red (FDC No. 2) (OH, N=N, OCH_3, OCH_3)

$(CH_3)_2N$–C_6H_4–N=N–C_6H_5

Butter Yellow

PROBLEM 12.16

Draw the structures of the starting materials for the diazonium coupling reaction used to prepare each of these dyes (the structures shown on page 425): (a) Methyl Orange, (b) Butter Yellow, (c) Para Red, and (d) Citrus Red.

12.6

Preparation of Amines

Two distinct approaches can be envisioned for synthesizing an amine. First, the carbon skeleton can be synthesized and then an —NH_2 group attached as a final step. This approach is illustrated in the discussion of alkylation in Section 12.6.1. The second approach involves attaching the nitrogen in an early step and then performing subsequent reactions to further elaborate the final structure. This approach is elaborated on in Sections 12.6.2 and 12.6.3.

12.6.1 Alkylation of Ammonia

Primary aliphatic amines can be effectively prepared from alkyl halides (primary or secondary only) by an S_N2 reaction with excess ammonia, as described in Section 12.4.1.

General reaction:

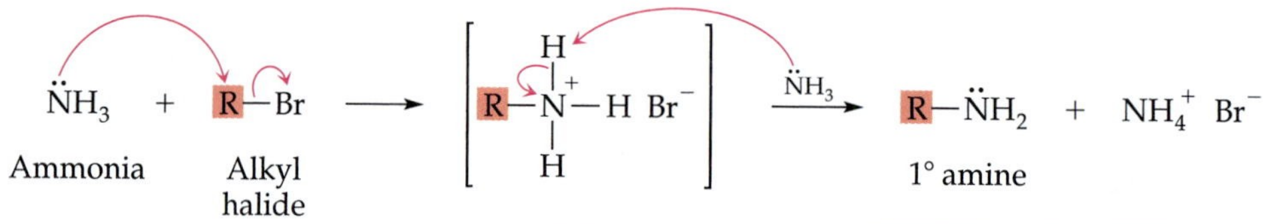

$$\ddot{N}H_3 + R{-}Br \longrightarrow \left[R{-}\overset{H}{\underset{H}{\overset{|}{\underset{|}{N^+}}}}{-}H\ Br^- \right] \xrightarrow{\ddot{N}H_3} R{-}\ddot{N}H_2 + NH_4^+\ Br^-$$

Ammonia — Alkyl halide — 1° amine

For example, benzylamine can be prepared from benzyl bromide and excess ammonia:

$$C_6H_5CH_2Br \xrightarrow{\text{excess } NH_3} C_6H_5CH_2NH_2$$

Benzyl bromide — Benzylamine

PROBLEM 12.17

Show the reactions necessary to prepare each amine from any hydrocarbon starting material. (*Hint:* Use retrosynthesis; more than one step is required.)

(a) cyclohexylamine
(b) aniline
(c) allylamine
(d) 3-methyl-2-butanamine

12.6.2 Reduction of Other Nitrogenous Compounds

Because amines contain the most highly reduced form of nitrogen bonded to at least one carbon atom, more highly oxidized forms of nitrogen can be reduced to amines. Although there are many such reactions, the three covered in this section are the most useful for elementary organic syntheses. In each, a nonnitrogenous compound is converted to a nitrogen-containing compound in an initial step, and the nitrogen-containing group is reduced in a second step. The only difference in the three processes is how the nitrogen atom is incorporated into the molecule; the reduction step is the same in all three cases.

Reduction of Azides

The **azide ion** (N_3^-), commercially available as sodium azide (NaN_3), is a very effective nucleophile. It reacts with primary or secondary alkyl halides in an S_N2 reaction to form alkyl azides, thereby forming the carbon-nitrogen bond required for an amine. Reduction of the alkyl azide with lithium aluminum hydride ($LiAlH_4$) results in displacement of molecular nitrogen (N_2) by a hydride anion and formation of a primary amine.

General reaction:

$$\underset{\text{Azide ion}}{\ddot{N}_3^-} + \underset{\text{Alkyl bromide}}{R{-}Br} \longrightarrow \underset{\text{Alkyl azide}}{R{-}\ddot{N}{=}\overset{+}{N}{=}\overset{-}{\ddot{N}}} \xrightarrow{LiAlH_4} \underset{1^\circ\ \text{amine}}{R{-}\ddot{N}H_2}$$

This two-step process accomplishes the same overall transformation as in the alkylation of ammonia—the conversion of R—Br to R—NH_2, with no change in the number of carbon atoms in the molecule. However, the azide substitution is usually much preferred because there is no risk of multiple alkylation reactions (see Section 12.4.1). For example, isopropyl bromide, $(CH_3)_2CH$—Br, can be converted to isopropylamine, $(CH_3)_2CH$—NH_2, in good yield:

$$\underset{\text{Isopropyl bromide}}{(CH_3)_2CH{-}Br} \xrightarrow{NaN_3} \underset{\text{Isopropyl azide}}{(CH_3)_2CH{-}N_3} \xrightarrow{LiAlH_4} \underset{\text{Isopropylamine}}{(CH_3)_2CH{-}NH_2}$$

Reduction of Nitriles

Cyanide ion (^-CN), available commercially as sodium cyanide (NaCN), is an effective nucleophile. Therefore, it can be used in an S_N2 reaction with primary or secondary alkyl halides to form compounds called **nitriles** (R—CN). Reduction of nitriles with lithium aluminum hydride produces primary amines (R—CH_2—NH_2).

General reaction:

$$\underset{\text{Cyanide ion}}{:\overset{-}{C}{\equiv}\ddot{N}} + \underset{\text{Alkyl bromide}}{R{-}Br} \longrightarrow \underset{\text{Nitrile}}{R{-}C{\equiv}\ddot{N}} \xrightarrow{LiAlH_4} \underset{1^\circ\ \text{amine}}{R{-}CH_2{-}\ddot{N}H_2}$$

In the reduction of azides, the number of carbons in the starting alkyl halide and the product amine is identical; but in the reduction of nitriles, the number of carbons in the product amine is one more than the number in the starting alkyl halide. In other words, a *homologation* occurs (a carbon is added to the parent chain). For example, ethyl bromide produces propylamine. It is a distinct advantage in synthesizing amines from alkyl halides to have available both methods, one that retains the same number of carbons (the azide method) and one that increases the number of carbons by one (the nitrile method).

$$R{-}Br \xrightarrow{NaN_3} RN_3 \text{ (Azide)} \xrightarrow{LiAlH_4} RNH_2 \text{ (1° amine)}$$

$$R{-}Br \text{ (Alkyl bromide)} \xrightarrow{NaCN} RCN \text{ (Nitrile)} \xrightarrow{LiAlH_4} RCH_2NH_2 \text{ (Homologated 1° amine)}$$

How to Solve a Problem

Convert cyclohexane to cyclohexylamine (C_6H_{11}—NH_2) and to cyclohexylmethanamine (C_6H_{11}—CH_2—NH_2).

Cyclohexane → Cyclohexylamine (NH_2)

Cyclohexane → Cyclohexylmethanamine (CH_2NH_2)

PROBLEM ANALYSIS

Both problems involve attaching groups to a cyclohexane ring. Recall that cycloalkanes are inert and undergo only one useful synthesis reaction—halogenation. Therefore, we should brominate first under radical conditions to form cyclohexyl bromide. Having a bromine on the ring opens the possibility of a myriad of substitution reactions, since bromide ion is a good leaving group.

To form cyclohexylamine from cyclohexyl bromide, the problem is simply attaching nitrogen; all the necessary carbon atoms are in place. Because we have an alkyl halide to work with, a substitution reaction with a nitrogen-containing species such as an azide ion can be used to form the carbon-nitrogen bond.

To form cyclohexylmethanamine from cyclohexyl bromide, we need to add a carbon as well as the nitrogen. We know that cyanide ion is a good nucleophile and attaches a carbon and a nitrogen in one step.

SOLUTION

Cyclohexane $\xrightarrow[h\nu]{Br_2}$ Cyclohexyl bromide (Br) $\xrightarrow{NaN_3}$ (N_3) $\xrightarrow{LiAlH_4}$ Cyclohexylamine (NH_2)

Cyclohexyl bromide $\xrightarrow{NaCN}$ (CN) $\xrightarrow{LiAlH_4}$ Cyclohexylmethanamine (CH_2NH_2)

PROBLEM 12.18

Synthesize each of the following amines from the designated alkyl bromide using reactions other than the alkylation of ammonia. (*Hint:* Use retrosynthesis.)

(a) 2-butanamine from 2-bromobutane

(b) cyclopropylamine from cyclopropyl bromide

(c) cyclopropylmethanamine from cyclopropyl bromide

(d) 2-methyl-1-butanamine from 2-bromobutane

(e) isopropylamine from isopropyl bromide

(f) 2-phenyl-1-ethanamine from benzyl bromide

Reduction of Amides

This process for the synthesis of amines involves converting one amine into another, more complex amine. The significant advantage of reducing amides is that this technique can be used to prepare secondary and tertiary amines, as well as primary amines. (Review Section 12.4.1 to see why secondary and tertiary amines are generally unavailable through alkylation.) The acylation reaction discussed in Section 12.4.2 converts an amine to an amide. This first step of the overall process establishes the carbon skeleton of the ultimate product (the correct number of carbon atoms and their arrangement) and incorporates the necessary nitrogen atom.

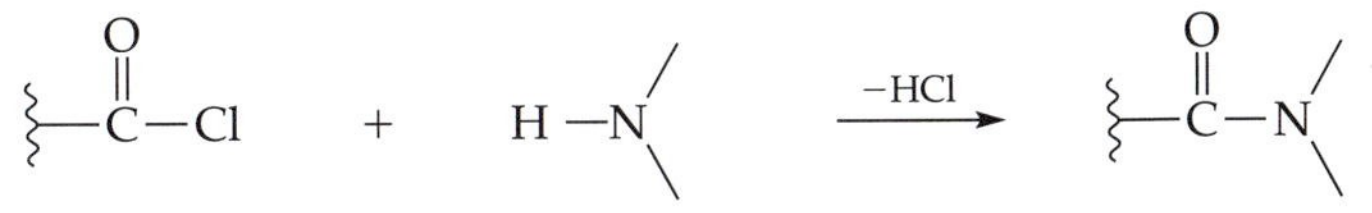

$$-\overset{O}{\overset{\|}{C}}-Cl \quad + \quad H-N\langle \quad \xrightarrow{-HCl} \quad -\overset{O}{\overset{\|}{C}}-N\langle$$

Acid chloride; 1° or 2° amine or ammonia; Amide

The second step of the synthesis is the reduction of the resulting amide with lithium aluminum hydride ($LiAlH_4$), a strong reducing agent. This reduction reaction converts the carbonyl group of the amide into a methylene group.

General reaction:

$$-\overset{O}{\overset{\|}{C}}-N\langle \quad + \quad LiAlH_4 \quad \longrightarrow \quad -CH_2-N\langle$$

Amide; Lithium aluminum hydride; Amine

If ammonia is used to form the amide, the resulting amine is primary. If a primary amine is used to form the amide, the resulting amine is secondary. If a secondary amine is used to form the amide, the resulting amine is tertiary. Thus, reduction of an amide is a reliable method for the conversion of a primary amine to a secondary amine or the conversion of a secondary amine to a tertiary amine.

$$\underset{\text{Acid chloride}}{R-CO-Cl} + \underset{\text{Ammonia}}{\ddot{N}H_3} \xrightarrow{-HCl} \underset{\text{Amide}}{R-CO-\ddot{N}H_2} \xrightarrow{LiAlH_4} \underset{1^\circ \text{ amine}}{R-CH_2-\ddot{N}H_2}$$

$$\underset{\text{Acid chloride}}{R-CO-Cl} + \underset{1^\circ \text{ amine}}{\ddot{N}H_2R'} \xrightarrow{-HCl} \underset{\text{Amide}}{R-CO-\ddot{N}HR'} \xrightarrow{LiAlH_4} \underset{2^\circ \text{ amine}}{R-CH_2-\ddot{N}HR'}$$

$$\underset{\text{Acid chloride}}{R-CO-Cl} + \underset{2^\circ \text{ amine}}{\ddot{N}HR'_2} \xrightarrow{-HCl} \underset{\text{Amide}}{R-CO-\ddot{N}R'_2} \xrightarrow{LiAlH_4} \underset{3^\circ \text{ amine}}{R-CH_2-\ddot{N}R'_2}$$

How to Solve a Problem

Convert methylamine to *N*-methylbenzylamine.

$$\underset{\text{Methylamine}}{CH_3-NH_2} \dashrightarrow \underset{N\text{-Methylbenzylamine}}{CH_3-NH-CH_2-C_6H_5}$$

PROBLEM ANALYSIS

First, we try to envisage how and where the structure of the starting material appears in the product. This highlights the group(s) that need(s) to be added in the synthesis. In this instance, it is clear that a benzyl group must be attached to the nitrogen of the starting amine. This is not normally accomplished by a simple alkylation with benzyl bromide (see Section 12.4.1 for the difficulty). However, we recognize that a benzyl group ($C_6H_5—CH_2—$) can be obtained from a benzoyl group ($C_6H_5—CO—$) via reduction and that a benzoyl group can be obtained from the corresponding acyl chloride, in this case, benzoyl chloride ($C_6H_5—CO—Cl$). Finally, we recognize that a benzoyl group can be attached to methylamine through a simple acylation reaction (see Section 12.4.2) to form an amide. Reduction of the amide yields the desired product.

SOLUTION

$$\underset{\text{Methylamine}}{CH_3-NH_2} \xrightarrow[\text{pyridine}]{C_6H_5-CO-Cl} C_6H_5-CO-NH-CH_3 \xrightarrow{LiAlH_4}$$

$$\underset{N\text{-Methylbenzylamine}}{C_6H_5-CH_2-NH-CH_3}$$

PROBLEM 12.19

Synthesize each of the following amines starting with a nitrogen-containing compound and using the amide reduction method: (a) *N*-ethylaniline, (b) 1-butanamine, (c) dipropylamine, and (d) triethylamine.

PROBLEM 12.20

Synthesize each of the following amines using any appropriate method, but starting with a non-nitrogenous compound: (a) cyclobutylmethanamine, (b) (*p*-chlorobenzyl)methylamine, (c) *N,N*-diethylcyclohexanamine, and (d) benzylamine.

12.6.3 Reduction of Nitroarenes

Aromatic amines, which are those with the amino group attached directly to an aromatic ring, cannot be formed by electrophilic aromatic substitution by an amino group on an aromatic ring because that group is not readily available as an electrophile. The nucleophilic substitution methods using azide or cyanide ions described earlier also do not apply because those ions do not bring about an S_N2 reaction on bromobenzene. Instead, aromatic amines are prepared by reduction of **nitroarenes,** aromatic rings with a nitro group (—NO_2) attached. Nitroarenes are readily available via direct nitration of aromatic compounds (see Section 9.3.3).

General reaction:

$$\underset{\text{Arene}}{Ar{-}H} \xrightarrow[H_2SO_4]{HNO_3} \underset{\text{Nitroarene}}{Ar{-}NO_2} \xrightarrow{Fe/HCl \text{ or } Pt/H_2} \underset{1^\circ \text{ aminoarene}}{Ar{-}NH_2}$$

Several reduction methods are available, the most common being a chemical method that uses iron in HCl and a catalytic method that uses hydrogen gas over a platinum catalyst. Having this choice of methods is critical because of the many cases in which a ring has two or more substituents that may be susceptible to reduction. For example, *p*-nitrostyrene can be reduced to *p*-aminostyrene by Fe/HCl; mild catalytic reduction produces *p*-ethylnitrobenzene, and more strenuous catalytic reduction produces *p*-ethylaniline. (Recall from Section 6.3 that alkenes are catalytically reduced to alkanes.)

$$\underset{p\text{-Nitrostyrene}}{O_2N{-}C_6H_4{-}CH{=}CH_2} \xrightarrow{Fe/HCl} \underset{p\text{-Aminostyrene}}{H_2N{-}C_6H_4{-}CH{=}CH_2}$$

$$\underset{p\text{-Nitrostyrene}}{O_2N{-}C_6H_4{-}CH{=}CH_2} \xrightarrow[\text{cold}]{Pt/H_2} \underset{p\text{-Ethylnitrobenzene}}{O_2N{-}C_6H_4{-}CH_2CH_3}$$

$$\underset{p\text{-Nitrostyrene}}{O_2N{-}C_6H_4{-}CH{=}CH_2} \xrightarrow[\text{heat}]{Pt/H_2} \underset{p\text{-Ethylaniline}}{H_2N{-}C_6H_4{-}CH_2CH_3}$$

PROBLEM 12.21

Synthesize each amine, starting with the given compound:

(a) *N,N*-dimethylaniline from benzene

(b) *N*-ethylaniline from aniline

(c) *p*-tolylmethanamine from toluene

(d) *N*-benzylcyclohexanamine from bromocyclohexane

CHEMISTRY AT WORK

IN THE BODY

Sulfa Drugs

Sulfonation of amines plays an important role in the pharmaceutical industry, especially in the formation of **sulfa drugs.** Sulfa drugs were among the earliest synthetic antibiotics used to treat bacterial infections.

The sulfonation of amines using aromatic sulfonyl chlorides (Ar—SO_2—Cl) to form **sulfonamides** is closely related to the acylation of amines to form amides (see Section 12.4.2):

$$R_2NH + Ar-SO_2-Cl \xrightarrow{NaOH} R_2N-SO_2-Ar + NaCl + H_2O$$

2° amine; Sulfonyl chloride; Sulfonamide

Sulfonamides are derivatives of sulfonic acids (Ar—SO_3H). Like the acylation reaction, the sulfonation reaction occurs only with primary and secondary amines, in which a hydrogen is replaced with a sulfonyl group (RSO_2—).

The parent compound for all sulfa drugs is sulfanilamide, derived from the sulfonation of ammonia:

$H_2N-C_6H_4-SO_2-NH_2$

Sulfanilamide

Sulfanilamide eventually was found to be too toxic for general use, but thousands of derivatives have been synthesized and tested. Most sulfa drugs used today have been derived from the sulfonation of primary amines rather than the sulfonation of ammonia. For example, sulfathiazole saved the lives of many people wounded in World War II; sulfapyridine provided the first successful treatment for pneumonia; and sulfadiazine is used to treat many different infections.

$H_2N-C_6H_4-SO_2-NH-R$

R-substituted sulfanilamide

R = thiazolyl (N, S): Sulfathiazole

R = pyridyl (N): Sulfapyridine

R = pyrimidinyl (N, N): Sulfadiazine

The antibiotic effect of sulfa drugs is due to their structural similarity to *p*-aminobenzoic acid (PABA) amide:

$H_2N-C_6H_4-CO-NH_2$

p-Aminobenzoic acid (PABA) amide

Antibiotics are antibacterial, and the sulfa drugs inhibit bacterial growth by interfering with the metabolic processes of bacteria. Thus, sulfa drugs are *antimetabolites.* The bacteria cannot distinguish between the sulfa drug and PABA amide because their sizes and electronic structures are very similar. The bacteria mistakenly incorporate the sulfa drug in place of PABA amide in the enzymatic synthesis of folic acid, essential for bacterial cell growth. The bacteria cease to grow and die.

The sulfa drug does not interfere with human metabolic processes because humans do not synthesize folic acid. Instead, we must obtain this essential vitamin from dietary sources, such as whole grains and deep-green vegetables.

12.7

Important and Interesting Amines

A wide variety of amines exist in nature, ranging from simple to complex, and many have become important because of their significant physiological effects. Simple amines include trimethylamine, which has a fishy odor, and putrescine [H_2N—$(CH_2)_4$—NH_2], which is found in decaying tissue. We will consider just a few important amines, beginning with the alkaloids.

12.7.1 Alkaloids

Many plants are rich sources of amines, so much so that the term **alkaloid** was adopted to describe such compounds (from the word *alkali* meaning "basic"). Alkaloids are usually obtained by grinding the whole plant or parts of it (for example, roots, leaves, or berries) and then extracting the organic material into an organic solvent, such as dichloromethane. The basic amines are extracted from the organic solvent using an acidic separation, as described in Section 12.3.2. The isolated amines can be purified and then used for their intended purpose. In some cases, such as nicotine from tobacco leaves or caffeine from coffee beans, the alkaloids are not isolated; they are ingested directly from the plant material and exert their physiological effects.

Delphinine
(A poisonous alkaloid from delphinium and larkspur plants, including *Delphium staphisagria L., Ranunculaceae*)

The delphinium, or larkspur, plant produces the poisonous alkaloid delphinine in its flowers—for no known biological function.

Interestingly, there is no apparent reason why plants produce some alkaloids, since these serve no known biological function for the plant. Others, however, have known functions, such as deterring attacks by insects or animals. Most alkaloids occur as a single enantiomer, and usually only that enantiomer is physiologically active. The natural occurrence of one pure enantiomer of an alkaloid makes possible the resolution (that is, the separation of a racemate into its two enantiomers) of certain chiral carboxylic acids (as we will see in Chapter 17).

Many alkaloids were first isolated by organic chemists as a result of knowledge obtained from primitive tribes, especially those in Asia and South and Central America. "Medicine men" often used extracts of indigenous plants for healing. As knowledge of this practice spread to the Western world, the plants were analyzed and the biologically active ingredient isolated and identified. Today, pharmaceutical companies test isolated compounds and obtain the necessary government approval to market the natural substance, a synthetic duplicate, or a synthetic compound of closely related structure as a prescription or over-the-counter drug. Ethnobotanists are constantly on the lookout for as yet undiscovered healing plants, especially in the diminishing rain forests, with the hope that the alkaloids involved might serve as cures, or as building blocks of cures, for modern diseases such as cancer, AIDS, and hepatitis.

Given the wide variety of chemical structures in alkaloids, a classification system based on the "parent" ring system evolved. Coniine, a member of the pyridine group, was the active ingredient in the Persian hemlock plant consumed by Socrates, which led to his death. Nicotine, shown in Section 12.1, is also a member of this group. It can be quite toxic in large doses, but in small doses (such as found in cigarettes), it stimulates the autonomic nervous system. After years of debate, there is now proof that nicotine is addictive.

Coniine

Socrates consumes hemlock while surrounded by his students.

Cocaine, obtained from the leaves of the coca shrub grown in the Andes Mountains, is a member of the tropane family of alkaloids. In small quantities, it is a stimulant and pain reliever, but continued use leads to addiction, depression, and even death. Synthetic substitutes for cocaine include procaine (Novocain), used in dentistry, and benzocaine, used in topical anesthetics such as sunburn ointments. Note that the synthetic anesthetics contain many of the structural features present in cocaine: an amine functional group, an ester group, and an aromatic ring.

Cocaine

Procaine (Novocain)

Benzocaine

The isoquinoline family of alkaloids includes probably the oldest-known pure alkaloid, morphine, obtained from the seeds of the opium poppy, *Papaver somniferum*. Crude extracts of poppy seeds were used for centuries as powerful painkillers, and the specific effect was later shown to be due to the very addictive morphine. Codeine, the *O*-methyl ether derivative of morphine, is also present in the same seeds, but it is a milder and less addictive analgesic, as well as an effective cough suppressant. Heroin is the diacetate ester of morphine; it does not occur naturally.

Because of the advantageous analgesic effects of these alkaloids, many similar compounds have been synthesized and tested in the hope that the addictive effects could be minimized. Such synthetic compounds are the active ingredients of Darvon and Demerol. All of these contain the basic morphine skeleton—an aromatic ring attached to a quaternary carbon, which is in turn attached to a two-carbon unit carrying a tertiary amine. Methadone, another member of this group, is used to treat opiate withdrawal symptoms.

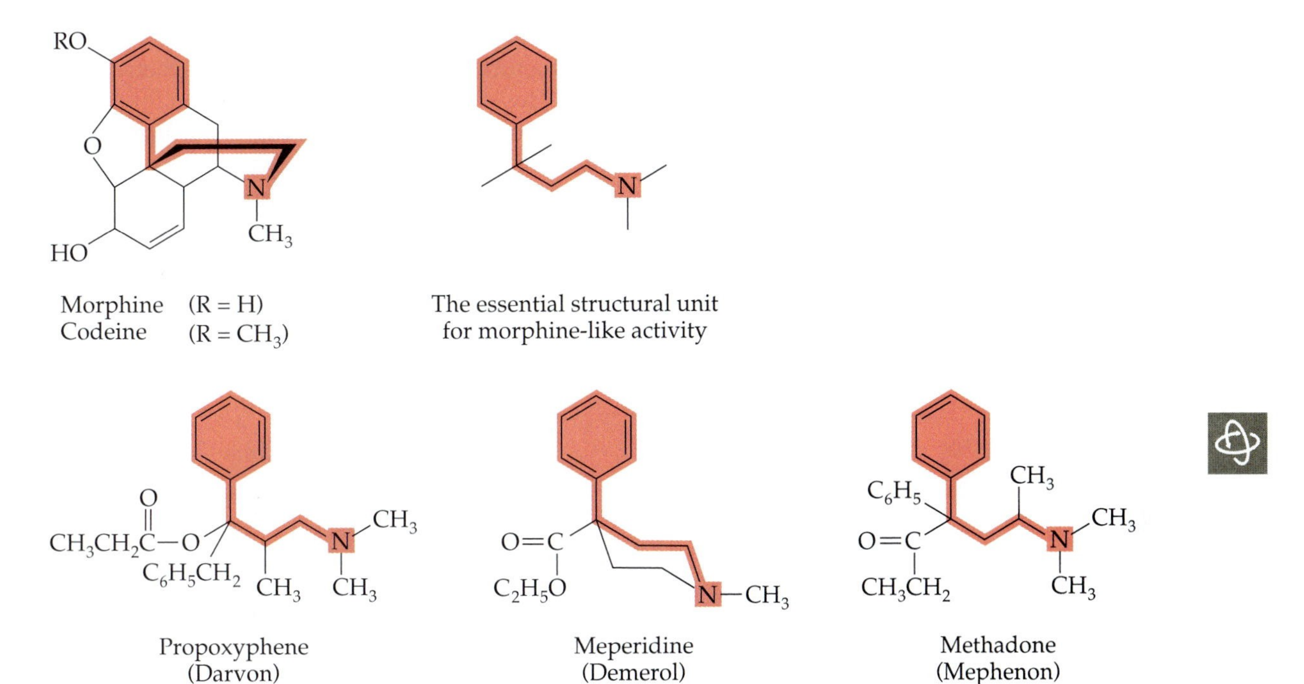

Morphine (R = H)
Codeine (R = CH_3)

The essential structural unit for morphine-like activity

Propoxyphene (Darvon)

Meperidine (Demerol)

Methadone (Mephenon)

These compounds illustrate a key fact about preparing synthetic drugs: It is often only necessary to incorporate just some of the original structural features of the naturally occurring compound. Considerable research is required to determine which part(s) of the molecule are essential for the desired physiological activity. On occasion, beneficial effects may be associated with one part of the molecule and undesirable side effects with another. Major research efforts, involving the synthesis and biological testing of hundreds or thousands of compounds, are undertaken to determine the right structural characteristics for a compound, so that a medicine that is safe and effective can eventually be marketed.

Some alkaloids have even more complex structures. One such medicinal alkaloid is quinine, obtained from the South American cinchona tree. Quinine was the first effective antimalarial drug. Reserpine, obtained from the roots of the Himalayan shrub *Rauwolfia serpentina,* was one of the early tranquilizers and is also an effective antihypertensive drug.

Quinine

Reserpine

The flowering heads of the cactus *Lophophora williamsi* are the source of peyote, which contains the hallucinogenic alkaloid mescaline.

12.7.2 Hormonal Amines

A number of hormonal amines are known, but we will consider just one family. Epinephrine, also known as adrenaline, is a stimulant of the sympathetic nervous system in humans. The physiologically active structural unit of epinephrine is a 2-phenylethylamine component ($C_6H_5CH_2CH_2NH_2$). This component is the foundation for a number of related compounds that have been synthesized or recognized. Ephedrine, originally extracted from the *ma-huang* plant in China, is used in decongestants because it shrinks the nasal membranes and reduces nasal excretions. Amphetamines, such as Benzedrine, are stimulants used as "stay-awake" medications. Mescaline is a hallucinogen found in peyote, a cactus used by some Native Americans in religious ceremonies.

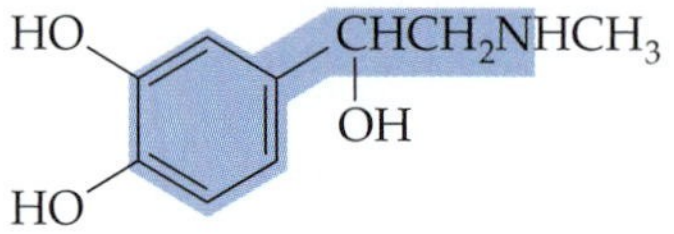

HO, HO, $CHCH_2NHCH_3$, OH

Epinephrine (adrenaline)

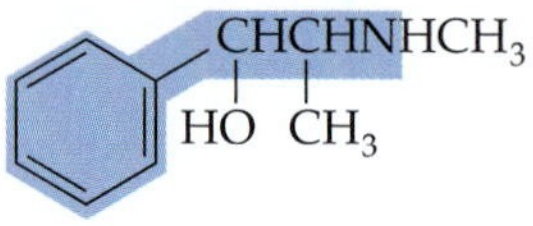

$CHCHNHCH_3$, HO, CH_3

Ephedrine

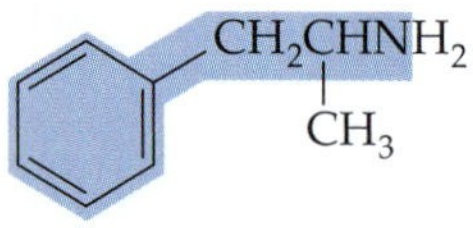

CH_2CHNH_2, CH_3

Benzedrine (amphetamine)

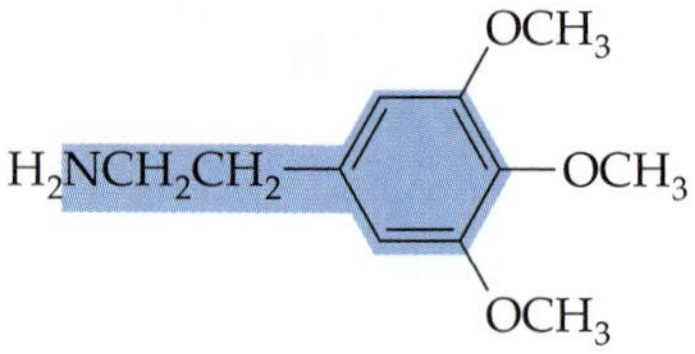

$H_2NCH_2CH_2$, OCH_3, OCH_3, OCH_3

Mescaline

Derivatives of 2-phenylethylamine are important in brain chemistry. Serotonin and dopamine have essential functions as neurotransmitters. Persons with Parkinson's disease have a deficiency of dopamine, and the disease occasionally can be treated with DOPA, a precursor of dopamine. Overproduction of dopamine in the brain is associated with schizophrenia. These kinds of compounds may also be implicated in Alzheimer's disease.

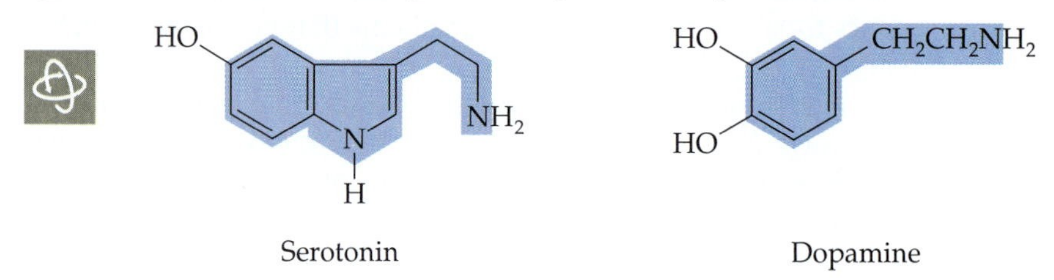

Serotonin

Dopamine

PROBLEM 12.22

Using your knowledge of amines and their synthesis, devise a synthesis of Benzedrine (structure shown above) starting with propylbenzene. (*Hint:* Use retrosynthesis.)

12.7.3 Other Interesting Amines

As noted earlier in this chapter, α-amino acids, which have the general structure R—CH(NH_2)COOH, are the building blocks of proteins, necessary for life itself. They will be discussed in some detail in Chapter 16.

New amine compounds that have potentially beneficial biological effects continue to be discovered. Some of these compounds come from very unusual sources. For example, in 1992, the relatively simple compound epibatidine was isolated in very small quantities from the mixture of chemicals secreted by an Ecuadorian frog, *Epipedobates tricolor* (see the introductory essay). In tests on mice, this compound proved to be 200 times as effective as morphine in blocking pain and appeared to function by a different mechanism than that of the opiates. It is also one of the rare naturally occurring compounds to contain chlorine. Several different syntheses of the compound have been reported, all as a result of high interest in the possibility that a long-sought pharmaceutical goal could be achieved—the availability of a nonsedating, nonopiate, and therefore nonaddicting but effective painkiller. Epibatidine itself cannot be used to treat humans because of side effects, but its discovery provides a new molecular structure on which to base the synthesis of similar compounds for biological testing. ABT-594 is one promising example of such a compound.

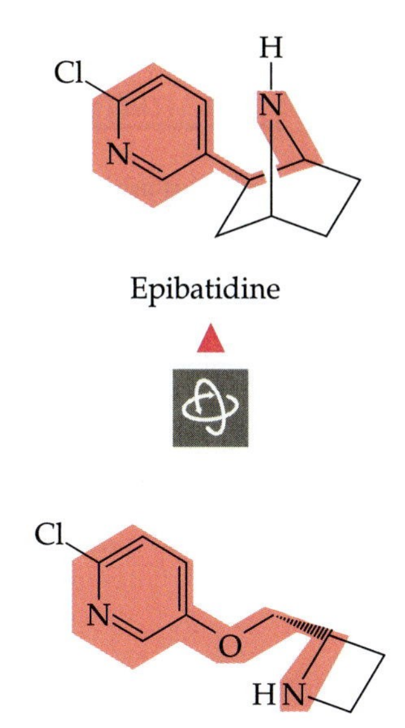

The search for useful, naturally occurring, and physiologically active compounds is a worldwide enterprise involving many chemists, biologists, and pharmaceutical companies. From this continuing effort undoubtedly will come new and effective compounds, many of them amines, to treat human diseases and otherwise improve the human condition.

Chapter Summary

Amines are organic compounds with an sp^3-hybridized nitrogen atom, which is trivalent. The nitrogen atom has an unshared pair of electrons and is bonded to one, two, or three alkyl and/or aryl groups. Amines are classified as primary (1°), secondary (2°), or tertiary (3°), depending on the number of alkyl or aryl groups attached to the nitrogen atom. The nitrogen atom may be part of an alicyclic system, a heterocyclic ring, or a heterocyclic aromatic ring. A nitrogen with four alkyl or aryl groups attached is tetravalent and is a **quaternary ammonium salt**. Amines have pyramidal geometry, but ammonium ions are tetrahedral.

The historical/common names for alkylamines are derived by naming the groups attached to nitrogen. The IUPAC system names alkylamines as **alkanamines,** using the standard stem names, suffixes, prefixes, and numbering system. Benzenoid amines are named as derivatives of aniline.

The outstanding properties of amines arise from the unshared pair of electrons on nitrogen. Amines are polar compounds and can serve as hydrogen-bond acceptors (if they have at least one hydrogen attached to nitrogen) or as hydrogen-bond donors. Amines are basic and are protonated by acids to form their **conjugate acids**, known as ammonium salts. The relative basicity of amines is reflected by the pK_a of their conjugate acids.

Amines are nucleophiles. They carry out S_N2 substitutions on primary and secondary alkyl halides (an **alkylation** reaction) to form ammonium salts that can be deprotonated to produce the higher amine. Primary and secondary amines undergo **acylation** by reaction with **acid (acyl) halides** to form **amides.**

Primary aromatic amines undergo **diazotization,** a reaction with nitrous acid to form **diazonium salts.** The nitrogen in such salts can be replaced by a variety of nucleophiles. This reaction is an effective means of producing aromatic iodides, fluorides, nitriles, and phenols. Reduction of the diazonium salt with hypophosphorous acid re-forms the starting arene.

Aromatic diazonium salts carry out a **coupling reaction** with activated aromatic rings to form **azo compounds,** which are used as dyes because of their long-wavelength absorption in the visible spectrum.

Primary aliphatic amines are prepared by reaction of an alkyl halide with excess ammonia. Alternatively, reaction of an alkyl halide with azide ion or cyanide ion, followed by $LiAlH_4$ reduction, produces primary amines selectively. Amines are also produced by $LiAlH_4$ reduction of amides, providing a means for converting a primary or secondary amine to a secondary or tertiary amine, respectively, by initial acylation followed by reduction. Aromatic amines are produced by nitration of an aromatic compound followed by catalytic or chemical reduction.

Summary of Reactions

1. Protonation of amines to form ammonium salts (Section 12.3.2)

$$R{-}NH_2 + HCl \longrightarrow R{-}\overset{+}{N}H_3\ Cl^-$$

1° alkyl amine — Alkylammonium chloride

2. Alkylation of amines (Section 12.4.1). Primary or secondary amines can be alkylated with primary or secondary alkyl halides to form more highly substituted amines. Tertiary amines form quaternary ammonium salts.

$$R{-}NH_2 + R'Br \xrightarrow{NaOH} RR'NH$$

1° amine — 1° or 2° alkyl bromide — Amine

$$R_3N + R'Br \xrightarrow{NaOH} R_3\overset{+}{N}R'\ Br^-$$

3° amine — 1° or 2° alkyl bromide — Quaternary ammonium bromide

3. Acylation of amines (Section 12.4.2). Amines with at least one hydrogen on nitrogen (that is, primary or secondary amines) form amides in which a hydrogen has been replaced by an acyl group.

$$R{-}NH_2 + R''COCl \xrightarrow{pyridine} R{-}NH{-}CO{-}R''$$

1° amine — Acyl chloride — Amide

4. Diazotization of primary aromatic amines (Section 12.5.1)

$$Ar{-}NH_2 \xrightarrow[HCl]{NaNO_2} ArN_2^+\ Cl^-$$

Aromatic amine — Aryl diazonium chloride

a. Displacement of nitrogen from a diazonium salt forms a substituted arene (Section 12.5.2).

$$ArN_2^+\ Cl^-$$

Aryl diazonium chloride

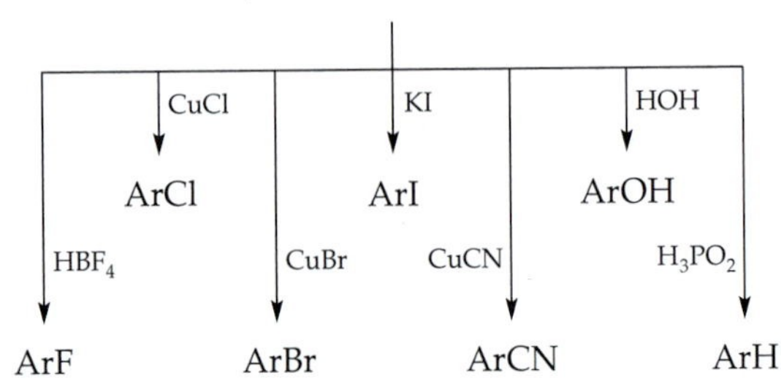

b. Coupling reaction of a diazonium salt forms an azo compound (Section 12.5.3).

$$ArN_2^+\ Cl^- + C_6H_5{-}X \longrightarrow Ar{-}N{=}N{-}C_6H_4{-}X$$

Aryl diazonium chloride — (X = OH, OR, or NR_2) — Azo compound

5. Preparation of primary alkyl amines by alkylation of ammonia (Section 12.6.1)

$$R{-}Br + NH_3\ (\text{excess}) \longrightarrow R{-}NH_2$$

1° or 2° alkyl bromide — 1° amine

6. Preparation of primary alkyl amines by reduction of azides or nitriles (Section 12.6.2)

$$R{-}Br \xrightarrow{NaN_3} R{-}N_3 \xrightarrow{LiAlH_4} R{-}NH_2$$

1° or 2° alkyl bromide — 1° amine

$$R{-}Br \xrightarrow{NaCN} R{-}CN \xrightarrow{LiAlH_4} R{-}CH_2NH_2$$

1° or 2° alkyl bromide — Homologated 1° amine

7. Preparation of primary, secondary, and tertiary amines by reduction of amides (Section 12.6.2)

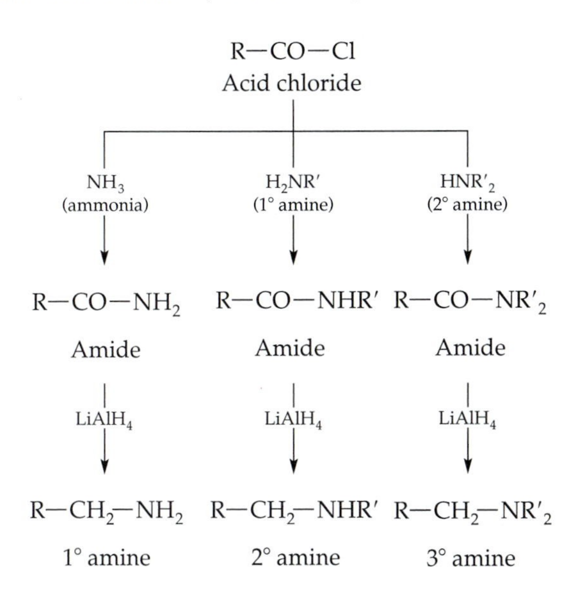

8. Preparation of primary aromatic amines by reduction of nitroarenes (Section 12.6.3)

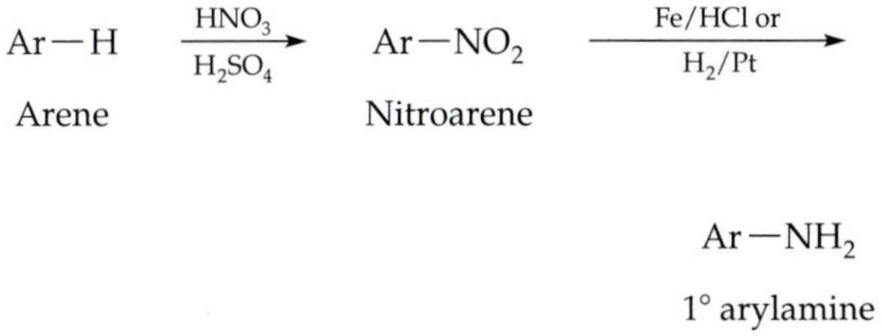

$$\underset{\text{Arene}}{Ar{-}H} \xrightarrow[H_2SO_4]{HNO_3} \underset{\text{Nitroarene}}{Ar{-}NO_2} \xrightarrow[H_2/Pt]{\text{Fe/HCl or}} \underset{1^\circ\ \text{arylamine}}{Ar{-}NH_2}$$

Additional Problems

Nomenclature

12.23 Draw the structural formula for each of the following compounds:

(a) *m*-bromoaniline
(b) benzylamine
(c) *N,N*-diisopropylaniline
(d) *trans*-4-(*N,N*-dimethylamino)-1-methylcyclohexane
(e) (*R*)-2-butanamine
(f) 2-methyl-3-pentanamine
(g) *cis*-1,2-cyclobutanediamine
(h) tetraethylammonium bromide
(i) *N,N,N*-trimethylanilinium chloride
(j) *p*-methoxy-*N*-methylaniline
(k) dimethylammonium chloride
(l) *p*-aminobenzoic acid
(m) 1,3-diaminobenzene
(n) 3-aminopyridine
(o) 2-(*N,N*-dimethylamino)pyrrole

12.24 For each of the compounds in Problem 12.23, indicate whether the amine is primary, secondary, tertiary, or quaternary.

12.25 Assign names to the following structures:

(a) $H_2N(CH_2)_6NH_2$

(b)  HO—(cyclohexane chair)—$N(CH_3)_2$

(c) 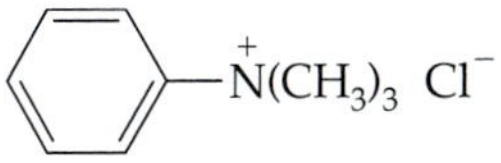 C_6H_5—$\overset{+}{N}(CH_3)_3\ Cl^-$

(d) 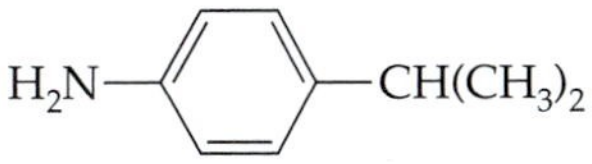 H_2N—C_6H_4—$CH(CH_3)_2$

(e) $CH_3CH_2NHCH_2CH(CH_3)_2$

(f) $(CH_3CH_2)_2NH$

(g) 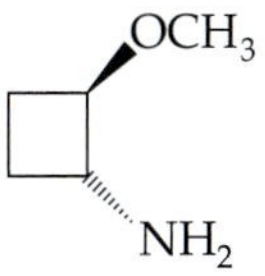 cyclobutane with OCH_3 and NH_2

(h) 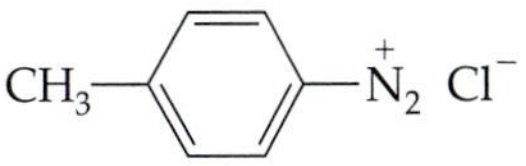 CH_3—C_6H_4—$\overset{+}{N_2}\ Cl^-$

(i) 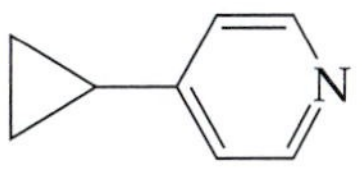 cyclopropyl—pyridine (N)

(j) $CH_3CH{=}CH{-}CH_2NH_2$

(k) $(CH_3CH_2CH_2)_4N^+ \ I^-$

(l) 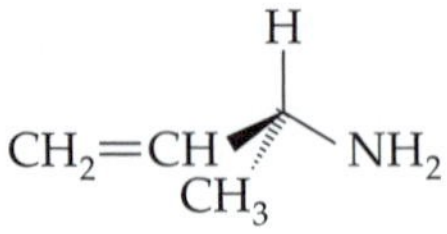 $CH_2{=}CH$–C(H)(CH_3)–NH_2

12.26 Many amines are known best by their historical names. A few examples are given here, preceded by the systematic names. Draw the structure of each amine, and indicate whether it is primary, secondary, or tertiary:

(a) *p*-methylaniline or *p*-aminotoluene, better known as *p*-toluidine

(b) 2-amino-1-phenylpropane, better known as amphetamine

(c) *p*-aminobenzenesulfonic acid, better known as sulfanilic acid

(d) azacyclohexane or hexahydropyridine, better known as piperidine

(e) azacyclopentane or tetrahydropyrrole, better known as pyrrolidine

(f) 2-(3-pyridyl)-*N*-methylpyrrolidine, better known as nicotine

Properties of Amines

12.27 Many medicinals (see Section 12.7) are large amines that are insoluble in water and are therefore administered as their hydrochloride salts, which are water-soluble and dissolve more readily in the bloodstream. Examples are the local anesthetics, which include benzocaine, procaine, and lidocaine. The structure of lidocaine, which is also important for treating ventricular arrhythmia, is shown. Draw the structure of its hydrochloride salt.

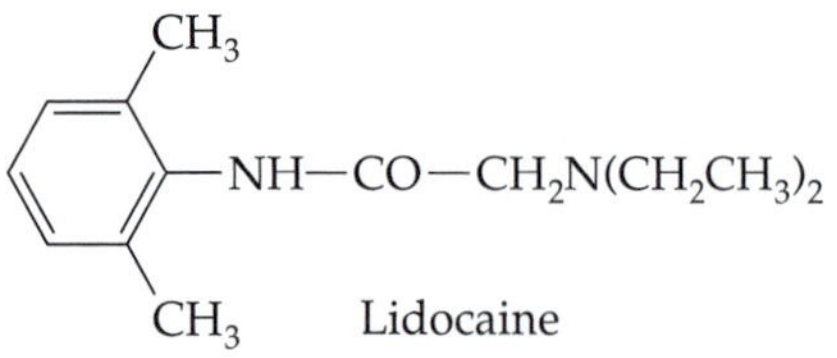

2,6-(CH_3)$_2C_6H_3$–$NH{-}CO{-}CH_2N(CH_2CH_3)_2$

Lidocaine

12.28 Explain why 1-butanamine is soluble in water, whereas pentane, of similar molecular weight, is not.

12.29 1-Propanol boils at 97°C, 1-propanamine boils at 49°C, and butane boils at 0°C. These compounds represent three different families, but their molecular weights are similar. Although you might expect their boiling points to be similar, clearly they are not. Explain this trend in boiling points.

12.30 Write an equation for each of the following reactions:

(a) aniline with hydrogen chloride

(b) diethylamine with sulfuric acid

(c) *N,N*-dimethylaniline with methyl iodide

(d) dimethylammonium chloride with sodium hydroxide

(e) triethylamine with ethyl iodide

12.31 Write acid-base equilibria for the following reactions, and show the direction in which each equilibrium lies:

(a) lidocaine hydrochloride (see Problem 12.27; pK_a ~10.7) with sodium bicarbonate (pK_a of H_2CO_3 ~6.4)

(b) lidocaine hydrochloride (see Problem 12.27; pK_a ~10.7) with sodium hydroxide (pK_a of H_2O ~15.7)

(c) phenol (pK_a ~10) with dimethylamine (pK_a of dimethylammonium chloride ~10.7)

(d) pyridine (pK_a of pyridinium chloride ~5.3) with acetic acid (pK_a ~4.8)

12.32 Indicate which member of each of the following pairs of compounds is more basic, and explain why:

(a) methylamine and dimethylamine

(b) aniline and *p*-fluoroaniline

(c) aniline and diphenylamine

(d) pyrrole and tetrahydropyrrole (pyrrolidine)

(e) aniline and acetanilide ($C_6H_5NH{-}COCH_3$)

12.33 Draw the resonance contributing structures for *p*-nitroaniline, and explain why it is a weaker base than aniline.

12.34 Develop a separation scheme to separate *N,N*-dimethylaniline from benzene when both are dissolved in an ether solution.

12.35 Draw the three-dimensional structures of the two enantiomers of *N*-methyl-2-pentanamine, and label them *R* and *S*.

12.36 How do you explain the fact that two isomeric quaternary ammonium salts are formed when 4-*t*-butyl-*N*-methylpiperidine is treated with benzyl chloride?

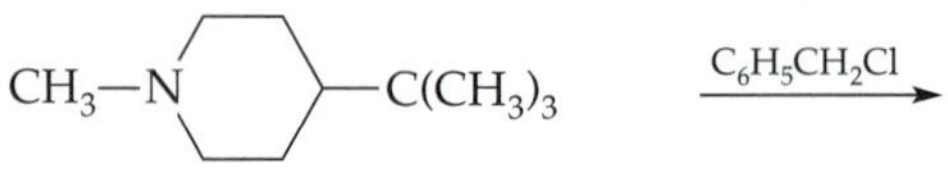

$CH_3{-}N$(piperidine ring)$-C(CH_3)_3 \xrightarrow{C_6H_5CH_2Cl}$ Two isomers

4-*t*-Butyl-*N*-methylpiperidine

12.37 Develop a separation scheme to separate aniline, phenol, and acetanilide dissolved in an ether solution.

Reactions of Amines

12.38 Show the product of each of the following reactions:

(a) aniline with excess methyl iodide

(b) benzyl bromide with excess ammonia

(c) (*R*)-2-bromopentane with excess ammonia

(d) trimethylamine with ethylene oxide (This reaction is a common preparation of choline; see p. 411. It was not studied, but you know the principles underlying it.)

12.39 Show the product expected from the reaction of benzylamine with each of the following reagents:

(a) hydrogen chloride

(b) excess methyl iodide

(c) acetyl chloride and pyridine

(d) tosyl chloride

12.40 Show the product of each of the following reactions:

(a) aniline with acetyl chloride and dilute NaOH

(b) the product of part (a) with $LiAlH_4$

(c) cyclohexylamine with benzoyl chloride (C_6H_5COCl) and pyridine

(d) *N*-methyl-*p*-chloroaniline with tosyl chloride

12.41 Show the structure of the diazonium salt that would result from treating each of the following amines with sodium nitrite and hydrochloric acid at 0°C:

(a) C_6H_5–NH_2

(b)

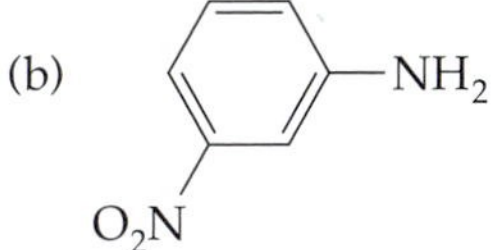

(c) N–H pyrrole with NH_2

(d)

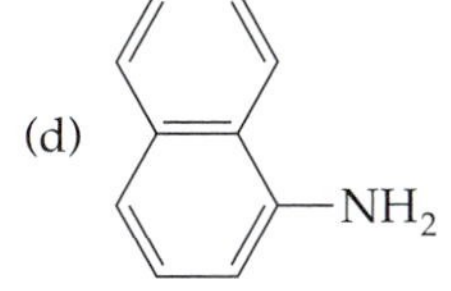

12.42 Show the products of the reaction of the diazonium salt prepared from *p*-toluidine (*p*-methylaniline) with each of the following reagents:

(a) CuCN (b) water at 50°C

(c) H_3PO_2 (d) KI

(e) CuCl (f) *N,N*-dimethylaniline

(g) phenol

12.43 Specify the reagent(s) needed to convert aniline into each compound:

(a) acetanilide (b) *N*-ethylaniline

(c) *p*-nitroaniline (d) iodobenzene

(e) phenol (f) benzonitrile (C_6H_5CN)

(g) benzene (h) fluorobenzene

(i) *N*-(*p*-chlorobenzyl)aniline

12.44 Show the product of the reaction of *p*-methoxyaniline first with sodium nitrite and hydrochloric acid and then with each of the following compounds:

(a) phenol

(b) anisole

(c) *N*-phenylpiperidine

(d) 2,6-dichloro-*N,N*-dimethylaniline

Preparation of Amines

12.45 Show the product of each of the following reactions:

(a) aniline with CH_3CH_2COCl and then $LiAlH_4$

(b) 2-butanamine with acetyl chloride and then $LiAlH_4$

(c) 1-phenyl-1-bromopropane with sodium azide and then $LiAlH_4$

(d) cyclopropylamine with benzoyl chloride and then $LiAlH_4$

(e) cyclohexyl bromide with sodium cyanide and then $LiAlH_4$

12.46 Show the reactions necessary to prepare each of the following amines from benzene, toluene, or any compound produced in a previous part (several steps may be required):

(a) aniline

(b) *p*-chloroaniline

(c) *m*-aminobenzoic acid

(d) *p*-aminobenzoic acid

(e) *m*-aminobenzenesulfonic acid

(f) dibenzylamine

12.47 Show how to prepare 1-pentanamine using the designated starting material(s):

(a) ammonia

(b) an amide

(c) an alkyl halide and sodium cyanide

(d) an alkyl halide and sodium azide

Mixed Problems

12.48 Describe how to accomplish the following synthesis by specifying the reagents needed. More than one reaction may be required for each conversion.

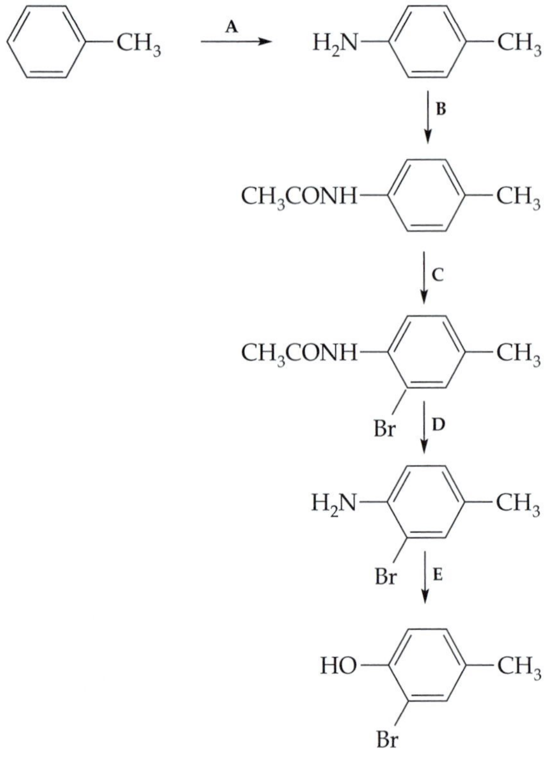

$C_6H_5CH_3$ —A→ H_2N–C_6H_4–CH_3 —B→ CH_3CONH–C_6H_4–CH_3 —C→ CH_3CONH–$C_6H_3(Br)$–CH_3 —D→ H_2N–$C_6H_3(Br)$–CH_3 —E→ HO–$C_6H_3(Br)$–CH_3

12.49 Show how to synthesize each of the following compounds, starting with benzene, toluene, or any compound produced in a previous part.

(a) *p*-cresol (*p*-methylphenol)

(b) *m*-dichlorobenzene

(c) *p*-fluorobenzoic acid

(d) *m*-bromoaniline

12.50 Specify a single chemical test that will distinguish between each of the following pairs of compounds, indicating what you would observe in each test:

(a) benzene and pyridine

(b) *p*-toluidine (*p*-methylaniline) and *N,N*-dimethyl-*p*-toluidine

(c) benzylamine and benzamide ($C_6H_5CONH_2$)

(d) 1,3-dimethylcyclohexane and 2,6-dimethylpiperidine

12.51 Indicate the simplest instrumental/spectroscopic technique that could be employed to distinguish between the pairs of compounds in Problem 12.50. Indicate what you would observe that would indicate the difference.

12.52 It is desired to synthesize a new sulfa drug from benzene, using the four major transformations shown. More than one step may be involved in each transformation. Show the reagents required to accomplish each step of this synthesis.

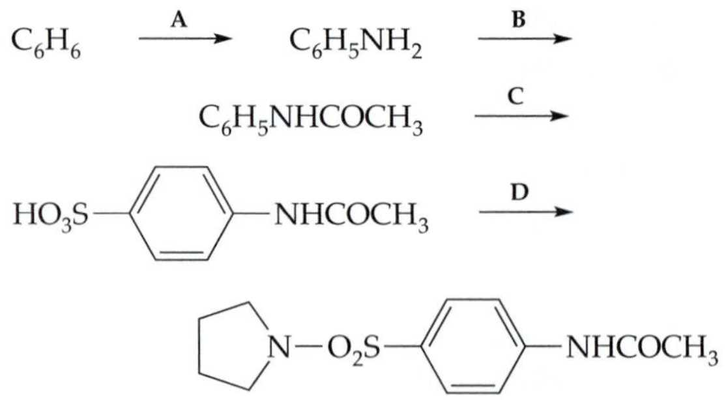

C_6H_6 —A→ $C_6H_5NH_2$ —B→ $C_6H_5NHCOCH_3$ —C→ HO_3S–C_6H_4–$NHCOCH_3$ —D→ (pyrrolidinyl)N–O_2S–C_6H_4–$NHCOCH_3$

12.53 Show the reactions needed to prepare *m*-chloroiodobenzene from benzene.

12.54 Show how to accomplish each conversion:

(a) anisole to *p*-iodoanisole

(b) 1-bromobutane to butylethylamine

(c) toluene to benzylamine

(d) bromocyclohexane to cyclohexylethylamine

(e) nitrobenzene to *m*-bromophenol

(f) toluene to *p*-iodobenzoic acid

12.55 Sunset Yellow (FDC No. 6) is a synthetic dye used to color margarine, among other uses. Show the two reactants and the reagents you would employ to prepare this dye using a diazonium coupling reaction.

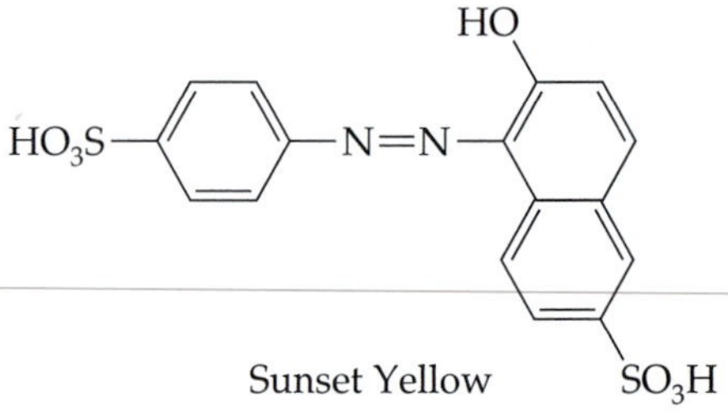

HO_3S–C_6H_4–N=N–(HO, SO_3H naphthalene)

Sunset Yellow

12.56 Propanil is a common herbicide. Show the reactions necessary to synthesize Propanil via the three transformations shown. More than one step may be required for each transformation.

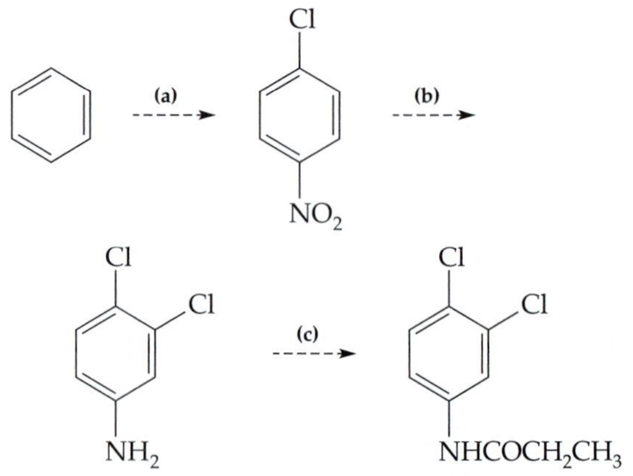

Benzene --(a)--> *p*-chloronitrobenzene (Cl, NO_2) --(b)--> 3,4-dichloroaniline (Cl, Cl, NH_2) --(c)--> Propanil (Cl, Cl, $NHCOCH_2CH_3$)

Propanil

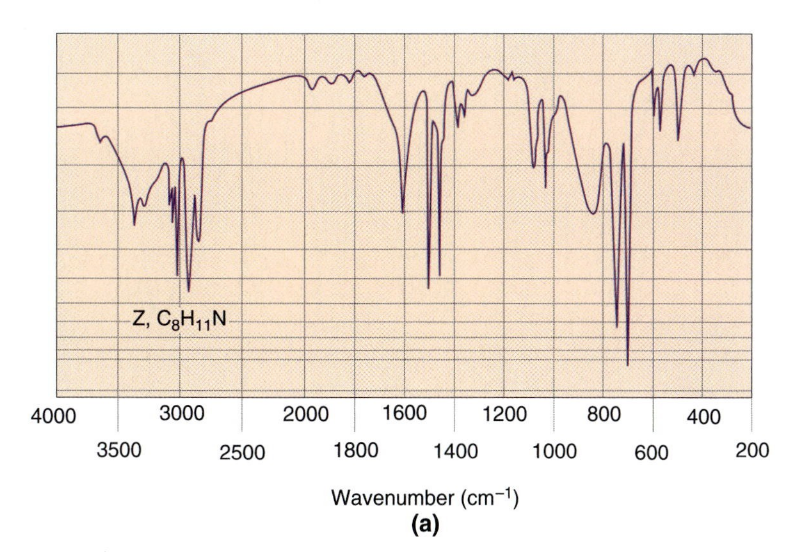

(5H)

Z, $C_8H_{11}N$

$C_8H_{11}N$

(2H + 2H)

(2H)

8.0 7.0 6.0 5.0 4.0 3.0 2.0 1.0 0

Chemical shift d (ppm)

(b)

FIGURE 12.5

(a) IR and (b) ^{1}H-NMR spectra of compound Z (Problem 12.58)

12.57 The local anesthetic lidocaine can be synthesized using 2,6-dimethylaniline and α-chloroacetyl chloride as starting materials. Show the reactions involved. (*Hint:* Use retrosynthesis.)

2,6-Dimethylaniline (CH_3, NH_2, CH_3) + $Cl{-}CO{-}CH_2Cl$ (α-Chloroacetyl chloride) --(several steps)--> Lidocaine (CH_3, $NHCOCH_2N(C_2H_5)_2$, CH_3)

12.58 Compound **X** (C_7H_7Br) shows ^{1}H-NMR peaks at 4.2 ppm (s, 2H) and 7.3 ppm (m, 5H). **X** reacts with sodium cyanide to yield **Y** (C_8H_7N), which has a similar ^{1}H-NMR spectrum. **Y** shows an IR peak at 2350 cm^{-1} and is insoluble in water. Reduction of **Y** with $LiAlH_4$ yields **Z** ($C_8H_{11}N$), which is insoluble in water but soluble in dilute acid and has the IR and ^{1}H-NMR spectra shown in Figure 12.5. What is the structure of **Z**?

12.59 Show the reactions necessary to complete the following transformations. [*Hint:* In parts (a) and (d), identify the new bonds that are easiest to form and plan to make them in the last steps of your synthesis.]

(a) → $CH_3CH_2CH_2CONH$

FIGURE 12.6

(a) IR and (b) ^{1}H-NMR spectra of an unknown compound (Problem 12.63).

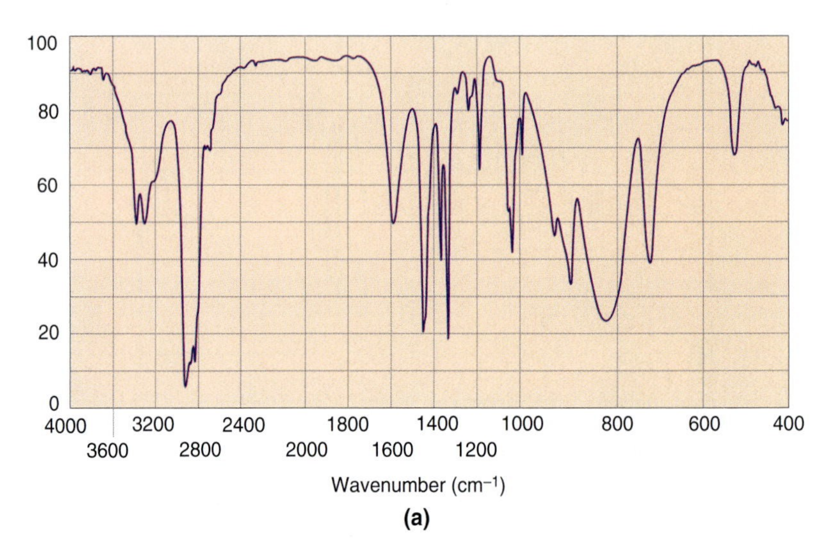

(a)

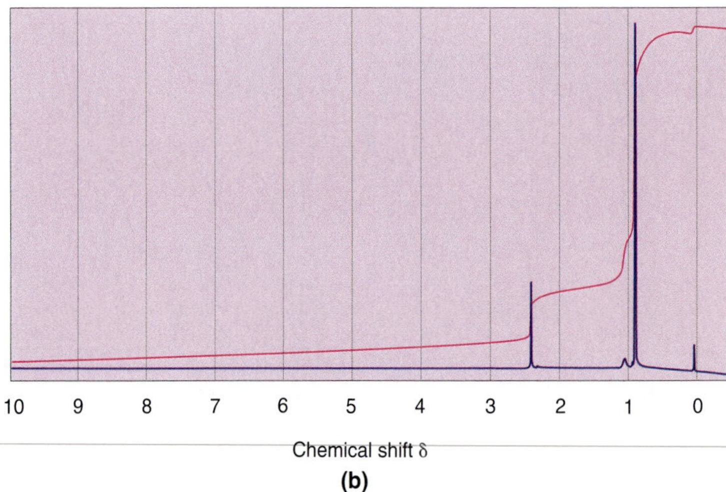

(b)

(b) Benzene ⟶ Benzene–CN

(c) Benzene–NO_2 ⟶ Br / Benzene–OH (meta)

(d) Benzene–CH_3 ⟶ $CH_3CONHCH_2CH_2$–Benzene

12.60 The mass spectrum of compound **A** shows a molecular ion at $m/z = 121$, a small peak at 122, and peaks at 106 and 77. **A** is insoluble in water but dissolves in dilute HCl. The ^{1}H-NMR spectrum shows a multiplet at 7.2 ppm, a singlet at 3.5 ppm, a singlet at 3.0 ppm, and a singlet at 2.1 ppm, with relative areas of 5:2:1:3. The IR spectrum shows a single peak near 3300 cm^{-1} and another peak at 1625 cm^{-1}. The UV-Vis spectrum shows a peak at 264 nm. Compound **A** reacts with tosyl chloride to produce a precipitate. Oxidation of **A** with hot potassium permanganate followed by removal of the MnO_2 precipitate and acidification of the colorless filtrate yields a colorless precipitate (**B**). **B** shows a very broad IR peak centered at 3300 cm^{-1} and sharp peaks at 1700 and 1625 cm^{-1}. Its ^{1}H-NMR spectrum shows a multiplet at 7.2 ppm and a singlet at 10.5 ppm, with relative areas of 5:1. What are the structures of **A** and **B**?

12.61 A compound with formula $C_8H_{11}N$ is insoluble in water but soluble in dilute HCl. Its mass spectrum shows peaks at $m/z = 121$, 106, 92, and 77. The IR spectrum has peaks at 1625 and 3300 cm^{-1}. There are peaks on the ^{1}H-NMR spectrum at 7.5 ppm (m, 5H), 3.2 ppm (q, 2H), and 1.0 ppm (t, 3H). What is the compound's structure?

CONCEPTUAL PROBLEM

Shades of Yellow

A textile designer was experimenting with different dyes. He found a substance that had a yellow color he liked, but the compound turned red in acid. Household ammonia restored the original yellow. The structure of the compound is shown here:

$(CH_3)_2N$—C_6H_4—N=N—C_6H_4—$SO_3^-\ Na^+$

- What is the molecular formula of this compound?
- Classify the amine group present in this compound.
- What feature of this compound causes it to be colored?
- What is the structure of this compound when it is dissolved in acidic solution?

The compound shown here is similar to both Sunset Yellow (see p. 369), currently used to give margarine its yellow color, and Butter Yellow (see p. 425), formerly used for that purpose.

- What does this suggest about the importance of the —$SO_3^-\ Na^+$ part of the structure in providing the yellow color to the dye?
- What is the approximate wavelength of UV-Vis absorption you would expect to see for the yellow dye?

12.62 Prolitane is an antidepressant drug for which there are several syntheses. Using benzene, butanoyl chloride ($CH_3CH_2CH_2COCl$), and pyrrolidine as starting materials, together with any other necessary reagents, devise a synthesis of Prolitane. (*Hint:* Use retrosynthesis.)

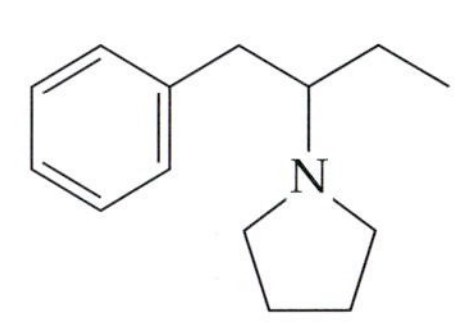

Prolitane

12.63 The mass spectrum of an unknown compound shows a molecular ion at $m/z = 87$. The IR and ^{1}H-NMR spectra are shown in Figure 12.6. The small singlet near 1.0 ppm disappears when the compound is shaken with D_2O. What is the compound's structure?

12.64 Phenacetin is a common analgesic used along with aspirin and caffeine in APC tablets, formerly used as pain relievers. Devise a synthesis starting with aniline. (*Hint:* Use retrosynthesis.)

CH_3CH_2O—C_6H_4—$NHCOCH_3$

Phenacetin

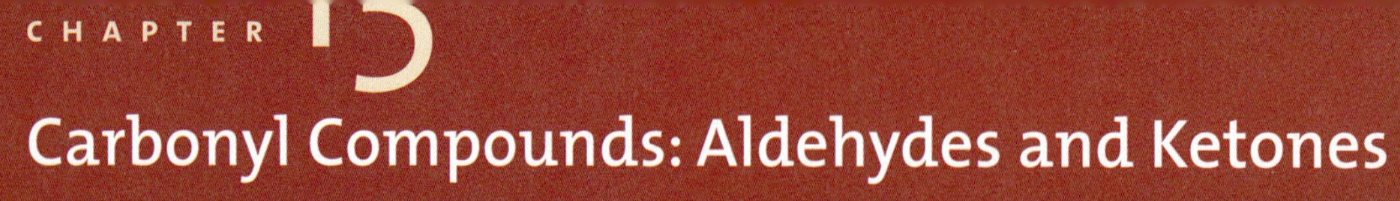

CHAPTER 15

Carbonyl Compounds: Aldehydes and Ketones

IN THIS STUDY of organic compounds, we have been concentrating on the chemistry of one family of compounds at a time. So far, each of these has had a functional group unique to it, but that is not the case with carbonyl compounds.

The term *carbonyl* is used two ways in organic chemistry. First, the **carbonyl group** is a functional group with a carbon double-bonded to oxygen (C=O), represented in condensed formulas as —CO—. The carbonyl group appears in two major groupings of families of compounds: (1) carbonyl compounds, including aldehydes (RCHO) and ketones (RCOR), and (2) carboxylic acids (RCOOH) and their derivatives, including acid halides (RCOCl), acid anhydrides [$(RCO)_2O$], esters (RCOOR′), and amides ($RCONH_2$). The chemistry of these two groupings of compounds varies slightly because of what is attached to the carbonyl group. Chapters 14 and 15 will describe carboxylic acids and their derivatives, while this chapter describes the chemistry of carbonyl compounds.

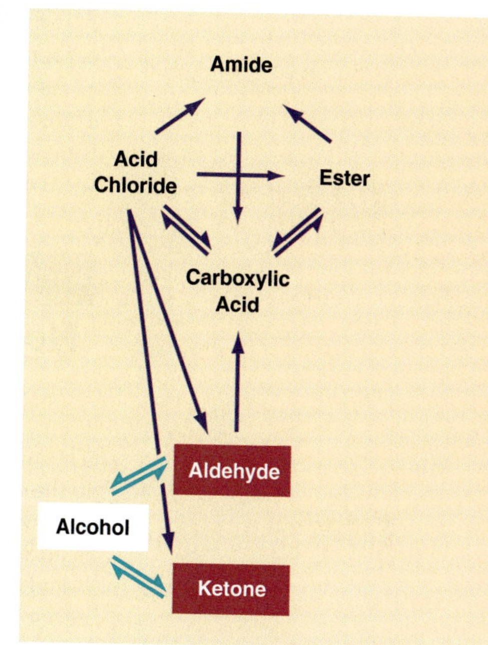

Aldehydes and ketones are both members of the carbonyl family of organic compounds.

The second use of the term *carbonyl* is to refer to the first grouping mentioned above, the **carbonyl compounds,** which consist of the two families aldehydes and ketones. These families are considered together because of the great similarity in their physical and chemical behavior; it is the carbonyl group that accounts for the distinctive chemistry of both. They can be characterized in this way:

- **Aldehydes** (RCHO and ArCHO) have at least one hydrogen attached to the carbonyl carbon:

$$R-\overset{O}{\overset{\|}{C}}-H \qquad Ar-\overset{O}{\overset{\|}{C}}-H$$

Aldehydes

- **Ketones** (R_2CO, Ar_2CO, or ArCOR) have two alkyl or aryl groups attached to the carbonyl carbon—and no hydrogen:

$$R-\overset{O}{\overset{\|}{C}}-R' \qquad R-\overset{O}{\overset{\|}{C}}-Ar \qquad Ar-\overset{O}{\overset{\|}{C}}-Ar'$$

Ketones

Ketones may be symmetrical (two identical groups attached to the carbonyl carbon) or unsymmetrical (two different groups attached to the carbonyl carbon). Also, the carbonyl carbon may be incorporated into a carbocyclic ring.

◀ Both sexes of the African civet cat produce civetone, a strong musky-smelling carbonyl compound. This substance has been used in perfumes for thousnds of years, and is now synthesized industrially.

Compounds with double bonds typically undergo addition reactions, as demonstrated by alkenes, which undergo addition to the carbon-carbon double bond (see Chapter 6). Addition to the carbon-oxygen double bond of the carbonyl group (see Sections 13.3–13.5) is different. Alkenes, with a relatively nonpolarized double bond, undergo mainly electrophilic addition; carbonyls, with a double bond polarized toward the more electronegative oxygen, undergo mainly nucleophilic addition at carbon.

$$>C{=}C< \; + \; E^+ \qquad\qquad >C{=}O \; + \; {}^-{:}Nuc$$

Addition to an alkene — Addition to a carbonyl group

Addition to a carbonyl group is an especially significant reaction because the use of carbon nucleophiles in attacking the carbonyl carbon results in the formation of new carbon-carbon bonds.

As we begin to consider carbonyl compounds, the right-hand portion of the Key to Transformations (inside front cover) becomes relevant. It summarizes the transformational chemistry among the families that have a carbon-oxygen double bond (families we will cover in Chapters 13 through 15).

13.1 Nomenclature of Carbonyls

13.1.1 IUPAC Nomenclature of Aldehydes

The aldehyde group is symbolized by —CHO, implying that the hydrogen is attached to the carbonyl carbon. Aldehydes are named using the appropriate alkane parent name for the number of carbons in the chain containing the aldehyde group. The suffix *-al* is added to the alkane name to indicate that the compound is an aldehyde. The carbonyl carbon is always carbon 1 (therefore, the number is not mentioned in the name). Other substituents, including hydroxyl, amino, and alkoxy groups and multiple carbon-carbon bonds, are designated by numerical positions relative to the aldehyde group. Typical names are propanal, 3-butenal, and 4-hydroxy-3-methylpentanal.

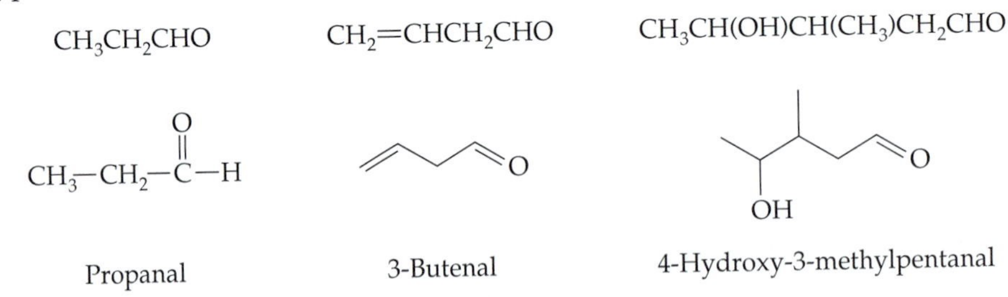

Propanal — 3-Butenal — 4-Hydroxy-3-methylpentanal

For cyclic aldehydes, the suffix *-carbaldehyde* is added to the ring name (for example, cyclobutanecarbaldehyde). The position of attachment of the carbonyl carbon to the ring is always numbered as carbon 1 of the ring, but the number 1 is not mentioned in the name. In the case of benzenoid aldehydes, the parent IUPAC name is benzenecarbaldehyde, but in practice this is almost never used. Instead, the historical name benzaldehyde is used, and substituents are designated by name and position.

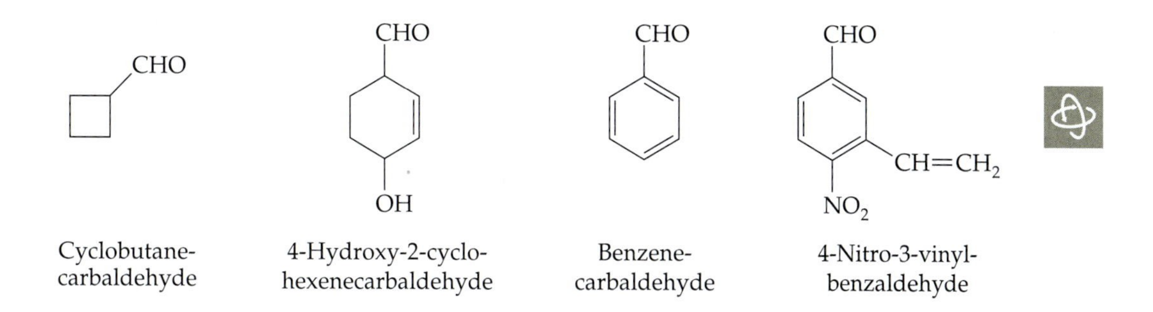

PROBLEM 13.1

Draw structures corresponding to the following names: (a) 2-methylbutanal, (b) 3,3-dimethyl-hexanal, (c) 2-bromocyclopropanecarbaldehyde, (d) *p*-nitrobenzaldehyde, and (e) 2-pentynal.

PROBLEM 13.2

Assign IUPAC names to the following aldehydes:

(a) $(CH_3)_2CHCHO$ (b) 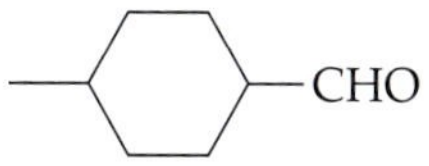 CHO (c) 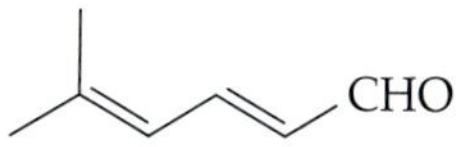 CHO

(d) 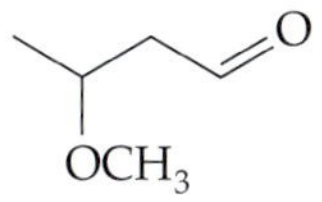 O, OCH_3 (e) 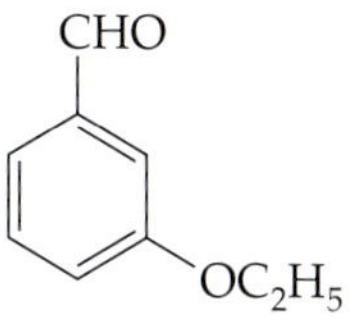 CHO, OC_2H_5 (f)  CHO

13.1.2 IUPAC Nomenclature of Ketones

Ketones are named using the alkane name for the longest straight chain of carbon atoms containing the ketone group and adding the suffix *-one* to the name. The position of the ketone must be indicated by a number, which is kept as small as possible. The positional designation of a ketone group takes precedence over other substituents—including multiple carbon-carbon bonds and hydroxyl, amino, and alkoxy groups—*except* aldehyde groups. Typical names are 2-butanone, 4-penten-2-one, and 4-phenyl-2-butanone. The same process is used to name cyclic ketones, but the number of the carbonyl carbon is usually not given, since it is designated as carbon 1.

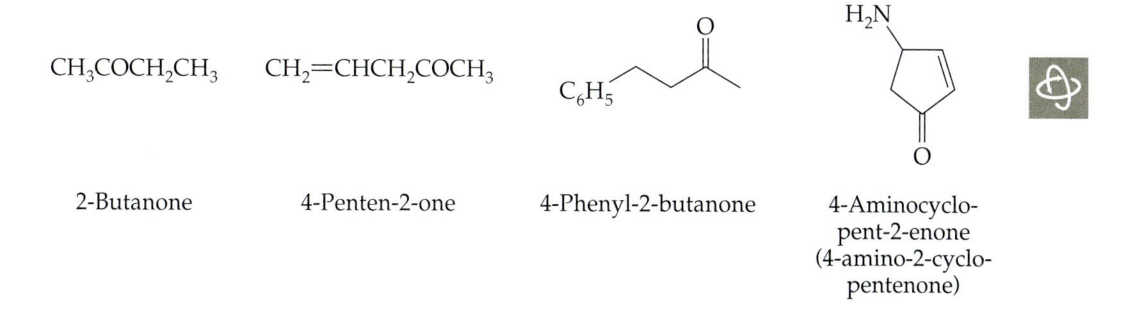

When the ketone group is not the highest-priority group (see Appendix A), it is treated as a substituent and referred to as an *oxo group*, located in the chain or ring by a position number. An example is 3-oxobutanal (CH_3COCH_2CHO).

How to Solve a Problem

Assign an IUPAC name to the following compound:

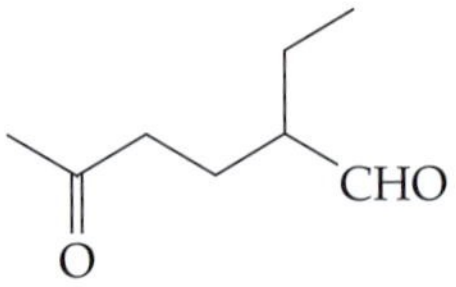

CHO

O

SOLUTION

There are two functional groups present. The aldehyde group has a higher priority than the ketone group. The longest continuous chain containing the aldehyde group has six carbons. Therefore, the name is derived from hexane, and the aldehyde family requires the suffix *-al*. Thus, the parent name is hexanal. There are two substituents: the oxo (ketone) group at carbon 5 and the ethyl group at carbon 2. The ethyl group takes alphabetical precedence in the name. Therefore, the name is 2-ethyl-5-oxohexanal.

PROBLEM 13.3

Draw structures corresponding to the following names: (a) 3-pentanone, (b) 2,2,6,6-tetra-methyl-cyclohex-3-enone, and (c) 1-hexen-3-one.

PROBLEM 13.4

Assign names to the following ketones:

(a) $CH_3CH(C_6H_5)COCH_3$ (b) O

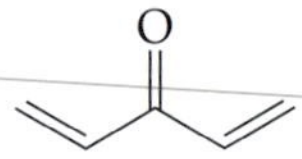

(c) =O F (d) O CH_3NH

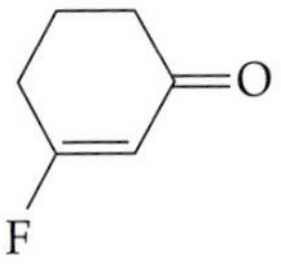

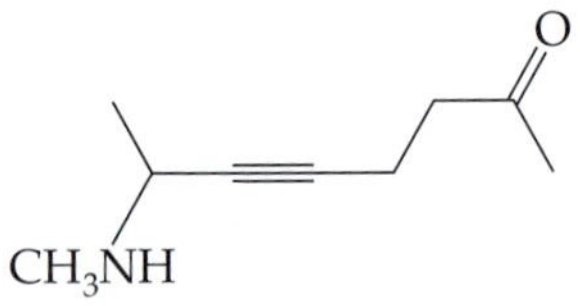

13.1.3 Historical/Common Names

The aldehyde functional group (—CHO) has been historically referred to as the **formyl group.** Occasionally, you will see *formyl* used as a substituent name—for example, cyclobutanecarbaldehyde (shown in Section 13.1.2) may be called formylcyclobutane.

Many aldehydes have historical names derived from a carboxylic acid name (because aldehydes are so readily oxidized to carboxylic acids; see Section 13.2.2). A few of these common names are widely used and should be learned: The common name formaldehyde (H_2CO) is derived from formic acid (HCOOH); acetaldehyde (CH_3CHO) is derived from acetic acid (CH_3COOH); and benzaldehyde (C_6H_5CHO) is derived from benzoic acid (C_6H_5COOH).

	H_2CO or CH_2O	CH_3CHO	C_6H_5CHO
Common name:	Formaldehyde	Acetaldehyde	Benzaldehyde
IUPAC name:	Methanal	Ethanal	Benzenecarbaldehyde

In many instances, the suffix *-al* has been attached to historical names to indicate the presence of an aldehyde group. Examples are retinal, the aldehyde form of vitamin A, and pyridoxal, the aldehyde form of vitamin B_6:

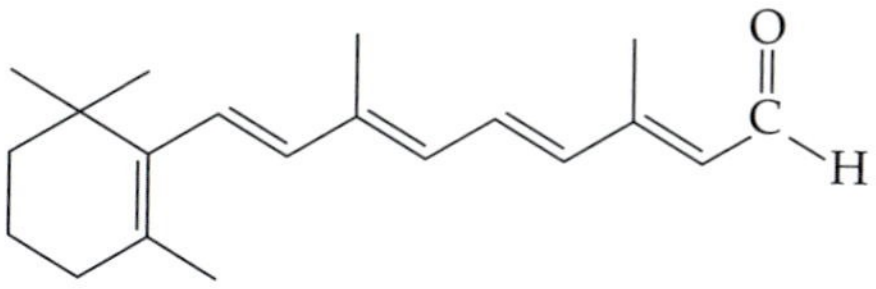

Retinal

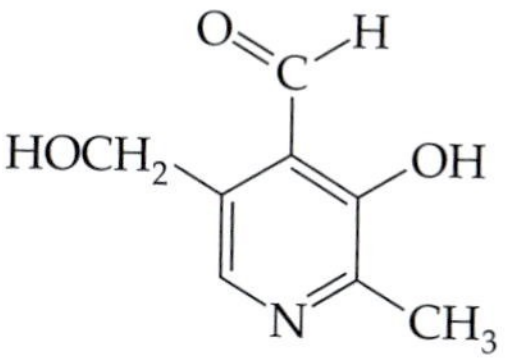

Pyridoxal

There are a few common ketone names that you should learn, including acetone, acetophenone, and benzophenone. In addition, many ketones are referred to by the names of the two groups attached to the carbonyl carbon followed by the word *ketone*. Examples are methyl ethyl ketone and dicyclopropyl ketone:

$CH_3{-}C({=}O){-}CH_3$ Acetone

$CH_3{-}C({=}O){-}C_6H_5$ Acetophenone

$C_6H_5{-}C({=}O){-}C_6H_5$ Benzophenone

$CH_3{-}C({=}O){-}C_2H_5$ Methyl ethyl ketone

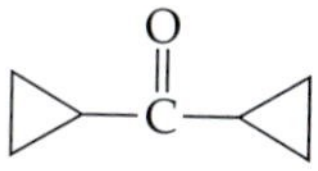

Dicyclopropyl ketone

Many ketones with complex structures have been given simple historically derived names in which the suffix *-one* has been attached to indicate the presence of a ketone group. Examples are testosterone (the male sex hormone; see Section 8.3.2), cortisone (the immune system medicinal; see Section 13.3.1), and civetone (the perfume ingredient excreted by the civet cat; see Section 13.7).

The group $R{-}C({=}O){-}$ (also written RCO—) is referred to generically as the **acyl group.** A specific acyl group is named after the carboxylic acid from which it is derived. For example, $CH_3C({=}O){-}$ is the acetyl group, which is derived from acetic acid, $CH_3C({=}O){-}OH$.

PROBLEM 13.5

Draw structures corresponding to the following names: (a) cyclobutyl methyl ketone, (b) *p*-nitroacetophenone, (c) *p*-*N*,*N*-dimethylaminobenzaldehyde, (d) diisopropyl ketone, (e) *p*-iodobenzophenone, and (f) diethyl ketone.

13.2

Structure and Properties of Carbonyls

13.2.1 Electronic Structure of the Carbonyl Group

The electronic structure of the carbonyl group is similar to that of the carbon-carbon double bond. Trigonally hybridized carbon is attached to trigonally hybridized oxygen through an sp^2-sp^2 sigma bond. The two remaining sp^2 orbitals of carbon are attached to alkyl or aryl groups or hydrogens. The other two sp^2 orbitals of oxygen remain as lone (unshared) pairs. Both carbon and oxygen also have an unhybridized $2p$ orbital containing one electron. Lateral overlap of these two orbitals produces a pi molecular orbital containing two electrons. Figure 13.1 shows the atomic orbital hybridization and the structure of the carbonyl group.

Note that the carbonyl carbon, the oxygen, and the two atoms bonded to the carbon through overlap with its sp^2 hybrid orbitals all lie in a single plane. For example, the three carbons and the oxygen of acetone lie in a plane (Figure 13.2).

In a C=C bond, the pi electron cloud is relatively equally shared by the two carbons. However, in a C=O bond, the sharing is unequal because the oxygen is more electronegative than the carbon, leading to polarization of the pi electron cloud toward oxygen. The carbonyl group can also be described as a resonance hybrid of two contributing structures, with the dipolar structure contributing much less than the nonpolar structure. The carbonyl group is often represented by a single structure with partial charges indicating the bond polarization.

$$\rangle C{=}\ddot{\underset{..}{O}} \longleftrightarrow \rangle \overset{+}{C}{-}\ddot{\underset{..}{O}}{:}^{-} \qquad\qquad \rangle \overset{\delta+}{C}{=}\overset{\delta-}{\ddot{\underset{..}{O}}}$$

Contributing structures for resonance stabilization of the carbonyl group

Bond polarization of the carbonyl group

13.2.2 Chemical Behavior of the Carbonyl Group

The distinctive electronic structure of the carbonyl group accounts for all of the specific reactions of aldehydes and ketones to be discussed in later sections. Four different kinds of chemical behavior lie behind these reactions: nucleophilic addition, protonation, α-hydrogen acidity, and oxidation of carbonyls.

FIGURE 13.1

Hybridization of atoms involved in a carbonyl bond and the structure of the carbonyl group.

	Ground state		Bonding orbitals
Carbon (Z = 6)	$1s^22s^22p^2$	(hybridization) →	$1s^2$ ↑ ↑ ↑ ↑ ; $2(sp^2)^3$ $2p$
Oxygen (Z = 8)	$1s^22s^22p^4$	(hybridization) →	$1s^2$ ↑↓ ↑↓ ↑ ↑ ; $2(sp^2)^5$ $2p$

$$\rangle C{=}\ddot{\underset{..}{O}}$$

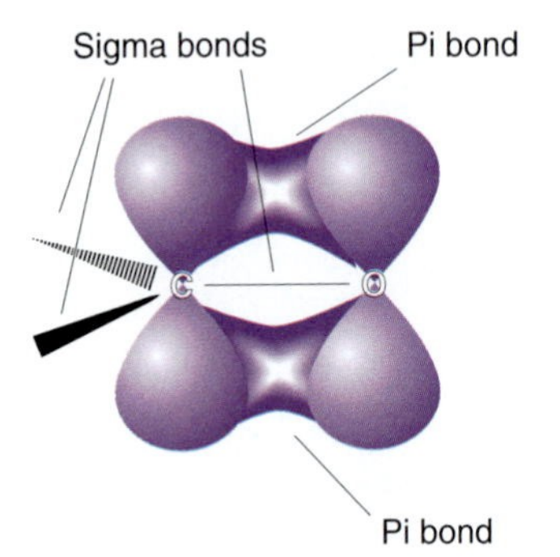

Nucleophilic Addition

The most characteristic reaction of the carbonyl group, as a result of its polarized structure, is nucleophilic addition by a wide variety of nucleophiles (:Nuc). The nucleophile attacks the electron-deficient carbonyl carbon from above or below the plane of the carbonyl group. Further, the intermediate resulting from such additions is an alkoxide, which is stable (see Section 4.2.3). Such reactions are completed by adding a proton source (even water) during work-up, so that the strongly basic alkoxide forms an alcohol as product.

$$C{=}\ddot{O}: \; + \; :Nuc \longrightarrow \left[C(\ddot{O}:^-)(Nuc) \right] \xrightarrow{H^+} C(\ddot{O}H)(Nuc)$$

Carbonyl — Alkoxide — Alcohol

The nucleophilic addition reactions discussed in the following sections all follow this general reaction mechanism and can be divided into three types:

1. Addition of the simple reagents hydride anion ($:H^-$) and cyanide anion ($:CN^-$) (Section 13.3)
2. Addition of amines (RNH_2) or alcohols (ROH) (Section 13.4)
3. Addition of carbanionic nucleophiles, such as organometallics (R—metal) or ylides (the Wittig reagent) (Section 13.5)

Protonation

The oxygen of a carbonyl group has a high electron density and therefore is susceptible to protonation. The positive charge resulting from such a reaction is stabilized by delocalization of the electrons of the double bond, which confers considerable positive character on the carbonyl carbon. Such a carbon is much more electron-deficient than is the carbon in an unprotonated carbonyl group. Therefore, protonation via acid catalysis is occasionally used to increase the susceptibility of the carbonyl group to attack by a nucleophile. This kind of reaction will be discussed in Section 13.4.2 (addition of alcohols).

$$C{=}\ddot{O}: \xrightarrow{H^+} \left[C{=}\overset{+}{\ddot{O}}{-}H \longleftrightarrow \overset{+}{C}{-}\ddot{O}{-}H \right] \xrightarrow{:Nuc} C(\ddot{O}H)(Nuc)$$

Carbonyl — Protonated carbonyl group — Alcohol

α-Hydrogen Acidity

The significant electron deficiency of the carbonyl carbon has the effect of making an **alpha (or α) hydrogen** (the hydrogen on an α-carbon, which is next to a carbonyl carbon) relatively acidic and therefore easy to remove with a moderate base. In addition, the resulting carbanion, called an **enolate anion,** is stabilized by resonance delocalization with the carbonyl group. This carbanion is a nucleophile, able to be used in nucleophilic substitutions and additions.

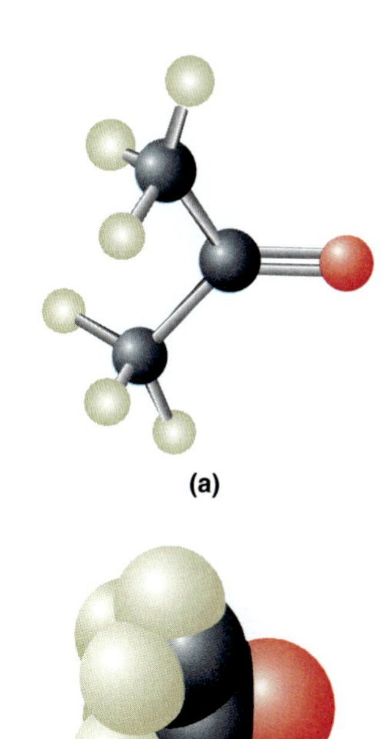

FIGURE 13.2

(a) Ball-and-stick and (b) space-filling models of acetone (CH_3COCH_3).

α-Hydrogen

Enolate anion

Oxidation of Carbonyls

Ketones are inert to the usual oxidation reactions. However, aldehydes are oxidized to carboxylic acids by a variety of oxidants.

$$\text{R—CHO} \xrightarrow{[O]} \text{R—COOH} \qquad \text{R}_2\text{CO} \xrightarrow{[O]} \text{No reaction}$$

Both strong oxidants such as potassium permanganate and Jones' reagent (sodium dichromate and sulfuric acid) and weak oxidants such as nitric acid will oxidize aldehydes.

$$\text{R—CHO} \xrightarrow{HNO_3} \text{R—COOH}$$

Simply standing in air causes some aldehydes to be oxidized. For example, benzaldehyde (C_6H_5CHO) is oxidized to benzoic acid (C_6H_5COOH) unless it is protected from oxygen.

As a result of their different susceptibilities to oxidation, aldehydes can be distinguished from ketones using a qualitative oxidation test (see Section 13.2.5).

Vanillin, the flavoring extract from vanilla beans, is not only an aldehyde—as a phenol it is the enol tautomer of a ketone.

13.2.3 Tautomerism of Carbonyls

The resonance contributing structures of the enolate anion show a negative charge in two locations: on the α-carbon and on oxygen. Reprotonation of an enolate anion can occur at either location. If the hydrogen attaches to the α-carbon, the result is the carbonyl structure, with a carbon-oxygen double bond, and called the **keto form.** (In this context, the term *keto form* applies to aldehydes and ketones.) If the hydrogen attaches instead to the oxygen, the result is the **enol form,** containing a carbon-carbon double bond (*enol* from the *-en* of an alkene and the *-ol* of an alcohol).

Keto form

Enolate anion

Enol form

The enol and keto forms of most carbonyl compounds exist in equilibrium, but the keto form is usually the overwhelmingly predominant component. For example, a vial of pure cyclohexanone exists as an equilibrium mixture of more than 99% in the keto form and less than 1% in the enol form. On the other hand, pentane-2,4-dione exists as an equilibrium mixture of 24% keto form and 76% enol form, mainly because of stabilization of the enol form by conjugation and hydrogen bonding.

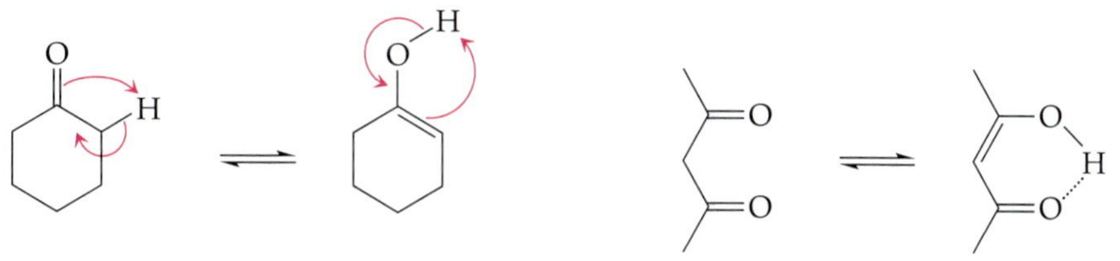

Keto (>99%) Enol (<1%)

Cyclohexanone

Keto (24%) Enol (76%)

Pentane-2,4-dione

The phenomenon of **tautomerism** is the equilibrium between the enol and keto forms of a compound, and it can occur spontaneously or upon acid or base catalysis. The resulting enol and ketone/aldehyde are constitutional isomers referred to as **tautomers,** which differ with respect to the location of an acidic hydrogen atom and a double bond. The process of **enolization** is the conversion of a keto form to an enol form; the process of **ketonization** is the conversion of an enol form to a keto form (even of an aldehyde). The electron flows (not the reaction mechanism) for the enolization and ketonization processes are shown above for cyclohexanone.

For most carbonyl compounds, the keto form is by far the predominant tautomer, and the enol form can be ignored in most chemical reactions. In those few reactions in which an enol is expected to form, the keto product is normally obtained as a result of rapid ketonization.

PROBLEM 13.6

Draw the enol form(s) of each of the following carbonyl compounds: (a) 2-propanone, (b) butanal, (c) cyclohexanone, and (d) 3-oxobutanal.

13.2.4 Physical Properties of Carbonyls

Carbonyl compounds are more soluble in water than comparably sized hydrocarbons and ethers, but not as soluble in water as alcohols. The carbonyl group is an effective electron donor in hydrogen bonding with water, enough so that C_3 and smaller carbonyl compounds are infinitely soluble in water. Acetone (CH_3COCH_3) and acetaldehyde (CH_3CHO) are totally miscible with water.

As a result of the polarity of the carbon-oxygen double bond, some dipole-dipole association occurs between molecules of liquid carbonyl compounds; this must be broken before vaporization can occur. Therefore, carbonyls are slightly higher-boiling than comparable hydrocarbons and ethers because of the extra energy required to dissociate clusters of their molecules.

13.2.5 Detection of Carbonyl Compounds

Spectroscopic Detection

The easiest way to detect the presence of a carbonyl group is by spectroscopy. The carbonyl group has a very characteristic strong and sharp absorption in the infrared spectrum at about 1720 cm^{-1}. The hydrogen of an aldehyde shows a characteristic moderate absorption at 2715 cm^{-1} (Figure 13.3a).

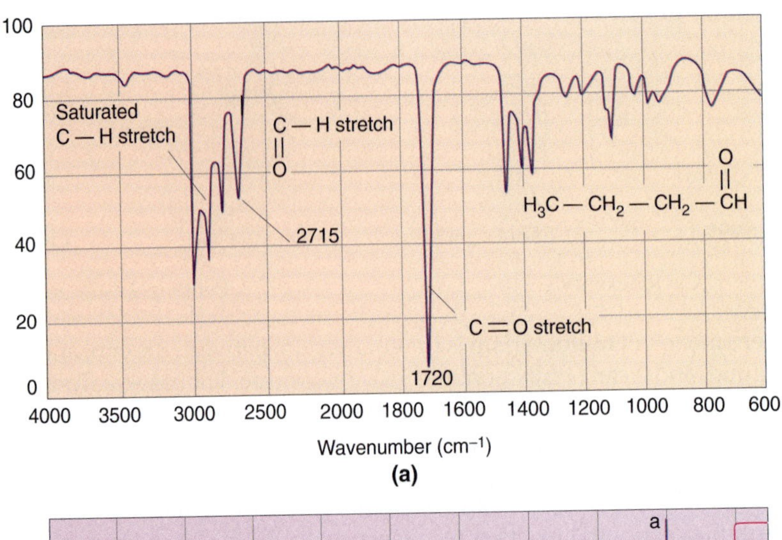

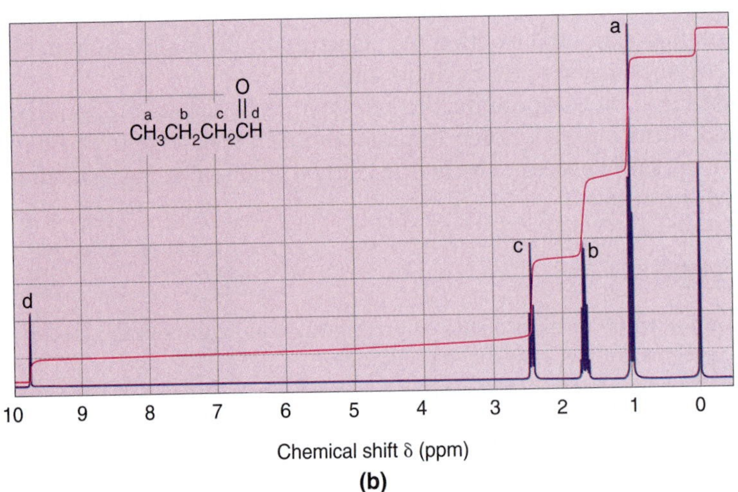

FIGURE 13.3

(a) IR spectrum and (b) ¹H-NMR spectrum of butanal.

NMR spectroscopy is also important. Because of its strong electron-withdrawing character (due to the positive polarity of the carbonyl carbon), the carbonyl group significantly deshields adjacent protons. Thus, whereas a normal methyl group shows a peak at about 1.0 ppm, a methyl group next to a carbonyl group shows a peak at about 2.5 ppm. The carbonyl hydrogen of an aldehyde is strongly deshielded and shows a very characteristic singlet at about 9 ppm (Figure 13.3b).

Chemical Detection

The presence of a carbonyl group is usually detected chemically by a color test involving reaction with a reagent called *2,4-dinitrophenylhydrazine (2,4-DNPH)*. As we will see in Section 13.4.1, this reagent reacts with the carbonyl group to form a crystalline *imine* that is bright orange in color, normally precipitates readily, and can be characterized by its melting point.

An aldehyde can be distinguished from a ketone by the *Tollens' test*, sometimes called the *silver mirror test*. This test relies on the fact that aldehydes can be easily oxidized to carboxylic acids, whereas ketones cannot be readily oxidized (review Sections 13.2.2 and 4.4):

$$\text{R—CHO} \xrightarrow{[O]} \text{R—COOH} \qquad \text{R}_2\text{CO} \xrightarrow{[O]} \text{No reaction}$$

A solution of silver nitrate in ammonia (Tollens' reagent) is added to the carbonyl compound in a test tube. If an aldehyde is present, the silver oxidizes it to a soluble carboxylate anion. The silver cation is reduced to metallic silver, which precipitates and coats the test tube with a "silver mirror." There is no reaction with a ketone.

$$\underset{\text{Aldehyde}}{R{-}CHO} + \underset{\text{Silver-ammonia complex ion}}{Ag(NH_3)_2^+} \longrightarrow \underset{\text{Carboxylate anion}}{R{-}COO^-} + \underset{\text{Metallic silver (silver mirror)}}{Ag^0\downarrow}$$

How to Solve a Problem

A compound with molecular formula $C_6H_{10}O$ does not react with metallic sodium or with bromine in CCl_4. However, it forms an orange precipitate with 2,4-DNPH and produces a silver mirror with Tollens' reagent. Draw a possible structure for the compound (there is more than one that fits).

PROBLEM ANALYSIS

First, the molecular formula should be "mined" for all possible information. It shows an HDI of 2, but there are no double or triple carbon-carbon bonds present because they would have reacted with bromine. Thus, the compound must contain one or two carbocyclic rings. However, we remember that a carbonyl group—possible because of the oxygen present—represents an HDI of 1 because of its double bond. These considerations indicate that $C_6H_{10}O$ must be either a carbocyclic compound with a carbonyl group or a bicyclic compound with a noncarbonyl oxygen-containing functional group.

Next, we try to deduce the nature of the oxygen functional group. We have studied only three oxygen functional groups to this point: alcohols, ethers, and carbonyls. We recall the characteristics of each of these: Alcohols are reactive, ethers are inert, and carbonyls are reactive. The absence of reaction with sodium indicates that the compound cannot be an alcohol (otherwise an alkoxide would have formed; see Section 4.2.3). The reaction with 2,4-DNPH indicates that the compound must contain a carbonyl group, and the positive Tollens' test indicates that it must be an aldehyde.

SOLUTION

$C_6H_{10}O$ must be an aldehyde containing a single carbocyclic ring. Since no information is provided to indicate the size of the ring, cyclopentanecarbaldehyde, among other compounds, is one possible structure.

CHO

Cyclopentanecarbaldehyde
($C_6H_{10}O$)

PROBLEM 13.7

Draw two other structures that fit the information provided in the preceding example problem. How could you distinguish among the three possible structures?

PROBLEM 13.8

A compound with molecular formula C_4H_8O shows an IR peak at 1700 cm^{-1}. The ^1H-NMR spectrum shows peaks at 2.6 ppm (s, 1.5H), 2.4 ppm (q, 1H), and 1.0 ppm (t, 1.5H). What is the compound's structure?

PROBLEM 13.9

A compound with the same molecular formula (C_4H_8O) as the compound in Problem 13.8 shows an IR peak at 1690 cm^{-1}. The NMR spectrum shows peaks at 0.9 ppm (d, 6H), 2.6 ppm (m, 1H), and 9.5 ppm (s, 1H). What is this compound's structure?

13.3

Addition to Carbonyls: Simple Nucleophiles

The general mechanism for the addition of nucleophiles to carbonyl compounds is shown here.

General nucleophilic addition mechanism:

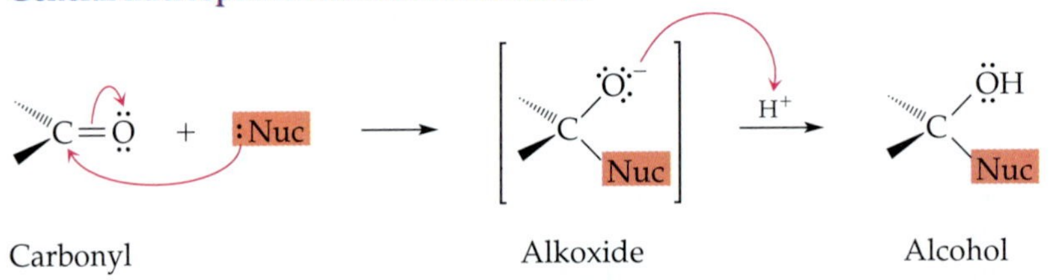

Carbonyl + :Nuc → [Alkoxide] —H^+→ Alcohol

Most of the addition reactions in this and the next two sections follow this general mechanism, with the major differences being in the nature of the nucleophile. The first step is the nucleophilic addition itself; the second step is the "work-up" of the reaction by protonation, usually in water.

13.3.1 Reduction of Carbonyls

Hydride Reduction to Alcohols

Carbonyl compounds are reduced to alcohols by reaction with sodium borohydride ($NaBH_4$), one of a family of *metal hydrides*. The term *hydride* implies that the hydrogen reacts not as a proton, but as an anion—that is, as :H^-, the hydride anion. Hydrogen serves as the anion in this case because the metal-hydrogen bond is polarized such that the metal (boron in this case) is positive, and, therefore, the hydrogen must be negative.

The borohydride reagent, in effect, *adds* a hydride anion (a very strong nucleophile) to the carbonyl carbon to form an alkoxide. Because there are four hydrogens on each boron atom, the reagent can reduce four equivalents of carbonyl compound. The intermediate alkoxide is converted to the alcohol simply by addition of water.

General reduction reaction:

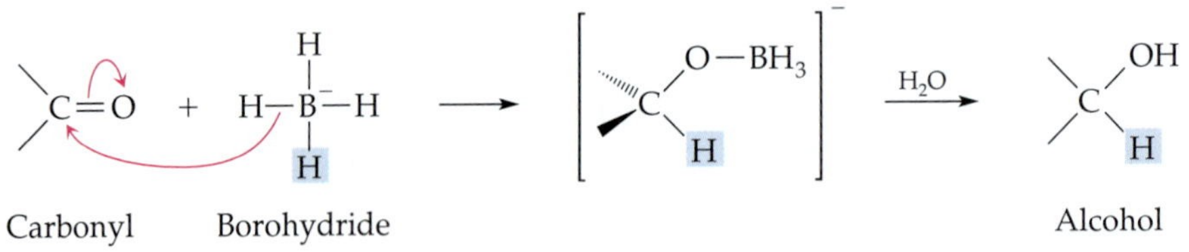

Carbonyl + Borohydride ($H-B^--H$) → [$C-O-BH_3$]$^-$ —H_2O→ Alcohol

Aldehydes form primary alcohols upon hydride reduction. This is the reverse of the reaction studied in Section 4.4, in which primary alcohols form aldehydes upon mild oxidation with pyridinium chlorochromate (PCC). Ketones form secondary alcohols upon hydride reduction. This is the reverse of the reaction studied in Section 4.4 in which secondary alcohols are oxidized to ketones with dichromate or permanganate reagents. These interconversions are represented in the Key to Transformations.

$$\underset{\text{Aldehyde}}{\text{RCHO}} \xrightarrow{\text{NaBH}_4} (\text{RCH}_2\text{O})_4\text{B}^- \ \text{Na}^+ \xrightarrow{\text{H}_2\text{O}} \underset{1^\circ \text{ alcohol}}{\text{RCH}_2\text{—OH}}$$

$$\underset{\text{Ketone}}{\text{R}_2\text{CO}} \xrightarrow{\text{NaBH}_4} (\text{R}_2\text{CHO})_4\text{B}^- \ \text{Na}^+ \xrightarrow{\text{H}_2\text{O}} \underset{2^\circ \text{ alcohol}}{\text{R}_2\text{CH—OH}}$$

The addition of the hydride ion to a carbonyl carbon can occur from above or below the plane of the carbonyl group. In the case of an unsymmetrical ketone, this means that two configurational isomers are produced, *R* and *S* enantiomers. Normally, they are produced in equal quantity, and the isolated product is thus the racemic secondary alcohol.

$$\text{RR}'\text{C=O} \xrightarrow{\text{NaBH}_4} \text{RR}'\text{C(H)OH} + \text{RR}'\text{C(H)OH}$$

Unsymmetrical ketone — Racemic secondary alcohol

PROBLEM 13.10

Show the product when each of the following is reduced with sodium borohydride:

(a) *p*-bromobenzaldehyde (b) isopropyl phenyl ketone (c) dicyclopropyl ketone

PROBLEM 13.11

Draw the structure of the carbonyl compound you would use to prepare each of the following alcohols by $NaBH_4$ reduction:

(a) ethanol (b) diphenylmethanol (c) 2-cyclopentenol

Reduction of a Carbonyl Group to a Methylene Group

A carbonyl group can also be reduced to a methylene group. However, this cannot be done using a metal hydride.

The Clemmensen reduction, usually employed in the reduction of an acyl aromatic (RCOAr) to an alkyl aromatic (RCH_2Ar) using zinc amalgam and HCl, has already been discussed in Section 9.3.6. This reaction is most effective when the carbonyl group is next to an aromatic ring, as it is in compounds produced via the Friedel-Crafts acylation. For example, acetophenone can be reduced to ethylbenzene:

$$\text{C}_6\text{H}_5\text{COCH}_3 \xrightarrow[\text{HCl}]{\text{Zn/Hg}} \text{C}_6\text{H}_5\text{CH}_2\text{CH}_3$$

Acetophenone — Ethylbenzene

A much milder and higher-yield method for reduction of all carbonyl compounds involves the desulfurization of a thioacetal; it will be described in Section 13.4.2.

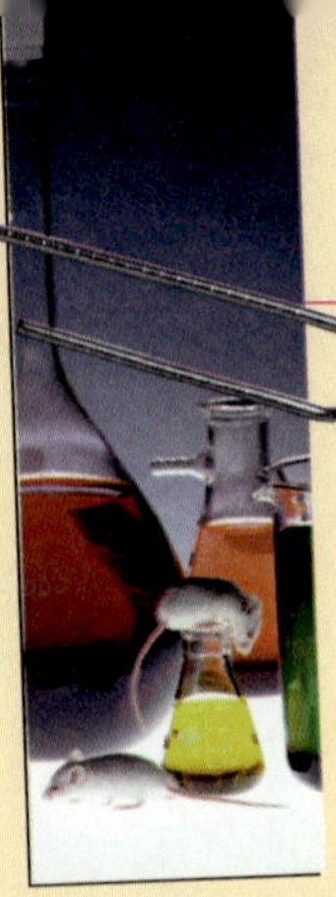

CHEMISTRY AT WORK

IN THE BODY

Cortisone and Hydrocortisone

Stress is a fact of life. On an emotional level, you deal with it every day; on a physiological level, so does your body. When you experience emotional or physical trauma, or even just the effects of an infection, your body sends out hormones to counteract the abnormality. One such hormone is hydrocortisone (cortisol), produced in the adrenal cortex located at the base of the skull. Hydrocortisone generally helps to keep your metabolism running smoothly by keeping your blood sugar level constant, even though your food intake is intermittent. Under stressful conditions, however, hydrocortisone causes stored proteins to break down and be released as amino acids into the bloodstream, providing building blocks for tissue repair. It also causes blood pressure to rise. At higher concentrations, hydrocortisone suppresses the activity of the immune system and can therefore act as an antiinflammatory agent. In this role, it has real value as a medicinal.

Hydrocortisone is a complex structure containing four rings, two carbonyl groups, three hydroxyl groups, one double bond, and seven stereocenters. It belongs to the steroid family of compounds (see Section 8.3.2) and is one of 128 possible stereoisomers. The body, as it turns out, has the means to produce the hormone hydrocortisone from the drug cortisone.

Cortisone is a medicinal widely used to treat inflammation, especially of the joints and especially resulting from injury; it is administered by injection or taken orally. Cortisone was first obtained by extraction from the brains of cattle, but is now readily available from pharmaceutical companies at relatively low cost because it is prepared by organic synthesis. It, like hydrocortisone, belongs to the steroid family. There are only six stereocenters in cortisone, so it is only one of 64 possible stereoisomers.

When cortisone enters the body, rapid stereospecific and enzymatic reduction of the carbonyl group at carbon 11 occurs to produce hydrocortisone. That is, that carbonyl group is reduced to an alcohol, the same overall process of reduction that occurs using $NaBH_4$ in the laboratory. Because cortisone is easier to synthesize than hydrocortisone, cortisone is applied as a medicinal, allowing the body to do the final chemical reduction to the active drug (hydrocortisone) with the correct stereochemistry at carbon 11.

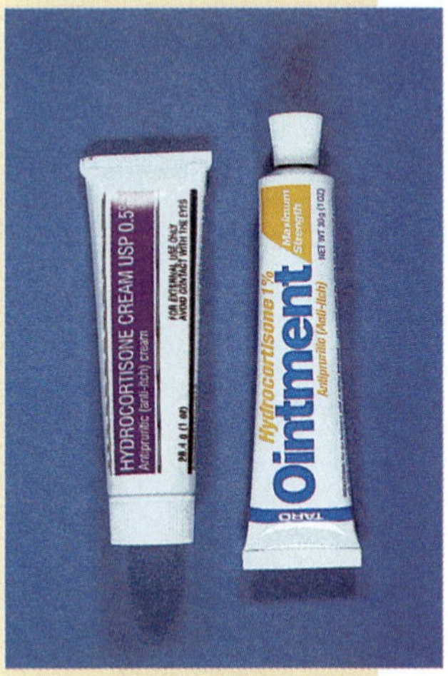

Cortisone in lower dosages is available over-the-counter as an ointment used to relieve itching and swelling.

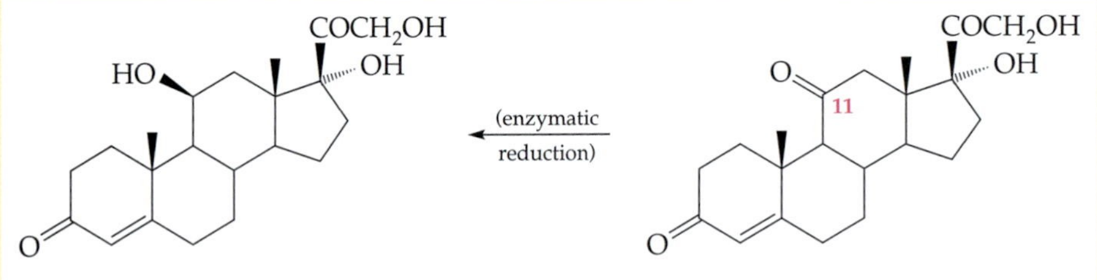

Hydrocortisone   Cortisone

PROBLEM 13.12

Show the reactions and reagents necessary to bring about each conversion:

(a) benzene to butylbenzene

(b) 2-phenylcyclopentanone to 1-phenylcyclopentene

13.3.2 Reaction of Carbonyls with Cyanide Ion

Carbonyl compounds undergo nucleophilic attack by the cyanide ion to yield an addition product called a **cyanohydrin.** The reaction is conducted using sodium cyanide at pH 10, so that the initially formed alkoxide anion can be protonated by the weak acid hydrogen cyanide present in the solution.

$$\text{C=O (Carbonyl group)} \xrightarrow{^{-}:\text{CN}} \text{Alkoxide anion} \xrightarrow[-\text{CN}^-]{\text{H—CN}} \text{Cyanohydrin (C(OH)CN)} \xrightarrow[\text{H}_2\text{O}]{\text{H}^+} \alpha\text{-Hydroxy acid (C(OH)COOH)}$$

Carbonyl group   Alkoxide anion   Cyanohydrin   α-Hydroxy acid

MOVIES
www.jbpub.com/organic-online

The importance of this simple addition reaction is that the cyano group (—CN) can be readily hydrolyzed to a carboxylic acid (—COOH), in particular to an α-hydroxy acid (the hydroxyl group is attached to the α-carbon). For example, acetaldehyde is converted to α-hydroxypropionic acid using the cyanide addition reaction. The α-hydroxy acids undergo reactions typical of alcohols and carboxylic acids.

$$CH_3CHO \xrightarrow[\text{HCN}]{\text{NaCN}} CH_3CH(OH)CN \xrightarrow[H_2O]{H^+} CH_3CH(OH)COOH$$

Acetaldehyde   Cyanohydrin   α-Hydroxypropionic acid

Note that cyanohydrin formation adds a carbon to the starting aldehyde chain. Thus, a one-carbon homologation has occurred, a very important process for synthesis. For example, phenylethanal can be homologated to 3-phenylpropanal using the cyanohydrin reaction. This conversion proceeds by dehydration of the α-hydroxy acid, hydrogenation of the resulting alkene, and reduction of the carboxylic acid to an aldehyde (via the acid chloride and another metal hydride, a reaction to be described in Section 13.6.2).

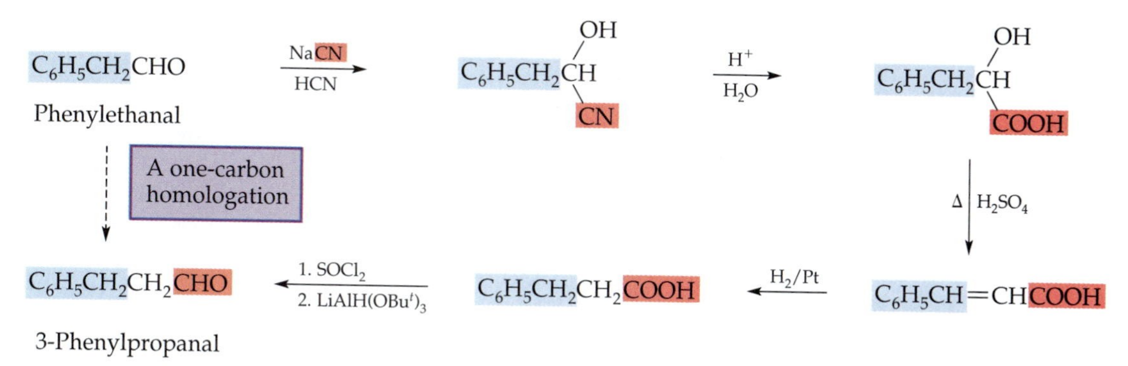

PROBLEM 13.13

Show the reactions and reagents necessary to complete each synthesis. (*Hint:* Apply your retrosynthetic skills.)

(a) cyclohexanone to 1-cyanocyclohexene

(b) benzaldehyde to 2-methoxy-2-phenylethanamine

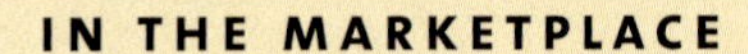

CHEMISTRY AT WORK

IN THE MARKETPLACE

Plexiglas

Plexiglas, a polymer of methyl methacrylate, is used for windshields in aircraft and "glass" walls around ice hockey rinks, as well as for other applications that require glasslike transparency and decidedly unglasslike strength. Plexiglas was the original bulletproof glass. Its strength and durability are also valuable in more mundane uses: Lucite paint uses the same polymer as a base. The raw material costs for producing poly(methyl methacrylate) are very low because all the ingredients are mass-produced very economically. The commercial preparation relies on the cyanohydrin reaction.

Acetone is converted to the cyanohydrin, which is then dehydrated. The unsaturated nitrile is hydrolyzed to the carboxylic acid and esterified with methanol in a single step to form the monomer, methyl methacrylate. This monomer is polymerized to the linear polymer, which is then formulated into the desired product.

To protect the spectators while allowing them a clear view of the action, Plexiglas surrounds hockey rinks—including this one at the University of North Dakota.

$$\underset{\text{Acetone}}{(CH_3)_2C{=}O} \xrightarrow[\text{HCN}]{\text{NaCN}} (CH_3)_2C(OH)(CN) \xrightarrow[\Delta]{H_2SO_4} CH_2{=}C(CH_3){-}CN \xrightarrow{CH_3OH/H_2SO_4} \underset{\text{Methyl methacrylate}}{CH_2{=}C(CH_3){-}COOCH_3}$$

$$\underset{\text{Methyl methacrylate}}{CH_2{=}C(CH_3){-}COOCH_3} \xrightarrow{\text{(polymerization)}} \underset{\text{Poly(methyl methacrylate)}}{\left(CH_2{-}C(CH_3)(COOCH_3) \right)_n}$$

13.4

Addition to Carbonyls: Primary Amines and Alcohols

13.4.1 Addition of Primary Amines

Primary amines, both aliphatic and aromatic, react with carbonyl compounds upon mild heating to form **imines,** compounds in which a C=N group has replaced the starting C=O group.

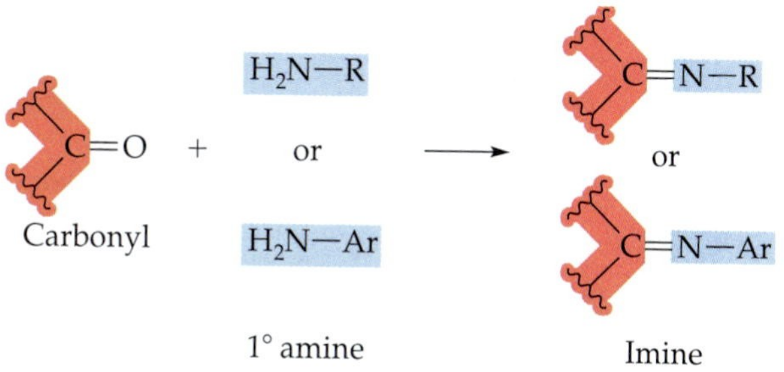

C=O + $H_2N—R$ or $H_2N—Ar$ → C=N—R or C=N—Ar

Carbonyl 1° amine Imine

For example, the amine aniline reacts with cyclohexanone to form an imine (*N*-phenyliminocyclohexane):

Cyclohexanone + Aniline $\xrightarrow[-H_2O]{\Delta}$ An imine (*N*-phenyliminocyclohexane)

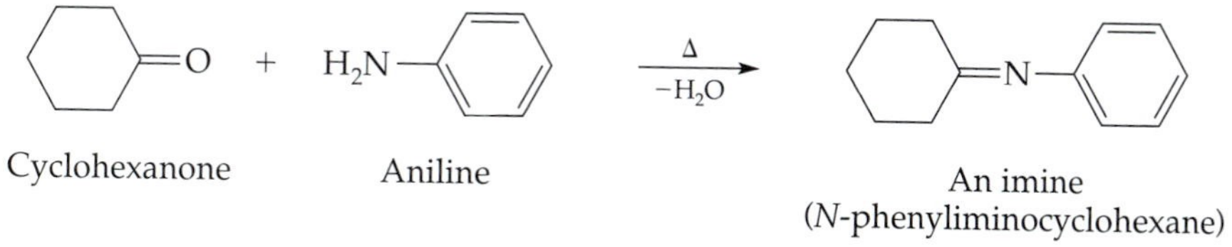

Although the product looks entirely different, this reaction is analogous to the hydride reduction and cyanide additions discussed in Sections 13.3.1 and 13.3.2. The first step is attack of the nucleophilic amine on the electron-deficient carbonyl carbon to form an alkoxide. The difference is in how the reaction is completed. Proton transfer forms the amino-alcohol, which is the standard adduct of nucleophilic addition but is normally not isolable. In this reaction, the alcohol-amine immediately loses water (dehydrates) upon heating to form the imine.

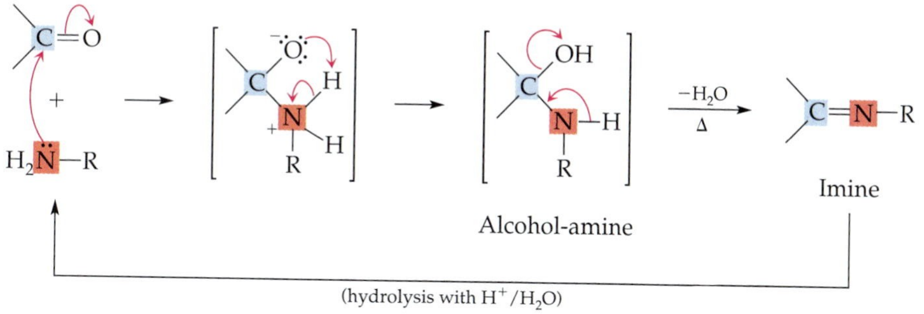

Alcohol-amine $\xrightarrow[\Delta]{-H_2O}$ Imine

(hydrolysis with H^+/H_2O)

Imines can be hydrolyzed to re-form the carbonyl compound and the amine using dilute acid and water; the reaction is simply the reverse of the formation process. Such reversibility is critical for biological reactions, such as the reactions of vitamins A and B_6 (as you will see in "Chemistry at Work" on page 465).

The imine formation reaction can be used to detect the presence of a carbonyl group in a compound (that is, it is a qualitative test for carbonyls) and to form crystalline derivatives of carbonyl compounds. This is an important application because many carbonyl compounds are liquids that are difficult to characterize chemically. The nitrogen of the primary amine can carry different substituents, leading to crystalline derivatives called oximes from hydroxylamine (H_2NOH), semicarbazones from semicarbazide ($N_2HNHCONH_2$), and 2,4-dinitrophenylhydrazones from 2,4-dinitrophenylhydrazine [$H_2NNHC_6H_3(NO_2)_2$-2,4].

General reactions:

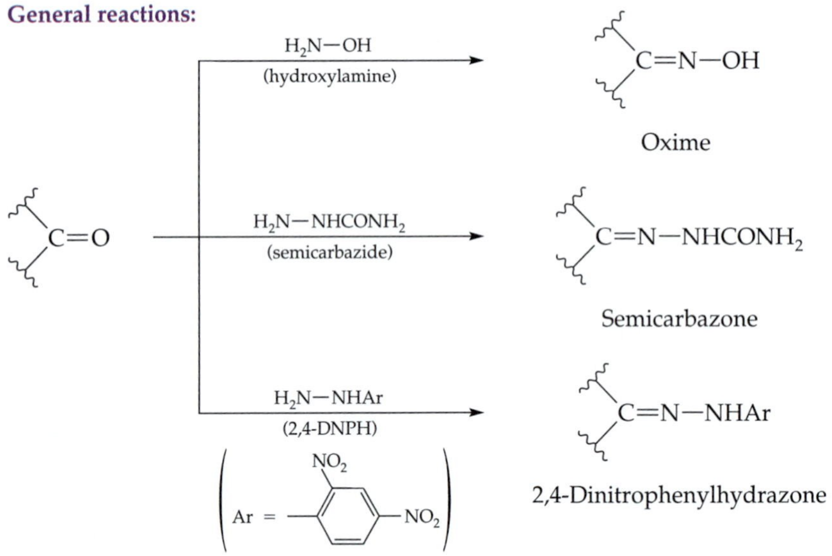

$$\ce{>C=O} \xrightarrow[\text{(hydroxylamine)}]{H_2N-OH} \ce{>C=N-OH}$$

Oxime

$$\ce{>C=O} \xrightarrow[\text{(semicarbazide)}]{H_2N-NHCONH_2} \ce{>C=N-NHCONH_2}$$

Semicarbazone

$$\ce{>C=O} \xrightarrow[\text{(2,4-DNPH)}]{H_2N-NHAr} \ce{>C=N-NHAr}$$

(Ar = 2,4-dinitrophenyl, $-C_6H_3(NO_2)_2$)

2,4-Dinitrophenylhydrazone

The 2,4-dinitrophenylhydrazones are especially useful and are formed readily from the reaction of 2,4-dinitrophenylhydrazine (2,4-DNPH) with a carbonyl compound. The fact that an unknown compound reacts with 2,4-DNPH and forms an orange-colored solid derivative indicates that it is an aldehyde or a ketone. The 2,4-dinitrophenylhydrazone derivative may be isolated and purified, and its melting point used to help identify the carbonyl compound.

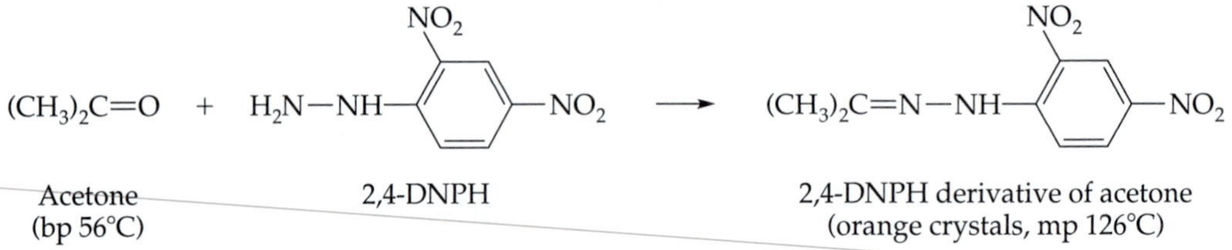

$$(CH_3)_2C{=}O + H_2N{-}NH{-}C_6H_3(NO_2)_2 \longrightarrow (CH_3)_2C{=}N{-}NH{-}C_6H_3(NO_2)_2$$

Acetone (bp 56°C) — 2,4-DNPH — 2,4-DNPH derivative of acetone (orange crystals, mp 126°C)

PROBLEM 13.14

Show the products of the following reactions:

(a) benzaldehyde with methylamine

(b) cyclohexanone with cyclohexylamine

(c) acetaldehyde with 2,4-DNPH

PROBLEM 13.15

Show the products of the acid-catalyzed hydrolysis of the following imines:

(a) $C_6H_5CH_2CH{=}NC_6H_5$ (b) 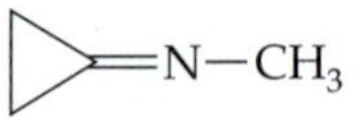 cyclopropylidene$=N-CH_3$ (c) 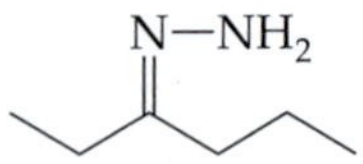 3-hexanone hydrazone ($C{=}N{-}NH_2$)

CHEMISTRY AT WORK

IN THE BODY

Imines and Vitamins A and B_6

As was noted in "Chemistry at Work" in Section 8.1.2, the chemical receptor of light in the eye is called *rhodopsin*. Rhodopsin is the imine formed between *cis*-retinal and the protein opsin. The nitrogen of the imine comes from the α-amino acid lysine, a part of opsin. When light strikes rhodopsin, it is isomerized to its *trans* form, which is then hydrolyzed to *trans*-retinal and opsin (cleaving the imine linkage). After the energy-consuming isomerization of retinal from the *trans* to the *cis* form, the *cis*-retinal reacts with opsin to re-form rhodopsin (the *cis*-imine), which is ready to absorb another photon of light.

Vitamin B_6, known as pyridoxal, is a coenzyme for the *transamination reaction*, a means for the body to retain nitrogen. This nitrogen is recycled in the synthesis of α-amino acids, which can then be converted into essential proteins. The steps of the reaction can be summarized as formation of an imine, tautomerization, and hydrolysis to carbonyl and amine.

By this process, pyridoxal is converted to pyridoxamine, and nitrogen is thereby "stored." The nitrogen-free keto acid continues to be metabolized to carbon dioxide and water, but the nitrogen atom of the amino acid has been retained in the body for "recycling." By reaction of pyridoxamine with a different keto acid to form an imine, the entire process can be reversed. This enables the pyridoxamine to transfer its nitrogen to a different carbonyl compound (to form pyridoxal and a different α-amino acid).

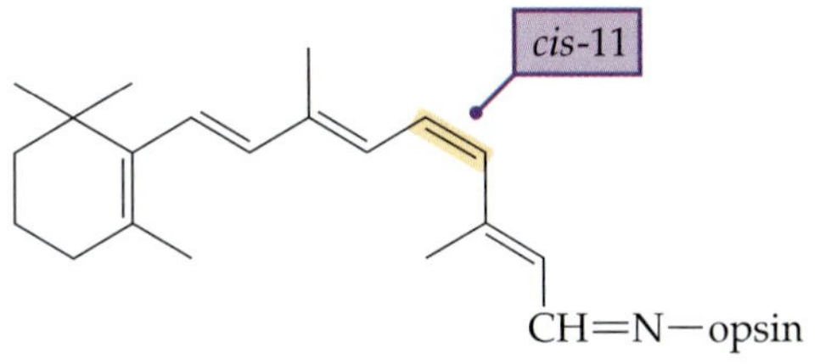
cis-11
CH=N—opsin

Rhodopsin

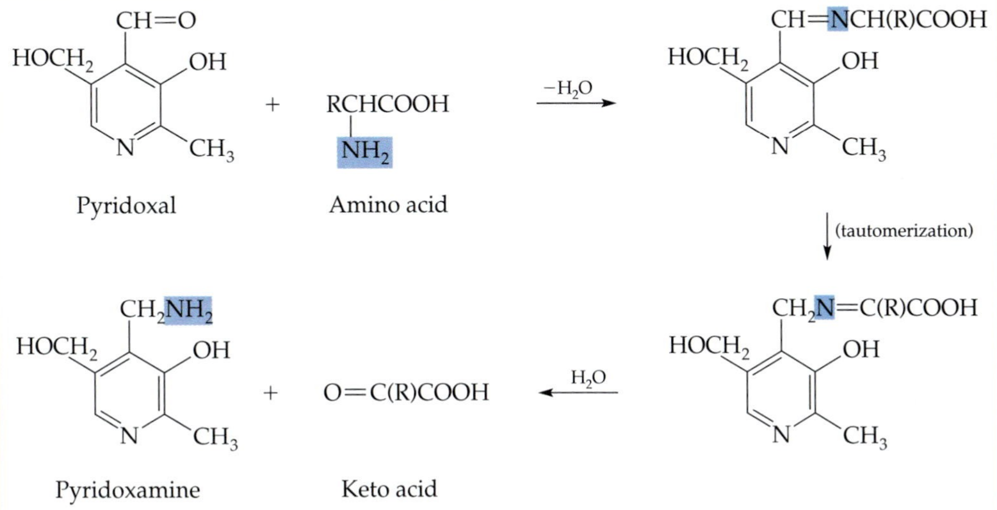
CH=O
HOCH2
OH
N
CH3
+
RCHCOOH
NH2
Pyridoxal
Amino acid
−H2O
CH=NCH(R)COOH
HOCH2
OH
N
CH3
(tautomerization)
CH2N=C(R)COOH
HOCH2
OH
N
CH3
H2O
CH2NH2
HOCH2
OH
N
CH3
+
O=C(R)COOH
Pyridoxamine
Keto acid

13.4.2 Addition of Alcohols and Thiols

Alcohols are weak nucleophiles (using an unshared electron pair on oxygen) and react with carbonyl compounds by nucleophilic attack at the carbonyl carbon. The reaction proceeds in a manner similar to the addition of primary amines (discussed in Section 13.4.1). However, while amines are sufficiently nucleophilic to react with carbonyls in the absence of a catalyst, alcohols are not. Therefore, acid catalysis is used to aid the reaction; protonation of the carbonyl oxygen makes the carbon more electron-deficient and more susceptible to nucleophilic attack (see Section 13.2.2).

Aldehydes and ketones react with alcohols to initially form **hemiacetals** ("half" acetals); continued reaction converts hemiacetals to **acetals.** Because all steps are equilibria, hemiacetals can seldom be isolated. In most attempts, they revert to the starting materials. Acetals, however, are easily isolated if water is removed as they are formed, preventing reversal of the reaction.

General reaction:

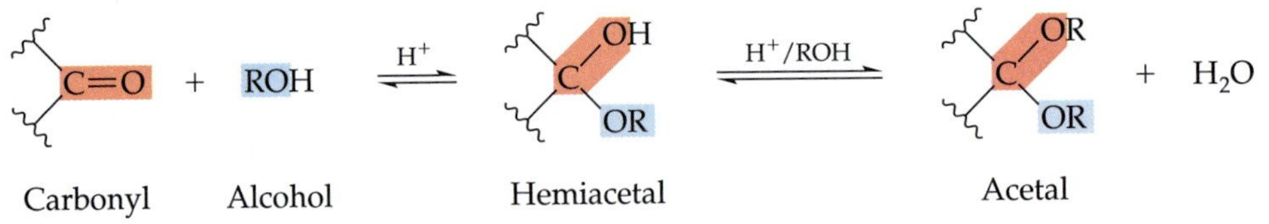

C=O + ROH ⇌ (H^+) C(OH)(OR) ⇌ (H^+/ROH) C(OR)(OR) + H_2O

Carbonyl Alcohol Hemiacetal Acetal

Frequently, a diol is used to form the acetal, most often ethanediol (ethylene glycol). In these cases, the second step is an *intramolecular* dehydration.

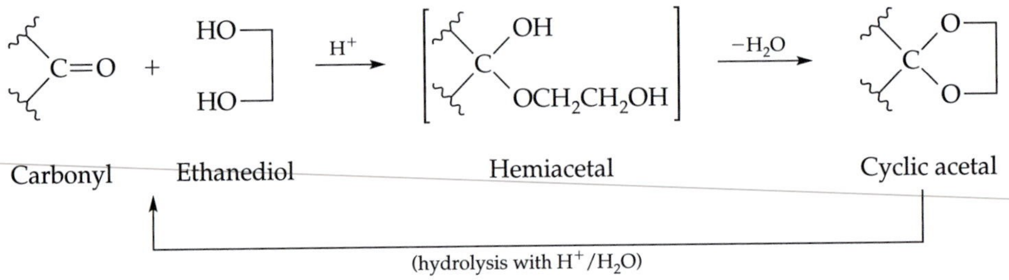

C=O + HO–CH₂CH₂–OH → (H^+) [C(OH)(OCH_2CH_2OH)] → ($-H_2O$) cyclic acetal

Carbonyl Ethanediol Hemiacetal Cyclic acetal

(hydrolysis with H^+/H_2O)

Acetals can be reconverted (hydrolyzed) to the starting alcohol and carbonyl compound by warming in water with acid catalysis, as shown above.

While acetals hydrolyze in acidic solution, they are perfectly stable in basic solution, making them good protecting groups for a carbonyl group in a complex compound. The acetal "masks" the carbonyl group while other reactions are conducted, and the carbonyl group can be restored by acid hydrolysis of the acetal. For example, to dehydrohalogenate an α-haloketone that also has a stereocenter adjacent to the carbonyl group (for example, 2-bromo-6-methylcyclohexanone) requires a strong base. Because a strong base could also racemize the carbon next to the carbonyl group (see Section 13.2.2), protection of the carbonyl as an acetal is essential. The risk of racemization vanishes when the carbonyl group disappears (the α-hydrogen is no longer acidic in an acetal), and so the stereochemistry of the α-methyl group is retained throughout the dehydrohalogenation.

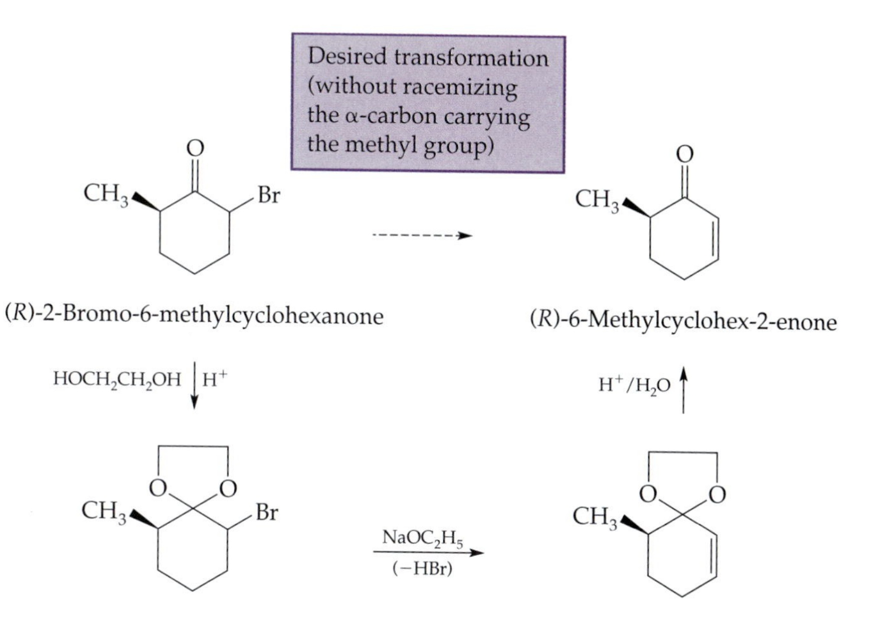

As was mentioned earlier, under normal reaction conditions, hemiacetals cannot be isolated. The exceptions are cyclic structures that arise because the alcohol group is in the appropriate proximity to the carbonyl group to spontaneously form a five- or six-membered ring. A simple example is 4-hydroxybutanal, which exists mainly in the hemiacetal form, even in the absence of acid catalysis:

$HOCH_2CH_2CH_2CHO$ ≡ HO–(chain)–CH=O ⇌ (cyclic) O–OH

4-Hydroxybutanal

Hemiacetal of 4-hydroxybutanal

More common examples of cyclic hemiacetals are the monosaccharide carbohydrates (see Section 4.7.2). For example, glucose exists mainly in its six-membered hemiacetal form:

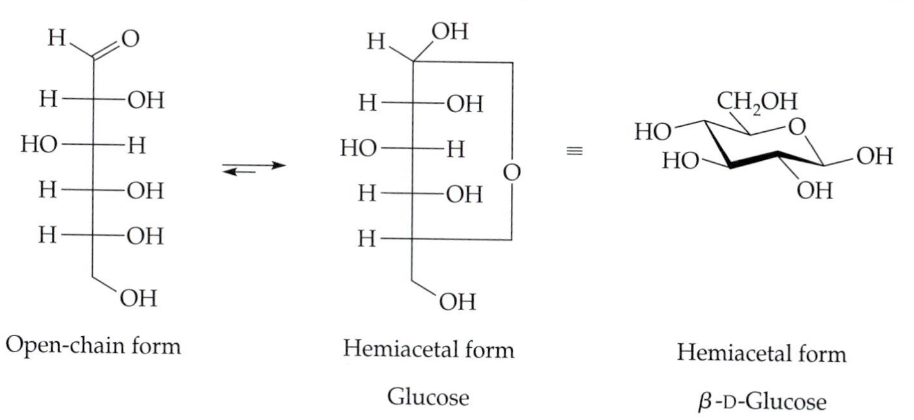

The mechanism of hemiacetal formation is straightforward, involving proton transfers and one nucleophilic attack, as shown in Figure 13.4. As was mentioned earlier, acid catalysis is essential because the alcohols are weaker nucleophiles than amines.

FIGURE 13.4

Mechanism of hemiacetal formation.

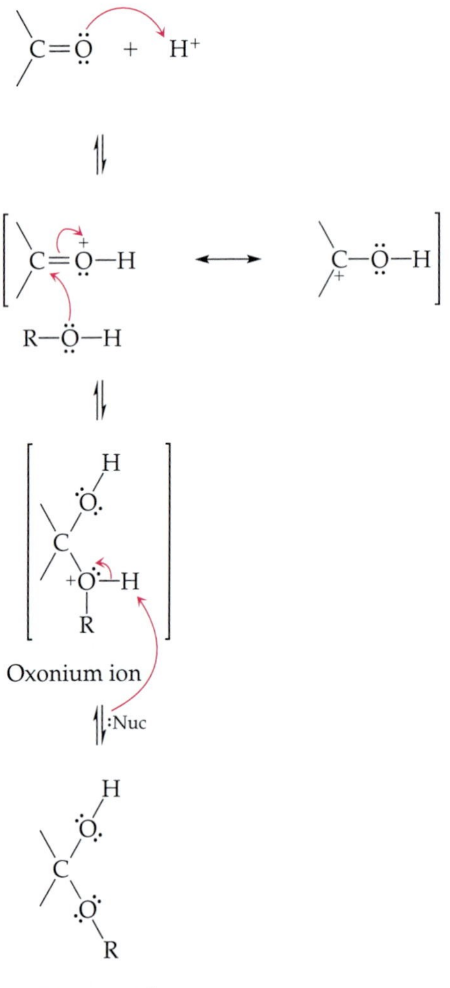

The carbonyl group is protonated, forming a resonance-stabilized cation. The cation has considerable positive charge on carbon, much more than in the unprotonated carbonyl group. This makes it more susceptible to attack by nucleophiles, especially weak ones like alcohols.

The nucleophilic alcohol attacks the electron-deficient carbon, and the pi electrons revert to oxygen. The donation of electrons by the alcohol's oxygen to form a bond means that it acquires a positive charge, becoming an oxonium ion.

Oxonium ion

The proton is removed from the trivalent oxygen, and the shared electrons revert to that oxygen, neutralizing the positive charge. Any nucleophile in the solution (a carbonyl group, the alcohol, etc.) will remove the proton in a simple acid-base reaction.

Hemiacetal

The mechanism of acetal formation involves an intramolecular dehydration between a hemiacetal and an alcohol, as shown in Figure 13.5. Normally, the water must be removed from the reaction as it is formed to isolate the acetal. The reaction can be reversed by treating an acetal with water and an acid catalyst, regenerating the carbonyl.

Carbonyl compounds react with thiols (the sulfur analog of alcohols) in exactly the same manner and by an analogous mechanism to form **thioacetals.** Most often, this reaction is accomplished using ethanedithiol to form a cyclic thioacetal.

General reaction:

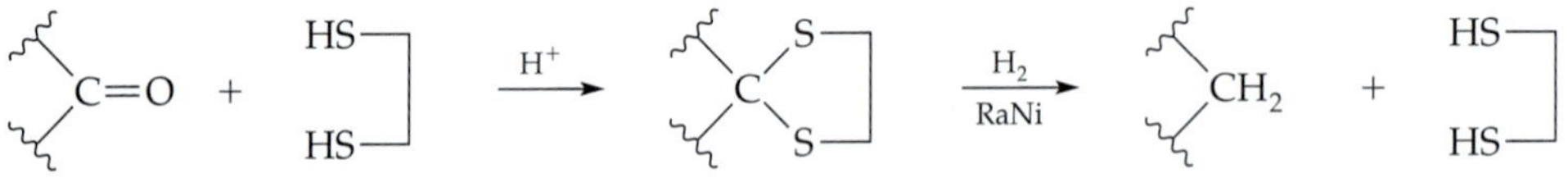

Carbonyl group + Ethanedithiol $\xrightarrow{H^+}$ Thioacetal $\xrightarrow[RaNi]{H_2}$ Methylene group (CH_2) + HS–CH₂CH₂–SH

The importance of this reaction is that cyclic thioacetals can be cleaved by molecular hydrogen over a Raney nickel catalyst (abbreviated RaNi), leaving a methylene group

Hemiacetal

Acetal

FIGURE 13.5

Mechanism of acetal formation.

where there once was a carbonyl group. This process of cleaving a bond with molecular hydrogen (H_2) and having a hydrogen add to each fragment is called *hydrogenolysis*. It is the best way to convert a carbonyl group to a methylene group. For example, cyclopentanone is converted to cyclopentane by this reaction:

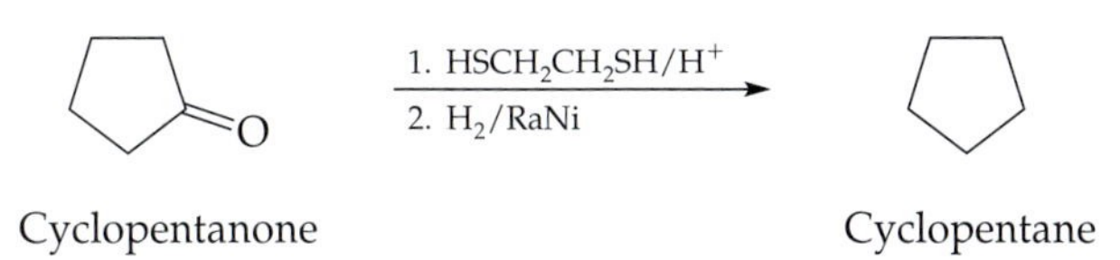

This conversion provides the opportunity to put a carbonyl group into a structure, use its electronic effect to influence other reactions, and then remove the carbonyl oxygen by the thioacetal hydrogenolysis reaction. However, the hydrogenolysis reaction cannot be employed if there is a carbon-carbon double or triple bond present, because hydrogenation of that bond will also occur.

How to Solve a Problem

Show the reactions necessary to convert the ribose portion (the C_5 carbohydrate; see Section 4.7.2) of *N*-acetyladenosine, a nucleoside used in DNA synthesis, into its 5-benzyl ether.

N-Acetyladenosine

5-Benzyl-*N*-acetyladenosine

PROBLEM ANALYSIS

The problem is to select only one of three hydroxyl groups and convert it into a benzyl ether. The solution has to lie in protecting the other two hydroxyl groups (on carbons 2 and 3) while we perform ether formation on the carbon 5 hydroxyl. This protection can be accomplished using acetal formation to "tie up" those hydroxyl groups. Note that those two hydroxyls are vicinal, just like the two hydroxyls of ethylene glycol. Therefore, we can react them with a carbonyl compound to form an acetal (called an *acetonide*), and then proceed with the Williamson synthesis. The final step is to remove the protecting group.

SOLUTION

First, we react the substrate with acetone and dilute acid, removing the water as it is formed, to produce the acetal. Then we react the acetal with metallic sodium to form the alkoxide at carbon 5. We then add benzyl bromide, and the S_N2 reaction produces the benzyl ether (the Williamson synthesis; see Section 5.1.2). Finally, we treat the product with dilute acid to hydrolyze the acetal back to acetone and free the two hydroxyl groups at carbons 2 and 3.

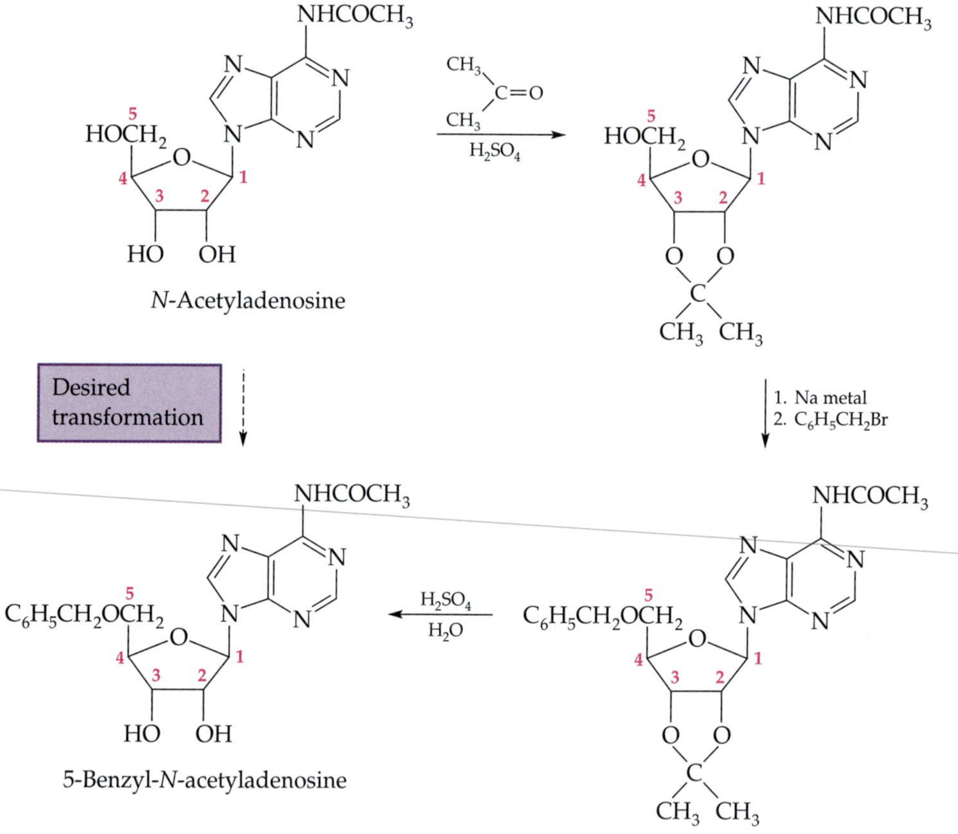

NHCOCH$_3$ N N N N HOCH$_2$ O HO OH 5 4 3 2 1

N-Acetyladenosine

$CH_3C(=O)CH_3$, H_2SO_4 →

NHCOCH$_3$ N N N N HOCH$_2$ O O O C CH$_3$ CH$_3$ 5 4 3 2 1

Desired transformation

1. Na metal
2. $C_6H_5CH_2Br$

NHCOCH$_3$ N N N N $C_6H_5CH_2OCH_2$ O O O C CH$_3$ CH$_3$ 5 4 3 2 1

← H_2SO_4, H_2O

NHCOCH$_3$ N N N N $C_6H_5CH_2OCH_2$ O HO OH 5 4 3 2 1

5-Benzyl-*N*-acetyladenosine

PROBLEM 13.16

Show the product of each of the following reactions:

(a) cyclohexanone and excess ethanol with acid catalysis
(b) benzaldehyde and ethylene dithiol with acid catalysis
(c) the dimethyl acetal of cyclohexanone with water and acid

PROBLEM 13.17

Show the reactions and reagents necessary to accomplish each conversion:

(a) benzaldehyde to toluene

(b) dicyclopentyl ketone to dicyclopentylmethane

13.5

Addition to Carbonyls: Carbanions

The reaction of carbanions with carbonyl compounds involves the attack of a carbon nucleophile on the carbonyl carbon, thereby forming a carbon-carbon bond. It is one of the most important reactions in organic chemistry because it permits the building of more complex structures from simpler structures. Carbanion reagents are very strong nucleophiles and react with almost any carbonyl compound.

The carbanions are usually formed as organometallic reagents, which must be generated in solution first and then allowed to react with an added carbonyl compound. One of the common organometallics is the Grignard reagent, an organomagnesium bromide (RMgBr or ArMgBr). The organolithium compounds (RLi and ArLi) also accomplish the reactions with carbonyl compounds described in this section; therefore, those reactions will not be separately detailed. (It is important to note that both aryl and alkyl bromides form Grignard reagents and organolithium compounds.) The third type of organometallic reagent discussed in this book, the organocuprate (R_2CuLi), is not sufficiently nucleophilic to add to carbonyl groups.

13.5.1 Addition of Grignard Reagents

As described in Section 3.5.1, Grignard reagents (RMgBr and ArMgBr) are prepared from alkyl or aryl bromides and metallic magnesium in ether solution. The metal-carbon bond in a Grignard reagent is quite ionic, being polarized such that the metal is the positive end and the carbon is the negative end. Thus, a Grignard reagent generally behaves as though it were a carbanion, a very strong nucleophile, which brings about nucleophilic addition to aldehydes and ketones, attaching the carbanion to the carbonyl carbon. The addition of Grignard reagents to carbonyls is also a very important method for synthesizing alcohols (see Section 4.5.2).

General reaction:

C=O + RMgBr —(1. ether solvent; 2. H^+/H_2O)→ C(OH)(R)

Carbonyl compound — Alcohol

For example, methylmagnesium bromide adds to acetophenone and after work-up (that is, quenching the reaction with dilute aqueous acid) yields 2-phenyl-2-propanol:

$$CH_3Br \xrightarrow[\text{ether}]{\text{Mg}} CH_3{-}MgBr$$

Methyl bromide → Methylmagnesium bromide

$$C_6H_5{-}\overset{O}{\overset{\|}{C}}{-}CH_3 + CH_3{-}MgBr \xrightarrow[\text{2. } H^+/H_2O]{\text{1. ether}} C_6H_5{-}\underset{CH_3}{\overset{OH}{C}}{-}CH_3$$

Acetophenone + Methylmagnesium bromide → 2-Phenyl-2-propanol

The end result is the addition of a methyl group to the carbonyl carbon and the conversion of the carbonyl group to an alcohol.

The mechanism of Grignard addition is consistent with the general nucleophilic addition mechanism presented at the beginning of Section 13.3. The reaction is conducted in ether in the absence of any moisture because the carbanion is sufficiently basic to extract a proton from water—or any other source of protons—and thus neutralize itself. The alkoxymagnesium salt is stable in solution, and protonation is required during workup to isolate the alcohol.

Carbonyl ($\delta+$ C=O $\delta-$) + Grignard reagent ($\delta-$ R—MgBr $\delta+$) $\xrightarrow{\text{ether}}$ [Alkoxymagnesium bromide ($^-$O: $^+$MgBr, C—R)] $\xrightarrow{H^+/H_2O}$ Alcohol (C—OH, R)

Carbonyl Grignard reagent Alkoxymagnesium bromide Alcohol

The importance of this reaction in synthesis is evident from the fact that primary, secondary, and tertiary alcohols can all be prepared by appropriate choice of the starting carbonyl compound. To synthesize an alcohol of any classification, you should apply the concept of retrosynthesis (thinking backward) to determine what options are available. Synthesis of any alcohol, RR′R″C—OH, can be brought about by adding any one of the three R groups as a carbanion to a carbonyl compound containing the other two R groups, as shown in Table 13.1. There are two limitations to this generalization:

1. If one of the R groups is hydrogen, it must be on the starting carbonyl compound because there is no Grignard reagent with the structure H—MgBr.
2. There cannot be any other functional groups present in the carbonyl compound or in the Grignard reagent with which the Grignard reagent will react, especially protic groups such as —OH and —NH_2. These groups can be protected in some instances, however.

It follows from the first limitation that there is only a single combination of reactants that can be used to prepare a primary alcohol: the Grignard reagent R—MgBr and formaldehyde, H_2CO. If the desired alcohol has one R group that is hydrogen, it is a secondary alcohol and there are two combinations of reactants that will produce it. If there are no hydrogens as R groups, then the desired alcohol is tertiary and there are three combinations that will produce it.

Application of these concepts is illustrated by the synthesis of cyclohexylmethanol from cyclohexane. Retrosynthetic analysis indicates that primary alcohols can be

TABLE 13.1

Retrosynthesis of an Alcohol from a Carbonyl Compound Using a Grignard Reagent

Target Alcohol		Carbonyl Compound	Carbanion	Grignard Reagent	Alkyl Halide
	⇗	RR′C=O	⁻:R″	R″—MgBr	R″—Br
R′—C(R)(R″)—OH	⟹	RR″C=O	⁻:R′	R′—MgBr	R′—Br
	⇘	R′R″C=O	⁻:R	R—MgBr	R—Br

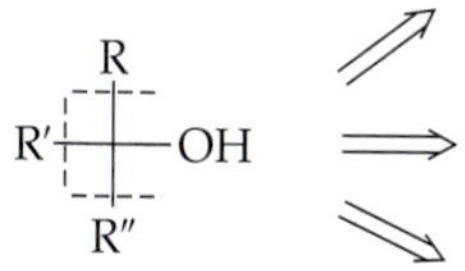

obtained from formaldehyde (H_2CO) and a Grignard reagent, and this particular alcohol requires cyclohexylmagnesium bromide.

$$C_6H_{11}-CH_2OH \Longrightarrow H_2C{=}O + C_6H_{11}-MgBr$$

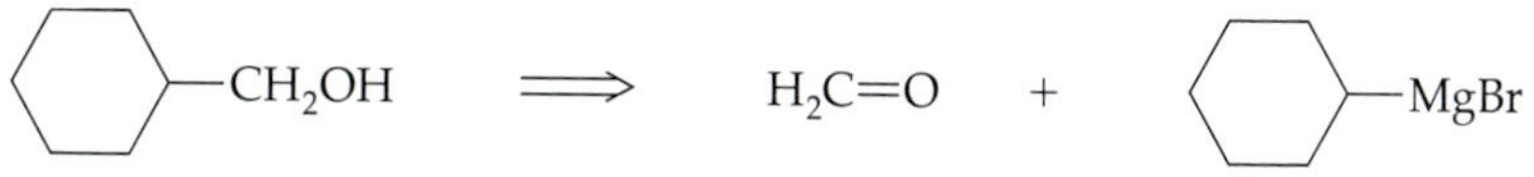

Cyclohexylmethanol — Formaldehyde — Cyclohexylmagnesium bromide

The Grignard reagent requires cyclohexyl bromide for its preparation. Since cyclohexane is the specified starting material, it must be converted to cyclohexyl bromide, a reaction effected by bromine under radical conditions. Therefore, the synthesis can be accomplished as follows:

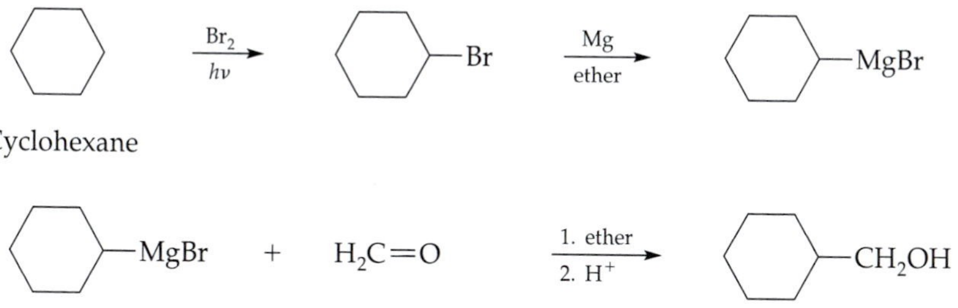

$$C_6H_{12} \xrightarrow[h\nu]{Br_2} C_6H_{11}-Br \xrightarrow[\text{ether}]{Mg} C_6H_{11}-MgBr$$

Cyclohexane

$$C_6H_{11}-MgBr + H_2C{=}O \xrightarrow[\text{2. }H^+]{\text{1. ether}} C_6H_{11}-CH_2OH$$

Cyclohexylmethanol

As another example, 2-phenyl-2-butanol can be prepared starting with 2-butanol:

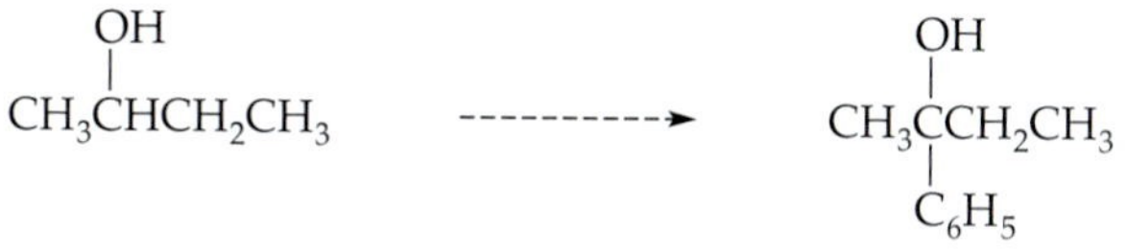

$$CH_3CH(OH)CH_2CH_3 \dashrightarrow CH_3C(OH)(C_6H_5)CH_2CH_3$$

2-Butanol — 2-Phenyl-2-butanol

Because the product is a tertiary alcohol (three nonhydrogen groups on the carbon carrying the hydroxyl group), two of the substituents can be obtained from a ketone and one can be obtained by adding a Grignard reagent. 2-Butanone—a ketone—can be obtained from the designated starting material—a secondary alcohol—by simple oxidation. Therefore, the synthesis is as follows:

$$\underset{\text{2-Butanol}}{CH_3\overset{OH}{\overset{|}{C}}HCH_2CH_3} \xrightarrow[H_2SO_4]{Na_2Cr_2O_7} \underset{\text{2-Butanone}}{CH_3\overset{O}{\overset{||}{C}}CH_2CH_3}$$

$$\underset{\text{Bromobenzene}}{C_6H_5Br} \xrightarrow[\text{ether}]{Mg} \underset{\text{Phenylmagnesium bromide}}{C_6H_5{-}MgBr}$$

$$CH_3\overset{O}{\overset{||}{C}}CH_2CH_3 + C_6H_5{-}MgBr \xrightarrow[2.\ H^+]{1.\ \text{ether}} \underset{\text{2-Phenyl-2-butanol}}{CH_3\overset{OH}{\overset{|}{C}}(C_6H_5)CH_2CH_3}$$

PROBLEM 13.18

Propose two additional routes for the synthesis of 2-phenyl-2-butanol, but use an organolithium compound rather than a Grignard reagent.

How to Solve a Problem

Convert 1-bromopropane to 1-bromobutane.

$$\underset{\text{1-Bromopropane}}{CH_3CH_2CH_2Br} \dashrightarrow \underset{\text{1-Bromobutane}}{CH_3CH_2CH_2CH_2Br}$$

PROBLEM ANALYSIS

The problem obviously is to add a single carbon to the propyl group (a one-carbon homologation) to form a butyl group. The only reactive part of 1-bromopropane is the bromine, so we focus our attention on replacing it with a carbon. We won't worry about getting the final bromine in place in 1-bromobutane until later. The bromine in 1-bromopropane can be used in any of three general ways: (1) It can be eliminated as HBr to form an alkene; (2) it can be substituted by a nucleophile in an S_N2 reaction; or (3) it can be converted to an organometallic reagent that will add to a carbonyl group. The shortest and most straightforward way to accomplish the overall conversion is to use the third approach. Formaldehyde is an obvious one-carbon substrate for a Grignard reaction. Therefore, a propyl magnesium Grignard reagent with formaldehyde will produce the four-carbon primary alcohol, and that alcohol's —OH group can then be converted to —Br.

SOLUTION

1-Bromopropane is converted to its Grignard reagent, to which is added formaldehyde. Workup with acid and water produces 1-butanol, which, upon heating with HBr, is converted to 1-bromobutane.

$$CH_3CH_2CH_2Br \xrightarrow[\text{ether}]{\text{Mg}} CH_3CH_2CH_2MgBr \xrightarrow[\text{2. } H^+]{\text{1. } CH_2O} CH_3CH_2CH_2CH_2OH \xrightarrow{\text{HBr}} CH_3CH_2CH_2CH_2Br$$

1-Bromopropane → 1-Bromobutane

PROBLEM 13.19

Show the reagents needed to bring about the following conversions, all of which should start with a Grignard addition reaction:

(a) acetophenone to 2-phenyl-2-pentanol
(b) cyclobutanone to 1-methylcyclobutene
(c) acetaldehyde to 3-methyl-3-hexanol

PROBLEM 13.20

Show how to accomplish the following syntheses, each of which involves using the Grignard reagent to build the carbon skeleton:

(a) benzaldehyde to 2-phenyl-2-propanol
(b) 2-butanol to 3-methyl-2-pentene
(c) cyclohexanone to 2-methylcyclohexanone

13.5.2 The Wittig Reaction

The **Wittig reaction** is the most effective means of producing alkenes with known regiochemistry and stereochemistry (discussion of the stereochemistry is beyond the scope of this book). (Recall that both dehydration of alcohols and dehydrohalogenation of alkyl halides produce alkenes with mixed regiochemistry—Zaitzev's rule.) Discovered in 1953, the Wittig reaction earned George Wittig a share of the 1979 Nobel Prize in chemistry.

The Wittig reaction requires the initial preparation of an organophosphorus reagent called an **ylide** (pronounced "ill-id") of the desired composition. The ylide is then reacted with a carbonyl compound to form an alkene and the by-product triphenylphosphine oxide. One double-bonded carbon of the alkene arises from the ylide and one from the carbonyl.

General reaction:

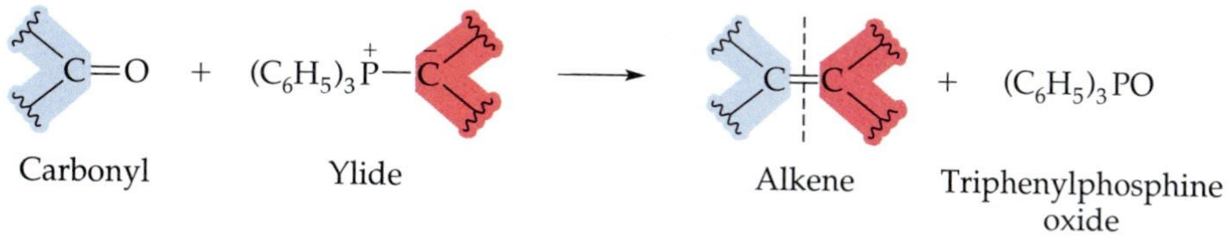

$$C{=}O + (C_6H_5)_3\overset{+}{P}{-}\overset{-}{C} \longrightarrow C{=}C + (C_6H_5)_3PO$$

Carbonyl, Ylide, Alkene, Triphenylphosphine oxide

The Wittig reaction occurs as though it were an attack of the ylide carbanion on the carbonyl carbon (as in the Grignard reaction). In fact, it actually appears to be a simultaneous (concerted) addition of both ends of the phosphorus-carbon bond of the ylide across the carbonyl double bond. This forms an intermediate called an *oxaphosphetane*,

which collapses (in a direction 90° to that by which it was formed) to produce an alkene and triphenylphosphine oxide. The driving force for the reaction is the formation of the highly stable triphenylphosphine oxide.

$$:\ddot{O}{=}C< \;+\; (C_6H_5)_3\overset{+}{P}{-}\overset{-}{C}< \longrightarrow \left[(C_6H_5)_3P{-}C{-}C{-}O\right] \longrightarrow (C_6H_5)_3P{=}O \;+\; >C{=}C<$$

Oxaphosphetane — Triphenylphosphine oxide — Alkene

For example, synthesis of methylenecyclohexane from cyclohexanone can be accomplished using phosphoniummethylide, while acetophenone and phosphoniumbenzylide produce 1,2-diphenylpropene:

$$\text{Cyclohexanone} \;+\; (C_6H_5)_3\overset{+}{P}{-}\overset{-}{C}H_2 \xrightarrow{-(C_6H_5)_3PO} \text{Methylenecyclohexane}$$

$$\underset{\text{Acetophenone}}{(C_6H_5)(CH_3)C{=}O} \;+\; (C_6H_5)_3\overset{+}{P}{-}\overset{-}{C}HC_6H_5 \xrightarrow{-(C_6H_5)_3PO} \underset{\text{1,2-Diphenylpropene}}{(C_6H_5)(CH_3)C{=}CHC_6H_5}$$

Ylides are prepared in two steps. The first step is the reaction of an alkyl bromide and triphenylphosphine in an S_N2 reaction (analogous to the alkylation of amines) to form a phosphonium salt. For example, alkylation of triphenylphosphine can be done using benzyl bromide:

$$\underset{\text{Triphenylphosphine}}{(C_6H_5)_3P:} \;+\; \underset{\text{Benzyl bromide}}{C_6H_5{-}CH_2{-}Br} \longrightarrow \underset{\text{Triphenylbenzylphosphonium bromide}}{(C_6H_5)_3\overset{+}{P}{-}CH_2C_6H_5\;\;Br^-}$$

Next, the phosphonium salt is treated with a very strong base, usually the organometallic butyllithium, which removes a single proton to produce the ylide:

$$(C_6H_5)_3\overset{+}{P}{-}CH_2C_6H_5\;\;Br^- \xrightarrow[-C_4H_{10},\;-LiBr]{\overset{\delta-}{C_4H_9}{-}\overset{\delta+}{Li}} \underset{\text{Ylide}}{(C_6H_5)_3\overset{+}{P}{-}\overset{-}{C}HC_6H_5}$$

In a Wittig synthesis, one carbon of the ultimate alkene product arises from the alkyl halide (any primary or secondary halide) used to make the ylide, and the other carbon arises from the carbonyl group. In principle, either carbon could be derived from either reagent, providing two synthetic routes to any alkene.

PROBLEM 13.21

Show the products of the following reactions:

(a) $(C_6H_5)_3\overset{+}{P}-CH_2C_6H_5 \xrightarrow[\text{2. } C_6H_5CHO]{\text{1. BuLi}}$

(b) $(C_6H_5)_3\overset{+}{P}-CH_2CH_3 \xrightarrow[\text{2. cyclohexenone}]{\text{1. BuLi}}$

PROBLEM 13.22

Devise a synthesis of each of the following alkenes using the Wittig reaction. (*Hint:* Use retrosynthesis.)

(a) 2-methyl-2-pentene
(b) methylenecyclobutane
(c) 1-phenyl-1,3-butadiene

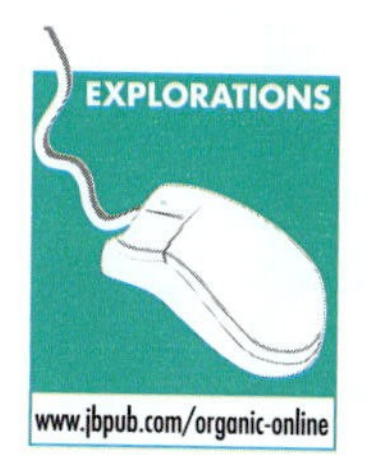

13.6

Preparation of Carbonyls

Aldehydes and ketones can be produced by many methods, several of which we have already considered. After briefly reviewing three of those, we will consider just two more. Both of these methods are effective and widely used and share certain similarities.

13.6.1 Review of Methods Presented Earlier

Oxidation of Alcohols

As described in Section 4.4, primary alcohols can be oxidized with pyridinium chlorochromate (PCC) to aldehydes. Ketones can be obtained from secondary alcohols by oxidation with sodium dichromate and sulfuric acid.

$$\underset{\text{1° alcohol}}{RCH_2-OH} \xrightarrow{\text{PCC}} \underset{\text{Aldehyde}}{RCHO}$$

$$\underset{\text{2° alcohol}}{R_2CH-OH} \xrightarrow[H_2SO_4]{Na_2Cr_2O_7} \underset{\text{Ketone}}{R_2C{=}O}$$

Friedel-Crafts Acylation

As described in Section 9.3.6, aromatic ketones (acyl benzenes) can be produced from the reaction of benzenoid compounds with acyl chlorides, which are derived from carboxylic acids.

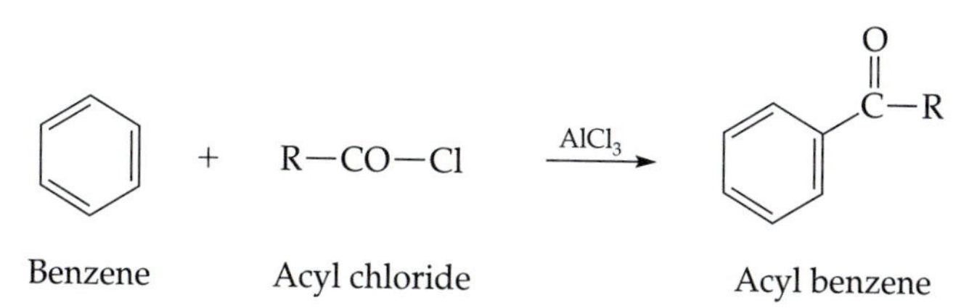

Green and yellow vegetables are primary dietary source of vitamin A precursor compounds.

CHEMISTRY AT WORK

The Commercial Synthesis of Vitamin A

The sale of vitamins and health supplements is big business in the United States. Although vitamins are naturally occurring, many of those on the market are synthesized in chemical manufacturing plants, using processes developed in the research laboratory. Vitamin A, needed for good vision as well as normal tooth and bone development, is an excellent example. The Wittig reaction figures in a key step in its synthesis.

The control of the stereochemistry of the double bonds is essential in the commercial synthesis of vitamin A (note that all the double bonds are *trans*). Forming double bonds regiospecifically and stereospecifically is difficult using the traditional elimination reactions (see Sections 6.5.1 and 6.5.2). The Wittig reaction presented a stereospecific and regiospecific route for alkene synthesis. As early as 1954, patents for the commercial production of vitamin A based on the Wittig reaction were filed by the BASF company in Germany. One of the commercial syntheses used shows how the necessary *trans* stereochemistry at the double bond between carbon 11 and carbon 12 is accomplished.

In order to function in the eye, vitamin A must be enzymatically converted to its aldehyde form (retinal) after ingestion (see Sections 8.1.2 and 13.4.1).

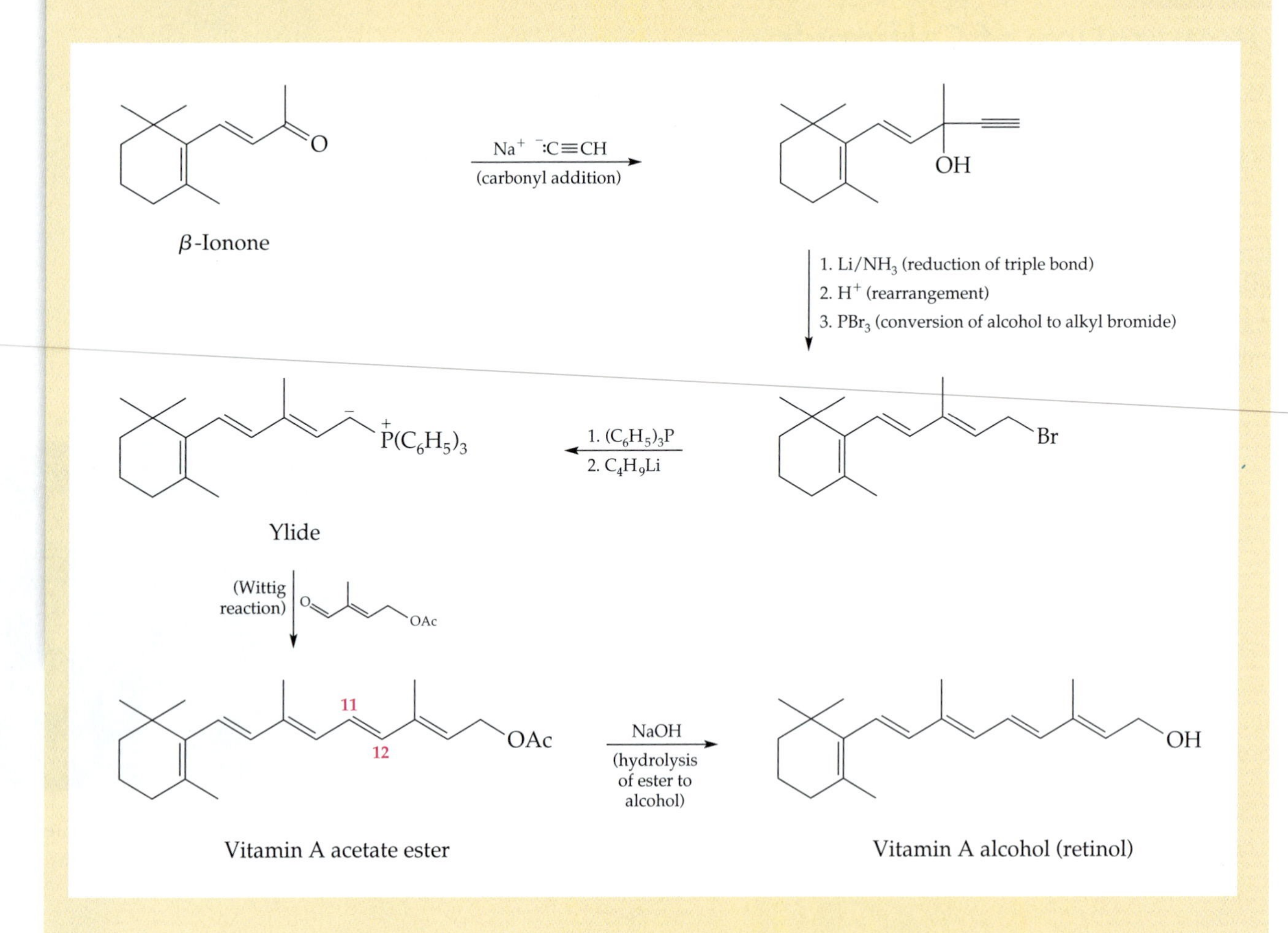

Ozonolysis of Alkenes

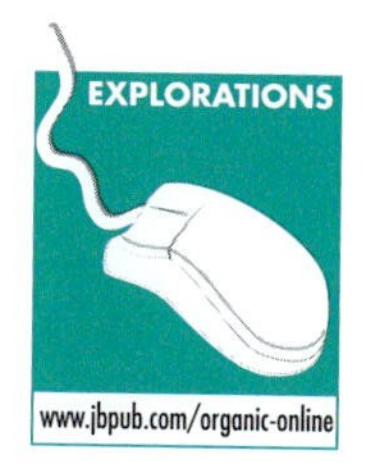

Although not generally used for synthesis, the cleavage of an alkene with ozone, as described in Section 6.5.3, produces carbonyl compounds. Recall that disubstituted double-bonded carbons become ketones and monosubstituted double-bonded carbons become aldehydes through ozonolysis.

$$\text{C=C} \xrightarrow[\text{2. Zn/H}_2\text{O}]{\text{1. O}_3} \text{C=O} + \text{O=C}$$

Alkene — Carbonyl compounds

13.6.2 Aldehydes from Acid Chlorides

Chapter 4 presented an oxidation-reduction hierarchy for certain families of compounds (see Table 4.1). In this hierarchy, aldehydes are below carboxylic acids; thus, aldehydes are easily oxidized to carboxylic acids (see Section 13.2.2), but carboxylic acids are difficult to reduce to aldehydes. This difficulty is circumvented by converting a carboxylic acid into the more reactive acid chloride, which can be readily reduced to an aldehyde.

General reaction:

$$\underset{\text{Carboxylic acid}}{\text{RCOOH}} \xrightarrow{\text{SOCl}_2} \underset{\text{Acid chloride}}{\text{RCOCl}} \xrightarrow{\text{LiAlH(OBu}^t)_3} \underset{\text{Aldehyde}}{\text{RCHO}}$$

Thionyl chloride ($SOCl_2$) is the reagent used to convert an alcohol into an alkyl chloride (see Section 4.3.2), and it performs the same function with a carboxylic acid, replacing the —OH group with —Cl. As we will see in Chapter 15, acid chlorides are very reactive and susceptible to attack at the carbonyl carbon by a nucleophile. Lithium tri-*t*-butoxyaluminum hydride [$LiAlH(OBu^t)_3$] is a metal hydride (a mild reducing agent) that displaces chloride with hydride to produce an aldehyde. The presence of the single hydride and its lower reactivity due to the presence of the three *t*-butoxy groups permit the formation of the aldehyde as the product; it is not reduced further (further reduction does occur if sodium borohydride is used; see Section 13.3.1).

We will see in Chapter 14 that there are many effective synthetic routes to carboxylic acids (see the Key to Transformations). The fact that such acids can readily be transformed into aldehydes by this reduction makes aldehydes readily synthesizable. For example, cyclohexanecarbaldehyde can be obtained by reduction of cyclohexanecarboxylic acid chloride, which itself is readily obtained from cyclohexane through bromination, formation of a Grignard reagent, and carboxylation with carbon dioxide—all of which will be described in Chapter 14—and conversion to the acid chloride with thionyl chloride, $SOCl_2$.

$$\text{C}_6\text{H}_{12} \xrightarrow[\text{peroxides}]{\text{Br}_2} \text{C}_6\text{H}_{11}\text{Br} \xrightarrow[\text{2. CO}_2\text{; 3. H}^+]{\text{1. Mg/ether}} \text{C}_6\text{H}_{11}\text{COOH} \xrightarrow[\text{2. LiAlH(OBu}^t)_3]{\text{1. SOCl}_2} \text{C}_6\text{H}_{11}\text{CHO}$$

PROBLEM 13.23

Show the reactants and reagents necessary for each synthesis:

(a) *p*-chlorobenzaldehyde from *p*-chlorobenzoic acid

(b) cyclobutanecarbaldehyde from cyclobutane

13.6.3 Ketones from Acid Chlorides

In Section 13.6.2, you saw that hydrogen replaces the chlorine of acid chlorides to form aldehydes; in a similar manner, alkyl groups can replace the chlorine to produce ketones.

$$\text{R—CO—Cl (Acid chloride)} \xrightarrow[-\text{Cl}^-]{:\text{H}} \text{R—CHO (Aldehyde)}$$

$$\text{R—CO—Cl (Acid chloride)} \xrightarrow[-\text{Cl}^-]{:\text{R}'} \text{R—CO—R}' \text{ (Ketone)}$$

Conceptually, this means that the alkyl group must be obtainable in the form of a carbanion reagent to bring about the displacement of chloride anion. The only effective means to prepare such carbanions is as organometallic reagents. However, in Section 13.5.1, you saw that Grignard reagents bring about addition to ketones. Therefore, if a Grignard (or organolithium) reagent were to react with an acid chloride, it would produce a ketone, but then the ketone would react immediately with additional Grignard reagent in the solution to form a tertiary alcohol. This problem is circumvented by using the weakest of the organometallic reagents discussed in Section 3.5.1, an *organocuprate*, which is too weak a nucleophile to add to a ketone.

$$\text{R}'\text{COOH (Carboxylic acid)} \xrightarrow{\text{SOCl}_2} \text{R}'\text{COCl (Acid chloride)} \xrightarrow{\text{RMgBr}} \left[\text{R}'\text{—C(=O)—R}\right] \xrightarrow{\text{RMgBr}} \text{R}'\text{—C(OH)(R)—R (3° alcohol)}$$

$$\text{R}'\text{COCl (Acid chloride)} \xrightarrow{\text{R}_2\text{CuLi}} \text{R}'\text{—C(=O)—R (Ketone)}$$

Using the organocuprate reaction, any ketone can, in principle, be prepared from a carboxylic acid. Further, there are always two approaches to any ketone, depending on which alkyl group comes from the acid and which from the organocuprate. One limitation is that the organocuprate cannot be aromatic. For example, isopropyl phenyl ketone can be prepared from benzoic acid and isopropyl bromide but *not* from bromobenzene and isobutyric acid:

$$C_6H_5COOH \xrightarrow{SOCl_2} C_6H_5COCl$$

Benzoic acid → Benzoyl chloride

$$(CH_3)_2CH{-}Br \xrightarrow{Li} (CH_3)_2CH{-}Li \xrightarrow{CuI} [(CH_3)_2CH]_2CuLi$$

Isopropyl bromide → Isopropyllithium → Lithium diisopropylcuprate

$$[(CH_3)_2CH]_2CuLi + C_6H_5COCl \longrightarrow C_6H_5{-}CO{-}CH(CH_3)_2$$

Lithium diisopropylcuprate + Benzoyl chloride → Isopropyl phenyl ketone

How to Solve a Problem

Devise a sequence of reactions to convert benzoic acid to 2-phenyl-2-butanol.

$$C_6H_5{-}COOH \dashrightarrow C_6H_5{-}C(OH)(CH_3){-}C_2H_5$$

Benzoic acid → 2-Phenyl-2-butanol

PROBLEM ANALYSIS

It is apparent that two new C—C bonds must be made, attaching a methyl group and an ethyl group to the carboxyl carbon of benzoic acid. First, we think of organometallics as a source of new carbon-carbon bonds. There is no direct reaction of organometallics with carboxylic acids (for one thing, the acid's proton would be abstracted by the Grignard reagent). However, tertiary alcohols can always be obtained by Grignard reaction with a ketone, and we can use retrosynthetic analysis to envision such a step. The necessary ketone can be obtained from benzoic acid using an organocuprate reagent with the acid chloride.

$$C_6H_5{-}C(OH)(CH_3){-}C_2H_5 \Longrightarrow C_6H_5{-}C(=O){-}CH_3 \Longrightarrow C_6H_5{-}C(=O){-}OH$$

2-Phenyl-2-butanol ⟹ Acetophenone ⟹ Benzoic acid

SOLUTION

Benzoic acid is converted to benzoyl chloride using thionyl chloride. Reaction of benzoyl chloride with lithium dimethylcuprate yields acetophenone. Reaction of acetophenone with ethylmagnesium bromide produces the desired 2-phenyl-2-butanol.

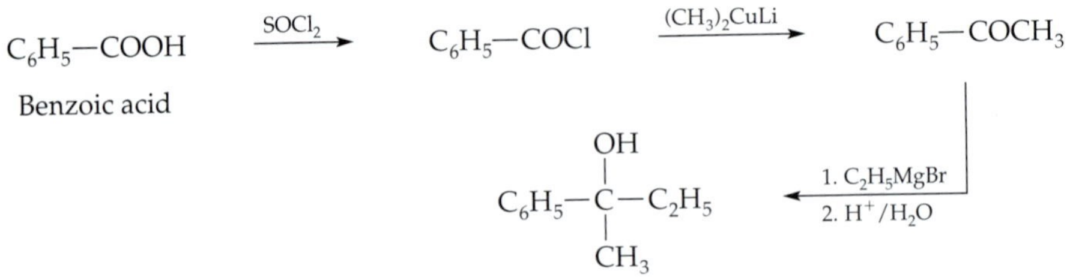

$$C_6H_5{-}COOH \xrightarrow{SOCl_2} C_6H_5{-}COCl \xrightarrow{(CH_3)_2CuLi} C_6H_5{-}COCH_3$$

Benzoic acid

$$C_6H_5{-}COCH_3 \xrightarrow[2.\ H^+/H_2O]{1.\ C_2H_5MgBr} C_6H_5{-}\underset{CH_3}{\overset{OH}{C}}{-}C_2H_5$$

2-Phenyl-2-butanol

PROBLEM 13.24

Show the reactions necessary to complete the following syntheses:

(a) benzoic acid to 3-phenyl-3-pentanol

(b) toluene to acetophenone (using the organocuprate reaction)

(c) propanal to diethyl ketone

13.7 Important and Interesting Carbonyl Compounds

The carbonyl group is important in numerous industrial processes. Formaldehyde (CH_2O) is produced commercially from the high-temperature catalytic oxidation of methanol (CH_3OH) in what is actually a dehydrogenation reaction. Formaldehyde is a gas (bp −21°C) that is used to produce polymers, such as urea-formaldehyde resins, for use in a wide variety of building and insulation materials. A solution of 37% formaldehyde in water/methanol is called *formalin*, and it is used as a disinfectant and a preservative for biological specimens.

Acetone (CH_3COCH_3) is an industrial chemical produced on a very large scale from propene or isopropyl alcohol or by oxidation of cumene (an industrial process that also produces phenol):

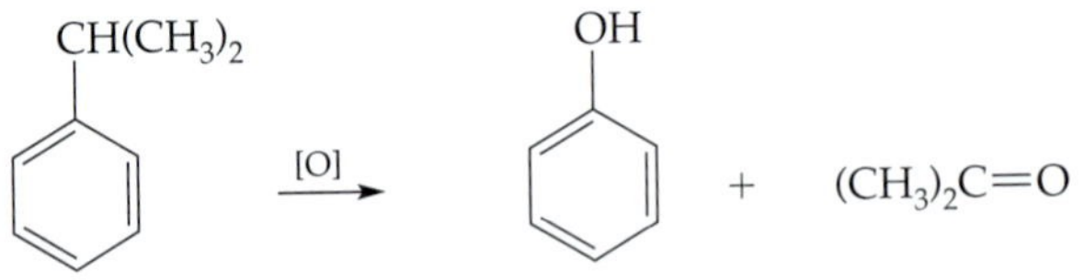

$$C_6H_5{-}CH(CH_3)_2 \xrightarrow{[O]} C_6H_5{-}OH + (CH_3)_2C{=}O$$

Cumene Phenol Acetone

Acetone is a superb organic solvent, but is also soluble in water. It is used to dissolve paints, dyes, resins, and many other organic materials.

Many aldehydes and ketones are flavorful compounds used as flavoring agents in cooking. Many are naturally occurring. Among these are a number of benzenoid compounds—for example, benzaldehyde, cinnamaldehyde, acetophenone, salicylaldehyde, vanillin, and piperonal:

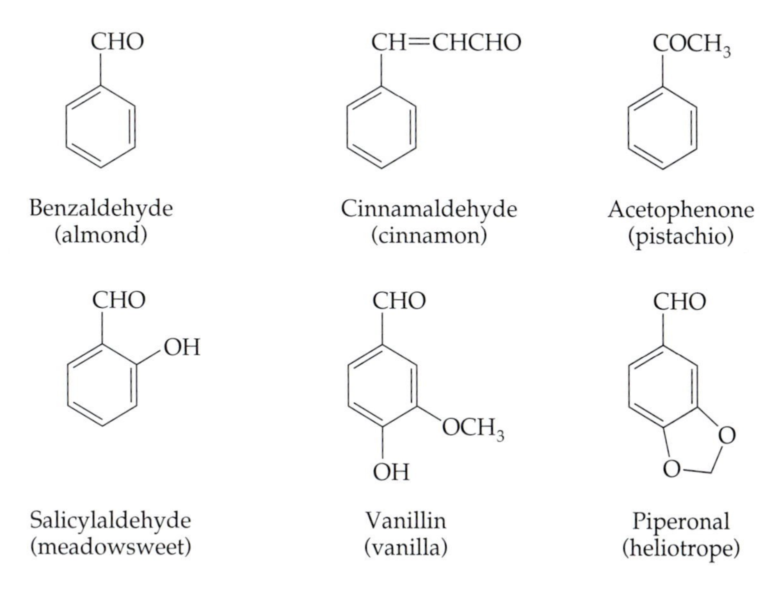

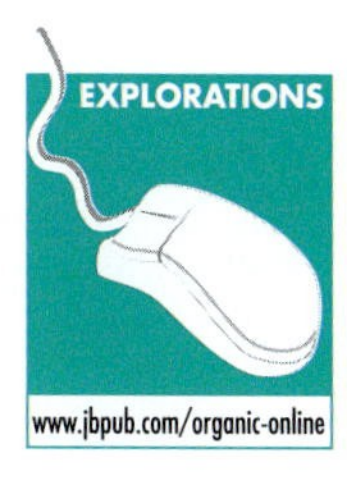

Some large macrocyclic ketones are used in the perfume industry; they include muscone (an excretion from the musk ox) and civetone (an excretion from the civet cat). Camphor is a bicyclic ketone used as a medicinal.

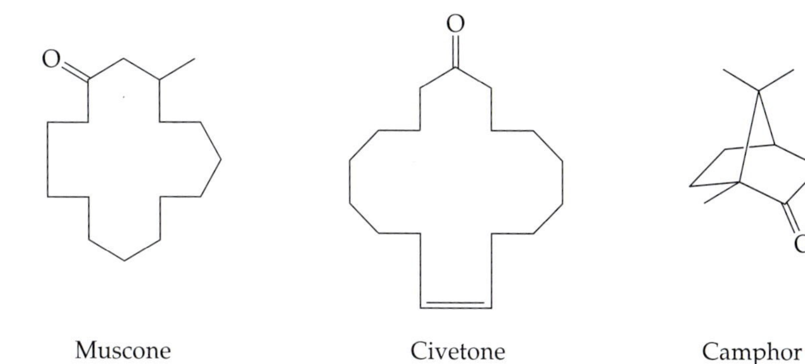

Virtually all monosaccharides (sugars) are carbonyl compounds; some are aldehydes and others ketones. However, most of the carbonyl groups are "masked" as hemiacetals or converted to acetals (glycosides), as in disaccharides, of which sucrose (table sugar) is an example. Two common monosaccharide sugars are glucose (an aldose) and fructose (a ketose), shown here in their open-chain and hemiacetal forms (see Section 4.7).

The perfume industry uses muscone as a perfume fixative to make scents last longer.

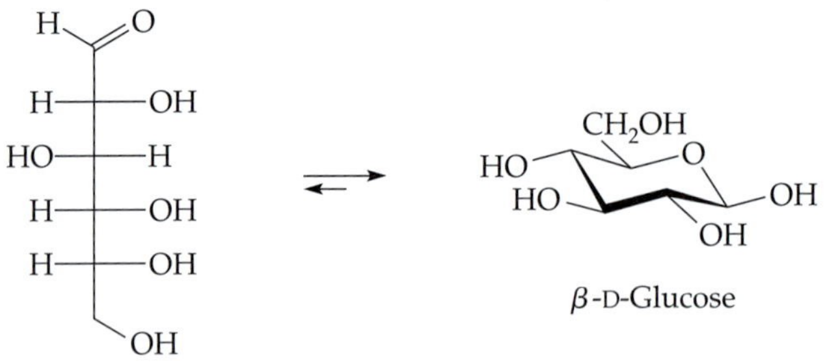

Open-chain aldose form

β-D-Glucose

Hemiacetal form

Glucose

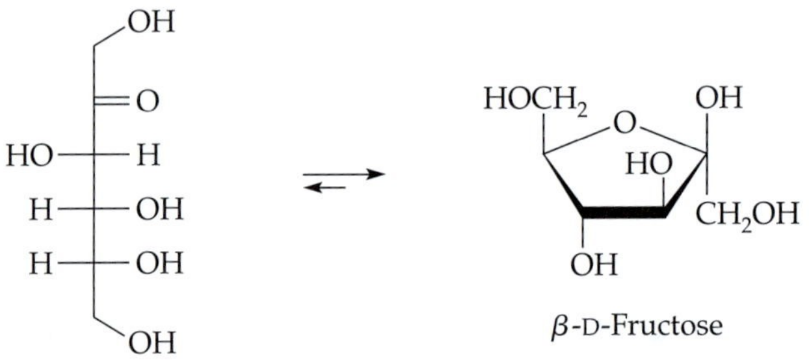

β-D-Fructose

Open-chain ketose form

Hemiacetal form

Fructose

Many important steroids are ketones. We have already encountered cortisone (see Section 13.3.1) and testosterone (see Section 8.3.2). Three other examples help to show the variety of biological effects of steroids: progesterone (a hormone excreted during pregnancy to suppress additional ovulation), norethindrone (the synthetic birth control steroid), and mifepristone (also known as RU-486, the "morning after" contraceptive). Note the suffix *-one* on each name, indicating a ketone.

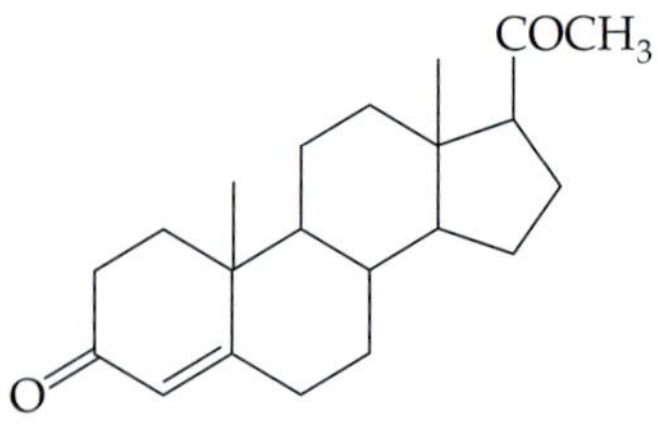

Progesterone

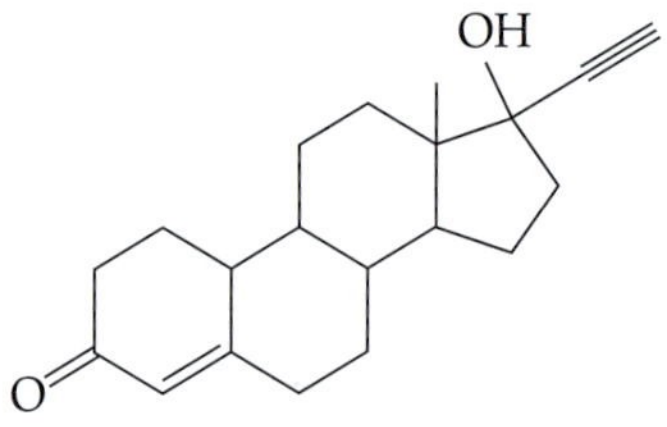

Norethindrone

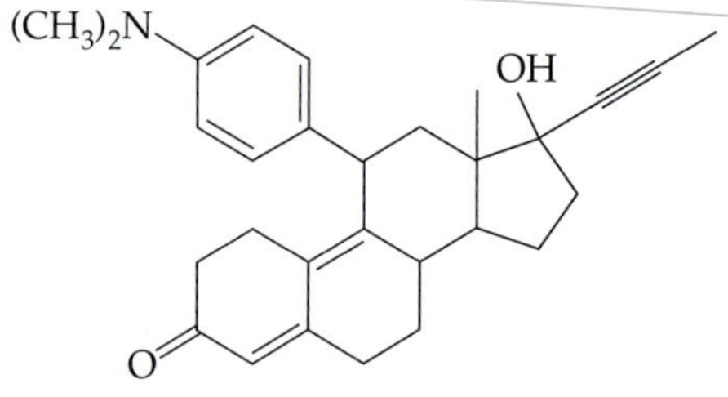

Mifepristone (RU-486)

The discovery of the hormonal effects of steroids, combined with their scarce availability from natural sources, led to much research. Initial attention focused on the total synthesis of steroids, but later the focus shifted to locating natural sources of steroidal compounds that could readily be converted into the desired steroids in the laboratory. One group of such steroidal compounds are the naturally occurring sapogenins found in desert plants. A specific example is hecogenin, which has been converted into human steroids, including cortisone:

Hecogenin

Desert plants are a natural source of the generic steroidal compounds like sapogenins that can be converted to specific steroids in the laboratory.

Finally, many pheromones contain carbonyl groups. We have encountered pheromones before. These chemicals are produced in small quantities by individual organisms and can be detected by other individuals of the same species. Pheromones function as repellents, alarm signals, sexual attractants, and so on. Here are some examples of pheromones containing carbonyl groups:

Alarm pheromone of the leaf-cutting ant

Sex attractant of the female cucumber beetle

Sex attractant of the female cockroach

Chapter Summary

Carbonyl compounds contain a **carbonyl group** (C=O) in one of two possible environments. In **aldehydes,** the carbonyl carbon is bonded to at least one hydrogen (the other bond may be to a hydrogen or to an alkyl or aryl group). In **ketones,** the carbonyl carbon is bonded to two carbons (alkyl or aryl groups). The carbonyl carbon atom is sp^2 hybridized and forms both a sigma bond and a pi bond with oxygen.

The carbon-oxygen carbonyl bond is polarized toward the oxygen, making the carbon electron-deficient. Thus, carbonyls (both aldehydes and ketones) undergo three general types of reactions: (1) *nucleophilic addition* by attack of a nucleophile at the carbonyl carbon; (2) protonation of the carbonyl oxygen to produce a resonance-stabilized cation; and (3) removal of a hydrogen from an **α-carbon** to produce a resonance-stabilized **enolate anion.** Aldehydes and ketones exist in equilibrium as isomeric **keto forms** and **enol forms,** a phenomenon known as **tautomerism.**

Aliphatic aldehydes, generically called *alkanals,* are named by replacing the final *-e* of the parent alkane name with the suffix *-al.* The carbonyl carbon is always carbon 1. Aliphatic ketones, generically called *alkanones,* are named by replacing the *-e* of the alkane name with the suffix *-one.* The location of the carbonyl group in ketones is indicated by a number.

Carbonyl compounds undergo reduction by sodium borohydride (which is, in effect, hydride addition) to form primary or secondary alcohols. Cyanide addition forms **cyanohydrins,** which can be hydrolyzed to α-hydroxy carboxylic acids. Reactions of carbonyls with primary amines form **imines,** some of which can be used to detect and characterize carbonyl groups. Addition reactions with alcohols under acid catalysis form **acetals,** which are useful as protecting groups. Use of ethanedithiol forms a dithioacetal that can be subjected to **hydrogenolysis,** converting a carbonyl group to a methylene group.

Grignard reagents and organolithium reagents, acting as carbanions, add to carbonyl compounds to form alcohols. The **Wittig reaction** of phosphonium ylides with carbonyl groups produces alkenes.

Aldehydes are prepared by reduction of acid chlorides with $LiAlH(OBu^t)_3$. Ketones are prepared by the reaction of lithium dialkylcuprates with acid chlorides and by the Friedel-Crafts acylation of arenes. Aldehydes and ketones are produced by oxidation of primary and secondary alcohols, respectively.

Summary of Reactions

1. Reduction of carbonyl compounds to produce primary or secondary alcohols (Section 13.3.1)

$$\text{C=O} + \text{NaBH}_4 \longrightarrow \text{C(OH)(H)}$$

Carbonyl compound + Sodium borohydride → Alcohol

2. Addition of cyanide ion to carbonyl compounds to produce cyanohydrins (Section 13.3.2)

$$\text{C=O} + \text{NaCN} \xrightarrow{\text{(pH 10)}} \text{C(OH)(CN)}$$

Carbonyl compound + Sodium cyanide → Cyanohydrin

3. Addition of primary amines to carbonyl compounds to produce imines (Section 13.4.1)

$$\text{C=O} + \text{H}_2\text{N—R} \text{ or } \text{H}_2\text{N—Ar} \longrightarrow \text{C=N—R} \text{ or } \text{C=N—Ar}$$

4. Addition of alcohols to carbonyl compounds to produce acetals (Section 13.4.2). The formation of a thioacetal followed by hydrogenolysis results in overall conversion of a carbonyl group to a methylene group.

$$\text{C=O} + 2\,\text{ROH} \xrightarrow{\text{H}_2\text{SO}_4} \text{C(OR)(OR)}$$

Carbonyl compound + Alcohol → Acetal

$$\text{C=O} + \text{HS—}\!\!\text{—SH} \xrightarrow{\text{H}_2\text{SO}_4} \text{C(S—)(S—)} \xrightarrow{\text{RaNi, H}_2} \text{C(H)(H)}$$

Carbonyl compound → Dithioacetal → Methylene group

5. Addition of Grignard reagents (or organolithium reagents) to carbonyl compounds to form alcohols (Section 13.5.1)

$$\text{C=O} + \text{RMgBr} \xrightarrow[\text{2. H}^+/\text{H}_2\text{O}]{\text{1. ether solvent}} \text{C(OH)(R)}$$

Carbonyl compound + Grignard reagent → Alcohol

6. Reaction of carbonyls with phosphonium ylides to form alkenes (the Wittig reaction) (Section 13.5.2)

$$\text{C=O} + \overset{-}{\text{C}}\text{—}\overset{+}{\text{P}}(\text{C}_6\text{H}_5)_3 \longrightarrow \text{C=C}$$

Carbonyl compound + Phosphonium ylide → Alkene

7. Conversion of acid chlorides to aldehydes with lithium tri-*t*-butoxyaluminum hydride (Section 13.6.2)

$$\text{R—}\overset{\text{O}}{\overset{\|}{\text{C}}}\text{—Cl} \xrightarrow{\text{LiAlH(OBu}^t)_3} \text{R—}\overset{\text{O}}{\overset{\|}{\text{C}}}\text{—H}$$

Acid chloride → Aldehyde

8. Conversion of acid chlorides to ketones using lithium dialkylcuprates (Section 13.6.3)

$$\text{R—}\overset{\text{O}}{\overset{\|}{\text{C}}}\text{—Cl} \xrightarrow{\text{LiCuR}'_2} \text{R—}\overset{\text{O}}{\overset{\|}{\text{C}}}\text{—R}'$$

Acid chloride → Ketone

Additional Problems

Nomenclature

13.25 Draw a structural formula corresponding to each of the following names:

(a) 2-pentanone

(b) (*E*)-2-methyl-2-pentenal

(c) 3-hydroxycyclopentanecarbaldehyde

(d) *p*-*N*,*N*-dimethylaminobenzaldehyde

(e) *p*-nitroacetophenone

(f) (*R*)-2-hydroxybutanal

(g) acetone

(h) ethyl isopropyl ketone

(i) benzophenone

(j) benzyl phenyl ketone

13.26 Supply a name for each of the following compounds:

(a) 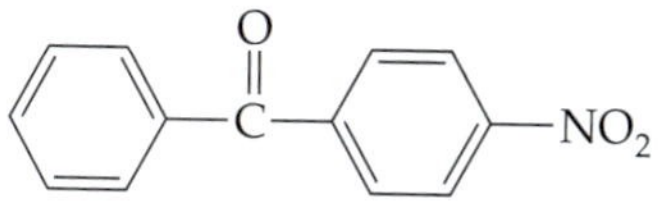 C_6H_5–C(=O)–C_6H_4–NO_2

(b) $(CH_3)_2CHCOCH_2CH_3$

(c) 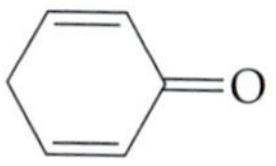 O

(d) $HOCH_2CH{=}CH{-}CH_2CHO$

(e) 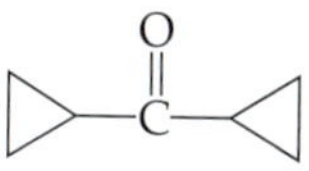 O ‖ C

(f) 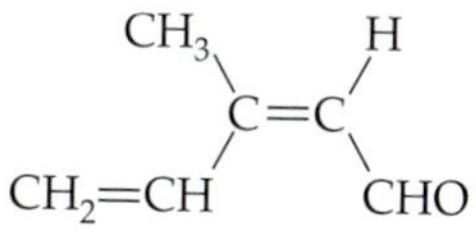 CH_3 H C=C $CH_2{=}CH$ CHO

(g) 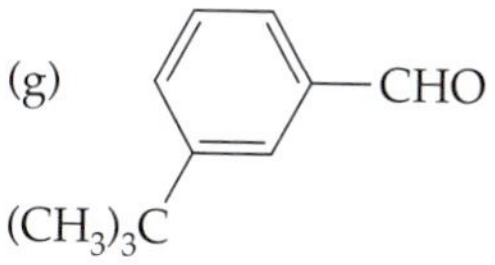 CHO $(CH_3)_3C$

(h) CHO

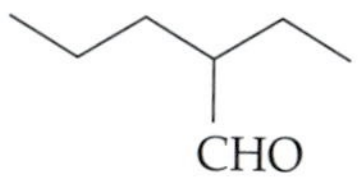

(i) 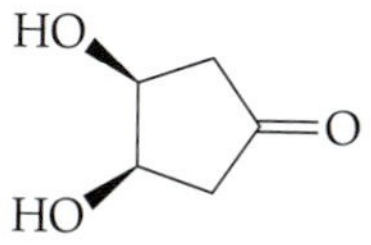 HO HO O

13.27 Draw a structural formula and supply an IUPAC name for each of the following compounds:

(a) allyl methyl ketone

(b) dibenzyl ketone

(c) phenylacetaldeyde

(d) *t*-butyl ethyl ketone

(e) divinyl ketone

13.28 What is wrong with each of the following names?

(a) 1-butanone

(b) cyclopentanal

(c) 2-phenylpropionaldehyde

(d) methylbenzaldehyde

(e) (*E*)-2-propenal

(f) 3-ketocyclohexanol

(g) 2,2-dimethyl-4-hexanone

Properties of Carbonyls

13.29 The three compounds pentane, 1-butanol, and butanal have similar molecular weights. Place them in order of increasing boiling point.

13.30 Draw the structure of an example of each of the following kinds of compounds:

(a) an acetal

(b) an enol

(c) a hemiacetal

(d) a 2,4-dinitrophenylhydrazone

(e) an imine

(f) a cyanohydrin

13.31 Explain why the α-hydrogen of an aldehyde is relatively acidic, and show the mechanism of its removal from propanal.

13.32 Show all possible enol forms for each of the following carbonyl compounds, and indicate which is expected to be the most stable (remember the sequence of alkene stability):

(a) isopropyl ethyl ketone

(b) $CH_3COCH_2COCH_3$

13.33 A compound is known to possess a carbonyl group. What single experiment (chemical or instrumental) would permit you to determine whether the compound is an aldehyde or a ketone?

13.34 An unknown compound shows a sharp and strong IR peak at 1720 cm^{-1}. The mass spectrum shows peaks at $m/z = 72$ ($M^{+\cdot}$), 57, and 43. The ^{1}H-NMR spectrum shows peaks at 9.5 ppm (s, 1H), 2.6 ppm (m, 1H), and 0.9 ppm (d, 6H). What is the compound's structure?

13.35 Show the mechanism by which (*R*)-2-methylbutanal is racemized (that is, converted into a mixture of equal parts of *R* and *S* enantiomers) when treated with aqueous sodium hydroxide.

Addition Reactions of Carbonyls

13.36 Show the product expected from reaction of acetophenone with each of the following reagents, followed by appropriate work-up:

(a) sodium borohydride
(b) methanol and catalytic sulfuric acid
(c) zinc amalgam and hydrochloric acid
(d) sodium cyanide at pH 10
(e) aniline
(f) ethanedithiol, then Raney nickel and hydrogen

13.37 Show the product expected from reaction of butanal with each of the following reagents, followed by appropriate work-up:
(a) silver nitrate in ammonia
(b) methylamine
(c) ethylene glycol and a trace of acid
(d) sodium cyanide at pH 10
(e) sodium borohydride
(f) nitric acid

13.38 Show how the cyanohydrin reaction can be used to homologate butanal and obtain pentanal.

13.39 Show the reactions necessary to convert phenylacetaldehyde to ethylbenzene.

13.40 Show the structure of the acetal formed from each of the following reactions, using H_2SO_4 as a catalyst:
(a) cyclohexanone and ethanol
(b) *cis*-1,2-cyclohexanediol and diethyl ketone
(c) benzaldehyde and ethanedithiol
(d) glucose (see Section 13.7) and methanol

13.41 Show the product(s) of the hydrolysis of each acetal with dilute acid:

(a) 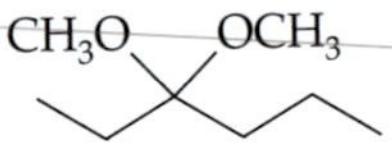 CH_3O OCH_3 (b) 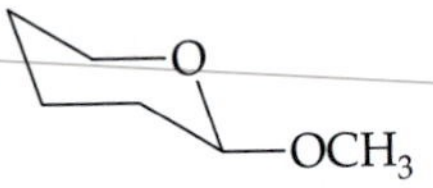 O OCH_3

(c) 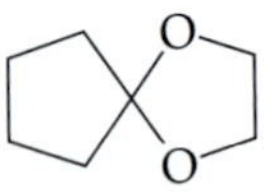 O O

13.42 Propenal can be converted to 2,3-dihydroxypropanal by reactions studied so far. Show how to accomplish this conversion, remembering that both the carbon-carbon double bond and the aldehyde group are susceptible to oxidation.

13.43 What chemical test can be employed to distinguish between the members of each pair of compounds? Show each "positive" reaction.
(a) benzyl alcohol and benzaldehyde
(b) cyclopentyl methyl ether and 2-hexanone
(c) methyl vinyl ketone and cyclobutanone
(d) pentanal and 3-pentanone

13.44 Show the product of the reaction of glucose (see Section 13.7) with each of the following reagents:
(a) sodium borohydride
(b) aniline
(c) sodium cyanide at pH 10
(d) ethanol and sulfuric acid

13.45 Explain the observation that when 4-hydroxypentanal is treated with methanol and catalytic amounts of sulfuric acid, 2-methoxy-5-methyltetrahydrofuran results.

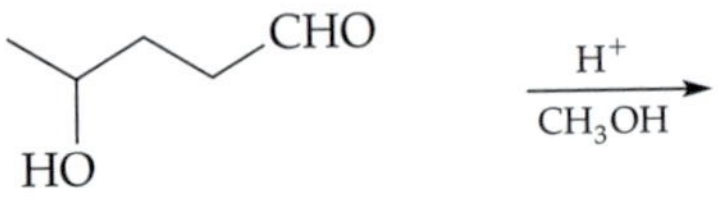

CHO HO $\xrightarrow[CH_3OH]{H^+}$

4-Hydroxypentanal

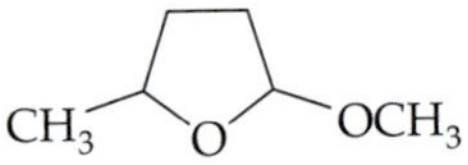

CH_3 O OCH_3

2-Methoxy-5-methyltetrahydrofuran

13.46 Show the product of the reaction of cyclohexanone with each of the following compounds:
(a) aniline
(b) phenylhydrazine ($C_6H_5NHNH_2$)
(c) hydroxylamine (H_2NOH)
(d) cyclohexylamine

13.47 Show the product of the reaction of phenylmagnesium bromide with each of the following compounds, followed by appropriate work-up with dilute acid and water:
(a) cyclohexanone
(b) benzaldehyde
(c) formaldehyde
(d) ethylene oxide
(e) *p*-methoxyacetophenone

13.48 Show the product of each reaction:
(a) ethyl bromide with triphenylphosphine
(b) the product of part (a) with butyllithium
(c) the product of part (b) with cyclohexanone

13.49 Indicate two combinations of reactants (ylide and carbonyl compound) that will produce each alkene:
(a) styrene
(b) 2-pentene
(c) 2,3-dimethyl-2-hexene
(d) 1-phenyl-2-(*p*-nitrophenyl)ethene

13.50 Show the reactants needed to synthesize each of the following alcohols using the Grignard reaction:
(a) cyclobutylmethanol
(b) *t*-butyl alcohol
(c) 2-phenyl-2-butanol

13.51 Show the product of the reaction of pentanal with each of the following reagents, followed by appropriate work-up:

(a) 2,4-dinitrophenylhydrazine
(b) sodium borohydride
(c) methyllithium
(d) nitric acid
(e) the product of part (d) with thionyl chloride
(f) the product of part (e) with benzene and aluminum chloride
(g) the product of part (e) with lithium diethylcuprate
(h) phenylmagnesium bromide
(i) methyltriphenylphosphonium bromide and butyllithium
(j) ethanedithiol and acid
(k) the product of part (j) with hydrogen and Raney nickel
(l) aniline
(m) potassium permanganate
(n) hydroxylamine
(o) sodium cyanide
(p) Jones' reagent (chromic acid)

Synthesis of Carbonyl Compounds

13.52 Show the reaction(s) necessary for the preparation of each carbonyl compound from the specified starting material:

(a) acetone from 2-propanol
(b) 2-methyl-2-butene from 2-propanol
(c) acetophenone from benzaldehyde
(d) cyclohexanone from cyclohexane
(e) dicyclohexyl ketone from cyclohexanecarbaldehyde
(f) 3-hexanone from 1-propanol
(g) cyclohexyl methyl ketone from acetaldehyde
(h) benzophenone from benzene
(i) 3-hexanol from 1-propanol

13.53 Show the reactions needed to convert propiophenone ($C_6H_5COCH_2CH_3$) into 1-phenyl-2-propanone.

Syntheses Using Carbonyls

13.54 Show two methods of converting 1-phenyl-2-propanone (phenylacetone) into 1-phenylpropane. (More than one step may be required.)

13.55 Identify the reagent(s) that are needed to complete each of the following transformations:

(a) $CH_3CH_2OH \xrightarrow{?} CH_3CH(OC_2H_5)_2$

(b) $C_6H_5CH_2Br \xrightarrow{?} (CH_3)_2CHCH(OH)CH_2C_6H_5$

(c) $C_6H_5COCH_2CH_3 \xrightarrow{?} C_6H_5C(=N{-}C_6H_5)CH_2CH_3$

(d) cyclohexanone ethylene ketal (1,4-dioxaspiro[4.5]decane) $\xrightarrow{?}$ cyclohexanone

13.56 Show how to accomplish the following transformations using any necessary reagents. (*Hint:* Think retrosynthetically.)

(a) acetone to 2-methyl-1-phenyl-1-propene
(b) cyclohexane to methylenecyclohexane
(c) 2-methylpropane to 2,2-dimethyl-1-propanol
(d) benzene to 1-phenylcyclohexanol

13.57 The Wittig reaction can be applied very effectively to synthesize conjugated dienes. Show two routes by which 1-phenyl-1,3-butadiene can be synthesized using the Wittig reaction.

13.58 Show all possible combinations of Grignard reagent and carbonyl compound that could be employed in the synthesis of each of the following compounds:

(a) $CH_3CHOHCH(CH_3)_2$
(b) $(CH_3)_3CCH_2OH$
(c) 4-ethyl-4-octanol

13.59 Show the reaction, if any, by which the alcohols in Problem 13.58 might be prepared by reduction of a carbonyl compound.

13.60 Starting with acetophenone and ethanol as the only starting materials, devise a synthesis for 2-phenyl-2-butene.

Mixed Problems

13.61 Indicate a chemical test that can be used to distinguish between the members of each of the following pairs of compounds. Describe the observation that constitutes a positive test.

(a) benzaldehyde and acetophenone
(b) cyclopentanone and cyclopentanol
(c) cyclopentanone and cyclopentene
(d) 3-pentanone and pentanal

13.62 Designate a spectroscopic technique that could be used in place of each test in Problem 13.61, and indicate the spectral differences between the compounds.

13.63 Employing 1-bromopropane as the starting material, show how each of the following alcohols can be prepared:

(a) 1-phenyl-1-butanol
(b) 1-butanol
(c) 2-butanol
(d) 2-pentanol

13.64 Gypsy moths have infested large areas of the northeastern United States, resulting in widespread deforestation. One of the control methods attempted has involved the use of the pheromone *disparlure*, the sex attractant of the female moth. The compound has been synthesized from two starting materials, first using a Wittig reaction to form a *cis*-alkene, which was then converted into disparlure, a *cis*-epoxide. Show the starting materials and reagents for these reactions.

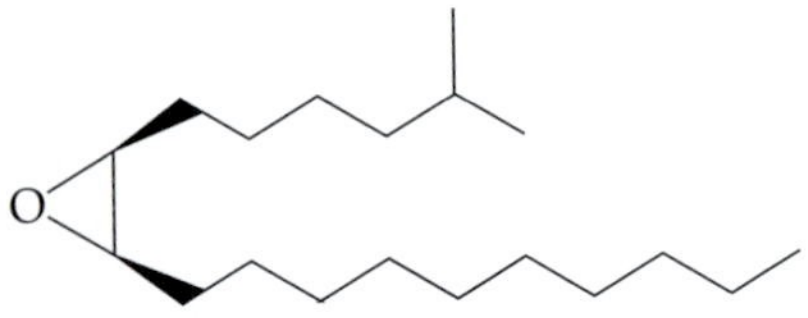

Disparlure

13.65 The following scheme for the synthesis of Valium has been published. Show the reagents necessary for each of the labeled steps:

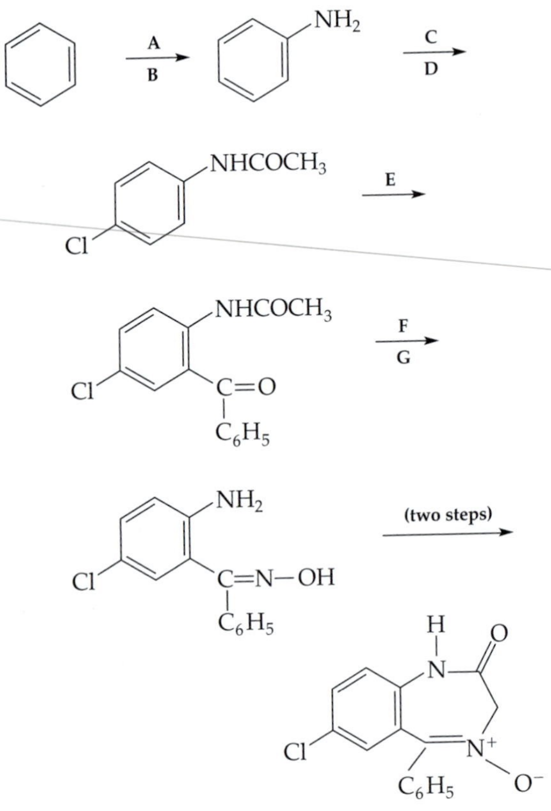

Benzene $\xrightarrow[B]{A}$ aniline (NH_2) $\xrightarrow[D]{C}$

Cl-substituted $NHCOCH_3$ compound $\xrightarrow{E}$

$NHCOCH_3$, Cl, $C=O$ with C_6H_5 $\xrightarrow[G]{F}$

NH_2, Cl, $C=N-OH$ with C_6H_5 $\xrightarrow{\text{(two steps)}}$

Valium (H, N, O, Cl, N^+, O^-, C_6H_5)

13.66 Propose a structure for a compound that has the molecular formula $C_7H_{14}O$. It showed an IR peak at 1720 cm^{-1}. On a ^1H-NMR spectrum, peaks were observed at 1.0 ppm (s, 9H), 2.5 ppm (s, 3H), and 2.6 ppm (s, 2H).

13.67 Draw structures for compounds **A–C** in the following synthetic sequence. Compound **B** gives a positive Tollens' test. Compound **C** shows no carbonyl or hydroxyl peaks in its IR spectrum.

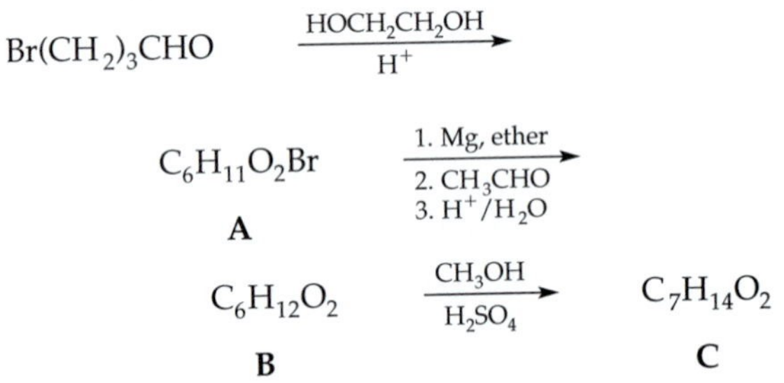

$$Br(CH_2)_3CHO \xrightarrow[H^+]{HOCH_2CH_2OH} \underset{\mathbf{A}}{C_6H_{11}O_2Br} \xrightarrow[\text{2. } CH_3CHO \text{ 3. } H^+/H_2O]{\text{1. Mg, ether}} \underset{\mathbf{B}}{C_6H_{12}O_2} \xrightarrow[H_2SO_4]{CH_3OH} \underset{\mathbf{C}}{C_7H_{14}O_2}$$

13.68 Glyceraldehyde is a very important biological compound whose structure serves as the stereochemical reference for carbohydrates (see Section 4.7) and α-amino acids and proteins (see Section 16.1.1). It can be synthesized from the readily available allyl alcohol. Indicate the reagents (**A–D**) that should be used for the steps of the synthesis.

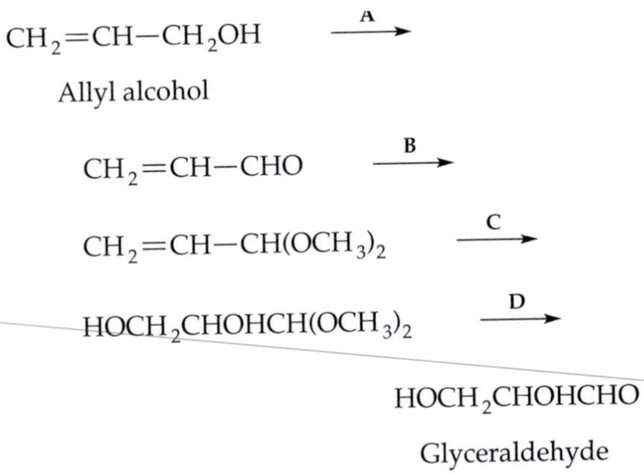

$CH_2{=}CH{-}CH_2OH$ (Allyl alcohol) $\xrightarrow{A}$

$CH_2{=}CH{-}CHO \xrightarrow{B}$

$CH_2{=}CH{-}CH(OCH_3)_2 \xrightarrow{C}$

$HOCH_2CHOHCH(OCH_3)_2 \xrightarrow{D}$

$HOCH_2CHOHCHO$
Glyceraldehyde

13.69 Identify the reagents needed to complete the following synthesis:

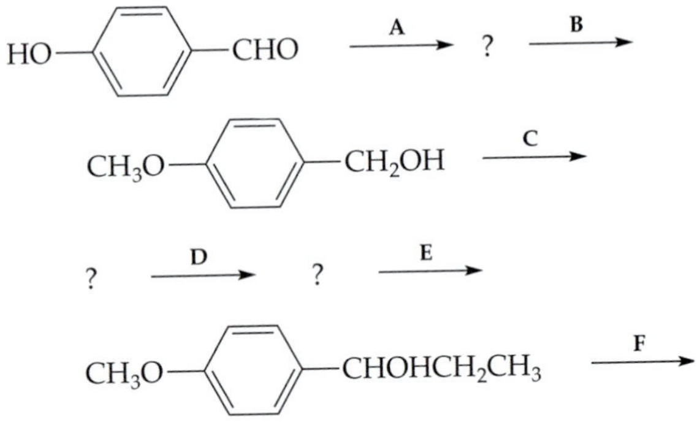

HO–C_6H_4–CHO $\xrightarrow{A}$? $\xrightarrow{B}$

CH_3O–C_6H_4–$CH_2OH \xrightarrow{C}$

? $\xrightarrow{D}$? $\xrightarrow{E}$

CH_3O–C_6H_4–$CHOHCH_2CH_3 \xrightarrow{F}$

Dollars and Scents

The cosmetics company you work for is interested in developing a line of perfumes. Some at the company think that perfumes made strictly from all-natural ingredients might add special appeal to the product line and thereby give the company an advantage in a fiercely competitive market. Others wonder if the cost of going "all natural" would be too prohibitive. It's your job to gather samples to assess scent quality and compare costs of natural versus synthetic ingredients.

You start with the fragrance of jasmine. One of your regular suppliers has some *jasmin absolut*, the essential oil extracted from jamine flowers, but it is extremely expensive. Three other suppliers offer you jasmine oil at just a fraction of what the first supplier wants, but you wonder if what they are selling is really the all-natural essential oil. You bring the four samples to a chemist you know and ask for an analysis.

The IR spectra of all four of the samples show a peak near 1690 cm^{-1}. After obtaining ^{1}H-NMR and mass spectra, the chemist identifies some of the compounds present in the four samples. Sample 1 consists of compound **A** and is fairly pure. Sample 2 is a mixture of compound **A** and its *trans* isomer. Sample 3 is compound **B**. Sample 4 is a complex mixture that includes compound **A** as a major component.

CH_3 ; O

A

C_6H_5—CH=C—C(=O)—H, with $CH_2CH_2CH_2CH_2CH_3$ on C

B

- What are the molecular formulas of the compounds shown?
- What does the presence of an IR peak at approximately 1690 cm^{-1} tell you about the structure of the compounds in each sample?
- The chemist has a vague memory that the active ingredient in jasmin absolut is a compound called *jasmone*. If this is true, which of the four samples is definitely a synthetic?
- Propose a synthesis of compound **A** from starting material **C**.

CH_3 ; OH

C

CH_3 ; C(=O)—H ; CH_3O ; OCH_3

D

Propose a synthesis of compound **A** and its *trans* isomer, as in sample 2, from starting material **D**. (*Hint:* Think Wittig.)

- The chemist also remembers that jasmin absolut does not include any isomers of jasmone. Which of the samples does this eliminate?
- Which sample is the natural sample?
- You recognize that compounds **A** and **B** have essentially identical fragrances. What does this tell you about the sense of smell?

Hundreds of thousands of jasmine blossoms are required to supply even this tiny amount of essential oil.

FIGURE 13.6

^{1}H-NMR spectra for two isomers with molecular formula $C_{10}H_{12}O$ (Problem 13.70).

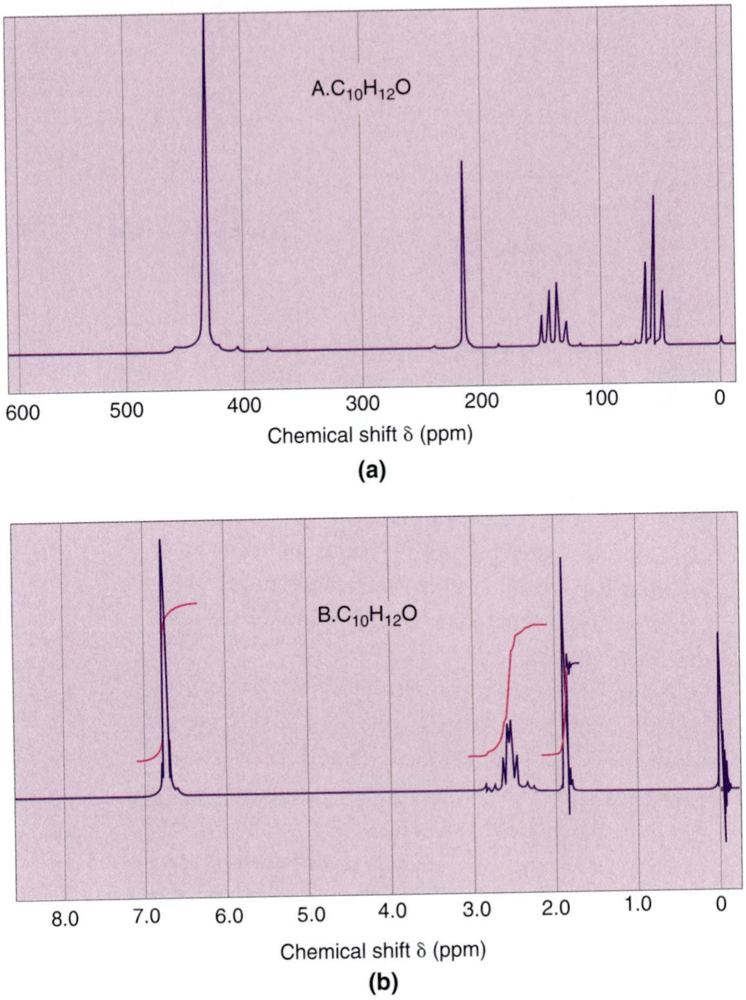
A.C10H12O
600 500 400 300 200 100 0
Chemical shift δ (ppm)
(a)
B.C10H12O
8.0 7.0 6.0 5.0 4.0 3.0 2.0 1.0 0
Chemical shift δ (ppm)
(b)

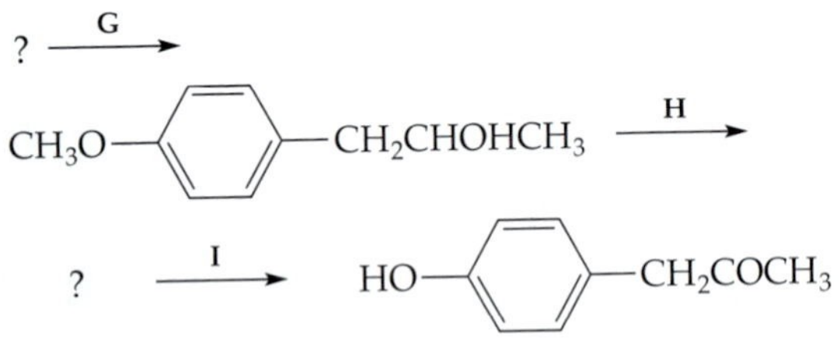
?
G
CH3O
CH2CHOHCH3
H
?
I
HO
CH2COCH3

13.70 Compounds **A** and **B** are isomers with molecular formula $C_{10}H_{12}O$; both show a strong IR peak at 1705 cm^{-1}. The ^{1}H-NMR spectrum for each is shown in Figure 13.6. Propose structures for **A** and **B.**

13.71 The preparation sold as Laetrile is obtained from the pits of apricots and has been promoted as a cancer cure, especially in Mexico. Its use for that purpose in the United States has not been approved by the FDA. The active ingredient, known as *amygdalin*, is a glycoside (acetal) of a cyanohydrin. (a) What is the absolute configuration of amygdalin at the cyanohydrin carbon (*R* or *S*)? (b) When amygdalin is hydrolyzed in dilute acid, the products are glucose, benzaldehyde, and HCN. Account for the formation of these products.

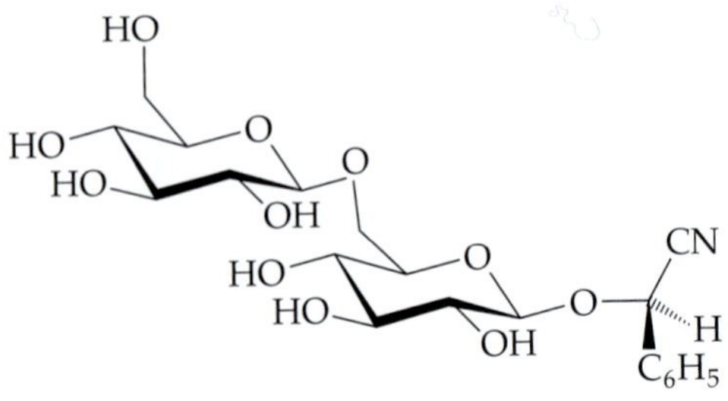
HO
HO
HO
O
O
OH
HO
HO
O
OH
O
CN
H
C6H5

Amygdalin (active ingredient in Laetrile)

CHAPTER 14

Carboxylic Acids

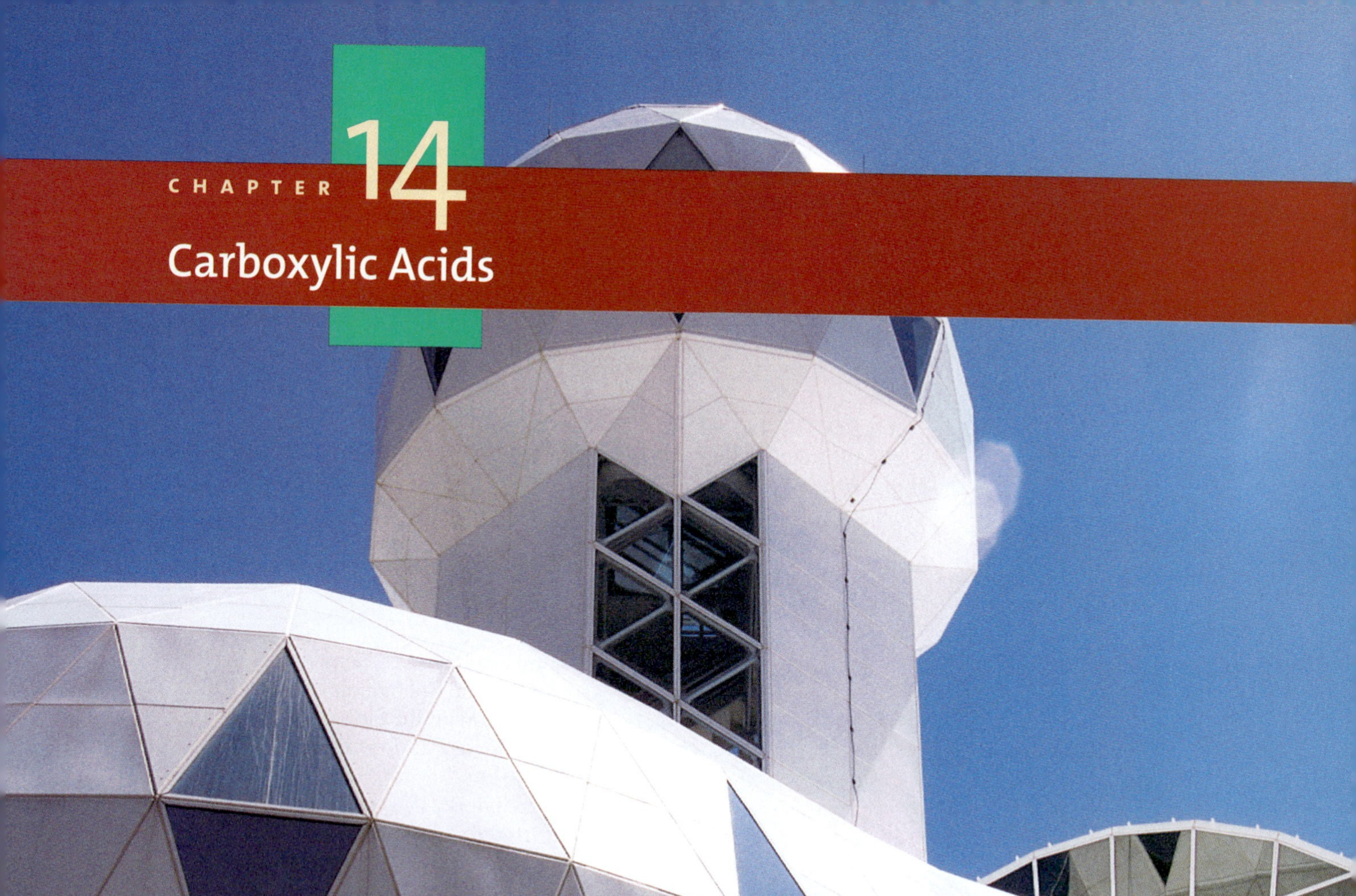

CARBOXYLIC ACIDS are a large family of organic compounds that contain the functional group called the **carboxyl group,** which can be represented in four ways:

$$-\overset{\overset{\displaystyle O}{\|}}{C}-OH \qquad -COOH \qquad -CO_2H$$

The carboxyl group

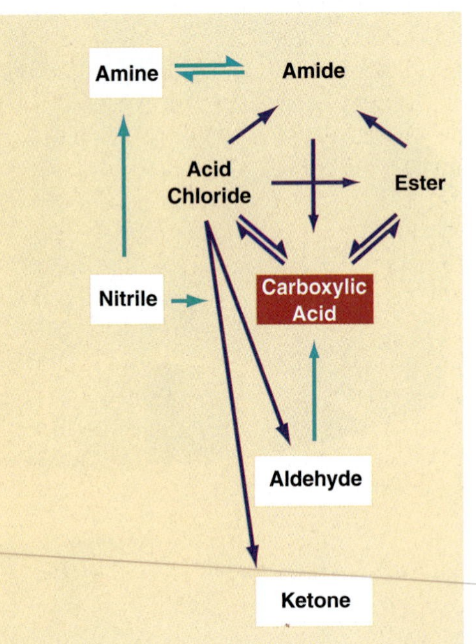

Carboxylic acids are organic acids.

Any aliphatic or aromatic compound can have a carboxyl group attached to it. The simplest carboxylic acid is formic acid (methanoic acid), HCOOH. The acid commonly encountered as an oxidation product in structure proofs is benzoic acid (C_6H_5—COOH). Fatty acids are long hydrocarbon chains with a carboxyl group at one end (R—COOH).

Earth has an oxidizing atmosphere (air is 21% oxygen), which eventually causes most chemicals to be converted to a higher oxidation state. As natural materials degrade upon exposure to the atmosphere, they are oxidized. For example, automobile tires gradually crack on the sidewalls because of oxidation of their polymer, and the end products of this degradation are carboxylic acids. Likewise, plant terpenes (see Section 8.3.2) are eventually oxidized to carboxylic acids. As you might expect, therefore, carboxylic acids are widespread in nature, in part because they represent the highest oxidation state possible for carbon while still bonded to an organic group. (The highest oxidation state of carbon occurs in carbon dioxide, CO_2.)

14.1 Nomenclature of Carboxylic Acids

14.1.1 IUPAC Nomenclature

The IUPAC nomenclature system requires naming the longest chain of carbon atoms containing the carboxyl group using the parent alkane name; the carbon of the carboxyl group is included in determining the parent alkane name. The alkane *-e* is dropped, and the suffix *-oic acid* is added to represent the functional group. Aliphatic carboxylic acids are known generically as **alkanoic acids** (see Figure 14.1). By definition, the carboxyl group must be at the beginning of the chain, and the carboxyl carbon is carbon 1. Thus, the three-carbon compound derived from propane and containing a carboxyl group, CH_3CH_2COOH, is named propanoic acid.

◀ Most industrial cleansers used today are composed of carboxylate anions.

Aliphatic carboxylic acids containing rings are named by adding the words *carboxylic acid* to the ring name, as in cyclobutanecarboxylic acid. (Note that the terminal *-e* of the alkane name is retained because the next letter in the name is not a vowel.)

CH_3CH_2COOH — Propanoic acid

$(CH_3)_2CHCH_2CH_2COOH$ — 4-Methylpentanoic acid

COOH — Cyclobutanecarboxylic acid

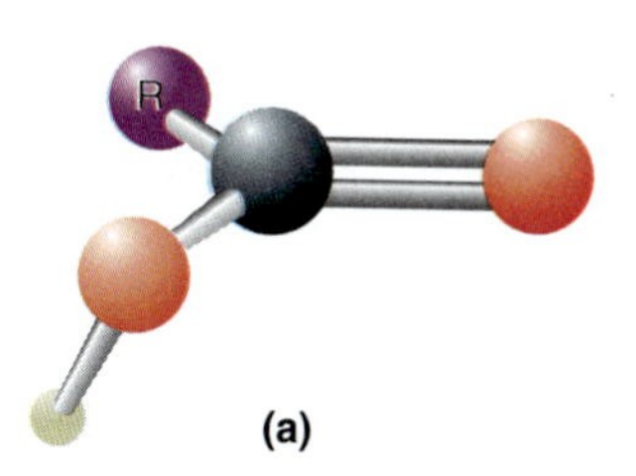

(a)

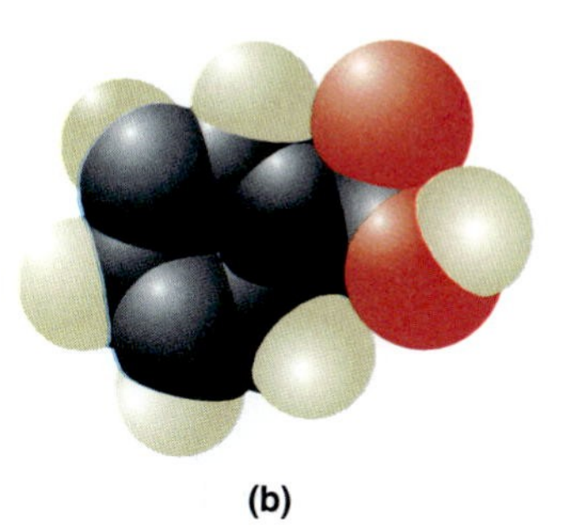
(b)

FIGURE 14.1

(a) Ball-and-stick model of an alkanoic acid. (b) Space-filling model of benzoic acid.

Benzenoid carboxylic acids may be named as derivatives of benzenecarboxylic acid (C_6H_5—COOH). However, this name is seldom used; instead, the historical name *benzoic acid* is used (see also Section 9.2.1).

COOH

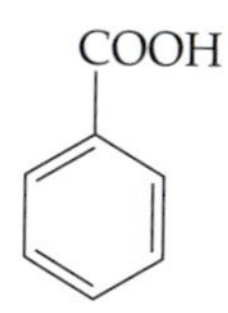

Benzoic acid

COOH, O_2N, OC_2H_5

3-Ethoxy-5-nitrobenzoic acid

Rules of Precedence

How should IUPAC names be derived for compounds containing more than one functional group? For example, should a compound containing a keto group and a carboxyl group be named as a ketocarboxylic acid or as a carboxyketone? Should a compound containing a double bond and a hydroxyl group be named as an alcohol or as an alkene? Now that all of the major functional groups have been introduced, it is appropriate to summarize the IUPAC rules for such circumstances. (For a fuller discussion, refer to Appendix A.)

- First, determine the order of precedence of the functional groups included in the compound (see Table 14.1). Use the carbon chain containing the functional group of highest precedence to establish the parent name and add the designated *suffix* for that group.
- Second, name all other functional groups of lower precedence using the designated *prefix*, indicating their positions by numbers. If needed, *-en-* or *-yn-* is inserted immediately before the highest-priority group's suffix, modifying the stem name from that of an alkane to that of an alkene or alkyne.

Application of these rules is reflected in the names 3-hydroxy-4-oxocyclohexanecarboxylic acid, 2-amino-3-butynoic acid, and (Z)-3-formyl-3-(4-methoxyphenyl)propenoic acid:

COOH

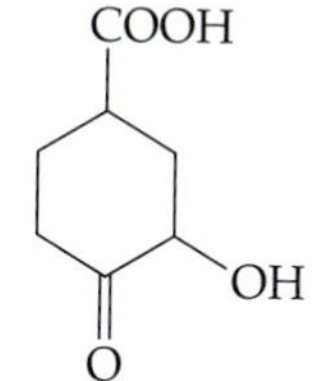

OH, O

3-Hydroxy-4-oxo-cyclohexanecarboxylic acid

NH_2, COOH

2-Amino-3-butynoic acid

CH_3O, CHO, COOH

(Z)-3-Formyl-3-(4-methoxyphenyl)-propenoic acid

The carboxylic acid, 3-methyl-3-hexanoic acid is one of the compounds associated with the odor of human perspiration.

TABLE 14.1

Precedence of Prefixes and Suffixes Used to Designate Functional Groups

Functional Group Name	Formula of Group	Suffix (when highest precedence)	Prefix (when lower precedence)
Carboxyl	—COOH	-oic acid or -carboxylic acid	carboxy-
Aldehyde	—CHO	-al or -carbaldehyde	formyl- or oxo-
Ketone	—CO—	-one	oxo-
Alcohol	—OH	-ol	hydroxy-
Amine	$—NH_2$	-amine	amino-
Ether	—OR	(none)	alkoxy-
Alkene	C=C	-ene	-en-
Alkyne	C≡C	-yne	-yn-

14.1.2 Historical/Common Names

$CH_3(CH_2)_{16}COOH$

Stearic acid
(octadecanoic acid)

Saturated fatty acid

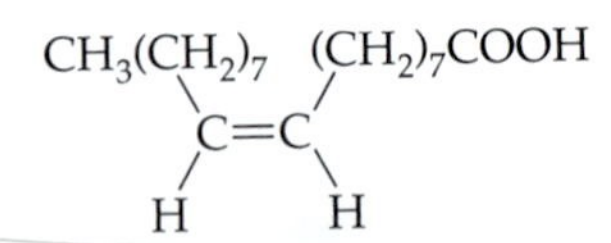

Oleic acid
(*cis*-9-octadecanoic acid)

Unsaturated fatty acid

Many carboxylic acids have very long histories and were assigned names long before their structures were known or IUPAC nomenclature existed. The names of the simplest carboxylic acids and of some common fatty acids, together with the source of each name, are given in Table 14.2. The historical names of the simple acids (the first four in Table 14.2) are widely used, so you should learn these now; you can learn the other names more gradually.

Long-chain carboxylic acids are often called *fatty acids*, because many of them were first obtained by hydrolysis of fats. Such naturally occurring acids always contain an even number of carbon atoms and may contain one or more double bonds, in which case they are called *unsaturated fatty acids*. Oleic acid is an unsaturated fatty acid. Lauric, myristic, palmitic, and stearic acids are the most common saturated fatty acids.

TABLE 14.2

IUPAC and Historical Names of Some Carboxylic Acids

Structure	IUPAC Name	Historical Name	Source of Name
HCOOH	Methanoic	Formic	*formica*, Latin for "ant"
CH_3COOH	Ethanoic	Acetic	*acetum*, Latin for "vinegar"
C_2H_5COOH	Propanoic	Propionic	*propion*, Greek for "fatty acid"
C_3H_7COOH	Butanoic	Butyric	*butyrum*, Latin for "butter"
$C_{11}H_{23}COOH$	Dodecanoic	Lauric	*laurus*, Latin for "laurel"
$C_{13}H_{27}COOH$	Tetradecanoic	Myristic	*myristikos*, Greek for "fragrant"
$C_{15}H_{31}COOH$	Hexadecanoic	Palmitic	*palma*, Latin for "palm tree"
$C_{17}H_{35}COOH$	Octadecanoic	Stearic	*stear*, Latin for "solid fat"

When a historical name is used for an aliphatic carboxylic acid, the position of a substituent is not indicated by a number; that is reserved for the IUPAC system. Instead, a Greek letter is employed. The position next to the the carboxyl is called the α (alpha) position. The next positions, in order, are called β (beta), γ (gamma), and δ (delta). Thus, β-hydroxybutyric acid is the historical name for 3-hydroxybutanoic acid:

$$\underset{\gamma}{CH_3}-\underset{\beta}{\overset{\overset{\displaystyle OH}{|}}{CH}}-\underset{\alpha}{CH_2}-COOH$$

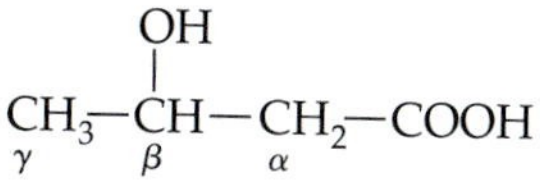

β-Hydroxybutyric acid
(3-hydroxybutanoic acid)

Aromatic carboxylic acids are usually referred to by their historical names. Benzoic acid is the simplest benzenoid carboxylic acid, while nicotinic acid derives its name from nicotine. Polynuclear aromatic hydrocarbon (PAH) carboxylic acids are named in a similar way—for example, 1-naphthoic acid (from naphthalene).

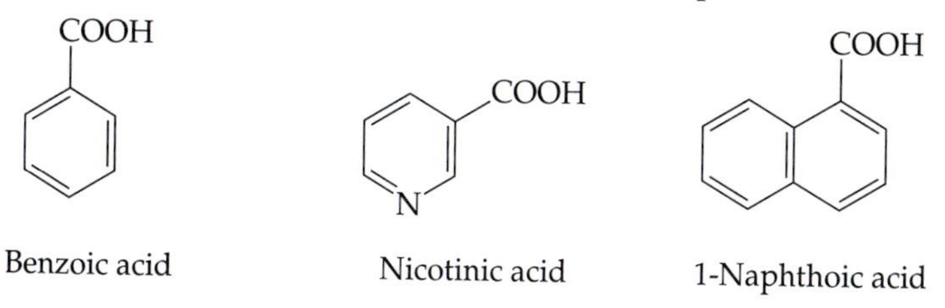

Benzoic acid Nicotinic acid 1-Naphthoic acid

Tobacco leaves, the primary ingredient in cigarettes and cigars, contains nicotine, from which nicotinic acid is derived.

How to Solve a Problem

Assign a IUPAC name to the following structure:

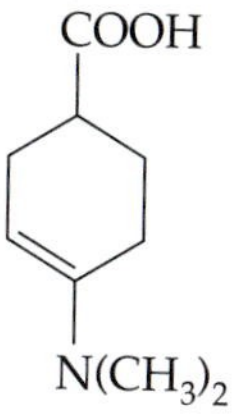

COOH

$N(CH_3)_2$

SOLUTION

The name is developed in a stepwise manner. Since the basic hydrocarbon structure is an aliphatic six-membered ring, the stem name is cyclohexane. There are three functional groups present: carboxylic acid, amine, and alkene. From Table 14.1, we can determine that the carboxyl group has the highest precedence. Therefore, the parent name is cyclohexanecarboxylic acid. The carbon carrying the carboxyl group is carbon 1. We indicate the presence of the carbon-carbon double bond by modifying the parent name from cyclohex*ane*carboxylic acid to cyclohex*ene*carboxylic acid, and we specify its location between carbons 3 and 4 of the ring by prefixing the name with a 3. So what we have to this point is 3-cyclohexenecarboxylic acid. The third functional group is an *N*,*N*-dimethylamino group attached to carbon 4. Therefore, the complete name is 4-(*N*,*N*-dimethylamino)-3-cyclohexenecarboxylic acid.

PROBLEM 14.1

Draw structures corresponding to the following names: (a) 2-methylpropanoic acid, (b) 2-butenoic acid, (c) 2,4-dichlorobenzoic acid, (d) α-hydroxypropionic acid, (e) 3-methoxy-4-oxopentanoic acid, and (f) β-ketobutyric acid.

PROBLEM 14.2

Assign names to the following structures:

(a) $(CH_3)_2CHCH_2COOH$

(b) 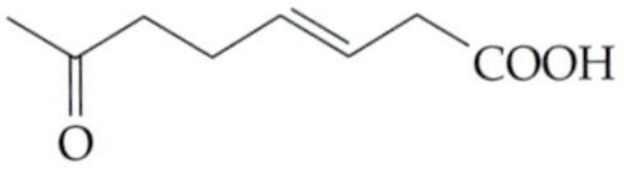 COOH, O

(c) 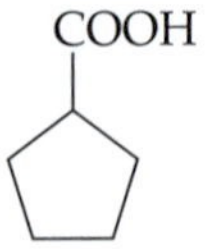 COOH

(d) 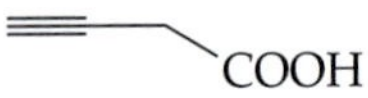 COOH

(e) $CH_3(CH_2)_9COOH$

(f) 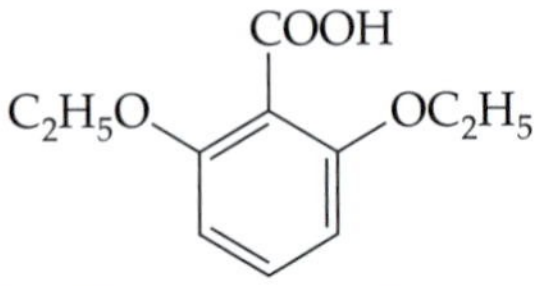 COOH, C_2H_5O, OC_2H_5

14.1.3 Dicarboxylic Acids

Dicarboxylic acids are named by adding the suffix *-dioic acid* to the alkane parent name, retaining the *-e*. The generic name is **alkanedioic acids.** Thus, ethanedioic acid, propanedioic acid, and butanedioic acid are the IUPAC names for the C_2, C_3, and C_4 diacids. However, the historical names are normally used for simple dicarboxylic acids. For the three compounds just named, the historical names are oxalic acid (from rhubarb), malonic acid (from apples), and succinic acid (from amber), respectively. Maleic acid is the C_4 *cis* unsaturated diacid; the *trans* isomer is fumaric acid.

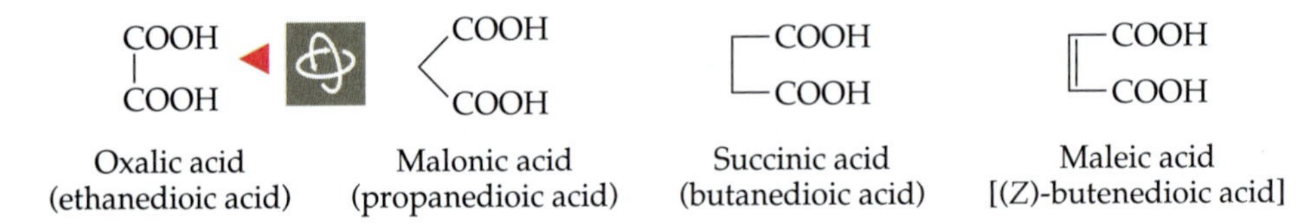

Oxalic acid (ethanedioic acid) Malonic acid (propanedioic acid) Succinic acid (butanedioic acid) Maleic acid [(Z)-butenedioic acid]

The three benzenoid dicarboxylic acids are phthalic acid, isophthalic acid, and terephthalic acid:

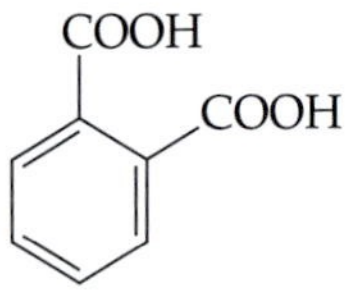

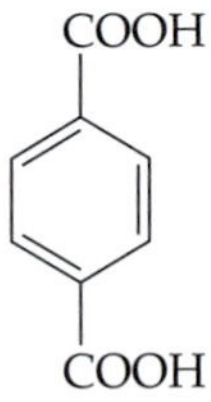

COOH, COOH — Phthalic acid

COOH, COOH — Isophthalic acid

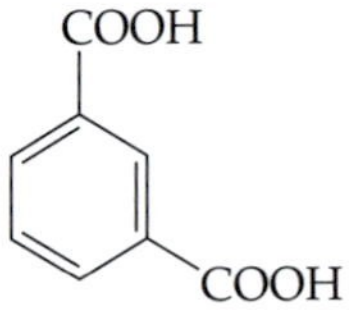

COOH, COOH — Terephthalic acid

14.2 Structure and Properties of Carboxylic Acids

14.2.1 Electronic Structure of Carboxylic Acids

The carboxyl group consists of a carbonyl group with a hydroxyl group attached to the carbon atom (Figure 14.1). The orbital description of the carbonyl group is identical to that for aldehydes and ketones (see Section 13.2.1). The electron-deficient character of the carbonyl carbon is satisfied to some extent by the ability of the hydroxyl oxygen to share electron density with the carbon, using its unshared electrons:

Resonance contributing structures for a carboxylic acid

One result of this delocalization is that the carbonyl group of carboxylic acids is not as electron-deficient as that in aldehydes or ketones. Therefore, carboxylic acids are not as susceptible as aldehydes and ketones to nucleophilic attack at the carbonyl carbon.

14.2.2 Acidity of Carboxylic Acids

The characteristic property of carboxylic acids, in contrast to most organic compounds, is that they are *acidic*—they readily give up a proton from the carboxyl group. The reason for this acidity is the stability of the **carboxylate anion** formed upon dissociation of the acid:

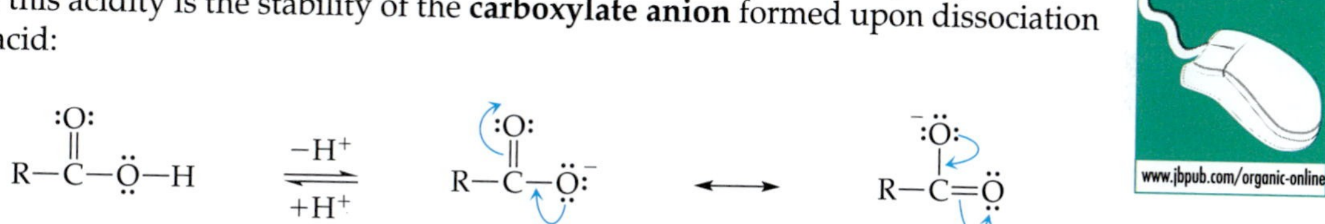

Carboxylic acid

Resonance contributing structures for the carboxylate anion

Two equivalent resonance contributing structures may be written for the carboxylate anion, meaning that the electron density is located equally on the two oxygen atoms.

Carboxylic acids are classified as weak acids (in contrast to the typical inorganic acids HCl, HNO_3, and H_2SO_4, which are strong acids). The pK_a values of carboxylic acids range from less than 1.0 to about 5.0, with most being closer to 5.0. Remember, the higher the pK_a value, the weaker the acid. In aqueous solution, carboxylic acids readily lose a proton to the familiar bases ammonia, sodium bicarbonate, sodium carbonate, and sodium hydroxide. Recall that the direction of any equilibrium can be predicted from the pK_a values of the two acids, as discussed in Section 1.3.1. The proton of acetic acid (a weak acid) is removed by sodium bicarbonate (a moderately weak base), forming the acetate anion:

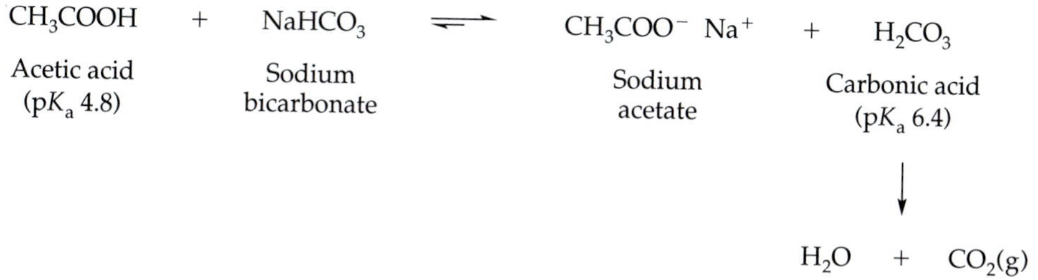

The carbonic acid formed immediately dissociates to water and carbon dioxide, which bubbles out of the solution. This visible reaction is used as a qualitative test to detect the presence of a carboxylic acid.

The acidity of a carboxylic acid can be affected (and therefore manipulated) by the nature of any substituents on the parent structure. Any substituent that further delocalizes the negative charge of the carboxylate anion further stabilizes that anion, facilitating its formation and lowering the pK_a of the carboxylic acid (it will have a higher acidity). In aliphatic carboxylic acids, the attachment of strong electron-withdrawing groups on the hydrocarbon chain increases the acidity by delocalizing the negative charge in the carboxylate anion. The magnitude of the effect is a function of the inductive effect (the electronegativity) of the substituent and its distance from the carboxyl carbon. The pK_a values for the following three series of compounds demonstrate the effect of increasing electronegativity and decreasing distance:

	CH_3COOH	ICH_2COOH	$BrCH_2COOH$	$ClCH_2COOH$	FCH_2COOH
pK_a	4.8	3.2	2.9	2.8	2.6

	$ClCH_2COOH$	$Cl_2CHCOOH$	Cl_3CCOOH
pK_a	2.8	1.5	0.7

	$ClCH_2CH_2CH_2COOH$	$CH_3CHClCH_2COOH$	$CH_3CH_2CHClCOOH$
pK_a	4.5	4.0	2.8

Increasing acid strength →

The same effects on acidity can be seen in benzenoid carboxylic acids. Again, the magnitude of the inductive effect of the substituent and the substituent's distance from the carboxyl carbon affect the pK_a. The methoxy group is relatively electron-donating, destabilizing the carboxylate anion; the chloro and nitro groups are electron-withdrawing, stabilizing the carboxylate anion by delocalizing the negative charge. (Recall the effects of substituents on aromatic rings, discussed in Section 9.4.)

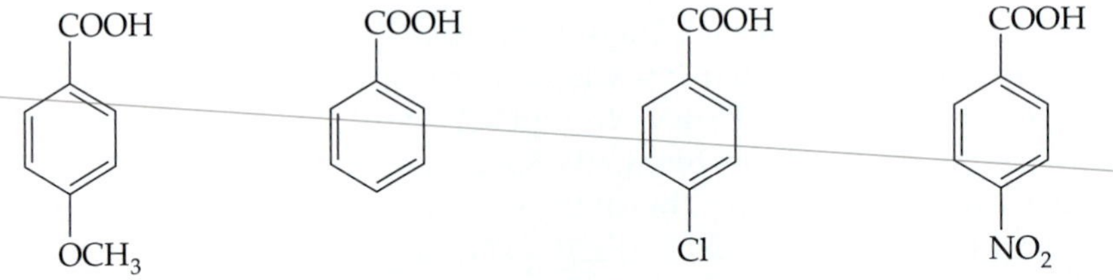

	COOH / OCH_3	COOH	COOH / Cl	COOH / NO_2
pK_a	4.5	4.2	4.0	3.4

Increasing acid strength →

PROBLEM 14.3

Place the members of each set of compounds in order of *decreasing* acidity:

(a) *o*-, *m*-, and *p*-fluorobenzoic acid

(b) cyclohexanecarboxylic acid, α-chlorocyclohexanecarboxylic acid, and β-chlorocyclohexanecarboxylic acid

(c) *p*-aminobenzoic acid and *p*-nitrobenzoic acid

(d) propanoic acid and 2,2-dibromopropanoic acid

14.2.3 Carboxylate Salts

A carboxylic acid reacts with a base in a typical acid-base reaction, resulting in removal of the acidic hydrogen and formation of a carboxylate anion. For example, sodium bicarbonate (a relatively weak base) reacts with carboxylic acids to produce a sodium carboxylate and carbonic acid.

General reaction:

$$\underset{\text{Carboxylic acid}}{RCOOH} + \underset{\text{Sodium bicarbonate}}{NaHCO_3} \longrightarrow \underset{\text{Sodium carboxylate}}{RCOO^- \ Na^+} + \underset{\text{Carbonic acid}}{H_2CO_3}$$

$$H_2CO_3 \longrightarrow H_2O + CO_2(g)$$

While higher carboxylic acids are insoluble in water, their sodium and potassium salts are generally water-soluble because of their ionic character. However, calcium, magnesium, iron, and other heavier-metal salts of carboxylic acids—those typically found in hard water—are generally insoluble in water.

Any carboxylate salt can be reconverted to the corresponding carboxylic acid by treatment with an acid stronger than the carboxylic acid. For example, potassium benzoate in aqueous solution is converted to benzoic acid by treatment with dilute hydrochloric acid. Once again, this is a straightforward acid-base reaction. In this instance, the benzoic acid precipitates from the solution.

$$\underset{\text{Potassium benzoate}}{C_6H_5COO^- \ K^+} + HCl \longrightarrow \underset{\text{Benzoic acid}}{C_6H_5COOH} + K^+ \ Cl^-$$

Carboxylate salts are named by first naming the cation involved, then replacing the suffix *-ic* of the acid name with the suffix *-ate*. Examples are sodium acetate from acetic acid, calcium stearate from stearic acid, magnesium propanoate from propanoic acid, and sodium benzoate from benzoic acid.

CH_3COONa	$[CH_3(CH_2)_{16}COO]_2Ca$	$[CH_3CH_2COO]_2Mg$	C_6H_5COONa
Sodium acetate	Calcium stearate	Magnesium propanoate	Sodium benzoate

Just as a reminder, the relative pK_a values of organic species encountered so far in this book are shown in Table 14.3. The pK_a of water is the reference point. Sodium hydroxide (pK_a of its conjugate acid, water, is 15.7) is strong enough to remove the proton from phenols and carboxylic acids. However, sodium bicarbonate (pK_a of its conjugate acid, carbonic acid, is 6.4) is only strong enough to remove the proton of carboxylic acids. This fact can be used to separate carboxylic acids from all other organic compounds, including phenols.

Consider an organic solution containing a variety of organic compounds, including a carboxylic acid. When treated with aqueous sodium bicarbonate, the carboxylic acid in the solution is converted to its salt (an ionic species). The carboxylate salt moves into the aqueous layer, which can be separated. Acidification of the aqueous layer with a stronger acid (usually dilute HCl) re-forms the carboxylic acid, which is

TABLE 14.3

Relative pK_a Values

Species	pK_a
Alkanes	~40
Amines	~33
Alcohols	~16
Water	15.7
Phenols	~10
Carboxylic acids	~5

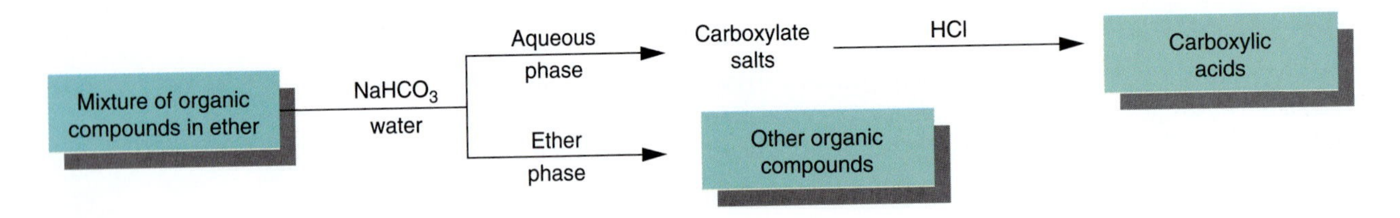

FIGURE 14.2

Separation scheme to isolate carboxylic acids.

usually insoluble in water at room temperature and can be isolated. A separation scheme illustrating these steps is shown in Figure 14.2.

PROBLEM 14.4

Design a separation scheme to show how a mixture of cyclohexanol, phenol, and benzoic acid dissolved in ether can be separated into the three pure components.

14.2.4 Physical Properties of Carboxylic Acids

Carboxylic acids are both hydrogen-bonding *donors* (they share electrons of their carbonyl oxygen with an electron-deficient hydrogen) and hydrogen-bonding *acceptors* (their acidic hydrogen shares electrons donated by electron donors). Thus, carboxylic acids of low molecular weight are infinitely soluble in water because of extensive hydrogen bonding with the water molecules. For instance, household vinegar is an aqueous solution of acetic acid:

Hydrogen bond

Acetic acid as a hydrogen-bond donor

Acetic acid as a hydrogen-bond acceptor

As the molecular weight of a carboxylic acid increases, so does the size of the hydrocarbon portion of the molecule, causing its *hydrophobicity* (repulsion for water) to increase. Eventually, this hydrophobicity outweighs the hydrophilicity of the carboxyl group, resulting in decreasing solubility. For example, hexanoic acid and benzoic acid are relatively insoluble in water at room temperature.

Carboxylic acids in their pure state exist as *dimers*, forming two intermolecular hydrogen bonds that create a dimeric structure:

Dimer of a carboxylic acid

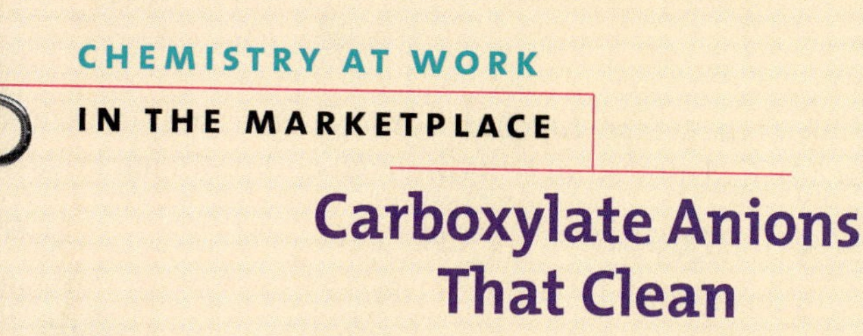

Carboxylate Anions That Clean

Soaps have been prepared for centuries by heating animal fats with solutions of lye (sodium hydroxide), a process called *saponification* (soap making). The actual reaction is the hydrolysis of esters, a reaction we will discuss in detail in Chapter 15. The result of saponification is the formation of sodium salts of a mixture of fatty acids (long-chain acids, generally C_{12}–C_{18}). A typical component of soap is sodium stearate.

Soap, like the detergents described in Section 9.3.5, is a *surfactant*, a surface-active agent that functions at the interface of two immiscible liquid phases (for example, water and oil). The long hydrocarbon chain of a soap is lipophilic ("fat-loving") and tends to dissolve most readily in organic solvents (recall that like dissolves like). In contrast, the carboxylate end of the soap molecule is ionic and therefore hydrophilic ("water-loving"); it dissolves in water.

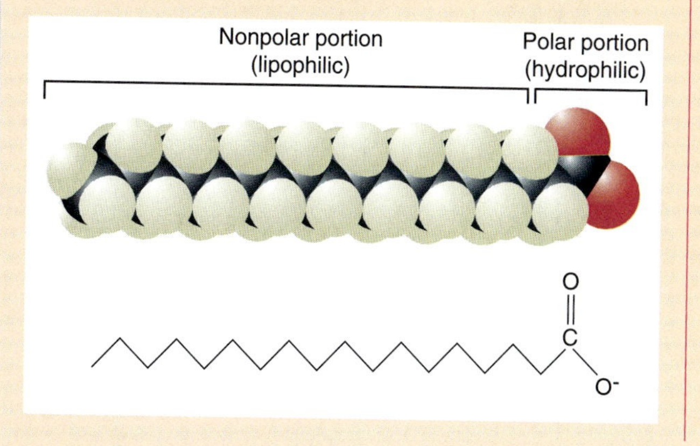

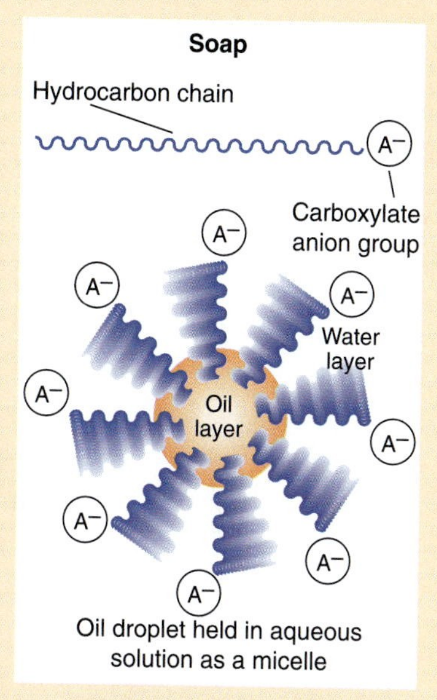

Oil droplet held in aqueous solution as a micelle

When a soap molecule is placed in a solution containing an organic phase and an aqueous phase, it tends to orient itself at the interface of the two phases: The lipophilic end is dissolved in the organic phase, and the hydrophilic end is dissolved in the aqueous phase. This results in a breaking down of the previously sharp line of division between the two immiscible phases. This effect can be seen by adding a drop of oil to a beaker of water. The oil drop remains well defined and clearly visible. But if a small amount of soap is added, with stirring, the drop appears to dissolve. The drop has broken down into tiny droplets (micelles) that have been surrounded by soap molecules. These micelles are so tiny that they are invisible to the naked eye. By this means, dirt is "solubilized" and can be removed from clothing by rinsing with water.

Soaps fell into disfavor because of two major problems: precipitation of metal salts in hard water and alkaline solutions (see Section 9.3.5). Most of the cleaning agents used today are not soaps, but *detergents*, which are also surfactants.

$$\text{(zigzag chain)}COO^-\ Na^+ \equiv CH_3(CH_2)_{16}COO^-\ Na^+$$

Sodium stearate (sodium octadecanoate)

These hydrogen bonds must be broken before vaporization can occur, and the bond breaking requires extra energy. As a result, carboxylic acids are higher-boiling than most compounds of comparable molecular weight. Further, for a given number of carbon atoms in the molecule, the boiling point decreases from the carboxylic acid to the alcohol to the aldehyde. This decrease in boiling point reflects the decreasing hydrogen-bonding ability of these three functional groups. For example, in the C_3 series, propanoic acid boils at 141°C, 1-propanol boils at 97°C, and propanal boils at 49°C.

CH_3CH_2COOH	$CH_3CH_2CH_2OH$	CH_3CH_2CHO
Propanoic acid (bp 141°C)	1-Propanol (bp 97°C)	Propanal (bp 49°C)

Decreasing hydrogen-bonding capability →

14.2.5 Detection of Carboxylic Acids

To detect the presence of a carboxylic acid, both spectroscopic and chemical means may be used. When analyzing a molecular formula for the hydrogen deficiency index (HDI), remember that the carboxyl group contains a double bond. For example, the molecular formula of propanoic acid, $C_3H_6O_2$, indicates an HDI of 1, showing the presence of the carbonyl double bond.

Spectroscopic Detection

Carboxylic acids can be readily detected in IR spectra because of the strong absorption of both the carbonyl group and the hydroxyl group (Figure 14.3a). The C=O stretch-

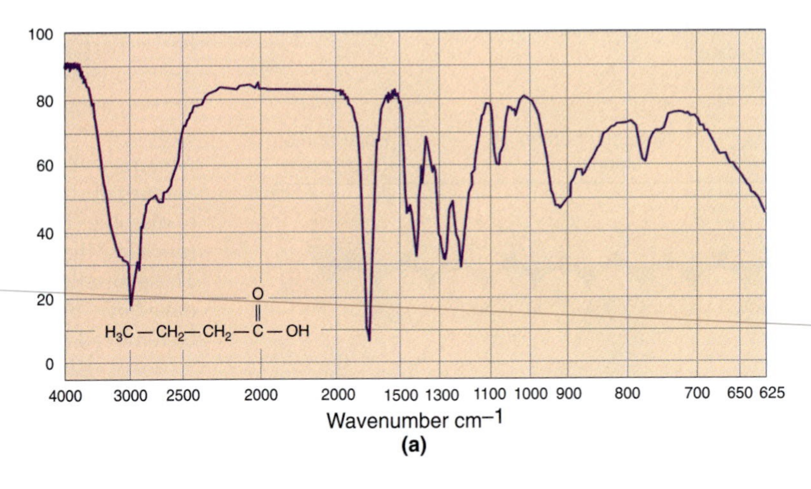

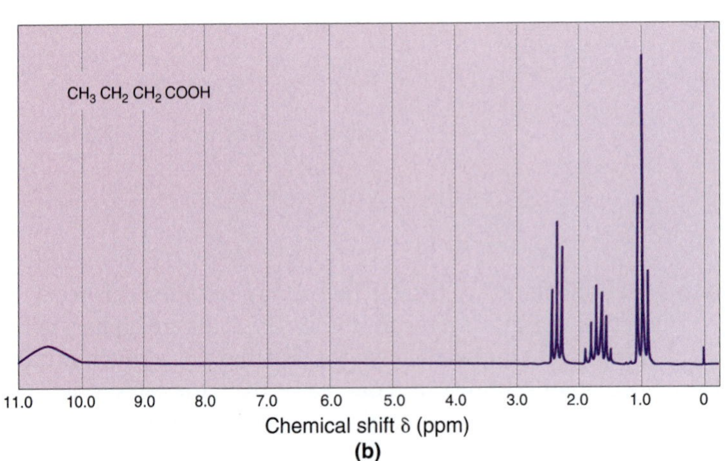

FIGURE 14.3

(a) IR and (b) ¹H-NMR spectra of butanoic acid.

ing appears strong near 1700 cm^{-1}. The O—H stretching absorption is very strong and very broad (much more so than for alcohols) and is centered near 3300 cm^{-1}.

The ^{1}H-NMR spectrum of a carboxylic acid is unique in that the carboxyl hydrogen is very strongly deshielded, because of the electronegativity of the —COO— entity. This hydrogen appears between 10 and 13 ppm, with no other hydrogens showing in the same region. Carboxyl groups also provide some deshielding of hydrogens on adjacent carbons (α-hydrogens), whose peaks appear at 2.0–2.5 ppm (Figure 14.3b)—just like the peaks for α-hydrogens of carbonyl groups (see Section 13.2.5).

Chemical Detection

The presence of a carboxylic acid that is insoluble in water is detected by its solubility in dilute sodium bicarbonate solution. The acid is converted to its carboxylate salt, which is soluble in water, and carbon dioxide bubbles out of the solution. For a water-soluble carboxylic acid, litmus paper detects the presence of the acid. You should recognize benzoic acid ($C_7H_6O_2$) in road map problems by its formula, given its frequent appearance as a degradation product. It is the very stable compound produced by strong oxidation of a benzenoid compound with an alkyl side chain (see Section 9.6.3).

How to Solve a Problem

The molecular formula of an unknown compound **X** is $C_8H_8O_2$. What is a possible structure for this compound? What experiments should we perform to determine the structure?

PROBLEM ANALYSIS

Our established procedure is to deduce all possible information from the molecular formula, develop a list of possible kinds of functional groups, use appropriate spectroscopic techniques to support or eliminate some of these possibilities, perform a minimal number of chemical tests to confirm or rule out our tentative conclusions, chemically convert the unknown compound into a compound whose structure is known and verifiable, and finally, review the evidence to ensure it fits our tentative structure.

SOLUTION

The molecular formula of **X** has an HDI of 5. Since the C : H ratio is 1, we suspect an aromatic compound, which accounts for 4 of the 5 units of unsaturation. Next, we focus on the oxygen atoms. With two oxygens present, a carboxylic acid is a possibility. However, we should also consider structures with two separate oxygen functions—for instance, a keto-aldehyde, a hydroxy-ketone, an oxy-aldehyde, and so on.

An IR spectrum distinguishes among most of these possibilities. Suppose the IR spectrum shows a broad, strong peak centered at about 3000 cm^{-1} and another peak near 1700 cm^{-1}. This is consistent with a carboxyl group (—COOH), and the carbonyl double bond would account for the remaining unit of unsaturation. If the IR spectrum also shows absorption near 1600 cm^{-1}, this is consistent with a benzenoid ring. To this point, **X** appears to be a benzenoid carboxylic acid, accounting for 7 of the 8 carbons in the molecular formula.

Next, we use the ^{1}H-NMR spectrum to confirm the IR information and to determine the kinds of hydrogens present. Suppose the ^{1}H-NMR spectrum shows the following peaks: 11.5 ppm (s, 1H), 7.2 ppm (m, 5H), and 2.1 ppm (s, 2H). Since the sum of the relative areas equals 8 and the formula includes eight hydrogens, these numbers are the actual numbers of hydrogens, not just their ratio. Therefore, the structure has a carboxyl group (one hydrogen), a benzene ring carrying one substituent (five of the possible six hydrogens are present), and a methylene group. The methylene group must connect the other two groups because the methylene group is unsplit and because the ring is monosubstituted.

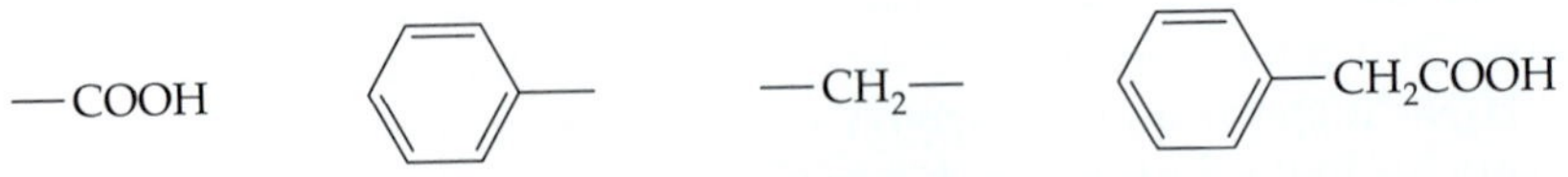

—COOH | (benzene ring)— | —CH_2— | (benzene ring)—CH_2COOH

Carboxyl | Phenyl | Methylene | Phenylacetic acid

The structure of **X** can only be that of phenylacetic acid. The compound is insoluble in water but dissolves when treated with sodium bicarbonate. Oxidation of phenylacetic acid with hot potassium permanganate and acidification of the subsequent solution provide a colorless precipitate with molecular formula $C_7H_6O_2$ and melting point 120–121°C, which must be benzoic acid. This confirms that there is only one substituent on the benzene ring.

PROBLEM 14.5

Compound **A,** $C_{10}H_{12}O_2$, is insoluble in water but dissolves in dilute sodium bicarbonate. The IR spectrum shows a peak at 1625 cm^{-1}, a strong peak at 1710 cm^{-1}, and a broad, strong peak centered at 3400 cm^{-1}. The ^1H-NMR spectrum shows several peaks as follows: 11.5 ppm (s, 1H), 7.2 ppm (m, 5H), 2.6 ppm (d, 2H), 2.2 ppm (m, 1H), and 1.0 ppm (d, 3H). Oxidation of **A** with hot potassium permanganate, followed by removal of the MnO_2 precipitate, produces a colorless filtrate. The filtrate, when acidified with HCl, produces a colorless precipitate (**B**), whose mass spectrum indicates a molecular weight of 122. What are the structures of **A** and **B**?

14.3 Reactions of Carboxylic Acids

14.3.1 Reduction of Carboxylic Acids

The carboxylic acid group can be reduced, but only by a powerful reducing reagent, such as lithium aluminum hydride. The product of such a reduction is a primary alcohol. (This is the reverse of the oxidation of primary alcohols to carboxylic acids, described in Section 4.4.)

General reaction:

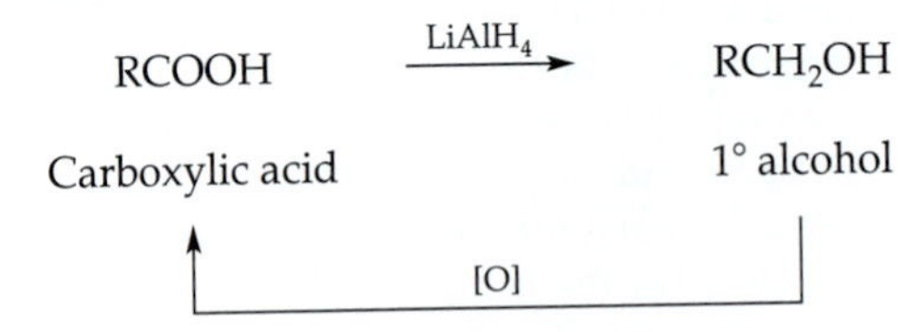

For example, benzoic acid produces benzyl alcohol, and butanoic acid is converted into 1-butanol:

$$C_6H_5COOH \xrightarrow{LiAlH_4} C_6H_5CH_2OH$$

Benzoic acid → Benzyl alcohol

$$CH_3CH_2CH_2COOH \xrightarrow{LiAlH_4} CH_3CH_2CH_2CH_2OH$$

Butanoic acid → 1-Butanol

CHEMISTRY AT WORK

IN THE ENVIRONMENT

Bee Pheromones

A colony of honeybees consists of hundreds of workers and just a single queen bee. Although both workers and queen are female, there is clearly a division of labor: The queen is the only bee able to reproduce; the worker bees gather food and build the hive. The difference between the queen and her "colonists" also extends to the chemical signals (pheromones) by which they communicate.

The major component of the queen's mandibular glands is (*E*)-9-oxo-2-decenoic acid. This compound serves as an attractant for the workers, keeping them swarming about the queen and precluding the development of another queen. If this pheromone weakens or disappears, the colony begins to nurture and rear a new queen bee.

The major component of a worker bee's mandibular gland is (*E*)-10-hydroxy-2-decenoic acid. This pheromone is mixed with the pollen collected by the workers.

It keeps the pollen from germinating so that it can be stored; pollen is the major protein food source for the hive.

The biosynthesis of each of these two major pheromones begins with the C_{18} fatty acid stearic acid.

$CH_3-(CH_2)_{16}-COOH$

Stearic acid

Queen bee pheromone
[(*E*)-9-oxo-2-decenoic acid]

Worker bee pheromone
[(*E*)-10-hydroxy-2-decenoic acid]

Although there are no limitations to this reaction in principle, there are significant limits to its application because of the strength of $LiAlH_4$; this is a much stronger reducing agent than $NaBH_4$ (sodium borohydride) or hydrogen plus a catalyst. $LiAlH_4$ also reduces aldehydes and ketones to alcohols (as does $NaBH_4$; see Section 13.3.1), and it reduces amides, nitriles, and azides to amines (see Section 12.6.2). For this reason, use of $LiAlH_4$ to reduce a carboxyl group is limited to compounds in which these other reducible groups are *not* present.

Because the carboxyl group is not reduced by the weaker reducing agent $NaBH_4$, that reagent can be used to selectively reduce aldehydes and ketones to alcohols (see Section 13.3.1) in the presence of a carboxyl group.

The two functional groups of *p*-acetylbenzoic acid can be reduced in all three possible ways by judicious use of the two metal hydrides:

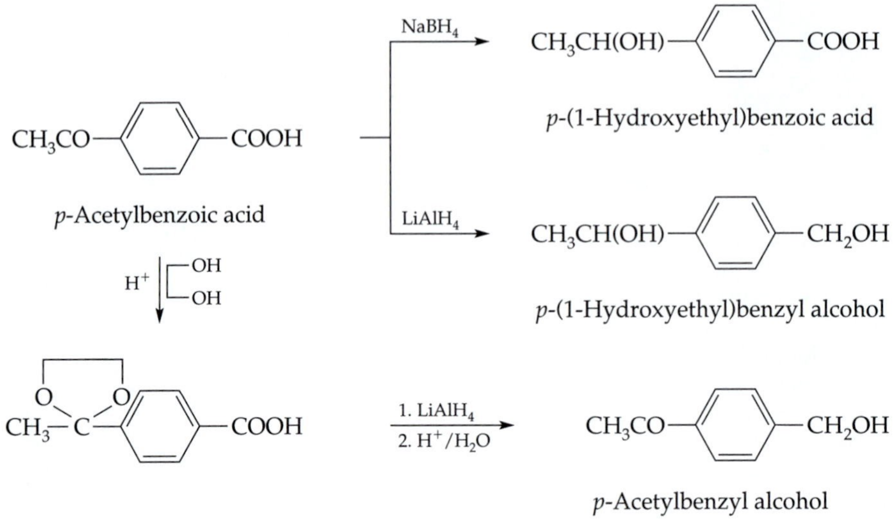

$CH_3CO-C_6H_4-COOH$ (*p*-Acetylbenzoic acid) $\xrightarrow{NaBH_4}$ $CH_3CH(OH)-C_6H_4-COOH$

p-(1-Hydroxyethyl)benzoic acid

$CH_3CO-C_6H_4-COOH$ $\xrightarrow{LiAlH_4}$ $CH_3CH(OH)-C_6H_4-CH_2OH$

p-(1-Hydroxyethyl)benzyl alcohol

p-Acetylbenzoic acid $\xrightarrow{H^+,\ HOCH_2CH_2OH}$ acetal ($CH_3-C(OCH_2CH_2O)-C_6H_4-COOH$) $\xrightarrow[2.\ H^+/H_2O]{1.\ LiAlH_4}$ $CH_3CO-C_6H_4-CH_2OH$

p-Acetylbenzyl alcohol

Sodium borohydride reduces only the ketone group. Lithium aluminum hydride reduces both functional groups. To reduce the carboxyl group but not the ketone group, the ketone group must be protected as its acetal. (Recall that acetals are inert to basic conditions, such as those created by $LiAlH_4$.) After the reduction, the ketone group is deprotected by hydrolysis in acidic solution.

PROBLEM 14.6

Show the product of reducing each of the following compounds with $LiAlH_4$:

(a) *p*-isopropylbenzoic acid
(b) *p*-carboxyacetanilide ($HOOCC_6H_4NHCOCH_3$)
(c) 1-cyclohexenecarboxylic acid

PROBLEM 14.7

What reagent(s) would you use for each of the following transformations?

(a) 3-oxopentanoic acid to 3-hydroxypentanoic acid
(b) 4-hexenoic acid to hexanoic acid
(c) 4-hexenoic acid to 4-hexen-1-ol
(d) butanedioic acid to 1,4-butanediol
(e) 3-oxopentanoic acid to 3-oxopentanal

14.3.2 Alpha Bromination of Alkanoic Acids

As indicated in the Key to Transformations, the attachment of a halogen to an alkane to produce an alkyl halide permits extensive chemistry to be accomplished—mainly because the halogen is so readily converted into other functional groups. As was noted in Chapters 3 and 6, halogens can be substituted by other nucleophiles (an S_N2 reaction) or can be eliminated by dehydrohalogenation (an E2 reaction). For this same reason—that the bromine can be converted to other groups—α-bromo carboxylic acids are also important synthetic intermediates. One very important application is their conversion to α-amino carboxylic acids (critical for protein synthesis, see Section 16.1.3). Further α-bromo carboxylic acids can be converted to α-hydroxy carboxylic acids and to α,β-unsaturated carboxylic acids. Recall that α-hydroxy carboxylic acids can also be obtained from aldehydes via the cyanohydrin synthesis (see Section 13.3.2).

$$R{-}CH_2{-}CH(Br){-}COOH \xrightarrow[(S_N2)]{\text{excess } NH_3} R{-}CH_2{-}CH(NH_2){-}COOH$$

α-Bromo carboxylic acid → α-Amino carboxylic acid

$$R{-}CH_2{-}CH(Br){-}COOH \xrightarrow[2.\ H^+]{1.\ NaOH} R{-}CH_2{-}CH(OH){-}COOH$$

α-Hydroxy carboxylic acid

$$R{-}CH_2{-}CH(Br){-}COOH \xrightarrow[2.\ H^+]{1.\ NaOC_2H_5} R{-}CH{=}CH{-}COOH$$

α,β-Unsaturated carboxylic acid

Although the processes available for the preparation of α-bromo carboxylic acids directly from carboxylic acids are somewhat exotic, one important method involves reaction of the carboxylic acid with phosphorus and bromine (known as the *Hell-Volhard-Zelinsky reaction*). The initial product is the α-bromo acid bromide; hydrolysis produces the desired α-bromo carboxylic acid.

General reaction:

$$RR'CH{-}COOH \xrightarrow{P,\ Br_2} \left[RR'C(Br){-}COBr\right] \xrightarrow{H_2O} RR'C(Br){-}COOH$$

Carboxylic acid → α-Bromo acid bromide → α-Bromo carboxylic acid

For example, propanoic acid is converted to 2-bromopropanoic acid (α-bromopropionic acid) using this reaction:

$$CH_3CH_2COOH \xrightarrow[2.\ H_2O]{1.\ P,\ Br_2} CH_3CHBrCOOH$$

Propanoic acid → 2-Bromopropanoic acid (α-bromopropionic acid)

How to Solve a Problem

Show the steps required to convert the readily available compound cyclopentanecarboxylic acid into 1-cyclopentenecarbaldehyde:

Cyclopentane—COOH - - - - → Cyclopentene—CHO

Cyclopentanecarboxylic acid → 1-Cyclopentenecarbaldehyde

PROBLEM ANALYSIS

First, we analyze the two structures to be sure we know what has to change. Here, a carboxyl group must be changed to an aldehyde group, and a double bond must be introduced. We consider each of these changes separately, using retrosynthesis.

We know from Section 13.6.2 that a standard synthesis of aldehydes is by reduction of a carboxylic acid chloride, which can be obtained from the carboxylic acid (check the Key to Transformations, on inside front cover).

R—CHO ⟹ R—COCl ⟹ R—COOH

Aldehyde   Acid chloride   Carboxylic acid

We know that an alkane cannot be converted directly into an alkene because alkanes are inert. There are really only four options available for alkene synthesis: two elimination reactions (dehydration of an alcohol and dehydrohalogenation of an alkyl halide), reduction of an alkyne, and the Wittig reaction (check the Key to Transformations). Alkyne reduction and the Wittig reaction are clearly inapplicable to this synthesis. Our challenge, then, is how to place a group on the starting carboxylic acid that can later be eliminated. That should get us thinking about brominating a carboxylic acid. The retrosynthesis looks like this:

COOH ⟹ COOH, Br ⟹ COOH

α,β-Unsaturated carboxylic acid α-Bromo carboxylic acid   Carboxylic acid

SOLUTION

The appropriate sequence for the two major conversions must be (1) introduce the double bond to the carboxylic acid and (2) reduce the acid to the aldehyde.

Bromination of the carboxylic acid, followed by dehydrohalogenation using sodium ethoxide, produces 1-cyclopentenecarboxylic acid. Conversion of the acid to its acid chloride and reduction with lithium tri(*t*-butoxy)aluminum hydride produces the desired product, 1-cyclopentenecarbaldehyde.

COOH —(1. P, Br_2; 2. H_2O)→ COOH, Br —(1. NaOEt, Δ; 2. H^+/H_2O)→ COOH

Cyclopentanecarboxylic acid   1-Cyclopentenecarboxylic acid

↓ $SOCl_2$

CHO ←($LiAlH(OBu^t)_3$)— COCl

1-Cyclopentenecarbaldehyde

PROBLEM 14.8

Show the reactions necessary to complete each synthesis:

(a) phenylethanoic acid to 2-hydroxy-2-phenylethanoic acid

(b) phenylethanoic acid to 2-amino-2-phenylethanoic acid

(c) phenylacetaldehyde to alanine (2-amino-3-phenylpropanoic acid)

(d) 3-phenylpropanoic acid to alanine [$C_6H_5CH_2CH(NH_2)COOH$]

(e) toluene to 3-phenylpropanoic acid

14.3.3 Decarboxylation of Carboxylic Acids

The reaction in which a carboxylic acid loses a molecule of carbon dioxide is called **decarboxylation:**

$$\underset{\text{Carboxylic acid}}{\text{R}-\overset{\overset{\displaystyle \text{O}}{\|}}{\text{C}}-\text{O}-\text{H}} \xrightarrow{\Delta} \underset{\text{Alkane}}{\text{R}-\text{H}} + CO_2$$

Although there is always considerable driving force toward formation of carbon dioxide, this reaction is seldom conducted with simple carboxylic acids in practice because of the strenuous conditions required.

There are two structural environments of carboxyl groups, however, that make decarboxylation quite easy. In both cases, there is a carbonyl group in the beta position relative to the carboxyl group (that is, at carbon 3).

General reaction:

$$\underset{\beta\text{-Carbonyl carboxylic acid}}{-\underset{\beta}{\overset{\overset{\displaystyle \text{O}}{\|}}{\text{C}}}-\underset{\alpha}{\text{CH}_2}-\overset{\overset{\displaystyle \text{O}}{\|}}{\text{C}}-\text{O}-\text{H}} \xrightarrow{-CO_2} \underset{\text{Carbonyl compound}}{-\underset{\beta}{\overset{\overset{\displaystyle \text{O}}{\|}}{\text{C}}}-\underset{\alpha}{\text{CH}_2}-\text{H}}$$

β-Keto carboxylic acids lose carbon dioxide readily upon heating to about 100°C:

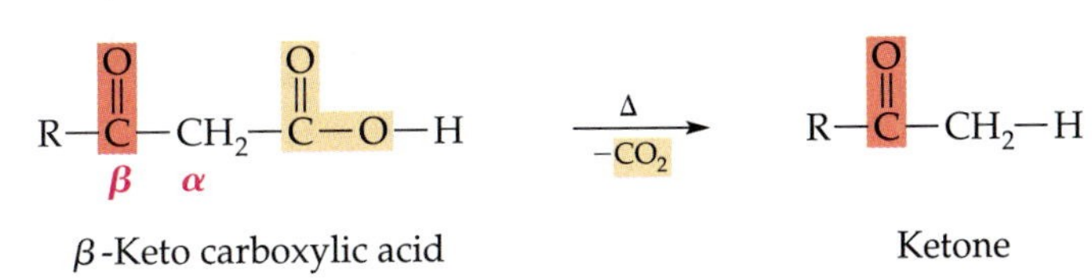

β-Keto carboxylic acid   Ketone

This reaction occurs via a six-membered transition state (the red curved arrows show this forming) and seems to be a concerted reaction. An enol is first formed by loss of carbon dioxide, but it quickly tautomerizes (ketonizes) to the more stable ketone.

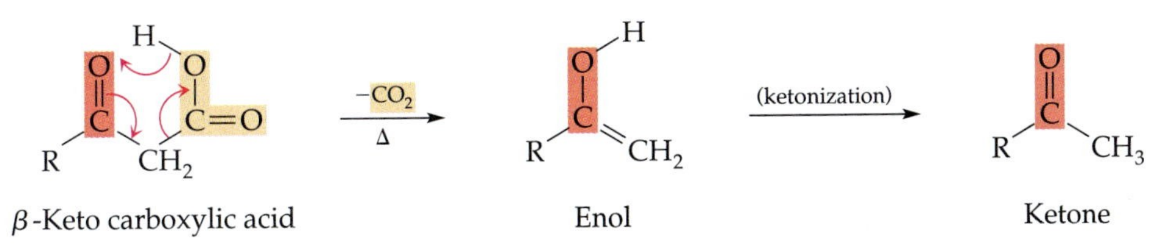

β-Keto carboxylic acid   Enol   Ketone

The second structural environment in which decarboxylation occurs readily is found in malonic acid and substituted malonic acids, more generally classifiable as 1,3-diacids:

$$\underset{\text{A substituted malonic acid}}{\text{H}-\text{O}-\underset{\beta}{\overset{\overset{\displaystyle \text{O}}{\|}}{\text{C}}}-\underset{\alpha}{\overset{\overset{\displaystyle \text{R}}{|}}{\text{CH}}}-\overset{\overset{\displaystyle \text{O}}{\|}}{\text{C}}-\text{O}-\text{H}} \xrightarrow[-CO_2]{\Delta} \underset{\text{Carboxylic acid}}{\text{H}-\text{O}-\overset{\overset{\displaystyle \text{O}}{\|}}{\text{C}}-\overset{\overset{\displaystyle \text{R}}{|}}{\text{CH}}-\text{H}}$$

Here again, there is a carbonyl group, which just happens to be part of another carboxyl group, located in a position beta to the carboxyl group that will leave as carbon dioxide. The mechanism is analogous to that for β-keto acids, proceeding through an enol intermediate that tautomerizes (ketonizes).

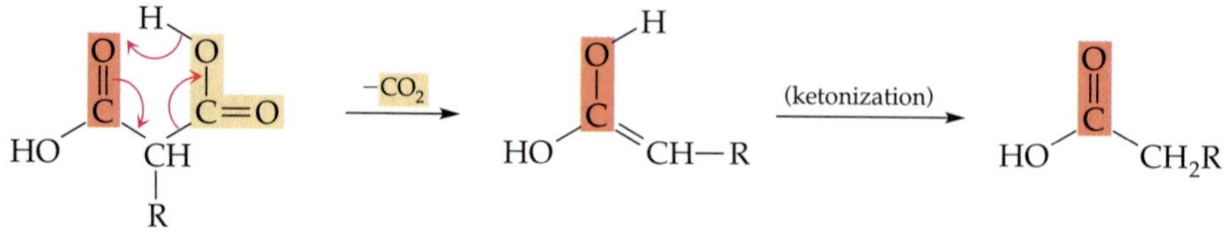

$-CO_2$ → (ketonization) →

A substituted malonic acid | Enol of a carboxylic acid | Carboxylic acid

Decarboxylation reactions play a very important role in human metabolism (see Chapter 19). The carbon dioxide that we exhale through our lungs arises, in part, from the decarboxylation of β-keto acids in metabolic cycles. Examples of these kinds of decarboxylations are (1) the conversion of acetoacetic acid (3-oxobutanoic acid) to acetone (propanone) and carbon dioxide, a reaction that occurs in some people who have a particular form of diabetes, and (2) the conversion of oxalosuccinic acid to α-ketoglutaric acid, a reaction that is part of the energy-producing Krebs cycle (see Section 19.2.1). In both instances, the carbon dioxide that is lost comes from a carboxyl group that is beta to another carbonyl group.

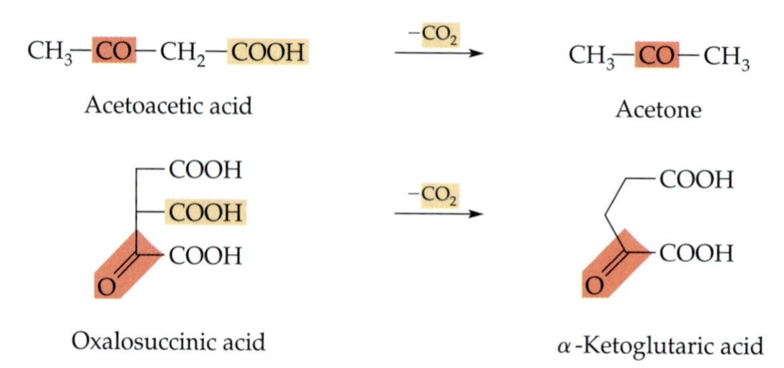

PROBLEM 14.9

What is the product of the decarboxylation of each of the following compounds?

(a) 2-isopropylpropanedioic acid
(b) 3-oxopentanoic acid
(c) diphenylmalonic acid
(d) 2-oxocyclohexanecarboxylic acid

PROBLEM 14.10

Show which carboxylic acid could be decarboxylated to produce each of the following compounds:

(a) diethylketone
(b) hexanoic acid
(c) phenylacetic acid
(d) 2,2-dimethylcyclobutanone

14.3.4 Conversion of Carboxylic Acids to Their Derivatives

One of the most important properties of carboxylic acids is their conversion into derivatives of carboxylic acids. **Derivatives** are compounds that can be formed from a certain type of starting material and, just as readily, reconverted to that starting material. The common structural feature among all of the derivatives of carboxylic acids is the **acyl group** (R—CO—). The interconversions involve changing only the substituent that is attached to the acyl group.

The upper right portion of the Key to Transformations (inside front cover) indicates the major interconversions involving carboxylic acid derivatives and the parent carboxylic acids. The interconversions between carboxylic acids and acid chlorides, esters, and amides are so important that they will be discussed separately in Chapter 15.

14.4 Preparation of Carboxylic Acids

We have already encountered some ways to prepare carboxylic acids via conversions from other functional groups; we will review these techniques in Section 14.4.3. First, we will consider two other major methods for the preparation of carboxylic acids. Both methods involve a one-carbon homologation step—that is, in the process of synthesizing the acid, one carbon is added to the starting compound.

14.4.1 Carboxylation of Organometallics

As we saw in Section 13.5.1, the Grignard reagent, a source of carbanions, brings about a nucleophilic attack on the carbonyl carbon of a carbonyl group. The end result is the addition of a carbanion "across" the carbonyl bond. This same reaction with a Grignard reagent, but this time using carbon dioxide (often as the solid, *dry ice*) in place of the carbonyl compound, produces a carboxylate anion. Acidification during work-up produces the carboxylic acid in high yield. Note that the carboxylic acid has one more carbon than the starting alkyl halide; this is a one-carbon homologation.

General reaction:

$$\underset{\text{Alkyl bromide}}{R{-}Br} \xrightarrow[\text{ether}]{Mg} \underset{\text{Grignard reagent}}{R{-}MgBr} \xrightarrow{CO_2} \underset{\text{Carboxylate salt}}{R{-}\overset{\overset{O}{\|}}{C}{-}O^-\ {}^+MgBr} \xrightarrow{HCl/H_2O} \underset{\text{Carboxylic acid}}{R{-}\overset{\overset{O}{\|}}{C}{-}OH}$$

The mechanism of **carboxylation** is analogous to that of addition to aldehydes and ketones, with the carbanionic center of the Grignard reagent attacking the electron-deficient carbonyl carbon. Addition of dilute acid as a proton source forms the carboxylic acid (in a simple acid-base reaction between carboxylate and HCl, which is a stronger acid).

$$\overset{\delta-}{R}{-}\overset{\delta+}{MgBr} + O{=}C{=}O \longrightarrow R{-}\overset{\overset{O}{\|}}{C}{-}O^-\ {}^+MgBr$$

PROBLEM 14.11

Using curved arrows, show the mechanism of the acid-base reaction between $CH_3COOMgBr$ and HCl, which is the work-up step of the carboxylation synthesis.

The Grignard reaction can be accomplished with any Grignard reagent, whether alkyl or aryl. Further, primary, secondary, and tertiary alkyl halides (usually bromides) all form Grignard reagents and bring about this reaction. This carboxylation reaction places a carboxyl group on a structure at a point where a bromine atom was located. Organolithium compounds also react in exactly the same way with carbon dioxide to produce carboxylic acids. For example, *p*-bromotoluene produces *p*-toluic acid, and allyl bromide produces 3-butenoic acid:

CH_3–C₆H₄–Br (*p*-Bromotoluene) → [1. Mg/ether; 2. CO_2; 3. HCl/H_2O] → CH_3–C₆H₄–COOH (*p*-Toluic acid)

Allyl bromide → [1. Mg/ether; 2. CO_2; 3. HCl/H_2O] → 3-Butenoic acid (COOH)

How to Solve a Problem

Show the reactions and reagents necessary for the conversion of cyclohexane to 1-cyclohexenecarboxylic acid:

Cyclohexane ⇢ 1-Cyclohexenecarboxylic acid (COOH)

PROBLEM ANALYSIS

The problem is twofold: adding a carbon to the cyclohexane ring and introducing a double bond into the ring. In any problem, we worry first about getting all of the carbons in place and only later about including the correct functional group. The reasoning behind this approach is that it is relatively easy to interconvert functional groups, as suggested by the Key to Transformations (inside front cover).

The synthesis of carboxylic acids using carboxylation of a Grignard reagent is a homologation. An alkyl bromide is the usual starting material. Retrosynthetic analysis indicates that the carboxyl group can be obtained via an alkyl bromide, and the bromide can be placed on the cyclohexane ring by radical bromination. Therefore, the first part of the problem is solved by producing cyclohexanecarboxylic acid.

Cyclohexanecarboxylic acid (COOH) ⟹ Bromocyclohexane (Br) ⟹ Cyclohexane

Once we have obtained cyclohexanecarboxylic acid, we need to introduce a double bond between carbons 1 and 2 of the ring. The only means of directly functionalizing an aliphatic carboxylic acid is by alpha bromination. Once a bromine is introduced at the alpha position, the compound can be dehydrohalogenated to produce the desired double bond, which is conjugated with that of the carboxyl group.

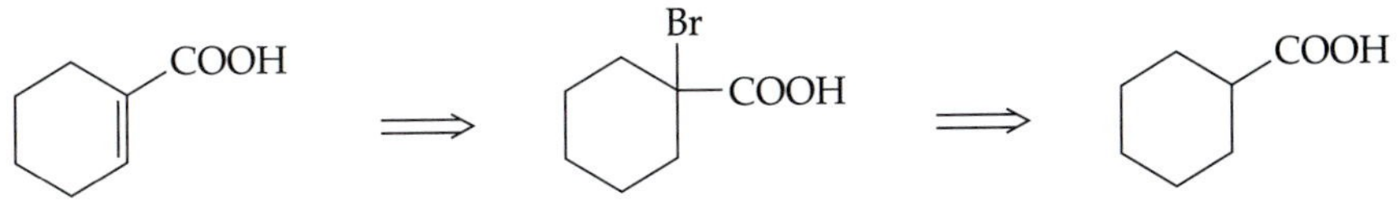

1-Cyclohexene-carboxylic acid ⟹ α-Bromocyclohexane-carboxylic acid ⟹ Cyclohexane-carboxylic acid

SOLUTION

Cyclohexane is brominated under radical conditions to form bromocyclohexane. Bromocyclohexane is converted to a Grignard reagent, which is reacted with carbon dioxide to produce cyclohexanecarboxylic acid. Reaction of the acid with bromine and phosphorus produces α-bromocyclohexanecarboxylic acid. When treated with hot sodium ethoxide, this acid undergoes an E2 elimination of HBr to produce the desired product, 1-cyclohexenecarboxylic acid.

Cyclohexane $\xrightarrow[\text{ROOR}]{Br_2}$ bromocyclohexane $\xrightarrow[3.\ H^+/H_2O]{1.\ Mg/\text{ether};\ 2.\ CO_2}$ cyclohexanecarboxylic acid $\xrightarrow{1.\ Br_2/P;\ 2.\ H_2O}$ α-bromocyclohexanecarboxylic acid $\xrightarrow{1.\ NaOC_2H_5;\ 2.\ H^+/H_2O}$ 1-Cyclohexenecarboxylic acid

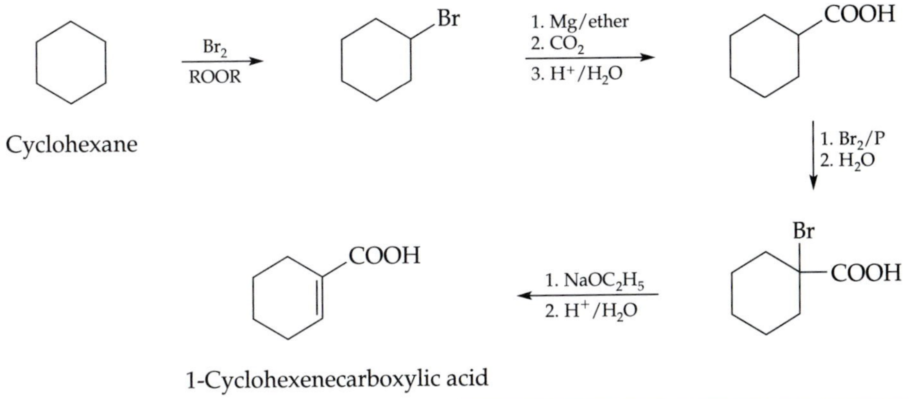

PROBLEM 14.12

Show the reactions for the preparation of each of the following acids using carboxylation: (a) *p*-isopropylbenzoic acid, (b) 2,2-dimethylpropanoic acid, and (c) cyclobutanecarboxylic acid.

PROBLEM 14.13

Show the reactions necessary for the conversion of 1-butene to (a) pentanoic acid and (b) 2-methylbutanoic acid.

14.4.2 Carboxylic Acids from Nitriles

EXPLORATIONS www.jbpub.com/organic-online

Nitriles (R—CN and Ar—CN) might be called organic cyanides, compounds in which the cyano group (—CN) is attached to an aliphatic (R) or aromatic group (Ar). The cyano group, containing a carbon-nitrogen triple bond, can be hydrolyzed in dilute aqueous acid to the carboxyl group. Thus, nitriles can be converted into carboxylic acids.

General reaction:

$$\underset{\text{Nitrile}}{R{-}C{\equiv}N} \xrightarrow[H_2O]{HCl} \underset{\text{Carboxylic acid}}{R{-}COOH} + \underset{\text{Ammonium chloride}}{NH_4Cl}$$

This reaction involves the simple protonation of the electronegative nitrogen, followed by attack of water at the electron-deficient carbon. An amide is formed as an intermediate, but it is not isolated under these conditions. Instead, it hydrolyzes to a carboxylic acid (see Section 15.4). The nitrogen atom is expelled as ammonia, which is protonated to the ammonium ion in the acidic solution. There are no limitations to this hydrolysis reaction, and it occurs in high yield.

The usefulness of this reaction clearly depends on the availability of nitriles, and these are readily available through three reactions introduced in earlier chapters and briefly reviewed here. Note that all three reactions involve a homologation.

S_N2 Substitution of Alkyl Halides with Cyanide

Reaction of an alkyl halide with sodium cyanide results in S_N2 displacement to form a nitrile (cyanide being an effective nucleophile; see Section 3.4). Hydrolysis produces the carboxylic acid. Therefore, in most structural environments, a carboxyl group can be placed where a bromine formerly was.

$$\underset{\text{Alkyl bromide}}{\text{R—Br}} \xrightarrow{\text{NaCN}} \underset{\text{Nitrile}}{\text{R—CN}} \xrightarrow{\text{HCl/H}_2\text{O}} \underset{\text{Carboxylic acid}}{\text{R—COOH}}$$

There is a major limitation to this synthesis of nitriles, however. Recall that S_N2 reactions are ineffective with tertiary halides because of competing elimination (E2) reactions. Thus, carboxylic acids in which the carboxyl group is attached to a tertiary carbon cannot be synthesized by this route. Remember, too, that S_N2 reactions are ineffective with aryl halides.

Formation of Cyanohydrins from Aldehydes

Reaction of an aldehyde with sodium cyanide at pH 10, resulting in a mixture of HCN and NaCN, brings about addition of hydrogen cyanide at the carbonyl group to form a cyanohydrin (see Section 13.3.2). Hydrolysis of the cyano group produces an α-hydroxy carboxylic acid.

$$\underset{\text{Aldehyde}}{\text{R—CHO}} \xrightarrow[\text{NaCN}]{\text{HCN}} \underset{\text{Cyanohydrin}}{\text{R—CH(OH)CN}} \xrightarrow{\text{HCl/H}_2\text{O}} \underset{\alpha\text{-Hydroxy carboxylic acid}}{\text{R—CH(OH)COOH}}$$

Conversion of Aromatic Amines into Aromatic Nitriles

Reaction of an aromatic amine with nitrous acid (HNO_2), formed from sodium nitrite and hydrochloric acid, produces a diazonium ion. Displacement of molecular nitrogen with cyanide ion produces an aromatic nitrile (see Section 12.5.2). Hydrolysis of the nitrile group produces an aromatic carboxylic acid.

$$\underset{\text{Aromatic amine}}{\text{Ar—NH}_2} \xrightarrow[\text{2. CuCN}]{\text{1. NaNO}_2\text{/HCl}} \underset{\text{Aromatic nitrile}}{\text{Ar—CN}} \xrightarrow{\text{HCl/H}_2\text{O}} \underset{\text{Aromatic carboxylic acid}}{\text{Ar—COOH}}$$

How to Solve a Problem

Show the reactions required to convert isopentyl bromide (3-methyl-1-bromobutane) to 4-methylpentanoic acid and to convert *t*-pentyl bromide (2-bromo-2-methylbutane) to 2,2-dimethylbutanoic acid. These carboxylic acids contain one carbon more than the starting material.

PROBLEM ANALYSIS AND SOLUTION

Both conversions require a homologation reaction, so hydrolysis of a nitrile and carboxylation are both candidate reactions. However, the nitrile reaction requires an S_N2 reaction of sodium cyanide with the alkyl bromide, which will not be effective with the tertiary halide *t*-pentyl bromide because of the competing elimination reaction. The nitrile route is effective with isopentyl bromide. The carboxylation process, however, will be effective with both halides.

Br —(1. Mg/ether; 2. CO_2; 3. HCl/H_2O)→ COOH

Isopentyl bromide → 4-Methylpentanoic acid

1. NaCN
2. HCl/H_2O

Br —(1. Mg/ether; 2. CO_2; 3. HCl/H_2O)→ COOH

t-Pentyl bromide → 2,2-Dimethylbutanoic acid

PROBLEM 14.14

Use the nitrile process to synthesize each of the following acids: (a) phenylacetic acid and (b) butanoic acid.

PROBLEM 14.15

Show the reactions necessary to bring about each conversion through a nitrile intermediate:

(a) cyclopropyl bromide to cyclopropanecarboxylic acid
(b) phenylacetaldehyde to β-phenylpropionic acid
(c) ethylbenzene to 2-phenylpropanoic acid
(d) bromobenzene to *p*-bromobenzoic acid

PROBLEM 14.16

Show the reactions and reagents necessary to transform (a) benzene to benzyl alcohol and (b) cyclopentane to cyclopentylmethanol. (*Hint:* Use a retrosynthesis involving a carboxylic acid.)

14.4.3 Other Syntheses of Carboxylic Acids

While studying other functional groups, we have encountered a number of reactions that result in the production of carboxylic acids. Some of these reactions are very useful for synthesizing carboxylic acids and so are reviewed here.

Oxidation of Alkyl Aromatics

Aromatic compounds carrying alkyl groups with at least one hydrogen on the alpha carbon undergo strong oxidation with potassium permanganate to produce aromatic acids (see Section 9.6.3). For example, cumene oxidizes to benzoic acid, and tetralin (tetrahydronaphthalene) to phthalic acid:

$$\text{Cumene } (C_6H_5\text{—}CH(CH_3)_2) \xrightarrow[2.\ H^+]{1.\ KMnO_4} \text{Benzoic acid } (C_6H_5\text{—}COOH)$$

$$\text{Tetralin} \xrightarrow[2.\ H^+]{1.\ KMnO_4} \text{Phthalic acid } (C_6H_4(COOH)_2)$$

Because this reaction involves very strong oxidizing conditions, it cannot be used if there are other groups on the aromatic ring that are either sensitive to oxidation themselves or that make the ring sensitive to oxidation. Prime examples of such groups are the amino and hydroxyl groups; however, these groups can be protected as their amides or esters, respectively. For example, *p*-toluidine is too sensitive to these reaction conditions to be oxidized directly (the ring is actually cleaved). However, use of an acetyl group to protect the amino group as the amide (see Section 12.4.2) allows the desired transformation to be brought about:

$$\underset{p\text{-Toluidine}}{CH_3\text{—}C_6H_4\text{—}NH_2} \xrightarrow{CH_3COCl} \underset{p\text{-Acetamidotoluene}}{CH_3\text{—}C_6H_4\text{—}NH\text{—}COCH_3}$$

$$\underset{p\text{-Acetamidotoluene}}{CH_3\text{—}C_6H_4\text{—}NH\text{—}COCH_3} \xrightarrow[2.\ H^+]{1.\ KMnO_4} \underset{p\text{-Acetamidobenzoic acid}}{HOOC\text{—}C_6H_4\text{—}NH\text{—}COCH_3} \xrightarrow{HCl/H_2O} \underset{p\text{-Aminobenzoic acid}}{HOOC\text{—}C_6H_4\text{—}NH_2}$$

Desired conversion: *p*-Toluidine → *p*-Aminobenzoic acid

Oxidation of Primary Alcohols

Primary alcohols can be oxidized to carboxylic acids using sodium dichromate and acid (see Section 4.4). In this synthesis of carboxylic acids, the number of carbon atoms remains the same. Because primary alcohols can be obtained from alkyl bromides by an S_N2 reaction, this process provides a means of converting an alkyl halide to a carboxylic acid *without homologation*. In contrast, the nitrile and carboxylation methods discussed earlier convert an alkyl halide to a carboxylic acid *with homologation*.

$$R\text{—}CH_2\text{—}Br \xrightarrow{NaOH} R\text{—}CH_2\text{—}OH \xrightarrow[H_2SO_4]{Na_2Cr_2O_7} R\text{—}COOH \quad \text{(no homologation)}$$

$$R\text{—}CH_2\text{—}Br \xrightarrow{NaCN} R\text{—}CH_2\text{—}CN \xrightarrow{HCl/H_2O} R\text{—}CH_2\text{—}COOH \quad \text{(homologation)}$$

$$R\text{—}CH_2\text{—}Br \xrightarrow[2.\ CO_2;\ 3.\ HCl/H_2O]{1.\ Li/ether} R\text{—}CH_2\text{—}COOH \quad \text{(homologation)}$$

PROBLEM 14.17

Show the reactions necessary to complete each of the following transformations:

(a) benzyl alcohol to phenylacetic acid
(b) ethylbenzene to 3-phenylpropanoic acid
(c) toluene to terephthalic acid

PROBLEM 14.18

Show the reactions necessary to complete each of the following conversions:

(a) toluene to *m*-bromobenzoic acid
(b) toluene to *p*-bromobenzoic acid
(c) cyclohexanone to cyclohexanecarboxylic acid
(d) isopropyl alcohol to 2-methylpropanoic acid

14.5

Important and Interesting Carboxylic Acids

Acetic acid is used industrially in very high volume, often to convert alcohols to acetate esters. (An ester is the result of dehydration between an alcohol and a carboxylic acid, to be described in Section 15.3.) A classic example of an acetate ester is cellulose acetate, one of the first clear films; it was used in making early motion pictures.

One of the better-known carboxylic acids is acetylsalicylic acid, otherwise known as *aspirin*. Now 100 years old, aspirin was the first mass-produced and mass-marketed drug. It is derived from salicylic acid, a precursor of which was originally obtained from the bark of the willow tree and used as an analgesic in the 18th century. However, its continued use induced side effects, apparently caused mainly by the phenolic hydroxyl group. These side effects were avoided by acetylating the phenolic hydroxyl group to form acetylsalicylic acid. The acetyl group is hydrolyzed upon digestion, and the active ingredient, salicylic acid, is formed.

The salts of a number of carboxylic acids are used as preservatives in foods such as these cookies.

Salicylic acid (benzene ring with OH and COOH) $\xrightarrow{\text{acetic anhydride}}$ Acetylsalicylic acid (aspirin) (benzene ring with $O-\overset{O}{\overset{\|}{C}}-CH_3$ and COOH)

The salts of a number of carboxylic acids are used as preservatives in food, especially baked goods. Prominent examples are sodium benzoate and calcium propanoate. Monosodium glutamate is used as a flavor enhancer and preservative. Nicotinic acid (pyridine-3-carboxylic acid) is also known as the vitamin niacin, which is added to many processed foods.

Terephthalic acid is a high-volume industrial chemical (35 billion pounds in 1995) made by the high-temperature catalytic oxidation of *p*-xylene (*p*-dimethylbenzene). It is esterified with ethylene glycol to make a polyester polymer, about 70% of which is spun to make the fiber sold as Dacron (Terylene in Great Britain), which is widely used in clothing. When fabricated into a film, the polymer is called Mylar and is used in the

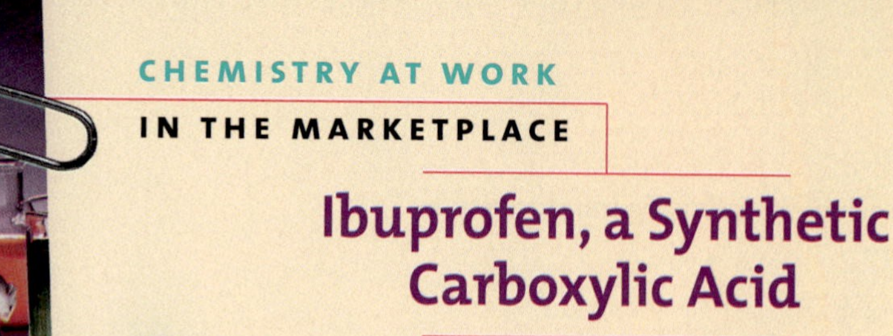

Ibuprofen, a Synthetic Carboxylic Acid

Aspirin is a very old and widely used analgesic, but a number of newer analgesics and anti-inflammatory agents have appeared on the market. One of these, commonly called *ibuprofen,* is the active ingredient in the over-the-counter medications tradenamed Motrin, Nuprin, and Advil. It is a carboxylic acid, and its chemical name is 2-(4-isobutylphenyl)propanoic acid.

Ibuprofen has a single stereocenter and therefore can exist in two enantiomeric forms. The commercial product is usually the racemate. However, only the *S* enantiomer is biologically active and has the desired medicinal effects. The *R* enantiomer is harmless, which is fortunate, since the two enantiomers are difficult to separate. The *S* enantiomer alone begins to take effect in only 12 minutes, whereas the racemate takes 38 minutes. Interestingly, and luckily, the body chemically converts the inactive *R* enantiomer into the active *S* enantiomer.

A number of commercial and laboratory syntheses for racemic ibuprofen have been published. Research continues to try to improve the efficiency of each industrial process, resulting in more advantageous economics and fewer by-products (which must be disposed of safely). In addition, new processes have been developed using proprietary metal catalysts to accomplish transformations not readily accomplished in the laboratory. Of particular interest are processes that will permit production and marketing of only the *S* enantiomer. Shown here is one of the older commercial processes, developed by the Boot Pure Drug Company, and a newer process, developed by the Hoechst Company. Both syntheses start with isobutylbenzene and use Friedel-Crafts acylation, but the Boot process requires six steps, while the Hoechst process requires only three.

Two synthetic routes from isobutylbenzene to ibuprofen:

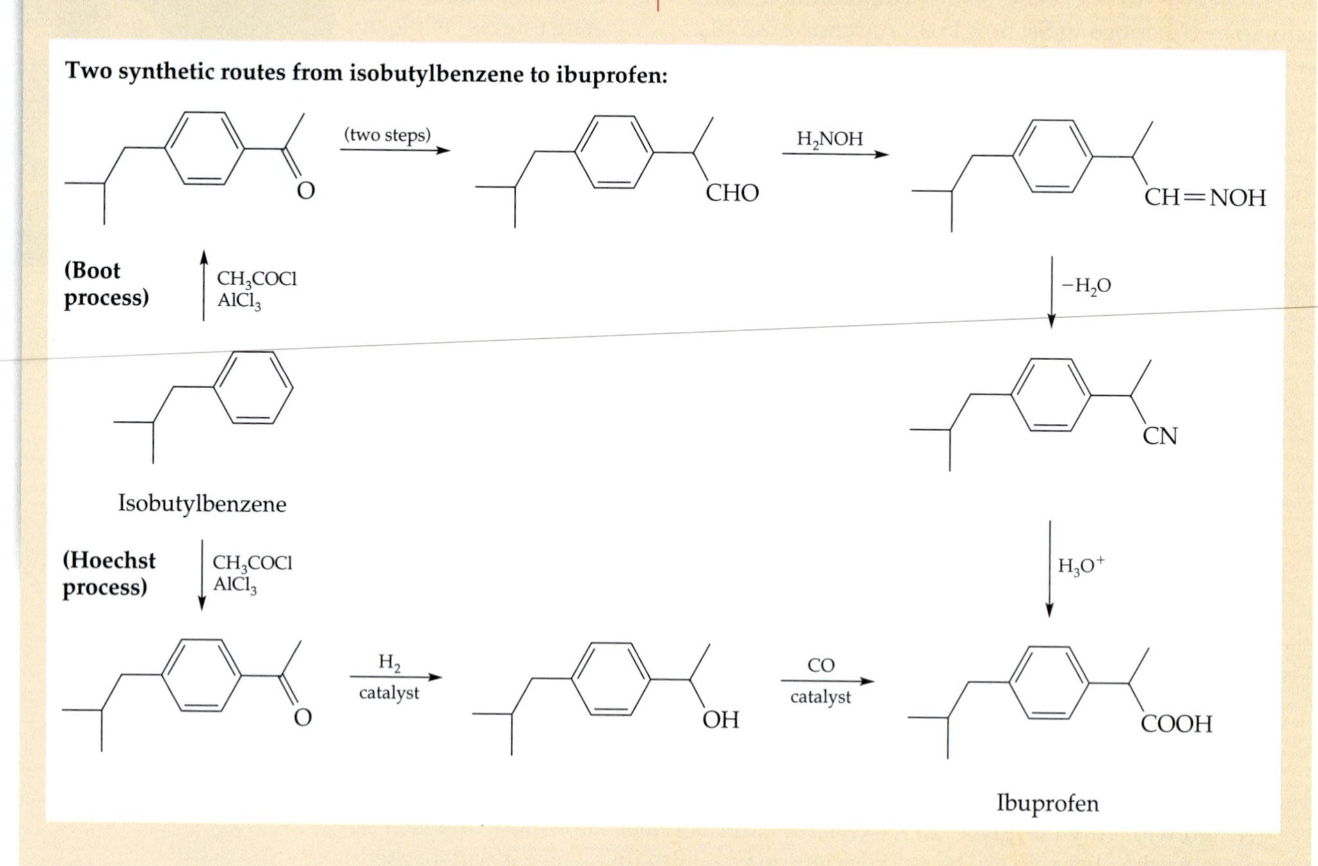

manufacture of magnetic tape. When the polymer is processed to form a resin (called *PET* for polyethylene terephthalate), it is used for bottling soft drinks, milk, and other foods. These products are recyclable.

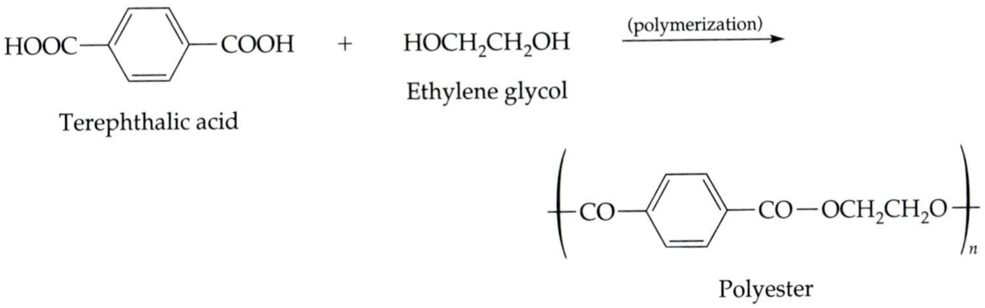

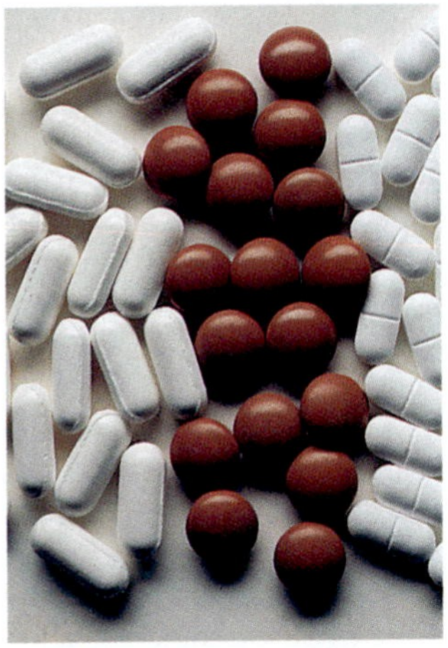

$$\text{HOOC}-\text{C}_6\text{H}_4-\text{COOH} + \text{HOCH}_2\text{CH}_2\text{OH} \xrightarrow{\text{(polymerization)}} \left[\text{CO}-\text{C}_6\text{H}_4-\text{CO}-\text{OCH}_2\text{CH}_2\text{O}\right]_n$$

Terephthalic acid

Ethylene glycol

Polyester

Acetominophen, ibuprofen and aspirin are examples of the varieties of painkillers available on the market. The active ingredient of ibuprofen (middle) is a carboxylic acid.

Carboxylic acids are very important in metabolic processes, as we will see in Chapter 19. The acyl group from carboxylic acids is a key reactant in the metabolism of fats, carbohydrates, and proteins.

Numerous biologically active compounds contain carboxyl groups. Some of the more interesting ones belong to a family of compounds called *prostaglandins,* which occur in trace amounts in virtually every cell. These compounds exhibit a wide array of regulatory effects on biological processes, including metabolism, and their overproduction is believed to be a major cause of pain and inflammation. All prostaglandins have the same overall structure, a C_{20} carboxylic acid containing a cyclopentane ring:

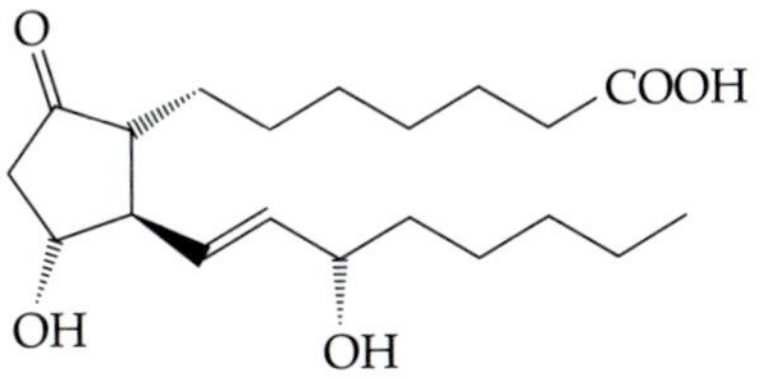

O
COOH
OH
OH

Prostaglandin E_1

The herbicide 2,4-D is also a carboxylic acid. It can be applied to plants in a water spray as its ammonium salt (water-soluble) or in a water emulsion as its ester (water-insoluble and therefore more resistant to being washed off by dew or rain). The synthetic sweetener aspartame, known commercially as NutraSweet, is a carboxylic acid. However, it also contains an amide, an ester, and an amine group.

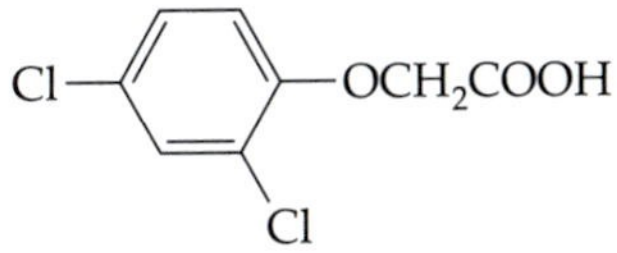

Cl
OCH_2COOH
Cl

2,4-D
(2,4-dichlorophenoxyacetic acid)

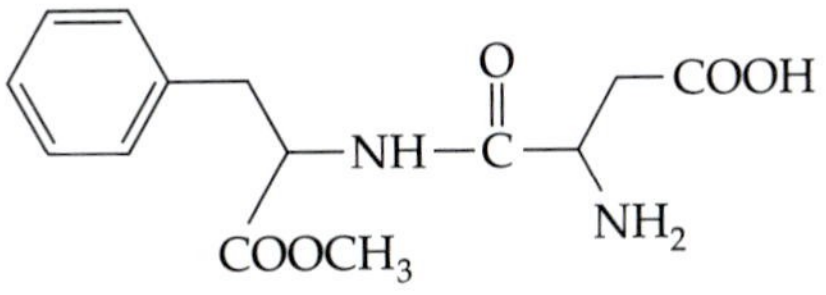

O
COOH
NH—C
$COOCH_3$
NH_2

Aspartame (NutraSweet)

Chapter Summary

Carboxylic acids contain the functional group called the **carboxyl group** (—COOH). The attachment of the electron-donating hydroxyl (—OH) group to the carbonyl group (C=O) increases the electron density at carbon; as a result, carboxylic acids do not undergo nucleophilic addition reactions like those of carbonyl compounds. The carboxyl group is a polar group, facilitating the solubility of lower-molecular-weight acids and serving as a hydrophilic center when attached to a lipophilic hydrocarbon.

IUPAC names of carboxylic acids are formed by naming the parent alkane (including the carboxyl carbon in the chain), dropping the *-e* ending, and adding the suffix *-oic acid*. The generic name is **alkanoic acid.** Dicarboxylic acids, containing two carboxyl groups in the parent chain, are called **alkanedioic acids** in general. Benzenoid carboxylic acids are named as derivatives of benzoic acid. Historical/common names are in widespread use for simple carboxylic and dicarboxylic acids.

The major property of carboxylic acids is their acidity, most having a pK_a of about 5. This acidity is due to the significant resonance stabilization of the **carboxylate anion** that results from removal of a proton. The pK_a values of carboxylic acids can be decreased (that is, their acidity increased) by the attachment of substituents that further delocalize the negative charge of the carboxylate anion.

The carboxyl group (—COOH) can be reduced by lithium aluminum hydride to a primary alcohol group (—CH_2OH). Reaction of an alkanoic acid with bromine and phosphorus results in bromination at the alpha carbon to produce α-bromo alkanoic acids. Carboxylic acids can be thermally **decarboxylated** when there is a carbonyl group beta to the carboxyl carbon.

Alkanoic acids are prepared by two major routes. Conversion of an alkyl or aryl bromide to a Grignard reagent or to an organolithium reagent, then **carboxylating** with carbon dioxide, produces a carboxylic acid. Substitution of primary or secondary alkyl halides with cyanide ion, then hydrolysis of the resulting **nitrile,** also produces carboxylic acids. Other routes to nitriles (aldehydes to cyanohydrins or aromatic amines to aromatic nitriles) can also be used to produce carboxylic acids. The oxidation of alkyl aromatics produces benzoic acids. The oxidation of primary alcohols also yields carboxylic acids.

Summary of Reactions

1. **Formation of carboxylate anions (Section 14.2.2).** Carboxylic acids are converted into carboxylate anions by reaction with bases as weak as sodium bicarbonate.

$$\underset{\text{Carboxylic acid}}{R-\overset{O}{\overset{\|}{C}}-O-H} + NaHCO_3 \longrightarrow \underset{\text{Sodium carboxylate}}{R-\overset{O}{\overset{\|}{C}}-O^- \; Na^+} + H_2O + CO_2$$

2. **Reduction of carboxylic acids (Section 14.3.1).** Lithium aluminum hydride is the only effective reagent for reducing carboxylic acids to primary alcohols.

$$\underset{\text{Carboxylic acid}}{R-\overset{O}{\overset{\|}{C}}-O-H} + LiAlH_4 \longrightarrow \underset{1^\circ\text{ alcohol}}{R-CH_2OH}$$

3. **Alpha bromination of alkanoic acids (Section 14.3.2).** Reaction of an alkanoic acid with phosphorus and bromine, followed by hydrolysis of the acid bromide, produces an α-bromocarboxylic acid.

$$\underset{\text{Carboxylic acid}}{R-CH_2-COOH} \xrightarrow[2.\ H_2O]{1.\ P,\ Br_2} \underset{\alpha\text{-Bromo carboxylic acid}}{R-CHBr-COOH}$$

4. **Decarboxylation of β-keto acids and malonic acids (Section 14.3.3).** Carboxylic acids that contain β-carbonyl groups lose carbon dioxide upon heating to about 100°C.

$$\underset{\beta\text{-Keto acid}}{R-\overset{O}{\overset{\|}{C}}-CH_2-\overset{O}{\overset{\|}{C}}-O-H} \xrightarrow{\Delta} \underset{\text{Ketone}}{R-\overset{O}{\overset{\|}{C}}-CH_3} + CO_2$$

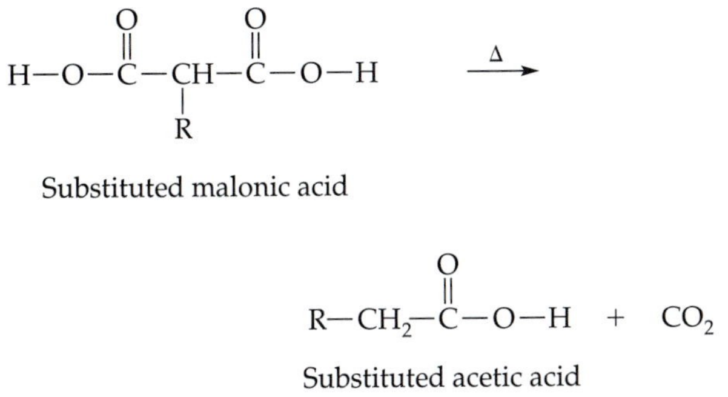

$$\text{H-O-}\overset{\text{O}}{\overset{\|}{\text{C}}}\text{-}\underset{\text{R}}{\underset{|}{\text{CH}}}\text{-}\overset{\text{O}}{\overset{\|}{\text{C}}}\text{-O-H} \xrightarrow{\Delta} \text{R-CH}_2\text{-}\overset{\text{O}}{\overset{\|}{\text{C}}}\text{-O-H} + CO_2$$

Substituted malonic acid → Substituted acetic acid

5. **Carboxylation of organometallics (Section 14.4.1).** Alkyl and aryl bromides are converted into Grignard reagents (or organolithium reagents), which add to the carbonyl group of carbon dioxide to form a carboxylate anion. Acidification during work-up produces the free acid.

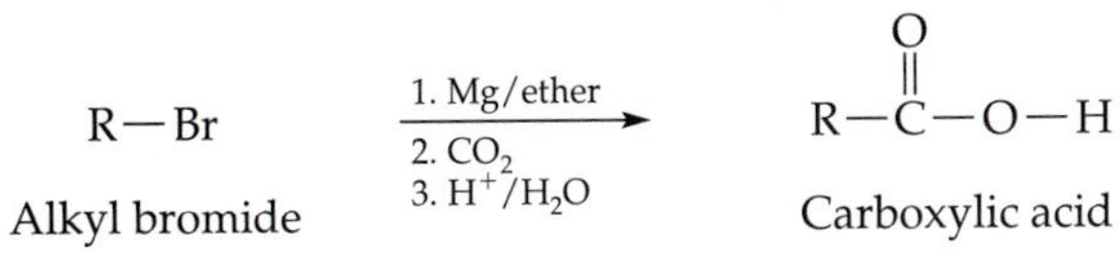

$$\text{R-Br} \xrightarrow[\text{2. } CO_2 \quad \text{3. } H^+/H_2O]{\text{1. Mg/ether}} \text{R-}\overset{\text{O}}{\overset{\|}{\text{C}}}\text{-O-H}$$

Alkyl bromide → Carboxylic acid

6. **Carboxylic acids from nitriles (Section 14.4.2).** Nitriles can be hydrolyzed to carboxylic acids. The two most prominent routes to nitriles are from alkyl halides and from aromatic amines.

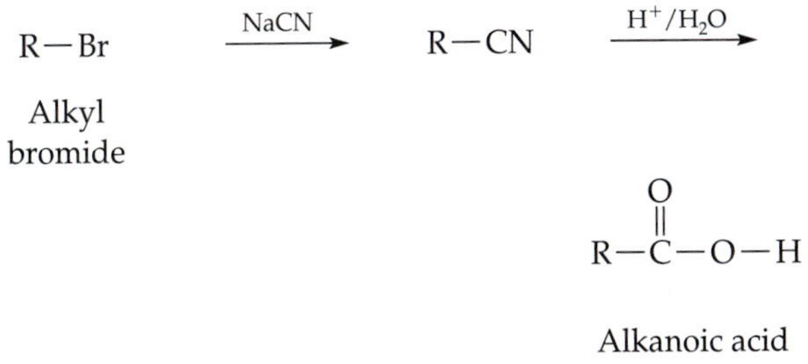

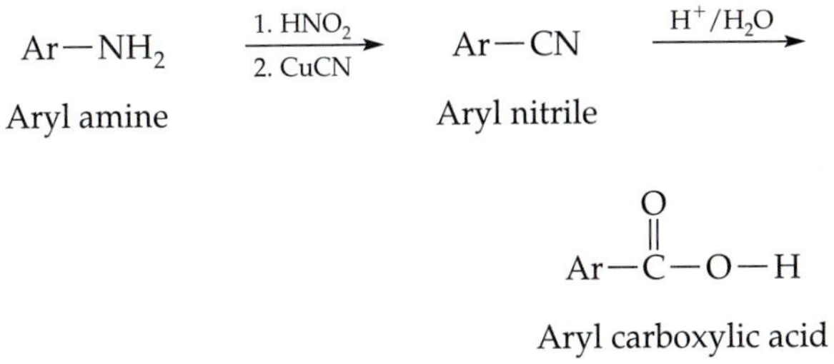

$$\text{R-Br} \xrightarrow{\text{NaCN}} \text{R-CN} \xrightarrow{H^+/H_2O} \text{R-}\overset{\text{O}}{\overset{\|}{\text{C}}}\text{-O-H}$$

Alkyl bromide → Alkanoic acid

$$\text{Ar-NH}_2 \xrightarrow[\text{2. CuCN}]{\text{1. } HNO_2} \text{Ar-CN} \xrightarrow{H^+/H_2O} \text{Ar-}\overset{\text{O}}{\overset{\|}{\text{C}}}\text{-O-H}$$

Aryl amine → Aryl nitrile → Aryl carboxylic acid

Additional Problems

Nomenclature

14.19 Draw the structural formula corresponding to each of the following names:

(a) 2-methylhexanoic acid
(b) (*E*)-3-pentenoic acid
(c) 2-oxobutanoic acid
(d) malonic acid
(e) 4-oxocyclohex-2-enecarboxylic acid
(f) 3,4-dinitrobenzoic acid
(g) 3-isopropylphthalic acid
(h) potassium propanoate
(i) *β*-aminobutyric acid
(j) butanedioic acid (succinic acid)
(k) 3,3-dimethylpentanedioic acid
(l) (*Z*)-2-methyl-2-hexenoic acid
(m) *cis*-butenedioic acid (maleic acid)
(n) *α*-chloroacetic acid
(o) sodium phenylacetate
(p) 3-hexynoic acid
(q) propenoic acid (acrylic acid)

14.20 What is wrong with each of the following names? Give the correct names.

(a) 1-carboxy-2-propanone
(b) *β*-methylbutanoic acid
(c) 3-chloro-4-bromobenzoic acid
(d) 3-carboxycyclohexanol
(e) 1,4-dicarboxybenzene
(f) 2-methylpropionic acid

14.21 Write the names corresponding to the following structures:

(a) 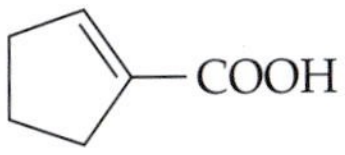 [cyclopentene ring with COOH on the double-bond carbon] COOH

(b) 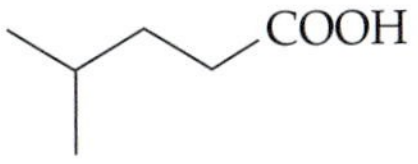 [branched chain] COOH

(c) 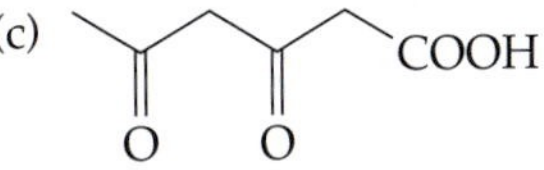 [chain with two C=O groups] COOH

(d) 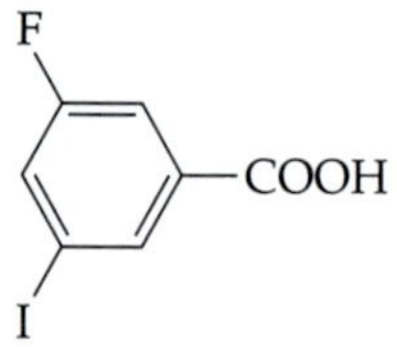 3-fluoro-5-iodobenzoic acid structure: F, COOH, I on benzene ring

(e) 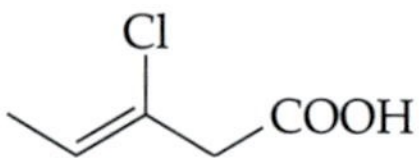 Cl, COOH

(f) $CH_3CHBrCH_2COOH$

(g)  HOOC—(cyclobutane)—COOH

(h) $CH_3CH(COOH)_2$

(i) 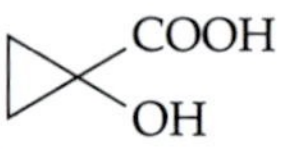 cyclopropane with COOH and OH

14.22 On food labels, you may see the following constituents listed as preservatives. Draw their structures:

(a) sodium benzoate

(b) calcium propionate (or propanoate)

(c) potassium sorbate (sorbic acid is $CH_3CH{=}CH{-}CH{=}CHCOOH$)

Properties

14.23 Rank each set of compounds in order of decreasing acidity:

(a) benzoic acid, *p*-nitrobenzoic acid

(b) propanoic acid, 2-fluoropropanoic acid, 2,2-difluoropropanoic acid

(c) acetic acid, ethanol, ethane

(d) benzoic acid, benzyl alcohol, phenol

(e) 2-chlorobutanoic acid, 3-chlorobutanoic acid

14.24 Place the sets of compounds in Problem 14.23 in order of decreasing pK_a.

14.25 Assign the correct pK_a value to each member of the following pairs of compounds:

(a) acetic acid and *p*-chlorobenzoic acid; pK_a values: 4.8 and 4.0

(b) propanoic acid and 3-chloropropanoic acid; pK_a values: 4.0 and 4.9

(c) *p*-nitrobenzoic acid and *p*-nitrophenol; pK_a values: 3.4 and 7.2

(d) benzoic acid and benzenesulfonic acid; pK_a values: 4.2 and 2.8

14.26 Complete each equation, and show the direction in which the equilibrium lies. (*Hint:* Remember pK_a values.)

(a) $CH_3COOH + NaHCO_3 \rightleftharpoons$

(b) 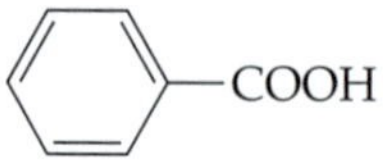 C_6H_5—COOH $+ (CH_3)_2NH \rightleftharpoons$

(c) $CH_3CH_2CH_2COO^- \ Na^+ + HCl \rightleftharpoons$

(d) $CH_3CH_2COOH + CH_3CH_2MgBr \rightleftharpoons$

(e) 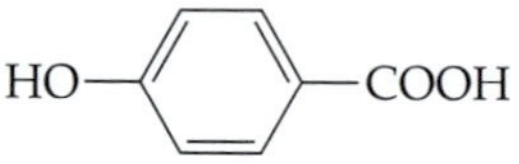 HO—C_6H_4—COOH $+ NaHCO_3 \rightleftharpoons$

(f) 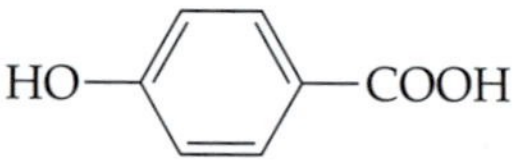 HO—C_6H_4—COOH $+ NaOH \rightleftharpoons$

(g) 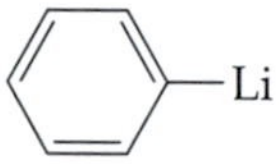 C_6H_5—Li $+ CH_3CH_2OH \rightleftharpoons$

14.27 Devise a separation scheme for isolating pyridine, benzoic acid, and *p*-cresol from a mixture of the three in an ether solvent.

14.28 A compound has a molecular formula $C_6H_{12}O_2$. Its IR and ^1H-NMR spectra are shown in Figure 14.4. What is the compound's structure?

Preparation of Carboxylic Acids

14.29 Show how benzoic acid can be prepared from each of the following starting materials, using any necessary reagents:

(a) toluene

(b) nitrobenzene

(c) bromobenzene

(d) benzyl alcohol

14.30 Show how 3-phenylpropanoic acid can be prepared from each of the following starting materials, using any necessary reagents:

(a) styrene

(b) phenylacetaldehyde

(c) toluene

(d) phenylethanoic acid

14.31 Show how to prepare pentanoic acid from each of the following compounds:

(a) 1-pentanol

(b) 1-pentene

(c) 1-propanol

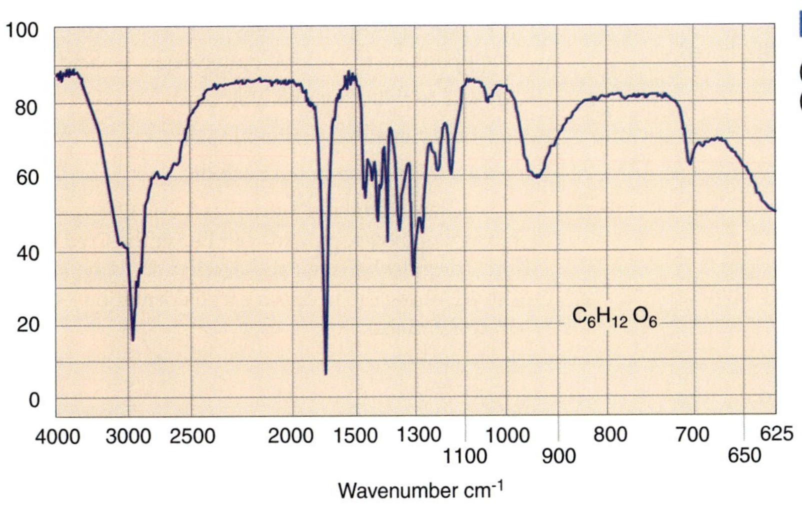

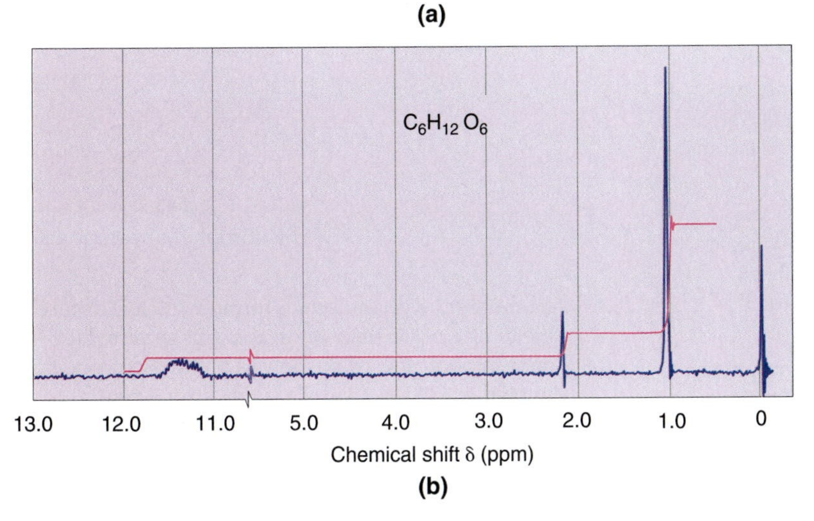

FIGURE 14.4

(a) IR and (b) ^{1}H-NMR spectra of $C_6H_{12}O_6$ (Problem 14.28).

14.32 Show how to convert ethylbenzene into *m*-bromobenzoic acid.

14.33 Show how to convert cyclohexanecarboxylic acid into each of the following compounds:

(a) 1-hydroxycyclohexanecarboxylic acid

(b) cyclohex-1-enecarboxylic acid

(c) 1-(*N,N*-dimethylamino)cyclohexanecarboxylic acid

14.34 Show how to convert pentanoic acid into 2-pentenoic acid.

14.35 Show how to convert cyclopentene into pentanedioic acid.

Reactions of Carboxylic Acids

14.36 Identify the product of decarboxylation of each of the following compounds:

(a) 5,5-dimethyl-2-oxocyclopentanecarboxylic acid

(b) diethylmalonic acid

(c) 3-methyl-2-oxobutanedioic acid

14.37 Show the carboxylic acid you would use as the starting material to prepare each of these compounds in a *single-step* reduction:

(a) 1-butanol

(b) benzyl alcohol

(c) cyclopropylmethanol

14.38 Draw the structure of the compound that can be decarboxylated to produce each of these compounds:

(a) diphenylacetic acid

(b) cyclohexanone

(c) acetone

(d) 2-methylpentanoic acid

14.39 Show the products after work-up of the reaction of 3-phenylbutanoic acid with bromine under the stated conditions:

(a) with $FeBr_3$

(b) with phosphorus

(c) with a peroxide

Mixed Problems

14.40 Show how to accomplish the following conversions. (*Hint:* Use retrosynthesis.)

(a) acetophenone to 2-phenylpropanoic acid

(b) cyclobutane to cyclobutanecarboxylic acid

(c) 4-oxocyclohexanecarboxylic acid to cyclohexanecarboxylic acid

(d) 4-oxocyclohexanecarboxylic acid to 4-oxocyclohexanecarbaldehyde

14.41 A compound with molecular formula $C_5H_8O_4$ shows three peaks in the ^{13}C-NMR spectrum and only two peaks in the ^1H-NMR spectrum, at 12.8 ppm (s, 1H) and 1.29 ppm (s, 3H). Propose a structure for the compound.

14.42 An isomer of the compound in Problem 14.41 shows four peaks in the ^{13}C-NMR spectrum and four peaks in the ^1H-NMR spectrum, at 12.7 ppm (s, 2H), 2.6 ppm (m, 1H), 2.4 ppm (d, 2H), and 0.9 ppm (d, 3H). Propose a structure for this isomer.

14.43 Show how to complete the following transformations (more than one step may be required):

(a) *t*-butyl alcohol to 2-methylpropanoic acid

(b) *t*-butyl alcohol to 2,2-dimethylpropanoic acid

(c) 1-propanol to propanoic acid

(d) 1-propanol to butanoic acid

(e) cyclopentane to cyclopentanecarboxylic acid

(f) anisole to *p*-methoxybenzoic acid

(g) 1,3-butadiene to 3-pentenoic acid

(h) *p*-diethylbenzene to terephthalic acid

14.44 The broadleaf herbicide known as 2,4-D (2,4-dichlorophenoxyacetic acid) can be synthesized from 2,4-dichloronitrobenzene and acetic acid using several steps. Show the necessary transformations. (*Hint:* Use retrosynthesis.)

Cl—(benzene ring, Cl)—NO_2 + CH_3COOH $\xrightarrow{\text{(several steps)}}$

Cl—(benzene ring, Cl)—OCH_2COOH

2,4-D
(2,4-dichlorophenoxyacetic acid)

14.45 A compound with molecular formula $C_8H_8O_3$ is insoluble in water but dissolves in aqueous sodium bicarbonate. The IR spectrum shows a broad, strong peak centered at 3300 cm^{-1} and a sharp, strong peak at 1710 cm^{-1}. The ^1H-NMR spectrum shows peaks at 10.5 ppm (s, 1H), 7.3 ppm (m, 4H), and 3.5 ppm (s, 3H). The decoupled ^{13}C-NMR spectrum shows six separate peaks. Propose a possible structure for the compound.

Getting Oriented

You are aboard the world-renowned train the *Orient Express*, en route to a conference of forensic pathologists in Budapest. As the train pulls into the station in Vienna, some of your fellow passengers disembark. As the conductor makes an accounting of the disembarking passengers, he realizes there is one too few—someone who should have left the train has remained behind. A search of each compartment in your car leads to the discovery of the body of a passenger. Austrian police are summoned, the car is detached from the train to remain in Vienna, with you aboard, and the rest of the train continues on to Budapest.

The initial police investigation reveals no obvious cause of death. However, a search of the compartment uncovers two vials in the medicine cabinet. One (sample A) contains only a few white tablets, and the other (sample B) is half-filled with white tablets. There are no labels on either of the vials. Most of the police department's forensic specialists have left for the conference you were to attend. Realizing that the forensic laboratory is shorthanded, you offer to provide some guidance to the remaining chemical technicians.

You first have a sample of each tablet burned on a spatula. Sample A does not burn, so you conclude that it is inorganic. Subsequent tests determine that these are simply salt tablets. This is consistent with the dead passenger's having recently been on a rigorous outdoor expedition in the Middle East. Sample B burns completely, so you conclude that it is organic. Since these tablets have a capital E stamped on them, you suspect they are Excedrin, a commercial analgesic. Referring to the *Merck Index*, you find that Excedrin contains acetaminophen, acetylsalicylic acid, and caffeine.

The police want to confirm your suspicion and ask you to design a scheme whereby the separate components of Excedrin can be isolated. They can then be identified by comparison of their physical properties with those given in the literature, including spectra. Your task, therefore, is to

- Design a separation scheme to isolate the three components of the Excedrin tablets, assuming they are first ground to a fine powder and all but the filler (starch) is dissolved in ethyl ether solvent.
- Indicate some important spectral peaks the technicians can look for in the IR and ^{1}H-NMR spectra of each of the pure compounds.

Alas, you must then rush off to Budapest, without learning the results of the investigation.

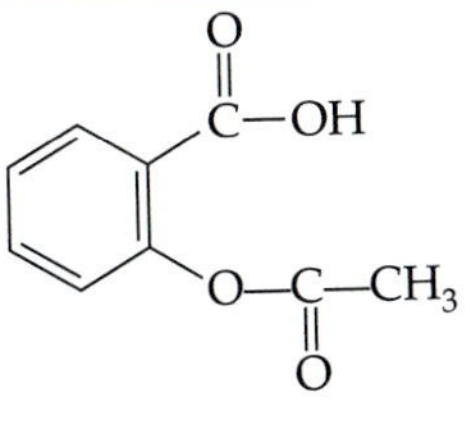

Aspirin

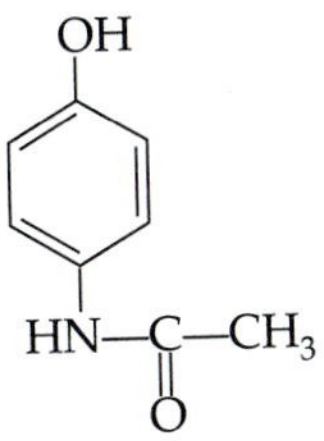

Acetaminophen

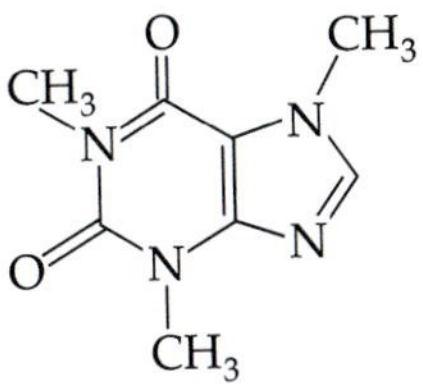

Caffeine

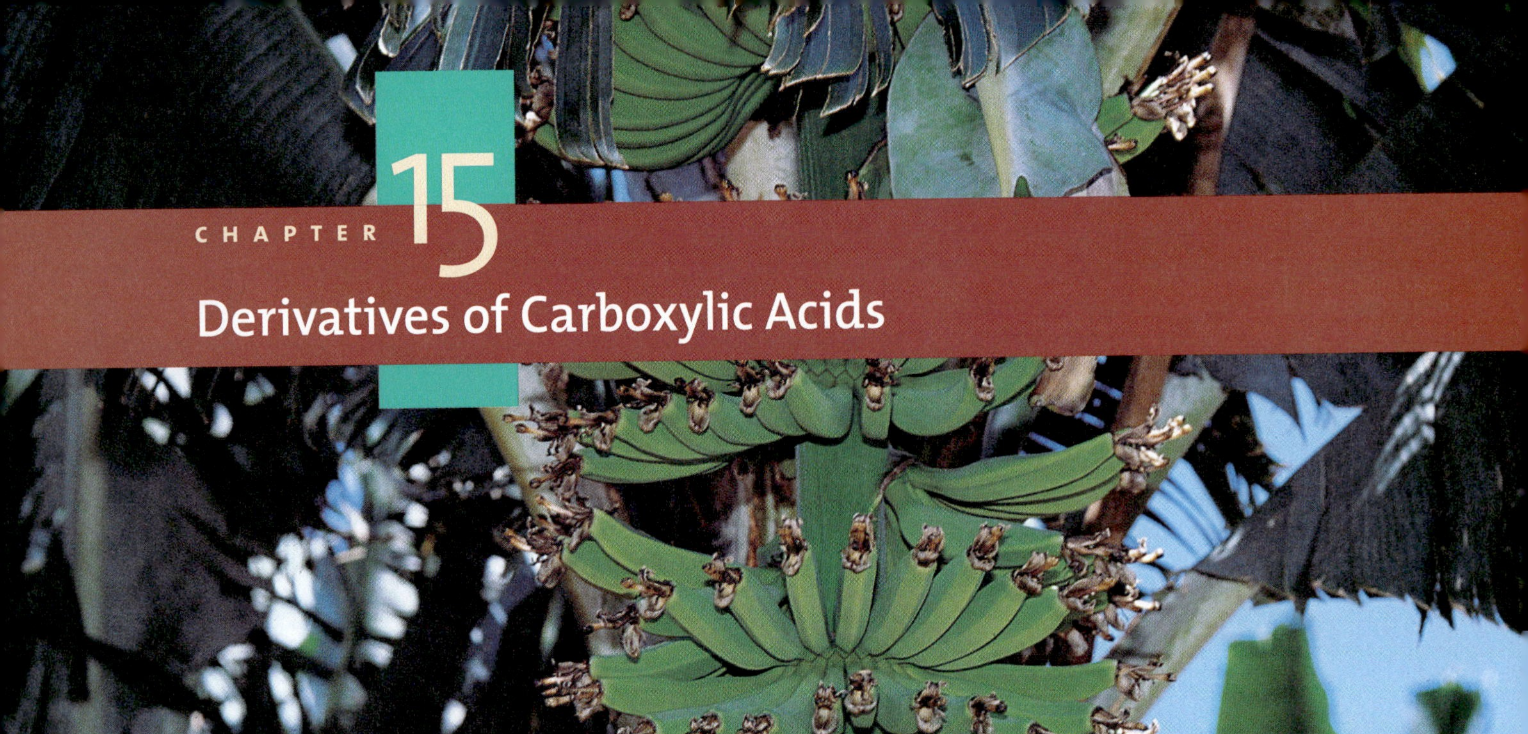

CHAPTER 15
Derivatives of Carboxylic Acids

CARBOXYLIC ACIDS $R-\overset{O}{\overset{\|}{C}}-O-H$ can be readily converted into four kinds of derivatives:

- acid halides, usually acid chlorides, $R-\overset{O}{\overset{\|}{C}}-Cl$
- acid anhydrides, $R-\overset{O}{\overset{\|}{C}}-O-\overset{O}{\overset{\|}{C}}-R$
- esters, $R-\overset{O}{\overset{\|}{C}}-O-R'$
- amides, $R-\underset{O}{\underset{\|}{C}}-NH_2$

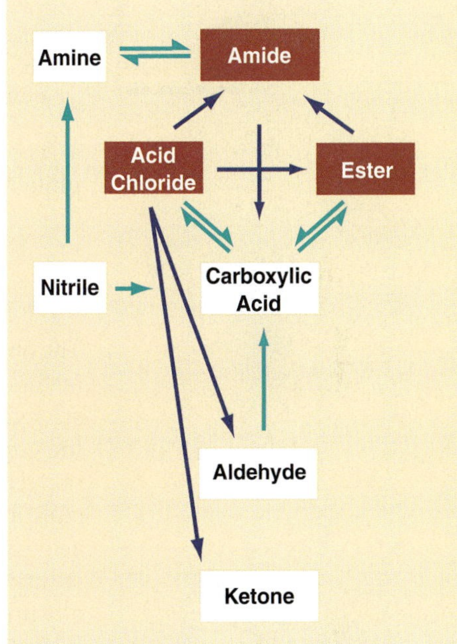

Derivatives of carboxylic acids are obtained from and can be reconverted to the parent carboxylic acids.

These derivatives can all be looked on as being derived from the **acyl group,** $R-\overset{O}{\overset{\|}{C}}-$. Interconversions among the different carboxylic acid derivatives simply involve changing the substituent to which the acyl group is attached. Each is called a **carboxylic acid derivative** because (1) each can be readily converted into its parent carboxylic acid and (2) the carbonyl group is present in each, but the hydroxyl group of the parent carboxylic acid has been replaced by another reactive heteroatom-containing entity (a chloro, carboxyl, alkoxy, or amino group).

Some of these derivatives can be interconverted as summarized in the diagram (a portion of the Key to Transformations, on the inside front cover). Note that acid anhydrides are omitted from this diagram because they are generally of minor importance, but see Section 15.2 for a brief discussion of their chemistry.

Nucleophilic attack at the carbonyl carbon is well known, as we saw in Chapter 13 with aldehydes and ketones (nucleophilic addition). With aldehydes and ketones, the result of reaction with a nucleophile is addition across the carbonyl bond to form first an intermediate alkoxide and then, after acidic work-up, an alcohol. However, when the carbonyl group carries a potential leaving group, as it does in the carboxylic acid derivatives, the end product of such a reaction is a new derivative. The carbonyl group has been re-formed, and the overall result is **nucleophilic acyl substitution.** The reaction is not simply an *addition* reaction, but an addition step followed by an ejection step (an *addition-ejection reaction,* more often referred to as an *addition-elimination reaction*).

◀ Many relatively simple esters, derivatives of carboxylic acids, have pleasant odors. The ester pentyl acetate is a component of the natural flavor and odor of bananas

$$R{-}C({=}O){-}R' + {:}Y \longrightarrow \left[R{-}C(O^-)(Y){-}R'\right] \xrightarrow{H^+} R{-}C(OH)(Y){-}R'$$

Aldehyde or ketone — Nucleophile — Alkoxide — **Addition** (Y′ = alkyl or aryl)

MOVIES
www.jbpub.com/organic-online

$$R{-}C({=}O){-}X + {:}Y \longrightarrow \left[R{-}C(O^-)(Y){-}X\right] \xrightarrow{-{:}X} R{-}C({=}O){-}Y$$

Carboxylic acid derivative — Nucleophile — Alkoxide — **Substitution** (X = Cl, OR, or NH_2)

The overall effect of the nucleophilic acyl substitution reaction is that one nucleophile has replaced another, converting one derivative into another.

$$R{-}CO{-}X + Y{:} \longrightarrow R{-}CO{-}Y + X{:}$$

Nucleophile — Leaving group

The obvious question is, which nucleophile will replace which other nucleophile? Three factors seem to be at work in these reactions, each complementing the other. First, the stronger nucleophile normally replaces the weaker nucleophile. Second, the better leaving group leaves from the alkoxide intermediate. Third, the most electron-deficient carbonyl carbon, that with the most electron-withdrawing and/or least electron-donating substituent attached, is the most easily attacked by a nucleophile. The end result is that acid chlorides are much more reactive (that is, more susceptible to nucleophilic acyl substitution) than esters, which are more reactive than amides. Carboxylic acids are virtually unreactive toward nucleophilic attack because of resonance delocalization from the hydroxyl to the carbonyl group (see Section 14.2.1) and proton removal by the nucleophile.

Relative susceptibility to nucleophilic attack at the carbonyl carbon:

RCOCl	>>	RCOOR′	>	$RCONH_2$	>>	RCOOH
Acid chloride		Ester		Amide		Acid

The flow chart in Figure 15.1 shows which interconversions among the derivatives are practical. We will discuss these reactions, called *alcoholysis*, *ammonolysis*, and *hydrolysis*, in some detail in the sections that follow. Of the many reactions possible for carboxylic acid derivatives, these select few provide an overview of the kinds of reactions undergone by these important compounds.

The most important carboxylic acid derivatives are the esters and the amides, which are naturally occurring and are important in both biological chemistry and industrial chemistry. Acid chlorides and acid anhydrides are less important because they are not naturally occurring, and they are reactive in the presence of moisture (hydrolysis), which means that they must either be prepared from their parent carboxylic acids shortly before use or be stored in the absence of moisture. Thus, they are used only as reactive reagents for the preparation of other compounds, mostly esters and amides.

We will briefly discuss acid chlorides and anhydrides first to illustrate their use in synthesis and then proceed to the more important esters and amides.

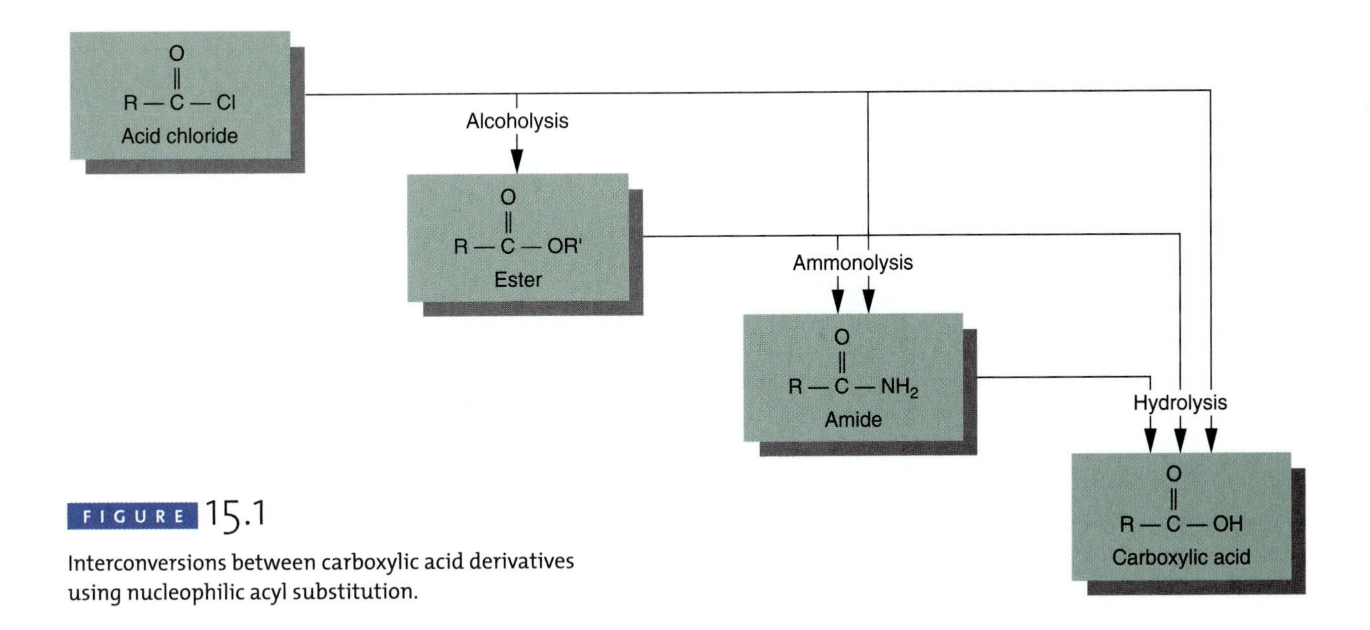

FIGURE 15.1

Interconversions between carboxylic acid derivatives using nucleophilic acyl substitution.

15.1

Acid Chlorides

Acid chlorides are compounds that have a chlorine atom attached to a carbonyl group. They are the most reactive of the carboxylic acid derivatives, mainly because chloride anion is such an effective leaving group. They must even be protected from moisture in the air because of their ready reaction with water. This reactivity has been utilized in a number of reactions that we have already discussed and will review briefly in this section.

Acid chlorides, also called *acyl chlorides*, are referred to as *acylating reagents* because they bring about **acylation reactions** (the attachment of an acyl group, RCO—, to another structure). There is no significant difference between the reactions of aromatic and aliphatic carboxylic acid chlorides. Acid chlorides may be looked on as the "reactive form" of carboxylic acids, which are themselves unreactive in nucleophilic acyl substitution reactions.

15.1.1 Nomenclature of Acid Chlorides

The group name for the halogen-containing compounds RCOX is *acid (or acyl) halide* (where X represents a halogen). However, acid chlorides (X = Cl) are the most prevalent members of the group and are represented by the general formula RCOCl or ArCOCl. Acid chlorides are named on the basis of the carboxylic acid from which they are derived, replacing the suffix *-ic* of the acid name with *-yl* and adding the word *chloride.* The same rule applies to both IUPAC and historical/common names. Thus, acetic or ethanoic acid becomes acetyl or ethanoyl chloride, benzoic acid becomes benzoyl chloride, and 2,2-dimethylpropanoic acid becomes 2,2-dimethylpropanoyl chloride:

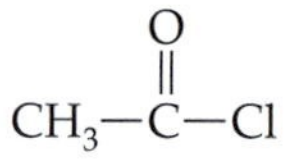

Acetyl chloride (ethanoyl chloride) — Benzoyl chloride 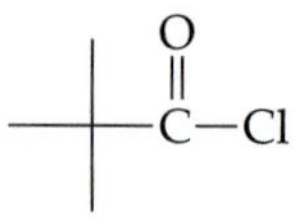 — 2,2-Dimethylpropanoyl chloride

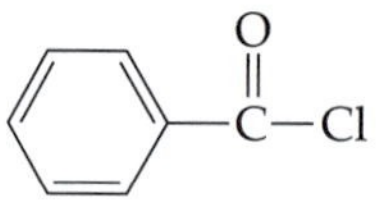

Acid chlorides derived from alicyclic carboxylic acids are named using *-carbonyl chloride* in place of *-carboxylic acid*. For example, cyclopentanecarboxylic acid yields the acid chloride cyclopentanecarbonyl chloride. Note also that sulfonic acids (RSO_3H) form analogous acid chlorides called sulfonyl chlorides (RSO_2Cl). For example, *p*-toluenesulfonic acid (TsOH) becomes *p*-toluenesulfonyl chloride (TsCl). These are very reactive sulfonation reagents, able to attach a sulfonyl group (RSO_2— or $ArSO_2$—) to other molecules.

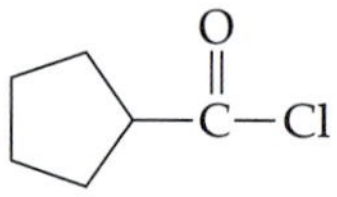

Cyclopentanecarbonyl chloride — *p*-Toluenesulfonyl chloride

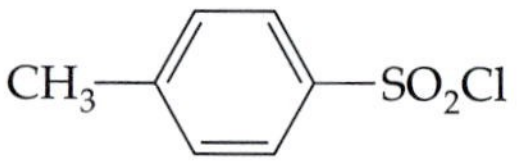

PROBLEM 15.1

Draw structures corresponding to the following names: (a) propanoyl chloride, (b) 3,5-dinitrobenzoyl chloride, and (c) 4-methylcyclohexanecarbonyl chloride.

PROBLEM 15.2

Assign names to the following compounds:

(a) 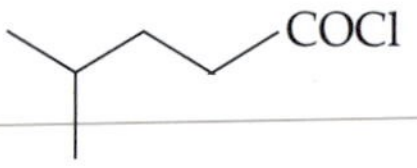 COCl (b) 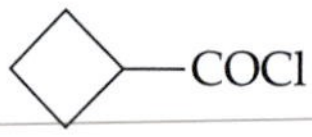 COCl (c) 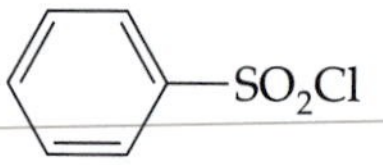 SO_2Cl

15.1.2 -Olysis Reactions of Acid Chlorides

Acid chlorides are the most reactive of the carboxylic acid derivatives. Their major use is conversion to other derivatives, as illustrated in Figure 15.1. They react at room temperature with nucleophiles—for example, with water to form carboxylic acids (*hydrolysis*), with alcohols to form esters (*alcoholysis*), and with ammonia and amines to form amides (*ammonolysis*). The mechanisms of these nucleophilic acyl substitution reactions are identical (see page 530). They involve initial addition via attack by the nucleophile (oxygen or nitrogen), followed by ejection of the chloride anion, and finally loss of a proton from the oxonium or ammonium ion. In each of these reactions, the overall effect is to replace the chlorine of the acid chloride with another nucleophile. Looked at another way, the outcome is acylation of the nucleophile—that is, replacement of a hydrogen in water, an alcohol, or an amine with the acyl group. For this reason, acid chlorides are referred to as *acylating reagents* (they acylate water, alcohols, or amines).

Hydrolysis of Acid Chlorides (Acylation of Water)

Acid chlorides react rapidly and exothermically with water at room temperature, so rapidly that acid chlorides, once prepared, must be kept in sealed vessels (desiccators) to prevent inadvertent exposure to water in the atmosphere. The reaction occurs by attack of the water molecule at the carbonyl carbon, with the chloride ion eventually leaving in a typical nucleophilic acyl substitution. This reaction is quite useless and is to be avoided because it simply reverses the normal synthesis of acid chlorides from carboxylic acids.

General hydrolysis reaction:

$$\underset{\text{Acid chloride}}{R{-}\overset{O}{\overset{\|}{C}}{-}Cl} + H_2O \longrightarrow \underset{\text{Carboxylic acid}}{R{-}\overset{O}{\overset{\|}{C}}{-}OH} + HCl$$

Alcoholysis of Acid Chlorides (Acylation of Alcohols)

Acid chlorides react rapidly at room temperature with all alcohols—primary, secondary, and tertiary—and with phenols. The product of the reaction is an **ester,** a derivative of a carboxylic acid in which the —OH of the carboxylic acid has been replaced by —OR or —OAr. This reaction is a standard laboratory means of preparing esters, but is seldom used on a commercial scale because the formation and handling of an acid chloride is expensive and there is a more economical means of forming esters on a large scale (see Section 15.3.4). The hydrogen chloride evolved in this reaction is trapped by adding an amine, usually pyridine.

General alcoholysis reaction:

$$\underset{\text{Acid chloride}}{R{-}\overset{O}{\overset{\|}{C}}{-}Cl} + \underset{\text{Alcohol}}{R'OH} \xrightarrow{\text{pyridine}} \underset{\text{Ester}}{R{-}\overset{O}{\overset{\|}{C}}{-}O{-}R'} + \text{HCl (trapped as pyridine·HCl)}$$

One of the major advantages of this reaction is that with an enantiomeric alcohol in which the stereocenter carries the —OH group, ester formation occurs with *retention of configuration* (that is, without racemization). This outcome is expected given the reaction mechanism—the carbon-oxygen bond in the alcohol is never broken, so the configuration of the stereocenter is retained:

$$R{-}\overset{O}{\overset{\|}{C}}{-}Cl + R'OH \longrightarrow \left[R{-}\underset{R'\overset{+}{O}H}{\underset{|}{\overset{O^-}{\overset{|}{C}}}}{-}Cl\right] \xrightarrow[2.\ -H^+]{1.\ -Cl^-} \underset{\text{Ester}}{R{-}\overset{O}{\overset{\|}{C}}{-}OR'}$$

For example, (*S*)-2-butanol reacts with acetyl chloride to produce (*S*)-2-butyl acetate. The configuration at the alcohol stereocenter is retained.

$$\underset{(S)\text{-2-Butanol}}{CH_3CH(OH)C_2H_5} \xrightarrow[\text{pyridine}]{CH_3COCl} \underset{(S)\text{-2-Butyl acetate}}{CH_3CH(C_2H_5){-}O{-}\overset{O}{\overset{\|}{C}}{-}CH_3}$$

Any alcohol can be acylated by this reaction. For example, acetyl chloride reacts with benzyl alcohol to form benzyl acetate:

$$\underset{\text{Acetyl chloride}}{CH_3COCl} + \underset{\text{Benzyl alcohol}}{C_6H_5CH_2OH} \xrightarrow{\text{pyridine}} \underset{\text{Benzyl acetate}}{CH_3COOCH_2C_6H_5}$$

PROBLEM 15.3

Show the products of the following reactions:

(a) cyclohexanol with benzoyl chloride and pyridine
(b) benzyl alcohol with cyclopropanecarbonyl chloride and pyridine
(c) 3,5-dichlorophenol with butanoyl chloride and pyridine

PROBLEM 15.4

What starting materials would you employ to produce each of the following compounds?

(a) $CH_3CH_2COOC_2H_5$ (b) 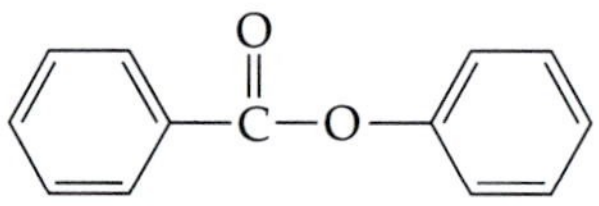 $C_6H_5-\overset{O}{\overset{\|}{C}}-O-C_6H_5$ (c) cyclobutane with two $-OOCC_6H_5$ groups

Ammonolysis of Acid Chlorides (Acylation of Amines)

Acid chlorides react with ammonia or any amine carrying at least one hydrogen on the nitrogen atom (primary or secondary amines) to form **amides.** In this case, the —Cl of the acid chloride is replaced by —NRR′ (R and R′ can be any alkyl groups, aryl groups, or hydrogens). This reaction can be thought of as the *ammonolysis* of an acyl chloride or as the *acylation* of an amine. This is the most effective means for producing amides, and it takes place rapidly at room temperature.

$$R-\overset{O}{\overset{\|}{C}}-Cl + \underset{\text{Ammonia}}{NH_3} \xrightarrow{-HCl} R-\overset{O}{\overset{\|}{C}}-NH_2$$

$$R-\overset{O}{\overset{\|}{C}}-Cl + \underset{1^\circ\ \text{amine}}{R'NH_2} \xrightarrow{-HCl} R-\overset{O}{\overset{\|}{C}}-NHR'$$

$$\underset{\text{Acid chloride}}{R-\overset{O}{\overset{\|}{C}}-Cl} + \underset{2^\circ\ \text{amine}}{R'_2NH} \xrightarrow{-HCl} \underset{\text{Amides}}{R-\overset{O}{\overset{\|}{C}}-NR'_2}$$

The mechanism of ammonolysis is the standard one for nucleophilic acyl substitution and involves addition of the nucleophilic nitrogen at the carbonyl carbon to form an alkoxide, followed by ejection of the chloride anion and, finally, loss of a proton from nitrogen. The last step explains why tertiary amines (R_3N) do not react with acid chlorides to form amides—there is no proton to lose to convert the ammonium salt to an amide.

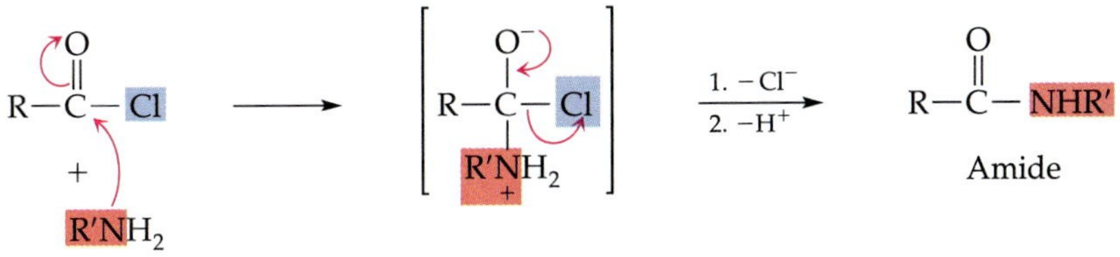

$$R{-}\overset{\overset{O}{\|}}{C}{-}Cl + R'NH_2 \longrightarrow \left[R{-}\overset{\overset{O^-}{|}}{\underset{\underset{R'\overset{+}{N}H_2}{|}}{C}}{-}Cl \right] \xrightarrow[2.\ -H^+]{1.\ -Cl^-} R{-}\overset{\overset{O}{\|}}{C}{-}NHR'$$

Amide

Any primary or secondary amine, and ammonia itself, can be acylated by this reaction. For example, *N*-methylaniline is converted to *N*-methylacetanilide with acetyl chloride:

$$C_6H_5NHCH_3 + CH_3COCl \longrightarrow C_6H_5N(CH_3)COCH_3$$

N-Methylaniline　Acetyl chloride　*N*-Methylacetanilide

PROBLEM 15.5

Show the products of the following reactions:

(a) cyclohexylamine with benzoyl chloride
(b) diethylamine with cyclopropanecarbonyl chloride
(c) 3,5-dichloroaniline with butanoyl chloride

PROBLEM 15.6

What starting materials would you use to produce each of the following compounds?

(a) $CH_3CH_2CON(C_2H_5)_2$　(b) 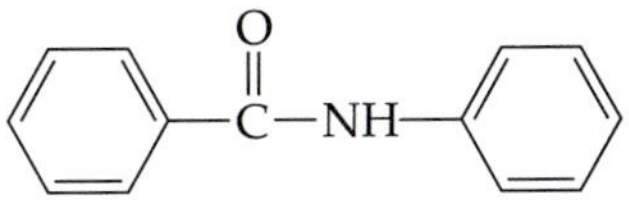 $C_6H_5{-}\overset{\overset{O}{\|}}{C}{-}NH{-}C_6H_5$　(c) cyclic compound with two $-NHCOC_6H_5$ groups

15.1.3 Review of Other Reactions of Acid Chlorides

Earlier, we have seen other reactions in which acid chlorides were used because of their reactivity. These reactions are reviewed briefly.

Reduction to Aldehydes

Reduction of an acid chloride with lithium tri(*t*-butoxy)aluminum hydride produces an aldehyde; a further reduction to an alcohol does not occur (see Section 13.6.2). The reaction occurs via attack by the nucleophilic hydride on the carbonyl carbon in a nucleophilic acyl substitution, eventually ejecting chloride.

$$R{-}COCl + LiAlH(OBu^t)_3 \longrightarrow R{-}CHO$$

Acid chloride　Aldehyde

Conversion to Ketones

Reaction of an acid chloride with the weakly nucleophilic organometallic reagent lithium dialkylcuprate produces a ketone (see Section 13.6.3). Primary, secondary, and tertiary alkyl bromides produce organocuprate reagents that react with any acyl chloride in a nucleophilic acyl substitution reaction.

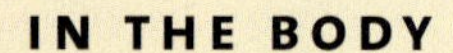

CHEMISTRY AT WORK

IN THE BODY

A Biological Equivalent of Acid Chlorides

The human body contains many carboxylic acids, esters, and amides. The transfer of acyl groups among various nucleophiles—such as water, alcohols, and amines—is a critical aspect of numerous biochemical processes, including metabolism. In a laboratory setting, acyl transfer, as in forming ethyl acetate from ethanol, can be accomplished using an acyl chloride. However, such a reactive reagent could not survive the aqueous environment of the human body.

In the human body, a compound called *coenzyme A*, a thiol, brings about most acyl transfers. One of its most important derivatives, *acetyl coenzyme A* (written as CH_3CO—S—CoA or acetyl-CoA), serves as an acetyl transfer reagent. Coenzyme A has a complex structure. The pantothenic acid portion is a B vitamin, and the adenine dinucleotide is a DNA nucleotide, but the "business end" of the molecule is the thiol group (—SH). Therefore, the coenzyme is frequently abbreviated as CoA—SH in chemical reactions.

Reaction of acetyl-CoA with choline, a compound important in nerve impulse transmission and containing a nucleophilic center in its hydroxyl group, results in the transfer of the acetyl group from acetyl-CoA to choline. The process is essentially one of *trans* esterification: conversion of a thioester (acetyl-CoA) into an ordinary ester (the acetate of choline) through acetyl transfer. The reaction occurs by the standard two-step mechanism of nucleophilic acyl substitution: addition of the nucleophile across the carbonyl group, followed by ejection of the leaving group (in this case, CoA—S^-).

Coenzyme A transfers many other acyl groups, especially in metabolic processes. In all instances, it is the thiolate anion (CoA—S^-) that is the good leaving group, corresponding to the chloride anion as the leaving group for acyl chlorides. However, the thiol group in CoA—SH is itself sufficiently nucleophilic to attack other sources of acyl groups to re-form reagents such as acetyl-CoA. (The critical role of CoA in human metabolism will be discussed further in Section 19.2.)

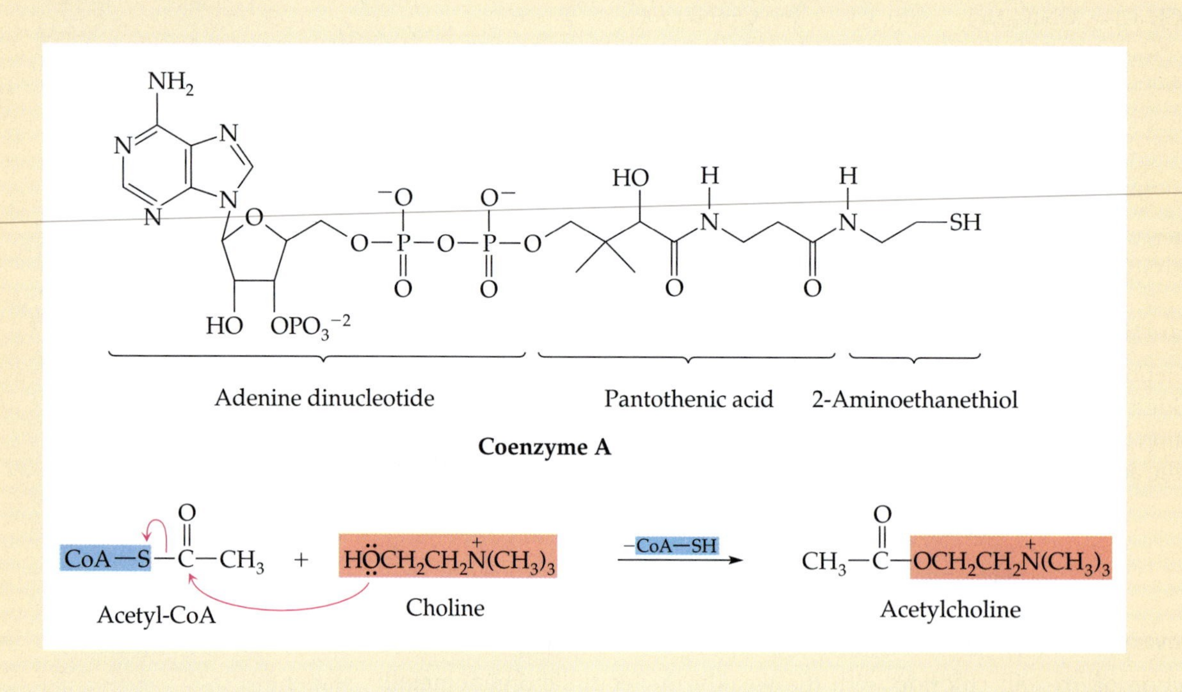

Coenzyme A

$$\underset{\text{Acid chloride}}{R\text{—}COCl} + R'_2CuLi \longrightarrow \underset{\text{Ketone}}{R\text{—}CO\text{—}R'}$$

Friedel-Crafts Acylation

When treated with an aluminum chloride catalyst, acid chlorides form acylium cations, which bring about electrophilic aromatic substitution on an aromatic ring (see Section 9.3.6).

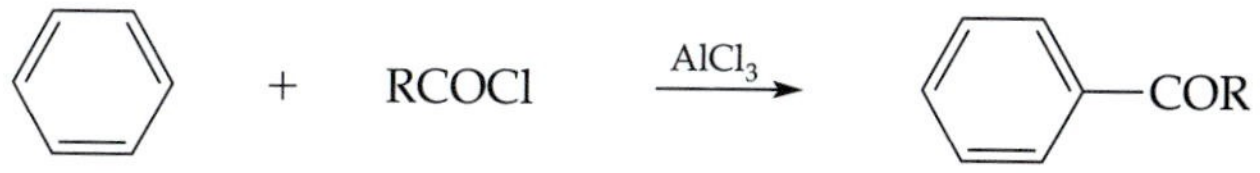

$$\underset{\text{Benzene}}{C_6H_6} + \underset{\text{Acid chloride}}{RCOCl} \xrightarrow{AlCl_3} \underset{\text{Acylbenzene}}{C_6H_5\text{—}COR}$$

PROBLEM 15.7

Show the products of the following reactions:

(a) propanoyl chloride with lithium diethylcuprate
(b) *p*-bromotoluene with acetyl chloride and aluminum chloride
(c) benzoyl chloride with lithium tri(*t*-butoxy)aluminum hydride

PROBLEM 15.8

What reagents would you use to accomplish each of the following conversions?

(a) C_6H_5COCl to $C_6H_5COCH_2CH(CH_3)_2$
(b) $CH_3CH_2CH_2Br$ to $CH_3CH_2CH_2COCH_3$
(c) C_2H_5COCl to $C_6H_5COCH_2CH_3$

PROBLEM 15.9

Explain why vapors of acetyl chloride should not be inhaled.

15.1.4 Preparation of Acid Chlorides

Acid chlorides are usually prepared by reaction of a carboxylic acid with thionyl chloride, $SOCl_2$, which is the double acid chloride of sulfurous acid, H_2SO_3. It is a very rapid and clean reaction that can be carried out at room temperature in a test tube. Both aliphatic and aromatic carboxylic acid chlorides are prepared by this reaction.

General reaction:

$$\underset{\text{Carboxylic acid}}{R\text{—}\overset{\overset{\displaystyle O}{\|}}{C}\text{—}OH} \xrightarrow{SOCl_2} \underset{\text{Acid chloride}}{R\text{—}\overset{\overset{\displaystyle O}{\|}}{C}\text{—}Cl} + HCl + SO_2$$

The reaction is analogous, both in outcome and in mechanism, to the reaction of thionyl chloride with an alcohol to form an alkyl chloride, in which chlorine has

CHEMISTRY AT WORK

IN THE BODY

Chemical Weapons

Warfare becomes *chemical* warfare when it involves the use of chemicals that inflict damage on military personnel and occasionally civilians. Some chemical weapons have the capacity to kill; others are used to immobilize their victims.

The first widespread use of chemical warfare agents occurred during World War I, when chlorine gas (Cl_2) and phosgene ($COCl_2$, a double acid chloride) were used by Germany against the Allies. Both of these reagents affected mainly the lungs, primarily through being hydrolyzed to hydrochloric acid. Mustard gas was also used during World War I. There were no significant chemical attacks during World War II, although both sides had large stockpiles of chemical weapons. It was not until the 1980s that chemical weapons again made the news, this time during the Iraq-Iran war. They were also a significant threat during the Persian Gulf War of 1991 and were used over several years by Iraq against its Kurdish minority. The terrorist attack on the Tokyo subway in 1995 involved the nerve gas Sarin.

Most nations have signed the Chemical Weapons Convention of 1993, and major programs are under way to dispose of stockpiled chemical munitions. The United States has spent a great deal of money building, testing, and operating chemical weapons disposal facilities.

At the moment, U.S. chemical weapons are stockpiled in "storage igloos". This igloo stores VX nerve agent.

One major category of chemical weapons, known collectively as "mustard gas," consists of compounds that burn exposed skin, including the tissue lining the lungs if the compound is inhaled. Sulfur mustard is a thioether, with two reactive halogens. Because it is rapidly absorbed by tissue, its effects are immediately felt, immobilizing, if not killing, anyone exposed.

$$ClCH_2CH_2{-}S{-}CH_2CH_2Cl$$

Sulfur mustard

replaced the hydroxyl group (see Section 4.3.2). In the final step, the chloride anion (a stronger nucleophile) replaces the chlorosulfite group (a better leaving group).

$$Cl{-}\overset{\overset{O}{\|}}{S}{-}Cl \;+\; R{-}\overset{\overset{O}{\|}}{C}{-}OH \xrightarrow[2.\ -H^+]{1.\ -Cl^-} \left[R{-}\overset{\overset{O}{\|}}{C}{-}OSOCl\right] \xrightarrow[2.\ -SO_2,\ -Cl^-]{1.\ Cl^-} R{-}\overset{\overset{O}{\|}}{C}{-}Cl$$

A second category of chemical weapons consists of "nerve gases," which may be designed either to kill or to temporarily incapacitate. Some nerve gases are designed to be persistent—they do not degrade rapidly and remain highly dangerous for a considerable time. Others are nonpersistent—they degrade in a short time, so that the area becomes safe for advancing troops. The compound called VX is a persistent nerve gas, whereas those known as Sarin (or GB) and Tabun (or GA) are nonpersistent. Sarin and Tabun, upon continued exposure to moisture in the atmosphere, hydrolyze to their corresponding nonlethal phosphoric acid derivatives.

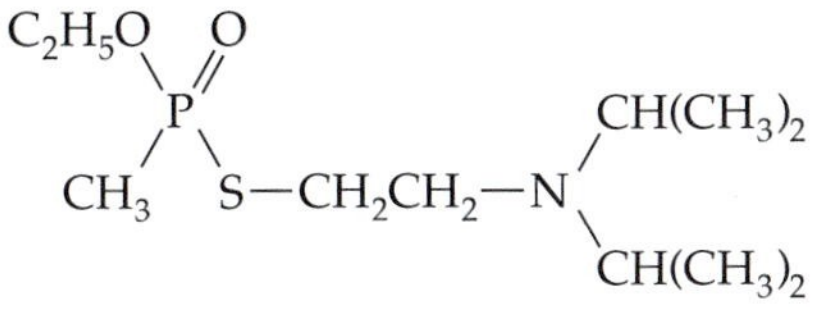

$$C_2H_5O-P(=O)(CH_3)-S-CH_2CH_2-N(CH(CH_3)_2)_2$$

VX

$$(CH_3)_2CHO-P(=O)(CH_3)F \qquad (CH_3)_2N-P(=O)(OC_2H_5)CN$$

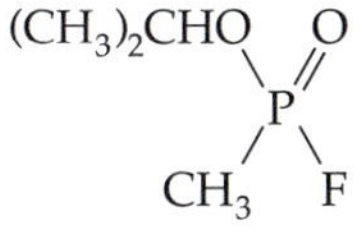

Sarin (GB) 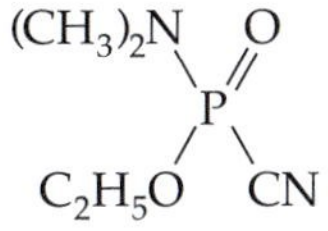 Tabun (GA)

The key to the effectiveness of most nerve gases is that they are chemically very reactive because they are acid halides or behave similarly to acid halides. Sarin is a phosphoric acid fluoride; the cyano group of Tabun is just as easily displaced as a fluoride anion. These reagents react with water, alcohols, and amines exactly as carboxylic acid chlorides do (see Section 15.1.2).

Nerve gases are targeted specifically at choline, which is produced in the transmission of nerve impulses (see "Chemistry at Work," page 411). Under normal conditions, choline is acetylated by the enzyme acetyl transferase, which contains acetyl-CoA, to form the ester acetylcholine. It is acetylcholine that moves across a nerve synapse to stimulate a receptor. Choline reacts with nerve gas in an alcoholysis reaction that esterifies its hydroxyl group, forming a phosphate ester. This blocks the hydroxyl group of choline and prevents its subsequent acetylation by the normal biological processes. If choline cannot be reconverted to acetylcholine, nerve transmission will cease, and muscular action, including breathing, will eventually stop.

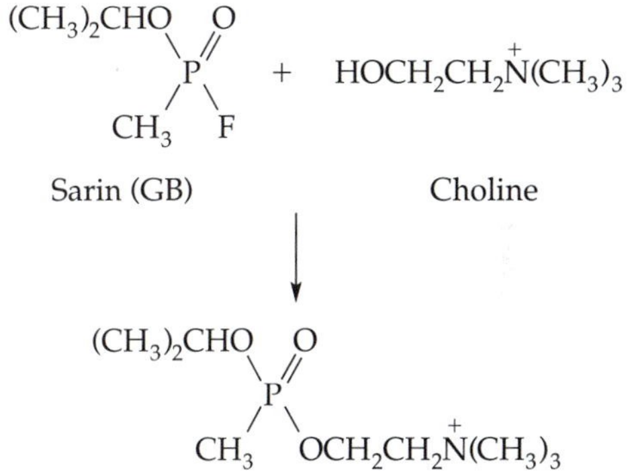

$$(CH_3)_2CHO-P(=O)(CH_3)F + HOCH_2CH_2\overset{+}{N}(CH_3)_3$$

Sarin (GB) Choline

$$\downarrow$$

$$(CH_3)_2CHO-P(=O)(CH_3)-OCH_2CH_2\overset{+}{N}(CH_3)_3$$

"Blocked" choline phosphate ester

For example, cyclopentanecarboxylic acid and *p*-nitrobenzoic acid are converted to their acid chlorides:

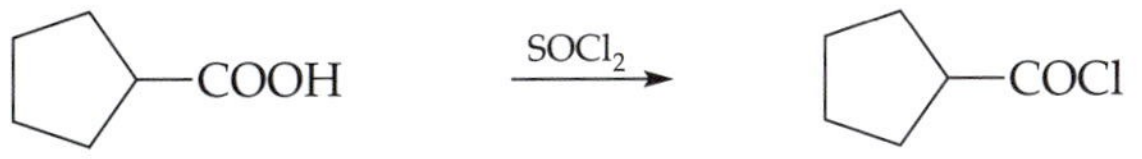

$$C_5H_9-COOH \xrightarrow{SOCl_2} C_5H_9-COCl$$

Cyclopentanecarboxylic acid Cyclopentanecarbonyl chloride

$$O_2N-C_6H_4-COOH \xrightarrow{SOCl_2} O_2N-C_6H_4-COCl$$

p-Nitrobenzoic acid *p*-Nitrobenzoyl chloride

15.2

Acid Anhydrides

Acid anhydrides result when the elements of water are removed from two carboxyl groups. With a monocarboxylic acid, the removal occurs between two molecules of the same acid:

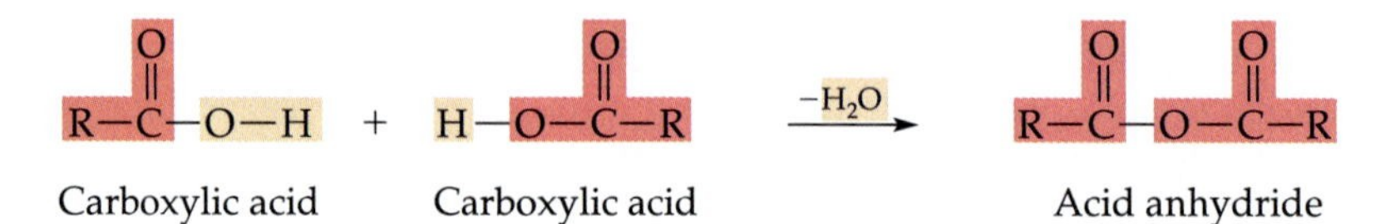

Carboxylic acid Carboxylic acid Acid anhydride

In the case of a dicarboxylic acid (such as succinic acid), the elements of water are removed from the two carboxyl groups in a single molecule to produce a cyclic anhydride:

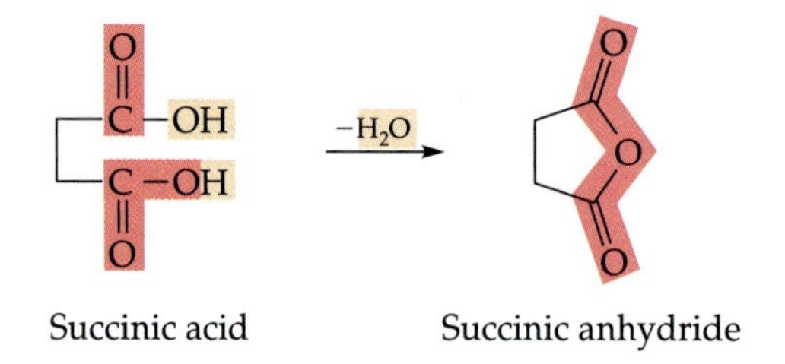

Succinic acid Succinic anhydride

Anhydrides are not naturally occurring. They have virtually no commercial end use. In a few instances, they are produced to serve as acylating reagents in the production of other compounds. Most acylation reactions that can be accomplished using an acid anhydride can also be accomplished using the comparable acid chloride. Because of their relative unimportance, we will look only briefly at acid anhydrides, and they are not included in the Key to Transformations.

15.2.1 Nomenclature of Acid Anhydrides

An acid anhydride is named using the name of the parent carboxylic acid from which it is formed, and replacing the word *acid* with the word *anhydride*. For example, acetic acid leads to acetic anhydride, succinic acid to succinic anhydride, and phthalic acid to phthalic anhydride.

$$CH_3-\overset{O}{\overset{\|}{C}}-O-\overset{O}{\overset{\|}{C}}-CH_3$$

or

$$CH_3CO-O-COCH_3$$

or

$$(CH_3CO_2)O$$

Acetic anhydride (ethanoic anhydride)

Succinic anhydride (butanedioic anhydride)

Phthalic anhydride

15.2.2 -Olysis Reactions of Acid Anhydrides

Acid anhydrides function exactly like acid chlorides in that they undergo nucleophilic acyl substitution reactions upon attack by a nucleophile. The only difference is that in the anhydride reactions, the leaving group is not the chloride anion but a carboxylate anion, which is almost as effective at leaving. The carboxylate anion is then protonated by proton transfer to form a carboxylic acid. Acid anhydrides are only slightly less reactive than acid chlorides.

Anhydrides react with alcohols or phenols to form an ester and a carboxylic acid (an alcoholysis), as shown for the conversion of salicylic acid to acetylsalicylic acid (aspirin):

Salicylic acid + $(CH_3CO)_2O$ (Acetic anhydride) ⟶ Acetylsalicylic acid (aspirin) + CH_3COOH (Acetic acid)

The mechanism of the reaction, involving nucleophilic attack by the alcohol or phenol at the carbonyl carbon in a standard addition, followed by ejection of the carboxylate anion, makes it evident why one equivalent of the carboxylic acid is ejected ("wasted") in each such reaction.

Acid anhydride —R'OH⟶ [tetrahedral intermediate] —1. $-RCOO^-$; 2. $-H^+$⟶ Ester

Thus, conversion of a carboxylic acid to an ester using an acid anhydride is only half as efficient as the same reaction using the acid chloride. However, acetic anhydride, because of its lower cost than acetyl chloride, is frequently chosen as the reagent to acetylate alcohols.

In the case of a cyclic anhydride, such as phthalic anhydride, the ester function and the carboxylic acid function remain in the same molecule:

Phthalic anhydride —CH_3OH⟶ Monomethyl phthalate

In a similar manner, an acid anhydride reacts with water to form two equivalents of the carboxylic acid. This *hydrolysis* reaction reverses the process by which anhydrides are formed (Section 15.2.3). A cyclic anhydride forms a dicarboxylic acid.

$$(RCO)_2O + H_2O \longrightarrow 2\,RCOOH$$

Acid anhydride — Acid

Amines also react with anhydrides in a similar *ammonolysis* reaction. For example, succinic anhydride reacts with methylamine to form *N*-methylsuccinamide:

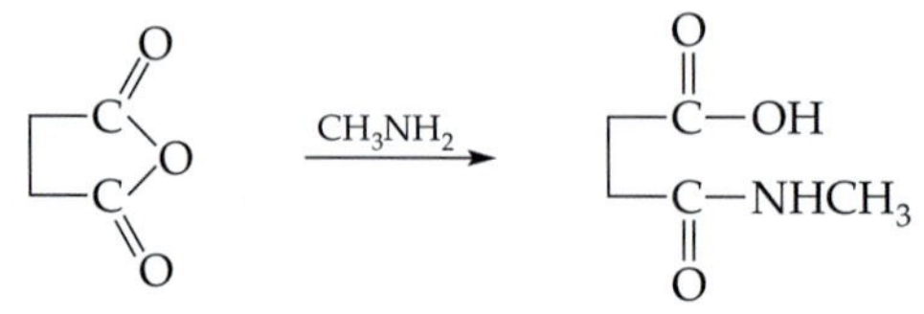

$$\text{Succinic anhydride} \xrightarrow{CH_3NH_2} HOOC{-}CH_2CH_2{-}CONHCH_3$$

Succinic anhydride — *N*-Methylsuccinamide

Acetic anhydride, because it is a more economical choice than acetyl chloride, is frequently used to acetylate amines, as in the formation of *N*-methylacetanilide from *N*-methylaniline.

PROBLEM 15.10

Show the products of the following reactions:

(a) phthalic anhydride and ethanol

(b) acetic anhydride and diphenylamine

(c) succinic anhydride and phenol

PROBLEM 15.11

Show the reagent necessary to accomplish each of the following conversions:

(a) 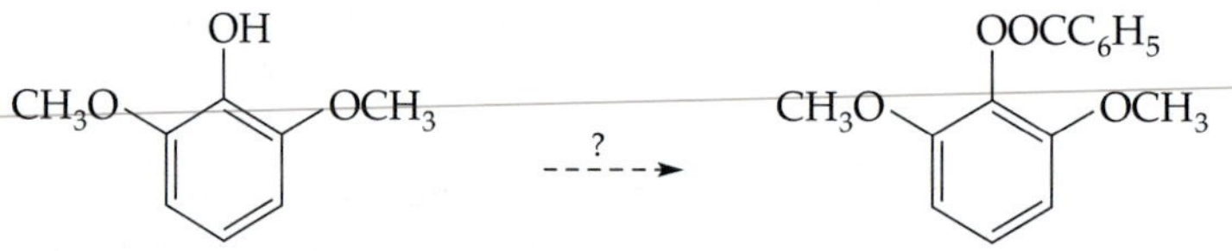 2,6-dimethoxyphenol (CH_3O, OH, OCH_3 on benzene ring) $\xrightarrow{?}$ (CH_3O, $OOCC_6H_5$, OCH_3 on benzene ring)

(b) $(CH_3CH_2CH_2)_2NH \xrightarrow{?} (CH_3CH_2CH_2)_2NCOCH_3$

15.2.3 Preparation of Acid Anhydrides

Acetic anhydride is a large-volume commercial reagent, so it is almost never prepared in the laboratory. We will consider the preparation of cyclic anhydrides because of their relative importance in the laboratory.

When a molecule contains two carboxyl groups in the appropriate proximity, most often located so as to form a five-membered ring, a molecule of water can be eliminated to form a cyclic anhydride. The reaction occurs readily, usually by simply heating to about 100°C. For example, succinic acid produces succinic anhydride, and phthalic acid produces phthalic anhydride:

$$\text{Succinic acid (HOOC–CH}_2\text{CH}_2\text{–COOH)} \xrightarrow[-H_2O]{100^\circ C} \text{Succinic anhydride}$$

Succinic acid Succinic anhydride

$$\text{Phthalic acid} \xrightarrow[-H_2O]{100^\circ C} \text{Phthalic anhydride}$$

Phthalic acid Phthalic anhydride

The two positional isomers of phthalic acid, the *meta* isomer (isophthalic acid) and the *para* isomer (terephthalic acid), have their carboxyl groups too far apart to form a cyclic anhydride. This fact is used to distinguish phthalic acid from the other two isomers; phthalic acid dehydrates upon heating to a cyclic anhydride, but the other two do not.

PROBLEM 15.12

Show the reactions and reagents necessary to accomplish each of the following syntheses:

(a) $HOOCCH_2CH_2COOH \xrightarrow{?} HOOCCH_2CH_2CONHC_6H_5$

(b) $C_6H_5COCH_3 \xrightarrow{?} C_6H_5\text{–}COOCH_3$

(c) $C_6H_6 \xrightarrow{?} C_6H_5\text{–}NHCOCH_3$

(d) Tetralin (1,2,3,4-tetrahydronaphthalene) $\xrightarrow{?}$ benzene ring bearing $COOCH_3$ and $COOH$ in ortho positions

15.3 Esters

Esters are important and widely occurring derivatives of carboxylic acids. They are compounds in which an acyl group is attached to an alkoxy group. In other words, the —OH group of a carboxylic acid has been replaced by an —OR group. The R group may be any alkyl or aryl group. An ester can be viewed as the product of an intermolecular dehydration between a carboxylic acid and an alcohol. In fact, some esters can be prepared in that manner, as we will see.

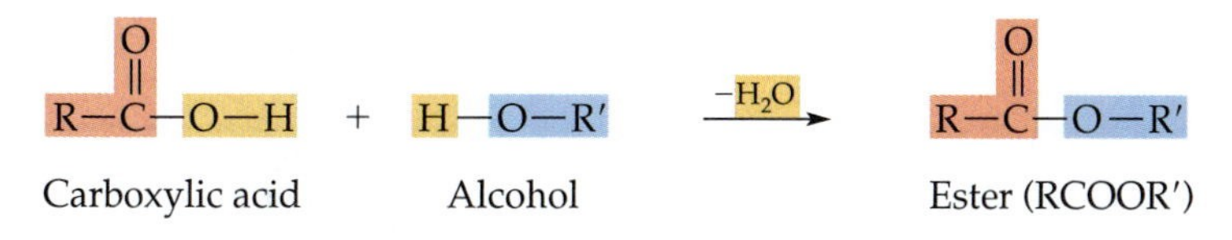

Esters occur naturally in fruits.

Esters are slightly polar compounds, owing to the presence of the polar carbonyl group. Because they have no hydrogen available for hydrogen bonding, they often have very low melting points, normally exist as liquids, and are insoluble in water. Many esters have pleasant odors, and some are used as flavorings; see Table 15.1 for some examples of esters that are used in this way. Many fruits and flowers contain complex mixtures of naturally occurring esters, which provide their particular flavors and odors.

Esters show strong carbonyl absorption in the infrared near 1750 cm^{-1}. The ester functional group also affects the ^{1}H-NMR shifts of attached —CH groups. Hydrogens alpha to the acyl group are deshielded to about 2.0 ppm (similar to α-hydrogens of ketones), while those adjacent to the alkoxy oxygen are more strongly deshielded to about 3.7 ppm (similar to those adjacent to the oxygen in alkyl ethers). For example, the isomeric esters ethyl acetate and methyl propanoate, indistinguishable from each other by infrared spectroscopy, show quite different NMR spectra (Figure 15.2).

Although we are discussing carboxylic acid esters—the products of intermolecular dehydration between an alcohol and a carboxylic acid—there are also esters of oxygen-containing inorganic acids. For example, sulfonic acids (RSO_3H, or RSO_2—OH) can be converted to sulfonate esters (RSO_3R', or RSO_2—OR′), and nitric acid (HNO_3, or $HONO_2$) can be converted to nitrate esters ($R'ONO_2$). One of the most important nitrate esters is nitroglycerine, the trinitrate ester of glycerol (1,2,3-propanetriol). It is a very unstable liquid that is highly explosive and shock-sensitive, as you learned in Chapter 9 (see "Chemistry at Work," p. 298). Nitroglycerine is also a rapid-acting blood vessel dilator, prescribed in pill form for those who suffer from angina, a painful heart condition.

15.3.1 Nomenclature of Esters

The names of esters are derived from the names of the carboxylic acids and alcohols from which the esters are derived. An ester name always has two separate words—for example, methyl acetate (or methyl ethanoate). The first word is the alkyl or aryl group derived from the alcohol component—in this case, methyl from methanol. The second word is derived from the carboxylic acid by replacing the suffix *-ic* with the

TABLE 15.1

Esters Used as Artificial Flavors

Compound	Chemical Name	Flavor
$(CH_3)_2CHCH_2CH_2OOCCH_3$	Isoamyl acetate	Banana
Benzene ring with $COOCH_3$ and OH (ortho)	Methyl salicylate	Wintergreen
$CH_3CH_2CH_2COOCH_3$	Methyl butyrate	Apple
$CH_3(CH_2)_6CH_2OOCCH_3$	Octyl acetate	Orange

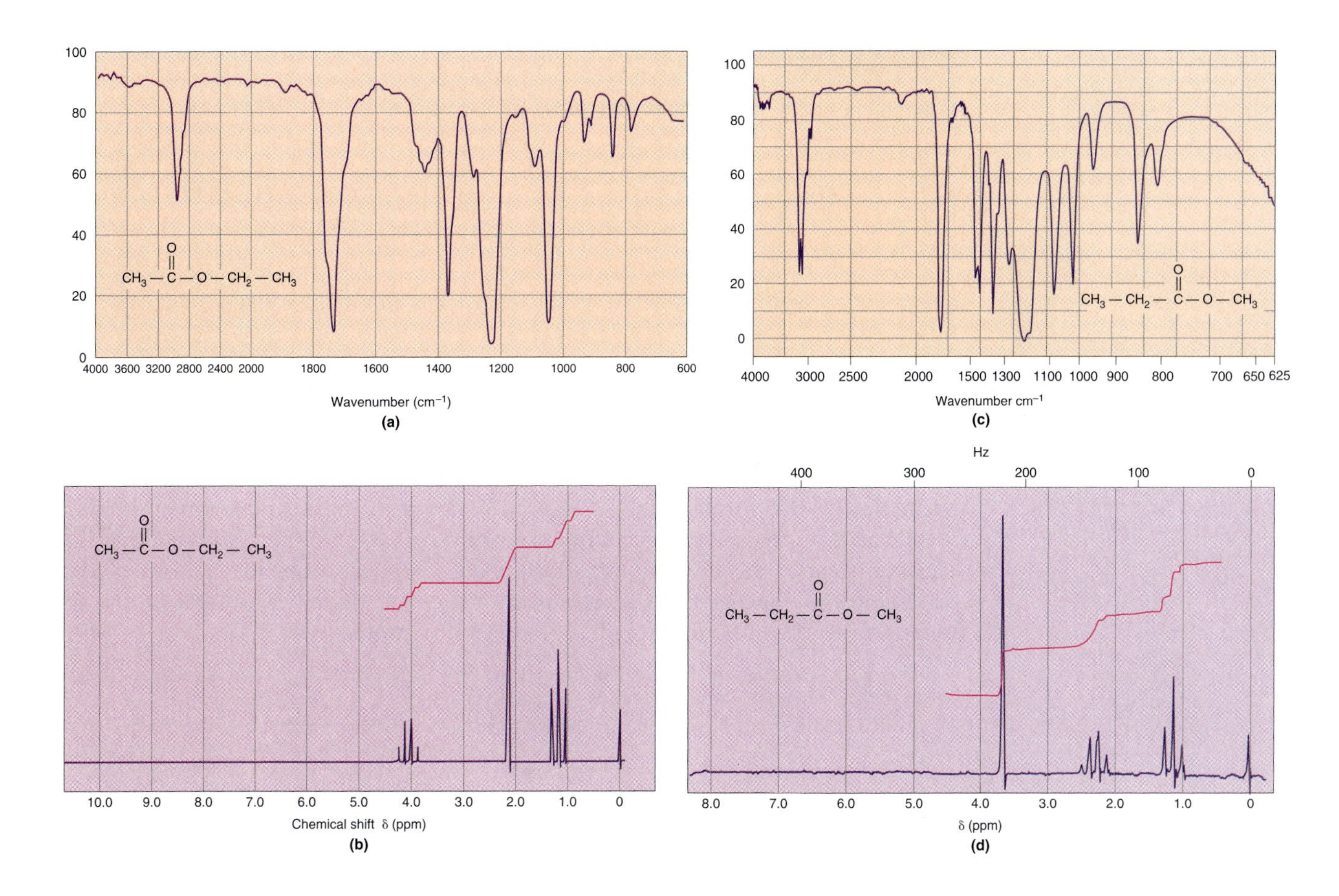

FIGURE 15.2

(a) IR and (b) ¹H-NMR spectra of ethyl acetate and (c) IR and (d) ¹H-NMR spectra of methyl propanoate.

suffix *-ate* (the same as for a carboxylate anion)—as in acetate from acetic. This method applies to both IUPAC names and common names. Other examples are 2-methylpropyl butanoate (isobutyl butyrate) and phenyl benzoate.

CH_3COOCH_3	$(CH_3)_2CHCH_2OOCCH_2CH_2CH_3$	$C_6H_5COOC_6H_5$
Methyl acetate (methyl ethanoate)	2-Methylpropyl butanoate (isobutyl butyrate)	Phenyl benzoate

PROBLEM 15.13

Draw a structure corresponding to each of the following names: (a) isopropyl benzoate, (b) phenyl (*E*)-2-butenoate, and (c) cyclopropyl cyclopropanecarboxylate.

PROBLEM 15.14

Assign names to the following compounds:

(a) cyclopentyl $-O-\overset{O}{\overset{\|}{C}}-C_2H_5$

(b) $CH_3CH_2CH_2\overset{O}{\overset{\|}{C}}-O-CH_2CH_2CH_3$

(c) cyclohexenyl $-O-\overset{O}{\overset{\|}{C}}-CH_3$

(d) $CH_3COCH_2COOCH_3$

15.3.2 -Olysis Reactions of Esters

As the Key to Transformations and Figure 15.1 suggest, esters can be converted to carboxylic acids (a hydrolysis reaction) and to amides (an ammonolysis reaction). Both reactions occur by the now familiar nucleophilic acyl substitution mechanism. Hydrolysis involves replacing the alkoxy group of the ester with a hydroxyl group, and ammonolysis involves replacing it with an amino group.

$$\underset{\text{Carboxylic acid}}{R\overset{O}{\overset{\|}{C}}-OH} \xleftarrow{\text{(hydrolysis)}} \underset{\text{Ester}}{R\overset{O}{\overset{\|}{C}}-OR'} \xrightarrow{\text{(ammonolysis)}} \underset{\text{Amide}}{R\overset{O}{\overset{\|}{C}}-NH_2}$$

Hydrolysis of Esters

Esters can be hydrolyzed to carboxylic acids using either acidic or basic conditions. Acid-catalyzed hydrolysis, less important than base-catalyzed hydrolysis, involves step-by-step reversal of the equilibrium reaction known as the *Fischer esterification* (described in Section 15.3.4). It is accomplished by adding excess water, forcing the reaction toward carboxylic acid and alcohol.

The most common method for the hydrolysis of esters involves use of aqueous base. This process is also referred to as *saponification* ("soap making"). The term comes from the early technique of making soap—the sodium salt of a fatty acid (see Sections 14.2.3 and 15.3.5). The hydrolysis reaction is conducted using aqueous sodium hydroxide, and it produces the alcohol and the sodium salt of the carboxylic acid. The acid salt is usually converted into the carboxylic acid by acidification with dilute mineral acid upon work-up.

General base-catalyzed hydrolysis (saponification):

$$\underset{\text{Ester}}{R\overset{O}{\overset{\|}{C}}-OR'} \xrightarrow{\text{NaOH}} \underset{\text{Alcohol}}{R'OH} + R\overset{O}{\overset{\|}{C}}-O^-\ Na^+ \xrightarrow[H_2O]{H^+} \underset{\text{Carboxylic acid}}{R\overset{O}{\overset{\|}{C}}-OH}$$

For example, methyl benzoate yields methanol and benzoic acid:

$$\underset{\text{Methyl benzoate}}{C_6H_5\overset{O}{\overset{\|}{C}}-OCH_3} \xrightarrow[2.\ H^+]{1.\ NaOH,\ \Delta} \underset{\text{Benzoic acid}}{C_6H_5\overset{O}{\overset{\|}{C}}-OH} + \underset{\text{Methanol}}{CH_3OH}$$

CHEMISTRY AT WORK

IN MEDICINE

Lactones and Macrolides

Cyclic esters, known as *lactones,* are compounds that arise from dehydration of a carboxylic acid group and an alcohol group located in the same molecule. The common name is derived from that of the parent hydroxy acid, with the position of the hydroxyl group indicated by a Greek letter. For example, β-hydroxybutyric acid gives β-butyrolactone:

β-Hydroxybutyric acid $\xrightarrow[-H_2O]{\Delta}$ β-Butyrolactone

Lactones are especially prone to form when a five- or six-membered ring results (γ- or δ-lactones, respectively), but can be difficult to synthesize otherwise.

γ-Butyrolactone γ-Valerolactone δ-Valerolactone

A common naturally occurring γ-lactone is ascorbic acid, also known as vitamin C. Found in citrus fruits, it is also synthesized commercially for the vitamin supplement market.

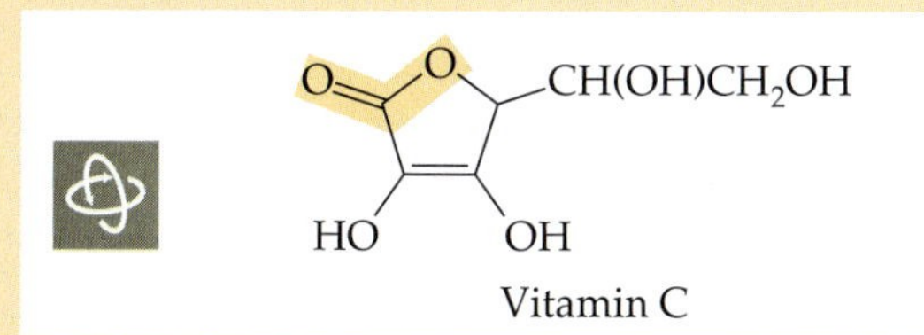

Vitamin C

The lactone coumarin, which was formerly used as a flavoring agent, is found in tonka beans, lavender oil, and sweet clover. It was discovered that when sweet clover is allowed to grow moldy, coumarin is converted into dicoumarol. Dicoumarol is an anticoagulant (it prevents blood clotting) and is used to treat heart disease.

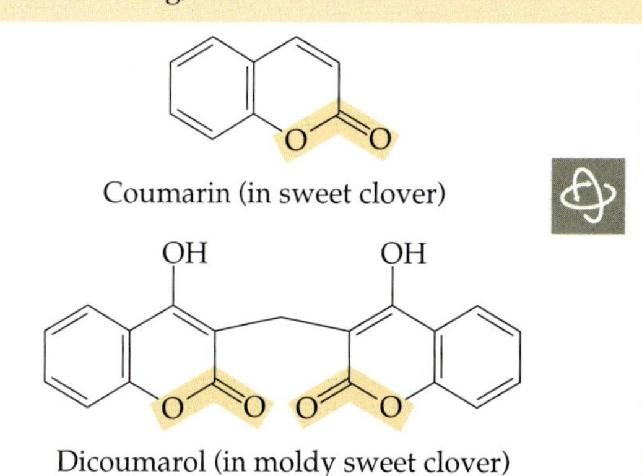

Coumarin (in sweet clover)

Dicoumarol (in moldy sweet clover)

Many large-ring lactones, called *macrolides,* exist in nature, and many have useful biological activity. The lactone of 15-hydroxypentadecanoic acid is obtained from the root of the plant *Angelica archangelica* and used in making perfumes. The 12-membered lactone antibiotic recifeiolide is obtained from a fungus.

Two other lactones have significant medical applications. Epothilone A, a natural bacterial product, was isolated in the 1980s, had its structure determined in 1996, and has been synthesized by two research groups. Its potential as an anticancer agent is greater than that of paclitaxel (taxol). The fungal metabolite mevastatin inhibits the enzyme that controls cholesterol biosynthesis. Because cholesterol is a major contributing factor to heart disease, the related compound pravastatin is being marketed as a prescription drug for the prevention of first heart attacks.

Recifeiolide Epothilone A Mevastatin

The mechanism of the base-catalyzed hydrolysis is a straightforward nucleophilic acyl substitution (two steps—addition then ejection), with hydroxide as the attacking nucleophile. This reaction is not reversible. A key piece of evidence for this mechanism is the observation that hydrolysis of an ester labeled with oxygen-18 at the alcoholic oxygen results in the ^{18}O appearing only in the product alcohol:

$$R-C(=O)-{}^{18}OCH_3 \xrightarrow[^{-}OH]{\text{(addition)}} R-C(O^-)(OH)-{}^{18}OCH_3 \xrightarrow{\text{(ejection)}} R-C(=O)-OH + {}^{-18}OCH_3$$

$$\longrightarrow R-C(=O)-O^- + CH_3{}^{18}OH \qquad R-C(=O)-O^- \xrightarrow{H^+} R-C(=O)-OH$$

Further evidence comes from the observation that hydrolysis of an ester of a chiral alcohol results in production of the alcohol with *retention of configuration* (the alcohol carbon-oxygen bond is not broken):

$$CH_3-C(=O)-O-CH(CH_3)C_2H_5 \xrightarrow[2.\ H^+]{1.\ NaOH,\ \Delta} HO-CH(CH_3)C_2H_5 + CH_3-C(=O)-OH$$

(*R*)-2-Butyl acetate → (*R*)-2-Butanol

The fact that the configuration of the alcohol is retained means that there was no attack by hydroxide on the alcoholic carbon of the ester. If there were, the configuration would have been inverted in what would be an S_N2 reaction.

It is important to reiterate that the mechanism for hydrolysis of sulfonate esters differs from that for hydrolysis of carboxylate esters in that attack by hydroxide occurs at the alcohol carbon, not at the sulfur atom (tosylate is an excellent leaving group). Therefore, hydrolysis of a sulfonate ester (such as a tosylate) of a chiral alcohol occurs with *inversion of configuration.* (An S_N2 attack occurs at the alkyl carbon, as described in Section 4.3.2.)

PROBLEM 15.15

Show the products when each of the following esters is hydrolyzed using aqueous sodium hydroxide:

(a) isopropyl benzoate
(b) phenyl 2-methylpropanoate
(c) (*R*)-2-butyl benzoate

PROBLEM 15.16

Show how to convert (*S*)-2-pentanol to (*R*)-2-pentanol using tosyl chloride as one of the reagents. (*Hint:* See Section 4.3.2.)

Ammonolysis of Esters

Esters react with ammonia or amines (primary or secondary only) to form amides, as suggested by the Key to Transformations and Figure 15.1.

CHEMISTRY AT WORK

IN AGRICULTURE

Persistent versus Nonpersistent Pesticides

As described in Sections 3.6.5 and 9.3.2, some chlorinated hydrocarbons, such as DDT, Lindane, and Aldrin, have been used as pesticides. In most cases, their use is now banned because they are "persistent pesticides"; they resist degradation under natural conditions and therefore "persist" in the environment. Their accumulation has led to many significant biological problems.

New families of compounds have been subsequently discovered that are effective insecticides but are "nonpersistent"; when exposed to the elements, they are chemically changed (degraded) into products that are nontoxic. One such insecticide that is widely used for agricultural purposes—mainly on cotton, rice, soybeans, and some vegetables—is methyl parathion. It is very effective in killing insects on contact. It is a "restricted-use pesticide," approved only for outdoor use in agricultural fields. Methyl parathion is a triester of phosphorothioic acid. It must be applied regularly because it degrades after exposure to air and water. This degradation results from hydrolysis of the three phosphate ester groups, with the products being the parent acid (phosphorothioic acid), *p*-nitrophenol, and methanol.

Recently, methyl parathion was used illegally by unlicensed applicators who sprayed it inside homes in parts of Mississippi, in order to control cockroaches. The inhabitants fell ill, and many had to be relocated while a major cleanup was undertaken. The people who became ill had been exposed to the insecticide for a fairly long time, because it fails to degrade rapidly indoors.

Degradation of methyl parathion:

$$(CH_3O)_2P(=S){-}O{-}C_6H_4{-}NO_2 \xrightarrow{\text{(hydrolysis)}} HO{-}C_6H_4{-}NO_2 + CH_3OH + (HO)_3PS$$

Methyl parathion

General ammonolysis reaction:

$$R{-}C(=O){-}OR' \xrightarrow{NH_3} R'OH + R{-}C(=O){-}NH_2$$

Ester → Alcohol + Amide

For example, ethyl benzoate and methylamine produce *N*-methylbenzamide and ethanol:

$$C_6H_5COOC_2H_5 \xrightarrow{CH_3NH_2} C_6H_5CONHCH_3 + C_2H_5OH$$

Ethyl benzoate → *N*-Methylbenzamide + Ethanol

The mechanism of the ammonolysis reaction is the familiar nucleophilic acyl substitution mechanism. Ammonia or the amine is the nucleophile adding to the carbonyl group, and then alkoxide ion is ejected:

$$\underset{\text{Ester}}{R{-}\overset{\displaystyle O}{\overset{\|}{C}}{-}OCH_3} \xrightarrow{NH_3} R{-}\overset{\displaystyle O^-}{\overset{|}{\underset{\underset{\displaystyle {}^+NH_3}{|}}{C}}}{-}OCH_3 \xrightarrow{-H^+} R{-}\overset{\displaystyle O^-}{\overset{|}{\underset{\underset{\displaystyle NH_2}{|}}{C}}}{-}OCH_3$$

$$CH_3OH \xleftarrow{H_2O} {}^-OCH_3 + \underset{\text{Amide}}{R{-}\overset{\displaystyle O}{\overset{\|}{C}}{-}NH_2}$$

The reaction of esters with amines provides a useful way to synthesize amides, (see Section 15.4.3).

PROBLEM 15.17

Show the products of the following reactions:

(a) phenyl benzoate with ammonia

(b) cyclohexyl acetate with dimethyl amine

(c) ethyl acetate with 2,4-dibromoaniline

15.3.3 Claisen Condensation of Esters

The Claisen condensation is the reaction by which your body produces fats and steroids. The steroid testosterone, in particular contributes to muscle development.

So far, the reactions we have considered for carboxylic acid derivatives have all occurred at the carbonyl group. However, recall the chemistry of aldehydes and ketones—in which their carbonyl group activates the α-hydrogens such that an enolate anion can be formed readily (see Section 13.2.2). In a similar manner, the carbonyl group of an ester has sufficient electron deficiency that it can stabilize an enolate anion.

When an ester is treated with a strong base, a proton is removed from the α-carbon to form a resonance-stabilized enolate anion of the ester. The base used in the reaction is the alkoxide anion corresponding to the alcohol group of the ester.

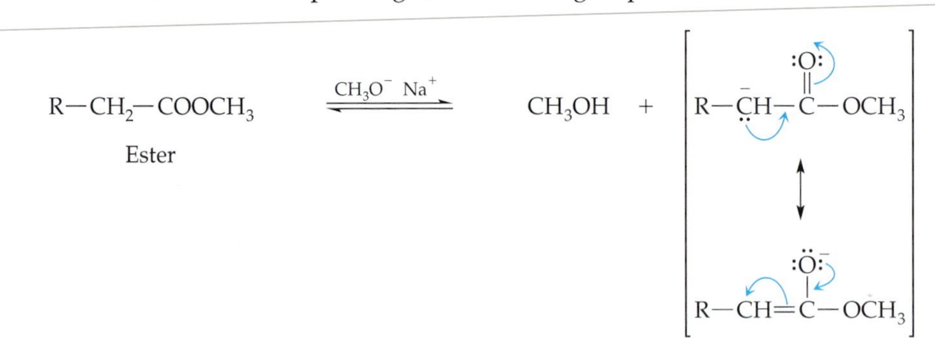

The **Claisen condensation** is a reaction that involves initial formation of an enolate anion from an ester. This anion then serves as a nucleophile in a nucleophilic acyl substitution reaction with another ester molecule. The enolate anion displaces that ester's alkoxy group. The overall effect is to join the α-carbon of one ester molecule to the carbonyl carbon of another, forming a new carbon-carbon bond. Thus, the product is always a β-keto ester.

General Claisen condensation:

$$2\text{R—CH}_2\text{—COOCH}_3 \xrightarrow{\text{CH}_3\text{ONa}} \text{R—CH}_2\text{—CO—CHR—COOCH}_3$$

Ester $\qquad\qquad$ β-Keto ester

For example, methyl 3-oxobutanoate is produced from methyl ethanoate:

$$\text{CH}_3\text{COOCH}_3 \xrightarrow{\text{CH}_3\text{ONa}} \text{CH}_3\text{COCH}_2\text{COOCH}_3$$

Methyl ethanoate (methyl acetate) $\qquad$ Methyl 3-oxobutanoate (methyl acetoacetate)

The mechanism of the Claisen condensation is no different from the other nucleophilic acyl substitutions we have considered. In this instance, an enolate anion has displaced an alkoxy leaving group, just as ammonia displaces an alkoxy leaving group to form an amide (see Section 15.3.2). The steps of the Claisen mechanism are outlined in Figure 15.3.

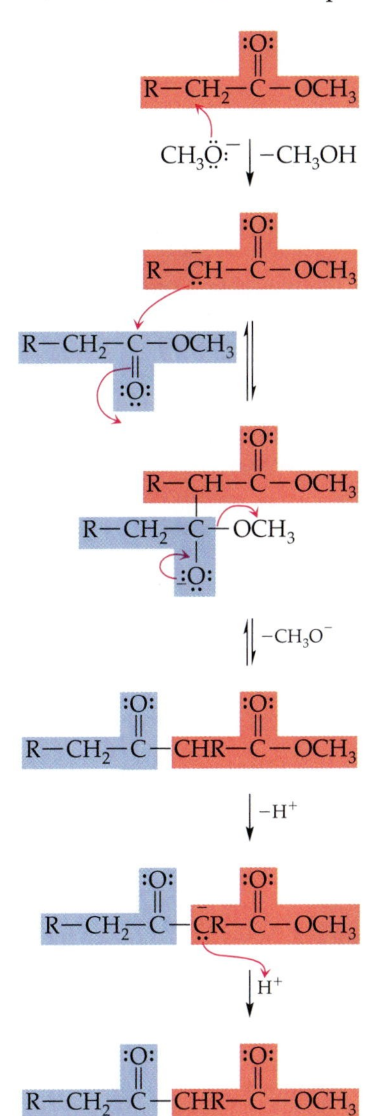

Methoxide anion acts as a base to remove an α-proton to form the stabilized enolate anion.

The nucleophilic enolate anion attacks the carbonyl group of a second ester molecule.

The adduct ejects a methoxide anion as the leaving group to form the β-keto ester.

The β-keto ester loses a proton to the methoxide anion. The resulting anion is a "double" enolate anion, stabilized by delocalization through both adjacent carbonyl groups.

Acidification regenerates the β-keto ester from the "double" enolate anion.

FIGURE 15.3

Mechanism of the Claisen condensation.

The key to the success of the Claisen condensation is that the product, a β-keto ester, has an especially acidic proton, which is removed to form a highly stable "double" enolate anion, stabilized by two adjacent carbonyl groups. It is the formation of this enolate anion that drives what is otherwise an unfavorable equilibrium. Therefore, only esters that have *two* α-hydrogens undergo the Claisen condensation. For example, $(CH_3)_2CHCOOCH_3$ will not undergo a Claisen condensation. The key to recognizing synthesis problems for which the Claisen condensation is potentially useful is to spot the β-carbonyl ester unit (—CO—C—$COOCH_3$).

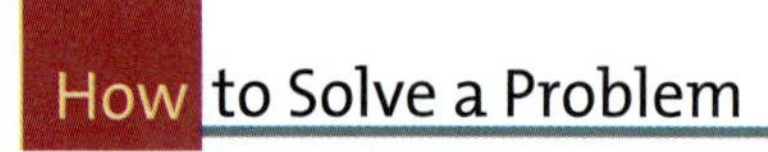

How to Solve a Problem

Outline a synthesis of 2-methyl-3-oxopentanoic acid from propanoic acid.

$$CH_3CH_2COOH \dashrightarrow CH_3CH_2COCH(CH_3)COOH$$

Propanoic acid → 2-Methyl-3-oxopentanoic acid

PROBLEM ANALYSIS

First, we inspect the product to determine if the reactant's structure seems to have been incorporated. In this instance, it appears that two molecules of the reactant's carbon skeleton have been connected—with the α-carbon of one unit bonded to the carbonyl carbon of the second unit. In other words, the —OH group has been displaced from one molecule of propanoic acid. Also, we recognize that the product is a β-keto acid, a structural unit obtainable via a Claisen condensation. We know, however, that the Claisen condensation occurs with esters, not with carboxylic acids.

Retrosynthetic analysis indicates that we could obtain an ester of the desired product using a Claisen condensation, and such an ester could be hydrolyzed to its carboxylic acid. Finally, to bring about a Claisen condensation, the propanoic acid must be converted to its ester.

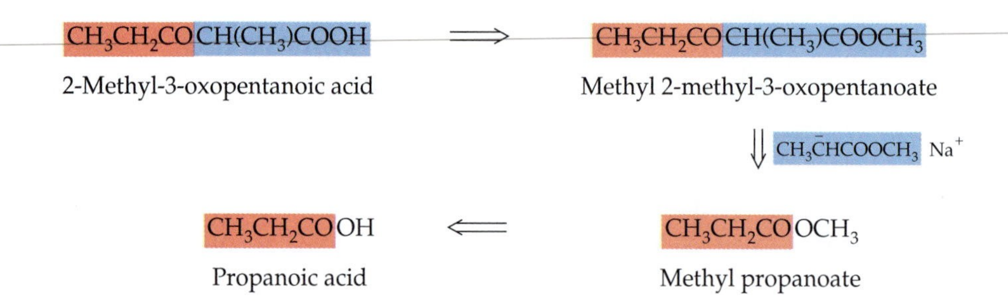

SOLUTION

We convert propanoic acid to its methyl ester by first forming the acid chloride (using thionyl chloride) and reacting it with methanol. Methyl propanoate is then treated with one equivalent of sodium methoxide (prepared from methanol and sodium metal) to bring about the Claisen condensation. Work-up with hydrochloric acid produces methyl 2-methyl-3-oxopentanoate, the β-keto ester. Hydrolysis of the β-keto ester with sodium hydroxide, followed by acidification, produces the desired product, 2-methyl-3-oxopentanoic acid.

$$CH_3CH_2COOH \xrightarrow{SOCl_2} CH_3CH_2COCl \xrightarrow{CH_3OH} CH_3CH_2COOCH_3$$

Propanoic acid → Methyl propanoate

$$CH_3CH_2COOCH_3 \xrightarrow{NaOCH_3} [CH_3\bar{C}HCOOCH_3\ Na^+] \xrightarrow{CH_3CH_2COOCH_3} [CH_3CH_2CO\bar{C}(CH_3)COOCH_3]$$

Methyl propanoate

$$[CH_3CH_2CO\bar{C}(CH_3)COOCH_3] \xrightarrow{HCl} CH_3CH_2COCH(CH_3)COOCH_3$$

Methyl 2-methyl-3-oxopentanoate

$$\xrightarrow[2.\ HCl]{1.\ NaOH} CH_3CH_2COCH(CH_3)COOH$$

2-Methyl-3-oxopentanoic acid

PROBLEM 15.18

Show the steps involved in the reaction of methyl phenylethanoate with sodium methoxide.

PROBLEM 15.19

Specify the starting materials needed to produce methyl 2-ethylhexanoate via a Claisen condensation.

15.3.4 Preparation of Esters

Esters are usually prepared by one of two methods. The first, the acylation of an alcohol (already described for acid chlorides and acid anhydrides), is easily carried out, especially on a small scale. When larger amounts of an ester are needed, the second method, the acid-catalyzed Fischer esterification, is used.

Alcoholysis of Acid Chlorides (Acylation of Alcohols)

As described in Sections 15.1.2 and 15.2.2, alcohols can be acylated by reaction with acid chlorides or acid anhydrides, forming esters in alcoholysis reactions. These reactions are successful with any carboxylic acid chloride and any alcohol, including phenols.

General acylation reaction:

$$R'OH \xrightarrow[\text{Acid chloride}]{RCOCl} \underset{\text{Ester}}{RCOOR'} \xleftarrow[\text{Acid anhydride}]{(RCO)_2O} R'OH$$

The two reactions are virtually identical, involving the usual nucleophilic acyl substitution mechanism initiated via attack by the nucleophilic alcohol on the carbonyl carbon. Usually, acetic anhydride is used to acetylate an alcohol, and 3,5-dinitrobenzoyl chloride is used to "derivatize" an alcohol:

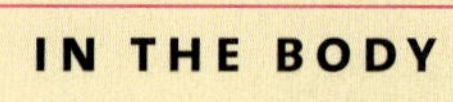

CHEMISTRY AT WORK

IN THE BODY

The Claisen Condensation and Biosynthesis

The Claisen condensation is a critical reaction in human metabolism. It is the reaction by which your body produces needed fats and steroids. As we will see in Section 19.2, the food you ingest is metabolized into smaller molecules, which are eventually further metabolized by oxidation to carbon dioxide and water. However, some of the simple compounds that result from metabolism are utilized by the body in the synthesis of needed chemicals, especially fats and proteins (a process called *biosynthesis* or *anabolism*).

One common product of metabolism is acetyl coenzyme A (CH_3—CO—S—CoA; see "Chemistry at Work," p. 536). This is an acetate ester of a thiol (rather than an alcohol), HS—CoA, and it behaves chemically like any other ester. It carries out two Claisen condensations resulting in the transformation of a two-carbon unit to a four-carbon unit, a β-keto ester (acetoacetyl-CoA).

The ketone group of the β-keto ester resulting from the Claisen condensation is reduced to a methylene group using two steps, resulting in a C_4 saturated ester, butyryl-CoA. Reaction of this C_4 ester with another acetyl-CoA enolate anion produces a C_6 unit via a Claisen condensation. This process continues, resulting in the addition of C_2 units (as enolate anions) until the needed fatty acid ester—for example, lauryl-CoA—is obtained. The fatty acid CoA ester then reacts with glycerol (propane-1,2,3-triol) to produce body fat (see Section 15.3.5). The repeated addition of C_2 units is why all naturally occurring fatty acids have an even number of carbons (see Section 14.1.2).

Acetoacetyl-CoA also serves as the starting point for the biosynthesis of 3-methyl-2-butenyl or 3-methyl-3-butenyl diphosphate, which has the carbon skeleton of isoprene [CH_2=C(CH_3)CH=CH_2; see Section 8.3.2]. This compound is a key intermediate in the biosynthesis of terpenes, cholesterol, and steroid hormones.

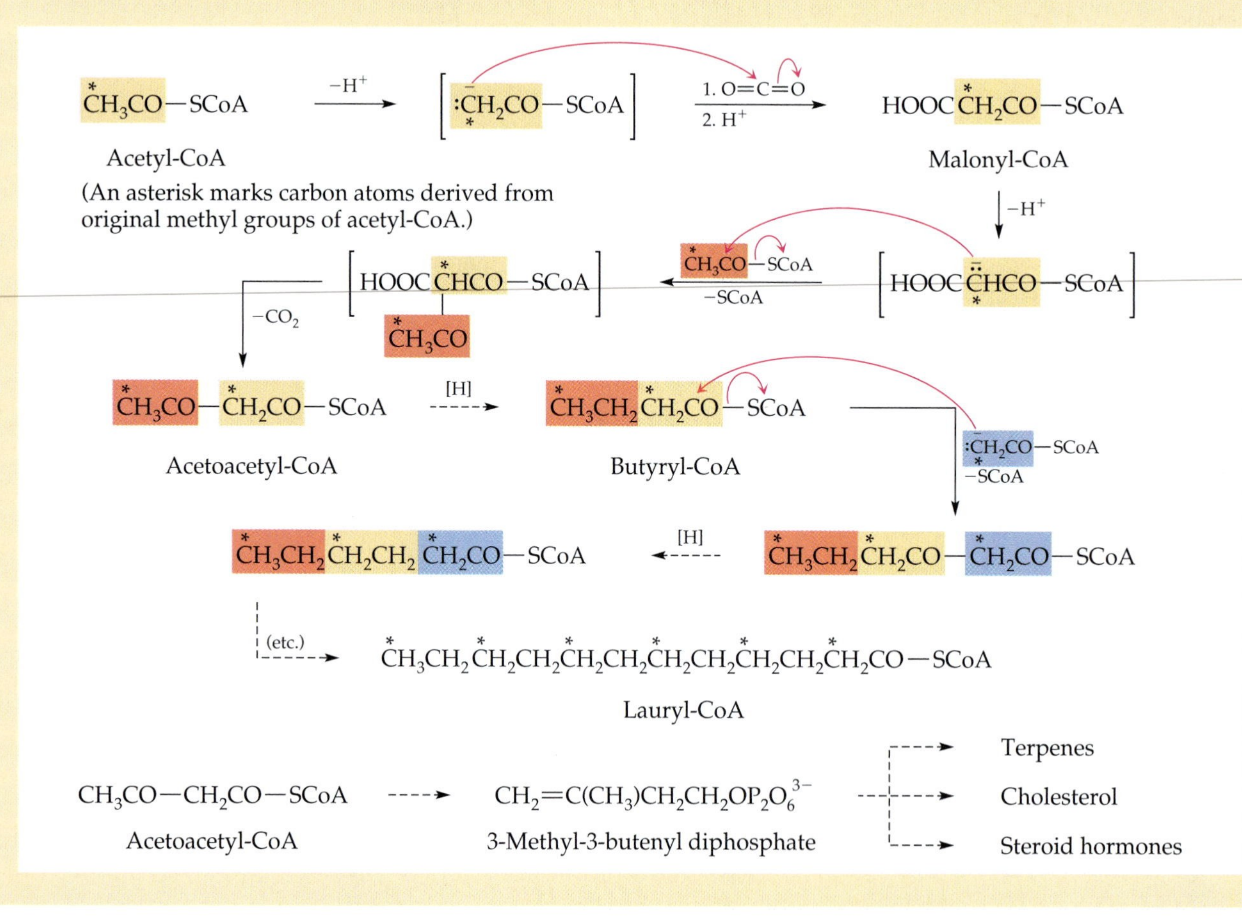

Cyclobutanol $\xrightarrow{(CH_3CO)_2O}$ Cyclobutyl acetate (cyclobutyl–$OOCCH_3$)

Cyclobutanol $\xrightarrow[\text{pyridine}]{\text{Cl–C(=O)–}C_6H_3(NO_2)_2}$ Cyclobutyl 3,5-dinitrobenzoate (cyclobutyl–OOC–$C_6H_3(NO_2)_2$)

Converting a carboxylic acid to its ester involves first converting the acid to its acid chloride with thionyl chloride and then adding an alcohol, usually methanol (in the presence of pyridine). For example, 1-naphthoic acid is converted to methyl 1-naphthoate in this manner:

1-Naphthoic acid ($C_{10}H_7COOH$) $\xrightarrow{SOCl_2}$ $C_{10}H_7COCl$ $\xrightarrow[\text{pyridine}]{CH_3OH}$ Methyl 1-naphthoate ($C_{10}H_7COOCH_3$)

How to Solve a Problem

Show the reactions and reagents necessary to convert 1-butanol into butyl butanoate.

PROBLEM ANALYSIS

Recognizing that an ester is formed from an acid and an alcohol, we can envision the product as containing two C_4 units joined through an ester linkage. Retrosynthetic analysis tells us that we can obtain the final product from butanoic acid and 1-butanol. We have 1-butanol available as a starting material, so the remaining problem is how to convert 1-butanol into butanoic acid. Since the number of carbons is identical in the two compounds, all that is required is a conversion of the primary alcohol into its corresponding carboxylic acid, a typical oxidation reaction. Therefore, the retrosynthetic scheme can be summarized as follows:

$$\underset{\text{Butyl butanoate}}{CH_3CH_2CH_2COOCH_2CH_2CH_2CH_3} \Longrightarrow \underset{\text{Butanoic acid}}{CH_3CH_2CH_2COOH} \Longrightarrow \underset{\text{1-Butanol}}{CH_3CH_2CH_2CH_2OH}$$

SOLUTION

1-Butanol is oxidized with sodium dichromate and sulfuric acid to butanoic acid. Conversion of this acid to its acid chloride with thionyl chloride, followed by reaction with 1-butanol, produces the desired product, butyl butanoate.

$$\underset{\text{1-Butanol}}{CH_3CH_2CH_2CH_2OH} \xrightarrow[H_2SO_4]{Na_2Cr_2O_7} \underset{\text{Butanoic acid}}{CH_3CH_2CH_2COOH} \xrightarrow{SOCl_2} \underset{\text{Butanoyl chloride}}{CH_3CH_2CH_2COCl}$$

$$\underset{\text{Butanoyl chloride}}{CH_3CH_2CH_2COCl} + \underset{\text{1-Butanol}}{CH_3CH_2CH_2CH_2OH} \xrightarrow{\text{pyridine}} \underset{\text{Butyl butanoate}}{CH_3CH_2CH_2COOCH_2CH_2CH_2CH_3}$$

PROBLEM 15.20

Show the product of each of the following reactions:

(a) propanol and acetic anhydride
(b) ethanol and succinic anhydride
(c) phenol and propanoyl chloride and pyridine
(d) cyclohexanol and 3,5-dinitrobenzoyl chloride and pyridine

PROBLEM 15.21

Show how to prepare each of the following esters from a carboxylic acid:

(a) isopropyl 4-oxo-2-pentenoate
(b) phenyl benzoate
(c) cyclopropyl cyclopropanecarboxylate

Fischer Esterification

Conversion of a carboxylic acid into an ester is also accomplished by heating the acid with a readily available alcohol, such as methanol or ethanol, in a process called the **Fischer esterification.** The alcohol is used in excess as the solvent. The process requires acid catalysis, usually supplied by adding a few drops of concentrated sulfuric acid. The reaction has an unfavorable equilibrium and so must be forced to the right. This can be accomplished by using excess alcohol (as the solvent) and/or by removing water as it is formed.

Fischer esterification:

$$\underset{\text{Carboxylic acid}}{RCOOH} + \underset{\text{Alcohol}}{R'OH} \overset{H^+}{\rightleftharpoons} \underset{\text{Ester}}{RCOOR'} + H_2O$$

Use of labeled oxygen atoms in the acid or alcohol has shown that the oxygen atom appearing in the water molecule ejected during the esterification comes from the acid, not from the alcohol.

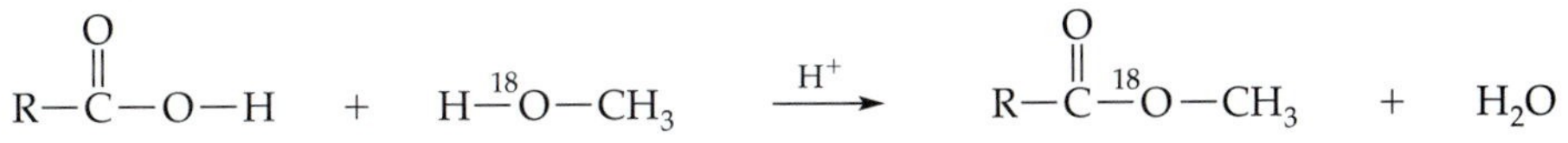

$$\underset{\text{Carboxylic acid}}{R-\overset{O}{\overset{\|}{C}}-O-H} + \underset{^{18}\text{O-Methanol}}{H-{}^{18}O-CH_3} \xrightarrow{H^+} \underset{\text{Methyl ester}}{R-\overset{O}{\overset{\|}{C}}-{}^{18}O-CH_3} + H_2O$$

The mechanism of the Fischer esterification involves attack by the nucleophile (alcohol) at the carbonyl carbon in a familiar nucleophilic acyl substitution. The only difference from other reactions we've considered that follow this mechanism is the acid catalysis. The carbonyl carbon of a carboxylic acid is not electron-deficient enough for a nucleophilic attack to occur. In the Fischer esterification, the acid catalyst protonates the carbonyl group, thereby making the carbonyl carbon sufficiently electron-deficient for the nucleophilic alcohol to attack. The mechanism is shown in Figure 15.4 using methanol.

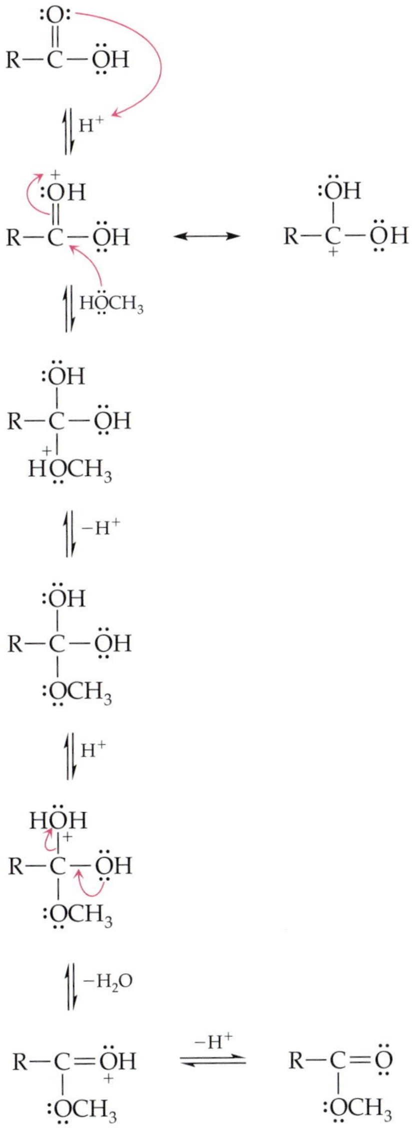

$$R-\overset{:O:}{\overset{\|}{C}}-\ddot{O}H \underset{}{\overset{H^+}{\rightleftharpoons}} R-\overset{:\overset{+}{O}H}{\overset{\|}{C}}-\ddot{O}H \longleftrightarrow R-\underset{+}{\overset{:\ddot{O}H}{\overset{|}{C}}}-\ddot{O}H$$

The carboxylic acid is protonated, and the resonance-stabilized cation has an electron-deficient carbon.

$$\overset{HOCH_3}{\rightleftharpoons} R-\underset{\underset{H\overset{+}{O}CH_3}{|}}{\overset{:\ddot{O}H}{\overset{|}{C}}}-\ddot{O}H$$

The carbonyl carbon is attacked by the nucleophilic alcohol. This is the addition step.

$$\overset{-H^+}{\rightleftharpoons} R-\underset{\underset{:\ddot{O}CH_3}{|}}{\overset{:\ddot{O}H}{\overset{|}{C}}}-\ddot{O}H$$

Loss of a proton from the oxonium cation produces a tetrahedral intermediate.

$$\overset{H^+}{\rightleftharpoons} R-\underset{\underset{:\ddot{O}CH_3}{|}}{\overset{H\ddot{O}H}{\overset{|+}{C}}}-\ddot{O}H$$

One of the hydroxyl groups is protonated to form an oxonium cation.

$$\overset{-H_2O}{\rightleftharpoons} R-\underset{\underset{:\ddot{O}CH_3}{|}}{C}=\overset{+}{\ddot{O}}H \overset{-H^+}{\rightleftharpoons} R-\underset{\underset{:\ddot{O}CH_3}{|}}{C}=\ddot{O}$$

The oxonium cation loses water. This is the ejection step.

A final deprotonation produces the ester.

FIGURE 15.4
Mechanism of the Fischer esterification.

The Fischer esterification can be accomplished using any carboxylic acid, but it is limited to primary or secondary alcohols because tertiary alcohols are prone to elimination (dehydration) under these conditions (see Section 6.6.1). Phenyl esters also cannot be prepared by this reaction because the phenolic hydroxyl group is insufficiently nucleophilic. Instead, phenyl esters must be prepared by the acylation of phenols with acid anhydrides or acid chlorides.

PROBLEM 15.22

Show the product of each of the following acid-catalyzed esterification reactions:

(a) ethanol and benzoic acid

(b) (*R*)-2-butanol and acetic acid

PROBLEM 15.23

Show the reagent(s) and conditions necessary to accomplish each conversion:

(a) isopropyl alcohol to isopropyl acetate

(b) phenol to phenyl benzoate

(c) cyclohexanol to cyclohexyl propanoate

(d) succinic anhydride to monophenyl succinate

15.3.5 Fats

Naturally occurring fats are esters formed from long-chain carboxylic acids, called **fatty acids,** and the triol known as glycerol (1,2,3-propanetriol). Fats are also called **triglycerides** and may be described as triacyl esters of glycerol. The three acyl groups in a triglyceride may be identical or different. Further, a particular fat (for example, olive oil or lard) may be a complex mixture of triglycerides.

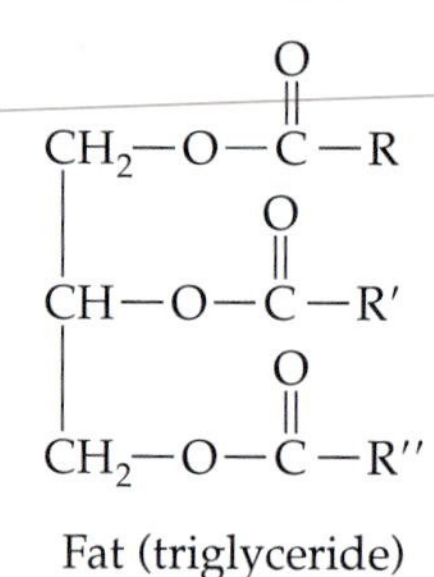

Fat (triglyceride)

Each fatty acid chain contains an even number of carbon atoms, owing to the mechanism by which plants and animals synthesize fats (see "Chemistry at Work," p. 554). Here are the historical and IUPAC names and structures of the most common saturated fatty acids:

C_{12}	lauric or dodecanoic acid	$CH_3(CH_2)_{10}COOH$
C_{14}	myristic or tetradecanoic acid	$CH_3(CH_2)_{12}COOH$
C_{16}	palmitic or hexadecanoic acid	$CH_3(CH_2)_{14}COOH$
C_{18}	stearic or octadecanoic acid	$CH_3(CH_2)_{16}COOH$

Beef tallow is a saturated fat. Canola oil, from the seeds of the canola plant, is an unsaturated fat.

There are also a number of unsaturated fatty acids. All of the double bonds are *cis* (Z), and there may be one or more double bonds present. Here are the historical and IUPAC names and structures of a few common unsaturated fatty acids:

oleic or 9-octadecenoic acid	$CH_3(CH_2)_7CH{=}CH(CH_2)_7COOH$
linoleic or 9,12-octadecadienoic acid	$CH_3(CH_2)_4CH{=}CH{-}CH_2{-}CH{=}CH(CH_2)_7COOH$
linolenic or 9,12,15-octadecatrienoic acid	$CH_3CH_2CH{=}CH{-}CH_2{-}CH{=}CH{-}CH_2{-}CH{=}CH(CH_2)_7COOH$

Animal fats are generally solid substances and have predominantly acyl groups from saturated fatty acids (that is, the hydrocarbon chains, R, have no double bonds). Lard and beef tallow are typical examples. The substances commonly called vegetable oils are actually fats as well. Vegetable fats differ from animal fats in having predominantly acyl groups from unsaturated fatty acids. (See Figure 15.5.) Vegetable oils are liquids at room temperature for two reasons: (1) Unsaturated compounds are normally lower-melting than saturated compounds (see Chapters 2 and 6), and (2) the *cis* geometry of the double bond means that the oil molecule has an overall shape that is irregular, which precludes compact crystal packing.

Vegetable oils such as canola oil, corn oil, soybean oil, olive oil, and palm oil have generally replaced animals fats in cooking largely because they are unsaturated. Medical recommendations stress the need to lower the amount of saturated fat in the diet. Many foods contain "partially hydrogenated" vegetable oils, oils that are partially saturated. Margarine is one such product. It is made by hydrogenating (saturating) *some* of the double bonds in a vegetable oil, just enough so that the saturated fat content is

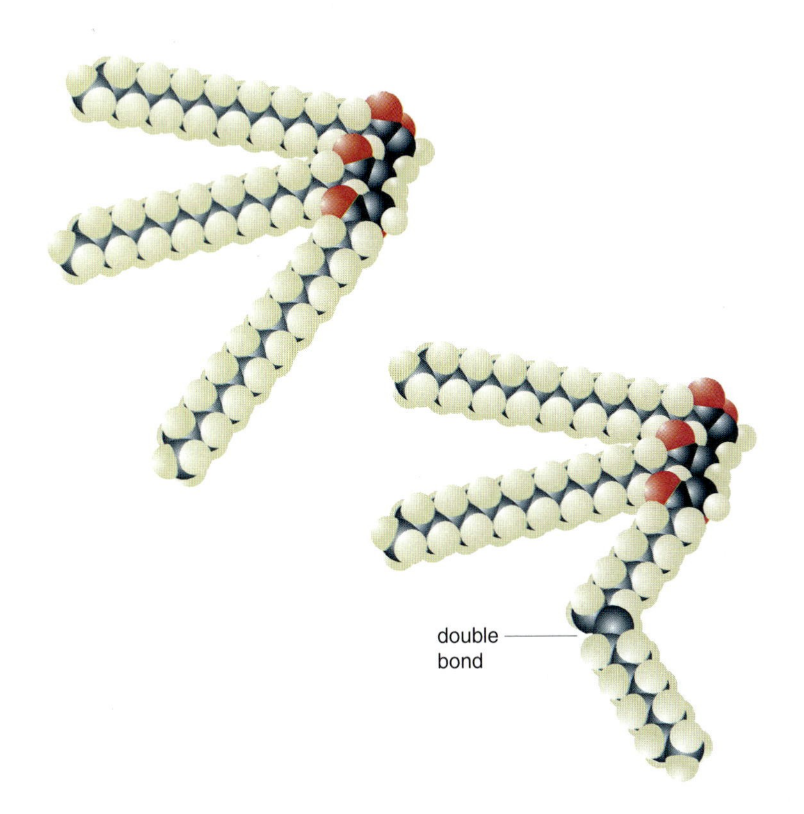

FIGURE 15.5

Space-filling models of saturated triglyceride (left) and unsaturated triglyceride (right).

CHEMISTRY AT WORK

IN THE MARKETPLACE

Olestra, the Fat Substitute

Americans have been diet-conscious for decades, but have also had a taste for fried foods and snacks—an obvious conflict! Most of the calories found in snack foods come from the fats used in the cooking process. Therefore, any product that could supply the appealing taste provided by fats but that also had a significantly lower caloric content might have "success" written all over it.

In 1996, Procter & Gamble received FDA approval for the introduction of a noncaloric edible fat called **Olestra** to be used in the preparation of "savory snack foods." Olestra is a fat in that it is an ester of long-chain fatty acids. However, in normal fats, the alcohol component of the ester is glycerol (1,2,3-propanetriol); in Olestra, the alcohol component is a common carbohydrate, sucrose (ordinary table sugar). Specifically, Olestra is mainly the octa-ester of sucrose, a synthetic material made from two naturally occurring components (sugar and fatty acids). There are other glycolipids (esters of fatty acids and sugars) in the human biological system, but none quite like Olestra.

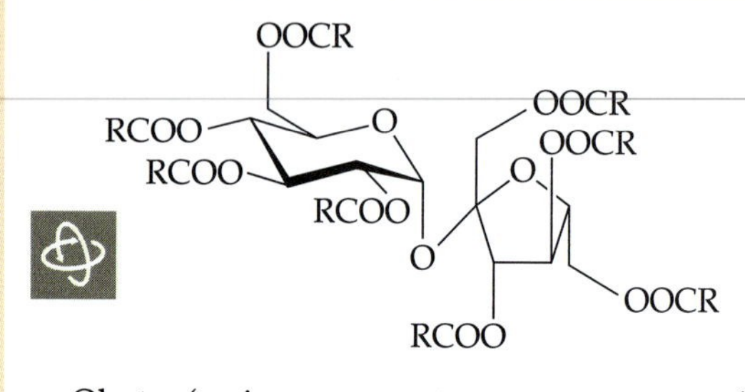

Olestra (main component, a sucrose octa-ester)

The key to the promise of Olestra is twofold. First, it is a fat-like compound that provides the rich taste characteristic of naturally occurring animal and vegetable fats, especially when fried. Second, Olestra is not metabolized in the body. Instead, it passes directly through the digestive system unhydrolyzed. In contrast, regular fats are hydrolyzed readily to glycerol and fatty acids. Thus, Olestra provides the desired taste but not the calories.

What prevents Olestra from being digested, its creators claim, is its size and shape, which prevents the usual digestive enzymes from carrying out hydrolysis. In early trials, Olestra produced some unpleasant side effects, but not sufficient to withhold the product from limited marketing and the opportunity to gain more information about it. It is now being marketed nationally under the trade name Olean in snack foods such as potato chips. The fatty acids for its synthesis are obtained by hydrolysis of naturally occurring fats, such as those found in soybean oil.

sufficient to raise the melting point and make the product semisolid. Thus, it can hold its shape, as in margarine "sticks" (See "Chemistry at Work," p. 192).

$$\text{Polyunsaturated fat} \xrightarrow[\text{catalyst}]{H_2} \text{Partially unsaturated fat (partially hydrogenated fat)}$$

Fats were at one time used to make soaps through the process known as *saponification* (soap making). As described in "Chemistry at Work," p. 503, and Section 15.3.2, soaps are surfactant agents, historically used for cleaning. While a few cleaning agents still use soaps as the surfactant, most cleaning agents now rely on detergents as more effective surfactants. Detergents are salts of aromatic and/or long-chain sulfonic acids, as described in "Chemistry at Work," p. 285.

$$\underset{\text{Soap (sodium carboxylate)}}{R{-}COO^{-}\ Na^{+}} \qquad \underset{\text{Detergent (sodium sulfonate)}}{R{-}SO_3^{-}\ Na^{+}}$$

PROBLEM 15.24

(a) Draw the structure of glyceryl tridecanoate.

(b) Name the following fat:

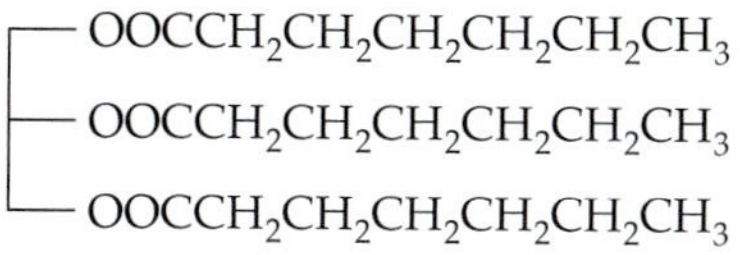

$$\begin{array}{l} -OOCCH_2CH_2CH_2CH_2CH_2CH_3 \\ -OOCCH_2CH_2CH_2CH_2CH_2CH_3 \\ -OOCCH_2CH_2CH_2CH_2CH_2CH_3 \end{array}$$

(c) Draw and name the products obtained from the saponification of glyceryl tridecanoate.

15.4 Amides

Amides are organic compounds in which an acyl group is attached to a nitrogen atom (R—CO—NR′R″). The nitrogen atom has two other sigma bonds, which may join it to hydrogens and/or alkyl or aryl groups. An amide may be considered the product of an intermolecular dehydration reaction between ammonia or a primary or a secondary amine and a carboxylic acid. However, amides are seldom actually prepared by such a reaction.

$$\underset{\text{Carboxylic acid}}{R{-}\overset{O}{\overset{\|}{C}}{-}OH} + \underset{\text{Amine}}{H{-}NR'R''} \xrightarrow{-H_2O} \underset{\text{Amide}}{RCONR'R''}$$

It is important to note that amides are neutral compounds. In other words, the basicity associated with amines (see Section 12.3.2)—which is due to the availability of unshared electrons on nitrogen—vanishes when an amine is converted into an amide. This change is due to the strong electron-withdrawing character of the acyl group of the amide (recall the electron deficiency of the carbonyl group; see Section 13.2.1). Two resonance contributing structures can be drawn for an amide, indicating the delocalization of the nitrogen electrons toward the carbonyl group:

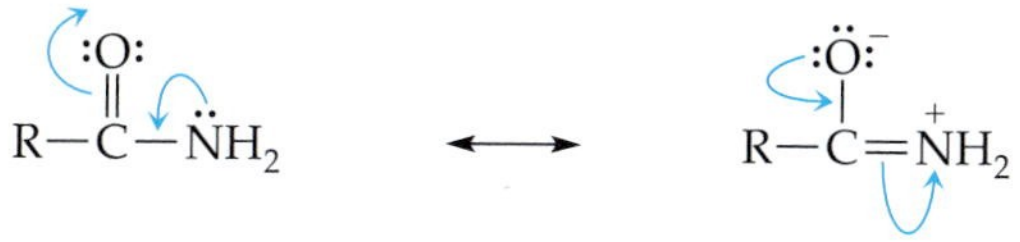

$$R{-}\overset{:O:}{\overset{\|}{C}}{-}\ddot{N}H_2 \longleftrightarrow R{-}\overset{:\ddot{O}:^{-}}{\overset{|}{C}}{=}\overset{+}{N}H_2$$

Resonance contributing structures for an amide

The result is that unlike amines, which dissolve in dilute aqueous acid as a result of protonation and formation of a soluble ammonium salt, amides are generally insoluble in water and are not protonated by dilute acid.

As the dipolar structure of one resonance contributor implies, amides are polar compounds. They can form strong hydrogen bonds with other compounds and can form dimers by intermolecular hydrogen bonding if there is an N—H bond present. The hydrogen-bonding ability of amides is very important in protein chemistry and in nucleic acid chemistry, as we will see in Chapters 16 and 19.

Amides show characteristic absorptions in the IR region (Figure 15.6a). The carbonyl peak is near 1650 cm^{-1}, a lower frequency than for other carbonyl groups because of the considerable single bond character. The key to the presence of an amide that contains an NH group is that the IR spectrum also shows N—H stretching absorption near 3300 cm^{-1}; an unsubstituted amide shows two such peaks, and a monosubstituted amide shows one. In the ^{1}H-NMR spectrum, the hydrogens alpha to an amide carbonyl group are deshielded and appear near 2.0 ppm (Figure 15.6b). The hydrogens on nitrogen, however, appear over a very wide range, from about 1 to 5 ppm, and so are not readily assignable based on only this type of spectrum; further spectral analysis may be necessary.

Amides occur naturally in many kinds of compounds, some of which will be illustrated in this chapter. However, the most important amides are proteins, those biomolecules that are so essential for life. The amide chemistry exhibited by proteins will be described in Chapter 16.

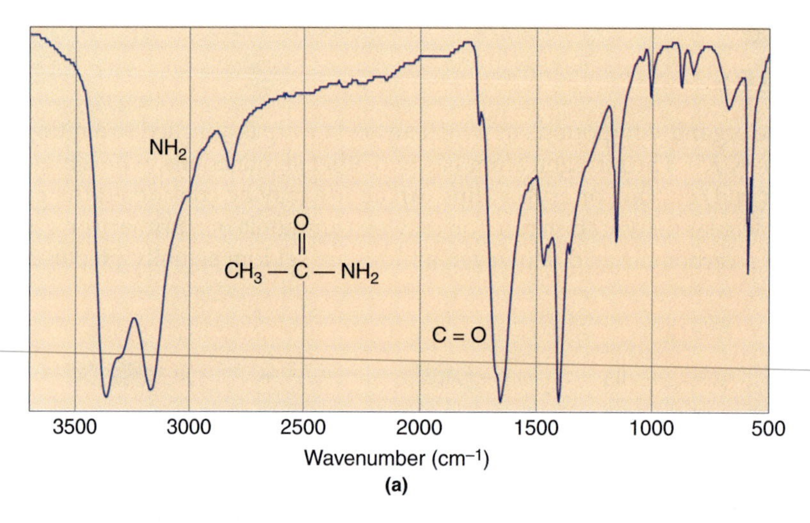

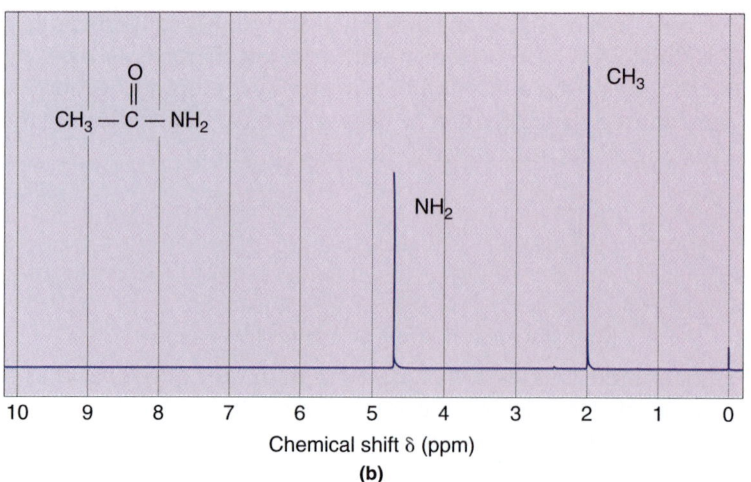

FIGURE 15.6

(a) IR and (b) ^{1}H-NMR spectra of acetamide.

15.4.1 Nomenclature of Amides

Amide names are based on those of the carboxylic acids from which the amides are derived. The suffix *-oic* of IUPAC names or the suffix *-ic* of historical names is replaced by the suffix *-amide*. Examples are benzamide (from benzoic acid), acetamide or ethanamide (from acetic acid or ethanoic acid), and 2-butenamide (from 2-butenoic acid).

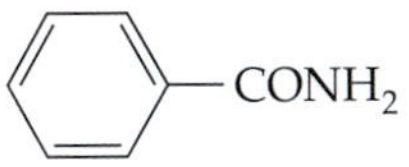

C_6H_5—$CONH_2$ Benzamide | CH_3CONH_2 Acetamide (ethanamide) | $CH_3CH{=}CHCONH_2$ 2-Butenamide

Amides are divided into three groups, depending on the substitution on nitrogen. Amides may be derived from ammonia, from primary amines (one nonhydrogen substituent on nitrogen), and from secondary amines (two substituents on nitrogen). The naming of substituted amides involves designating each nitrogen substituent with the prefix *N*-. Examples are *N*-phenylacetamide (best known by the historical name acetanilide), *N,N*-dimethylcyclopropanecarboxamide, and *N*-methyl-*N*-phenylpropanamide.

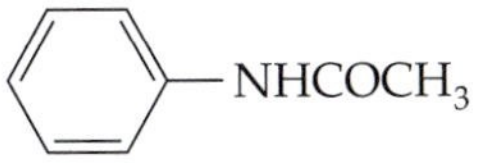

C_6H_5—$NHCOCH_3$ *N*-Phenylacetamide (acetanilide) | cyclopropyl—$CON(CH_3)_2$ *N,N*-Dimethylcyclopropanecarboxamide | $CH_3CH_2CON(CH_3)C_6H_5$ *N*-Methyl-*N*-phenylpropanamide

PROBLEM 15.25

Draw the structures of the following compounds: (a) hexanamide, (b) *N*-(*p*-nitrophenyl)acetamide, (c) *N*-methyl-3-hydroxybutanamide, (d) *N,N*-diphenylcyclohexanecarboxamide, and (e) *N*-benzylpentanamide.

PROBLEM 15.26

Assign names to the following compounds:

(a) $C_6H_5CONHC_6H_5$ (b) 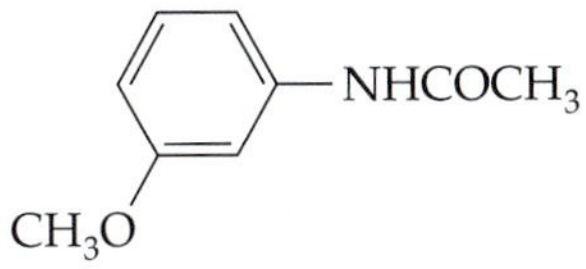 benzene ring bearing $NHCOCH_3$ with CH_3O meta (c) $CH_3CH_2CON(CH_3)_2$

15.4.2 Reactions of Amides

Hydrolysis of Amides

The amide bond can be hydrolyzed (cleaved with the addition of water) to yield either ammonia or an amine plus a carboxylic acid, as suggested by the Key to Transformations.

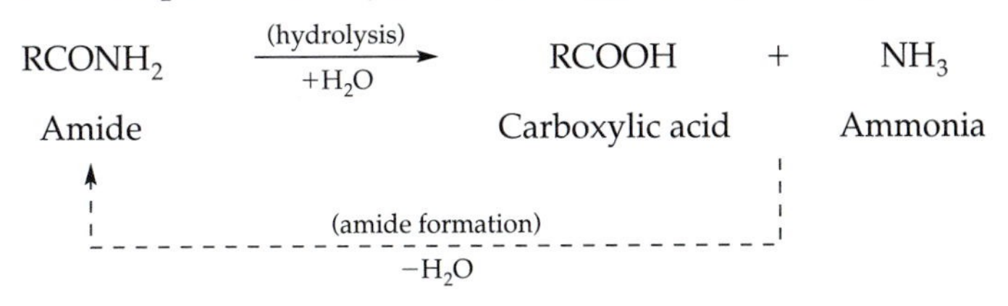

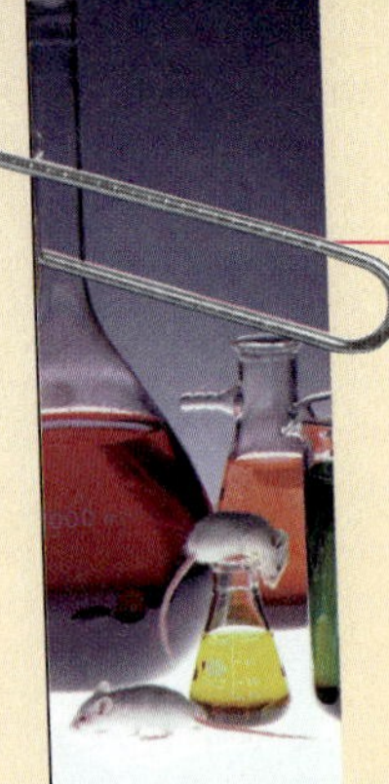

CHEMISTRY AT WORK

IN THE MARKETPLACE

Urea

Urea, the major component of urine, is the double amide of carbonic acid. It is an end product of protein metabolism, and its excretion serves to remove excess nitrogen from the body (humans excrete about 30 g of urea per day).

$$H_2N{-}\overset{\overset{\large O}{\|}}{C}{-}NH_2 \qquad HO{-}\overset{\overset{\large O}{\|}}{C}{-}OH \qquad Cl{-}\overset{\overset{\large O}{\|}}{C}{-}Cl$$

Urea — Carbonic acid — Phosgene

Urea can be prepared in the laboratory by reacting phosgene (the double acid chloride of carbonic acid) with ammonia in an ammonolysis reaction. However, urea is made commercially—and that means in billions of pounds per year—by heating carbon dioxide with ammonia.

Ammonia is an active growth agent consumed by plants as a source of nitrogen. Urea is the primary component in many fertilizers for two reasons. First, it is efficient because it is 47% nitrogen by weight. Second, urea is a "slow-release" fertilizer, releasing ammonia into the soil slowly as the urea gradually hydrolyzes and reverts to ammonia and carbon dioxide:

$$\underset{\text{Urea}}{H_2N{-}\overset{\overset{\large O}{\|}}{C}{-}NH_2} \xrightarrow{H_2O} 2\,NH_3 + CO_2$$

Farmers spread millions of tons of urea-based fertilizers to enhance crop growth.

In "rapid-release" fertilizers, such as ammonium nitrate and liquid ammonia, the nitrogen is immediately available to plants. This produces an immediate growth "spurt," but has no long-term fertilizing effect.

Urea is also important in the polymer resin industry. Reaction of urea with formaldehyde produces urea-formaldehyde resins through addition of the —NH group of urea across the carbonyl group of formaldehyde, followed by dehydration to form a three-dimensional network.

Adding small amounts of such resins to cotton fabrics produces the crease resistance so heavily advertised. Particle board used in home construction is made by heating wood chips and sawdust with urea-formaldehyde resins. The formaldehyde present in some homes appears to arise from slow decomposition of these resins.

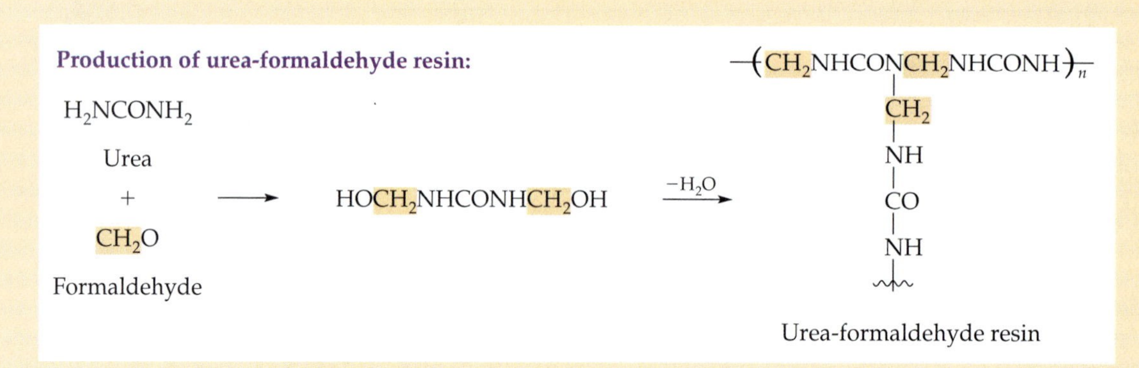

Hydrolysis is the reversal of the formal (not practical) amide-forming process and requires rather strenuous conditions. Hydrolysis can be accomplished using a basic solution (aqueous sodium hydroxide), but it can also be accomplished, and generally more readily, using acid catalysis.

General hydrolysis reaction:

$$\underset{\text{Amide}}{RCONH_2} \xrightarrow[H_2O]{H^+} \underset{\text{Carboxylic acid}}{RCOOH} + NH_4^+$$

Hydrolysis of an amide proceeds by the standard nucleophilic acyl substitution mechanism. Acid catalysis makes hydrolysis occur more readily because protonation of the carbonyl group facilitates attack by the nucleophile—in this case, water. In the acidic solution, the amine or ammonia formed in the hydrolysis is protonated, resulting in an ammonium salt. The amine can be recovered, if necessary, by making the solution basic.

Amide $R{-}C({=}O){-}NH_2$ $\xrightarrow{H^+}$ $R{-}C({=}\overset{+}{O}H){-}NH_2$ $\xrightarrow{H_2O}$ $R{-}C(OH)(NH_2)(H_2O^+)$ (proton migration) $\rightarrow$ $R{-}C(OH)(HO)(\overset{+}{N}H_3)$ $\xrightarrow{-NH_3}$ $R{-}C({=}\overset{+}{O}H){-}HO$ $\xrightarrow{-H^+}$ $R{-}C({=}O){-}OH$ Carboxylic acid

Examples of amide hydrolysis are cleavage of *N*-phenylbenzamide and cleavage of *N,N*-dimethylacetamide. As we will see in Chapters 16 and 19, the hydrolysis reaction is central to the metabolism of proteins, which involves cleavage to a mixture of α-amino acids.

$$\underset{N\text{-Phenylbenzamide}}{C_6H_5CONHC_6H_5} \xrightarrow[H_2O]{HCl} \underset{\text{Benzoic acid}}{C_6H_5COOH} + C_6H_5\overset{+}{N}H_3 \xrightarrow{NaOH} \underset{\text{Aniline}}{C_6H_5NH_2}$$

$$\underset{N,N\text{-Dimethylacetamide}}{CH_3CON(CH_3)_2} \xrightarrow[H_2O]{HCl} \underset{\text{Acetic acid}}{CH_3COOH} + \overset{+}{N}H_2(CH_3)_2 \xrightarrow{NaOH} \underset{\text{Dimethylamine}}{NH(CH_3)_2}$$

Reduction of Amides

The carbonyl group of an amide can be reduced to a methylene group using lithium aluminum hydride, as described in Section 12.6.2. By this means, amides can be converted into primary amines, secondary amines, or tertiary amines, depending on the nature of the starting amide.

$$\underset{\text{Amine}}{R'_2NH} \xrightarrow{RCOCl} \underset{\text{Amide}}{RCONR'_2} \xrightarrow{LiAlH_4} \underset{\text{Amine}}{RCH_2NR'_2}$$

How to Solve a Problem

Show the reactions necessary to convert benzene to benzylamine:

Benzene ⇢ Benzylamine (C_6H_5–CH_2NH_2)

PROBLEM ANALYSIS

We need to realize that in many instances there will be more than one reasonable synthesis. Also, some syntheses may be inordinately lengthy, which will significantly reduce the overall yield of the product. In this synthesis, the key problem is to establish a new carbon-carbon bond to the ring. Then the challenge is to place a nitrogen on the benzylic carbon.

An amine of this sort can be obtained by two possible reductions: (1) of a nitrile or (2) of an amide. (1) The nitrile can only be obtained through the diazonium reaction of aniline. Aniline is obtained by reduction of nitrobenzene, which can be obtained from benzene by direct nitration. (2) The amide can be obtained from benzoic acid, which can be obtained via carboxylation of a Grignard reagent prepared from bromobenzene. Bromobenzene can be obtained by direct bromination of benzene. This retrosynthetic analysis yields the following schemes:

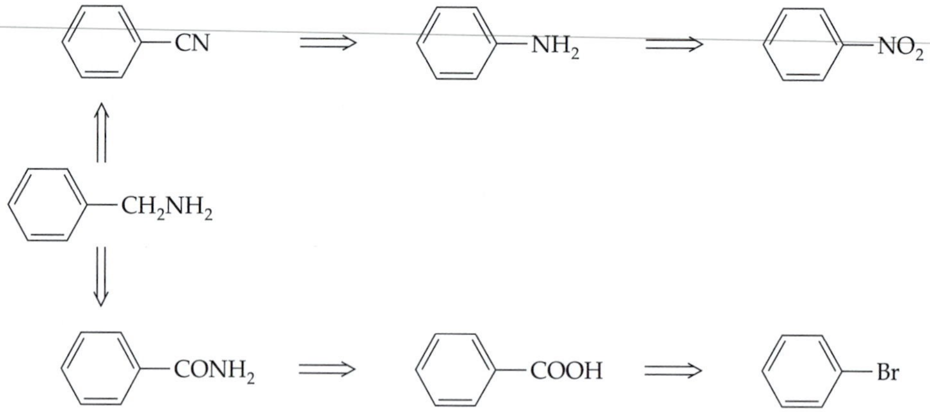

$$C_6H_5CH_2NH_2 \Longrightarrow C_6H_5CN \Longrightarrow C_6H_5NH_2 \Longrightarrow C_6H_5NO_2$$

$$C_6H_5CH_2NH_2 \Longrightarrow C_6H_5CONH_2 \Longrightarrow C_6H_5COOH \Longrightarrow C_6H_5Br$$

SOLUTIONS

The amide route is probably more efficient because of better reaction yields, but both are good routes. Bromination of benzene produces bromobenzene. Conversion to its Grignard reagent with magnesium, followed by carboxylation, produces benzoic acid. Treatment with thionyl chloride, then ammonia, produces benzamide. Reduction with lithium aluminum hydride yields the desired benzylamine.

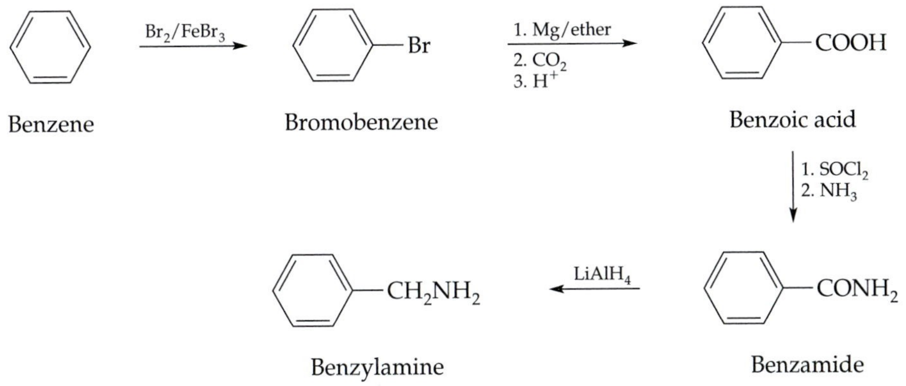

$$\text{Benzene} \xrightarrow{Br_2/FeBr_3} \text{Bromobenzene} \xrightarrow[\text{2. } CO_2 \quad \text{3. } H^+]{\text{1. Mg/ether}} \text{Benzoic acid (}C_6H_5COOH\text{)}$$

$$\text{Benzoic acid} \xrightarrow[\text{2. } NH_3]{\text{1. } SOCl_2} \text{Benzamide (}C_6H_5CONH_2\text{)} \xrightarrow{LiAlH_4} \text{Benzylamine (}C_6H_5CH_2NH_2\text{)}$$

PROBLEM 15.27

Show the reagents necessary for each step of the alternate synthesis of benzylamine from benzene, using the nitrobenzene route outlined in the preceding example problem.

PROBLEM 15.28

Show the product of each of the following reactions:

(a) *N*-ethylbenzamide heated with dilute sulfuric acid
(b) *N,N*-diphenylacetamide with lithium aluminum hydride
(c) cyclopropanecarboxamide with lithium aluminum hydride

15.4.3 Preparation of Amides

Both methods of amide preparation have been described in earlier sections but are briefly reviewed here.

Ammonolysis of Acid Chlorides

The most effective means of converting a carboxylic acid into its amide derivative is to first convert it into the reactive acid chloride, which is then treated with the appropriate amine.

General ammonolysis reaction:

$$\underset{\text{Carboxylic acid}}{RCOOH} \xrightarrow[\text{2. } NH_3]{\text{1. } SOCl_2} \underset{\text{Amide}}{RCONH_2}$$

We considered this reaction of amines with acid chlorides in detail in Section 15.1.2 as the ammonolysis of acid chlorides and in Section 12.4.2 as the acylation of amines. If an amine is to be acetylated, to form an acetamide derivative, acetic anhydride may be used in place of acetyl chloride, although one equivalent of acetic acid will be a byproduct (see Section 15.2.2).

$$\underset{\text{Amine}}{R_2NH} + \underset{\text{Acetic anhydride}}{(CH_3CO)_2O} \longrightarrow \underset{\text{Amide}}{CH_3CONR_2} + \underset{\text{Acetic acid}}{CH_3COOH}$$

CHEMISTRY AT WORK

IN THE MARKETPLACE

Lactams and Antibiotics

In this day and age, when relief from bacterial infection is usually just a doctor's visit away, it is hard to imagine that in the early part of the twentieth century, such infections were a leading cause of death. Pure chance and an open window changed all that in 1928, when an airborne mold blew into the laboratory of Alexander Fleming in London's St. Mary's Hospital. The mold, which Fleming identified as *Penicillium notatum,* landed in an open petri dish, multiplied, and killed the bacteria that the dish contained. The antibiotic penicillin was discovered.

A penicillin is a member of a family of drugs that all have a four-membered cyclic amide fused to a five-membered thiazole ring. Cyclic amides, known as *lactams,* are formed by intramolecular dehydration of an amino acid. (Note the similarity to cyclic esters, called lactones; see "Chemistry at Work," p. 547.) Lactams are classified as β-lactams, γ-lactams, δ-lactams, and so on, depending on the location of the amino group in the parent amino acid. For example, ϵ-aminocaproic acid produces ϵ-caprolactam, a starting material for the commercial production of nylon-6.

Lactams became of special interest when penicillin was discovered. There are various penicillins that differ in the acyl group attached to the amide nitrogen. Penicillin G was the first to be used in humans and proved to be an enormously effective antibiotic. However, bacterial strains gradually developed resistance to some early forms of penicillin, so newer penicillin derivatives were developed, including ampicillin, amoxicillin, and penicillin V. Moreover, some people are allergic to some penicillins.

R—CONH ... S ... N ... O ... COOH

Penicillin G	R = $C_6H_5CH_2$—
Penicillin V	R = $C_6H_5OCH_2$—
Ampicillin	R = $C_6H_5CH(NH_2)$—
Amoxicillin	R = $(p)HOC_6H_4CH(NH_2)$—

Penicillins have been synthesized in the laboratory, but commercial synthesis currently relies on microbial fermentation. In the search for new antibiotics in the penicillin family, it is necessary to first synthesize new compounds in the laboratory and then subject them to

Ammonolysis of Esters

Amides can also be produced by heating esters with amines. We discussed this reaction in Section 15.3.2. In this ammonolysis of an ester, the more nucleophilic —NH_2 group replaces the less nucleophilic —OR group via the usual nucleophilic acyl substitution mechanism (see Section 15.3.2 and Figure 15.1).

General ammonolysis reaction:

As an example, ethyl benzoate reacts with aniline to produce *N*-phenylbenzamide; and in a commercially important reaction, urea (instead of an amine), reacts with dimethyl malonate in a double amide-forming reaction to produce barbituric acid. Barbituric acid is the parent compound of the barbiturate family of drugs, which includes the prescription drug phenobarbital, used as an anticonvulsant, hypnotic, and sedative.

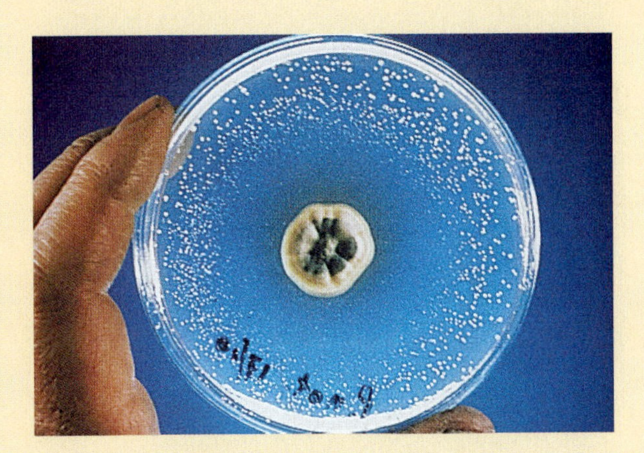

The *Penicillium* mold in the center of this petri dish is inhibiting the growth of the white *Staphylococcus* colonies around the perimeter.

testing for safety and effectiveness. Once an effective compound has been discovered and proved, it is often synthesized microbially to allow particular structures to be incorporated. For example, penicillin V is produced by adding 2-phenoxyethanol to a *Penicillium* mold; the microbe oxidizes this compound and incorporates the resulting phenoxyacetyl group ($C_6H_5OCH_2CO$—) into the amide portion of the molecule.

A newer family of very effective lactam antibiotics is the cephalosporins, first isolated from a fungus. Once again, a number of structural variants are on the market, with variations in the acyl group (RCO—) and thiazine side chain (R′).

R—CONH ... S ... N ... O ... R′ ... COOH

Cephalothin	R =	$C_4H_3SCH_2$—
	R′ =	CH_3COOCH_2—
Cephalexin	R =	$C_6H_5CH(NH_2)$—
	R′ =	CH_3—
Cefaclor	R =	$C_6H_5CH(NH_2)$—
	R′ =	Cl—

Production of ϵ-caprolactam:

$H_2N(CH_2)_5COOH \xrightarrow{-H_2O}$ ϵ-Caprolactam (ring positions ϵ, δ, γ, β, α; H–N, =O) $\dashrightarrow$ $-\!\!(NH(CH_2)_5CO)\!\!-_n$

ϵ-Aminocaproic acid — ϵ-Caprolactam — Nylon-6

Ethyl benzoate ($C_6H_5COOC_2H_5$) + Aniline ($C_6H_5NH_2$) $\xrightarrow{-C_2H_5OH}$ $C_6H_5CONHC_6H_5$

Ethyl benzoate — Aniline — *N*-Phenylbenzamide

$CH_2(COOCH_3)_2$ + $(H_2N)_2C{=}O$ $\xrightarrow{-(2\,CH_3OH)}$ Barbituric acid

Dimethyl malonate — Urea — Barbituric acid — Phenobarbital (C_6H_5, C_2H_5)

How to Solve a Problem

Show how to convert ethyl benzoate to benzamide and, separately, how to convert benzamide back to ethyl benzoate.

PROBLEM ANALYSIS

The problem is to convert derivatives of benzoic acid into each other:

$$C_6H_5{-}COOC_2H_5 \rightleftarrows C_6H_5{-}CONH_2$$

Ethyl benzoate ⇄ Benzamide

Keeping in mind the hierarchy of carboxylic acid derivative conversions (see Figure 15.1), we recall that esters can be converted to amides by nucleophilic acyl substitution. Therefore, conversion of ethyl benzoate to benzamide requires only reaction with ammonia for the substitution to occur.

The same hierarchy of derivative interconversions tells us that we cannot convert an amide into an ester by a simple nucleophilic acyl substitution with alkoxide. Therefore, we must consider alternative routes presented by Figure 15.1. The only reasonable way to get from an amide to an ester is to convert the amide to the parent carboxylic acid first. At that point, we have two choices: (1) conversion of the acid to the ester using Fischer esterification (heating with ethanol and acid catalyst) or (2) conversion of the acid to its acid chloride, followed by alcoholysis with ethanol.

SOLUTION

Reaction of ethyl benzoate with ammonia accomplishes the nucleophilic acyl substitution directly, forming benzamide and expelling ethanol:

$$\underset{\text{Ethyl benzoate}}{C_6H_5{-}COOC_2H_5} \xrightarrow{NH_3} \underset{\text{Benzamide}}{C_6H_5{-}CONH_2} + \underset{\text{Ethanol}}{C_2H_5OH}$$

Hydrolysis of benzamide with dilute acid produces benzoic acid. Reacting the acid with thionyl chloride produces benzoyl chloride, which, when reacted with ethanol, produces ethyl benzoate. Alternatively, reaction of benzoic acid with ethanol (which also serves as the solvent) in the presence of a few drops of concentrated sulfuric acid gives the same product:

$$\underset{\text{Benzamide}}{C_6H_5{-}CONH_2} \xrightarrow{H^+/H_2O} \underset{\text{Benzoic acid}}{C_6H_5{-}COOH} \xrightarrow{SOCl_2} \underset{\text{Benzoyl chloride}}{C_6H_5{-}COCl} \xrightarrow{C_2H_5OH} \underset{\text{Ethyl benzoate}}{C_6H_5{-}COOC_2H_5}$$

$$\underset{\text{Benzoic acid}}{C_6H_5{-}COOH} \xrightarrow[H_2SO_4]{C_2H_5OH} \underset{\text{Ethyl benzoate}}{C_6H_5{-}COOC_2H_5}$$

PROBLEM 15.29

Show the reactions for preparation of each of the following amides starting with two different compounds—a carboxylic acid and one of its derivatives:

(a) *N*-methylbenzamide

(b) *N,N*-diethylcyclohexanecarboxamide

Chapter Summary

There are four key **carboxylic acid derivatives** in which the —OH group of the acid has been replaced by another functional substituent. **Acid chlorides,** which have —Cl in place of the —OH group, and **acid anhydrides,** which have —OOCR (carboxylate) in place of the —OH group, are very reactive derivatives. They are mostly used as reagents to attach an acyl group to other groups (that is, to carry out an **acylation reaction**). Acid chlorides are prepared from carboxylic acids by reaction with thionyl chloride. Cyclic anhydrides (such as those with five-membered rings) are prepared by eliminating the elements of water from two carboxyl groups of a dicarboxylic acid, through heating.

Esters contain an alkoxy group (—OR) in place of the —OH group of a carboxylic acid. **Amides** contain an amino group (—NH_2, —NHR, or NRR′) in place of the —OH group.

The usual reaction of a carboxylic acid derivative is **nucleophilic acyl substitution.** Using this reaction and the appropriate nucleophile, any carboxylic acid derivative can be converted into a derivative lying to its right in the following sequence.

RCOCl	RCOOR′	$RCONH_2$	RCOOH
Acid chloride	Ester	Amide	Acid

Acid chlorides and acid anhydrides are converted to esters by *alcoholysis* with any alcohol or phenol. Acid chlorides, acid anhydrides, and esters are converted into amides by *ammonolysis* with ammonia or a primary or secondary amine. Acid chlorides, acid anhydrides, esters, and amides are converted into the parent carboxylic acid by *hydrolysis* with water or with aqueous base.

Carboxylic acids can be converted into esters and amides by first converting them to acid chlorides and then conducting an alcoholysis or ammonolysis reaction. Esters can also be prepared from carboxylic acids by reaction with excess alcohol in the presence of an acid catalyst (**Fischer esterification**).

Esters with two α-hydrogens undergo a **Claisen condensation** in the presence of an alkoxide base; the result is the nucleophilic acyl substitution of an alkoxy group by the enolate anion of the ester. A β-keto ester is the product.

Summary of Reactions

1. Preparation of acid chlorides (Section 15.1.4)

$$\underset{\text{Carboxylic acid}}{\text{RCOOH}} \xrightarrow[-SO_2,\ -HCl]{SOCl_2} \underset{\text{Acid chloride}}{\text{RCOCl}}$$

2. Preparation of cyclic anhydrides (Section 15.2.3)

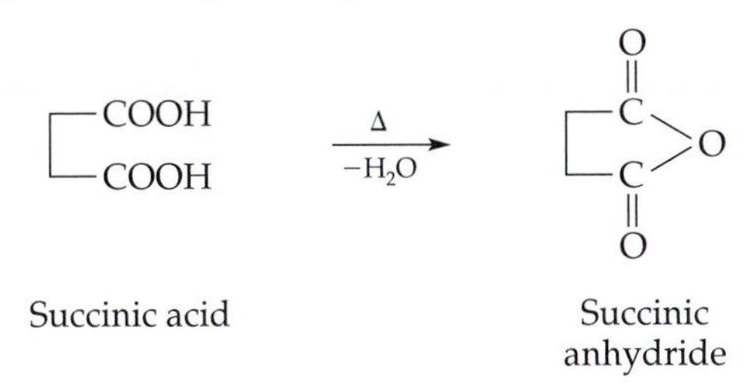

3. Hydrolysis of carboxylic acid derivatives

$$\underset{\text{Acid chloride}}{\text{RCOCl}} \xrightarrow[-HCl]{H_2O} \underset{\text{Carboxylic acid}}{\text{RCOOH}}$$

$$\underset{\text{Acid anhydride}}{(RCO)_2O} \xrightarrow{H_2O} \underset{\text{Carboxylic acid}}{2\ \text{RCOOH}}$$

$$\underset{\text{Ester}}{\text{RCOOR}'} \xrightarrow[2.\ H^+,\ -R'OH]{1.\ NaOH} \underset{\text{Carboxylic acid}}{\text{RCOOH}}$$

$$\underset{\text{Amide}}{RCONH_2} \xrightarrow[-NH_4^+]{H^+/H_2O} \underset{\text{Carboxylic acid}}{\text{RCOOH}}$$

4. Ammonolysis of carboxylic acid derivatives

$$\underset{\text{Acid chloride}}{RCOCl} \xrightarrow[-NH_4Cl]{NH_3} \underset{\text{Amide}}{RCONH_2}$$

$$\underset{\text{Acid anhydride}}{(RCO)_2O} \xrightarrow[-RCOOH]{NH_3} \underset{\text{Amide}}{RCONH_2}$$

$$\underset{\text{Ester}}{RCOOR'} \xrightarrow[-R'OH]{NH_3} \underset{\text{Amide}}{RCONH_2}$$

5. Alcoholysis of carboxylic acid derivatives

$$\underset{\text{Acid chloride}}{RCOCl} \xrightarrow[\text{pyridine}]{R'OH} \underset{\text{Ester}}{RCOOR'}$$

$$\underset{\text{Acid anhydride}}{(RCO)_2O} \xrightarrow[-RCOOH]{R'OH} \underset{\text{Ester}}{RCOOR'}$$

6. Preparation of esters by Fischer esterification (Section 15.3.4)

$$\underset{\text{Carboxylic acid}}{RCOOH} + \underset{\text{Alcohol}}{R'OH} \xrightarrow[-H_2O]{H_2SO_4} \underset{\text{Ester}}{RCOOR'}$$

7. Claisen condensation of esters (Section 15.3.3)

$$\underset{\text{Ester}}{2\ RCH_2COOR'} \xrightarrow[2.\ H^+]{1.\ NaOR'} \underset{\beta\text{-Keto ester}}{RCH_2COCH(R)COOR'}$$

Additional Problems

Nomenclature

15.30 Write names corresponding to the following structures:

(a) 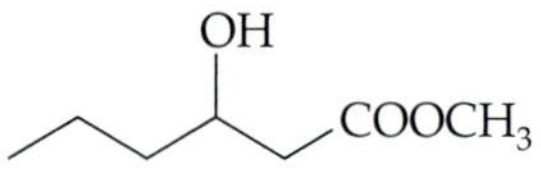 $CH_3CH_2CH_2CH(OH)CH_2COOCH_3$

(b) 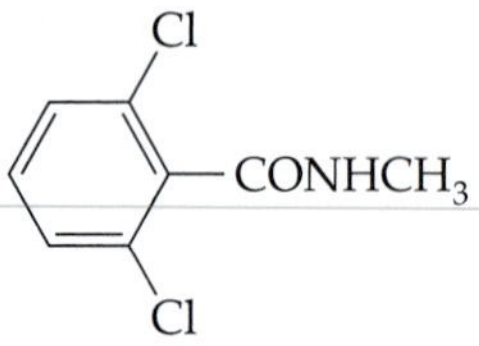 2,6-dichlorophenyl–$CONHCH_3$

(c) 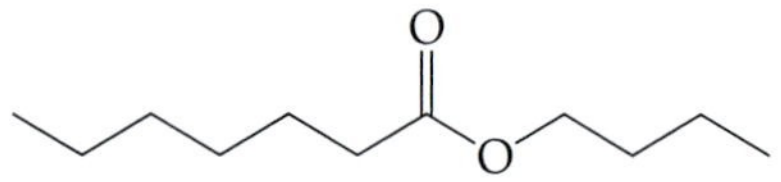 $CH_3(CH_2)_5C(=O)O(CH_2)_3CH_3$

(d) 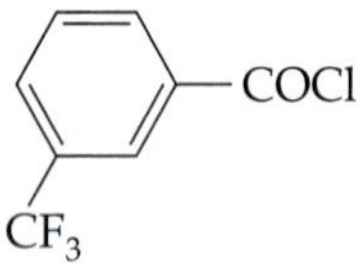 3-(CF_3)phenyl–$COCl$

(e) 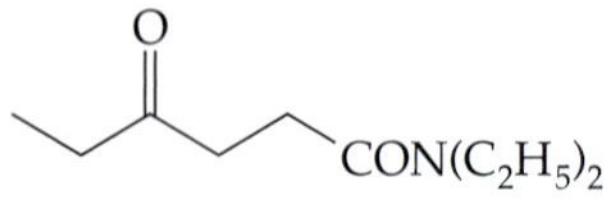 $CH_3CH_2C(=O)CH_2CH_2CON(C_2H_5)_2$

(f) 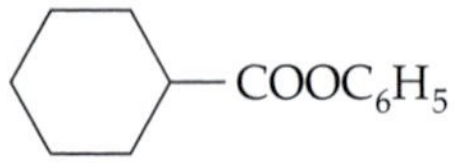 cyclohexyl–$COOC_6H_5$ (g) $(C_6H_5CO)_2O$

(h) 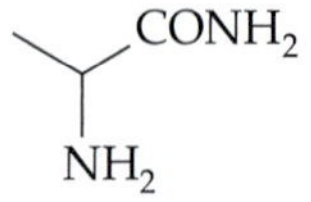 $(CH_3)_2C(NH_2)CONH_2$ (i)  $CH_2(COOC_2H_5)_2$

15.31 Draw the structural formula corresponding to each of the following names:

(a) 2-methylhexanoyl chloride
(b) propanoic anhydride
(c) *p*-nitrobenzamide
(d) isopropyl 2-propenoate
(e) diphenyl succinate
(f) 2,4-dichlorobenzoyl chloride
(g) cyclopentyl benzoate
(h) *N*-ethyl-2-oxo-3-(*E*)-pentenamide
(i) acetanilide (*N*-phenylacetamide)
(j) phthalic anhydride
(k) dimethyl malonate
(l) tosyl chloride (*p*-toluenesulfonyl chloride)

Reactions

15.32 Show the product(s) of the reaction of benzoyl chloride with each of the following compounds:

(a) piperidine
(b) cyclohexanol
(c) ammonia
(d) isopropyl alcohol
(e) water
(f) benzene and aluminum chloride
(g) lithium tri(*t*-butoxy)aluminum hydride
(h) lithium diethylcuprate
(i) aniline
(j) phenol

15.33 Show the product of each reaction of benzamide:

(a) heating with dilute acid

(b) with lithium aluminum hydride

15.34 Show the product obtained from the reaction of glycerol with excess acetyl chloride.

15.35 Show the product of the following reactions:

(a) isopropyl alcohol and acetic acid with H_2SO_4 catalyst

(b) cyclohexanol and acetic anhydride

(c) (*S*)-2-butanol and benzoyl chloride

(d) *t*-butylbenzoate heated with aqueous sodium hydroxide

(e) ethyl benzoate with dimethylamine

(f) (*S*)-2-butyl acetate with aqueous sodium hydroxide

(g) (*S*)-2-butyl tosylate with aqueous sodium hydroxide

(h) glycerol triacetate with aqueous sodium hydroxide

(i) ethyl acetate with sodium ethoxide, followed by acidification

15.36 Show the product of each step of the following conversion:

(a) methyl propanoate with sodium methoxide

(b) acidification of the solution from (a) with HCl

(c) heating the product of (b) with aqueous sodium hydroxide

(d) treatment of the solution from (c) with cold HCl

(e) heating the product of (d) until gas evolution ceases

15.37 Show the products of the following condensation reactions, with acidification as the work-up:

(a) ethyl acetate with sodium ethoxide

(b) ethyl propanoate with sodium ethoxide

15.38 Show the final product obtained by applying the same steps outlined in Problem 15.36 to each of the following sets of reactants:

(a) ethyl acetate with NaOEt

(b) ethyl butanoate with NaOMe

15.39 Show the product(s) of the reaction of benzoic anhydride with each of the following compounds:

(a) dimethylamine

(b) phenol

(c) isopropyl alcohol

(d) water

Syntheses

15.40 Show how to accomplish the following transformations:

(a) (*E*)-2-pentenoic acid to (*E*)-2-pentenamide

(b) benzoic acid to phenyl benzoate

(c) propanoic acid to 2-aminopropanamide

15.41 Show how to accomplish the following syntheses:

(a) benzene to ethyl benzoate

(b) bromobenzene to *p*-nitrobenzamide

(c) cyclohexane to methyl cyclohexanecarboxylate

(d) cyclohexene to dimethyl hexanedioate

15.42 Show two different routes for the conversion of benzoic acid to cyclohexyl benzoate.

15.43 Show the starting materials you would employ to synthesize the following ketones, using the Claisen condensation as one of the steps. (*Hint:* Apply retrosynthesis.)

(a) $CH_3CH_2CH_2COCH_2CH_2CH_3$

(b) diphenylacetone (1,3-diphenyl-2-propanone)

Mixed Problems

15.44 Aspirin is formed when salicylic acid (*o*-hydroxybenzoic acid) is acetylated. Show the reaction and the structure of the product.

15.45 When salicylic acid is converted to its methyl ester, the product is called oil of wintergreen. Show this conversion and the structure of the product.

15.46 The active ingredient in the bug repellent sold as Off is *N,N*-diethyl-*m*-toluamide (*N,N*-diethyl-*m*-methylbenzamide). Show how it could be prepared from *m*-toluic acid (*m*-methylbenzoic acid).

15.47 Propanil is an herbicide that is an aromatic amide and can be synthesized starting with benzene and propanoic acid. 2,4-Dichloronitrobenzene is an intermediate in the synthesis. Show the reactions (and products of each step) needed to accomplish the total synthesis of Propanil.

(a) (three steps) → Cl, Cl, NO_2 → (b) (two steps) → Cl, Cl, $NHCOCH_2CH_3$

Propanil

15.48 The artificial sweetener sold as NutraSweet is a compound called aspartame. It is not recommended for use under hydrolytic conditions (for baking or in hot drinks for extended periods of time) because it readily hydrolyzes. Show all the products you would expect from *complete* acid-catalyzed hydrolysis of aspartame.

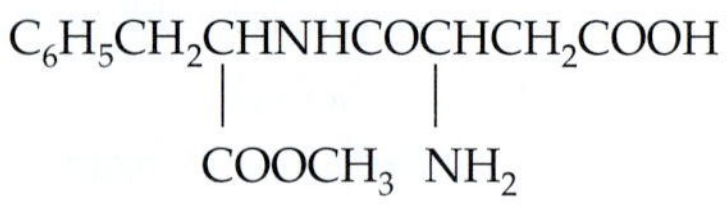

$C_6H_5CH_2CH(COOCH_3)NHCOCH(NH_2)CH_2COOH$

Aspartame

15.49 The local anesthetic benzocaine, used in many sunburn ointments, can be readily synthesized from toluene. Show the reactions necessary for this synthesis, with *p*-toluidine and *p*-acetamidobenzoic acid as intermediates:

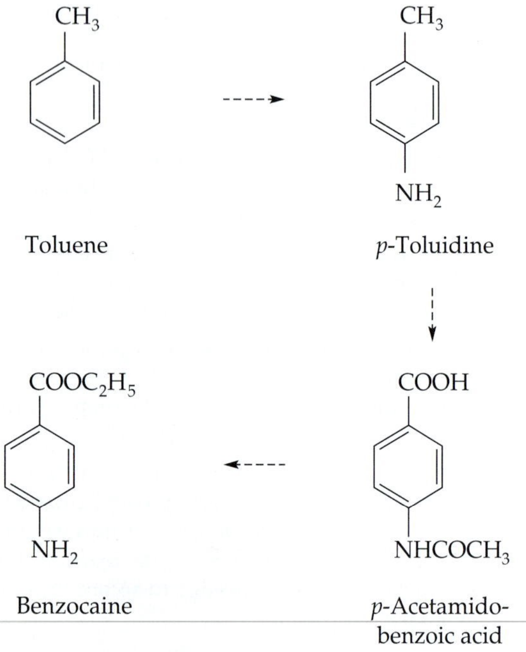

Toluene (CH_3) → *p*-Toluidine (CH_3, NH_2) → *p*-Acetamidobenzoic acid (COOH, $NHCOCH_3$) → Benzocaine ($COOC_2H_5$, NH_2)

15.50 Procaine (Novocain) is a local anesthetic used frequently by dentists. Using the compounds shown below as starting material and intermediates, show the reactions necessary to complete a synthesis of procaine.

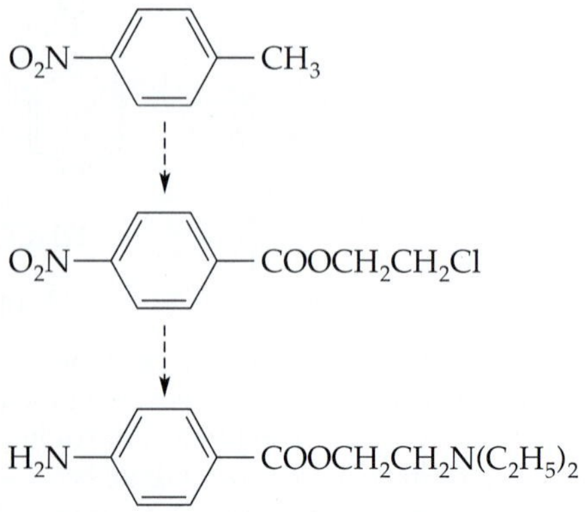

$O_2N–C_6H_4–CH_3$

↓

$O_2N–C_6H_4–COOCH_2CH_2Cl$

↓

$H_2N–C_6H_4–COOCH_2CH_2N(C_2H_5)_2$

Procaine (Novocain)

15.51 Atropine is a poisonous alkaloid obtained from the nightshade plant. In very low doses, it acts as a muscle relaxant and is used regularly by ophthalmologists to dilate the pupil of the human eye. When it is hydrolyzed, atropine produces two fragments: the amino-alcohol called tropine and tropic acid [$C_6H_5CH(CH_2OH)COOH$]. Tropine is not an enantiomer—its mirror image is superimposable. It can be dehydrated by phosphoric acid to tropidene. Upon ozonolysis, tropidene produces a dialdehyde, *N*-methyl-2-formyl-5-(formylmethyl)pyrrolidine, shown here. What is the structure of atropine?

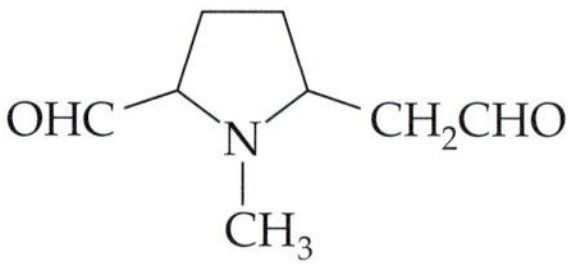

OHC–(pyrrolidine, N–CH_3)–CH_2CHO

15.52 The compound tetracaine is used as a spinal anesthetic. It can be synthesized from *p*-toluidine, using $HOCH_2CH_2N(CH_3)_2$ as a reagent in the final step. Show the steps (a) from *p*-toluidine to *N*-butyl-*p*-toluidine and (b) from *N*-butyl-*p*-aminobenzoic acid to tetracaine.

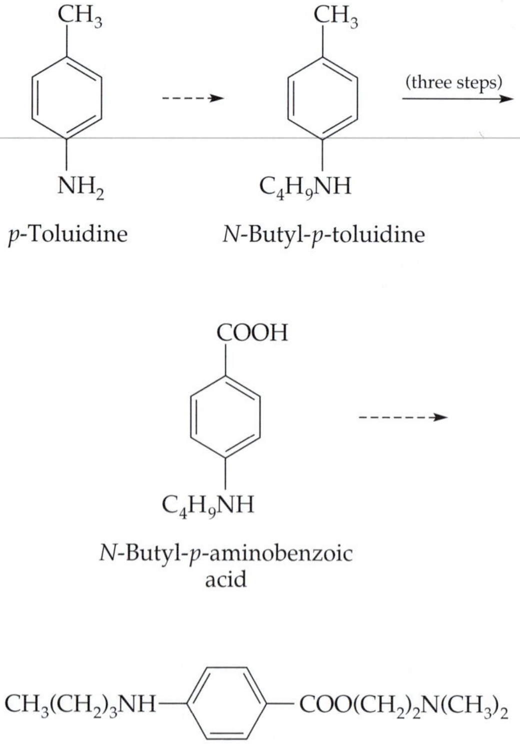

p-Toluidine (CH_3, NH_2) → *N*-Butyl-*p*-toluidine (CH_3, C_4H_9NH) —(three steps)→

N-Butyl-*p*-aminobenzoic acid (COOH, C_4H_9NH) →

$CH_3(CH_2)_3NH–C_6H_4–COO(CH_2)_2N(CH_3)_2$

Tetracaine

CONCEPTUAL PROBLEM

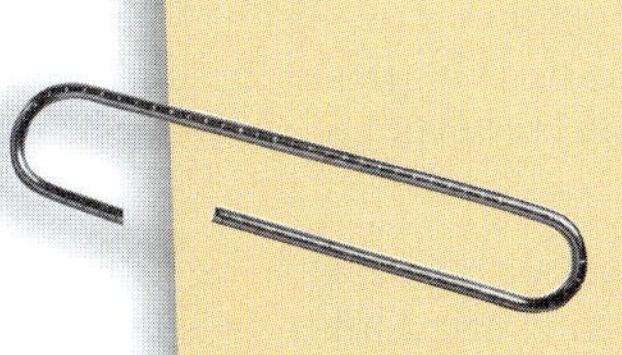

Fat-Free But Not Necessarily Calorie-Free

The science reporter for the local newspaper is doing a story about the zero-calorie fat substitute Olestra. She wants to get some background information from a food scientist, and so she calls you. You arrange to meet over lunch. She brings with her a chemist's report that shows three chemical structures:

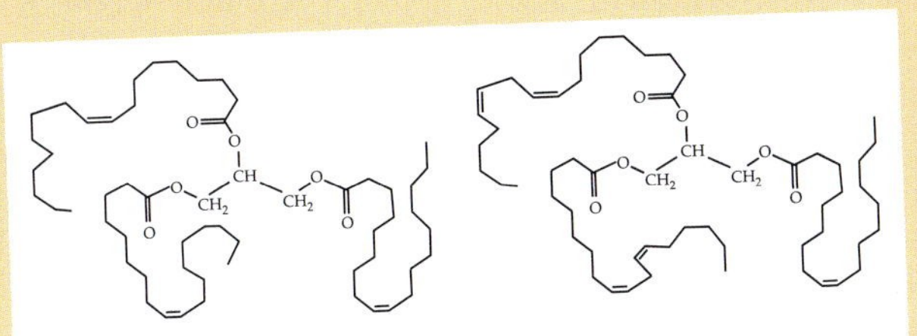

Fat from olive oil

Fat from corn oil

Molecule of Olestra

She can see the structural similarities between the fats from olive oil and corn oil. And it's clear that the Olestra molecule is quite different. She asks you how a chemist would describe the differences.

The reporter recalls an earlier story she did on the cancer-fighting drug paclitaxel. In that case, at least, the sheer size and complexity of the molecule made it very difficult to synthesize. She knows that a molecule of Olestra is about the same size, yet Proctor & Gamble can make Olestra by the ton. What's the difference?

Digestive enzymes easily break up fats into glycerol and fatty acids, but they don't break up Olestra into sucrose and fatty acids, even though both fats and Olestra are esters. That seems odd. Why should it happen that way?

Olestra products are required to be fortified with vitamins A, D, E, and K because Olestra makes it hard for the body to absorb these fat-soluble vitamins. Obviously, ordinary fats don't cause that problem. Why?

It seems contradictory to say that Olestra is fat-free and has no calories when in fact it is made from fat and sugar. How do you explain this?

The reporter has come up with a catchy title for her piece, "Olestra: A Polyester You Can Eat." She asks how you like it. Do you rain on her parade? Is it important that she be technically accurate?

Every snacker's dream, every dieter's nightmare—an unending supply of potato chips. The makers of Olestra think they have the answer: a zero-calorie fat substitute. Chips made with Olestra can be labeled fat-free, but they are not calorie-free. Only the calories coming from the replaced fat are missing; each chip still has 70 calories coming from other ingredients.

15.53 A medicinal compound (**A**) with the formula $C_{13}H_{20}O_2N_2$ is insoluble in water and dilute sodium hydroxide but soluble in dilute HCl. The IR spectrum shows peaks at 3350 and 3300 cm^{-1}, as well as at 1730 cm^{-1}. The ^1H-NMR spectrum shows peaks at 7.3 ppm (m, 2H), 2.5 ppm (q, 2H), 2.3 ppm (t, 1H), 2.1 ppm (t, 1H), 1.7 ppm (s, 1H), and 0.9 ppm (t, 3H). When hydrolyzed with dilute sodium hydroxide, **A** slowly dissolves. Extraction of the aqueous basic solution with ether, followed by evaporation of the ether, produces a colorless liquid, compound **B.** The remaining aqueous layer is acidified to a neutral pH and produces a colorless precipitate, compound **C.**

Compound **B,** whose molecular formula is $C_6H_{15}ON$, is insoluble in base but dissolves in acid. Further, it gives off a gas when treated with metallic sodium. Its IR spectrum shows a strong sharp peak at 3300 cm^{-1}. The ^1H-NMR spectrum shows peaks at 3.2 ppm (s, 1H), 2.5 ppm (q, 4H), 2.3 ppm (t, 2H), 2.1 ppm (t, 2H), and 0.9 ppm (t, 6H). By comparison of spectra, **B** is found to be identical to a compound prepared by reaction of diethylamine with ethylene oxide.

Compound **C,** whose molecular formula is $C_7H_7O_2N$, is insoluble in water but dissolves in both dilute base and dilute acid. Its ^1H-NMR spectrum shows peaks at 11.5 ppm (s, 1H), 7.3 ppm (m, 4H), and 1.7 ppm (s, 2H). Compound **C** shows five different peaks in the decoupled ^{13}C-NMR spectrum. Propose structures for compounds **A, B,** and **C.**

15.54 Two sample vials are known to contain isomeric compounds (**X** and **Y**) with formula $C_9H_{10}O_2$. The IR spectra of the two samples are different in the fingerprint region, but both contain a strong peak near 1675 cm^{-1}, a moderate peak at 1600 cm^{-1}, and no peak near 3300 cm^{-1}. Separately heating each sample with dilute sodium hydroxide, then acidifying the resultant solution, produces precipitates that are shown to be carboxylic acids by IR and ^1H-NMR spectra. Methanol can also be detected in both hydrolysis solutions. The ^1H-NMR spectrum of **X** shows peaks at 7.3 ppm (m, 4H), 3.5 ppm (s, 3H), and 2.4 ppm (s, 3H). The ^{13}C-NMR spectrum indicates the presence of seven different kinds of carbon. The ^1H-NMR spectrum of **Y** shows peaks at 7.2 ppm (m, 5H), 3.5 ppm (s, 3H), and 3.6 ppm (s, 2H). Propose structures for compounds **X** and **Y.**

15.55 A compound of formula $C_5H_8O_4$ shows a peak at 1740 cm^{-1} in its IR spectrum and two singlets at 2.5 ppm (1H) and 3.4 ppm (3H) in its ^1H-NMR spectrum. What is its structure?

15.56 Compound **A** is a pleasantly odiferous liquid with the formula $C_{11}H_{14}O_2$. It is insoluble in water, dilute acid, and dilute base. Its most obvious IR peak is at 1690 cm^{-1}. It does not react with Jones' reagent (chromic acid), with acetyl chloride, or with 2,4-dinitrophenylhydrazine. Compound **A** can be hydrolyzed by heating with dilute sodium hydroxide. Work-up of the reaction mixture leads to what is obviously a carboxylic acid (**B**) and a low-molecular-weight alcohol (**C**). Compound **C** is oxidized by Jones' reagent and produces acetone. Compound **B** ($C_8H_8O_2$) shows a broad peak in the IR spectrum centered at 3300 cm^{-1} and a sharp peak at 1720 cm^{-1}. The ^1H-NMR spectrum of **B** shows peaks at 11.3 ppm (s, 1H), 7.3 ppm (m, 5H), and 3.6 ppm (s, 2H). Propose structures for compounds **A, B,** and **C.**

15.57 Compound **D** ($C_{11}H_{14}O_2$), an isomer of **A** from Problem 15.56, is also insoluble in water and in dilute base. It shows a sharp peak in the IR spectrum at 1700 cm^{-1}. Its ^1H-NMR spectrum has the following peaks: 7.2 ppm (m, 4H), 4.0 ppm (q, 2H), 3.5 ppm (s, 2H), 2.3 ppm (s, 3H), and 1.1 ppm (t, 3H). There is no evidence that hydroxyl, ketone, or aldehyde groups are present. Compound **D** hydrolyzes in dilute base to produce, after appropriate work-up, a carboxylic acid (**E,** $C_9H_{10}O_2$) and an alcohol (**F**). The ^1H-NMR spectrum of **E** shows peaks at 10.5 ppm (s, 1H), 7.2 ppm (m, 4H), 3.5 ppm (s, 2H), and 2.3 ppm (s, 3H). Strong oxidation of **E** with hot potassium permanganate yields another carboxylic acid (**G,** $C_8H_6O_4$). The ^1H-NMR spectrum of **G** shows peaks at 10.7 ppm (s, 1H) and 7.3 ppm (m, 2H). Heating **G** leads to the formation of a new substance (**H,** $C_8H_4O_3$), which can be reconverted to **G** with dilute acid and water. Propose structures for compounds **D–H.**

15.58 Propose a structure for a compound that has the formula $C_6H_{12}O_2$, that is insoluble in water and dilute base, and whose ^1H-NMR spectrum shows two singlets at 3.5 ppm (1H) and 0.9 ppm (3H).

CHAPTER 16

Proteins

SECTION 15.4 FOCUSED on the chemistry of amides, one of the derivatives of carboxylic acids. From there, it is a small step to the chemistry of the polyamides called *proteins,* one of the primary constituents of all living matter. Proteins make up about half the dry weight of cells and are vital to the functioning of all living systems, both plant and animal. This chapter will introduce the structure and functions of proteins.

$$\mathrm{R{-}\underset{\displaystyle NH_2}{\underset{|}{CH}}{-}COOH}$$

α-Amino acid

Proteins are long-chain polyamides formed from **α-amino acids,** alkanoic acids carrying an amine group on the carbon alpha to the carboxyl carbon. The amide bond of a protein is formed when the amine group of one α-amino acid connects to the carboxyl group of another α-amino acid. Before we begin discussing the proteins themselves, we need to look at the building blocks of proteins—the α-amino acids.

The amide bond between two α-amino acids is called a **peptide bond.** A compound containing two α-amino acids joined by a peptide bond is called a *dipeptide,* and compounds containing more α-amino acids are called *tripeptides, tetrapeptides,* and so on. Compounds containing large numbers of α-amino acids, usually more than 10, are called *polypeptides.* In a peptide chain, one end has a free amino group, and it is called the N-terminus. The other end has a free carboxyl group, and it is called the C-terminus. By convention, the N-terminus is always written at the left end of the chain.

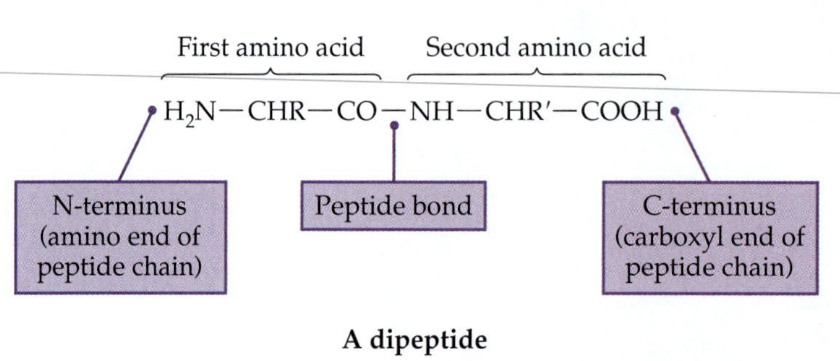

A dipeptide

Proteins may be quite small polypeptides, such as the natural hormone oxytocin, with only 9 α-amino acids; oxytocin is used to stimulate uterine contractions to induce labor during childbirth. Insulin is a polypeptide of moderate size, with 51 α-amino acids. On the large side is the protein globin (of hemoglobin), with 574 α-amino acids.

◀ When Mark McGwire broke Roger Maris' homerun record, he not only made headlines for his feat, but also popularized the use of the amino acid Creatine. Creatine is a naturally occuring substance that promotes muscle development.

16.1

α-Amino Acids

16.1.1 Structure of α-Amino Acids

A total of 22 different α-amino acids are found in proteins; 20 of these are found free in the human body and are used in protein synthesis. The other two, hydroxyproline and cystine, are formed by conversion from proline and cysteine, two α-amino acids in protein chains. The human body can synthesize all but 8 of the 20 α-amino acids; these 8 are therefore called *essential amino acids* because they must be obtained from the diet.

All α-amino acids except glycine (α-aminoacetic acid) have a stereocenter and exist in proteins of higher organisms only as the single enantiomer with the L configuration. (The L configuration is the opposite of the D configuration of carbohydrates, which relates them to D-glyceraldehyde; see Section 4.7.1.) The historical L designation implies an *S* configuration for all α-amino acids except cysteine, which is *R*.

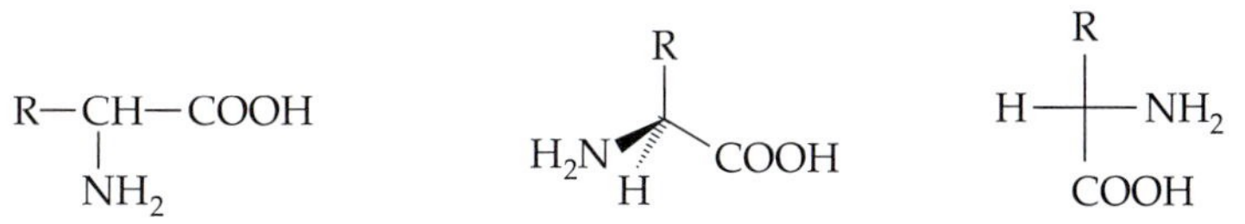

R—CH—COOH with NH_2 on CH

α-Amino acid

L-α-amino acids

Table 16.1 gives the formulas of the 22 α-amino acids found in proteins, together with their names, abbreviations, and an indication of which are essential. They have been grouped into categories based on the nature of their side chains. The three-letter abbreviations given in the table are widely used to represent the α-amino acids in proteins and peptides. This system saves a tremendous amount of space and effort and makes the structures easier to read. For example, the tetrapeptide formed from glycine, alanine, serine, and histidine is written as

Gly—Ala—Ser—His

rather than as

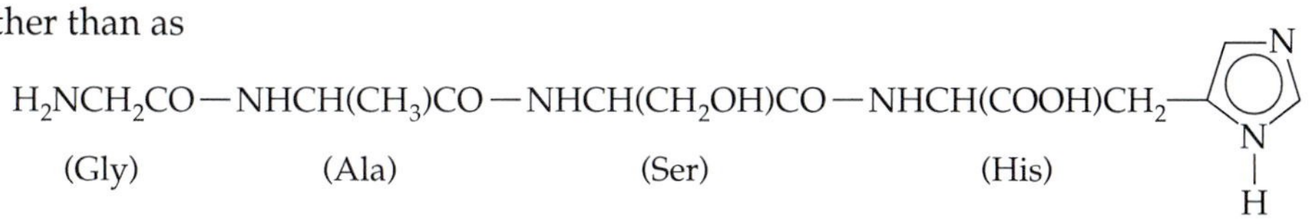

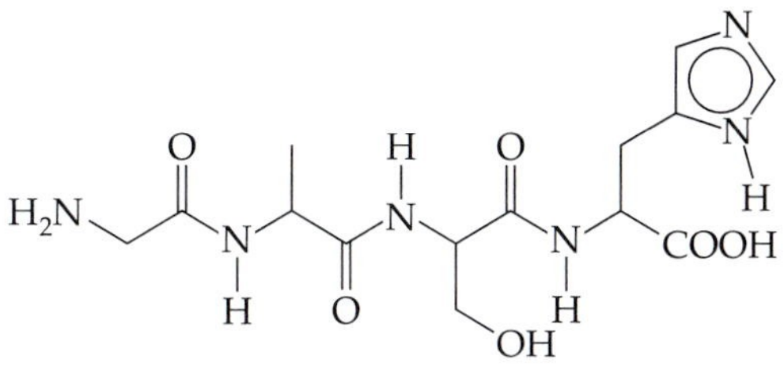

$H_2NCH_2CO—NHCH(CH_3)CO—NHCH(CH_2OH)CO—NHCH(COOH)CH_2$—(imidazolyl)

(Gly) (Ala) (Ser) (His)

(Gly) (Ala) (Ser) (His)

The bond connecting each pair of α-amino acids is an amide (peptide) bond. According to this representation, glycine (the leftmost α-amino acid) has its amino group free and histidine has its carboxyl group free. Therefore, for this tetrapeptide, glycine is the *N-terminal* α-amino acid and histidine is the *C-terminal* α-amino acid.

TABLE 16.1

L-α-Amino Acids [$RCH(NH_2)COOH$] Found in Proteins

Structure of R—		Name	Abbreviation
Alkyl Side Chain	—H	Glycine	Gly
	$—CH_3$	Alanine	Ala
	$—CH(CH_3)_2$	Valine	Val*
	$—CH_2CH(CH_3)_2$	Leucine	Leu*
	$—CH(CH_3)CH_2CH_3$	Isoleucine	Ile*
	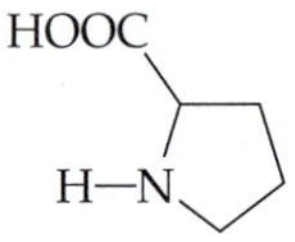 HOOC / H—N (pyrrolidine ring) (complete structure)	Proline	Pro
Side Chain Containing Aromatic Ring	$—CH_2—C_6H_5$	Phenylalanine	Phe*
	$—CH_2—C_6H_4—OH(p)$	Tyrosine	Tyr
	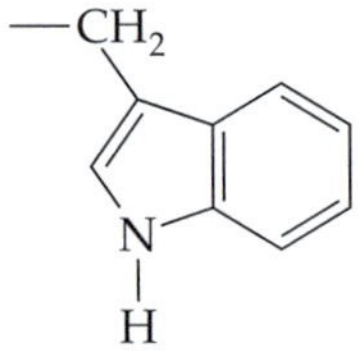 $—CH_2$ (indole ring, N—H)	Tryptophan	Trp*
Polar Side Chain	$—CH_2OH$	Serine	Ser
	$—CH(CH_3)OH$	Threonine	Thr*
	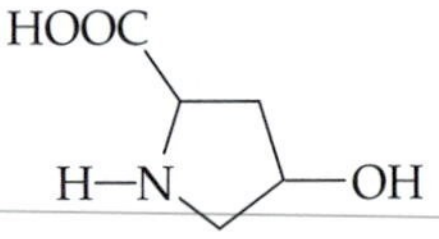 HOOC / H—N (pyrrolidine ring) —OH (complete structure)	Hydroxyproline	Hyp
	$—CH_2SH$	Cysteine	Cys
	$—CH_2S—SCH_2—$	Cystine	Cys-Cys
	$—CH_2CH_2SCH_3$	Methionine	Met*
	$—CH_2CONH_2$	Asparagine	Asn
	$—CH_2CH_2CONH_2$	Glutamine	Gln
Carboxyl Side Chain	$—CH_2COOH$	Aspartic acid	Asp
	$—CH_2CH_2COOH$	Glutamic acid	Glu
Amine Side Chain	$—CH_2CH_2CH_2CH_2NH_2$	Lysine	Lys*
	$—CH_2CH_2CH_2NHC(=NH)NH_2$	Arginine	Arg
	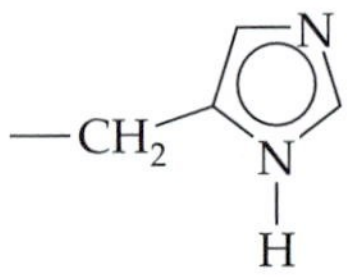 $—CH_2$ (imidazole ring, N, N—H)	Histidine	His

* = essential

PROBLEM 16.1

A few α-amino acids of the "unnatural" D configuration have been detected in bacterial proteins. Draw the three-dimensional structure for D-alanine.

16.1.2 Properties of α-Amino Acids

α-Amino acids contain two functional groups that we have studied: the amino group (see Chapter 12) and the carboxyl group (see Chapter 14). The most unique property of any α-amino acid results from the fact that a single molecule contains both an *acidic* carboxyl group and a *basic* amino group. (To keep things simple for this discussion, we will ignore those α-amino acids that have two amino groups or two carboxyl groups.) The result is that an α-amino acid can act as either an acid or a base; such compounds are **amphoteric.** In the solid state, α-amino acids exist as dipolar ions called **zwitterions** (a German term meaning "hermaphrodite ions," ions with both positive and negative character), in which the acidic carboxyl group has protonated the basic amino group. The zwitterion can be considered an *internal* salt (consisting of an ammonium cation and a carboxylate anion).

$$\underset{\text{Cation}}{\text{R—CH}(\text{NH}_3^+)\text{—COOH}} \underset{+\text{H}^+}{\overset{-\text{H}^+}{\rightleftharpoons}} \underset{\text{Zwitterion}}{\text{R—CH}(\text{NH}_3^+)\text{—COO}^-} \underset{+\text{H}^+}{\overset{-\text{H}^+}{\rightleftharpoons}} \underset{\text{Anion}}{\text{R—CH}(\text{NH}_2)\text{—COO}^-}$$

Increasing pH shifts equilibrium in this direction →

← Decreasing pH shifts equilibrium in this direction

In aqueous solution, α-amino acids exist in an equilibrium among three forms: the zwitterion, a cationic form, and an anionic form. The pH of a solution containing an α-amino acid determines the form in which it exists. The pK_a of the carboxyl group is about 2, and the pK_a of the conjugate acid of the amino group (that is, of the ammonium cation) is about 9. In an acid solution (pH ~1), the α-amino acid is entirely in its cationic form; the strong acid present has protonated the amino group and the carboxylate anion. However, if the pH is gradually raised (for example, by the gradual addition of sodium hydroxide), the most acidic proton is lost first (the carboxyl proton, pK_a ~2). When the pH reaches what is called the *isoelectric point,* the concentration of the zwitterion is at its maximum. (The isoelectric point, abbreviated pI, is the average of the pK_a values of the two acidic groups—usually about 6.) If the pH continues to rise, the second proton (the ammonium proton, pK_a ~9) is lost and the α-amino acid exists entirely in its anionic form.

α-Amino acids undergo the reactions characteristic of both amines and carboxylic acids. To do so, the amino and carboxyl groups must be "free"—that is, un-ionized. Therefore, the pH of a solution is extremely important when considering the reactions of α-amino acids. For example, for an α-amino acid to be acetylated by acetyl chloride requires that the amino group bring about a nucleophilic acyl substitution on the acetyl chloride. This can only happen if the amino group is free to use its unshared elec-

FIGURE 16.1

Reactions of α-amino acids requiring free amino or carboxyl groups.

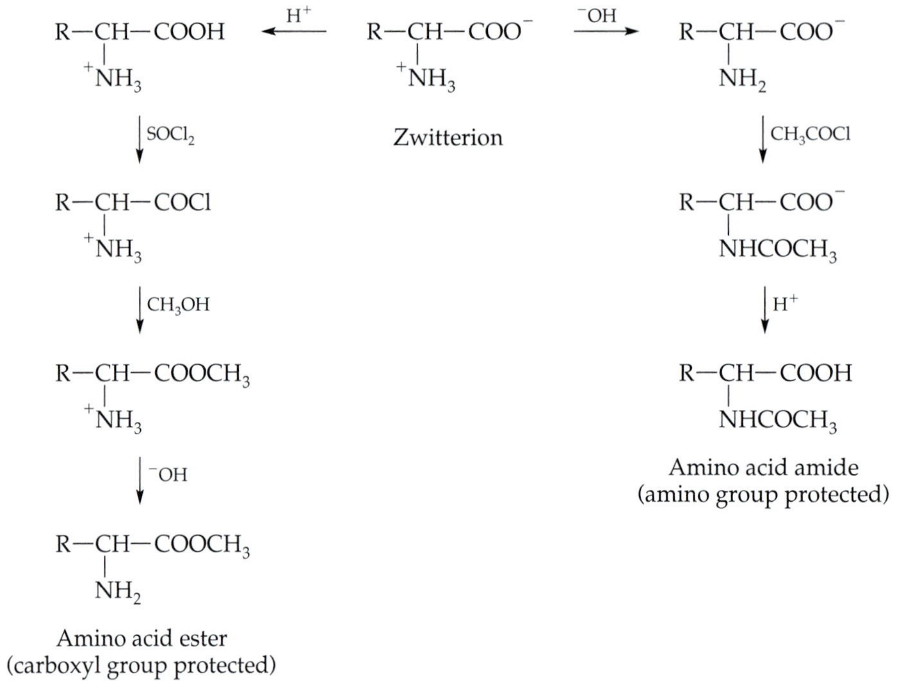

$$R{-}CH({}^{+}NH_3){-}COOH \xleftarrow{H^+} R{-}CH({}^{+}NH_3){-}COO^- \xrightarrow{^-OH} R{-}CH(NH_2){-}COO^-$$

Zwitterion

$$R{-}CH({}^{+}NH_3){-}COOH \xrightarrow{SOCl_2} R{-}CH({}^{+}NH_3){-}COCl \xrightarrow{CH_3OH} R{-}CH({}^{+}NH_3){-}COOCH_3 \xrightarrow{^-OH} R{-}CH(NH_2){-}COOCH_3$$

Amino acid ester (carboxyl group protected)

$$R{-}CH(NH_2){-}COO^- \xrightarrow{CH_3COCl} R{-}CH(NHCOCH_3){-}COO^- \xrightarrow{H^+} R{-}CH(NHCOCH_3){-}COOH$$

Amino acid amide (amino group protected)

trons—in other words, if it is not protonated. Thus, such a reaction requires the anionic form of the amino acid, which exists only in basic solution (Figure 16.1). Conversely, converting the carboxyl group of an amino acid to an ester group by alcoholysis of the corresponding acid chloride derivative requires that the carboxyl group be intact and not ionized into the carboxylate anion. In other words, the carboxyl group must be protonated, which only occurs in acidic solution (see Figure 16.1).

Forming peptides from individual α-amino acids involves "doing chemistry" on one functional group to the exclusion of the other. This is accomplished by *protecting* the amino or carboxyl group—that is, converting it into a derivative that will temporarily block its normal behavior, but from which it can be recovered. The amino group is converted to an amide and is no longer nucleophilic (recall that amides are neutral). The carboxyl group is converted to an ester and is therefore no longer acidic. The ability to protect (and deprotect) the functional groups in α-amino acids is essential to peptide synthesis.

16.1.3 Syntheses of α-Amino Acids

α-Amino acids can be obtained by ammonolysis of α-bromocarboxylic acids (available from carboxylic acids by bromination). (See Sections 12.6.1 and 14.3.2.)

General alkylation reaction:

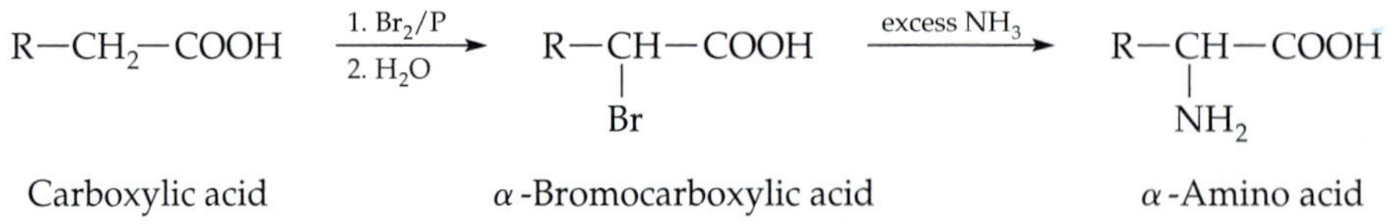

$$R{-}CH_2{-}COOH \xrightarrow[2.\ H_2O]{1.\ Br_2/P} R{-}CH(Br){-}COOH \xrightarrow{\text{excess } NH_3} R{-}CH(NH_2){-}COOH$$

Carboxylic acid — α-Bromocarboxylic acid — α-Amino acid

A number of other methods are also available, but we will consider only one more, the **Strecker synthesis.** Reaction of an aldehyde with ammonia and hydrogen cyanide

(HCN) produces an α-aminonitrile. The cyano group is then hydrolyzed to a carboxyl group, producing an α-amino acid. The reaction probably proceeds via imine formation (see Section 13.4.1), followed by addition of HCN, which is exactly analogous to the cyanohydrin-forming reaction of aldehydes (see Section 13.3.2).

General Strecker synthesis:

$$\underset{\text{Aldehyde}}{R{-}CHO} \xrightarrow{NH_3} \underset{\text{Imine}}{[R{-}CH{=}NH]} \xrightarrow{HCN} \underset{\alpha\text{-Aminonitrile}}{R{-}\underset{NH_2}{\underset{|}{C}}HCN} \xrightarrow{HCl/H_2O} \underset{\alpha\text{-Amino acid}}{R{-}\underset{NH_2}{\underset{|}{C}}HCOOH}$$

The two reactions just considered produce an α-amino acid in racemic form. To be used in peptide synthesis, the L enantiomer must be obtained. There are two approaches to this problem. The first is to accomplish a resolution of the racemate into the two enantiomers, keeping the L-amino acid. This is a laborious process but it has been carried out for many years in the laboratory and in industry. The second approach is to bring about a stereoselective synthesis of the L enantiomer. Such syntheses have been developed and commercialized only in recent years. Both approaches will be described in Chapter 17.

PROBLEM 16.2

Devise two syntheses for α-aminophenylacetic acid: (a) starting with toluene and (b) starting with benzaldehyde.

16.1.4 Analysis for α-Amino Acids

Determining the structure of a protein depends on being able to analyze the protein for all 22 possible α-amino acids. Acid hydrolysis of a protein—that is, the acid-catalyzed cleavage of all the peptide (amide) bonds as described in Section 15.4.2—produces a mixture containing all of the α-amino acids that are present in the protein. For example, the tripeptide Phe-Gly-Ser is hydrolyzed to phenylalanine, glycine, and serine:

$$C_6H_5CH_2\underset{NH_2}{\underset{|}{C}}H\overset{O}{\overset{\|}{C}}{-}NHCH_2\overset{O}{\overset{\|}{C}}{-}NHCH(CH_2OH)COOH$$

Phe-Gly-Ser

$$\downarrow H^+/H_2O$$

$$\underset{\text{Phenylalanine}}{C_6H_5CH_2\underset{NH_2}{\underset{|}{C}}HCOOH} + \underset{\text{Glycine}}{H_2NCH_2COOH} + \underset{\text{Serine}}{H_2NCH(CH_2OH)COOH}$$

Determining which α-amino acids are present in a hydrolysis mixture and in what relative amounts is accomplished using an instrument called an automatic amino acid analyzer (AAAA). Each α-amino acid is detected using a colorimetric procedure. The result is a graph like that shown in Figure 16.2.

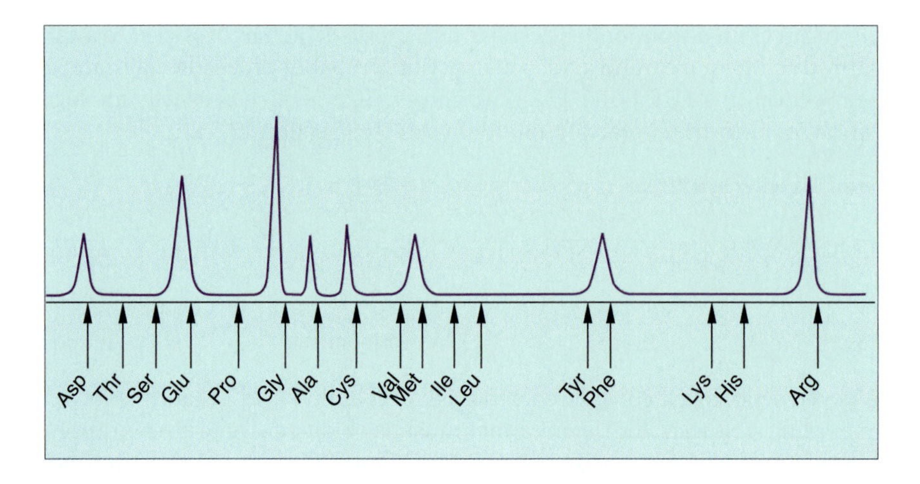

FIGURE 16.2

Graph produced by an automatic amino acid analyzer (AAAA).

For example, analysis of the B chain of bovine insulin reveals a composition of 16 different α-amino acids forming a chain of 30 units (subscripts indicate multiple occurrences):

$$Ala_2, Arg, Asn, Cys_2, Gln, Glu_2, Gly_3, His_2, Leu_4, Lys, Phe_3, Pro, Ser, Thr, Tyr_2, Val_3$$

In general, when the number of α-amino acids in a protein is given, it means the *total number* of α-amino acids in the protein, not how many *different* α-amino acids are present.

After the α-amino acids present in a protein and their relative amounts have been determined, the most difficult step remains, that of determining the sequence of their connections. As we will see in Section 16.3.1, techniques have been devised to determine the sequence.

PROBLEM 16.3

A small peptide is subjected to automatic amino acid analysis. It produces peaks for only three α-amino acids: alanine (Ala), histidine (His), and proline (Pro), in the ratio 1:1:1. The peptide is shown to have a molecular weight of 323. Draw all possible structures that should be considered for this peptide, using the abbreviations for the α-amino acids.

16.2 Classification of Proteins

Many different kinds of proteins are known, encompassing an amazing variety of functions and roles in living systems. Each protein contains a unique sequence of α-amino acids, and the human body alone contains about 100,000 different proteins. Yet, this complexity is derived from the simplicity of just 22 α-amino acids linked together. Before looking more closely at protein structure, we will consider the different kinds of proteins. Several different classification systems are used to describe proteins.

16.2.1 Classification by Composition

The first classification system for proteins depends on their composition. **Simple proteins** are those consisting of only α-amino acids. Examples include insulin (the hormone that controls the level of blood sugar) and blood serum albumin. Far more

TABLE 16.2

Conjugated Proteins

Conjugated Protein	Example	Nonprotein Component
Lipoprotein	Plasma lipoprotein	Cholesterol
Glycoprotein	γ-Globulin	Carbohydrate
	Interferon	Carbohydrate
Phosphoprotein	Casein	Phosphate ester
Metalloprotein	Hemoglobin	Heme
Nucleoprotein	DNA	Deoxyribonucleic acid
	RNA	Ribonucleic acid
	Virus	DNA or RNA

numerous are the **conjugated proteins,** compounds in which a protein has a nonprotein compound attached to it. For example, glycoproteins, which are a major component of cell membranes, contain a carbohydrate fragment (*glyco* means "sugar"). In some conjugated proteins, such as lipoproteins (*lipo* means "fat") and nucleoproteins (*nucleo* refers to a nucleic acid, DNA or RNA), the nonprotein component is the major constituent. Table 16.2 lists some conjugated proteins.

16.2.2 Classification by Shape

A second classification system for proteins is based on their three-dimensional shapes. Proteins that consist of long, relatively continuous chains of side-by-side polypeptides are known as **fibrous proteins.** These proteins, which resemble a multiwound fiber bundle, are very strong. They are also insoluble in water and body fluids, mainly because the majority of their R groups exposed on their surface are nonpolar (in other words, they are hydrophobic). These two characteristics make fibrous proteins ideally suited as components of the body's structural material. Examples include collagen (found in tendons, muscles, and hooves), keratin (found in hair, skin, and fingernails), and actin and myosin (the key muscle proteins).

Proteins that are coiled into compact, approximately spherical shapes are called **globular proteins.** These proteins are soluble in water and body fluids, mainly because they contain α-amino acids with polar R groups (such as amino and carboxyl groups) exposed on their surface—groups that are known to be hydrophilic. Globular proteins can generally move about within cells, unlike the fibrous proteins that are incorporated into the body's structure. Examples of globular proteins include enzymes, transport proteins such as hemoglobin, and hormones such as insulin. The proteins listed in Table 16.3 are organized by shape as well as by function.

16.2.3 Classification by Function

A third way to classify proteins is by their biological functions. The diversity of protein function is extraordinary, as suggested by the list in Table 16.3.

TABLE 16.3

Biological Functions of Some Proteins

Type of Protein	Example	Function
Fibrous		
Structural	Collagen	In tendons
Surface	Keratin	In fingernails
Contractile	Myosin	In muscle
Globular		
Protection	Antibodies	Fight infection
Enzymes	Chymotrypsin	Catalysis
Transport	Hemoglobin	Transport oxygen and carbon dioxide
Hormones	Insulin	Control blood sugar
Storage	Casein	Store nutrients
Defense	Venom	Repel attackers

16.3

Structure of Proteins

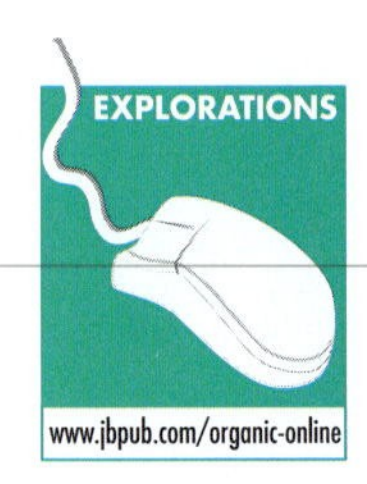

The structure of proteins is much more complex than that of any of the families of organic compounds we have considered. This structural complexity arises from the fact that proteins are such large molecules, often referred to as **macromolecules.** A protein may consist of just a single strand, or it may include several strands. A protein may have just a few α-amino acids, or it may have thousands:

- Oxytocin is a single-stranded protein, consisting of 9 α-amino acids.

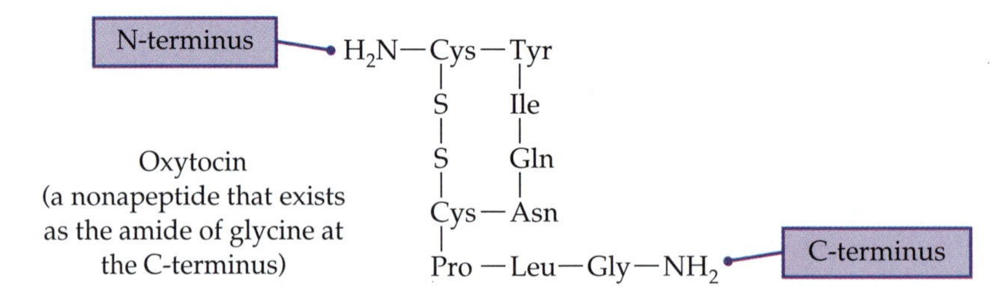

Oxytocin
(a nonapeptide that exists as the amide of glycine at the C-terminus)

- Insulin is a two-chain protein, consisting of 51 α-amino acids.
- Myoglobin is a single-chain protein, consisting of 153 α-amino acids.
- Gamma globulin is a four-chain protein, consisting of 1320 α-amino acids.

Proteins provide a classic example of how chemical structure, and ultimately three-dimensional shape, determines biological function. There are four levels of protein structure that are important in understanding how proteins function. The simplest two-dimensional description of connectivity is called the primary structure. The three-

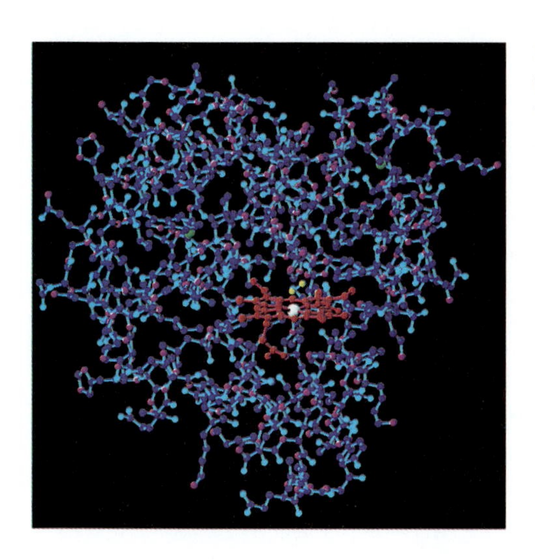

Myoglobin is the oxygen carrying protein that is used by muscles as a rapidly available source of oxygen.

dimensional gross structure—specifically, the secondary, tertiary, and quaternary structure—describes how protein molecules are aggregated in order to carry out their specific functions.

16.3.1 Primary Structure of Proteins

The most basic aspect of protein structure is the **primary structure.** This is, quite simply, the specific sequence in which the α-amino acids are joined together by peptide bonds. Primary structure is represented using the three-letter abbreviations for the α-amino acids (see Table 16.1). For example, the primary structure of human insulin, a two-chain protein, is shown in Figure 16.3. Note that the two chains of insulin, the A chain and B chain, are held together by two disulfide (—S—S—) linkages. These result from oxidation of —SH groups on the α-amino acid cysteine (Cys), as described for thiols in Section 4.1.2.

Primary structure is determined through a process known as **sequence analysis**—determination of the actual sequence in which the various α-amino acids are connected to one another. The two general approaches to sequence analysis provide complementary information.

FIGURE 16.3

Human insulin (an A chain of 21 amino acids and a B chain of 30 amino acids).

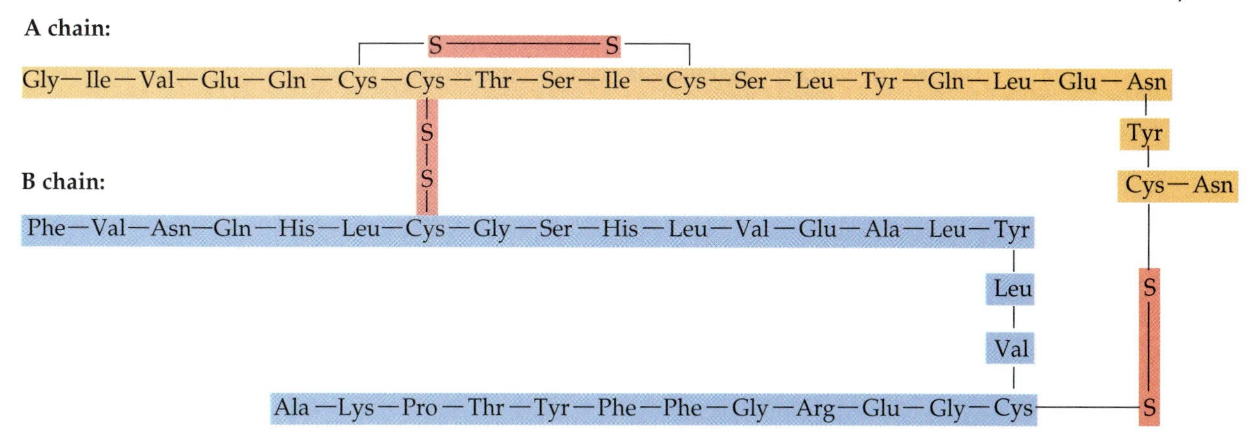

The first approach is **end-group analysis**—determining which α-amino acids are at the N-terminus and C-terminus of the chain. End-group analysis relies on "tagging" the N-terminal or C-terminal α-amino acid in the peptide. Then, upon hydrolysis of the peptide, the terminal α-amino acid can be readily identified through its "tag."

N-terminal analysis is done by reacting a polypeptide with phenylisothiocyanate, to which the free amino group adds. Treatment with dry hydrogen chloride cleaves the first peptide bond, leaving the remainder of the polypeptide chain intact and releasing a fragment that rearranges into the phenylthiohydantoin derivative of the N-terminal amino acid. The phenylthiohydantoin is identified by comparison with the known phenylthiohydantoins of the α-amino acids. The process can be repeated up to about 20 times with the remaining peptide fragment (which has a different N-terminal amino acid exposed each time) to determine the sequence of several α-amino acids from the N-terminus of the peptide chain. This process of "clipping off" α-amino acids one at a time from the N-terminus is called the **Edman degradation.**

$C_6H_5{-}N{=}C{=}S$

Phenylisothiocyanate

+

$H_2N{-}CHRCO{-}NH{-}peptide$

Polypeptide chain

$\longrightarrow$ $C_6H_5{-}NH{-}CS{-}NH{-}CHRCO{-}NH{-}peptide$

A thiourea

↓ HCl

α-Amino acid fragment

A phenylthiohydantoin (ring: $C_6H_5{-}N$, C=S, NH, C(R), C=O) + $H_2N{-}peptide$

Polypeptide with one less α-amino acid

C-terminal analysis is done using the enzyme carboxypeptidase, which is known to cleave peptides at the terminal carboxyl end, releasing the free α-amino acid. Because the enzyme continues to attack the carboxyl end of the shortened peptide chain and release amino acids, it is necessary to monitor the rate at which free α-amino acids appear in the solution to detect their sequence of appearance.

For example, the two types of end-group analysis can be used to determine that Gly is the N-terminus α-amino acid and Arg is the C-terminus α-amino acid in a peptide known to contain Gly, Lys, Ala, Tyr, Ile, and Arg:

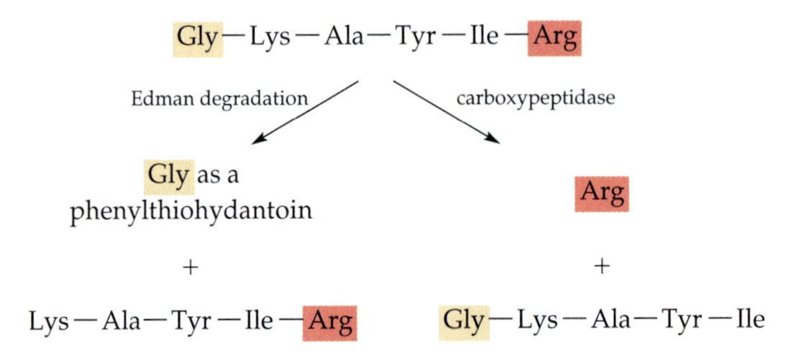

The second approach to sequence analysis is **fragment analysis,** which employs various means of cleaving a peptide into smaller peptide fragments, each of which may then be analyzed by end-group analysis. One technique is to undertake partial acid-catalyzed hydrolysis of a peptide. Such a cleavage occurs randomly, not at any predictable peptide linkage. A more systematic fragment analysis technique employs enzymes that are known to cleave only certain peptide linkages. For example, the enzyme trypsin only cleaves a peptide bond in which the carboxyl group is part of lysine (Lys) or arginine (Arg). Chymotrypsin only cleaves peptide bonds at the carboxyl groups of phenylalanine (Phe), tyrosine (Tyr), and tryptophan (Trp). Based on the overlapping information that can be obtained from knowing these cleavages have occurred, it is possible to determine the exact amino acid sequence of very large proteins, as demonstrated in the following example problem.

How to Solve a Problem

Determine the α-amino acid sequence and structure of a hypothetical polypeptide that produces the following experimental results.

(1) AAAA indicates that the following nine α-amino acids are present, all in the same amount: Ala, Arg, Gly, His, Ile, Leu, Lys, Phe, and Tyr. Based on the molecular weight of the peptide, it is deduced that only one of each α-amino acid is present.

(2) N-terminal analysis by Edman degradation indicates Gly at that end.

(3) C-terminal analysis using carboxypeptidase indicates His at that end.

(4) Hydrolysis with trypsin produces three fragments: a dipeptide, a tripeptide, and a tetrapeptide. By Edman degradation, these fragments are shown to be

Dipeptide: Gly—Lys
Tripeptide: Phe—Leu—His
Tetrapeptide: Ala—Tyr—Ile—Arg

(5) Hydrolysis with chymotrypsin also produces three fragments: a dipeptide, a tripeptide, and a tetrapeptide. By Edman degradation, these fragments are shown to be

Dipeptide: Leu—His
Tripeptide: Ile—Arg—Phe
Tetrapeptide: Gly—Lys—Ala—Tyr

SOLUTION

We arrange the known fragments from left to right, starting with a Gly fragment, known to be the N-terminus. The overlapping amino acid sequences become apparent, leading to a final overall sequence for the nonapeptide.

Fragments: Gly—Lys—Ala—Tyr
Ala—Tyr—Ile—Arg
Ile—Arg—Phe
Phe—Leu—His

Final result: Gly—Lys—Ala—Tyr—Ile—Arg—Phe—Leu—His

In retrospect, we can see that the enzymatic cleavages of the polypeptide took place at these points:

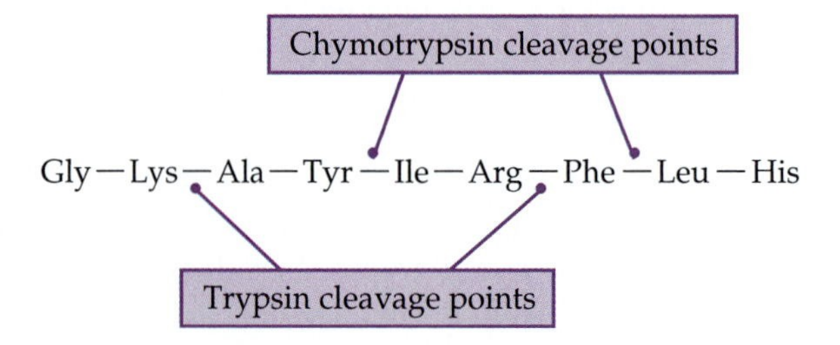

PROBLEM 16.4

For the tetrapeptide His-Lys-Phe-Leu, identify the following:

(a) the α-amino acid obtained from Edman degradation
(b) the fragment(s) obtained via trypsin cleavage

PROBLEM 16.5

Show the structure of a peptide that fits the following information:

(a) AAAA shows Arg, Gly, Ile, Phe, Pro_2, and Val.
(b) Partial hydrolysis produces three fragments: Arg-Gly-Pro, Pro-Phe-Ile-Val, and Pro-Pro.

16.3.2 Secondary Structure of Proteins

How a protein chain is arranged or oriented in three-dimensional space is, in large part, described by what is called the **secondary structure.** The secondary structure of proteins is determined, in part, by two aspects of organic chemistry we studied in previous chapters—resonance and hydrogen bonding.

Resonance delocalization within the amide (peptide) group causes restricted rotation about the nitrogen-carbon bond, because it acquires partial double bond character:

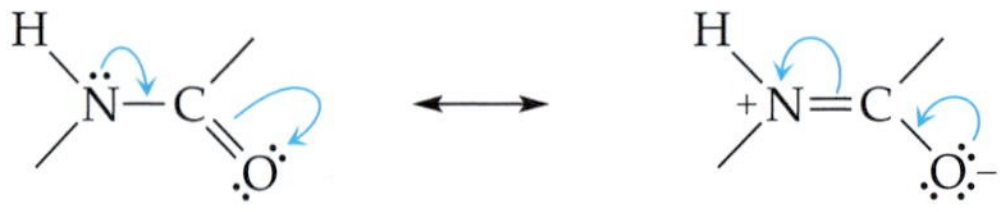

Resonance contributing structures for an amide

The NH—C=O grouping lies in a plane with bond angles of about 120°. The carbon atoms to which the nitrogen and the carbonyl carbon are attached (the α-carbons of both amino acids) must also lie in the same plane. Therefore, the four atoms of the peptide backbone (—C—N—C—C—) lie in a plane, but there are two variants:

Planar structures for the peptide backbone

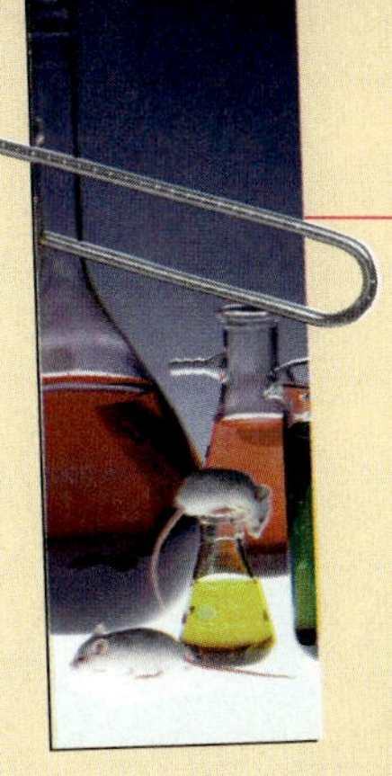

CHEMISTRY AT WORK

IN THE MARKETPLACE

Nylon and Other Polyamides

It has been known for a long time that wool and silk are proteins; in time, it was discovered that proteins are polyamides. In 1935, the first commercial synthetic polyamide, called Nylon, was produced by the DuPont Company, formulated into a fiber, and marketed as nylon stockings. A shopping frenzy resulted.

There are several different nylon polymers. Nylon-6,6 is made from the C_6 dicarboxylic acid called adipic acid (hexanedioic acid) and the C_6 diamine called hexamethylenediamine (1,6-diaminohexane). Heating a mixture of the two causes loss of a water molecule between an amino group and a carboxyl group, forming an amide group.

$$\underset{\text{Hexanedioic acid}}{HOOC(CH_2)_4COOH} + \underset{\text{1,6-Diaminohexane}}{H_2N(CH_2)_6NH_2} \xrightarrow[\Delta]{-H_2O} \underset{\text{Nylon-6,6}}{\left[OC(CH_2)_4CONH(CH_2)_6NH\right]_n}$$

The reaction repeats itself to form a long-chain polyamide polymer that is employed in a number of manufacturing processes. It can be spun into fibers to make products as diverse as nylon stockings and long-wear carpet. Alternatively, it can be molded into rigid materials such as lightweight gears and automobile parts.

Nylon-6 was originally produced by heating ϵ-aminocaproic acid and forming the polyamide chain by eliminating water between identical molecules. It was later found feasible to obtain the same product by ring opening and polymerization of the internal amide, ϵ-caprolactam (see Sections 18.3.4 and 18.4.1).

ϵ-Caprolactam (cyclic amide: C=O, NH in seven-membered ring) —(rearrangement) Δ→ Nylon-6

ϵ-Aminocaproic acid $HOOC(CH_2)_5NH_2$ —($-H_2O$) Δ→ Nylon-6

$$\underset{\text{Nylon-6}}{\left[OC(CH_2)_5NH\right]_n}$$

More recently, extremely strong polyamide compounds called *aramids* have been prepared by replacing the aliphatic sections of nylon with aromatic rings. For example, *p*-phenylenediamine (1,4-diaminobenzene) and terephthalic acid (the same diacid used in making Dacron) produce the polyamide sold as Kevlar. It is used instead of steel to make radial tire cord and also in crash helmets and bulletproof vests.

Production of Kevlar:

$$\underset{\text{Terephthalic acid}}{HOOC{-}C_6H_4{-}COOH} + \underset{p\text{-Phenylenediamine}}{H_2N{-}C_6H_4{-}NH_2} \xrightarrow[-H_2O]{\Delta} \underset{\text{Kevlar}}{\left[OC{-}C_6H_4{-}CONH{-}C_6H_4{-}NH\right]_n}$$

There is free rotation about the other single bonds in a peptide. Therefore, these are the rotational options for the peptide:

Restricted rotation

Free rotation | Free rotation

—NH—CH(R)—C(=O)—NH—CH(R)—C(=O)—

These four backbone atoms lie in the same plane

(a)

The secondary structure of a protein is generally held in place by hydrogen bonding between the —NH group of one amide and the C=O group of another amide. Because a protein chain contains a large number of peptide bonds, there is ample opportunity for extensive hydrogen bonding. Such hydrogen bonding may occur between peptide linkages of two separate chains or between peptide linkages in sufficiently close proximity within a single chain.

(b)

N—H ⋯⋯ Ö=C

Hydrogen bond between —NH group of one amide and C=O group of another amide

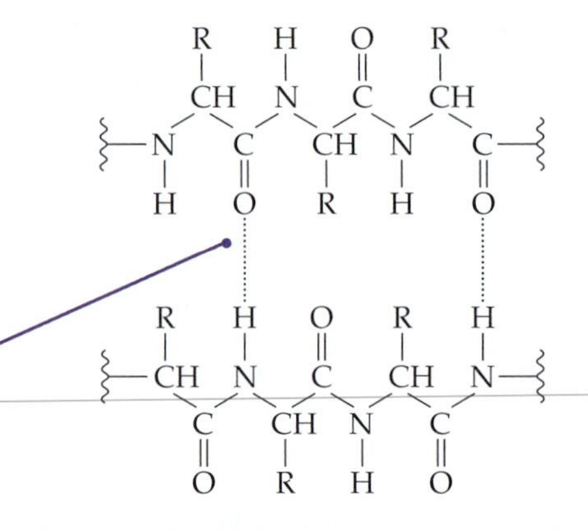

Hydrogen bonds between two protein chains

There are two well-known kinds of secondary structure. The first kind is the **α-helix,** in which a protein chain is coiled about a central axis. You can envision the helix by wrapping a colored string of wool or a pipe cleaner around a tall drinking glass, left to right, starting at the bottom and working your way up the glass. The helix is alpha because the coiling is to the right as it moves up the glass, just as in a right-threaded screw (Figure 16.4a). Alternatively, you can examine a typical coiled computer or telephone cord (Figure 16.4b). Your particular cord may be alpha (coiling to the right) or beta (coiling to the left).

The α-helix in a protein (Figure 16.4c) is held in place by hydrogen bonding that occurs between the hydrogens of —NH groups and the oxygens of C=O groups that are oriented toward each other on the *inside of the helix.* It takes 3.6 amino acids to complete one turn of the helix to enable such hydrogen bonding. The R groups of the amino acids are oriented on the *outside* of the helix, pointed away from the central axis.

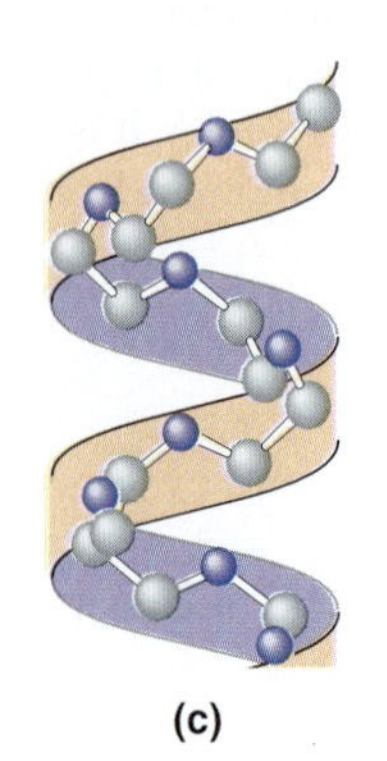

(c)

FIGURE 16.4

Three examples of an α-helix.

The proteins α-keratin (found in hair and wool) and myosin and actin (found in muscle) have helical secondary structures. When any of these proteins is stretched, the helix becomes elongated as hydrogen bonds break; upon relaxation, the hydrogen bonds re-

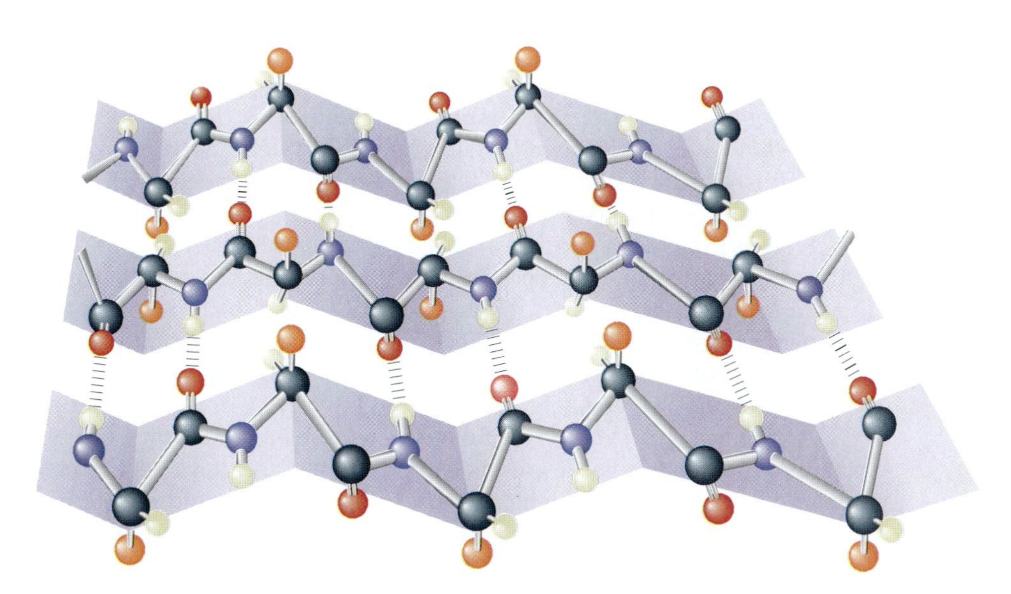

FIGURE 16.5

The pleated sheet secondary structure of proteins.

form and the α-helical shape returns. This is much like a coiled spring being stretched and then assuming its original shape when released. Even though each hydrogen bond in an α-helix is relatively weak (about 5 kcal/mol), the combined energy of many of these bonds provides the stabilizing energy for the helical structure.

The second kind of secondary structure for proteins is the **pleated sheet.** In this structure, two protein chains are lined up parallel to each other but running in opposite directions—that is, the N-terminus of one chain is near the C-terminus of the other. This puts the —NH group of one chain opposite the C═O group of the other chain, permitting hydrogen bonding that holds the sheet together. The pleated sheet looks like a piece of paper that has been folded back on itself many times (like a folded road map) and then stretched out, but not to its full length. This creates a hill-and-valley pattern. A planar amide group and its two attached α-carbons lie in the plane of each pleat, with the tetrahedral α-carbons at the edges, or creases, of each pleat. This means that the angles between the folds of the pleat are about 109.5°, corresponding to the geometry of the α-carbons. The R groups of each α-amino acid project from the creases of the pleated sheet, alternately pointing above or below the overall plane of the sheet (Figure 16.5).

The pleated sheet structure forces the R groups of the α-amino acids in the two chains to be in relatively close proximity (on the same side of the sheet), causing some intergroup repulsion. Therefore, pleated sheet structures are stable only when the majority of the α-amino acids have very small R groups or no R groups (glycine). A common example is fibroin, the protein of silk, which consists mainly of glycine and alanine. This protein does not stretch significantly. "Dragline silk," which forms the major framework of a spider web, is 42% glycine (R = H) and 25% alanine (R = CH_3) and exists mainly as a pleated sheet.

"Dragline Silk", which forms the framework of a spider web, is composed of fibroin. Fibroin consists mainly of glycine and alanine.

Finally, just to complicate matters, some large proteins have structural regions that are pleated, others that are helical, and still others that are neither. For example, about 25% of the α-amino acids in lysozyme, an enzyme with 129 amino acids, are in three separate helical regions, while about 7% exist in two pleated regions (Figure 16.6).

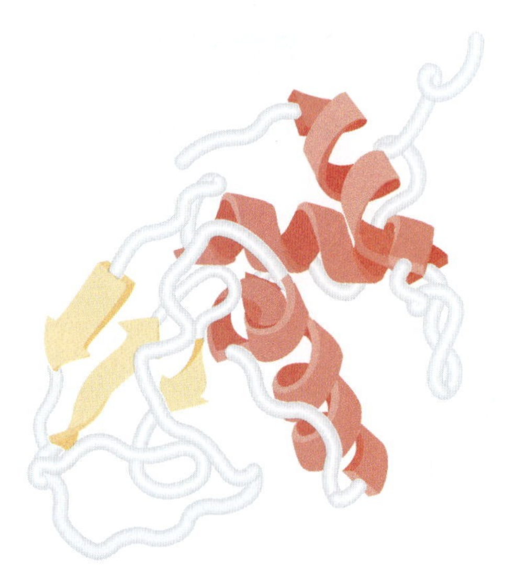

FIGURE 16.6

Helical, pleated sheet, and other regions of secondary structure of a lysosome.

16.3.3 Tertiary Structure of Proteins

The overall shape of a protein molecule is defined by the **tertiary structure**—how the protein chains, whether pleated or helical, are folded. The tertiary structure determines whether a protein is elongated (and thus fibrous) or folded (and thus globular).

Fibrous proteins (Figure 16.7a) are elongated in their overall shape, and they are insoluble in water because the majority of their R groups are nonpolar (in other words, hydrophobic). Examples of such proteins are the structural proteins keratin (found in hair, feathers, claws, and fingernails) and collagen (found in hooves, tendons, cartilage, and blood vessels). Globular proteins (Figure 16.7b) tend to form globules—spherical masses. Such proteins are generally water-soluble because they contain amino acids with polar R groups (such as amino and carboxyl groups). Examples of globular proteins are enzymes, antibodies, and myoglobin, all of which are soluble in body fluids and therefore transportable.

The tertiary structure of a protein brings into close proximity α-amino acids that are very far apart in the primary structure, thereby creating chemically active sites. The tertiary structure is held together through a number of chemical interactions with which you are now familiar: hydrogen bonding, oxidation of cysteine units [$HSCH_2CH(NH—)CHCO—$] to cystine units to form disulfide (—S—S—) linkages (as in insulin—see Figure 16.3), and electrostatic attraction between carboxylate groups (as in aspartic acid and glutamic acid) and ammonium groups (as in arginine, lysine, and histidine). Such interactions are represented in Figure 16.8.

When an egg is cooked, protein denaturation is irreversible.

A protein must be in its proper shape (tertiary structure) to fulfill its designated biological function. When a protein is *denatured,* such as by treatment with acid or alcohol or by heating, it unfolds and becomes biologically inactive—its tertiary structure has been destroyed. In most cases, protein denaturation is irreversible, as when an egg is cooked. However, with some proteins (such as some enzymes), renaturation may occur, with the protein folding itself back into its tertiary structure. After refolding, a process that has recently been observed using high-speed laser techniques, the protein's biological function is reactivated.

16.3.4 Quaternary Structure of Proteins

Certain proteins, usually globular proteins, further organize into aggregates, or assemblies. The spatial orientation of these aggregates is described as the **quaternary**

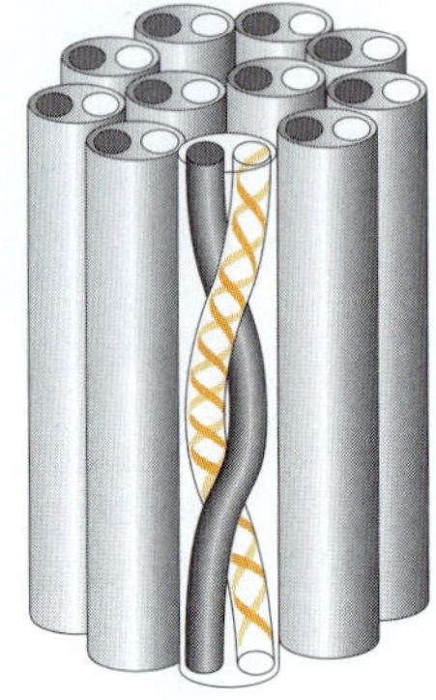

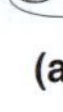

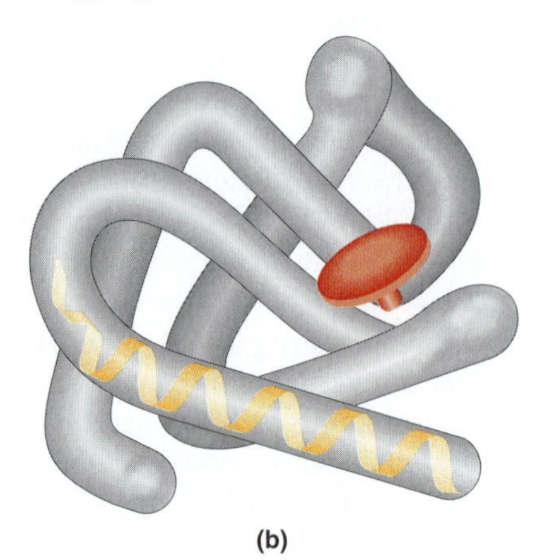

FIGURE 16.7

Schematic representation of (a) a fibrous protein and (b) a globular protein.

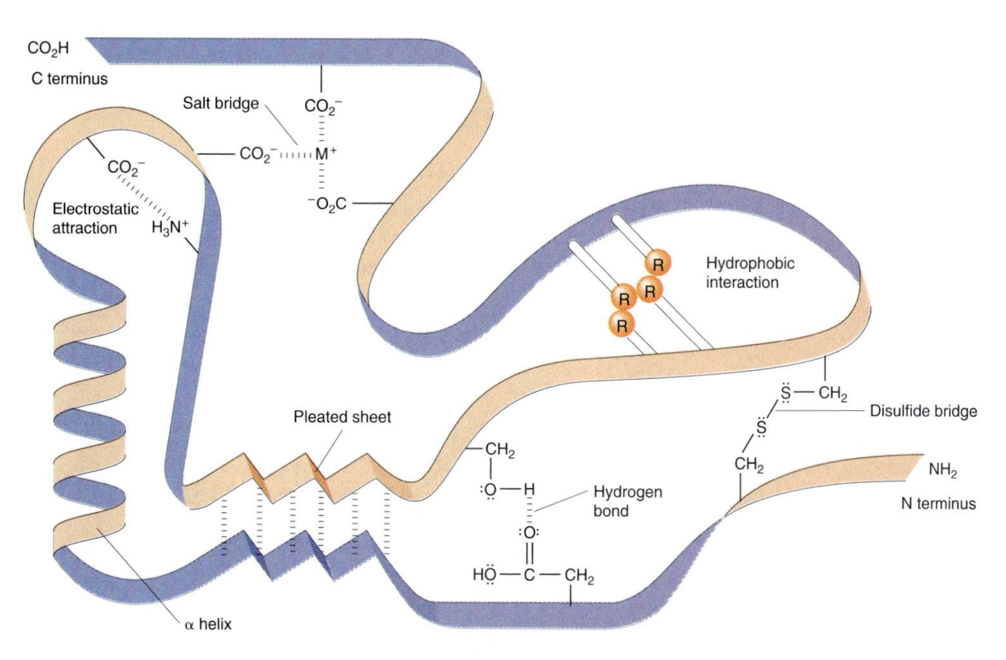

FIGURE 16.8

A representation of a globular protein showing the chemical interactions that produce the tertiary structure.

structure of the protein. A classic example is hemoglobin, the substance in red blood cells responsible for transporting oxygen from the lungs and carbon dioxide to the lungs. Hemoglobin (molecular weight about 64,500) consists of four individual protein chains, two α chains of 141 α-amino acids each, and two β chains of 146 α-amino acids each. Each of these chains has a globular tertiary structure (Figure 16.9a). In hemoglobin, each protein chain is bonded to a heme molecule containing an iron atom in the center of a porphyrin ring structure (Figure 16.9b). It is the iron atom that binds to the oxygen or carbon dioxide during transport in the bloodstream. It is also the iron atom that will complex even more strongly and irreversibly with carbon monoxide or cyanide ion, making exposure to these substances so deadly. (Because the iron is fully complexed by these poisons, the heme cannot then transport the oxygen needed to support life.)

Although the heme unit is the critical transport unit of hemoglobin, the importance of the protein portion (the globin) cannot be overstated. A mutated form of globin, in which only one α-amino acid out of 146 in the β chain is abnormal, is the cause of sickle cell anemia. The naturally occurring glutamic acid is replaced by valine in the mutated form.

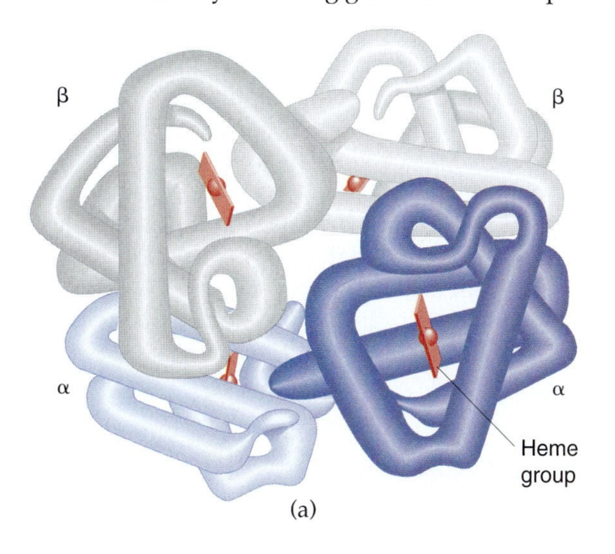

(a)

CH_3 CH_2CH_2COOH

$CH_2{=}CH$ N CH_2CH_2COOH

N—Fe—N

CH_3 N CH_3

$CH_2{=}CH$ CH_3

(b)

FIGURE 16.9

(a) A representation of hemoglobin, showing the assemblage of four globular proteins. (b) A heme molecule.

16.4

Synthesis of Peptides and Proteins

The synthesis of peptides and proteins presents a very difficult challenge because of the identical bifunctional nature of each α-amino acid reactant. The task involves creating many identical structural linkages (peptide bonds). Protein synthesis was accomplished by first discovering how to synthesize small peptides.

16.4.1 Solution Synthesis

To synthesize a dipeptide, two amino acids need to be joined by a peptide (amide) bond. But which amino group and which carboxyl group will be involved? For example, the coupling of glycine with alanine by forming a peptide bond could produce four different dipeptide products:

$$\underset{\text{Gly}}{H_2NCH_2COOH} + \underset{\text{Ala}}{H_2NCH(CH_3)COOH} \dashrightarrow \text{Gly—Ala} + \text{Ala—Gly} + \text{Gly—Gly} + \text{Ala—Ala}$$

The solution to this problem is to protect (block) those functional groups that are *not* to be involved in forming a new peptide bond and to do so with protecting groups that can be readily removed. However, the protecting groups must be removable using reactions that will not cleave the newly formed peptide bond, which is susceptible mainly to hydrolysis. The modern peptide synthesis process, therefore, has five distinct steps (shown in Figure 16.10 for the synthesis of the dipeptide Gly-Ala):

Step 1. Protect the amino group of the α-amino acid glycine (**A1**) that will be the N-terminus. Convert the glycine to an amide (**A2**).

Step 2. Protect the carboxyl group of the α-amino acid alanine (**B1**) that will be the C-terminus. Convert the alanine to an ester (**B2**).

Step 3. Now there is only a single free amino group and a single free carboxyl group that can be joined to form the peptide linkage. Couple the free amino group of **B2** with the free carboxyl group of **A2**, forming the new peptide bond.

Step 4. Remove the protecting group on the N-terminus.

Step 5. Remove the protecting group on the C-terminus.

There are several possibilities for the condensation reaction of step 3. All involve converting the free carboxyl group of **A2** into a form that is susceptible to nucleophilic acyl substitution by the nucleophilic free amino group of **B2.** Forming an acid chloride from **A2** seems obvious and was an early approach. Then ammonolysis with the free amino group of **B2** produces the peptide (amide). However, the most effective reagent for the condensation reaction is *N,N*-dicyclohexylcarbodiimide, abbreviated DCC. (The nature of this condensation reaction is beyond the scope of this book.) Treatment of any primary amine with any carboxylic acid in the presence of DCC produces the expected amide.

$$C_6H_{11}\text{—}N\text{=}C\text{=}N\text{—}C_6H_{11}$$

N,N-Dicyclohexylcarbodiimide (DCC)

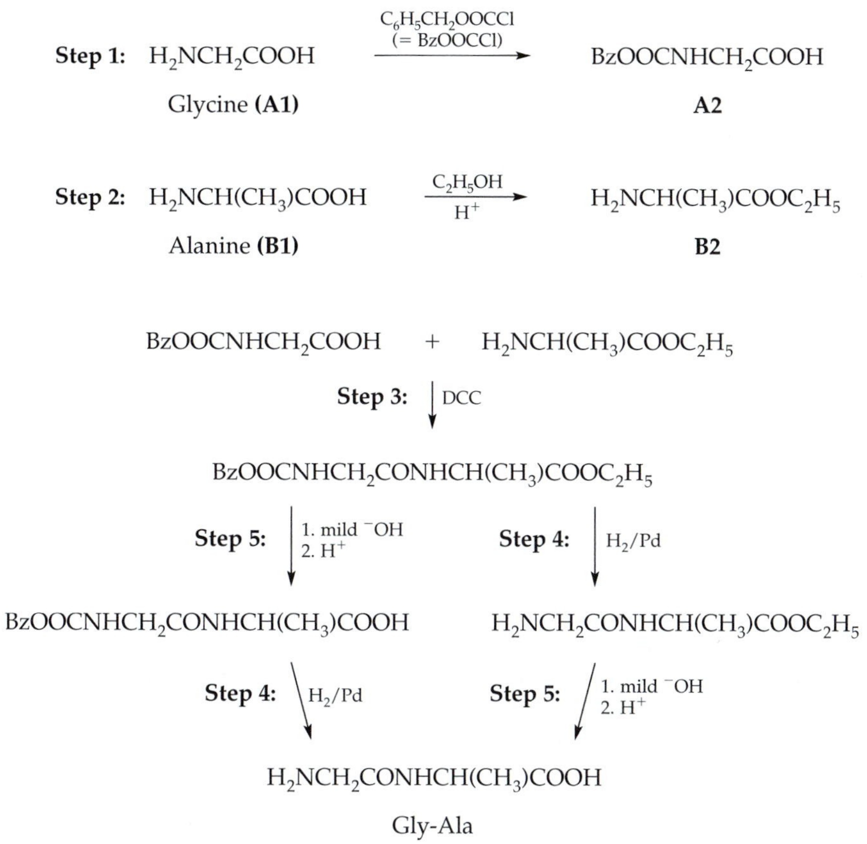

$$\textbf{Step 1:}\quad \underset{\text{Glycine }\mathbf{(A1)}}{H_2NCH_2COOH} \xrightarrow[\;]{C_6H_5CH_2OOCCl\ (= BzOOCCl)} \underset{\mathbf{A2}}{BzOOCNHCH_2COOH}$$

$$\textbf{Step 2:}\quad \underset{\text{Alanine }\mathbf{(B1)}}{H_2NCH(CH_3)COOH} \xrightarrow[H^+]{C_2H_5OH} \underset{\mathbf{B2}}{H_2NCH(CH_3)COOC_2H_5}$$

$$BzOOCNHCH_2COOH + H_2NCH(CH_3)COOC_2H_5$$

Step 3: ↓ DCC

$$BzOOCNHCH_2CONHCH(CH_3)COOC_2H_5$$

Step 5: ↓ 1. mild ^-OH 2. H^+ — **Step 4:** ↓ H_2/Pd

$$BzOOCNHCH_2CONHCH(CH_3)COOH \qquad H_2NCH_2CONHCH(CH_3)COOC_2H_5$$

Step 4: ↘ H_2/Pd — **Step 5:** ↙ 1. mild ^-OH 2. H^+

$$\underset{\text{Gly-Ala}}{H_2NCH_2CONHCH(CH_3)COOH}$$

FIGURE 16.10

Synthesis of the dipeptide Gly-Ala.

Note that the two protecting groups, the benzyloxycarbonyl group (BzOOC—, or $C_6H_5CH_2OOC$—) and the ethyl group (C_2H_5—), are removed separately (steps 4 and 5). This separate deprotection of the terminal amino or carboxyl group of the dipeptide permits continued building to a tripeptide, if desired, using a third amino acid (appropriately protected) and bringing about its condensation reaction with the dipeptide. The process may be repeated as many times as necessary. However, if the dipeptide Gly-Ala is the desired final product, a benzyl ester (from benzyl alcohol) rather than the ethyl ester can be prepared in step 2. This benzyl ester will also be cleaved by the hydrogenolysis reaction of step 4, producing the desired dipeptide in a single cleavage step.

$$BzOOCNHCH_2CONHCH(CH_3)COOBz \xrightarrow{H_2/Pd} \underset{\text{Gly-Ala}}{H_2NCH_2CONHCH(CH_3)COOH}$$

Numerous protecting groups for amino and carboxyl functions have been developed. One often used for amino groups is benzyloxycarbonyl chloride, an acid chloride with which any free amino group reacts in an ammonolysis reaction to form an amide called a carbamate (see Section 15.4.3). The advantage of this reagent is that it can be removed by hydrogenolysis of the benzyl-oxygen bond. This unique cleavage with hydrogen and a metal catalyst is better than hydrolysis because it does not affect a normal peptide bond. The resulting *N*-carboxyl group spontaneously loses carbon dioxide (analogous to evolution of carbon dioxide by carbonic acid).

Protection of amino group:

$$C_6H_5CH_2{-}O{-}\overset{\overset{\displaystyle O}{\|}}{C}{-}Cl \quad + \quad RNH_2 \longrightarrow C_6H_5CH_2{-}O{-}\overset{\overset{\displaystyle O}{\|}}{C}{-}NHR$$

Benzyloxycarbonyl chloride — Amine — Protected amine

Deprotection of amino group:

$$C_6H_5CH_2{-}O{-}\overset{\overset{\displaystyle O}{\|}}{C}{-}NHR \xrightarrow{H_2/Pd} C_6H_5CH_3 \quad + \quad \left[HO{-}\overset{\overset{\displaystyle O}{\|}}{C}{-}NHR\right]$$

Protected amine — Toluene — *N*-carboxylic acid (a carbamic acid)

$$\downarrow$$

$$CO_2(g) \quad + \quad H_2NR$$

Amine

Carboxyl groups can be protected as simple ethyl or methyl esters, formed using acid catalysis (see Section 15.3.4). These ester groups can be cleaved by base-catalyzed hydrolysis under much milder conditions than needed to cleave any amide (peptide) bond. Alternatively, a carboxyl group can be protected as a benzyl ester, formed by esterification with benzyl alcohol. The advantage of this protecting group is that it, too, can be removed by hydrogenolysis of the benzyl-oxygen bond.

Protection of carboxyl group:

$$C_6H_5CH_2OH \quad + \quad RCOOH \xrightarrow{H^+} RCOOCH_2C_6H_5$$

Benzyl alcohol — Carboxylic acid — Benzyl ester

Deprotection of carboxyl group:

$$RCOOCH_2C_6H_5 \xrightarrow{H_2/Pd} C_6H_5CH_3 \quad + \quad RCOOH$$

Benzyl ester — Toluene — Carboxylic acid

PROBLEM 16.6

Starting with valine and phenylalanine, design a synthesis of the dipeptide Val-Phe.

16.4.2 Solid-Phase Synthesis

A significant advance in the synthesis of polypeptides was developed in 1962 by R. B. Merrifield, who received the Nobel Prize in chemistry in 1984 for this process. It is called *solid-phase synthesis* and involves attaching the desired C-terminal α-amino acid to a solid polymer through a benzylic ester linkage and then sequentially condensing the free amino group to an individual N-protected amino acid. Thus, the polypeptide chain grows from the solid polymer support, with coupling always occurring at the N-terminus. The reactions occur in solution at the surface of the support, and the growing peptide chain is held to the insoluble resin as by-products are washed away. The final step is cleavage of the ester linkage binding the peptide chain to the resin by hydrogenolysis of the benzyl-oxygen bond.

In these reactions, the amino groups are protected by a *t*-butyloxycarbonyl group, $(CH_3)_3COOC$— (abbreviated Boc). The protection is accomplished by *t*-butyloxycarbonyl chloride, $(CH_3)_3COOCCl$, which can be removed by a number of nonaqueous acids that will not cleave the benzyl ester linkage holding the peptide chain to the resin. The solid-phase synthesis is illustrated in Figure 16.11 for the condensation of three α-amino acids (**A1, A2,** and **A3**) into a tripeptide (**A3-A2-A1**). Note that **A3,** the last amino acid added, is the N-terminal amino acid.

Solution and solid-phase syntheses of polypeptides involve "growing" a peptide chain by adding one amino acid at a time, a process sometimes called **linear synthesis.** This is a slow process, and even if the yield of each condensation reaction is 90%, synthesis of a decapeptide will have an overall yield of only 39%. The alternative approach is called **convergent synthesis,** in which small peptides are synthesized and then condensed into larger peptides, which are finally condensed to produce the end product. The yields of such an approach are considerably higher, and several teams can be working on the synthesis of various small peptides simultaneously.

Small-protein synthesis has become rather routine. Even large proteins can be synthesized using the solid-phase technique. A stellar example is Merrifield's 1969 synthesis of the enzyme ribonuclease, a protein of 124 α-amino acids. The synthesis

FIGURE 16.11

Solid-phase synthesis of a tripeptide (A3-A2-A1).

$$\underset{\textbf{A1}}{\text{Boc—NHCHRCOO}^-} + \text{ClCH}_2\text{C}_6\text{H}_4\text{—resin}$$

The carboxylate anion of the first amino acid (**A1**) effects S_N2 displacement of chlorine to form a benzylic ester linkage to the solid resin.

$$\downarrow$$

$$\text{Boc—NHCHRCOOCH}_2\text{C}_6\text{H}_4\text{—resin}$$

$$\downarrow \text{CF}_3\text{COOH}$$

The protecting group Boc is removed to free the amino group of **A1**.

$$\text{H}_2\text{NCHRCOOCH}_2\text{C}_6\text{H}_4\text{—resin}$$

$$\text{DCC} \downarrow \text{Boc—NHCHR'COOH} \quad \textbf{(A2)}$$

The free amino group of **A1** is coupled to the free carboxyl group of the second protected amino acid (**A2**).

$$\text{Boc—NHCHR'CO—NHCHRCOOCH}_2\text{C}_6\text{H}_4\text{—resin}$$

$$\downarrow \text{CF}_3\text{COOH}$$

The protecting group Boc is removed to free the amino group of **A2**.

$$\text{H}_2\text{NCHR'CO—NHCHRCOOCH}_2\text{C}_6\text{H}_4\text{—resin}$$

$$\text{DCC} \downarrow \text{Boc—NHCHR''COOH} \quad \textbf{(A3)}$$

The free amino group of **A2** is coupled to the free carboxyl group of the third protected amino acid (**A3**).

$$\text{Boc—NHCHR''CO—NHCHR'CO—NHCHRCOOCH}_2\text{C}_6\text{H}_4\text{—resin}$$

$$\downarrow \text{CF}_3\text{COOH}$$

The protecting group Boc is removed to free the amino group of **A3**.

$$\text{H}_2\text{NCHR''CO—NHCHR'CO—NHCHRCOOCH}_2\text{C}_6\text{H}_4\text{—resin}$$

$$\downarrow \text{H}_2/\text{Pd}$$

The benzyl-oxygen bond joining the growing peptide to the resin is cleaved by hydrogenolysis. The desired tripeptide (**A3-A2-A1**) is isolated.

$$\underset{\textbf{A3-A2-A1}}{\text{H}_2\text{NCHR''CO—NHCHR'CO—NHCHRCOOH}}$$

required 369 chemical reactions and 11,931 individual operations, all of which were accomplished by an automated protein synthesizer.

Today, many proteins are prepared by biosynthesis, using the appropriate "piece" of DNA containing the "code" for the synthesis of that particular protein (see Section 19.1.4). This fragment of DNA is simply inserted into a bacterium in a growing medium containing the necessary α-amino acids, and the bacterium is then induced to produce the protein enzymatically. For example, strains of *E. coli* can be induced to produce human insulin. This process is much more rapid and economical than sequential laboratory synthesis.

Stepwise protein synthesis is still very important, however, because it permits the synthesis of "designer" proteins and natural proteins in which one or more α-amino acids at specific known positions are replaced. This process allows the biological testing of the effect of such α-amino acid substitutions, which occasionally occur naturally as a result of genetic mutations.

16.5 Enzymes

16.5.1 Biological Catalysts

Some reactions in the body are catalyzed by enzymes called *oxidoreductases*. Enzymes are effective catalysts, but even the best catalyst cannot function well when one of the reactants (in this case, oxygen) is in short supply.

An important function of some globular proteins is to serve as biological catalysts known as **enzymes.** All chemical reactions occurring in living systems are catalyzed by enzymes.

As we have seen in earlier chapters, experimental conditions such as heat, pressure, specific solvents, and acid or base catalysis can be used to facilitate chemical reactions in the laboratory. However, these conditions are not feasible in living cells, where all chemical reactions must occur under identical conditions: at body temperature (37°C), in aqueous solution, and usually near neutral pH. Enzymes are extremely effective chemical catalysts, able to exert their effects under these cellular conditions. In fact, some have the ability to increase the rate of a reaction to 10^{12} times the noncatalyzed rate. For example, hydrolysis of a fat (an ester) to a carboxylic acid and an alcohol in aqueous solution is extremely slow, but this hydrolysis occurs rapidly when a lipase (hydrolytic enzyme) is added.

Enzymes are globular proteins with a hydrophilic exterior (the R groups are polar; see Table 16.1), making them water-soluble. Enzymes catalyze a reaction by first forming an *enzyme-substrate complex* with the specific compound (the substrate) on which the reaction will occur:

$$\text{Enzyme} + \text{Substrate} \longrightarrow \left[\begin{array}{c}\text{Enzyme-substrate}\\ \text{complex}\end{array}\right] \longrightarrow \text{Enzyme} + \text{Product(s)}$$

This complexation occurs at what is called the **active site** of the enzyme—the region of the protein containing the specific α-amino acids that will bind the substrate.

An enzyme binds only to a very specific substrate, having a particular structure and stereochemistry. This is the basis for what is called the *lock-and-key concept* of enzyme catalysis (Figure 16.12). The enzyme with its specific shape and constitution is the lock, and the only acceptable substrate is the key, making for a perfect "fit." This specificity is exemplified by the fact that different enzymes are required to hydrolyze very similar disaccharides (see Section 4.7.3). For example, the enzyme maltase catalyzes the

Enzyme Substrate Enzyme-substrate complex

FIGURE 16.12

Lock-and-key relationship between an enzyme and its substrate.

hydrolysis of maltose (an acetal formed from two glucose units joined by a 1,4-α-linkage) into two molecules of glucose. But maltase will not catalyze the hydrolysis of cellobiose (also formed from two glucose units but joined by a 1,4-β-linkage) or of sucrose (a glucose and fructose unit joined by a 1,2-α-linkage). Further, since enzymes are chiral reagents (remember, proteins consist of L-amino acids), *the reactions they catalyze are stereospecific* (see Section 17.4).

As an example of enzyme function and active site geometry, let's consider the chymotrypsin-catalyzed hydrolysis of proteins at the carboxyl group of the α-amino acids phenylalanine, tyrosine, and tryptophan. Chymotrypsin is a protease, an enzyme that hydrolyzes proteins at a peptide linkage, producing a free carboxyl group and a free amino group. The overall shape of the enzyme is similar to that of a cloverleaf—it is folded so that amino acids at positions 195 (serine), 57 (histidine), and 102 (aspartic acid) are near one another to form the active site (Figure 16.13).

The hydroxyl group of serine serves as a nucleophile to carry out a nucleophilic acyl substitution at the peptide linkage. The serine concurrently transfers its proton to histidine. The resulting tetrahedral oxyanion then re-forms the carbonyl group by breaking the carbon-nitrogen peptide bond and ejecting the nitrogen-containing fragment (an amino end of the remaining polypeptide). The carboxyl-containing portion of the original protein is later freed from serine by hydrolysis of the serine ester functional group (see Section 15.3.2).

The overall result is the cleavage of a protein (R′NHCOR) into two smaller polypeptides, one with a new N-terminus ($R'NH_2$) and the other with a new C-terminus (RCOOH). The only substance consumed in the process is water, so the process is a hydrolysis reaction analogous to that for any amide (see Section 15.4.2). The key to the efficiency (rapid rate) of the reaction is that the enzyme binds the substrate in the exact orientation needed for the nucleophilic hydroxyl group of serine to add across the carbonyl bond and initiate the nucleophilic acyl substitution.

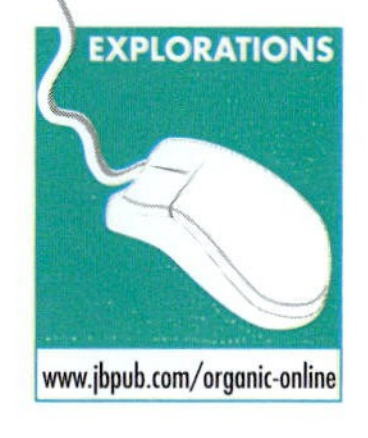

A number of **competitive inhibitors** of enzyme function have been discovered. These are compounds that compete with the normal substrate for complexation at the active site of the enzyme. For example, sulfanilamide, one of the sulfa drugs, is a competitive inhibitor of the bacterial enzyme that incorporates *p*-aminobenzoic acid into folic acid, an essential constituent for bacterial function (see "Chemistry at Work" p. 432). The competitive inhibition results from the close similarity of sulfanilamide to *p*-aminobenzoic acid amide in regard to chemical structure and functional group characteristics. The drug literally "fools" the enzyme into binding with it, thus denying the bacterium its essential *p*-aminobenzoic acid amide.

H_2N–C_6H_4–SO_2NH_2 H_2N–C_6H_4–$CONH_2$

Sulfanilamide *p*-Aminobenzoic acid amide

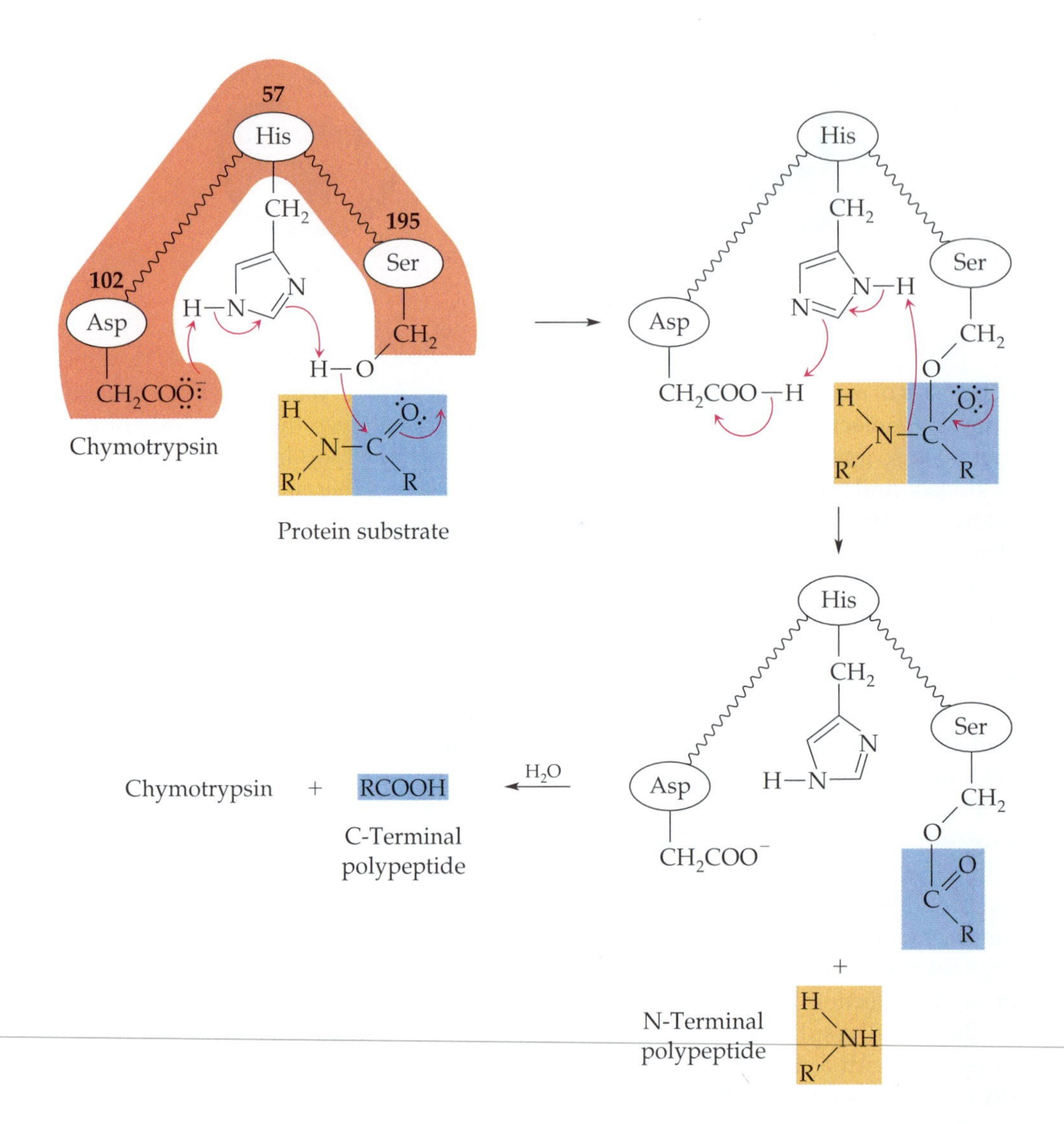

FIGURE 16.13

Chymotrypsin-catalyzed hydrolysis of a protein.

Enzymes are categorized based on the kind of chemical reaction they catalyze. Examples are hydrolases (hydrolysis), reductases (reduction), oxidases (oxidation), isomerases (isomerization), and proteases (hydrolysis of proteins). Note that all enzyme names end in *-ase*. Many enzymes are named using two words, the first being the substrate and the second being the name of the chemical process. For example, glucose-6-phosphatase attaches a phosphate group to carbon 6 of glucose. DNA polymerase converts the nucleotides that make up DNA (monomers) into the nucleic acid (a polymer). There are, however, many common enzymes that have been assigned simple names, such as the metabolic proteases trypsin, chymotrypsin, and pepsin.

16.5.2 Coenzymes

Most enzymes are simply proteins, and in many the catalytic function is exerted by the protein. However, some enzymes are conjugate proteins with a nonprotein component that is required for enzymatic function. These nonprotein entities, called **coenzymes,** are usually complex organic molecules. In such enzymes, the role of the protein

CHEMISTRY AT WORK

IN THE MARKETPLACE

Commercial Applications of Enzymes

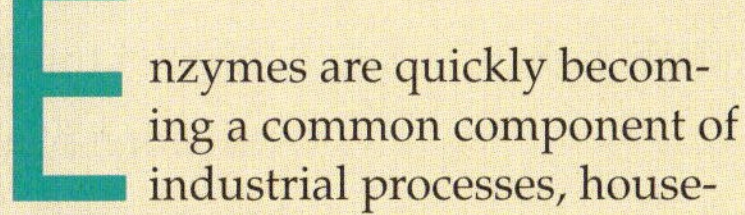

Enzymes are quickly becoming a common component of industrial processes, household products, and medical applications. Why? Because they offer a very efficient and stereospecific means of accomplishing needed chemical reactions under mild conditions.

High-fructose syrup, equivalent in sweetness to sucrose and used in many products (especially as the sweetener for soft-drinks), is prepared from cornstarch in quantities of several billion pounds per year. It is made in large plants using three enzymes: The enzyme α-amylase partially breaks down the cornstarch (a polymer of α-D-glucose) into dextrins (small polyglucose fragments). Then glucoamylase depolymerizes the dextrins into the monomer D-glucose. Finally, glucose isomerase converts D-glucose into D-fructose.

Another commercial application of enzymes is the addition of industrially prepared enzymes to many detergents. The enzymes improve the detergents' cleaning effectiveness by breaking down protein-based dirt particles. Still another commercial enzyme is papain, a component of some meat tenderizers that breaks down the proteins in the meat. There are also a number of medical applications of enzymes, including enzyme replacement therapies.

One reason for the increased use of enzymes commercially is availability; genetic engineering now enables the economical large-scale production of enzymes using bacterial fermentation processes. Previously, all such enzymes had to be obtained by extraction from natural sources.

Production of D-fructose:

Starch (poly-α-D-glucose) $\xrightarrow{\alpha\text{-amylase}}$ Dextrin (6–10 D-glucose units) $\xrightarrow{\text{glucoamylase}}$ D-Glucose $\xrightarrow{\text{glucose isomerase}}$ D-Fructose

component is to identify, attract, and hold the substrate (the lock-and-key concept), but the actual catalyzed reaction occurs between the substrate and the coenzyme. The coenzyme might be called the "business end" of the enzyme.

Many coenzymes are derived from vitamins, especially the B vitamins. That is why these vitamins are essential dietary components. Their absence in the diet is associated with many diseases (for example, scurvy, beriberi, pellagra). These diseases have become relatively rare, at least in developed countries, because of wider availability of vitamin supplements and better nutrition. A vitamin is frequently coupled with additional structural groups to form the complete coenzyme, but the functional part of the coenzyme is the vitamin portion. Several coenzymes are listed in Table 16.4.

The functioning of a coenzyme can be illustrated by the oxidation in body cells of an alcohol group (in lactic acid) to a carbonyl group (in pyruvic acid) by <u>n</u>icotinamide <u>a</u>denine <u>d</u>inucleotide (NAD^+).

Rennin, an enzyme found in the stomach of cows, coagulates milk into cheese.

Oxidation of lactic acid to pyruvic acid:

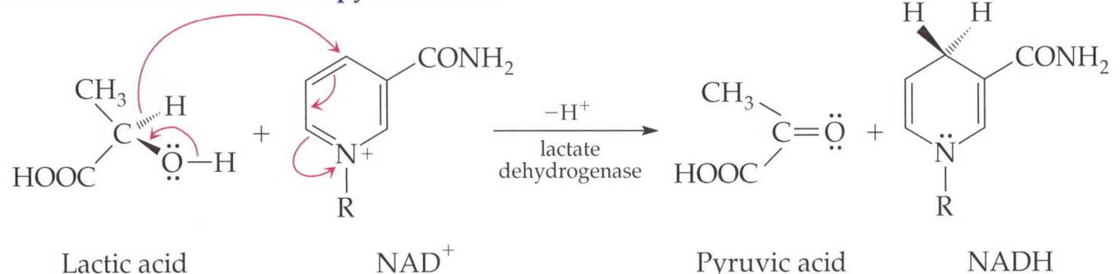

TABLE 16.4

Vitamin Components of Coenzymes

Vitamin	Coenzyme	Function
Niacin	NAD^+	Oxidation of alcohols
Pyridoxine (B_6)	Pyridoxal phosphate	Nitrogen transfer and storage
Riboflavin (B_2)	$FADH_2$	Reduction of alkenes to alkanes
Pantothenic acid	Coenzyme A	Acyl transfer
Thiamine (B_1)	Thiamine pyrophosphate	Decarboxylation of α-keto acids
Folic acid (B_c)	Tetrahydrofolic acid	One-carbon transfer

As the alcohol is oxidized, the coenzyme NAD^+ is reduced to NADH. A proton and two electrons are transferred from the alcohol to the pyridinium ring, accomplishing its reduction. Recall that oxidation of a secondary alcohol to a ketone is a standard chemical process, usually accomplished using sodium dichromate and sulfuric acid in the laboratory (see Section 4.4). In that case, Cr(VI) is reduced to Cr(III).

One of the key differences between coenzymes and laboratory reagents is that coenzymes are always recycled. In other words, they are two-way catalysts. For example, those that are oxidized also reduce. After bringing about the usual reaction, the coenzyme is regenerated. Therefore, once NAD^+ has been reduced to NADH by the alcohol (accompanied by oxidation of the alcohol to a carbonyl), NADH is oxidized back to NAD^+ by serving as a reducing agent. NADH reduces another carbonyl group to an alcohol (similar to the $NaBH_4$ reduction of a carbonyl group; see Section 13.3.1). This reduction process is important in the biosynthesis of fats. The sequential reduction of a coenzyme followed by its oxidation is called a *cycle*. These cycles are represented by converging arrows.

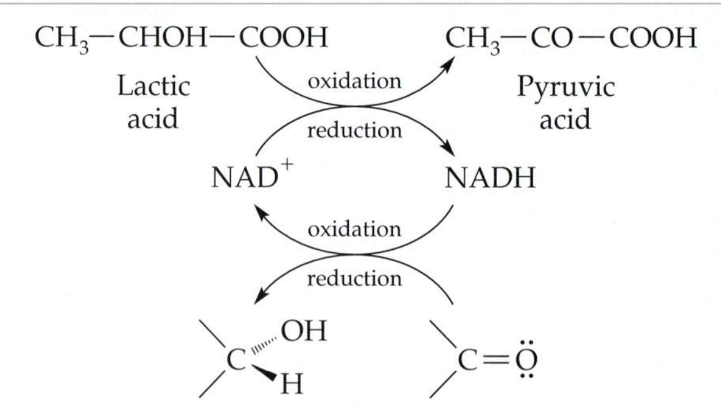

Recycling of NADH to NAD^+

What is really happening is that electrons (and the protons that follow them) are being passed between compounds, flowing through the coenzyme as an intermediate and a catalyst. (Remember that oxidation of a compound means that the compound loses electrons, whereas reduction means that it gains electrons.) A coenzyme that is able to bring about oxidations and reductions, depending on which form it is in, is known as a *redox coenzyme*. Several enzymes are involved sequentially in oxidation reactions, the electrons eventually being transferred to molecular oxygen to produce the final product of oxidation, water.

Chapter Summary

Proteins are polyamides, **macromolecules** in which α-amino acids are linked through an amide bond between the —NH_2 group of one α-amino acid and the —COOH group of another. There are 22 **α-amino acids** found in proteins, 14 of which are synthesized by human metabolic systems, and 8 of which (*the essential amino acids*) are not. All α-amino acids except glycine have at least one stereocenter, are chiral, and are of the L configuration (in contrast to the D-configuration found in carbohydrates). α-Amino acids are **amphoteric** compounds and exist as dipolar ions called **zwitterions.** Racemic α-amino acids are synthesized using the **Strecker synthesis** (reaction between an aldehyde, HCN, and ammonia) or by ammonolysis of α-bromocarboxylic acids.

The amide bond between two α-amino acids is called a **peptide bond,** and a compound with two or more α-amino acids connected is called a **peptide.** Proteins are therefore **polypeptides,** usually containing from 10 up to hundreds of α-amino acids.

Proteins are classified several ways: by composition (**simple** or **conjugated proteins**), by shape (**fibrous** or **globular proteins**), and by biological function. The structure of proteins involves four levels: **primary structure** (the composition and sequence of the α-amino acids), **secondary structure** (α-helix or pleated sheet), **tertiary structure** (elongated/fibrous or folded/globular), and **quaternary structure** (assemblies of globular proteins).

The structure of peptides or proteins is determined by two processes. The first process determines the composition of the peptide using an instrument called the Automatic Amino Acid Analyzer. From this analysis can be deduced how many units of each α-amino acid are present in the peptide. The second process is **sequence analysis,** which determines the sequence in which the constituent α-amino acids are connected through peptide bonds. **End-group analysis** and **fragment analysis** together permit the determination of the structure of any protein.

Peptide synthesis, and therefore protein synthesis, involves connecting two α-amino acids together by a peptide bond. This process can be repeated many times to synthesize small peptides. **Linear synthesis** involves synthesizing a peptide one α-amino acid at a time. **Convergent synthesis** involves synthesizing several small peptides and finally joining them together.

Enzymes are globular proteins that function as biological catalysts. Enzymes conduct very specific chemical reactions on particular substrates. Each enzyme contains an **active site,** a region of usually nonconnected α-amino acids in the enzyme that holds the substrate while the chemical reaction occurs. Some enzymes are pure protein, while others have associated with them a nonprotein portion called a **coenzyme.** Frequently, the coenzymes are vitamins, which are essential dietary constituents.

Summary of Reactions

1. Synthesis of α-amino acids (Section 16.1.3)

a. Ammonolysis of α-bromocarboxylic acids

$$\underset{\text{Carboxylic acid}}{R{-}CH_2{-}COOH} \xrightarrow[2.\ H_2O]{1.\ Br_2/P} \underset{\alpha\text{-Bromocarboxylic acid}}{R{-}\underset{|}{\overset{}{C}}H(Br){-}COOH} \xrightarrow{NH_3} \underset{\alpha\text{-Amino acid}}{R{-}CH(NH_2){-}COOH}$$

b. The Strecker synthesis

$$\underset{\text{Aldehyde}}{R{-}CHO} \xrightarrow[HCN]{NH_3} \underset{\alpha\text{-Aminonitrile}}{R{-}CH(NH_2)CN} \xrightarrow{HCl/H_2O} \underset{\alpha\text{-Amino acid}}{R{-}CH(NH_2)COOH}$$

2. **Edman degradation of a peptide (Section 16.3.1).** The N-terminus α-amino acid can be removed selectively through reaction with phenylisothiocyanate followed by HCl cleavage.

$$H_2N{-}CHRCO{-}NH{-}peptide \xrightarrow[2.\ HCl]{1.\ C_6H_5{-}N{=}C{=}S}$$

Polypeptide

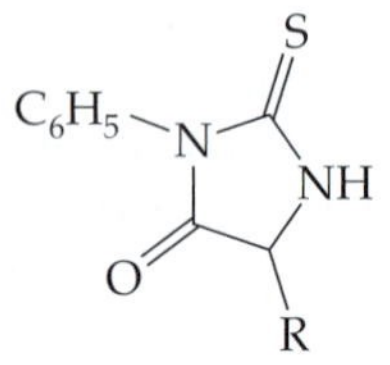

S, C_6H_5, N, NH, O, R + $H_2N{-}peptide$

A phenylthiohydantoin — Polypeptide with one less α-amino acid

3. **Synthesis of peptides (Section 16.4.1).** Two protected α-amino acids or two protected peptides can be condensed using *N,N*-dicyclohexylcarbodiimide (DCC). The protecting groups can be removed without cleavage of the new peptide bond.

$BzOOCNHCHRCOOH + H_2NCH(R')COOC_2H_5$

N-Protected amino acid — C-Protected amino acid

$BzOOCNHCHRCONHCH(R')COOC_2H_5$

↓ (removal of protecting groups)

$H_2NCHRCONHCH(R')COOH$

A dipeptide

Additional Problems

α-Amino Acids

16.7 Proteins are occasionally detected using their UV spectra. Which of the α-amino acids in Table 16.1 would you expect to absorb UV radiation?

16.8 Show the following reactions of phenylalanine:

(a) with dilute NaOH

(b) with dilute HCl

(c) with benzoyl chloride and pyridine

(d) with acetic anhydride

(e) the product of (d) with thionyl chloride, then ethanol

16.9 Show the reactions necessary to accomplish each conversion:

(a) acetaldehyde to alanine

(b) 3-phenylpropanoic acid to phenylalanine

(c) succinic acid to aspartic acid

16.10 6-Aminohexanoic acid can be converted to lysine using alpha bromination. However, this reaction cannot be conducted in the presence of an amino group. With this thought in mind, design a synthesis of lysine from 6-aminohexanoic acid.

16.11 Which of the α-amino acids in Table 16.1 have more than one stereocenter?

16.12 Draw the structure of each of the following α-amino acids at the given pH values:

(a) glycine at pH 2 and 13

(b) aspartic acid at pH 2 and 13

(c) lysine at pH 2 and 13

16.13 Write the equilibrium between the nonpolar and the zwitterionic forms of (a) alanine and (b) proline.

16.14 In its acidic form, tyrosine shows three pK_a values: 2.2, 9.1, and 10.1. Write equations showing how this form of tyrosine changes as the pH of a solution is gradually raised from 1 to 13.

16.15 Protonated alanine has pK_a values of 2.3 and 9.4, compared to the pK_a value of 4.9 for propanoic acid. Account for this observation.

16.16 Show the reaction of serine with excess acetic anhydride.

Peptides

16.17 The pain felt when a wasp stings is related to the release by the blood plasma globulins of a nonapeptide called bradykinin. Its composition is Arg_2, Gly, Phe_2, Pro_3, and Ser. Fragment analysis produced the following peptides: Phe-Ser, Pro-Gly-Phe, Pro-Pro, Ser-Pro-Phe, Phe-Arg, and Arg-Pro. End-group analysis showed that both terminal amino acids are Arg. What is the structure of the nonapeptide bradykinin?

CONCEPTUAL PROBLEM

A Question of Identity

You are a chemical technician, newly employed by a biotechnology start-up company. You are asked to start a synthesis of a tripeptide, Gly-Lys-Tyr, which is needed for one of the company's hot new research projects. The individual amino acids (Gly, Lys, and Tyr) are delivered to your lab in three bottles on a tray. However, much to your dismay, the labels are lying on the bottom of the tray! Not knowing which label belongs to which bottle, and not wishing to appear incompetent on your first day on the job, you decide to see if you can figure out which bottle contains which amino acid.

- What are some of the ways that you could use to uniquely identify at least one of the amino acids?

Another task that falls to you is to supervise two premedical students working in your lab during the summer. They are studying a protein that has been found to catalyze part of the process of oxidizing fatty acid molecules. In preliminary experiments, the students found that the protein was soluble in water and consisted only of amino acids. You ask them to do two things:

- Classify this protein by composition, shape, and function.
- Name three amino acids that might be more than usually abundant on the outer surface of this protein.

16.18 Write the condensed structure for each of the following peptides:

(a) Gly-Ser

(b) Ala-Phe-Val

(c) Lys-Gly-Cys

16.19 Show the reaction and products for the hydrolysis of Gly-Ala using dilute HCl.

16.20 How many different tripeptides can be formed by coupling one unit each of Gly, Ala, and Phe?

16.21 Show the reactions needed to synthesize the dipeptide Ala-Phe using protecting groups.

16.22 As we will see in Section 19.1.4, the biosynthesis of proteins, which occurs in the cell under the influence of RNA, involves the reaction between the free amino group of one α-amino acid and the carboxyl group (in the form of an ester) of an adjacent α-amino acid, thus forming a peptide bond. This reaction is identical to the laboratory one by which esters react with amines to form amides (see Section 15.3.2). Show (a) the dipeptide product you would expect from the reaction of the two appropriately protected α-amino acids shown here and (b) the deprotected peptide.

$C_6H_5CH_2OOCNHCH(CH_3)COOCH_3$

N-Protected methyl ester of alanine

+

$H_2NCH(CH_2C_6H_5)COOCH_2C_6H_5$

C-Protected phenylalanine

↓

Dipeptide

Proteins

16.23 Define each of the following terms and give an example:

(a) peptide bond

(b) essential amino acid

(c) zwitterion

(d) amphoteric compound

(e) tripeptide

(f) L configuration of α-amino acids

16.24 Proteins can be classified three different ways: by composition, by shape, and by function. Describe these classifications.

16.25 What is meant by the primary structure of a protein?

16.26 What is it that holds proteins in their secondary structure, and how are these secondary structures described?

16.27 In hydrogen bonding between proteins, what are the electron donors and the electron acceptors? Draw the partial structure of a protein that shows one such hydrogen bond.

16.28 From the list of α-amino acids in Table 16.1, identify (a) several amino acids whose R groups should be on the outer surface in fibrous proteins and (b) several amino acids whose R groups should be on the outer surface in globular proteins. Explain your choices.

16.29 Do you expect enzymes to be fibrous or globular proteins? Explain.

16.30 What is meant by the denaturation of a protein?

16.31 What is meant by the tertiary and quaternary structures of hemoglobin?

16.32 What is an enzyme, and what is its role?

16.33 What is a coenzyme, and what is its role?

16.34 What is the active site of an enzyme?

16.35 Why are most enzymatic reactions stereospecific?

CHAPTER 17

Optical Activity

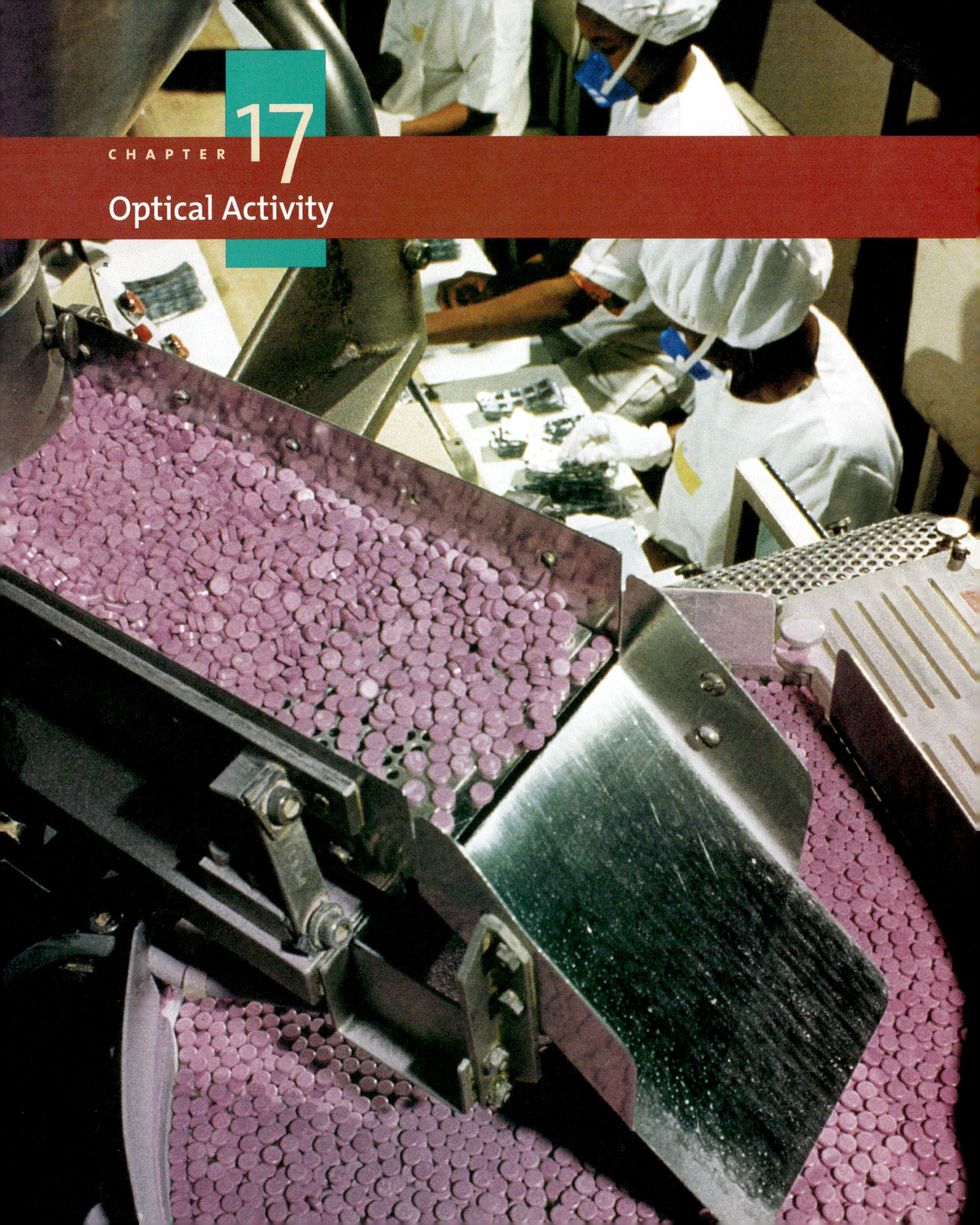

YOU HAVE BEEN introduced to all of the essential functional groups of organic chemistry. This book has stressed the importance of the three-dimensional shape of molecules and introduced several kinds of isomerism. You have seen how shape and orientation affect chemical reactions. Now that you have experience with the different families of organic compounds, this chapter revisits the subject of isomerism. First, it briefly summarizes all the different types of isomerism you have studied. Then, and more importantly, it will show how the phenomenon known as *optical activity* is related to enantiomers.

17.1 Isomerism Revisited

The kinds of isomerism important to organic chemistry are outlined in Figure 17.1. A more detailed chart is included on the inside back cover of this book. As a refresher, here are brief definitions of the different kinds of isomers.

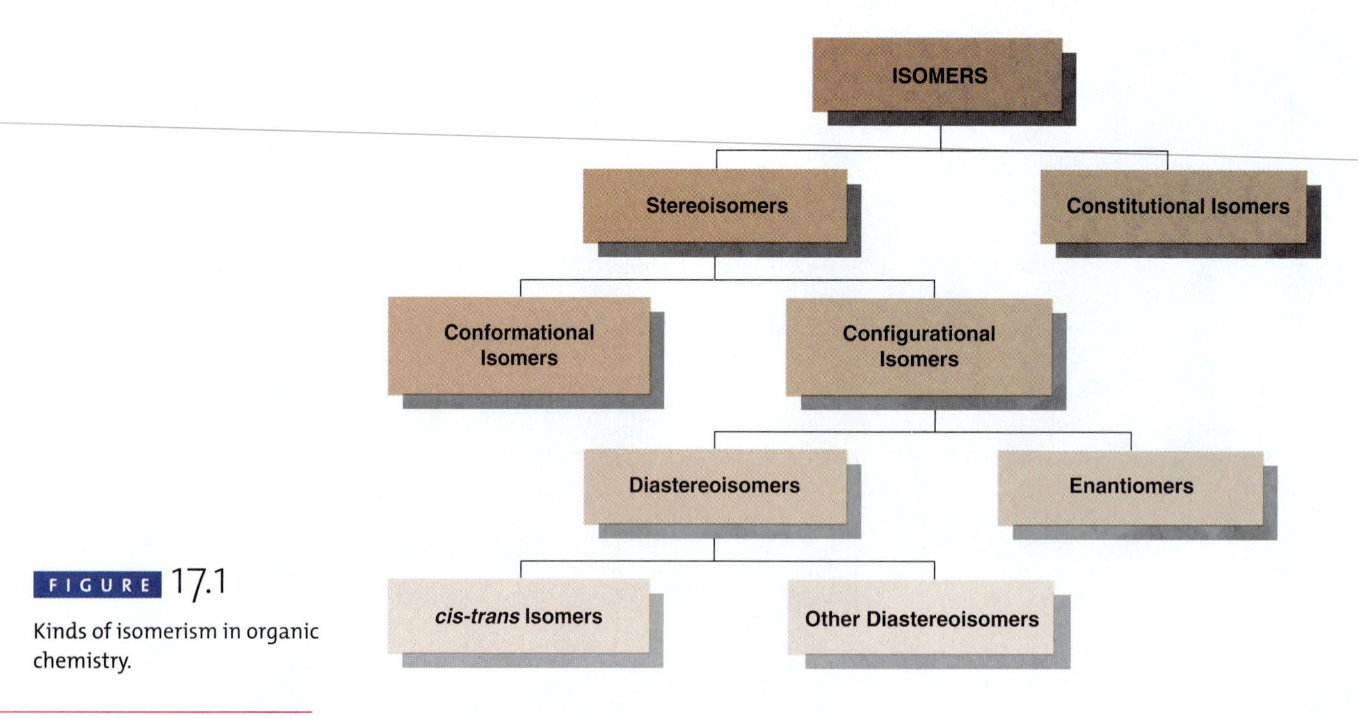

FIGURE 17.1

Kinds of isomerism in organic chemistry.

◀ Drug tablets, once synthesized are carefully packaged to exclude air and moisture and prevent further reaction.

Isomers are compounds that share the same molecular formula but have different structural formulas. There are two major categories of isomers: constitutional isomers and stereoisomers.

Constitutional isomers are compounds that share the same molecular formula but have different bond connectivities (see Section 2.2.1). This difference can be discerned using two-dimensional drawings. Constitutional isomers cannot be directly interconverted, but they can be readily separated based on their different physical properties. There are three major categories of constitutional isomers.

- *Skeletal isomers* are constitutional isomers with different carbon skeletons, such as butane and 2-methylpropane, both C_4H_{10}:

$CH_3-CH_2-CH_2-CH_3$ Butane

$CH_3-CH(CH_3)-CH_3$ 2-Methylpropane

- *Positional isomers* are constitutional isomers with substituents attached at different positions on the same carbon skeleton, such as *o*-chlorotoluene and *m*-chlorotoluene, both C_7H_7Cl:

o-Chlorotoluene *p*-Chlorotoluene

- *Functional group isomers* are constitutional isomers with different functional groups, such as ethanol and methyl ether, both C_2H_6O:

CH_3-CH_2-O-H Ethanol

CH_3-O-CH_3 Methyl ether

Stereoisomers are compounds that share the same molecular formula and the same bond connectivities (in contrast to constitutional isomers), but have different spatial arrangements of atoms or groups. The difference between stereoisomers is best discerned by using drawings that show three-dimensionality or by using models. There are two broad categories of stereoisomers: conformational isomers and configurational isomers.

Conformational isomers are compounds that share the same molecular formula and the same bond connectivities, but have different orientations of their atoms in space (see Section 2.2.2) and can be interconverted by rotation about one or more sigma bonds. Normally, two conformational isomers cannot be separated because of the low energy barrier for their interconversion. Examples are (1) the axial and equatorial isomers of methylcyclohexane and (2) the *anti* and *gauche* conformations of butane:

Axial ⇌ Equatorial

Methylcyclohexane

Anti ⇌ *Gauche*

Butane

Configurational isomers are compounds that share the same molecular formula and the same bond connectivities, but have different orientations of their atoms in space and cannot be interconverted except by breaking bonds—a process that rarely occurs because of the high energy required. There are two categories of configurational isomers: enantiomers and diastereoisomers.

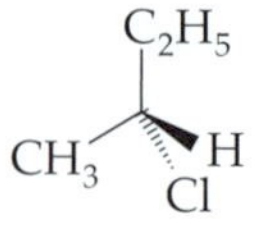

C_2H_5 / CH_3 / H / Cl

(*R*)-2-Chlorobutane

C_2H_5 / H / CH_3 / Cl

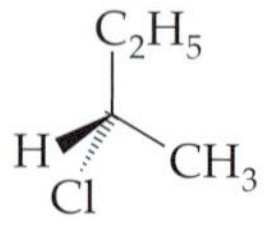

(*S*)-2-Chlorobutane

Enantiomers are compounds that share the same molecular formula and the same connectivities, but are *chiral* (see Section 3.3). That is, they are related to one another in the same way as an object and its mirror image. Enantiomers cannot be separated by normal physical processes and cannot be readily and directly interconverted. All enantiomers studied in this book contain one or more stereocenters, usually a carbon atom with four different groups attached, and have opposite configurations at the stereocenters. For example, (*R*)-2-chlorobutane and (*S*)-2-chlorobutane are a pair of enantiomers.

Diastereoisomers are compounds that share the same molecular formula and the same bond connectivities, but have different orientations in space. They do not have an object/mirror-image relationship with each other. Diastereoisomers can be subdivided into *cis-trans* isomers and other diastereoisomers.

Cis-trans isomers are compounds that share the same molecular formula and the same bond connectivities, but have their substituents attached differently with respect to a fixed plane of reference (see Sections 2.5.2 and 6.2.2). Examples of structures having planes of reference are cycloalkane rings, where groups may be attached to the same (*cis*) or opposite (*trans*) sides, and alkenes, where groups may be attached to the same (*cis*) or opposite (*trans*) sides of the rigid double bond. Specific examples are *cis*- and *trans*-1,2-dimethylcyclopropane and *cis*- and *trans*-1,2-dichloroethene:

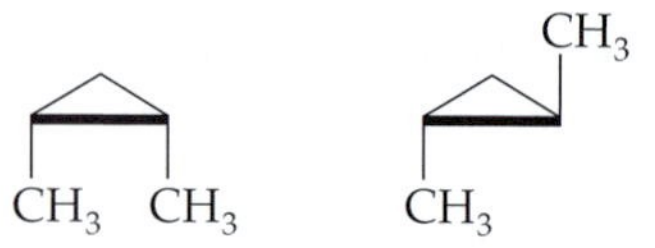

CH_3 CH_3 — *cis*; CH_3 CH_3 — *trans*

1,2-Dimethylcyclopropane

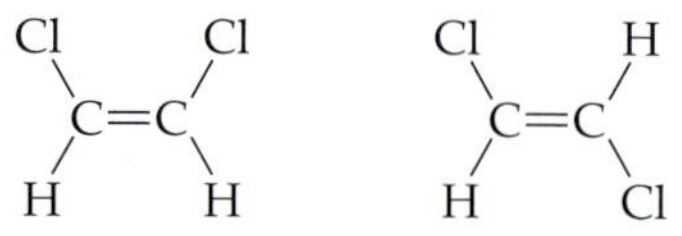

Cl, Cl, C=C, H, H — *cis*; Cl, H, C=C, H, Cl — *trans*

1,2-Dichloroethene

Other diastereoisomers share the same molecular formula and have the same bond connectivities. However, they have more than one stereocenter and the same configuration at one or more stereocenters. They are not enantiomers. Such isomers will be discussed in detail in Section 17.3.1.

17.2 Enantiomers and Optical Activity

17.2.1 Enantiomers Revisited

Enantiomers are a pair of configurational isomers for which the three-dimensional structure of one is the mirror image of the other. The two enantiomers are not identical (their structural formulas are not superimposable) and are therefore said to be *chiral*, a term derived from the Greek word for "handedness." The property of chirality is associated with many three-dimensional objects, including the human hand. Your left hand is not identical to your right hand, but the relationship they share is that of an object to its image, as seen in a mirror.

As you learned in Chapter 3, it is possible to predict whether an organic compound is chiral simply by drawing its structure in two dimensions, but the ultimate test is the test of superimposability. You can test for superimposability by drawing a structure and its mirror image or by making molecular models. For example, for 2-chlorobutane ($CH_3CHClCH_2CH_3$), you can draw structural formula I, representing the object, and then structural formula II, its mirror image:

Mirror

I	II
(*R*)-2-Chlorobutane	(*S*)-2-Chlorobutane

Envisioning the manipulation of II in three-dimensional space makes clear that II is different from I. No matter what orientation it has, II cannot be superimposed on I. Therefore, 2-chlorobutane is chiral, and I and II are a pair of configurational isomers called enantiomers. As described in Section 3.3.2, the configurations of the two enantiomers about the central carbon are opposite and are identified by the letters *R* and *S*.

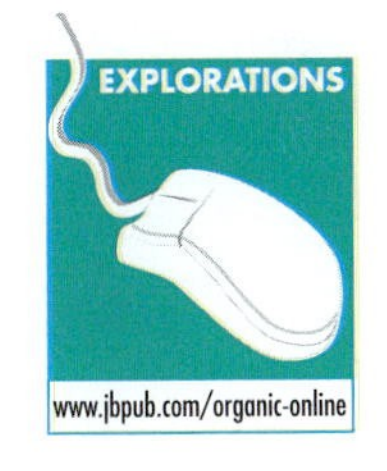

That 2-chlorobutane is chiral can be predicted without drawing three-dimensional structures because it has a **stereocenter,** a carbon atom with four different groups attached. (Note that other tetrahedral central atoms, such as tetrahedral nitrogen, silicon, or tetrahedral phosphorus, can also serve as stereocenters.) Any compound with a single stereocenter exists as two enantiomers. Another predictor for chirality is the absence of a plane of symmetry in the compound. You cannot draw a plane through 2-chlorobutane that would create two mirror-image halves.

Now, apply the three tests for chirality to 3-chloropentane:

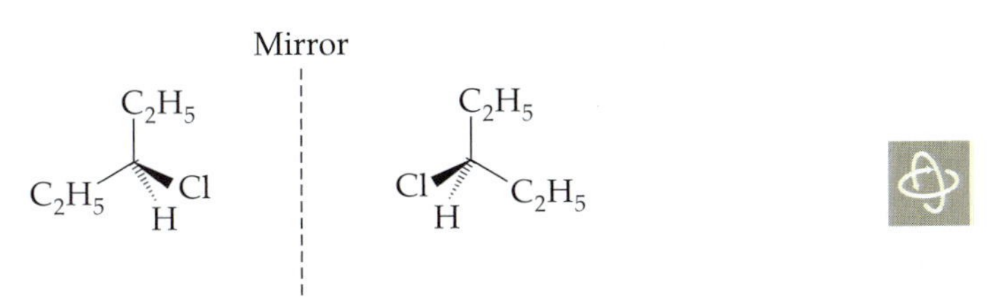

3-Chloropentane

Examination reveals that 3-chloropentane has no stereocenter and does have a plane of symmetry; therefore, it should not be chiral and should not exist as two enantiomers. The plane of symmetry, in this case, is the imaginary plane through the hydrogen, the central carbon, and the chlorine. By testing for superimposability, you can determine that the image is superimposable on the object. So, by definition, 3-chloropentane is not chiral; therefore, it is *achiral*. There are no configurational isomers of 3-chloropentane.

A pair of enantiomers together in equal proportion is referred to as a **racemate,** or a **racemic mixture.** As we will see in Section 17.2.3, a racemate is also referred to as a *dl* or a (±) pair. Under normal conditions, a chemical reaction that leads to the formation of a chiral compound with a stereocenter from an achiral compound produces a racemic mixture—that is, a 50/50 mixture of the *R* and *S* enantiomers.

A unique property of individual enantiomers is their ability to rotate a beam of plane-polarized light—a property called *optical activity*. The means of detecting this property will be described in Section 17.2.2.

PROBLEM 17.1

Indicate which of the following compounds has a stereocenter and mark the stereocenter with an asterisk:

(a) isopropyl benzoate
(b) *N*,*N*-dimethyl 2-methylbutanamide
(c) 3-phenyl-2-butanone
(d) 2-vinylcyclohexanone
(e) 1-methylcyclopentanecarbaldehyde
(f) 1,3-dibromocyclohexane
(g) 1,4-dichlorocyclohexane

PROBLEM 17.2

How many stereocenters are there in each of the following compounds?

(a) 2-aminopropanoic acid
(b) 2,5-dimethyl-3-hexanone
(c) 2,3-diphenylbutane
(d) sorbitol (1,2,3,4,5,6-hexahydroxyhexane)

17.2.2 Determination of Optical Activity

Plane-Polarized Light

Ordinary light (regardless of its color or wavelength) consists of waves vibrating in all directions perpendicular to the direction of propagation. Passing a beam of light through certain polarizing materials, such as a Nicol prism (the mineral calcite) or a sheet of Polaroid, blocks light waves except those that are vibrating in a particular plane. The emerging beam of light has the distinctive property of being *plane-polarized* and can be used to measure the optical activity of a compound (Figure 17.2). When plane-polarized light is passed through an optically active substance, the plane of vibration rotates by a number of degrees (0–360°) that is specific to that substance and that can be physically measured. The rotation also has a direction: appearing either clockwise (+) or counterclockwise (−) to the observer.

If a beam of plane-polarized light is directed at a second sheet of Polaroid, it passes through if the polarizing orientations of the two sheets are identical. However, if the

FIGURE 17.2

A light beam at three stages: normal, plane-polarized, and rotated plane-polarized.

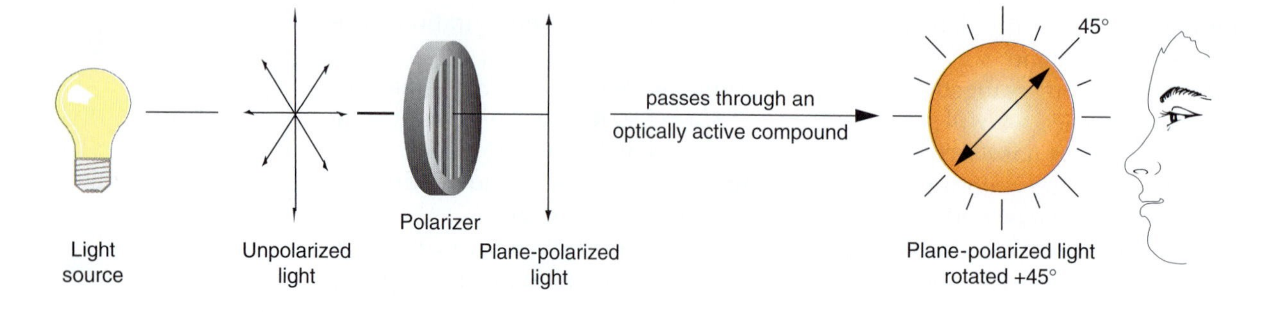

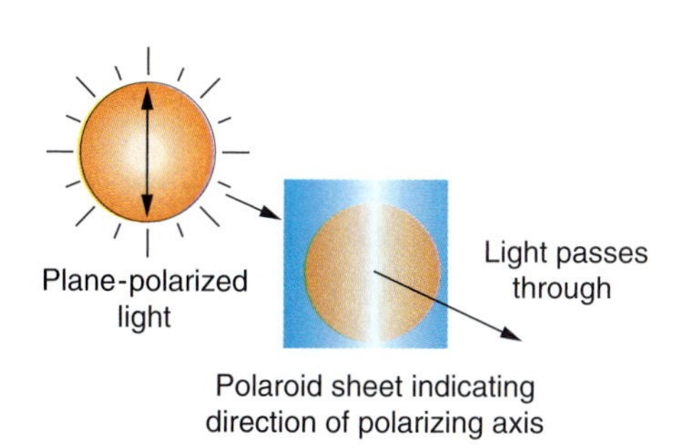

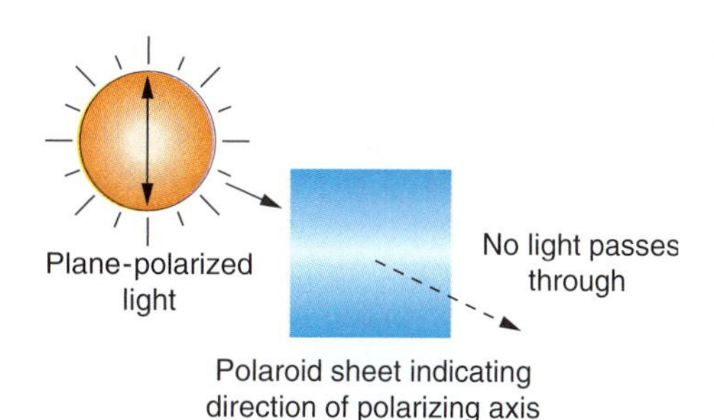

FIGURE 17.3

Effect of a sheet of Polaroid on the transmission of plane-polarized light.

second sheet is at 90° to the first, no light passes through (Figure 17.3). (You can see this effect by holding one pair of Polaroid sunglasses at a 90° angle to a second pair.) As the second sheet of Polaroid is rotated gradually from 90° to 0°, in reference to the first sheet, an increasing amount of polarized light passes through. When the polarizing axis of the second sheet of Polaroid is perfectly aligned with the axis of the plane-polarized light (that of the first sheet), all the light is transmitted through the second sheet (see Figure 17.3). A setup similar to the one just described is incorporated into the polarimeter, an instrument for measuring the extent of rotation of plane-polarized light by optically active compounds.

The Polarimeter

A **polarimeter** is an instrument used to convert ordinary light into plane-polarized light, pass the polarized light through a sample, and measure the extent of rotation of the plane-polarized light. It consists of a polarizing crystal (a prism), a sealed tube to hold the sample, and an analyzing crystal (Figure 17.4).

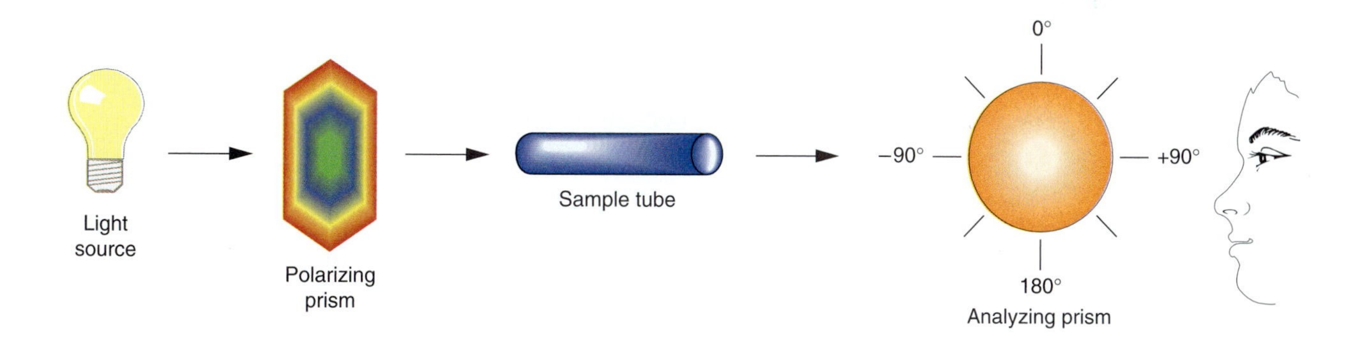

FIGURE 17.4

Schematic of a polarimeter.

The light source used in a polarimeter is usually a sodium lamp, which emits a consistent yellow light (just like that of sodium vapor street lamps) at 589 nm (the so-called D line of sodium). The eyepiece of the polarimeter contains the analyzing prism. The eyepiece is rotated as much as needed so that the light from the lamp can be clearly seen. At this point, the angle of rotation of the analyzer matches that of the plane-polarized light passing through the sample. This rotation, known as the **observed rotation, $[\alpha]_{obs}$,** depends on the substance, the temperature, the wavelength of light used (usually 589 nm), the solvent used to dissolve the sample, the concentration of the sample, and even the length of the sample tube.

Given all the variables involved, chemists use a standardized set of conditions (temperature, wavelength, tube length) on which to base their comparisons. The result is the **specific rotation, $[\alpha]_D^t$,** which is defined as follows, where t is the temperature of the sample (usually near room temperature, ~20°C) and D is the D line of sodium:

$$[\alpha]_D^t = \frac{[\alpha]_{obs}}{\text{length of tube (dm)} \times \text{concentration of sample (g/mL)}}$$

The specific rotation is a characteristic physical property of every enantiomer. The specific rotation of a sample of a compound can be compared with that known for the pure compound to identify the sample and determine its purity. The **optical purity** of a sample is that percentage of the sample made up of that single enantiomer. A pure sample has 100% optical purity; the calculated specific rotation is equal to that of the pure reference sample.

Optical purity can also be used to measure concentration. For example, pure sucrose (table sugar) has a specific rotation of +66.4° in water at 20°C using the D line of sodium. In the sugar industry, the concentration of sugar can be determined at any time during the refining process (which takes place in water) by quickly determining the optical activity of the solution. If a rotation of +33.2° is observed in a sample, the solution must be half as concentrated as the reference solution—that is, there are half as many sugar molecules available to rotate the plane-polarized light.

PROBLEM 17.3

Calculate the specific rotation of a solution of 5 g of nicotine in 100 mL of ethanol at 20°C taken in a 1 dm tube with $[\alpha]_{obs} = -8.45°$.

17.2.3 Optical Activity of Enantiomers

Because two enantiomers are isomers, they are different compounds and can be expected to have different properties. However, in contrast to other kinds of isomers, the major physical and chemical properties of a pair of enantiomers are identical. In other words, two enantiomers have the same boiling point, melting point, refractive index, solubility, spectra peaks, and so on. Thus, it is impossible to separate a pair of enantiomers (a racemate) using the normal chemical and physical separation techniques, such as distillation, crystallization from different solvents, or acid or base extraction. (Section 17.2.4 will discuss the one way to separate a pair of enantiomers, using a process called *resolution*.) The key difference in physical properties between a pair of enantiomers is the direction in which they rotate plane-polarized light.

Because of the unique property of enantiomers to rotate plane-polarized light, they are said to be **optically active.** Each enantiomer of a pair causes an optical rotation of equal magnitude *but opposite direction*. For example, (*R*)-2-butanol has a rotation of −13.9°, while (*S*)-2-butanol has a rotation of +13.9°. A positive value indicates a clockwise rotation; a negative value indicates a counterclockwise rotation. Therefore, the complete description of these two enantiomers is *R*-(−)-2-butanol and *S*-(+)-2-butanol.

Compounds that rotate the plane of light to the right (the clockwise or plus direction) are called *dextrorotatory* (*d*), while those that rotate light to the left (the counterclockwise or minus direction) are called *levorotatory* (*l*). (Note that the prefixes *d*- and *l*-, which refer to direction of rotation, are unrelated to the prefixes D- and L-, which refer to the relative configurations of enantiomers of carbohydrates and α-amino acids.)

An enantiomer rotates plane-polarized light, but a racemate, being a 50/50 mixture of the two enantiomers, is optically inactive. Half the molecules present in a racemic mixture rotate the light clockwise a certain number of degrees, and the other half rotate the light the same number of degrees in the opposite direction. The net effect is no rotation at all. Thus, a racemic compound is designated as *dl* or ±—for example, *dl*-2-butanol or (±)-2-butanol. A racemate behaves chemically and physically as though it were a single compound with its own characteristic physical and chemical behavior.

Remember that two enantiomers, in addition to rotating a beam of plane-polarized light in opposite directions, have opposite absolute configurations (the orientation of their substituents around the stereocenter carbon; see Section 3.3.2). Only in recent years, with the advent of X-ray crystallography, has it been possible to determine the absolute configuration of any particular enantiomer. For example, the *d* and *l* forms of lactic acid have been known for nearly two centuries, but only fairly recently has it been proved that *d*-lactic acid has the *S* configuration and *l*-lactic acid has the *R* configuration.

COOH, HO, CH_3, H — (*R*)-(−)-Lactic acid, $[\alpha]_D = -2.6°$

COOH, CH_3, OH, H — (*S*)-(+)-Lactic acid, $[\alpha]_D = +2.6°$

$\xrightarrow{\text{NaOH}}$

$COO^- \; Na^+$, CH_3, OH, H — (*S*)-(−)-Sodium lactate, $[\alpha]_D = -13.5°$

Note that there is no relationship between the *direction of rotation* exhibited by an enantiomer (+ or −) and its *absolute configuration* (*R* or *S*). Proof that there is no relationship between configuration and rotation can be found in the fact that *S*-(+)-lactic acid is converted into its sodium salt by reaction with sodium hydroxide solution, with no change in its configuration. However, there is a change in the magnitude and direction of the optical rotation.

PROBLEM 17.4

Indicate which of the following compounds are optically active and which are optically inactive: (a) *dl*-2-chloropentane, (b) (*R*)-2-phenyl-3-hexanone, (c) methyl (±)-2-(*N*-methylamino)propanoate, (d) potassium 4-nitrobenzoate, and (e) (*S*)-2-methylcyclohexanone.

17.2.4 Resolution of Racemates

Because most physical properties of two enantiomers are identical, there is no obvious physical means of separating a racemic mixture into the two enantiomers. A chemical process was eventually developed for separating a *dl* mixture into *d* and *l* enantiomers. It is called **resolution,** and it depends on (1) the concept of converting enantiomers into

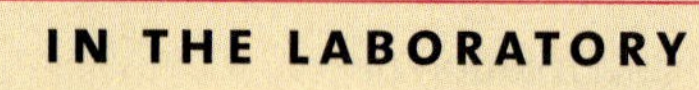

CHEMISTRY AT WORK

IN THE LABORATORY

The Mutarotation of Glucose

When D-glucose is crystallized from water, two different forms can be obtained. Colorless crystals with a melting point of 146°C are obtained by cooling an aqueous solution; these are pure α-D-glucopyranose. However, if an aqueous solution is slowly evaporated at 98°C, crystals with a melting point of 150°C are obtained; these are pure β-D-glucopyranose. These two forms are the *anomers* of D-glucose, in which the only difference is the configuration of the hydroxyl group on carbon 1 of the hemiacetal form of D-glucose (see Section 4.7.2). They are diastereoisomers (as will become evident in Section 17.3.1) and therefore have different physical properties.

Both of these anomers are optically active. Pure α-D-glucose has a specific rotation of +112°, whereas pure β-D-glucose has a specific rotation of +19°. Interestingly, when *either* of these isomers is dissolved in pure water and the solution is allowed to stand, the optical rotation of the solution gradually changes to +53°. This gradual change of an optical rotation to an equilibrium value is known as *mutarotation.*

What is occurring in this mutarotation is the establishment of an equilibrium between the two anomeric (α- and β-) forms of D-glucopyranose (and only trace amounts of the open-chain form and furanose forms). From the value of the specific rotation at equilibrium, the composition of the equilibrium mixture can be determined to be about 64% β-D-glucose and 36% α-D-glucose, the more stable equatorial β anomer predominating.

α-D-Glucopyranose (mp 146°C, $[\alpha]_D = +112°$) ⇌ Open-chain form of D-glucose ⇌ β-D-Glucopyranose (mp 150°C, $[\alpha]_D = +19°$)

diastereoisomers and (2) the availability of naturally occurring enantiomers, especially alkaloids (see Section 12.7.1), for use as **resolving agents.**

In a resolution process, a racemate (*R* and *S* enantiomers in equal proportion) is reacted with an enantiomeric resolving agent (B) to form a mixture of two products (*R*—B and *S*—B) (Figure 17.5). The two products of the reaction are no longer enantiomers; they are no longer mirror images of each other. (The *R* and *S* portions of *R*—B and *S*—B have opposite configurations, but the B portion has the same configuration in both products; therefore, the two products are not mirror images of each other.) They are, by definition, diastereoisomers—configurational isomers that have two stereocenters and that do not have a mirror-image relationship.

Diastereoisomers have different chemical and physical properties (different solubilities and melting points, for example) and therefore can be separated by the usual

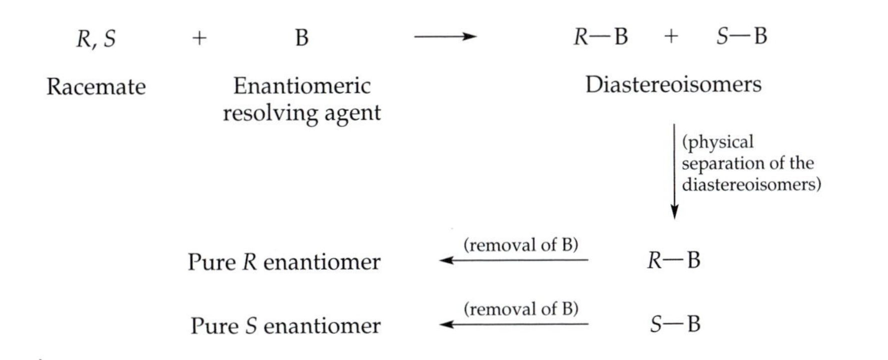

FIGURE 17.5 Resolution of a racemate (*R*, *S* = enantiomers; B = resolving agent).

means, such as crystallization. Once separated, the diastereoisomers are separately reconverted (usually by hydrolysis or another simple reaction) into their starting materials. The resolving agent B is removed, and the result is separated enantiomers, *R* and *S* (see Figure 17.5).

As a specific example, racemic mandelic acid (2-hydroxy-2-phenylethanoic acid) has been resolved using the natural alkaloid (+)-cinchonine as the resolving agent. This alkaloid exists in the dextrorotatory form ($[\alpha]_D = +228°$) in the bark of cinchona trees found in South America. The first step is the reaction between the resolving agent and the racemate to form a mixture of two diastereoisomers. The carboxylic acid protonates the most basic nitrogen atom of the alkaloid in an acid-base reaction to form two diastereoisomeric ammonium salts:

HO H N N + HO H COOH → HO H COO^- BH^+

(+)-Cinchonine (±)-Mandelic acid Diastereoisomeric cinchonium salts of mandelic acid (B = cinchonine)

The resolution process involves the separation of the two salts by selective crystallization: They have different solubilities. The separated salts are then each treated with dilute hydrochloric acid to regenerate the enantiomeric mandelic acids, protonating cinchonine in the process. The properties of the various reactants and products are shown in Figure 17.6, where B represents cinchonine and AH represents mandelic acid.

A variety of chemical reactions and reagents can be used in the resolution process. The only requirements are that the diastereoisomers be readily separable and that the cleavage reaction for regenerating the separated enantiomers not racemize any stereocenters. Esterification is another frequently used resolving reaction because of the existence of naturally occurring enantiomeric alcohols and carboxylic acids that can be used as resolving agents.

FIGURE 17.6

Resolution of racemic mandelic acid.

AH	+	B	⟶	$A^- \; BH^+$
(±)-Mandelic acid (mp 118–120°C, $[\alpha]_D = 0°$)		(+)-Cinchonine (mp 255°C, $[\alpha]_D = +228°$)		Mixture of diastereoisomeric salts

↓ (fractional crystallization)

(*S*)-(−)-Mandelic acid (mp 133–135°C, $[\alpha]_D = -158°$)	$\xleftarrow{HCl^*}$	Salt I (mp 176–178°C, $[\alpha]_D = +154°$)
(*R*)-(+)-Mandelic acid (mp 133–134°C, $[\alpha]_D = +157°$)	$\xleftarrow{HCl^*}$	Salt II (mp 165°C, $[\alpha]_D = +92°$)

* The cinchonine is transformed into its hydrochloride $BH^+ \; Cl^-$.

PROBLEM 17.5

Suggest a kind of resolving agent (that is, a family of compounds) that could be used in the resolution of each of the following compounds, and indicate the general structure of the diastereoisomers that would be formed:

(a) 2-butanol

(b) α-aminophenylacetic acid

(c) 2-methylbutanal

(d) 1-amino-2-vinylcyclohexane

Although the resolution process is frequently laborious, modern chemical technology has permitted resolutions to be carried out even on an industrial scale. For example, the α-amino acid lysine, $H_2N(CH_2)_4CHNH_2COOH$, used as a feed supplement for cattle, is produced on a multi-ton scale in its enantiomeric form by a resolution process.

Although it is difficult to synthesize individual enantiomers of compounds, it is essential to be able to do so because almost all biological reactions occur between enantiomers. Recall that all but one of the 22 naturally occurring α-amino acids are chiral and that all major carbohydrates are chiral. Also, all enzymes are chiral. Although enantiomers have identical physical properties, except for the direction of rotation of plane-polarized light, they react differently with enantiomeric chemical reagents. Therefore, if a racemic mixture is administered to a biological system, it is likely that the system will react differently with each enantiomer. Fortunately, nature has provided a ready source of many enantiomers that can be used in resolutions or as starting materials for the synthesis of enantiomeric products. By these means, chemists can produce enantiomeric chemicals.

With the discovery that some drugs previously sold as racemates pose a public health risk because of differing biological effects exhibited by the two enantiomers, the FDA has begun to require, in some instances, that only the active enantiomer of a drug be sold in the United States (see "Chemistry at Work," p. 87). The common analgesic ibuprofen, also used as a nonsteroidal anti-inflammatory drug (NSAID), is sold as its racemate in products such as Motrin, Nuprin, and Advil. The (*R*)-(−) enantiomer is biologically inactive, with all of the desired activity due to the (*S*)-(+) enantiomer. A process has now been developed that allows production of only the (*S*)-(+) enantiomer. This enantiomerically pure drug is already on the market in Europe and is expected to be available soon in the United States.

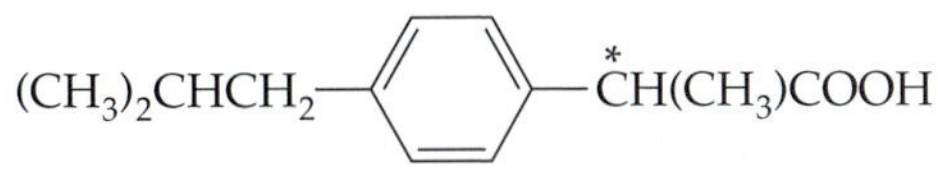

$(CH_3)_2CHCH_2$–C_6H_4–$\overset{*}{C}H(CH_3)COOH$

Ibuprofen
(* = stereocenter)

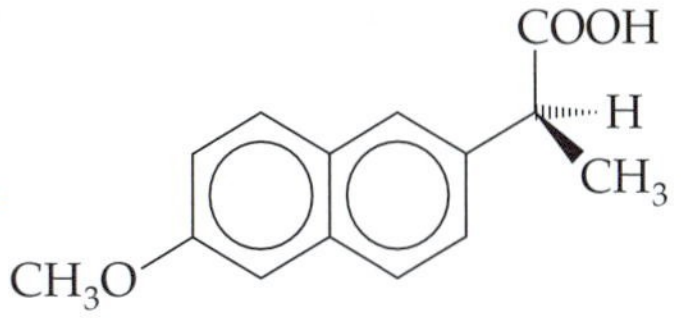

CH_3O COOH H CH_3

(*S*)-(+)-Naproxen

A newer anti-inflammatory drug is naproxen, which is 2-(6-methoxy-2-naphthyl)propanoic acid, a chiral compound. Only the (*S*)-(+) enantiomer is used because the (*R*)-(−) enantiomer is reported to be a liver toxin. The sodium salt of naproxen is marketed as Aleve.

Overall, chiral drugs should be more biologically efficient and safer than the racemic drugs they replace. However, they will undoubtedly be more expensive because of the resolution processes involved or the stereospecific or stereoselective syntheses that must be developed (see Section 17.4.5). Chiral drugs are almost certain to be the wave of the future in pharmaceuticals.

17.3

Multiple Stereocenters

To this point, the discussion of chiral compounds has focused on those that have a single stereocenter. However, many organic compounds have more than one stereocenter. For such compounds, the obvious question is: How many isomers may exist for this compound? The general rule is that *the maximum possible number of configurational isomers is 2^n, where n equals the number of stereocenters.* Thus, for a compound with a single stereocenter, there are $2^1 = 2$ isomers—a pair of enantiomers. With two stereocenters present (and they need not be adjacent), there is a maximum of $2^2 = 4$ isomers.

One of the major challenges in determining the structure of any natural product is, in addition to determining its structural formula, to determine its *stereoformula*—the exact stereochemistry at each stereocenter. In the case of cholesterol, which contains eight stereocenters, there are 256 possible isomers ($2^8 = 256$), of which cholesterol itself is only one. In the case of glucose, there are five stereocenters, and glucose itself is one of 32 possible isomers.

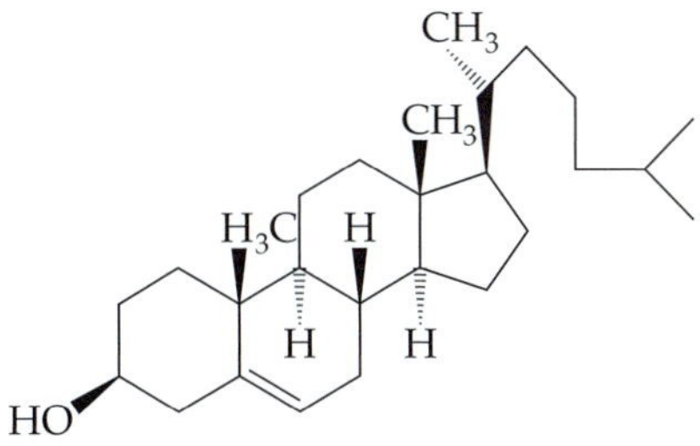

CH_3 CH_3 H_3C H H H HO

Cholesterol
(mp 149°C, $[\alpha]_D = -39.5°$)

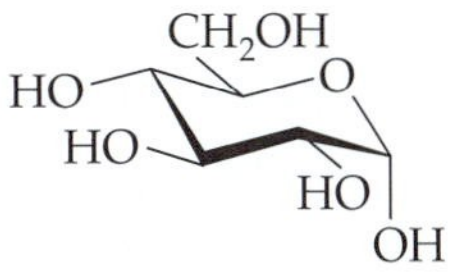

CH_2OH O HO HO HO OH

α-D-Glucopyranose
(mp 146°C, $[\alpha]_D = +112°$)

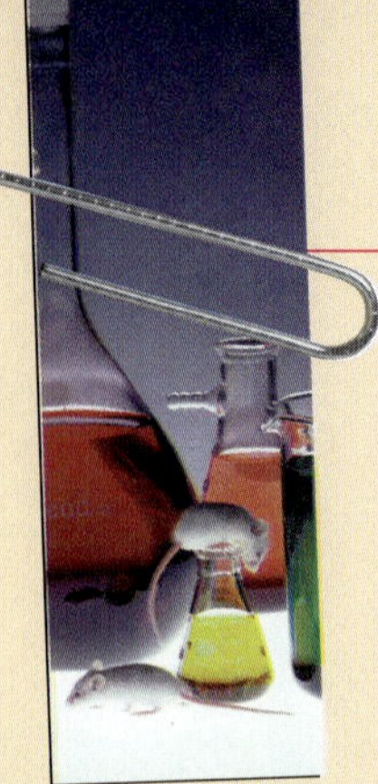

CHEMISTRY AT WORK

IN THE LABORATORY

The Discovery of Enantiomers

Believe it or not, a lot of important chemical discoveries were made in and around the breweries, wineries, and distilleries of Europe in the nineteenth century. One such occurrence was Louis Pasteur's discovery of enantiomers, which played a key role in advancing knowledge of stereochemistry.

The fact that certain substances were optically active had been known since 1812, when French physicist Jean Baptiste Biot discovered that quartz crystals could rotate a beam of plane-polarized light. He later determined that many naturally occurring substances, including turpentine, cane sugar, and camphor, had this property of being optically active.

In 1848, Louis Pasteur was working with two forms of tartaric acid that were found crystallized in casks as a by-product of the wine-making process. These acids seemed to be chemically identical except for one thing. The form known as tartaric acid was dextrorotatory; the other, first known as "racemic acid" (*racemus* is the Latin word for "grapes") but later known as paratartaric acid, was optically inactive.

Working with a microscope, Pasteur observed that crystals of the optically active sodium ammonium salt of tartaric acid appeared not to be symmetrical. Because this salt was optically active and dextrorotatory in solution as well, he realized that optical activity was not just a property of the crystal shape, but of the molecule itself.

Then Pasteur examined the optically inactive salt of paratartaric acid under the microscope. He found that there were two distinct types of crystals present, one the mirror image of the other. He painstakingly separated the crystals under the microscope, using a pair of tweezers, and then dissolved each type in water. One set of crystals was dextrorotatory; it exhibited a specific rotation identical to that of the tartaric acid salt derivative. The other set of crystals was levorotatory; it gave the opposite rotation. When mixed in a 50/50 ratio, a solution of the two crystal forms was optically inactive.

COOH / H—C—OH / HO—C—H / COOH

(2*R*,3*R*)-(+)-Tartaric acid

$[\alpha]_D = +12.5°$

COOH / HO—C—H / H—C—OH / COOH

(2*S*,3*S*)-(−)-Tartaric acid

$[\alpha]_D = -12.5°$

Pasteur concluded that tartaric acid existed in two physical orientations—what are now referred to as enantiomers—one the mirror image of the other. Pasteur's racemic or paratartaric acid was in fact a racemate, *dl*-tartaric acid. The term *racemic* has since been used to describe not a specific compound but an equal mixture of two enantiomers of a chiral compound.

Pasteur did not know what it was about tartaric acid that made it optically active. The concepts of tetrahedral carbon, stereocenters, and chirality did not arise until 26 years later, in 1874, when they were proposed independently by chemists Jacobus van't Hoff and Joseph Le Bel. That chiral molecular structure is the cause of optical activity is now a known fact.

Pasteur's discovery was aided by a bit of luck: Few compounds crystallize as conveniently in recognizable enantiomeric forms as do the tartaric acid salts. The modern process for the separation (resolution) of racemates into enantiomers can be a laborious one, as described in Section 17.2.4.

17.3.1 Acyclic Compounds

To determine the three-dimensional structure of each possible enantiomer of a given acyclic compound, you could draw the three-dimensional structure of every possible structural variation. This practice is limited to smaller molecules because keeping track of the orientations of many groups in three-dimensional space is difficult.

Consider the compound 2-bromo-3-chlorobutane. It has two stereocenters, so there should be four isomers. Satisfy yourself that these drawings do, in fact, represent four different structures, by making models, if necessary, and testing for superimposability:

CH_3	CH_3	CH_3	CH_3
H–C–Br	Br–C–H	H–C–Br	Br–C–H
H–C–Cl	Cl–C–H	Cl–C–H	H–C–Cl
CH_3	CH_3	CH_3	CH_3
I	II	III	IV
(2*S*,3*R*)	(2*R*,3*S*)	(2*S*,3*S*)	(2*R*,3*R*)

I and II: Enantiomers of each other; diastereoisomers of III and IV

III and IV: Enantiomers of each other; diastereoisomers of I and II

Isomers I and II are a pair of enantiomers—note the object/mirror-image relationship of their structural formulas. You can arrive at this same conclusion by determining the configuration of each carbon: I is 2*S*, 3*R* and II is 2*R*, 3*S* (remember that *S* is the mirror image of *R*; see Section 3.3.2 for how to assign absolute configurations). Isomers III and IV are also a pair of enantiomers, with the following configurations: III is 2*S*, 3*S* and IV is 2*R*, 3*R*. Therefore, if 2-bromo-3-chlorobutane were synthesized in such a way as to produce all possible configurational isomers in a single process, the mixture would be separable by normal methods into two optically inactive racemates: I and II, *and* III and IV. Each of the racemates could then be resolved into its two enantiomers, both of which are optically active.

In considering the stereochemistry of compounds with more than one stereocenter, we have come to the last entry on the chart of isomers (see Figure 17.1): the "other diastereoisomers" defined in Section 17.1. Whereas isomers I and II of 2-bromo-3-chlorobutane are enantiomers of each other, and isomers III and IV are enantiomers of each other, the relationship of I to III and IV is that of a diastereoisomer. In other words, they have the same constitutional formulas, they are configurational isomers, they are chiral compounds, but they are not enantiomers of each other. Likewise, the relationship of II to III and IV, the relationship of III to I and II, and the relationship of IV to I and II are all diastereoisomeric relationships. These relationships can be summarized as follows:

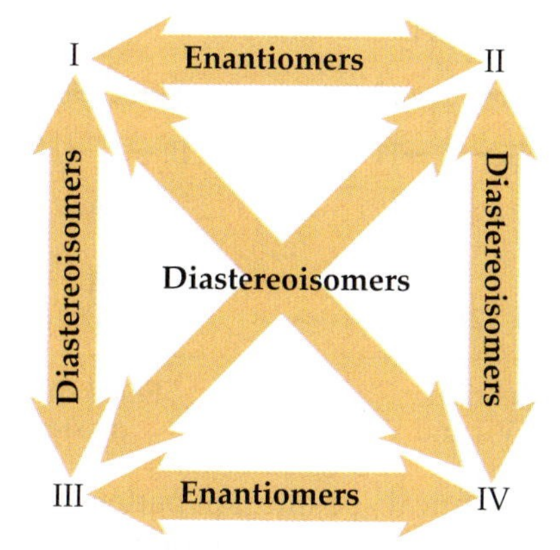

Compounds with a diastereoisomeric relationship usually have different physical and chemical properties and can therefore be separated by the usual physical methods. Note also that diastereoisomers such as I and III have one carbon in which the configuration is the same; here carbon 2 has the *S* configuration in both cases:

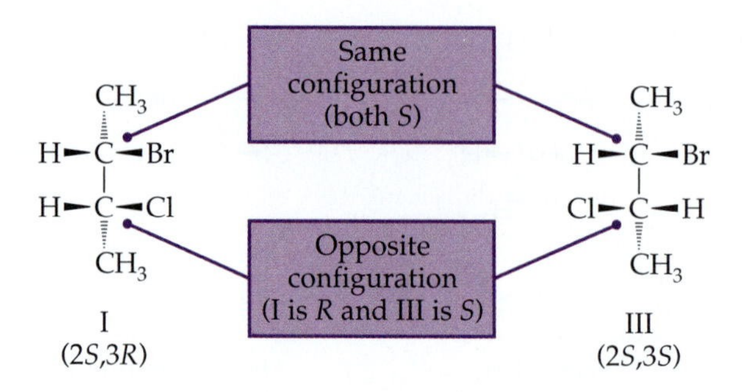

PROBLEM 17.6

D-Fructose, the sugar also known as levulose and used as the major sweetener in soft drinks, is a 2-ketohexose. How many configurational isomers do the open-chain and furanose forms of D-fructose have?

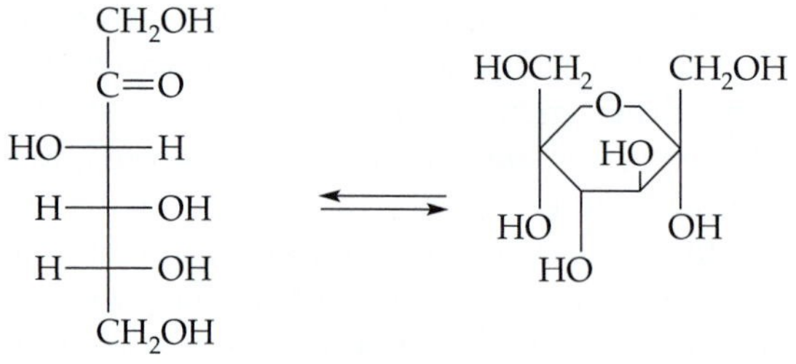

D-Fructose

An additional factor is introduced when the two stereocenters are identically substituted, as in 2,3-dibromobutane. Once again, the general rule predicts that since there are two stereocenters, there should be a *maximum* of four isomers:

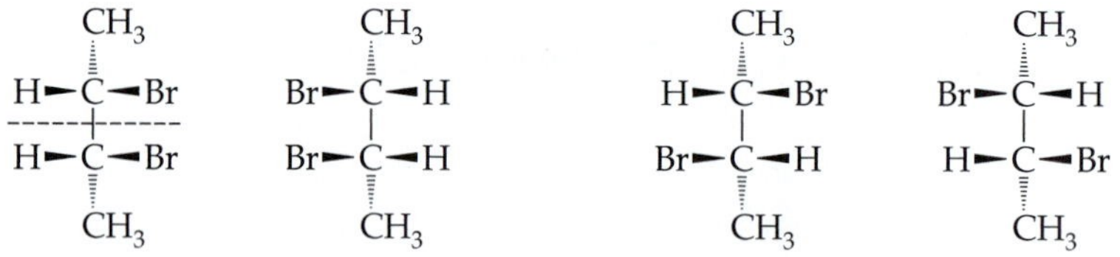

V (2*S*,3*R*) VI (2*R*,3*S*) VII (2*S*, 3*S*) VIII (2*R*,3*R*)

V and VI are identical; diastereoisomers of VII and VIII

Enantiomers of each other; diastereoisomers of V

In this case, because of the identical substitution pattern, V and VI are identical compounds, despite the two different *R-S* configurations. V can be superimposed on VI and therefore represents a single *achiral* compound. Therefore, there are only three configurational isomers of 2,3-dibromobutane, not the maximum number predicted, which is four.

We can predict that isomer V is achiral even without drawing its mirror image, VI, because it has a plane of symmetry (shown as a dashed line). A structure has a plane of symmetry if it can be "sliced in half" by a plane such that one half is the mirror image of the other. Any molecule with a plane of symmetry cannot be chiral; therefore, isomer V is not an enantiomer. Isomers that have stereocenters but are not chiral are called ***meso* isomers;** such an isomer is said to be the *meso form.* The plane of symmetry of *meso*-2,3-dibromobutane is more clearly seen in Figure 17.7; the top half of the molecule is the reflection of the bottom half.

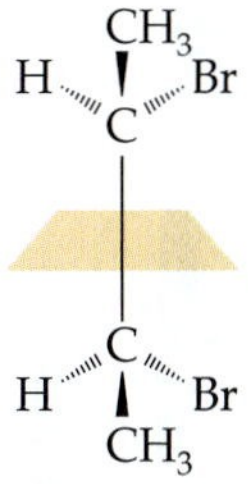

A *meso* form is not optically active. The relationship between the *meso* form of a compound and the other enantiomers is that of diastereoisomers. If 2,3-dibromobutane were to be synthesized by a reaction that produced all possible configurational isomers, two optically inactive compounds could be isolated from the reaction. The *meso* form (V) could be separated from the racemate (VII and VIII) by normal physical and chemical means. However, the racemate could then be separated into its optically active enantiomers by resolution. Thus, a total of three isomers would be produced by the reaction.

The lesson to be learned from this example is that the general rule for predicting the number of configurational isomers (2^n) that can exist for compounds having one or more stereocenters predicts the *maximum possible number* of isomers, not the actual number. There may be fewer isomers if any of the isomers are found to have a plane of symmetry and are therefore not chiral; that is, they are *meso* forms.

CH_3, H, Br, C — Plane of symmetry — C, H, Br, CH_3

FIGURE 17.7

meso-2,3-Dibromobutane and its plane of symmetry.

PROBLEM 17.7

Indicate which of the following compounds have a *meso* form, and draw each: (a) 2,3-butanediamine, (b) 2,4-dibromopentane, (c) 2,3-dibromopentane, and (d) cyclopropane-1,2-diol.

17.3.2 Fischer Projections

As you can imagine, as the number of stereocenters in an acyclic compound increases, it becomes increasingly difficult to test for superimposability. It is hard to keep straight which groups are in the front or in the back while drawing structural formulas and envisioning all possible configurations in three dimensions. Emil Fischer, in addition to solving the dilemma of how to determine relative configurations of long-chain carbohydrates and amino acids (see "Chemistry at Work," p. 626), also came up with a simple means of portraying these configurations (see Section 4.7.2). A **Fischer projection** is a two-dimensional drawing representing a three-dimensional structure and drawn according to the following rules:

- The longest carbon chain is oriented vertically, with the most highly oxidized group at the top.
- Horizontal lines represent bonds coming toward the viewer, out of the plane of the paper.
- Vertical lines represent bonds extending away from the viewer, behind the plane of the paper.
- The intersection of two lines represents a tetrahedral carbon.

Fischer projections can be used to test for superimposability as long as you follow one cardinal rule: You can rotate the drawing in the plane of the paper, but you cannot rotate it out of that plane. Consider, for example, the Fischer projection and the corresponding structural formulas indicating three-dimensionality for 2-bromobutane:

CHEMISTRY AT WORK

IN THE MARKETPLACE

Relative and Absolute Configurations

In the early days of organic chemistry—in fact, up until 1951—the absolute configurations of carbohydrates and α-amino acids were not known. Though the absolute configurations could not be determined, chemists could determine the relative configurations.

As a starting point, the configuration of the simple compound glyceraldehyde [$HOCH_2CH(OH)CHO$], with a single stereocenter, was used as a reference standard because it was viewed as the simplest carbohydrate and it existed as two known enantiomers. Although the actual configuration of (+)-glyceraldehyde was not then known, Emil Fischer arbitrarily assigned to (+)-glyceraldehyde a configuration that had the —OH group of the stereocenter to the right in the drawn formula. The symbol D was used for the configuration of (+)-glyceraldehyde, and L for that of (−)-glyceraldehyde.

CHO
H—+—OH
CH_2OH
D-(+)-Glyceraldehyde

CHO
HO—+—H
CH_2OH
L-(−)-Glyceraldehyde

Other compounds could be related configurationally to D-(+)-glyceraldehyde by means of chemical reactions that were known not to break the C—OH bond; thus, the stereocenter *and* the compound's configuration remained intact. It turned out that most natural sugars had the D configuration and most α-amino acids had the L configuration. Compounds that had the same configuration as D-glyceraldehyde were said to have the same relative configuration.

As an example of how relative configurations were determined, a series of chemical transformations proved that the biologically important compound (−)-lactic acid has the same configuration as D-(+)-glyceraldehyde. What is important here is that none of the reactions used affected the configuration at the single stereocenter. Using reactions that would not affect the position of the hydroxyl group on carbon 2, chemists showed that (+)-isoserine is related to both D-(+)-glyceraldehyde and (−)-lactic acid, proving that all three have the same configuration at the stereocenter, symbolized as D, but now known to be *R*.

In a similar manner, the major sugars, such as glucose, were carefully degraded starting from the top (carbon 1) and going down the chain, until only a single stereocenter was left (that which had been carbon 5), leaving glyceraldehyde as the product. Because the resulting product was identical to the known D-(+)-glyceraldehyde, it was proved that carbon 5 of glucose had the D configuration. The configuration of the other three stereocenters (carbons 2 through 4) could then be deduced by controlled stereospecific degradation reactions.

1 CHO
H—2—OH
HO—3—H
H—4—OH
H—5—OH
CH_2OH
Open-chain form of D-(+)-glucose

(degradation to remove carbons 1–3 and oxidation of carbon 4) →

CHO
H—+—OH
CH_2OH
D-(+)-Glyceraldehyde

In 1951, Dutch chemist Bijvoet used X-ray crystallography to determine that the absolute configuration of (+)-tartaric acid was *S*. Since (−)-tartaric acid had been previously related to D-(+)-glyceraldehyde, this proved that D-(+)-glyceraldehyde had the *R* configuration. The myriad of compounds that had been related to D-(+)-glyceraldehyde must, by extension, have the *R* configuration.

For historical reasons, including the massive amount of published literature on the relative configurations of many compounds, the symbols D and L are still used with carbohydrates and α-amino acids, but they normally mean *R* and *S*, respectively.

CHO / H—+—OH / CH_2OH (D-(+)-Glyceraldehyde) ← HNO_3 — COOH / H—+—OH / CH_2OH ← HNO_2, H_2O — COOH / H—+—OH / CH_2NH_2 ((+)-Isoserine) — HNO_2, HBr → COOH / H—+—OH / CH_2Br — Zn/H^+ → COOH / H—+—OH / CH_3 ((−)-Lactic acid)

Mirror

Structural formula — Fischer projection — (R)-2-Bromobutane | Fischer projection — Structural formula — (S)-2-Bromobutane

The two enantiomers of 2-bromobutane

You already know how to determine the configuration of a compound from its three-dimensional structure (see Section 3.3.2). Now you need to learn how to determine configuration from a Fischer projection. The key is to remember that the horizontal bonds are coming up out of the plane of the paper. Using (*R*)-2-bromobutane as the example (Figure 17.8), you should view the stereocenter from below the paper and behind the bromine; that way you can be looking at the back side of the carbon of the C—H bond. With the C—H bond pointed away from you, you can see that the priority of groups (**a** = Br, **b** = C_2H_5, **c** = CH_3) runs in a clockwise direction.

(rotate 109° about the C—CH_3 axis) — Clockwise

Fischer projection

FIGURE 17.8 Views of (*R*)-2-bromobutane.

The stereochemistry of 2,3-dibromobutane can be readily analyzed using Fischer projections:

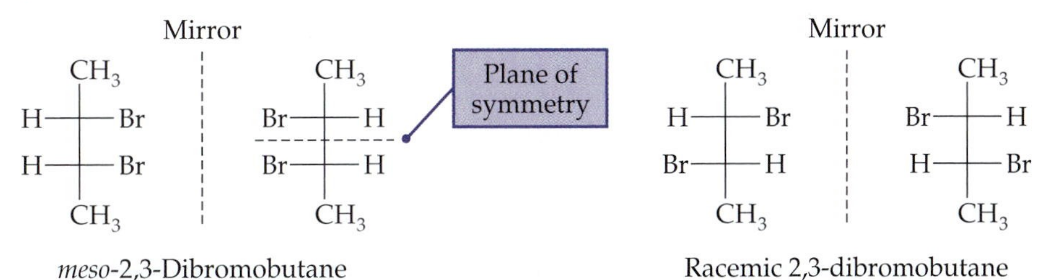

You can see that the *meso* form has a plane of symmetry; therefore, it is not chiral and is not an enantiomer. You can verify this by noting that the mirror image of *meso*-2,3-dibromobutane, when rotated 180° in the plane of the paper, is identical to the object. Using similar tests, you can see that the racemic form is chiral; there are two enantiomers. When the right-hand drawing is rotated 180° in the plane of the paper, it is not superimposable on the left-hand drawing. During these manipulations to test for superimposability, you do not have to worry about the three-dimensional orientation of the structure; the appropriate stereochemistry is maintained.

How to Solve a Problem

(a) Draw the two Fischer projections for 2-pentanol (object and mirror image), and determine whether the compound is chiral.

(b) Determine the absolute configurations of the stereocenters from the Fischer projections.
(c) Draw the three-dimensional structure corresponding to each Fischer projection, and assign its configuration.

SOLUTION

The structure of 2-pentanol produces the Fischer projection labeled **A** (remember that the longest carbon chain is drawn vertically). This can be shortened to **B** because there are no stereocenters in the top three carbons. Then the mirror image of **B** is drawn; this is **C.**

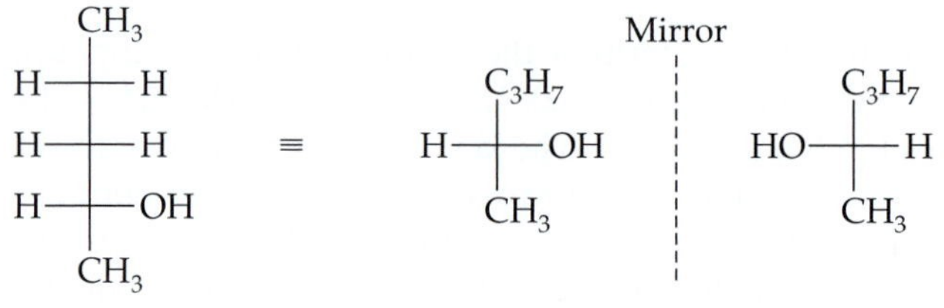

CH_3

H—+—H

H—+—H ≡ H—+—OH (top C_3H_7, bottom CH_3) | Mirror | HO—+—H (top C_3H_7, bottom CH_3)

H—+—OH

CH_3

A **B** **C**

(a) 2-Pentanol is chiral because the mirror image (**C**) is not superimposable on the object (**B**); this could be predicted because carbon 2 is a stereocenter. This conclusion is verified by the fact that when we rotate structure **C** 180° in the plane of the paper and superimpose it on **B,** the two are not identical. Therefore, they are enantiomers.

(b) The absolute configuration of **B** is determined by viewing **B** along the C—H axis from the right rear of the molecule (behind OH). From this perspective, the priority sequence of groups (**a** = OH, **b** = C_3H_7, **c** = CH_3) can be seen to run clockwise, so the configuration of **B** is *R*. The same approach with **C** indicates that the priority sequence of groups runs counterclockwise, so its configuration is *S*.

(c) The three-dimensional structures of **B** and **C** are as follows:

C_3H_7 / H—C—OH / CH_3 C_3H_7 / HO—C—H / CH_3

B **C**

(*R*)-2-Pentanol (*S*)-2-Pentanol

PROBLEM 17.8

Show the structures of the following enantiomers using Fischer projections:

(a) 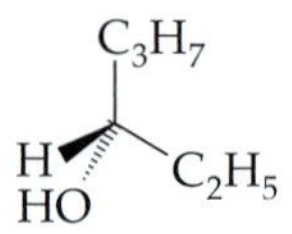 C_3H_7 / H, HO, C_2H_5 (b) (*R*)-2-chloropentane (c) (*S*)-alanine [$CH_3CH(NH_2)COOH$]

Fischer projections are especially helpful when carbohydrates are drawn in their open-chain form because there may be several stereocenters involved. For example, the open-chain form of aldohexose has four stereocenters, leading to a maximum of 16 isomers, of which glucose is only one. From the Fischer projection, it is obvious that D-(+)-glucose is an enantiomer; it is nonsuperimposable on its mirror image. (Also, it clearly does not have a plane of symmetry.) The carbon 4 epimer of glucose is known as galactose, a configurational isomer of glucose.

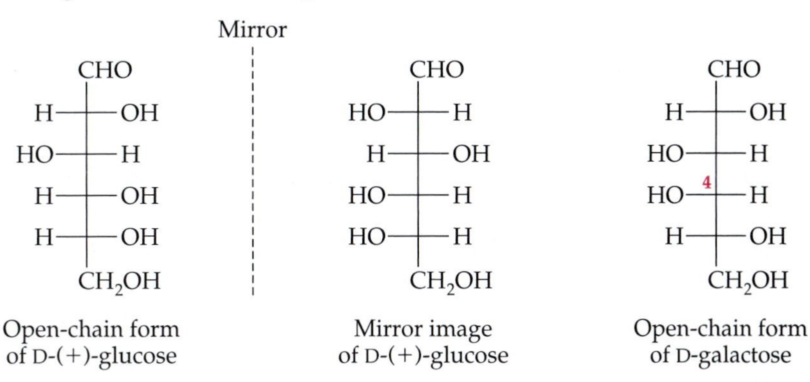

Open-chain form of D-(+)-glucose

Mirror image of D-(+)-glucose

Open-chain form of D-galactose

PROBLEM 17.9

Write the full IUPAC name, including a specification of configuration, for each compound represented by its Fischer projection:

(a) 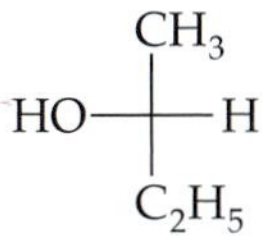 Fischer projection: CH_3 (top), HO (left), H (right), C_2H_5 (bottom)

(b) 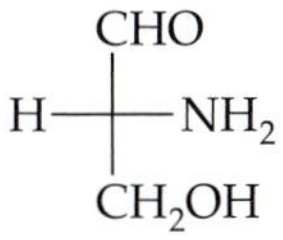 Fischer projection: CHO (top), H (left), NH_2 (right), CH_2OH (bottom)

PROBLEM 17.10

Label each stereocenter in D-(+)-glucose (shown above) with its absolute configuration (*R* or *S*).

PROBLEM 17.11

(a) Draw Fischer projections for all the configurational isomers of the open-chain form of the aldotetrose 2,3,4-trihydroxybutanal [$HOCH_2CH(OH)CH(OH)CHO$]. Label each as an enantiomer, a diastereoisomer, and so forth. Name each isomer (including labeling each stereocenter as the *R* or *S* configuration), and indicate whether each is an enantiomer or a *meso* form. Indicate which isomers are optically active.

(b) Repeat the process described in part (a) for 1,2,3,4-butanetetrol [$HOCH_2CH(OH)CH(OH)CH_2OH$].

17.3.3 Cyclic Compounds

The stereochemistry of cyclic compounds containing multiple stereocenters can be analyzed using the same approach as for acyclic compounds. As always, the ultimate test for chirality is drawing the isomer and testing for the superimposability of its mirror image. For example, 1,2-dimethylcyclopropane has two stereocenters (each with four different groups attached), so there are a *maximum* of four configurational isomers ($2^2 = 4$). However, note that the *cis* isomer has a plane of symmetry and is therefore an optically inactive *meso* form. The *trans* isomer exists as a racemate made up of a pair of enantiomers:

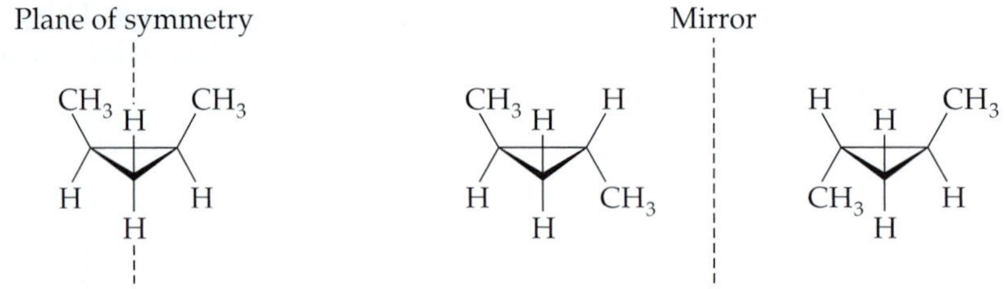

Plane of symmetry

Mirror

cis-1,2-Dimethylcyclopropane (*meso*)

trans-(±)-1,2-Dimethylcyclopropane (racemate)

This is an instance of having to consider two types of configurational isomers simultaneously—*cis-trans* isomers and enantiomers.

The stereochemistry of cyclohexane rings can be analyzed using the conformational structures, but it is easier to use planar drawings. For example, *cis*-1,2-dichlorocyclohexane has two stereocenters but also a plane of symmetry, making it a *meso* isomer. However, *trans*-1,2-dichlorocyclohexane exists as a racemate:

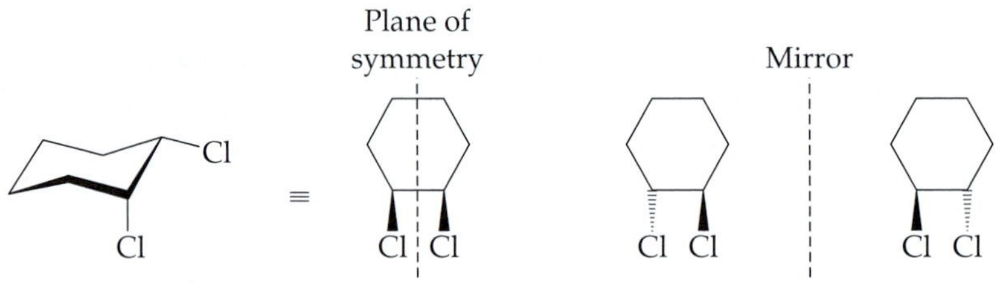

Plane of symmetry

Mirror

cis-1,2-Dichlorocyclohexane (*meso*)

trans-1,2-Dichlorocyclohexane (racemate)

PROBLEM 17.12

A total of four dibromocyclopropanes exist. Draw and fully name each. Identify the kinds of isomeric relationships that exist between the compounds.

PROBLEM 17.13

Draw the configurational isomers of each of the following compounds:

(a) 1,2-dibromocyclobutane

(b) 1,3-dichlorocyclohexane

(c) 1-bromo-4-methylcyclohexane

17.4 Chemistry of Enantiomers

We can now apply the knowledge of functional group chemistry obtained in earlier chapters to chiral molecules to understand how enantiomeric products are produced. Much of modern organic chemistry, especially that dealing with actual or potential medicinals, involves working with or producing enantiomers that are necessary for specific biological functions. This section briefly looks at the nature of reactions that involve or produce enantiomers.

CHEMISTRY AT WORK

IN THE BODY

Enantiomers and Odor

A classic case of the different effects of enantiomers on a chiral environment is the different odors exhibited by some enantiomeric pairs. Odors are detected by the olfactory sensors in the nose, and these sensors are believed to contain specific chemical receptor sites. The receptor sites are specific configurations of compounds with which volatile odiferous compounds interact. The receptor sites are believed to be chiral; so one enantiomer of an odiferous compound "fits" that particular site and therefore produces a chemical interaction that is then transferred by the nervous system to the brain. The enantiomer with the opposite configuration does not fit that particular site, but may fit another receptor site and produce a different odor sensation.

The odor exhibited by spearmint oil is caused by the monoterpene compound *R*-(−)-carvone. Its enantiomer is *S*-(+)-carvone, which is found in caraway seed oil. Very different odors are produced by olfactory interaction with these two enantiomeric but otherwise identical terpenes.

Another example from the same monoterpene family of compounds (C_{10} compounds) is the pair of hydroxyaldehydes known as the citronellals. The two hydroxydihydrocitronellals exhibit different odors: a mint odor from the (*R*)-(+) enantiomer and a lily-of-the-valley odor from the (*S*)-(−) enantiomer.

Lily of the Valley.

(*R*)-(−)-Carvone (spearmint)

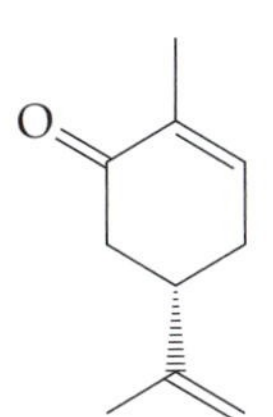

(*S*)-(+)-Carvone (caraway)

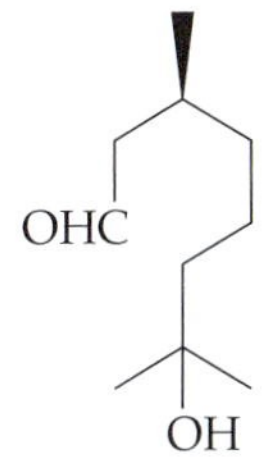

(*S*)-(−)-Hydroxydihydrocitronellal (lily-of-the-valley)

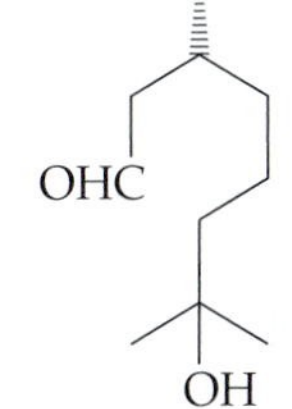

(*R*)-(+)-Hydroxydihydrocitronellal (mint)

17.4.1 Review of Reactions at Stereocenters: Inversion, Retention, and Racemization

Predicting the course of a reaction at a stereocenter relies on knowledge of the mechanism of the reaction. For example, recall that S_N2 reactions involve back-side attack at a carbon (see Section 3.4.1). Thus, such reactions result in stereospecific *inversion* of configuration at the stereocenter.

C_2H_5, CH_3, H, Br —[$NaOCH_3$ (S_N2)]→ C_2H_5, CH_3O, H, CH_3

(*S*)-2-Bromobutane → (*R*)-2-Methoxybutane

Reactions producing carbocations at the stereocenter, such as S_N1 reactions (see Section 3.4.3), are not stereospecific and result in *racemization,* due to the symmetry of the carbocation intermediate.

(S)-1-Bromo-1-phenylethane $\xrightarrow[(S_N1)]{CH_3OH}$ Racemic 1-methoxy-1-phenylethane

There are also certain conversions that can be accomplished stereospecifically with *retention* of configuration at the stereocenter. For example, recall that an alcohol with the —OH group attached to a stereocenter can be stereospecifically converted to an alkyl halide, either with overall inversion of configuration or with overall retention of configuration (this via two inversions; see Section 4.3.2):

R—OH (*S* configuration) $\xrightarrow[\text{(retention)}]{SOCl_2}$ [R—OSOCl] (*S* configuration) $\xrightarrow[\text{(inversion)}]{Cl^-}$ R—Cl (*R* configuration) **(overall inversion)**

R—OH (*S* configuration) $\xrightarrow[\text{(retention)}]{TsCl}$ R—OTs (*S* configuration) $\xrightarrow[\text{(inversion)}]{NaOH}$ R—OH (*R* configuration) $\xrightarrow[\text{(inversion)}]{SOCl_2}$ R—Cl (*S* configuration) **(overall retention)**

Therefore, many reactions at stereocenters can be stereospecific. In fact, chemists try to avoid nonstereospecific reactions with enantiomers because of the racemization that will result.

17.4.2 Reaction of Achiral Compounds with Achiral Reagents

When an achiral compound reacts with an achiral reagent to produce a compound with a new stereocenter, a racemic mixture always results. There is nothing in the reactant or reagent that induces the formation of one enantiomer in preference to another. For example, addition of hydrochloric acid to 2-butene produces racemic 2-chlorobutane because the intermediate carbocation has a plane of symmetry (the two carbons and the hydrogen attached to the sp^2-hybridized carbon of the carbocation lie in the same plane). The attacking chloride anion cannot distinguish between the two "sides" of the plane; it attacks from both above and below the carbocation plane. Thus, a 50/50 mixture of both products results, which is, by definition, a racemic mixture.

$CH_3-CH=CH-CH_3$ (2-Butene) $\xrightarrow{H^+}$ [Carbocation] $\xrightarrow{:Cl^-}$ Racemic 2-chlorobutane

PROBLEM 17.14

Show all the possible products (both major and minor) that could be obtained from the reaction of 1-methylcyclopentene with dilute aqueous sulfuric acid (addition of H_2O). Assign a full name and stereochemistry to each product.

17.4.3 Reaction of Achiral Compounds with Chiral Reagents

In contrast with the situation described in Section 17.4.2 (achiral compounds, achiral reagents, and racemic products), use of a chiral reagent with an achiral compound can produce a chiral product. A classic example is the reduction of pyruvic acid (achiral) to lactic acid (chiral), a reaction that occurs as a late step in carbohydrate metabolism in muscle tissue, when there is a deficiency of oxygen (this occurs when lactic acid has accumulated during a period of strenuous exercise).

When pyruvic acid is reduced in the laboratory with sodium borohydride, an achiral reagent, racemic lactic acid results. The hydride ion can approach the planar carbonyl group from either above or below the plane, leading to production of both enantiomers in equal amounts. By contrast, enzymatic reduction of pyruvic acid with NADH (nicotinamide adenine dinucleotide) produces only (*S*)-(+)-lactic acid.

CH_3, HOOC–C=O (Pyruvic acid) —$NaBH_4$→ CH_3, HOOC, OH, H–C + CH_3, HOOC, H, OH–C (Racemic lactic acid)

Pyruvic acid —NADH→ CH_3, HOOC, H, OH–C ((*S*)-(+)-Lactic acid)

Enzymes are chiral reagents, with chirality arising mainly from the fact that the protein component of an enzyme consists of L-amino acids. The chiral (asymmetrical) surface of the enzyme binds with the substrate. The delivery of the reducing electrons and the proton (that is, effectively a hydride ion) occurs preferentially from one side of the carbonyl group, resulting in the enantiomeric *S*-(+)-lactic acid product.

Chemists have adapted many normally achiral reagents to chiral chemistry by attaching a chiral group to the reagent, thereby making the entire reagent chiral. The chiral group does not change the basic nature of the reagent but simply provides a chiral environment for the reaction. This usually results in the production of an excess of one of the two possible enantiomers, called an **enantiomeric excess (ee)** and expressed as a percentage.

17.4.4 Reaction of Chiral Compounds with Achiral Reagents

When a chiral substrate reacts with an achiral reagent at a site other than the stereocenter, the chirality of the substrate can influence the stereochemistry of the reaction. This generally leads to the predominant production of one possible configurational isomer over the other. In other words, a chiral substrate produces an excess of one of the

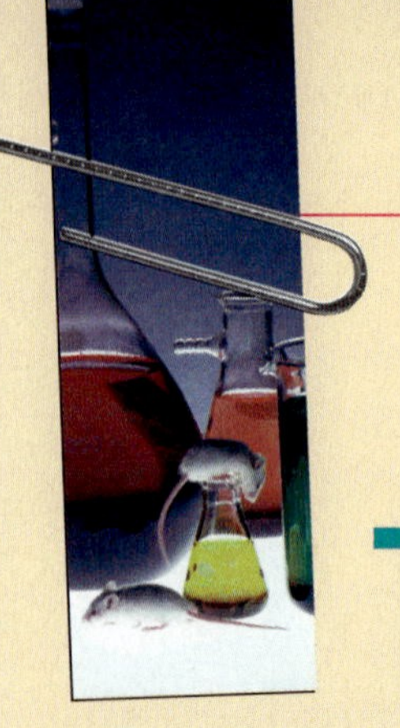

CHEMISTRY AT WORK

IN THE MARKETPLACE

Synthesis of Enantiomers

The synthesis of enantiomers is extremely important in biological chemistry, especially with potential medicinals. There are really only two options for accomplishing such syntheses. The usual approach for many years was to design a synthesis starting with achiral substrates and eventually perform a resolution to obtain the desired enantiomer. This resolution could be performed early in the synthetic sequence or even at the final stage.

The second, and now more common, approach is to start with naturally occurring enantiomers and perform only stereospecific reactions on them. In effect, nature has already performed the initial resolution. For example, synthesis of the alkaloid (+)-lycoricidine (with four stereocenters) was first done in 1976 but yielded only a racemic mixture. By 1993, there had been four syntheses of the desired enantiomer, one of them starting with optically active L-arabinose and retaining the specific configurations of several stereocenters throughout the synthetic sequence.

An interesting approach to the preparation of an enantiomeric medicinal is illustrated by the commercial synthesis of the internal ammonium salt carnitine, which is involved in fatty acid metabolism. Only the (*R*)-(−) enantiomer is biologically active. In fact, the (*S*)-(+) enantiomer inhibits the action of the (*R*)-(−) enantiomer. The racemate of carnitinamide is manufactured in multiton quantities, and the *R* enantiomer is separated by resolution as a diastereoisomeric ammonium salt, using an enantiomeric carboxylic acid as the resolving agent. However, this means that half of the product of the production process, the *S* enantiomer, is useless. Recently, however, a synthetic scheme has been devised to rescue the previously discarded *S* enantiomer and convert it into the *R* enantiomer through a sequence of stereospecific reactions.

L-Arabinose

(9 steps)

$(CH_3)_3\overset{+}{N}$–CH(OH)–$CONH_2$ (Cl^-) —(i-BuOH, HCl)→ $(CH_3)_3\overset{+}{N}$–…–$COOBu^i$ —($(CH_3SO_2)_2O$)→ $(CH_3)_3\overset{+}{N}$–…(OSO_2CH_3)–$COOBu^i$ —(HCl/H_2O)→ $(CH_3)_3\overset{+}{N}$–…(OSO_2CH_3)–COOH —($NaHCO_3$ (S_N2))→ β-lactone —($NaHCO_3$, H_2O)→ $(CH_3)_3\overset{+}{N}$–…(OH)–COO^-

(*S*)-(+)-Carnitinamide

(*R*)-(−)-Carnitine

two possible diastereoisomers. For example, the addition of hydrogen to a cyclohexene ring carrying a chiral group can occur from either side of the double bond in a *syn* addition (see Section 6.3). Two diastereoisomers are produced, but the chiral group causes the addition to occur preferentially (sometimes almost exclusively) from one side of the ring to produce an excess of one diastereoisomer (*R,R* or *R,S*).

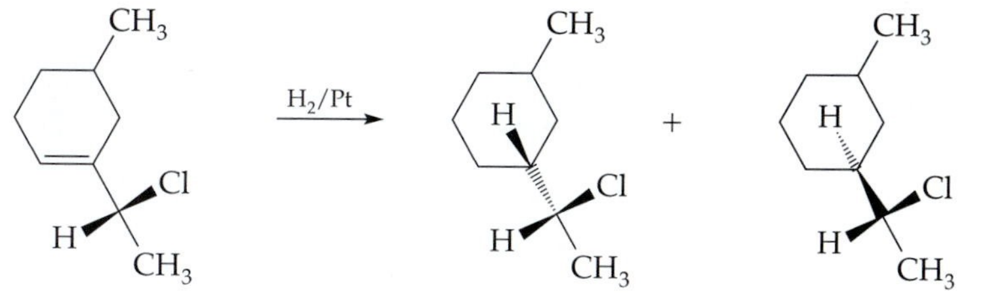

H_2/Pt

R enantiomer — *R,R* diastereoisomer + *R,S* diastereoisomer

Such observations have led to the regular use of a group of compounds called **chiral auxiliaries,** which are chiral compounds that are temporarily attached to a substrate. They make the substrate chiral while a reaction is executed with an achiral reagent. This results in an excess of one diastereoisomer, called a **diastereoisomeric excess (de),** from which the chiral auxiliary is subsequently removed. The advantage is that an excess of one enantiomer can be produced without going through a resolution. For example, when a chiral auxiliary is first attached to the hydroxyl group, reduction of achiral (*E*)-2-methyl-2-buten-1-ol results in an enantiomeric excess of 2-methyl-1-butanol after the chiral auxiliary is removed by hydrolysis.

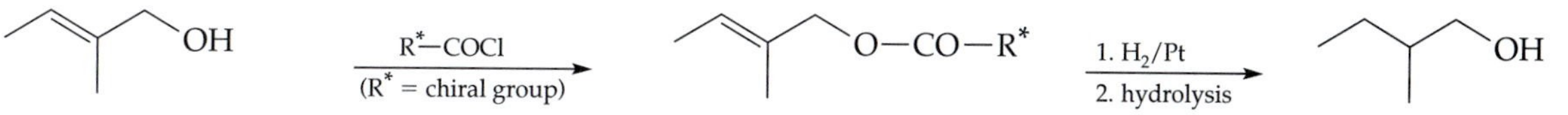

R^*—COCl (R^* = chiral group); 1. H_2/Pt 2. hydrolysis

(Z)-2-Methyl-2-buten-1-ol → O—CO—R^* → 2-Methyl-1-butanol

The specific rotation of pure *d*-2-methyl-1-butanol is +3.8°. Thus, if the rotation of the sample prepared by the preceding reaction is +3.0°, then the enantiomeric excess, or optical purity, of the sample is $(3.0 \div 3.8) \times 100 = 79\%$. This means that 79% of the mixture is the *d* form and 21% of the mixture is the *dl* form (a racemate). Put another way, the *d* portion of the mixture is 79% + 10.5% (half of the *dl*) = 89.5%, while 10.5% (half of the racemate) is the *l* form.

PROBLEM 17.15

Reduction of 2-butanone to 2-butanol with a reagent that carries a chiral auxiliary produces an alcohol with $[\alpha]_{obs} = +10.0°$. Pure (+)-2-butanol has $[\alpha]_{obs} = +13.9°$. What is the percentage of *d* and *l* forms present in the product mixture?

PROBLEM 17.16

Show and fully name the products that would result from the sodium borohydride reduction of (*R*)-1,3,4-trihydroxy-2-butanone. Indicate their stereochemical relationships.

PROBLEM 17.17

(*E*)-2,2-Dimethyl-3-hexene is brominated with *N*-bromosuccinimide (recall allylic bromination; see Section 10.2.3). Show the products that result, name each, and indicate the kind of isomeric relationships that exist between them. Will the product mixture be optically active?

CHEMISTRY AT WORK

IN THE UNIVERSE

Chirality and Life

Concerning the origins of life on earth, some interesting chemical questions arise. Why are the carbohydrate components of DNA and RNA (the genetic material of all known living organisms) only of the *R* (D) configuration? Why are the α-amino acids that make up proteins in most forms of life only L-amino acids (that is, of the *S* configuration)? The reason why these compounds, which are the very chemical essence of life, have such uniform chirality (homochirality) but opposite chirality is not known, but it is the subject of wide speculation and some research. A related question is, Which came first, life or homochirality?

Living systems produce molecules as enantiomers, while compounds produced in the laboratory are normally racemates. If life started from inanimate (nonliving) materials, how did the original racemic compounds evolve to enantiomers? In 1953, it was demonstrated that passing an electric charge through a mixture of methane, ammonia, and water—constituents presumed to have been present on prebiotic earth—produces some α-amino acids. However, because they were produced from achiral starting materials, these were racemates. How could such racemates evolve into enantiomers necessary for life on earth? Was there a resolution? How?

Some scientists wonder if the first living organisms on earth were of extraterrestrial origin and brought chirality with them. Perhaps life itself was not brought to earth, but prebiotic enantiomers were. One class of meteorites that have landed on earth, known as the carbonaceous chondrites, contain some complex organic compounds, including some α-amino acids. An example is the Murchison meteorite. These α-amino acids have been shown to be enantiomers and may be of extraterrestrial origin. A NASA space mission is being designed to send a small spacecraft to rendezvous with the comet Wirtanen and sample it to determine (1) if there is organic material in the comet, (2) if there are compounds present that might be considered prebiotic, and (3) if such compounds are enantiomeric or racemic.

Few of the many meteorites that streak through the night sky survive to be examined for their content of organic compounds.

On the question of which came first on earth, microorganisms or homochirality, some argue that homochirality came first—genetic material (DNA) could not copy itself if it were racemic. However, others argue that homochirality is an artifact of life—that it came about through a natural selection process and was not a precursor of life. Watch for more on this debate in your favorite science magazine.

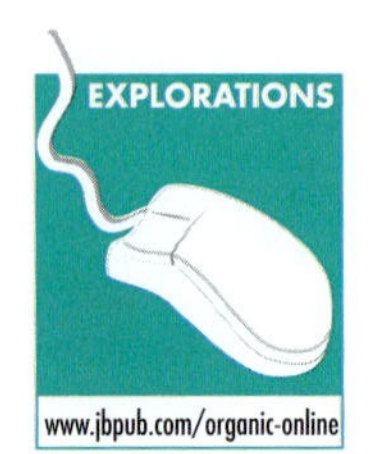

17.5

Stereochemistry of Biological Processes

Most of the chemical reactions occurring in metabolic processes involve chiral enzymes, so most of these reactions produce enantiomeric products. Most of the active ingredients in biological systems, such as vitamins, hormones, regulating agents, and coenzymes, are also enantiomeric. So are all proteins and carbohydrates. In other words,

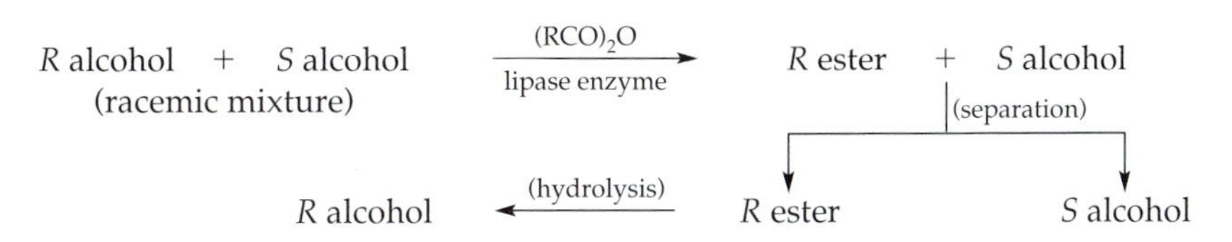

FIGURE 17.9

Use of a lipase enzyme to accomplish a stereoselective reaction, resulting in resolution of a racemic alcohol.

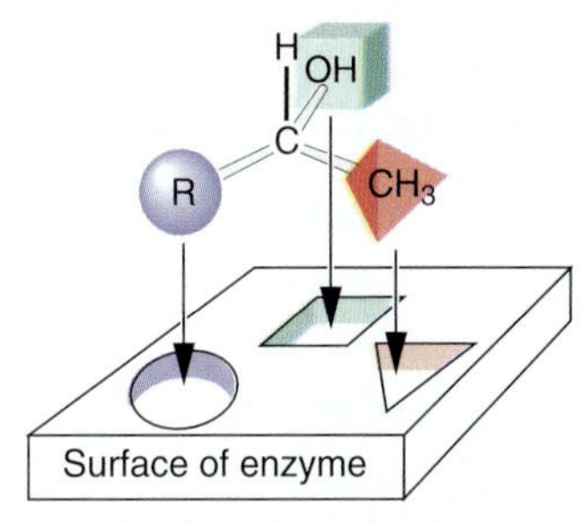

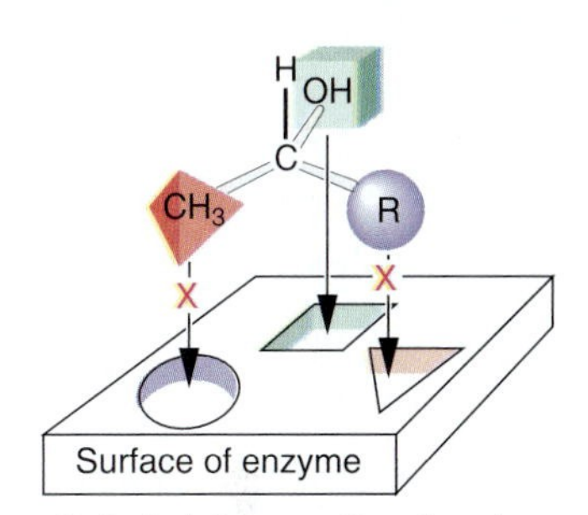

FIGURE 17.10

Drawing showing how the *R* alcohol "fits" the enzyme's active site allowing that alcohol to be esterified. The *S* alcohol does not fit.

life itself depends on enantiomers (chiral compounds). This means that when drugs are designed to have particular effects on biological systems, they exert those effects through an enantiomeric form, even if a racemate is administered. For example, ibuprofen (an analgesic and NSAID found in Motrin, Nuprin, and Advil) is sold as the racemate, but its biological effect is exerted by the *S* enantiomer only.

The chirality of enzymes, as found in bacteria and other sources, can be used to accomplish organic syntheses of enantiomers through their stereospecific catalytic effects. In fact, biological reagents are now widely used for industrial chemical processes. For example, lipases are a family of enzymes that catalyze esterification reactions. Reaction of an achiral acid anhydride with a racemic alcohol in the presence of a lipase produces mainly the ester from one enantiomer, leaving the other enantiomeric alcohol unchanged and recoverable. Separation and hydrolysis of the ester produces the other enantiomeric alcohol, as illustrated in Figure 17.9. This is a roundabout but very efficient resolution of the racemic alcohol.

The chirality of enzymes and their selectivity in reacting with one enantiomer of a substrate but not the other can be illustrated using the preceding lipase-catalyzed esterification. The chiral enzyme surface binds with the *R* alcohol but not with the *S* alcohol, as illustrated in Figure 17.10.

Chapter Summary

A number of different kinds of **isomers** are possible for many organic compounds. Enantiomers are a biologically important category of isomers.

An optically inactive **racemic mixture** can be resolved into the two optically active enantiomers of opposite configurations. This **resolution** is accomplished using a **resolving agent,** a single enantiomer. The only difference in physical properties between two enantiomers is their optical activity, and the characteristic property of enantiomers is that they are **optically active;** that is, they rotate a beam of plane-polarized light. The magnitude and direction of that rotation comprise the **observed rotation** that can be measured using a **polarimeter.** The **specific rotation** is a characteristic and reproducible physical property of each enantiomer calculated from its observed rotation. There is no relation between the absolute configuration (*R* or *S*) of an enantiomer and the direction in which it rotates plane-polarized light.

Configurational isomers with multiple **stereocenters** have a maximum of 2^n possible isomers, where n is the number of stereocenters present. **Fischer projections** are used to represent the configurations of such isomers in two dimensions. ***Meso* isomers** are those that have multiple stereocenters but contain a plane of symmetry. Therefore, they are not chiral and do not rotate plane-polarized light. Diastereoisomers contain more than one stereocenter, and the configuration of at least one of those stereocenters (but not all) must be the same in both isomers.

The use of a chiral reagent or the attachment of a **chiral auxiliary** to a substrate permits the production of an **enantiomeric excess** of one enantiomer in a reaction. In organic syntheses, enantiomers can be produced either by carrying out a resolution of the product (or an intermediate) or by commencing a stereospecific synthesis with a single enantiomer of starting material and using only reactions that are stereospecific.

Additional Problems

Termininology

17.18 Define the following terms used in organic chemistry:

(a) isomers

(b) stereocenter

(c) chiral molecule

(d) enantiomers

(e) racemic mixture

(f) configurational isomers

(g) conformational isomers

(h) resolution

(i) levorotatory

(j) diastereoisomers

(k) plane-polarized light

(l) constitutional isomers

(m) specific rotation

(n) dextrorotatory

(o) racemate

(p) *meso* isomer

(q) observed rotation

(r) *cis-trans* isomers

17.19 Draw structures for the following compounds:

(a) (*Z*)-3-methyl-2-heptene

(b) (*E*)-2-chloro-2-pentene

(c) 3-bromo-6-methylocta-(3*E*,5*Z*)-diene

17.20 Assign an *E* or *Z* configuration to each of the following alkenes:

(a) 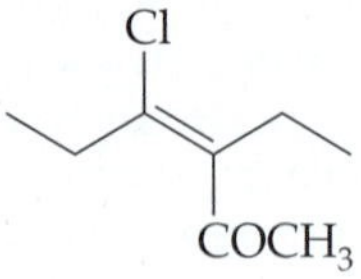

(b) 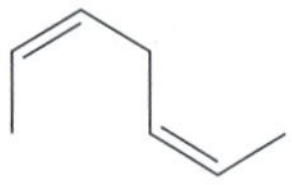

(c) 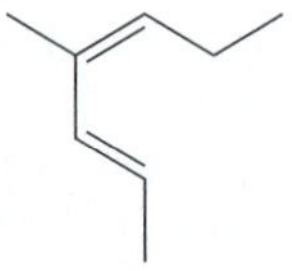

17.21 Many pheromones are effective in only a single configuration. Draw the structure of the following pheromones:

(a) the sex attractant of the winter moth, (1,3*E*,6*E*,9*E*)-nonadecatetraene

(b) the sex attractant of the coddling moth, 3-ethyl-7-methyl-(2*Z*,6*E*)-decadien-1-ol

(c) the sex attractant of the honeybee, (*E*)-9-oxo-2-decenoic acid

(d) the "juvenile hormone" of the silkworm moth, the carbon 10 and carbon 11 epoxide of methyl (2*E*,6*E*,10*Z*)-tridecatrienoate

Chirality and Configurations

17.22 Mark with an asterisk each stereocenter in the following compounds:

(a) 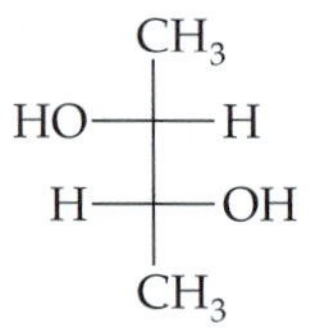

(b) 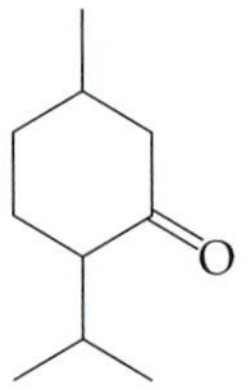

Menthone

(c) $CH_3CHOHCH_2CH_2CHBrCH(CH_3)_2$

(d) $CH_3CHOHCHNH_2COOH$

Threonine

(e) 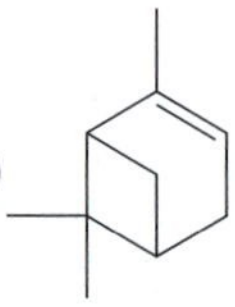

α-Pinene

(f) 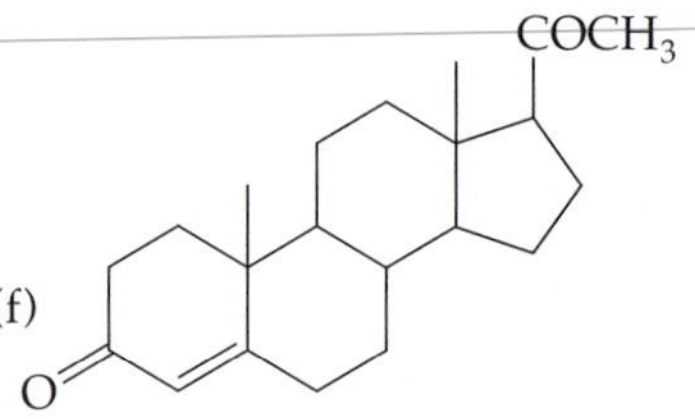

Progesterone

17.23 Draw the other enantiomer (mirror image) of each of the following compounds:

(a) (*R*)-2-iodopentane

(b) (2*R*,3*R*)-dibromobutane

(c) (*S*)-3-penten-2-ol

17.24 Indicate which of the following compounds can exist as racemates:

(a) *cis*-1,2-dichlorocyclohexane

(b) 2-methylpiperidine

(c) *trans*-1,4-dimethylcyclohexane

(d) *cis*-1,3-divinylcyclohexane

(e) 1-phenyl-1-bromo-2-pentanone

(f) methyl cyclohex-2-en-4-onecarboxylate

(g) 2-butyl benzoate

(h) 3-methylcyclohexyl acetate

(i) 4-(*N,N*-dimethylamino)-2-pentyne

17.25 The C_4 carbohydrates with an aldehyde group are called *aldotetroses* (IUPAC name, 2,3,4-trihydroxybutanal). Draw each of the configurational isomers, and label each as a D or L isomer. Indicate in parentheses the configuration (*R* or *S*) at carbon 2 and carbon 3 for each isomer.

17.26 Which of the following compounds are optically active?

(a) methylcyclohexane

(b) (*S*)-2-hexanol

(c) (*E*)-2-hexene

(d) *dl*-2-aminobutanoic acid

(e) (±)-2-butyl acetate

(f) D-glyceraldehyde

(g) L-phenylalanine

(h) (*R*)-3-hydroxypentanoic acid

17.27 The two carvones shown are responsible for the flavor of caraway seeds and spearmint oil, respectively. Assign an *R* or *S* configuration to each stereocenter.

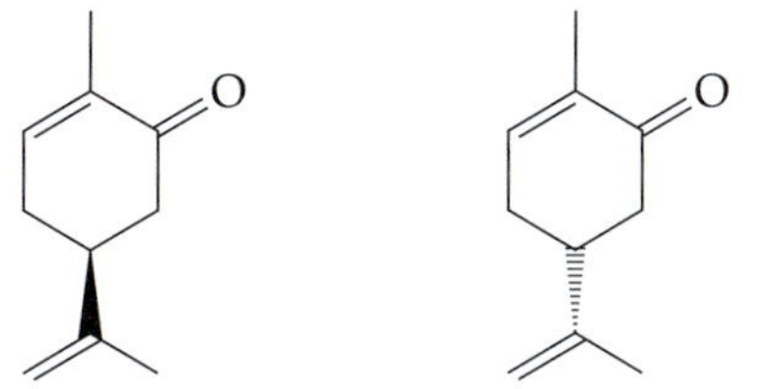

(+)-Carvone ($[\alpha]_D = +62.5°$) (caraway)

(−)-Carvone ($[\alpha]_D = -62.5°$) (spearmint)

17.28 Assign the *R* or *S* configuration to each Fischer projection:

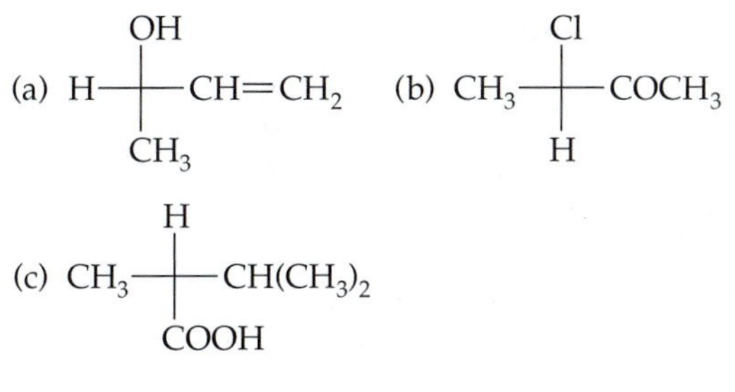

(a) top: OH; left: H; right: $CH{=}CH_2$; bottom: CH_3

(b) top: Cl; left: CH_3; right: $COCH_3$; bottom: H

(c) top: H; left: CH_3; right: $CH(CH_3)_2$; bottom: COOH

17.29 Assign the *R* or *S* configuration to each stereocenter of the following compounds:

(a) COOH; H—C—OH; H—C—OH; COOH

(b) CH_3; H—C—OH; Cl—C—H; COOH

(c) COOH; HO—C—H; H—C—OH; COOH

(d) $COCH_3$; H—C—OH; H—C—OH; CH_3

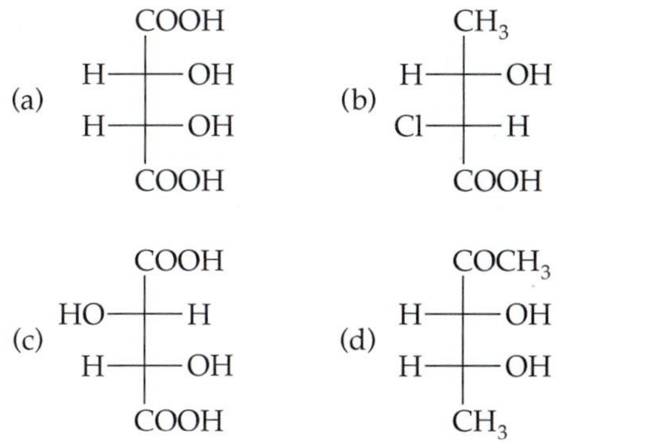

17.30 Suggest the kind of resolving agent (family of compounds) that could be employed to resolve each of the following compounds, and indicate the kind of compound the diastereoisomer will be:

(a) 2-butanol

(b) 2-aminopentane

(c) 2-phenylbutanoic acid

(d) 3-methyl-2-pentanone

(e) ibuprofen [2-(4-isobutylphenyl)propanoic acid]

(f) lactic acid (2-hydroxypropanoic acid)

(g) *p*-(2-butyl)aniline

17.31 Indicate which of compounds **B–D** have the same configuration as compound **A** and which are enantiomers of **A**:

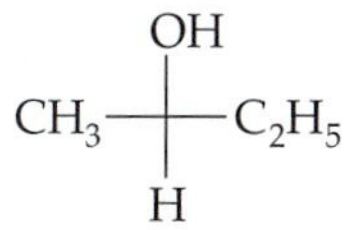

A: top: OH; left: CH_3; right: C_2H_5; bottom: H

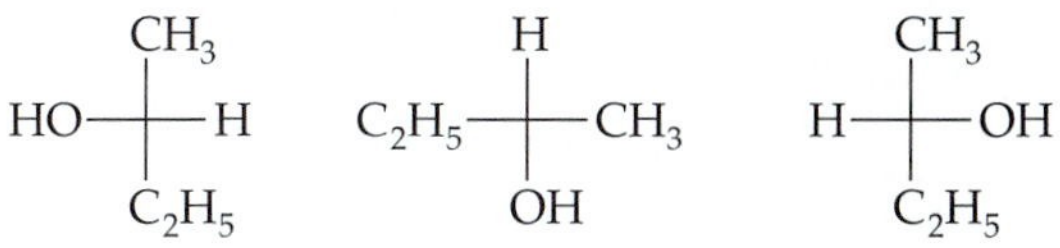

B: top: CH_3; left: HO; right: H; bottom: C_2H_5

C: top: H; left: C_2H_5; right: CH_3; bottom: OH

D: top: CH_3; left: H; right: OH; bottom: C_2H_5

Multiple Stereocenters

17.32 Two possible configurations for a compound with three stereocenters are *R,S,R* and the mirror image *S,R,S*. How many optical isomers in total are possible for the compound? What are the configurations of the other possible isomers for the compound?

17.33 For a saturated compound with five stereocenters, what is the maximum possible number of configurational isomers?

17.34 For each of the following compounds, determine whether there is a plane of symmetry:

(a) *cis*-1,2-dibromocyclobutane

(b) *trans*-1,2-dibromocyclopropane

(c) *trans*-1,4-dimethylcyclohexane

(d) *cis*-1,3-dichlorocyclohexane

(e) (2*R*,3*S*)-butanediol

17.35 Draw all of the possible configurational isomers for 2,3-dichloropentane, label each isomer as to configuration of the stereocenter(s), and indicate the relationship of the isomers to each other.

17.36 Repeat Problem 17.35 for 2,4-dichloropentane.

17.37 Tartaric acid (2,3-dihydroxysuccinic acid) exists in two naturally occurring forms, both of which are optically inactive. One form melts at 206°C, and the other melts at 140°C. The 206°C form can be resolved into two enantiomers, each with a melting point of 170°C. The 140°C form cannot be resolved.

(a) What term can be used to describe the 206°C form?

(b) What term can be used to describe the 140°C form?

(c) What is the relationship between the 170°C enantiomers and the 140°C form?

(d) Draw a Fischer projection for the 140°C form. Write its full IUPAC name, including the absolute configuration of each stereocenter.

(e) Draw Fischer projections for the 170°C enantiomers. Write their full IUPAC names, including the absolute configuration of each stereocenter.

Reactions of Chiral Compounds

17.38 Predict the stereochemical outcome of each of the following reactions by indicating the name and configuration of the stereoisomer produced. Briefly explain each answer.

(a) (*R*)-2-butyl acetate with dilute sodium hydroxide

(b) (*S*)-2-pentyl tosylate with sodium hydroxide

(c) (*R*)-1-iodo-1-phenylethane with methanol

(d) (*R*)-2-bromopentane with sodium methoxide

(e) 1-butene with dilute sulfuric acid

(f) (*R*)-2-butanol with thionyl chloride and pyridine

(g) styrene with HCl

(h) 2-methyl-2-butene with diborane then hydrogen peroxide

(i) cyclohexene with cold potassium permanganate

(j) cyclopentene oxide with sodium methoxide

17.39 Show the reactions that would be necessary to obtain (*R*)-3-methyl-2-butanol from 2-methyl-2-butene.

17.40 Explain why laboratory reduction (for example, with $NaBH_4$) of cortisone to hydrocortisone usually produces two isomers, whereas in the body the same reaction produces only a single isomer.

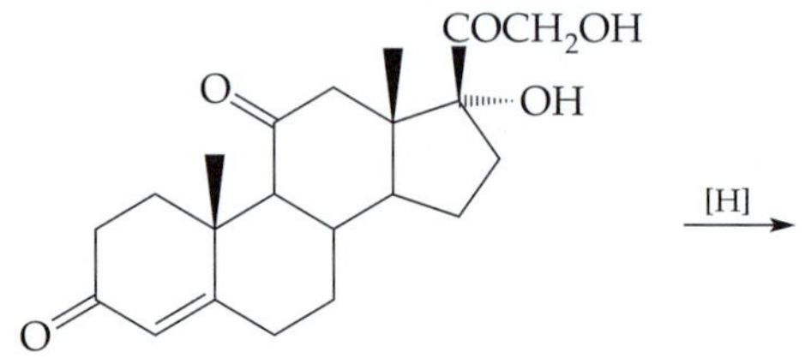

$COCH_2OH$ OH O O [H] →

Cortisone

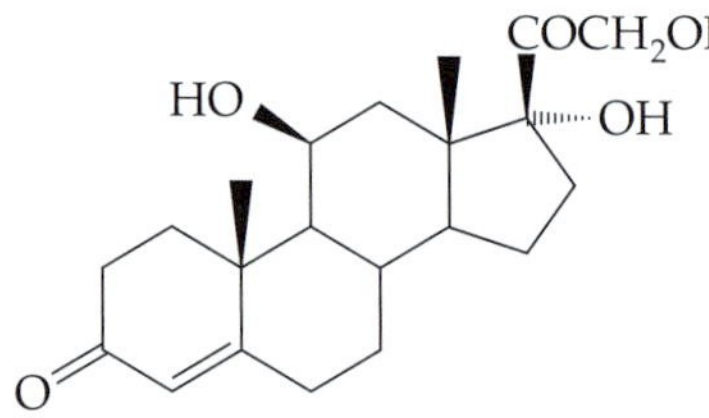

$COCH_2OH$ HO OH O

Hydrocortisone

17.41 Predict the products from the ionic addition of HBr to (*S*)-3-bromo-1-butene, and describe the stereochemistry that results. If there is more than one product, describe their relationship to each other.

17.42 Explain why the acid-catalyzed hydration of styrene produces a racemic mixture.

17.43 Show the products, including stereochemistry, from each of the following reactions with *E*-3-methyl-2-hexene:

(a) hydrogen over a platinum catalyst

(b) cold potassium permanganate

(c) water and catalytic amounts of sulfuric acid

(d) bromine in CCl_4

Mixed Problems

17.44 For each of the following pairs of compounds, indicate what the structural/stereochemical relationship is between the two:

(a) $CH_3CH(OH)CH_2CH_3$

and

$CH_3CH_2CH_2CH_2OH$

(b) 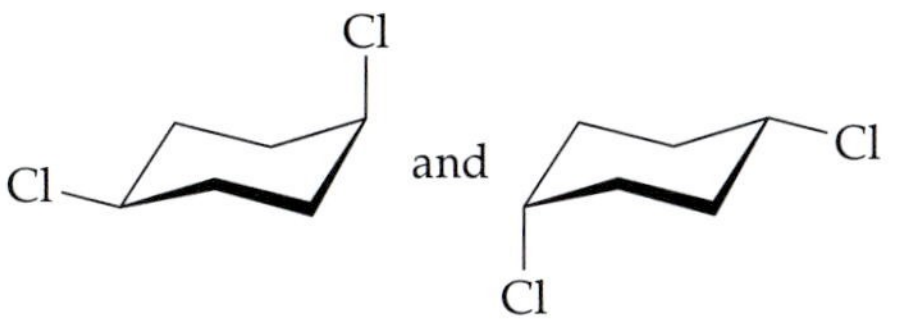 and

(c) 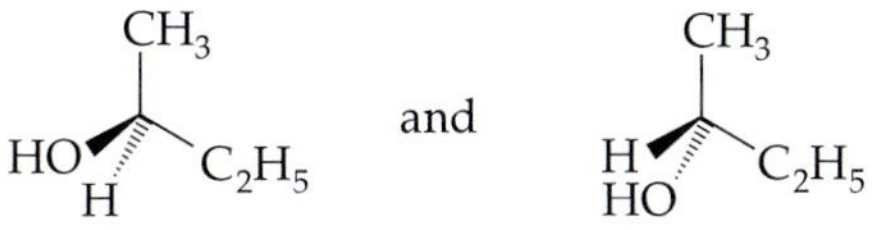 and

(d) 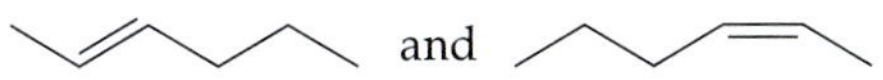 and

(e) 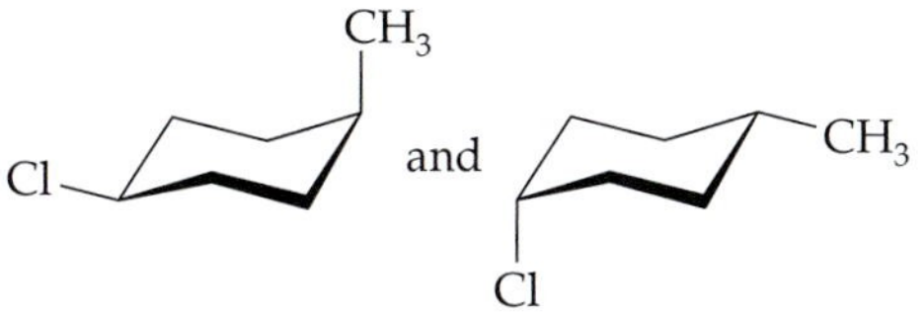 and

(f) 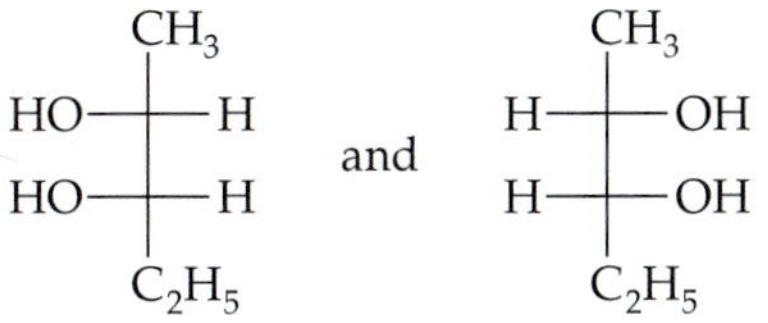 and

(g) 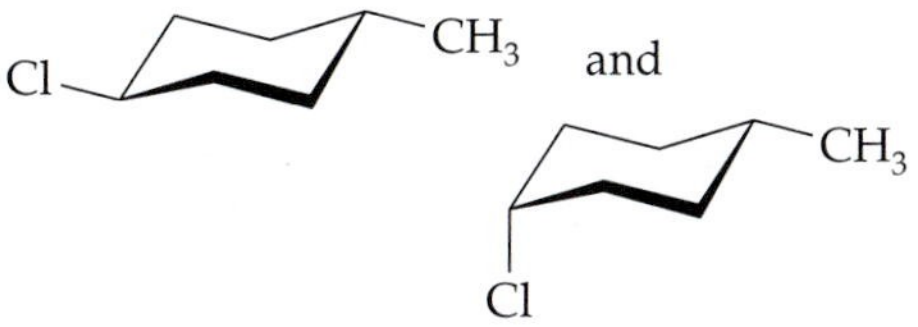 and

(h) 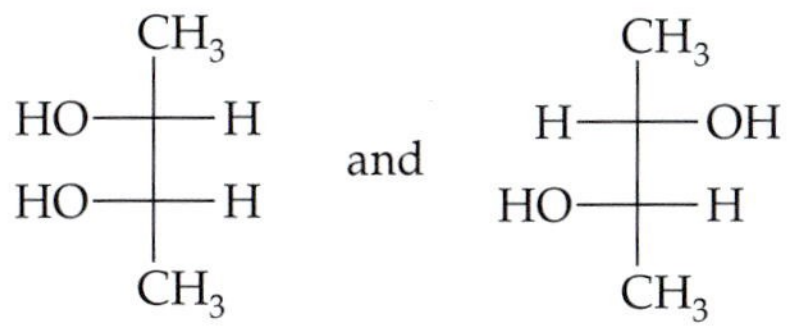 and

17.45 How many configurational isomers, and what kind, exist for each of the following compounds?

(a) 4-hexen-2-ol

(b) 3-cyclopentenecarboxylic acid

(c) 2-chloro-2-butenoic acid

(d) glycerol (propane-1,2,3-triol)

(e) lactic acid (2-hydroxypropanoic acid)

(f) 2,3-dimethylsuccinic acid

17.46 Draw the structure for an unsaturated compound with the formula C_5H_9Br that fits each description:

(a) shows *cis-trans* isomerism but no optical activity

(b) shows no *cis-trans* isomerism but is optically active

(c) shows no *cis-trans* isomerism and no optical activity

(d) shows *cis-trans* isomerism and is optically active

17.47 Compound **A,** with formula C_8H_{14}, is optically active. Catalytic reduction yields optically inactive **B,** C_8H_{18}. Ozonolysis of **A** produces two products, one of which is acetic acid. The other (compound **C**) is an optically active carboxylic acid, $C_6H_{12}O_2$. What is the structure of each compound?

17.48 For each of the following reactions, show the product(s) expected, indicate how many configurational isomers are possible, and indicate whether or not the product(s) will be optically active:

(a) 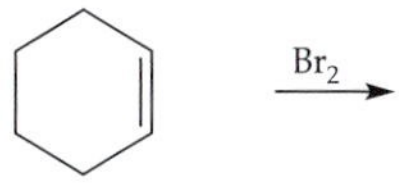 $\xrightarrow{Br_2}$

(b) 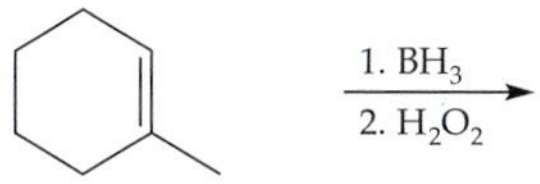 $\xrightarrow[2.\ H_2O_2]{1.\ BH_3}$

(c) 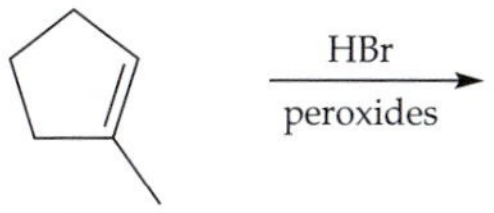 $\xrightarrow[\text{peroxides}]{HBr}$

(d)  $\xrightarrow{H_2O/H_2SO_4}$

(e) 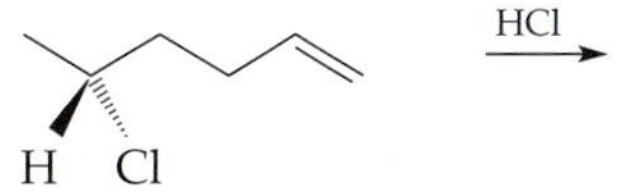 $\xrightarrow{HCl}$

(*R* enantiomer)

(f) $CH_3CH_2CH{=}CHCH_2CH_3$ $\xrightarrow{\text{peracetic acid}}$

(a mixture of *cis* and *trans*)

(g) CH_3CH_2CHO $\xrightarrow[2.\ H^+/H_2O]{1.\ CH_3MgBr}$

17.49 The two D-aldotetroses are known as erythrose and threose. Oxidation of either with nitric acid produces a tartaric acid (2,3-dihydroxysuccinic acid). Threose produces an optically active tartaric acid, while erythrose produces an optically inactive (and unresolvable) tartaric acid. From this information, deduce the structures of threose and erthyrose, and assign *R* or *S* configurations to their stereocenters.

17.50 D-Galactose, an optically active aldohexose monosaccharide, is coupled to glucose in the disaccharide lactose (see Section 4.7.3), has four stereocenters in its open-chain form, and is a carbon-4 epimer of glucose. It can be oxidized to the corresponding C_6 dicarboxylic acid with the modest oxidizing agent nitric acid. The dicarboxylic acid is optically inactive. Explain.

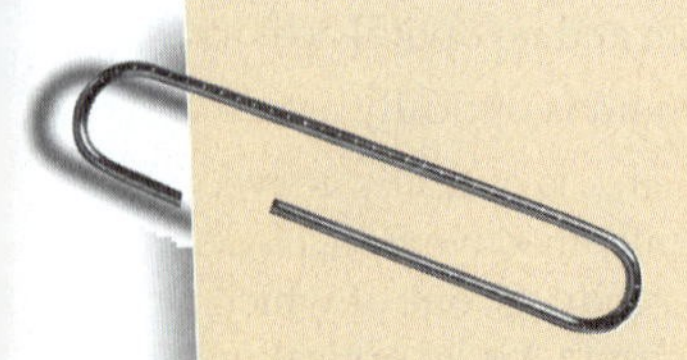

CONCEPTUAL PROBLEM

Drugs

As a pharmacist, you read with interest that thalidomide has recently been approved by the Food and Drug Administration for use in the treatment of Hansen's disease (more commonly known as leprosy). This announcement comes almost 40 years after the drug was banned by that agency because it caused birth defects in babies born to women who had used it to combat morning sickness during early pregnancy. What's of particular interest to you are all the new regulations associated with the drug, which will be marketed as Thalomid. The manufacturer of the drug has developed a System for Thalidomide Education and Prescribing Safety (STEPS) program. Only physicians who register with the program may prescribe Thalomid. The patients, both male and female, must also register, as well as complying with mandatory contraceptive measures and filling out surveys. The prescriptions will be for no more than a 28-day supply, with no automatic refills.

This is quite a contrast to the drug's introduction in 1958. It was touted as nontoxic, and there was even some pressure to sell it over the counter, with no prescription needed. It was its use to control morning sickness that revealed its one terrible and tragic property. It is a powerful teratogen, a substance that causes birth defects. The structure of thalidomide is

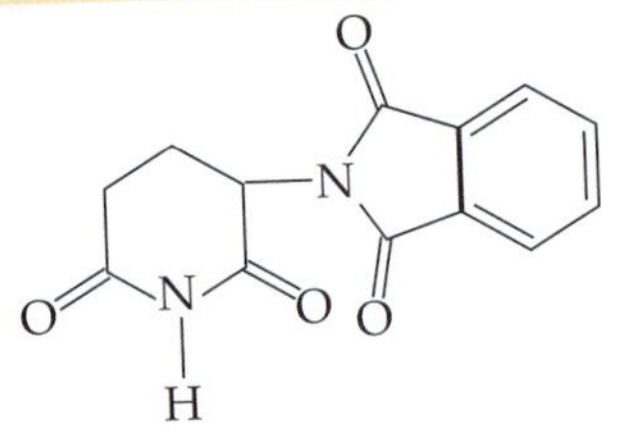

O, N, O, N, O, O, H

Thalidomide

- A thalidomide molecule contains one stereocenter. Where is it in the structure shown above?
- Thalidomide was manufactured from achiral materials and carefully purified, but no resolution was performed. What was therefore present in the thalidomide prescribed for morning sickness?
- Draw structures of the two enantiomers of thalidomide, and identify the structures you have drawn as *R* and *S* isomers.
- More recent research has shown that the *R* isomer is active against leprosy, and the *S* isomer is the teratogen. What does this tell you about the point of interaction of thalidomide with the body's biochemical systems?

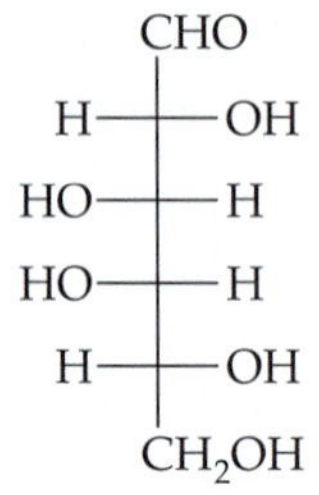

CHO
H—OH
HO—H
HO—H
H—OH
CH_2OH

Open-chain form of D-galactose

17.51 Considering the possible mechanisms for the following two preparations of 2-ethoxy-1-phenylpropane from (*d*)-1-phenyl-2-propanol, explain the observed optical rotations in terms of the stereochemistry of each reaction:

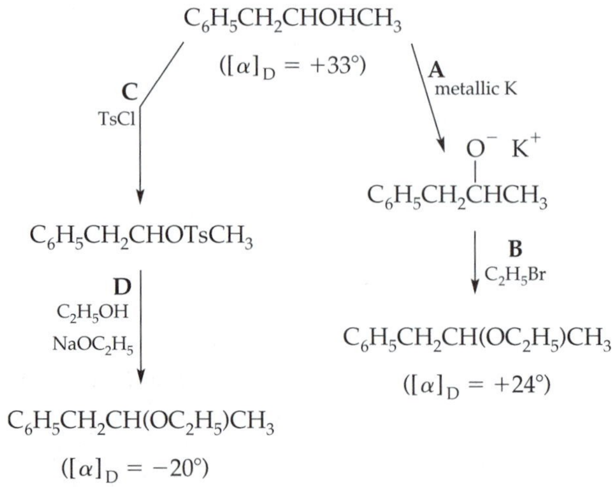

$C_6H_5CH_2CHOHCH_3$ ($[\alpha]_D = +33°$)

Route C (TsCl) → $C_6H_5CH_2CHOTsCH_3$ → D (C_2H_5OH, $NaOC_2H_5$) → $C_6H_5CH_2CH(OC_2H_5)CH_3$ ($[\alpha]_D = -20°$)

Route A (metallic K) → $C_6H_5CH_2CH(O^- K^+)CH_3$ → B (C_2H_5Br) → $C_6H_5CH_2CH(OC_2H_5)CH_3$ ($[\alpha]_D = +24°$)

CHAPTER 18

Synthetic Polymers

HUMANS HAVE USED natural polymers to advantage throughout history: cellulose (wood and straw) for shelter, starch (corn and potatoes) for food, and protein (hides and hair) for warmth. With the advent of synthetic polymers in the mid-twentieth century, the uses of polymers multiplied immeasurably: vinyl siding and foam insulation for shelter; Nylon, Orlon, and polyester fabrics for clothing; plastic automobile and aircraft parts as well as synthetic tires for transportation; polyethylene bags and foam packaging; lucite and latex paints; videotape and compact disks; and synthetic heart valves and prosthetics. Most of these synthetic polymers are organic polymers. One of the great successes of the chemical industry is the ability to discover and economically manufacture polymers with a wide variety of physical properties, specifically designed for particular functions.

Human beings have utilized natural polymers for shelter thoughout history. Today, we utilize many synthetic polymers such as vinyl siding and foam insulation.

Chemists are continually searching for new polymers that have interesting and beneficial properties. Of particular interest are polymers, or mixtures of polymers, that might serve medical needs. For example, researchers continue to seek more stable polymers for use in artificial hearts and heart components (such as valves). Another major target of research is a material that can serve in optical storage devices, using light signals instead of electronic signals to encode information for computers and related devices. Such a polymer is intended to replace the electronically and magnetically based devices currently in use. Finally, the search continues for polymeric materials that have the necessary functional properties for their intended uses but that are also more biodegradable than the polymers now used. Striking a balance between the stability needed for a useful product lifetime and the degree of instability needed to facilitate decomposition in landfills or elsewhere is a difficult chemical puzzle. Recycling represents only a partial solution. (See "Chemistry at Work," p. 665.)

18.1 Introduction to Polymer Chemistry

The chemistry covered so far in this book has focused primarily on molecules containing no more than 30 carbon atoms, with relatively low molecular weights (less than 500). These molecules, which generally have no more than a few functional groups, are those on which most organic chemistry is conducted, whether in the laboratory or in nature. However, there are many examples of both naturally occurring and synthetic **macromolecules,** compounds that have hundreds, even thousands, of carbon atoms and molecular weights in the millions. A **polymer** (*poly* means "many") is one kind of macromolecule made up of a large number of smaller molecules called **monomers** (*mono*

◀ Most plastic products begin their lives as colored chips. They are then melted and molded into their final shapes.

means "single") that have been chemically joined together. (This description of polymers applies to both natural and synthetic ones.)

In most polymers, the repeating unit is the same throughout (monomer-A):

$$\text{A} \longrightarrow \text{(etc.)—A—A—A—A—A—(etc.)}$$

Monomer Polymer

For example, the monomer of the naturally occurring polymer cellulose is the monosaccharide β-D-glucose, and the monomer of natural rubber is isoprene (2-methyl-1,3-butadiene). The monomer of the synthetic polymer tradenamed Teflon is tetrafluoroethylene.

$$\text{—(}CH_2\text{—}C(CH_3)\text{=}CH\text{—}CH_2\text{)}_n\text{—} \qquad \text{—(}CF_2\text{—}CF_2\text{)}_n\text{—}$$

Cellulose (poly-β-D-glucose) Natural rubber (polyisoprene) Teflon (polytetrafluoroethylene)

As shown above, the repeating unit of a polymer is represented by drawing the structural formula of the monomer with an open bond at both ends, bisected by parentheses. The small letter *n* indicates that an unspecified but large number of such repeating units are joined together. The polymer molecules may be of uniform size and the value of *n* may be known. However, most polymers are mixtures of chains containing different numbers of monomer units; in this case, *n* varies and is known only as an average.

Some polymers contain two or more different kinds of repeating units (monomers), in which case they are referred to as **copolymers.** The sequence of the monomers in a copolymer may be random or regular:

$$\text{A} + \text{B} \longrightarrow \text{(etc.)—A—B—A—B—(etc.)}$$

Two monomers Regular copolymer

The synthetic plastic film sold as Saran Wrap is a copolymer of chloroethene and 1,1-dichloroethene; ABS plastic is a synthetic copolymer of acrylonitrile (CH_2=CHCN), butadiene (CH_2=CH—CH=CH_2), and styrene (C_6H_5CH=CH_2). (We will consider their structures later in this chapter.) Proteins are natural copolymers of up to 22 different α-amino acids (see Section 16.2.1); DNA and RNA are natural copolymers of eight different nucleotides (see Section 19.1).

PROBLEM 18.1

Based on your knowledge of proteins from Chapter 16, draw a generalized structural formula for a protein macromolecule.

18.1.1 Classification of Polymers by Structure

Polymeric structures fall into two broad categories: linear polymers and branched polymers. A **linear polymer** is a long chain of atoms formed by repeating units of a monomer joined end to end. There are no connections between such chains; they are like a bunch of separate pieces of rope. Two monomer combinations are possible for linear polymers, as illustrated in Figure 18.1. The first type of linear polymer (Figure 18.1a) is formed from a single monomer that has two different functional groups (one

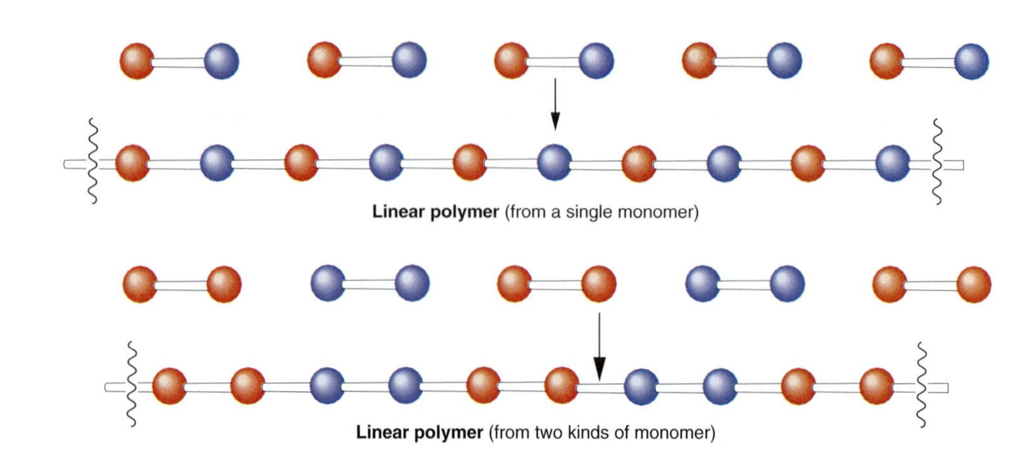

FIGURE 18.1

Representation of the formation of two types of linear polymers.

at each end). Nylon-6 is of this type. The second type of linear polymer (Figure 18.1b) is formed from two different monomers, each of which has the same functional group at both ends. Nylon-6,6 and Dacron are examples of this type, as we will see.

In **branched, or cross-linked, polymers,** the polymer chains are joined together in a two- or even three-dimensional network. Bonds can form between linear polymer chains (Figure 18.2a), a type of cross-linking that occurs in the vulcanization of natural rubber (see Section 18.3.6). A network can form if a monomer has three functional groups, permitting chains to grow in three directions simultaneously (Figure 18.2b). Starch is a heavily branched polymer of α-D-glucose in which some glycoside linkages are formed from carbon 1 to carbon 4, some to carbon 6, and some to both carbons.

18.1.2 Classification of Polymers by Formation

Polymers are also classified into two categories based on the chemistry used in their synthesis—the process of polymerization. In **addition polymers** (also called *chain-growth polymers*), monomers are joined together end to end, and all atoms present in the

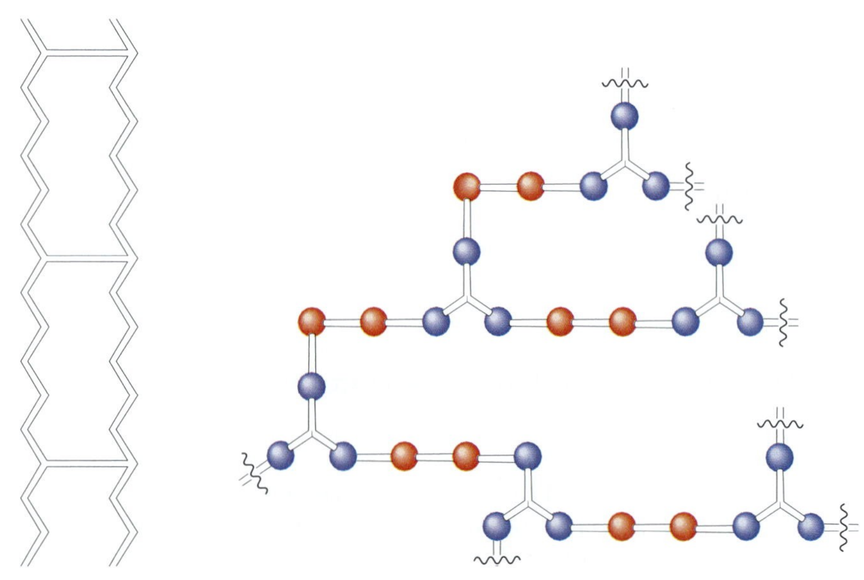

FIGURE 18.2

Representation of two types of branched polymers.

monomer are retained in the polymer. These polymers are invariably formed from monomers that are unsaturated compounds; a pi bond is converted into two sigma bonds. (Recall one addition reaction of alkenes in which hydrogen chloride and ethylene add to form ethyl chloride, and all atoms are retained in the product.) For example, as we will see, polyethylene is formed by the addition of ethylene monomers to each other.

A second type of polymerization results from the reaction of two functional groups that *condense* to form a new functional group. A small molecule (frequently water) is usually ejected in the process. These polymers are called **condensation polymers** (or *step-growth polymers*). An example is a polyester that is formed from an alcohol and a carboxylic acid; water is eliminated in ester formation (see the Fischer esterification in Section 15.3.4).

We will discuss both types of polymerization in greater detail in Sections 18.3 and 18.4.

18.2

Physical Properties of Polymers

In the 1930s, chemists began to synthesize polymers, trying to create and mass-produce materials with specific properties. As you might expect, the properties of polymers are directly related to their size, shape, composition, and uniformity. Any synthetic polymer is a mixture of molecules of different sizes and molecular weights, depending on the number of monomer repeating units incorporated. For example, a batch of polyethylene may be made up of thousands of polyethylene chains, with different chains incorporating different numbers of monomer units. Chemists are constantly searching for ways to make polymers more homogeneous because their properties will then be more consistent and predictable. Melting point and degree of crystallinity are two of the most important physical properties of any polymer because they are often determining factors in how the polymer can be used. We will discuss these properties on the basis of the general concept of polymer chains, without considering any particular polymer's chemical structure.

Polymers that can be synthesized with a high degree of structural regularity (for example, linearity), so that their chains can pack tightly into an organized and regular crystal lattice, are higher-melting. For example, linear polyethylene (Figure 18.3a) is highly crystalline and melts near 135°C, whereas branched polyethylene is amorphous and melts at a much lower temperature, which somewhat limits its applications.

Other polymers, such as natural rubber, with a melting point of 30°C, are amorphous solids because the randomly coiled shape of the chains precludes their packing into regular crystal patterns (Figure 18.3b). This property of natural rubber makes it unsatisfactory for use in tires because the tires would lose their shape and eventually melt. Nylon is a synthetic polymer that has some regions that are crystalline and other regions that are amorphous (Figure 18.3c). Its semicrystalline nature makes Nylon suitable for casting in molds; its high melting point of 265°C enables it to be used in applications where high temperature stability is required. In 1995, Nylon was even used to fabricate the intake manifold for some Chrysler automobiles. This manifold of synthetic polymer has greater strength, lower weight, and a smoother internal flow surface than does an aluminum manifold.

The kinds of properties a polymer exhibits determine the functions it can have. There are three broad categories of polymers based on bulk properties:

1. **Elastomers** are amorphous polymers with a coiled molecular shape that gives them elastic properties. The synthetic rubbers used for vehicle tires and the natural rubber used for rubber bands are two examples.

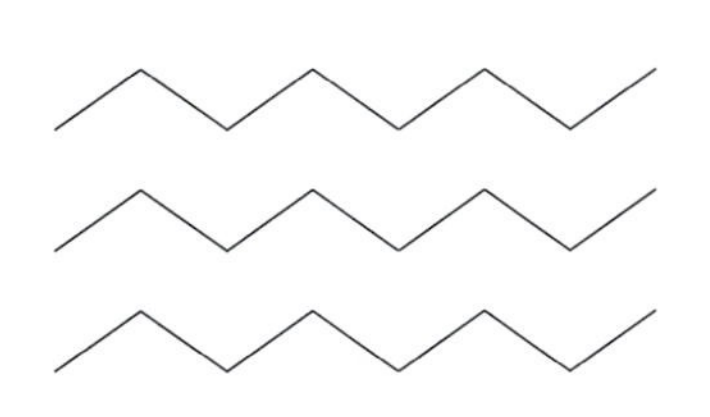

Linear polyethylene (crystalline)
(a)

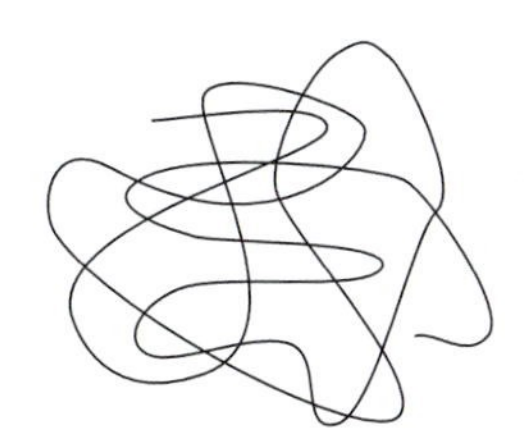

Rubber (amorphous)
(b)

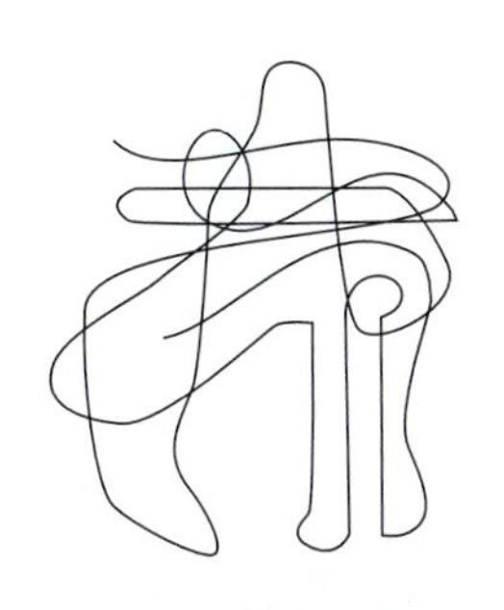

Nylon (semicrystalline)
(c)

FIGURE 18.3

Representations of (a) crystalline, (b) amorphous, and (c) semicrystalline polymers. (Parallel straight lines represent crystalline areas, and curved lines represent amorphous areas.)

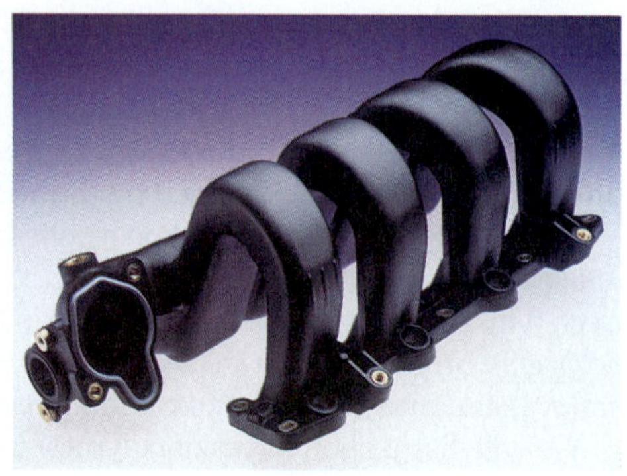

Natural rubber, an amorphous polymer, is made from the milky sap of rubber plants. Because it is a semicrystalline polymer, nylon can be molded into rigid structures, such as air intake manifolds for cars.

2. **Fibers** are made from semicrystalline polymers. The melted polymer is drawn through a small orifice, creating a fiber with increased alignment of the polymer molecules. Nylon is a good example.
3. **Plastics** are rigid synthetics that are made from highly crystalline polymers, such as high-density polyethylene (HDPE), polypropylene, and aramids. Very high degrees of cross-linking in a polymer also provide rigidity, as is the case with the plastic sold as Bakelite.

Synthetic polymers are generally produced by chemical companies in bulk quantities, often in the form of granules that are supplied to product manufacturers. Fabricators process the granules in one of three ways, related to the three categories based on properties. The bulk polymer may be heated and drawn into a fiber by forcing it through a small orifice as it cools, a process called *spinning*. This fiber may then be woven into fabrics ranging from nylon stockings to clothing to carpeting. Alternatively, the polymer may be heated to its melting point and poured into a casting for molding into various shapes, producing items such as tires, gears, bottles, and the entire dashboard panels of some automobiles.

Various additives may be included in the final processing of a polymer, and they significantly affect the properties of the final product. For example, poly(vinyl chloride), or PVC, is well known as the rigid plastic piping used in household plumbing. However, processing the same polymer with dibutyl phthalate added as a plasticizer lowers the polymer's melting point and serves as a lubricant between its chains, thereby adding flexibility. This enables the product to be used as the soft vinyl found on automobile tops, in raincoats, and in shoes.

Polymers are drawn into fibers by forcing it through a small orifice as it cools. This fiber can then be woven into different types of fabrics.

Some polymers, called *resins*, are classified as **thermosetting polymers,** which means that once cast into a mold, they cannot be remelted and remolded. Bakelite resin is a very highly cross-linked (and therefore very rigid) polymer formed from formaldehyde and phenol. Bakelite is used in electrical fixtures (plugs and outlets), household appliances, dishes, and bowling balls. Epoxy resins and epoxy glues are other examples of thermosetting polymers (see Section 5.2). Such polymers cannot be recycled.

Thermoplastic polymers can be remelted and reshaped, and some are suitable for recycling. Common examples are the PET plastics [poly(ethylene terephthalate) polyesters] used in soft-drink bottles and other containers and certain nylons.

18.3 Addition Polymers

18.3.1 Mechanisms of Polymeric Addition Reactions

Addition, or chain-growth, polymers are formed via addition reactions of alkenes, whose mechanisms were presented in Chapters 6 and 8. The approach is to first add a reagent to an alkene monomer to produce a reactive intermediate. The reactive intermediate then adds to another molecule of monomer. In other words, the idea is to initiate a **chain reaction,** a reaction that once started will continue (in theory, at least) until all of the monomer is consumed.

$$\underset{\text{Alkene monomer}}{CH_2{=}CH(X)} \xrightarrow[\text{polymerization)}]{\text{(addition}} \underset{\text{Addition polymer}}{\left(CH_2{-}CH(X) \right)_n}$$

Addition polymerization can be initiated under cationic conditions, radical conditions, and anionic conditions. The principle is the same in each case: An **initiator** is used to bring about the initial addition to the alkene—the *initiation step.* This forms a reactive intermediate—carbocation, radical, or carbanion. The reactive intermediate carries out an addition on another molecule of alkene—the *propagation step.* This process repeats itself over and over.

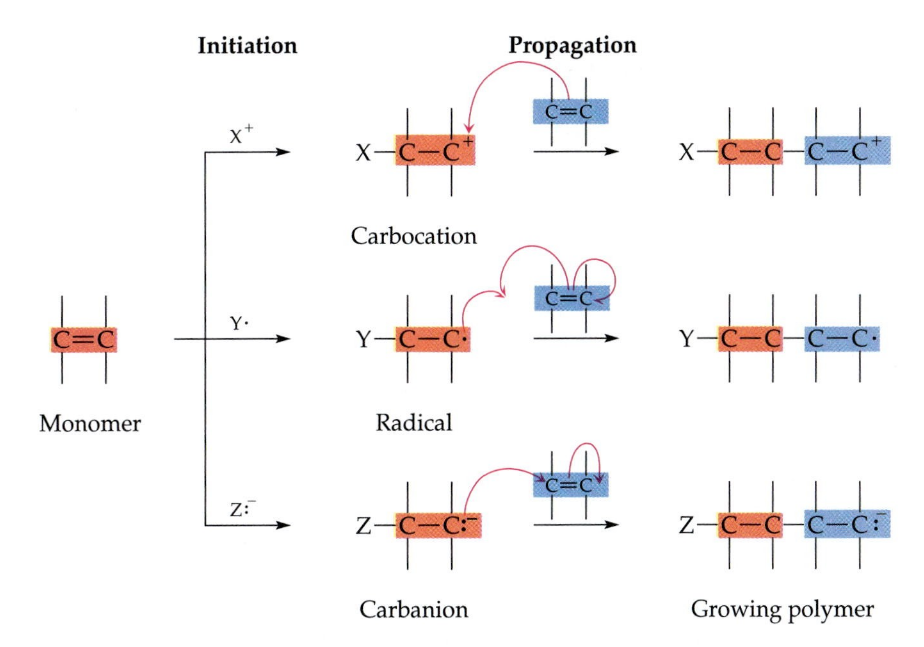

The net effect of addition polymerization is that the pi bond of the monomer is broken and two new sigma bonds are formed in an overall exothermic process. The initiator remains attached to one end of the polymer, but in view of the huge size of the polymer molecule, its presence is so insignificant that it does not affect the overall properties of the final product. The growing polymer chain is eventually terminated by one of several possible reactions, as we will see. Again, the exact structure of the terminal group is immaterial with respect to the bulk properties of a polymer.

Which mechanism is appropriate for preparing an addition polymer depends on the ability of the alkene monomer to form the reactive intermediate. (Recall the earlier discussions of the stability of carbocations, carbanions, and radicals.) Radical polymerization is effective with almost any alkene. Cationic polymerization is generally limited to those alkenes that form stable tertiary carbocations in the initiation step. Anionic polymerization is usually effective only when the carbanionic intermediate is stabilized by attached electron-withdrawing groups.

PROBLEM 18.2

Show the initiation step, the structure of the reactive intermediate, and the structure of the polymer resulting from each of the following polymerizations. Explain the regiochemistry you have shown.

(a) styrene ($C_6H_5CH{=}CH_2$) and R·
(b) isobutylene [$(CH_3)_2C{=}CH_2$] and H^+
(c) acrylonitrile ($CH_2{=}CHCN$) and $:R^-$

18.3.2 Radical Polymerization

Radical polymerization begins with the addition to the monomer of a small amount of a radical initiator (see Section 10.1.1). Typically, a peroxide such as benzoyl peroxide is added, which, upon heating, forms two benzoyloxy radicals. An example is the polymerization of styrene (Figure 18.4).

The nature of the termination reaction is relatively unimportant because of the minute ratio of terminal groups (two per chain) to the number of monomer units (thousands). Shown in Figure 18.4 are two termination reactions: elimination to form an alkene end group or hydrogen-atom abstraction to form an alkane end group. A third kind of termination reaction is the combination of two growing radicals to form a longer chain. It is also possible to add *chain terminators* at some point during the polymerization process to stop chain growth. For example, thiols (RSH) are effective in terminating a radical chain reaction. They readily give up a hydrogen atom, producing a thiol radical that is quite unreactive and unable to start a new chain:

$$C_6H_5COO{-}\left(CH_2\underset{\substack{|\\ C_6H_5}}{CH}\right)_n{-}CH_2{-}\dot{C}HC_6H_5 \xrightarrow[-R-S\cdot]{H-S-R} C_6H_5COO{-}\left(CH_2\underset{\substack{|\\ C_6H_5}}{CH}\right)_n{-}CH_2{-}CH_2C_6H_5$$

Polystyrene

One additional and very important reaction frequently occurs in radical polymerization—but not at the end of a chain. In this reaction, known as **chain transfer,** a growing chain radical abstracts a hydrogen atom from another chain, thereby producing a new radical within that chain's interior. Growth occurs at the new radical site by reaction with additional monomer. This process leads to branching of the chain, as shown in Figure 18.5 for polystyrene. This property may or may not be desirable. Chain branching leads to more amorphous polymers.

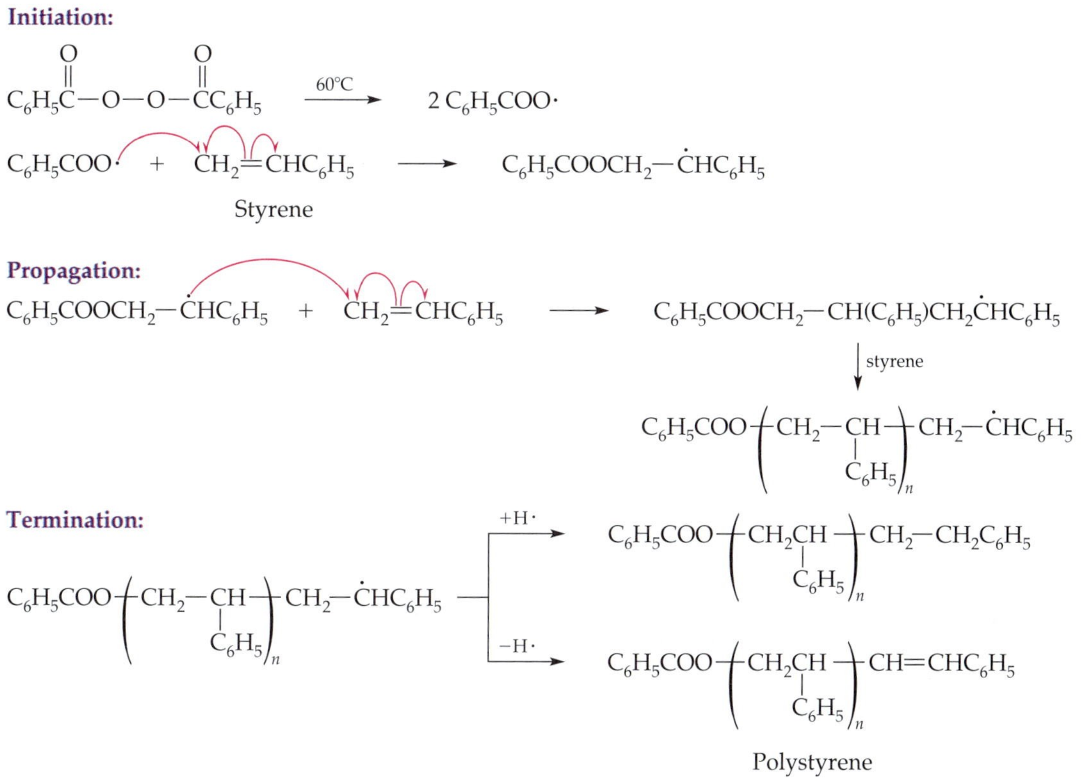

Initiation:

$$C_6H_5\overset{O}{\overset{\|}{C}}-O-O-\overset{O}{\overset{\|}{C}}C_6H_5 \xrightarrow{60^\circ C} 2\,C_6H_5COO\cdot$$

$$C_6H_5COO\cdot + \underset{\text{Styrene}}{CH_2{=}CHC_6H_5} \longrightarrow C_6H_5COOCH_2-\dot{C}HC_6H_5$$

Propagation:

$$C_6H_5COOCH_2-\dot{C}HC_6H_5 + CH_2{=}CHC_6H_5 \longrightarrow C_6H_5COOCH_2-CH(C_6H_5)CH_2\dot{C}HC_6H_5$$

$$\downarrow \text{styrene}$$

$$C_6H_5COO\left(CH_2-\underset{C_6H_5}{\underset{|}{C}H}\right)_n CH_2-\dot{C}HC_6H_5$$

Termination:

$$C_6H_5COO\left(CH_2-\underset{C_6H_5}{\underset{|}{C}H}\right)_n CH_2-\dot{C}HC_6H_5 \xrightarrow{+H\cdot} C_6H_5COO\left(CH_2\underset{C_6H_5}{\underset{|}{C}H}\right)_n CH_2-CH_2C_6H_5$$

$$C_6H_5COO\left(CH_2-\underset{C_6H_5}{\underset{|}{C}H}\right)_n CH_2-\dot{C}HC_6H_5 \xrightarrow{-H\cdot} C_6H_5COO\left(CH_2\underset{C_6H_5}{\underset{|}{C}H}\right)_n CH{=}CHC_6H_5$$

Polystyrene

FIGURE 18.4

The mechanism of radical polymerization of styrene to form polystyrene.

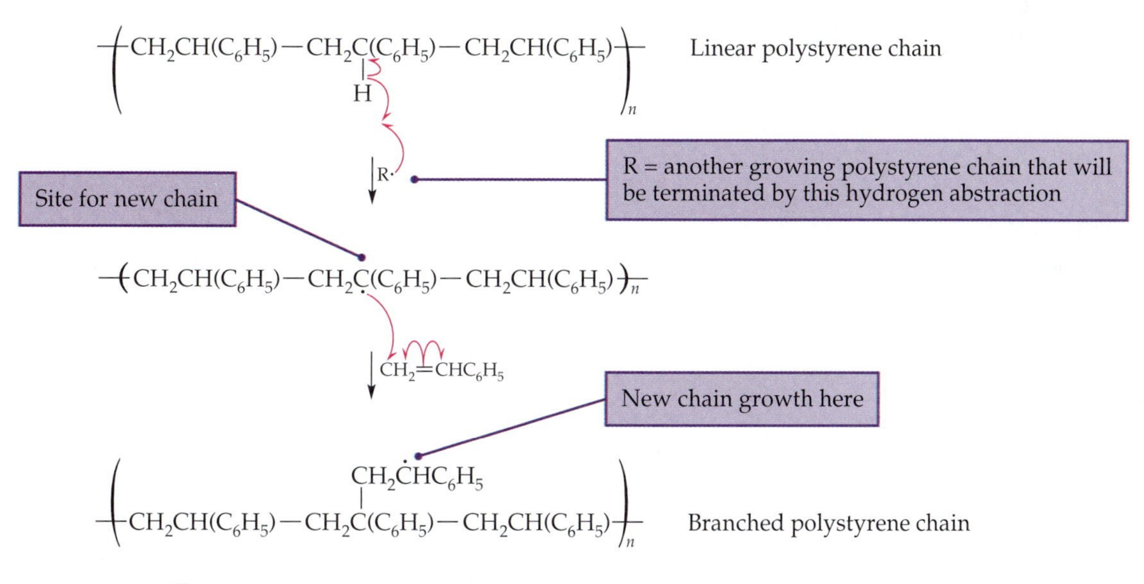

FIGURE 18.5

Chain-transfer reaction leading to chain branching.

Now let's consider the preparation and uses of a sampling of radical polymerization products.

Polystyrene

Polystyrene is produced in very large quantities by the radical polymerization of styrene (vinyl benzene):

$$C_6H_5CH{=}CH_2 \xrightarrow[\text{initiator}]{\text{radical}} \left(\begin{matrix} CH-CH_2 \\ | \\ C_6H_5 \end{matrix}\right)_n$$

Styrene → Polystyrene

The average molecular weight of polystyrene is about 2 million. It is molded to form the casings for many household appliances, including radios, television sets, and computers.

Styrofoam is prepared by dissolving small amounts of a low-boiling liquid (such as pentane) in the melted polystyrene. This low-boiling additive vaporizes. Then, as the polymer cools, the vapor is trapped as bubbles, creating "holes." Recently, carbon dioxide has been introduced as the "blowing agent" to create the bubbles. Styrofoam is a very lightweight and effective insulating material used as insulation board in construction and molded to form ice chests, foam cups, refrigerator insulation, and other insulating items. Because it can be molded into many shapes, it is often used as foam packaging surrounding fragile products—objects made of glass or delicate electronics, like audio equipment or computers.

Styrene is frequently used in copolymers. Polymerization with acrylonitrile produces a strong, clear material used in automobile headlight lenses. Copolymerization of styrene with butadiene and acrylonitrile produces ABS plastics.

$$\underset{\text{Styrene}}{CH_2{=}CHC_6H_5} + \underset{\text{Acrylonitrile}}{CH_2{=}CHCN} \longrightarrow \left(\begin{matrix} CH_2-CH-CH_2-CH \\ \quad\;\; | \qquad\quad\;\; | \\ \quad C_6H_5 \qquad CN \end{matrix}\right)_n$$

Styrene-acrylonitrile copolymer

The white packing material that is often found protecting computers and other electrical equipment, is made of polystyrene.

$$\underset{\text{Acrylonitrile (A)}}{CH_2{=}CHCN} + \underset{\text{Butadiene (B)}}{CH_2{=}CH-CH{=}CH_2} + \underset{\text{Styrene (S)}}{C_6H_5CH{=}CH_2}$$

$$\downarrow$$

$$\left(\begin{matrix} CH_2-CH-CH_2-CH{=}CH-CH_2-CH_2-CH \\ | \qquad\qquad\qquad\qquad\qquad\qquad | \\ CN \qquad\qquad\qquad\qquad\qquad C_6H_5 \end{matrix}\right)_n$$

ABS copolymer

Polystyrene can be cross-linked by adding a small amount of divinylbenzene during polymerization. The main chains can continue to grow, but the divinylbenzene allows the connection of two parallel chains, producing a much more rigid polymer than styrene itself.

$$\text{Styrene} + \text{Divinylbenzene} \longrightarrow$$

Styrene Divinylbenzene

$$-\!\!\left(\text{CH}-\text{CH}_2-\text{CH}-\text{CH}_2-\text{CH}-\text{CH}_2\right)_x\!\!-$$
$$-\!\!\left(\text{CH}-\text{CH}_2-\text{CH}-\text{CH}_2-\text{CH}-\text{CH}_2\right)_y\!\!-$$

Copolymer of styrene and divinylbenzene (cross-linked)

PROBLEM 18.3

The ion-exchange resin used in water softeners is the sodium salt of polystyrenesulfonic acid prepared from polystyrene. Show the reaction and reagent(s) necessary for this preparation.

Polyethylene

Polymerization of ethylene produces a polymer by the same radical mechanism as for polystyrene.

$$\underset{\text{Ethylene}}{CH_2{=}CH_2} \longrightarrow \underset{\text{Polyethylene}}{-\!\!\left(CH_2-CH_2\right)_n\!\!-}$$

Polyethylene should be a linear polymer, a long chain of methylene groups. However, in most instances, considerable branching occurs, probably through chain transfer reactions. This branching produces an amorphous polymer that melts well below 100°C; it is known as low-density polyethylene (LDPE). Products employing LDPE include films, packaging (such as plastic bags for dry-cleaned clothing and fresh produce), and protective heavy sheeting. Polyethylene can be produced in a wide range of molecular weights, allowing for a wide range of uses. See Section 18.3.5 for the preparation of high-density polyethylene (HDPE), which has little chain branching and is therefore more crystalline and higher-melting. This makes it suitable for use in rigid products, especially containers and bottles.

Poly(vinyl chloride)

The radical polymerization of vinyl chloride (chloroethene) produces poly(vinyl chloride), or PVC. A major use for PVC is in plastic piping for household and industrial plumbing. Vinyl chloride and 1,1-dichloroethene react to form a copolymer that can be formed into clear sheets. One well-known product of this process is sold as Saran Wrap.

$$\underset{\text{Vinyl chloride}}{CH_2{=}CHCl} \longrightarrow \underset{\text{Poly(vinyl chloride) (PVC)}}{\left(-CH_2-\underset{Cl}{\underset{|}{CH}}-\right)_n}$$

$$\underset{\text{1,1-Dichloroethene}}{CH_2{=}CCl_2} + \underset{\text{Vinyl chloride}}{CH_2{=}CHCl} \longrightarrow \underset{\text{Saran Wrap}}{\left(-CH_2-CCl_2-CH_2-\underset{Cl}{\underset{|}{CH}}-\right)_n}$$

Saran Wrap, a brand of clear plastic wrap, is a copolymer that has been extruded into thin sheets.

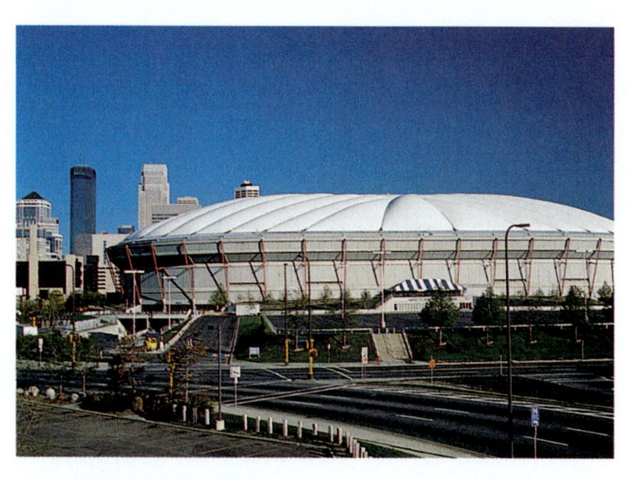

Teflon is used as the roof fabric on some covered sports stadiums such as Humphrey Metrodome in Minnesota.

Teflon

Radical polymerization of tetrafluoroethylene produces a linear chain of difluoromethylene groups—the polymer sold as **Teflon.** The carbon-fluorine bonds are very strong, so the material is inert. The unshared electrons of fluorine are nonpolarizable, leading to minimal electrostatic attractive forces between chains. Thus, Teflon is a slippery ("nonstick") material because of the low level of interaction between the chains, allowing them to slide past each other. Teflon is used to coat frying pans and is also used as the roof fabric on some covered sports stadiums.

$$\underset{\text{Tetrafluoroethylene}}{CF_2{=}CF_2} \longrightarrow \underset{\text{Teflon}}{\left(CF_2{-}CF_2 \right)_n}$$

Teflon is also the fabric used to make Gore-Tex products, including shoes and clothing that "breathe" yet are waterproof. The very small pores in the polymer permit water vapor (and therefore perspiration) to escape through the fabric but are too small for liquid water to penetrate.

Polyacrylonitrile

The radical polymerization of acrylonitrile ($CH_2{=}CH{-}CN$) produces a very tough material that is used mainly as a fiber for materials ranging from clothing to the artificial playing surfaces in many athletic stadiums. The generic term is acrylic fiber; trade names include Orlon and Acrilan.

$$\underset{\text{Acrylonitrile}}{CH_2{=}CHCN} \longrightarrow \underset{\text{Orlon, Acrilan}}{\left(CH_2{-}\underset{\displaystyle CN}{\underset{|}{CH}} \right)_n}$$

Poly(methyl methacrylate)

The radical polymerization of the ester methyl methacrylate (methyl 2-methylpropenoate) produces a polymer that is used to make Plexiglas and Lucite. Plexiglas has excellent transparency properties and also very high strength. Thus, it can be used in place of glass, especially in windshields, around hockey rinks, and as bulletproof "glass" (see "Chemistry at Work," p. 462).

$$\underset{\text{Methyl methacrylate}}{CH_2{=}C(CH_3)COOCH_3} \longrightarrow \underset{\text{Poly(methyl methacrylate)}}{\left(CH_2{-}\overset{\displaystyle CH_3}{\overset{|}{\underset{\displaystyle COOCH_3}{\underset{|}{C}}}} \right)_n}$$

How to Solve a Problem

The compound known as isoprene (2-methyl-1,3-butadiene; see Section 8.4.1) is polymerized under radical conditions to a polymer similar to natural rubber. Without being concerned about the stereochemistry involved, show the mechanism of this polymerization and account for the structure of the product.

$$CH_2{=}C(CH_3){-}CH{=}CH_2 \dashrightarrow \left(CH_2{-}C(CH_3){=}CH{-}CH_2\right)_n$$

Isoprene — Polyisoprene

PROBLEM ANALYSIS AND SOLUTION

Recall that compounds such as isoprene are known as *dienes*, but they react initially like any alkene, undergoing addition reactions at the methylene terminus. Addition of a radical initiator to isoprene leads to addition at a terminal carbon to form an allylic (resonance-stabilized) radical intermediate:

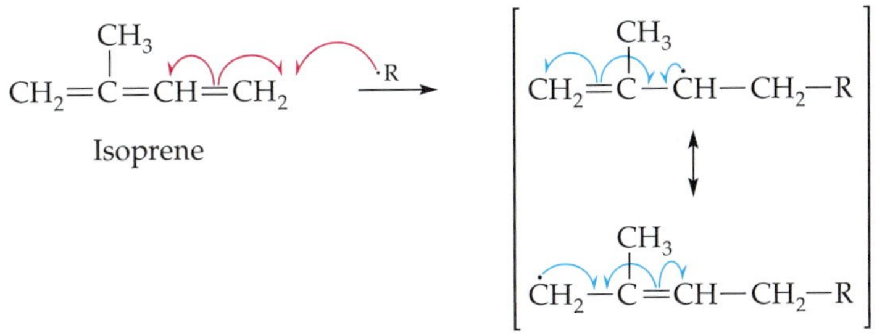

$$CH_2{=}C(CH_3){-}CH{=}CH_2 \xrightarrow{\cdot R} \left[CH_2{=}C(CH_3){-}\dot{C}H{-}CH_2{-}R \longleftrightarrow \dot{C}H_2{-}C(CH_3){=}CH{-}CH_2{-}R \right]$$

Isoprene

This intermediate adds to another molecule of isoprene in the same manner as the initiator (·R) did to start the growing chain. Repetition of this step many times produces polyisoprene:

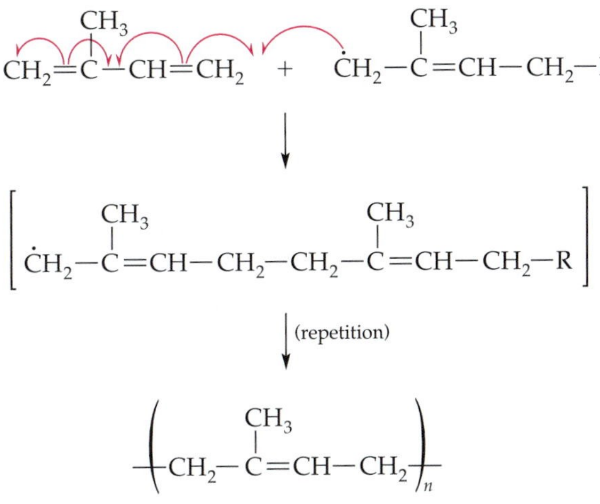

$$CH_2{=}C(CH_3){-}CH{=}CH_2 + \dot{C}H_2{-}C(CH_3){=}CH{-}CH_2{-}R$$

$$\downarrow$$

$$\left[\dot{C}H_2{-}C(CH_3){=}CH{-}CH_2{-}CH_2{-}C(CH_3){=}CH{-}CH_2{-}R\right]$$

$$\downarrow \text{(repetition)}$$

$$\left(CH_2{-}C(CH_3){=}CH{-}CH_2\right)_n$$

Polyisoprene

Therefore, the mechanism involves radical addition to an alkene, which happens to be a conjugated diene, so the propagating radical of the chain reaction is an allylic radical.

PROBLEM 18.4

Show the monomer(s) needed to prepare each of the following addition polymers:

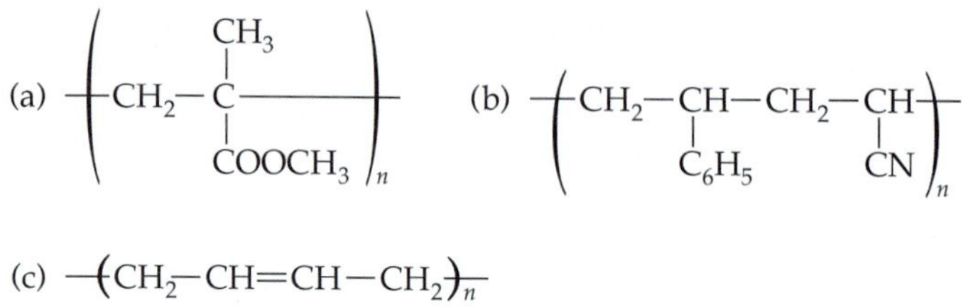

(a) $\left(CH_2-\underset{COOCH_3}{\overset{CH_3}{C}} \right)_n$ (b) $\left(CH_2-\underset{C_6H_5}{CH}-CH_2-\underset{CN}{CH} \right)_n$

(c) $(CH_2-CH{=}CH-CH_2)_n$

18.3.3 Cationic Polymerization

For alkenes, the use of **cationic polymerization** (involving formation of a carbocation as the propagating reactive intermediate) is practical when the initial carbocation forms readily because of its stability. For example, isobutylene reacts with an electrophile, such as concentrated sulfuric acid or, more usually, boron trifluoride, to form a *t*-butyl cation (which also is an electrophile). (Recall electrophilic addition, discussed in Section 6.4.) This carbocation then adds to another molecule of isobutylene, forming yet another tertiary carbocation, and so on. Continuation of this process leads to the polymer polyisobutylene.

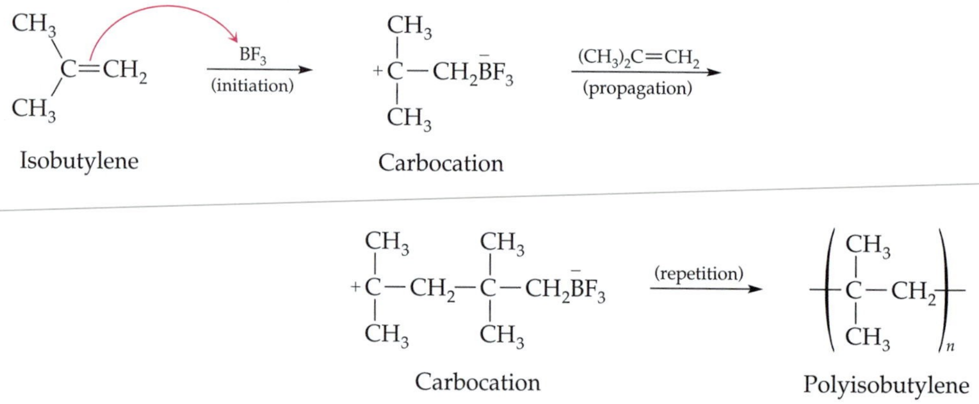

$$\underset{\text{Isobutylene}}{(CH_3)_2C{=}CH_2} \xrightarrow[\text{(initiation)}]{BF_3} \underset{\text{Carbocation}}{{}^{+}\underset{CH_3}{\overset{CH_3}{C}}-CH_2\bar{B}F_3} \xrightarrow[\text{(propagation)}]{(CH_3)_2C{=}CH_2}$$

$$\underset{\text{Carbocation}}{{}^{+}\underset{CH_3}{\overset{CH_3}{C}}-CH_2-\underset{CH_3}{\overset{CH_3}{C}}-CH_2\bar{B}F_3} \xrightarrow{\text{(repetition)}} \underset{\text{Polyisobutylene}}{\left(\underset{CH_3}{\overset{CH_3}{C}}-CH_2 \right)_n}$$

The termination step occurs when the growing polymer loses a proton to form an alkene end group, reacts with a nucleophile (such as water) added to the solution to produce an alcohol end group, or acquires a hydride anion from another chain to produce an alkane end group.

Low-molecular-weight polyisobutylene is typically used as an adhesive for paper products (paper labels and pressure-sensitive tapes), while the high-molecular-weight polymer is called *butyl rubber* and is used in inner tubes for rubber tires.

PROBLEM 18.5

Show the initiation and propagation steps for the acid-catalyzed polymerization of styrene. Draw the final polymer structure, and explain the regiochemistry you have shown.

18.3.4 Anionic Polymerization

Anions can also be used to initiate polymerization reactions of some alkenes. In **anionic polymerization,** the initiator must be a strong nucleophile—usually a carbanion such as a Grignard reagent or an organolithium reagent. The addition of this anion to the alkene is a somewhat unusual reaction because alkenes normally add electrophiles, not nucleophiles. Therefore, this polymerization process works best when the intermediate carbanion has attached to it groups that can stabilize it through resonance delocalization, such as ester and cyano groups. The mechanism of anionic polymerization is illustrated here for the preparation of polyacrylonitrile:

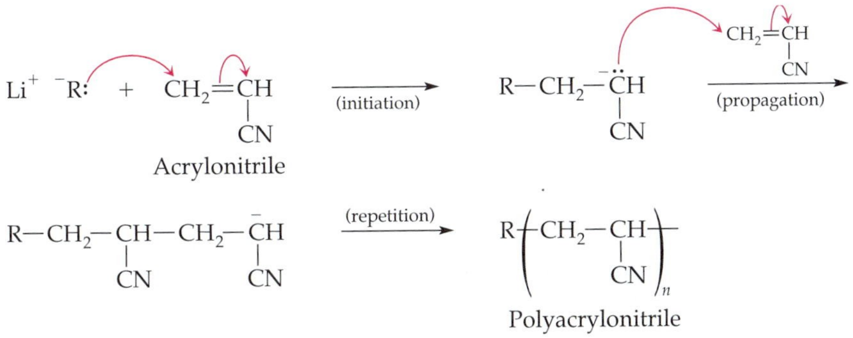

$$Li^+ \ {}^-R{:} + CH_2{=}CH(CN) \xrightarrow{\text{(initiation)}} R{-}CH_2{-}\ddot{C}H^-(CN) \xrightarrow[\text{(propagation)}]{CH_2=CH(CN)}$$

Acrylonitrile

$$R{-}CH_2{-}CH(CN){-}CH_2{-}\bar{C}H(CN) \xrightarrow{\text{(repetition)}} R{+}CH_2{-}CH(CN){+}_n$$

Polyacrylonitrile

One unique characteristic of anionic polymerization is that the carbanion at the end of the chain cannot neutralize itself for termination, in contrast to a radical or a carbocation. The polymer remains in solution as a reactive carbanion until a proton source, usually water, is added to "cap" the end of the chain. Thus, such a polymer is sometimes called a "living polymer." Anionic polymerization provides the opportunity to prepare a polymer by allowing one monomer to be consumed completely and then adding a different monomer, which enables the carbanionic chain to resume growing. Out of this approach has come what are known as **block copolymers,** polymers with two distinct segments in their chains, consisting of two different monomers. For example, styrene can be polymerized, and then acrylonitrile can be added to produce a block copolymer (Figure 18.6).

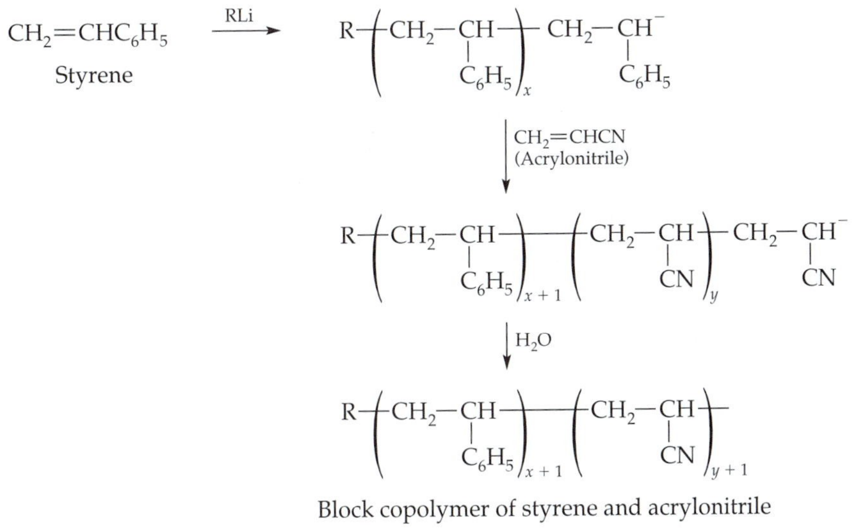

$$\underset{\text{Styrene}}{CH_2{=}CHC_6H_5} \xrightarrow{RLi} R{+}CH_2{-}CH(C_6H_5){+}_x\,CH_2{-}CH^-(C_6H_5)$$

$$\downarrow CH_2{=}CHCN \text{ (Acrylonitrile)}$$

$$R{+}CH_2{-}CH(C_6H_5){+}_{x+1}{+}CH_2{-}CH(CN){+}_y\,CH_2{-}CH^-(CN)$$

$$\downarrow H_2O$$

$$R{+}CH_2{-}CH(C_6H_5){+}_{x+1}{+}CH_2{-}CH(CN){+}_{y+1}$$

Block copolymer of styrene and acrylonitrile

FIGURE 18.6 The formation of a block copolymer of styrene and acrylonitrile.

PROBLEM 18.6

Show the mechanism of anionic polymerization of methyl acrylate ($CH_2{=}CH{-}COOCH_3$), and account for the regiochemistry of the product.

18.3.5 Isotactic Polymers

A major advance in polymer chemistry occurred when it was discovered by Karl Ziegler and Giulio Natta in 1953 that a complex metal-based catalysis system produces an addition polymer called an *isotactic polymer,* in which the stereochemical configuration is the same at each stereocenter.

Applying the Ziegler-Natta process to propylene produces isotactic polypropylene. Every other carbon is a stereocenter (the polymer is chiral), and all the stereocenters have the same configuration. The methyl groups are on the same "side" of the polymer chain, leading to a very regular structure that can pack tightly as a crystal. Further, there is no chain transfer, so the polymer is purely linear. This crystalline polypropylene can be molded into very strong products, including many automobile parts, computer casings, and audio equipment. In contrast, radical polymerization of propylene produces *atactic* polypropylene, which has a random mixture of stereocenter configurations and is an amorphous polymer.

CH_3 H CH_3 H CH_3 H CH_3 H n

Isotactic polypropylene

Ziegler and Natta shared the 1963 Nobel Prize in chemistry for their discovery, which can be applied to the polymerization of any monosubstituted alkene. The catalysis system is a complex mixture of a trialkylaluminum compound and titanium tetrachloride, which is not soluble in the reaction medium. The polymerization process therefore occurs at the surface of the catalyst (this is known as a *heterogeneous process*). The full details of the mechanism of the reaction are still unknown, but it undoubtedly involves coordination of the alkene to a metal site in a specific spatial orientation.

The same process is applied to the polymerization of ethylene (although there are no stereocenters in polyethylene). The result is absolutely linear polyethylene, with none of the branching that usually results from radical polymerization. Polyethylene prepared by this process is highly crystalline and therefore much harder and stronger than that produced by radical polymerization. (It is high-density polyethylene or HDPE, rather than LDPE.) HDPE is used in many hard-sided products, such as piping, toys, and housewares.

18.3.6 Rubber

Natural rubber is *cis*-polyisoprene, a linear polymer of isoprene (2-methyl-1,3-butadiene) with a molecular weight of over 100,000. It is produced by the rubber tree, *Hevea brasiliensis,* originally found in Brazil but now grown mainly in Southeast Asia. The double bonds all have *cis* stereochemistry, which results in a highly coiled structure and an amorphous polymer. Natural rubber can be stretched easily into a more ordered structure (the chains become more linear), but returns rapidly to the coiled structure upon release, giving it the highly desirable property of elasticity. Upon gentle heating in the absence of air, natural rubber depolymerizes to produce isoprene:

$$\left(\!-CH_2(CH_3)C{=}CH(H)CH_2-\!\right)_n \xrightarrow{\Delta} CH_2{=}C(CH_3){-}CH{=}CH_2$$

Natural rubber (*cis*-polyisoprene) → Isoprene

There were many early attempts to produce "synthetic natural rubber" by radical or cationic polymerization of isoprene, but they were unsuccessful for two reasons. First, most attempts resulted in mainly *trans*-polyisoprene (recall that *trans* is the more stable alkene configuration), a much less amorphous polymer. Interestingly, a natural product closely related to natural rubber, *gutta percha* (produced by trees of the *Dichopsis* family), is *trans*-polyisoprene. Gutta percha is of lower molecular weight and is less coiled (more crystalline) than natural rubber (the *cis* isomer), and is therefore less flexible. Gutta percha is used commercially for "balata" golf ball covers and electrical insulation.

The second reason for the early failures to obtain pure *cis*-polyisoprene (natural rubber) is that the polymerization resulted in considerable branching of the polymer chains. This occurred because 1,2-addition occurred along with the desired 1,4-addition (see Section 8.2) in the propagation step of the chain reaction.

$$CH_2{=}C(CH_3){-}CH{=}CH_2 \xrightarrow{\cdot R} R{-}CH_2{-}\dot{C}(CH_3){-}CH{=}CH_2 \longleftrightarrow R{-}CH_2{-}C(CH_3){=}CH{-}\dot{C}H_2$$

(1,2-addition of isoprene) → Branched polyisoprene

(1,4-addition of isoprene) → Linear polyisoprene

In 1955, it was found that application of the Ziegler-Natta polymerization process (the process used to produce isotactic polymers, see Section 18.3.5) to isoprene produced either *cis*- or *trans*-polyisoprene, depending on the particular metal catalyst employed. Thus, synthetic natural rubber, identical in all respects to that produced by the rubber tree, became readily available.

$$CH_2{=}C(CH_3){-}CH{=}CH_2 \xrightarrow{\text{(Ziegler-Natta catalysis)}} \left(\!-CH_2(CH_3)C{=}CH(H)CH_2-\!\right)_n$$

Isoprene → Natural rubber (*cis*-polyisoprene)

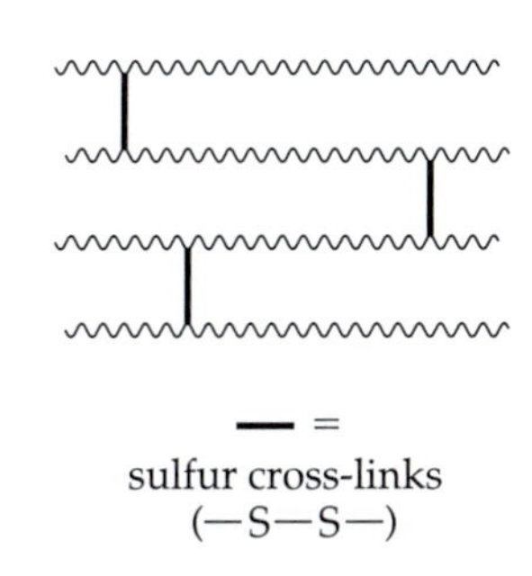

FIGURE 18.7 Schematic representation of hydrocarbon chains of rubber that are cross-linked by sulfur through the vulcanization process discovered by Charles Goodyear.

Natural rubber is not suitable for use by itself in vehicle tires. In 1839, Charles Goodyear discovered the process of *vulcanization*, which makes natural rubber much harder and less affected by heat or cold, yet still somewhat elastic. In other words, it makes natural rubber suitable for use in vehicle tires. This process, still used today in only slightly modified form, involves heating rubber with sulfur and other ingredients. The major change is that the hydrocarbon chains in rubber become cross-linked (Figure 18.7). The cross-linked chains stretch but do not slide past each other and tear, as occurs in natural rubber.

A major component of long-wear automobile and truck tires is the synthetic rubber called SBR (styrene-butadiene rubber), a copolymer of styrene and 1,3-butadiene. Originally called BuNa-S rubber when discovered in Germany, SBR is produced in twice the volume of natural rubber by polymerizing a mixture of 1,3-butadiene and styrene:

$$\underset{\text{1,3-Butadiene}}{CH_2{=}CH{-}CH{=}CH_2} + \underset{\text{Styrene}}{C_6H_5{-}CH{=}CH_2} \xrightarrow{\text{(polymerization)}} \underset{\text{SBR (synthetic rubber)}}{\left(CH_2{-}CH{=}CH{-}CH_2{-}CH_2{-}\underset{C_6H_5}{CH} \right)_n}$$

The tread wear of tires has increased enormously over the past five decades, thanks to polymer chemistry. Tires used to wear out after a few hundred miles. Now it is not uncommon to see tires give 75,000 miles of use. Modern tires are manufactured from mixtures of polymeric materials, employing some natural rubber along with synthetics. The tread itself is mainly SBR and natural rubber. The tire cords may be polyester polymers (see Section 18.4.2), and the belting in radial tires may be steel wire (in steel-belted tires) or polyamide (aramid) fiber (see Section 18.4.1).

PROBLEM 18.7

Show the mechanism of the radical polymerization of 1,3-butadiene to form a linear polymer.

18.4 Condensation Polymers

Condensation polymers are formed from the reaction between two functional groups to form a new functional group. In the process, a small molecule (usually water) is ejected. An example is the reaction between an acid and an alcohol to form an ester and eliminate water (Fischer esterification; see Section 15.3.4), which has been adapted to produce polyesters. In contrast to addition reactions, condensation reactions are not chain reactions. Instead, condensation polymers grow one condensation reaction at a time, in a stepwise fashion, with each reaction occurring totally independent of any other. (Thus, these polymers are also referred to as *step-growth polymers*.)

As illustrated in Figure 18.8, condensation reactions are of two types. The first involves two different monomers, each with the same functional group at both ends. The second involves a single monomer carrying both of the reacting functional groups, one at each end.

In condensation polymerization, chain growth does not occur only by the monomer reacting with the growing polymer, although that does happen to some extent. It is also possible for two *oligomers*—polymer segments containing several monomer units already joined—to react together. In this way, the chain grows by several monomer units all at once. This means that controlling the chain length is somewhat more complex than in addition polymerization.

18.4.1 Polyamides

Polyamides were first investigated for use as synthetic fibers because two natural fibers, silk and wool, were found to be polyamides. (In fact, they are both polypeptides—pro-

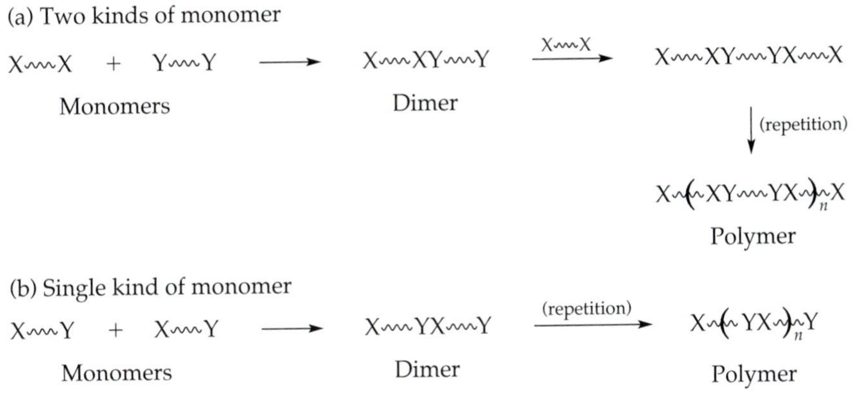

(a) Two kinds of monomer

X〰X + Y〰Y ⟶ X〰XY〰Y —(X〰X)⟶ X〰XY〰YX〰X

Monomers Dimer

↓ (repetition)

X〰(〰XY〰YX〰)$_n$〰X

Polymer

(b) Single kind of monomer

X〰Y + X〰Y ⟶ X〰YX〰Y —(repetition)⟶ X〰(〰YX〰)$_n$〰Y

Monomers Dimer Polymer

X and Y = two different functional groups (for example, X = COOH and Y = NH_2)
XY and YX = functional groups resulting from reaction of X with Y (for example, the amide group)

FIGURE 18.8

Representation of two kinds of condensation reactions.

teins; also recall that cotton, the other major natural fiber, is derived from cellulose, a polysaccharide.) Silk is produced by the silkworm as a polypeptide composed of a random sequence of only two α-amino acids, glycine (H_2NCH_2COOH) and alanine [$H_2NCH(CH_3)COOH$]. Dragline silk, produced by spiders for building the frames of their webs, also consists mainly of glycine (42%) and alanine (25%). Its high strength is due to the crystallinity of the polyalanine segments. Wool, grown like hair by many animals, is also a polypeptide, but one that incorporates many different α-amino acids—most importantly, cysteine [$H_2NCH(CH_2SH)COOH$]. In wool, the peptide chains are extensively cross-linked via oxidation of cysteine units on two chains to form a sulfur-sulfur bond as a link.

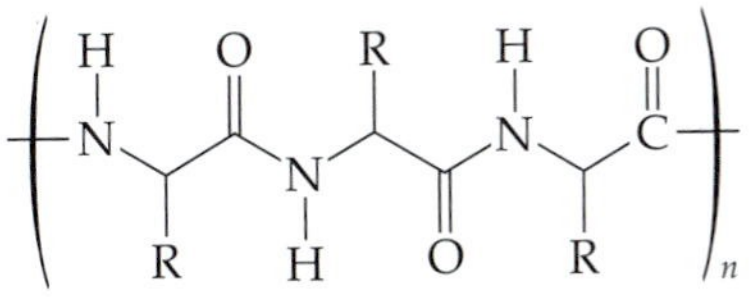

Silk
($R = CH_3$ or H
in random sequence)

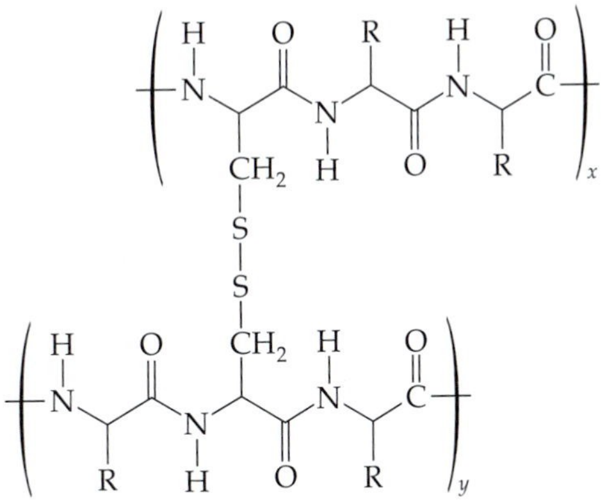

Wool

A polyamide is usually formed by the reaction of a dicarboxylic acid with a diamine; water is ejected upon formation of each amide group. Nylon-6,6 is formed by reaction of hexanedioic acid with 1,6-hexanediamine (the 6,6 designation indicates that the two components of this nylon each contain six carbon atoms):

$$HOOC(CH_2)_4COOH + H_2N(CH_2)_6NH_2 \xrightarrow[-H_2O]{\Delta} -\!\!\left(OC(CH_2)_4CONH(CH_2)_6NH\right)_n\!\!-$$

Hexanedioic acid 1,6-Hexanediamine Nylon-6,6

Synthetic polymers have had a great impact on our marketplace. This kayaker is surrounded by polymers from his clothing to his helmet to the kayak he is in.

A semicrystalline polymer with a molecular weight of about 10,000, Nylon-6,6 makes a very strong fiber when drawn. It was the first commercial polymer used in fabrics, and it stole the show at the 1939 World's Fair in New York, when nylon stockings were exhibited. "Nylons," as they are commonly called, became available nationwide in 1940. However, they were withdrawn from the market during World War II (from 1941 through 1946), when available manufacturing space was refitted to produce parachutes. The impact of the introduction of synthetic fabrics into the marketplace is evidenced by the near riots that occurred after the war, when nylon stockings were reintroduced. Silk stockings all but vanished from that point on as a mass-market commodity.

Polyamides that contain aromatic rings are called *aramids*. They are much stronger and more rigid than Nylon-6,6 because their nearly planar shape (due to the benzene rings) permits close crystal packing. For example, Kevlar is produced from terephthalic acid and 1,4-diaminobenzene. Kevlar is used in bulletproof vests, crash helmets, and some radial tires (in cords replacing the steel wires of steel-belted radial tires).

$$HOOC-C_6H_4-COOH \text{ (Terephthalic acid)} + H_2N-C_6H_4-NH_2 \text{ (}p\text{-Phenylenediamine (1,4-diaminobenzene))} \xrightarrow{-H_2O} \left(OC-C_6H_4-CONH-C_6H_4-NH\right)_n \text{ (Kevlar)}$$

Nylon-6 is also a polyamide. It was originally prepared by a condensation reaction of 6-aminohexanoic acid. In this monomer, the two reactive functional groups are incorporated in a single molecule. However, production now relies on a ring-opening reaction, initiated by attack of water at the carbonyl carbon of ϵ-caprolactam.

$$H_2N(CH_2)_5COOH \text{ (}\epsilon\text{-Aminocaproic acid (6-aminohexanoic acid))} \xrightarrow{-H_2O} \left(NH(CH_2)_5CO\right)_n \text{ (Nylon-6)}$$

$$\epsilon\text{-Caprolactam} \xrightarrow{\text{trace of } H_2O} \left(NH(CH_2)_5CO\right)_n \text{ (Nylon-6)}$$

PROBLEM 18.8

(a) Show and name the monomers that would be needed to produce Nylon-4,4. Also draw the structure of the resulting polymer.

(b) Show the monomer that would be needed to produce Nylon-4. Also draw the structure of the resulting polymer.

18.4.2 Polyesters

As the name implies, polyesters contain repeating ester functional groups. The most common polyester is formed from terephthalic acid and ethylene glycol and is called poly(ethylene terephthalate). When spun into a fiber and made into fabrics, it is sold as Dacron, Terylene, or just polyester. When molded into "plastic" bottles (such as soft-drink bottles), it is known as PET. When processed into a thin film, the polymer is sold as Mylar, which is used to make magnetic tapes (audio, video, and computer).

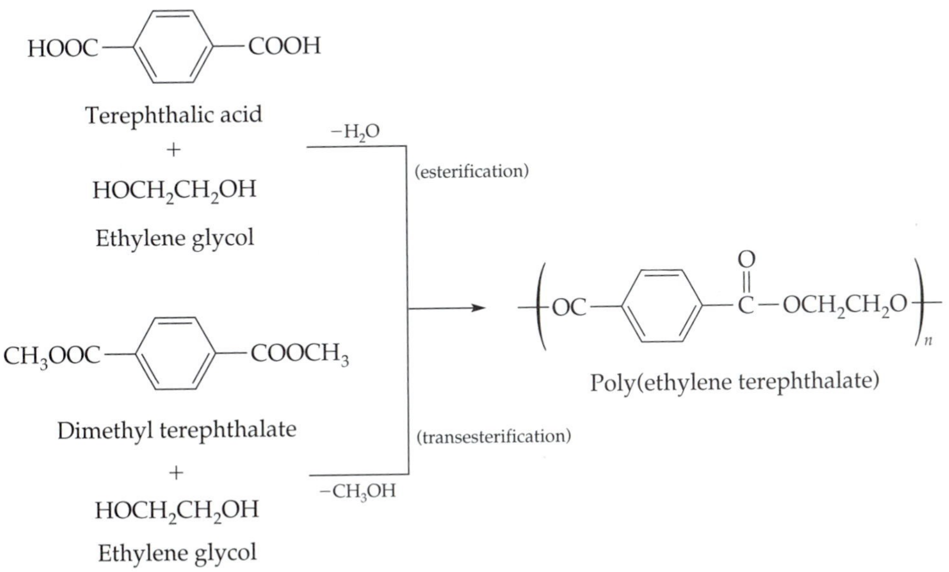

HOOC–C_6H_4–COOH
Terephthalic acid
+
$HOCH_2CH_2OH$
Ethylene glycol

$-H_2O$ (esterification)

CH_3OOC–C_6H_4–$COOCH_3$
Dimethyl terephthalate
+
$HOCH_2CH_2OH$
Ethylene glycol

$-CH_3OH$ (transesterification)

$\rightarrow$ $\left(OC–C_6H_4–\overset{O}{\overset{\|}{C}}–OCH_2CH_2O \right)_n$
Poly(ethylene terephthalate)

Industrial production of poly(ethylene terephthalate) is accomplished by preparing dimethyl terephthalate and reacting it with ethylene glycol to bring about a transesterification reaction (an ester exchange reaction). The lower-boiling methanol is recovered and recycled by distillation. The Fischer esterification mechanism operates in this process, with the glycol as the initial nucleophile (see Section 15.3.4).

Although, technically speaking, the polymers made from carbonic acid and alcohols are polyesters, they are referred to as *polycarbonates*. Reaction of bisphenol A with phosgene (the double acid chloride of carbonic acid) or ester exchange of bisphenol A with diethyl carbonate produces the polymer known as Lexan, a very strong material used for crash helmets and bulletproof vests. Polycarbonates are also used to produce shatter-resistant eyeglasses and clear automobile headlight lenses.

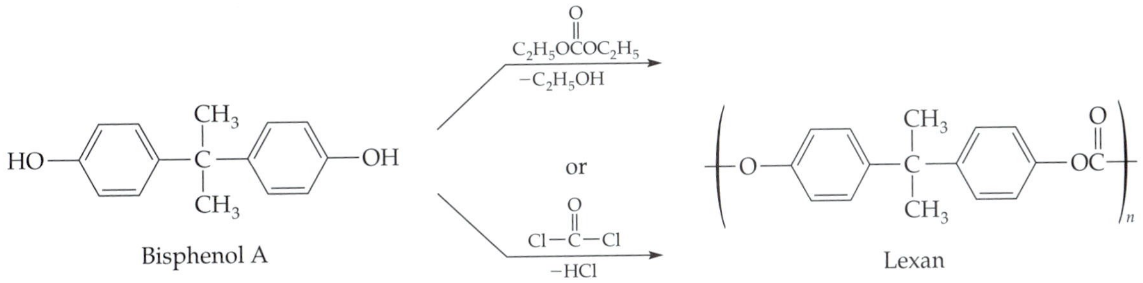

HO–C_6H_4–$C(CH_3)_2$–C_6H_4–OH
Bisphenol A

$C_2H_5O\overset{O}{\overset{\|}{C}}OC_2H_5$, $-C_2H_5OH$
or
$Cl–\overset{O}{\overset{\|}{C}}–Cl$, $-HCl$

$\rightarrow$ $\left(O–C_6H_4–C(CH_3)_2–C_6H_4–O\overset{O}{\overset{\|}{C}} \right)_n$
Lexan

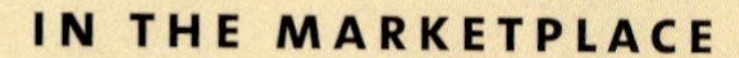

CHEMISTRY AT WORK

IN THE MARKETPLACE

Plastics Degradation and Recycling

Since the invasion of plastics into everyday life, the amount of solid waste generated in this high-consuming society has multiplied tremendously. Although not necessarily a cause-and-effect relationship, waste disposal has also become a much more pressing social issue. This has led to a reversal of sorts for plastics manufacturers: Durability, in some instances, is out, and degradability (decomposition) and recyclability are in. A delicate balance between the three characteristics is desirable.

The issue of waste degradation applies not only to plastics. In fact, nonplastic wastes persist much longer in landfills than previously recognized. Nonetheless, partially in response to government requirements, the plastics industry has followed two approaches in producing degradable products.

The first approach is to make some plastics *photodegradable* (in spite of the fact that materials in landfills are not generally exposed to the light needed to promote photodegradation). This can be accomplished in the formulation steps (that is, during extrusion, molding, and so on) by addition of certain additives to the pure polymer resin. The additives absorb ultraviolet light and so initiate photoreactions that result in chain fracture (usually radical reactions). The additives must absorb radiation of a wavelength shorter than 315 nm, the lowest wavelength transmitted by glass. (If additives absorbed higher-wavelength radiation, photodegradation could result from household and commercial lighting, shortening the useful lifetime of products still in use.) For example, the plastic rings made for six-packs of beverages are produced from polyethylene containing about 1% carbon monoxide to speed photodegradation. Some polymers

$$\underset{\substack{\text{3-Hydroxybutanoic acid}\\(\beta\text{-hydroxybutyric acid})}}{CH_3\overset{\overset{OH}{|}}{C}HCH_2COOH} \xrightarrow{\text{(polymerization)}} \underset{\text{Polyhydroxybutyrate (PHB)}}{\left(\overset{\overset{CH_3}{|}}{C}HCH_2COO\right)_n}$$

18.4.3 Polyurethanes

Urethanes, also called carbamates, are derivatives of carbonic acid (HO—CO—OH) containing an amide functional group and an ester functional group (RNHCOOR′). (Recall that urea is the diamide of carbonic acid.) Urethanes are, theoretically, obtainable from phosgene (the double acid chloride of carbonic acid, Cl—CO—Cl) by stepwise reaction with an alcohol and an amine (that is, alcoholysis followed by ammonolysis; see Section 15.1.2). In practice, they are obtained from a family of compounds called *isocyanates* (R—N=C=O) by reaction with alcohols. This reaction involves nucleophilic attack of the alcohol at the carbonyl carbon but results in addition across the C=N bond to form the urethane. The carbonyl group remains intact.

$$\underset{\text{Phosgene}}{Cl-\overset{\overset{O}{\|}}{C}-Cl} \xrightarrow[-HCl]{ROH} \left[RO-\overset{\overset{O}{\|}}{C}-Cl\right] \xrightarrow[-HCl]{R'NH_2} \underset{\substack{\text{Urethane}\\\text{(carbamate)}}}{RO-\overset{\overset{O}{\|}}{C}-NHR'}$$

$$\underset{\text{Isocyanate}}{R'-N=C=O} \xrightarrow{R\ddot{O}H} \underset{\substack{\text{Urethane}\\\text{(carbamate)}}}{RO-\overset{\overset{O}{\|}}{C}-NHR'}$$

are produced containing functional groups that facilitate the absorption of light and the photodegradation process.

The second approach is to find a way to make synthetic polymers, which are normally resistant to biodegradation, in some way biodegradable. This can be done by designing new polymers (so-called bioplastics) or by reformulating existing polymers so that they will be biodegradable by the microorganisms typically found in landfills. For example, *polyhydroxybutyrate*, or PHB, which is synthesized by the bacterium *Alicaligenes eutrophus* as an energy reserve, is now being produced commercially. PHB can be molded and made into films. Under aerobic conditions, it is metabolized to carbon dioxide and water; under anaerobic conditions, typical in landfills, it degrades to methane and water.

Biodegradability is also important in some medical uses of polymers. For example, poly(lactic acid) is used in implants intended to serve only temporarily, until damaged tissue repairs itself. Eventually, the polymer hydrolyzes to lactic acid, which degrades to carbon dioxide and water via normal metabolic processes in the body (see Section 19.2).

The other option available for stemming the tide of plastics in landfills is recycling. Recycling has developed into a major commercial enterprise and has contributed to a significant decline in the amount of plastic deposited in landfills. Plastics are not all alike, and so one of the biggest challenges that recyclers face is achieving homogeneity in the material collected. The success of widespread commercial and residential plastics recycling depends largely on labeling plastics with codes that facilitate easy sorting by the consumer. Plastic beverage containers, which are polyethyleneterephthalate (PET; see Section 18.4.2), are coded △ Plastic milk containers, which are high-density polyethylene (HDPE; see Section 18.3.5), are coded △. The separate collection of these plastics, along with removal of their labels and caps, results in a fairly homogeneous material that can be remelted and reextruded into new products. These two types of plastics alone account for about 40% of the total plastics used for packaging in the United States.

Recycled plastic pellets. Recycling plastic offers a good alternative for stemming the tide of plastics in landfills.

$$\left[\begin{array}{c}CH_3\\ |\\ CHCOO\end{array}\right]_n \xrightarrow{H_2O} CH_3CH(OH)COOH \xrightarrow{\text{(metabolism)}} CO_2 + H_2O$$

Poly(lactic acid) — Lactic acid

The reaction of isocyanates with alcohols to form urethanes has been adapted to polymerization in which a diisocyanate is reacted with a diol, producing a polyurethane. The most common polymerization reaction uses ethylene glycol and toluene-2,4-diisocyanate:

$$HOCH_2CH_2OH \text{ (Ethylene glycol)} + OCN\text{–}C_6H_3(CH_3)\text{–}NCO \text{ (Toluene-2,4-diisocyanate)} \longrightarrow \left[OCNH\text{–}C_6H_3(CH_3)\text{–}NHCOOCH_2CH_2O\right]_n \text{ (Polyurethane)}$$

Other preparations use a diol that is already a small polymeric unit.

Polyurethanes are the major source of soft (elastomeric) foams, like those used for pillows, seat cushions, and some packaging. Such products require a polymer with some cross-linking, which is achieved by using excess isocyanate monomer and allowing the amide functional group in the polymer chains to add across the isocyanate groups. The presence of excess isocyanate is also important in producing the bubbles essential for a foam. A small amount of water is added near the end of the polymerization process. The water reacts with free isocyanate groups, initially forming a carbamic acid that produces gaseous carbon dioxide, which expands in the warm melted polymer, forming bubbles. Thus, the weight-to-volume ratio is greatly reduced, producing a lightweight and soft foam.

$$\underset{\text{Isocyanate}}{R{-}N{=}C{=}O} \xrightarrow{H_2O} \underset{\text{Carbamic acid}}{\left[RNH{-}\overset{\overset{O}{\parallel}}{C}{-}OH\right]} \longrightarrow RNH_2 + CO_2(g)$$

PROBLEM 18.9

(a) Show the polymer that would result from the reaction of succinoyl chloride ($ClOCCH_2CH_2COCl$) with 1,4-dihydroxybenzene (hydroquinone).

(b) Show the polymer that would result from the reaction of 1,3-propanediol with benzene-1,4-diisocyanate.

Chapter Summary

Polymers are **macromolecules** consisting of large numbers of small molecules that have been chemically joined together. The small molecule that is the repeating unit is called the **monomer.** Polymers may have thousands of repeating units and molecular weights that can be in the millions. Some polymers incorporate two or more kinds of monomers in a regular or random sequence and are called **copolymers.**

Polymer chains may be **linear** or **branched (cross-linked).** Polymers that have a high degree of structural regularity are crystalline (individual polymer molecules can pack tightly together in a crystal lattice) and higher-melting. Those with less regularity in their structure are amorphous and lower-melting. Polymers may serve as **elastomers, fibers,** or **plastics,** depending on their molecular structure. Polymers also may be **thermosetting** or **thermoplastic.**

One major class of polymers is **addition polymers** (also called *chain-growth polymers*). The addition of an **initiator** across a carbon-carbon double bond results in a reactive intermediate that adds across the double bond of another monomer molecule in a **chain reaction.** The initiator may be a radical (**radical polymerization**), a cation (**cationic polymerization**), or an anion (**anionic polymerization**). Radical polymerization can lead to **chain transfer** and resultant branching. Some examples of addition polymers are polyethylene, polystyrene, poly(vinyl chloride), polytetrafluoroethylene (Teflon), polyacrylonitrile, polyethers, and poly(methyl methacrylate). Anionic polymerization results in "living polymers" that can be used to produce **block copolymers.** Radical polymerization usually yields *atactic polymers*, which have a random mixture of stereocenter configurations. Complex metal-based catalysis systems result in the formation of *isotactic polymers*, those with the same configuration at all stereocenters; the regularity of their structures results in high melting points. Dienes can also be polymerized by addition polymerization. Natural rubber is *cis*-polyisoprene, while gutta percha is *trans*-polyisoprene.

The second major class of polymers is **condensation polymers.** They are formed from the reaction between two functional groups to form a new functional group. In the process, a small molecule (usually water) is ejected, as in ester or amide formation. Condensation polymers are also referred to as *step-growth polymers*. Examples of condensation polymers are *polyamides* (such as Nylon), *polyesters* (such as Dacron), *polycarbonates*, and *polyurethanes*.

Additional Problems

Terminology

18.10 Define the following terms used in organic chemistry:

(a) linear polymer (b) cross-linked polymer
(c) addition polymer (d) condensation polymer

18.11 For the following partial polymer structures, first indicate which are simple polymers and which are copolymers, and then indicate the monomer units involved:

(a) $\xi{-}CH_2CHClCH_2CHClCH_2CHCl{-}\xi$

(b) $\text{—CH}_2\text{CHClCCl}_2\text{CCl}_2\text{CH}_2\text{CHCl—}$

(c) $\text{—CH}_2\text{OCH}_2\text{OCH}_2\text{OCH}_2\text{O—}$

(d) $\text{—OCH}_2\text{CH}_2\text{OOCCH}_2\text{CH}_2\text{CH}_2\text{CH}_2\text{CO—}$

(e) 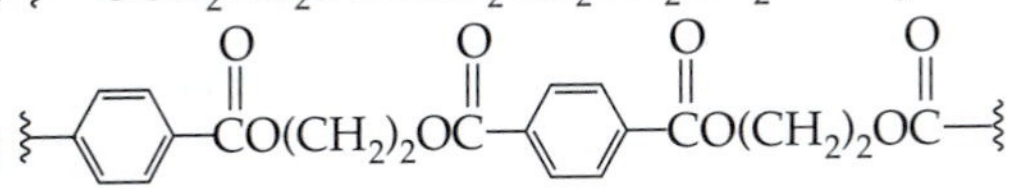 $\text{—C}_6\text{H}_4\text{—C(=O)O(CH}_2)_2\text{OC(=O)—C}_6\text{H}_4\text{—C(=O)O(CH}_2)_2\text{OC(=O)—}$

(f) $\text{—CH}_2\text{CCl=CHCH}_2\text{CH}_2\text{CCl=CHCH}_2\text{—}$

18.12 Explain the following terms used in polymer chemistry:

(a) plastic (b) fiber
(c) elastomer (d) thermoplastic
(e) thermosetting (f) crystalline
(g) amorphous

18.13 The two stereoisomeric forms of polyisoprene have very different properties. Natural rubber, which contains *Z* double bonds, is an elastomer; gutta percha, which contains *E* double bonds, is a rigid material. Account for the difference in properties.

18.14 The discovery of how to produce isotactic polymers was a major advance in polymer chemistry. Explain what an isotactic polymer is, and indicate what structural feature the monomer must have for such a polymer to result.

18.15 Can polyisobutylene be isotactic? Explain.

Addition Polymers

18.16 Show a three-monomer segment of each of the following:

(a) polypropylene
(b) poly(vinyl acetate)
(c) polytetrachlorethene

18.17 Show the initiation step for an addition polymerization of ethylene that involves each of the following:

(a) radical polymerization
(b) anionic polymerization
(c) cationic polymerization

18.18 Show the initiation and propagation steps for the radical polymerization of vinyl chloride.

18.19 Explain why cationic polymerization of styrene always occurs "head to tail" (that is, with the methylene group of one unit attached to the methinyl group of another unit) and never "head to head" or "tail to tail."

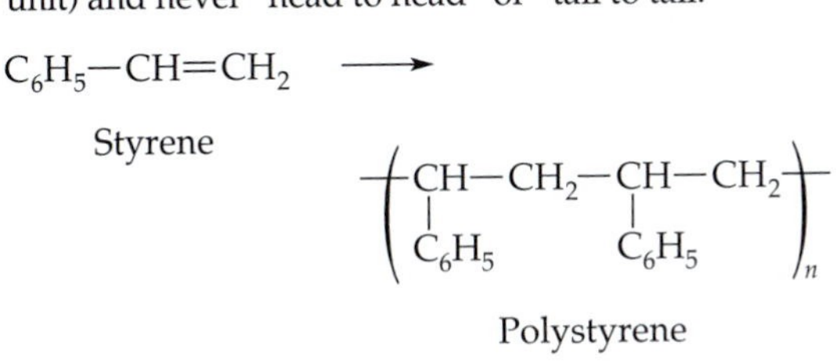

$$\underset{\text{Styrene}}{C_6H_5\text{—CH=CH}_2} \longrightarrow \underset{\text{Polystyrene}}{\left(\text{CH(C}_6\text{H}_5)\text{—CH}_2\text{—CH(C}_6\text{H}_5)\text{—CH}_2\right)_n}$$

18.20 Show the mechanism for the radical polymerization of chloroprene ($CH_2\text{=CCl—CH=}CH_2$) to neoprene, which involves 1,4-addition.

18.21 High-density polyethylene (HDPE) can be used to produce rigid materials. What structural feature of the polymer gives HDPE its properties?

18.22 Gutta percha is the relatively crystalline polymer of isoprene ($CH_2\text{=}CCH_3\text{—CH=}CH_2$) with the remaining double bond in the *E* configuration. It results from the 1,4-addition polymerization of isoprene. What change, if any, would you expect in the polymer's properties if the polymerization included substantial 1,2-addition as well as 1,4-addition?

18.23 Draw the repeating unit from the copolymerization of styrene and vinyl chloride.

18.24 Polyisobutylene, used in butyl rubber for inner tubes, results from the polymerization of isobutylene (2-methylpropene), usually by a cationic process.

(a) Show the mechanism of the formation of polyisobutylene using acid catalysis, and draw the structure of a segment of the polymer.

(b) Is polyisobutylene an addition polymer or a condensation polymer?

(c) Is this polymerization head-to-tail or head-to-head?

(d) What is it that makes this polymerization mechanism effective whereas anionic polymerization is not effective?

18.25 In the process of chain transfer (Section 18.3.2), as a new radical approaches a polystyrene chain, there are two different kinds of hydrogens available for removal. Which hydrogen will be abstracted, and why?

18.26 Draw a three-monomer segment of isotactic poly(vinyl chloride).

Condensation Polymers

18.27 (a) Draw the structure of a repeating unit of a hypothetical polyamide formed from succinic acid and 1,3-propanediamine.

(b) Devise a possible synthesis of 1,3-propanediamine from a nonnitrogenous starting material.

18.28 Formaldehyde can be polymerized into polymethylene ether ($\text{(—CH}_2\text{—O—CH}_2\text{—O—)}_n$, also called *paraformaldehyde,* by treatment with an acid. (*Hint:* How should a proton interact with a carbonyl group?) Show a mechanism by which this polymerization occurs. The polymer can be "unzipped" by reversing the synthesis unless the hydroxyl end group is capped by esterification. Show such an esterification reaction using acetyl chloride.

18.29 Show the repeating unit of a condensation polymer formed from malonic acid and ethylene glycol.

18.30 A polymer called Carbowax, which is poly(ethylene glycol), can be formed by reacting ethylene oxide with hydroxide anion as initiator. Show the reaction involved and the product. (*Hint:* Remember the typical ring-opening behavior of epoxides.)

18.31 The structure of polyamide fiber known as Qiana is shown. Draw the structures of the monomers involved.

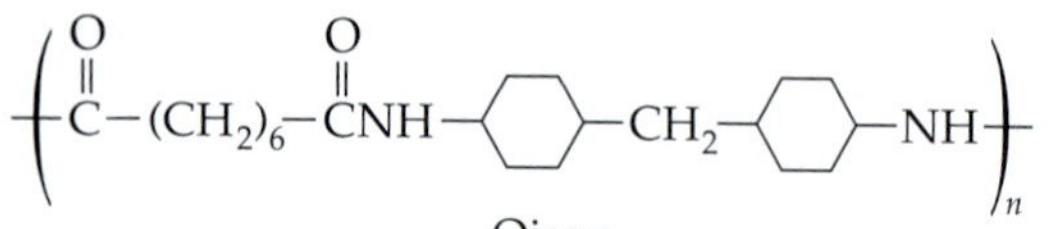

$$\left[-\overset{O}{\overset{\|}{C}}-(CH_2)_6-\overset{O}{\overset{\|}{C}}NH-\langle\bigcirc\rangle-CH_2-\langle\bigcirc\rangle-NH- \right]_n$$

Qiana

18.32 Nomex is a very strong polyamide fiber made from *m*-diaminobenzene and isophthalic acid (the *meta* isomer). Show the repeating unit of the polymer.

18.33 Draw the monomers necessary for the production of the commerical polyester fiber known as Kodel.

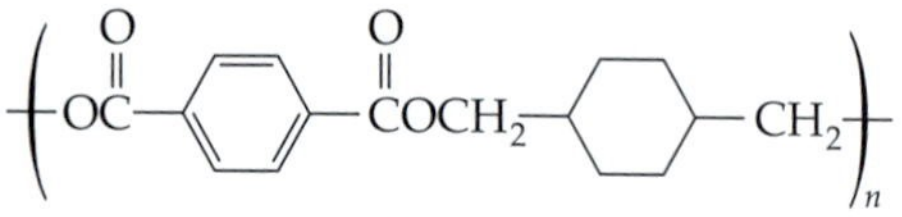

$$\left[-O\overset{O}{\overset{\|}{C}}-\langle\bigcirc\rangle-\overset{O}{\overset{\|}{C}}OCH_2-\langle\bigcirc\rangle-CH_2- \right]_n$$

Kodel

Mixed Problems

18.34 Polystyrene can be converted into a number of different polymers by using normal aromatic substitution reactions. Show the products of these reactions of a two-monomer segment of polystyrene:

(a) with fuming sulfuric acid

(b) the product of part (a) with sodium hydroxide (*Note:* This results in a cation-exchange resin of the type used in water softeners.)

(c) with HNO_3/H_2SO_4

(d) the product of part (c) with Fe/HCl

(e) the product of part (d) with excess methyl iodide (*Note:* This results in an anion-exchange resin.)

18.35 Why is styrene especially well-suited for cationic polymerization, whereas propene is less useful for such polymerization?

18.36 Show how the structure of natural rubber is proved by the fact that only 4-oxopentanal is obtained upon ozonolysis.

18.37 The monomer used to prepare poly(methyl methacrylate), used in Plexiglas and Lucite paints, is methyl methacrylate. It can be prepared from acetone, HCN, and methanol, three economical starting materials. Show the reactions necessary for the synthesis of the monomer.

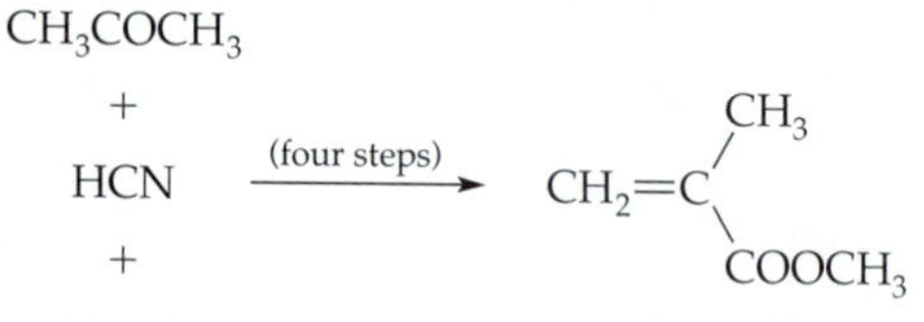

$$CH_3COCH_3 + HCN + CH_3OH \xrightarrow{\text{(four steps)}} CH_2{=}C(CH_3)COOCH_3$$

Methyl methacrylate

18.38 Show the mechanism by which tetrahydrofuran can be polymerized by acid catalysis to the polyether used in the manufacture of the stretch fabric Lycra.

$$\text{Tetrahydrofuran} \xrightarrow{H^+} HO{-}\!\left[(CH_2)_4{-}O\right]_n\!{-}$$

Tetrahydrofuran — Polyether

18.39 One of the monomers used in the preparation of Nylon-6,6 is 1,6-hexanediamine (hexamethylenediamine). It can be synthesized from 1,3-butadiene. Show how to accomplish this synthesis. (*Hint:* Apply retrosynthesis and remember how amines are prepared.)

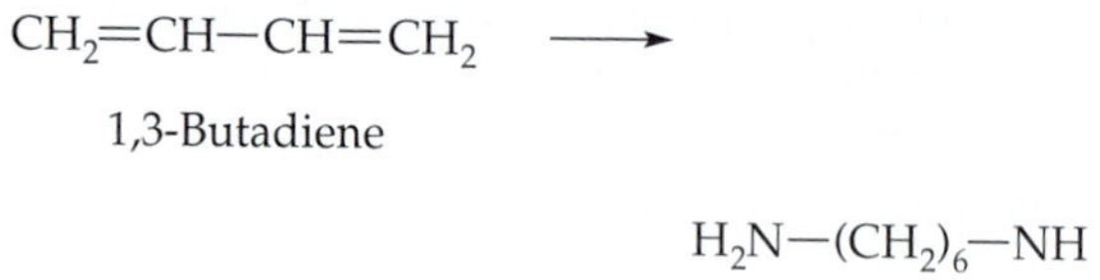

$$CH_2{=}CH{-}CH{=}CH_2 \longrightarrow$$

1,3-Butadiene

$$H_2N{-}(CH_2)_6{-}NH$$

1,6-Hexanediamine

18.40 A highly cross-linked thermosetting resin is called Glyptal. It is formed from the reaction of glycerol (1,2,3-propanetriol) with phthalic anhydride. Without necessarily drawing a structure, describe what kind of polymer Glyptal is. (*Hint:* Remember what anhydrides do with nucleophiles, including alcohols.)

18.41 Poly(vinyl alcohol) is a useful polymer, but vinyl alcohol is not an appropriate starting material for producing it. Explain. Instead, poly(vinyl alcohol) is made from poly(vinyl acetate), which is prepared from vinyl acetate ($CH_2{=}CH{-}OOCCH_3$). Draw a segment of poly(vinyl acetate), and show the reaction necessary to convert it into poly(vinyl alcohol).

18.42 The stronger versions of Superglue result from the anionic polymerization of methyl 2-cyanopropenoate. This polymerization is accomplished by adding a small amount of base to the monomer. Show the mechanism of the polymerization using a generalized base, $:B^-$, and explain why this monomer is so readily polymerized at room temperature by base.

CONCEPTUAL PROBLEM

Know Your Polymers

As a materials scientist working for an auto manufacturer, you are considering three polymer samples for possible use in new models. Unfortunately, the shipment of samples was in a warehouse in a coastal city when a hurricane hit, and the labels came off all the samples when the warehouse flooded. You decide to examine the samples anyway.

Sample 1 is a hard solid. When it is heated strongly, even in the absence of air, it decomposes rather than melting.

- Give two different classifications for this polymer based on its bulk properties.
- What feature of its molecular structure is responsible for its physical properties?

Sample 2 is stretchable and flexible, and it returns to its original shape when released.

- Classify this polymer based on its bulk properties.
- What structural feature is responsible for its physical properties?
- Where in a car is this material most likely to be used?

Sample 3 consists of long filaments. When heated in the presence of acid, the filaments decompose, absorbing water to produce the following two compounds:

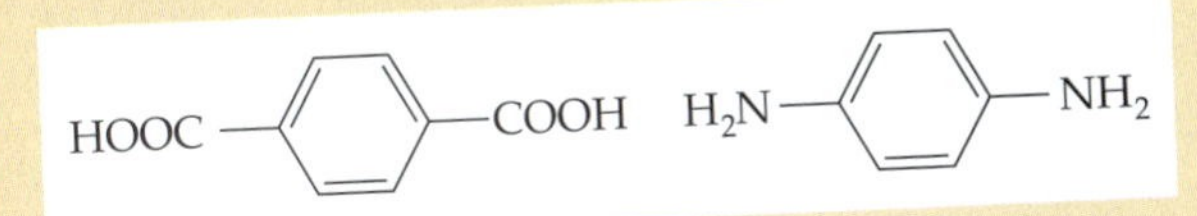

- Classify this polymer based on its bulk properties.
- Classify this polymer based on the type of reaction used to produce it.
- Classify the polymer based on the functional groups present.
- Identity the specific monomer used.
- Where in the car is this material most likely to be used?

CHAPTER 19

Introduction to Biological Chemistry

VIRTUALLY ALL of the biological functions of living systems—plants, animals, microorganisms—have a chemical basis. Chemistry drives the mechanisms of nerve impulse transmission, digestion, energy storage, intraspecies communication, muscle contraction, replication of genetic information—the list could go on and on. The basis of biological functioning is chemistry, and almost always organic chemistry. The chemical basis of biological functions and activities is studied by scientists in the fields of chemistry, biological chemistry, biology, genetics, biochemistry, microbiology, biomedicine, pharmacology, and physiology. Some amazing compounds and unique chemical control mechanisms have been discovered through the study of the chemistry of living organisms.

This concluding chapter considers some applications of organic chemistry to biological systems. You should watch for specific reactions and relate them to the Key to Transformations (inside front cover) and material covered in earlier chapters.

A host of very interesting topics could be covered under the heading "Biological Chemistry," more than enough for another book. Rather than providing superficial coverage of a large number of topics, this chapter instead presents a description of three topics that illustrate the fascinating application of chemistry to living systems: nucleic acids (the genetic code), the chemistry of metabolism, and the chemistry of medicinals (chemotherapy). Further coverage of these and similar topics can be found in more advanced textbooks and in courses in advanced organic chemistry and biochemistry. And remember, we have already discussed the elementary chemistry of carbohydrates (see Section 4.7), fats (see Section 15.3.5), and proteins (see Section 16.3).

19.1 Nucleic Acids: DNA and RNA

Nucleic acids are the very "stuff of life" in that they contain and transmit the genetic information for life processes to develop in the cell, to be maintained, and to be passed on to offspring. In particular, nucleic acids are responsible for the cell's biosynthesis of all of the approximately 100,000 necessary body proteins. If such syntheses go awry, for whatever reason, the result is usually a genetic defect, a disease, or a body malfunction.

19.1.1 The Primary Structure of Nucleic Acids

Section 9.7.3 introduced the structures of DNA and RNA and their component parts, as part of the discussion on the heterocyclic bases (purines and pyrimidines). The primary structures of the two nucleic acids DNA (deoxyribonucleic acid) and RNA (ribonucleic acid) consist of long-chain polymers of nucleotide units. A **nucleotide** con-

◀ DNA Fingerprinting: These electrophoresis gels are used to separate fragments of DNA into bands, which can then be used to identify the owner of that DNA.

tains three components: a heterocyclic base (B), a C_5 ribofuranose sugar, or pentose (R), and a molecule of phosphoric acid (as, phosphate, P). The polymer has a *backbone* made up of an alternating sequence of sugar (R) and phosphate (P) units, with a base (B) attached to each sugar. A fragment containing just a base (B) attached to a sugar (R) is called a **nucleoside.** Nucleic acids can be represented schematically as follows:

B = heterocyclic base
R = sugar (ribose or deoxyribose)
P = phosphate

—R—P—R—P—R—P—R—P— (each R bearing a B)
Nucleic acid

R—P (R bearing a B)
Nucleotide

R (bearing a B)
Nucleoside

Although it is composed of thousands of nucleotide units, DNA contains only four heterocyclic bases: **adenine, guanine, cytosine,** and **thymine,** abbreviated **A, G, C,** and **T,** respectively.

Adenine   Guanine   Cytosine   Thymine

The four heterocyclic bases incorporated in DNA

Adenine and guanine are derivatives of the heterocycle purine, while cytosine and thymine are derivatives of pyrimidine (see Section 9.7.3). RNA also contains only four heterocyclic bases, three of which are identical to those found in DNA (A, G, and C). The fourth RNA base is **uracil (U),** a demethylated derivative of thymine.

Adenine   Guanine   Cytosine   Uracil

The four heterocyclic bases incorporated in RNA

A segment of a single strand of typical DNA is shown in Figure 19.1, including the structural formula of the sugar-phosphate backbone and two bases. Note that the connecting linkages in the backbone are to carbons 5′ and 3′ of the sugar.

Nucleic acids are hydrolyzed by enzymes or in the laboratory to their three components:

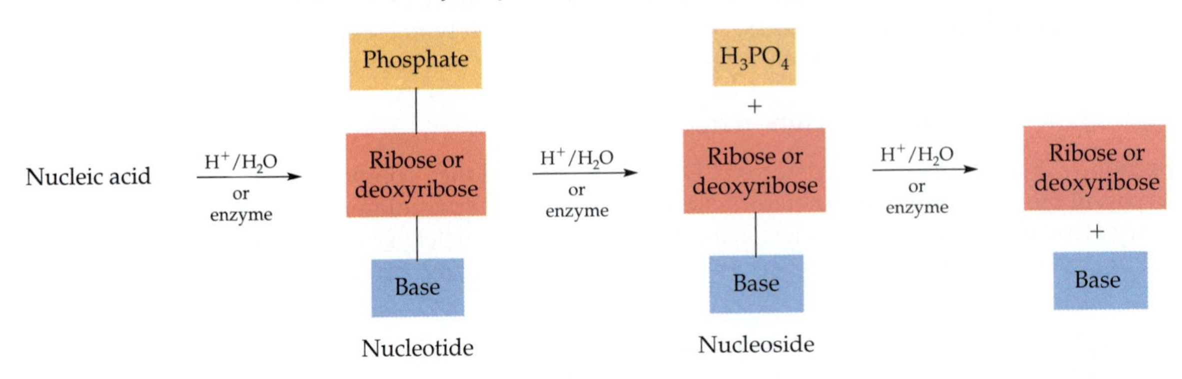

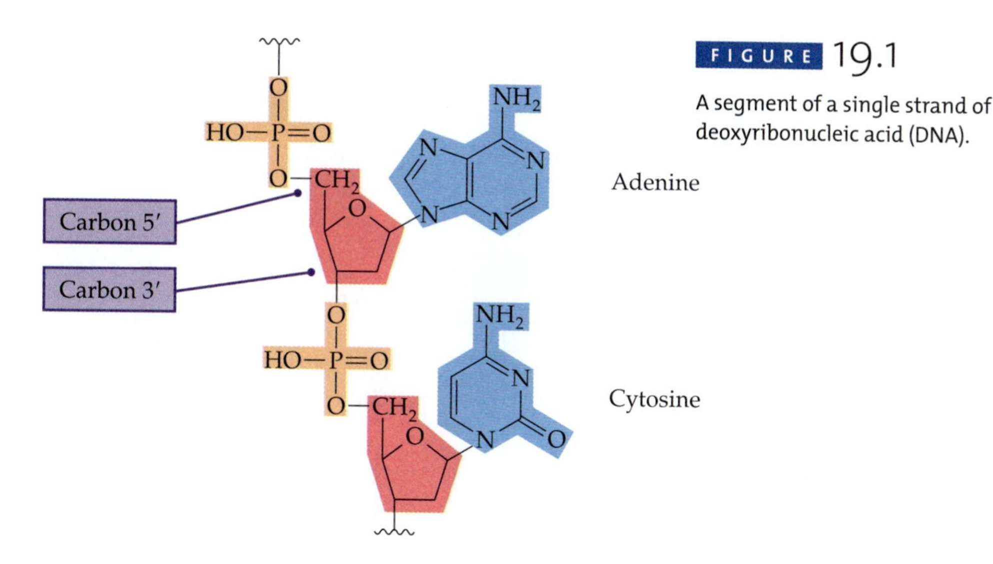

FIGURE 19.1

A segment of a single strand of deoxyribonucleic acid (DNA).

Recall that a sugar normally has an oxygen bonded to every carbon, most often in the form of an —OH group. The sugars in DNA and RNA are C_5 sugars, generically known as pentoses and specifically members of the ribose family. In DNA, the 2′ carbon of ribose is deoxygenated. Therefore, the sugar is known as 2′-deoxyribose (providing the D of DNA). In RNA, the sugar is normal ribose, with all hydroxyl groups present.

From this description, it should be clear that there are only two fundamental differences between the primary structures of DNA and RNA: The two sugars are different (by one oxygen atom), and one of the four bases is different (by one methyl group).

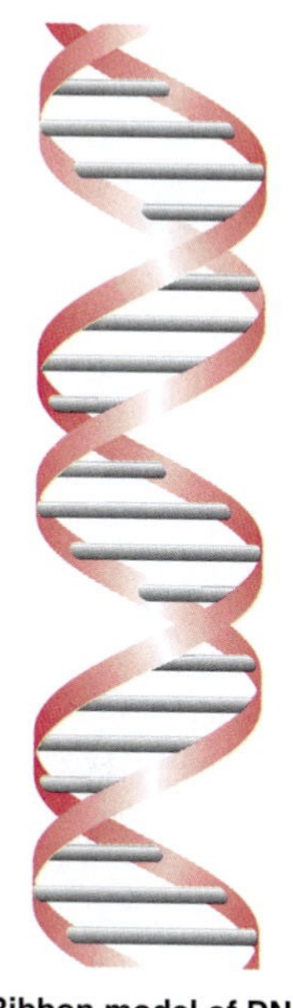

Ribbon model of DNA
(a)

19.1.2 The Secondary Structure of DNA

DNA exists in the cell as two parallel strands of nucleic acid coiled about each other in a **double helix.** This term became an important part of the language of science when the structure was first proposed by James Watson and Francis Crick in 1953; their work was later recognized with the 1962 Nobel Prize in physiology and medicine (shared with Maurice Wilkins). The double helix is the secondary structure of DNA and can be envisioned as two pieces of ribbon, each representing the sugar-phosphate backbone of a single strand of nucleic acid, wrapped around a hollow glass cylinder (such as a tall drinking glass) in an alpha direction. The two strands of DNA are oriented in opposite directions, so that one strand ends in a free hydroxyl group on carbon 3′ and the other ends in a free hydroxyl group on carbon 5′. The bases project from the backbone into the *interior* of the helix. (Recall that the α-helix protein structure has the side chains oriented to the *exterior* of the helix.) Figure 19.2 shows two representations of a section of DNA.

The two strands in the DNA double helix are held in position by a very precise pattern of hydrogen bonding between sets of two bases, known as **base pairs.** One base pair is adenine and thymine (A-T), and the other is guanine and cytosine (G-C). Thus, the two strands of DNA are not identical, but instead are complementary in a way that allows maximum hydrogen bonding. In each base pair, one base is a purine derivative (adenine or guanine) and the other base is a pyrimidine derivative (cytosine or thymine). Wherever there is an adenine in one strand, a thymine is located opposite

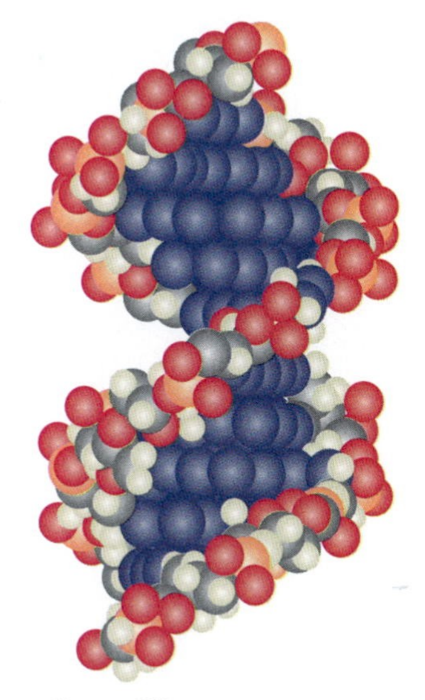

Space-filling model of DNA
(b)

FIGURE 19.2

(a) Ribbon model and (b) space-filling model of DNA.

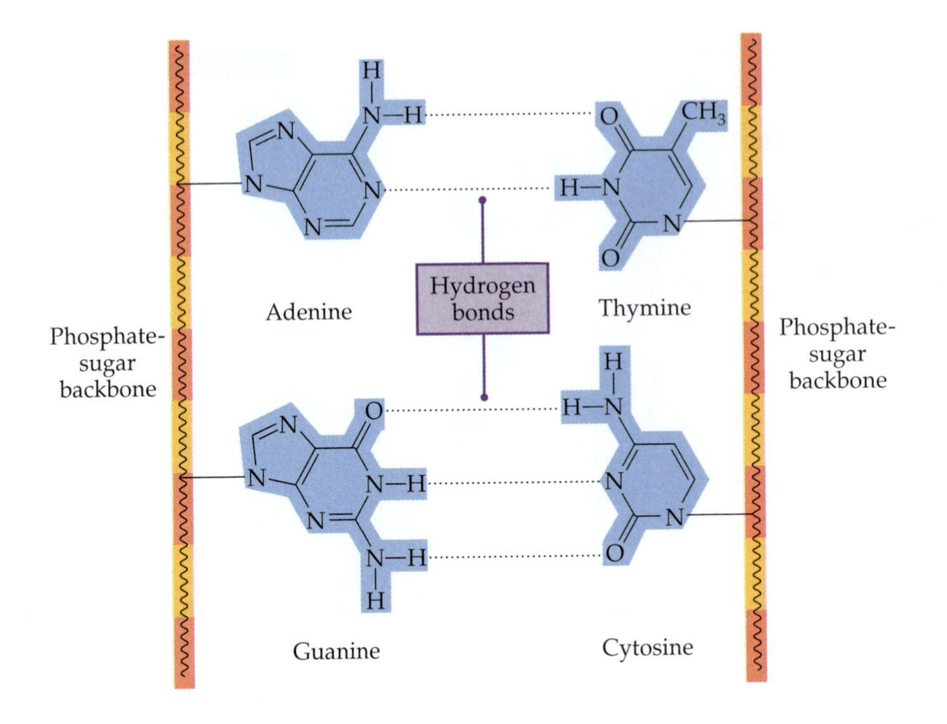

FIGURE 19.3

Hydrogen bonding of complementary base pairs in DNA.

in the other strand: an A-T base pair. Wherever there is a guanine in one strand, a cytosine is located opposite in the other strand: a G-C base pair (Figure 19.3). The aggregate energy of the many hydrogen bonds within the helix is sufficient to maintain the helical structure.

The DNA helix is about 20 Å in diameter (1 Å = 10^{-10} m). The vertical distance between base pairs in a strand of DNA is about 3.4 Å, and 10 base pairs are required for one complete turn of the helix.

Human DNA is located in the nucleus of each cell, within 23 pairs of chromosomes (46 chromosomes in all). Each chromosome is made up of one large DNA molecule of about 200 million nucleotides. The chromosome itself is about 5 μm long, yet if the two DNA strands were uncoiled and laid out in a line, they would measure about 5 cm. All together, a single human cell contains about 6 billion base pairs in its DNA. Each strand of DNA has numerous segments of nucleotides called **genes,** which are associated with specific chemical and biological functions. These genes are not necessarily contiguous along the DNA strand.

The human *genome* (the sum total of the approximately 100,000 human genes) contains about 3 billion base pairs. A major international project is currently under way to determine its complete structure, a process known as *sequencing* the human genome. The mind-boggling magnitude of this effort can be appreciated by the fact that a Nobel Prize was awarded to Frederick Sanger (his second) in 1980 for the sequencing of a virus chromosome that contained a "mere" 5375 nucleotides.

The DNA located in cell nuclei is the sole repository of all the genetic information of an organism. As we will see, this information is stored in the sequence of DNA's base pairs. For this vital information to be used—for protein synthesis, for example—it must be transferred to other sites in the body where biological and chemical functions are carried out—usually outside the cell nucleus. Three information-transfer processes occur:

1. **Replication**—the process whereby DNA reproduces itself

2. **Transcription**—the process whereby DNA passes its information to RNA for eventual use outside the cell nucleus
3. **Translation**—the process of decoding the information contained in RNA so that it can be used in protein synthesis

Keep in mind that this discussion of secondary structure pertains to DNA only. RNA does not exist as a double helix but as a single strand of nucleotides. We will look more closely at RNA a bit later.

19.1.3 Replication of DNA

Replication of DNA is necessary to life; it is occurring all the time in any organism. Through the process of cell division, it enables an organism to grow and to renew itself. It also enables life to continue from one generation to the next by passing on genetic information to offspring through specialized germ cells. Because the genetic information is contained in a precise sequence of base pairs in DNA, this information must be replicated without error. It is estimated that errors do occur—about once in every 10 billion replications. However, a number of enzymes exist specifically to recognize and correct errors in DNA. But even those errors that do persist do not necessarily spell genetic disaster. Some of them, called *mutations*, provide for the gradual evolution of a species.

The replication process for DNA is enzyme-catalyzed. It starts with the gradual uncoiling of the double helix, which means exposure of the bases of each of the two complementary strands of nucleic acid, ordinarily protected within the interior of the helix. The cell nucleus contains within it many unattached nucleotides that can form new complementary base pairs, by hydrogen bonding with the exposed bases on both strands. Through catalysis by the enzyme DNA polymerase, the new nucleotides are joined together to form a new strand of nucleic acid (that is, the sugar-phosphate backbones connected to the bases are joined). Thus, where there had been an A-T base pair in the helix, replication produces two base pairs, A-T and T-A, as two new helices begin to form. This process continues until the two original strands of nucleic acid in the double helix have totally separated and each of those strands has a new complementary strand hydrogen-bonded to it. Thus, two identical DNA double helices exist where there had been one before. As the cell divides, one DNA double helix goes with each cell nucleus; the DNA is thus passed on. Figure 19.4 is a representation of the replication process when partially completed.

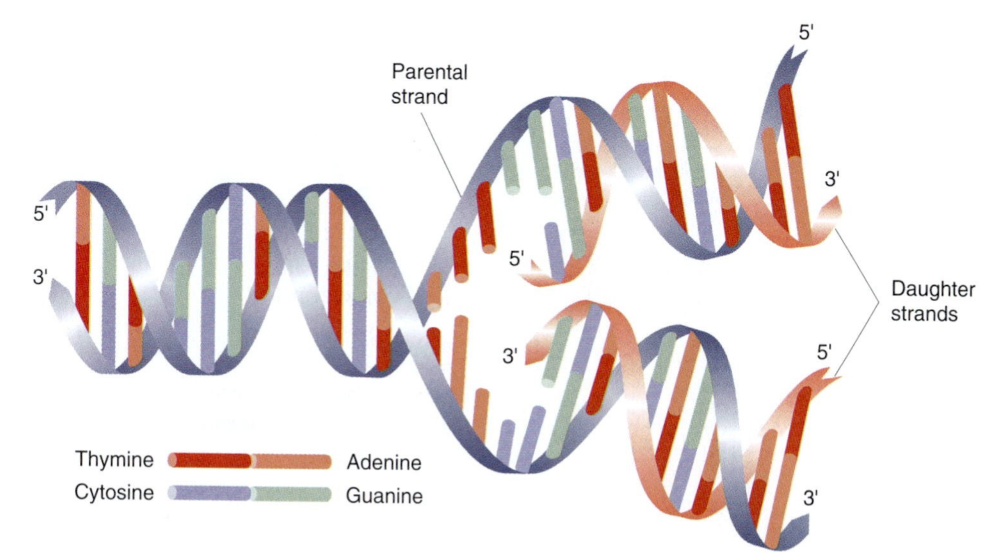

FIGURE 19.4 Representation of partially replicated DNA.

19.1.4 Protein Biosynthesis

It is estimated that there are about 100,000 different proteins in the human body. These proteins mediate all the body's chemical and biological activity. Each protein, which is a specific sequence of α-amino acids, is synthesized in the cytoplasm of a cell. The exact sequence of α-amino acids in each protein is encoded in a gene, a particular sequence of nucleotides in the DNA located in the cell nucleus. Altogether, the process of protein synthesis in the cells (which is called *biosynthesis* as opposed to laboratory synthesis) involves the gene and its DNA, messenger RNA (mRNA), transfer RNA (tRNA), and the storehouse of α-amino acids located in the cytoplasm.

Recall that RNA (ribonucleic acid) is chemically similar to DNA, with two exceptions: The sugar in RNA is ribose rather than 2′-deoxyribose, and the pyrimidine base in RNA is uracil (U) rather than thymine (it also forms a base pair with adenine, A-U). RNA molecules are also much smaller than DNA molecules—RNA has many fewer nucleotides—and are single-stranded, not a double helix.

Transcription

The process of information transfer from DNA (the genes) to **messenger RNA (mRNA)** occurs in the nuclei of cells and is called *transcription*. The DNA becomes partially uncoiled to expose the section of nucleotides that makes up a particular gene. One strand of the DNA helix is known as the *informational strand*, and the other is known as the *template strand*. Individual ribonucleotides in the nucleus arrange themselves in a complementary fashion (A-U, T-A, C-G, and G-C) by hydrogen bonding along the template strand. They are then connected to each other (polymerized) to form a strand of messenger RNA, a process similar to DNA replication. The mRNA then dissociates from the template strand, the DNA recoils into its double helix, and the mRNA migrates through the wall of the cell nucleus to enter the cytoplasm. This transcription process creates mRNA molecules whose sequence of bases is an exact copy of the sequence of bases in the informational strand of DNA (except that uracil has replaced thymine) (Figure 19.5).

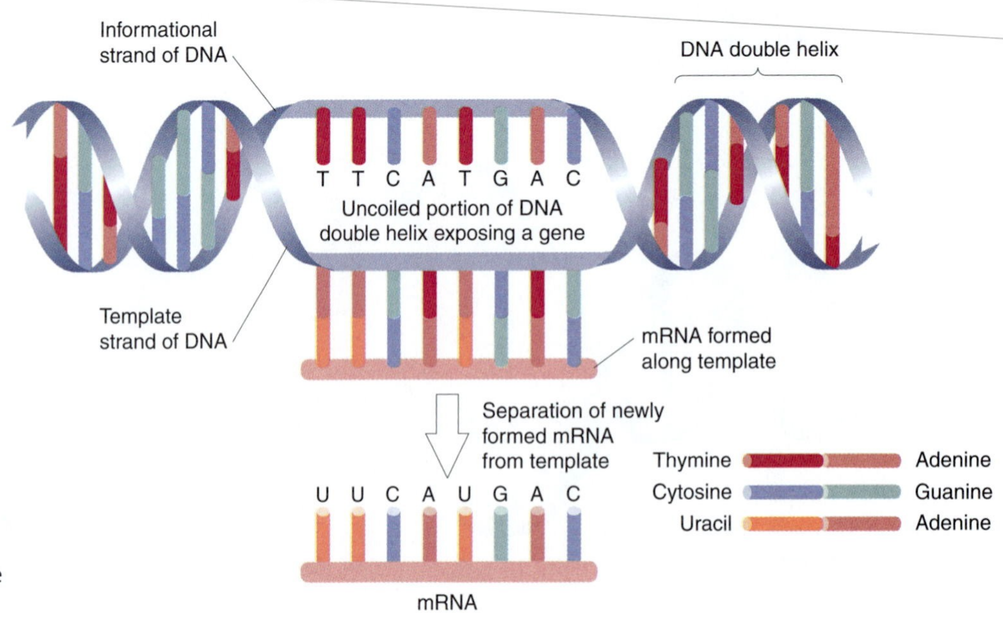

FIGURE 19.5

Transcription of base sequence from DNA to mRNA.

Translation

After entering the cytoplasm of the cell, mRNA becomes associated with small particles called *ribosomes,* which contain about 60% RNA and 40% protein. Each sequence of three contiguous bases (a triplet) on the mRNA makes up what is called a **codon.** Smaller ribonucleotides in the ribosome called **transfer RNA (tRNA)** include a specific sequence of three bases that make up an **anticodon**—a triplet that is complementary to a codon. It is the codon-anticodon complementarity that brings each tRNA to the mRNA chain to bind (hydrogen-bond) at a specific site. This process of codon-anticodon recognition is called *translation.*

Each tRNA has attached to it a single α-amino acid. Thus, each tRNA acts as a *carrier* for that particular α-amino acid. The tRNA structures are well known. There are 64 known tRNAs, and each tRNA contains about 70 to 100 ribonucleotides. Included in each nucleotide chain is the specific sequence of three bases that make up the tRNA anticodon that recognizes the complementary codon of mRNA and forms hydrogen bonds with it. At the end of the tRNA chain is a 3′-hydroxyl group of a ribose. (Recall that both the 3′- and 5′-hydroxyl groups of all but the end ribose units are bonded to phosphate to form the backbone of RNA.) This terminal 3′-hydroxyl group is bonded through an ester linkage to the carboxyl group of a specific α-amino acid:

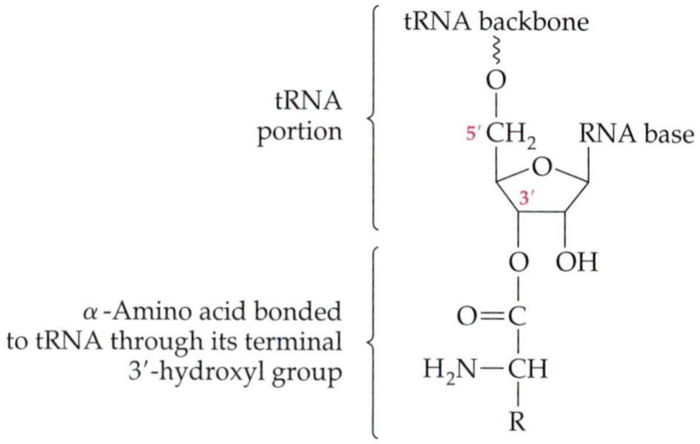

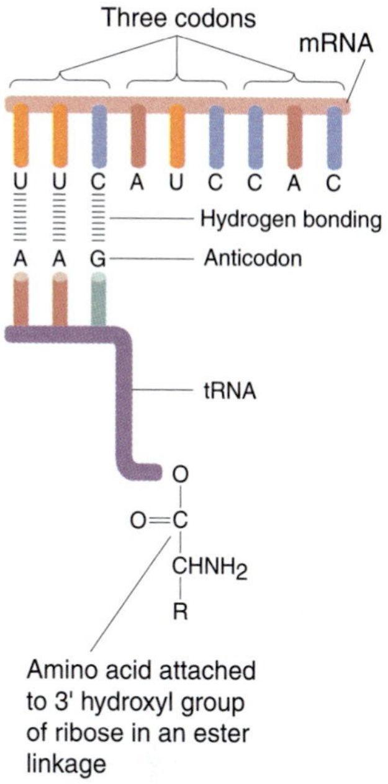

tRNA portion: tRNA backbone — O — 5′ CH_2 — ribose (3′) — RNA base; OH

α-Amino acid bonded to tRNA through its terminal 3′-hydroxyl group: O — $O{=}C$ — $H_2N{-}CH$ — R

Three codons / mRNA: U U C A U C C A C — Hydrogen bonding — Anticodon: A A G — tRNA — O — $O{=}C$ — $CHNH_2$ — R — Amino acid attached to 3′ hydroxyl group of ribose in an ester linkage

FIGURE 19.6 The translation process by which tRNAs bring α-amino acids to mRNA for protein synthesis.

The tRNAs sequentially form hydrogen bonds with the mRNA, and the α-amino acids borne to the site are thereby in position to be connected to each other by peptide-forming enzymes (Figure 19.6).

As the amino acids come into position to be connected to one another, a peptide-forming reaction occurs, catalyzed by an enzyme (Figure 19.7). This "connecting" reaction is a familiar one, that of a nucleophilic amine reacting with an ester, displacing alkoxide

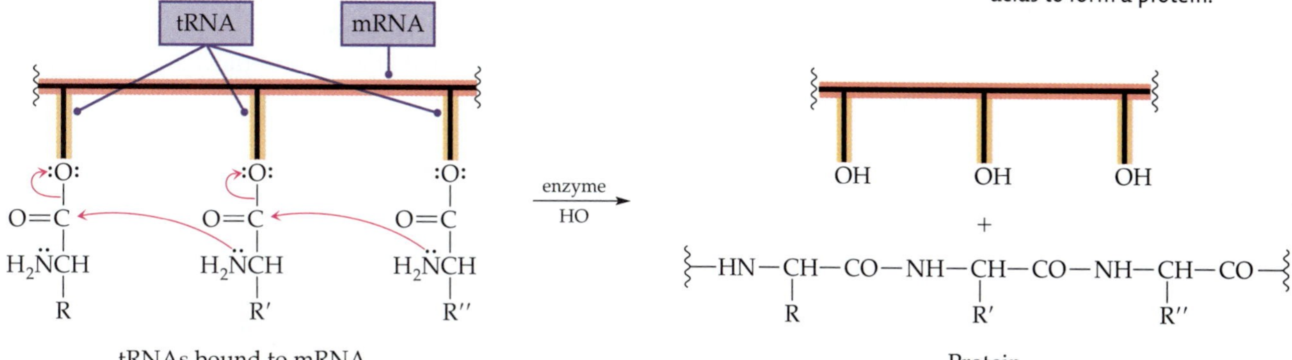

FIGURE 19.7 Reaction of tRNA-carried amino acids to form a protein.

to form an amide (a nucleophilic acyl substitution reaction; see Sections 15.3.2 and 15.4.3). By a series of these reactions transforming ester linkages into amide (peptide) linkages, the protein is gradually "zipped together" with its precise sequence of α-amino acids.

The total process of information transfer from DNA to final α-amino acid sequence in a protein therefore occurs through several steps, with great preciseness. For example, a sequence of A-G-T in the uncoiled template strand of DNA produces a codon of U-C-A in mRNA, which in turn is recognized by a tRNA whose anticodon is A-G-U. This tRNA carries the α-amino acid serine, and the mRNA codon position ensures a specific location for serine in the protein being synthesized. The overall sequence for protein synthesis can be summarized like this:

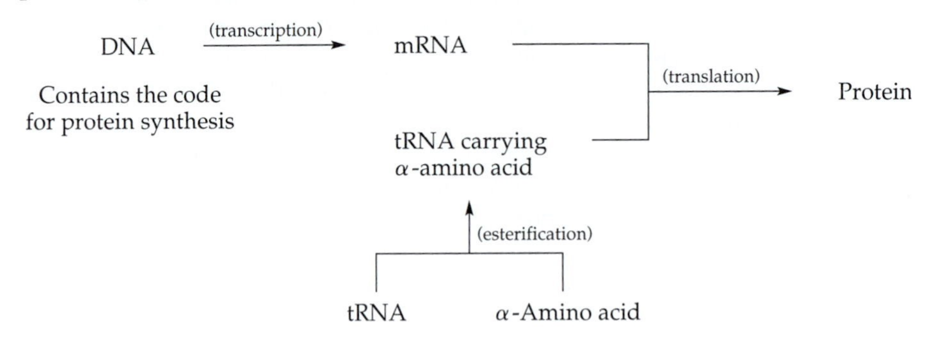

19.1.5 The Genetic Code

You can see from the description of the transcription and translation processes that the genetic code consists of a series of triplets (or triads) that ultimately specify the incorporation of specific α-amino acids into proteins in a particular sequence. The genetic code results in the biosynthesis of the proteins in the human body, each of which has a unique sequence of α-amino acids. Each particular sequence of triplets of nucleic acid bases (triads) is stored in DNA, then *transcribed* to mRNA, and finally *translated* by tRNA to create a specific sequence of α-amino acids for biosynthesis into a particular protein.

Because there are only four different bases in DNA (A,T, C, and G), and because a sequence of three bases is required for a particular codon on mRNA, there are $4^3 = 64$ possible sequences. Of these 64 possible codons, 61 are known to be associated with specific α-amino acids. The other three codons are termination (stop) codes, providing the signal that the protein chain is complete. Since there are only 20 different α-amino acids incorporated into human proteins, it is clear that some of these amino acids must be encoded by more than one triad. For example, while tryptophan is encoded by only one triad (U-G-G), other α-amino acids are encoded by up to six different triads. Table 19.1 shows all possible triads and the α-amino acids they encode. You could, from this table, construct an inverse table that would list the α-amino acids and all the codons that apply to each.

How to Solve a Problem

Using Table 19.1, indicate the triplet code in a template strand of DNA that would ultimately result in formation of the peptide Val-Phe:

$$(CH_3)_2CHCH(NH_2)CO{-}NHCH(CH_2C_6H_5)COOH$$

Val-Phe

TABLE 19.1

mRNA Codons for α-Amino Acids

		Third Base			
First Base	**Second Base**	**A**	**U**	**G**	**C**
A	A	Lys	Asn	Lys	Asn
	U	Ile	Ile	Met	Ile
	G	Arg	Ser	Arg	Ser
	C	Thr	Thr	Thr	Thr
U	A	Stop	Tyr	Stop	Tyr
	U	Leu	Phe	Leu	Phe
	G	Stop	Cys	Trp	Cys
	C	Ser	Ser	Ser	Ser
G	A	Glu	Asp	Glu	Asp
	U	Val	Val	Val	Val
	G	Gly	Gly	Gly	Gly
	C	Ala	Ala	Ala	Ala
C	A	Gln	His	Gln	His
	U	Leu	Leu	Leu	Leu
	G	Arg	Arg	Arg	Arg
	C	Pro	Pro	Pro	Pro

PROBLEM ANALYSIS

The solution of this problem involves three discrete steps: (1) determining the codon in the mRNA that encodes for the desired α-amino acid; (2) determining the complementary bases that must be in the template strand of DNA; and (3) determining the complementary bases that must be in the informational strand of DNA. (We can use Figure 19.5 to assist in the second and third steps.)

SOLUTION

Val (valine) is encoded by G-U-A (as well as by G-U-U, G-U-G, and G-U-C) in mRNA. The complementary informational strand of DNA is C-A-T. The template strand of DNA is therefore G-T-A. Note that this “double” complementarity results in G-U-A (mRNA) arising from G-T-A (template DNA). These triads are the same except that uracil (U) replaces thymine (T).

In a similar manner, we determine that Phe is encoded by U-U-U in mRNA. Therefore, the informational strand of DNA must have A-A-A, and the template strand of DNA must have T-T-T.

Therefore, the two triads on the template strand of DNA that would lead to Val-Phe are G-T-A and T-T-T.

19.1.6 Gene Therapy

Armed with the ability to determine the nucleotide sequence of genes, using processes similar to those applied to sequencing proteins (see Section 16.2.2), the Human Genome Project was launched in 1985. This is a coordinated worldwide effort to "map" all of the human genes—a gargantuan task to determine the nucleotide sequence of every human gene. (A total of about 3 billion nucleotides are present in human DNA.) As part of this effort, mapping projects are also under way for a number of much simpler organisms. A major accomplishment was reported in 1995—the complete genome map was produced for a very simple bacterium, *Mycoplasma genitalium*, consisting of only 580,070 base pairs. A total of 470 genes and their functions—including replication, transcription, translation, DNA repair, cellular transport, and energy metabolism—were identified. In 1996, the complete genome of the yeast *Saccharomyces cerevisiae*, containing 12,068,000 bases and 6340 genes, was elucidated.

About 4000 diseases are known to be genetically inherited. As human genes are sequenced (identified), scientists come closer to being able to distinguish between normal genes and those that are abnormal, thus pinpointing the causes of various diseases. There has already been progress in this direction. For example, in human chromosome 7 (containing about 15 million base pairs), the gene responsible for cystic fibrosis has been identified. That gene encompasses about 2.4 million base pairs. Also, sites on chromosomes 1, 14, and 21, among other locations, have been associated with the onset of Alzheimer's disease. Defective genes have also been identified for several other diseases, including Lou Gehrig's disease, Huntington's disease, congenital glaucoma, and some forms of multiple sclerosis.

Genetic mutations occur constantly by the alteration of DNA or the incorporation of the "wrong" base into a DNA sequence. In many instances, DNA repair enzymes correct these errors, but not always. Mutations can be caused by "natural" errors in nucleotide polymerization, but also by external forces, such as high-energy radiation and cancer-causing agents. It is known, for example, that UV radiation leads to the dimerization of some DNA bases, which prevents the uncoiling of a double helix.

Once a mutated gene is identified and the defect in that gene is located, a form of *gene therapy* may be undertaken. The defective piece of DNA may be replaced by a "good" piece of DNA. In other words, patients are administered segments of DNA designed to replace or supplement the segments carrying the genetic defect. For example, sickle cell anemia is known to be caused by incorporation of the α-amino acid valine in place of the normal glutamine in the protein globin. The triggering event is the mutation in the sixth codon of the β-globin gene, a mutation in which the base adenine (A) is replaced by thymine (T). There has been some success in correcting this mutation in the laboratory.

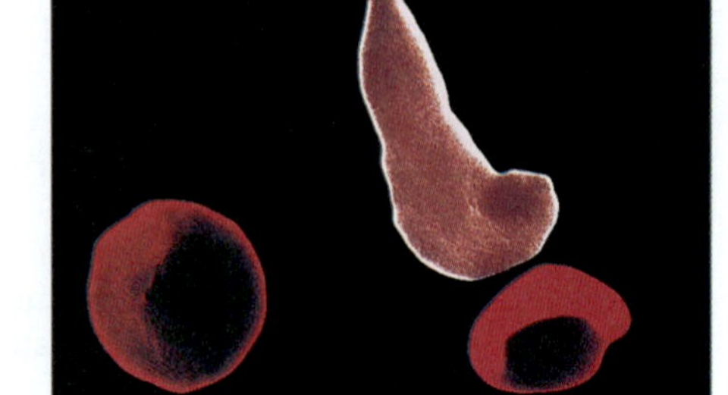

A mutation that affects the structure of the hemoglobin protein causes normally round red blood cells to acquire this pointed, sickle shape (center).

19.1.7 Genetic Engineering

Genetically engineered corn resists damage by the European corn borer.

The discovery that the polymerase chain reaction (PCR) is initiated by the enzyme *DNA polymerase* created the opportunity to "mass replicate" DNA. For example, a tiny sample of DNA-containing material—a single hair, a drop of semen, or a drop of dried blood (either recent or ancient)—can be subjected to the PCR to produce sufficient DNA to determine its source by analysis. Because the genetic "fingerprint" of each individual is different, this process permits DNA analysis to be employed in forensic efforts and determination of biological relationship. The mass replication of DNA also means that specially "engineered" DNA can be produced in quantity.

Genetic engineering involves the laboratory synthesis of small DNA segments (polynucleotides) that have the potential for incorporation into DNA. Most often, such DNA is incorporated into a microorganism, with the express purpose of causing the organism to synthesize specific desired proteins. However, synthetic DNA segments can also be incorporated into human DNA. The synthesis of polynucleotide segments is extremely complex and laborious, with problems analogous to those faced in protein synthesis (see Section 16.2.4). The process has become automated, however, just as for proteins, with the development of a "gene machine." As a result, it is not unusual to synthesize polynucleotide sequences of a few hundred base pairs for use in genetic engineering.

Once a desired polynucleotide has been synthesized, whether in the laboratory or using enzymatic processes, it is "spliced" into the DNA of an organism. This added piece of DNA creates new characteristics in that organism, implemented by the proteins synthesized by the inserted piece of polynucleotide. Many of the major accomplishments associated with genetic engineering have occurred in agriculture, where this gene-splicing approach has produced so-called *transgenic crops*. Here are some examples:

- A tomato marketed as the Flavr Savr Tomato contains a new gene that causes it to resist softening and ripen slower, enabling it to be picked ripe, rather than being picked green.
- Squash have been genetically engineered to resist attack by viruses.
- Cotton and soybeans have been developed that carry resistance to widely applied herbicides such as Round-Up and Bromoxynil.
- Insect resistance has been introduced to several plant species, including cotton, potatoes, and corn, using *Bacillus thuringiensis* genes. The new protein inhibits attack by European corn borers, which cause great crop damage around the world each year.

19.2 Metabolism

The human body relies on external sources for energy and chemical raw materials. Nutrients are converted into needed chemicals and energy via metabolic processes. **Metabolism** involves two different types of processes. **Catabolism** describes those reactions that break down complex materials into simpler materials and energy. **Anabolism** describes those reactions by which the body converts simple materials into needed complex materials like proteins (a process also called *biosynthesis*).

This section describes a few examples of metabolic chemistry. But first, it is essential to understand the role of a key component of energy-related processes—the compound called **ATP** (adenosine triphosphate).

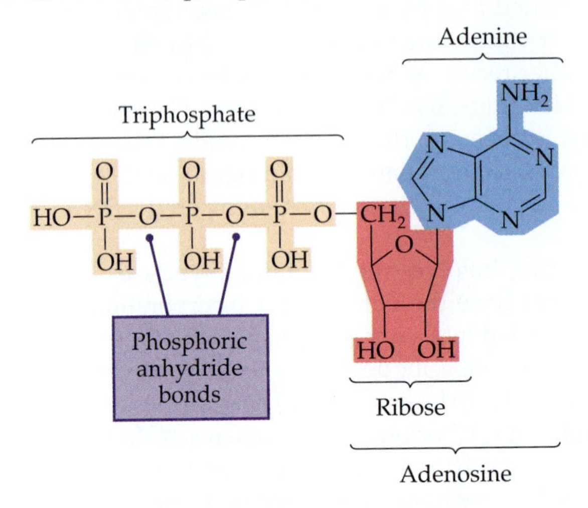

Adenosine triphosphate (ATP)

Note that ATP contains a triphosphorylated nucleoside (adenosine = adenine (blue) + ribose (red)). The key feature of ATP is that it contains two phosphate-phosphate bonds. These are often called *high-energy* bonds: They require considerable energy to form, and they release considerable energy when hydrolyzed to phosphoric acid. Thus, these bonds appear to store energy. They are phosphoric anhydride bonds, analogous to carboxylic anhydride bonds (see Section 15.2.2), which are also rapidly and exothermically hydrolyzed.

$$\mathrm{HO{-}P(=O)(OH){-}O{-}P(=O)(OH){-}OH} \xrightarrow{H_2O} 2\ \mathrm{HO{-}P(=O)(OH){-}OH} + 7\ \text{kcal/mol}$$

Diphosphate (phosphoric acid anhydride) — Phosphoric acid

$$\mathrm{R{-}C(=O){-}O{-}C(=O){-}R} \xrightarrow{H_2O} 2\ \mathrm{R{-}C(=O){-}OH} + 12\ \text{kcal/mol}$$

Carboxylic acid anhydride — Carboxylic acid

ATP is produced using the energy released from certain enzyme-catalyzed reactions, as we will see later. The starting compound for this production is adenosine monophosphate (AMP), which is a nucleotide found in the cytoplasm of cells. If one phosphate-phosphate bond is added by incorporating a second phosphate anion, forming a phosphoric anhydride bond, AMP becomes adenosine diphosphate (ADP). When a second phosphate-phosphate bond is created, ADP is converted to ATP. Conversely, when ATP serves as an energy source, first one and then the second phosphate-phosphate bond is broken, and inorganic phosphate ions (PO_4^{3-}) are released.

When you eat, you take in the sources of energy for your body: carbohydrates, proteins, and fats. These materials are synthesized in nature from other starting materials and energy (see Section 19.2.4). The original energy source is the sun. Through a long series of chemical reactions (photosynthesis) starting with carbon dioxide and water, solar energy is converted into carbohydrates, initially glucose. Animals ingest carbohydrates to produce fats and proteins. Humans ingest carbohydrates, proteins, and fats and catabolize them to release their energy. The overall energy flow in carbohydrate biosynthesis and catabolism, from sunlight energy to human energy, can be depicted as follows:

$$6\,CO_2 + 6\,H_2O + 670\text{ kcal} \xrightarrow{\text{(photosynthesis)}} \underset{\text{Glucose}}{C_6H_{12}O_6} + 6\,O_2$$

$$\underset{\text{Glucose}}{C_6H_{12}O_6} + 6\,O_2 \xrightarrow{\text{(catabolism)}} 6\,CO_2 + 6\,H_2O + 670\text{ kcal}$$

The following sections will briefly summarize the chemical reactions involved in several catabolic and anabolic processes. These reactions have all been described in earlier chapters for simple compounds, so you should be able to recognize them.

19.2.1 Catabolism of Carbohydrates

The overall pathways involved in carbohydrate metabolism are illustrated in Figure 19.8. Humans' major source of carbohydrates is starch, which is hydrolyzed (acid hydrolysis of acetals and ketals) in the digestive system to its monomer glucose (see Section 4.7). Other carbohydrate sources include glucose itself, fructose, sucrose, maltose, and lactose. The glucose is absorbed from the stomach and intestines into the bloodstream and carried throughout the body to sites where energy is needed. Excess glucose is repolymerized to a compound called *glycogen,* which is stored mainly in the liver but also in the muscles. Glycogen is very rapidly hydrolyzed to glucose on demand and can be transported in the bloodstream to cells for catabolism by their enzyme systems.

The catabolism of glucose occurs in two major phases. The first phase involves conversion of glucose into two C_3 fragments (pyruvic acid), each of which loses CO_2 to form acetyl-CoA. The second phase involves the overall conversion of acetyl-CoA to CO_2 in what is known as the **Krebs cycle,** or tricarboxylic acid (TCA) cycle. Because of the complexity of the reactions involved, we will look at each of these phases separately. Many details have been omitted, including many of the enzyme systems that are involved.

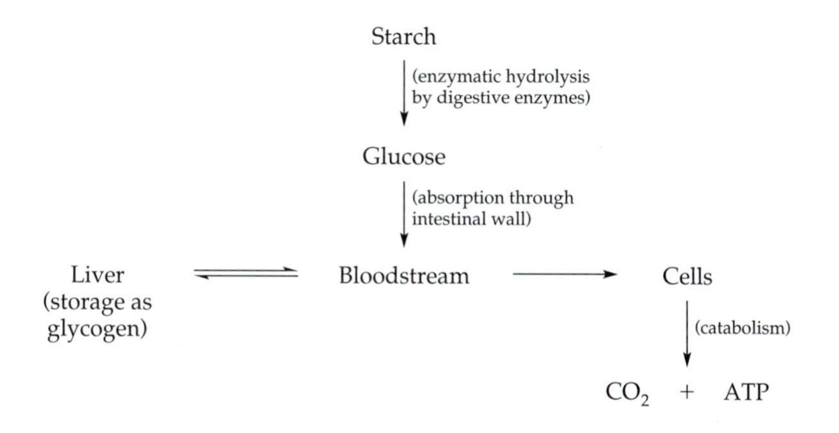

FIGURE 19.8
Source and fate of glucose.

Catabolism to Pyruvic Acid

Glucose is phosphorylated at carbon 6 by ATP and is then isomerized to fructose-6-phosphate via an ene-diol intermediate [—C(OH)=C(OH)—]. Fructose-6-phosphate is then phosphorylated at carbon 1 by a second molecule of ATP to produce fructose-1,6-diphosphate. (Phosphorylations are esterification reactions of phosphoric acid and an alcohol; see Section 15.3.) By cleavage between carbons 3 and 4, two C_3 fragments are eventually produced; then the dihydroxyacetone monophosphate isomerizes by enolization then ketonization (see Section 13.2.3) to glyceraldehyde-3-phosphate. As a result of this entire sequence, two ATP units are consumed, glucose is cleaved to two C_3 molecules, and no energy is produced (Figure 19.9).

Next, glyceraldehyde-3-phosphate undergoes addition at its carbonyl group by the anion of phosphoric acid (analogous to cyanohydrin formation by aldehydes; see Section 13.3.2). The new alcohol at carbon 1 of the adduct is oxidized by NAD^+ (see Section 16.5.2) to the corresponding glyceric acid derivative, resulting in a high-energy anhydride bond between phosphoric acid and the carboxylic acid. This phosphate group is then transferred to ADP to form ATP, and the energy of the phosphoric anhydride bond is now stored in ATP. Next, the phosphate on carbon 3 migrates to carbon 2 (an internal transesterification), and dehydration (see Section 6.6.1) occurs to produce an enol-phosphate. This also contains a high-energy phosphate bond, which is transferred to another molecule of ADP, forming ATP. This step releases the enol, which rapidly tautomerizes to pyruvic acid (see Section 13.2.3). In this overall sequence, then, the two C_3 units have produced a total of four units of ATP, which store energy (Figure 19.10).

The final sequence in this phase of glucose catabolism involves the decarboxylation of pyruvic acid to form acetic acid. This decarboxylation is a critical step in catabolism and involves three major enzyme systems that depend on the presence of vitamins as coenzymes: thiamine for the decarboxylase, niacin for the NAD^+ oxidation, and

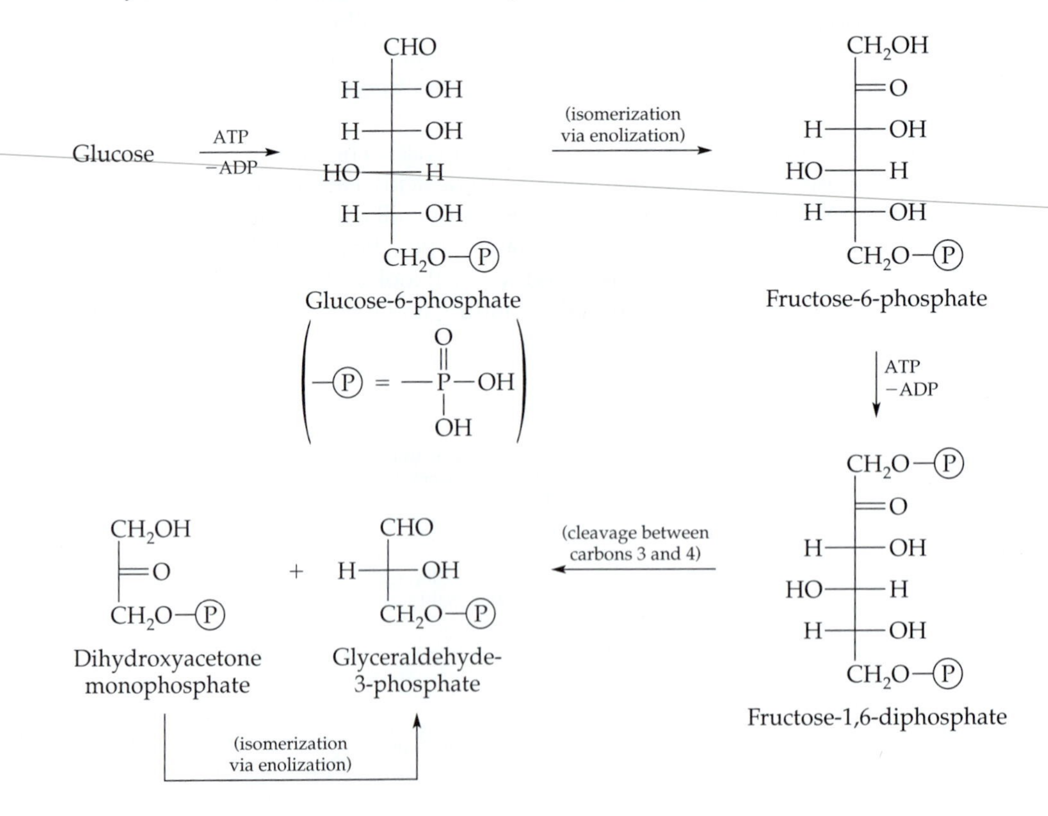

FIGURE 19.9

Metabolism of glucose to C_3 compounds.

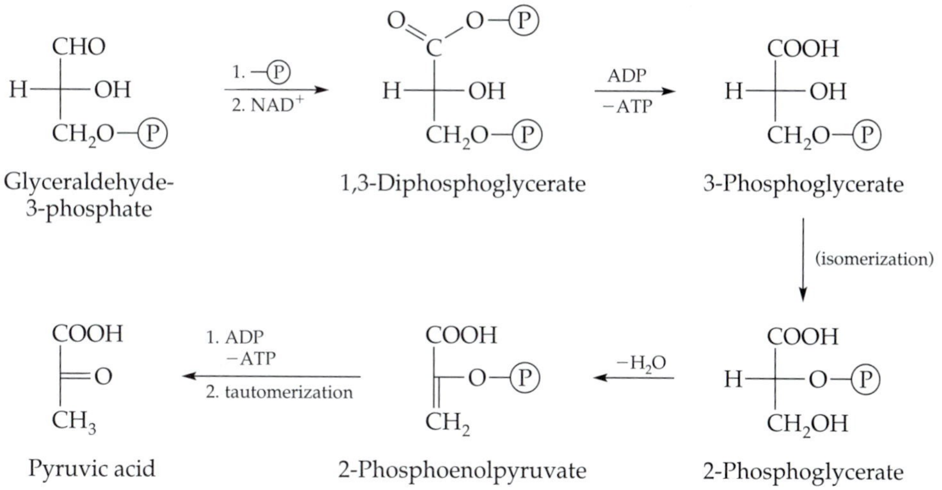

Glyceraldehyde-3-phosphate →(1. –Ⓟ; 2. NAD^+)→ 1,3-Diphosphoglycerate →(ADP, –ATP)→ 3-Phosphoglycerate →(isomerization)→ 2-Phosphoglycerate →($-H_2O$)→ 2-Phosphoenolpyruvate →(1. ADP, –ATP; 2. tautomerization)→ Pyruvic acid

FIGURE 19.10

Metabolism of glyceraldehyde-3-phosphate to pyruvic acid.

riboflavin for coenzyme-A. The end result is a thioester linkage of CoA with an acetyl group, with carbon dioxide being expelled. The product is acetyl-CoA, a key compound in all metabolic processes (see "Chemistry at Work," p. 536).

$$\underset{\text{Pyruvic acid}}{CH_3{-}CO{-}COOH} \xrightarrow[\text{CoA-SH, NAD}^+]{\text{thiamine}} CO_2(g) + \underset{\text{Acetyl-CoA}}{CH_3CO{-}S{-}CoA}$$

So far, each glucose molecule has been degraded to two molecules of CO_2 and two molecules of acetyl-CoA and has produced a net of two units of ATP:

$$\underset{\text{Glucose}}{C_6H_{12}O_6} \xrightarrow{\text{(many steps)}} \underset{\text{Acetyl-CoA}}{2\,CH_3COSCoA} + 2\,CO_2 + 2\,ATP$$

This entire process, called the **Embden-Meyerhof pathway,** is an anaerobic process, for which oxygen is not required. Acetyl-CoA next enters the Krebs cycle.

In the absence of sufficient oxygen, as in prolonged exercise, the pyruvic acid anion is reduced temporarily by NADH (see Section 16.5.2) to lactate anion:

$$\underset{\text{Pyruvate anion}}{CH_3{-}C(=O){-}C(=O){-}O^-} + NADH + H^+ \rightleftharpoons \underset{\text{Lactate anion}}{CH_3{-}CH(OH){-}C(=O){-}O^-} + NAD^+$$

The lactate anion is stored in the muscle tissue and bloodstream until the oxygen supply is adequate. When oxygen again becomes available, the lactate is reoxidized to pyruvate. This storage of lactate is responsible for muscle fatigue.

The Krebs Cycle

The second phase of glucose catabolism involves the conversion of the acetyl-CoA into CO_2 and energy (ATP) (Figure 19.11). This phase is a "cycle" because the starting compound that reacts with acetyl-CoA is ultimately re-created, and two equivalents of CO_2 are evolved. It may appear as though the C_2 acetyl unit is directly converted to CO_2, but this is not the case, as we will see. The individual reactions involved in this interesting cycle have been described in earlier chapters and should be recognizable to you. Discovery of this cycle led to the 1953 Nobel Prize for physiology or medicine for Hans Krebs.

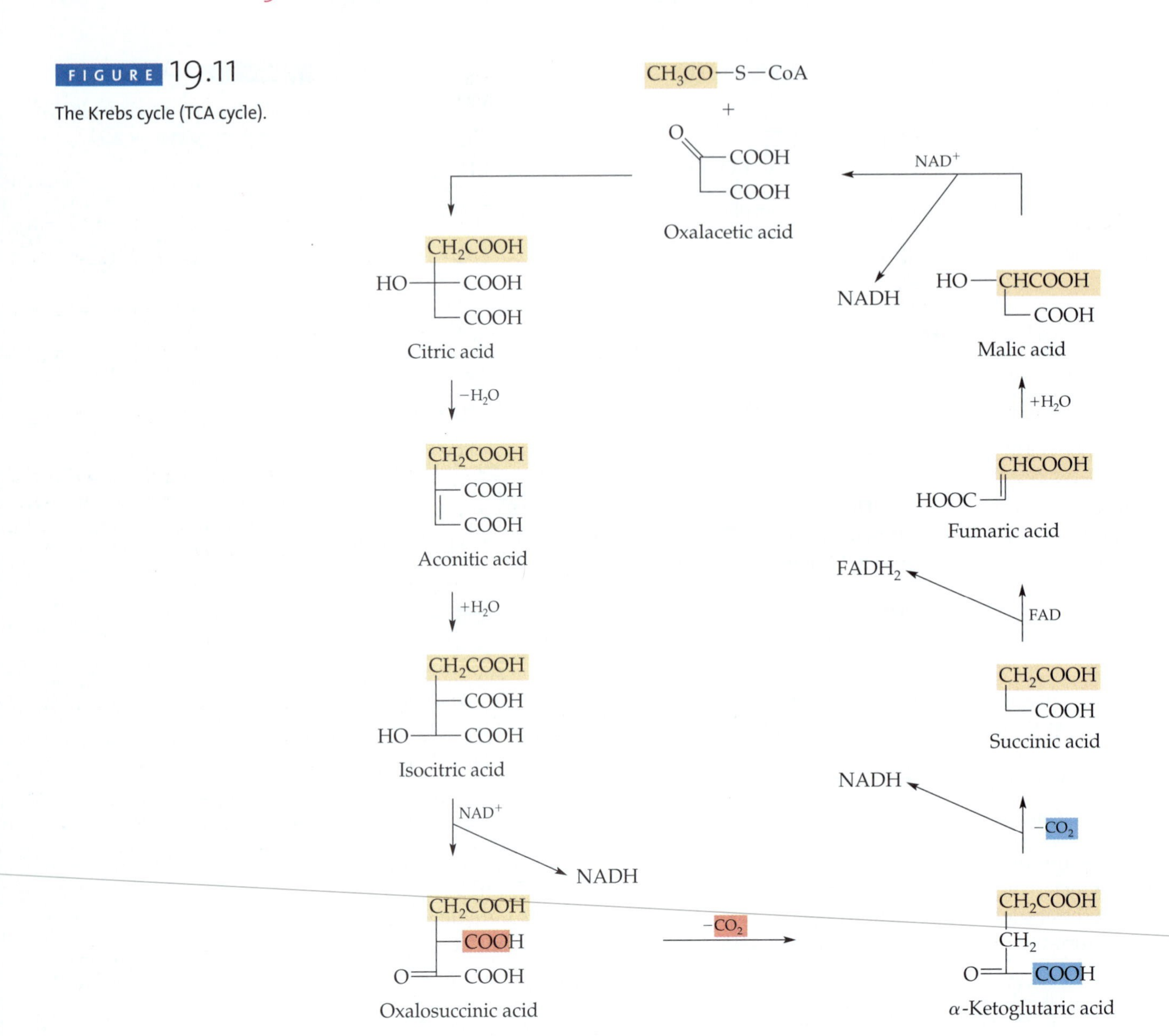

FIGURE 19.11

The Krebs cycle (TCA cycle).

Acetyl-CoA reacts with oxalacetic acid to form citric acid. In this reaction, the methyl group of acetyl-CoA loses a proton to form a stabilized carbanion (see Section 15.3.3), which adds across the carbonyl bond of oxalacetic acid (recall carbanion additions in Section 13.5). Dehydration of the citric acid, followed by rehydration with the opposite regiochemistry, yields isocitric acid (see Sections 6.4.2 and 6.6.1). Oxidation of isocitric acid by NAD^+ produces oxalosuccinic acid (a β-keto acid), which loses carbon dioxide (decarboxylates) readily to form α-ketoglutaric acid (decarboxylation was discussed in Section 14.3.3). Thiamine decarboxylates this α-keto acid to form succinic acid (analogous to pyruvic acid decarboxylating to acetic acid). Finally, succinic acid is converted to the starting oxalacetic acid by oxidation (dehydrogenation) to fumaric acid, hydration of the fumaric acid, and oxidation of the alcohol group to a ketone group. Figure 19.11 shows this entire cycle, with the two carbons of the entering acetyl group identifiable throughout.

The Krebs cycle involves the input of a C_2 unit (acetyl-CoA) and the output of two carbon atoms in two molecules of CO_2. In the process, four equivalents of reduced coenzymes (NADH and $FADH_2$) are produced. Through an extensive series of reactions, these coenzymes transfer their electrons and hydrogen to oxygen in the cell, reducing the oxygen to water and in the process producing 16 ATP molecules (equivalent to about 112 kcal/mol).

The final balanced equation for the catabolism of 1 molecule of glucose is the production of 6 molecules of carbon dioxide and 34 molecules of ATP:

$$\underset{\text{Glucose}}{C_6H_{12}O_6} \xrightarrow{\text{(overall)}} 6\,CO_2 + 34\text{ ATP}$$

It can be seen that the bulk of the energy (32 of 34 ATPs) is produced in the Krebs cycle and that oxygen is essential for the ultimate release of that energy. Thus, any event that restricts the supply of oxygen to the cells results in a rapid decrease in energy production (chemical energy and thermal energy) needed for body processes. For this reason, traumatic events such as carbon monoxide poisoning (which prevents hemoglobin from transporting oxygen) and heart attack (which diminishes the circulation of blood carrying oxygen to the cells) quickly result in tissue and organ damage, which frequently leads to death.

19.2.2 Catabolism of Fats

Fats are the most energy-rich nutrient per unit of weight. For example, 1 gram of palmitic acid (the C_{16} fatty acid) produces about 9.4 kcal when combusted with oxygen, but 1 gram of glucose produces only about 3.7 kcal. However, fat catabolism is a much slower process than carbohydrate catabolism. For this reason, people who need large amounts of energy over a short period of time, like marathon runners, prepare themselves by eating large quantities of carbohydrates beforehand; they "carb up" and store large amounts of glycogen in their bodies.

The initial step in the catabolism of fats is the hydrolysis of the triglyceride into glycerol and fatty acids (see Section 15.3.5). This is accomplished by enzymes in the digestive tract. The fatty acids then enter the bloodstream. If there is no immediate demand for energy, the fatty acids are reconverted to fats and may be deposited as adipose tissue. When there is a demand for energy, lipase enzymes in the adipose tissue hydrolyze the fats back to fatty acids, which are then transported to sites of catabolism. The fatty acid is esterified by coenzyme-A, which serves as a carrying vehicle throughout the catabolic processes, requiring a unit of ATP to form the thioester, acyl-CoA.

$$\underset{\text{Fat}}{\begin{array}{l} -O-\overset{O}{\overset{\|}{C}}-R \\ -O-\overset{O}{\overset{\|}{C}}-R \\ -O-\overset{O}{\overset{\|}{C}}-R \end{array}} \xrightarrow{\text{digestive enzymes}} \underset{\text{Fatty acid}}{R-\overset{O}{\overset{\|}{C}}-OH} \xrightarrow[\text{ATP}]{\text{CoA–SH}} \underset{\text{Acyl-CoA}}{R-\overset{O}{\overset{\|}{C}}-S-CoA}$$

The acyl-CoA is subject to a series of enzymatic conversions that result in the fatty acid chain being shortened by two carbons at a time; the cleavage product at each stage is acetyl-CoA. The first step in the cleavage is dehydrogenation by a dehydrogenase enzyme, with flavin adenine dinucleotide (FAD) as the coenzyme. This produces an α,β-unsaturated thioester. Addition of water to the carbon-carbon double bond of this thioester is followed by oxidation of the alcohol by NAD^+ to a β-keto thioester. Finally, a reverse Claisen condensation (see Section 15.3.3) takes place, with the thiolate anion of CoA acting as the nucleophile in acyl substitution by attacking the β-carbonyl group to expel the carbanion of acetyl-CoA. The products are acetyl-CoA and a shortened acyl-CoA. The overall sequence is illustrated here using palmitic acid as the fatty acid:

$$\underset{\text{Palmityl-CoA}}{CH_3(CH_2)_{12}CH_2CH_2CO{-}S{-}CoA} \xrightarrow[(-H_2)]{FAD} CH_3(CH_2)_{12}CH{=}CHCO{-}S{-}CoA$$

$$\xrightarrow{+H_2O} CH_3(CH_2)_{12}CHOHCH_2CO{-}S{-}CoA \xrightarrow[(-H_2)]{NAD^+} CH_3(CH_2)_{12}COCH_2CO{-}S{-}CoA$$

$$\xrightarrow{CoA{-}S^-} \underset{\text{Myristyl-CoA}}{CH_3(CH_2)_{12}CO{-}S{-}CoA} + \underset{\text{Acetyl-CoA}}{CH_3CO{-}S{-}CoA}$$

This process results in the fatty acid chain being degraded two carbons at a time. In the case of palmitic acid, the overall result is eight units of acetyl-CoA. All fatty acids have an even number of carbon atoms, thus only acetyl-CoA is produced as the final product of this process. The acetyl-CoA is catabolized to CO_2 through the Krebs cycle (described earlier).

In fatty acid catabolism, ATP is not produced directly, but results from the eventual oxidation of $FADH_2$ and NADH by oxygen through a long electron-transfer process. The 16 carbons of palmitic acid eventually produce about 130 net units of ATP. Put another way, fatty acids produce a net of about 8 ATPs per carbon, while glucose produces about 6 ATPs per carbon. Thus, fats are a more productive source of energy than carbohydrates.

19.2.3 Catabolism of Proteins

The catabolism of proteins is not a major energy source, but it is the *only* source of nitrogen for the body. Nitrogen is required for the biosynthesis of α-amino acids, proteins, nucleic acids, porphyrins, and many other essential nitrogen-containing compounds.

The catabolism of proteins begins when they enter the digestive system, where proteolytic enzymes hydrolyze them to α-amino acids. Once absorbed into the bloodstream, the α-amino acids enter the body's "amino acid pool" (Figure 19.12) and can have several different fates:

- They can be biosynthesized directly into new proteins in the cell (see Section 19.1.4). This is especially important for the eight essential amino acids, which the body cannot synthesize and which must be supplied through the diet (see Section 16.1.1).
- They can be incorporated into other essential nitrogenous compounds, such as nucleic acids and hemoglobin.
- They can be deaminated to α-keto acids, which are then catabolized by processes similar to fat catabolism and eventually enter the Krebs cycle. In this case, the nitrogen is stored for later reuse in biosynthesis.

The *deamination* of an α-amino acid involves an enzyme whose coenzyme is derived from pyridoxal phosphate; pyridoxal is vitamin B_6 (Figure 19.13). The enzyme is

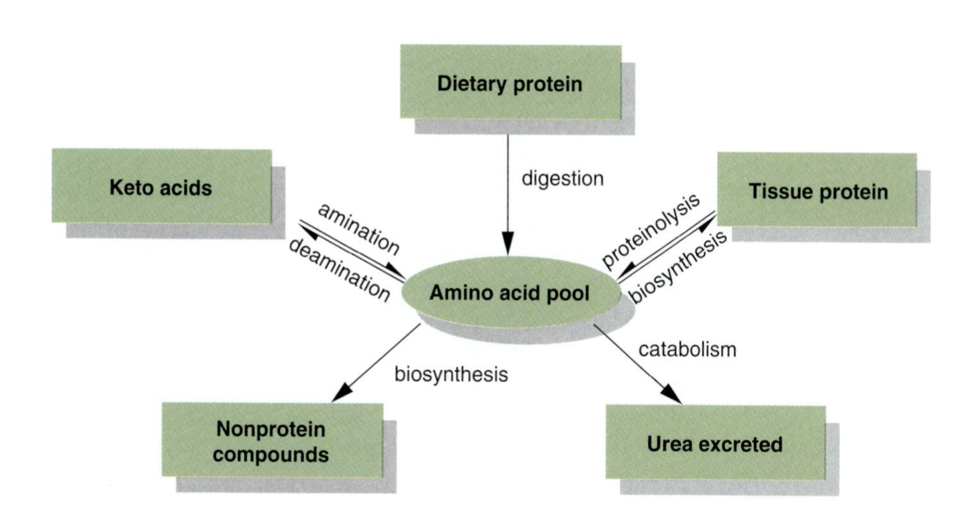

FIGURE 19.12 The α-amino acid pool.

bound to its coenzyme by an imine linkage (see Section 13.4.1). Reaction of the enzyme with an α-amino acid involves imine exchange, in which the amino group of the amino acid attacks the carbon end of the imine. The new imine, using the α-amino acid amino group, undergoes tautomerization to yet another imine by nucleophilic removal of the proton from the α-carbon, an electron shift toward nitrogen, and proton attachment at the benzylic carbon (a tautomerization process; see Section 13.2.3). Finally,

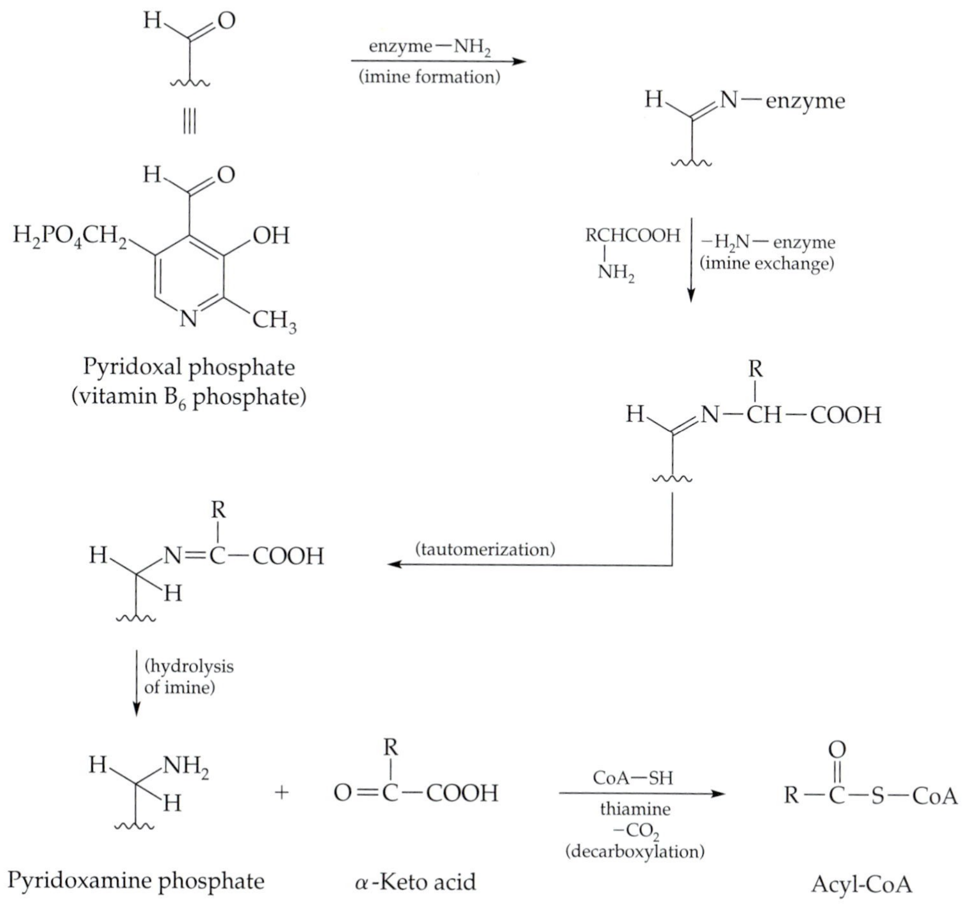

H–C(=O)– ≡ pyridoxal phosphate ($H_2PO_4CH_2$, OH, CH_3, N ring)

Pyridoxal phosphate (vitamin B_6 phosphate)

enzyme—NH_2 (imine formation) → H–C(=N—enzyme)–

RCH(NH_2)COOH, $-H_2N$—enzyme (imine exchange) → H–C(=N—CH(R)—COOH)–

(tautomerization) → H–CH(–N=C(R)—COOH)–

(hydrolysis of imine) → H–CH(NH_2)– + O=C(R)—COOH

Pyridoxamine phosphate + α-Keto acid

CoA—SH, thiamine, $-CO_2$ (decarboxylation) → R—C(=O)—S—CoA

Acyl-CoA

FIGURE 19.13 Deamination of α-amino acids (a transamination sequence).

hydrolysis of the new imine bond produces the α-keto acid and pyridoxamine phosphate. The α-keto acid then undergoes the fatty acid catabolism by loss of CO_2 and the formation of acyl-CoA. This overall movement of an amino group from an α-amino acid to pyridoxal is called **transamination.**

The entire reaction sequence shown in Figure 19.13 is reversible. It is this process that makes 12 α-amino acids *nonessential* in the diet; the body can synthesize them. The pyridoxamine serves as a nitrogen storage compound, extracting nitrogen from α-amino acids. This nitrogen is subsequently recycled as it is transferred to α-keto acids to form α-amino acids needed in protein synthesis (another transamination).

The body stores all of the nitrogen it needs. If excess protein is ingested, the excess nitrogen is excreted in the form of urea ($H_2N—CO—NH_2$). The deamination process occurs in the liver. It involves a series of reactions called the *urea cycle* and consumes energy in the form of ATP. If liver function is impaired, the urea cycle does not function, and toxic ammonia is formed instead.

19.2.4 Biosynthesis (Anabolism)

The various processes by which essential chemicals in biological systems are synthesized are well known. The details of these processes are beyond the scope of this book, but we will briefly consider the overall anabolic process.

Nitrogen fixation by leguminous plants is a critical link in the entire food chain because it is the major process by which elemental nitrogen enters the food chain (Figure 19.14). The nitrogen incorporated in such plants, in the form of plant protein, is digested by animals who use the resulting α-amino acids and nitrogen to form animal protein. For humans, animal protein is the major source of nitrogen and the essential α-amino acids. Some plant protein, such as from soybeans, can also serve as a source of nitrogen and essential α-amino acids.

The process of photosynthesis to produce carbohydrates is well understood. It occurs in plants and involves the capture of energy from sunlight and carbon dioxide to produce carbohydrates and give off oxygen. Plant carbohydrates are a major source of human energy, as shown in Section 19.2.1.

The third major food source for humans is fat, obtained from both animals and plants. Plant fats are biosynthesized from acetyl-CoA by the reversal of the fat catabolism process

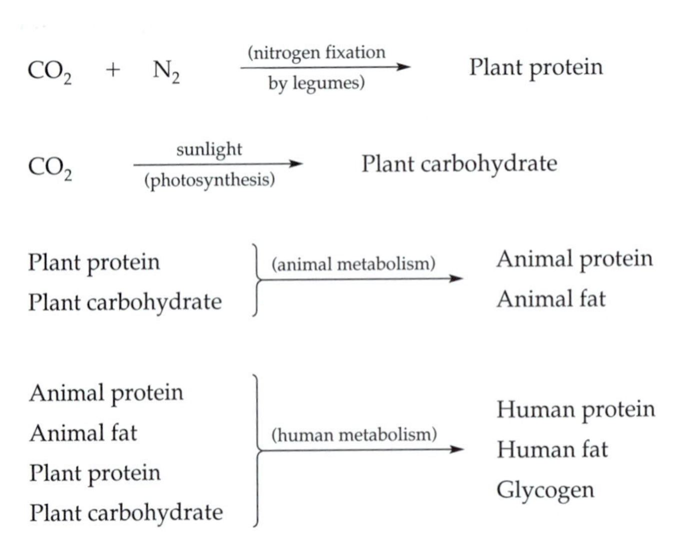

FIGURE 19.14 Sources of proteins, carbohydrates, and fats.

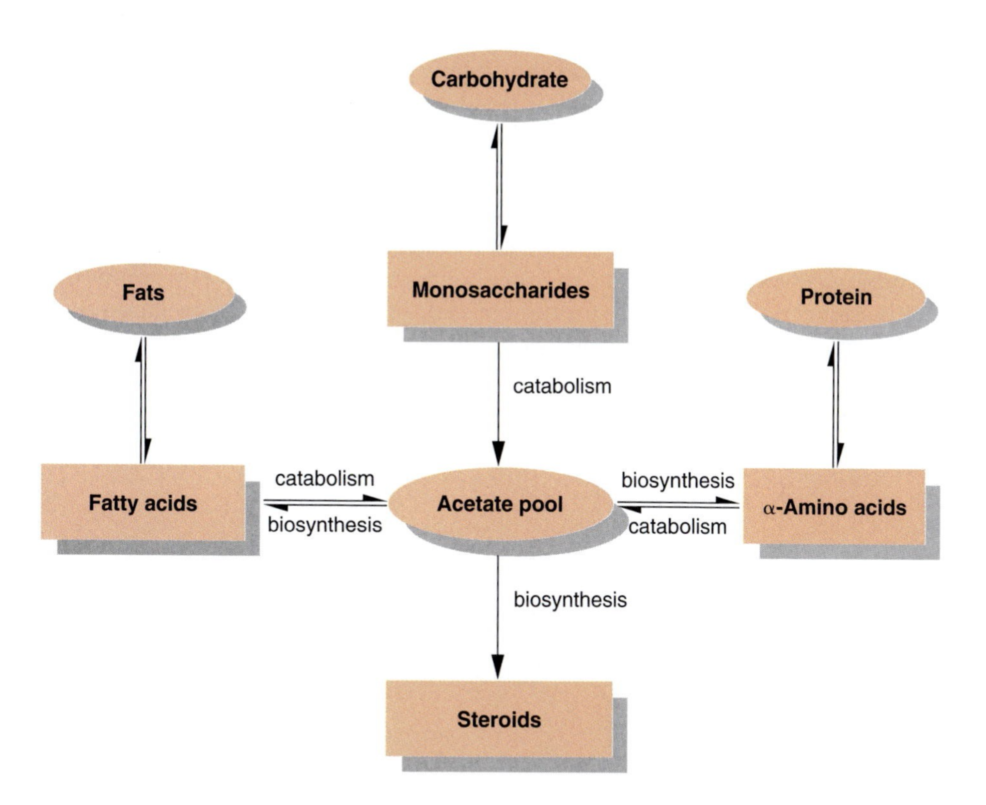

FIGURE 19.15

The "acetate pool": input and output.

(see Section 19.2.2). Animal fat is biosynthesized by the same process. The "acetate pool," essentially acetyl-CoA, is created in humans by catabolism of carbohydrates, fats, and proteins (Figure 19.15). This pool serves as a source of biosynthesized fatty acids and α-amino acids, the starting point for fat and protein biosynthesis. The acetate pool is also the starting point for the synthesis of essential steroids, such as cholesterol, sex hormones, and adrenocortical steroids (see Section 8.4.2).

19.3 The Chemical Basis of Therapies

With all that is known about the chemistry of biological function, it has become possible to devise treatments for many of the diseases and abnormalities that ravage the world's human population. In most instances, these treatments include the administration of a medicinal (drug or pharmaceutical) that has been designed not only to have the beneficial effect desired, but also not to have detrimental side effects. The balancing of beneficial and detrimental effects is frequently a difficult choice. Most medicinals are the result of organic synthesis in the laboratories of pharmaceutical companies. Usually, this means a totally synthetic process (a *total synthesis*); occasionally, the synthesis starts with a natural product, which is then converted to the final product by subsequent synthetic steps (a *semisynthesis*).

Human disease states generally arise from one of four circumstances: (1) the shortage of an essential chemical; (2) the overproduction of a chemical; (3) the malignant growth of cells; or (4) invasion by an external organism. This section looks briefly at examples of each type of disease state and the chemical therapies employed. Recall that birth defects are not true disease states, but rather abnormalities that more and more frequently are being treated by gene therapy (see Section 19.1.6).

19.3.1 Treatment of Chemical Shortage

Many of the diseases that have existed throughout human history are caused by the underavailability of chemicals essential to the body. Such diseases are now relatively rare in developed societies because their symptoms are recognized early, preventive steps are often taken, and nutritional practices have improved immensely. Diseases readily treatable by simple additions to the diet include scurvy, caused by lack of vitamin C (ascorbic acid, an important antioxidant); beriberi, caused by lack of vitamin B_1 (thiamine is essential as a coenzyme for catabolism); and goiter, caused by lack of iodide ion (to form the hormone thyroxine from the α-amino acid tyrosine).

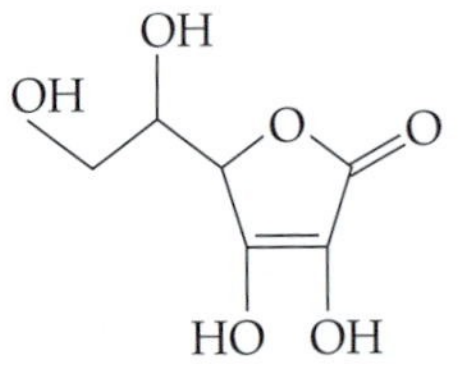

Ascorbic acid
(absence causes scurvy)

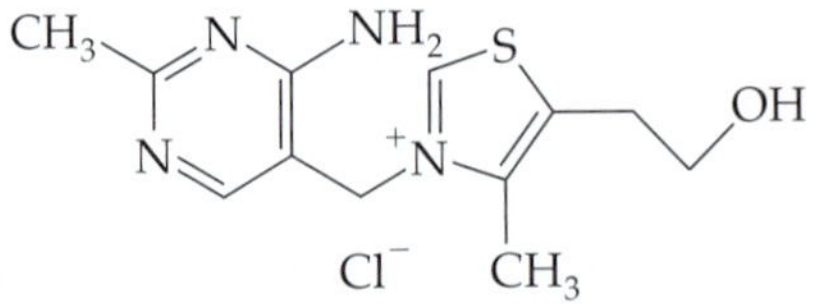

Thiamine
(absence causes berberi)

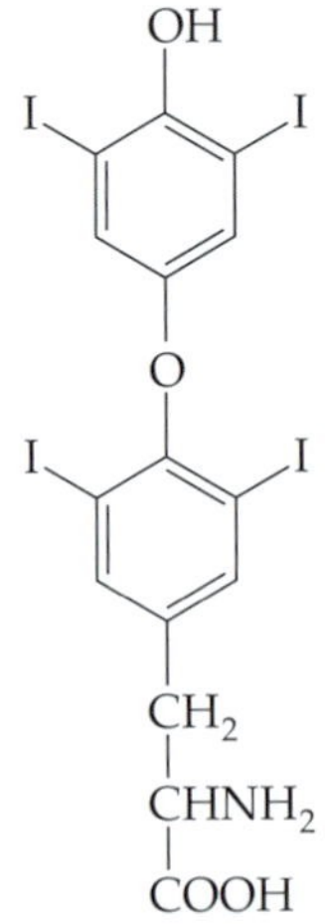

Thyroxine
(absence causes goiter)

Diabetes is caused by the inadequate production of the hormone insulin in the pancreas. Insulin is essential for the conversion of glycogen to glucose (in order to maintain the correct level of glucose in the blood) and for catabolism. Insulin is a protein with 51 amino acids (see Figure 16.3) and is available through a number of sources. However, it usually cannot be taken orally because, as a protein, it will be hydrolyzed by the proteolytic enzymes in the digestive system and thus destroyed before it can be absorbed. Therefore, it normally is administered by subcutaneous injection.

19.3.2 Treatment of Chemical Overproduction

The α-amino acid histidine undergoes enzymatic decarboxylation to form histamine, which, under normal circumstances, is degraded by the enzyme histaminase.

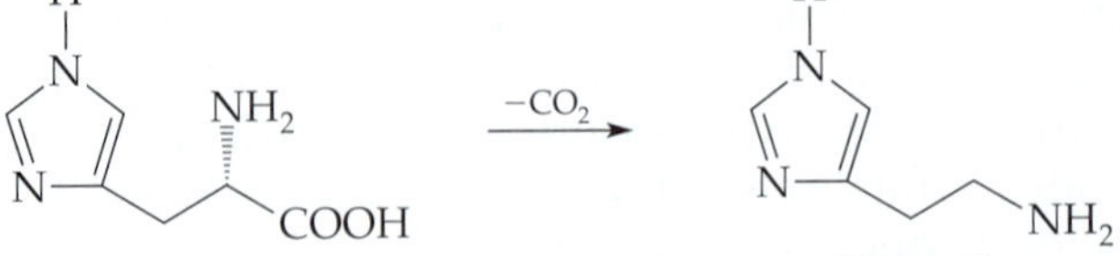

L-Histidine $\xrightarrow{-CO_2}$ Histamine

However, as a consequence of inflammation or allergic reactions, the concentration of histamine may rise too high, causing unpleasant effects such as vasodilation. Antihistamines are compounds that block the effect of histamine by complexing at the vasodilation receptor site, thereby excluding histamine. Examples of antihistamines are the products marketed as Benadryl and Dimetapp.

$(C_6H_5)_2CHOCH_2CH_2N(CH_3)_2$

Benadryl

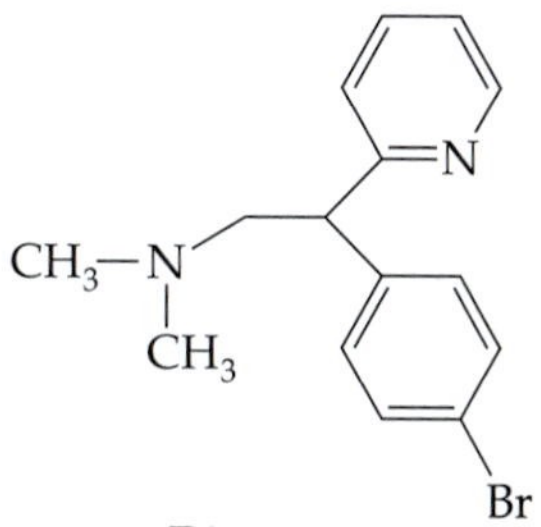

Dimetapp

The overproduction of stomach acid (and the enzyme pepsin) is also triggered by excess histamine, but by its binding at a different receptor site. For many years, the common treatment for stomach acid was to add a base (an antacid) to lower the acidity by neutralizing it. There are many such antacids on the market, but they treat only the symptoms of the problem. More recent research has revealed the nature of the receptor site for histamine that causes the release of excess stomach acid and excess pepsin. Medicinals have been designed to fit and occupy (block) the receptor site and prevent histamine from associating with it. These synthetic blockers include cimetidine (marketed as Tagamet), and ranitidine (marketed as Zantac).

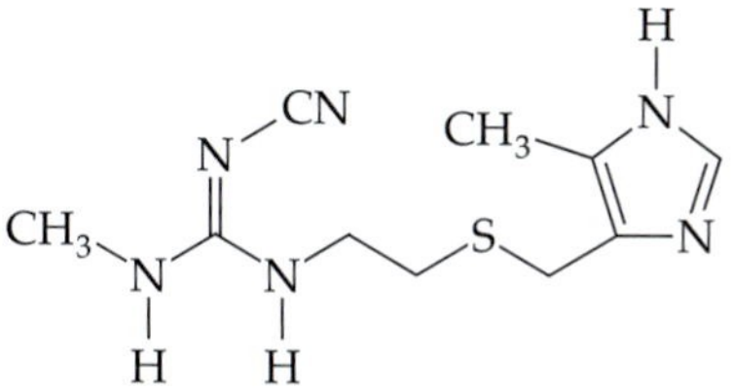

Cimetidine
(Tagamet)

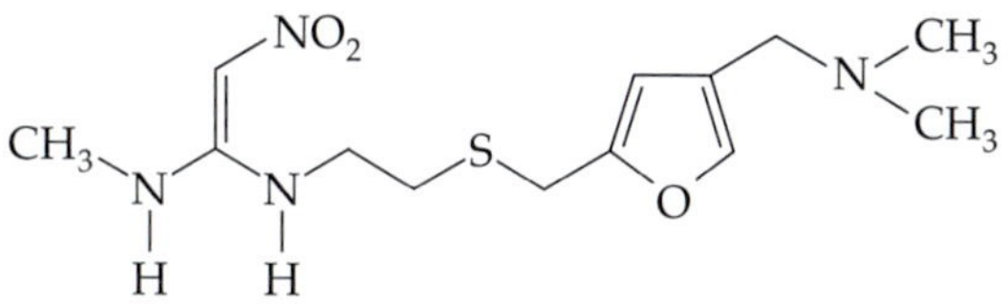

Ranitidine
(Zantac)

19.3.3 Treatment of Malignant Growth

Cancer is the malignant (unregulated) growth of cells. Its natural causes are as yet unknown, but many chemicals are known to be cancer-causing (carcinogenic)—including simple compounds such as benzene and complex compounds such as benzo[a]pyrene, found in coal tar and cigarette smoke. In addition, high-energy radiation, such as gamma rays, and excess ultraviolet radiation are also known to cause cancer. Because cell division involves replication of nucleic acids, much research has been targeted at finding chemicals that will stop rapid DNA replication. However, if the chemical is too effective, it will stop all cell division (both normal and abnormal), and that will kill the patient. Therefore, a delicate balance of effectiveness is required for an anticancer medicinal to be approved for use on humans.

One approach to cancer treatment has been the synthesis of **antimetabolites,** compounds that are similar to those needed for biosynthesis but different enough to block the process. In essence, the secret is to find a chemical that will "fool" an enzyme into reacting with it by mistake, with the result that further catalysis by that enzyme is precluded.

The drug 5-fluorouracil is an effective cancer antimetabolite that disrupts the incorporation of pyrimidine bases into RNA (recall that uracil is one of the four RNA bases). After being administered, the antimetabolite is converted to its nucleotide, which reacts with an enzyme that normally converts a uracil molecule to a thymine molecule by methylating carbon 5. However, the antimetabolite has that position blocked by a fluorine atom, and so it remains complexed with the enzyme, precluding production of the thymine nucleotide, an essential component in nucleic acid biosynthesis. Therefore, cell division is inhibited.

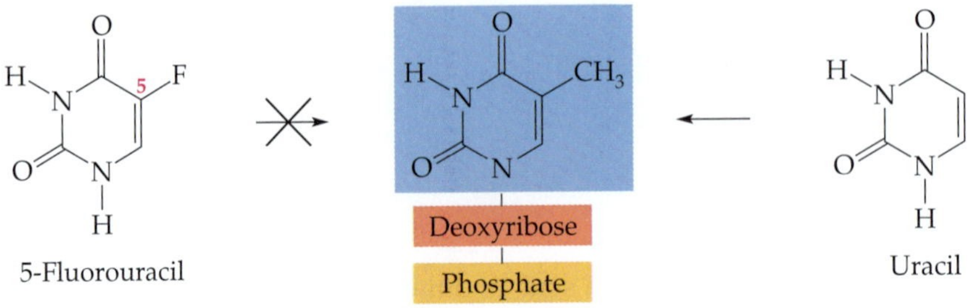

5-Fluorouracil

Deoxyribose

Phosphate

Thymine nucleotide

Uracil

Arabinosylcytosine is another antimetabolite, effective with some leukemias and lymphomas. It works because the ribose unit of cytosine nucleoside has been replaced by arabinose, an epimer at carbon 2. Incorporation of this antimetabolite inhibits DNA polymerase.

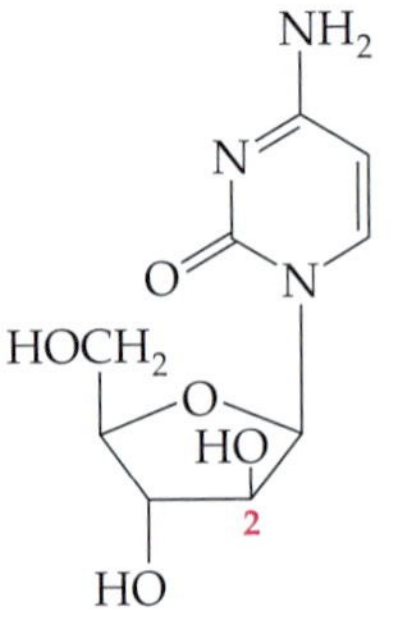

Arabinofuranosylcytosine

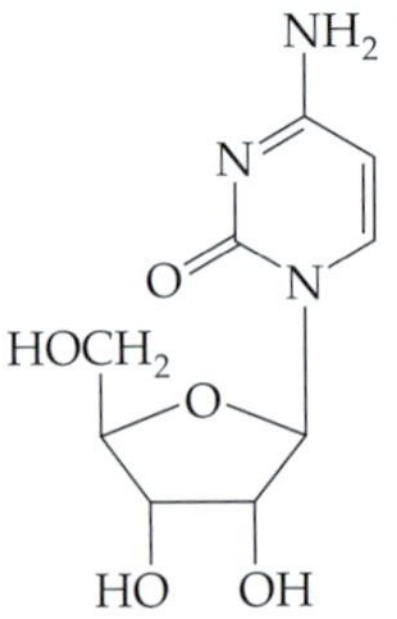

Cytosine nucleoside

The antimetabolite methotrexate is effective in treating some uterine cancers and leukemias by blocking the synthesis of purine bases. Folic acid is converted into tetrahydrofolic acid, a cofactor for an enzyme that brings about the transfer of single carbons for the biosynthesis of purine bases. Methotrexate is similar enough in structure to folic acid that it is incorporated in the redox enzyme. However, it cannot be reduced by that enzyme and remains attached to it. Thus, it blocks the functioning of that enzyme and ultimately inhibits nucleic acid synthesis and cell division.

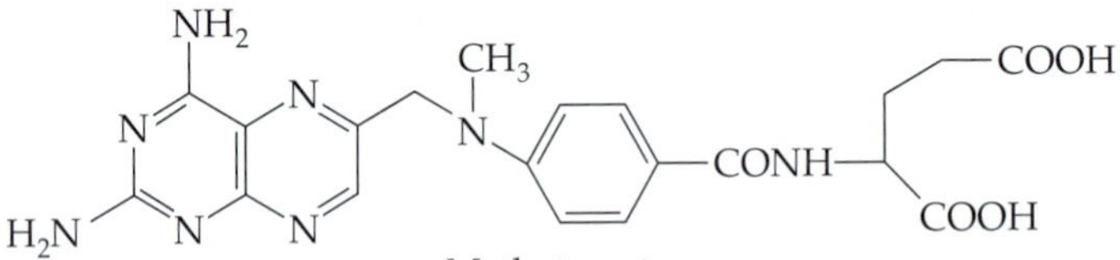

Methotrexate

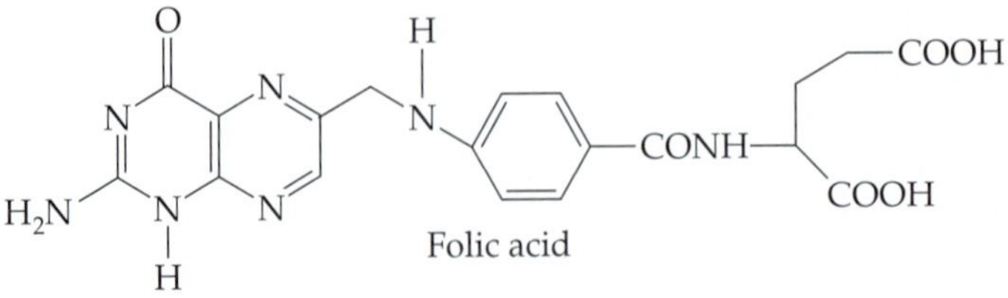

Folic acid

A relatively new class of compounds called *ene-diynes* have very promising antitumor activity (see "Chemistry at Work," p. 236). The common structural feature of these ene-diynes is the conjugated double and triple bond system. The chemical details of its action involve what is called a *Bergman cyclization* to form a new benzene ring that is also a diradical. (Recall the high reactivity of radicals; see Chapter 10.)

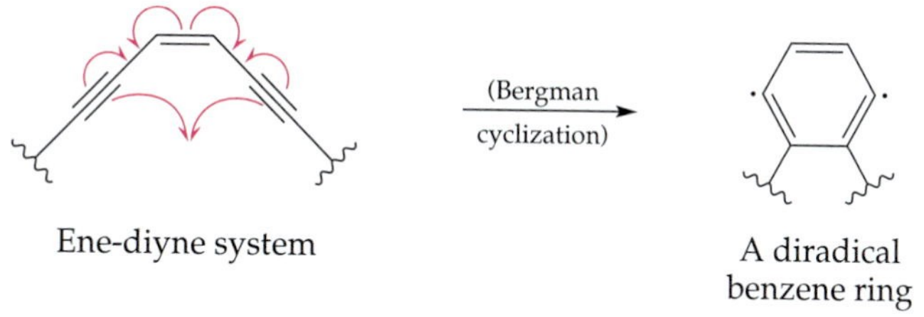

(Bergman cyclization)

Ene-diyne system

A diradical benzene ring

This cyclization process was discovered by Robert Bergman long before the discovery of the biologically active ene-diynes. It is the resonance stabilization of the product benzenoid ring which drives the cyclization.

Three examples of ene-diynes are dynemicin A, calicheamicin, and neocarzinostatin. They were discovered in minute quantities as fermentation products of bacteria and as products of a mold taken from soil. They all have the property of reacting with the two strands of the DNA double helix and "cutting" them. Of course, this precludes further DNA replication and cell division. The natural ene-diynes are not sufficiently selective to single out cancer cells for DNA cutting, so research is continuing using semisynthesis to find related compounds that will be more selective. In an outstanding feat, dynemicin A was totally synthesized, and so derivative compounds are now available by synthesis.

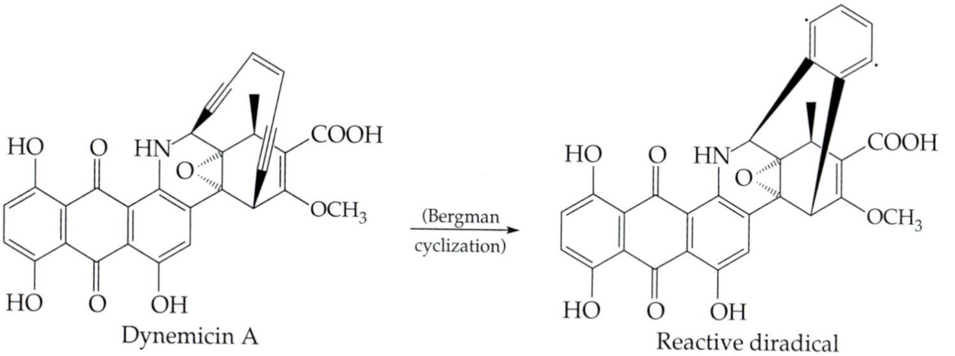

HO O HN COOH O OCH_3 HO O OH

Dynemicin A

(Bergman cyclization)

HO O HN COOH O OCH_3 HO O OH

Reactive diradical

The diradical resulting from the Bergman cyclization is very reactive and is the actual DNA cleaving agent, acting at specific points along the DNA chain to abstract a hydrogen atom from carbon 5′ of deoxyribose. The resulting carbon radical is converted to an aldehyde, thereby dissociating carbon 5′ from the nucleic acid backbone. Thus, the DNA strand is "cut," and further replication of the normal DNA is impossible.

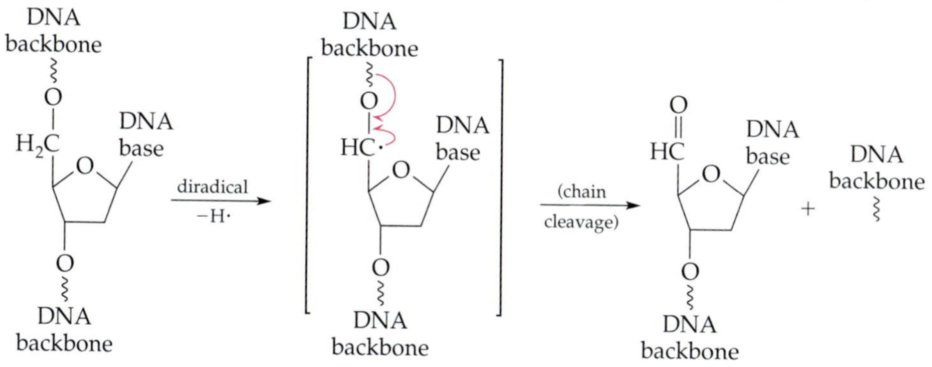

DNA backbone — O — H_2C — DNA base — O — DNA backbone

diradical, $-H\cdot$

[DNA backbone — O — $HC\cdot$ — DNA base — O — DNA backbone]

(chain cleavage)

O = HC — DNA base — O — DNA backbone + DNA backbone

Another recently discovered complex chemical, paclitaxel (formerly called taxol), has been found to have potentially beneficial use in treating ovarian, lung, and breast cancers. This chemical was discovered in the bark of Pacific yew trees in very small quantities, but harvesting the bark killed the rare trees. A total synthesis was eventually accomplished, but more recently, a related compound has been isolated from the renewable needles of another *Taxus* species and from other sources as well. This compound has been converted in the laboratory to paclitaxel and related compounds in sufficient quantities for testing (a semisynthesis). The biosynthetic pathway to paclitaxel is also becoming clear, providing the potential for producing larger quantities via a cell culture process.

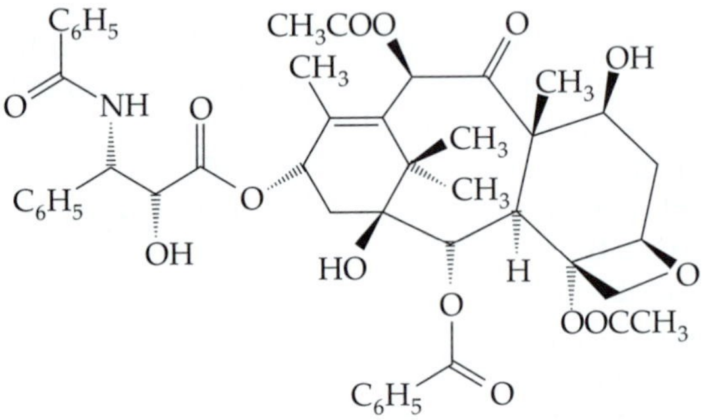

Paclitaxel (taxol)

19.3.4 Treatment of Infections

The human body is subject to infection from two kinds of foreign organisms: bacteria and viruses. The immune system, when operating normally, is capable of successfully fighting such invaders through the synthesis of appropriate antibodies. In some instances, however, external treatments are also needed.

Antibiotics

Bacterial infections are usually treated with antibiotics. Fortunately, there are enough differences between the metabolic processes of bacteria and those of mammalian cells to target bacterial cells selectively. Of several different kinds of antibiotics and many different modes of action, we will consider only a couple of examples here.

Among the earliest synthetic antibiotics were the sulfa drugs (see "Chemistry at Work," p. 432), all based on the structure of sulfanilamide (p-$H_2N—C_6H_4—SO_2NH_2$); sulfathiazole is one example. Sulfa drugs are antimetabolites because they stop the bacterial synthesis of folic acid by "fooling" the enzyme, which complexes to the sulfanilamide because of its similarity to *p*-aminobenzoic acid amide. (Can you spot the PABA unit in folic acid, shown on page 694?) Since folic acid is essential for bacterial growth, the bacterium ceases to multiply. Sulfa drugs cannot inhibit synthesis of folic acid in the human body because folic acid is not synthesized there. Instead, it is obtained through the diet as a vitamin.

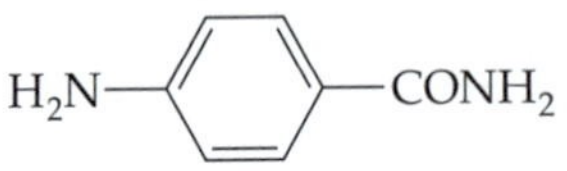

p-Aminobenzoic acid amide 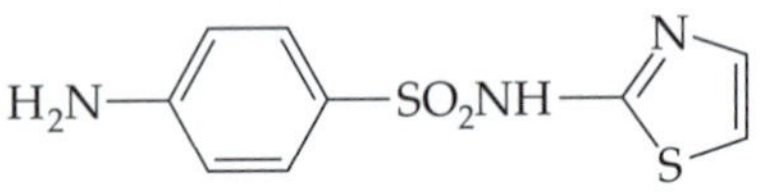   Sulfathiazole

The classic antibiotic is penicillin, discovered serendipitously in a fungus by Alexander Fleming in 1928 ("Chemistry at Work," p. 568). It was 13 years before penicillin G was introduced to medical practice, but once in use, its success was amazing, especially during World War II. Subsequently, a number of semisynthetic derivatives (such as ampicillin and amoxicillin) were marketed, and these are especially useful for those allergic to penicillin G. What distinguishes one penicillin from another is a variation in the R group. A newer family of antibiotics are the cephalosporins, obtained from another fungus. Again, a number of semisynthetic structural variants (such as cephalothin and cefaclor) are on the market. The penicillins and cephalosporins, both being β-lactams, act in the same manner, inhibiting the formation of bacterial cell walls.

Penicillin G
($R = C_6H_5CH_2—$)

Cephalothin
($R = C_4H_3SCH_2—$;
$R' = —CH_2OOCCH_3$)

Antiviral Agents

Viral infections (including the common cold, herpes, influenza, and HIV) are caused by viruses, which infect the host by entry into its cells. Curing viral infections is much more difficult than curing bacterial infections because of the difficulty in selectively targeting the virus without damaging the host.

Viruses consist of an outer core of protein surrounding an inner core of nucleic acid. The nucleic acid carries the information necessary for the virus to replicate. There are two kinds of viruses: those whose inner core is DNA (such as herpes viruses) and those whose inner core is RNA (for example, the flu virus and the human immunodeficiency virus, HIV, responsible for AIDS). A virus attaches itself to the host cell, enters the cell, sheds its protein coat, and finally enters the nucleus of the cell. In that environment, the virus uses the available bases, sugars, and phosphate to replicate itself, just like the host cell DNA does. The replicated viruses then leave the nucleus and the cell to infect other cells. In effect, the virus commandeers the host cell's nucleic acid replication capability to multiply itself. Antiviral agents have been designed to attack the viral replication process at several points.

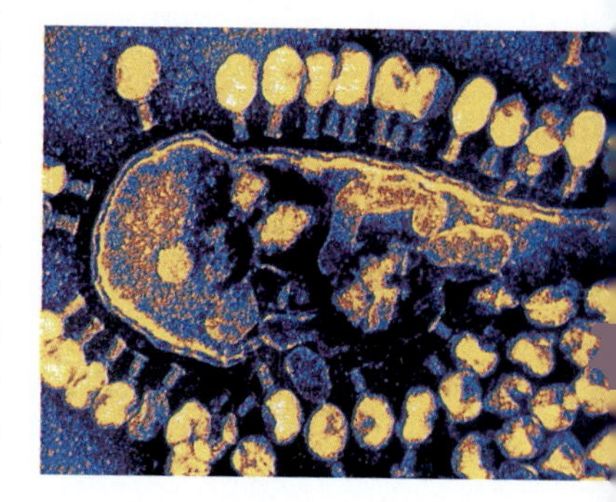

Here large numbers of viruses (yellow) have attached themselves to a bacterial cell. An effective antiviral agent would keep them from replicating within the cell.

The herpes virus can be treated using acyclovir (ACV), which is structurally related to 2-deoxyguanosine, an essential ingredient for DNA synthesis. The enzyme DNA polymerase forms an enzyme-substrate complex with ACV, but there is no 3′-hydroxyl group to be phosphorylated for chain growth. (Recall that the DNA backbone includes phosphate groups attached to the 5′- and 3′-hydroxyl groups of ribose.) Thus, chain growth is terminated and the enzyme remains blocked, thereby terminating viral replication.

Acyclovir

2′-Deoxyguanosine

HIV causes AIDS (acquired immune deficiency syndrome). This virus operates by destroying the body's immune system, thus making the body susceptible to a wide variety of diseases, including those normally held in check by a functioning immune system. HIV is classified as a *retrovirus* because its nucleic acid is RNA rather than DNA. Thus, when it enters a host immune cell (called a *T4 lymphocyte*), it migrates to the nucleus, where an enzyme called *reverse transcriptase* decodes the RNA and synthesizes DNA. (Recall the transcription process from DNA to RNA; see Section 19.1.3.) The viral DNA seems to remain inactive for varying periods of time in the cell nucleus, but then, for unknown reasons, suddenly becomes active. This is why a person can be HIV-positive but not yet have AIDS. When replication is activated, the viral DNA seems to take over the nucleic acid synthetic machinery of the cell. Thus, it is able, through normal replication and transcription processes, to produce large amounts of viral RNA in a very short period of time and infect most T4 cells. The T4 cell machinery that controls the immune system is virtually shut down, making the patient susceptible to many different diseases.

The entire HIV process is not yet fully understood. One of the problems with treating for HIV is the great speed with which it replicates. This means that HIV can readily mutate and thereby acquire resistance to drugs. Thus, constant drug therapy is needed to keep the viral concentration as low as possible.

The first major treatment for AIDS was a series of drugs known as *RTIs—reverse transcriptase inhibitors*. The first of these was the drug known as zidovudine or azidothymidine (AZT). AZT is an analog of 2′-deoxythymidine, a nucleoside necessary for the production of DNA in human cells. The similar structure of AZT fools the enzyme reverse transcriptase into forming an enzyme-substrate complex. However, the absence of a 3′-hydroxyl group prevents the growth of a nucleic acid chain and blocks the functioning of the enzyme. Fortunately, AZT binds more strongly to reverse transcriptase than to human DNA polymerase. Other RTIs are often administered in conjunction with AZT. These include ddI (didanosine: 2′,3′-dideoxyinosine); ddC (zalcitabine: 2′,3′-dideoxycytidine); 3TC (lamivudine, or 2′,3′-dideoxy-3′- thiacytidine); and d4T (2′,3′-didehydro-3′deoxythymidine, tradenamed Zerit). Note the different DNA bases incorporated in these drugs (the T in AZT and d4T represents the base thymine and the C in ddC represents the base cytosine) as well as the relatively slight variations in the deoxyribose portion of the structures:

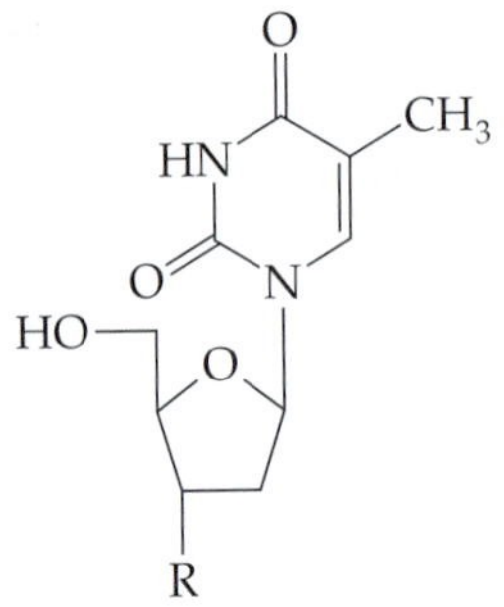

AZT, R = N_3
(azidothymidine, zidovudine)
2′-Deoxythymidine, R = OH

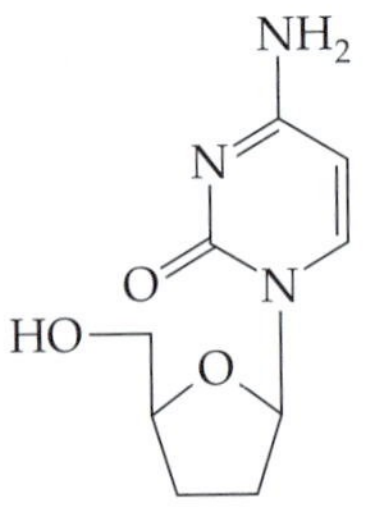

ddC
(zalcitabine)

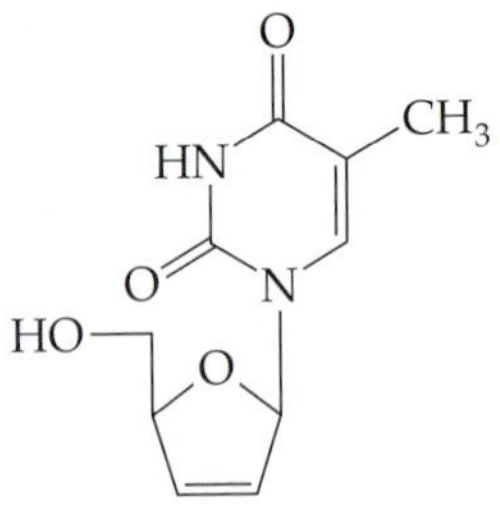

d4T
(Zerit, stavudine)

A newer direction in HIV treatment involves the use of *protease inhibitors*. Inhibition of the essential HIV enzyme protease has been attained with several candidate drugs, two of which are indinavir and ritonavir. These drugs affect the protein portion of the virus, not the nucleic acid portion. They inhibit the ability of the virus to organize its proteins prior to carrying out additional cell infections. Indinavir is a totally synthetic drug that is manufactured in multi-ton quantities as a single enantiomer with five chiral centers:

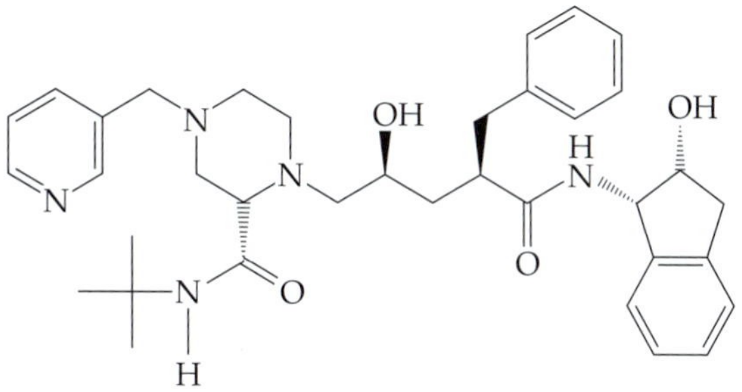

Indinavir

Protease inhibitors have been shown to lead to a dramatic lowering of the concentration of HIV in the bloodstream.

Chapter Summary

The nucleic acids, deoxyribonucleic acid (DNA) and ribonucleic acid (RNA), are polymers of nucleotides. A **nucleotide** consists of three compounds joined together, a heterocyclic base, a C_5 sugar, and a phosphate group. DNA exists as a **double helix,** a double-stranded α-helix held together by hydrogen bonding between complementary structured **base pairs.** RNA exists as a single strand of polynucleotide.

DNA is **replicated** by becoming gradually uncoiled; complementary nucleotides are polymerized along its exposed surfaces. Thus, one double helix becomes two. Protein synthesis is accomplished by a DNA helix becoming partially uncoiled, exposing a segment of the informational strand and the complementary template strand. **Messenger RNA (mRNA)** is formed along the template strand (**transcription**). The mRNA migrates from the nucleus to the cytoplasm of the cell, where small nucleotides called **transfer RNAs (tRNAs)** arrange themselves along the mRNA (**translation**). Each tRNA unit carries a specific α-amino acid, and once held in place along the mRNA strand, the α-amino acids are joined together to form a protein.

A **gene** is a section of DNA in a chromosome that carries the information necessary to accomplish synthesis of a specific protein. The genetic code is the sequence of triplets (or triads) of nucleic acid bases in DNA that gives rise to the incorporation of specific α-amino acids in a particular order in proteins. A triad sequence in mRNA is called a **codon;** and a triad sequence in tRNA is called an **anticodon.**

The body relies on **metabolism** for energy and biosynthesis. The **catabolism** of carbohydrates produces **ATP** (adenosine triphosphate), a molecule that stores energy. Carbohydrates are catabolized first to acetyl-CoA via the **Embden-Meyerhof pathway.** Acetyl-CoA then enters the **Krebs cycle,** which produces carbon dioxide, water, and energy. Fats are catabolized by hydrolysis to fatty acids; then cleavage of acyl-CoA produces units of acetyl-CoA. Proteins are catabolized to α-keto acids, many of which also are ultimately converted to acetyl-CoA. The nitrogen from α-amino acid catabolism is stored by **transamination** with vitamin B_6 (pyridoxal).

Diseases caused by chemical shortages are treated by administering chemical additives or by ensuring the supply of the necessary chemicals, such as vitamins, in the diet. Overproduction of chemicals—for example, the excess production of histamine—is treated by blocking the receptor sites of enzymes, which either produce the oversupply or interact with the oversupplied chemicals. Treatment of malignant growth involves the use of **antimetabolites** to block enzyme systems essential for rapid growth or of chemical agents that selectively destroy the DNA used in such growth. Treatment of invasions by bacteria occurs through the use of antibiotics, including derivatives of the sulfa drugs and the β-lactams, penicillin and cephalosporin. Treatment of invasions by viruses relies on chemicals designed to block the viral replication and transmission processes by interfering with nucleic acid replication or protein synthesis.

Problems

Nucleic Acids

19.1 Adenine and guanine are purines; cytosine, thymine, and uracil are pyrimidines. Explain what that means.

19.2 Caffeine is a member of the purine family. What structural features of caffeine preclude its being incorporated mistakenly as a DNA base in place of adenine or guanine?

Caffeine

Adenine

Guanine

19.3 Although the bases cytosine, thymine, and uracil are called pyrimidine bases, their structures as written do not look like that of pyrimidine, which is a heterocyclic aromatic compound (see Section 9.7.3). Write equilibria showing how cytosine, thymine, and uracil can tautomerize to pyrimidine-like structures.

Pyrimidine

19.4 What is a nucleotide? Draw an example.

19.5 What is a nucleoside? Draw an example.

19.6 Nucleosides are stable in water and dilute base, but they are hydrolyzed to the purine or pyrimidine base and ribose in dilute acid. Explain why this is so, and write the hydrolysis reaction for adenosine (adenine-ribose). (*Hint:* Think of acetals.)

19.7 What are the essential differences between DNA and RNA?

19.8 What is meant by a base pair? Give an example.

19.9 What is a gene?

19.10 Define the following terms:

(a) transcription

(b) translation

(c) replication

(d) codon

(e) anticodon

19.11 What is the role of tRNA, and where does it fulfill this role?

19.12 Why is the genetic code called a triplet code?

19.13 What is genetic engineering?

19.14 What is the so-called PCR process?

19.15 What is meant by the term *human genome*?

19.16 Show the product of the reaction between methyl acetate and ethylamine, a reaction that is identical to the peptide-forming reaction that occurs in protein biosynthesis.

19.17 If a segment of DNA has the base sequence A-C-G-A-T, what is the base sequence of the corresponding mRNA?

19.18 For the example given in Problem 19.17, what is the anticodon sequence in the tRNA for the first triplet?

19.19 Using Table 19.1, show the peptide that would be produced from a gene that has an informational strand of DNA consisting of the following sequence of bases:

TAG—CAA—GTG—TGT—TGA

19.20 If the adenine in a T-A-T triad in the informational strand of DNA were replaced by guanine through a mutation, what change would result in the protein synthesized by the DNA?

19.21 If a segment of an informational strand of DNA has the sequence of bases A-T-G, what is the sequence of bases it will produce in the complementary template strand of DNA, in mRNA, and in tRNA?

19.22 A tRNA segment has alanine attached to it. Write the sequence of bases in the informational strand of DNA that was translated into this segment.

19.23 If a peptide has the amino acid sequence Ala-Gly-Phe-Tyr-Trp, what was the base sequence in the original informational strand of DNA.

Metabolism

19.24 Show the tautomerization that isomerizes dihydroxyacetone monophosphate to glyceraldehyde-3-phosphate.

19.25 (a) Write a mechanism (using curved arrows) for the reaction of phosphate anion with glyceraldehyde-3-phosphate. (*Hint:* Recall the typical nucleophilic addition undergone by aldehydes.)

CONCEPTUAL PROBLEM

As a horticulturist, you are asked to assess the qualities of a new hybrid tomato plant. The plant's genes have reportedly been "engineered," through recombinant DNA techniques, to make it produce less acidic fruit. To test the success of the genetic engineering, you plant a test plot of the hybrid tomato and one of a wild-type variety of tomato. Both plots are cared for in exactly the same way, receiving the same amounts of water, sunlight, and fertilizer. At the end of the growing season, the hybrid plants not only produce a sweeter fruit—as advertised—but they are significantly larger and healthier than the wild-type plants.

- What conclusions can you draw about this case of genetic engineering in particular?
- What conclusions can you draw about the ramifications of recombinant DNA techniques in general?

Genetically altered tomatoes.

(b) Show the reaction (not the mechanism) by which the product of (a) can be converted into 1,3-diphosphoglycerate. (Note that the overall process can be viewed as oxidation of the aldehyde group in the starting material to a carboxylic acid group, which appears in the form of an anhydride.)

19.26 If pyruvic acid could somehow be decarboxylated, what would be the product? What reaction would then be necessary to produce acetic acid, which could be esterified by CoA?

19.27 Write a generalized mechanism (using curved arrows) to account for the first step of the Krebs cycle, the reaction of acetyl-CoA with oxalacetic acid in the presence of a base to produce citric acid. (*Hint:* Recall the two major chemical properties of an ester: It undergoes nucleophilic acyl substitution and the α-hydrogen is acidic.)

19.28 The catabolism of fats involves a reverse Claisen condensation to produce acetyl-CoA. A Claisen condensation is involved in the anabolism (biosynthesis) of fats in the body. Show the mechanism of this forward step (using curved arrows) for the reaction of acetyl-CoA with myristyl-CoA using $CoA\text{-}S^-$ as the catalyst.

19.29 The reaction sequence for deamination of an α-amino acid to an α-keto acid by pyridoxal phosphate has a tautomeric isomerization step. Using partial structures, show the mechanism for this tautomerization using a weak acid catalyst in water.

Chemical Basis of Therapies

19.30 What does the term *antibiotic* mean?

19.31 What is an antimetabolite?

19.32 What is the concept underlying the development of successful antihistamines?

19.33 Devise a total synthesis of sulfathiazole from benzene involving three transformations: (1) benzene to acetanilide, (2) acetanilide to sulfanilic acid ($H_2NC_6H_4SO_3H$-*p*), and (3) sulfanilic acid to sulfathiazole, using 2-aminothiazole. (*Hint:* Use retrosynthesis.)

H_2N–C_6H_4–SO_2NH–(thiazol-2-yl) Sulfathiazole H_2N–(thiazol-2-yl) 2-Aminothiazole

19.34 How does a retrovirus differ from other viruses?

19.35 What is the difference between AIDS and HIV?

19.36 What is the key conceptual difference between designing medicinals to fight bacterial diseases and designing medicinals to fight viral diseases?

Appendix A

Nomenclature of Organic Compounds: A Summary

In the early years of organic chemistry, the name for a new organic compound was chosen by the individual who discovered the compound. Many names were based on the compound's sources—formic acid (HCOOH), for example, was obtained from red ants and named after the Latin word for "ants," *formica*. But as the number of compounds grew, so grew the need for a systematic means by which a chemist could know the name of a compound or determine the structure of a compound from its name. Finally, the International Union of Pure and Applied Chemistry (IUPAC, pronounced "I-you-pac") developed a nomenclature system that allowed anyone to derive a structure from a name, and vice versa. The application of this nomenclature to the various functional groups is described in the book and is more effectively learned gradually as you work through the chapters. For convenience, however, this appendix summarizes in one place the essential IUPAC nomenclature rules.

IUPAC names contain up to four parts: a *stem* to indicate the number of carbons involved in the chain or ring; a *suffix* to indicate the family of compounds to which the compound belongs (the stem and the suffix become the *parent name*); *prefixes* to indicate the presence of any substituents; and *numbers*, where appropriate, to indicate the locations of the substituents and certain functional groups.

A.1 Compounds without a Functional Group

Unbranched Alkanes

Alkanes are considered the parent compounds on which the IUPAC nomenclature system is based. Once you understand the rules for naming alkanes, the naming of compounds containing functional groups follows in a systematic and evolutionary manner.

An alkane is named by identifying and naming the longest continuous chain of carbon atoms. Then the suffix to indicate that the compound is an alkane (*-ane*) is added. The parent names for various-length carbon chains are listed in Table A.1 for alkanes up to C_{20}. For C_1–C_4 carbon chains, the historical names (methane, ethane, propane, and butane) were adopted by IUPAC, but for all longer chains, the Latin or Greek term for the number is employed (for example, *penta-* for a five-carbon chain). In accord with this system, the compound $CH_3CH_2CH_2CH_2CH_3$ is pentane, and $CH_3(CH_2)_{10}CH_3$ is dodecane.

Alkyl Groups

The term *alkyl group* refers to an alkane with a hydrogen atom removed, usually from a terminal carbon. The name of the alkyl group is derived by removing the suffix *-ane* from the parent alkane name and replacing it with the suffix *-yl*. The names of the simple alkyl groups are shown in Table A.2.

Branched Alkanes

Alkanes may have alkyl groups attached to the parent chain. In that case, all of the carbons in the molecule are not in a single continuous chain. These compounds are known as *branched alkanes*. The group comprising a branch may be simple, such as ethyl

TABLE A.1

IUPAC Parent Names of Unbranched Alkanes

Number of Carbon Atoms	Parent Name	Number of Carbon Atoms	Parent Name
1	Methane	11	Undecane
2	Ethane	12	Dodecane
3	Propane	13	Tridecane
4	Butane	14	Tetradecane
5	Pentane	15	Pentadecane
6	Hexane	16	Hexadecane
7	Heptane	17	Heptadecane
8	Octane	18	Octadecane
9	Nonane	19	Nonadecane
10	Decane	20	Icosane

or methyl, or more complex with branching of its own (in that case, it may be referred to as a side chain, to be described shortly).

Branched alkanes are named by following these rules:

1. Identify and name the longest continuous chain of carbons (this is the parent name) using the suffix *-ane.*
2. Number the chain from the end closest to the substituent (the point of branching).
3. Designate the location of the substituent by the number of the carbon on the chain to which it is attached. The chain is numbered from the end that will provide the lowest possible number for a substituent.
4. Name the substituent as an alkyl group.
5. To actually write the name, remember that the number precedes the prefix and is connected to it by a hyphen. There is no hyphen or space between the prefix and the stem name.

TABLE A.2

Simple Alkyl Group Names

Number of Carbons	Parent Alkane Name	Alkyl Group Name	Alkyl Group
1	Methane	Methyl	$CH_3—$
2	Ethane	Ethyl	$CH_3CH_2—$
3	Propane	Propyl	$CH_3CH_2CH_2—$
4	Butane	Butyl	$CH_3CH_2CH_2CH_2—$
5	Pentane	Pentyl	$CH_3CH_2CH_2CH_2CH_2—$

For example, the following compound is named 2-methylhexane, *not* 5-methylhexane:

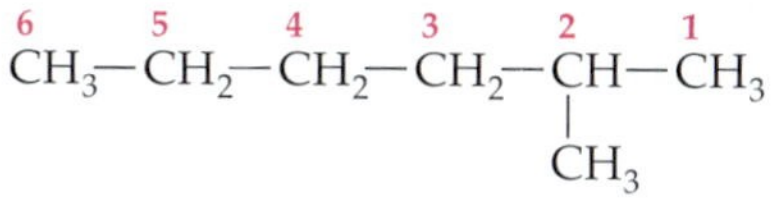

$$\overset{6}{CH_3}-\overset{5}{CH_2}-\overset{4}{CH_2}-\overset{3}{CH_2}-\overset{2}{\underset{\underset{CH_3}{|}}{CH}}-\overset{1}{CH_3}$$

2-Methylhexane

Multiple alkyl group substituents are treated by naming each in alphabetical order and assigning a location number to each. If two or more substituents are identical, they are named together using the prefix *di-* or *tri-* (a prefix does not determine the alphabetical sequence), but there must be a number assigned for each individual substituent (these numbers are separated by commas). Thus, the following compounds are 4-ethyl-3-methylheptane (*not* 3-methyl-4-ethylheptane) and 2,2-dimethylbutane (*not* 3,3-dimethylbutane):

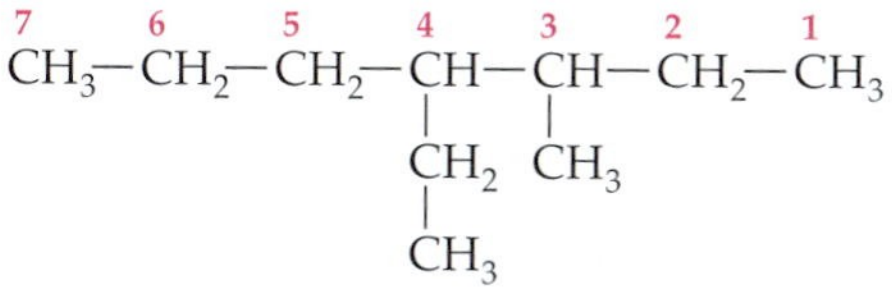

$$\overset{7}{CH_3}-\overset{6}{CH_2}-\overset{5}{CH_2}-\overset{4}{\underset{\underset{\underset{\underset{CH_3}{|}}{CH_2}}{|}}{CH}}-\overset{3}{\underset{\underset{CH_3}{|}}{CH}}-\overset{2}{CH_2}-\overset{1}{CH_3}$$

4-Ethyl-3-methylheptane

$$\overset{4}{CH_3}-\overset{3}{CH_2}-\overset{\overset{CH_3}{2|}}{\underset{\underset{CH_3}{|}}{C}}-\overset{1}{CH_3}$$

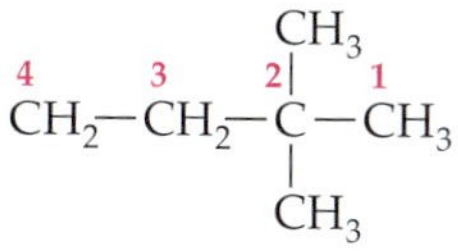

2,2-Dimethylbutane

Branched Side Chains

Compounds frequently carry branched side chains, but they are usually simple. Here are some common branched alkyl groups:

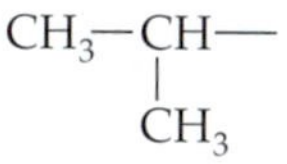

$$CH_3-\underset{\underset{CH_3}{|}}{CH}-$$

Isopropyl

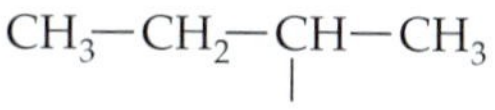

$$CH_3-CH_2-\underset{|}{CH}-CH_3$$

sec-Butyl

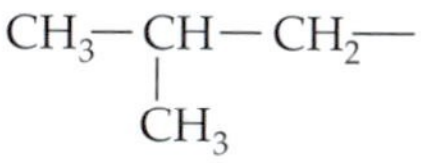

$$CH_3-\underset{\underset{CH_3}{|}}{CH}-CH_2-$$

Isobutyl

$$CH_3-\overset{\overset{CH_3}{|}}{\underset{\underset{CH_3}{|}}{C}}-$$

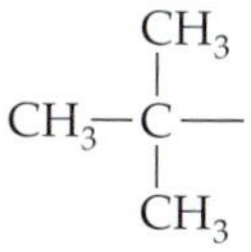

tert-Butyl
(*t*-butyl)

Other Substituents

Certain functional groups (called *subordinate groups*) are named as substituents on the parent chain rather than by using a suffix (discussed below). Included are the following: fluoro- (—F), chloro- (—Cl), bromo- (—Br), iodo- (—I), nitro- (—NO_2), azido- (—N_3), diazo- (—N_2^+), and alkoxy- (—OR). Thus, the following compound is 4-bromo-4-isopropyl-3-methoxyheptane:

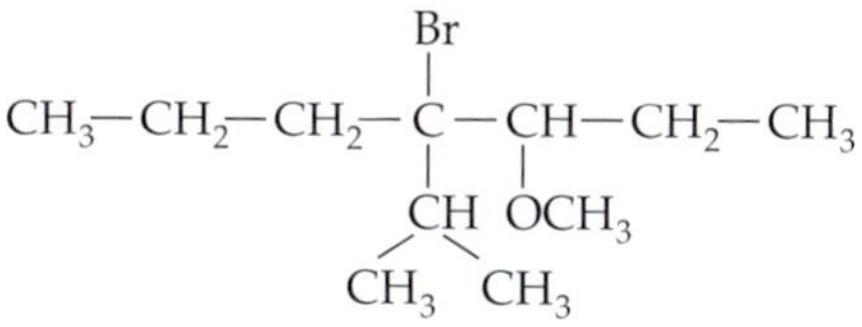

$$CH_3-CH_2-CH_2-\overset{\overset{Br}{|}}{\underset{\underset{CH_3\ \ CH_3}{\underset{/\ \ \backslash}{CH}}}{\underset{|}{C}}}-\underset{\underset{OCH_3}{|}}{CH}-CH_2-CH_3$$

4-Bromo-4-isopropyl-3-methoxyheptane

Cycloalkanes

Cycloalkanes are named following the same general rules as for alkanes except that the parent ring is named as a cycloalkane, with the prefix *cyclo-* attached to the stem name. The stem name indicates the number of carbons in the ring (its size), as in the four-carbon cyclobutane and the six-carbon cyclohexane. Any substituents are located on the ring by numbers, and the numbering sequence around the ring is chosen to keep the numbers as low as possible. The following compounds are named 1-chloro-2-nitrocyclobutane (*not* 1-chloro-4-nitrocyclobutane) and 1,1-diethyl-3-fluorocyclohexane (*not* 1-fluoro-3,3-diethylcyclohexane):

1-Chloro-2-nitrocyclobutane

1,1-Diethyl-3-fluorocyclohexane

When there is a single substituent on a ring, the numerical location indicator is not necessary, such as in isopropylcyclopropane.

A.2 Compounds Containing Functional Groups

Major functional groups are called *principal groups* (in contrast to the subordinate groups mentioned earlier). The presence of a principal functional group is reflected by a unique suffix on the compound name and a number indicating the position of the functional group. The functional group must be included in the chain from which the parent name is derived. Examples of such parent names are propanoic acid (CH_3CH_2COOH) and 1-butanol ($CH_3CH_2CH_2CH_2OH$). Other substituents are located on such a parent compound in the manner described earlier (for example, 3-chloropropanoic acid, $ClCH_2CH_2COOH$). Table A.3 lists the suffixes for the common functional groups.

Note that the location of five of the common functional groups listed in the table must be indicated by a number. The parent chain is numbered so as to keep this number as low as possible, and then any other substituents are assigned numbers. The location of an aldehyde or carboxylic acid group need not be indicated by a number because these can only be at the end of a chain (the carbon atom in either of these groups is automatically considered to be carbon 1). The number assigned to a double bond or a triple bond in alkenes or alkynes identifies the lowest possible numbered carbon; the multiple bond extends between that carbon and the next higher numbered carbon.

A.3 Polyfunctional Compound Nomenclature

Here are the rules for determining the nomenclature of compounds that contain more than one principal functional group:

1. Determine the order of precedence of the principal functional groups in the compound, using Table A.4.
2. Choose the highest-precedence group and use its designated suffix in conjunction with the appropriate stem (which indicates the number of carbons in the longest continuous chain containing that functional group) to derive the parent name.
3. If appropriate for that functional group (see Table A.3), assign a number for its location, keeping this number as low as possible.

TABLE A.3

Selected Functional Groups

Group Name	Structure	IUPAC Suffix	Example
Alkene	$>C=C<$	-ene	$CH_3CH_2CH=CH_2$ 1-Butene
Alkyne	$-C\equiv C-$	-yne	$CH_3C\equiv CCH_3$ 2-Butyne
Alcohol	$-OH$	-ol	$CH_3CH(OH)CH_3$ 2-Propanol
Amine	$-NH_2$	-amine	$CH_3CH_2CH_2NH_2$ 1-Propanamine
Aldehyde	$-C(=O)-H$	-al	CH_3CH_2CHO Propanal
Ketone	$-C(=O)-$	-one	$CH_3CH_2COCH_3$ 2-Butanone
Carboxylic acid	$-C(=O)-OH$	-oic acid	CH_3CH_2COOH Propanoic acid

TABLE A.4

Decreasing Order of Functional Group Precedence for IUPAC Nomenclature

Functional Group Name	Formula	Suffix (when highest precedence)	Prefix (when lower precedence)
Carboxyl	$-COOH$	-oic acid or -carboxylic acid	carboxy-
Aldehyde	$-CHO$	-al or -carbaldehyde	formyl- oxo-
Ketone	$-CO-$	-one	oxo-
Alcohol	$-OH$	-ol	hydroxy-
Amine	$-NH_2$	-amine	amino-
Ether	$-OR$	none	alkoxy-
Alkene	$C=C$	-ene	en-
Alkyne	$C\equiv C$	-yne	yn-

4. Name all other functional groups of lower precedence using their designated prefixes.
5. Assign location numbers for these functional groups using the chain numbering already determined in step (3).
6. For compounds containing both a double and a triple bond, assign numbers to produce the lowest possible total number. If the alternative sets of numbers are equivalent, use the lower number for the double bond.

Here are three examples illustrating the application of these rules:

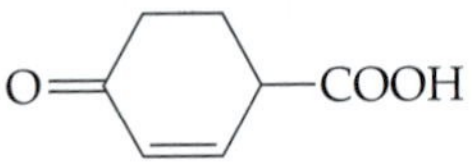

O= —COOH

4-Oxo-2-cyclohexene-carboxylic acid

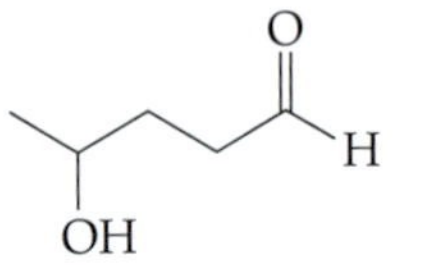

O H OH

4-Hydroxypentanal

1-Hepten-4-yne

A.4 Benzenoid Compounds

The benzene ring is considered a parent structure just like an alkane chain. Alkyl groups, halogens, and alkoxy, azido, diazo, and nitro groups are considered substituents on the parent and are therefore named using prefixes. Examples are isopropylbenzene, nitrobenzene, and azidobenzene. Principal functional groups on benzene rings are incorporated into parent names, as in benzoic acid (C_6H_5COOH), benzaldehyde (C_6H_5CHO), acetophenone ($C_6H_5COCH_3$), phenol (C_6H_5OH), and aniline ($C_6H_5NH_2$).

When two groups are attached to a benzene ring, their locational relationship is described by the prefix *ortho-*, *meta-*, or *para-*, usually using only the first letter of each. When there are three or more substituents on the benzene ring, their relative locations are indicated by numbers assigned to the six ring locations, with location number 1 being reserved for the highest-precedence group.

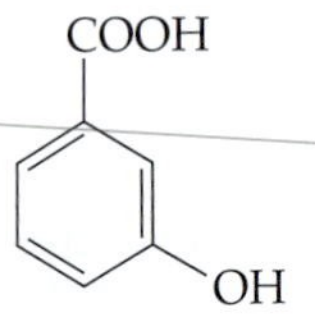

COOH OH

m-Hydroxybenzoic acid

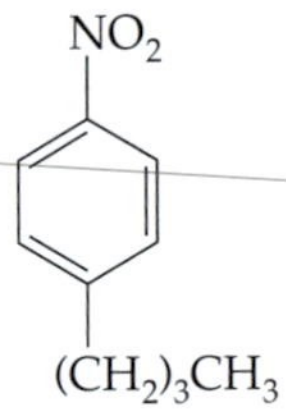

NO_2 $(CH_2)_3CH_3$

p-Butylnitrobenzene

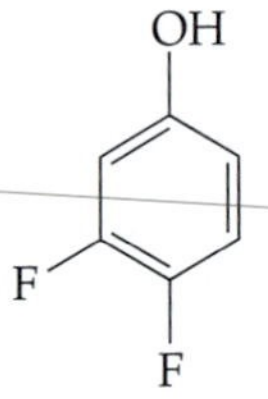

OH F F

3,4-Difluorophenol

When a benzene ring is attached to an alkane chain, it may be considered a substituent on that chain, as in 3-phenyl-1-propanol ($C_6H_5CH_2CH_2CH_2OH$).

Parentheses are occasionally used when naming complex substituents if their absence would lead to ambiguity. Here are a few examples:

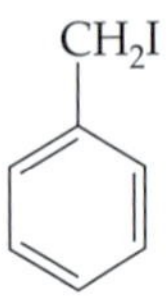

CH_2I

(Iodomethyl)benzene

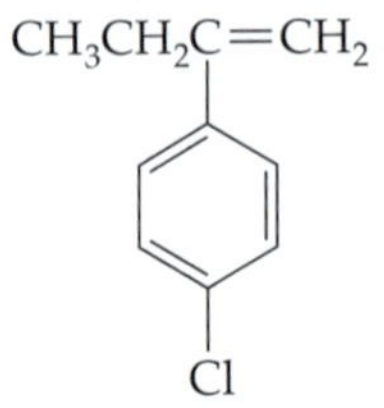

$CH_3CH_2C{=}CH_2$ Cl

2-(*p*-Chlorophenyl)-1-butene

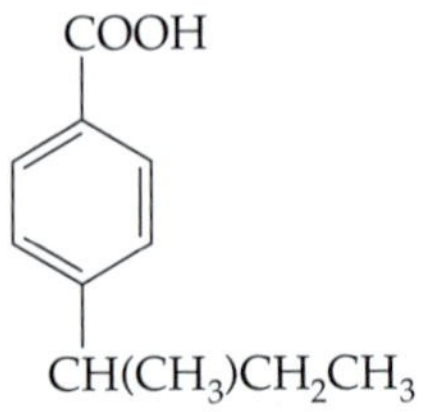

COOH $CH(CH_3)CH_2CH_3$

p-(1-Methylpropyl)benzoic acid

Appendix B

Group Assignments in Infrared Spectra

Type of Bond	Group	Family of Compounds	Range (cm^{-1})
Single Bonds	C—H	Alkanes	2850–3300
	=C—H	Alkenes, aromatics	3000–3100
	≡C—H	Alkynes	3300–3320
	O—H	Alcohols	3200–3600
	N—H	Amines	3300–3500
Double Bonds	C=C	Alkenes, aromatics	1600–1680
	C=O	Carbonyls	1680–1750
		Aldehydes, ketones	1710–1750
		Carboxylic acids	1700–1725
		Esters, amides	1680–1750
	C=N	Imines	1500–1650
Triple Bonds	C≡C	Alkynes	2100–2200
	C≡N	Nitriles	2200–2300

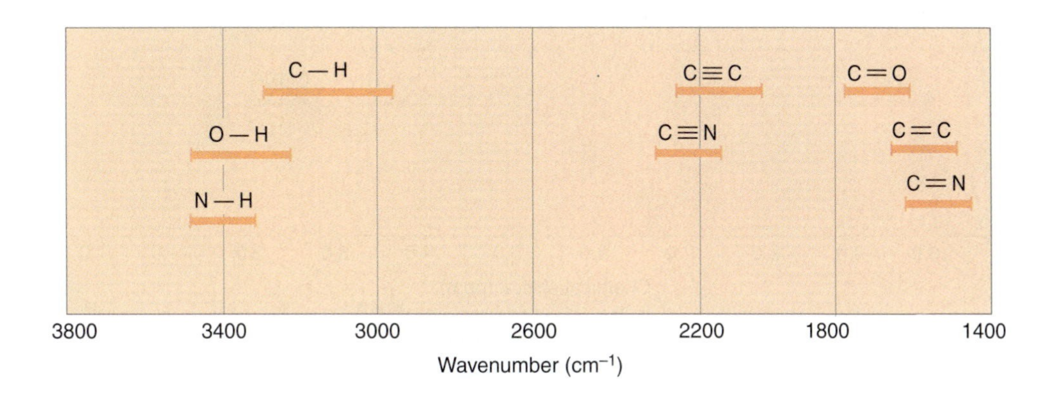

Appendix C

Group Assignments in ^{1}H-NMR Spectra

Type of Hydrogen	Chemical Shift (ppm)	Type of Hydrogen	Chemical Shift (ppm)
C—CH_3	0.8–1.0	—CH—O—	3.3–4.0
C—CH_2	1.0–1.4	—CH—N—	2.2–2.9
C—CH	1.0–1.6	O═C—CH	2.0–2.6
C═C—C—H	1.6–1.9	O═C—H	9.5–9.7
Ar—C—H	2.2–2.8	O═C—O—H	10–13
C═C—H	4.6–5.7	Halogen—C—H	3.1–4.1
C≡C—H	2.5–2.7	O—H	0.5–6.0*
Ar—H	6.5–8.5	N—H	0.6–3.0*

*The chemical shifts of hydroxyl and amino hydrogens are variable and cannot be predicted with accuracy. The best advice is to identify those peaks by elimination or by shaking the sample with D_2O (deuterium will replace the hydrogen, and the peak will disappear because deuterium does not show up in NMR spectra).

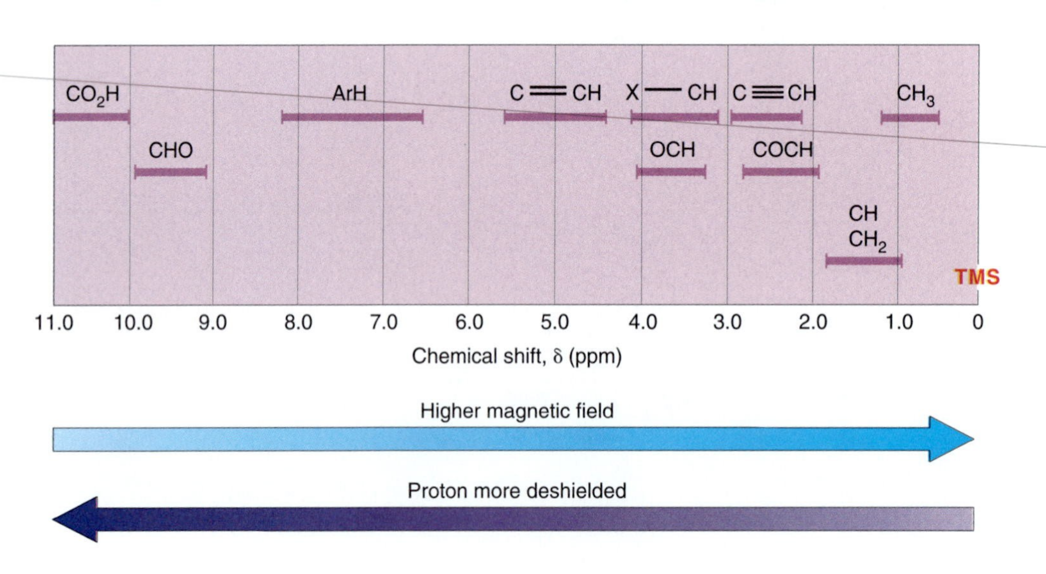

Appendix D
pK_a Values of Some Important Compounds

Name	General Structure	Typical pK_a Value*	Conjugate Base That Results	Typical Reagent for Proton Removal
Organic Compounds				
Sulfonic acid	RSO_3H	<1	RSO_3^-	NH_3
Carboxylic acid	RCOOH	3–5	$RCOO^-$	$NaHCO_3$
Alkyl ammonium salt	RNH_3^+	9–11	RNH_2	NaOH
Aryl ammonium salt	$ArNH_3^+$	4–5	$ArNH_2$	$NaHCO_3$
Phenol	ArOH	9–10	ArO^-	NaOH
Alcohol	ROH	16–19	RO^-	Na metal
Aldehyde	RCH_2CHO	14–16	RCH^-CHO	NaOH
Ester	RCH_2COOR'	20–25	RCH^-COOR'	$NaOCH_3$
Alkyne	RCCH	25	RCC^-	$NaNH_2$
Alkane	R_3CH	40–60	R_3C^-	None
Typical Reagents				
Carbonic acid	H_2CO_3	6.4	HCO_3^-	NaOH
Methanol	CH_3OH	15.5	CH_3O^-	Na metal
Water	H_2O	15.7	HO^-	$NaOCH_3$
t-Butyl alcohol	$(CH_3)_3COH$	18	$(CH_3)_3CO^-$	K metal
Ammonia	NH_3	33	NH_2^-	Na metal

*These are pK_a values for typical unsubstituted compounds. Recall that substitution may change these values by several pK_a units.

Index/Glossary

Key terms in the text appear here in **bold** followed by the definition.

mainly to the *ortho* and *para* positions on a benzene ring, 292
***Ortho* position** position 2 on a benzene ring, 273
Oxalacetic acid, 686
Oxalic acid, 498
Oxalosuccinic acid, 686
Oxaphosphetane, 475
Oxazole, 317
Oxidase, 602
Oxidation an increase in the oxygen content of an organic species or a decrease in its hydrogen content
of alcohols, 138, 477
of aldehydes, 454
of alkanes, 69
of alkenes, 203
of alkynes, 241
of aromatic side chains, 309
Oxidation states, 138
Oximes a kind of imine ($R_2C{=}NOH$) resulting from reaction of a carbonyl compound with hydroxylamine (H_2NOH), 463
Oxiranes (epoxides) three-membered cyclic ethers, 168
Oxo group, 450
(*E*)-9-Oxo-2-decenoic acid, 507
Oxonium ion an oxygen atom with three groups covalently attached, making the oxygen atom positively charged, 131, 170
Oxyacetylene torch, 242
Oxybenzone, 370
Oxygen-18, 548
Oxygenates, 176
Oxytocin, 586
Ozone layer, 344
Ozonide, 205
Ozonolysis of an alkene the reaction of an alkene with ozone to form an ozonide, which, after treatment with zinc and water (a reduction), results in two carbonyl compounds from the cleavage of the double bond, 205

PABA (*p*-aminobenzoic acid), 371
PABA amide, 432, 696
Paclitaxel, 5, 9, 696
Palmitic acid, catabolism of, 688
***Para* position** position 4 on a benzene ring, 273
Para Red, 425
Paraffins, 55
Paratartaric acid, 622
Partially hydrogenated vegetable oil, 192
Pasteur, L., 622
Pauli exclusion principle the principle that a maximum of two electrons, of opposing spin, may occupy an orbital, 10
PCBs, 281
PCC, 139
Peak a position (that is, wavelength or frequency) in a spectrum at which absorption of radiation has occurred; may be sharp or broad, 364
Penicillin, 6, 568, 696
1,4-Pentadiene, 184
Pentaerythritol nitrate (PETN), 298
2-Pentene, 183
3-Penten-1-ol, 184
Peptide bond an amide bond connecting two α-amino acids, 417, 578
Peptides compounds in which two or more α-amino acids are connected through amide (peptide) bond(s), 578
hydrolysis of, 583
protecting groups in, 597
synthesis of, 596
Peracids compounds of general formula RCO_3H, formed from a carboxylic acid by reaction with hydrogen peroxide, 168, 203
Perchloroethane, 78
Perchloroethylene, 111
Perfluorodecalin, 76, 113
Period, in periodic table, 11
Peroxides compounds containing an "extra" oxygen atom involved in an oxygen-oxygen sigma bond, 330
Peroxy acid, 168
Peroxy radical, 334
Persian hemlock, 434
Persistent pesticide a pesticide that is not readily decomposed upon exposure to the elements and therefore remains effective for a considerable time, 280, 549
PET crystalline poly(ethylene terephthalate), 520, 649, 663, 665
PETN, 298
Petroleum, 57
Peyote, 436
Phenanthrene, 313
Phenobarbitol, 568
Phenolic ethers
cleavage of, 304
synthesis of, 304
Phenols aromatic compounds with one or more hydroxyl groups attached to the ring, 300, 305
acidity of, 300
O-alkylation of, 304
deprotection of, 304
synthesis of, 304, 419
substitution of, 303
Phenoxy radical, 342
Phenyl group the C_6H_5— group, 273
Phenyl isothiocyanate, 588
Phenylalanine, 154
Phenylketonuria (PKU), 154
Phenylthiohydantoins, 588
Pheromones compounds that are the basis of chemical communication, usually between members of the same species; examples are alarm pheromones and sex pheromones, 245, 485
of bees, 507
Phosgene, 538, 564, 663
Phosphonium ylide, 475
Phosphorus tribromide, 135
Photodegradable plastics, 664
Photon a packet of energy emitted by electromagnetic radiation, described by the relationship $E = h\nu$, 364
Photoresist technology, 424
Photosynthesis, 690
Phthalic acid, 309, 498
Phthalic anhydride, 543
Pi (π) bond a bond formed by the overlap of two parallel *p* orbitals, 186
Pi (π) orbital a molecular orbital formed by the overlap of two parallel *p* orbitals, 186, 235
Picric acid, 302
α-Pinene, 261
Piperidine, 317, 404
Piperonal, 482
pK_a a measure of the strength of an acid, the negative log of the acid dissociation constant (K_a), 32
of alcohols, 126
of carboxylic acids, 500
of phenols, 300
pK_b, 408
Plane of symmetry an imaginary plane through a compound such that one half is the mirror image of the other half, 81, 627
Plane-polarized light light in which all the waves vibrate in the same plane, 614
Plant hormone, 222
Plastics rigid materials fabricated from highly crystalline polymers, 648
biodegradable, 665
photodegradable, 664
recycling of, 665
Pleated sheet one of two secondary structures of proteins, in which two chains of polypeptide are aligned parallel to each other with the N-terminus adjacent to the C-terminus, 593
Plexiglas a glass-like product fabricated from poly(methyl methacrylate), 462, 655
Poison ivy, 305
Polar molecule, 29
Polarimeter an instrument for passing plane-polarized light through a solution of a compound and measuring the extent and direction of rotation of that light, 615
Polarizability the ease with which a pair of

Credits

(continuation of the copyright page)

TABLE OF CONTENTS

p. x: courtesy of Michele Browner, Roche Bioscience; p. xi: ©Sinclair Stammers/Science Photo Library/Photo Researchers, Inc.; p. xii: (top) ©Enderson, Colorado College, (bottom) ©Mitch Kezar/PHOTOTAKE/PNI; p. xiii: ©P. Dumas/Eurelios/Science Photo Library/Photo Researchers, Inc.; p. xiv: ©Nigel Cattlin/Holt Studios International/Photo Researchers, Inc.; p. xv: courtesy of Robert Williams, the Hubble Deep Field Team, and NASA; inset (NASA). p. xvi: (top) NASA/GSFC; (bottom) ©James Prince/Science Source/Photo Researchers, Inc.; p. xvii: ©1998 PhotoDisk, Inc.; p. xviii: ©1998 PhotoDisk, Inc.; p. xix: ©Michael W. Nelson/Stock South/PNI; p. xx: ©Geoff Tompkinson/Science Photo Library/Photo Researchers, Inc.; p. xxi: ©Dr. Gopal/Science Photo Library/Photo Researchers, Inc.

INTRODUCTORY ESSAY

p. 2: (left) ©Earl Roberge/Photo Researchers, Inc.; (middle) ©David M. Stone/PHOTO/NATS, Inc.; (right) ©Robert Rathe/Stock Boston/PNI; p. 3: ©Peter Menzel/Stock Boston/PNI; p. 4: ©1998 PhotoDisc, Inc.; p. 5: ©GJLP/CNRI/PHOTOTAKE; p. 6: ©1998 PhotoDisc, Inc.

CHEMISTRY AT WORK

photo of glassware appearing in upper left of all boxes: ©1998 PhotoDisc, Inc.

SPECTRA

Figs. 11.20, 11.42, 11.44b, 14.3, 14.4, 15.6 ©Sadtler Research Laboratories, Division of Bio-Rad Laboratories, Inc. (1981-1993). Permission for the publication herein of Sadtler Standard Spectra® has been granted, and all rights are reserved, by Sadtler Research Laboratories, Division of Bio-Rad Laboratories, Inc.

CHAPTER 1

p. 8: ©Matthew McVay/Tony Stone Images/PNI; p. 23: ©Michael Iscaro/Black Star/PNI; p. 28: (right) courtesy of Michele Browner, Roche Bioscience; p. 30: ©Charles D. Winters/Photo Researchers, Inc.; p. 32: ©Charles D. Winters/Photo Researchers, Inc.; p. 38: ©Peter K. Ziminski/Visuals Unlimited.

CHAPTER 2

p. 40: ©Ken Graham/Tony Stone Images; p. 43: ©Paul Silverman/Fundamental Photographs, NYC; p. 55: ©Ben Phillips/PHOTO/NATS, Inc.; p. 56: ©Joe Sohm/Tony Stone Images; p. 58: (right) ©Diane Schiumo/Fundamental Photographs, NYC; p. 68: (right) ©Sinclair Stammers/Science Photo Library/Photo Researchers, Inc.; p. 70: ©Christopher Morris/Black Star; p. 74: ©Arnulf Husmo/Tony Stones Images.

CHAPTER 3

p. 75: ©Rosenfeld Images Ltd/Science Photo Library/Photo Researchers, Inc.; p. 81: ©1998 PhotoDisc, Inc.; p. 87: (right) ©Will & Deni McIntyre/Photo Researchers, Inc.; p. 93: (top and bottom) ©1998 PhotoDisc, Inc.; p. 110: (top) ©Bruno Barbey/Magnum Photos/PNI, (bottom) ©Richard Magna/Fundamental Photographs, NYC; p. 111: NASA/GSFC; p. 112: ©Enderson, Colorado College; p. 113: ©Paul Shambroom/Science Source/Photo Researchers, Inc.; p. 117: ©TSM/Robert Essex, 1998.

CHAPTER 4

p. 119: ©Dan McCoy/Rainbow/PNI; p. 124: (right) ©Renee Lynn/Tony Stone Images; p. 130: (bottom) ©Gerhard Stief/Photo Researchers, Inc.; p. 141: (right) ©Dennis MacDonald/PhotoEdit/PNI; p. 144: (top) ©Ken Kaminsky/PHOTOTAKE, (bottom) ©Mitch Kezar/PHOTOTAKE/PNI; p. 150: ©Deborah M. Crowell/PHOTO/NATS, Inc.; p. 152: (left) ©Mark Tomalty/Masterfile; (right) ©Junebug Clark/Photo Researchers, Inc.; p. 153 ©R. Hamilton Smith/Tony Stone Images/PNI; p. 159: ©Charlie Ott/Photo Researchers, Inc.

CHAPTER 5

p. 160: ©Arthur V. Mauritius GMBH/PHOTOTAKE; p. 167: ©1998 PhotoDisc, Inc.; p. 174: ©TMS/Peter Fisher, 1998; p. 179: ©Gwyn M. Kibbe/Stock Boston.

CHAPTER 6

p. 181: ©TSM/Charles Krebs, 1998; p. 184: ©Randy G. Taylor/Leo de Wys; p. 192: (right) ©Richard Megna/Fundamental Photographs, NYC; p. 205: ©Scott Camazine/Photo Researchers, Inc.; p. 206: ©Steven P. Parker/Photo Researchers, Inc.; p. 222: ©P. Dumas/Eurelios/Science Photo Library/Photo Researchers, Inc.; p. 299: ©Christopher Brown/Stock Boston.

CHAPTER 7

p. 230: ©Art Wolfe/AllStock/PNI; p. 242: ©Charles D. Winters/Timeframe Photography/Photo Researchers, Inc.;p. 245: (bottom) ©Nigel Cattlin/Holt Studios International/Photo Researchers, Inc.; p. 249: ©Nigel Cattlin/Holt Studios International/Photo Researchers, Inc.

CHAPTER 8

p. 250: ©Bill Horsman/Stock Boston; p. 261: ©Thomas R. Fletcher/Stock Boston; p. 263: ©Lee Landau/PHOTO/NATS, Inc.; p. 264: ©Michael P. Gadomski/Photo Researchers, Inc.

CHAPTER 9

p. 267: courtesy of Robert Williams, the Hubble Deep Field Team, and NASA; (inset) NASA; p. 268: ©1998 PhotoDisk, Inc.; p. 280: ©Gordon Roberts/Holt Studios/Photo Researchers, Inc.; p. 283: ©1998 PhotoDisk, Inc.; p. 298: ©Jim Zipp/Photo Researchers, Inc.; p. 300: ©Don Johnston/PHOTO/NATS, Inc.; p. 305: ©John Dudak/PHOTOTAKE/PNI; p. 314: ©Ken Eward/Science Source/Photo Researchers, Inc.; p. 315: ©American Cancer Society/PHOTOTAKE; p. 318: ©Frederica Georgia/Photo Researchers, Inc.; p. 320: ©Peter Menzel/Stock Boston/PNI; p. 327: ©Iscar Sosa/Black Star.

CHAPTER 10

p. 328: ©Will & Deni McIntyre/Photo Researchers, Inc.; p. 333: ©TSM/Chris Jones, 1998; p. 334: NASA; p. 341: © Scott Camazine & Sue Trainor/Photo Researchers, Inc.; p. 344: NASA/GFSC.

CHAPTER 11

p. 350: ©Roy Ohms/Masterfile; p. 357: ©James Holmes/Oxford Centre for Molecular Sciences/Science Photo Library/Photo Researchers, Inc.; p. 365: ©Tom Pantages; p. 366: ©Gregory Dimijian/Photo Researchers, Inc.; p. 367: ©Kristen Brochmann/Fundamental Photographs, NYC; p. 376: © Young-Wolff/PhotoEdit/PNI; p. 377: ©James Prince/Science Source/Photo Researchers, Inc.; p. 387: ©Laurence Dutton/Tony Stone Images; p. 400: Maximiliam Stock Ltd/Science Photo Library/Photo Researchers, Inc.

CHAPTER 12

p. 401: ©Jeff Corwin/Tony Stone Images; p. 403: ©Mark Gibson Photography; p. 417 ©1998 PhotoDisc, Inc.; p. 424: © Rosenfeld Images/PhotoResearchers, Inc.; p. 433: ©Mark Gibson Photography; p. 434: ©Archive Photo; p. 436: ©R. Konig/JACANA/Photo Researchers, Inc.; p. 445: ©1998 PhotoDisc, Inc.

CHAPTER 13

p. 446: ©Anthony Bannister/Photo Researchers, Inc.; p. 454: ©1998 PhotoDisc, Inc.; p. 460 ©Leonard Lessin FBPA; p. 462: courtesy of University of North Dakota; p. 477: ©Gary D. McMichael/Photo Researchers, Inc.; p. 483: ©Fletcher & Baylis/Photo Researchers, Inc.; p. 485: ©Ellis Herwig/Stock Boston; p. 492: ©Klaus Guldbrandsen/Science Photo Library/Photo Researchers, Inc.

CHAPTER 14

p. 493: ©Cassey Cohen/PhotoEdit; p. 495: ©Bob Daemmrich/Stock Boston; p. 497: ©1998 PhotoDisc, Inc.; p. 507: © Scott Camazine/Photo Researchers, Inc.; p. 519: Charles Gurton/UniPhoto; p. 521: ©Will & Deni McIntyre/Photo Researchers, Inc.; p. 527: ©1998 PhotoDisc, Inc.

CHAPTER 15

p. 528: ©Nigel Cattlin/Photo Researchers, Inc.; p. 538: U.S. Army; p. 544: ©D. Young-Wolff MR/PhotoEdit; p. 550: ©Mark E. Gibson; p. 558: (top) ©Benelux/Photo Researchers, Inc.; (bottom) ©Mark E. Gibson; p. 560: ©Tom Pantages; p. 564: ©Michael W. Nelson/Stock South/PNI; p. 569: ©James Webb/PHOTOTAKE; p. 575: ©David Joel/Tony Stone Images.

CHAPTER 16

p. 577: ©AP WideWorld; p. 587: ©Ken Eward/Science Source/Photo Researchers, Inc.; p. 593: ©John Dudak/PHOTOTAKE; p. 594: ©Stephen Frisch/Stock Boston; p. 600 ©Karen Su/Stock Boston, p. 603: ©David Joel/Tony Stone Images.

CHAPTER 17

p. 609: ©Geoff Tompkinson/Science Photo Library/Photo Researchers, Inc.; p. 631: ©Jerry Howard/Stock Boston/PNI; p. 636: ©Denis Milon.

CHAPTER 18

p. 643: ©Bob Krist/Tony Stone Images; p. 644: (top) ©Hubertus Kanus/Photo Researchers, Inc.; p. 644: (bottom) ©Peter Southwick/Stock Boston; p. 648: (upper left) ©1998 PhotoDisc, Inc.; (upper right) ©DuPont Zytel Nylon 66; (lower left) ©Keith Wood/Tony Stone Images; p. 652: ©John Colett/Stock Boston; p. 653: ©George Haling/Photo Researchers, Inc.; p. 654: ©ChromoSohm/Sohm/Photo Researchers, Inc.; p. 662: ©1998 Photo Disc, Inc.; p. 665: ©Hank Morgan/Science Source/Photo Researchers, Inc.; p. 669: ©Mark Richards/Photo Edit.

CHAPTER 19

p. 670: ©James Holmes/Science Photo Library/Photo Researchers, Inc.; p. 680: ©Dr. Gopal Murti/Science Photo Library/Photo Researchers, Inc.; p. 681: ©Scott Camazine/Photo Researchers, Inc.; p. 697: ©Lee D. Simons/Science Stock/Photo Researchers, Inc.; p. 701: ©Gary Wagner/Stock Boston.